Fascinating Life Sciences

This interdisciplinary series brings together the most essential and captivating topics in the life sciences. They range from the plant sciences to zoology, from the microbiome to macrobiome, and from basic biology to biotechnology. The series not only highlights fascinating research; it also discusses major challenges associated with the life sciences and related disciplines and outlines future research directions. Individual volumes provide in-depth information, are richly illustrated with photographs, illustrations, and maps, and feature suggestions for further reading or glossaries where appropriate.

Interested researchers in all areas of the life sciences, as well as biology enthusiasts, will find the series' interdisciplinary focus and highly readable volumes especially appealing.

More information about this series at http://www.springer.com/series/15408

Valentí Rull • Ana Carolina Carnaval
Editors

Neotropical Diversification: Patterns and Processes

Springer

Editors
Valentí Rull
Laboratory of Paleoecology
Institute of Earth Sciences Jaume Almera (CSIC)
Barcelona, Spain

Ana Carolina Carnaval
Department of Biology
City College of New York
New York, NY, USA

Biology Program, Graduate Center of CUNY
New York, NY, USA

ISSN 2509-6745 ISSN 2509-6753 (electronic)
Fascinating Life Sciences
ISBN 978-3-030-31166-7 ISBN 978-3-030-31167-4 (eBook)
https://doi.org/10.1007/978-3-030-31167-4

This Springer imprint is published by the registered company Springer Nature Switzerland AG.
The registered company address is: Gewerbestrasse 11, 6330 Cham, Switzerland

Acknowledgements

The editors are very grateful to all the authors who contributed to this book for their dedication and hard work leading to high-quality chapters. We also wish to thank the reviewers for their time and expertise (in alphabetic order): Alexandre Antonelli, Chrsitine Bacon, John M. Bates, João M. Capurucho, Rosane Collevatti, Guarino Colli, Chris Dick, José A. Diniz Filho, Leandro Duarte, Joost F. Duivervoorden, José M. Fernández Palacios, Stephen Hopper, Marie-Pierre Ledru, Peter Linder, Francis Mayle, Fabian Michelangeli, Cristina Miyaki, Toby Pennington, Ivan Prates, Renato Recoder, Marcelo Reginato, Camila Ribas, Ignacio de la Riva, Miguel T. Rodrigues, Fernando Silveira, Beryl Simpson and Ángel Viloria. We greatly acknowledge the invitation of Lars Koerner, from Springer, for editing/writing the book and Sivachandran Ravanan for his constant support in editorial and production tasks.

Contents

Chapter 1
Introduction

Valentí Rull and Ana Carolina Carnaval

This book addresses the colossal biodiversity of the Neotropics and provides a range of potential explanations for its origin and maintenance. In general, we use the term "Neotropics" as synonym of "tropical America"—i.e., the part of this continent situated between the tropics of Cancer and Capricorn—rather than of the Neotropical biogeographical realm, which includes also the southern part of South America and excludes the Mexican highlands. However, the geographical distribution of the taxonomic groups studied in some chapters makes necessary to consider certain extratropical areas for a thorough study.

The main differential feature of this book with respect to other compilations and reviews on the same subject is that it is not attached to any particular idea on how the Neotropical biodiversity has originated. Instead, it considers that there is no a single explanation applicable to any Neotropical region or taxonomic group. We (the editors) are convinced that different environmental drivers acting at multiple spatio-temporal scales have led to the complex environmental-evolutionary-ecological interactions that resulted in current biogeographical patterns (Rull 2011), and this compilation is a testimony of that.

For this reason, we have invited scholars who are working on the subject regardless their personal views on Neotropical diversification. Each author has been free to choose the topic of his/her paper and to develop it according to their own preferences, without any formal or conceptual constraint besides the editorial author's instructions. As a result, the book gathers as many views as possible on the

V. Rull (✉)
Laboratory of Paleoecology, Institute of Earth Sciences Jaume Almera (ICTJA-CSIC), Barcelona, Spain
e-mail: vrull@ictja.csic.es

A. C. Carnaval
Department of Biology, City College of New York, New York, NY, USA

Biology Program, Graduate Center of CUNY, New York, NY, USA

V. Rull, A. C. Carnaval (eds.), *Neotropical Diversification: Patterns and Processes*, Fascinating Life Sciences, https://doi.org/10.1007/978-3-030-31167-4_1

causes of the extant Neotropical biodiversity, according to the awareness of the editors and the availability of the contributors.

It could be argued that this could create confusion, rather than help the reader to acquire a sound view of the status of the problem. We do not agree with that opinion, for several reasons. After more than three centuries of study, it is clear that the origin of Neotropical biodiversity and of large-scale diversity patterns (e.g., the latitudinal diversity gradients) are not simple problems with a single explanation (Brown 2014). Any attempt to reduce global macroecological patterns to a particular cause runs the risk of being inaccurate or unrealistic. It may be that a general theory on Neotropical biodiversity is just a human expectation or construct, and that explanations should be explored case by case, biome by biome, or region by region (Carnaval et al. 2014). Syntheses such as the presented here may be a good starting point to test these ideas. Avoiding the potential pitfalls of focusing the narrative on one or few sensationalist messages, regardless the soundness of its empirical support—a not unusual practice in current scientific literature—we created space for multiple views and data, allowing the contributing scholars to build and share their own perspective on Neotropical diversity and diversification. We consider this to be more useful for those readers interested in figuring out the true state of the art.

1 Book Plan

The book has been subdivided into three main parts: (I) Neotropical diversity and diversification, (II) regional biodiversity patterns and diversification processes, (III) taxon-based studies at the local and regional scales. An epilogue links the topics discussed here with the conservation of Neotropical biodiversity. The first part examines diversity patterns and diversification processes at a Neotropical scale using multitaxon approaches or broad taxonomic categories, such as plants, diatoms or birds, among others. The second part gathers chapters focused on more specific but large-scale Neotropical regions. Diversity patterns and diversification processes have been most intensely studied in the Amazon, and this region is often used as the preferred model for discussing the origin and maintenance of Neotropical biodiversity. In Part II, however, we ensured to include areas that go beyond the Amazon basin, considering, for example, the Atlantic rainforests, the Mexican forests, and the Caribbean, a region that is rarely contemplated in studies about Neotropical diversification. The third section of the book includes studies of particular taxonomic groups (amphibians, reptiles, flightless birds, angiosdperm families) on specific Neotropical areas such as the Guayanan and the Andean highlands, or selected river catchments. Below, we briefly summarize the content of chapters of the main book sections.

Part I: Neotropical Diversity and Diversification

This part is composed of eight chapters.

Rull introduces the focus area of this book, and discusses Neotropical diversification from historical and conceptual perspectives. Four historical steps are recognized, namely the discovery of the latitudinal biodiversity gradients, the biogeographic phase, the paleoecological phase, and the phylogeographical phase. Conceptually, this chapter explores the main handicaps that have hindered the progress towards an explanation for the comparatively high Neotropical biodiversity, with emphasis on the shifting from one paradigm to another, the geographical and taxonomic extrapolation of particular case studies, the selection of biased evidence to support different hypotheses, the identification of Pleistocene diversification with the refugial hypothesis, and the inference of diversification mechanisms from solely diversification timing. Some solutions are proposed to circumvent these drawbacks.

Baker et al. follow Rull's opening by revisiting the Haffer's refuge hypothesis in light of the paleoclimatic evidence gathered during the last 50 years. Their message to us is that the role of Quaternary climatic changes on Neotropical diversification should be reevaluated. The authors emphasize two main features that should be considered in addition to the glacial-interglacial cyclicity, namely the existence of an east-west precipitation dipole due to precessional cycles of ~20-kyr period, and the occurrence of phases of anomalously high rainfall of ~1500-y recurrence consistent with cold North-Atlantic stadials (Dansgaard-Oeschger cycles). The combination of these climatic modes conform a much more complex scenario of climatic change in time and space than originally proposed by the refuge hypothesis.

The biological significance of climatic transitions, this time in space, is the focus of the next chapter. Nascimento et al. investigate the effect of the Tropic of Capricorn in the distribution of Neotropical communities, particularly comparing the changes observed in the composition of biological communities in geographical space. By cleaning and analyzing nearly 15 million records of 124,066 species, all available online, they show that only in the cis-Andean region one can observe a latitudinal gradient in the degree of species turnover. Contrary to previous taxon-specific analyses, this chapter suggests that the tropical-subtropical transition does not influence the rate at which turnover indexes change in South America—at least not when data from multiple groups are combined. Although the data indicate that the replacement of species is largely responsible for changes in community composition, the Tropic of Capricorn does not seem to represent a common marker of concerted changes in community turnover at this large scale.

Villalobos et al. also focus on large-scale patterns of biodiversity in geographic space, but they utilize new methods in the field of macroecology to incorporate evolutionary thinking into studies of species ranges. They illustrate these novel approaches with data from the Furnariides, a model group of Neotropical birds. Their contribution shows that multiple metrics now enable the inclusion of speciation and extinction processes into data analyses and provide coherent results and insights. The chapter ends with a discussing of ways to expand the field by incorporating lineage dispersal over evolutionary time as a major contributor to geographic diversity gradients, and the promise of simulation studies in historical inference.

Similar to Nascimento et al., Maestri and collaborators combine data from thousands of species and explore possible correlates of biodiversity patterns at a continental scale. Yet, they specifically explore how the spatial distribution of evolutionary history compare across major groups of organisms. For that, they combine phylogenetic data and occurrence information from 1100 + species of vertebrates whose evolutionary histories are primarily confined to the Neotropics. They find that the spatial patterns of phylogenetic dissimilarity between areas are very similar across groups, despite the fact that they have different ecological requirements and ages. Their analysis also suggests that biogeographical relationships across regions explain phylogenetic similarity patterns better than present-day environments.

Nores discusses the biogeographical patterns observed in the birds of tropical and subtropical South America. Using range data, he introduces and discusses major geological and climatic events in the Neogene and the Pleistocene that are believed to have impacted these fauna, including the uplift of the Andes (splitting pan-Amazonia into a trans- and a cis-Andean regions, and leading to the formation of large Amazonian rivers), and the closing of the Isthmus of Panama (allowing for faunal interchange, increase in diversity, and new radiations). He argues that Amazonia and other tropical forests remained predominantly forested over time, even during glacial periods, and claims that marine transgressions in the Tertiary and lower Quaternary, tied to Amazonian rivers, were important by fragmenting species ranges and leading to speciation. Nores ends this chapter alerting us to the significant drop in species diversity and abundance recently observed in South America, particularly in protected areas of Amazonia, the lower Eastern Andean slopes, and the Pacific Forest—and likely tied to anthropogenic climate change.

Brower and Garzón-Orduña also explore the history of another widespread group of animals—Neotropical butterflies—but through the use of molecular data. Because the fossil record is too scarce to study the age of origin of these species, the authors estimate time of speciation through phylogenetic trees calibrated with relaxed molecular clocks. Brower and Garzón-Orduña compile clock-based age estimates for six butterfly families, at a Neotropical level, containing ~800 species in ~100 genera. The authors conclude that crown ages, that is, the ages of the initial diversification of the genera, range from Oligocene to Miocene, whereas species pairs diverged mostly during the Pleistocene. The analysis suggest that Neotropical butterflies are relatively young, and that Pleistocene climate changes likely played a significant role, whereas biogeographically significant geological events in the Miocene only set the stage for their subsequent evolution.

While studies on the diversity and community composition of macroorganisms are abundant, this is not the case for microorganisms, especially in the tropics, where this type of surveys are particularly scarce. Benito and Fritz fill this gap in our book, reviewing the diversity and distribution of Neotropical diatoms using a newly created database for tropical South America, which also includes geoclimatic and limnological parameters, as environmental drivers of community change. Using island biogeography and metacommunity approaches, the authors identify several aspects that may influence biogeographical patterns in diatoms, notably differences in regional floras, latitudinal gradients in species richness, dispersal ability, and environmental filtering. They also emphasize the need for incorporating diatoms into macroecological frameworks for studying community assembly in the Neotropics.

Part II: Regional Biodiversity Patterns and Diversification Processes
This part has 11 chapters that are sorted geographically.

Cracraft et al. revisit Amazonian diversification in relation to the origin of modern landscape features modeled by tectonic activity, drainage evolution, and climatic change. These authors observe that, although current Amazon landscapes were shaped during the last 4–5 million years, much of bird, mammal and butterfly diversity, at species level, originated during the last 1–2 million years. Therefore, the more significant landscape changes for the evolution of the current Amazon biota should have occurred over the last 1–2 million years. Cracraft et al. conclude that many mysteries still remain in relation to the origin of current Amazonian biodiversity, and that further research should be transdisciplinary and integrative, thus avoiding simplistic explanations. Phylogeographic analyses of individual clades should be replaced by among-clade studies that consider a wider spatial context and the differential action of landscape changes on different clades.

To address this question at a smaller geographic scale, Capurucho and collaborators focus on an under-explored region of Amazonia: the White Sand Ecosystems (WSE). These are nutrient-poor, quartz-rich sandy soils patchily distributed in areas of low elevation areas. Gathering data on community composition and the phylogeny of birds and plants, the authors demonstrate that the WSE present a unique assemblage of species. Although these areas pre-date the Quaternary, the molecular data suggest that their Pleistocene history has been dynamic. In the face of anthropogenic change, the authors emphasize that the WSE's low resilience to disturbance, as inferred through their historical analysis, ought to be incorporated into conservation planning.

Many traditional studies of the origin of Amazonian biodiversity, like Cracraft et al. and Capurucho and collaborators', have focused on the action of natural drivers such as tectonic or climatic changes—but largely disregarded the contribution of human activities on local biodiversity patterns. Montoya et al. analyze the potential impact of pre-Columbian anthropogenic activity on today's diversity of Amazonian plants, based on a review of the recent literature. Pre-Columbian human influence began with plant domestication and possibly megafaunal extinction, and remained until European contact, as shown by several types of evidence, including landscape transformations (i.e., savanna expansions) and changes in forest composition and soil properties. However, most studies developed to date are of local nature and are unable to account for the high spatial heterogeneity of the Amazon region. The authors conclude that more transdisciplinary interaction is needed for a proper understanding of human contribution to the Amazonian biodiversity, with emphasis on the last few millennia.

Azevedo and collaborators focus on the history of another key Neotropical ecosystem: the South American savannas. By combining data from the fossil record, paleoenvironmental proxies, and phylogenetic information of plants and animals, the authors claim that the phenomenal diversity of the South American Neotropical savannas did not emerge from a long history of stability. Paleoenvironmental data indicate that appropriate environmental conditions for the occurrence of savannas were in place since the middle Miocene, in agreement with phylogenetic analyses that recovered that as the beginning of the diversification phase of C4 grasses. However, the review finds that woody plants diversified only a few millions year later. This time lag, say the authors, may indicate that trees adapted to the fire-prone savannas reached ecological dominance only after the emergence of open habitats and their associated fauna.

Colevatti et al. review the biogeographic patterns and the diversification timing of Angiosperms of the South American Dry Diagonal (SADD), an open vegetation belt extending from northeastern Brazil to northern Argentina that includes several xeric biomes. The main goal is to unravel whether the Neogene tectonic/paleogeographic events or the Quaternary climatic changes were the main responsible for biotic diversification. Despite the low proportion of species studied, the authors highlight some regular patterns, notably that major lineage divergences occurred during the Pliocene but most intraspecific divergence took place during the Early and Middle Pleistocene. This suggests that Quaternary climate changes played a significant role but the authors emphasize that species responded differently to these environmental shifts and challenge the hypothesis of glacial refugia and the Last Glacial Maximum (LGM) on the diversification of the SADD Angiosperms.

Guedes et al. discuss the role of the major Non-Andean South American mountain ranges as biodiversity cradles and museums, emphasizing the Serra do Mar Range, the Mantiqueira Mountains, the Espinhaço Mountains, the Northeastern Highlands, the Central Brazilian Highlands, and the Pantepui region. Data published to date show that the Neogene and the Quaternary were important periods for the diversification for many terrestrial groups, resulting in the high endemism that today characterizes the non-Andean mountains. They also discuss how biotic interchange among the mountain ranges themselves, and the Andes, resulted in the patterns observed today.

Peres et al. zooms into the patterns of diversity and endemism of the ecosystem that encompass three of the mountain ranges discussed by Guedes and collaborators: the Brazilian Atlantic rainforest. Their review indicates that higher species richness is often found in the topographically complex coast of Rio de Janeiro and São Paulo, including the Mar and Mantiqueira ranges discussed by Guedes, though not exclusively (e.g. coastal Bahia). Data from multiple groups of plants and animals demonstrate striking differences in the composition of communities in the northern vs. the southern half of the forest, although the exact point of community turnover varies among groups. Several areas of unique taxonomic composition and endemism have been consistently identified across plants and animals, and those defined by organisms with low dispersal ability, such as terrestrial invertebrates, are nested within the areas identified by groups of higher ability to disperse, like birds. By contrasting regional diversity patterns and the processes that led to them, the authors argue that both climate and landscape changed acted in combination to lead to the astonishing diversity of this hotspot. They also argue that while the spatial patterns of species and genetic diversity are similar across plants and animals, the processes that led to them—and their timing—are widely different.

Silveira et al. argue that the study of Neotropical diversity has largely focused on relatively recent geological environments such as Amazonia or the Andes, whereas older and extremely nutrient-poor settings have been traditionally overlooked, despite their outstanding biodiversity and endemism patterns. This is the case of the so called Old Climatically-Buffered Infertile Landscapes (OCBILs) and, specifically, of the Brazilian "campos rupestres", which Silveira et al. use to propose a theoretical diversification framework including mechanistic explanations that link ecological and evolutionary processes to vegetation patterns and functional traits. This example is proposed as a model to demonstrate how the OCBIL theory and its predictions can help understanding the complex history of Neotropical plant diversification and inform their conservation.

Rull and Vegas-Vilarrúbia focus on another area included within the OCBIL category, as is the biogeographical province of Pantepui, formed by the assemblage of flat summits of the Guiana table mountains (tepuis), situated in the confluence of the Amazon and the Orinoco basins, between approximately 1500 and 3000 m elevation. Pantetui is one of the few pristine areas remaining on Earth and has been considered a natural laboratory for the study of Neotropical biodiversity. The diversity and endemism patterns of Pantepui are outstanding and its biota and ecosystems are unparalleled across the planet. The authors use the most recent update on the biodiversity of Pantepui, including algae, bryophytes, vascular plants, aquatic insects, butterflies, scorpions, land snails, amphibians, reptiles, birds, mammals and vertebrate parasites, to take the first steps towards a biogeographical, ecological and evolutionary synthesis on the origin of Pantepui biota and ecosystems. Direct (notably fire) and indirect (global warming) threats to Pantepui biodiversity are also analyzed.

Reginato and Michelangeli explore the data emerging from rich, well-curated, and digitized plant collections from Eastern Brazil to advance on the review provided by Peres and collaborators and identify areas of unique community composition (or bioregions). To enable analyses across groups with striking different dispersal abilities, diversity and endemism, they provide three levels of grouping—and the small, intermediate, and large spatial scales. The intermediate-level analysis splits E Brazil into 10 bioregions, which cluster into 5 super-bioregions at the larger scale, and which can be further subdivided into 23 sub-regions, at the smaller scale. High diversity and endemism is flagged both in the mountain ranges of the Serra do Mar and Mantiqueira discussed by Guedes and collaborators, but also in the campos rupestres system introduced by Silveira et al., and lowland areas of Bahia, as described by Peres et al. The rasters of the inferred bioregions are available for download and use by the scientific community at large.

Figueroa-Rangel and collaborators transport us north to Central America. Two major types of Mexican forests have been recognized: the Neartic forests, in the north, and the Neotropical forests, in the south, separated by a transition zone called the Trans-Mexican Volcanic Belt (TMVB). Figueroa-Rangel et al. review the long-term plant diversity changes along an altitudinal transect of the TMVB and the

Neotropical forests, ranging from sea level to the highlands (~3900 m). For that, they use a variety of diversity indices applied to Holocene pollen records. The results reveal that Neotropical forests are more diverse than forests with Holarctic affinities, and that Holocene climate change has been decisive for the shaping of these diversity patterns.

Finally, Roncal et al. discuss an important but often overlooked Neotropical biodiversity hotspot: the Caribbean archipelago. The origins of the high diversity observed in this area is very poorly understood. Roncal et al. use a combination of time-calibrated phylogenetic studies and biogeographical inferences to test the main hypotheses erected to explain Caribbean plant diversity, namely Late Cretaceous vicariance, Oligocene GAARlandia colonization, recurrent Cenozoic long-distance colonization, and in situ speciation. The authors conclude that plants repeatedly colonized the Caribbean islands during the last 60 million years (roughly, the Cenozoic) by long-distance dispersal from the American mainland, especially from Central and South America. They also raise doubts on the existence of GAARlandia, a hypothetical emerged land connecting the present Greater Antilles (Cuba, Hispaniola, Puerto Rico, Jamaica), which would have promoted plant migration among them.

Part III: Taxon-Based Studies at the Local and Regional Scales
This part has eight chapters, each devoted to a unique taxonomic group and its diversification history in a key Neotropical region.

Señaris and Rojas-Runjaic describe and discuss diversity patterns of the herpetofauna (amphibians and reptiles) of the Venezuelan Guayana (also Guiana), one of the largest wilderness areas on the planet. They document how species richness decreases, and endemism increases, with elevation: 34% (amphibians) and 22% (reptiles) of the local fauna is endemic to upland areas, whereas widespread species, or forms with an Amazonian-Guianan distribution, dominate the lowland fauna. This distribution pattern makes conservation particularly difficult in the region, especially in the face of climate change—a topic discussed by the authors.

Exploring a similar topic, Cadena and Céspedes zoom into the processes leading to species turnover along elevational gradients, but they use the Andes as a model system. Combining phylogenetic and occurrence data from a group of passerine birds, they find that most species have small elevational ranges and that those taxa that are more closely related to each other have similar elevational distributions across mountains. Conversely, those species that replace each other along a mountain are distantly related. Based on this pattern, the authors propose that speciation happened mostly in allopatry, across mountains, instead of not along environmental gradients within mountains—suggesting that species accumulation in montane environments largely reflects colonization instead of in situ divergence.

Leubert and collaborators also study patterns of species distributions, this time with the aim of understanding niche evolution in South America. Specifically, they use data from ten plant clades to study a well-documented disjunction: that between the southern Andes (SA) and southern Brazil (SB). These areas tend to have higher precipitation relative to their intervening areas, where no species of the disjunct clades occur. Comparing divergence times between SA and SB, obtained from phylogenetic studies, they show a wide range of split times across groups—from 27 to 1 Ma. This, they argue, demonstrates that SA-SB disjunctions reflect different clade histories as opposed to a single explanation behind the common pattern observed—much like Peres et al.'s findings for coastal Brazil. Comparing the realized climatic niches of species across SA and SB, however, they find that they do not overlap—suggesting that niche evolution may have occurred across multiple groups. This hypothesis, however, remains to be tested.

Moraes and collaborators transport us to a very distinct environment to investigate the potential role of another environmental barrier on the distribution of local diversity—the Tapajós River, in Amazonia. Using molecular data from 16 species of amphibians and reptiles, sampled from across the Tapajós river basin, the authors find that most groups show spatial patterns of genetic diversity that are consistent with the present-day location of the river: they observe high levels of genetic divergence across river banks. However, much like Leubert et al., the range of riverine-associated divergence times is wide and continuous—suggesting a dynamic history and multiple events of vicariance or colonization.

Recoder and Rodrigues focus on another Neotropical river system to test how changes in geomorphology and climate may have impacted local patterns of diversification—the São Francisco. Located within the domain of the semi-arid Caatinga, the São Francisco river is known for its extensive paleodunes and for the spectacularly diverse sand-dwelling fauna endemic to the dune region. The authors compile existing and new data from local reptiles to ask whether a paleolacustrine vicariant diversification hypothesis, once proposed to explain the origin of this endemic fauna, agrees with current patterns of diversification. That hypothesis suggested that the river flowed into an interior lake during the last glacial maximum, but later made its way out into the Atlantic, isolating populations on opposite margins. The phylogenetic data gathered here show that this model is not sufficient to explain all pattern observed: divergence times differ across species, and some splits pre-date the Pleistocene. Again, Neotropical history is complex and cannot be attributed to a single event.

Ignacio de La Riva calls attention to the diversitication processes impacting high-Andean direct-developing frogs, particularly focusing on species occurring from Colombia to Bolivia. Through his review, he clarifies how similar life histories and ecological niches are associated with a particular morphology: local frogs are small-bodied, have short legs, and simple digital tips, likely reflecting adaptions to live in humid mossed habitats of the cloud forests and paramos. According to the author, new studies indicate that this is a convergent morphology across distantly related groups, and that the limited information available about these species render inferences about their evolutionary history, and the history of these unique Neotropical high-elevation areas, tentative at best.

Olave and collaborators also discuss morphological evolution, but are able to incorporate extensive molecular data available for the Liolaemini—the most diverse lizard group of the southern half of South America—in their chapter. Combining molecular and morphological data, they identify different patterns of diversification among the three genera that compose the clade. They propose that the richness of one of the genera, the species-rich *Liolaemus*, is associated with low extinction rates over time—while the opposite is observed in *Phymaturus*. Based on their analyses, they also argue that natural selection is tied to body size evolution and net diversification in this group, but notice that the trajectories and processes are different across the genera studied.

In the last chapter of Part III, Vargas and Dick focus on a group of 10 genera (ca. 232 species) of Lechytidaceae—the third most abundant family of trees in Amazon forests. Using a time-callibrated phylogeny of more than 100 species, they discuss the diversification history of this important group. Their results suggest that these plants dispersed from the Paleotropics long after the Gondwana break-up, and that most clades diversified in the Miocene. The analysis places the Guayana floristic region as ancestral to several clades, agreeing with Señaris and Rojas-Runjaic about the evolutionary relevance of that area. The results also suggest that this clade diversified through a pattern of steady accumulation of species, as opposed to well-marked pulses in response to specific geomorphological or climatic events.

Epilogue

Closing our book, Carnaval compiles several conservation-relevant messages discussed throughout this volume. Although not the main goal of this contribution, the author argues that one can and should leverage the large, heterogeneous and collective body of knowledge presented in these chapters to inform and guide conservation in the tropical Americas. She emphasizes some of the messages that we are about to hear in the chapters to come, particularly those about the status and patterns of diversity in the Neotropics, along with the authors' insights about the ecological and evolutionary process, and the biological mechanisms, that underscore them. This understanding, she argues, may allow us to better anticipate the behavior of the Neotropical biota as we move forward—a step which will require us not only to integrate data and insight across the biological and physical sciences, as we have done here, but also to reach out to the social sciences. But that is a topic for another book.

References

Brown JH (2014) Why are there so many species in the tropics? J Biogeogr 41:8–22

Carnaval AC, Waltari E, Rodrigues MT et al (2014) Prediction of phylogeographic endemism in an environmental complex biome. Proc R Soc B 281:20141461

Rull V (2011) Neotropical biodiversity: timing and potential drivers. Trends Ecol Evol 26:508–513

Part I
Neotropical Diversity and Diversification

Chapter 2
Neotropical Diversification: Historical Overview and Conceptual Insights

Valentí Rull

Abstract Studying the causes of biological diversification and the main environmental drivers involved is useful not only for the progress of fundamental science but also to inform conservation practices. Unraveling the origin and maintenance of the comparatively high Neotropical biodiversity is important to understand the global latitudinal biodiversity gradients (LBGs), which is one of the more general and conspicuous biogeographical patterns on Earth. This chapter reviews the historical development of the study of Neotropical diversification, in order to highlight the influence of methodological progress and to identify the conceptual developments that have appeared through history. Four main steps are recognized and analyzed, namely the discovery of the LBGs by pioneer naturalists, the first biogeographic studies, the inception of paleoecology and the recent revolution of molecular phylogeography. This historical account ends with an update of the current state of the study of Neotropical diversification and the main conceptual handicaps that are believed to slow progress towards a general theory on this topic. Among these constraints, emphasis is placed on (1) the shifting from one paradigm to another, (2) the extrapolation from particular case studies to the whole Neotropics, (3) the selection of biased evidence to support either one or another hypothesis, (4) the assumption that Pleistocene diversification equals to refuge diversification, and (5) the straightforward inference of diversification drivers from diversification timing. The main corollary is that the attainment of a general theory on Neotropical diversification is being delayed by conceptual, rather than methodological causes. Some solutions are proposed based on the Chamberlin's multiple-working-hypotheses scheme and a conceptual research framework to address the problem from this perspective is suggested.

Keywords Latitudinal biodiversity gradients · Diversification drivers · Historical developments · Conceptual insights · Biogeography · Paleoecology · Phylogeography

---

V. Rull (✉)
Institute of Earth Sciences Jaume Almera (ICTJA-CSIC), Barcelona, Spain
e-mail: vrull@ictja.csic.es

V. Rull, A. C. Carnaval (eds.), *Neotropical Diversification: Patterns and Processes*, Fascinating Life Sciences, https://doi.org/10.1007/978-3-030-31167-4_2

1 Introduction

This paper reviews the historical development of research on the origin of Neotropical biodiversity under a conceptual perspective, in an attempt to identify the main factors that have hindered the attainment of a holistic theory on this subject. The Neotropics, or the American tropics, are among the most biodiverse regions in the world. A synthetic perspective on the origin and maintenance of such diversity would be relevant to understand the global Latitudinal Biodiversity Gradients (LBGs), one of the most pervasive global macroecological patterns on Earth (Lomolino et al. 2010; Cox et al. 2016). In addition, understanding how extant biodiversity and its major geographical patterns have been generated and sustained is needed for better informed conservation and restoration practices (Winter et al. 2013; Dietl et al. 2015).

The review emphasizes the environmental drivers and the evolutionary processes that have contributed to the origination of Neotropical species, ecological mechanisms favoring species' coexistence and biodiversity maintenance will be addressed when appropriate but are treated in more depth elsewhere (Wright 2002; Mittelbach et al. 2007; Cannon and Lerdau 2015; Usinowicz et al. 2017). Following the definitions of Pielou (1975), the measure of diversity employed in this paper is species diversity, that is, the number of species in a given geographical context. Other measures such as ecological diversity, which also considers the relative abundance of each species, are not analyzed. Species richness is here considered at local (α-diversity) and regional (γ-diversity) levels; diversity trends along environmental gradients or across regions or ecosystems/biomes (β-diversity) and other spatial patterns (Anderson et al. 2011; Wiegand et al. 2017) are not under scrutiny.

Technical or specialized jargon and concepts have been avoided or appropriately explained, due to the transdisciplinary character of the topic, the variety of research fields and methodological approaches involved and the general interest that the topic may have for a wide scientific audience, including advanced students. However, basic knowledge on general disciplines (as for example geology, evolution, ecology, biogeography or genetics, among others) is required. It should also be stressed that, given the great amount of literature available on Neotropical diversification and the continued publication of numerous papers on this subject, the literature provided here is by no means exhaustive but a selection aimed to be, at least, representative. The use of papers on diversification studies of particular taxonomic groups and/or geographical regions has been kept to a minimum; emphasis is placed on general studies, revisions, meta-analyses and other synthetic views that address the problem of Neotropical diversity from more general perspectives. It should also be stressed that evidence-based studies and models are preferred. Top-down modelling approaches using flawed biological assumptions, as for example species' equivalence—i.e., the purported irrelevance of specific biological traits and particular niche features—to favor previously established stochastic processes (e.g., Colwell and Hurt 1994; Hubbell 2001; Scheffer et al. 2018), are not analyzed. This paper is mainly a personal approach to the problem and uses data, concepts and arguments already published by the author elsewhere, in which case the corresponding references are provided in order to avoid self-plagiarism and eventual double-publication claims (Gutiérrez and Block 2013; Rozensweig and Schnitzer 2013). This is the only reason why own author's citations are overrepresented with respect to others.

The paper has been subdivided into four sections for more clarity. In order to provide a worldwide perspective, the first part briefly summarizes the issue of the LBGs and their potential causes. The second part emphasizes the comparatively higher Neotropical biodiversity with respect to other tropical areas and how this has been framed in global conservation strategies. The third section provides a historical account on the development of hypotheses about Neotropical diversification and the nature of the evidence used to propose and to test them. The fourth section emphasizes several conceptual and methodological constraints that have slowed the progress towards a synthesis on the origin and maintenance of Neotropical diversity and proposes some potential solutions. Finally, some hints are provided on the possible future research trends on Neotropical diversification, with emphasis on the mentioned conceptual constraints.

2 The Latitudinal Biodiversity Gradients

The occurrence of Earth's biodiversity gradients from the species-rich tropics to the almost barren poles is a classical biogeographical topic since the late eighteenth century, when the first biogeographical explorations took place (Lomolino et al. 2010). Johann Forster, who travelled around the world with the famous Captain James Cook, was the first to note the latitudinal trends in plant diversity attributing this fact to the increased intensity of heat towards the tropics (Forster 1778). Since then, the LBGs have been recognized in many terrestrial and marine organisms, as well as in total biodiversity patterns, and is now considered a first-order biogeographical pattern (Rozenzweig 1995; Willig et al. 2003; Hildebrand 2004; Krug et al. 2009; Yasurara et al. 2012). The LBGs do not seem to be only a present-day-only feature; paleontological studies have shown that similar patterns were already present during several Paleozoic and Cenozoic phases, especially those characterized by global cooling or 'icehouse' phases. Contrarily, the Mesozoic 'greenhouse' phase seems to have been characterized by reverse patterns characterized by depleted tropical biodiversity and extra-tropical richness peaks (Mannion et al. 2014). Understanding the environmental drivers and the ecological and evolutionary processes and mechanisms involved in the generation and maintenance of the LBGs is still challenging (Mittelbach et al. 2007; Lomolino et al. 2010; Brown 2014).

In general terms, the LBGs are considered to be generated and maintained by coupled ecological and evolutionary processes. Geographical differences in origination (speciation), extinction and dispersal (*s. l.*) over evolutionary timescales seem to have been essential (Krug et al. 2009; Mannion et al. 2014). Ecological mechanisms related to predation, competition, colonization ability, mating behavior, germination, differential growth or habitat heterogeneity, among others, would have been responsible for minimizing extinction and favoring species coexistence thus contributing to the maintenance of a high biodiversity (Wright 2002; Mittelbach et al. 2007; Stein et al. 2014; Cannon and Lerdau 2015; Usinowicz et al. 2017). Stebbins (1974) asked whether the tropics are more diverse due to higher speciation rates (cradle hypothesis) or lower extinction rates (museum hypothesis). Others consider that net diversification—the balance between speciation and extinction—is higher in the tropics (Mittelbach et al. 2007). In addition to speciation and extinction, dispersal has also been considered crucial to explain the LBG (Jablonski et al. 2006). Whatever the case, the environmental drivers and the ecological and evolutionary processes involved in speciation, extinction and dispersal trends in both time and space are the object of intense debate.

Some hypotheses favor the dominant action of single factors of worldwide extent, as for example the energy-gradient hypothesis, according to which, the increasing solar energy from the poles to the equator would be responsible for increased tropical productivity thus promoting higher diversity (Connell and Orias 1964; MacArthur 1965). Others emphasize the combined action of environmental and ecological factors by assuming that higher tropical diversity is due to more stable climates, which favor ecological interaction and niche diversification, whereas lower extra-tropical diversity is due to the depleting action of harsh climates, notably the Pleistocene glaciations (Fischer 1960; Pianka 1966; Slobodkin and Sanders 1969). The time-area hypothesis proposes that the tropics have accumulated more species than the temperate zones because they are bigger and have remained in its present geographical location for more time (Farrell et al. 1992; Ricklefs and Schluter 1993). The niche-conservatism hypothesis considers that tropics are both a cradle and a museum and, as a consequence, their older species largely retain their ancestral traits and niches (Wiens and Donoghue 2004). The out-of-the-tropics hypothesis contends that the tropics are more diverse because most taxa originated in the tropics and migrated toward higher latitudes but still remaining in the tropics (Jablonski et al. 2006; Jansson et al. 2013). At local and regional scales, migration—the dispersal across hospitable terrains (Pielou 1979)—can be important, either as species' input (immigration) or output (emigration) (Gaston and Spicer 2005).

Until recently, these and other hypotheses about the origin and maintenance of the LBGs were tested using mainly present-day biogeographical evidence (sometimes coupled with paleontological and paleoecological data) and theoretical modelling. Yet, the recent development of molecular phylogenetic tools has revolutionized the field. Empirical testing using these new methods has supported a number of the existing hypotheses or combination of them (Brown 2014; Kerhoff et al. 2014; Antonelli et al. 2015; Fine 2015; Schluter 2016) but the debate continues. According to Hurlbert and Stegen (2014) and Jablonski et al. (2017), the tendency to focus on single causal factors and processes is blocking progress towards a general LBG theory, which needs a more synthetic and integrative approach. Such synthesis should consider habitat and niche features, ecological interactions, evolutionary trends and biogeographical shifts, as well as their variability along environmental gradients and across temporal scales (Pontarp et al. 2019). Whittaker et al. (2001) and Willis and Whittaker (2002) question the possibility of attributing global

Table 2.1 A hierarchical framework for processes influencing biodiversity (simplified from Willis and Whittaker 2002)

Spatial scale	Species richness	Environmental variables	Temporal scale
Local	Within communities and habitat patches	Fine-scale biotic and abiotic interactions	1–100 years
Landscape	Between communities	Soils, elevation, peninsula effect	100–1000 years
Regional	Within continents	Radiation budget, water availability, area, latitude	Last 10,000 years
Continental	Across continents	Pleistocene glacial/interglacial cycles, mountain-building events, ardification	Last 1–10 million years
Global	Between continents and geographical realms	Continental drift, sea-level changes	Last 10–100 million years

biodiversity patterns to a single causal explanation and propose a hierarchical framework that considers the more influential processes and their corresponding spatio-temporal scales (Table 2.1).

3 The Neotropics

In the American continents, the LBG is particularly apparent (Fig. 2.1). This paper is concerned with the tropical part of these continents, the ecozone known as the Neotropics (Schultz 2005), lying between the tropics of Cancer and Capricorn (Fig. 2.2). Therefore, the term 'Neotropical' used here refers to the Neotropics and should not be confused with the Neotropical biogeographic realm, encompassing not only the tropical Americas but also the whole South America (Lomolino et al. 2010). Defined in this way, the Neotropics range from central Mexico and the Greater Antilles, to the north, to the southern end of Bolivia, to the south. The Neotropics exhibits a complex topography, ranging from the sea level to above 5000 elevation (Fig. 2.2) and encompasses a varied range of biomes, from the driest deserts to the more humid forests on earth (Fig. 2.3).

The amazing biodiversity of the Neotropics, as compared to the rest of the planet, is well illustrated by higher plants, which have been considered as surrogates of general biological diversity. For example, it has been estimated that Amazonian forests contain more tree species diversity in a square kilometre than do all the temperate forests of Europe, North America, and Asia combined (Usinowicz et al. 2017). The Netropics has also been considered the more species-rich tropical region, with ca. 100,000 species of seed plants, almost 40% of the world's total (Antonelli and Sanmartín 2011). These patterns are similar for many animal species including amphibians, mammals, birds, butterflies and reptiles, among others (Antonelli et al. 2015). Therefore, the Neotropics could be considered the most biodiverse region of

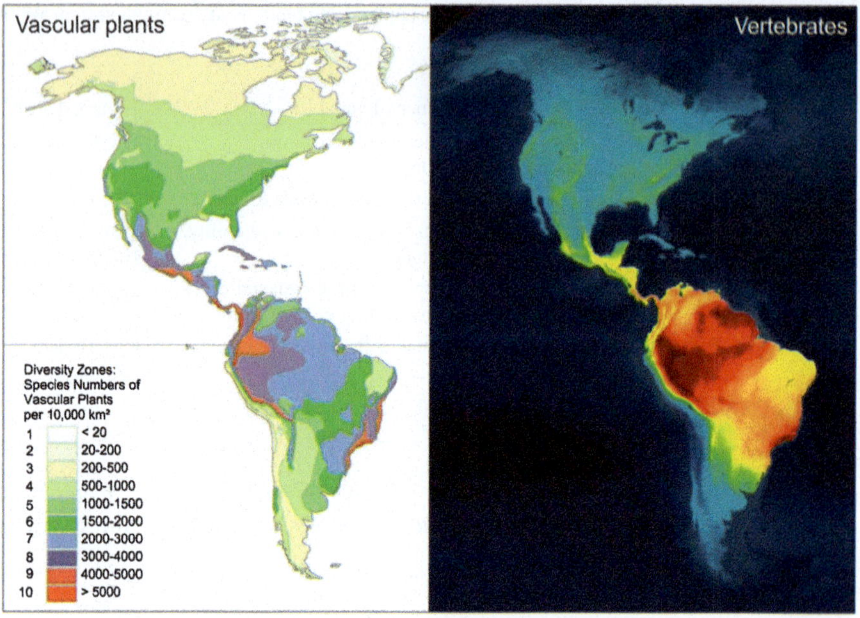

Fig. 2.1 Latitudinal distribution of species richness among vascular plants (left) and vertebrates (right) across the American continents. Vascular plant biodiversity is expressed in species number per unit area (adapted from Mutke and Bathlott 2005) and vertebrates as the total number of species, which is maximum at the brown side of the color spectrum and minimum at the blue side of the color spectrum (adapted from Mannion et al. 2014)

the world and, therefore, a well suited place to study the drivers, processes and mechanisms involved in the generation of the LBGs.

The Neotropics is also a keystone region for biodiversity conservation as manifested, for example, in the occurrence of six of the 25 world's biodiversity hotspots considered as conservation priorities, namely the Caribbean, the Mesoamerican, the Atlantic forest, the Cerrado, the Tropical Andes and the Chocó/Darién/Western Ecuador areas (Fig. 2.4). A hotspot has been defined as a region holding exceptional biodiversity levels, mainly of vascular plants and vertebrates, and suffering exceptional rates of habitat loss (Myers et al. 2000). The Amazon rainforests are not included in this classification because, despite their amazing biodiversity, they have not been considered to be under exceptional risk of habitat loss. However, given the current rates of rainforest destruction in Amazonia, this region might eventually be included in the list. If so, the whole Neotropics would be considered a mega-biodiversity hotspot. In addition, the Neotropics encompasses 6 of the 17 world's countries defined as megadiverse, namely Brazil, Colombia, Ecuador, Mexico, Peru and Venezuela (Fig. 2.5). As a whole, these 17 countries account for ~70% of the total world's biodiversity and bear at least 5000 endemic species, thus deserving special conservation priorities (Mittermeier et al. 1997).

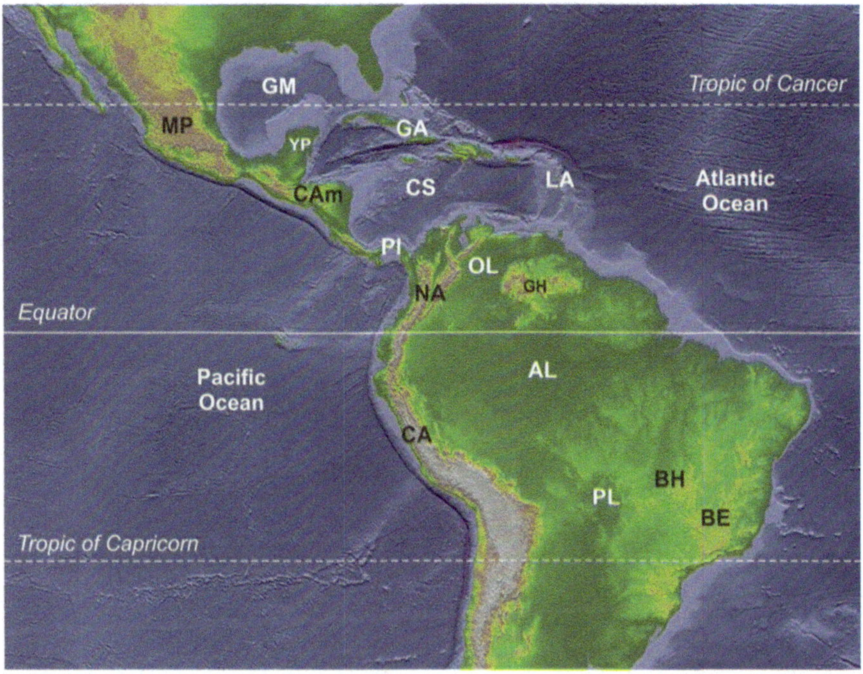

Fig. 2.2 Topographic/bathymetric map of the Neotropics with the main physiographical regions. *AL* Amazon Lowlands, *BE* Brazil East Coast Range, *BH* Brazilian Highlands, *CA* Central Andes, *Cam* Central America, *CS* Caribbean Sea, *GA* Greater Antilles, *GH* Guiana/Guayana Highlands, *GM* Gulf of Mexico, *LA* Lesser Antilles, *MP* Mexican Plateau, *NA* Northern Andes, *OL* Orinoco Lowlands, *PI* Panamá Isthmus, *PL* Pantanal, *YP* Yucatán Peninsula. Base map: NOAA National center for Environmental Information (https://ngdc.noaa.gov/mgg/topo/globega2.html)

Therefore, although the purely academic study of the ecological and evolutionary causes of the high Neotropical biodiversity makes sense by itself, it also has fundamental practical significance in the face of direct (e.g., deforestation and other forms of habitat destruction) and indirect (notably anthropogenic global warming) threats resulting from human activities. This makes of Neotropical biodiversity research a central activity, from basic and applied perspectives, towards a more sustainable use of natural resources.

4 Historical Outline

For more clarity, the historical account on the study of the origin of Neotropical biodiversity has been subdivided into four main periods: (1) the discovery of the American LBG by pioneer naturalists, (2) the first attempts to explain this LBG using present-day biogeographical patterns, (3) the incoming of paleoecological

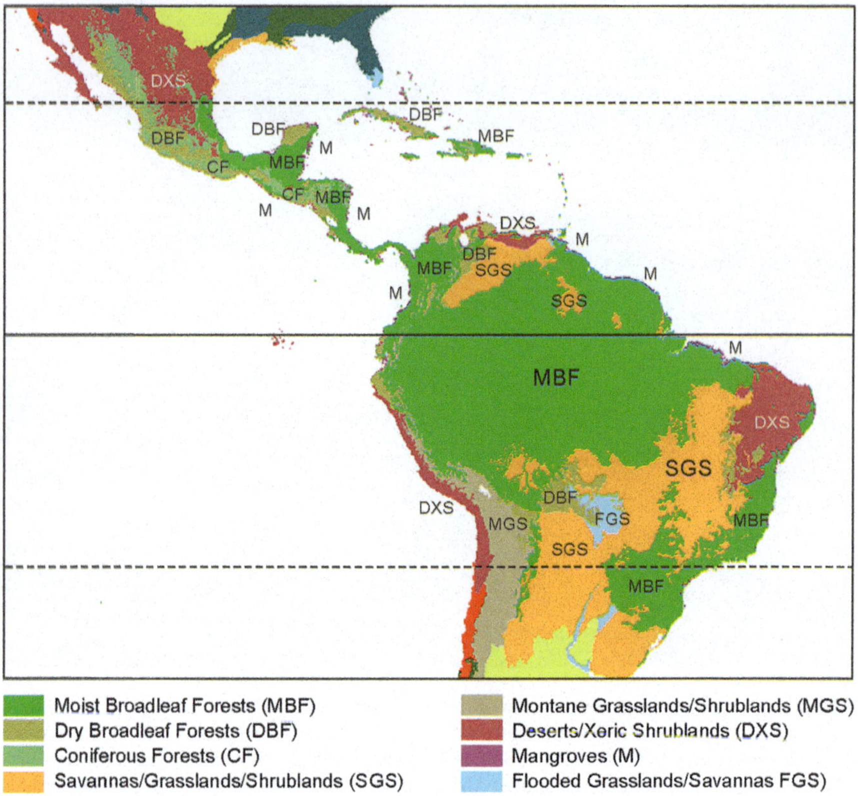

Fig. 2.3 Map of the Neotropics and its major biomes. Redrawn from Olson et al. (2001)

evidence to the investigations and (4) the appearance and the rapid development of molecular phylogenetic and phylogeographic methods. This does not mean that these approaches have replaced one another through history, as a number of them have eventually been combined to obtain a broader picture. This has been especially true in the last years, when some synthetic approaches have been attempted, which is briefly discussed at the end of this section.

4.1 The Beginnings

Few decades after the discovery of the general LBGs by Johann R Forster (1778), between 1799 and 1804, Alexander von Humboldt did similar observations in the New World. He noted an increase not only in the number of plant species but also in "structure, grace of form, and mixture of colors, as also in perpetual youth and vigor of organic life" (von Humboldt 1850) towards the tropics, and he attributed this fact to climatic constraints (Hawkins 2001). Humboldt also realized that the latitudinal

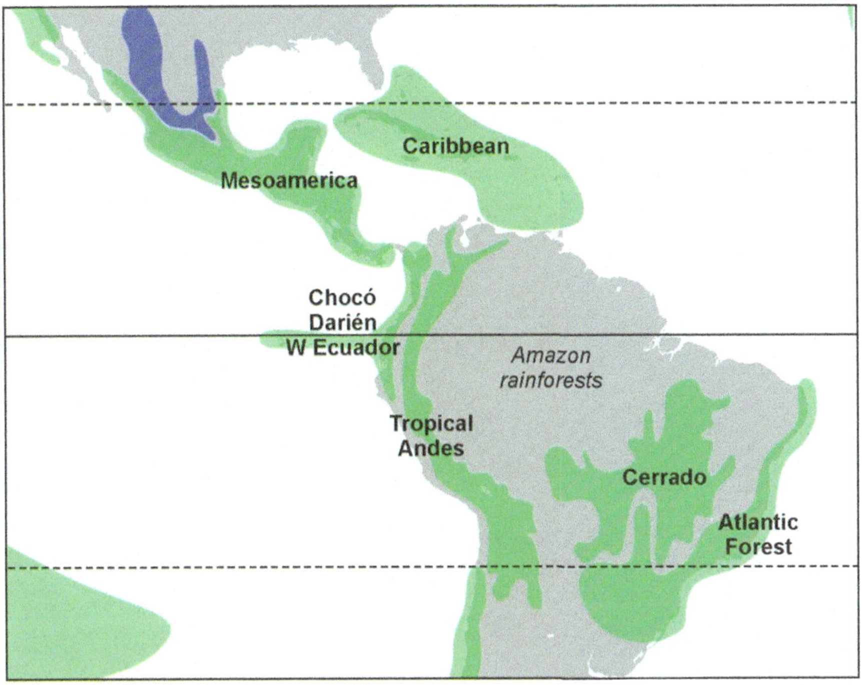

Fig. 2.4 Biodiversity hotspots (green areas) of the Neotropics, according to Myers et al. (2000). The Amazon rainforests are also indicated although they have not yet been considered a biodiversity hotspot (see text). Modified from an original map of Wikimedia Commons (https://en.wikipedia.org/wiki/Biodiversity_hotspot), licensed by Creative Commons 3.0

floristic gradients could be observed at more local scales in the form of elevational gradients across mountains (Lomolino et al. 2010). Charles Darwin (1839) and Alfred R Wallace (1853) also highlighted the luxuriance of plant and animal life in the Neotropics, especially in Amazonia. Wallace (1878) pointed out that these pioneer observations not only revealed a major biogeographical pattern, but also implicitly hypothesized about its potential causes, with emphasis on the increasing climatic severity from the tropics to the poles and the resulting intensification of the struggle for the existence (Lomolino et al. 2010).

Until the 1970s, the Neotropics—and the tropics, in general—were considered climatically stable areas that have been free from the influence of Pleistocene glaciations. In this context, diversification theories considering only rather static latitudinal energy gradients like the present ones dominated the scene. This view changed when widespread evidence of Pleistocene climatic shifts in tropical areas started to accumulate (Damuth and Fairbridge 1970; Emiliani 1971; Ab'Saber 1982). In the Neotropics, these environmental changes were considered to have fostered biotic diversification, mainly through sea-level changes, temperature-driven elevational migrations of montane ecological belts, and intense moisture changes in the lowlands (Simpson 1971). The first general theories about Neotropical

Fig. 2.5 Megadiverse countries (green areas, yellow text) of the Neotropics according to Mittermeier et al. (1997). Sovereign states from the Lesser Antilles: Antigua & Barbuda (AB), Barbados (B), Dominica (D), Grenada (G), Saint Kitts and Nevis (SK), Santa Lucia (SL), Saint Vincent and the Grenadines (SV), Trinidad and Tobago (TT)

diversification considering environmental instability appeared in the third quarter of the twentieth century and emphasized the role of speciation, extinction and dispersal under the influence of climate changes and topographic and paleogeographic shifts derived from continental drift.

4.2 *Biogeographical Inferences*

The first hypotheses consisted of evolutionary inferences from extant biogeographical patterns. Based on the present-day distribution of lowland Amazon rainforests' avifauna and the assumption of severe climatic aridity during the Plesitocene glaciations, Haffer (1969) proposed that these forests would have been fragmented into isolated patches, or refugia, in a sea of non-forested vegetation, which would have favored speciation by vicariance. More humid interglacial phases would have caused forest re-expansion and coalescence, thus promoting sympatric speciation and hybridization along secondary contact zones. The same model was proposed shortly after by Vanzolini and Williams (1970), while working on lizards. These

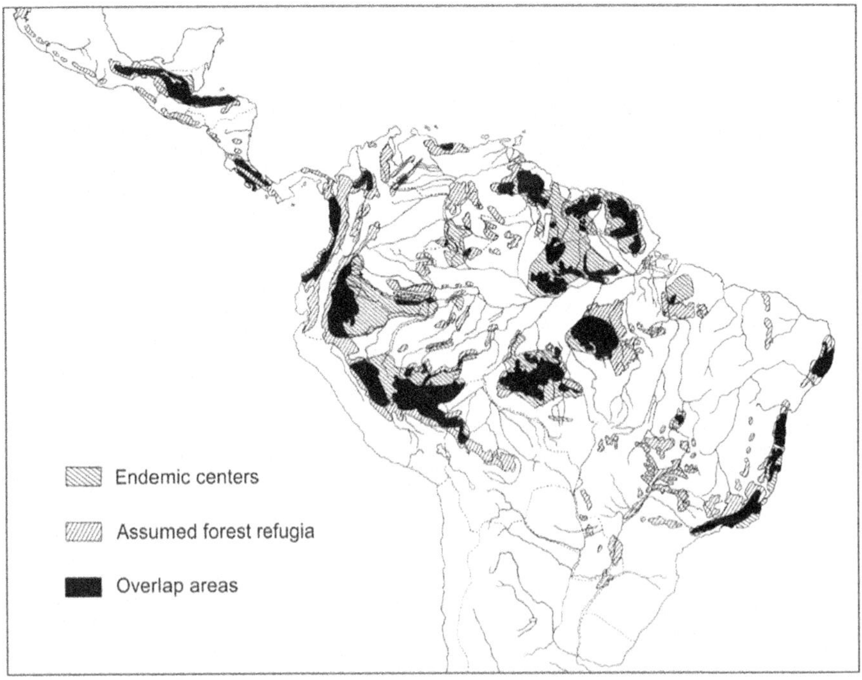

Fig. 2.6 Overlapping patterns of assumed glacial forest refugial areas, as deduced from paleoclimatic, pedological and geomorphological evidence, and present-day endemism centers of plants, butterflies and birds. Redrawn from Whitmore and Prance (1987)

trends would have repeated several times during the Pleistocene and, although eventual extinctions were not dismissed, this recurrence would have led to forest faunas significantly more diverse than the pre-Pleistocene ones. Based on the diversity and endemism patterns of Amazonian birds and the modern distribution of precipitation, Haffer (1969) identified nine areas as potential candidates for Pleistocene glacial refugia. According to this author, the refuge hypothesis would be applicable to other animal groups such as insects, amphibians, reptiles and mammals. Further biogeographical analyses on these and other groups of organisms (e.g. butterflies and some plant families) were considered to be consistent with the refuge hypothesis predictions and this diversification model became paradigmatic (Prance 1982; Whitmore and Prance 1987). The combination of paleoeclimatic, pedological, geomorphological and biogeographical evidence resulted in the mapping of the potential distribution of the presumed refugia for the lowland Neotropics (Fig. 2.6). A similar diversification scenario was proposed for the avifauna of the Andes, where interglacial upward migration of mountain belts would have led to the isolation of high-mountain biomes thus favoring vicariance, whereas glacial downward migration would have fostered coalescence and gene flow (Simpson 1971). Again, the main supporting evidence were present-day biogeographical patterns.

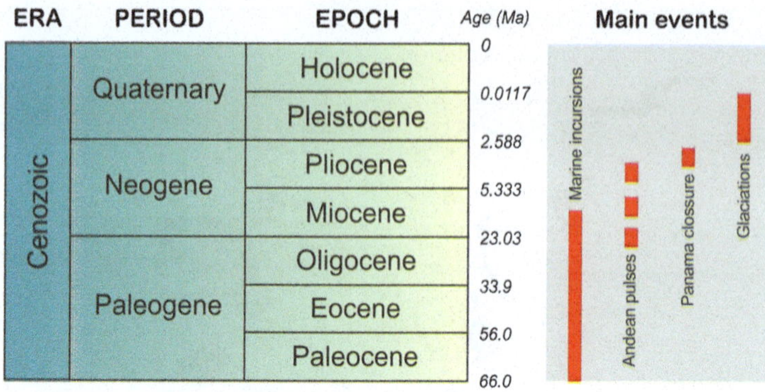

Fig. 2.7 Geochronological units of the International Stratigraphic Chart (Cohen et al. 2013, updated 2015), and the main tectonic and climatic events discussed in the text, according to Hoorn et al. (2010). The first peaks of north-central Andean building occurred at the Oligocene/Miocene transition (ca. 23 Ma) but the most intense pulses took place in the Middle Miocene (ca. 12 Ma) and the Early Pliocene (ca. 4.5 Ma). The closure of the Panama Isthmus occurred during the Pliocene (ca. 3.5 Ma)

According to Raven and Axelrod (1974), Pleistocene climate changes would have been important in the shaping of present-day Neotropical biodiversity, but pre-Pleistocene paleogeographic changes linked to continental drift were considered to be essential for the formation of modern biotas. These authors discussed the topic using the worldwide distribution of angiosperms and vertebrates, and their migrational possibilities during the Cenozoic, in the context of a changing physical context characterized by (1) dramatic changes in land-sea geographical patterns, (2) the uplift of new cordilleras and (3) other consequences of plate tectonics potentially responsible for the waxing and waning of migration pathways and barriers. The debate between the relevance of either Pleistocene climate changes or pre-Pleistocene (mostly Cenozoic) tectonic events on the shaping of Neotropical biodiversity started with Gentry (1982), who explicitly raised the dilemma in the title of its influential paper. Using the biogeographical patterns of the extant angiosperm flora, mainly at genera and family levels, this author concluded that the Pliocene closure of the Panama Isthmus and the Mio-Pliocene Andean uplift were the most significant events for the origin of the high Neotropical diversity (Fig. 2.7). The Panama Isthmus (Fig. 2.2) would have provided the pathway for northern temperate taxa to colonize South America in the so called Great American Biotic Interchange (GABI) (Marshall et al. 1982). The Andean orogeny would have created new montane habitats that allowed northern biotas to persist and migrate southward. According to Gentry (1982), "The evolutionary phenomena associated with the Andean uplift account for almost half of the total Neotropical flora and are thus largely responsible for the excess floristic richness of the Neotropics" (p. 589). Pleistocene climate changes, although discussed, were not considered to have the same importance as the mentioned tectonically-driven processes (Fig. 2.7). This author considered that co-evolutionary relationships with pollinators played a prominent role in Neotropical plant evolution.

4.3 The Inception of Paleoecology

In the late 1980s, Quaternary paleoecological evidence was added to the debate. After an extensive literature review, Colinvaux (1987) did not find enough evidence to support glacial aridity during the Last Glacial Maximum (LGM), occurred ca. 21,000 years ago. In addition, further palynological analyses documented the continuity of lowland Amazon forests during the LGM, which contradicted the predictions of the refuge hypothesis concerning forest fragmentation (Colinvaux et al. 1996). Instead, these forests would have changed their composition due to downward migration of montane species caused by cooling, moderate precipitation reduction and atmospheric CO_2 depletion (Bush 1994). Acting together, these phenomena could have provided the physical basis to explain the observed biogeographical patterns without the need for the refuge hypothesis (Colinvaux et al. 2000). This was called the disturbance-vicariance hypothesis. However, this view was not shared by all paleoecologists working on the Neotropics and the initial Haffer's idea of potential glacial forest refugia on current Amazon areas of high precipitation remained (van der Hammen and Hooghiemstra 2000). A debate settled that still persists nowadays between the defenders and the detractors of the refuge hypothesis.

The proponents of the disturbance-vicariance hypothesis vividly combated the refuge hypothesis and emphasized the importance of the Andean uplift and the stability of the Amazon forests through the Cenozoic (Colinvaux and De Oliveira 2001). These authors noted that the modern Amazon basin was formed during the Miocene, when the Andean uplift caused a major shift in the drainage patterns of northern South America. Indeed, until the early Miocene, most of the drainage of the present western Amazon region was directed to the north but in the late Miocene, the uplift that shaped the modern Andes drastically changed this pattern and formed the huge Amazon and Orinoco river systems draining to the Atlantic Ocean (Hoorn et al. 1995). In addition, some preliminary palynological data suggested that many taxa of the modern Amazon forests were already present in the middle Miocene and climates were warm tropical (Hoorn 1994). This, combined with their own evidence of forest continuity during the last glaciation, led Colinvaux and De Oliveira (2001) to propose that present-like Amazon rainforests had been present and stable since the middle Miocene, in agreement with the museum hypothesis of the high Neotropical biodiversity, as a consequence of reduced extinction due to high environmental stability. The refuge hypothesis was considered useful to explain modern biogeographical patterns in temperate areas but unlikely for the Neotropics (Willis and Whittaker 2000; Bush and De Oliverira 2006). The Neogene origin of modern Amazon rainforests was considered to be supported by palynological and paleobotanical data (Jaramillo et al. 2010).

Besides the refuge hypothesis, the disturbance-vicariance hypothesis and the hypotheses emphasizing tectonically-driven processes such as the Andean uplift or the closure of the Panama Isthmus, other existing proposals for Neotropical diversification involving Pleistocene and Neogene drivers and processes were summarized by Haffer (1997). According to the river hypothesis, the ancestral widespread and uniform populations of Amazonian animals were fragmented and isolated by the progressive development of the Amazon fluvial network during the Neogene and the early Quaternary, thus favoring allopatric speciation. A variant is the river-refuge hypothesis, according to which forest refugia would have been separated by non-forested vegetation and also by rivers, or by both, depending on the case. The gradient hypothesis predicted parapatric speciation across step environmental gradients and was assumed to be active in marginal populations even in the absence of physical disconnection. Nores (1999) observed a consistent pattern of high-diversity spots in areas above 100 m elevation in the Amazon lowlands and hypothesized that sea-level increases of this magnitude would have isolated these areas thus favoring vicariance. A variant of the refuge hypothesis is the vanishing refuge model, according to which, refugial areas should not necessarily be constant through time and may experience habitat changes thus providing new adaptive opportunities (Damasceno et al. 2014). Another possibility suggested by Rull et al. (1989) contends that Neotropical biota would have survived the cold and possibly arid glacial climates in small but diverse and widespread clusters, or microrefugia, under special local microclimatic conditions (see also Rull 2009, 2010).

4.4 The Molecular Revolution

In the last decades, the methodological refinement of molecular DNA phylogenic methods has revolutionized the study of biological diversification by providing phylogenetic trees for extant species that can be calibrated in time units and mapped, thus providing spatiotemporal evidence on speciation events. In the Neotropics, the first studies of this type found a contrasting pattern between the Andes and the Amazon lowlands. In the Andes, many endemic bird species were of recent origin, probably due to the habitat complexity created by the relatively recent uplift. In the Amazon lowlands, on the contrary, many endemisms were Neogene phyletic relicts, which was considered to support the hypothesis of the long-standing Neotropical stability (Fjeldsă 1994). Similar patterns found in other animal taxa (amphibia, reptiles and small mammals) were used to question the refuge hypothesis for the lowland Amazon rainforest faunas (Moritz et al. 2000). In other biomes, however, elements of both Neogene and Pleistocene diversification were recognized (Pennington et al. 2004; Carnaval and Bates 2007). During the 1990s and the early 2000s, molecular phylogenetic and phylogeographic studies on particular groups of Neotropical organisms proliferated and a variety of diversification drivers—i.e., glacial aridity, temperature and CO_2 depletion, Andean uplift or the Panama closure—were proposed according to the case and the geographical area

(Rull 2008 and references therein). In spite of this dramatic increase in the amount of evidence to unravel the origin of Neotropical diversity, the polarization between the Pleistocene refuge hypothesis and the Neogene tectonic hypotheses continued. A synthetic effort seamed thus pertinent to organize and understand this body of knowledge in the search for potential regularities.

A first meta-analysis was performed on >1400 species (105 genera and other supra-specific lineages) with dated molecular phylogenies encompassing the whole Neotropics—a unusual feature in previous literature, mainly concerned with the Amazon basin—covering a wide range of terrestrial and marine organism groups (amphibians, arachnids, birds, corals, echinoderms, fishes, insects, mammals, mollusks, reptiles and plants) (Rull 2008). These species originated in a continual manner since the Eocene-Oligocene until the Pleistocene and no diversification bursts were observed at any particular time (Fig. 2.8). Nearly the half of these species emerged during the Neogene, while the other half were of Pleistocene origin (Fig. 2.9). This suggested that extant Neotropical biodiversity originated through a diversity of drivers and evolutionary mechanisms. Therefore, it was proposed that the dual simplistic Neogene-Pleistocene controversy did not make sense as the current Neotropical biodiversity was likely the result of a complex interplay of ecological and evolutionary processes across spatial and temporal scales, initiated by Neogene tectonic events and continued under the action of Pleistocene climatic changes (Rull 2011a).

A further meta-analysis on almost 190 Amazon genera of plants and animals (insects, fishes, amphibians and birds) with dated phylogenies found that most of these genera emerged during the Neogene (Fig. 2.10) and concluded that current Neotropical biodiversity patterns originated before the Pleistocene and that the Mio-Pliocene Andean uplift was a paramount diversification driver (Hoorn et al. 2010). This study used the dating of the initial diversification event of each genus represented by the crown node or the node of the crown group, which is the monophyletic clade that contains all extant members of the genus (Quental and Marshall 2010). Several types of geological, paleontological, paleoecological and paleoclimatic evidence was provided in support of this Neogene hypothesis (Hoorn and Wesselingh 2010). According to this view, the Andean uplift not only changed drastically the drainage and climatic patterns of northern South America, but also created new montane environments and a complex topography that provided new opportunities for in situ speciation and a dispersal barrier promoting vicariance at both sides of the cordillera (Antonelli et al. 2010; Hoorn et al. 2013). Pleistocene diversification was explicitly dismissed (Antonelli and Sanmartín 2011). The same team of authors argued that Miocene diversification was fundamental for the origin of extant Neotropical biodiversity and contended that "the Quaternary was a time of distribution shifts, but can no longer be considered a time of diversification in Amazonia", rather, "the Quaternary should possibly be considered as a period of net loss of biodiversity" (Wesselingh et al. 2010, p .421). This return to the Neogene paradigm was further debated on the basis of conceptual and methodological features (Hoorn et al. 2011; Rull 2011b), which will be discussed later on. Rull (2014), using recent molecular phylogeographic evidence of intense Pleistocene

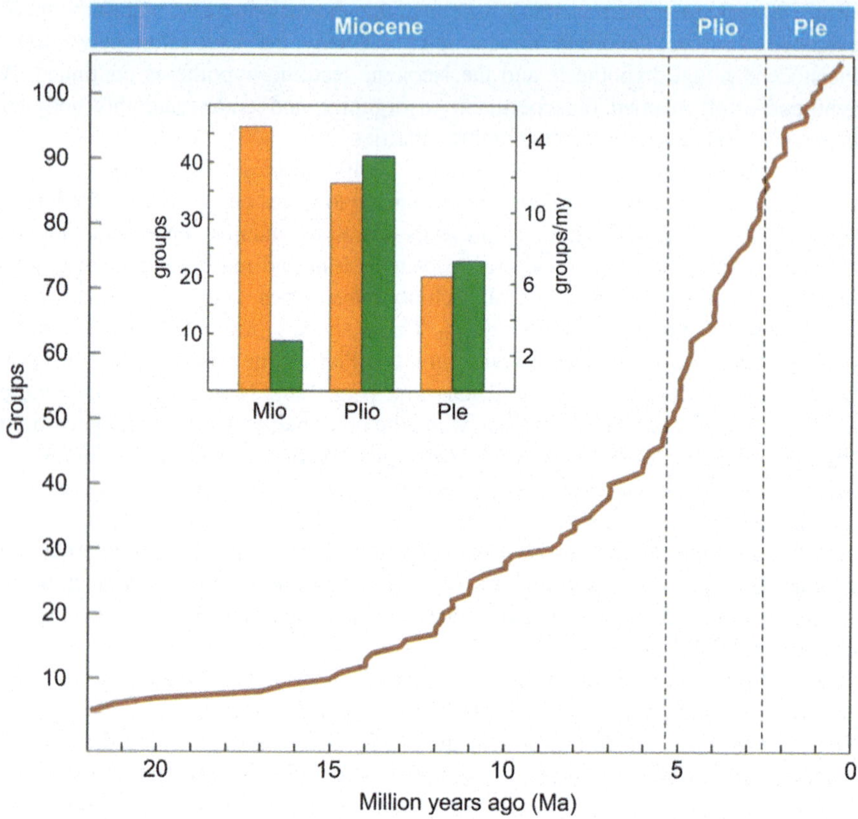

Fig. 2.8 Continuous Neotropical diversification trends since the Miocene until the Pleistocene, as shown by a meta-analysis of >1400 species belonging to 105 groups (genera and other supra-specific lineages). The brown solid line represents the age of the initial speciation event for each group sorted chronologically. The column plot displays the amount of these groups that initiated their speciation in each epoch, in both absolute numbers (orange bars, left scale) and rates, expressed as groups per million years (green bars, right scale). *Mio* Miocene, *Plio* Pliocene, *Ple* Pleistocene. Redrawn from Rull (2013)

diversification amongst many Andean lineages, insisted on the combined action of mountain building and climatic change in the shaping of current Neotropical biodiversity. The potential effect of Pleistocene climate changes were illustrated for two extra-Andean mountain complexes such as the Atlantic Forest, in the south-western Brazilian coasts (Fig. 2.3), and the Guayana Highlands (Fig. 2.2), between the Orinoco and the Amazon basins (Rull 2004, 2005; Carnaval and Bates 2007; Carnaval and Moritz 2008).

Another meta-analysis (Turchetto-Zolet et al. 2013) of molecular phylogeographies of ca. 480 South American species of algae, plants, invertebrates, fishes, amphibians, reptiles, birds and mammals including many tropical lineages, reported again similar percentages of Pleistocene and pre-Pleistocene diversification

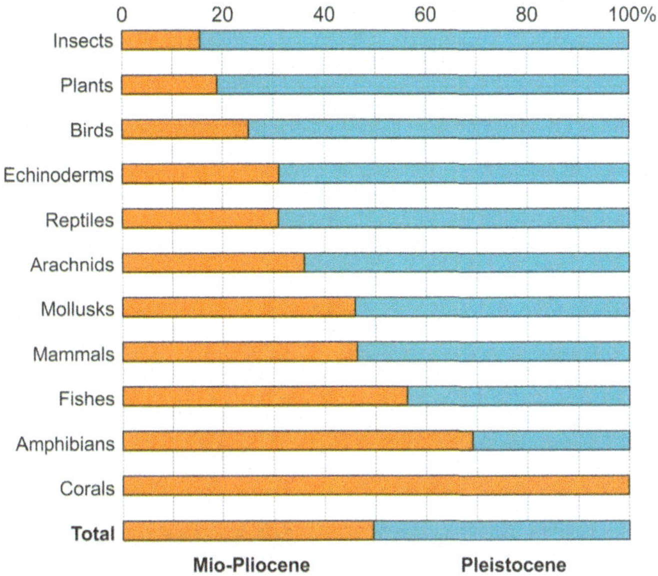

Fig. 2.9 Percentages of species originated either during Pleistocene (blues bars) or pre-Pleistocene (orange bars) times using the same data set as in Fig. 2.8. Redrawn from Rull (2008)

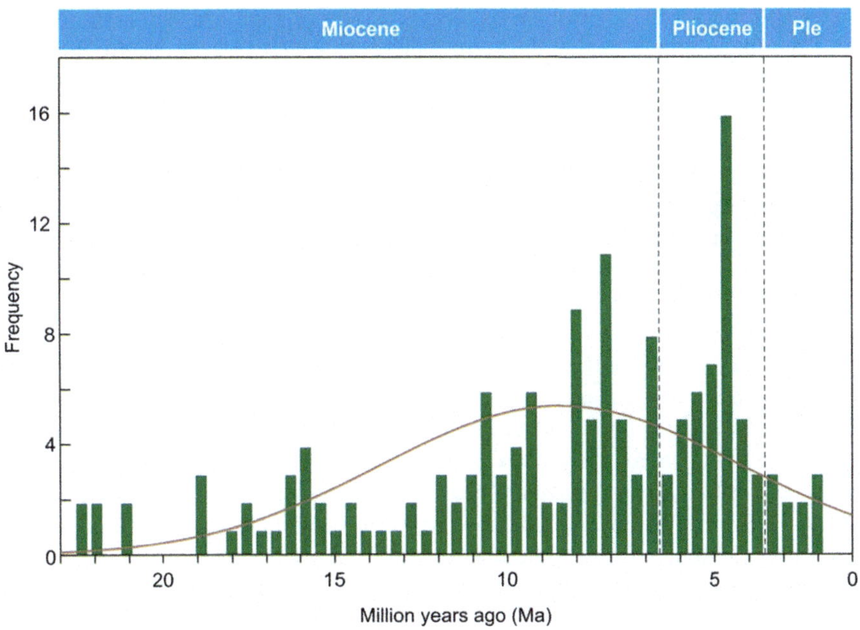

Fig. 2.10 Histogram and normalized density function of the ages of initial diversification events (crown nodes) of 189 Amazon living genera including plants, insects, fishes, amphibians and birds. Redrawn from Rull (2011a) after raw data from Hoorn et al. (2010)

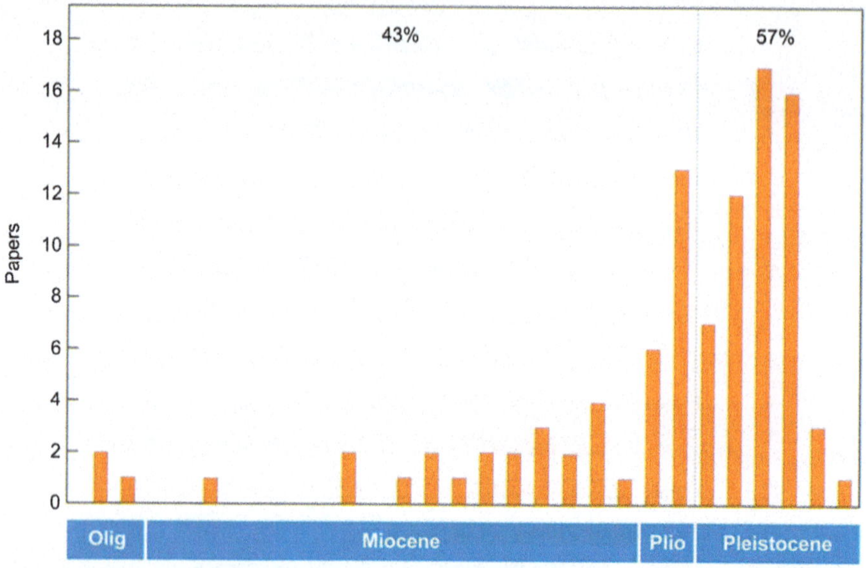

Fig. 2.11 Timing of lineage divergence of South American taxa based on phylogeographic studies published until 2011. Note that diversification amounts (expressed in percentages) are similar for Pleistocene and pre-Pleistocene times. Redrawn from Turchetto-Zolet et al. (2013)

thus reinforcing the idea that the current diversity patterns of the Neotropics were shaped by the combination of Pleistocene climatic oscillations and Mio-Pliocene tectonic events (Fig. 2.11). Moreover, the idea of the Pleistocene as a time of extinction was not supported by empirical evidence. Estimating extinction using molecular phylogenetic data is problematic as it deals with the genome of living organisms (e.g., Quental and Marshall 2010; Pyron and Burbrink 2013). Extinction trends through time are better approached using the paleoecological record. Using this type of evidence, Willis and Bhagwat (2009) highlighted that, in plants, there is a single extinction documented for the entire Pleistocene, whereas in animals, the more significant extinctions recorded are the disappearance of large mammals during the Late Pleistocene-Early Holocene, which have been related to climate changes and human activities. Palaeoecological evidence has shown that, rather than extinction, Pleistocene glacial-interglacial changes have promoted large scale migrations and range shifts, which have been decisive for the shaping of present-day biogeographical patterns and the assembly of extant communities (Willis and Bhagwat 2009). This, combined with the mounting molecular phylogenetic evidence of Pleistocene speciation, suggested that the Pleistocene was a time of net diversification in the Neotropics thus supporting the view of a continuous diversification trend since the Neogene to the Pleistocene (Rull 2012a).

Molecular phylogenetic results have also been used to suggest inter-continental long-distance dispersal, a process not contemplated in the classical studies previously mentioned about the role of migration in the shaping of Neotropical diversity

(Raven and Axelrod 1974; Gentry 1982). For example, Cody et al. (2010) found evidence that plants would have been able to cross the Isthmus of Panama region well before its closure, which suggested that the GABI started earlier for plants likely due to the possibility of long-distance dispersal across marine barriers (but see also Carrillo et al. 2015). A fascinating idea is the possibility of trans-oceanic dispersal between Africa and South America, which is also supported by a number of molecular plant phylogenies and would have been important for the assembly of present-day Amazonian rainforest flora (Pennington and Dick 2004; Weeks et al. 2014). Antonelli et al. (2015) contended that the Neotropics has been an engine for global plant diversity, as species' emigration from the American tropics has largely exceeded immigration during the Cenozoic. This would support, in part, the out-of-the-tropics hypothesis for the occurrence of LBGs. Dispersal ability has been proposed as an important diversification mechanism even within the Neotropics. For example, Smith et al. (2014) argued that landscape change is not needed to explain the current biogeographical patterns of Neotropical birds at species level, which is more consistent with a model that combines persistence and dispersal across the landscape matrix followed by allopatric speciation. Similar results were obtained by Dexter et al. (2017) for several genera of Amazonian trees and Prates et al. (2016a, 2017) for lizards, although in this last case, dispersal seem to have been triggered by Pleistocene climate changes.

The amount of molecular phylogenetic studies on Neotropical taxa has greatly increased since the beginning of this century (Fig. 2.12) thus providing new

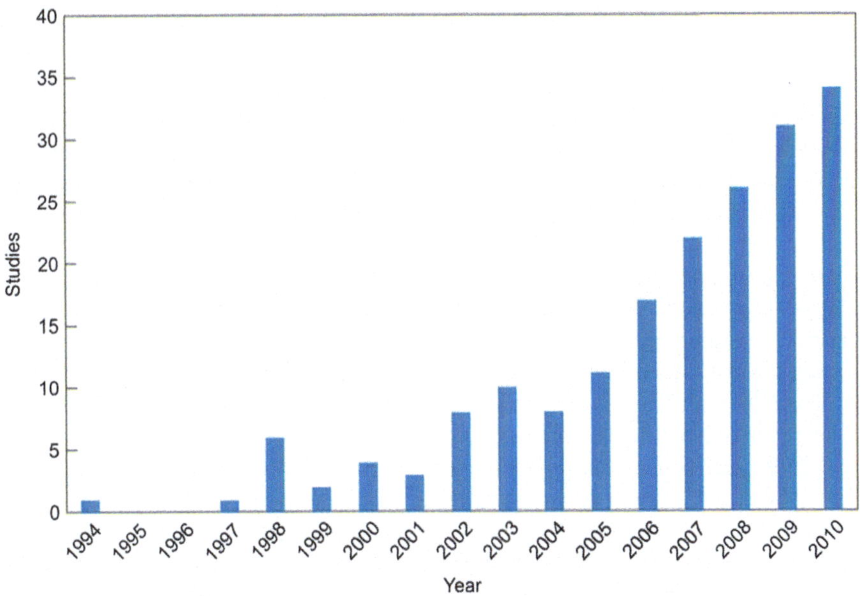

Fig. 2.12 Number of phylogeographical studies published between 1994 and 2010 on South American organisms, most of them from tropical areas. Redrawn from Turchetto-Zolet (2013)

opportunities for synthetic approaches. Some attempts in this direction have been made in the last years. For example, Hughes et al. (2013) recognized the complexity of Neotropical diversification but insisted on the Neogene paradigm by arguing that a cluster of geological and climatic events occurred during this period—i.e., the Late Miocene global cooling, the latest Andean uplift occurred 9–10 Ma and the formation of the Panama landbridge ca. 3.5 Ma (Fig. 2.7)—coincided with the preponderance of Mid to Late Miocene crown ages of species-rich Neotropical clades thus suggesting that the Miocene was "a pivotal time for the establishment and diversification of the modern Neotropical flora and that a very large fraction of extant Neotropical plant species diversity has arisen within the last ten million years" (p. 13). These authors suggested that niche conservatism and high speciation rates, rather than low extinction rates, would have been fundamental for Neotropical diversification (museum hypothesis). Pleistocene diversification was not addressed.

A cursory examination of the great amount of studies on Neotropical diversification suggests that different taxonomic groups would have been submitted to different drivers, processes and mechanisms across time and space, which might lead some to believe that there is no a general explanation for the high Neotropical biodiversity. The identification of potential general patterns would require a comprehensive Neotropical database of molecular phylogenetic studies at species level and a reanalysis of older phylogenetic trees using the latest-generation dating methods (Rull 2011a). An attempt in this direction is a recent synthesis based on a very large worldwide plant database (ca. 22,600 species) and updated methods of chronological calibration found that the comparatively higher Neotropical diversity was associated with higher speciation and extinction rates than in tropical Asia and Africa (Antonelli et al. 2015). This was interpreted in terms of a higher species turnover and shorter average species longevity, which was preliminary attributed to the considerable landscape dynamics that has affected the Neotropics since the Miocene, as well as to differences in biome size, niche space and climatic history, with respect to other tropical areas. Using a similar approach, the same research team performed a meta-analysis on 4450 species of varied taxonomic categories and biomes from the whole Neotropics and concluded that Amazonia was the main source of Neotropical biodiversity, as >2800 lineages originated in this region and dispersed to other Neotropical regions, especially to Mesoamerica (Antonelli et al. 2018). Again, biotic turnover among regions was considered the main diversification factor. According to these authors, such biotic interchange would have been active during the last >60 million years with a general increase towards the present.

Together with these recent synthetic attempts, more and more case studies on a wide range of Neotropical organisms subjected to a variety of diversification drivers and ecological constraints appear almost every week in specialized journals and edited books. A number of these studies insist on the Neogene-Pleistocene controversy and try to make generalizations while others do not but this debate is still alive. A general impression is that the Neogene paradigm has experienced a revitalization whereas the Pleistocene paradigm has been comparatively underrated, some even consider that "the effects of Quaternary climatic changes on Neotropical rain forest vegetation remain something as a black hole in Neotropical biogeography" (Cavers and Dick 2013, p. 616). This could be due to the difficulty of demonstrating a causal relationship between past fragmentation and extant diversity patterns and also to the unfeasibility of attributing speciation events to particular Pleistocene shifts using the current tools for dating phylogenetic trees (Pennington and Dick 2010). A further effort is needed for data storing, processing and integration of the great amount of information on Neotropical diversification that is produced every year. But merely the sum of all the available information will not suffice to attain a holistic perspective and a change of strategy seems necessary. Hughes et al. (2013), suggested to study diversification patterns separately in different biomes to include not only geological and climatic drivers but also the ecological constraints involved in the assembly of present-day biotas. Examples of such procedure on the assembly of a wide range of Neotropical forested and non-forested biomes are Simon et al. (2009), Pennington et al. (2010), Särkinen et al. (2012), Erickson et al. (2014), Willis et al. (2014) and Honorio Coronado et al. (2015), among others.

5 Conceptual and Methodological Insights

The history of research on Neotropical diversification has been more than a succession of hypotheses and theories trying to explain new evidence provided by novel methodological tools and the improvement of existing ones. There is a conceptual background behind all of the debates and discussions mentioned above that is worth to be highlighted. Such theoretical developments have not necessarily proceeded parallel to the methodological improvements; some concepts and views have maintained through time, others have been replaced and others have been revitalized with the advent of new evidence that was previously unavailable. This section briefly discusses the more relevant conceptual developments, with emphasis on those that have hindered the advancement towards a synthetic theory of Neotropical diversification. Such account might seem irrelevant as many of the topics discussed would be considered by some as obvious, even trivial issues. However, a number of anachronistic views and fruitless debates still persist in the study of Neotropical diversification that should be revised if a truly holistic diversification framework is to be achieved. The main flawed approaches discussed here are: (1) the shifting from one paradigm to another, (2) the extrapolation from particular case studies to the whole Neotropics, (3) the selection of different lines of evidence—notably different taxonomic categories—to support either one or another hypothesis, (4) the implicit assumption that Pleistocene diversification equals to refuge diversification and (5) the straightforward inference of diversification drivers from diversification timing (Rull 2012a, 2013, 2015). These drawbacks are briefly discussed and several potential solutions are suggested.

5.1 Paradigm Shifting

As discussed in the historical section, the biogeographically-based Pleistocene refuge hypothesis was paradigmatic until the late 1980s (Whitmore and Prance 1987), when paleoecological data started to be used to reject such explanation in favor of the hypothesis of an assumed stability and continuity of the Amazon rainforests since the middle Miocene until the present (Colinvaux and De Oliveira 2001). Further molecular phylogenetic studies were used to dismiss Pleistocene diversification, supporting the Neogene hypothesis, and advocated for tectonically-driven events such as the Andean orogeny or the closure of the Panama Isthmus as the main diversification drivers (Hoorn et al. 2010). This shift from the refuge hypothesis paradigm to the Neogene paradigm was radical and many scholars considered that these hypotheses were not only opposed but also excluding. It is not unusual that the defenders of either one or another of these hypotheses highlight the evidence needed to support their preferred explanation ignoring contradictory evidence. In the late twentieth century, the refuge hypothesis was the favorite explanation but in the last decades, the Neogene hypothesis seems to be preferred. Shifting from one paradigm to another is unlikely to result in progress towards a better understanding of Neotropical diversification (Rull 2013). In light of the available evidence, the continued debate between the Pleistocene and the Neogene paradigms as excluding explanations has been considered rather simplistic, outdated and sterile (Rull 2008).

5.2 Unwarranted Generalizations

Another flawed approach is to use case studies of particular geographical regions and/or groups of organisms to try to develop general theories on Neotropical diversification. There are some taxonomic groups whose diversification patterns are consistent with a particular type of diversification driver, as for example the refuge hypothesis (Haffer 1969), the microrefuge hypothesis (Rull and Montoya 2014), the vanishing refuge hypothesis (Damasceno et al. 2014), the disturbance-vicariance hypothesis (Noonan and Gaucher 2005), the river hypothesis (Ribas et al. 2011), the Andean uplift (Hughes and Eastwood 2006) or the closure of the Panama isthmus (Cody et al. 2010), among others. But this does not mean that each of these drivers can be extrapolated to the whole Neotropics. The high physiographic, environmental and biotic heterogeneity of the Neotropics, together with the long history of paleogeographic reorganizations and paleoclimatic changes (Hoorn and Wesselingh 2010), precludes any generalization from selected case studies. In this context, it is noteworthy that the Amazon lowlands, especially the rainforests they support, have been the object of the majority of Neotropical diversification studies and there is the danger of considering these environments as representatives of the whole Neotropics. For example, multidisciplinary studies including GIS modelling, paleoecology and molecular phylogenetics (Carnaval and Moritz 2008; Carnaval et al. 2009, 2014; Porto et al. 2013) seem to be consistent with the existence of a complex patterns of Pleistocene forest refugia in the so called Atlantic forest, situated at the south-eastern Brazilian coasts (Fig. 2.3). Therefore, any extrapolation from the Amazon or the Atlantic forest models to the whole Neotropics would be unwarranted. The tendency to infer general trends from particular examples was more frequent in earlier times and has notably decreased with time and the flourishing of diversification studies but, unfortunately, it has not been yet completely eradicated. It should also be highlighted that most studies on Neotropical diversification have been carried out on terrestrial organisms and the validity of the conclusions obtained for marine environments is not granted.

5.3 *Taxonomic Resolution*

Choosing the appropriate taxonomic category is essential in diversification studies. Moritz (1994) defined the Evolutionary Significant Unit (ESU) as the lower taxonomic unit needed to represent the actual genetic variability associated with a distinct evolutionary potential and considered the ESU as the preferred category for conservation purposes (Moritz 2002). Usually, ESUs coincide with species or sub-specific categories, which are the commonly used to define current biodiversity patterns. More recent approaches based on Bayesian phylogenetics and phylogeography (BPP) methods have been used for species delimitation as an alternative to ESUs (Fujita and Leaché 2011). Both ESUs and BPP methods emphasize the suitability of specific and sub-specific categories for biodiversity assessment. Supra-specific categories, as for example genera, are unreliable biodiversity descriptors as they may contain a wide range of internal diversity, from one to thousand or more species (Rull 2012a). Therefore, unraveling the origin and maintenance of current Neotropical biodiversity should be preferably addressed using species and sub-specific units, that is, ESUs. Paleontological and paleoecological studies have important limitations in this sense, as identification at species level is rare. However, molecular phylogenetic studies are able to attain specific and sub-specific resolution thus rightly reproducing true genetic variability relevant for evolutionary potential. In spite of this, a number of phylogenetic and phylogeographic studies on the origin of Neotropical diversity are performed using genera and other supra-specific categories (e.g., Hoorn et al. 2010). Using species or genera makes a great difference in the final results obtained, as the second underestimate recent diversification and, therefore, the evolutionary origin of extant ESUs. A good example of using either one or another taxonomic category can be found in the Gimnosperm order of the Cycadales, commonly called cycads.

The cycads have been traditionally considered as 'living fossils', as their extinct representatives with almost the same morphological characteristics as the living ones are present in 280 million years old Permian rocks. However, a recent study has revealed that most of the ca. 300 extant cycad species originated during the last

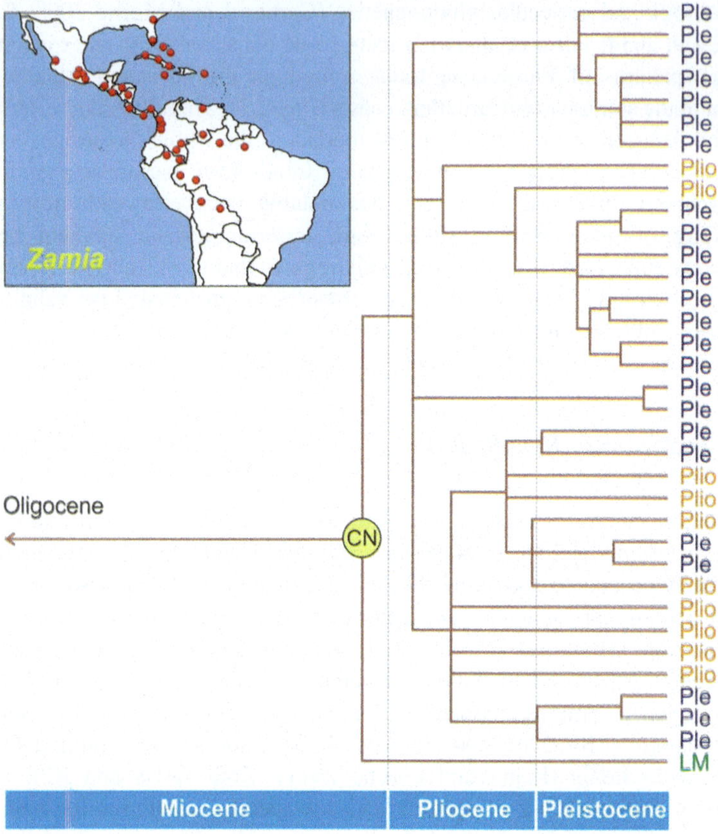

Fig. 2.13 Time-constrained phylogenetic tree for 35 species of the cycad genus *Zamia*, which diverged from *Microcycas* in the Oligocene. *CN* Crown node, *LM* Late Miocene, *Plio* Pliocene, *Ple* Pleistocene. Redrawn from Nagalingum et al. (2011). Geographical distribution according to Whitelock (2002)

12 million years (Nagalingum et al. 2011). The more diverse Neotropical genus of cycads is *Zamia*, with almost 80 species. Nagalingum et al. (2011) studied 35 of these species and found that this genus separated from *Microzamia* during the Oligocene and started to diversify in the Late Miocene (Fig. 2.13). This would suggest that some event occurred during the Late Miocene would have been the responsible for the diversification of this genus. But a closer examination reveals that only one species, the oldest one, is of Miocene origin, whereas 10 species originated during Pliocene and the remaining 24 emerged in the Pleistocene (Rull 2012b). Therefore, further Pliocene and Pleistocene events would have been involved for the origin of most studied species. The lesson is that using crown dating alone, the Neogene paradigm would be favored (e.g., Hoorn et al. 2010; Hughes et al. 2013), whereas using species dating, the idea of a complex diversification process involving Miocene, Pliocene and Pleistocene drivers and processes seems more realistic (Rull

2011a, 2012a). In summary, crown dating is useful to identify the initial diversification event and its possible causes, whereas species dating is needed to understand the origin of the full range of extant diversity, at the ESU level, and to infer the potential drivers involved (Rull 2013).

Taxonomic resolution is also important in paleoecology. For example, the palynological evidence used to support to the Neogene origin of modern Amazon rainforests (Jaramillo et al. 2010) is weak as is not based on actual species but on fossil species with assumed taxonomic affinities with extant families and genera. Besides the reliability of such affinities, the fragmentary nature and the low taxonomic resolution of the fossil record is not enough to firmly assess that modern Amazon forests originated in the Neogene and remained constant through time (see also Prates et al. 2015).

5.4 Pleistocene Diversification and Refugia

There is also a tendency, among a number of researchers, to assume, explicitly or not, that Pleistocene diversification should have followed the model of the refuge hypothesis. Indeed, many defenders of this hypothesis implicitly accept that demonstrating Pleistocene diversification supports refuge diversification (e.g., Garzón-Orduña et al. 2014), whereas many critics believe that dismissing Pleistocene diversification automatically rejects diversification on glacial refugia (e.g., Hoorn et al. 2010). However, as has been explained in the historical section, there are other diversification models for the Pleistocene as for example the disturbance-vicariance hypothesis, the river hypothesis and its variants, the gradient hypothesis, the vanishing refuge hypothesis or the microrefugia hypothesis, among others (Haffer 1997; Nores 1999; Rull 2009; Antonelli and Sanmartín 2011; Antonelli et al. 2010; Damasceno et al. 2014; Rull and Montoya 2014). Therefore, when Pleistocene diversification is supported by empirical data, a careful examination of the potential mechanisms involved should be addressed, rather than relying only in the refuge model.

5.5 Diversification Timing and Drivers

It is also common to deduce eventual diversification drivers and mechanisms simply from the chronological coincidence between the age of diversification of a given lineage and its species and the occurrence of some specific or singular geological or environmental event. This is the case, for example, of the studies that observe a diversification event at the time of the Andean uplift or the occurrence of Pleistocene glaciations and automatically link the biodiversity increases with these phenomena. Examples are numerous as this is a usual procedure. In essence, this is not a flawed procedure but it should only be considered the first step of a more in-depth research needed to identify the ecological and evolutionary processes and mechanisms unequivocally relating the geological/environmental shift with the diversification trends observed. For example, the Pliocene was a time of intense environmental shifts and paleogeographic reorganizations including the latest Andean uplift, the closure of the Panama Isthmus and the onset of the global cooling that ended in Pleistocene glaciations. These events occurred in a narrow temporal window between ca. 4 and ca. 3 million years before present (Coates et al. 1992; Mora et al. 2010; McKay et al. 2012), a time interval that could be accommodated within the statistical error of phylogenetic tree dating. Therefore, the increasing speciation trend occurred during the Pliocene (Figs. 2.8, 2.10 and 2.11) could be attributed to any of these events or a combination of them but the exact drivers, processes and mechanisms involved cannot be evaluated using solely this chronological coincidence.

5.6 Potential Solutions

An obvious solution is to avoid the above shortcomings but this seems to be more difficult than expected given their persistence in time. Some conceptual tools addressed to provide new comprehensive theoretical frameworks may be helpful. A first step could be the adoption of an open-minded attitude able to address the full spatio-temporal complexity of the problem. One possibility is the use of the multiple working hypotheses framework proposed more than a century ago by Thomas Chamberlin (1890). According to this author, we humans are prone to attribute a phenomenon to a single cause and tend to erect premature or precipitate explanations turning into ruling theories that direct research. Indeed, when a ruling theory is established, most efforts are oriented to finding evidence to either support or reject it, depending on individual interests. This is what happened with the refuge theory for Neotropical diversification. Alternatively, a potential explanation for empirical observations can be considered as a working hypothesis to be tested, rather than a ruling theory. According to Chamberlin (1890), however, the risk of a working hypothesis becoming a controlling idea remains high, as it is occurring at present with the Neogene paradigm for Neotropical diversification. The multiple working hypotheses framework seeks to explore any explanation possible for the observed phenomena and to develop every testable hypothesis in relation to the possible causes. This approach promotes thoroughness, suggests lines of inquiry that might otherwise be overlooked and develops the habit of parallel and complex thought (Chamberlin 1890). This approach would seem trivial but it is surprisingly unusual in the study of Neotropical diversification (Rull 2013), where the Pleistocene and the Neogene ruling theories are still alive and remain as excluding options for many, especially those who ignore or explicitly dismiss Pleistocene diversification (e.g., Antonelli and Sanmartín 2011; Wesselingh et al. 2010; Hughes et al. 2013).

Using the multiple hypotheses approach requires at least six conceptual steps to relate the extant biodiversity patterns with their corresponding speciation drivers, processes and mechanisms (Rull 2015) (Fig. 2.14). First, the research question

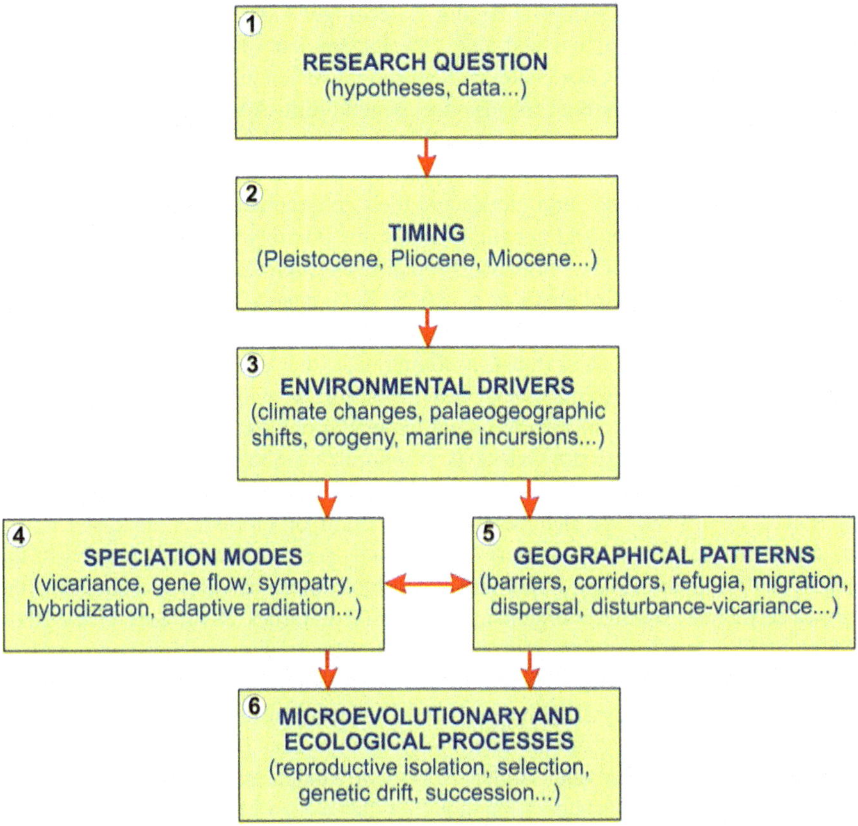

Fig. 2.14 Schematic view of the conceptual steps discussed in the text for making inferences about the origin of biodiversity using paleoecological and phylogenetic evidence. Redrawn from Rull (2015)

should be clear, in order to focus the type of study to be developed and the more suitable empirical evidence to be seek. The second step is to use empirical data to estimate the speciation timing and establish the more likely chronological correlations with the candidate environmental drivers. In order to avoid circularity, it is essential that the corresponding phylogenies have been dated with events different from these drivers (Baker et al. 2014). For example, we cannot calibrate a phylogenetic tree with the age of the closure of the Isthmus of Panama and then try to test the influence of this event on speciation. The next step is critical and is the analysis of the potential speciation patterns and processes, that is, the more likely speciation modes and their geographical expression. Using the same example, the closure of the Panama Isthmus reversed the evolutionary trends in terrestrial and marine organisms leading to allopatric speciation in the sea and gene flow in the land. Finally, speciation is not instantaneous but proceeds through a variety of microevolutionary mechanisms involving genotypic changes at the population level occurring within the community that, in turn, is changing via ecological succession. For example, in the case of speciation across and along the Panama Isthmus, ecological processes and genetic mechanisms associated with reproductive isolation and gene flow should have been more relevant that, for instance, genetic drift. Addressing all these points with empirical data in each single study would be difficult but any suggestion about potential relationships between speciation timing and causes should be evaluated in light of the possible speciation modes and their geographical expression, as well as the ecological processes and genetic mechanisms possibly involved. Some models trying to integrate all these elements to explain Neotropical diversity are in progress (Graham et al. 2014; Collevatti et al. 2015; Beheregaray et al. 2015) but studies across different spatial, temporal and biological scales are still poorly addressed.

Any conceptual change must accommodate the continued methodological improvements in the disciplines that provide empirical evidence for data bases and meta-analyses, notably paleoecology and molecular phylogenetics. One of the most decisive paleoecological drawbacks is taxonomic resolution as expressed in the difficulty of attaining the species level in fossil identification (Rull 2014). Some advances in this venture include the introduction of biomarker analysis, as for example DNA analysis of well-preserved fossils (i.e., ancient DNA) and also the study of free sedimentary DNA and other molecules that allow identifying the presence of non-fossilized organisms (Brown and Barnes 2015; Key et al. 2017; Marx 2017). Another recent innovation is the use of Fourier transform infrared spectroscopy (FTIR), a promising technique to increase taxonomic resolution, especially in palynology (Jardine et al. 2017). Regarding molecular phylogenetics, constant progresses are made in dating phylogenetic trees (Laurin 2012; Pyron and Burbrink 2013; Hipsley and Müller 2014; van Tuinen and Torres 2015), which makes necessary the reconsideration of earlier data using old dating methods—as for example those assuming constant mutation rates and extrapolating them among different taxonomic groups—in order to make them comparable with latest studies using updated dating methods. Ideally, data bases should contain the raw phylogenetic information needed for dating, rather than solely the ages obtained when this information was produced. In this way, the ages of species and the clades they form could be constantly updated and homogenized. It would be especially useful to increase dating resolution and minimize statistical errors, in order to attain a higher precision in the dating of diversification events. This would contribute to enhance the reliability of chronological coincidences between speciation and geological/environmental events thus improving the formulation and testing of causal hypotheses, especially in more recent times. The possibility of reliably estimating extinction using molecular phylogenetics would be important to obtain a more complete picture of this parameter, which estimation currently relies on the fragmentary and taxonomically imprecise fossil record. New approaches using Bayesian methods combined with biogeographic and paleontological models are under way (Sanmartín and Meseguer 2016). Finally, it should be stressed that increasing the number, quality and interdisciplinarity of raw data to be stored in databases is not enough for fruitful collaboration among the different disciplines that participate in the study of Neotropical diversification.

6 Future Directions

Usually, a section with this title is attempted to discuss potential methodological developments useful for raising new hypotheses and/or improving and enhancing the empirical basis for their testing. In the case of Neotropical diversification, methodological research is undoubtedly very important and some hints have already been provided in the former section on potential solutions. However, as it has been seen in the historical section, methodological progress alone has not been enough thus far to develop a general theory on the origin and maintenance of Neotropical biodiversity and this is not expected to occur in the near future without a change in the current conceptual framework of most studies. In other words, progress towards a general theory of Neotropical biodiversity would be hampered by conceptual, rather than methodological constraints. In this sense, two aspects seem to be of paramount importance, namely the need for truly multidisciplinary approaches and the abandonment of premeditated ruling theories. Rather than pursuing evidence for either supporting or rejecting biased and incomplete hypotheses, an open-minded multiple working hypotheses framework including evolutionary, ecological, biogeographical and geological considerations, among others, is urgently needed. Recent synthetic attempts using phylogenetic and phylogeographical evidence are strongly biased towards the Neogene paradigm and ignore or explicitly dismiss the abundant body of evidence on Pleistocene diversification. This seems to be a reaction to the past attachment of many studies to the refuge hypothesis, which has been erroneously considered by many as a synonym of Pleistocene speciation.

The multidisciplinary nature of the team works addressing these aspects of the problem should not be restricted to the combination of different research fields but also of different points of view on the problem. Research teams should be open to approaches different from the paradigms they defend; otherwise, the origin and maintenance of Neotropical diversity will remain enigmatic, no matter the money invested, the amount of people involved and the variety of research disciplines participating in its study. The Neotropical mystery will hardly be solved by big projects with the participation of large and multidisciplinary teams, whose researchers are engaged in a single paradigm or ruling theory. Figure 2.14 is an attempt to provide a basis for a conceptual framework based on the multiple working hypotheses idea. It is hoped that this could be improved, perhaps even deeply modified, in the future in order to include all the available possibilities in the progress towards a holistic theory of Neotropical diversification.

7 Final Remarks

The idea of a general theory of Neotropical diversification, as well as for the origin of the LBGs, is underlying most studies on this topic. However, it could be asked whether such general theory really exists or it is just a human expectancy.

Biodiversity gradients are complex features and the fact that we describe this pattern with a single term does not mean that it is due to a single factor. As mentioned above, different diversification drivers and mechanisms have produced different diversity patterns across regions and taxonomic groups. Does this mean that there is no a general explanation for the high Neotropical biodiversity and we should go case by case, biome by biome or region by region? (Prates et al. 2016b). It is hard to escape the idea that a general pattern should have a general explanation. The question is whether it is possible to accommodate all spatial heterogeneity, ecological disparity and temporal variability into a single conceptual framework. One of the more general approaches is the Stebbins' (1974) cradle vs. museum model and its variants, based on the occurrence of latitudinal gradients in speciation and extinction rates. However, such a general approach is primarily descriptive and does not provide causal explanations. The same is true for other general approaches as for example the energy-gradient hypothesis or the climate-stability hypothesis (McGlone 1996). On the contrary, specific case studies on particular taxonomic groups and/or geographical regions suggest fairly detailed, although very heterogeneous, causal diversification relationships. Thus, there seems to be an inverse relationship between generality and explanatory power that would prevent to attain a satisfactory explanation of general validity.

But progresses towards a unifying theory with sufficient explanatory power are worth to be done. Among the options available to date, a combination of the hierarchical framework of Willis and Whittaker (2002) (Table 2.1) and the multiple hypothesis framework of Chamberlin (1890), as expressed in Fig. 2.14, might be useful to address the issue of Neotropical diversification. Under such approach, the clear definition of the research question and its taxonomic, spatial and temporal scope appears as a fundamental premise to progress in the right direction. Alternative ways of thinking could also help. For example, the usual question: why the Neotropics is so diverse?, which has directed the research to date, implicitly assumes that lower extra-tropical diversity is the norm and the comparatively higher tropical diversity requires an explanation. Such fallacy is a cultural bias resulting from the extrapolation of ecological and evolutionary principles obtained in temperate regions to the tropics, whose study started later and was initially conducted by scholars from temperate countries (Raby 2017). In the reverse situation, lower extra-tropical diversity would have been the unusual feature to be explained.

Acknowledgements The author is very grateful to Ana Carnaval, Peter Linder and Beryl Simpson who critically reviewed the first draft of the manuscript and provided valuable comments and suggestions that improved and enriched the original text.

References

Ab'Saber AN (1982) The paleoclimate and paleoecology of Brazilian Amazonia. In: Prance G (ed) Biological diversification in the tropics. Columbia University Press, New York, pp 41–59

Anderson MJ, Crist TO, Chase JM et al (2011) Navigating the multiple meanings of β diversity: a roadmap for the practicing ecologist. Ecol Lett 14:19–28

Antonelli A, Sanmartín I (2011) Why are there so many plants species in the Neotropics? Taxon 60:403–414

Antonelli A, Quijada-Mascareñas A, Crawford AJ et al (2010) Molecular studies and phylogeography of Amazonian tetrapods and their relation to geological and climatic models. In: Hoorn C, Wesselingh F (eds) Amazonia: landscape and species evolution. A look into the past. Wiley-Blackwell, Chichester, pp 386–404

Antonelli A, Zizka A, Silvestro D et al (2015) An engine for global plant diversity: highest evolutionary turnover and emigration in the American tropics. Front Genet 6:130. https://doi.org/10.3389/fgene.2015.00130

Antonelli A, Zizka A, Antunes F et al (2018) Amazonia is the primary source of Neotropical biodiversity. Proc Natl Acad Sci U S A 115:6034–6039

Baker PA, Fritz SC, Dick CW et al (2014) The emerging field of geogenomics: constraining geological problems with genetic data. Earth Sci Rev 135:38–47

Beheregaray LB, Cook GM, Chao NL, Landguth EL (2015) Ecological speciation in the tropics: insights from comparative genetic studies in Amazonia. Front Genet 5:477. https://doi.org/10.3389/fgene.2014.00477

Brown JH (2014) Why are there so many species in the tropics? J Biogeogr 41:8–22

Brown TA, Barnes IM (2015) The current and future applications of ancient DNA in Quaternary science. J Quat Sci 30:144–153

Bush MB (1994) Amazonian speciation: a necessarily complex model. J Biogeogr 21:5–17

Bush MB, De Oliverira PE (2006) The rise and fall of the Refugial Hypothesis of Amazon speciation: a paleoecological perspective. Biota Neotrop 6:1. https://doi.org/10.1590/S1676-06032006000100002

Cannon CH, Lerdau M (2015) Variable mating behaviors and the maintenance of tropical biodiversity. Front Genet 6:183. https://doi.org/10.3389/fgene.2015.00183

Carnaval AC, Bates JM (2007) Amphibian DNA shows marked genetic structure and tracks Pleistocene climate change in northeastern Brazil. Evolution 61:2942–2957

Carnaval AC, Moritz C (2008) Historical climate modelling predicts patterns of current biodiversity in the Brazilian Atlantic forest. J Biogeogr 35:1187–1201

Carnaval AC, Hickerson MJ, Haddad CFB et al (2009) Stability predicts genetic diversity in the Brazilian Atlantic fortest hotspot. Science 323:785–789

Carnaval AC, Waltari E, Rodrigues MT et al (2014) Prediction of phylogeographic endemism in an environmental complex biome. Proc R Soc B 281:20141461

Carrillo JD, Forasiepi A, Jaramillo C, Sánchez-Villagra MR (2015) Neotropical mammal diversity and the Great American Biotic Interchange: spatial and temporal variation in Souht Americs'a fossil record. Front Genet 6:451. https://doi.org/10.3389/fgene.2014.00451

Cavers S, Dick CW (2013) Phylogeography of neotropical trees. J Biogeogr 40:615–617

Chamberlin TC (1890) The method of multiple working hypotheses. Science (old series) 15:92–96 (reprinted 1965, 148:754–759)

Coates AG, Jackson JB, Collins LS et al (1992) Clossure of the isthmus of Panama: the near-shore marine record of Costa Rica and western Panama. Geol Soc Am Bull 104:814–828

Cody S, Richardson JE, Rull V, Ellis C, Pennington RT (2010) The Great American Biotic Interchange revisited. Ecography 33:326–332

Cohen KM, Finney SC, Gibbard PL, Fan J-X (2013) The ICS international stratigraphic chart. Episodes 36:199–204

Colinvaux PA (1987) Amazon diversity in light of the paleoecological record. Quat Sci Rev 6:93–114

Colinvaux PA, De Oliveira PE (2001) Amazon plant diversity and climate through the Cenozoic. Palaeogeogr Palaeoclimatol Palaeoecol 166:51–63

Colinvaux PA, De Oliveira PE, Moreno JE, Miller MC, Bush MB (1996) A long pollen record from lowland Amazonia: forest and cooling in glacial times. Science 274:85–88

Colinvaux PA, De Oliveira PE, Bush MB (2000) Amazonian and neotropical plant communities on glacial time-scales: the failure of the aridity and refuge hypothesis. Quat Sci Rev 19:141–169
Collevatti RG, Terribile LC, Diniz-Filho JAF, Lima-Ribeiro MS (2015) Multimodel inference in comparative phylogeography: an integrative approach based on multiple lines of evidence. Front Genet 6:31. https://doi.org/10.3389/fgene.2015.00031
Colwell RK, Hurt GC (1994) Non-biological gradients in species richness and a spurious Rapoport effect. Am Nat 144:57–595
Connell JH, Orias E (1964) The ecological regulation of species diversity. Am Nat 98:399–414
Cox CB, Moore PD, Ladle R (2016) Biogeography: an ecological and evolutionary approach. Wiley-Blackwell, Chichester
Damasceno R, Strangas ML, Carnaval AC et al (2014) Revisiting the vanishing refuge model of diversification. Front Genet 5:353. https://doi.org/10.3389/fgene.2014.00353
Damuth JE, Fairbridge RW (1970) Equatorial Atlantic deep-sea arkosic sands and ice-age aridity in tropical South America. Bull Geol Soc Am 81:189–206
Darwin C (1839) Journal of the researches into the geology and natural history of various countries visited by HMS Beagle from 1832 to 1836. Henry Colburn, London
Dexter KG, Lavin M, Torke BM et al (2017) Dispersal assembly of rain forest tree communities across the Amazon basin. Proc Natl Acad Sci U S A 114:2645–2650
Dietl GP, Kidwell SM, Brenner M et al (2015) Conservation paleobiology: leveraging knowledge of the past to inform conservation and restoration. Annu Rev Earth Planet Sci 43:79–103
Emiliani C (1971) The amplitude of Pleistocene climatic cycles at low latitudes and the isotopic composition of glacial ice. In: Turekian KK (ed) The Late Cenozoic glacial ages. Yale University Press, New Haven, pp 183–197
Erickson DL, Jones FA, Swenson NG et al (2014) Comparative evolutionary diversity and phylogenetic structure across multiple forest dynamics plots: a mega-phylogeny approach. Front Genet 5:358. https://doi.org/10.3389/fgene.2014.00358
Farrell BD, Mitter C, Futuyma DJ (1992) Diversification at the insect-plant interface. Bioscience 42:34–42
Fine PVA (2015) Ecological and evolutionary drivers of geographic variation in species diversity. Annu Rev Ecol Evol Syst 46:369–392
Fischer AG (1960) Latitudinal variation in organic diversity. Evolution 14:64–81
Fjeldsă J (1994) Geographical patterns dor relict and young species of birds in Afriva and South America and implications for conservation priorities. Biodivers Conserv 3:207–226
Forster JR (1778) Observations made during a voyage made round the world, on physical geography, natural history and ethic philosophy. G Robinson, London
Fujita MK, Leaché AD (2011) A coalescent perspective on delimiting and naming species: a reply to Bauer et al. Proc R Soc B 278(1705):493–495. https://doi.org/10.1098/rspb.2010.1864
Garzón-Orduña JJ, Benetti-Longhini JE, Brower AVZ (2014) Timing the diversification of the Amazionian biota: butterfly divergences are consistent with Pleistocene refugia. J Biogeogr 41:1631–1638
Gaston KJ, Spicer JI (2005) Biodiversity, an introduction. Blackwell, Oxford
Gentry AH (1982) Neotropical floristic diversity: phytogeographical connections between Central and South America, Pleistocene climatic fluctuations, or an accident of the Andean orogeny? Ann Mo Bot Gard 69:557–593
Graham CH, Carnaval AC, Cadena CD et al (2014) The origin and maintenance of montane diversity: integrating evolutionary and ecological processes. Ecography 37:1–9
Gutiérrez RJ, Block B (2013) Self-plagiarism in publishing. J Wildl Manag 77:1487–1488
Haffer J (1969) Speciation in Amazonian forest birds. Science 165:131–137
Haffer J (1997) Alternative models of vertebrate speciation in Amazonia: an overview. Biodivers Conserv 6:451–476
Hawkins BA (2001) Ecology's oldest pattern? Trends Ecol Evol 16:470
Hildebrand H (2004) On the generality of the latitudinal diversity gradient. Am Nat 163:192–211

Hipsley CA, Müller J (2014) Beyond fossil calibrations: realities of molecular clock practices in evolutionary biology. Front Genet 5:138. https://doi.org/10.3389/fgene.2014.00138

Honorio Coronado EN, Dexter KG, Pennington RT et al (2015) Phylogenetic diversity of Amazonian tree communities. Divers Distrib 21:1295–1307

Hoorn C (1994) An environmental reconstruction of the paleo-Amazon river system (Middle-Late Miocene, NW Amazonia). Palaeogeogr Palaeoclimatol Palaeoecol 112:187–238

Hoorn C, Wesselingh F (2010) Amazonia: landscape and species evolution. A look into the past. Wiley-Blackwell, Chichester

Hoorn C, Guerrero J, Sarmiento GA, Lorente MA (1995) Andean tectonics as a cause for changing drainage patterns in Miocene northern South America. Geology 23:237–241

Hoorn C, Wesselingh FP, ter steege H et al (2010) Amazonia through time: Andean uplift, climate change, landscape evolution, and biodiversity. Science 330:927–931

Hoorn C, Wesselingh FP, ter Steege H et al (2011) Origins of biodiversity-response. Science 331:399–400

Hoorn C, Mosbrugger V, Mulch A, Antonelli A (2013) Biodiversity from mountain building. Nat Geosci 6:154

Hubbell SP (2001) The unified neutral theory of biodiversity and biogeography. Princeton University Press, Princeton

Hughes C, Eastwood R (2006) Island radiation on a continental scale: exceptional rates of plant diversification after uplift of the Andes. Proc Natl Acad Sci U S A 103:10334–10339

Hughes CE, Pennington RT, Antonelli A (2013) Neotropical plant evolution: assembling the big picture. Bot J Linn Soc 171:1–18

Hurlbert AH, Stegen JC (2014) On the processes generating latitudinal richness gradients: identifying diagnostic patterns and predictions. Front Genet 5:420. https://doi.org/10.3389/fgene.2014.00420

Jablonski D, Roy K, Valentine JW (2006) Out of the tropics: evolutionary dynamics of the latitudinal diversity gradient. Science 314:102–106

Jablonski D, Huang S, Roy K, Valentine JW (2017) Shaping the latitudinal diversity gradient: new perspectives from a synthesis of paleobiology and biogeography. Am Nat 189:1–12

Jansson R, Rodríguez-Castañeda G, Harding LE (2013) What can multiple phylogenies say about the latitudinal diversity gradient? A new look at the tropical conservatism, out of the tropics, and diversification rate hypotheses. Evolution 67:1741–1755

Jaramillo C, Hoorn C, Silva SAF et al (2010) The origin of the modern Amazon rainforest: implications of the palynological and paleobotanical record. In: Hoorn C, Wesselingh F (eds) Amazonia: landscape and species evolution. A look into the past. Wiley-Blackwell, Chichester, pp 317–334

Jardine PE, Abernethy FAJ, Lomax BH et al (2017) Shedding light on sporopollenin chemistry, with reference to UV reconstructions. Rev Palaeobot Palynol 238:1–6

Kerhoff AJ, Moriarty PE, Weiser MD (2014) The latitudinal richness gradient in New World woody angiospermns is consistent with the tropical conservatism hypothesis. Proc Natl Acad Sci U S A 111:8125–8130

Key FM, Posth C, Krause J et al (2017) Mining metagenomic data sets for ancient DNA: recommended protocols for authentication. Trends Genet 33:508–520

Krug AZ, Jablonski D, Valentine JW, Roy K (2009) Generation of earth's first-order biodiversity pattern. Astrobiology 9:113–124

Laurin M (2012) Recent progresses in paleontological methods for dating the tree of life. Front Genet 3:130. https://doi.org/10.3389/fgene.2012.00130

Lomolino MV, Riddle BR, Brown JH, Whittaker RJ (2010) Biogeography. Sinauer, Sunderland

MacArthur RH (1965) Patterns of species diversity. Biol Rev 40:510–533

Mannion PD, Upchurch P, Benson RBJ, Goswani A (2014) The latitudinal diversity gradient through deep time. Trends Ecol Evol 29:42–50

Marshall LG, Webb SD, Sepkoski JJ, Raup DM (1982) Mammalian evolution and the great American biotic interchange. Science 215:1351–1357

Marx V (2017) Genetics: new tales from ancient DNA. Nat Methods 14:771–774
McGlone MS (1996) When history matters: scale, time, climate and tree diversity. Glob Ecol Biogeogr Lett 5:309–314
McKay R, Naish T, Carter L et al (2012) Antarctic and Southern Ocean influences on Late Pleistocene global cooling. Proc Natl Acad Sci U S A 109:6423–6428
Mittelbach GG, Schemske D, Cornell HV et al (2007) Evolution and the latitudinal diversity gradient: speciation, extinction and biogeography. Ecol Lett 10:315–331
Mittermeier RA, Mittermeier CG, Robles P (1997) Megadiversity: earth's biologically wealthiest nations. CEMEX, Mexico
Mora A, Baby P, Roddaz M et al (2010) tectonic history of the Andes and sub-Adean zones: implications for the development of the Amazon drainage basin. In: Hoorn C, Wesselingh F (eds) Amazonia: landscape and species evolution. A look into the past. Wiley-Blackwell, Chichester, pp 39–60
Moritz C (1994) Defining 'evolutionarily significant units' for conservation purposes. Trends Ecol Evol 9:373–375
Moritz C (2002) Strategies to protect biological diversity and the evolutionary processes that sustain it. Syst Biol 51:238–254
Moritz C, Patton JL, Schneider CJ, Smith TB (2000) Diversification of rainforest faunas: an integrated molecular approach. Annu Rev Ecol Syst 31:533–563
Mutke J, Bathlott W (2005) Patterns of vascular plant diversity at continental to global scales. Biol Skr 55:521–531
Myers N, Mittermeier RA, Mittermeier CG, da Fonseca GAB, Kent J (2000) Biodiversity hotspots for conservation priorities. Nature 403:853–858
Nagalingum NS, Marshall CR, Quental TB et al (2011) Recent synchrononous radiation of a living fossil. Science 334:796–799
Noonan BP, Gaucher P (2005) Phylogeography and demography of Guyanan harlequin toads (*Atelopus*): diversification within a refuge. Mol Ecol 14:3017–3031
Nores M (1999) An alternative hypothesis for the origin of Amazonian bird diversity. J Biogeogr 26:475–485
Olson DM, Dinerstein E, Wikramanayake ED et al (2001) Terrestrial ecoregions of the world: a new map of life on Earth. Bioscience 51:933–938
Pennington RT, Dick CW (2004) The role immigrants in the assembly of South American rainforest flora. Philos Trans R Soc Lond 359:1611–1622
Pennington RT, Dick CW (2010) Diversification of the Amazon flora and its relation to key geological and environmental events: a molecular perspective. In: Hoorn C, Wesselingh F (eds) Amazonia: landscape and species evolution. A look into the past. Wiley-Blackwell, Chichester, pp 373–385
Pennington RT, Prado LM, Pendry CA, Pell SK, Butterwoth CA (2004) Historical climate change and speciation: neotropical seasonally dry forest plants show patterns of both Tertiary and Quaternary diversification. Philos Trans R Soc Lond 359:515–537
Pennington RT, Lavin M, Särkinen T et al (2010) Contrasting plant diversification histories within the Andean biodiversity hotspot. Proc Natl Acad Sci U S A 107:13783–13787
Pianka ER (1966) Latitudinal gradients in species diversity: a review of concepts. Am Nat 100:33–46
Pielou EC (1975) Ecological diversity. Wiley, New York
Pielou EC (1979) Biogeography. Wiley, New York
Pontarp M, Bunnefeld L, Cabral JS et al (2019) The latitudinal diversity gradient: novel understanding through mechanistic eco-evolutionary models. Trends Ecol Evol 34:211–223
Porto TJ, Carnaval AC, da Rocha PLB (2013) Evaluating forest refugia models using species distribution models, model filling and inclusion: a case study with 14 Brazilian species. J Biogeogr 19:330–340
Prance GT (1982) Biological diversification in the tropics. Columbia University Press, New York

Prates I, Rodrigues MT, Melo-Sampaio PR, Carnaval AC (2015) Phylogenetic relationships of Amazonian anole lizards (*Dactyloa*): taxonomic implications, new insights about phenotypic evolution and the timing of diversification. Mol Phylogenet Evol 82:258–268

Prates I, Rivera D, Rodrigues MT, Carnaval AC (2016a) A mid-Pleistocene rainforest corridor enabled synchronous invasions of the Atlantic Forest by Amazonian aanole lizards. Mol Ecol 25:5174–5186

Prates I, Xue AT, Brown JL, Alvarado-Serrano DF et al (2016b) Inferring responses to climate dynamics from historical demography in neotropical forest lizards. Proc Natl Acad Sci U S A 113:7978–1985

Prates I, Melo-Sampaio PR, Drummond LO et al (2017) Biogeographic links between southern Atlantic Forest and western South America: rediscovery, re-description, and phylogenetic relationships of two rare montane anole lizards from Brasil. Mol Phylogenet Evol 113:49–58

Pyron RA, Burbrink FT (2013) Phylogenetic estimates of speciation and extinction rates for testing ecological and evolutionary hypotheses. Trends Ecol Evol 28:729–736

Quental TB, Marshall CR (2010) Diversity dynamics, molecular phylogenies need the fossil record. Trends Ecol Evol 25:434–441

Raby M (2017) The colonial origins of tropical field stations. Am Sci 105:216–223

Raven PH, Axelrod DI (1974) Angiosperm biogeography and past continental movements. Ann Mo Bot Gard 61:539–673

Ribas CC, Aleixo A, Nogueira ACR et al (2011) A palaeobiogeographic model for biotic diversification within Amazonia over the past three million years. Proc R Soc B 279:681–689

Ricklefs RE, Schluter D (1993) Species diversity: regional and historical influences. In: Ricklefs RE, Schluter D (eds) Ecological communities: historical and geographical perspectives. University Chicago Press, Chicago, pp 350–363

Rozensweig M, Schnitzer AE (2013) Self-plagiarism: perspectives for librarians. Coll Res Libr 74:492–494

Rozenzweig ML (1995) Species diversity in space and time. Cambridge University Press, Cambridge

Rull V (2004) Biogeography of the 'Lost World': a palaeoecological perspective. Earth-Sci Rev 67:125–137

Rull V (2005) Biotic diversification in the Guayana Highlands. J Biogeogr 32:921–927

Rull V (2008) Speciation timing and neotropical biodiversity: the Tertary-Quaternary debate in the light of molecular phylogenetic evidence. Mol Ecol 17:2722–2729

Rull V (2009) Microrefugia. J Biogeogr 36:481–484

Rull V (2010) On microrefugia and cryptic refugia. J Biogeogr 37:1623–1627

Rull V (2011a) Neotropical biodiversity: timing and potential drivers. Trends Ecol Evol 26:508–513

Rull V (2011b) Origins of biodiversity. Science 331:398–399

Rull V (2012a) Palaeobiodiversity and taxonomic resolution: linking past trends with present patterns. J Biogeogr 39:1005–1006

Rull V (2012b) Cycad diversification and tropical biodiversity. Collect Bot 31:103–106

Rull V (2013) Some problems in the study of the origin of neotropical biodiversity using palaeoecological and molecular phylogenetic evidence. Syst Biodivers 11:415–423

Rull V (2014) Biodiversity, mountains and climate change. Collect Bot 33:75–79

Rull V (2015) Pleistocene speciation is not refuge speciation. J Biogeogr 42:602–609

Rull V, Montoya E (2014) *Mauritia flexuosa* palm swamp communities: natural or human-made? A palynological study of the Gran Sabana region (northern South America) within a neotropical context. Quat Sci Rev 99:17–33

Rull V, Schubert C, Aravena R (1989) Palynological studies in the Venezuelan Guayana Shield: preliminary results. Curr Res Pleistocene 5:54–56

Sanmartín I, Meseguer AS (2016) Extinction in phylogenetics and biogeography: from timetrees to patterns of biotic assemblage. Front Genet 7:35. https://doi.org/10.3389/fgene.2016.00035

Särkinen T, Pennington RT, Lavin M et al (2012) Evolutionary islands in the Andes: persistence and isolation explain high endemism in Adean dry tropical forests. J Biogeogr 39:884–900
Scheffer M, van Nes EH, Vergnon R (2018) Toward a unifying theory of biodiversity. Proc Natl Acad Sci U S A 115:639–641
Schluter D (2016) Speciation, ecological opportunity, and latitude. Am Nat 187:1–18
Schultz J (2005) The ecozones of the world: the ecological divisions of the biosphere. Springer, Berlin
Simon MF, Grether R, de Queiroz LP et al (2009) Recent assembly of the Cerrado, a neotropical plant diversity hotspot, by in situ evolution and adaptations to fire. Proc Natl Acad Sci U S A 106:20359–20364
Simpson B (1971) Pleistocene changes in the flora and fauna of South America. Science 173:771–780
Slobodkin LB, Sanders HL (1969) On the contribution of environmental predictability to species diversity. In: Woodwell GM, Smith HH (eds) Ecological systems. Brookhaven Symp Biol 22:82–93
Smith BT, McCormack JE, Cuervo AM et al (2014) The drivers of tropical speciation. Nature 515:406–409
Stebbins GL (1974) Flowering plants: evolution above the species level. Belknap Press, Cambridge
Stein A, Gerstner K, Kreft H (2014) Environmental heterogeneity as a universal driver of species richness across taxa, biomes and spatial scales. Ecol Lett 17:866–880
Turchetto-Zolet AC, Pinheiro F, Salgueiro F, Palma-Silva C (2013) Phylogeographical patterns shed light on evolutionary process in South America. Mol Ecol 22:1193–1213
Usinowicz J, Chang-Yang CH, Chen YY et al (2017) Temporal coexistence mechanisms contribute to the latitudinal gradient in forest diversity. Nature 550:105–108. https://doi.org/10.1038/nature24038
van der Hammen T, Hooghiemstra H (2000) Neogene and Quaternary history of vegetation, climate, and plant diversity in Amazonia. Quat Sci Rev 19:725–742
van Tuinen M, Torres CR (2015) Potential for bias and low precision in molecular divergence time estimation of the Canopy of Life: an example from aquatic bird families. Front Genet 6:203. https://doi.org/10.3389/fgene.2015.00203
Vanzolini PE, Williams EE (1970) South American anoles: the geographical differentiation and evolution of the *Anolis chrysolepis* species group (Sauria, Iguanidae). Arq Zool São Paulo 19:1–298
von Humboldt A (1850) Views of nature: or contemplation on the sublime phenomena of creation. Harrison & Sons, London
Wallace AR (1853) A narrative of travels on the Amazon and Rio Negro. Dover, New York
Wallace AR (1878) Tropical nature and other essays. Macmillan, New York
Weeks A, Zapata F, Pell SK et al (2014) To move or to evolve: contrasting patterns of intercontinental connectivity and climatic niche evolution in "Therebintaceae" (Anacardiaceae and Burseraceae). Front Genet 5:409. https://doi.org/10.3389/fgene.2004.00409
Wesselingh F, Hoorn C, Kroonenberg SB et al (2010) On the origin of Amazonian landscapes and biodiversity: a synthesis. In: Hoorn C, Wesselingh F (eds) Amazonia: landscape and species evolution. A look into the past. Wiley-Blackwell, Chichester, pp 421–431
Whitelock LM (2002) The cycads. Timber Press, Portland
Whitmore TC, Prance GT (1987) Biogeography and Quaternary history in tropical Latin America. Oxford University Press, New York
Whittaker RJ, Willis KJ, Field R (2001) Scale and species richness: towards a general, hierarchical theory of species diversity. J Biogeogr 28:453–470
Wiegand T, Uriarte M, Kraft NJB et al (2017) Spatially explicit metrics of species diversity, functional diversity, and phylogenetic diversity: insights into plant community assembly processes. Annu Rev Ecol Evol Syst 48:329–351
Wiens JJ, Donoghue MJ (2004) Historical biogeography, ecology and species richness. Trends Ecol Evol 19:639644

Willig MR, Kaufman DM, Stevens RD (2003) Latitudinal gradients of biodiversity: pattern, process, scales and synthesis. Annu Rev Ecol Evol Syst 34:273–309
Willis KJ, Bhagwat SL (2009) Biodiversity and climate change. Science 326:806–807
Willis KJ, Whittaker RJ (2000) The refugial debate. Science 287:1406–1407
Willis KJ, Whittaker RJ (2002) Species diversity-scale matters. Science 295:1245–1247
Willis CG, Franzone BF, Xi Z, Davis CG (2014) The establishment of central American migratory corridors and the biogeographic origins of seasonally dry tropical forests in Mexico. Front Genet 5:433. https://doi.org/10.3389/fgene.2014.00433
Winter M, Devictor V, Schweiger O (2013) Phylogenetic diversity and nature conservation: where are we? Trends Ecol Evol 28:199–204
Wright JS (2002) Plant diversity in tropical forests: a review of mechanisms of species coexistence. Oecologia 130:1–14
Yasurara M, Hunt G, Dowsett HJ et al (2012) Latitudinal species diversity gradient of marine zooplankton for the last three million years. Ecol Lett 15:1174–1179

Chapter 3
Beyond Refugia: New Insights on Quaternary Climate Variation and the Evolution of Biotic Diversity in Tropical South America

Paul A. Baker, Sherilyn C. Fritz, David S. Battisti, Christopher W. Dick, Oscar M. Vargas, Gregory P. Asner, Roberta E. Martin, Alexander Wheatley, and Ivan Prates

Abstract Haffer's (Science 165: 131–137, 1969) Pleistocene refuge theory has provided motivation for 50 years of investigation into the connections between climate, biome dynamics, and neotropical speciation, although aspects of the original theory are not supported by subsequent studies. Recent advances in paleoclimatology suggest the need for reevaluating the role of Quaternary climate on evolutionary history in tropical South America. In addition to the many repeated large-amplitude climate changes associated with Pleistocene glacial-interglacial stages (~40 kyr and 100 kyr cyclicity), we highlight two aspects of Quaternary climate change in tropical South America: (1) an east-west precipitation dipole, induced by solar radiation changes associated with Earth's precessional variations

P. A. Baker (✉) · A. Wheatley
Division of Earth and Ocean Sciences, Duke University, Durham, NC, USA
e-mail: pbaker@duke.edu

S. C. Fritz
Department of Earth and Atmospheric Sciences, School of Biological Sciences, University of Nebraska-Lincoln, Lincoln, NE, USA
e-mail: sfritz2@unl.edu

D. S. Battisti
Department of Atmospheric Sciences, University of Washington, Seattle, WA, USA

C. W. Dick · O. M. Vargas
Department of Ecology & Evolutionary Biology, University of Michigan, Ann Arbor, MI, USA

G. P. Asner · R. E. Martin
Center for Global Discovery and Conservation Science, Arizona State University, Tucson, AZ, USA

I. Prates
Department of Vertebrate Zoology, National Museum of Natural History, Smithsonian Institution, Washington, DC, USA

V. Rull, A. C. Carnaval (eds.), *Neotropical Diversification: Patterns and Processes*, Fascinating Life Sciences, https://doi.org/10.1007/978-3-030-31167-4_3

(~20 kyr cyclicity); and (2) periods of anomalously high precipitation that persisted for centuries-to-millennia (return frequencies ~1500 years) congruent with cold "Heinrich events" and cold Dansgaard-Oeschger "stadials" of the North Atlantic region. The spatial footprint of precipitation increase due to this North Atlantic forcing extended across almost all of tropical South America south of the equator. Combined, these three climate modes present a picture of climate change with different spatial and temporal patterns than envisioned in the original Pleistocene refuge theory.

Responding to these climate changes, biomes expanded and contracted and became respectively connected and disjunct. Biome change undoubtedly influenced biotic diversification, but the nature of diversification likely was more complex than envisioned by the original Pleistocene refuge theory. In the lowlands, intermittent forest expansion and contraction led to species dispersal and subsequent isolation, promoting lineage diversification. These pulses of climate-driven biotic interchange profoundly altered the composition of regional species pools and triggered new evolutionary radiations. In the special case of the tropical Andean forests adjacent to the Amazon lowlands, new phylogenetic data provide abundant evidence for rapid biotic diversification during the Pleistocene. During warm interglacials and interstadials, lowland taxa dispersed upslope. Isolation in these disjunct climate refugia led to extinction for some taxa and speciation for others.

Keywords Refugia · Tropical South America · Quaternary · Paleoclimate · Phylogenetics · Geogenomics

1 Introduction

Publication of Haffer's (1969) Pleistocene refuge theory encouraged researchers to seek connections between climate-biome dynamics and neotropical diversification. Based on centers of modern endemism identified respectively for birds and lizards, Haffer (1969) and Vanzolini and Williams (1970) independently proposed that pulses of forest contraction and expansion in the Amazon basin, driven by Pleistocene glacial-interglacial climate cycles, forced population isolation and divergence, promoting allopatric speciation in forest refugia. This model assumed large-magnitude drying of the Amazon during the global ice ages that produced the contraction of wet tropical forest into isolated patches with replacement of intervening forest by vast expanses of savanna. In recent years, the Pleistocene refuge theory for Amazonia has been rejected by some who found little evidence either for significant Pleistocene aridity or for forest fragmentation (Colinveaux et al. 1996; Bush et al. 2004). The contention was also made that crown-group ages in some groups of organisms date to the Neogene, in disagreement with the temporal framework of diversification implied in the Pleistocene refuge theory (Moritz et al. 2000; Hoorn et al. 2010; Prates et al. 2015). Yet phylogenetic studies across a wide range of plant and animal taxa increasingly point to the fact that many neotropical

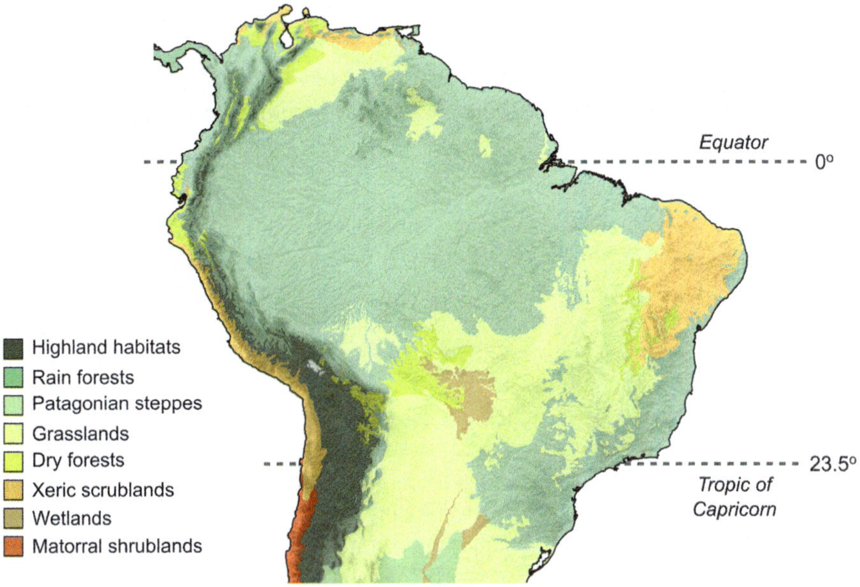

Fig. 3.1 Biome map of tropical South America depicting the distribution of highland habitats (paramo and montane forests in the Andes and Guiana Shield region), rain forests (including Amazonia, Chocó, and the Brazilian Atlantic Forest), Patagonian steppes, grasslands (including the Cerrado, Chaco, Pampas, and Venezuelan Llanos), dry forests, xeric scrublands (including the Brazilian Caatinga), wetlands (including the Brazilian Pantanal), and Chilean Matorral scrublands

sister species did actually diverge in the Pleistocene (Richardson et al. 2001; Hughes and Eastwood 2006; Madriñán et al. 2013; Garzón-Orduña et al. 2014; Koenen et al. 2015; Byrne et al. 2016), opening the possibility of a role for Quaternary climate variation in the diversification of these biota (Rangel et al. 2018; Wheatley et al. 2019, but see Rull and Carnaval 2019 and Vargas and Dick 2019).

In this contribution, we first review major advances in our knowledge of the nature of Quaternary paleoclimate variation at scales relevant to community compositional change and biotic diversification in tropical South America. We address if, and how, Quaternary climate variation affected forest distribution (Fig. 3.1) and composition, to the limited extent that we know either. Finally, we briefly discuss the evidence from biogeographic and phylogenetic studies relevant to the question of how regional taxa may have evolved during, and in response to, periods of Quaternary climate change.

2 Climate Variability During the Quaternary

2.1 Was Amazonia Drier During the Last Glacial Maximum (and Other Glacial Intervals of the Quaternary) Than It Is Today?

We start our discussion of Quaternary climate variability by addressing the first question that has been oft revisited, but never definitively answered. Earth's global climate was stunningly different only 20,000 years ago, at the Last Glacial Maximum (LGM), compared to today. Atmospheric CO_2 was reduced to ~180 ppmv; large ice sheets were present across boreal continental regions, particularly in North America; global eustatic sea level dropped 120 m below present; and global air temperatures averaged ca. 5 °C colder than modern. In tropical South America, Andean snow lines lowered approximately 1 km and glacial erosion greatly increased the sediment load of Andean rivers and sediment delivery to inland basins (Fritz et al. 2007) and to the Atlantic Ocean. The Amazon continental shelf was exposed subaerially and likely became vegetation covered (Leite et al. 2016), and the Amazon river incised its course for nearly 300 km across the shelf, delivering much of its sediment load to the Amazon deep-sea fan (Nace et al. 2014). These changes during the LGM were only the latest manifestation of some 50 prior glacial stages that occurred over the past 2.6 million years: lower-amplitude warm-cold cycles every 40 kyr from 2.6 to ca. 0.8 Ma and higher-amplitude cycles every 100 kyr since (Lisiecki and Raymo 2005).

A common and persistent misconception about Amazon paleoclimate is that the LGM (and earlier glacial stages) was drier than present throughout all of Amazonia. Whereas thermodynamics informs that ca. 35% less water vapor can be held in saturated air due to the ca. 5 °C air temperature lowering deduced for the LGM in tropical South America (Stute et al. 1995), atmospheric dynamics can compensate for lower water vapor content by increasing low-level winds bringing moisture from its Atlantic source into the Amazon. Such an increase could have been brought about both by a southward shift of the zonal mean position of the western Atlantic Inter-tropical Convergence Zone (ITCZ) during the LGM (Black et al. 1999; Peterson et al. 2000; Baker et al. 2001a) and the posited strengthening of the northeast Trades (McGee et al. 2018). Summer insolation over South America was at a maximum during the LGM and intensified the South American summer monsoon (SASM) (Baker et al. 2001a, b; Cruz et al. 2005). Together, these three factors (ITCZ, trade winds, insolation) led to a generally high rate of precipitation, perhaps similar to modern, during the LGM in the SASM-region, i.e. the tropical central Andean region, the western Amazon, and subtropical South America. That conclusion is based on multiple lacustrine (e.g., Baker et al. 2001a, b; Baker and Fritz 2015) and speleothem (e.g., Cruz et al. 2005; Cheng et al. 2013) records of LGM climate from sites around the western and southern periphery of the Amazon (Fig. 3.2), from which we infer that western Amazon precipitation was high (i.e. comparable to modern) during the LGM (Table 3.1).

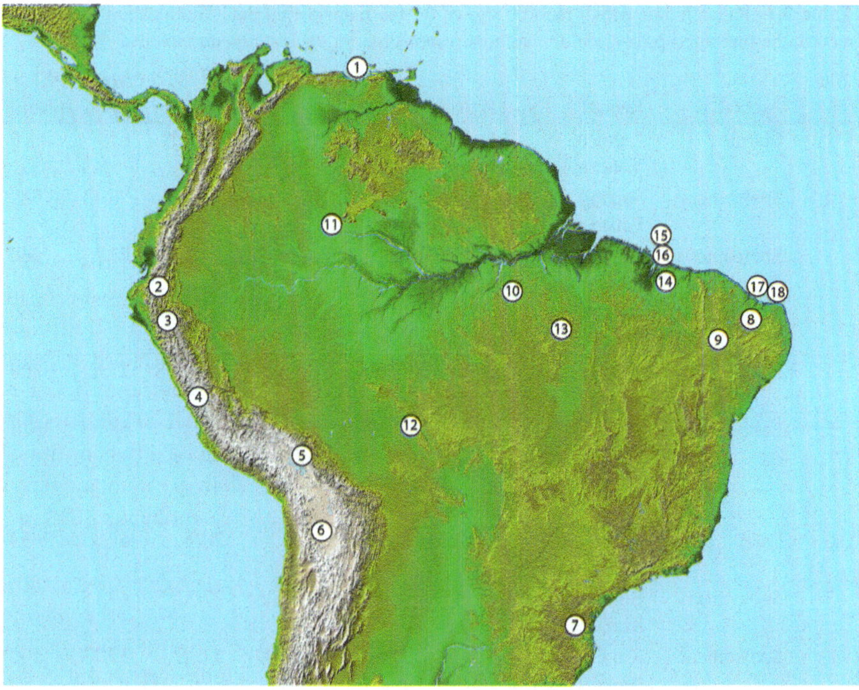

Fig. 3.2 Location of sites discussed in the text superimposed on a digital elevation model of South America. See Table 3.1 for site key and additional site details

Fewer paleoclimate records of the LGM exist from locations within the central and eastern parts of the Amazon basin and northeastern Brazil. Yet within the last decade, new speleothem records from northeastern Brazil (Cruz et al. 2009) and the eastern Amazon (Wang et al. 2017) indicate significant spatial variation of precipitation at orbital time scales across tropical South America. Specifically, Cruz et al. (2009) uncovered a precipitation dipole between the western Amazon/tropical central Andes and northeastern Brazil, varying with precessional (20 kyr) periodicity (Fig. 3.3). This finding was reinforced by a speleothem record from eastern Amazonia (Wang et al. 2017). Thus, in the western pole (western Amazonia, the tropical central Andes, the subtropics), precipitation increased during periods of increased summer insolation, while in the eastern pole (northeastern Brazil, the eastern Amazon), precipitation decreased during the same periods of higher summer insolation.

The spatial footprint of this precipitation dipole is not well defined, because of the limited number of paleoclimate sites. Moreover, the longest speleothem record from the eastern dipole region (Wang et al. 2017) only extends back to 45,000 years before present. Although no terrestrial paleoclimate records that are presently available can confirm its long-term persistence, the existence of the east-west precipitation dipole is a robust feature of many different climate models (Fig. 3.4) forced by

Table 3.1 Paleoclimate proxy records shown on the map in Fig. 3.2. The location and type of record, the measured proxy, and the authors responsible for each study are included

Site #	Archive	Record	Proxies	References	Lat.	Long °W	Elev. (m)
1	Marine	Caraico, Venezuela	Color	Peterson et al. (2000)	10.5°N	65	0
2	Speleothem	Santiago, Ecuador	$\delta^{18}O$	Mosblech et al. (2012)	3.02°S	78.13	980
3	Speleothem	Diamante, Perú	$\delta^{18}O$	Cheng et al. (2013)	5.73°S	77.5	960
	Speleothem	El Condor, Perú	$\delta^{18}O$	Cheng et al. (2013)	5.93°S	77.3	860
4	Speleothem	Pacupahuain, Perú	$\delta^{18}O$	Kanner et al. (2012)	11.24°S	75.82	3800
5	Lake	Titicaca, Bolivia/Perú	$\delta^{13}C$	Baker et al. (2001a, b), Fritz et al. (2007, 2010)	16°S	68.5	3810
6	Lake	Uyuni, Bolivia	Gamma	Baker et al. (2001a, b)	20°S	68	3653
7	Speleothem	Botuverå, Brazil	$\delta^{18}O$	Cruz et al. (2005)	27.22°S	49.15	230
8	Speleothem	Rio Grande, Brazil	$\delta^{18}O$	Cruz et al. (2009)	5.6°S	37.73	100
9	Groundwater	Maranhao, Brazil	Noble gas	Stute et al. (1995)	7°S	41.5	400
10	Speleothem	Paraíso, Brazil	$\delta^{18}O$	Wang et al. (2017)	4.07°S	55.45	60
11	Lake	Hill of Six Lakes, Brazil	Pollen	Bush et al. (2004)	0.30°N	66.67	75
12	Lake	Bella Viata, Bolivia	Pollen	Punyasena et al. (2008), Burbridge et al. (2004)	13.62°S	61.55	225
	Lake	Champlain, Bolivia	Pollen	Punyasena et al. (2008), Burbridge et al. (2004)	14.47°S	61.07	225
13	Lake	Carajas, Brazil	Pollen	Absy et al. (1991), Hermanowski et al. (2012), Reis et al. (2017)	6.4°S	50.42	730
14	Lake	Cacó, Brazil	Pollen	Ledru et al. (2006)	2.97°S	43.42	0
15	Marine	CDH-86, Brazil	XRF	Nace et al. (2014)	0.33°S	44.21	0
16	Marine	GeoB-16,205, Brazil	Pollen	Bouimetarhan et al. (2018)	1.21°S	43.05	0

(continued)

Table 3.1 (continued)

Site #	Archive	Record	Proxies	References	Lat.	Long °W	Elev. (m)
17	Marine	GeoB-3912, Brazil	XRF	Arz et al. (1998)	3.67°S	37.43	0
	Marine	GeoB-3104, Brazil	Pollen	Behling et al. (2000)	3.67°S	37.43	0
18	Marine	GeoB-3910, Brazil	Pollen	Dupont et al. (2010)	4.25°S	36.34	0

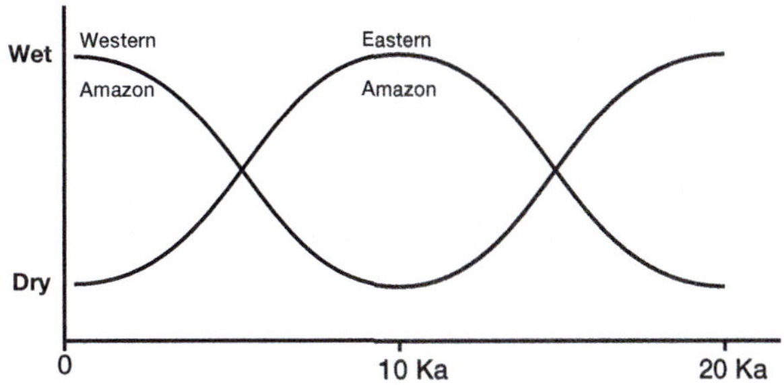

Fig. 3.3 Sketch of the temporal evolution of precipitation of the eastern and western Amazon through one precession (20 kyr) cycle

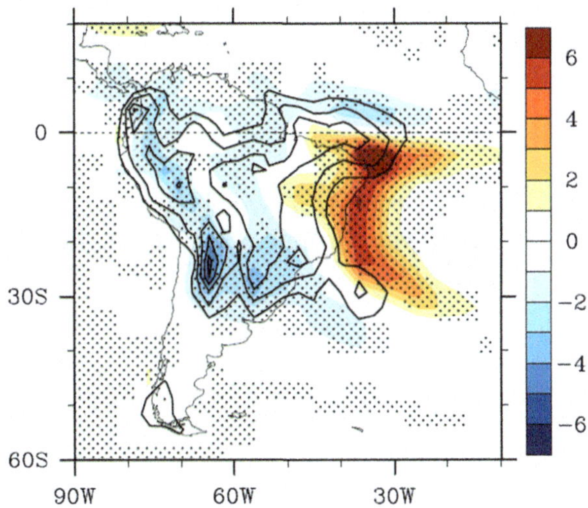

Fig. 3.4 A map of the east to west precipitation dipole in tropical South America on precession time scales (20 kyr) that shows the difference in DJF precipitation (mm d^{-1}) observed between low summer insolation (218 Ka) and high summer insolation (207 Ka) experiments. From Liu and Battisti (2015)

processionally induced changes in insolation (Prado et al. 2013; Liu and Battisti 2015).

Wang et al. (2017) calculated on the basis of their speleothem isotopic record that precipitation during the LGM was 58% of modern in the eastern Amazon region, supporting the idea of a drier eastern Amazon at this time. However, they failed to adjust speleothem oxygen isotopic values for the estimated 5 °C cooling at the LGM (Stute et al. 1995) and changes in seawater $\delta^{18}O$ (Schrag et al. 2002). When their speleothem record is corrected for these two effects (by subtracting ~2‰ from the LGM speleothem $\delta^{18}O$ value, see Baker and Fritz 2015), LGM $\delta^{18}O$ values nearly match modern $\delta^{18}O$ values in the same speleothem, from which we conclude that LGM precipitation was nearly equal to modern levels. That is, LGM precipitation in the eastern Amazon was similar to modern precipitation in that region, although relatively low compared to contemporaneous precipitation levels from regions farther west. The corrected record indicates that it was the early-to-mid Holocene time period (~9000–5000 years before present) that had exceptional levels of precipitation. This period was evidently far wetter-than-modern in the eastern Amazon, while far drier-than-modern in the western Amazon (Punyasena et al. 2008) and tropical central Andes (Baker et al. 2001a; Cross et al. 2001).

In conclusion, data and models agree that there were large-amplitude fluctuations in precipitation amount on precessional time scales for the past 40,000 years in the near-equatorial regions, with drier conditions in the east accompanied by wetter conditions in the west, and vice versa. Furthermore, models suggest that that this east-west precipitation dipole existed throughout the entire Quaternary and prior to that. However, the exact spatial footprint of the dipole, the magnitude of precipitation change, and its impact on the forest and associated biota, all remain to be determined. Whereas it is evident from the data that spatially variable regions of climate change and forest expansion/contraction (e.g. Cheng et al. 2013) did not resemble the patterns envisioned by Haffer (1969), it is expected that climate-driven variation of forest biomes during the Quaternary had a profound influence on biotic dispersal, gene flow, and divergence (see below).

2.2 Short-Term (Millennial) Extreme Precipitation Events: Were They Sufficient to Establish or Destroy Forested Habitat?

Some of the intervals of most extreme precipitation change in tropical South America during the late Quaternary are not tied to gradual variations in insolation driven by orbital forcing but instead were shorter-duration anomalies, persisting for several hundred to a few thousand years. This "millennial variability" is associated with large changes in sea surface temperatures in the North Atlantic region.

One example of such millennial variability, predominantly present during the last glacial stage, is Dansgaard-Oeschger (D-O) cyclicity. D-O cycles were first identified in the oxygen isotopic record of Greenland ice cores and determined to be warm-cold oscillations of air temperature (Dansgaard et al. 1993). Approximately 25 D-O cycles occurred between 90 and 10 Ka (Rahmstorf 2002). D-O cyclicity is also present in Antarctic ice cores (Steig and Alley 2002), where temperatures are anti-phased with Greenland (the "bipolar seesaw") and amplitudes are of lower magnitude (see Pedro et al. 2018 and references therein). D-O cyclicity is recorded in sea-surface temperature proxies from North Atlantic sediment and is widespread elsewhere in the Northern Hemisphere (Voelker 2002). D-O cyclicity affected precipitation across the Neotropics: cold periods in the North Atlantic are associated with an apparent intensification of the SASM and increased precipitation everywhere in tropical South America south of the equator (Fritz et al. 2010; Kanner et al. 2012; Cheng et al. 2013; Wang et al. 2017), with the possible exception of the Pacific coast. How this persistent millennial variation during glacial stages impacted biotic communities in the heart of Amazonia remains to be addressed. The regional instability associated with the occurrence of at least a couple dozen pulses of alternating increased and decreased precipitation over such a short period (i.e., ~80 ky) may have led to recurrent extirpation of forest organisms in northeastern Brazil. This dynamic may explain a pattern of lower species richness in eastern, as compared to the climatically more stable western, Amazonia (e.g., Jenkins et al. 2013).

Heinrich events are a second type of millennial climate variability during glacial stages, closely related to D-O stadials ("stadials" are the cold phases of the D-O cycles). Heinrich events appear to occur only during the most extreme D-O stadials. Heinrich events are characterized by the massive discharge of icebergs, primarily derived from the Laurentide Ice Sheet, into the North Atlantic Ocean (Broecker 1994). Subsequent melting of the icebergs significantly decreased sea-surface salinity of the North Atlantic, possibly shutting down the Atlantic meridional overturning circulation and increasing the abundance of sea ice in the North Atlantic. Although both the causes and the consequences of Heinrich events are still debated, there is no doubt that large-scale climate impacts, synchronous with the Heinrich events, were felt in many far-field regions (Hemming 2004; Vellinga and Wood 2002). Paleoceanographic records from the Brazilian continental margin (Arz et al. 1998; Nace et al. 2014) document large increases of river runoff in northeastern Brazil synchronous with Heinrich events. On the Altiplano of Bolivia, Heinrich Event 1 (H1) brought about flooding and major expansion of now dry lakes (Sylvestre et al. 1999; Baker et al. 2001b). And Heinrich events coincide with the most negative $\delta^{18}O$ values (indicating peak wet conditions) in speleothem records from the central Andes of Peru (Kanner et al. 2012), the western Amazon (Mosblech et al. 2012; Cheng et al. 2013), and the eastern Amazon (Wang et al. 2017).

Thus, D-O stadials and Heinrich events had similar impacts on tropical South American climate (Fig. 3.5), although Heinrich events were apparently associated with more extreme climates than were the D-O stadials (Zhang et al. 2017), consistent with the fact that the former represented more extreme cold conditions in the North Atlantic region. Paleoclimate data show that wet climates associated with these North Atlantic cold events occurred simultaneously across all of tropical South America, south of the equator, in both eastern and western regions. Some of these wet events lasted as long as 5000 years (Nace et al. 2014), although most were shorter in duration. It seems likely that these North Atlantic warm-cold swings, associated respectively with large amplitude dry-wet swings in tropical South America, brought about profound changes in forest composition and extent. Ecological studies have found that the recovery of rain forest following disturbance happens quickly, within only a couple hundred years (Liebsch et al. 2008). Moreover, vertebrate species can diffuse through thousands of kilometers of suitable

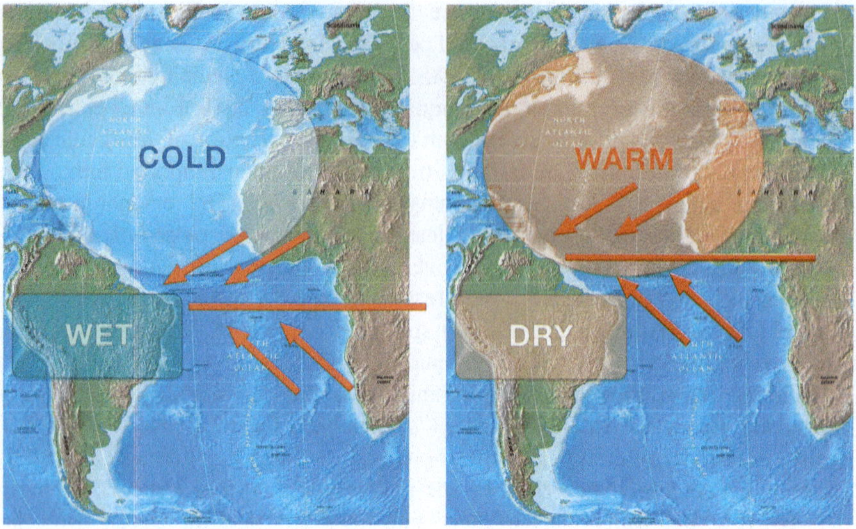

Fig. 3.5 A sketch of the impact of millennial cold and warm variability in the North Atlantic region on the position of the ITCZ and trade winds and on precipitation throughout tropical South America. See text for additional details

habitat within only a few decades, as suggested by ecological studies of amphibian, mammal, and bird species (Phillips et al. 2007; Stodart and Parer 1988; Wehtje 2003). As a result, the temporal scale of the climatic shifts implicated in D-O stadials and Heinrich events certainly was sufficient to lead to pronounced changes in habitat and species distributions in northern South America.

3 Biological Responses to Quaternary Climate Variation

3.1 Biome Expansions and Contractions in Response to Orbital and Millennial Climate Change: Was Pleistocene Climate Variability Responsible for Speciation?

Paleoecological data compiled over the last few decades suggest that tropical forest composition was dynamic in response to changes in climate, but there remains little evidence of large-scale shifts in the geographic distribution of major Amazonian biomes on glacial to interglacial time scales as envisioned by Haffer (Bush et al. 2004). The apparent persistence of wet tropical forest in the western to central Amazon (e.g., Cheng et al. 2013) has major implications for evolutionary studies of other organisms associated with forest settings. For instance, some animal populations may have remained stable even when faced with temporal turnover of tree species, as long as the structural forest environment remained similar over time. Yet, the paucity of sites in present-day Amazon forest and of records that date back more than ~60,000 years means that we still have a very incomplete picture of variability in the distribution and species composition of regional biomes through time (Fig. 3.1). While the available information points to stability of wet tropical forests in western to central Amazonia, there is clear evidence of biome expansion and contraction near rainforest-savanna ecotones in the southern and eastern Amazon regions (Absy et al. 1991; Burbridge et al. 2004; Hermanowski et al. 2012; Reis et al. 2017).

Pollen data also suggest vegetation responses to the large wet millennial events of the late-Quaternary in regions marginal to, but outside of, Amazonia. In areas of northeastern Brazil now occupied by semi-arid Caatinga, humid gallery forests apparently expanded during Heinrich 1 and the Younger Dryas events (Behling et al. 2000; Ledru et al. 2006; Dupont et al. 2010; Bouimetarhan et al. 2018). This forest expansion may have been sufficiently extensive to have produced a landscape mosaic with wet forest corridors connecting the Atlantic and eastern Amazon forests (Cheng et al. 2013; Bouimetarhan et al. 2018). However, phylogenetic evidence of dispersal events through gallery forests in the open habitats of central and northern Brazil is currently lacking.

Phylogenetic patterns in distinct organisms suggest links between Quaternary climate variation and population divergence. Patterns of genetic structure in species that colonized the Atlantic Forest from Amazonia indicate population genetic differentiation following forest contraction and biome separation (Dal-Vechio et al. 2018; Prates et al. 2018). These climate-driven events of population divergence, inferred to have happened recently, provide a mechanism to explain speciation and new evolutionary radiations also at deeper timescales. For example, a pattern of sister relationships between species and clades from Amazonia and the Atlantic Forest in several bird and small mammal groups (Costa 2003; Batalha-Filho et al. 2013) is consistent with the hypothesis that expansion of open and dry habitats (i.e., Caatinga dry forest and Cerrado savanna) following intervals of wet forest expansion favored speciation of rainforest organisms.

Lastly, climatic stability may contribute to the accumulation of biodiversity at regional spatial scales. In an example from the Atlantic Forest region, Carnaval et al. (2009, 2014) tested whether vertebrate phylogenetic endemism was related to forest persistence over the last 120 kyr. Their analysis, using climate model output, suggests that forest stability was a necessary factor for maintaining high levels of lineage endemism. However, different factors best explain patterns of endemism in the northern and southern regions of the Atlantic Forest; whereas lineage endemism is better predicted by climatic stability through time in the north, spatial climatic heterogeneity was a more important predictor of lineage endemism in the south. Such examples of integration of geologic and genetic approaches (geogenomics, *sensu* Baker et al. 2014) provide a potentially powerful means of testing models of how climatic history has affected evolutionary history and shaped current spatial patterns of biodiversity.

3.2 *Shifts in Community Composition Through Climate-Mediated Dispersal*

Studies of diversification in Amazonian organisms have often emphasized in situ speciation, but the composition of local species pools is also strongly affected by migration between regions (Wiens 2004; Antonelli et al. 2018), and several examples of climate-mediated dispersal during the Quaternary have been proposed for Amazonia. For instance, a stepping-stone like corridor of dry vegetation formations may have favored migration of dry-adapted tree species (Bush 1994) and viperid snakes (Quijada-Mascareñas et al. 2007) through an otherwise wet forest matrix. Similarly, climate-driven geomorphic processes that affected the distribution of sandy soils may have created corridors for dispersal of dry-adapted taxa (D'Apolito et al. 2017); alternatively, drier conditions may have permitted dispersal of white sand forest specialists within a former non-analog forest matrix (Capurucho et al. 2013). The patchy distribution and low species endemism of the physiognomically distinctive white sand flora across the Amazon basin (Adeney et al. 2016) support the latter idea. In addition, pollen records suggest that cold-adapted upland tree taxa (e.g. *Podocarpus*) descended to the lowlands and expanded through the Amazon forest in response to past climate cooling (Colinveaux et al. 1996), creating assemblages that have no modern analog (Bush et al. 2004; Reis et al. 2017). These episodes of range expansion, followed by subsequent extinction within a wetter Amazon, could explain plant disjunctions between the Guiana Shield and Andean slopes (Berry and Riina 2005).

Climate-driven opportunities for dispersal may have produced large-scale biogeographic interactions between Amazonia and other South American biomes, with pronounced effects on the composition of regional assemblages. For instance, reconstructions of population history on the basis of genetic data from vertebrate taxa point to the establishment of rainforest corridors connecting eastern Amazonia with the northern Atlantic Forest in northeastern Brazil during the Pleistocene (Batalha-Filho et al. 2013; Dal-Vechio et al. 2018; Prates et al. 2016a, b, 2018). This finding is consistent with reconstructed pulses of increased precipitation on the basis of speleothem records, as discussed above (Cheng et al. 2013). Thus, climate-driven habitat shifts may have affected biotic composition and associated gene pools in Amazonia and adjacent regions by favoring dispersal.

3.3 *Perspectives on Climate, Topography, Soils, and Diversification in the Quaternary*

An increasing number of studies support a direct link between Pleistocene climatic variability, dispersal, and speciation in tropical South America. Several groups of lowland rain forest and Andean taxa diversified during the Pleistocene (e.g., Richardson et al. 2001; Kay et al. 2005; Hughes and Eastwood 2006; Lavin 2006; Erkens et al. 2007; Pouchon et al. 2018), and patterns of phylogenetic structure in birds, mammals, and reptiles support pulses of forest expansion that are spatially congruent across taxa, such as in present-day northeastern Brazil (Batalha-Filho et al. 2013; Costa 2003; Dal-Vechio et al. 2018; Prates et al. 2016b). The timing of population divergence and speciation for many taxa, as well as the inferred routes of range shifts, match expectations derived from the known variability of Pleistocene climate. These large-amplitude climate shifts extended not only into the Amazon basin but also into higher elevations in the tropical Andes.

Pleistocene climate variability superimposed upon pre-existing Andean topography provides a mechanism for driving speciation in high relief terrains. It has been suggested for plants (Gentry 1982) and frogs (Santos et al. 2009) and recently modeled (Rangel et al. 2018; Wheatley et al. 2019) that a significant proportion of the taxonomic diversification of tropical South America has arisen in the tropical Andes. Although Neogene and earlier history may explain some of this diversity (Luebert and Wiegend 2014; Antonelli et al. 2009), many Andean diversification events are recent, overlapping with the timing of Quaternary climatic dynamism (van der Hammen and Cleef 1986; Madriñán et al. 2013). Under interglacial or interstadial warm conditions, montane forest and alpine taxa colonize upslope, where populations may be disjunct from each other because of topographic isolation—valleys form genetic barriers whose taxon-specific effective porosity is related to their width, depth, and the physiological tolerances of different organisms (e.g., Wiens 2004). Subsequent cooling can reconnect isolated populations as they disperse downhill, encouraging genetic exchange between formerly disjunct populations, but also introducing newly divergent species back into the lowlands. Mountain tops thus serve as real climate refugia promoting allopatric speciation boosted by biotic interactions and ecological divergence (Lagomarsino et al. 2016; Vargas and Simpson 2019).

The Andes are only the most extreme of the mountains that circumscribe the Amazon. Lower-elevation ranges, such as the Guianan tablelands and Roraima to the north, the Serra do Divisor to the west, the Serra dos Gradaus and Serra do Cachimbo to the south, virtually surround the central Amazon basin. Pleistocene climate variability is known to have strongly impacted the biota of at least some of these regions (e.g., Reis et al. 2017), and new species may have arisen in these regions by climate-driven dispersal followed by isolation on mountain tops, a mechanism similar to that proposed for the Andes. Moreover, novel taxa originating in these high elevation habitats adjacent to Amazonia may have dispersed downward, providing lineages to central Amazonia (Santos et al. 2009).

Another important layer needed to improve our understanding of how Quaternary climate change affected neotropical biotas is the synergistic influence of geological substrate and soils. For example, although it has been posited that seasonally dry forests are currently restricted into their "refugia" and were more broadly distributed during the LGM (Pennington et al. 2000), the absence of rich, high pH soils between contemporary dry forests would have hindered their broad expansion. Although the distribution of soils has been poorly mapped in the Andes-Amazon region, the functional response of the tree canopy to spatially-varying substrate has been resolved at local scales and mapped over large biogeographic areas (Asner et al. 2015), providing new tools for assessing potential edaphic and nutrient influences on biodiversity and its origins.

Climate variability may also help to explain attributes of ecological communities. This is the case, for instance, in the observation that many tree species in the Amazon region are rare. It is possible that the low abundances of these species represent population declines in response to climatic shifts, while more favorable climates in the past may have sustained larger populations. Similarly, periodic mixing and population changes associated with spatio-temporal climate variability may explain why Amazon tree communities are widely dispersed and appear to have little community phylogenetic structure (Dexter et al. 2017). Lastly, climate variability may also help to explain how drought-tolerant and drought-intolerant plant species come to reside in the same forests (Esquivel-Muelbert et al. 2019).

4 Conclusions

Although Haffer (1969) posited large climate and biome variation during the Quaternary, at the time of his publication he could not marshal supporting paleoclimatic or paleobiotic information. Paleoclimatic studies during the intervening decades have fully established that there was very large, extrinsically forced, coherent, climate variation during the Quaternary across the Amazon, northeastern Brazil, subtropical Brazil, and the tropical Andes. It is now clear that the Amazon basin as a whole was not uniformly drier-than-modern during Pleistocene glacial intervals. For instance, the western Amazon, the central Andes, and the southeastern sub-tropics were similarly wet as at present during the LGM, whereas the eastern Amazon/northeast Brazil region was similarly dry as today. Past wet and dry intervals of the Quaternary occurred on both orbital and millennial timescales, and the spatial footprints of wet and dry regions also varied, both east-west and north-south. In particular, large precipitation and effective moisture increases occurred coherently on millennial timescales across all of tropical South America, south of the equator, concurrent with cold "Heinrich events" and D-O stadials of the North Atlantic region. In short, Quaternary climate variation was neither monolithic nor necessarily contemporaneous solely with global glacial-interglacial cycles; rather the pacing and spatial footprint of past climates varied on both orbital and sub-orbital timescales.

Genetic evidence from a range of taxa is consistent with the hypothesis that alternately wet and dry conditions during the Quaternary brought about pulses of forest expansion and contraction. Biogeographic investigations of forest taxa recovered population changes and speciation events within the temporal and spatial scales implicated in climatic change associated with the precession cycles, Heinrich events, and D-O stadials. These pulses of climate-driven habitat shifts had dramatic effects on the composition of regional species pools and likely contributed to diversification. For instance, changes in habitat distribution led to connections and biogeographic exchange among presently distinct neotropical biomes, such as Amazonia and the Atlantic Forest. Moreover, forest contraction concomitant with the expansion of open and dry regions (such as the Cerrado and Caatinga) led to separation of major forested regions; isolated in these "mega-refugia", forest populations diverged and became new species, and species became new clades. The genetic evidence also supports that Quaternary climate variation promoted explosive radiations in the regions of high topographic relief surrounding central Amazonia, which subsequently provided lineages and increased the diversity of lowland forests. New biological and paleoenvironmental data convincingly support that Quaternary climate variation played a major role in evolutionary diversification in tropical South America. The responses of co-distributed organisms to these climate dynamics may have been mediated, to some extent, by distinct ecological tolerances and their underlying physiological and life history traits (Prates et al. 2016a, b).

Acknowledgements Writing of the manuscript was supported by a NASA workshop grant (NASA 15-BIODIV15-0013) to SF, as well as by NSF EAR-1338694 to PB, SF, DB, and CD. We are grateful to the Charles Darwin Foundation, Galapagos for their support during the writing of an initial version of this manuscript. John Megahan, University of Michigan, assisted with graphics.

References

Absy ML, Cleef A, Fornier M, Servant M, Siffedine A, Da Silva MF, Soubies F, Suguio K, Turcq B, Van der Hammen T (1991) Mise en evidence de quatre phases d'ouverture de la foret dense dans le sud-est de L'Amazonie au cours des 60,000 dernieres annees. Premiere comparaison avec d'autres regions tropicales. C R Acad Sci II 312:673–678

Adeney JM, Christensen NL, Vicentini A, Cohn-Haft M (2016) White-sand ecosystems in Amazonia. Biotropica 48:7–23

Antonelli A, Nylander JAA, Persson C, Sanmartin I (2009) Tracing the impact of the Andean uplift on Neotropical plant evolution. Proc Natl Acad Sci U S A 106:9749–9754. https://doi.org/10.1073/pnas.0811421106

Antonelli A, Zizka A, Carvalho FA, Scharn R, Bacon CD, Silvestro D, Condamine FL (2018) Amazonia is the primary source of Neotropical biodiversity. Proc Natl Acad Sci U S A 115:6034–6039. https://doi.org/10.1073/pnas.1713819115

Arz HW, Patzold J, Wefer G (1998) Correlated millennial-scale changes in surface hydrography and terrigenous sediment yield inferred from last-Glacial marine deposits off Northeastern Brazil. Quat Res 50:157–166

Asner GP, Anderson CB, Martin RE, Tupayachi R, Knapp DE, Sinca F (2015) Landscape biogeochemistry reflected in shifting distributions of chemical traits in the Amazon forest canopy. Nat Geosci 8:567–573

Baker PA, Fritz SC (2015) Nature and causes of Quaternary climate variation of tropical South America. Quat Sci Rev 124:31–47

Baker PA, Seltzer GO, Fritz SC, Dunbar RB, Grove M, Tapia P, Cross S, Rowe H, Broda J (2001a) The history of South American tropical climate for the past 25,000 years. Science 291:640–643

Baker PA, Rigsby CA, Seltzer GO, Fritz SC, Lowenstein T, Bacher N, Veliz Y (2001b) Tropical climate changes at millennial and orbital timescales revealed by deep drilling on the South American Altiplano. Nature 409:698–701

Baker PA, Fritz SC, Dick CW, Eckert AJ, Horton BK, Manzoni S, Ribas CC, Garzione CN, Battisti DS (2014) The emerging field of geogenomics: constraining geologic problems with genetic data. Earth Sci Rev 135:38–47

Batalha-Filho H, Fjelsa J, Fabre P-H, Yumi C (2013) Connections between the Atlantic and the Amazonian forest avifaunas represent distinct historical events. J Ornithol 154:41–50

Behling H, Arz HW, Patzold J, Wefer G (2000) Late Quaternary vegetational and climate dynamics in northeastern Brazil, inferences from marine core GeoB 3104-1. Quat Sci Rev 19:981–994

Berry PE, Riina R (2005) Insights into the diversity of the Pantepui Flora and the biogeographic complexity of the Guayana Shield. Biologiske Skrifter 5:145–167

Black DE, Peterson LT, Overpeck JT, Kaplan A, Evans MN, Kashgarian M (1999) Eight centuries of North Atlantic Ocean atmosphere variability. Science 286:1709–1713

Bouimetarhan I, Chiessi C, Gonzalez-Arango CG, Dupont L, Voigt I, Prange M, Zonneveld K (2018) Intermittent development of forest corridors in northeastern Brazil during the last deglaciation: climatic and ecologic evidence. Quat Sci Rev 192:86–96

Broecker WS (1994) Massive iceberg discharges as triggers for global climate change. Nature 372:421–424

Burbridge RE, Mayle FE, Killeen TJ (2004) Fifty-thousand year vegetation and climate history of Noel Kempff Mercado National Park, Bolivian Amazon. Quat Res 61:215–230

Bush MB (1994) Amazonian speciation: a necessarily complex model. J Biogeogr 21:5–18

Bush MB, De Oliveira PE, Colinvaux PA, Miller MC, Moreno JE (2004) Amazonian paleoecological histories: one hill, three watersheds. Palaeogeogr Palaeoclimatol Palaeoecol 214:359–393

Byrne H, Rylands AB, Carneiro JC, Alfaro JWL, Bertuol F, Silva MNF, Messias M, Groves CP, Mittermeier RA, Farias I, Hrbek T, Schneider H, Sampaio I, Boubli JP (2016) Phylogenetic relationships of the New World titi monkeys (Callicebus): first appraisal of taxonomy based on molecular evidence. Front Zool 13:10. https://doi.org/10.1186/s12983-016-0142-4

Capurucho JMG, Cornelius C, Borges SH, Cohn-Haft M, Aleixo A, Metzger JP, Ribas CC (2013) Combining phylogeography and landscape genetics of *Xenoplpo atronitens* (Aves: Pipridae), a white sand campina specialist, to understand Pleistocene landscape evolution in Amazonia. Biol J Linn Soc 110:60–76

Carnaval AC, Hickerson MJ, Haddad CFB, Rodrigues MT, Moritz C (2009) Stability predicts genetic diversity in the Brazilian Atlantic forest hotspot. Science 323:785–789

Carnaval AC, Waltari E, Rodrigues MT, Rosauer D, VanDerWal J, Damasceno R, Prates I, Strangas M, Spanos Z, Rivera D, Pie MR, Firkowski CR, Bornschein MR, Ribero LF, Moritz C (2014) Prediction of phylogeographic endemism in an environmentally complex biome. Proc R Soc B 281:20141461

Cheng H, Sinha A, Cruz FW, Wang X, Edwards LR, d'Horta FM, Ribas CC, Vuille M, Stott LD, Auler AS (2013) Climate change patterns in Amazonia and biodiversity. Nat Commun 4:1411

Colinveaux PA, DeOlivera PE, Moreno JE, Miller MC, Bush MB (1996) A long pollen record from lowland Amazonia: forest cooling in Glacial times. Science 274:85–88

Costa LP (2003) The historical bridge between the Amazon and the Atlantic Forest of Brazil: a study of molecular phylogeography with small mammals. J Biogeogr 30:71–86

Cross S, Baker P, Seltzer G, Fritz S, Dunbar R (2001) Late Quaternary climate and hydrology of tropical South America inferred from an isotopic and chemical model of Lake Titicaca, Bolivia and Peru. Quat Res 56:1–9

Cruz FWJ, Burns SJ, Karmann I, Sharp WD, Vuille M, Cardoso AO, Ferrari JA, Silva Dias PL, Viana OJ (2005) Insolation-driven changes in atmospheric circulation over the past 116,000 years in subtropical Brazil. Nature 434:63–66

Cruz FW, Vuille M, Burns SJ, Wang X, Cheng H, Werner M, Edwards RL, Karmann I, Auler AS, Nguyen H (2009) Orbitally driven east-west anti-phasing of South American precipitation. Nat Geosci 2:210–214

D'Apolito CD, Absy ML, Latrubesse EM (2017) The movement of pre-adapted cool taxa in north-central Amazonia during the last glacial. Quat Sci Rev 169:1–12

Dal-Vechio F, Prates I, Grazziotin FG, Zaher H, Rodrigues MT (2018) Phylogeography and historical demography of the arboreal pit viper *Bothrops bilineatus* (Serpentes, Crotalinae) reveal multiple connections between Amazonian and Atlantic rainforests. J Biogeogr 45:2415–2426

Dansgaard W, Johnsen SJ, Clausen HB, Dahl-Jensen D, Gunderstrup NS, Hammer CU, Hvidberg CS, Steffensen JP, Sveinbjirnsdottir AE, Jouzel J, Bond G (1993) Evidence for general instability of past climate from a 250-year ice core record. Nature 364:218–220

Dexter KG, Lavin M, Torke BM, Twyford AD, Kursar TA, Coley PD, Drake C, Hollands R, Pennington RT (2017) Dispersal assembly of rain forest tree communities across the Amazon basin. Proc Natl Acad Sci U S A 114:2645–2650. https://doi.org/10.1073/pnas.1613655114

Dupont L, Schlutz F, Ewah CT, Jennerjahn TC, Pal A, Behling H (2010) Two-step vegetation response to enhanced precipitation in Northeast Brazil during Heinrich event 1. Glob Chang Biol 16:1647–1660

Erkens RHJ, Chatrou LW, Maas JW (2007) A rapid diversification of rainforest trees (*Guatteria*; Annonaceae) following dispersal from Central into South America. Mol Phylogenet Evol 44:399–411. https://doi.org/10.1016/j.ympev.2007.02.017

Esquivel-Muelbert A, Baker TR, Dexter KG (2019) Compositional response of Amazon forests to climate change. Glob Chang Biol 25:39–56

Fritz SC, Baker PA, Seltzer GO, Ballantyne A, Tapia P, Cheng H, Edwards L (2007) Quaternary glaciation and hydrologic variation in the South American tropics as reconstructed from the Lake Titicaca drilling project. Quat Res 68:410–420

Fritz SC, Baker PA, Ekdahl E, Seltzer GO, Stevens LR (2010) Millennial-scale climate variability during the Last Glacial period in the tropical Andes. Quat Sci Rev 29:1017–1024

Garzón-Orduña IJ, Benetti Longhini JE, Brower AVZ (2014) Timing the diversification of the Amazonian biota: butterfly divergences are consistent with Pleistocene refugia. J Biogeogr 41:1631–1638

Gentry A (1982) Neotropical floristic diversity: phytogeographical connections between Central and South America. Pleistocene climatic fluctuations or an accident of Andean orogeny? Ann Mo Bot Gard 69:557–593

Haffer J (1969) Speciation in Amazonian forest birds. Science 165:131–137

Hemming SR (2004) Heinrich events: massive late Pleistocene detritus layers of the North Atlantic and their global climate imprint. Rev Geophys 42(1):RG1005

Hermanowski B, Lima da Costa M, Carvalho AT, Behling H (2012) Palaeoenvironmental dynamics and underlying climatic changes in southeast Amazonia (Serra Sul dos Carajas, Brazil) during the late Pleistocene and Holocene. Palaeogeogr Palaeoclimatol Palaeoecol 365–366:227–246

Hoorn C, Wesselingh FP, ter Steege H, Bermudez MA, Mora A, Sevink J, Sanmartin I, Sanchez-Meseguer A, Anderson CL, Figueiredo JP, Jaramillo C, Riff D, Negri FR, Hooghiemstra H, Lundberg J, Stadler T, Sarkinen T, Antonelli A (2010) Amazonia through time: Andean uplift, climate change, landscape evolution, and biodiversity. Science 330:927–931

Hughes C, Eastwood R (2006) Island radiation on a continental scale: exceptional rates of plant diversification after uplift of the Andes. Proc Natl Acad Sci 103:10334–10339

Jenkins CN, Pimm SL, Joppa LN (2013) Global patterns of terrestrial vertebrate diversity and conservation. Proc Natl Acad Sci 110(28):E2602–E2610

Kanner LC, Burns SJ, Cheng H, Edwards RL (2012) High-latitude forcing of the South American Summer Monsoon during the last Glacial. Science 335:570–573

Kay KM, Reeves PA, Olmstead RG, Schemske DW (2005) Rapid speciation and the evolution of hummingbird pollination in neotropical *Costus* subgenus *Costus* (Costaceae): evidence from nrDNA its and ETS sequences. Am J Bot 92:1899–1910. https://doi.org/10.3732/ajb.92.11.1899

Koenen EJM, Clarkson JJ, Pennington TD, Chatrou LW (2015) Recently evolved diversity and convergent radiations of rainforest mahoganies (Meliaceae) shed new light on the origins of rainforest hyperdiversity. New Phytol 207:327–339. https://doi.org/10.1111/nph.13490

Lagomarsino LP, Condamine FL, Antonelli A, Mulch A, Davis CC (2016) The abiotic and biotic drivers of rapid diversification in Andean bellflowers (Campanulaceae). New Phytol 210:1430–1442. https://doi.org/10.1111/nph.13920

Lavin M (2006) Floristic and geographical stability of discontinuous seasonally dry tropical forests explains patterns of plant phylogeny and endemism. In: Pennington RT, Lewis GP, Ratter JA (eds) Neotropical savannas and seasonally dry forests: plant diversity, biogeography and conservation. CRC Press, Boca Raton, FL, pp 433–448

Ledru MP, Ceccantini G, Gouveia S, Lopez-Saez JA, Pessenda L, Ribero AS (2006) Millennial-scale climatic and vegetation changes in a northern Cerrado (northeast Brazil) since the Last Glacial Maximum. Quat Sci Rev 25:1110–1126

Leite YLR, Costa LP, Loss AC, Rocha RG, Batalha-Filho H, Bastos AC, Quaresma VS, Fagundes V, Paresque R, Passamani M, Pardini R (2016) Neotropical forest expansion during last glacial period challenges refuge hypothesis. Proc Natl Acad Sci U S A 113:1008–1013

Liebsch D, Marques MC, Goldenberg R (2008) How long does the Atlantic Rain Forest take to recover after a disturbance? Changes in species composition and ecological features during secondary succession. Biol Conserv 141:1717–1725

Lisiecki LE, Raymo ME (2005) A Pliocene-Pleistocene stack of 57 globally distributed benthic δ18O records. Palaeoceanography 20:PA1003

Liu X, Battisti DS (2015) The influence of orbital forcing of tropical insolation on the climate and isotopic composition of precipitation in South America. J Clim 28:4841–4862

Luebert F, Wiegend W (2014) Phylogenetic insights into Andean plant diversification. Front Ecol Evol 2:article 27. https://doi.org/10.3389/fevo.2014.00027

Madriñán S, Cortés AJ, Richardson JE (2013) Páramo is the world's fastest evolving and coolest biodiversity hotspot. Front Genet 4:192

McGee D, Moreno-Chamarro E, Green B, Marshall J, Galbraith E, Bradtmiller L (2018) Hemispherically asymmetric trade wind changes as signatures of past ITCZ shifts. Quat Sci Rev 180:214–228

Moritz C, Patton JL, Schneider CJ, Smith TB (2000) Diversification of rainforest faunas: an integrated molecular approach. Annu Rev Ecol Syst 31:533–563

Mosblech NAS, Bush MB, Gosling WD, Hodell DA, Thomas L, van Calsteren P, Correa-Metrio A, Valencia BG, Curtis J, van Woesik R (2012) North Atlantic forcing of Amazonian precipitation during the last ice age. Nat Geosci 5:817–820

Nace TE, Baker PA, Dwyer GS, Silva CG, Rigsby CA, Burns SJ, Giosan L, Ott-Bliesner B, Liu Z, Zhu J (2014) The role of North Brazil Current transport in the paleoclimate of the Brazilian Nordeste margin and paleoceanography of the western tropical Atlantic during the late Quaternary. Palaeogeogr Palaeoclimatol Palaeoecol 415:3–13

Pedro JB, Jochum M, Buizert C, He F, Barker S, Rasmussen SO (2018) Beyond the bipolar seesaw: toward a process understanding of interhemispheric coupling. Quat Sci Rev 192:27–46

Pennington RT, Prado DE, Pendry CA (2000) Neotropical seasonally dry forests and Quaternary vegetation changes. J Biogeogr 27:261–273

Peterson LC, Haug GH, Hughen KA, Rohl U (2000) Rapid changes in the hydrologic cycle of the tropical Atlantic during the Last Glacial. Science 290:1947–1951

Phillips BL, Brown GP, Greenlees M, Webb JK, Shine R (2007) Rapid expansion of the cane toad (*Bufo marinus*) invasion front in tropical Australia. Austral Ecol 32:169–176

Pouchon C, Fernández A, Nassar JM, Boyer F, Aubert S, Lavergne S, Mavárez J (2018) Phylogenomic analysis of the explosive adaptive radiation of the Espeletia complex (Asteraceae) in the Tropical Andes. Syst Biol 67:1041–1060

Prado LF, Wainer I, Chiessi CM (2013) Mid-Holocene PMIP3/CMIP5 model results: intercomparison for the South American monsoon system. The Holocene 23:1915–1920

Prates I, Rodrigues MT, Melo-Sampaio PR, Carnaval AC (2015) Phylogenetic relationships of Amazonian anole lizards (*Dactyloa*): taxonomic implications, new insights about phenotypic evolution and the timing of diversification. Mol Phylogenet Evol 82:258–268

Prates I, Xue AT, Brown JL, Alvarado-Serrano DF, Rodrigues MT, Hickerson MJ, Carnaval AC (2016a) Inferring responses to climate dynamics from historical demography in neotropical forest lizards. Proc Natl Acad Sci 113:7978–7985

Prates I, Rivera D, Rodrigues MT, Carnaval AC (2016b) A mid-Pleistocene rainforest corridor enabled synchronous invasions of the Atlantic Forest by Amazonian anole lizards. Mol Ecol 25:5174–5186

Prates I, Penna A, Rodrigues MT, Carnaval AC (2018) Local adaptation in mainland anole lizards: integrating population history and genome-environment associations. Ecol Evol 8:11932–11944

Punyasena SW, Mayle FE, McElwain JC (2008) Quantitative estimates of glacial and Holocene temperature and precipitation change in lowland Amazonian Bolivia. Geology 36:667–670

Quijada-Mascareñas A, Ferguson JE, Pook CE, Wuster W (2007) Phylogeographic patterns of Trans-Amazonian vicariants and Amazonian biogeography: the Neotropical rattlesnake (*Crotalus durissus* complex) as an example. J Biogeogr 34:1296–1312

Rahmstorf S (2002) Ocean circulation and climate during the past 120,000 years. Nature 419:207

Rangel TF, Edwards NR, Holden PB, Diniz-Filho JAF, Gosling WD, Coelho MTP, Cassemiro FAS, Rahbek C, Colwell RK (2018) Modeling the ecology and evolution of biodiversity: biogeographic cradles, museums, and graves. Science 361:244

Reis LS, Guimaraes J, Souza-Filho P, Sahoo PK, Figueiredo M, Giannini T (2017) Environmental and vegetation changes in southeastern Amazonia during the late Pleistocene and Holocene. Quat Int 449:83–105

Richardson JE, Pennington RT, Pennington TD, Hollingsworth PM (2001) Rapid diversification of a species-rich genus of neotropical rain forest trees. Science 293:2242–2245

Rull V, Carnaval AC (2019) Introduction. In: Rull V, Carnaval A (eds) Neotropical diversification. Springer, Heidelberg

Santos JC, Coloma LA, Summers K, Caldwell JP, Ree R, Cannatella DC (2009) Amazonian amphibian diversity is primarily derived from late Miocene Andean lineages. PLoS Biol 7(3): e1000056

Schrag DP, Adkins JF, McIntyre K, Alexander JL, Hodell DA, Charles CD, McManus JF (2002) The oxygen isotopic composition of seawater during the Last Glacial Maximum. Quat Sci Rev 21:331–342

Steig E, Alley R (2002) Phase relationships between Antarctic and Greenland climate records. Ann Glaciol 35:451–456

Stodart E, Parer I (1988) Colonisation of Australia by the rabbit *Oryctolagus cuniculus* (L.). CSIRO, Canberra

Stute M, Forster M, Frischkorn H, Serejo CP, Broecker WS (1995) Cooling of tropical Brazil during the last glacial maximum. Science 269:379–383

Sylvestre F, Servant M, Servant-Vildary S, Causse C, Fournier M, Ybert JP (1999) Lake-level chronology on the southern Bolivian Altiplano (18–23°S) during late-Glacial time and the early Holocene. Quat Res 51:54–66

van der Hammen T, Cleef AM (1986) Development of the high Andean páramo flora and vegetation. In: Vuilleumier F, Monasterio M (eds) High altitude tropical biogeography. Oxford University Press, New York, pp 153–201

Vanzolini P, Williams E (1970) South American anoles: the geographic differentiation and evolution of the anolis Chrysolepis species group (Sauria, Iguanidae). Arquivos de Zoologia 19:125–298

Vargas OM, Dick CW (2019) Diversification history of Neotropical Lecythidaceae, an ecologically dominant tree family of Amazon rain forest. In: Rull V, Carnaval A (eds) Neotropical diversification. Springer, Heidelberg

Vargas OM, Simpson BB (2019) Allopatric speciation versus parapatric ecological divergence in high Andean plants (Asteraceae: *Piofontia*). bioRxiv:868216. https://doi.org/10.1101/868216

Vellinga M, Wood RA (2002) Global climatic impacts of a collapse of the Atlantic thermohaline circulation. Clim Chang 54:251–267

Voelker AH (2002) Global distribution of centennial-scale records for Marine Isotope Stage (MIS) 3: a database. Quat Sci Rev 21:1185–1212

Wang X, Edwards RL, Auler AS, Cheng H, Kong X, Wang Y, Cruz FW, Dorale JA, Chiang H-W (2017) Hydroclimate changes across the Amazon lowlands over the past 45,000 years. Nature 541:204–207

Wehtje W (2003) The range expansion of the great-tailed grackle (*Quiscalus mexicanus* Gmelin) in North America since 1880. J Biogeogr 30:1593–1607

Wheatley A et al (2019) Modeling the origins of biodiversity in the tropical Andes (manuscript in preparation)

Wiens JJ (2004) Speciation and ecology revisited: phylogenetic niche conservatism and the origin of species. Evolution 58:193–197

Zhang Y, Chiessi CM, Mulitza S, Sawakuchi AO, Haggi C, Zabel M, Portioho-Ramos C, Schefuß E, Crivellari S, Wefer G (2017) Different precipitation patterns across tropical South America during Heinrich and Dansgaard-Oeschger stadials. Quat Sci Rev 177:1–9

Chapter 4
Tropical and Subtropical South America: A Study of Community Turnover Across Environmental Gradients

João Claudio Sousa Nascimento, Millke Jasmine Arminini Morales, Wendy Yohana Arroyo-Pérez, Juliana José, Mônica Paiva Quast, and Vera Nisaka Solferini

Abstract The Tropic of Capricorn, at 23°S, delimits the Tropical and Subtropical regions of the Southern hemisphere and marks a transition of diverse climatic variables that affect the distribution of biodiversity in the South American continent. To study the effect of the Tropic in the distribution of species, and to explore how community composition changes southwards, we analyze patterns of beta-diversity across the Neotropical region. Using locality data from GBIF, we show a latitudinal gradient in the degree of species turnover, but only in the cis-Andean region. Although previous taxon-centric analyses detected breakpoints in the components of beta-diversity at the tropical-subtropical transition, we were unable to identify them at large scale and level of data aggregation studied here: there was no single latitude at which a change in the behavior of beta-diversity could be detected. Although the replacement of species is the main component responsible for the changes in the composition of the communities in South America, the Tropic of Capricorn does not seem to represent a threshold in community turnover at this large scale.

Keywords Beta-diversity · Latitudinal gradient · Neotropics · GBIF · Break zones

---

J. C. S. Nascimento · M. J. A. Morales · W. Y. Arroyo-Pérez · M. P. Quast
Department of Genetics, Evolution, Microbiology and Immunology, Institute of Biology, University of Campinas, Campinas, Brazil

J. José
Laboratory of Genomics, Department of Genetics, Evolution, Microbiology and Immunology, Institute of Biology, University of Campinas, Campinas, Brazil

V. N. Solferini (✉)
Department of Genetics, Evolution, Microbiology and Immunology, Institute of Biology, University of Campinas, Campinas, Brazil

Center for Computational Engineering and Sciences, University of Campinas, Campinas, Brazil
e-mail: veras@unicamp.br

V. Rull, A. C. Carnaval (eds.), *Neotropical Diversification: Patterns and Processes*, Fascinating Life Sciences, https://doi.org/10.1007/978-3-030-31167-4_4

1 Introduction

At a global scale, the boundaries between tropical and subtropical areas are defined by an imaginary latitudinal line at 23°26′S: the Tropic of Capricorn. This old-established line marks the southern limit where the sunlight can reach the surface of the Earth in a perpendicular position; its Northern counterpart is the Tropic of Cancer.

These imaginary lines also work as a general reference for global climate. The different insolation north and south of the tropic lines result in differential heating in land and sea surfaces, creating a pattern of air circulation throughout the globe. At the Equator, as a consequence of the higher temperatures, humid air moves upwards and forms massive rain clouds; when reaching the troposphere, the now drier air moves towards the poles. During this process, heat loss increases the density of the air column, causing a descending flow around latitudes 20–30° in both hemispheres. These high-pressure zones inhibit the formation of clouds, promoting the establishment of local deserts. Reaching the surface, the air column divides itself into two branches: one moving to the pole and the other to the Equator. Due to Earth's rotation, the air flows to the Equator from east to west and to the Poles in the opposite direction. In both hemispheres, around latitude 60°, the flow from the tropic line zone meets the winds from the poles in a convergent zone where weather fronts and cyclones are formed. Together with these pressure areas, differences in lightning and in the relief contribute to characterize the climate of a region (Reboita et al. 2012).

We are interested in how these climatic shifts impact community composition. In community ecology, studies of compositional turnover focus on beta-diversity to understand the mechanisms that structure communities across spatial, environmental, and temporal gradients (Cadotte and Davies 2016). The term beta-diversity was introduced by Whittaker (1960) as *the extent of change in community composition, or degree of community differentiation, in relation to a complex-gradient of environment, or a pattern of environments.* Usually, beta-diversity is applied to compare species composition of different biotas, which can be done through dissimilarity indices such as Jaccard's and Sørensen's (Jost 2007). However, differences in species composition among sites can be generated by two different phenomena: (1) spatial turnover, or the replacement of species by others from site to site and (2) nestedness, when the biota of one site is a subset of the richest site (Baselga 2010).

Beta-diversity has been shown to increase towards lower latitudes, and some have suggested this to be a general pattern in ecology (Soininen et al. 2007). In a global meta-analysis, Soininen et al. (2018) detected a significant latitudinal gradient in beta-diversity for both the total beta-diversity and the turnover component. Possible explanations for this observation invoke Rapoport's rule, which asserts that species ranges increase from low to high latitudes (implying lower turnover rates at higher latitudes; Qian 2009), or also the narrower climatic niches of tropical species, resulting in limited dispersal capability and range (Janzen 1967; Qian 2009). However, while several taxon-specific studies have reported a latitudinal gradient in beta-diversity (Koleff et al. 2003; Qian 2009; Qian et al. 2009; Qian and Ricklefs 2007), others were unable to detect this trend (Rodríguez and Arita 2004), and some even questioned its existence (Kraft et al. 2011).

Moreover, it has been suggested that the components of beta-diversity behave differently across latitudinal gradients, and that there are breakpoints where a change in the prevalent component of beta-diversity, turnover or nestedness, can be detected (Baselga et al. 2012; Castro-Insua et al. 2016). The identification of these latitudinal thresholds may be a first step to successfully unveil the potential drivers of beta-diversity (Castro-Insua et al. 2016). However, the studies that have identified the location of these breaks sampled just a few vertebrate groups; the generality has not been established. In this chapter, we explore general patterns of beta-diversity in South America using a spatial method to calculate and partition the dissimilarity index of beta-diversity with the aim of investigating any effect of the tropical-subtropical transition in the structure of its communities.

2 South America: General Descriptions of Geographical Landscape and Climate

Most of South America is located right under the Equator, having the Tropic of Capricorn crossing its southern portion. Overall, the continent occupies an area of roughly 18 million km^2. In the north-south direction, the continent extends over 8000 km and is connected to Central and North America by the Isthmus of Panama. The widest portion of continent is found just south of the Equator, extending east to west for 5000 km (Orme 2007). It sits over the South American tectonic plate; in the western coast of the continent, where this plate meets the Nazca plate, a long chain of mountains, the Andes cordillera, reaches heights over 6000 m.

Just by its placement in the globe, from 13°N to 56°S and hence covering about 70° in latitude, South America is expected to harbor distinct climate zones. Adding complexity to this basic pattern, the Andes cordillera impacts air circulation and rain distribution (Garreaud 2009); in addition, the somewhat recent closure of the Panama isthmus has altered sea-water circulation and also influenced the continent's climate pattern. Regional details of the relief, and a mosaic of soils that differ in age and composition, form the backstage of a continent that is one of the most biodiverse places in the world (Trewick 2017).

South America has several climatic regimes according to its astronomical location, general circulation, surface features, altitude and exposure—as well as the climatic elements (temperature, rainfall, atmospheric pressures, winds, relative humidity and cloudiness) correlated with them (Eidt 1969). According to Köppen's classification system, which uses precipitation and temperature data, the continent presents four major types of climate: a tropical rain climate, a temperate climate, an arid climate (steppe, desert) and a tundra climate (Peel et al. 2007). Variations of those main types are also present, but depend on the seasonality of climatic elements

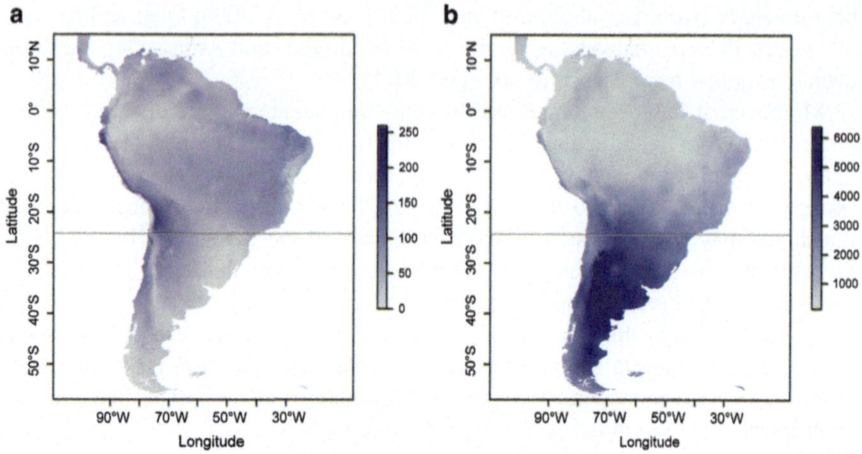

Fig. 4.1 Maps of precipitation (**a**) and temperature (**b**) seasonality, obtained from the WorldClim database (Fick and Hijmans 2017; http://www.worldclim.org/). Horizontal line indicates the Tropic of Capricorn. Temperature seasonality values are the standard deviation multiplied by 100 of monthly temperature estimates. Precipitation seasonality values are the coefficient of variation of monthly precipitation

(Eidt 1969). Temperatures in South America are the highest in the tropical region, with an annual average above 25 °C, and mean temperature decreases with increased latitude.

While precipitation changes spatially and temporally in South America, precipitation seasonality is not clearly delimited by the Tropic (Fig. 4.1a). In the eastern portion of the continent, areas north of the Tropic have higher seasonality relative to southern sites. Yet, much higher values are observed in the west, along the Andean region. One of the major climatic effects of the Andes cordillera is to restrict the entrance of the South Pacific anticyclone winds, causing an elongated area of rain shadow that extends east of the Andes, from southern Bolivia to Argentina. It also produces extraordinarily high precipitation on the windward slopes of the Pacific coasts of Colombia and southern Chile. An example of the rainfall variability over the Andes is the contrast observed at 18°–23°S, encompassing the Atacama Desert in the east, and the Chaco wetlands in the west (Garreaud 2000; Garreaud and Aceituno 2001).

In contrast, the location of the Tropic is more strongly linked to changes in temperature seasonality (Fig. 4.1b). As expected for subtropical climates, the southern portion of the continent presents the highest level of temperature seasonality. However, contrasting patterns are still observed between eastern and western South America. In the west coast, seasonal temperature changes are less pronounced throughout the year.

3 Biodiversity in South America

Diversity can be studied at various spatial scales (from local to broad geographical regions), levels of organization (from biomes to genes), or functional levels (from interspecific relationships to the regulation of gene expression). Common to these studies is the drive to investigate the origins of diversity and the mechanisms underlying its structure and maintenance.

Since the first naturalists arrived in South America, nearly 200 years ago, a great effort has been directed to making inventories of the continent's intriguing great number of species. Together with species' identification, the geographical location of collected specimens made it possible to recognize distinct biomes or phytogeographic regions, comprised by a very distinct set of abiotic conditions and a particular set of organisms. Since the very beginning, the aim of biogeographers has been to recognize spatial patterns of species' structure that might indicate similar tolerances to similar environmental conditions (Brown and Lomolino 1998).

The complex biogeographical history of South America began with the westward drift of the South America tectonic plate away from the African plate. For about 100 million years, this land mass was surrounded by water (Bigarella 1970). However, the biota had never been isolated, and a dynamic exchange of lineages with other mainland areas has been shown by data from diverse taxonomic groups. Later, the Andean uplift and the closure of Panama isthmus provided new opportunities for the increase of diversity. The Andean uplift was fundamental to shape Amazonian landscapes and ecosystem, having a striking effect on regional climate. Sedimentary and paleontological records, dated molecular phylogenies, and present-day species distributions reveal the link between the high species richness observed in northern South America and the changes induced by Andean mountain building (Hoorn et al. 2010). Likewise, the emergence of the Isthmus of Panama has greatly impacted South American diversity. The Great American Biotic Interchange, through several distinct pulses of dispersal which were made possible by the emergence of the isthmus, is considered one of the largest episodes of biological migration between previously disconnected landmasses (Bacon et al. 2015).

Spatial changes in the rate of species turnover have been reported for South America. Buckley and Jetz (2008), for instance, show a change in amphibian turnover near the Tropic of Capricorn, likely related to higher environment turnover in higher latitudes. Castro-Insua et al. (2016) identified the existence of latitudinal breakpoints in the components of beta-diversity that are generally concordant with the location of the tropical-subtropical transition. Using vertebrate data, they noticed that the community composition in South America are due to species replacement from the equatorial region to 20–25°S; from this point towards higher latitudes, however, the changes in community composition are mainly associated with species loss (Castro-Insua et al. 2016). This suggests that distinct ecological and historical processes associated with turnover and nestedness drive community composition. Data from riparian forest communities provide another example, as they differ strikingly depending on whether one samples a tropical or a subtropical site in southern Brazil (Giehl and Jarenkow 2012). Specifically, subtropical localities have fewer species and higher phylogenetic clustering relative to the tropical sites, suggesting that only a subset of tropical lineages is able to occupy the subtropical climates and diversify in this region (Giehl and Jarenkow 2012). Lower temperatures and precipitation seasonality have also been used to explain changes in community composition by limiting dispersal (Dobrovolski et al. 2012).

Phylogeographical studies in South America also provide clues about the processes happening at the tropical/subtropical transition, and demonstrate that responses to former landscape and climatic shifts have impacted the current distribution of biodiversity (Turchetto-Zolet et al. 2013). By looking at spatial patterns of genetic diversity within species, phylogeographers identify patterns of spatial turnover in co-distributed species, and recognize phylogeographic break zones (Marske et al. 2013). By focusing on historical demography, they also provide clues about how variations in the physical and biotic environment may have influenced the distribution and history of species in the past (Kumar and Kumar 2018). Phylogeographic studies conducted in South America suggest that Quaternary climate cycles impacted the subtropical region of South America and influenced local diversification. For instance, data suggest that *Nephila sexpunctata*, a spider endemic to Neotropical dry habitats, has experienced strong population and spatial expansion after the Last Glacial Maximum; today, analyses of its genetic structure demonstrate the existence of a subtropical lineage restricted to the Chaco (Bartoleti et al. 2017). The widespread bee *Bombus pauloensis*, which occupies multiple geographical and climate regions, is today structured into three main lineages—with one of them occurring in the subtropics. This subtropical lineage is thought to have undergone a southern expansion in the Last Interglacial period (Françoso et al. 2016), tracking climate.

4 Does the Tropical-Subtropical Environmental Transition Influence Community Composition in South America?

To discern the mechanisms underlying beta-diversity patterns, Baselga (2010) proposed partitioning the total dissimilarity (β_{sor}) in dissimilarity due to spatial turnover, which can be measured with Simpson dissimilarity index (β_{sim}), and a nestedness-resultant component (β_{nes}).

We investigated the general pattern of beta-diversity along South America, with the specific aim of analyzing the extent to which the tropical-subtropical transition might structure community organization in the continent. For that, we downloaded all species occurrence records available for the South American continent, from the Global Biodiversity Information Facility (GBIF.org 2019, last accessed 02/27/2019), obtaining roughly 15 million records of 124,066 species. We removed duplicated rows and masked localities to exclude sampling errors, also removing occurrences located in the ocean. As proposed by Baselga (2010), we partitioned the original Sørensen pairwise dissimilarity in two components, estimating its components of nestedness (β_{nes}) and spatial turnover (β_{sim}). The spatial turnover component was measured with the Simpson dissimilarity index, and the resultant component was attributed to nestedness.

We constructed a presence-and-absence matrix in a $1^\circ \times 1^\circ$ gridded map of South America and estimated a dissimilarity index and its components for each cell, calculated as the mean value of dissimilarity between the focal cell and each of its eight neighboring cells. We then analyzed the relationship between latitude and dissimilarity measures for the general South America continent, as well as for the Andean and Cis-Andean regions separately. We followed the definition of Morrone (2014) in delimiting the Andean region and separating it from the Cis-Andean region (Neotropical region in Morrone 2014). To test the impact of poorly sampled areas on the final results, we performed additional analyses by excluding all cells with less than 1000, 5000 and 10,000 records. All analyses were conducted in the packages betapart (Baselga and Orme 2012), letsR (Vilela and Villalobos 2015) and CommEvol (Melo 2013), in the software R (R Development Core Team 2019).

To test if regression models adjusted to the Andean vs. the Cis-Andean datasets differed for each index, we estimated probability density functions for the difference between parameters, using 1000 bootstrap iterations. In each iteration, one sample from each dataset was generated, a regression model was adjusted to each one, and the differences between slopes and intercepts was annotated. The p-value was estimated as two times (to account for a bilateral alternative hypothesis) of the proportion of iterations in which the difference between parameters was higher (or lower) than zero. This analysis was conducted in the base package of R (R Core Team 2018).

Our results demonstrate that spatial turnover (β_{sim}) is the main component of total beta-diversity in South America, being observed in the great majority of cells (Fig. 4.2). Nestedness, in contrast, is low across most of the continent. This is in general agreement with the global-scale meta-analyses conducted by Soininen et al. (2018).

Our maps reveal higher levels of total dissimilarity (β_{sor}) and Simpson dissimilarity (β_{sim}) in the Amazon basin and in the Cerrado, as well as along the Andean cordillera (Fig. 4.2a, b). This is in disagreement with previous studies, which identified the Andean region as that of higher turnover in the Neotropics (Buckley and Jetz 2008; Gaston et al. 2007; Maestri and Patterson 2016). We cannot rule out the possibility that our failure to recover these results may reflect biases in the spatial coverage of the data we used here, particularly in the Andes. Our analysis suggests that the nestedness-resultant component (β_{nes}) is uniformly distributed along the continent, with few areas of high values (Fig. 4.2c).

We find support for a relationship between latitude and community dissimilarity in South America, but this relationship differs between the Andean and Cis-Andean regions. In fact, the Sørensen beta-diversity presented a significant difference between Andean and Cis-Andean intercepts and slopes (Table 4.1). Outside of the Andes, we detect a linear negative relationship between latitude and the Sørensen dissimilarity index (Fig. 4.3a, blue), similarly to Soininen et al. (2018). The same trend is observed when the turnover component is analyzed separately (Fig. 4.3b),

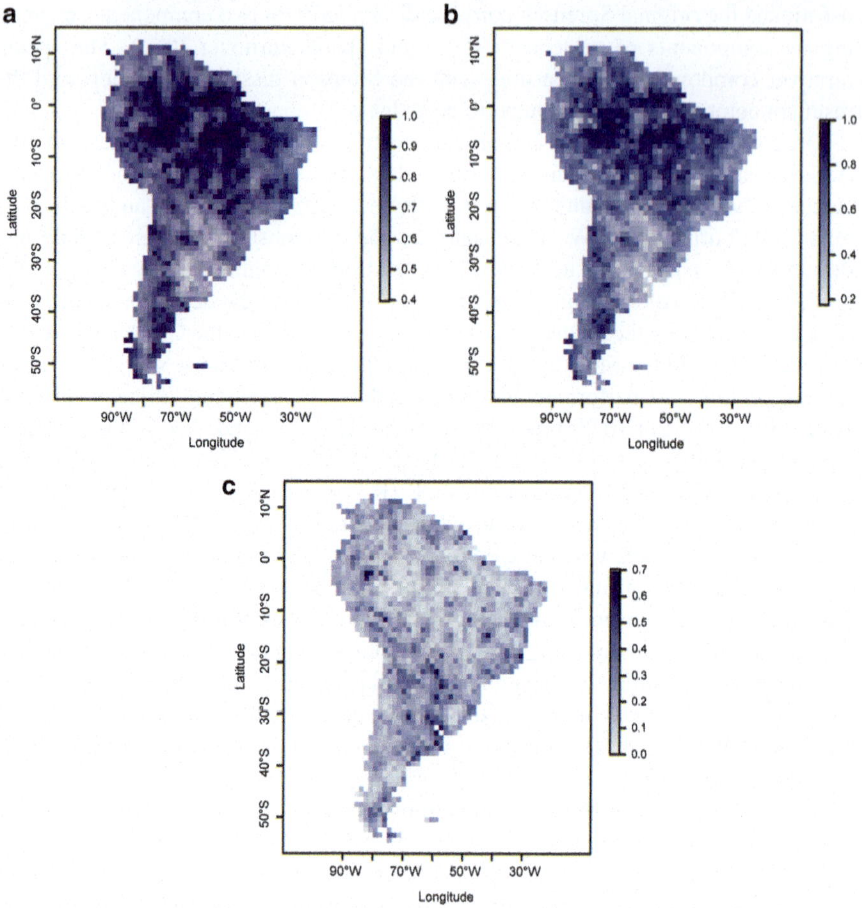

Fig. 4.2 Distribution of beta diversity in South America, calculated for each 1° × 1° cell as the mean of pairwise dissimilarity among all of its eight adjacent cells. (**a**) Sørensen dissimilarity index (β_{sor}); (**b**) Simpson dissimilarity index (β_{sim}); (**c**) Nestedness-resultant component (β_{nes})

Table 4.1 Intercepts and slopes for beta diversity estimates Sørensen dissimilarity index (β_{sor}), Simpson dissimilarity index (β_{sim}) and nestedness-resultant component (β_{nes}), in the Andean and Cis-Andean regions

	β_{sor}	β_{sim}	β_{nes}
Intercept			
Andean	0.7323	0.5741	0.2050
Cis-Andean	0.8781	0.7575	0.1372
	p-value = 0	p-value = 0	p-value = 0
Slope			
Andean	0.0009	0.0011	−0.0005
Cis-Andean	−0.0065	−0.0082	0.0038
	p-value = 0	p-value = 0	p-value = 0

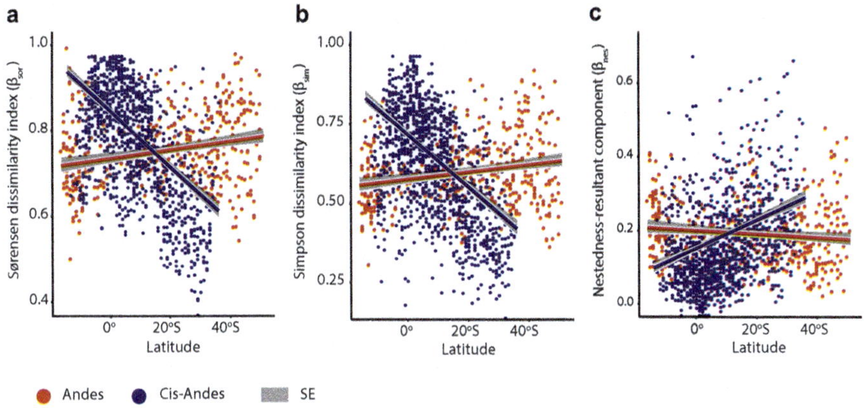

Fig. 4.3 Distribution of pairwise dissimilarity measures from 10°N to 57°S. Andean and Cis-Andean region are delineated. (**a**) Sørensen dissimilarity index (β_{sor}); (**b**) Simpson dissimilarity index (β_{sim}); (**c**) Nestedness-resultant component (β_{nes}). *SE* standard error

but not when the nestedness-resultant component is plotted (Fig. 4.3c). For the latter, β_{nes} increases with higher latitudes, again similarly to Soininen et al. (2018).

In the Andean region, however, the patterns are inverted. When all GBIF data are used in a single analysis, a linear positive relationship is observed between latitude and Sørensen dissimilarity index (Fig. 4.3a, red). The nestedness-resultant component (β_{nes}) showed no significant relationship with latitude (Fig. 4.3c). Moreover, all components showed smaller variation across the latitudinal gradient in the Andean region relative to the Cis-Andean region, suggesting that the processes that shape the latitudinal gradient of beta-diversity differ across these two areas.

To explore whether spatial biases in the availability of data may explain the patterns observed in the Andean region, we re-ran our analyses with smaller and more densely sampled points, as described previously. For the Cis-Andean region, these analyses returned similar results regarding the patterns of total beta-diversity and turnover, although recovering a lack of relationship between latitude and the nestedness component of beta-diversity (Fig. 4.4, blue). Nevertheless, the patterns inferred for the Andean region (Fig. 4.4, red) differ from those observed when all data are utilized. Particularly, the relationships between latitude and beta-diversity were no longer significant in the reduced datasets. This increases our confidence in the existence of a gradient of total beta-diversity for the Cis-Andean region, but points to a limited ability to draw conclusions about the Andean region given the data explored here.

Importantly, the data suggest that the tropical-subtropical transition does not influence the rate at which beta-diversity indexes change in South America. A piecewise regression identified a breakpoint at 0°S for the turnover component for the South America region. For the Cis-Andean region, we identified a breakpoint at 6°S. For the Andean region, we identified a breakpoint at 44°S, tough the regressions fit weakly to the data (results no shown). These findings contrast with previous

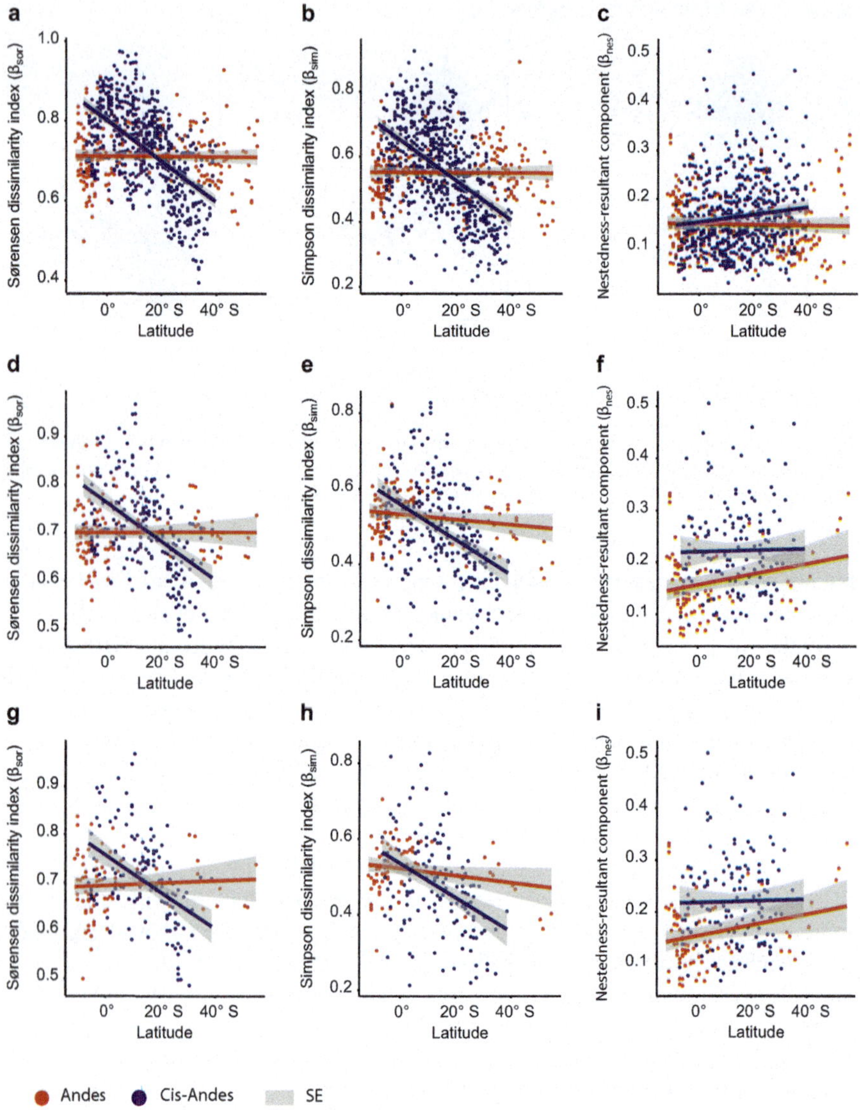

Fig. 4.4 Distribution of pairwise dissimilarity measures from 10°N to 57°S, after removing poor sampled cells. Andean and Cis-Andean region are delineated. Samplings for cells with more than 1000 localities are in plots (**a**), (**b**) and (**c**) for, respectively, Sørensen dissimilarity index (β_{sor}), Simpson dissimilarity index (β_{sim}) and nestedness-resultant component (β_{nes}). Samplings for cells with more than 5000 localities are in plots (**d**), (**e**) and (**f**) for, respectively, Sørensen dissimilarity index (β_{sor}), Simpson dissimilarity index (β_{sim}) and nestedness-resultant component (β_{nes}). Samplings for cells with more than 10,000 localities are in plots (**g**), (**h**) and (**i**) for, respectively, Sørensen dissimilarity index (β_{sor}), Simpson dissimilarity index (β_{sim}) and nestedness-resultant component (β_{nes}). *SE* standard error

studies that showed the turnover rates to change at latitudes close to 20°–30°S (Castro-Insua et al. 2016) or 41°S (Soininen et al. 2018). However, because taxonomic groups differ in the ways they respond to geographical clines (Dobrovolski et al. 2012; Qian 2009; Rodríguez and Arita 2004), it is possible that specific patterns of each group may be masked and lost when we aggregate a diverse array of species, like we did here. Further analyses of smaller scope are needed to test whether differences in the location of breakpoints are observed across taxonomic groups.

The data utilized here represent various groups and sources of uncertainty. Even considering its numerous sources, GBIF is not a complete database. Lack of sampling and taxonomic errors exist for various regions and groups. We sought to circumvent some of these drawbacks by eliminating poorly sampled cells; the results still point to a general pattern in the structuring and composition of South American communities. The main result points to the presence of a latitudinal gradient of total beta diversity and turnover, and a constant latitudinal rate of changes east of the Andes. Although previous taxon-centric analyses detected the existence of breakpoints in the components of beta-diversity at the tropical-subtropical transition, we were unable to identify them at large scale and level of data aggregation studied here: there was no single latitude at which a change in the behavior of beta-diversity could be detected. Although the replacement of species is the main component in the changes of the composition of the communities in South America, the Tropic of Capricorn does not represent a threshold in community turnover at this large scale.

References

Bacon CD, Silvestro D, Jaramillo C, Smith BT, Chakrabarty P, Antonelli A (2015) Biological evidence supports an early and complex emergence of the Isthmus of Panama. Proc Natl Acad Sci U S A 112(19):6110–6115

Bartoleti LFM, Peres EA, Sobral-Souza T, Fontes FHM, Silva MJ, Solferini VN (2017) Phylogeography of the dry vegetation endemic species *Nephila sexpunctata* (Araneae: Araneidae) suggests recent expansion of the Neotropical Dry Diagonal. J Biogeogr 44 (9):2007–2020

Baselga A (2010) Partitioning the turnover and nestedness components of beta diversity. Glob Ecol Biogeogr 19(1):134–143

Baselga A, Orme CDL (2012) Betapart: an R package for the study of beta diversity. Methods Ecol Evol 3(5):808–812

Baselga A, Gómez-Rodríguez C, Lobo JM (2012) Historical legacies in world amphibian diversity revealed by the turnover and nestedness components of beta diversity. PLoS One 7(2):e32341

Bigarella JJ (1970) Continental drift and paleocurrent analysis (A comparison between Africa and South America). In Second international Gondwana symposium, South Africa, pp 73–97

Brown J, Lomolino M (1998) Biogeography. Sinauer Associates, Sunderland, MA

Buckley LB, Jetz W (2008) Linking global turnover of species and environments. Proc Natl Acad Sci U S A 105(46):17836–17841

Cadotte MW, Davies TJ (2016) Phylogenetic diversity across space and time. Phylogenies. In: Phylogenies in ecology: a guide to concepts and methods. Princeton University Press, Princeton, pp 143–171

Castro-Insua A, Gómez-Rodríguez C, Baselga A (2016) Break the pattern: breakpoints in beta diversity of vertebrates are general across clades and suggest common historical causes. Glob Ecol Biogeogr 25(11):1279–1283

Dobrovolski R, Melo AS, Cassemiro FA, Diniz-Filho JAF (2012) Climatic history and dispersal ability explain the relative importance of turnover and nestedness components of beta diversity. Glob Ecol Biogeogr 21(2):191–197

Eidt RC (1969) The climatology of south America. In: Biogeography and ecology in South America. Springer, Berlin, pp 54–81

Fick SE, Hijmans RJ (2017) WorldClim 2: new 1-km spatial resolution climate surfaces for global land areas. Int J Climatol 37(12):4302–4315

Françoso E, Zuntini AR, Carnaval AC, Arias MC (2016) Comparative phylogeography in the Atlantic forest and Brazilian savannas: pleistocene fluctuations and dispersal shape spatial patterns in two bumblebees. BMC Evol Biol 16(1):267

Garreaud R (2000) Intraseasonal variability of moisture and rainfall over the South American Altiplano. Mon Weather Rev 128(9):3337–3346

Garreaud RD (2009) The Andes climate and weather. Adv Geosci 22:3–11

Garreaud R, Aceituno P (2001) Interannual rainfall variability over the South American Altiplano. J Clim 14(12):2779–2789

Gaston KJ, Davies RG, Orme CDL, Olson VA, Thomas GH, Ding T-S, Rasmussen PC, Lennon JJ, Bennett PM, Owens IP (2007) Spatial turnover in the global avifauna. Proc R Soc B Biol Sci 274(1618):1567–1574

GBIF.org (2019) (27th February 2019) GBIF occurrence download. https://doi.org/10.15468/dl.7nlqaq

Giehl ELH, Jarenkow JA (2012) Niche conservatism and the differences in species richness at the transition of tropical and subtropical climates in South America. Ecography 35(10):933–943

Hoorn C, Wesselingh F, Ter Steege H, Bermudez M, Mora A, Sevink J, Sanmartín I, Sanchez-Meseguer A, Anderson C, Figueiredo J (2010) Amazonia through time: Andean uplift, climate change, landscape evolution, and biodiversity. Science 330(6006):927–931

Janzen DH (1967) Why mountain passes are higher in the tropics. Am Nat 101(919):233–249

Jost L (2007) Partitioning diversity into independent alpha and beta components. Ecology 88(10):2427–2439

Koleff P, Lennon JJ, Gaston KJ (2003) Are there latitudinal gradients in species turnover? Glob Ecol Biogeogr 12(6):483–498

Kraft NJ, Comita LS, Chase JM, Sanders NJ, Swenson NG, Crist TO, Stegen JC, Vellend M, Boyle B, Anderson MJ (2011) Disentangling the drivers of β diversity along latitudinal and elevational gradients. Science 333(6050):1755–1758

Kumar R, Kumar V (2018) A review of phylogeography: biotic and abiotic factors. Geol Ecol Landsc 2(4):268–274

Maestri R, Patterson BD (2016) Patterns of species richness and turnover for the South American rodent fauna. PLoS One 11(3):e0151895

Marske KA, Rahbek C, Nogués-Bravo D (2013) Phylogeography: spanning the ecology-evolution continuum. Ecography 36(11):1169–1181

Melo A (2013) CommEcol: community ecology analyses. http://commecol.r-forge.r-project.org

Morrone JJ (2014) Biogeographical regionalisation of the Neotropical region. Zootaxa 3782(1):1–110

Orme AR (2007) The tectonic framework of South America. Phys Geogr S Am 7:3

Peel MC, Finlayson BL, McMahon TA (2007) Updated world map of the Köppen-Geiger climate classification. Hydrol Earth Syst Sci Discuss 4(2):439–473

Qian H (2009) Beta diversity in relation to dispersal ability for vascular plants in North America. Glob Ecol Biogeogr 18(3):327–332

Qian H, Ricklefs RE (2007) A latitudinal gradient in large-scale beta diversity for vascular plants in North America. Ecol Lett 10(8):737–744

Qian H, Badgley C, Fox DL (2009) The latitudinal gradient of beta diversity in relation to climate and topography for mammals in North America. Glob Ecol Biogeogr 18(1):111–122

R Development Core Team (2019) R: a language and environment for statistical computing. R Foundation for statistical computing, Vienna, Austria

Reboita MS, Krusche N, Ambrizzi T, Rocha RP (2012) Entendendo o Tempo e o Clima na América do Sul. Terrae Didatica 8(1):34–50

Rodríguez PT, Arita H (2004) Beta diversity and latitude in North American mammals: testing the hypothesis of covariation. Ecography 27(5):547–556

Soininen J, Lennon JJ, Hillebrand H (2007) A multivariate analysis of beta diversity across organisms and environments. Ecology 88(11):2830–2838

Soininen J, Heino J, Wang J (2018) A meta-analysis of nestedness and turnover components of beta diversity across organisms and ecosystems. Glob Ecol Biogeogr 27(1):96–109

Trewick S (2017) Plate tectonics in biogeography. In International encyclopedia of geography: people, the earth, environment and technology. Wiley. https://doi.org/10.1002/9781118786352.wbieg0638

Turchetto-Zolet A, Pinheiro F, Salgueiro F, Palma-Silva C (2013) Phylogeographical patterns shed light on evolutionary process in South America. Mol Ecol 22(5):1193–1213

Vilela B, Villalobos F (2015) letsR: a new R package for data handling and analysis in macroecology. Methods Ecol Evol 6(10):1229–1234

Whittaker RH (1960) Vegetation of the Siskiyou mountains, Oregon and California. Ecol Monogr 30(3):279–338

Chapter 5
Evolutionary Macroecology and the Geographical Patterns of Neotropical Diversification

Fabricio Villalobos, Jesús N. Pinto-Ledezma, and José Alexandre Felizola Diniz-Filho

Abstract Over the last decades, the geographical distribution of species, as well as its associated patterns have been at the core of the macroecology research program. Gradients in geographic range size and shape, as well as range overlap (species richness), reveal broad-scale patterns that may help to infer underlying ecological processes, mainly related to climatic and environmental variation. However, it is clear now that evolutionary processes are at least equally important, demanding the inclusion of an evolutionary dimension to better understand such patterns. In this review, we discuss recent macroecological approaches to study evolutionary patterns at the geographical scale, and exemplify some of these approaches with data from a model group of Neotropical birds, the Furnariides.

Keywords Community phylogenetics · Latitudinal diversity gradients · Macroevolution · Niche conservatism · Species diversity · Speciation rates

---

F. Villalobos
Laboratorio de Macroecología Evolutiva, Red de Biología Evolutiva, Instituto de Ecología, A.C., Xalapa, Veracruz, Mexico

J. N. Pinto-Ledezma
Department of Ecology, Evolution and Behavior, University of Minnesota, Saint Paul, MN, USA

Museo de História Natural Noel Kempff Mercado, Universidad Autónoma Gabriel René Moreno, Santa Cruz de la Sierra, Bolivia

J. A. F. Diniz-Filho (✉)
Departamento de Ecologia, Instituto de Ciências Biologicas, Universidade Federal de Goiás, Goiânia, GO, Brazil

V. Rull, A. C. Carnaval (eds.), *Neotropical Diversification: Patterns and Processes*, Fascinating Life Sciences, https://doi.org/10.1007/978-3-030-31167-4_5

1 Introduction

Since the eighteenth century, broad-scale patterns of diversity called the attention of naturalists. Recognizing that tropical regions have higher species richness relative to temperate areas, Alexander von Humboldt was the first one to propose it to emerge from climatic differences (Hawkins 2001). This ubiquitous pattern has since then been known as the Latitudinal Diversity Gradient (LDG) and, although the global distribution of biodiversity is indeed far more complex than a simple unidirectional gradient (Hawkins and Diniz-Filho 2004), the difference in species richness between temperate and tropical regions tends to capture the most evident facet of the distribution of life on Earth: its geographic heterogeneity.

Early explanations for the LDG in the 1950s and 1960s followed von Humboldt's tradition and focused on the strong correlations observed between diversity (i.e., species richness) and components of current environmental variation—especially combinations of temperature and precipitation (Simpson 1964; Pianka 1966; O'Brien 2006; Hawkins et al. 2003; Brown 2012). These high correlations suggested a causal explanation, and spurred the development of hypotheses that aimed to identify the mechanisms affecting species distributions and hence driving geographical patterns (e.g., Currie et al. 2004). Although these diversity-environment correlations suggested "pure ecological explanations" that involved population-level processes tied to dispersal and aggregation of tropical organisms, it quickly became clear that deep-time evolutionary processes should also be taken into account to explain the LDG (Rohdes 1992; Ricklefs 2004). In fact, as early as 1937, Theodosius Dobzhansky had proposed that diversity gradients should be explained by an interaction between ecological and evolutionary mechanisms, in which evolution would drive the dimensions of the niche—the set of biotic and abiotic factors that allow a species to exist indefinitely—that would allow different patterns of niche packing throughout environmental gradients. Today, it is consensus that the LDG should be explained not only by current climatic factors, but also by the long-term dynamics of such climatic factors and by events happening throughout the evolution of the species (Fine 2015).

In the last 40 years, several papers have synthetized knowledge about LDG and discussed its potential causes at distinct spatial and temporal scales (e.g., Pianka 1966; Rohde 1992; Currie et al. 2004; Hillebrand 2004; Mittelbach et al. 2007; Fine 2015). Explanations gradually shifted from purely ecological mechanisms driving population-level processes to deep time evolutionary patterns and processes (Schemske and Mittelbach 2017). Currie et al. (2004) and Mittelbach et al. (2007), for instance, proposed two ways to couple evolutionary with ecological processes to generate the LDG. The first focused on the different ways by which speciation and extinction rates (and their balance; the diversification rate) vary geographically (see Schluter and Pennel 2017). The second emphasized how different regions are occupied at different moments in a clade's history, leading to different times for speciation-extinction events and creating gradients in the accumulated number of species. More recently, these two possibilities (speciation-extinction dynamics and time for diversity accumulation) have been considered as evolutionary mechanisms to explain the LDG (Fine 2015). Indeed, both mechanisms are expressed in two of the most prominent hypotheses currently posed to explain the LDG: the tropical niche conservatism (Wiens and Donoghue 2004) and the out-of-the-tropics (Jablonski et al. 2006) hypotheses.

The tropical niche conservatism hypothesis (TNC) posits that most clades originated in the tropics and have occupied these regions for a longer time period compared to the fewer and more recent clades occupying temperate areas; this would have led to the accumulation of more species in the tropics relative to the temperate zone, resulting in the observed LDG (Wiens and Donoghue 2004). The TNC hypothesis does not explicitly predict differences in speciation and extinction rates, but focuses simply on the time available for speciation, tied to a limited dispersal from tropical to temperate regions. Alternatively, the out-of-the-tropics hypothesis (OTT) also postulates a tropical origin for most clades—but with additional higher dispersal out of (rather than into) the tropics, as well as comparatively higher and lower speciation and extinction rates, respectively, in these regions (Jablonski et al. 2006). Testing the predictions of both LDG hypotheses, whether related to time for speciation or evolutionary rates, requires comparisons between tropical and temperate regions in a spatial context. However, beyond more general difficulties related to scale effects, and considering that evolutionary (speciation-extinction) dynamics is a property of the clade, one of the main problems hindering our understanding of diversity gradients is that these dynamics are to be measured and evaluated in geographical space (Hawkins et al. 2006). It is still challenging to properly combine these two dimensions—space and time—in an analytical sense.

Our goal here is to review the application of alternative methodological approaches, under a macroecological perspective, to investigate evolutionary patterns and processes associated with the LDG in the Neotropics. To achieve this, we discuss the rationale behind these approaches, and their application in published studies that focused on or included the Neotropical region and clades. To illustrate their use, we apply them to available data from a model clade of passerine birds: the infraorder Furnariides. The Furnariides clade constitutes the largest continental radiation of vertebrates endemic to the Neotropics, comprising nearly 51% of suboscines species and 11% of all Passeriformes (Ricklefs 2002; Claramunt 2010). Furnariides present high levels of ecological diversity (e.g. morphology and life histories), and are distributed in nearly all major terrestrial habitats across the Neotropics. Highest concentrations of Furnariides species are found in forest habitats in the Amazon and the Atlantic Forest, whereas lowest species concentration can be found in open habitats such us the Andean and Patagonian deserts, and the Llanos of Venezuela and Colombia (Pinto-Ledezma et al. 2017). Furnariides thus show the classic latitudinal diversity gradient (Fig. 5.1a), and their ecological and evolutionary patterns make them an ideal system to review the use of macroecological approaches to study Neotropical diversification. All data used here to map species' geographic distributions (Furnariides range maps; see Pinto-Ledezma et al. 2017) come from Ridgely et al. (2012); the phylogenetic hypothesis inferred for the group follows Pinto-Ledezma et al. (2019). This phylogeny is a maximum clade credibility tree

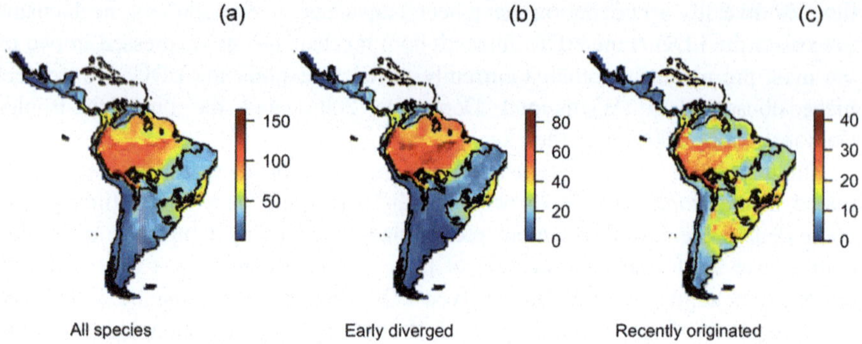

Fig. 5.1 Geographic patterns of species richness for different sets of species of the Furnariides clade across the Neotropics; (**a**) all species, (**b**) early-diverged species and (**c**) recently-originated species. Map lines correspond to the division between forest and open habitats. Notice that to estimate the number of early-diverged and recently-originated species we ranked each species by calculating the number of nodes separating a species from the root of the tree (Root distance metric, RD). Then, using the 35% and 65% percentiles of RD across all species we defined, respectively, early-diverged species as those with few nodes from the root of the tree and recently-originated species as those with many nodes, yielding a balanced number of early-diverged (283) and recently-originated (236) species

obtained from a posterior distribution of trees built under a Bayesian framework and based on nine molecular markers (Pinto-Ledezma et al. 2019).

2 Mapping Phylogenetic History in Geographic Space

At broad spatial scales, diversity gradients are usually described from species' geographic ranges and their overlap (Gotelli et al. 2009). The standard methodology for doing so is to overlay a grid of cells onto the study area, wherein species richness (alpha diversity) is estimated by counting the (co)occurrence of species within each grid-cell. As a basic property describing species presence or absence within a cell, geographic ranges can be further used to describe diversity gradients beyond species richness. For instance, different spatial patterns can be derived by summarizing attributes of the species known (or assumed) to be present in each cell (Hawkins et al. 2017). This "assemblage-based" approach can be used to create a raster of a particular trait (e.g. body size) that is summarized across species, as commonly done in studies of ecogeographical rules (e.g. Bergmann's Rule; Olalla-Tárraga et al. 2010).

Following the current interest on evolutionary drivers of diversity patterns (Schemske and Mittelbach 2017), assemblage-based approaches now also include the creation of "evolutionary variables" from the species' phylogenetic information (Hawkins et al. 2017). Different phylogenetic metrics can be used to represent the evolutionary development of assemblages, such as their mean root distance (MRD;

Kerr and Currie 1999), relative phylogenetic diversity (rPD, Davies and Buckley 2011) or diversification rate (DR, Jetz et al. 2012). Mean root distance was one of the first evolutionary variable to be mapped (Kerr and Currie 1999), and it continues to be widely used despite its dependence on tree topology instead of branch lengths (Hawkins et al. 2012; Kennedy et al. 2014; Pinto-Ledezma et al. 2017). MRD quantifies the average distance of species as the number of nodes separating each taxon (one tip of the tree) to the root of the phylogeny, allowing the classification of assemblages along a continuum that spans from early-diverged to recently-originated assemblages (Pinto-Ledezma et al. 2017). Patterns in MRD have been useful to test hypotheses on diversity gradients related to evolutionary time, such as the Tropical Niche Conservatism (TNC) hypothesis, which posits that, for clades originated in the tropics, most lineages persist, diversify, and accumulate within the tropics—whereas few lineages are able to establish outside the ancestral conditions (e.g. temperate regions; Wiens and Donoghue 2004). Such pattern of lineage accumulation is expected to be discernable in the distribution of early-diverged ("old/basal") and recently-derived ("young/derived") species, with the former accumulating in the ancestral environment, and the latter doing so in different environments (Hawkins et al. 2006). Hawkins et al. (2006), for instance, explicitly used a "deconstruction" approach based on the distribution of root distance among species, and obtained richness for "basal" and "derived" clades for Neotropical birds. As expected under the TNC, current species richness is better explained by basal richness than by derived richness (Hawkins et al. 2006).

Kennedy et al. (2014) used a similar approach and showed different diversity patterns between the two suborders of passerine birds (oscines and suboscines). They used a measure related to the MRD, called the diversification rate (DR; Jetz et al. 2012), to rank species along the old-young continuum. The DR represents the number of diversification events along the history of a lineage (Jetz et al. 2012), and is highly correlated with the species root distances (Pinto-Ledezma et al. 2017). Kennedy et al. (2014) found that data from both oscines and subsocines support the TNC hypothesis, with young species preferentially accumulating in temperate regions, but young and old species doing so in the tropics. However, the spatial patterns in DR differed between these suborders, with oscines showing a southward pattern and suboscines a more tropical-temperate gradient. Such findings support the inferred biogeographic histories of both suborders, with a New World colonization from the north for the oscines, and colonization from the south for the suboscines. Later, Kennedy et al. (2016) expanded their approach to 10 large clades of passerine birds and confirmed the role of spatiotemporal differences in diversification in the establishment of current diversity gradients in the New World.

Claims about TNC influencing birds' diversity gradients (Hawkins et al. 2006; Jetz et al. 2012; Kennedy et al. 2014, 2016) can also be extended to our study group, the Furnariides. Furnariids present similar patterns to those from the abovementioned studies, with high species co-occurrence in tropical areas being dominated by early-diverged species, and low species co-occurrence in subtropical to temperate areas, which are mainly dominated by recently-derived species (Fig. 5.1b, c). MRD and DR reveal a similar pattern (Fig. 5.2a, b). Furnariides

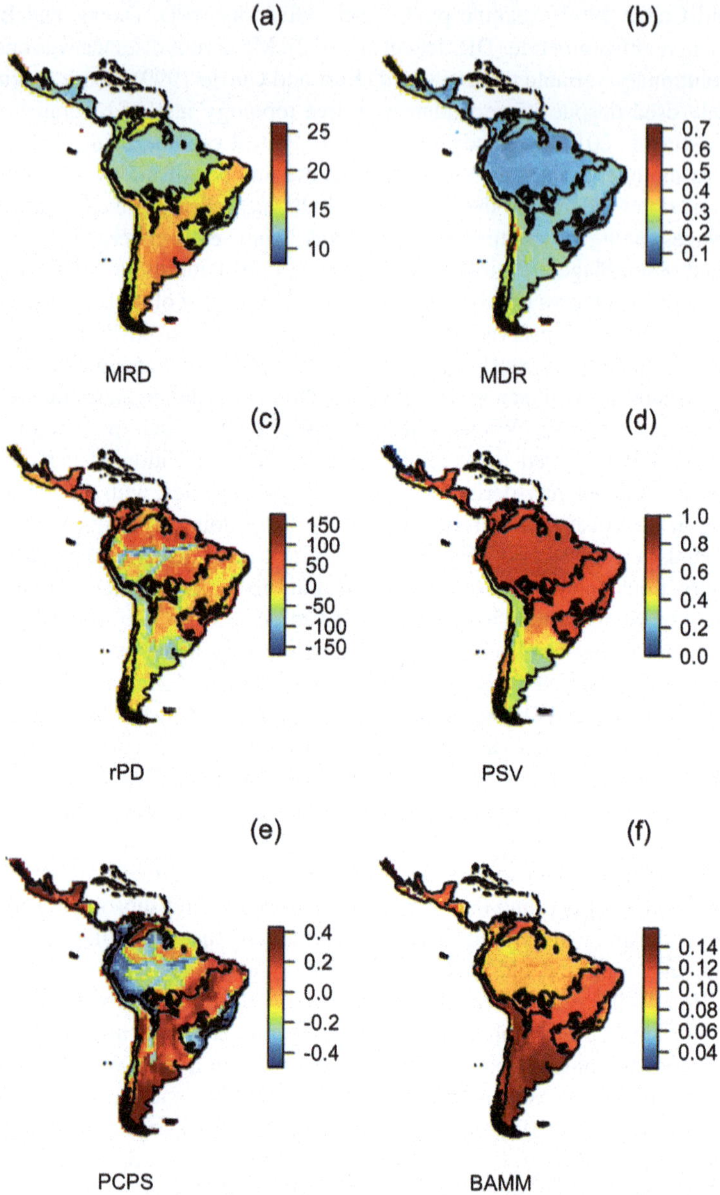

Fig. 5.2 Geographic patterns of evolutionary variables of Furnariides assemblages; (**a**) Mean Root Distance (MRD), (**b**) Mean Diversification Rate (MDR), (**c**) relative Phylogenetic Diversity (rPD), (**d**) Phylogenetic Species Variability (PSV), (**e**) first axis of the Principal Coordinates of Phylogenetic Structure (PCPS), (**f**) BAMM-derived speciation rates (BAMM). Map lines correspond to the division between forest and open habitats

patterns further show that both evolutionary variables, MRD and DR, are highly correlated ($r = 0.69$, $p < 0.001$) and thus may be used interchangeably when describing evolutionary patterns at the geographic scale (see also Pinto-Ledezma et al. 2017).

Another evolutionary variable commonly used is the phylogenetic diversity of assemblages. Phylogenetic diversity (PD) describes the sum of the branch lengths (estimated from a known phylogeny) that connect all species within an assemblage, and thus is highly correlated with the number of species considered (Faith 2013). Considering the relationship between species richness and PD, Davies and Buckley (2011) evaluated the spatial patterns of mammalian relative PD (rPD; the residual PD after controlling for species richness) to infer patterns of diversification and migration. They considered low rPD regions as evidence of rapid speciation and low immigration, and high rPD regions as those of low speciation rates and frequent long-distance immigration. Such different regions were clearly highlighted by a New vs Old World pattern, where the New World (specially the Neotropics) presented low rPD and the Old World (especially the Afrotropics) showed high rPD, consistent with the history of mammalian origin and dispersal (Springer et al. 2011; see also Fritz and Rahbek 2012; Volkamp et al. 2017 for a similar approach in amphibians and birds, respectively). Similarly, in Furnariides, mapping rPD shows that tropical areas (e.g., the Chocó ecoregion) tend to present low speciation rates, whereas temperate areas show mid- to high speciation rates (Fig. 5.2c). That being said, one can also identify tropical areas (e.g, the tropical Andes in Bolivia and Peru) consistent with high speciation rates (Fig. 5.2c), which may be linked to environmental variability (Morales-Barbero et al. 2018).

3 Expanding the Community Phylogenetics Framework to Macroecology

The development of a community phylogenetics (CP) framework (Webb et al. 2002) that explicitly integrates ecology and evolution renovated the interest in combining both disciplines to study diversity patterns (Cavender-Bares et al. 2012). The basis of such framework relies on studying the phylogenetic structure of species (i.e. their genealogical relationships) inhabiting an assemblage, under the assumption that species interact via traits that are not randomly distributed in the phylogeny (Vamosi et al. 2009). Accordingly, CP metrics represent an assemblage's phylogenetic structure in terms of overdispersion (i.e. species being more distantly related, on average, than expected by chance), clustering (i.e. species being more closely related, on average, than expected by chance) or randomness (Webb et al. 2002). A plethora of such metrics have been developed (Miller et al. 2017a), with recent proposals trying to clarify and guide the user based on their mathematical properties (Tucker et al. 2016).

The field of community phylogenetics has primarily focused on local-scale diversity patterns (e.g. within and among sampling plots; Cardillo 2011). However, about a decade ago, studies began to apply a CP approach under a macroecological perspective by evaluating the phylogenetic structure of assemblages within large geographical regions (Algar et al. 2009; Cardillo et al. 2008; Cardillo and Meijaard 2010; Cooper et al. 2008). The reasoning behind such an expansion of CP to macroecology is that the former allows a phylogenetically explicit evaluation of evolutionary hypotheses that seek to explain large-scale diversity patterns (Cardillo 2011; Tucker et al. 2016). In a macroecological context, the phylogenetic structure of an assemblage can be interpreted in terms of actual evolutionary processes such as geographic speciation and phylogenetic niche conservatism (Kissling et al. 2012; Villalobos et al. 2013).

One of the first applications of CP to macroecology, in which species' distributions were used to map phylogenetic structure, evaluated the Tropical Niche Conservatism (TNC) hypothesis as an explanation for the LDG of New World treefrogs (Hylidae; Algar et al. 2009). Under the TNC hypothesis, Algar et al. expected to find a correlation between species richness and phylogenetic structure (using the Phylogenetic Species Variability index; PSV, Helmus et al. 2007), as well as between each of these variables and a common set of explanatory climatic variables. However, they were unable to detect a correlation between species richness and phylogenetic structure, and found each one to be correlated with a unique climatic variable, thus not supporting TNC as an explanation for the LDG in treefrogs (Algar et al. 2009). Differently from Algar et al. (2009), however, studies of other biological groups—including Furnariides (Pinto-Ledezma et al. 2017)—were able to apply a similar approach and found support for TNC as one of the explanations for diversity gradients in the Neotropics. Indeed, there is broad correspondence between Furnariides species richness (Fig. 5.1a) and phylogenetic structure (PSV; Fig. 5.2d), with more related species co-occurring in species-poor subtropical/temperate areas, and more distantly related species co-occurring in species-rich tropical areas.

Another community phylogenetics tool that has been recently expanded to a biogeographical scale is the analysis of Principal Coordinates of Phylogenetic Structure (PCPS; Duarte et al. 2016). Given a map in which each grid-cell is associated with a list of species, and given a phylogenetic hypothesis relating all species, this method extracts eigenvectors from a matrix expressing the phylogeny-weighted species composition (i.e. species presence considering their phylogenetic similarity) for each cell (Duarte et al. 2016). These eigenvectors reveal orthogonal phylogenetic gradients across the studied region. Maestri et al. (2016) recently applied PCPS to model the body size variation (Bergmann's rule) among sigmodontine rodents in South America, with the first PCPS being used as an explanatory variable revealing assemblage phylogenetic structure. By doing so, Maestri et al. (2016) found that the combined influenced of both phylogenetic composition and environmental factors is more important than each variable alone in driving Bergmann's rule in sigmodontine rodents. When we applied the PCPS approach to our Furnariides dataset, the spatial gradient derived from the first PCPS

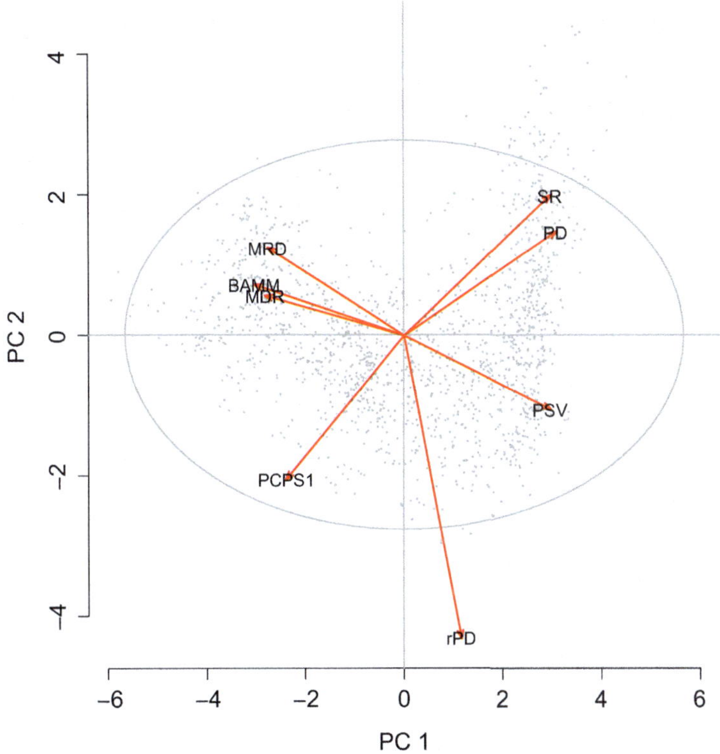

Fig. 5.3 Principal Component Analysis (PCA) biplot showing the ordination of Furnariides assemblages according to the grouping of their evolutionary variables. *SR* species richness, *PD* phylogenetic diversity; the rest of the acronyms correspond to those of Fig. 5.2

axis showed a clear distinction in phylogenetic structure between open and forest habitat assemblages—similar to those found for the other metrics (MRD, DR, PSV; Fig. 5.2e) and confirming our previous findings of differential phylogenetic structure between these habitats (Pinto-Ledezma et al. 2017, 2019).

Our analyses of the Furnariides in the Neotropics show that the alternative methods used to detect assemblage phylogenetic structure (Fig. 5.2) are, indeed, quite similar. This similarity is evident when we use a Principal Component Analysis (PCA) on the spatial correlation matrix among the maps of the six phylogenetic metrics and species richness. In this analysis, the loadings of the PCA were used to express the similarity in geographic space of the first two PCs, which accounted for 83% of the variation among the maps. In general, the alternative methods produced quite similar results (i.e., along the first axes of the PCA), with the exception of rPD—which showed mixed patterns between open and forest habitats (Fig. 5.3). As previously discussed, all methods tend to show the high level of "derivedness" of the Furnariides assemblages in open habitats.

Significant phylogenetic structure (clustering or overdispersion), the basis of the CP framework, depends on the scale of the sampling pool from which assemblages are composed, both spatially and phylogenetically (Cavender-Bares et al. 2009; Kissling et al. 2012; Graham et al. 2018). Indeed, evaluating assemblage phylogenetic structure in relation to sampling pools from different spatial extents can inform about the critical scales at which assembly processes act (Kissling et al. 2012). For instance, by studying a collection of palm assemblages across the world at the global, hemispheric, and biogeographic realm scales, Kissling et al. (2012) found that Neotropical assemblages show significant phylogenetic clustering within the first two spatial scales, but not the third. They interpreted such pattern as a strong signal of in situ diversification and geographic isolation. Considering different phylogenetic scales (a hierarchy of mutually nested clades; e.g. large clades containing subclades as in taxonomic families containing genera; Graham et al. 2018), can also lead to different findings regarding the phylogenetic structure of assemblages. For instance, in New World leaf-nose bats (Phyllostomidae), which are a major component of the Neotropical mammal fauna, assemblages showed high PSV values (thus phylogenetic overdispersion) towards the equator, which may be related to higher diversification in this region (Villalobos et al. 2013; Arita et al. 2014). However, another study showed opposite patterns when considering a higher-level taxon, the bat superfamily Noctilionoidea (Stevens et al. 2013), with lower PSV values and thus phylogenetic clustering towards the equator. Such opposite patterns across phylogenetic scales highlight the abovementioned need to consider the effect of scale (Graham et al. 2018).

Finally, another relevant scale component in a study of phylogenetic structure patterns is the unit of analysis—whether sites (site-based) or species (species-based). A site-based approach is most commonly used in community phylogenetics studies, whether at local, regional or global scales (Cardillo 2011; Kissling et al. 2012). However, considering individual assemblages as the study unit (site-based) is not necessarily equivalent to considering complete species' ranges (species-based) as such unit; especially when the former approach does not consider all sites where a particular species occurs (Villalobos et al. 2013). Given that the geographical coexistence among species (dictated by the size, shape and location of their whole ranges; Gotelli et al. 2009) ultimately determines large-scale biodiversity patterns, we should expect causal processes to be discernable at the level of species' ranges (Villalobos et al. 2013; Barnagaud et al. 2014). Thus, a species-based approach can be at least complementary to a site-based approach to evaluate evolutionary processes at biogeographic scales (Villalobos et al. 2017). So far, only one study has applied the phylogenetic field approach for a Neotropical clade (bat family Phyllostomidae; Villalobos et al. 2013), but studies at global and regional scales for birds (Barnagaud et al. 2014; Miller et al. 2017b) and mammals (Villalobos et al. 2017) have shown the potential of such approach in revealing interesting patterns and their potential evolutionary causes.

4 New Methods to Study Diversification at Macroecological Scales

In the last few years, there has also been an impressive increase in the availability of phylogenetic comparative methods (PCMs) to study evolutionary processes (Garamzegi 2014; Cooper et al. 2016)—mainly in terms of diversification rates (i.e. speciation minus extinction rates; Morlon 2014). While most diversification models provide estimates on speciation and extinction rates for the whole phylogeny (ies) under study, Rabosky (2014) introduced BAMM (Bayesian Analysis of Macroevolutionary Mixtures) to evaluate diversification rates at any point in time (including each of the individual species) across a time-calibrated phylogenetic tree. Given known limitations for accurately estimating extinction rates, BAMM-derived rates commonly focused solely on speciation rates (Rabosky et al. 2015).

Because BAMM allows us to estimate rates as a species-level attribute, they can be rasterized and mapped in geographic space by assigning rate values to the geographic distribution of the species, as explained above. Yet so far, few studies have mapped rates to evaluate, for instance, the presence of important areas of diversification (Perez-Escobar et al. 2017), or the museum/cradle role of diversity refugia (e.g. Morales-Barbero et al. 2018). For example, Perez-Escobar et al. (2017) mapped speciation rates of Neotropical orchid lineages and were able to identify speciation hotspots in different regions such as the Northern Andes and Central America. Their findings support geological events, such as the Andean uplift, in driving evolutionary processes of megadiverse groups like Neotropical orchids (see also Lagomarsino et al. 2016 for similar findings in bellflowers).

Contrary to those orchid and bellflower studies, our Furnariides maps do not flag the existence of speciation hotspot areas (Fig. 5.1f). Instead, they suggest that subtropical and temperate savannas, as a continuum, represent diversification arenas for this bird clade, supporting our previous findings using different approaches (Pinto-Ledezma et al. 2017). It is important to note, however, that mapping processes such as speciation rates also come with several caveats. For instance, assuming that speciation processes have occurred at sites where species are currently distributed can be particularly risky, especially given the dynamics of species ranges (Liow and Stenseth 2007).

All of the aforementioned approaches are based on mapping phylogenetic information of multiple species onto space, either by considering local inventories, or by rasterizing their geographic distributions. An alternative approach is to use the distributional information on multiple species and "map them" onto a phylogeny, by defining a "geographic trait" of species whose trait states depend on the particular location of each species within a geographic domain (e.g. tropical vs temperate distribution) (Goldberg et al. 2011; Jablonski et al. 2017). Indeed, recent phylogenetic comparative methods allow speciation and extinction rates to depend on species' traits. These methods aim to disentangle the relationship between trait evolution and diversification rates, collectively referred as state-dependent speciation and extinction (SSE) methods (O'Meara and Beaulieu 2016). One such method is the geographic state speciation and extinction (GeoSSE) model (Goldberg et al. 2011), which considers a species' geographic range as its trait, and includes spatial dynamics (i.e. dispersal rates) over two regions on the diversification process (Ramiadantsoa et al. 2017). Considering spatial dynamics, particularly range expansion and contraction (Goldberg et al. 2011), the GeoSSE model allows the inclusion of a third fundamental process responsible for the change in species numbers at a biogeographic scale, along with speciation and extinction: dispersal (Ricklefs 2004). In doing so, the GeoSSE model permits more complex exploration of evolutionary hypotheses.

Considering the tropical and temperate regions, for instance, recent studies have applied the GeoSSE model to evaluate predictions related to the Latitudinal Diversity Gradient in terms of speciation, extinction and dispersal rates. Based on existing interpretations of the LDG, species-rich areas (e.g. the tropics) would be expected to show higher speciation and/or dispersal rates relative to species-poor areas (temperate regions; Schluter and Pennel 2017). Yet, while some of studies have found support for higher rates in the tropics (mammals, birds, insects and flowering plants, Jansson et al. 2013; mammals, Rolland et al. 2014), others have found the opposite (reptiles, Pyron 2014). Such differences in macroevolutionary rates can also be observed between adjacent environments and habitats (Schluter and Pennel 2017), especially within species-rich regions such as the Neotropics (Pinto-Ledezma et al. 2017). Our previous work on Furnariides (Pinto-Ledezma et al. 2017) evaluated such rate differences between species-rich forest habitats and species-poor open habitats within the Neotropics, revealing a counterintuitive pattern where the latter habitats showed higher rates than the former habitats. This led us to suggest that the highest species richness in forest habitats depends upon the high diversification and rapid species colonization from open to forest habitats, as well as the stability of forest habitats through time, to allow for local species accumulation (Pinto-Ledezma et al. 2017).

Concerns on the use of SSE models have been raised recently, mainly related to interpretation of results and inflated Type I errors (Rabosky and Goldberg 2015; O'Meara and Beaulieu 2016; Alves et al. 2017). However, the few Neotropical studies published so far have applied a battery of methods to evaluate the robustness of their results (e.g. GeoSSE, BAMM, diversity-dependent models), showing that different methods were in fact consistent (Lagomarsino et al. 2016; Perez-Escobar et al. 2017; Pinto-Ledezma et al. 2017). Thus, despite well-founded concerns on SSE models, they appear to be useful exploratory tools.

5 Perspectives and Concluding Remarks

Our review discussed the advances and caveats of the distinct macroecological approaches currently available to investigate evolutionary processes associated with spatial gradients of species richness. Given that the multi-scale data needed to test these evolutionary processes have only been available in the last few years, or decades at most (e.g. megaphylogenies, deep-time paleoclimatic data), several research avenues may still be explored.

Moreover, the analytical methods discussed here are not exhaustive. In addition to speciation and extinction processes, one may want to further focus on lineage dispersal over evolutionary time as a major contributor to geographic diversity gradients. To this aim, parametric biogeography has gained prominence, including methods implemented by software such as DIVA, DEC, LAGRANGE and BioGeoBEARS (see Ronquist and Sanmartín 2011; Albert and Antonelli 2017, and references therein). Parametric biogeography methods rely on the assumption that closely related taxa must have a similar geographic history (Ronquist and Sanmartín 2011), and are thus useful to disentangle the relative roles of cladogenetic and anagenetic processes driving diversity gradients. Different studies have used these approaches to evaluate dispersal as a driver of large scale patterns of species diversity, and to infer the ancestral areas (i.e. centres of origin) of clades (Wiens et al. 2011; Pyron 2014; Barker et al. 2015; Claramunt and Cracraft 2015; Dupin et al. 2017), or to support previous findings (Rolland et al. 2015). Integration with the overall research program on LDG is still missing, though, which opens an important research avenue (Albert and Antonelli 2017).

Another relevant framework to aid studies of spatial gradients is the simulation of biogeographical and macroecological patterns. Conceptually, one major difficulty in understanding the causes behind diversity patterns at large spatial and temporal scales is the inability to apply experimental approaches, leaving statistical analyses as an only option (Gotelli et al. 2009). Recently, however, macroecological approaches have bridged such difficulty by applying computer simulation models as an alternative to purely statistical analyses, allowing the explicit consideration of evolutionary dynamics in a spatial domain under changing environments. Rangel et al. (2007), for instance, applied a simulation model to evaluate the relative contribution of evolutionary mechanisms such as speciation and niche conservatism or evolution in creating the LDG for South American birds. They found that a Neotropical origin of species and strong conservatism of their niches generated an emergent diversity gradient that closely resembled the observed one (Rangel et al. 2007; see also Rahbek et al. 2007). While the complexity of the parameter space that the simulation models must explore may pose a computational challenge, more efficient programming and high-performance computers are overcoming these shortfalls (Cabral et al. 2017). A more complex model that couples empirical paleoclimatic data, dispersal, and adaptation to changing environments, while allowing for further parametrization regarding group-specific ecology and life history, has just been released (Rangel et al. 2018).

In summary, our review demonstrates that we now have multiple coherent metrics for capturing distinct components of evolutionary factors of diversity gradients. We have shown here that combining such evolutionary components into the study of diversity gradients—from species distributions and ages, to their phylogenetic structure and macroevolutionary rates—is not only possible, but also required for a comprehensive understanding of the most conspicuous pattern of biodiversity: its geographical variation. This is an exciting time for evolutionary macroecology.

Acknowledgments Work by FV and JAFDF has been supported by several grants from Conselho Nacional de Desenvolvimento Científico e Tecnológico (CNPq), Brazil. FV is currently supported by CONACYT, Mexico. JNP-L. is supported by the University of Minnesota College of Biological Sciences' Grand Challenges in Biology Postdoctoral Program. This chapter is also developed in the context of National Institutes for Science and Technology (INCT) in Ecology, Evolution and Biodiversity Conservation, supported by MCTIC/CNPq (proc. 465610/2014-5) and FAPEG (proc. 201810267000023).

References

Albert JS, Antonelli A (2017) Society for the study of systematic biology symposium: frontiers in parametric biogeography. Syst Biol 66:125–127

Algar AC, Kerr JT, Currie DJ (2009) Evolutionary constraints on regional faunas: whom, but not how many. Ecol Lett 12:c57–c65

Alves DMCC, Diniz-Filho JAF, Villalobos F (2017) Geographical diversification and the effect of model and data inadequacies: the bat diversity gradient as a case study. Biol J Linn Soc 121:894–906

Arita HT, Vargas-Barón J, Villalobos F (2014) Latitudinal gradients of genus richness and endemism and the diversification of New World bats. Ecography 37:1024–1033

Barker FK, Burns KJ, Klicka J et al (2015) New insights into New World biogeography: an integrated view from the phylogeny of blackbirds, cardinals, sparrows, tanagers, warblers, and allies. Auk 132:333–348

Barnagaud JY, Kissling WD, Sandel B et al (2014) Ecological traits influence the phylogenetic structure of bird species co-occurrences worldwide. Ecol Lett 17:811–820

Brown JH (2012) Why are there so many species in the tropics? J Biogeogr 41:8–22

Cabral JS, Valente L, Hartig F (2017) Mechanistic simulation models in macroecology and biogeography: state-of-art and prospects. Ecography 40:267–280

Cardillo M (2011) Phylogenetic structure of mammal assemblages at large geographical scales: linking phylogenetic community ecology with macroecology. Proc R Soc B 366:2545–2553

Cardillo M, Meijaard E (2010) Phylogeny and co-occurrence of mammal species of southeast Asian islands. Glob Ecol Biogeogr 19:465–474

Cardillo M, Gittleman JL, Purvis A (2008) Global patterns in the phylogenetic structure of island mammal assemblages. Proc R Soc B 275:1549–1556

Cavender-Bares J, Kozak HK, Fine PVA et al (2009) The merging of community ecology and phylogenetic biology. Ecol Lett 12:693–715

Cavender-Bares J, Ackerly DD, Kozak HK (2012) Integrating ecology and phylogenetics: the footprint of history in modern-day communities. Ecology 91:s1s3

Claramunt S (2010) Discovering exceptional diversifications at continental scales: the case of the endemic families of neotropical suboscine passerines. Evolution 64:2004–2019

Claramunt S, Cracraft J (2015) A new time tree reveals Earth history's imprint on the evolution of modern birds. Sci Adv 1:e1501005

Cooper N, Rodríguez J, Purvis A (2008) A common tendency for phylogenetic overdispersion in mammalian assemblages. Proc R Soc B 275:2031–2037

Cooper N, Thomas GH, FitzJohn RG (2016) Shedding light on the 'dark side' of phylogenetic comparative methods. Methods Ecol Evol 7:693–699

Currie DJ, Mittelbach GG, Cornell HV et al (2004) Predictions and tests of climate-based hypotheses of broad-scale variation in taxonomic richness. Ecol Lett 7:1121–1134

Davies TJ, Buckley LB (2011) Phylogenetic diversity as a window into the evolutionary and biogeographic histories of present-day richness gradients for mammals. Philos Trans R Soc B 366:2414–2425

Dobzhansky TH (1937) Genetics and the origin of species. Columbia University Press, New York
Duarte LD, Debastiani VJ, Freitas AV, Pillar VD (2016) Dissecting phylogenetic fuzzy weighting: theory and application in metacommunity phylogenetics. Methods Ecol Evol 7:937–946
Dupin J, Matzke NJ, Särkinen T et al (2017) Bayesian estimation of the global biogeographical history of the Solanaceae. J Biogeogr 44:887–899
Faith DP (2013) Biodiversity and evolutionary history: useful extensions of the PD phylogenetic diversity assessment framework. Ann NY Acad Sci 1298:69–89
Fine PVA (2015) Ecological and evolutionary drivers of geographic variation in species diversity. Ann Rev Ecol Evol Syst 46:369–392
Fritz SA, Rahbek C (2012) Global patterns of amphibian phylogenetic diversity. J Biogeogr 39:1373–1382
Garamzegi LZ (ed) (2014) Modern phylogenetic comparative methods and their application in evolutionary biology: concepts and practice. Springer, Berlin
Goldberg EE, Lancaster LT, Ree RH (2011) Phylogenetic inference of reciprocal effects between geographic range evolution and diversification. Syst Biol 60:451–465
Gotelli NJ, Anderson MJ, Arita HT et al (2009) Patterns and causes of species richness: a general simulation model for macroecology. Ecol Lett 12:873–886
Graham CH, Storch D, Machac A (2018) Phylogenetic scale in ecology and evolution. Glob Ecol Biogeogr 27:175–187
Hawkins BA (2001) Ecology's oldest pattern. Trends Ecol Evol 16:170
Hawkins BA, Diniz-Filho JAF (2004) 'Latitude' and geographic patterns in species richness. Ecography 27:268–272
Hawkins BA, Field R, Cornell HV et al (2003) Energy, water, and broad-scale geographic patterns of species richness. Ecology 84:3105–3117
Hawkins BA, Diniz-Filho JAF, Jaramillo CA et al (2006) Post-Eocene climate change, niche conservatism, and the latitudinal diversity gradient of New World birds. J Biogeogr 33:770–780
Hawkins BA, McCain CM, Davies TJ et al (2012) Different evolutionary histories underlie congruent species richness gradients of birds and mammals. J Biogeogr 39:825–841
Hawkins BA, Leroy B, Rodríguez MA et al (2017) Structural bias in aggregated species-level variables driven by repeated species co-occurrences: a pervasive problem in community and assemblage data. J Biogeogr 44:1199–1211
Helmus MR, Bland TJ, Williams CK et al (2007) Phylogenetic measures of biodiversity. Am Nat 169:e68–e83
Hillebrand H (2004) On the generality of the latitudinal diversity gradient. Am Nat 163:192–211
Jablonski D, Roy K, Valetine JW (2006) Out of the tropics: evolutionary dynamics of the latitudinal diversity gradient. Science 314:102–106
Jablonski D, Huang S, Roy K et al (2017) Shaping the latitudinal diversity gradients: new perspectives from a synthesis of paleobiology and biogeography. Am Nat 189:1–12
Jansson R, Rodríguez-Castañeda G, Harding LE (2013) What can multiple phylogenies say about the latitudinal diversity gradient? a new look at the tropical conservatism, out of the tropics, and diversification rate hypotheses. Evolution 67(6):1741–1755
Jetz W, Thomas GH, Joy JB et al (2012) The global diversity of birds in space and time. Nature 491:444–448
Kennedy JD, Wang Z, Weir JT et al (2014) Into and out of the tropics: the generation of the latitudinal gradient among New World passerine birds. J Biogeogr 41:1746–1757
Kennedy JD, Price TD, Fjeldså J et al (2016) Historical limits on species co-occurrence determine variation in clade richness among New World passerine birds. J Biogeogr 44:736–747
Kerr JT, Currie DJ (1999) The relative importance of evolutionary and environmental controls on broad-scale patterns of species richness in North America. Ecoscience 6:329–337
Kissling WD, Dorman CF, Groeneveld J et al (2012) Towards novel approaches to modelling biotic interactions in multispecies assemblages at large spatial extents. J Biogeogr 39:2163–2178
Lagomarsino LP, Condamine FL, Antonelli A et al (2016) The abiotic and biotic drivers of rapid diversification in Andean bellflowers (Campanulaceae). New Phytol 210:1430–1442

Liow LH, Stenseth NC (2007) The rise and fall of species: implications for macroevolutionary and macroecological studies. Proc R Soc B 274:2745–2752

Maestri R, Luza AL, Barros LD, Hartz SM, Ferrari A, Freitas TR, Duarte LD (2016) Geographical variation of body size in sigmodontine rodents depends on both environment and phylogenetic composition of communities. J Biogeogr 43:1192–1202

Miller ET, Farine DR, Trisos CH (2017a) Phylogenetic community structure metrics and null models: a review with new methods and software. Ecography 40:461–477

Miller ET, Wagner SK, Harmon LJ et al (2017b) Radiating despite a lack of character: ecological divergence among closely related, morphologically similar honeyeaters (Aves: Meliphagidae) co-occurring in arid Australian environments. Am Nat 189:E14–E30

Mittelbach GG, Schemske DW, Cornell HV et al (2007) Evolution and the latitudinal diversity gradient: speciation, extinction and biogeography. Ecol Lett 10:315–331

Morales-Barbero J, Martinez PA, Ferrer-Castán D et al (2018) Quaternary refugia are associated with higher speciation rates in mammalian faunas of the western Palaearctic. Ecography 41:607–621

Morlon H (2014) Phylogenetic approaches for studying diversification. Ecol Lett 17:508–525

O'Brien EM (2006) Biological relativity to water–energy dynamics. J Biogeogr 33:1868–1888

O'Meara BC, Beaulieu JM (2016) Past, future, and present of state-dependent models of diversification. Am J Bot 103:792–795

Olalla-Tárraga MA, Bini LM, Diniz-Filho JAF et al (2010) Cross-species and assemblage-based approaches to Bergmann's rule and the biogeography of body size in Plethodon salamanders of eastern North America. Ecography 33:362–368

Perez-Escobar OA, Chomicki G, Condamine FL et al (2017) Recent origin and rapid speciation of Neotropical orchids in the world's richest plant biodiversity hotspot. New Phytol 215:891–905

Pianka ER (1966) Latitudinal gradients in species diversity: a review of concepts. Am Nat 100:33–46

Pinto-Ledezma JN, Simon L, Diniz-Filho JAF et al (2017) The geographic diversification of Furnariides: the role of forest versus open habitats in driving species richness gradients. J Biogeogr 44:1683–1693

Pinto Ledezma JN, Jahn AE, Cueto VR, Diniz-Filho JAF, Villalobos F (2019) Drivers of phylogenetic assemblage structure of the furnariides, a widespread clade of lowland neotropical birds. Am Nat 193(2):E41–E56

Pyron RA (2014) Biogeographic analysis reveals ancient continental vicariance and recent oceanic dispersal in amphibians. Syst Biol 63:779–797

Rabosky DL (2014) Automatic detection of key innovations, rate shifts, and diversity dependence on phylogenetic trees. PLoS One 9:e89543

Rabosky DL, Goldberg EE (2015) Model inadequacy and mistaken inference of trait-dependent speciation. Syst Biol 64:340–355

Rabosky DL, Title PO, Huang H (2015) Minimal effects of latitude on present-day speciation rates in New World birds. Proc R Soc B 282:20142889

Rahbek C, Gotelli N, Colwell RK, Entsminger GL, Rangel TFLVB, Graves GR (2007) Predicting continental-scale patterns of bird species richness with spatially explicit models. Proc R Soc Lond B 274:165–174

Ramiadantsoa T, Sirén J, Hanski I (2017) Phylogenetic comparative method for geographical radiation. Ann Zool Fenn 54:237–257

Rangel TFL, Diniz-Filho JAF, Colwell RK (2007) Species richness and evolutionary niche dynamics: a spatial pattern–oriented simulation experiment. Am Nat 170(4):602–616

Rangel TFR, Edwards RR, Holden PB et al (2018) Modeling the ecology and evolution of biodiversity: biogeographical cradles, museums, and graves. Science 361:eaar5452

Ricklefs RE (2002) Splendid isolation: historical ecology of the South American passerine fauna. J Avian Biol 33:207–211

Ricklefs RE (2004) A comprehensive framework for global patterns in biodiversity. Ecol Lett 7:1–15

Ridgely RS Allnutt TF, Brooks T, McNicol DK, Mehl-man DW, Young BE, Zook JR, BirdLife Inter- national (2012) Digital distribution maps of the birds of the Western Hemisphere, version 5.0. In BirdLife Interna- tional and NatureServe (2012) Bird species distribution maps of the world. Version 2.0. BirdLife International, Cambridge, UK and Nature Serve, Arlington, USA

Rohde K (1992) Latitudinal gradients in species diversity: the search for the primary cause. Oikos 65:514–527

Rolland J, Condamine FL, Jiguet F et al (2014) Faster speciation and reduced extinction in the tropics contribute to the mammalian latitudinal diversity gradient. PLoS Biol 12:e1001775

Rolland J, Condamine FL, Beeravolu CR et al (2015) Dispersal is a major driver of the latitudinal diversity gradient of Carnivora. Glob Ecol Biogeogr 24:1059–1071

Ronquist F, Sanmartín I (2011) Phylogenetic methods in biogeography. Annu Rev Ecol Evol Syst 42:441–464

Schemske DW, Mittelbach GG (2017) Latitudinal gradients in species diversity: reflections on Pianka's 1966 article and look forward. Am Nat 189:599–603

Schluter D, Pennel MW (2017) Speciation gradients and the distribution of biodiversity. Nature 546:48–55

Simpson GG (1964) Species densities of North American recent mammals. Syst Zool 13:57–73

Springer MS, Meredith RW, Janecka JE et al (2011) The historical biogeography of Mammalia. Philos Trans R Soc B 366:2478–2502

Stevens RD, Tello JS, Gavilanez MM (2013) Stronger tests of mechanisms underlying geographic gradients of biodiversity: insights from the dimensionality of biodiversity. PLoS One 8:e56853

Tucker CM, Cadotte MW, Carvalho SB et al (2016) A guide to phylogenetic metrics for conservation, community ecology and macroecology. Biol Rev 92:698–715

Vamosi SM, Heard SB, Vamosi JC et al (2009) Emerging patterns in comparative analysis of phylogenetic community structure. Mol Ecol 18:572–592

Villalobos F, Rangel TF, Diniz-Filho JAF (2013) Phylogenetic fields of species: cross-species patterns of phylogenetic structure and geographical coexistence. Proc R Soc B 280:20122570

Villalobos F, Olalla-Tarraga MA, Cianciaruso MV, Rangel TF, Diniz-Filho JAF (2017) Global patterns of mammalian co-occurrence: phylogenetic and body size structure within species ranges. J Biogeogr 44:136–146

Volkamp A, Baker DJ, Stephens PA et al (2017) Global patterns in the divergence between phylogenetic diversity and species richness in terrestrial birds. J Biogeogr 44:709–721

Webb CO, Ackerly DD, McPeek MA et al (2002) Phylogenies and community ecology. Annu Rev Ecol Evol Syst 33:475–505

Wiens JJ, Donoghue MJ (2004) Historical biogeography, ecology and species richness. Trends Ecol Evol 19:639–644

Wiens JJ, Pyron RA, Moen DS (2011) Phylogenetic origins of local-scale diversity patterns and the causes of Amazonian megadiversity. Ecol Lett 14:643–652

Chapter 6
Evolutionary Imprints on Species Distribution Patterns Across the Neotropics

Renan Maestri and Leandro Duarte

Abstract Species diversity is unevenly distributed across the Neotropical region. Distinct biological groups have notably similar patterns of species richness, with peaks in richness usually associated with tropical regions. However, how diversification occurred across space and time for each group, leading to the current diversity patterns, are less understood. Differences among clades can be expected from their distinct ecologies and evolutionary histories. We mapped the occurrence of 1100 + species of Neotropical vertebrates belonging to seven clades with evolutionary histories primarily confined to the Neotropics. For each clade, we analyzed the spatial distribution of phylogenetic turnover of species assemblages, and contrasted the observed patterns among different clades. We also investigated which environmental or biogeographic factors were most associated with phylogenetic turnover, using permutational regression on dissimilarity matrices. We discovered that spatial patterns of phylogenetic turnover between sites are strikingly similar for distinct vertebrate groups, and cannot be predicted from within-lineage relatedness—a surprising result given the differences in ecology and history of colonization among clades. For most groups, biogeographic units predicted phylogenetic turnover better than environmental variables. This suggests that vicariant and dispersal events that took place during the settlement of each vertebrate clade may have been similar in mode and place, even if the timing was different. The fact that environmental variables alone were less predictive of phylogenetic turnover for most clades suggests that current climatic and topographical features are less determinant of phylogenetic lineage distribution than former, biogeographic processes.

Keywords Biogeographical patterns · Historical diversification · Phylogenetic beta diversity · Phylogenetic composition · Phylogenetic fuzzy weighting · Principal coordinates of phylogenetic structure · Tetrapods

---

R. Maestri · L. Duarte (✉)
Departamento de Ecologia, Universidade Federal do Rio Grande do Sul, Porto Alegre, RS, Brazil

V. Rull, A. C. Carnaval (eds.), *Neotropical Diversification: Patterns and Processes*, Fascinating Life Sciences, https://doi.org/10.1007/978-3-030-31167-4_6

1 Introduction

The complex climatic and geological history of the Neotropical region gave rise to one of the richest species diversity in the world (Antonelli and Sanmartín 2011). Alternating periods of isolation and biotic exchanges (Simpson 1980; Patterson and Costa 2012), repeated fluctuations in the sea level and changes in river systems (Hoorn 1993; Hoorn et al. 1995), the rise and modification of mountain chains (Garzione et al. 2008; Hoorn et al. 2010), and climatic variation in time and space (e.g., Rull 2005; Goin et al. 2012), all shaped the diversity patterns found in the Neotropics today. In this region, as in others, diversification—representing the balance between speciation and extinction—is the result of vicariance, dispersal, and ecological specialization processes acting over different times and with different intensities, depending on the taxa of focus (Antonelli et al. 2018a). Several hypotheses have been proposed to explain species distribution patterns, and the latitudinal diversity gradient, that are observed in clades primarily distributed in the Neotropics (Hawkins et al. 2003; Rull 2011). Yet, while macroecological studies have asked why some regions are richer in species than others, and sought to identify correlates of diversity gradients within and across taxa (Diniz-Filho et al. 2007; Fergnani and Ruggiero 2017), the spatial patterns of the historical processes that ultimately gave rise to current diversity patterns in the Neotropics (Antonelli et al. 2009, 2018c; Fine et al. 2014) are less explored.

We are particularly interested in understanding and comparing how different evolutionary lineages occupied the geographical space, and diversified in the Neotropics. Species diversity patterns are known to be fairly similar for several groups of Neotropical macro-organisms. For instance, the familiar relationship between latitude and species diversity holds from plants to vertebrates (Ramos Pereira and Palmeirim 2013; Kerkhoff et al. 2014). Nonetheless, it is unknown if the patterns of colonization and diversification in the Neotropics were also similar across clades. How do groups differ in the relative importance of in situ diversification versus repeated colonization of Neotropical biomes? One can hypothesize multiple diversification trajectories for a group. One clade may have achieved high richness in Amazonia, for instance, via independent diversification of species belonging to distantly related sub-clades. In contrast, another group may have achieved the same diversity pattern via diversification within a single Amazonian sub-clade, giving rise to multiple species through an evolutionary radiation. Patterns of colonization and diversification are therefore theoretically independent of species diversity patterns, but ultimately contribute to determining them.

Our goal is to compare how different evolutionary lineages colonized and diversified within the Neotropics, and to investigate the environmental and biogeographic correlates of these diversification processes. For that, we combine information about regional diversity patterns and the distribution of phylogenetic composition across space, which we quantify by mapping phylogenetic beta diversity (Graham and Fine 2008). Phylogenetic composition expresses the representativeness of different clades across a set of local species assemblages (Peres-Neto et al. 2012; Duarte et al. 2016), enabling the quantification of phylogenetic turnover across space. Here, we refer to phylogenetic composition as a measure of phylogenetic beta diversity (Duarte et al. 2016), and define it as the pairwise dissimilarity between species assemblages, computed through phylogenetic fuzzy weighting (Pillar and Duarte 2010). Hereafter, we use the term phylogenetic turnover to denote differences in phylogenetic composition across assemblages.

Phylogenetic turnover provides clues about how lineages have colonized and diversified within a region. For example, differences in phylogenetic composition between sites is often explained by habitat differentiation driving ecological niche divergence between clades, leading to a pattern in which particular clades occupy particular ecological regions (e.g., niche-based differentiation—Soberón 2007). Within this scenario, ecological adaptation followed by niche conservatism can account for observations of ecologically similar regions being colonized by species that belong to particular clades (e.g., species of a given clade occupying tropical forests; Cavender-Bares et al. 2009). Alternatively, historical processes linked to a common geological event, and similar vicariant and dispersal events, can cause distinct clades to occupy particular regions through processes that are not related to niche differentiation (Ficetola et al. 2017; Mazel et al. 2017). Most commonly, however, ecological and biogeographical historic factors are expected to interact (Duarte et al. 2014b; Maestri et al. 2016), and the combination of their relative importance helps to infer dispersal processes from distribution patterns within clades. Here, we used biogeographical variables to represent historical events of dispersal and vicariance, and current climatic and topographical variables as environmental descriptors of phylogenetic turnover across the Neotropics. Our reasoning is that major diversification events can be linked with key biogeographical splits (e.g., in the colonization of unoccupied biogeographical regions following by in situ diversification) or mediated by major climatic and topographic variables (Antonelli et al. 2018b). Such events provide an explanation for how species become distributed as they currently are, which will be reflected in the current phylogenetic turnover patterns (Graham and Fine 2008; Mazel et al. 2017).

For this, we gathered locality and phylogenetic data from Neotropical amphibians, reptiles, birds and non-volant mammals. By sampling widely across the Tree of Life, we included groups with contrasting histories of colonization and diversification. Among terrestrial vertebrates, for instance, amphibian groups such as the Hylidae have a long history of occupation of the Neotropics, dating back ~70 million years as estimated from molecular phylogenies (Pyron 2014). Sigmodontine rodents, on the other hand, colonized the South American continent around just ten million years ago (Steppan and Schenk 2017). Moreover, clades of amphibians, reptiles, mammals and birds have colonized the Neotropics from different routes and in different times (Faivovich et al. 2005; Weir et al. 2009; Antonelli et al. 2010; Jansa et al. 2014; Maestri et al. 2019). These historical differences, together with the intrinsic biological distinctiveness of each group, account for their current species distribution patterns. By exploring and comparing the spatial distribution of phylogenetic turnover within these groups, we seek to answer three major questions: (1) do these major clades of terrestrial vertebrates show similar spatial patterns of phylogenetic turnover across the Neotropics?, (2) is the dissimilarity in patterns greater for distantly related groups (e.g., a mammalian clade relative to an amphibian clade) than for closely related groups (e.g., marsupials relative to rodents)?, and (3) is the phylogenetic turnover of sites better predicted by major climatic and topographic variables, or by biogeographic history?

2 Methods

2.1 Data

We mapped species occurrences within seven vertebrate clades whose evolutionary histories were primarily centered in the Neotropical region. Clades were selected to include representatives of the major terrestrial vertebrate groups (amphibians, reptiles, birds and non-volant mammals), representing independent, well-sampled, widely distributed, and largely monophyletic radiations in the Neotropics. Range maps were gathered from IUCN (2008) for clades belonging to two amphibian families (the Hylidae and the Leptodactylidae), and to one mammal family (Didelphidae). Maestri et al. (2019) was used to gather ranges from species belonging to a second mammal family (Sigmodontinae). From BirdLife International (2016), we extracted ranges for two bird families (Tyrannidae and Furnariidae). Finally, ranges of species belonging to one reptile clade, the genus *Bothrops*, were obtained from Fenker et al. (2014). Matrices of presence/absence of species were constructed for each clade over an equal area grid of 110 km in Behrman projection, with cells distributed across the Neotropical region *sensu stricto* (Morrone 2014a) without islands, totaling 1382 cells.

For each clade, a phylogenetic hypothesis was obtained from the literature. Amphibian groups followed the consensus phylogenetic hypothesis of Pyron (2014), which is one of the most comprehensive phylogenies to date. To represent phylogenetic relationships for the genus *Bothrops*, we used the squamate tree of Tonini et al. (2016). Didelphidae relationships followed the phylogeny of Amador and Giannini (2016), the most inclusive hypothesis to date, and the Sigmodontinae phylogeny was taken from the 11-genes tree presented by Maestri et al. (2017). Tyrannidae phylogenetic relationships followed the BirdTree project (Jetz et al. 2012), and a Furnariidae tree was taken from Derryberry et al. (2011). In most cases, those trees were built with both nuclear and mitochondrial genes.

Presence/absence matrices were then pruned, so that only those species represented in the available phylogeny were kept for subsequent analyses. The resultant datasets consisted of 225 species of Hylidae (~48% of the total number of species in the clade), 69 species of Leptodactylidae (47%), 27 species of *Bothrops* (63%), 67 species of Didelphidae (61%), 260 species of Sigmodontinae (65%), 265 species of Tyrannidae (66%), and 223 species of Furnariidae (76%). Occurrence maps for each group covered all cells of the grid, over the entire Neotropical region *sensu strictu*.

2.2 Phylogenetic Turnover

Variation in phylogenetic composition among grid cells (i.e. phylogenetic turnover) was measured with the phylogenetic fuzzy-weighting method of Pillar and Duarte (2010). This method accounts for phylogenetic distances and tree imbalance, and show greater statistical performance—higher power and increased effect sizes in statistical analyses—than other metrics of phylogenetic beta-diversity (Duarte et al. 2014a, 2016). But what is phylogenetic fuzzy-weighting? For each clade, the phylogenetic distances among species were converted to phylogenetic similarities and later standardized within columns, resulting in a matrix **Q** depicting degrees of phylogenetic belonging of each species to every other species (Duarte et al. 2016). The pairwise degree of phylogenetic belonging of species i to species j captures, in a single value, the amount of phylogenetic covariance between i and j, and also the rate of diversification between i and the ancestral node δ connecting i to j. If the path linking species i to δ shows a higher rate of diversification than the path connecting j to δ, species i will show lower degree of belonging to j than j to i. In other words, matrix **Q** is a matrix that expresses, simultaneously, symmetric phylogenetic covariances among species and also asymmetric diversification trajectories connecting them—which implies that the topological arrangement of the phylogenetic tree is also captured. As any other current estimates of diversification, caution is needed in the interpretation of those rates as truly capturing net diversification, especially when extinction was high (Title and Rabosky 2018). The method is described in detail in Duarte et al. (2016). Matrix **Q** was then multiplied by a matrix that describes species occurrences (presence/absence) in the cells, resulting in matrix **P**, which describes phylogeny-weighted species composition (or simply phylogenetic composition) for a set of assemblages. Differences in phylogenetic composition among cells express phylogenetic turnover.

To map phylogenetic turnover across the cells, we performed Principal Coordinates Analysis on the matrix **P** of each clade. Eigenvectors from this analysis capture gradients of phylogenetic composition, which express phylogenetic beta diversity patterns across grid cells (Duarte et al. 2016)—the so-called Principal Coordinates of Phylogenetic Structure, or simply PCPS (Duarte 2011). Phylogenetic composition (**P**) and the PCPS have found widespread use in studies of phylogenetic structuring of metacommunities (e.g. Duarte et al. 2012; Gianuca et al. 2013; Carlucci et al. 2016; Maestri et al. 2016; Prieto et al. 2017), and in studies that predict spatial variation in traits (Maestri 2017; Duarte et al. 2018; Maestri et al. 2018). Since the method captures the differential frequency of species belonging to different clades across grid cells, it is useful to understand patterns of lineage occupation in space. Both **P** matrices and PCPS were computed in the R statistical environment (R Core Team 2018) using the package *PCPS* (Debastiani and Duarte 2014). Phylogenetic mapping of species scores on the PCPS axis was achieved with the R package *phytools* (Revell 2012).

2.3 Environmental and Biogeographic Predictors of Phylogenetic Turnover

We also explored if and to which extent the spatial patterns of phylogenetic turnover observed within each clade correlate to the environments experienced by Neotropical species, or to their biogeographical history. As potential environmental predictors of variation in phylogenetic turnover between grid cells, we used four climatic variables (mean annual temperature, temperature seasonality, mean annual precipitation and precipitation seasonality) and two topographical variables (mean altitude and altitudinal range variation). Climatic variables were taken from the WorldClim2 database (Fick and Hijmans 2017), and topographic variables were taken from NASA (http://neo.sci.gsfc.nasa.gov/).

As potential biogeographic predictors of phylogenetic turnover, we used the biogeographic classification of the Neotropical region into domains, as proposed by Morrone (2014b). For that, we used the shape-file formatted map of Löwenberg-Neto (2014). Morrone's biogeographic domains are expert-defined polygons (see a discussion in Antonelli 2018) informed from the geographic distributions of multiple taxa (plants, insects, arachnids and vertebrates) across the Neotropics. They are thought to reflect geological instances (e.g., collision of tectonic plates, marine transgressions, mountain uplift) and vicariant and dispersal events (Löwenberg-Neto 2014; Morrone 2014a, b).

We are aware that the distinction between environmental and biogeographical classifications is somewhat artificial, since broad-scale biogeographical regions overlap in some extent with broad-scale climatic areas. For that reason, we also quantified, in our models, the joint contributions of domains and environmental variables relative to their exclusive contribution to explain phylogenetic turnover (see below).

Questions 1 and 2: Similarity in the Phylogenetic Turnover Patterns Among Clades

To compare the spatial patterns of phylogenetic turnover across major clades, we investigated the correlation of the matrices of dissimilarity (**P**) among clades with a Procrustes superimposition (Peres-Neto and Jackson 2001), as implemented in the R package *vegan* (Oksanen et al. 2019). The Procrustes superimposition rotates each matrix to maximize similarity with every other matrix, by minimizing the sum of square differences between them. With that, we estimated a residual sum of squares that we interpreted as a correlation-like metric, using row permutations to approximate a significance value (Jackson 1995). To investigate the correlation among the major axis of phylogenetic turnover, we also calculated Pearson correlations among the PCPS 1 of each clade.

Question 3: Associations Between Phylogenetic Turnover and Biogeographical and Environmental Predictors

To investigate the exclusive and joint contributions of biogeographical and environmental variables as predictors of phylogenetic turnover pairwise square-rooted Bray-

Curtis dissimilarities between cells were computed for each matrix **P**, and taken as response variables in regressions on distance matrices (Mcardle and Anderson 2001). These regressions are hereafter called 'ADONIS'. ADONIS is a method akin to both distance-based Analysis of Variance—PERMANOVA (Pillar and Orlóci 1996; Anderson 2001), and AMOVA (Excoffier et al. 1992). Predictor variables included the six environmental variables, the seven biogeographic domains, or a combination of both. The statistical significance of ADONIS was calculated through two null models, one built by freely shuffling cells across the spatial gradient (site shuffle), and another built by shuffling species position across the phylogenetic tree (taxa shuffle; Duarte et al. 2016). ADONIS was carried out using the function 'matrix.p.sig' of *PCPS* package, which implements null models based on function 'adonis2' of the *vegan* package (Oksanen et al. 2019).

To account for pure spatial autocorrelation in the models above, we computed spatial filters based on Principal Coordinates of Neighbor Matrices (Borcard and Legendre 2002), using the geographic coordinates of each cell in the grid, using the function 'pcnm' of the R package *vegan* (Oksanen et al. 2019). All spatial filters not correlated to both environmental and biogeographical predictors were considered as a potential source of pure spatial autocorrelation in phylogenetic turnover across the cells (Duarte et al. 2014b). To quantify spatial autocorrelation in phylogenetic turnover for each clade, we performed the two ADONIS-based null model tests described above.

3 Results

Questions 1 and 2: Similarity in the Phylogenetic Turnover Patterns Among Clades

Overall, PCPS 1 or 2 showed highly similar spatial patterns of phylogenetic turnover among vertebrate clades (Fig. 6.1). Notably, within all Neotropical groups, the Amazonian region houses clades that are phylogenetically distant from all others. The Procrustes alignment of matrices of phylogenetic dissimilarities among groups confirmed the general trends observed in graphical spatial patterns (Table 6.1).

Correlations among matrices did not follow phylogenetic relatedness among vertebrate clades. For example, the pattern of phylogenetic turnover for frogs of the family Hylidae is more similar to the patterns observed in Sigmodontinae (mammals) and Furnariidae (birds) than to that of amphibians of the Leptodactylidae family. Similarly, the patterns within the Sigmodontinae are more similar to those of birds (Tyrannidae and Furnariidae) than to that of mammals of the family Didelphidae (Table 6.1).

Question 3: Associations Between Phylogenetic Turnover and Biogeographical and Environmental Predictors

Environmental variables and biogeographic dominion influenced phylogenetic turnover across cells in all major clades, as demonstrated by the low *P*-values obtained

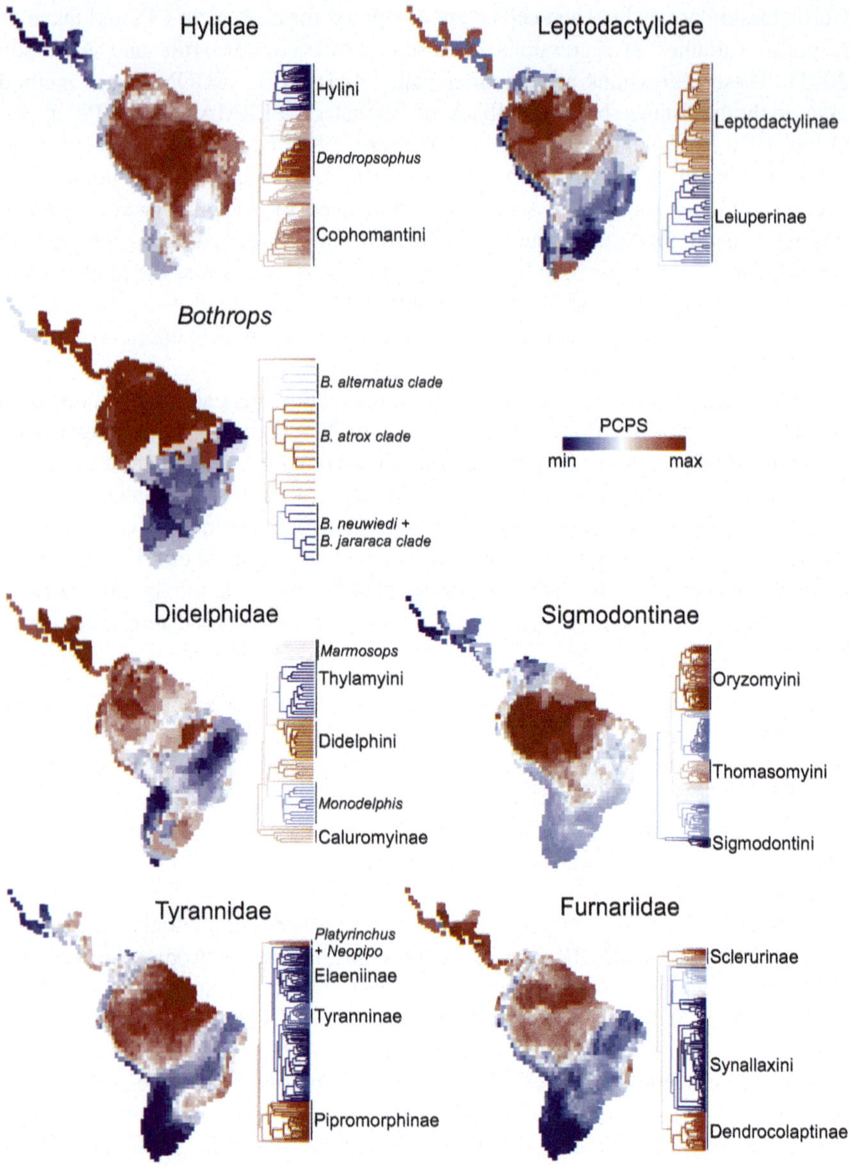

Fig. 6.1 Patterns of phylogenetic turnover for terrestrial vertebrates across the Neotropics. Each map represents the gradient of phylogenetic composition (a PCPS, see main text) most associated with biogeographical dominions. Usually, those same PCPS captured the largest variation in phylogenetic dissimilarity (PCPS 1), except for Hylidae and Sigmodontinae that are represented by the PCPS 2. Next to each map, a phylogeny of the corresponding clade is represented, with terminal colors corresponding to species scores on the PCPS. Maps are provided in Behrmann equal-area projection

Table 6.1 Procrustes correlations among phylogenetic dissimilarity matrices (i.e. phylogenetic turnover) of different clades are shown in the lower diagonal

	Hylidae	Leptodactylidae	*Bothrops*	Didelphidae	Sigmodontinae	Tyrannidae	Furnariidae
Hylidae		0.209	0.263	0.263	0.419	0.269	0.402
Leptodactylidae	0.377		0.505	0.285	0.362	0.533	0.562
Bothrops	0.387	0.42		0.635	0.673	0.679	0.759
Didelphidae	0.381	0.364	0.502		0.650	0.449	0.630
Sigmodontinae	0.53	0.392	0.401	0.425		0.394	0.769
Tyrannidae	0.395	0.459	0.511	0.43	0.403		0.750
Furnariidae	0.468	0.465	0.515	0.443	0.593	0.587	

All Procrustes correlations were higher than expected under the null hypotheses of no association ($P < 0.001$). Pairwise Pearson correlations among the PCPS 1 of each clade are shown in the upper diagonal

Table 6.2 Permutational regression on distance matrices (ADONIS) of phylogenetic dissimilarities between grid cells distributed across the Neotropics (i.e. phylogenetic turnover) on environmental variables, biogeographic dominion, and spatial autocorrelation

Clade	Number of species	R^2	F	P_{site}	P_{taxa}
Hylidae	225				
Biogeographic dominion		0.268	83.91	0.002	0.008
Environment		0.185	52.06	0.002	0.322
Biogeography +. Environment		0.382	70.59	0.002	0.010
Spatial autocorrelation		0.016	2.74	0.002	0.102
Leptodactylidae	69				
Biogeographic dominion		0.251	76.76	0.002	0.596
Environment		0.213	62.16	0.002	0.392
Biogeography +. Environment		0.342	59.26	0.002	0.766
Spatial autocorrelation		0.036	3.64	0.002	0.002
Bothrops	27				
Biogeographic dominion		0.299	97.89	0.002	0.408
Environment		0.274	86.43	0.002	0.164
Biogeography +. Environment		0.422	83.24	0.002	0.230
Spatial autocorrelation		0.006	4.09	0.004	0.282
Didelphidae	67				
Biogeographic dominion		0.255	78.25	0.002	0.190
Environment		0.207	59.71	0.002	0.368
Biogeography +. Environment		0.323	54.41	0.002	0.518
Spatial autocorrelation		0.021	2.73	0.002	0.070
Sigmodontinae	260				
Biogeographic dominion		0.358	127.52	0.002	0.016
Environment		0.157	42.79	0.002	0.624
Biogeography +. Environment		0.416	81.30	0.002	0.032
Spatial autocorrelation		0.002	2.20	0.068	0.060
Tyrannidae	265				
Biogeographic dominion		0.556	287.25	0.002	0.002
Environment		0.291	94.06	0.002	0.082
Biogeography +. Environment		0.657	218.94	0.002	0.002
Spatial autocorrelation		0.004	5.13	0.002	0.466
Furnariidae	223				
Biogeographic dominion		0.465	198.87	0.002	0.002
Environment		0.449	186.62	0.002	0.002
Biogeography +. Environment		0.616	182.99	0.002	0.002
Spatial autocorrelation		0.007	3.13	0.002	0.304

P-values for site and taxa shuffle null models were obtained by permutation tests (499 runs)

relative to the site-shuffle null model (Table 6.2). Nonetheless, the *P*-values obtained relative to the taxa-shuffle model indicate that this association is truly mediated by phylogenetic history within the Hylidae frogs, Sigmodontinae mammals, and Tyrannidae and Furnariidae birds (all showing low *P*-values for taxa-shuffle; see

Table 6.2). We cannot claim the same for Leptodactylidae, Didelphidae, and *Bothrops*, given the high *P*-values obtained when species were shuffled across the tree to recalculated phylogenetic composition (i.e., after controlling for the effect of species composition on phylogenetic composition).

Among the groups with significant associations, biogeographic domain was the major driver of phylogenetic turnover within Hylidae, Sigmodontinae and Tyrannidae: in these groups, environmental variables showed negligible influence on the distribution of phylogenetic lineages. In Furnariidae, however, the distribution of lineages was found to be influenced by both environmental conditions of sites and biogeographic domain (Table 6.2). Irrespectively, the elevated values of R^2 suggest a stronger influence of biogeographic domains in comparison to environmental variables on the phylogenetic turnover of all these groups.

We detected some signal of spatial autocorrelation in the dissimilarity matrices of all lineages except Sigmodontinae (Table 6.2). It was nonetheless very low ($R^2 < 0.04$), and influenced only the distribution of species, independently of phylogenetic relationships among them. The only exception to this general trend was Leptodactylidae; for this group, while spatial autocorrelation was low ($R^2 = 0.034$), it was the only predictor for which the null hypothesis of no association with phylogenetic relatedness among species was rejected ($P_{\text{taxa}} = 0.002$, Table 6.2).

4 Discussion

Phylogenetic turnover of three out of the seven clades of terrestrial vertebrates analyzed here responded significantly only to biogeographic regions, and for all clades the biogeographic regions had a stronger influence than current environmental variables. This suggests that the history of dispersal and vicariance of each clade is a stronger determinant of its phylogenetic turnover over the Neotropics relative to present-day environmental conditions. This conclusion stems from the fact that most groups show strong phylogenetic turnover between the Amazonian region and other areas of the southeastern South America—a spatial pattern that is consistently observed across groups (Procrustes correlations varying from 0.36 to 0.59).

Our analysis indicated that the two domains trespassing the vast Amazonian region (the Boreal Brazilian and the South Brazilian) have been colonized by particular clades of each vertebrate group, forming a distinction between these Amazonian regions and those biogeographic regions in southeastern South America. Consistent differences in species composition along this Northwestern-Southeastern axis was also recovered in a study considering 36 taxa including insects, arachnids, and plants (Morrone 2014a). A similar pattern classified the Northwestern and Southeastern regions as distinct zoogeographical regions within the Neotropical realm, according to species distributions and phylogenetic turnover of more than 20,000 vertebrate species (Holt et al. 2013). Such distinction could have occurred via vicariant or dispersal processes, starting with the formation of the Madeira and

Mamoré rivers in the Cretaceous, and also influenced by marine transgressions in Amazonia during the Miocene (Nihei and De Carvalho 2007; Jaramillo et al. 2017).

Although similar broad-scale responses to historical processes appear to drive the consistent patterns of species distribution observed across clades, neither the age of the clades nor the phylogenetic relatedness among them seem to predict within-clade phylogenetic turnover in spatial distribution patterns. Older clades had similar phylogenetic turnover patterns to younger ones. For instance, Hylidae's (67.61 Ma) spatial patterns of phylogenetic turnover was most strongly correlated with that of the Sigmodontinae (12.65 Ma). Indeed, phylogenetically distantly related vertebrate clades tended to show more similar spatial distribution patterns to each other than more closely related ones (e.g., Sigmodontinae was as strongly correlated with bird clades and *Bothrops* than with Didelphidae). Therefore, regardless of the deeper phylogenetic relationships among the vertebrate clades and the timing of colonization, the same biogeographic regions may have acted as independent arenas for in situ diversification of each clade.

Despite the fact that multiple groups showed a distinction between the lineages occupying Amazonia vs. those in southeastern South America, for some clades (e.g., Tyrannidae and Furnariidae), the Amazonian species belonged to the earliest divergent nodes. For others (e.g., Sigmodontinae), the region was occupied by species that belonged to the late divergent nodes (Fig. 6.1). The order in which the domains were occupied during each clade's radiation underlies these differences, and future studies may investigate the evolutionary history of each clade in further detail to better understand the interplay between diversification processes and phylogenetic turnover patterns for different clades.

Environmental conditions seem less important than biogeographic units to explain phylogenetic turnover for each clade. Although ecological variables influenced species distribution across the Neotropics for most clades, such influence was not often mediated by phylogenetic relatedness among species ($P > 0.05$ for taxa shuffle in Table 6.1). This finding suggests that climate and topography may have a more recent effect on the distribution of species, while domains may capture deeper patterns of phylogenetic distribution (see also Duarte et al. 2014a; Ficetola et al. 2017). For the Furnariidae, however, the influence of environmental features on species distributions was actually found to be mediated by phylogenetic relatedness among them, suggesting a strong role of climatic and topographic variables on clade distribution along their evolutionary histories. Such interplay might explain similar phylogenetic composition patterns for Furnariidae across climatically similar but historically and geographically distinct Amazonian and Atlantic forest regions. A similar result was also found for Tyrannidae, albeit with lower probabilistic support (Table 6.2, Fig. 6.1).

5 Concluding Remarks

In this chapter, we showed the existence of congruent spatial patterns of phylogenetic turnover across Neotropical terrestrial vertebrate clades, regardless of the degree of phylogenetic relatedness among them. Biogeographic units were better predictors of phylogenetic turnover between sites than environmental variables, suggesting that similar histories of dispersal and vicariance shaped the phylogenetic distribution of species inside each clade, yielding the large-scale phylogenetic beta diversity patterns we observe nowadays. This congruence is largely driven by the phylogenetic history of clades distributed in the Amazonian region, given their phylogenetic distinctiveness relative to those occupying other biogeographical areas.

Acknowledgments We thank Lucila Amador for providing a phylogeny for Didelphidae and Cristiano Nogueira and Jéssica Fenker for data on *Bothrops*. Alexandre Antonelli, Ana Carolina Carnaval, and one anonymous reviewer provided extensive feedbacks that improved previous versions of this chapter. RM was supported by a CNPq postdoctoral fellowship (150391/2017-0) during the initial preparation of this chapter. LD research activities have been supported by CNPq Productivity Fellowship (grant 307886/2015-8) and developed in the context of National Institutes for Science and Technology (INCT) in Ecology, Evolution and Biodiversity Conservation, supported by MCTIC/CNPq (proc. 465610/2014-5) and FAPEG (proc. 201810267000023).

References

Amador LI, Giannini NP (2016) Phylogeny and evolution of body mass in didelphid marsupials (Marsupialia: Didelphimorphia: Didelphidae). Org Divers Evol 16:641–657. https://doi.org/10.1007/s13127-015-0259-x

Anderson MJ (2001) A new method for non parametric multivariate analysis of variance. Austral Ecol 26:32–46. https://doi.org/10.1111/j.1442-9993.2001.01070.pp.x

Antonelli A (2018) Neotropical biogeography: regionalisation and evolution. Q Rev Biol 93:266

Antonelli A, Sanmartín I (2011) Why are there so many plant species in the Neotropics? Taxon 60:403–414. https://doi.org/10.2307/41317138

Antonelli A, Nylander JAA, Persson C, Sanmartin I (2009) Tracing the impact of the Andean uplift on Neotropical plant evolution. Proc Natl Acad Sci U S A 106:9749–9754. https://doi.org/10.1073/pnas.0811421106

Antonelli A, Quijada-Masareñas A, Crawford AJ et al (2010) Molecular studies and phylogeography of Amazonian tetrapods and their relation to geological and climatic models. In: Hoorn C, Wesselingh FP (eds) Amazonia, landscape and species evolution, 1st edn. Blackwell Publishing, Oxford

Antonelli A, Ariza M, Albert J et al (2018a) Conceptual and empirical advances in Neotropical biodiversity research. PeerJ 6:1–53. https://doi.org/10.7717/peerj.5644

Antonelli A, Kissling WD, Flantua SGA et al (2018b) Geological and climatic influences on mountain biodiversity. Nat Geosci 11:718–725

Antonelli A, Zizka A, Carvalho FA et al (2018c) Amazonia is the primary source of Neotropical biodiversity. Proc Natl Acad Sci U S A 115(23):6034–6039. https://doi.org/10.1073/pnas.1713819115

BirdLife International and Handbook of the Birds of the World (2016) Bird species distribution maps of the world. Version 6.0. http://datazone.birdlife.org/species/requestdis

Borcard D, Legendre P (2002) All-scale spatial analysis of ecological data by means of principal coordinates of neighbour matrices. Ecol Model 153:51–68

Carlucci MB, Seger GDS, Sheil D et al (2016) Phylogenetic composition and structure of tree communities shed light on historical processes influencing tropical rainforest diversity. Ecography (Cop) 40:521–530. https://doi.org/10.1111/ecog.02104

Cavender-Bares J, Kozak KH, Fine PVA, Kembel SW (2009) The merging of community ecology and phylogenetic biology. Ecol Lett 12:693–715. https://doi.org/10.1111/j.1461-0248.2009.01314.x

Debastiani VJ, Duarte LDS (2014) PCPS – an R-package for exploring phylogenetic eigenvectors across metacommunities. Front Biogeogr 6:144–148

Derryberry EP, Claramunt S, Derryberry G et al (2011) Lineage diversification and morphological evolution in a large-scale continental radiation: the neotropical ovenbirds and woodcreepers (aves: furnariidae). Evolution (NY) 65:2973–2986. https://doi.org/10.1111/j.1558-5646.2011.01374.x

Diniz-Filho JAF, Rangel TFLVB, Bini LM, Hawkins BA (2007) Macroevolutionary dynamics in environmental space and the latitudinal diversity gradient in New World birds. Proc R Soc B Biol Sci 274:43–52. https://doi.org/10.1098/rspb.2006.3712

Duarte LDS (2011) Phylogenetic habitat filtering influences forest nucleation in grasslands. Oikos 120:208–215. https://doi.org/10.1111/j.1600-0706.2010.18898.x

Duarte LDS, Prieto PV, Pillar VD (2012) Assessing spatial and environmental drivers of phylogenetic structure in Brazilian Araucaria forests. Ecography (Cop) 35:952–960. https://doi.org/10.1111/j.1600-0587.2011.07193.x

Duarte LDS, Bergamin RS, Marcilio-Silva V et al (2014a) Phylobetadiversity among forest types in the Brazilian Atlantic Forest complex. PLoS One 9:1–10. https://doi.org/10.1371/journal.pone.0105043

Duarte LDS, Both C, Debastiani VJ et al (2014b) Climate effects on amphibian distributions depend on phylogenetic resolution and the biogeographical history of taxa. Glob Ecol Biogeogr 23:213–222. https://doi.org/10.1111/geb.12089

Duarte LDS, Debastiani VJ, Freitas AVLL, Pillar VD (2016) Dissecting phylogenetic fuzzy weighting: theory and application in metacommunity phylogenetics. Methods Ecol Evol 7:937–946. https://doi.org/10.1111/2041-210X.12547

Duarte LDS, Debastiani VJ, Carlucci MB, Diniz-Filho JAF (2018) Analyzing community-weighted trait means across environmental gradients: should phylogeny stay or should it go? Ecology 99 (2):385–398. https://doi.org/10.1002/ecy.2081

Excoffier L, Smouse PE, Quattro JM (1992) Analysis of molecular variance inferred from metric distances among DNA haplotypes: application to human mitochondrial DNA restriction data. Genetics 131:479–491

Faivovich J, Haddad CFB, Garcia PCA et al (2005) Systematic review of the frog family Hylidae, with special reference to Hylinae: phylogenetic analysis and taxonomic revision. Bull Am Mus Nat Hist 294:1–240. https://doi.org/10.1206/0003-0090(2005)294[0001:SROTFF]2.0.CO;2

Fenker J, Tedeschi LG, Pyron RA, Nogueira C d C (2014) Phylogenetic diversity, habitat loss and conservation in South American pitvipers (Crotalinae: Bothrops and Bothrocophias). Divers Distrib 20:1108–1119. https://doi.org/10.1111/ddi.12217

Fergnani PN, Ruggiero A (2017) The latitudinal diversity gradient in South American mammals revisited using a regional analysis approach: the importance of climate at extra-tropical latitudes and history towards the tropics. PLoS One 12:1–19. https://doi.org/10.1371/journal.pone.0128264.Funding

Ficetola GF, Mazel F, Thuiller W (2017) Global determinants of zoogeographical boundaries. Nat Ecol Evol 1:89. https://doi.org/10.1038/s41559-017-0089

Fick S, Hijmans R (2017) Worldclim 2: new 1-km spatial resolution climate surfaces for global land areas. Int J Climatol 37(12):4302–4315. https://doi.org/10.1002/joc.5086

Fine PVA, Zapata F, Daly DC (2014) Investigating processes of neotropical rain forest tree diversification by examining the evolution and historical biogeography of the protieae (Burseraceae). Evolution (NY) 68:1988–2004. https://doi.org/10.1111/evo.12414
Garzione CN, Hoke GD, Libarkin JC et al (2008) Rise of the Andes. Science 320:1304–1307. https://doi.org/10.1126/science.1148615
Gianuca AT, Dias RA, Debastiani VJ, Duarte LDS (2013) Habitat filtering influences the phylogenetic structure of avian communities across a coastal gradient in southern Brazil. Austral Ecol 39:29–38
Goin FJ, Gelfo JN, Chornogubsky L et al (2012) Origins, radiations, and distribution of South American mammals: from greenhouse to icehouse worlds. In: Patterson BD, Costa LP (eds) Bones, clones and biomes: the history and geography of recent Neotropical mammals. The University of Chicago Press, Chicago
Graham CH, Fine PVA (2008) Phylogenetic beta diversity: linking ecological and evolutionary processes across space in time. Ecol Lett 11:1265–1277
Hawkins BA, Porter EE, Felizola Diniz-Filho JA (2003) Productivity and history as predictors of the latitudinal diversity gradient of terrestrial birds. Ecology 84:1608–1623. https://doi.org/10.1890/0012-9658(2003)084[1608:PAHAPO]2.0.CO;2
Holt BG, Lessard J-P, Borregaard MK et al (2013) An update of Wallace's zoogeographic regions of the world. Science (80–) 339:74–78
Hoorn C (1993) Marine incursions and the influence of Andean tectonics on the Miocene depositional history of northwestern Amazonia: results of a palynostratigraphic study. Palaeogeogr Palaeoclimatol Palaeoecol 105:267–309
Hoorn C, Guerrero J, Sarmiento GA, Lorente MA (1995) Andean tectonics as a cause for changing drainage patterns in Miocene northern South America. Geology 23:237–240
Hoorn C, Wesselingh FP, Ter Steege H et al (2010) Amazonia through time: Andean uplift, climate change, landscape evolution, and biodiversity. Science (80–) 330:927–931. https://doi.org/10.1126/science.1194585
IUCN (2008) IUCN redlist of threatened species. www.iucnredlist.org. Accessed 2 Jan 2018
Jackson D (1995) PROTEST: a Procrustean randomization test of community environment concordance. Écoscience 2:297–303
Jansa SA, Barker FK, Voss RS (2014) The early diversification history of didelphid marsupials: a window into South America' s "Splendid Isolation". Evolution (NY) 68:684–695. https://doi.org/10.1111/evo.12290
Jaramillo C, Romero I, Apolito CD et al (2017) Miocene flooding events of western Amazonia. Sci Adv 3:e1601693
Jetz W, Thomas GH, Joy JB et al (2012) The global diversity of birds in space and time. Nature 491:444–448. https://doi.org/10.1038/nature11631
Kerkhoff AJ, Moriarty PE, Weiser MD (2014) The latitudinal species richness gradient in New World woody angiosperms is consistent with the tropical conservatism hypothesis. Proc Natl Acad Sci U S A 111:8125–8130. https://doi.org/10.1073/pnas.1308932111
Löwenberg-Neto P (2014) Andean region: a shapefile of Morrone's (2014) biogeographical regionalisation. Zootaxa 3802:300–300. https://doi.org/10.11646/zootaxa.3985.4.9
Maestri R (2017) Using phylogenetic clade composition to understand biogeographical variation in functional traits. Front Biogeogr 9.3:e34435. https://doi.org/10.21425/F59334435
Maestri R, Luza AL, Barros LD et al (2016) Geographical variation of body size in sigmodontine rodents depends on both environment and phylogenetic composition of communities. J Biogeogr 43:1–11. https://doi.org/10.1111/jbi.12718
Maestri R, Monteiro LR, Fornel R et al (2017) The ecology of a continental evolutionary radiation: Is the radiation of sigmodontine rodents adaptive? Evolution (NY) 71:610–632. https://doi.org/10.1111/evo.13155
Maestri R, Monteiro LR, Fornel R et al (2018) Geometric morphometrics meets metacommunity ecology: environment and lineage distribution affects spatial variation in shape. Ecography (Cop) 41:90–100. https://doi.org/10.1111/ecog.03001

Maestri R, Upham NS, Patterson BD (2019) Tracing the diversification history of a Neogene rodent invasion into South America. Ecography (Cop) 42(4):683–695. https://doi.org/10.1111/ecog.04102

Mazel F, Wüest RO, Lessard J-P et al (2017) Global patterns of β-diversity along the phylogenetic time-scale: the role of climate and plate tectonics. Glob Ecol Biogeogr 26:1211–1221. https://doi.org/10.1111/geb.12632

Mcardle BH, Anderson MJ (2001) Fitting multivariate models to community data: a comment on distance-based redundancy analysis. Ecology 82:290–297. https://doi.org/10.2307/2680104

Morrone JJ (2014a) Cladistic biogeography of the Neotropical region: identifying the main events in the diversification of the terrestrial biota. Cladistics 30:202–214

Morrone JJ (2014b) Biogeographical regionalisation of the Neotropical region. Zootaxa 3782:1–110

Nihei SS, De Carvalho CJB (2007) Systematics and biogeography of Polietina Schnabl & Dziedzicki (Diptera, Muscidae): neotropical area relationships and Amazonia as a composite area. Syst Entomol 32:477–501. https://doi.org/10.1111/j.1365-3113.2006.00376.x

Oksanen J, Blanchet FG, Friendly M, Kindt R, Legendre P, McGlinn D, Minchin PR, O'Hara RB, Simpson GL, Solymos P, Stevens MHH, Szoecs E, Wagner H (2019) vegan: community ecology package. R package version 2:5–6. https://CRAN.R-project.org/package=vegan

Patterson BD, Costa LP (eds) (2012) Bones, clones and biomes: the history and geography of recent Neotropical mammals. The University of Chicago Press, Chicago

Peres-Neto PR, Jackson DA (2001) How well do multivariate data sets match? The advantages of a procrustean superimposition approach over the Mantel test. Oecologia 129:169–178. https://doi.org/10.1007/s004420100720

Peres-Neto P-R, Leibold M, Dray S (2012) Assessing the effects of spatial contingency and environmental filtering on metacommunity phylogenetics. Ecology 93:14–30. https://doi.org/10.1890/11-0494.1

Pillar VD, Duarte L d S (2010) A framework for metacommunity analysis of phylogenetic structure. Ecol Lett 13:587–596. https://doi.org/10.1111/j.1461-0248.2010.01456.x

Pillar VDP, Orlóci L (1996) On randomization testing in vegetation science: multifactor comparisons of relevé groups. J Veg Sci 7:585–592. https://doi.org/10.2307/3236308

Prieto PV, Seger GDS, Sánchez-Tapia A et al (2017) Secondary succession and fire disturbance promote dominance of a late-diverging tree lineage in a lowland Neotropical forest. Plant Ecol Divers 10:311–322. https://doi.org/10.1080/17550874.2017.1379568

Pyron RA (2014) Biogeographic analysis reveals ancient continental vicariance and recent oceanic dispersal in amphibians. Syst Biol 63:779–797. https://doi.org/10.1093/sysbio/syu042

R Core Team (2018) R: a language and environment for statistical computing

Ramos Pereira MJ, Palmeirim JM (2013) Latitudinal diversity gradients in New World bats: are they a consequence of niche conservatism? PLoS One 8:e69245. https://doi.org/10.1371/journal.pone.0069245

Revell LJ (2012) Phytools: an R package for phylogenetic comparative biology (and other things). Methods Ecol Evol 3:217–223. https://doi.org/10.1111/j.2041-210X.2011.00169.x

Rull V (2005) Biotic diversification in the Guayana Highlands: a proposal. J Biogeogr 32:921–927. https://doi.org/10.1111/j.1365-2699.2005.01252.x

Rull V (2011) Neotropical biodiversity: timing and potential drivers. Trends Ecol Evol 26:508–513. https://doi.org/10.1016/j.tree.2011.05.011

Simpson GG (1980) Splendid isolation: the curious history of South American mammals. Yale University Press, New Haven

Soberón J (2007) Grinnellian and Eltonian niches and geographic distributions of species. Ecol Lett 10:1115–1123. https://doi.org/10.1111/j.1461-0248.2007.01107.x

Steppan SJ, Schenk JJ (2017) Muroid rodent phylogenetics: 900-species tree reveals increasing diversification rates. PLoS One 12:e0183070. https://doi.org/10.1371/journal.pone.0183070

Title PO, Rabosky DL (2018) Tip rates, phylogenies, and diversification: what are we estimating, and how good are the estimates? bioRxiv. doi: https://doi.org/10.1101/369124

Tonini JFR, Beard KH, Ferreira RB et al (2016) Fully-sampled phylogenies of squamates reveal evolutionary patterns in threat status. Biol Conserv 204:23–31. https://doi.org/10.1016/j.biocon.2016.03.039

Weir JT, Bermingham E, Schluter D (2009) The Great American Biotic Interchange in birds. Proc Natl Acad Sci U S A 106:21737–21742. https://doi.org/10.1073/pnas.0903811106

Chapter 7
Diatom Diversity and Biogeography Across Tropical South America

Xavier Benito and Sherilyn C. Fritz

Abstract Determining the mechanisms of community assembly forms the foundation of biogeography and community ecology. Studies of the biodiversity of Neotropical macro-organisms have recognized the roles of environmental, spatial, and historical factors in structuring communities at different spatial and temporal scales, but the influences on microorganisms are not well studied. Diatoms (unicellular siliceous algae) are a very species-rich group of algae, disperse widely, and respond rapidly and sensitively to environmental variation. In this chapter, we use a newly constructed database of tropical South American diatoms (8°N–30°S and 58–79°W) to analyze the distribution and diversity patterns of species across lentic and lotic environments, including predictors that describe local (limnology) and regional (geo-climatic) factors. We provide a conceptual framework that examines community responses (species composition and diversity) to environmental and spatial factors by applying paleolimnology and macroecological approaches (Theory of Island of Biogeography and metacommunity theory). Diatom biogeographic patterns emerge for (1) identifying regionally distinct floras with a significant association with geographic groups of lakes, (2) determining latitudinal gradients of species richness and community structure, and (3) estimating the joint contribution of environmental filtering and dispersal dynamics in affecting diatom community composition across time. We emphasize the potential for incorporating diatom ecology and paleolimnology into a macroecology framework to further develop multiple, interdisciplinary hypotheses on the mechanisms driving microbial diversity and community assembly in the Neotropics.

X. Benito (✉)
National Socio-Environmental Synthesis Center (SESYNC), University of Maryland, Annapolis, MD, USA

Department of Earth and Atmospheric Sciences, School of Biological Sciences, University of Nebraska–Lincoln, Lincoln, NE, USA
e-mail: xbenitogranell@sesync.org

S. C. Fritz
Department of Earth and Atmospheric Sciences, School of Biological Sciences, University of Nebraska–Lincoln, Lincoln, NE, USA
e-mail: sfritz2@unl.edu

V. Rull, A. C. Carnaval (eds.), *Neotropical Diversification: Patterns and Processes*, Fascinating Life Sciences, https://doi.org/10.1007/978-3-030-31167-4_7

Keywords Amazon · Andes · Latitudinal diversity · Diatoms · Lakes · Microbial biogeography · Metacommunities

1 Introduction

The overlap between biogeography and community ecology is rapidly expanding, particularly for questions that examine the effects of spatio-temporal gradients and historical legacies on contemporary biotas (Jenkins and Ricklefs 2011). Both disciplines are spatial in nature, but whereas environmental filtering can act at varied spatial scales (Soininen 2007), more controversy exists about how historical and regional factors influence local communities (Heino et al. 2016). Biogeography and community ecology generally assume that most local assemblages are at non-equilibrium relative to the processes involved in environmental filtering, in which local abiotic and biotic factors shape ecological communities (Ricklefs and Jenkins 2011). In other words, regional and historical factors often override local environmental conditions in explaining species abundance and composition, even at small spatial scales (Ricklefs 1987). As a consequence, two paradigms coexist to explain community assembly processes: the 'niche paradigm' (i.e. environmental gradients are the major influence, thus emphasizing species autecology), and the 'island paradigm' (i.e. assembly rules are governed by dispersal-based mechanisms) (Antonelli et al. 2018; Weiher and Keddy 2001). Patterns in multiple biological groups in the Neotropics have been evaluated, including plants (Costa et al. 2005), vertebrates (Valdujo et al. 2013), invertebrates (Gonçalves-Souza et al. 2014), zooplankton (Declerck et al. 2011), and diatoms (Wetzel et al. 2012).

Macroecological patterns of microorganisms are poorly known in the Neotropics. Diatoms are siliceous unicellular eukaryotic algae distributed in nearly all types of aquatic systems, and several features make them effective model microorganisms for exploring microbial distribution at large spatial scales: (1) they are the most diverse algal group in rivers and lakes (Mann 1999), (2) extensive characterization of species (both molecular and morphological data) from a wide range of aquatic ecosystems have been made at varied time scales (Burge et al. 2018), and (3) species-environment relationships are reasonably well documented for many groups, and many species respond directly and sensitively to environmental change (Smol and Stoermer 2010). Despite these characteristics and recent advances in analyzing distribution patterns at sub-continental scales (Jüttner et al. 2010; Potapova and Charles 2002; Soininen et al. 2016; Verleyen et al. 2009; Vijver and Beyens 2007; Vyverman et al. 2007), biogeographic patterns in diatoms remain understudied, likely because of: (1) the difficulty of morphological delimitation for the most widespread and ecologically relevant genera (e.g. *Gomphonema*, *Nitzschia*, *Fragilaria*) (Round et al. 1990); (2) the high degree of cryptic taxon variation (Mann and Droop 1996); and (3) the lack of spatially explicit inter-calibrated diatom datasets (Vyverman et al. 2007). Hence, the data available for diatoms preclude sound assessments of whether microbial ecological patterns converge with biogeographic regionalization schemes for macro-organisms (e.g. realms, regions, ecoregions) (Barberán et al. 2014; Vilhena and Antonelli 2015). Biogeography integrates geological and climatic history with the composition, diversity, and distribution of biota by delineating distinct biogeographic realms (Cox et al. 2005). Because diatoms contribute to ecosystem structure and functioning as one of the important components at the base of the food chain in freshwater ecosystems (lakes, rivers, ponds), it is timely to evaluate biogeographic diversity patterns in diatom communities.

A traditional view in microbial biogeography is that unicellular organisms lack biogeographic patterns, because of large population densities and highly efficient dispersal (Finlay 2002). However, this generality has inspired a growing body of research that seeks to understand the relative role of local environmental variables, biotic interactions, and dispersal-related factors in structuring microbial communities (Jenkins and Medley 2011). For instance, it has been shown that dispersal abilities and body size (Bie et al. 2012; Jamoneau et al. 2018), habitat connectivity (Vyverman et al. 2007), and ecosystem size (Reche et al. 2005) may have shaped modern global microbial distributions. In the specific case of diatoms, recent findings indicate that species richness and assemblage structure are not only determined by local environmental characteristics but also are driven by spatial factors operating at broader scales (Pajunen et al. 2016; Pinseel et al. 2017; Verleyen et al. 2009). Indeed, empirical and conceptual findings suggest that these two driving forces (i.e. species sorting and dispersal related factors) are not mutually exclusive and may jointly shape diatom communities according to a hierarchical environmental filtering model (Poff 1997). This model predicts that species from a regional pool must pass through a series of "nested" filters to join a local community. For instance, precipitation may influence catchment productivity (Wetzel 2001), which affects biogeochemical fluxes from terrestrial to aquatic systems, and latitudinal gradients of solar radiation have strong but complex influences on multiple niche-related variables, such as water temperature, conductivity, and resource supply (Passy 2010; Pinel-Alloul et al. 2013). Furthermore, landscape-related factors, such as topography, aquatic connectivity, and ecoregion type may also affect diatoms, because they integrate ecological factors that influence species composition and diversity directly or indirectly (Pajunen et al. 2017).

In this chapter, we synthesize the earlier literature and recent studies (Benito et al. 2018a, b) on the distribution and diversity of diatoms in tropical South America. In doing so, we present a conceptual framework (Fig. 7.1) for examining patterns and processes that maintain and generate diatom diversity in tropical South America, with a special focus on the applicability of macroecological approaches (Theory of Island of Biogeography and metacommunity theory). We also demonstrate the use of different numerical approaches for examining diversity and community assembly. We use a newly created tropical South American diatom database (8°N–30°S and 58–79°W) available in the Dryad repository (Benito et al. 2018a, b) and GitHub with a georeferenced map of the inventory of the diatom sites (https://github.com/xbenitogranell/diatoms-biogeography-southamerica) (Fig. 7.2). The database comprises

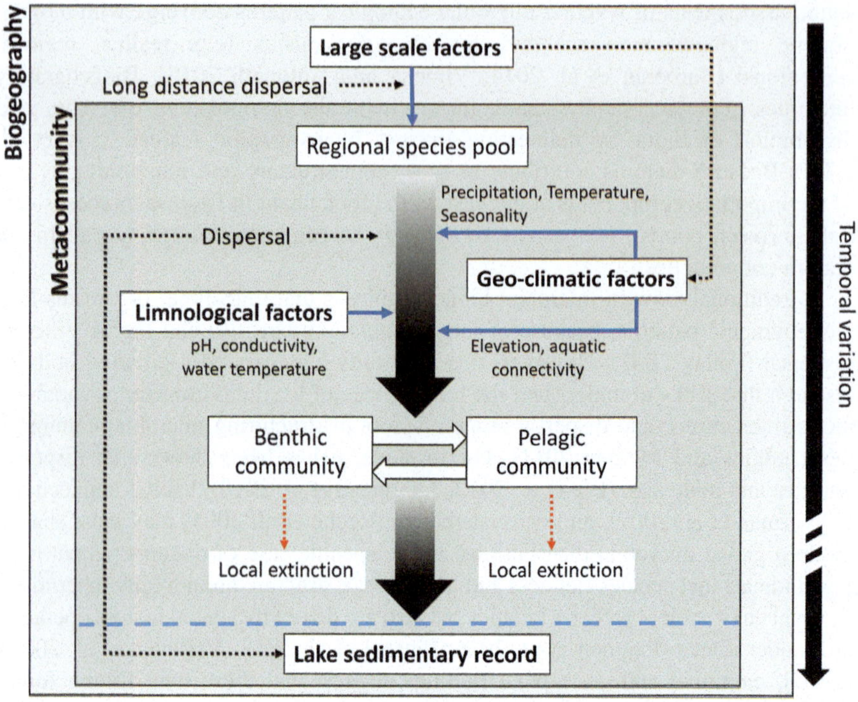

Fig. 7.1 Conceptual diagram of the relationships among processes affecting diatom community responses (community composition and species richness). The relationships indicate both the direct (blue solid lines) and indirect (dashed lines) effects of different groups of variables (i.e. unique and shared effects extracted from variance partitioning analysis, see Fig. 7.5). Note the hierarchically organized niche- and dispersal-based assembly processes that interact and drive diatom communities at biogeography and metacommunity spatial scales; paleolimnology (lake sedimentary records) can provide long-term data on changes in diatom macroecological patterns

published and unpublished studies from lentic and lotic environments sampled by different authors for different purposes (e.g. paleoclimatic reconstructions, taxonomy, biodiversity). Diatom samples correspond to multiple habitats (e.g. sediment surface, periphyton, and plankton) and cover the period 1978–2017. Measures to standardize sampling efforts included transformation of site-by-species abundance matrices to site-by-species presence-absence matrices. Taxonomic harmonization consisted of aggregating varieties of species, scanning the data for taxonomic synonyms, updating diatom taxonomic nomenclature according Guiry and Guiry (2017), and lumping species complexes (e.g., *Achnanthidium minutissimum*, *Sellaphora pupula*, *Discostella stelligera*, *Ulnaria ulna*) from the database entries.

Below, we first review the literature on diatom distribution and the associated limnological habitats in tropical South America. We then examine the influence of different environmental factors, from regional to local spatial scales, on benthic diatom species diversity in riverine and lacustrine environments using linear regression models. Subsequently, we estimate the ecological controls of latitudinal-related

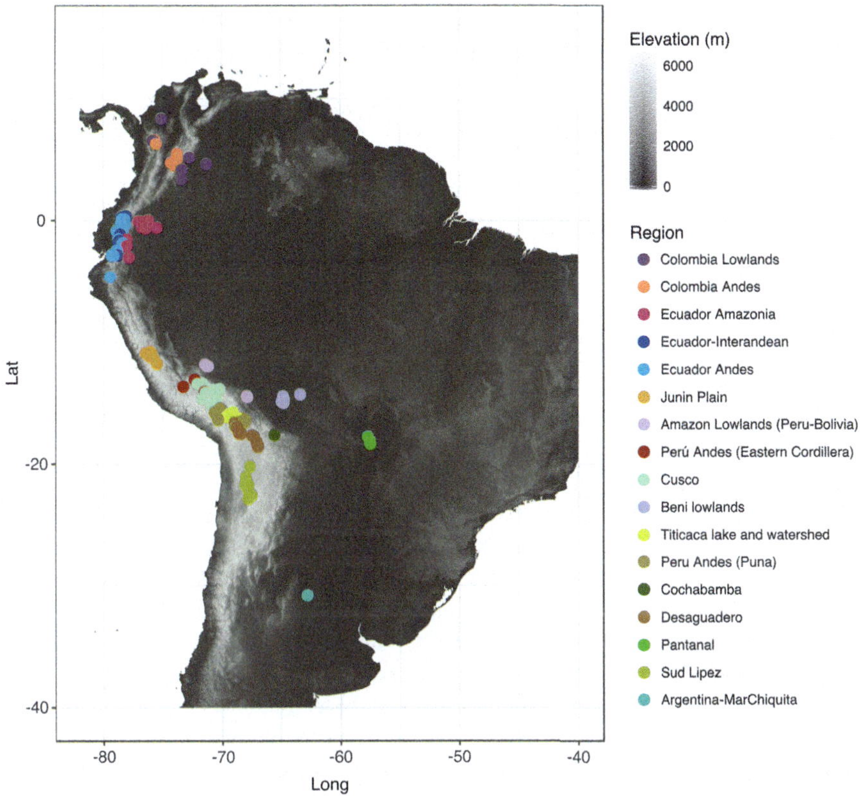

Fig. 7.2 Map of tropical South America showing the location of the study lakes (n = 195) grouped by diatom datasets distributed across the Andes and the Andean foreland plains

factors on lake diatom community composition using variance partitioning. These relationships indicate both the direct and indirect effects of different groups of variables (i.e. unique and shared fraction extracted from variance partitioning analysis and regression models). Finally, we try to integrate the different factors that operate at multiple spatial and temporal scales: large scale (comprising historic factors), regional scale (comprising dispersal and geo-climatic factors), and local environment (comprising limnological factors). The ultimate goal is to compare the commonalities in ecoregion patterning among groups of macro-organisms (vegetation) and diatoms in the Neotropics, and to recognize the importance of hierarchically organized abiotic drivers that interact and drive the flux of energy and materials and ultimately shape the functioning of aquatic ecosystems. We propose that integrating multi-site data on diatom distribution will provide new opportunities for addressing research questions that are multidisciplinary by nature, such as the effects of global environmental change in Neotropical freshwaters and diversification patterns and processes in the microbial world.

2 Lakes and Streams of Tropical South America: Natural Laboratories for the Study of Diatom Diversity and Ecology

The tropical Andes and adjoining Amazonian lowlands are a mega-diverse region, with high levels of provincialism and endemism of both macro and micro-organisms, at least partially linked to strong orographic gradients (Kessler 2002; Metzeltin and Lange-Bertalot 2007; Morales et al. 2014; Särkinen et al. 2012). The geographic diversity (latitude, longitude, altitude), coupled with diverse physiographic regions and geologic substrates produce notable climatic and limnological variability that can be used to examine the factors that structure the species composition and diversity of diatom assemblages. Yet, the relative importance of major correlates of diatom community composition have long been focused on local environmental variables (Montoya-Moreno et al. 2012; Morales and Vis 2007; Sala et al. 2008; Servant-Vildary and Roux 1990; Tapia et al. 2006), without a consideration of spatial factors (Benito et al. 2018a). Biogeographic patterns also need to be interpreted in the context of historical influences (Cox et al. 2005). For example, in tropical South America, the uplift of the Andes, paleoclimate variation, and the geomorphic evolution of large river systems created habitat heterogeneity and complex gradients of water and energy balance that played a crucial role in the development of biodiversity (Baker and Fritz 2015; Flantua and Hooghiemstra 2018). However, integrated hypotheses about biogeographic patterns in diatoms are still poorly developed and largely untested, partly because of the challenge in unravelling the many direct and indirect connections among ecological processes operating at different spatial and temporal scales (Antonelli et al. 2009; Luebert and Weigend 2014).

The lakes and streams of tropical South America are effective systems for examining biogeography and biodiversity patterns of diatoms for several reasons. First, they are well-defined ecosystems many of which have persisted over long geological time-scales, allowing evolutionary processes (speciation, extinction) to affect diatoms (Fritz et al. 2012; Spanbauer et al. 2018). Second, despite contrasts in the climatic (Baker and Fritz 2015) and evolutionary history of various regions (Hughes and Eastwood 2006), broadly similar ecoregions and communities are present due to the persistent topographic profile and orientation of the Andes throughout much of the Neogene (Luebert and Weigend 2014). Third, they are in a topographically diverse landscape characterized by notable spatial variability in climate (Killeen et al. 2007; Baker and Fritz 2015). Fourth, they are characterized by a large number of topographically closed basins, forming a hydrological continuum at the watershed scale in which biological communities are directly and indirectly sustained by the physical template (Colinvaux and Steinitz-Kannan 1980; Särkinen et al. 2012). Finally, many spatially derived environmental datasets of climatic and topographic parameters are available to use for testing hypothesis about spatial influences on biotic communities (e.g. Flantua et al. 2015).

Despite being more densely populated than the adjacent Amazon forelands, Andean freshwaters are often referred to as pristine ecosystems, suggesting that climatic controls are the major drivers of their dynamics (Sarmiento 2002; Micheluti et al. 2015). However, paleoecological and archaeological data indicate that for millennia humans have been an integral part of the Andes, shaping a very dynamic cultural landscape (Jantz and Behling 2012). Currently, fire, land-use changes, cattle farming, and fish stocking are widespread across many Andean lake-basins and affect the composition and productivity of lacustrine biological communities (Aguilera et al. 2006; Jampel 2016; Van Colen et al. 2017). For instance, Van Colen et al. (2018) found that intensive cattle grazing in high-elevation Ecuadorean lakes led to shifts in diatom community composition. In the Andes, sampling near human settlements with easy access (e.g. towns, roads, hiking trails, etc.) may affect diatom species richness estimates, because low-diversity pollutant-resistant diatom taxa commonly are associated with human-impacted landscapes (Morales et al. 2012). Thus, lakes located in the inter Andean valley of Ecuador (2900–2000 m), which have higher nutrient concentrations and conductivity ranges, primarily driven by higher human density and land-use, show distinct recent changes in diatom community composition, whereas lakes located the Andean highlands (>3500 m) show very limited limnological change (Benito et al. 2019). Future studies that seek to analyze diatom community responses require understanding complex landscape dynamics over space and time, including diverse anthropogenic influences, as well as past historical legacies associated with varied types and magnitudes of environmental change on geological time scales (Fritz et al. 2018).

The early diatom floristic studies in tropical South America date from the nineteenth century by Ehrenberg (1843, 1873). In the twentieth century, Frenguelli (1939, 1942) in Argentina, Hustedt (1927) in Chile, and Manguin (1964) in Peru, made important contributions to documenting the South American diatom flora, particularly in the Andes. More recently, major taxonomic monographs have been published for the Andes (Rumrich et al. 2000), adjacent Amazon lowlands (Metzeltin and Lange-Bertalot 2007), and Uruguay (Metzeltin et al. 2005). Despite the great variety of aquatic ecosystems for many disciplines that use diatoms as a central component (diversity, ecology, paleoclimatology) (Ekdahl et al. 2008; Fritz et al. 2007, 2012; Servant-Vildary et al. 2002; Tapia et al. 2006; Velez 2003), the study of South American diatom diversity has never been thoroughly investigated to evaluate the mechanisms contributing to maintaining diatom community structure at sub-continental to continental scales.

Several more recent studies have documented the composition of the diatom flora of tropical South America at different spatial scales. Benito et al. (2018a), analyzed a subset of benthic diatom samples from the tropical South American diatom database and found a total of 1086 taxa in 163 samples, representing 109 genera. *Pinnularia* and *Nitzschia* were the richest diatom genera, with 102 species (8.6% each of the total number of taxa), followed by *Navicula* (8.4%), *Eunotia* (7.5%), and *Gomphonema* (7%). A total of 204 diatom taxa (19% of the total number of taxa) were not assigned to described diatoms based on existing literature, remaining therefore as unknown species with potential to be new taxa for the science. Several authors have suggested that the number of new diatom taxa in the region will likely increase due to revision of misidentified original diatom material (Morales et al. 2014). Indeed, some authors have challenged the view that Andean diatoms are cosmopolitan, arguing that a higher number of endemic taxa likely exist than those currently known (Maidana et al. 2011; Morales et al. 2012). Rumrich et al. (2000) identified 888 diatom taxa from 350 samples randomly distributed from Venezuela to Argentina along the Andes, documenting a relatively high proportion (42%) of cosmopolitan taxa (i.e. reported in European and North American temperate regions), whereas a much smaller proportion (9.5%) were new or could not be identified with available literature. In contrast, Morales et al. (2012) found 228 taxa in a single sample from an Andean Bolivian stream, where 24% may be endemic taxa. Moreover, recent advances in taxonomy (e.g. molecular markers) will lead to the discovery of taxa showing more restricted geographical distributions and narrower ecological characteristics (Vijver and Beyens 2007). As shown by several authors, the prevailing diatom species concept arguably is based on coarse taxonomic classification and minimizes true diversity, endemism, and geographical distribution of diatom communities (Mann 1999; Mann and Droop 1996; Sabbe et al. 2001). Improved classification of diatoms using modern species concepts and new molecular tools will be essential for future use of diatoms as bioindicators in tropical South American freshwaters and for taxonomic comparisons of difficult taxa (i.e., *Fragilaria* and its recently revised genera) or among tropical regions.

According to Fourtanier and Kociolek (1999), more than >900 diatom genera are cosmopolitan. Although the South American diatom flora lacks taxonomically calibrated regional data sets, and most early studies analyzed a limited number of samples using old and geographically scattered floras (Morales et al. 2008), some notable observations of taxa showing high levels of provinciality and endemism exist. For instance, the genus *Halorundia* is endemic to southern Chile, described by Diaz (2006) to accommodate *Achnanthes speciosa* Hustedt (1927). In the Bolivian and Chilean Altiplano, the genera *Amphora* and *Navicula* have a significant number of endemic species restricted to shallow hypersaline lakes (Blanco et al. 2013; Frenguelli 1939; Hustedt 1927; Patrick 1961; Sala and Maidana 2003; Servant-Vildary 1986). Also, high-elevation saline environments from Argentina contain *Navicula* species restricted to small geographic areas (Maidana and Seeligmann 2006; Maidana et al. 2011). In the Peruvian Andes, a survey from 10 freshwater lakes revealed that 32% of the diatom species identified were unknown outside South America (Manguin and Manguin 1964). More recent studies of diatom material from lakes and streams of the Lake Titicaca and Desaguadero watershed described new benthic fragilarioid and planktic taxa (Morales et al. 2012; Tapia et al. 2004). A relatively high proportion of endemic species belong to the genera *Actinella* and *Eunotia*, mostly restricted to the Brazilian Amazon in acidic freshwater habitats (Wetzel et al. 2010). In the Amazon region of Ecuador, the diatom flora of seasonally flooded lakes are characterized by endemic *Pinnularia* taxa (Oliveira and Steinitz-Kannan 1992). Finally, in lakes of the Galapagos islands, diatom endemism has been described from early regional expeditions to contemporary times (Cleve 1881; Patrick 1970; Seddon et al. 2011, 2014), although the characteristics of oceanic islands cannot be directly compared with continental islands, such as high-elevation lakes.

3 Effects of Local Environments, Climate, and Spatial Factors: Biogeography Meets Community Ecology

A prior examination of biogeographical patterns of diatom communities that analyzed patterns of species richness showed that geographic regions within Ecuador had more shared diatom species than within other regions of the central Andes and adjacent Amazon lowlands of Bolivia and Peru (Fig. 7.3), despite similar geographic distances between highlands and lowlands. Diatom floras are regionally distinctive, with a relatively low averaged proportion of shared species of diatoms among all study regions (*ca.* 21%) and high levels of compositional turnover (beta-diversity) (Benito et al. 2018a). Major differences between highlands and lowlands, and among Andean regions, suggest the relevance of physiography and the altitudinal gradients that influence both regional and local climate (Killeen et al. 2007; Valencia et al. 2016). For example, dispersal may be inhibited in some regions, because they are cold and arid, with a low density and small number of suitable aquatic environments, such as Sud Lipez with its shallow hypersaline lakes in flat terrain of the Altiplano plateau. In contrast, humid regions in more topographically rugged landscapes (e.g. northern Altiplano and Peruvian Andes) have a higher number of freshwater systems connected by rivers, streams, and wetlands. Moreover, many of these landscapes have changed their spatial extent as a result of long-term climate change. In the northern Andes, for instance, the Páramo habitat shifted altitudinally over a gradient of 1500 m, and its aerial extent fluctuated between isolated patches and more extensive areas during much of the late-Quaternary (Flantua and Hooghiemstra 2018).

The importance of aquatic connectivity in shaping species composition and richness is a fundamental component of the Theory of Island Biogeography: species number at equilibrium (colonization versus extinction rates) is a function of habitat isolation and size (MacArthur and Wilson 1967). Linear models of diatom patterns in tropical South America (Table 7.1) indicate a significant increase in diatom species richness with increased water connectivity, suggesting that a decrease in the remoteness of the system increases species number. The importance of spatial influences is further supported by variance partitioning results of a larger number of lakes; the results indicate the dominant effect of broad-scale factors related to dispersal (e.g. geographic distance) relative to factors operating at the local scale (e.g. limnological variables) in determining diatom species composition (Fig. 7.4). The absence of environmental variables difficult to measure and/or unmeasured in the field may underestimate the influence of local environmental factors in diatom species composition, and thus we cannot discard the possibility that the observed spatial variation in diatom composition was strongly influenced by some local

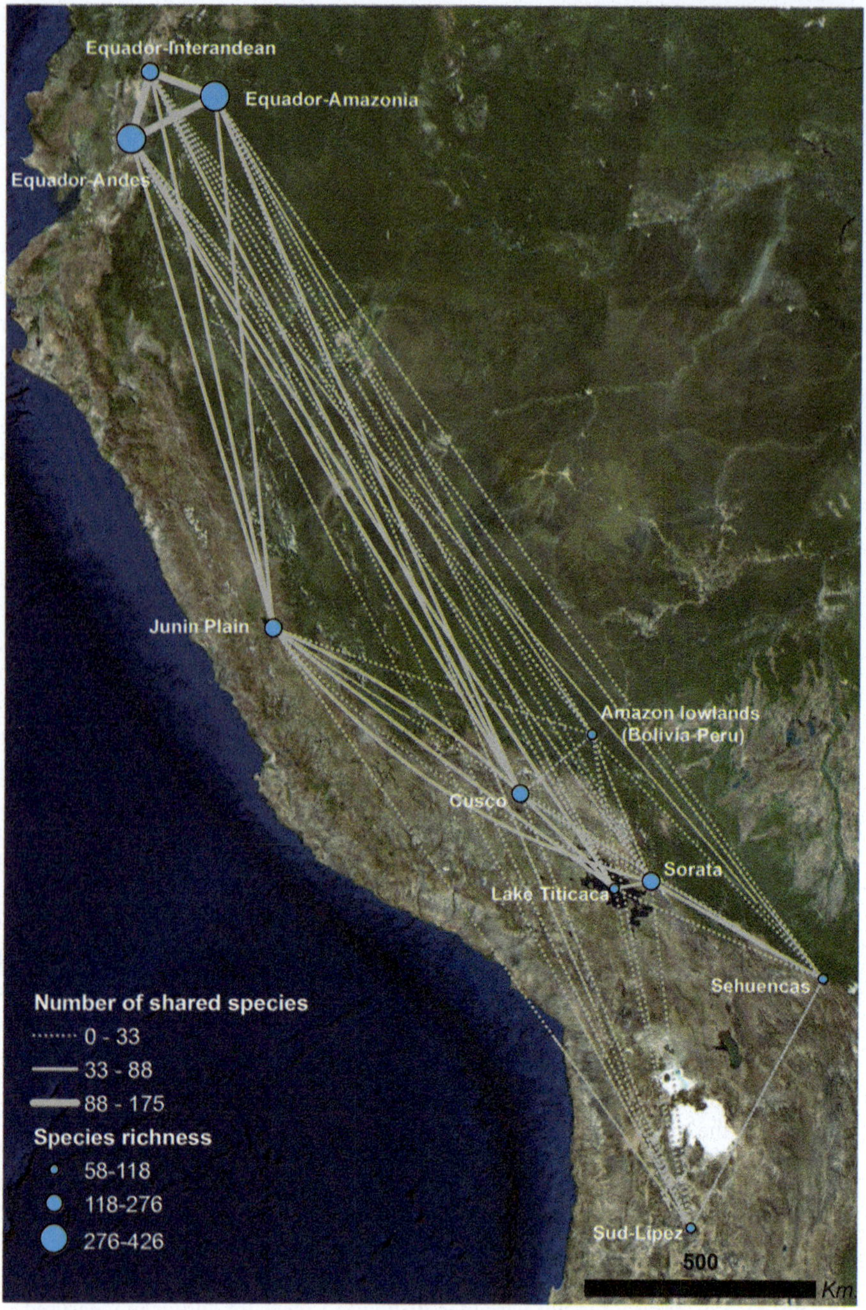

Fig. 7.3 Flow map showing the number of shared diatom species among all regions as indicated by proportional line widths. The total diatom richness (alpha diversity) of each study region is shown by proportional circles. Diatom species richness (alpha diversity) was estimated for each study region using the Chao 2 estimator, a nonparametric asymptotic estimator in order to account for differences in sampling effort between study regions (reprinted with permission from Benito et al. 2018a)

Table 7.1 Results of the Generalized Linear Mixed Model (GLMM) analysis of diatom species richness as a function of local (limnological) and regional (geo-climatic) predictors

Variable	Model-averaged coefficient	Adjusted SE	Variable importance
Intercept	3.11	0.34	
Conductivity (μS/cm)	−0.12	0.04	1.00
Connectivity	0.17	0.03	1.00
(number water bodies in a 50 km² grid cell)			
Ecoregion	0.43	0.20	0.80
Elevation (m)	−0.35	0.10	1.00
% aquatic habitat	0.19	0.05	1.00
(% freshwater habitat in a 50 km² grid cell)			
MAT (°C)	−0.21	0.04	1.00
pH	0.06	0.02	1.00
Precipitation seasonality	−0.34	0.09	1.00
(coefficient of variation; mm)			
Temperature seasonality	0.21	0.07	1.00
(standard deviation; °C)			

Only significant variables selected across all candidate models are shown. Model-averaged coefficients are averaged by model weights. Higher positive or negative coefficients indicate a stronger influence. Variable importance represents the selection probability of each independent variable

N candidate models (ΔAICc < 7) $= 27$

Variables excluded from all candidate models: Longitude, MAP, waterbody size and water temperature

Latitude was initially excluded from the full GLMM due to its high collinearity effects (Variance Inflation Factor > 10)

predictors not included in the study (Pajunen et al. 2016; Soininen 2007). In addition, it is also apparent that the relative importance of spatial (dispersal) and environmental (niche) processes vary not only with spatial scale, but also with time. For instance, Bramburger et al. (2008) experimentally demonstrated that dispersal-based processes assemble local diatom communities in a set of inter-connected Indonesian lakes, but as colonization progressed, their importance was replaced by niche-based processes related to deterministic and competition-driven processes. In our studies, the time span of diatom collection (30 years) in tropical South American lakes exerted a strong control on community composition (Benito et al. 2018b). However, empirical and observational studies are still lacking to understand how these processes interact to maintain and generate diatom diversity at different temporal scales.

Distinct latitudinal patterns of diatom species richness and composition also are apparent (Figs. 7.3 and 7.4). The latitudinal gradient of biodiversity, which commonly shows increasing diversity towards the equator, is one of the best documented patterns on Earth, with consistent patterns across a wide range of taxonomic groups (e.g. plants, invertebrates, mammals; Hillebrand 2004). In the case of diatoms,

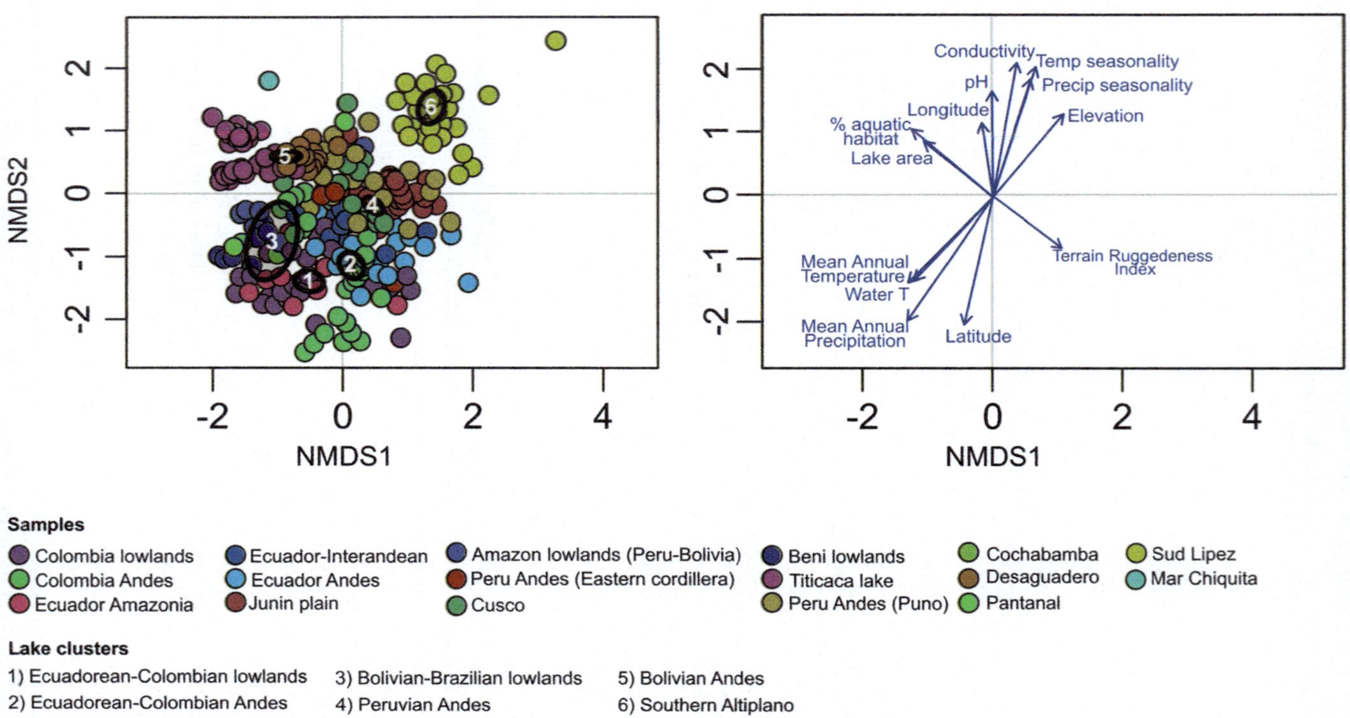

Fig. 7.4 Non-metric multidimensional scaling (NMDS) based on Chao dissimilarity diatom matrix. Left: Site scores labeled by regions; lake clusters are identified using Principal Component Analysis (PCA) from lake environmental data and using site scores as input to group samples with cluster analysis. Right: environmental variable fitting showing the selected variables; the length of each vector is proportional to the correlation between variables and NMDS axes. Lake clusters are arranged from low to high latitude. Modified from Benito et al. (2018b)

low-latitude regions of the Andes and Amazon of Ecuador contain the richest diatom floras and exhibit the highest beta diversity (i.e. turnover) (Fig. 7.3). Negative latitudinal diatom diversity gradients have been found in some studies of other regions (Vyverman et al. 2007), but others have identified either a reverse or no latitudinal diversity gradient (Hillebrand and Azovsky 2001; Soininen et al. 2016). In tropical South America, after modelling diatom species richness using linear regressions, climatic and regional factors that vary with latitude (e.g. precipitation seasonality, mean annual temperature, ecoregion) had the strongest relationships with large-scale diversity patterns (Benito et al. 2018a). The metacommunity theory, which recognizes a set of local ecological communities that are connected by the dispersal of potentially interacting species, provides a flexible framework for the integration of regional (e.g. dispersal, climate) and local (e.g. biotic interactions, limnology) drivers of species composition (Leibold et al. 2004). However, hypotheses that integrate metacommunity processes into the latitudinal gradient paradigm are mostly untested today. Jocque et al. (2010) proposed that variability in climate with latitude mediates shifts between habitat specialization and dispersal at the metacommunity level, generating gradients of species richness, dispersal ability, endemism. Benito et al. (2018b) found that species sorting and dispersal dynamics jointly drive regional assembly of diatom metacommunities, and that prevalence of ecological guild variation [as proxy of differential dispersal capabilities of different taxa (Wetzel et al. 2012; Jamoneau et al. 2018)] in response to higher regionally structured environment toward the equator might explain the formation of the latitudinal diatom metacommunity gradient.

At the metacommunity scale, geographical distances limit the dispersal of organisms, thus imposing spatially structured variability in species composition independent of environmental variation. Despite the notorious challenge in explicitly measuring dispersal for diatoms, because they can disperse passively by wind and water currents or actively by animal vectors (Kristiansen 1996), such a 'distance effect' has been documented in neotropical lotic and lentic communities. For instance, in the Amazon Negro River basin, dissimilarity of benthic diatom community composition increased with geographic distance, which led different tributaries of the Negro River to harbor different diatom communities (Wetzel et al. 2012). In closed-basin Andean lakes of Ecuador, Steinitz-Kannan (1979) demonstrated an increased plankton dissimilarity with increasing geographic distance. The decay of similarity with distance may be related to historical legacies in micro-organisms, because modern communities are influenced by past (or recent) dispersal limitation (Martiny et al. 2006; Verleyen et al. 2009; Vilmi et al. 2017; Vyverman et al. 2007).

The term diatom metacommunities can be used to define regional diatom communities of tropical South America (Fig. 7.4). We classified sets of lakes according to their environmental similarity, resulting in six lake clusters arranged following a latitudinal gradient and used these clusters to investigate ecological drivers of diatom community structure within lake clusters (see Benito et al. (2018b) for further information on statistical analysis). Our ongoing investigation suggests that geo-climatic factors related to climate and landscape configuration are important mechanisms structuring benthic diatom metacommunities, as indicated by variance

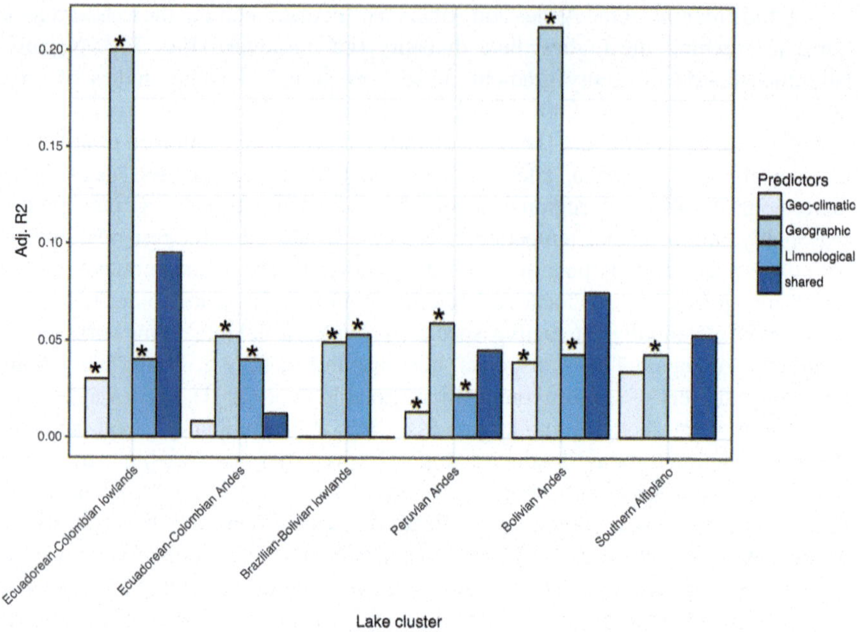

Fig. 7.5 Variance partitioning results of geo-climatic, geographic, and limnological variables on diatom species composition data across the six lake clusters. Lake clusters were identified using Principal Component Analysis (PCA) from lake environmental data and using site scores as input to group samples with cluster analysis (see Benito et al. 2018b for further information on statistical analysis). Asterisks denote the statistical significance ($p < 0.05$) of the pure fractions of geo-climatic, geographic or limnological variables. Lake clusters are arranged from low to high latitude

partitioning (Fig. 7.5). Although our linear regressions corrected for spatial autocorrelation and environmental collinearity, this observation does not exclude the relevance of local environmental factors that are nested within regional-scale factors (Poff 1997). Hence, we hypothesized that geo-climatic effects might be also partially manifested via local environmental variables. Accordingly, and despite the limited number of limnological variables, we found significant effects between limnological factors and diatom community composition in groups of lakes after factoring out geographical distances, as indicated by partial Mantel tests results (Benito et al. 2018a). This is the case in the Peruvian Andes near Cusco and in the Junin Plain and in the Andean and inter-Andean Ecuadorean lakes (Fig. 7.4). In contrast, diatom metacommunities from Lake Titicaca, the Desaguadero watershed, Peruvian Amazon lowlands, and Sud Lipez in the Bolivian Altiplano showed clear regional separation in species composition, suggesting community similarity due to spatially structured environments. Indeed, there was a only a weak association between limnological and diatom matrices in these regions but a strong correlation with the variable ‘ecoregion’, except for the Sud Lipez region. The latter could reflect, at least in part, the influence of extreme limnological conditions in the southern part of the

Bolivian Altiplano, which is characterized by hypersaline shallow lakes as a result of basin geology and high evaporation rates (Sylvestre et al. 2001). The diatom flora here consists of low-diversity benthic communities, with the highest proportion of endemic taxa across all regions analyzed (Benito et al. 2018a; Blanco et al. 2013; Servant-Vildary et al. 2002). Nevertheless, the heterogeneous water chemical data available in the diatom database may underestimate the local environmental influences on species composition and hence future studies are needed. If modern diatom-environment datasets are expanded, they can improve palaeoenvironmental reconstructions using downcore diatom assemblages, including for evaluating nutrient changes driven by recent variation in land-use and climate (Barta et al. 2018).

Metacommunity processes are also expected to operate within-lakes, thus at very small spatial scales. Diatom community responses, including species composition and species richness, differ among lake habitats, thereby highlighting the relevance of species sorting due to substrate type (e.g. mud, plants, rocks), and dispersal between different lake habitats (pelagic versus benthic communities) as a result of the fluid aquatic environment (Fig. 7.1). Local environmental conditions differ among these lake habitats, especially nutrient cycling and environmental stability (Vadeboncoeur et al. 2006), which can act as a strong environmental filter for both stream and lake diatoms (Passy 2007; Pla-Rabés and Catalan 2018; Yang and Flower 2012). Unfortunately, our diatom database from tropical South America suffers from a paucity of nutrient data, and most of the samples correspond to benthic habitats (sediment surface and periphyton) and very few from the pelagic zone. The general conditions of lake habitats and their influence on diatom metacommunity structure can be inferred from ancillary studies. For instance, surface sediment accumulates diatoms transported from multiple habitats acting as a source, but also harbors site-specific communities acting as sink (Bramburger et al. 2008). Hence, surface sediment samples often show the highest species richness in lake ecosystems, because they mix taxa from multiple parts of the lake system (Yang and Flower 2012). Nonetheless, it is the diversity found in the peripheral communities that eventually determines between-lake diversity of diatoms. The area of peripheral (benthic) habitats may also introduce differences in species richness due to species-area relationships (Bolgovics et al. 2016). Habitats with higher diatom species richness often cover more benthic area (Pla-Rabés and Catalan 2018; Stone and Fritz 2004). The area covered by benthic communities in the peripheral lake zones, therefore, may determine rates of colonization and extinction in pelagic communities (Colinvaux and Steinitz-Kannan 1980) (Fig. 7.1).

In the tropical Andes and Amazon, the sparse geographic coverage of long-term biological monitoring data limits an evaluation of the extent of the impacts of different types of environmental change on lake ecosystems. Lake sediments record environmental signals for metacommunity research at timescales and spatial ranges not available in the observational record and can help to fill this gap (Fig. 7.1) (Winegardner et al. 2017). Thus, metacommunity dynamics of diatom assemblages can be incorporated into a paleolimnogical framework for biodiversity studies and analysis of multiple stressors (Gregory-Eaves and Beisner 2011; Benito et al. 2019). Given the newly created diatom database for tropical South America (https://github.

com/xbenitogranell/diatoms-biogeography-southamerica), future studies comparing past and modern diatom assemblage structuring can strengthen inferences about which lake ecosystems are most sensitive to local (e.g., agriculture, fish stocking) versus regional (e.g., climate warming, land-use) stressors as environmental change intensifies in tropical regions.

4 Conclusions and Implications

There is clear potential for new studies in tropical South America to address research questions related to the taxonomy, ecology, and biogeography of diatoms, as well as other algal groups. Future work will benefit from harmonized taxonomic data sets using the existing inventories of the tropical South American diatom database to understand diatom diversity patterns and drivers in mountain and lowland settings. There is also a potential to evaluate colonization and extinction events using phylogeograhic approaches and molecular data preserved in lake sediments (Burge et al. 2018), as well as species diversification linked to geologic and climatic histories (Valencia et al. 2016). For example, one recent study suggests that diatom communities have changed in relative isolation during periods of unfavourable climatic conditions in a lake basin acting as microrefugia (de Novaes Nascimento et al. 2018). A broader range of basins within a spatially structure landscape could provide tests of the universality of this observation.

Every new study dealing with the geographic distribution of microorganisms in general, and diatoms in particular, contributes to an evaluation of the degree of endemism and cosmopolitanism of different groups (Vanormelingen et al. 2008; Williams 2011). Prioritization of regions that contribute disproportionately to maintaining regional diversity is only possible when the geographic distribution of species can be reliably determined. Microorganisms are often neglected in spatial ecological studies, despite their role in ecosystem functioning and structure (Casamayor 2017). Consequently, diatoms can be more widely and better used to support conservation planning and assessment of ecosystem services. However, the quest to understand whether or not South American tropical freshwaters hold a highly diverse reservoir of diatom communities with evolutionary potential (Spanbauer et al. 2018) is in its infancy. In other mountainous lake regions, including the Alps (Monchamp et al. 2018), Pyrenees (Casamayor 2017), Himalayas (Liu et al. 2006), and the African Highlands (Taylor et al. 2009) studies suggest that idiosyncratic microbial populations may be severely impacted by climate change and land-use practices. As more regional diatom studies appear in tropical South America, the perspectives of conservation biology that consider habitat conservation may also be applied in this region.

In this chapter, we have explored how the combination of climate, topography, and landscape configuration created large-scale ecoregion patterning of diatom communities, partly influenced by local limnological variables, in patterns similar to what has been described for the macroscopic flora of the Andean Páramo and Puna

(Hughes and Eastwood 2006). This suggests the potential to unify the conceptual frameworks of microbial ecology and macroecology to better understand mechanistic drivers of niche, stochastic, and historical processes (Martiny et al. 2006). Diatoms can contribute to this approach given the influence of geophysical and climatic gradients on freshwater systems of tropical South America. The topographic complexity, the influence of geodiversity as a surrogate for diatom diversity patterns, combined with palaeotopographic reconstructions (Flantua and Hooghiemstra 2018) and phylogenetic data (Spanbauer et al. 2018) can provide a powerful platform to develop multiple, interdisciplinary hypotheses on processes driving Neotropical diversification in the microbial world.

References

Aguilera X, Declerck S, De Meester L, Maldonado M, Ollevier F (2006) Tropical high Andes lakes: a limnological survey and an assessment of exotic rainbow trout (*Oncorhynchus mykiss*). Limnologica 36:258–268

Antonelli A, Nylander JA, Persson C, Sanmartín I (2009) Tracing the impact of the Andean uplift on Neotropical plant evolution. Proc Natl Acad Sci U S A 106:9749–9754

Antonelli A, Ariza M, Albert J, Andermann T, Azevedo J, Bacon C, Faurby S, Guedes T, Hoorn C, Lohmann LG (2018) Conceptual and empirical advances in Neotropical biodiversity research. PeerJ 6:e5644

Baker P, Fritz SC (2015) Nature and causes of quaternary climate variation of tropical South America. Quat Sci Rev 124:31–47

Barberán A, Casamayor EO, Fierer N (2014) The microbial contribution to macroecology. Front Microbiol 5:1–8

Barta B, Mouillet C, Espinosa R, Andino P, Jacobsen D, Christoffersen KS (2018) Glacial-fed and p K.S. lake ecosystems in the tropical high Andes. Hydrobiologia 813:19–32

Benito X, Fritz SC, Steinitz-Kannan M, Tapia PM, Kelly MA, Lowell TV (2018a) Geo-climatic factors drive diatom community distribution in tropical South American freshwaters. J Ecol 106 (4):1660–1672. https://doi.org/10.5061/dryad.ck7pt

Benito X, Fritz SC, Steinitz-Kannan M, Vélez MI, McGlue MM (2018b) Lake regionalization and diatom metacommunity structuring in tropical South America. Ecol Evol 8(16):7865–7878. https://doi.org/10.5061/dryad.6jk8h77

Benito X, Feitl M, Fritz SC, Mosquera PV, Schneider T, Hampel H, Quevedo L, Steinitz-Kannan M (2019) Identifying temporal and spatial patterns of diatom community change in the tropical Andes over the last ~150 years. J Biogeogr 46(8):1889–1900

Bie T, Meester L, Brendonck L, Martens K, Goddeeris B, Ercken D, Hampel H, Denys L, Vanhecke L, Gucht K, Wichelen J, Vyverman W, Declerck SAJ (2012) Body size and dispersal mode as key traits determining metacommunity structure of aquatic organisms. Ecol Lett 15:740–747

Blanco S, Álvarez-Blanco I, Cejudo-Figueiras C, De Godos I, Bécares E, Muñoz R, Guzman HO, Vargas VA, Soto R (2013) New diatom taxa from high-altitude Andean saline lakes. Diatom Res 28:13–27

Bolgovics Á, Ács É, Várbíró G, Görgényi J, Borics G (2016) Species area relationship (SAR) for benthic diatoms: a study on aquatic islands. Hydrobiologia 764:91–102

Bramburger AJ, Hamilton PB, Hehanussa PE, Haffner GD (2008) Processes regulating the community composition and relative abundance of taxa in the diatom communities of the Malili Lakes, Sulawesi Island, Indonesia. Hydrobiologia 615:215–224

Burge DR, Edlund MB, Frisch D (2018) Paleolimnology and resurrection ecology: the future of reconstructing the past. Evol Appl 11(1):42–59. https://doi.org/10.1111/eva.12556

Casamayor EO (2017) Towards a microbial conservation perspective in high mountain lakes. In: High mountain conservation in a changing world. Springer, Cham, pp 157–180

Cleve PT (1881) On some new and little known diatoms. Kongl, Boktryckeriet

Colinvaux P, Steinitz-Kannan M (1980) Species richness and area in Galapagos and Andean lakes: equilibrium phytoplankton communities and a paradox of the zooplankton. In: Kerfoot WC (ed) Evolution and ecology of zooplankton communities. The University Press, England, pp 697–711

Costa FR, Magnusson WE, Luizao RC (2005) Mesoscale distribution patterns of Amazonian understorey herbs in relation to topography, soil and watersheds. J Ecol 93:863–878

Cox C, Moore P, Marquardt W, Demaree R (2005) Biogeography: an ecological and evolutionary approach, 6th edn. Blackwell, Oxford

Declerck SAJ, Coronel JS, Legendre P, Brendonck L (2011) Scale dependency of processes structuring metacommunities of cladocerans in temporary pools of High-Andes wetlands. Ecography 34:296–305

de Novaes Nascimento M, Laurenzi AG, Valencia BG, Van R, Bush M (2018) A 12,700-year history of paleolimnological change from an Andean microrefugium. The Holocene 29 (2):231–243. https://doi.org/10.1177/0959683618810400

Diaz CA (2006) A new monoraphid diatom genus: Haloroundia Diaz & Maidana. Nova Hedwig Beih 130:177

Ehrenberg CG (1843) Verbreitung und Einfluss des mikroskopischen Lebens in Süd-und Nord-Amerika. Königlichen Akademie der Wissenschaften

Ehrenberg CG (1873) Mikrogeologische Studien über das kleinste Leben der Meeres-Tiefgründe aller Zonen und dessen geologischen Einfluss. Buchdr. der K. Akademie der Wissenschaften, in Kommission bei F. Dümmler

Ekdahl EJ, Fritz SC, Baker PA, Rigsby CA, Coley K (2008) Holocene multidecadal- to millennial-scale hydrologic variability on the South American Altiplano. The Holocene 18:867–876

Finlay BJ (2002) Global dispersal of free-living microbial eukaryote species. Science 296:1061–1063

Flantua SG, Hooghiemstra H (2018) Historical connectivity and mountain biodiversity. In: Hoorn C, Perrigo A, Antonelli A (eds) Mountains, climate and biodiversity. Wiley, Chichester

Flantua SG, Hooghiemstra H, Grimm EC, Behling H, Bush MB, González-Arango C, Gosling WD, Ledru M-P, Lozano-García S, Maldonado A (2015) Updated site compilation of the Latin American pollen database. Rev Palaeobot Palynol 223:104–115

Fourtanier E, Kociolek J (1999) Catalogue of the diatom genera. Diatom Res 14:1–90

Frenguelli J (1939) Diatomeas del lago Titicaca. Notas del Museo de la Plata 4:175–198

Frenguelli J (1942) Diatomeas del Neuquén (Patagonia). Instituto del Museo de la Universidad de La Plata (Sección Botánica) 5:73–219

Fritz SC, Baker PA, Seltzer GO, Ballantyne A, Tapia P, Cheng H, Edwards RL (2007) Quaternary glaciation and hydrologic variation in the South American tropics as reconstructed from the Lake Titicaca drilling project. Quat Res 68:410–420

Fritz SC, Baker PA, Tapia P, Spanbauer T, Westover K (2012) Evolution of the Lake Titicaca basin and its diatom flora over the last ~370,000 years. Palaeogeogr Palaeoclimatol Palaeoecol 317–318:93–103

Fritz SC, Benito X, Steinitz-Kannan M (2018) Long-term and regional perspectives on recent change in lacustrine diatom communities in the tropical Andes. J Paleolimnol 61(2):251–262. https://doi.org/10.1007/s10933-018-0056-6

Gonçalves-Souza T, Romero GQ, Cottenie K (2014) Metacommunity versus biogeography: a case study of two groups of Neotropical vegetation-dwelling arthropods. PLoS One 9:e115137
Gregory-Eaves I, Beisner BE (2011) Palaeolimnological insights for biodiversity science: an emerging field. Freshw Biol 56:2653–2661
Guiry MD, Guiry GM (2017) AlgaeBase. World-Wide Electronic Publication, National University of Ireland, Galway
Heino J, Soininen J, Alahuhta J, Lappalainen J, Virtanen R (2016) Metacommunity ecology meets biogeography: effects of geographical region, spatial dynamics and environmental filtering on community structure in aquatic organisms. Oecologia 183:121–137
Hillebrand H (2004) On the generality of the latitudinal diversity gradient. Am Nat 163:192–211
Hillebrand H, Azovsky AI (2001) Body size determines the strength of the latitudinal diversity gradient. Ecography 24:251–256
Hughes C, Eastwood R (2006) Island radiation on a continental scale: exceptional rates of plant diversification after uplift of the Andes. Proc Natl Acad Sci U S A 103:1334–1339
Hustedt F (1927) Fossile Bacillariaceen aus dem Loa-Becken in der Atacama-Wüste. Chile Archiv fur Hydrobiologie 18:224–251
Jamoneau A, Passy SI, Soininen J, Leboucher T, Tison-Rosebery J (2018) Beta diversity of diatom species and ecological guilds: response to environmental and spatial mechanisms along the stream watercourse. Freshw Biol 63:62–73
Jampel C (2016) Cattle-based livelihoods, changes in the taskscape, and human–bear conflict in the Ecuadorian Andes. Geoforum 69:84–93
Jantz N, Behling H (2012) A Holocene environmental record reflecting vegetation, climate, and fire variability at the Pantz, of Quimsacocha, southwestern Ecuadorian Andes. Veg Hist Archaeobotany 21:169–185
Jenkins D, Medley K (2011) Microbes as a test of biogeographic principles. In: Biogeography of microscopic organisms: is everything small everywhere? Cambridge University Press, Cambridge, pp 309–320
Jenkins DG, Ricklefs RE (2011) Biogeography and ecology: two views of one world. Philos Trans R Soc Lond B Biol Sci 366:2331–2335
Jocque M, Field R, Brendonck L, Meester LD (2010) Climatic control of dispersal–ecological specialization trade-offs: a metacommunity process at the heart of the latitudinal diversity gradient? Glob Ecol Biogeogr 19:244–252
Jüttner I, Chimonides PDJ, Ormerod SJ, Cox EJ (2010) Ecology and biogeography of Himalayan diatoms: distribution along gradients of altitude, stream habitat and water chemistry. Fundam Appl Limnol/Archiv für Hydrobiologie 177:293–311
Kessler M (2002) The elevational gradient of Andean plant endemism: varying influences of taxon-specific traits and topography at different taxonomic levels. J Biogeogr 29:1159–1165
Killeen TJ, Douglas M, Consiglio T, Jørgensen PM, Mejia J (2007) Dry spots and wet spots in the Andean hotspot. J Biogeogr 34:1357–1373
Kristiansen J (1996) Biogeography of freshwater algae - conclusions and perspectives. Hydrobiologia 336:159–161
Leibold MA, Holyoak M, Mouquet N, Amarasekare P, Chase JM, Hoopes MF, Holt RD, Shurin JB, Law R, Tilman D, Loreau M, Gonzalez A (2004) The metacommunity concept: a framework for multi-scale community ecology. Ecol Lett 7:601–613
Liu Y, Yao T, Jiao N, Kang S, Zeng Y, Huang S (2006) Microbial community structure in moraine lakes and glacial meltwaters, Mount Everest. FEMS Microbiol Lett 265:98–105
Luebert F, Weigend M (2014) Phylogenetic insights into Andean plant diversification. Front Ecol Evol 2:27
MacArthur R, Wilson EO (1967) The theory of island biogeography. Princeton University Press, Princeton, NJ
Maidana N, Seeligmann C (2006) Diatomeas (Bacillariophyceae) de ambientes acuáticos de altura de la Provincia de Catamarca, Argentina II. Boletín de la Sociedad Argentina de Botanica 41:1–13

Maidana NI, Seeligmann C, Morales MR (2011) El género Navicula sensu stricto (Bacillariophyceae) en humedales de altura de Jujuy, Argentina. Boletín de la Sociedad Argentina de Botanica 46:13–29

Manguin E (1964) Contribution à la connaissance des diatomées des Andes du Pérou. Edition du Muséum

Mann DG (1999) The species concept in diatoms. Phycologia 38:437–495

Mann DG, Droop SJM (1996) Biodiversity, biogeography and conservation of diatoms. In: Biogeography of freshwater algae. Springer, Dordrecht, pp 19–32

Martiny JBH, Bohannan BJM, Brown JH, Colwell RK, Fuhrman JA, Green JL, Horner-Devine MC, Kane M, Krumins JA, Kuske CR, Morin PJ, Naeem S, Øvreås L, Reysenbach A-L, Smith VH, Staley JT (2006) Microbial biogeography: putting microorganisms on the map. Nat Rev Microbiol 4:102–112

Metzeltin D, Lange-Bertalot H (2007) Tropical diatoms of South America II. In: Iconographia Diatomologica 18. A.R.G. Gantner Verlag K.G., Königstein

Metzeltin D, Lange-Bertalot H, Garcia-Rodriguez F (2005) Diatoms of Uruguay. Iconographia Diatomologica 15:1–736

Michelutti N, Wolfe A, Cooke C, Hobbs W, Vuille M (2015) Climate change forces new ecological states in tropical Andean lakes. PLoS One 10:e0115338

Monchamp ME, Spaak P, Domaizon I, Dubois N, Bouffard D, Pomati F (2018) Homogenization of lake cyanobacterial communities over a century of climate change and eutrophication. Nat Ecol Evol 2:317

Montoya-Moreno Y, Sala S, Vouilloud A, Aguirre N (2012) Periphytic diatoms (bacillariophyta) of ayapel flood plain. Colombia. I. Caldasia 34:457–474

Morales E, Vis M (2007) Epilithic diatoms (Bacillariophyceae) from cloud forest and alpine streams in Bolivia, South America. Proc Acad Nat Sci Phila 156:123–155

Morales EA, Fernández E, Fernandez CE, Lizarro D, Alcoreza P (2008) Algal studies in Bolivia: a compilation and preliminary analysis of existing psychological literature. Gayana. Botánica 65:93–108

Morales EA, Novais MH, Chávez G, Hoffmann L, Ector L (2012) Diatoms (Bacillariophyceae) from the Bolivian Altiplano: three new araphid species from the Desaguadero River draining Lake Titicaca. Fottea 12:41–58

Morales EA, Wetzel CE, Rivera SF, Van de Vijver B, Ector L (2014) Current taxonomic studies on the diatom flora (Bacillariophyceae) of the Bolivian Altiplano, South America, with possible consequences for palaeoecological assessments. J Micropalaentol 33:1–9

Oliveira PD, Steinitz-Kannan M (1992) The diatom flora (Bacillariophyceae) of the Cuyabeno Faunistic Reserve, Ecuadorian Amazonia. Nova Hedwigia 54:515–552

Pajunen V, Luoto M, Soininen J (2016) Climate is an important driver for stream diatom distributions. Glob Ecol Biogeogr 25:198–206

Pajunen V, Luoto M, Soininen J (2017) Unravelling direct and indirect effects of hierarchical factors driving microbial stream communities. J Biogeogr 44:2376–2385

Passy SI (2007) Diatom ecological guilds display distinct and predictable behavior along nutrient and disturbance gradients in running waters. Aquat Bot 86:171–178

Passy SI (2010) A distinct latitudinal gradient of diatom diversity is linked to resource supply. Ecology 91:36–41

Patrick R (1961) Diatoms (Bacillariophyceae) from the alimentary tract of *Phoenicoparrus jamesi* (Sclater). Postilla 49:43–50

Patrick R (1970) diatom flora of some lakes of the Galapagos Islands. Nova Hedwigia, Beihefte

Pinel-Alloul B, André A, Legendre P, Cardille JA, Patalas K, Salki A (2013) Large-scale geographic patterns of diversity and community structure of pelagic crustacean zooplankton in Canadian lakes. Glob Ecol Biogeogr 22:784–795

Pinseel E, de Vijver BV, Kavan J, Verleyen E, Kopalová K (2017) Diversity, ecology and community structure of the freshwater littoral diatom flora from Petuniabukta (Spitsbergen). Polar Biol 40:533–551

Pla-Rabés S, Catalan J (2018) Diatom species variation between lake habitats: implications for interpretation of paleolimnological records. J Paleolimnol 60(2):1–19
Poff NL (1997) Landscape filters and species traits: towards mechanistic understanding and prediction in stream ecology. J N Am Benthol Soc 16:391–409
Potapova MG, Charles DF (2002) Benthic diatoms in USA rivers: distributions along spatial and environmental gradients. J Biogeogr 29:167–187
Reche I, Pulido-Villena E, Morales-Baquero R, Casamayor EO (2005) Does ecosystem size determine aquatic bacterial richness? Ecology 86:1715–1722
Ricklefs RE (1987) Community diversity: relative roles of local and regional processes. Science 235:167–171
Ricklefs RE, Jenkins DG (2011) Biogeography and ecology: towards the integration of two disciplines. Philos Trans R Soc Lond B Biol Sci 366:2438–2448
Round FE, Crawford RM, Mann DG (1990) Diatoms: biology and morphology of the genera. Cambridge University Press, Cambridge
Rumrich U, Lange-Bertalot H, Rumrich M (2000) Diatoms of the Andes, from Venezuela to Patagonia/Tierra del Fuego, and two additional contributions. In: Iconographia diatomologica, vol 9. ARG Gartner Verlag KG, Königstein
Sabbe K, Vanhoutte K, Lowe RL, Bergey EA, Biggs BJ, Francoeur S, Hodgson D, Vyverman W (2001) Six new Actinella (Bacillariophyta) species from Papua New Guinea, Australia and New Zealand: further evidence for widespread diatom endemism in the Australasian region. Eur J Phycol 36:321–340
Sala SE, Maidana NI (2003) Morphology and taxonomy of Amphora atacamae Frenguelli (Bacillariophyceae). Diatom Res 18:69–78
Sala SE, Ramírez JJ, Plata Y (2008) Diatoms from lentic and lotic systems in antioquia, chocó and santander departments in Colombia. Revista de Biologia Tropical 56:1159–1178
Särkinen T, Pennington R, Lavin M (2012) Evolutionary islands in the Andes: persistence and isolation explain high endemism in Andean dry tropical forests. J Biogeogr 39:884–900
Sarmiento FO (2002) Anthropogenic change in the landscapes of highland Ecuador. Geogr Rev 92:213–234
Seddon AW, Froyd CA, Witkowski A (2011) Diatoms (Bacillariophyta) of isolated islands I: new taxa in the genus Navicula sensu stricto from the Galapagos islands. J Phycol 47:861–879
Seddon AW, Witkowski A, Froyd CA, KurzydłOwski KJ, Grzonka J, Willis KJ (2014) Diatoms from isolated islands II: Pseudostaurosira diablarum, a new species from a mangrove ecosystem in the Galápagos Islands. Diatom Res 29:201–211
Servant-Vildary S (1986) Les diatomées actuelles des Andes de Bolivie (Taxonomie, écologie). Cahiers de Micropaléontologie 1:99–124
Servant-Vildary S, Roux M (1990) Multivariate analysis of diatoms and water chemistry in Bolivian saline lakes. Hydrobiologia 197:267–290
Servant-Vildary S, Risacher F, Roux M (2002) Diatom based transfer function for estimating the chemical composition of fossil water. Calibration based on salt lakes of the Lipez area in the southwestern. Carnets de Geologie/Notebooks on Geology, Maintenon
Smol JP, Stoermer EF (2010) The diatoms: applications for the environmental and earth sciences. Cambridge University Press, Cambridge
Soininen J (2007) Environmental and spatial control of freshwater diatoms – a review. Diatom Res 22:473–490
Soininen J, Jamoneau A, Rosebery J, Passy SI (2016) Global patterns and drivers of species and trait composition in diatoms. Glob Ecol Biogeogr 25:940–950
Spanbauer TL, Fritz SC, Baker PA (2018) Punctuated changes in the morphology of an endemic diatom from Lake Titicaca. Paleobiology 44:89–100
Steinitz-Kannan M (1979) Comparative limnology of Ecuadorian Lakes: a study of species number and composition of plankton communities of the Galapagos Islands and Equatorial Andes. Ph. D. Dissertation, The Ohio State University, Ohio

Stone JR, Fritz SC (2004) Three-dimensional modeling of lacustrine diatom habitat areas: improving paleolimnological interpretation of planktic: benthic ratios. Limnol Oceanogr 49:1540–1548
Sylvestre F, Servant-Vildary S, Roux M (2001) Diatom-based ionic concentration and salinity models from the south Bolivian Altiplano (15–23°S). J Paleolimnol 25:279–295
Tapia PM, Theriot EC, Fritz SC, Cruces F, Rivera P (2004) Distribution and morphometry analysis of Cyclostephanos andinus comb. nov., a planktonic diatom from the central Andes. Diatom Res 19:311–327
Tapia PM, Fritz SC, Seltzer GO, Rodbell DT, Metiever S (2006) Contemporary distribution and late-quaternary stratigraphy of diatoms in the Junin Plain, central Andes, Peru. Boletin Sociedad Geologica Peru 101:19–42
Taylor RG, Mileham L, Tindimugaya C, Mwebembezi L (2009) Recent glacial recession and its impact on alpine riverflow in the Rwenzori Mountains of Uganda. J Afr Earth Sci 55:205–213
Vadeboncoeur Y, Kalff J, Christoffersen K, Jeppesen E (2006) Substratum as a driver of variation in periphyton chlorophyll and productivity in lakes. J N Am Benthol Soc 25:379–392
Valdujo PH, Carnaval ACO, Graham CH (2013) Environmental correlates of anuran beta diversity in the Brazilian Cerrado. Ecography 36:708–717
Valencia BG, Matthews-Bird F, Urrego DH, Williams JJ, Gosling WD, Bush M (2016) Andean microrefugia: testing the Holocene to predict the Anthropocene. New Phytol 212:510–522
Van Colen WR, Mosquera P, Vanderstukken M, Goiris K, Carrasco M-C, Decaestecker E, Alonso M, Leon-Tamariz F, Muylaert K (2017) Limnology and trophic status of glacial lakes in the tropical Andes (Cajas National Park, Ecuador). Freshw Biol 62:458–473
Van Colen W, Mosquera PV, Hampel H, Muylaert K (2018) Link between cattle and the trophic status of tropical high mountain lakes in p K., grasslands in Ecuador. Lakes Reserv Res Manag 23(4):303–311. https://doi.org/10.1111/lre.12237
Vanormelingen P, Verleyen E, Vyverman W (2008) The diversity and distribution of diatoms: from cosmopolitanism to narrow endemism. Biodivers Conserv 17:393–405
Velez MA (2003) A contribution of diatom analysis to Lateglacial and Holocene environmental reconstructions of Colombian lowland and montane ecosystems. Ph.D. Dissertation, IBED, Universiteit van Amsterdam, Amsterdam
Verleyen E, Vyverman W, Sterken M, Hodgson DA, De Wever A, Juggins S, Van De Vijver B, Jones VJ, Vanormelingen P, Roberts D, Flower R, Kilroy C, Souffreau C, Sabbe K (2009) The importance of dispersal related and local factors in shaping the taxonomic structure of diatom metacommunities. Oikos 118:1239–1249
Vijver BV, Beyens L (2007) Biogeography and ecology of freshwater diatoms in Subantarctica: a review. J Biogeogr 26:993–1000
Vilhena DA, Antonelli A (2015) A network approach for identifying and delimiting biogeographical regions. Nat Commun 6:6848
Vilmi A, Karjalainen SM, Heino J (2017) Ecological uniqueness of stream and lake diatom communities shows different macroecological patterns. Divers Distrib 23:1042–1053
Vyverman W, Verleyen E, Sabbe K, Vanhoutte K, Sterken M, Hodgson DA, Mann DG, Juggins S, Van de Vijver B, Jones V, Flower R, Roberts D, Chepurnov VA, Kilroy C, Vanormelingen P, Wever AD (2007) Historical processes constrain patterns in global diatom diversity. Ecology 88:1924–1931
Weiher E, Keddy P (2001) Ecological assembly rules: perspectives, advances, retreats. Cambridge University Press, Cambridge
Wetzel RG (2001) Lake and river ecosystems. Limnology 37:490–525
Wetzel CE, Ector L, Hoffmann L, Bicudo DDC (2010) Colonial planktonic Eunotia (Bacillariophyceae) from Brazilian Amazon: taxonomy and biogeographical considerations on the *E. asterionelloides* species complex. Nova Hedwigia 91:49–86
Wetzel CE, de Bicudo DC, Ector L, Lobo EA, Soininen J, Landeiro VL, Bini LM (2012) Distance decay of similarity in neotropical diatom communities. PLoS One 7:10–11

Williams DM (2011) Historical biogeography, microbial endemism and the role of classification: everything is endemic. In: Biogeography of microscopic organisms: is everything small everywhere? Cambridge University Press, Cambridge, pp 11–31
Winegardner AK, Legendre P, Beisner BE, Gregory-Eaves I (2017) Diatom diversity patterns over the past c. 150 years across the conterminous United States of America: identifying mechanisms behind beta diversity. Glob Ecol Biogeogr 26:1303–1315
Yang H, Flower RJ (2012) Effects of light and substrate on the benthic diatoms in an oligotrophic lake: a comparison between natural and artificial substrates. J Phycol 48:1166–1177

Chapter 8
Avian Diversity in Humid Tropical and Subtropical South American Forests, with a Discussion About Their Related Climatic and Geological Underpinnings

Manuel Nores

Abstract The tropical and subtropical regions of South America occupy ca. 14,250,000 km^2, of which 8,000,000 km^2 are covered by humid forests. Among these, Amazonia, the Atlantic Forest, the forests of the Eastern Andean Slopes, and the Pacific Forest stand out. About 1014 species of birds (324 non-passerines and 690 passerines) are endemic to these forests. Their lineages are believed to have been impacted by five major geological and climatic events in the Neogene and the Pleistocene: the uplift of the Andes, the formation of the Isthmus of Panama, glaciations, sea-level rise, and alternating period of differing humidity. In this contribution, inspired by an analysis of bird distribution patterns inferred from superimposed species ranges, I discuss possible processes that may have influenced biological diversification in tropical and subtropical South America. Data demonstrate that the Andean uplift divided the original pan-Amazonia into a trans-Andean region (the Pacific forest) and a cis-Andean region (present-day Amazonia), and, about 10–7 Ma, induced the formation of large Amazonian rivers. There is evidence that Amazonia and other tropical forests remained predominantly forested over time, even during the cooler and drier glacial periods. Marine transgressions during the upper Tertiary and lower Quaternary, and the formation of the Amazonian rivers following the rise of the Andes, likely fragmented the forest and operated as mechanisms to generate new species. Moreover, the closing of the Panama Isthmus and connections between Amazonia and the Atlantic forest, and the southern Andean and the Paranaense/Atlantic forests, led to the migration of taxa from other biogeographic regions—substantially increasing diversity in these biomes. Biogeographical patterns can promote insight and lead to the generation of hypotheses about the processes that have shaped present-day patterns of avian biodiversity in the region, but testing the latter will require multidisciplinary studies that integrate geology,

Electronic supplementary material The online version of this chapter (https://doi.org/10.1007/978-3-030-31167-4_8) contains supplementary material, which is available to authorized users.

M. Nores (✉)
Centro de Zoología Aplicada, Universidad Nacional de Córdoba, Córdoba, Argentina

V. Rull, A. C. Carnaval (eds.), *Neotropical Diversification: Patterns and Processes*, Fascinating Life Sciences, https://doi.org/10.1007/978-3-030-31167-4_8

paleoecology, molecular phylogenetics, and phylogeography. Moreover, despite the historical processes responsible for the extraordinary richness of the tropical and subtropical avifauna, available data point out to a serious problem affecting bird populations in humid forests of South America—especially in Amazonia, the lower Eastern Andean Slopes, and the Pacific Forest: a severe reduction in the diversity of birds. This drop in diversity is appreciable not only in disturbed habitats, but also in national parks and reserves, suggesting that climate change may be involved.

Keywords Forest birds · South America · Biogeography · Diversity · Status · Amazonia · Pacific forest · Eastern Andean Slopes · Atlantic forest · Isthmus of Panama · Sea level rise

1 Introduction

The tropical and subtropical regions of South America are considerable in size, reaching about 14,250,000 km^2, and extending from northern South America to approximately 30°S (Catamarca, in Argentina, and Rio Grande do Sul, in Brazil; Fig. 8.1), where the range of the continuous forests ends. In montane areas, the subtropical zone rises to 2300–2500 m above sea level (Ridgely and Tudor 1989). Approximately 57% of this region (ca. 8,000,000 km^2 as per satellite images) is covered by humid forests. These are mostly lowland (terra firme) forest, but cloud forests, mountain forests, gallery forests, flooded forests (varzea, igapó, mangroves), and white-sand forests are also present (Hueck and Seibert 1972, pers. obs.)

The origin of the great species richness observed in the forested areas of tropical and subtropical South America, particularly in Amazonia, has long intrigued biologists and naturalists. Moritz et al. (2000) stated that molecular approaches indicates that, in most cases, the divergence of extant sister taxa in rainforests predates the Pleistocene. Aleixo and Rossetti (2007) reviewed the molecular systematics of 18 Amazonian avian lineages and found that half diversified predominantly during the Pliocene. Lastly, of the 1400+ Neotropical forest species whose origins have been studied through molecular phylogenies, 49% were estimated to have diversified in the Quaternary. The other half is linked to a Neogene origin, or earlier still (Rull 2011). These time estimates are not surprising: at a global scale, the Neogene was a period in which much of the configuration of the modern world was formed (Potter and Szatmari 2009; Hoorn et al. 2010; Rull 2011).

In this contribution, I review some of the hypotheses and processes associated with the origins of diversity within a particularly relevant and well-studied group of tropical and subtropical forest animals of South America: birds. I specifically focus on patterns of avifauna diversity, and their underlying processes, in four focal forest areas: Amazonia, the Pacific Forest, the forests of the Eastern Andean Slopes, and the Atlantic Forest.

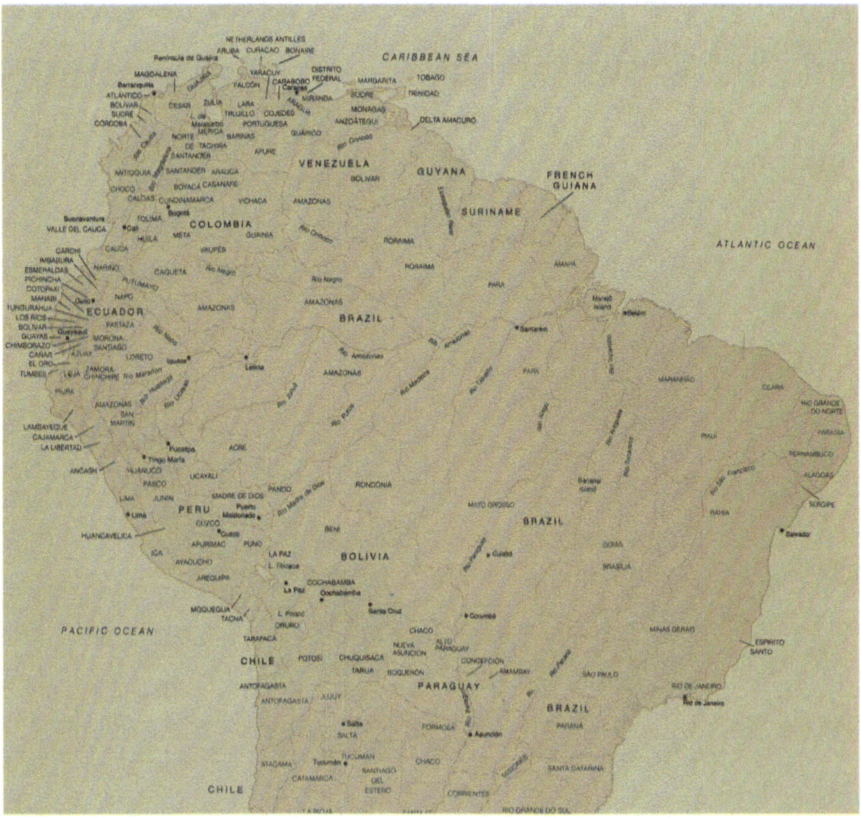

Fig. 8.1 Map of tropical and subtropical South America (adapted from Ridgely and Tudor 1989)

For each focal area, I illustrate general patterns of bird diversity by superimposing species range information available in the Literature and data repositories (Davis 1972; Ridgely 1976; Ridgely and Tudor 1989, 1994; del Hoyo et al. 1992–2013; Hilty and Brown 1986; Ridgely and Greenfield 2001; Hilty 2003; Stotz et al. 1996; Rodríguez-Mata et al. 2006; Schulenberg et al. 2007; https://neotropical.birds.cornell.edu and https://www.xeno-canto.org), as well as my own records, per domain (Müller 1973; Haffer 1974). Although this approach is a simplistic way to analyze the distribution of species in an area, particular given the rise of molecular studies (Moritz et al. 2000; Rull 2006), it can be helpful to locate centers of origin and dispersal, especially in habitats with latitudinal arrangement, such as the Atlantic Forest, the Pacific Forest and the forest of the eastern slope of the Andes. Another of the advantages of the method is that it allows the analysis of all species of a given biome. To avoid deficient sampling and problems related to modification of habitats, I assumed that the presence of at least two endemic species is necessary to consider a given region as an area of endemism (Silva et al. 2004). Nomenclature and order follow Remsen et al. (2018).

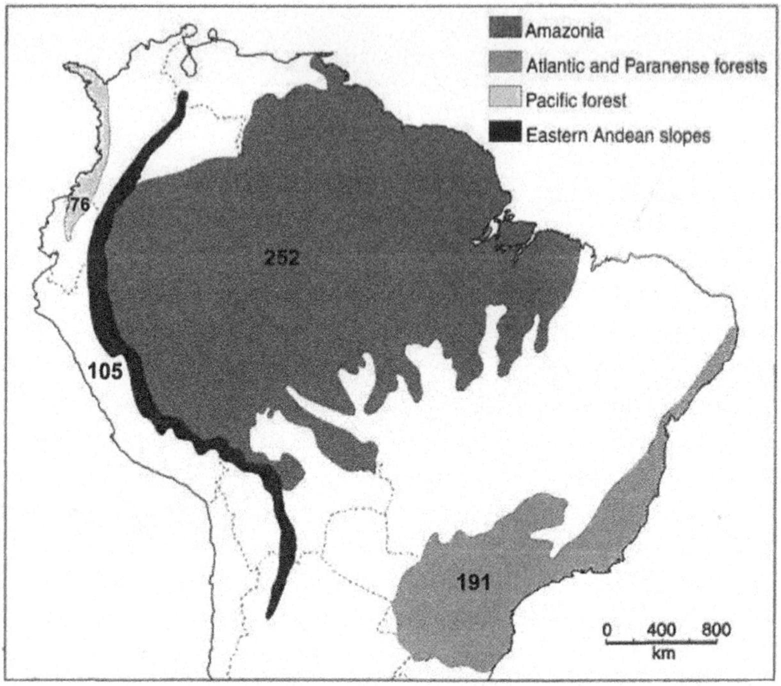

Fig. 8.2 Sketch map of the main biomes of the humid tropics and subtropics of South America, with numbers of endemic bird species of each of them

Together, these datasets indicate that about 1014 species (323 non-passerines and 691 passerines) are endemic to the tropical and subtropical forests of South America. Of them, approximately 252 are endemic to Amazonia, ca.191 are endemic to the Atlantic forest (of which 19 occur from 400/500 m upwards), ca. 105 are endemic to the forest of Eastern Andean Slopes of the Andes (between 500 and 2500 m), and about 76 species are endemic to the Pacific Forest of western Colombia, Ecuador and eastern Panama (Chocó Center), of which 36 occur only from 400/500 m upwards (Fig. 8.2). A smaller number of species exist in the foothills and adjoining lowlands of the Western and Central Andes of Colombia (Nechí Center), in the Tepuis, in the Santa Marta Mts., in the Venezuelan Coastal Range, and in the Paria Peninsula. The remaining endemic species occur in more than a domain.

I then review five geologic and climatic events that took place in the Neogene and Pleistocene, which are frequently referred to in discussions of the biogeography of tropical America: the uplift of the Andes, the formation of the Isthmus of Panama, glaciations, sea-level rise, and alternating periods of differing humidity. I discuss them under the light of available biological data, and relate them to the diversity patterns documented in each focal area. Lastly, I close my discussion with novel data and observations on the status and abundances of birds in the target areas.

2 Target Forests in Tropical and Subtropical South America

2.1 *Amazonia*

Amazonia constitutes the most extensive tract of tropical rainforest on the planet (Richards 1952), and is considered the richest ecosystem in the world (Colinvaux 1997). Insofar as birds are concerned, it the most important habitat in the Neotropical region (Amadon 1973; Haffer 1974, 1978). Haffer's (1978) definition of Amazonia encompasses not only the basin of the Amazon River but also that of the Orinoco River and the Guiana lowland forest along the Atlantic coast, which are continuous with the forest of the Amazon basin.

For the purpose of this chapter, I consider as Amazonian only those bird species that are restricted to the lowland that extends from the Atlantic coast to the foot of the Andes and of the Tepuis, up to about 500 m (Nores 2011), and also that of the Orinoco Basin and the Guiana lowland (Haffer 1978). The western boundary of this Amazonia definition is therefore an imaginary line that passes along the sites where the Andes rise abruptly out of the plain (Fig. 8.3). So defined, Amazonia covers an area of about 6 million km^2 and is in contact with the forest of the Eastern Andean Slopes for about 3000 km. Approximately 252 species of birds are strictly endemic to Amazonia, and 337 are not strictly endemic (this includes Amazonian species occurring at up to 700 m on the eastern slopes of the Andes, or that inhabit the lowlands up to about 500 m, but that are occasionally or locally recorded at higher altitudes; Electronic Supplementary Materials Appendix 1).

Amazonia extends for about 4000 km, from the Atlantic coast to the foot of the Andes. The slope is gentle and almost imperceptible from sea level to approximately 500 m altitude, but the eastern Andes rise abruptly out of the plain at about 500 m. At that point, the relief becomes very steep and there is a change in the vegetation, produced mainly by a cloud effect. Likely as a result of this, a large percentage of the 284 species of Amazonian birds that reach the foot of the Andes (88.35% in Colombia, 67.26% in Ecuador and 51.30% in Peru) do not extend more than 500 m up the Eastern Andean Slopes (Table 8.1). In the opposite direction, of 140 species of birds that inhabit the eastern slopes of the Andes, only 23 reach the foot of the Andes (500 m). Of these, 18 (78%) do not occur at an altitude lower than 500–400 m (Nores 2011).

2.2 *The Pacific Forest*

The Pacific Forest extends west of the Andes for approximately 1000 km, from the north of Colombia to the north of Ecuador, with some species reaching the east of Panama and northwestern Peru (Chapman 1917, 1926; Haffer 1967; Cracraft 1985).

Fig. 8.3 The suggested western boundary of Amazonia (from Nores 2011)

Table 8.1 Altitudes reached by bird species inhabiting western Amazonia (from Nores 2011)

Altitude (m)	Colombia (%)	Ecuador (%)	Peru (%)
<500–600	167 (88.35)	113 (67.26)	138 (51.30)
Between 700 and 1000	6 (2.07)	35 (20.83)	68 (25.28)
>1000	16 (10.58)	20 (11.90)	63 (23.42)

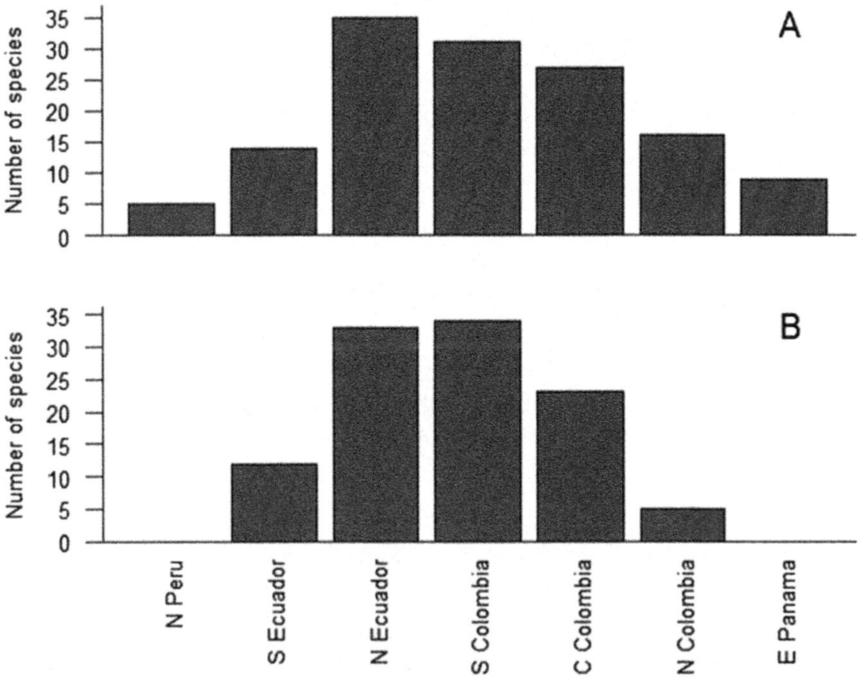

Fig. 8.4 Numbers of species of birds in the many regions of the Pacific Forest. (**a**) Based on superimposed ranges of 40 endemic species of the "Chocó Center". (**b**) Based on superimposed ranges of 36 endemic species of the "Pacific Slopes Center"

Its climate is tropical, and it is the most humid zone of South America, with up to 10,000 mm of rainfall per year. Analysis of the distribution of birds shows that ca. 76 species are endemic to this region (Electronic Supplementary Materials Appendix 2), of which 36 live above 400–500 m (Appendix 1). Although some elements of the low elevation zone blend into those of the upper zone, each has its unique species, and for this reason I have called them here the "Chocó Center" and the "Pacific Slopes Center", respectively. Superimposed ranges of the 40 species of the "Chocó Center" show that there is a peak in species richness in northern Ecuador and southern Colombia, with decreasing species richness to the south and to the north (Fig. 8.4a). A similar pattern is seen when the ranges of the 36 species living from 400/500 m upwards are superimposed (Fig. 8.4b). The decline of biodiversity

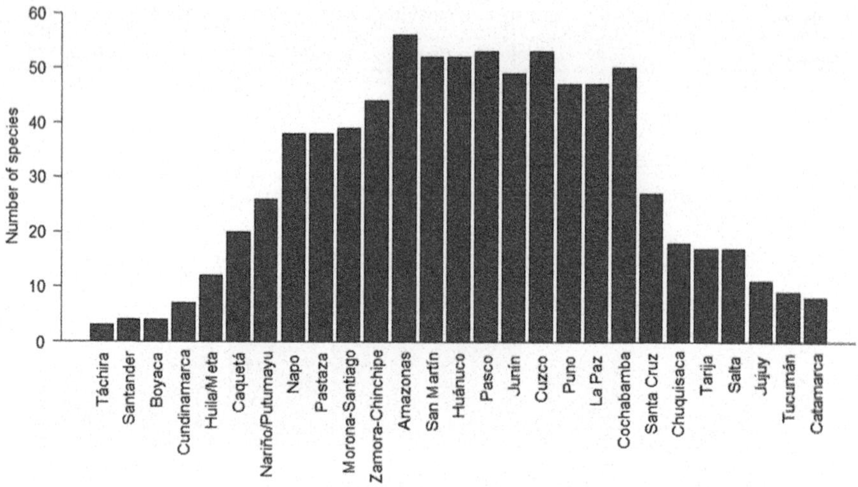

Fig. 8.5 Numbers of species of birds in the many regions of the Eastern Slopes of the Andes, based on superimposed ranges of 105 species endemic to the forest

with latitude is also evident here, both in the "Chocó and the Pacific Slopes centers". Bird species are concentrated around the Equator and decrease towards the poles.

2.3 *The Forest of the Eastern Andean Slopes*

The forest of the Eastern Andean Slopes extends for approximately 5000 km, from Táchira in Venezuela to Catamarca in Argentina, and from 500 m of elevation upwards (Nores 2011), covering a latitudinal range of about 40°. The austral sector in Bolivia and Argentina is known as the Yungas, and in their southernmost part (in Argentina), extends for the eastern slopes of the Sierras Pampeanas. Its climate is mostly tropical, with rainfall exceeding 3000 mm in eastern Ecuador and western Peru, and the dominant vegetation is composed of cloud and montane forests (Hueck and Seibert 1972; Cabrera and Willink 1973).

An analysis of the distribution of local birds shows that about 105 species are endemic to this biome (Electronic Supplementary Materials Appendix 3), occurring between 500 and 2500 m. Superimposed ranges of these species show that they are concentrated in Peru and northern Bolivia, decreasing gently to the north and more steeply to the south (Fig. 8.5). Seven species are endemic to northern Peru (Amazonas, Cajamarca and San Martín): *Xenoglaux loweryi, Picumnus steindachneri, Herpsilochmus parkeri, Grallaricula ochraceifrons, Henicorhina leucoptera, Thlypopsis inornata* and *Wetmorethraupis sterrhopteron*. Four are restricted to northern Bolivia (La Paz and Cochabamba): *Hemitriccus spodiops, Lipaugus uropygialis, Myrmotherula grisea* and *Stilpnia meyerdeschauenseei*. One species is endemic to southern Ecuador (Morona-Santiago and Zamora-Chinchipe):

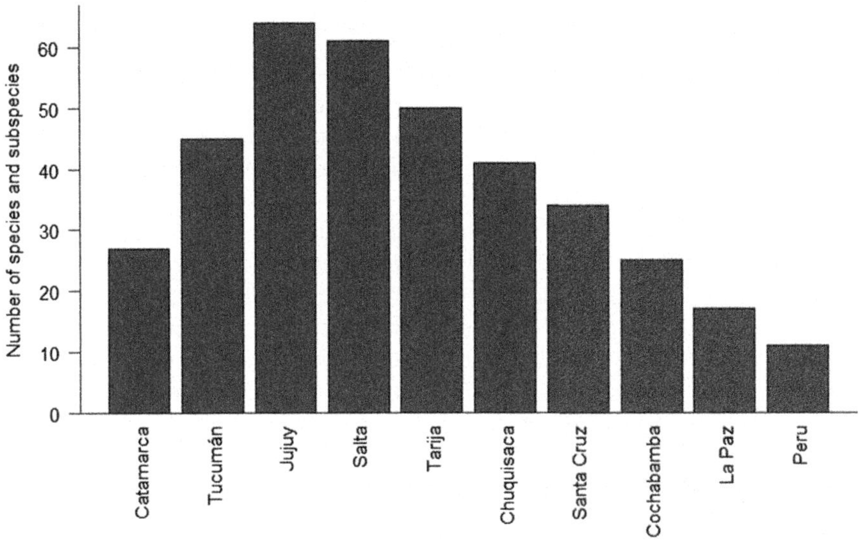

Fig. 8.6 Numbers of species and subspecies of birds in the many regions of the southern Andes (Yungas of northern Argentina and Bolivia), based on superimposed ranges of 65 endemic species and subspecies

Pyrrhura albipectus, one to Colombia (Cundinamarca): *Grallaria kaestneri,* one to southern Venezuela (Táchira): *Grallaria chthonia*; and three to northwestern Argentina: *Scytalopus superciliaris*, *Cinclus schulzii* and *Atlapetes citrinellus*. Following the criterion of two or more endemic species needed to recognize a center of endemism (Silva et al. 2004), one can flag Amazonas/Cajamarca/San Martin (Northern Peru Center), Cochabamba/La Paz (Northern Bolivia Center), and northwestern Argentina (Northwestern Argentina Center) as centers. A decline in bird richness with latitude is not so evident here. Bird species are concentrated south of the Equator and decrease not only to the south, but also towards the north.

Superimposed ranges of the 65 species and subspecies that inhabit the southern part of this biome (Yungas of Argentina and Bolivia) show that the greatest species richness is located in Jujuy Province, Argentina, decreasing steeply to the south and to the north (Fig. 8.6). This reinforces the idea that the Yungas of northwestern Argentina, more precisely the Calilegua mountain, Jujuy, can be considered a center of endemism, and add three species to this center: *Penelope dabbenei*, *Amazona tucumana* and *Phacellodomos maculipectus* and 35 subspecies.

2.4 The Atlantic Forest

The Atlantic Forest Domain (Ab'Saber 1977) extends along 3000 km and covers a latitudinal range of 25°, from the northeastern to the southern regions of Brazil, from

Rio Grande do Norte to Rio Grande do Sul. Numerous species of birds extending west into the Paranaense Forest of northeastern Argentina and eastern Paraguay. The northern section occupies a thin coastal strip, while in the south it extends from the coast to as far as 400 km inland. The band narrows considerably at the latitude of São Francisco River, where it is 25 km wide (up to 12 km upstream of the township of Penedo) and then expands again, being 50 km wide at the latitude of Aracaju in the state of Sergipe (pers. obs).

Most studies agree that there is a historical separation between the northern and southern parts of the Atlantic Forest, whose limit more or less coincide with the Doce River Valley (northern Espírito Santo State; Finschii and Pirani 2009). Silva et al. (2004) identified four areas of endemism for passerine birds in the Atlantic Forest: Pernambuco, Central Bahia, Coastal Bahia and Serra do Mar.

About 191 species of birds are endemic to this forest (Electronic Supplementary Materials Appendix 4), 19 of which occur only from 400 m and upwards (Appendix 2), in the area of endemism refered to by Stattersfield et al. (1998) as the "Atlantic Forest Mountains". Superimposed ranges of the 172 remaining species of the "Atlantic Forest Lowlands" show that they concentrate in Espírito Santo, Rio de Janeiro and São Paulo, decreasing gently to the south and more steeply to the north (Fig. 8.7a). A similar pattern is seen when the ranges of the 19 species living above 400 m are superimposed (Fig. 8.7b). A remarkable fact is that the northern states of Alagoas and Pernambuco, despite having a low number of species, have eight endemic species and 21 endemic subspecies (forming the "Pernambuco Center"; Pinto 1940, 1978; Silveira et al. 2003). The concentration of species in Espírito Santo, Rio do Janeiro, São Paulo and east of Minas Gerais seems to be mainly related to habitat quality, since to the south the forest becomes subtropical (the tropic of Capricorn crosses through the center of São Paulo) and to the north it narrows from the center of the state of Bahia. The wide tropical zone is at the southern part of the Atlantic Forest (Finschii and Pirani 2009; Sobral-Souza et al. 2015). However, the south of Bahia, which is wide and is in the tropical zone, experiences an unexpected marked decrease in species richness (Fig. 8.7a). The decline of richness of birds with latitude is evident here to the south, but a reverse latitudinal gradient is produced to the north, where species richness decreases towards the Equator, maybe because of the narrowing of the forest band.

Several Atlantic Forest birds have disjunct forest distributions, and are found in other domains. Seventy-nine species of birds are shared between the Amazonian and Atlantic forests, for instance. The superimposed ranges of these species in the Atlantic Forest show a quite different pattern (Fig. 8.7c) from those of the endemic Atlantic Forest birds (Fig. 8.7a, b). Bahia and Alagoas show a large number of species, while Rio de Janeiro, São Paulo and the other southern states lost importance. Disjunctions are also seen across other forest systems. One example is that observed between the southern Yungas and the Paranaense/Atlantic forests (Nores 1992; Fig. 8.8), which is discussed below.

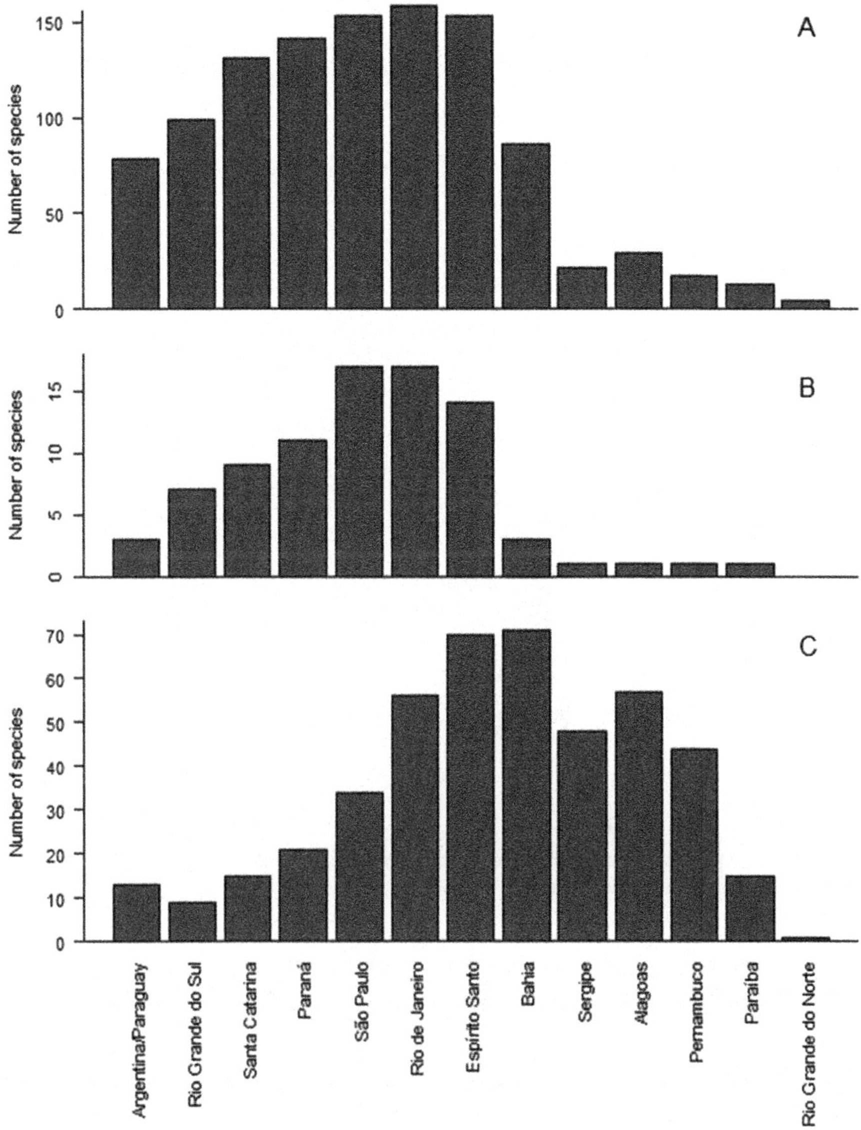

Fig. 8.7 Numbers of species of birds in the many regions of the Atlantic Forest. (**a**) Superimposed ranges of 172 endemic species of the "Atlantic Forest Lowlands Center". (**b**) superimposed ranges of 19 endemic species of the "Atlantic Forest Mountains Center". (**c**) superimposed ranges of 79 species disjunctly distributed between the Atlantic Forest and Amazonia

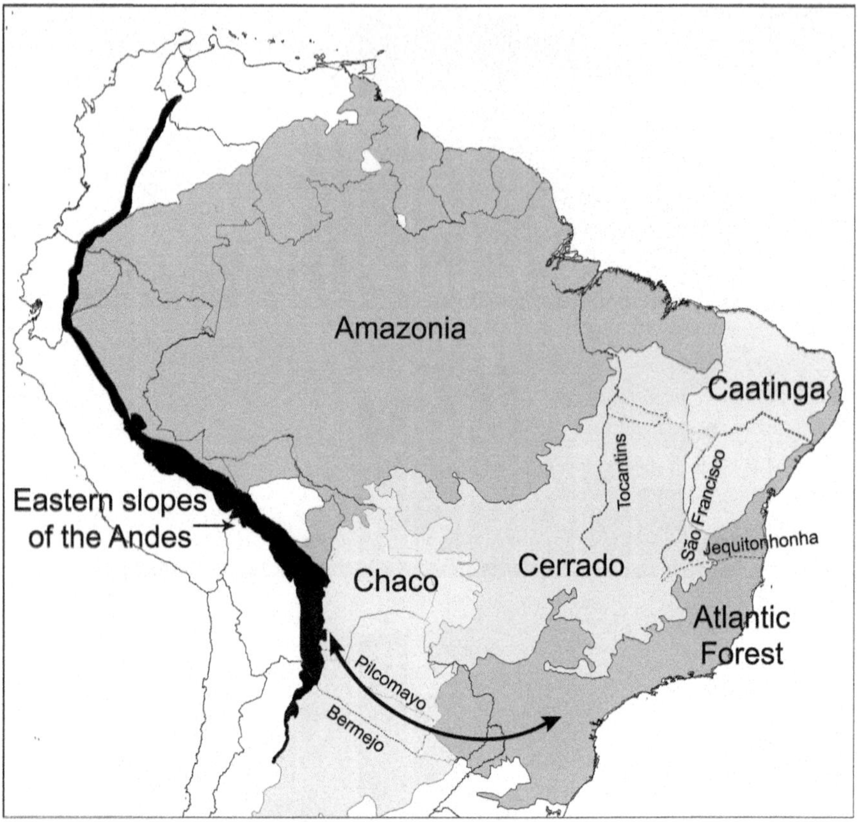

Fig. 8.8 Hypotetical connections between the Atlantic Forest and Amazonia and between the Atlantic Forest and the Andean forest

3 Main Historical Processes Impacting Diversity in Tropical and Subtropical South America

To discuss the relative roles of the many processes tied to the diversification of the South American avifauna, I review them below.

3.1 The Andean Uplift

The Andes make up the largest orogenic system developed by the subduction of an oceanic crust along a continental margin (Ramos 2009). Although the Andean uplift was generally assumed to reflect a gradual rise, recent studies indicate that their elevation remained relatively stable for about tens of millions of years, followed by rapid (1–4 million years) uplift of 1.5 km or more (Garzione et al. 2008). The first uplift of the Eastern and Western cordilleras in Colombia started in the Paleocene, approximately 60 Ma (Hammen 1961; Simpson 1975). Intensified mountain building in the northern Andes occurred at about 23 Ma, yet the most intense uplift happened in the middle Miocene and early Pliocene (12–4.5 Ma; Potter and Szatmari 2009; Mora et al. 2010). Present-day altitudes were reached ca. 2.7 Ma (Gregory-Wodzicki 2000; Garzione et al. 2008).

These movements had profound consequences on the geography of the remaining South American continent (Antonelli et al. 2009; Antonelli and Sanmartín 2011). For one, they changed the original course of the Amazon system: instead of flowing northwestwards, it began to flow to the Atlantic side, as observed today (Hoorn 1993, 1997). The Andean uplift also had climatological implications, by forming the only barrier to atmospheric circulation in the Southern Hemisphere (Gregory-Wodzicki 2000). It is also believed to have played an important role in the diversification of Neotropical organisms by creating new high elevation habitats, and by isolating lowland organisms on either side of the mountains (Brumfield and Edwards 2007).

Prior to Andean influence, the area known today as Amazonia was thought to have been part of a much larger "pan-Amazonian" region, which, before the late Miocene (until 10 Ma), included the area of the present Amazon, Orinoco, and Magdalena drainage basins (Hoorn et al. 2010). It has been proposed that the coastal lowland of Colombia and Ecuador was submerged before the uplift of the Andes (Harrington 1962; Mégard 1992), but that the sea withdrew from most of that area in the Paleocene. The uplift of the Andes, concomitant with the emergence of forest cover on the emergent land, is believed to have resulted in several trans-Andean populations of Amazonian origin (Chapman 1917; Brumfield and Capparella 1996).

The Andean Uplift and Its Perceived Impact on Bird Diversification Then and Now

The role of the Andean uplift as a driver of avian diversification has been an issue of debate. Generally, the Andes have been seen as promoters of a build-up of biodiversity in lowland Neotropical faunas through both vicariance-based speciation (during uplift) and dispersal-based speciation (following uplift; Weir and Price 2011). Using bird distribution data from Colombia and Ecuador, Chapman (1917) suggested that the initial uplift of the Andes led to the early divergence of some trans-Andean birds (a.k.a the Andean Uplift hypothesis). This initial separation process of the trans-Andean populations of birds was proposed to have begun when the northern Andes had an altitude a little over 1000 m (probably in the middle Miocene ~15 Ma, Gregory-Wodzicki 2000), representing a barrier for some of the lowland taxa. Given data on bird distribution patterns and chronological data, Nores (2004) suggested that endemic bird genera of the Pacific rainforest (*Androdon, Damophila, Xenornis, Sipia, Pittasoma, Sapayoa, Carpodectes, Aphanotriccus, Chrysothlypis*), none of which occur, presently, at elevations higher than 1100–1200 m, may have originated from Amazonian species that were isolated early by the rising of the Andes. The last uplift of the Andes, between 5 and

2.7 Ma (Gregory-Wodzicki 2000), would have separated populations of lowland species that also occur higher altitudes in the Andean slopes.

Yet, molecular evidence suggests a more complex process. Using protein electrophoretic data from 14 taxonomically diverse species groups of birds, for instance, Brumfield and Capparella (1996) showed that some taxa such as *Microrhopias quixensis* and *Tityra semifasciata* exhibit little across-Andes genetic divergence, supporting gene flow despite the final uplift of the Eastern Andean Cordillera approximately 2–3 Ma. More recently, genomic-scale data analyzed by Smith et al. (2014) identified multiple (9–29) divergence events across the Andes, dating to the Pleistocene. Those authors proposed that the Andean uplift did not have a direct effect on diversification via synchronic vicariance across groups, but rather played an indirect role in divergence by acting as a semi-permeable barrier to post-uplift dispersal.

3.2 The Formation of the Isthmus of Panama

Another geological process frequently brought up in discussions of avian diversification is the formation of the Isthmus of Panama. During the Paleocene, a wide gap is thought to have separated South America from North America and part of Central America—a gap that is now occupied by Nicaragua, Costa Rica and Panama (Malfait and Dinkelman 1972). Geological and benthic foraminifera data indicate that a volcanic archipelago initiated the filling of this gap, establishing itself from the continental area southwards to Central Panama in the middle Miocene (~15 Ma; Coates and Obando 1996; Collins 1996; Collins et al. 1996). Benthic foraminipher data suggest that the isthmus was already formed by the end of the Miocene, when only three main marine corridors connected the Pacific and Caribbean waters: the San Carlos basin (between northern Costa Rica and southern Nicaragua), the Panama Canal Basin, and the Atrato basin in northwestern Colombia (Collins et al. 1996). Based on geological, coral reef, and benthic foraminifer information, the final closure of the Isthmus is thought to have occurred in the middle Pliocene, ~3.5 Ma (Collins 1996, 1999; Keigwin 1982; Coates et al. 1992; Hartley 2003; Johnson and Weckstein 2011; Leigh et al. 2013).

Yet these dates and processes have been contested, particularly in the light of biological analyses based on molecular dating. Smith and Clicka (2010), for example, compiled 64 avian phylogenetic studies and applied a relaxed molecular clock to estimate the timing of trans-isthmus diversification events. They reported that 24% of 135 trans-Isthmus diversification events identified between North and South Americas occurred prior to 4 Ma. Other, new geological and chronological data suggest that the Central America seaway disappeared by about 15 Ma (Montes et al. 2012; Coates et al. 2013). Bacon et al. (2015), based on molecular phylogenetic work, found significant waves of dispersal of terrestrial organisms at approximately ca. 20 and 6 Ma. Yet, based on a reanalysis of geological, paleontological and molecular records, O'Dea et al. (2016) suggested that the formation of the Isthmus of

Panama would have occurred around 2.8 Ma, stating that the evidence used to support an older isthmus is inconclusive. Their dates have been nonetheless been contested by Erkens and Hoorn (2016).

Regardless of the date, there is consensus among researchers regarding the impact of the final closing of the Isthmus: it enabled a considerable interchange of the (previously isolated) biota of these continents, a phenomenon referred to as the "Great American Biotic Interchange" (GABI; Marshall 1988; Marshall et al. 1982; Vuilleumier 1985; Webb 1991; Weir et al. 2009; Smith and Clicka 2010; Johnson and Weckstein 2011; Leigh et al. 2013). Although frequently referred to in biogeographical studies of animals, its impact on plant species has been debated. Cody et al. (2010), for instance, suggested that perhaps the closure of Isthmus may have been less critical for plant migration given that plants dispersed more effectively than animals. Based on genetic evidence, Smith et al. (2014) argued that avian diversification across the Isthmus of Panama were not synchronous, suggesting instead 1–7 distinct divergence events.

The Isthmus and Its Perceived Impact on Bird Dispersal and Diversification The closing of the Isthmus is expected to have impacted the distribution and diversification of many tropical birds (Vuilleumier 1985; Weir et al. 2009; Smith and Clicka 2010). However, because of the incompleteness of the fossil record, and the ability of many bird species to fly over water barriers (Weir et al. 2009), it is difficult to discuss the role of the isthmus for this group of animals.

To assess the role of the Isthmus in the dispersal of bird species from North and Central America to the then newly emerged zone and South America, I compiled information about the distribution of non-migrant species that have poor ability to disperse over the sea, and which are present from Honduras (that is, the southernmost part of Central America before the formation of the Isthmus) southwards. For that, I built a species list using geographical range information provided in Davis (1972), Ridgely (1976), Ridgely and Tudor (1989, 1994), del Hoyo et al. (1992–2013), Neotropical Birds (Cornell Lab), and Xenocanto, and determined the number of species that occur in Honduras, Nicaragua, Costa Rica, western Panama, eastern Panama and Colombia.

Of the ca. 200 species that occur in Honduras, 171 (85.5%) also occur in Nicaragua; 131 (66.5%) in Costa Rica; 85 (42.5%) in western Panama; 53 (26.5%) eastern Panama and 44 (22%) arrive to South America, in the Pacific Forest, in northwestern Colombia Fig. 8.9). These percentages are similar between non-passerines and passerines, and involve 13 and 15 families, respectively. In addition, there are 71 species endemic to the Isthmus, especially in Costa Rica and western Panama (Chiriquí Center, Brown 1975), which may have differentiated *in loco*, after the formation. When focusing south of the Isthmus, the number of species seemingly crossing north is relatively smaller. Of about 258 species present in northwestern Colombia, 165 (64%) also occur in eastern Panama, 149 (57.7%) in western Panama; 92 (35.6%) in Costa Rica; 30 (11.6 %) in Nicaragua and only 17 (6.6%) are found in Honduras (Fig. 8.9). However, while one may argue that these data are consistent with a hypothesis that the direction of traffic of birds across

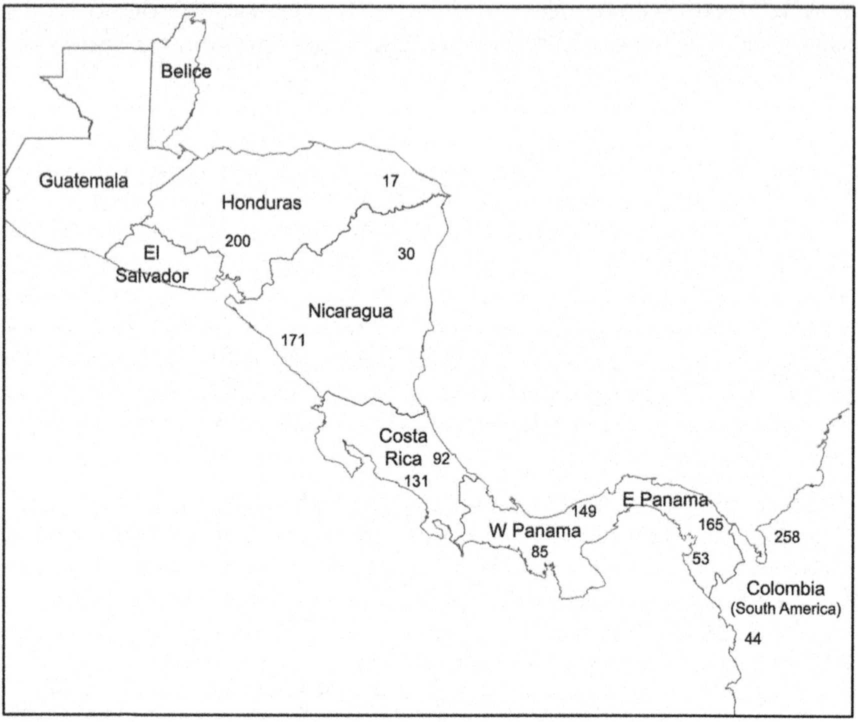

Fig. 8.9 Number of birds species shared between Honduras and countries to the south (placed on the western part of each country), and those shared between Colombia and countries to the north (placed on the eastern part of each country)

the isthmus was principally north to south, molecular data and phylogenetic analyses are needed to test them appropriately. In fact, available molecular data published to date suggest that the direction of traffic of birds across the Isthmus was primarily south to north (Weir et al. 2009).

There are also about 200 non-migrant species that have poor ability to disperse over the sea, and which have continuous distribution from South America to north of Central America or Mexico (Davis 1972; Ridgely 1976; Ridgely and Tudor 1989, 1994; del Hoyo et al. 1992–2013; Neotropical Birds (Cornell Lab) and Xenocanto). These species probably dispersed via the Isthmus. Together, the range data suggest that about 535 species of birds may have used the Isthmus for dispersal, of which c. 260 would have participated in the "Great American Biotic Interchange", i.e. dispersed all the way from Honduras to South America or the other way round.

3.3 Glaciations

Four to six glacial periods of about 100,000 years each, separated by interglacial periods of 100,000–300,000 years, are generally recognized worldwide (Flint 1957; Ericson et al. 1964; Ericson and Wolling 1968; Pielou 1979; Gibbard and van Kolfschoten 2005; Balco and Rovey 2010; Bacon et al. 2015). The most direct effect of these glaciations was ice accumulation; wherever the ground was covered by ice, it is assumed that all (or most) forms of life was momentary lost (Pielou 1979). Geological and biological data indicate that glaciations gave rise to glacial tongues that isolated animal populations, compressed and lowered life zones in the mountains, led to the formation of glacial lakes and climatic cooling, lowered sea level, decrease in rainfall, and reduced carbon dioxide levels (Flint 1957; Haffer 1969, 1974; Pielou 1979; Colinvaux 1993; Mayle et al. 2004).

The impact of these glacial periods in the tropics has been a matter of debate. It was generally believed that the tropics have been characterized by stable long-term temperatures, and that the tropical oceans and adjacent land areas were only 2 °C (or less) cooler than now (Climap 1976; Gates 1976). Yet more recent studies suggest that the tropics have experienced decrease in temperature similar to those occurring outside the tropics (Liu and Colinvaux 1985; Webb et al. 1997; Bush et al. 2011; Vonhof and Kaandorp 2010). Several authors agree that the Amazon Basin was ca. 5 °C cooler than today at the last glacial maximum (Stute et al. 1995; Mayle et al. 2004; Bush et al. 2011; Wang et al. 2017), that the carbon dioxide level was concurrently reduced, and that the relative abundances and distribution of many species (particularly plants) were shifted in the ice-age Amazon (Colinvaux 1997; Mayle et al. 2004). Colinvaux et al. (1996) indicated that glacial age forests were comparable to modern forests, but also included species now restricted to higher elevations by temperature. However, Pennington et al. (2000) argued that pollen spectra from Amazonia, previously interpreted as a continuous rainforest signal with intrusion of some Andean elements, could be interpreted as a predominantly seasonally dry forest signal.

The effect of glaciations on birds would have been mostly indirect, and have been related to periods of aridity (Haffer 1969, 1974; Vanzolini and Williams 1970; Prance 1973) or to the combination of reduced temperatures, precipitation and atmospheric CO2 concentrations (Colinavaux 1993), that would have taken place during the glacial maximum.

The Glaciations and Their Perceived Impact on Bird Dispersal and Diversification

I hypothesize that the glaciations particularly impacted the distribution and diversification of the avifauna in the southern part of the Eastern Andean Slopes. Six species (*Penelope dabbenei, Amazona tucumana, Phacellodomus maculipectus, Scytalopus superciliaris*, *Cinclus schulzii* and *Atlapetes citrinellus*) and 35 subspecies are endemic to the Northwestern Argentina Center, in this forest. It is not clear what processes led to the formation of this center: one possibility may be that the Calilegua Mountains were once a climatic refugium. The Yungas of northwestern

Argentina and southern Bolivia are much less wet than the forests of the northern Yungas and that the rest of the Eastern Andean Slopes, and may have been eliminated with moderate decrease in rainfall during the glacial maximum. However, the Calilegua Mountains, which are now apparently more humid and subjected to a longer rainy season relative to their surrounding areas, may have remained wet over time.

3.4 Sea-Level Rise

Sea-level rise has been associated with bird diversification in Amazonia and the Pacific Forest (Nores 1999, 2004). While sea-level changes undoubtedly occurred in the past (Flint 1957; Fairbridge 1961; Tanner 1968; Hallam 1971; Hays and Pitman 1973; Haq et al. 1987; Hoorn 1993; Räsänen et al. 1995; Kerr 1996; Naish 1997; Lovejoy et al. 2006; Dwyer and Chandler 2009; Rohling et al. 2009; Bush et al. 2011; Raymo et al. 2011; Hannah 2015), the levels reached by the sea, and the timing of high sea-level events, are a matter of debate. As pointed out by Lambeck et al. (2002), amplitudes of sea-level rise during the early interglacials are difficult to understand because several of the sites where records are preserved were also subject to tectonic uplift.

Independent sources of data suggest that sea levels reached ca. 100 m on some occasions. An episode of sea-level rise of about 100 m during the late Tertiary is shown in the eustatic curves of Haq et al. (1987), and it is corroborated by analyses of oxygen-isotope composition of carbonate sediments (Kerr 1996). This event is thought to have had an estimated duration of ~800,000 year, and would have occurred in the lower Pliocene (~5 Ma) or (more probably) in the middle Pliocene (~3 Ma), when the Earth was significantly warmer than today, and Antactica experimented maximum deglatiation (Barrett et al. 1992; Barron 1996; Dowsett et al. 1996; Ashworth and Kuschel 2003; Ashworth and Preece 2003; Ashworth and Thompson 2003). Bush et al. (2011) argued that the rise of the Andes caused forebasin subsidence in western Amazonia in the Miocene or Pliocene. Although the sea-level was inferred to be only 30–60 m higher than at present, the net effect of the subsidence was thought to correspond to ca. 100 m rise in relative sea-level. Earlier studies also indicated sea level increases of ~100 m in the Tertiary (Zeuner 1945; Fittkau 1974).

Reports of high sea levels also exist for the Late Pliocene and Pleistocene: Kaufman and Brigham-Grette (1993) recognized five high sea level events in Nome coastal plain, Alaska. According to Raymo et al. (2011), those levels would have risen up to 60 m above present. Several marine features and beach deposits occurring along many coasts of the world, also support the hypothesis of sea level rise of about 100 m during the Quaternary (Flint's 1957; Fairbridge 1961; Ward and Jessup 1965; Klammer 1971; Idnurm and Cook 1980). That being said, other studies have also indicated lower levels (Dwyer and Chandler 2009; Rohling et al. 2009; Raymo et al. 2011, 2018; Dutton et al. 2015).

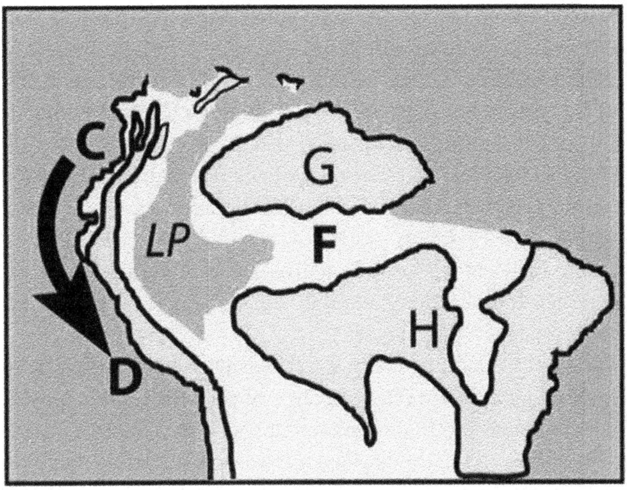

Fig. 8.10 Middle Miocene paleogeographic map of South America, showing Lake Pebas (LP) in Western Amazonia connected to the Caribbean marine incursion. G: Guiana Shield; H: Southeastern South America; F: Amazonia; C: Northern Andes; D: Central Andes (from Antonelli et al. 2009)

While studying the Pebas formation in Western Amazonia, Räsänen et al. (1995) suggested the existence of late Miocene marine incursions not only from the Atlantic Ocean (via the Amazon River), but also from the Caribbean and from the present-day Paraná drainage. However, they were unclear on whether the Pebasian and Paranaense embayment were synchronous or were ever interconnected. Recent studies indicate that Lake Pebas, which extended up to 1,000,000 km^2 (Fig. 8.10) and persisted for c. 17 million years until 7 Ma, would have been connected to the Caribbean Ocean and possibly to the Pacific Ocean through the Cuenca basin, but not to the Amazon or Paraná Rivers (Wesselingh et al. 2002; Vonhof et al. 2003; Wesselingh and Salo 2006; Hoorn et al. 2010; Fig. 8.10).

Studies of continental ice coverage, and melting, are particularly relevant to inferences of sea level rise. The Antarctic Ice Sheet contains 25–30 $\times$ 10^6 km^3 of ice (Drewry et al. 1982; Vaughan 2000; Ingólfsson 2004), which is potentially equivalent to 57–66 m of sea-level (Denton et al. 1991; Vaughan 2000); the melting of all ice in Antarctica and Greenland would raise sea level approximately 75 m (Hannah 2015). However, there is significant debate about the extent of the Antarctic ice sheet during the period of Pliocene warmth. Some argue that extensive ice covered the continent in the middle Miocene (~14 Ma), persisting more or less continually to the present (Kennett and Shackleton 1976; Burckle et al. 1996; Marchant and Denton 1996; Raymo et al. 2011).

Other studies, however, have suggested that Antarctica was deglaciated in the middle Pliocene (~3 Ma). In interior East Antactica, for instance, marine diatoms found in glacial deposits (Barron 1996), tied to the presence of *Nothofagus* trees, suggest that the middle Pliocene climatic conditions were significantly warmer than those of today, and that the region must have been ice-free and below sea-level (Barrett et al. 1992; Dowsett et al. 1996).

Antartic biogeographic sudies carried out by Ashworth and colleagues in the Meyer Desert Formation have also allowed to found a fossil of a higher fly (Diptera: Cyclorrhapha; Ashworth and Thompson 2003), freshwater molluscs (Ashworth and Preece 2003), a fish species, listroderine weevils (Curculionidae), and other insects inhabiting a shrub and herb tundra with cushion plants, *Ranunculus*, and patches of dwarf *Nothofagus* and *Podocarpus* (Ashworth and Kuschel 2003). According to these authors, the landscape was sufficiently stable to contain a lake in the middle Pliocene, being ice-free for long enough during the summers to support algae, freshwater molluscs and a fish species.

Raymo et al. (2011) estimated a middle Pliocene worldwide sea-level up to 40 m above present, arguing that an eustatic change of such magnitude must have required complete deglaciation of the West Antarctica and Greenland ice sheets, and significant loss of ice in East Antarctica. Given evidence that East Antarctica (which is twice the size of West Antarctica) suffered extensive deglaciation in the mid-Pliocene (Barrett et al. 1992; Barron 1996; Dowsett et al. 1996; Ashworth and Kuschel 2003; Ashworth and Preece 2003; Ashworth and Thompson 2003), it seems reasonable to infer that sea-level must have been considerably higher than 40 m in the middle Pliocene. Present-day Antarctic ice would have established in the late Pliocene, about the same time as the first northern ice sheets developed (Shackleton and Opdyke 1977; Barrett et al. 1992).

Proposed Impacts of Sea-Level Rise in Avian Diversification

Although the Amazon's high diversity was likely a consequence of several factors that operated during the Cenozoic (Haffer 1967; Cracraft and Prum 1988; Bush 1994; Brumfield and Capparella 1996; Rull 2011; Brumfield 2012), I have proposed that marine transgressions during the upper Tertiary and lower Quaternary were probably the landscape changes with most impact on the Amazon biota (Nores 1999, 2000, 2013). Under a scenario of sea-level rise of about 100 m, the Amazonian forest was likely fragmented into numerous true islands and archipelagos (Fig. 8.11), isolating and differentiating populations of birds and other animals and plants. Webb (1995) and Bates (2001), based on inferences of late Miocene marine incursions suggested by Räsänen et al. (1995), proposed that the Amazonian forests were split into three parts: (1) the Guiana Shield, (2) the Brazilian Shield (Belém and Pará), and (3) along the base of the eastern slope of the Andes (Inambari and Napo). However, this hypothesis does not seem feasible because the connection of the Pebas Lake with the Amazon and Paraná rivers would not have existed by then (Wesselingh et al. 2002; Vonhof et al. 2003; Wesselingh and Salo 2006; Lovejoy et al. 1998; Antonelli et al. 2009; Fig. 8.10).

I also believe that marine transgressions impacted the avifauna of other Neotropical forests. For instance, I have proposed that marine transgressions during the upper Tertiary and lower Quaternary may have isolated the Chocó Center of the Pacific forest, leaving a large rainforest area in the north of Ecuador and in the south of Colombia, and a narrow strip in the center and north of Colombia (Nores 2004;

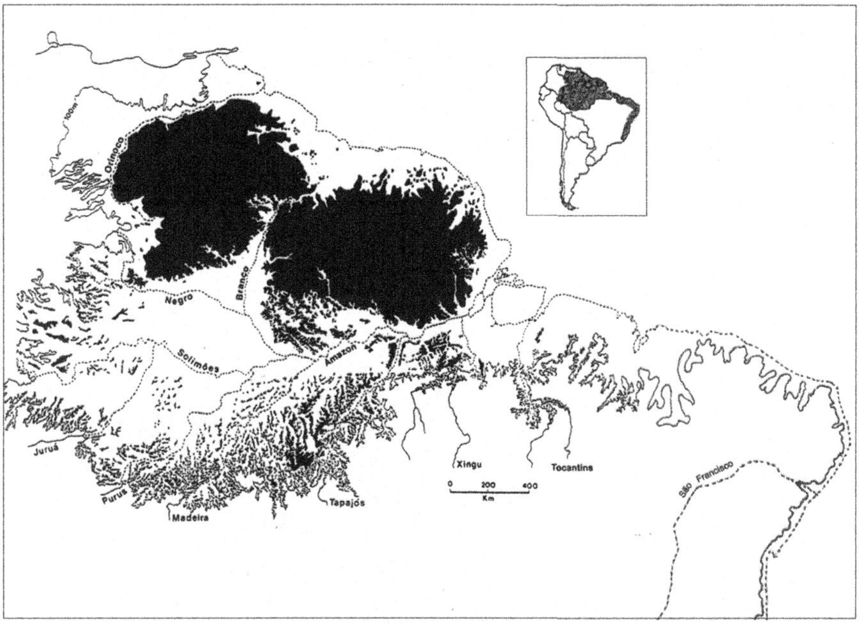

Fig. 8.11 Proposed location of islands and archipelagos (in black) in Amazonian lowland during a sea-level rise of 100 m and the marine transgression from the Atlantic Ocean via the São Francisco River (modified from Nores 1999)

Fig. 8.12). It remains to be tested whether this may have resulted in the current concentration of species in the south of the "Chocó Center". In the Atlantic forest, I wonder if the existence of eight species endemic in Alagoas and Pernambuco (Pinto 1940, 1978; Silveira et al. 2003) may reflect sea-level rise during the middle Pliocene, which would have produced a marine transgression from the Atlantic Ocean via the São Francisco River (Fig. 8.11). This would have interrupted the coastal band and isolated populations of these species north of the transgression. As the band is 25 km wide in this region, and a maximum altitude of 25–30 m, moderate sea-level rises during the Quaternary would have been enough to interrupt it and promote differentiation of the endemic subspecies.

3.5 *Alternating Arid and Humid Periods*

Several authors have suggested the occurrence of arid periods in tropical America during the glacial maximum (Hammen 1961, 1974; Van der Hammen and Absy 1994; Haffer 1969, 1974, 1978; Absy et al. 1991; Clapperton 1993; Hooghiemstra 1997; Hewitt 2000; Latrubesse 2002; D'Apolito et al. 2013). Van der Hammen and Absy (1994) suggested a 40% reduction of annual rainfall, which would have

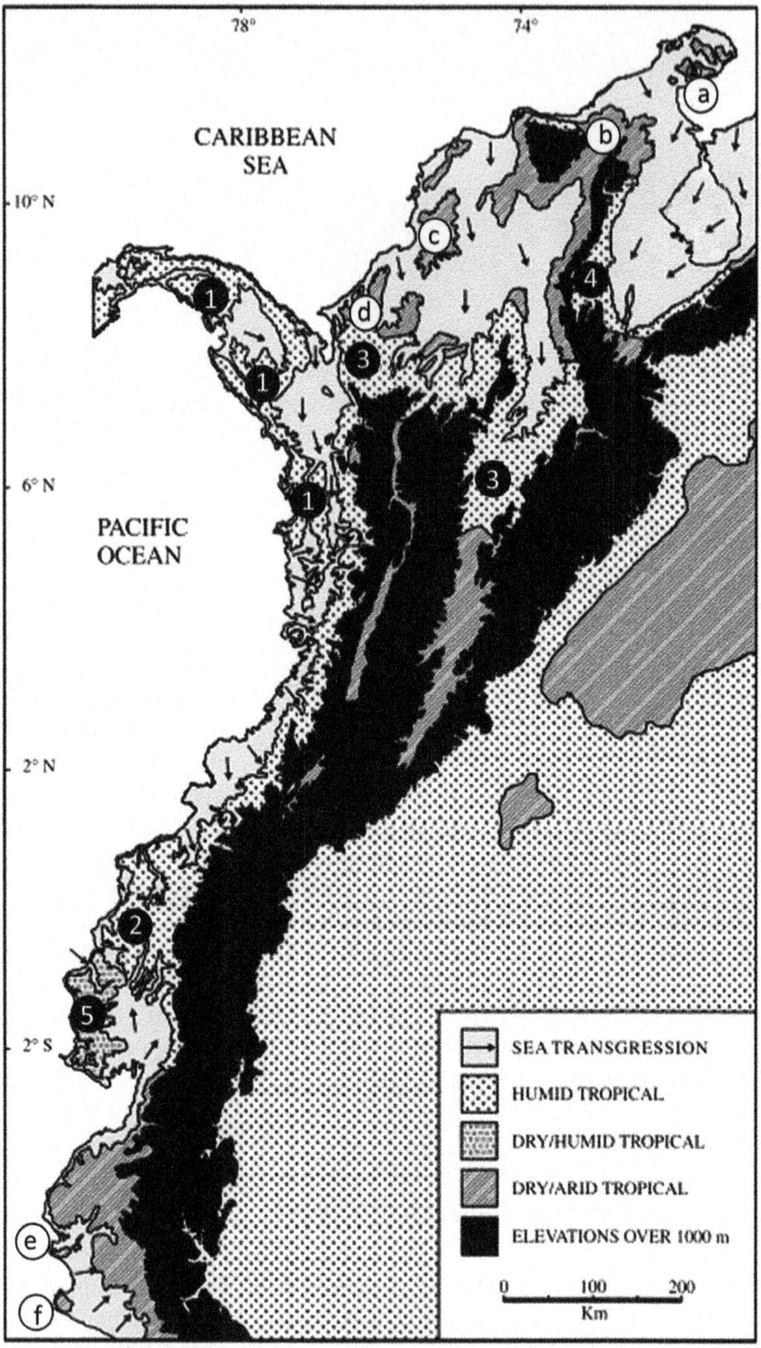

Fig. 8.12 Proposed distribution of land and sea in northwestern South America during a sea-level rise of about 100 m. Numbers indicate humid areas that would have been isolated by the sea: (1) Darién, (2) Chocó, (3) Nechí, (4) Catatumbo, (5) Tumbesian; letters indicate arid areas (from Nores 2004)

produced fragmentation of the rainforest into several blocks. Hewitt (2000) also noted that conditions during the last ice age were colder and drier in the tropics, extending deserts and savannah while reducing rainforest. Interestingly, however, Hewitt (2000) cited the work of Colinvaux and Bush (Colinvaux et al. 1996; Colinvaux 1997) to support that idea—but those authors are some of the main opponents of the hypothesis of Amazonian aridization during the glacial maximum.

Using speleothems from Paraiso Cave in eastern Amazonia, Wang et al. (2017) proposed that precipitation in the region, during the Last Glacial Maximum (around 21,000 years ago), was about 58% of modern levels—reaching 142% during the mid- Holocene (about 6000 years ago). They suggested that the Amazon was drier during the last glacial period, with much less recycling of water and probably reduced plant transpiration, although the rainforest persisted throughout this time. Cheng et al. (2013) pointed out that speleothem oxygen isotope records from the last glacial period imply a modest increase in rainfall in western Amazonia but a significant drying in eastern Amazonia, where forests may have been more susceptible to fragmentation. However, this magnitude of cyclic tropical aridity is controversial. Several authors argue that a reduction in Amazonian precipitation did occur during the glacial periods, but that it was likely smaller than 20% and hence insufficient to eliminate rainforest coverage from most of the Amazonian lowland (Colinvaux 1987, 1989; Bush 1994; Bush and de Oliveira 2006; Bush et al. 2011; Haberle 1997; Hoorn 1997; Hoorn et al. 2010; Piperno 1997; Colinvaux et al. 2000, 2001; Knapp and Mallet 2003; Rull 2015).

Alternating climatic conditions have also been associated with diversification processes in the Atlantic Forest: Carnaval et al. (2009), for instance, proposed that the coastal southern Atlantic forest was more climatically unstable relative to the central region, which may served as a large climatic refugium for lowland species in the late Pleistocene.

It is also possible that an alternation between dry and humid peridos have contributed to the diversity of the Pacific rainforest. For instance, it has been suggested that several bird species would have originated following the immigration of Amazonian elements towards the Pacific rainforest, through the Caribbean lowland (Haffer 1967). At present there are several species of forest birds (*Glaucis hirsutus, Myiopagis gaimardii, Tyrannulus elatus, Rhynchocyclus olivaceus, Piprites chloris, Ceratopipra erythrocephala, Cantorchilus leucotis, Microcerculus marginatus, Tersina viridis, Cacicus cela, Psarocolius decumanus, Sporophila angolensis*) that have continuous distribution from the Amazonia to the Pacific, through the Caribbean lowland. Because most of these species also occur in dry forest, the Caribbean lowland likely does not act as a barrier for these species. Obligatory rainforest species, on the contrary, could have crossed during more humid periods than today, when dry forests were replaced by rainforests (Haffer 1967). It may be that species such as *Trogon viridis, Myrmotherula brachyura, M. surinamensis, Pyriglena leuconota, Myiopagis caniceps, Tolmomyias assimilis, Cnipodectes subbrunneus, Myiobius atricaudus, Querula purpurata, Cryptopipo holochlora, Lepidothrix coronata, Manacus manacus* have originated following immigration of Amazonian elements through low passes in the Andes of southern

Ecuador. The ranges of these species include the Amazonia and the Pacific rainforest, but not the Caribbean lowland. The deep valleys of this region are mostly dry today, but were likely occupied by forest during humid climatic periods of the past (Haffer 1967).

Alternating Climates and Forest Disjunctions

Paleoecological records dating from interglacial phases indicate the existence of periods of higher humidity and temperature, and more extensive forest coverage, relative to today (Hammen 1974; Markgraf 1985; De Oliveira et al. 1999; Bueno et al. 2016; Wang et al. 2017). The last of them has been proposed to have occurred about 7500 year BP (Markgraf 1985) or 6000 years ago (Wang et al. 2017). During those moister interglacial periods, currently disjunct regions such as Amazonia and the Atlantic Forest, and the Yungas and Atlantic/Parana forests, may have been connected (Chapman 1926; Smith 1962; Vanzolini 1968; Haffer 1985; Nores 1992; Por 1992; Willis 1992; Sick 1993; Silva 1995; Silva 1996; de Vivo 1997; Oliveira et al. 1999; Behling et al. 2000; Costa 2003; Auler et al. 2004; Wang et al. 2004, 2017; Vilela et al. 2009; Patel et al. 2011; Batalha-Filho et al. 2013; Cheng et al. 2013; Sobral-Souza 2015; Thomé et al. 2016; Trujillo-Arias et al. 2017; Vergara et al. 2017). It is assumed that these periods of connectivity allowed an important degree of avian interchange between Amazonian, Andean, and the Atlantic forests—resulting in diversity increase in these domains.

Batalha-Filho et al. (2013) explored this aspect in depth, in birds with disjunct Atlantic and Amazonian forest distribution. Combining phylogenetic data with distributional data of suboscines living in Amazonia and the Atlantic Forest, those authors suggested three historical connections between these domains, two young ones in the north, and an old one in the south. Unlike me, however, those authors did not take the spatial accumulation of diversity into consideration while tracing their inferred pathway of connectivity and divergence. In their map (Fig. 8.1b), for instance, the highest concentration of species is in the Espírito Santo State, while the arrow points to a place between the present states of Alagoas and Sergipe.

I hypothesize that the connections between Amazonia and the Atlantic Forest happened largely via southern Bahia and Alagoas, through gallery forests. Specifically, I hypothesize that three rivers may have played an important role in these connections: the Tocantins, the São Francisco and the Jequitinhonha (Fig. 8.8). In its final stretch, the Tocantins River crosses ca. 400 km of Amazonian forest. Seventy-six (95%) of the 79 bird species with disjunct distributions occur in the forest area along the final stretch of the Tocantins River. From the Tocantins, I hypothesize that those species may have reached the Atlantic Forest via the São Francisco (whose mouth is located between the states of Alagoas and Sergipe) and the Jequitinhonha (mouth in southern Bahia). This could have happened assuming that the Tocantins and the São Francisco were once connected to each other by smaller-order rivers, such as do Sono, Manuel Alves, Preto, and Grande. This proposed role of rivers as promoters of interchange of species between Amazonia and the Atlantic Forest is not new. In predominantly non-forest regions (Caatinga, Cerrado, Chaco and Pampas), long-distance forest expansion is thought to have occurred generally along rivers and streams, via gallery forests which may have allowed the dispersal of rainforest species (Hueck and Seibert 1972; Darrieu 1986; Redford and da Fonseca 1986; Nores 1992; Nores et al. 2005; Oliveira-Filho and Ratter 1995; Silva 1996; Johnson et al. 1999).

Another connection of the Atlantic Forest, but in this case with the Andean forest (Yungas of Bolivia and north of Argentina), may have occurred via the Bermejo and Pilcomayo Rivers (Nores 1992; Fig. 8.8). At present, two patterns of bird distribution support this hypothesis. The first is the disjunct distribution of species in the Yungas and the Paranaense/Atlantic forests, despite being separated by 450 km of xerophytic Chaco woodland. The second involves several zones of secondary contact located in the Chaco lowland, where woodland and grassland species and subspecies interact (Short 1975). Moreover, the presence of dry riverbeds with relict forest patches, and the occurrence of forest trees in the upper part of the channels of the Bermejo and the Pilcomayo Rivers, indicate that the area used to be more humid.

Gallery forests are present in zones of 900 mm rainfall or higher in subtropical and temperate South America. A notable example is the Uruguay river, which has a continuous gallery forest (750 km long) that extends from the southern border of the Paranaense forest (1600 mm of rainfall) to its mouth, and then 100 km more along the Río de la Plata to Punta Lara, where the rainfall is about 900 mm (Nores et al. 2005). In the Bermejo and Pilcomayo rivers, gallery forests penetrate up to 200 km into the Chaco, which also matches the 900 mm isohyet (Nores 1992). As the lowest rainfall in this area is 600–700 mm, it may be assumed that the gallery forest was continuous between the Yungas and the Paranaense/Atlantic forests, via the Bermejo and Pilcomayo rivers, under 300 mm more rain in a former, wetter period (Wang et al. 2017). Based on species distribution patterns, I propose that 33 species of birds were involved in this connection (Appendix 3). Studying the disjunct distribution of the haverstmen (*Discocyrtus dilatatus*) in this area, by combining two independent approaches (paleodistributional reconstruction, using Species Distribution Modeling, and phylogeographic analyses based on the cytochrome oxidase subunit I molecular marker), Vergara et al. (2017) found support for the existence of this corridor joining the Atlantic Forest with the Yungas, calling it the "Nores' paleobridge". In contrast, based on the study of one bird species: *Arremon flavirostris*, Trujillo-Arias et al. (2017) suggested that a connection between the tropical Andes and the Atlantic forest may have existed through a main Cerrado corridor, and perhaps also through a Chaco corridor, ca. 150 km to the south of the Bermejo River.

4 Hypotheses to Explain Rainforest Biodiversity

To explain the high diversity of birds and other animals and plants of the humid forests, several hypotheses have been proposed, summarized and discussed in the literature (Haffer 1997, 2001; Nores 2000, 2013; Aleixo 2004; Rull 2011). The earliest one argued that the rainforest had remained largely stable during most of the Cenozoic, leading to a gradual accumulation of species (Richards 1952; Darlington

1957; Fisher 1960; Sanders 1968, 1969; Schwabe 1969). This *Stability-time hypothesis*, as it is known, was proposed by Sanders (1968, 1969), who observed high diversity of mud-dwelling animals in the deep-sea floor, despite the darkness, cold, and low productivity of the environment. Sanders suggested that the Amazon forest and other tropical forests should be thought of as analogous to the deep sea: places with a stable supply of annual moisture and warmth, in which extinction should be rare. This hypothesis has been criticized based on species richness data from marine benthic fauna (Abele and Walters 1979; Thistle 1983), and given evidence that the tropics experienced environmental shifts similar to those occurring in extra-tropical regions (Colinvaux et al. 1996; Bush et al. 2004; Rull 2011). However, those criticisms fail to get to the core Sanders' idea: that Amazonia and other tropical forests have remained mostly forested over time (Baker et al. 2001; Mayle et al. 2004; Bush and de Oliveira 2006; Bush et al. 2011), leading to a gradual accumulation of species.

Since the 1960s, however, there has been a significant change in the rainforest climatic stability-time concept. Multiple studies proposed the existence of periods of arid climate in Amazonia during the glacial maximum, which fragmented the forest, and that the intervening regions were covered by savannah (Hammen 1961, 1974; Haffer 1969, 1974, 2001; Simpson and Haffer 1978; Absy et al. 1991; Clapperton 1993; Hooghiemstra 1997; Hewitt 2000; D'Apolito et al. 2013). This has been the basis for the *Refuge hypothesis* (Haffer 1969; Vanzolini and Williams 1970; Prance 1973; Haffer and Prance 2001), which assumed that a 40% reduction in annual rainfall (Van der Hammen and Absy 1994) would have fragmented the rainforest. These fragments, presumably isolated by arid conditions, have been proposed to function as refugia for populations of birds, as well as other animals and plants, which were believed to have undergone divergent evolution during periods of geographic isolation.

As this hypothesis became more widely known, the idea of rainforest stability was practically abandoned. Still, the former was still controversial—for several authors raised the alternative that although a reduction in Amazonian precipitation did occur during glacial periods, it may not have been sufficient to eliminate the rainforest from most of the Amazonian lowland (Colinvaux 1987, 1989; Bush 1994; Bush et al. 2011; Haberle 1997; Hoorn 1997; Piperno 1997; Colinvaux et al. 2000, 2001; Nores 2000; Pennington et al. 2000). In other words, Amazonia would have remained mostly forested over time, even during the cooler and drier glacial periods (Cowling et al. 2001; Mayle et al. 2004; Bush and de Oliveira 2006; Bush et al. 2011; Bush and Metcalfe 2012). Critics of the refuge hypothesis also brought up that forest fragmentation resulting from a substantial reduction of rainfall would have had a double negative effect on the biota of the refugia: area reduction and impoverishment of the vegetation structure, which may have been an instance of loss rather than gain in biodiversity (Nores 2000; Cheng et al. 2013).

Colinvaux (1993) proposed that the causes of vegetational changes in the Pleistocene were cooling and reduced carbon dioxide, rather than aridity. The combination of reduced temperatures, precipitation, and atmospheric CO2 concentrations resulted in forests structurally and floristically quite different from those of today

(Colinvaux 1993; Mayle et al. 2004). Cold-adapted Andean taxa mixed with rainforest taxa in central areas, dry forest species and lianas probably became important in the more seasonal southern Amazon forests, and savannahs expanded in forest-savannah ecotones (Bonaccorso et al. 2006; Bush and de Oliveira 2006; Hoorn et al. 2010). The *Disturbance-vicariance hypothesis*, as this hypothesis was named (Colinvaux 1993), proposes that lowland animal species presumably lived in the Amazonian bottomland 100–300 m above present sea level, where temperatures were roughly 6 °C lower than at present. These regions constituted a dynamic borderland between the upland and lowland, where species distribution and abundance oscillated throughout the Pleistocene as a result of invasion and counterinvasion. Many interactions between the invaders were competitive, thus favoring species isolation.

A small variation of this hypothesis is that of Cowling et al. (2001), which is based on paleovegetation modeling simulations of the lowland Amazon basin. It considered that glacial cooling was probably responsible for maintaining glacial forest cover by reducing photorespiration and decreasing evapotranspiration, and argued that heterogeneous variations in glacial vegetation leaf area index may have promoted allopatric speciation by geographically isolating species in the forest understory.

Another hypothesis commonly debated in the context of the Amazonian biota is the *Riverine barrier hypothesis*. Although rivers have been used by naturalists to explain various distribution patterns in Amazonia since the nineteenth century (Wallace; Sclater and Salvin; Hellmayr), it was only in the 1960s that it was formalized as a hypothesis (Sick 1967). At ~10 Ma, Andean sediments reached the Atlantic coast through the Amazon drainage system, with the Amazon River becoming fully established at ~7 Ma (Figueiredo et al. 2009, 2010) or in the early Pliocene (~5 Ma) (Latrubesse et al. 2010). Sick (1967) proposed that large Amazonian rivers, formed as a consequence of the rising of the Andes, produced isolation and differentiation of lowland forest animals by fragmenting their ranges. Indeed, it has been argued that the high biodiversity of the Amazonian rainforest is correlated with its large number of rivers (Capparella 1991; Ayres and Clutton-Brock 1992; Naka et al. 2012; Ribas et al. 2012; Pomara et al. 2014; Thom and Aleixo 2015). Nevertheless, while multiple studies (Haffer 1969, 1974; Patton et al. 1994; Gascon et al. 1996, 1998, 2000; Hayes and Sewlal 2004; Weir et al. 2015) propose that Amazonian rivers may be effective barriers to the dispersal of nonvolant or nonaquatic animals (e.g. monkeys), or even understory birds, this does not imply that the formation of large rivers led to speciation per se. Weir et al. (2015) pointed out that co-distributed taxa endemic to adjacent interfluves in Amazonia, whose large geographic ranges are generally separated by rivers, come into geographic contact in the headwater regions where they hybridize. These results, they add, support the key argument that geographic contact and gene flow in headwater regions should prevent speciation.

The lack of evidence for the *Refuge hypothesis,* and the apparent failings of the *Riverine barrier hypothesis* in view of gene flow in headwater regions, are addressed by a hybrid of these hypotheses—the *River-refuge hypothesis* (Capparella 1991;

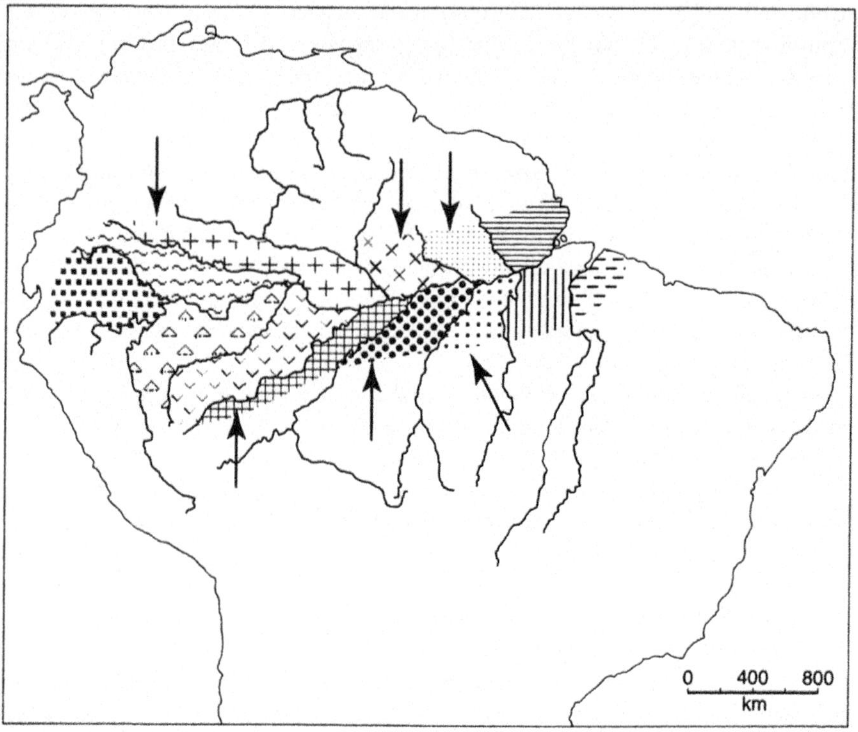

Fig. 8.13 Schematic representation of the River-refuge hypothesis, showing semi-refuges resulting from the combination of retraction of wet forest towards the center of Amazonia and to wider river sections (from Haffer1997)

Ayres and Clutton-Brock 1992; Haffer 1997; Weir et al. 2015). This hypothesis acknowledges that rivers in headwater regions fail to provide barriers strong enough to promote speciation, and builds on paleo-pollen records that indicate periods of past retraction of wet forest at the edges of Amazonia towards its center, which does not require an important reduction of rainfall as required by the *Refuge hypothesis*. Rivers are wider towards the center of Amazonia and, in combination with the retraction of wet forest out of headwater regions, are believed to provide sufficient barriers to promote speciation under this model (Weir et al. 2015).

The *River-refuge hypothesis* seems a plausible model of speciation in a case of forest retraction as schematized by Haffer (1997; Fig. 8.13). However, when that scheme is compared with the Amazonian precipitation map, showing 20% reduction of effective rainfall during the glacial maximum (Bush 1994; Fig. 8.14), this hypothesis appears less convincing. Most of the semi-refugia proposed either fall in areas expected to be semi-arid, or in equally or less humid areas relative to the headwater regions (Nores 2000).

Lastly, the *Marine transgression (=island) hypothesis* is also relevant to discussions of biological diversification in Amazonia. It proposes that Amazonia and other

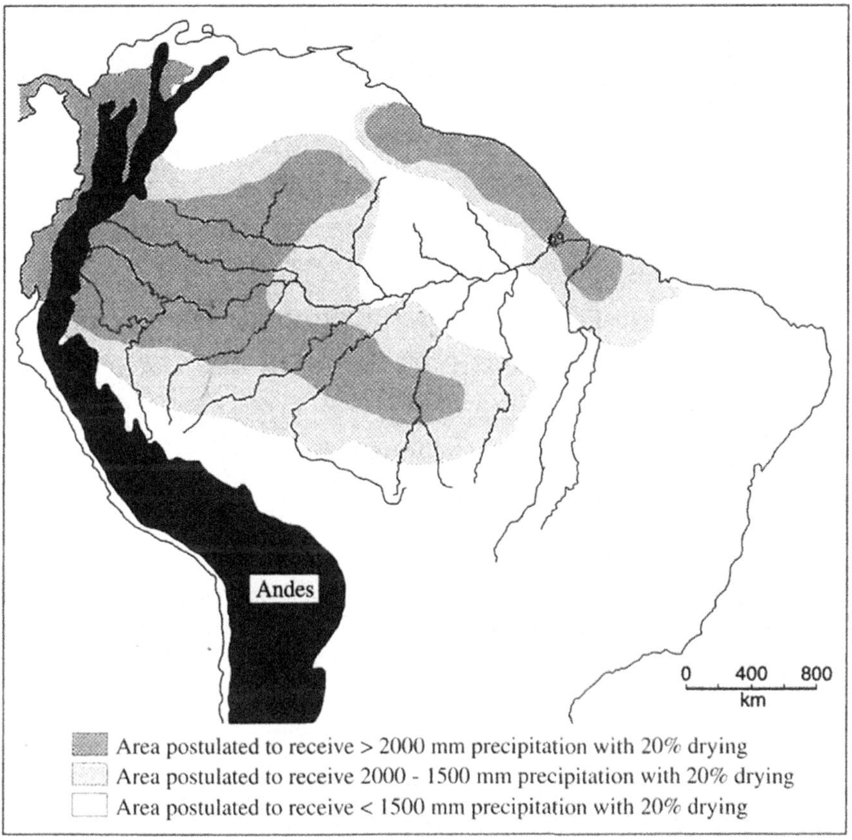

Fig. 8.14 Sketch map showing postulated annual precipitation on the Amazon basin following 20% drying during the glacial maximum (adapted from Bush 1994)

tropical forests became fragmented into numerous true islands and archipelagos during the rise of about 100 m in sea-level in the upper Tertiary and lower Quaternary, allowing animal and plant species to differentiate thanks to geographical isolation (Nores 1999, 2000, 2004; see also Bates 2001). Sea-level rise would also have isolated areas on the Pacific coast of Colombia, Ecuador and Panama, constituting the Darién, Chocó, Nechí, Catatumbo, and north Tumbesian areas of endemism (Nores 2004), also isolating populations on both sides of the Tocantins and São Francisco Rivers (this chapter, Figs. 8.11 and 8.12). This hypothesis proposes that the once isolated forest re-expanded as sea-level lowered, allowing birds and other organisms to expand their ranges according to their dispersal ability. Although segregation via sea-level rise would have isolated taxa of similar age, certain variables of the islands where the species would have evolved (size, distance to source and interisland distance) may have modified the impact of fragmentation (Nores 1999). The geographical coincidence between the areas of endemism and the areas that would have been isolated by a sea-level rise of about 100 m (Nores 1999, 2004) suggests that sea-level rises could have played an important role in the differentiation of forest birds (see also Willis and Whittaker 2000; Grau et al. 2005; Ribas et al. 2005; Peterson et al. 2010).

Because it has been proposed that sea level rise in Amazonia was unable to reach as much as 100 m (Dwyer and Chandler 2009; Rohling et al. 2009; Raymo et al. 2011, 2018), and as a Brazilian gazetteer is now available with altitudes of central and western Amazonia, I verified how forest connectivity would have been impacted by lower sea level values. With a sea level rise of about 50 m, marine transgressions via the Amazon River could have penetrated approximately to the 66°W, the mouth of the Juruá River, isolating two main regions of Amazonia: the Guiana shield and the Brazilian shield. Sea water would also have entered via the tributaries of the Amazon River, isolating areas on both sides of them. It is proposed that most of the islands shown in Fig. 8.11 would have formed, but land masses would have been closer together, forming larger islands. With a sea level rise of 70–75 m, transgressions would have reached to 69°W (eastern Colombia), resulting on a scenario similar to that produced by a sea level rise of 100 m (Nores 2013).

Insights from Molecular Studies

In the last few decades, phylogenetic and phylogeographic studies have contributed to the general discussion of former changes in Neotropical forests (Harrison 1991; Patton et al. 1994, 1996, 2000; Hall and Harvey 2002; Marks et al. 2002; Ribas et al. 2005, 2012, 2018; Brumfield and Edwards 2007; Rull 2011; Cabanne et al. 2008; Bonaccorso et al. 2011; Derryberry et al. 2011; Fernandes et al. 2012, 2014; d'Horta et al. 2013; Patel et al. 2011; Sousa-Neves et al. 2013; Smith et al. 2014; Harvey and Brumfield 2015; Thom and Aleixo 2015; Weir et al. 2015; Matos et al. 2016; Choueri et al. 2017; Ferreira et al. 2017; Schultz et al. 2017; to cite a few).While investigating whether species consist of one or more independently evolving units, and studying their relationships (Zink and Barrowclough 2008), phylogeography has had crucial implications for the understanding of bird diversification in the Neotropics, including estimates the age of extant species and clades (Rull 2011). However, methodological challenges do exist. For instance, the widespread use of linked haplotypic markers (e.g. mitochondrial DNA), which has limited inference power particularly regarding the timing of divergences and demographic changes, led authors to suggest that inferences without corroboration from additional evidence are unwarranted (e.g. Zink and Barrowclough 2008). Peterson (2006, 2009) also documented wide variation regarding molecular clock calibration. Moreover, large-scale extrapolations or attempts to clarify complex biogeographic patterns based on phylogeographic studies of one or two species that are restricted to particular regions or habitat types should be interpreted with care.

These data must be integrated with the analyses of biogeographic patterns, paleoecological (e.g. fossil pollen), and geological data to test the hypotheses and the impacts of the events discussed in this chapter. They also should be discussed under the light of multiple alternative hypotheses, instead of just one. Maldonado-Coelho et al. (2013), for instance, carried out a molecular analysis of the two subspecies of the White-backed Fire-eye (*Pyriglena leuconota*) occurring on opposite sides of the Tocantins River, and concluded that the river was probably the historical barrier promoting population divergence in these birds. However, it may be that it was not just the river, per se, that drove speciation. Despite being ca. 17 km wide at the mouth, the Tocantins narrows upstream, and at only 70–80 km it separates into many arms that are 2–3 km wide, with numerous islands in between, and continues to become more and more narrow. Maldonado-Coelho et al. (2013) did not take in account the fact that marine transgressions during sea-level rises could have penetrated, via the Tocantins River, to the border of the rainforest (Fig. 8.11), significantly expanding the width of the river and promoting population divergence, as observed in *Pyriglena leuconota* and other species and subspecies representative of the Belém and Pará centers of endemism (Nores 1999).

One of the most striking results of molecular analysis is that of Smith et al. (2014), who employed multiple markers and species to find out that tumultuous changes of the South American landscape may not have led to marked pulses in speciation, despite the existence of a large amount of evidence suggesting the opposite. Their study proposes that dispersal and differentiation on a matrix shaped by large-scale landscape events would have been a major driver of avian speciation in lowland Neotropical rainforests. Furthermore, Claramunt et al. (2011) clarified that the relationship between dispersal ability and speciation can be positive or negative, because dispersal can favor speciation (by permitting expansion across barriers) or inhibit speciation (by maintaining gene flow among populations). As historical demography and biogeographic hypothesis testing enters a new age of genomic level sampling, hopefully many other exciting results will appear in the near future.

5 Present-Day Trends in Bird Diversity and Abundance in Tropical and Subtropical South America

Unfortunately bird populations in humid tropical and subtropical forests of South America seem to be changing drastically, to the extent that almost all sites that I have studied in the last years, especially in Amazonia, the lower Eastern Andean Slopes (up to 700 m) and the Pacific Forest, severely lack birds.

To evaluate the severity of this problem and compare bird richness and abundance across sites, I performed one-day (12 h) censuses in several humid forest locations, including rainforest, cloud forest, and seasonal forest. I used available trails (usually one) and surveyed birds from sunrise to sunset. The observations were comparable in South America and Central America, consisting of the recording all the birds observed (including those that were not possible to identify), from sunrise to sunset, while walked along trails. When unidentified birds were considered to be a different species from those already recorded (for reasons of color, size, flight type,

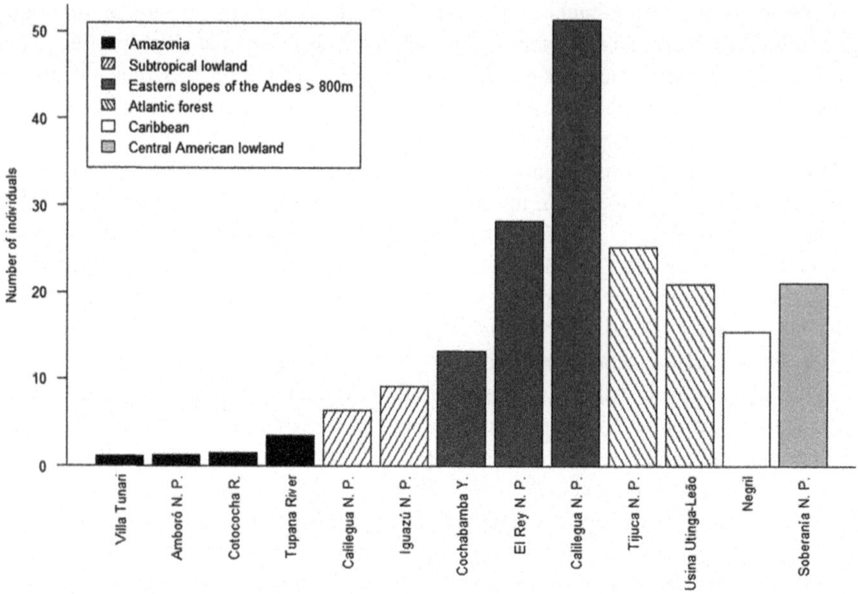

Fig. 8.15 Number of individuals observed per watching hour in humid forest sites in South and Central America. The very high value of Calilegua National Park at 800 m is somewhat overestimated because the census was carried out only during the first 3 h in the morning, when bird activity was highest

etc.), they were added to the species number and to the individuals' number. The censuses did not include open areas. In Fig. 8.15, I graph the number of individuals counted in the censuses, to which I have added two more: one carried out in Negril (Jamaica) on November 3, 2012, and another in Usina Utinga-Leão (state of Alagoas, Brazil) on June 12, 2018. It is striking to observe the low number of birds observed in Amazonia, ranging from 1.16 to 3.5 per watching hour. It is also remarkable that this low diversity and activity were observed not only in disturbed habitats, but also in well-preserved national parks and reserves, suggesting that climate change may be involved (Nores 2009, 2013). Given the rate at which humans have been impacting our planet, and hence modifying natural patterns of diversity distribution, it may be time to more effectively incorporate anthropogenic activity in new biogeographical and macroecological analyses inspired by field data.

Appendices

Appendix 1: Endemic Birds of the "Pacific Slopes Center"

Odontophorus melanonotus, Heliangelus strophianus, Heliodoxa imperatrix, Coeligena wilsoni, Urosticte benjamini, Aglaiocercus coelestis, Eriocnemis mirabilis, Calliphlox mitchellii, Haplophaedia lugens, Megascops columbianus, Glaucidium nubicola, Semnornis ramphastinus, Andigena laminirostris, Micrastur plumbeus, Sipia nigricauda, Grallaria flavotincta, Scytalopus vicinior, Pseudocolaptes johnsoni, Thripadectes ignobilis, Margarornis stellatus, Pipreola jucunda, Cephalopterus penduliger, Machaeropterus deliciosus, Cyanolyca pulchra, Entomodestes coracinus, Diglossa indigotica, Anisognathus notabilis, Chlorochrysa phoenicotis, Bangsia edwardsi, B. aureocincta, Ixothraupis rufigula, Chlorothraupis stolzmanni, Oreothraupis arremonops, Chlorospingus semifuscus, C. flavovirens, Chlorophonia flavirostris.

Appendix 2: Endemic Birds of the "Atlantic Forest Mountains Center"

Piculus aurulentus, Dysithamnus xanthopterus, Drymophila rubricollis, D. genei, D. ochropyga, Myrmoderus loricatus, Merulaxis ater, Anabazenops fuscus, Cranioleuca pallida, Phylloscartes difficilis, Carpornis cucullata, Tijuca atra, T. condita, Ilicura militaris, Sporophila frontalis, Castanozoster thoracicus, Tangara cyanocephala, T. desmaresti, T. cyanoventris.

Appendix 3: Species that Would Have Been Involved in the Corridor Joining the Atlantic Forest with the Yungas

Crypturellus obsoletus, Penelope obscura, Lurocalis semitorquatus, Antrostomus sericocaudatus, Lophornis chalybeus, Tigrisoma fasciatum, Parabuteo leucorrhous, Aegolius harrisii, Batara cinerea, Thamnophilus ruficapillus, Herpsilochmus rufimarginatus, Chamaeza campanisona, Xenops minutus, Lochmias nematura, Philydor rufum, Phylloscartes ventralis, Phyllomyias burmeisteri, Elaenia obscura, Poecilotriccus plumbeiceps, Ramphotrigon megacephalum, Pachyramphus castaneus, Oxyruncus cristatus, Piprites chloris, Turdus albicollis, Trichothraupis melanops, Cissopis leverianus, Pipraeidea melanonota, Arremon flavirostris, Habia rubica, Myiothlypis rivularis, Cacicus chrysopterus, Euphonia cyanocephala, Chlorophonia cyanea.

Literature Cited

Ab'Saber AN (1977) Os dominios morfoclimáticos da América do Sul. Geomorfologia 53:1–23

Abele LG, Walters K (1979) Marine benthic diversity: a critique and alternative explanation. J Biogeogr 6:115–126

Absy ML, Cleef A, Fournier M, Martin M, Servant M, Sifeddine A, Ferreira da Silva M, Soubies F, Suguio K, Turcq B, Van der Hammen T (1991) Mise en évidence de quatre phases d'ouverture de la forêt dense dans le sud-est de l'Amazonie au cours des 60000 dernières années. Première comparaison avec d'autres régions tropicales. CR Acad Sci Paris Sér II 312:673–678

Aleixo A (2004) Historical diversification of a terra-firme forest bird superspecies: a phylogeographic perspective on the role of different hypotheses of Amazonian diversification. Evolution 58:1301–1317

Aleixo A, Rossetti DF (2007) Avian gene trees, landscape evolution, and geology: towards a modern synthesis of Amazonian historical biogeography? J Ornithol 148(Suppl 2):S443–S453

Amadon D (1973) Birds of the Congo and Amazon: a comparison. In: Meggers BJ et al (eds) Tropical forest ecosystems in Africa and South America: a comparative review. Smithsonian Institution Press, Washington, DC, pp 167–267

Antonelli A, Sanmartín I (2011) Why are there so many plant species in the Neotropics? Taxon 60:403–414

Antonelli A, Nylander JAA, Persson C, Sanmartín I (2009) Tracing the impact of the Andean uplift on Neotropical plant evolution. Proc Natl Acad Sci U S A 106:9749–9754

Ashworth AC, Kuschel G (2003) Fossil weevils (Coleoptera: Curculionidae) from latitude 85°S Antarctica. Palaeogeogr Palaeoclimatol Palaeoecol 191:191–202

Ashworth AC, Preece RC (2003) The first freshwater molluscs from Antarctica. J Molluscan Stud 69:89–92

Ashworth AC, Thompson FC (2003) A fly in the biogeographic ointment. Nature 423:136–136

Auler AS, Smart PL (2001) Late Quaternary paleoclimate in semiarid northeastern Brazil from U-series dating of travertine and water-table speleothems. Quat Res 55:159–167

Auler AS, Wang AX, Edwards RL, Cheng H, Cristalli PS, Smart PL, Richards DA (2004) Quaternary ecological and geomorphic changes associated with rainfall events in presently semi-arid northeastern Brazil. J Quat Sci 19:693–701

Ayres JMC, Clutton-Brock TH (1992) River boundaries and species range size in Amazonian primates. Am Nat 140:531–537

Bacon CD, Silvestro D, Jaramillo C, Smith BT, Chakrabarty P, Antonelli A (2015) Biological evidence supports an early and complex emergence of the Isthmus of Panama. Proc Natl Acad Sci U S A 112:6110–6115

Baker PA, Seltzer GO, Fritz SC, Dunbar RB, Grove MJ, Tapia PM, Cross SL, Rowe HD, Broda JP (2001) The history of South American tropical precipitation for the past 25,000 years. Science 291:640–643

Balco G, Rovey CW (2010) Absolute chronology for major Pleistocene advances of the Laurentide Ice Sheet. Geol Soc Am 38:795–798

Barrett PJ, Adams CJ, Mcintosh WC, Swisher CC, Wilson GS (1992) Geochronological evidence supporting Antarctic deglaciation three million years ago. Nature 359:816–818

Barron JA (1996) Diatom constraints on the position of the Antarctic polar front in the middle part of the Pliocene. Mar Micropaleontol 27:195–204

Batalha-Filho H, Fjeldså J, Fabre PH, Miyaki CY (2013) Connections between the Atlantic and Amazonian forest avifaunas represent distinct historical events. J Ornithol 154:41–50

Bates JM (2001) Avian diversification in Amazonia: evidence for historical complexity and a vicariance model for a basic pattern of diversification. In: Viera IMAD, Incao JM, Silva C, Oreneds D (eds) Diversidade Biológica e Cultural da Amazônia. Museu Paraense Emilio Goeldi, Belém, pp 119–138

Behling H, Arz HW, Pätzold J, Wefer G (2000) Late Quaternary vegetational and climate dynamics in northeastern Brazil, inferences from marine core GeoB 3104-1. Quat Sci Rev 19:981–994

Bonaccorso E, Koch I, Peterson AT (2006) Pleistocene fragmentation of Amazon species' ranges. Divers Distrib 12:157–164

Bonaccorso E, Guayasamin JM, Peterson AT, Navarro-Sigüenza AG (2011) Molecular phylogeny and systematics of Neotropical toucanets in the genus Aulacorhynchus (Aves, Ramphastidae). Zool Scr 40:346–349

Brown KS Jr (1975) Geographical patterns of evolution in Neotropical Lepidoptera. Systematics and derivation of known and new Heliconiini (Nymphalidae: Nymphalinae). J Entomol 44:201–242

Brumfield RT (2012) Inferring the origins of lowland Neotropical birds. Auk 129:367–376

Brumfield RT, Capparella AP (1996) Historical diversification of birds in northwestern South America: a molecular perspective on the role of vicariant events. Evolution 50:1607–1624

Brumfield RT, Edwards SV (2007) Evolution into and out of the Andes: a Bayesian analysis of historical diversification in *Thamnophilus antshrikes*. Evolution 61:346–367

Bueno ML, Toby Pennington R, Dexter KG, Yoshino Kamino LH, Pontara V, Neves DM, Ratter JA, de Oliveira-Filho AT (2016) Effects of Quaternary climatic fluctuations on the distribution of Neotropical savannah tree species. Ecography 39:1–12

Burckle LH, Mortlock R, Rudolph S (1996) No evidence for extreme, long term warming in early Pliocene sediments of the Southern Ocean. Mar Micropaleontol 27:215–226

Bush MB (1994) Amazonian speciation: a necessarily complex model. J Biogeogr 21:5–17

Bush MB, Colinvaux PA (1990) A long record of climatic and vegetation change in lowland Panama. J Veg Sci 1:105–119

Bush MB, de Oliveira PE (2006) The rise and fall of the Refugial Hypothesis of Amazonian Speciation: a paleoecological perspective. Biota Neotrop 6. Online version

Bush MB, Metcalfe SE (2012) Latin America and the Caribbean. In: Metcalfe SE, Nash DJ (eds) Quaternary environmental change in the tropics. John Wiley and Sons, New York, pp 263–311

Bush MB, Silman MR, Urrego DH (2004) 48,000 years of climate and forest change from a biodiversity hotspot. Science 303:827–829

Bush MB, Gosling WD, Colinvaux PA (2011) Climate and vegetation change in the lowlands of the Amazon Basin. In: Bush MB, Flenley JR, Gosling WD (eds) Tropical rainforest responses to climatic change. Springer, Berlin, pp 61–84

Cabanne GS, d'Horta FM, Sari EHR, Miyaki CM (2008) Nuclear and mitochondrial phylogeography of the Atlantic Forest endemic *Xiphorhynchus fuscus* (Aves Dendrocolaptidae): biogeography and systematics implications. Mol Phylogenet Evol 49:760–773

Cabrera AL, Willink A (1973) Biogeografía de América Latina. Monografía 13, Serie de Biología. OEA, Washington, DC

Capparella A (1991) Neotropical avian diversity and riverine barriers. In: Bell BD (ed) Acta XX Congressus Internationalis Ornithologici. New Zealand Ornithological Congress Trust Board, Wellington, pp 307–316

Carnaval AC, Hickerson MJ, Haddad CFB, Rodrigues MT, Moritz C (2009) Stability ptredicts genetic diversity in the Brazilian Atlantic forest hotspot. Science 323:785–789

Chapman FM (1917) The distribution of bird-life in Colombia. Bull Am Mus Nat Hist 36:1–729

Chapman FM (1926) The distribution of bird-life in Ecuador. Bull Am Mus Nat Hist 55:1–784

Cheng H, Sinha A, Cruz FW, Wang X, Edwards RL, d'Horta FM, Ribas CC, Vuille M, Stott LD, Auler AS (2013) Climate change patterns in Amazonia and biodiversity. Nat Comms 4:1411. https://doi.org/10.1038/ncomms2415

Choueri EL, Gubili C, Borges SH, Thom G, Sawakuchi AO, Soares EAA, Ribas CC (2017) Phylogeography and population dynamics of Antbirds (Thamnophilidae) from Amazonian fluvial islands. J Biogeogr 44(10):2284–2294. https://doi.org/10.1111/jbi.13042

Clapperton CM (1993) Nature of environmental changes in South America at the last glacial maximum. Palaeogeogr Palaeoclimatol Palaeoecol 101:189–208

Claramunt S, Derryberry EP, Remsen JV Jr, Brumfield RT (2011) High dispersal ability inhibits speciation in a continental radiation of passerine birds. Proc R Soc B 279:1567–1574

CLIMAP (1976) The surface of the ice-age Earth. Science 191:1131–1136

Coates AG, Obando JA (1996) The geologic evolution of the Central American Isthmus. In: Jackson JBC, Budd AF, Coates AG (eds) Evolution and environment in tropical America. University of Chicago Press, Chicago, pp 21–56

Coates AG, Jackson JBC, Collins LS, Cronin TM, Dowset HJ, Bybell LM, Jung P, Obando J (1992) Closure of the Isthmus of Panama: the near-shores marine records of Costa Rica and western Panama. Geol Soc Am Bull 104:814–828
Coates AG, Stallard RF, Robert F (2013) How old is the Isthmus of Panama? Bull Mar Sci 89:801–813
Cody S, Richardson JE, Rull V, Ellis C, Toby Pennington R (2010) The Great American Biotic Interchange revisited. Ecography 33:326–332
Colinvaux PA (1987) Amazon diversity in light of the paleoecological record. Quat Sci Rev 6:93–114
Colinvaux PA (1989) The past and future Amazon. Sci Am 260:102–108
Colinvaux PA (1993) Pleistocene biogeography and diversity in tropical forests of South America. In: Goldblatt P (ed) Biological relationships between Africa and South America. Yale University Press, New Haven, CT, pp 473–499
Colinvaux PA (1997) An arid Amazon? Trends Ecol Evol 12:318–319
Colinvaux PA, De Oliveira PE, Moreno JE, Miller MC, Bush MB (1996) A long pollen record from lowland Amazonia: forest and cooling in glacial times. Science 274:85–88
Colinvaux PA, de Oliveira PE, Bush MB (2000) Amazonian and Neotropical plant communities on glacial time-scales: the failure of the aridity and refuge hypotheses. Quat Sci Rev 19:141–169
Colinvaux PA, Irion G, Räsänen ME, Bush MB, Nunes de Melo JAS (2001) A paradigm to be dicarded: geological and palaeoecological data falsify the Haffer and Prance refuge hypothesis of Amazonian speciation. Amazoniana 16:609–646
Collins LS (1996) Environmental changes in Caribbean shallow waters relative to the closing tropical American seaway. In: Jackson JBC, Budd AF, Coates AG (eds) Evolution and environment in tropical America. University of Chicago Press, Chicago, pp 130–167
Collins LS (1999) The Miocene to recent diversity of Caribbean benthic foraminifera from the Central American Isthmus. Bull Am Paleontol 357:91–107
Collins LS, Coates AG, Berggren WA, Aubri MP, Zhang J (1996) The late Miocene Panama isthmian strait. Geology 24:687–690
Costa LP (2003) The historical bridge between the Amazon and the Atlantic Forest of Brazil: a study of molecular phylogeography with small mammals. J Biogeogr 30:71–86
Cowling SA, Maslin MA, Sykes MT (2001) Paleovegetation simulations of lowland Amazonia and implications for Neotropical allopatry and speciation. Quat Res 55:140–149
Cracraft J (1985) Historical biogeography and patterns of differentiation within the South American avifauna: areas of endemism. Ornithol Monogr 36:49–84
Cracraft J, Prum RO (1988) Patterns and processes of diversification: speciation and historical congruence in some Neotropical birds. Evolution 42:603–620
D'Apolito C, Absy ML, Latrubesse EM (2013) The Hill of Six Lakes revisited: new data and re-evaluation of a key Pleistocene Amazon site. Quat Sci Rev 76:140–155
d'Horta FM, Cuervo AM, Ribas CC, Brumfield RT, Miyaki CY (2013) Phylogeny and comparative phylogeography of *Sclerurus*(Aves: Furnariidae) reveal constant and cryptic diversification in an old radiation of rain forest understorey specialists. J Biogeogr 40:37–49
Darlington PJ (1957) Zoogeography. The geographical distribution of animal. Wiley, New York
Darrieu CA (1986) Estudios sobre la avifauna de Corrientes. III. Nuevos registros de aves Passeriformes (Dendrocolaptidae, Furnariidae, Formicariidae, Cotingidae y Pipridae) y consideraciones sobre su distribución geográfica. Hist Nat 6:93–99
Davis LI (1972) A field guide to the Mexico and Central America. Texas University Press, Austin
De Oliveira PE, Barreto AMF, Suguio K (1999) Late Pleistocene/Holocene climatic and vegetational history of the Brazilian caatinga: the fossil dunes of the middle São Francisco River. Palaeogeogr Palaeoclimatol Palaeoecol 152:319–337
de Vivo M (1997) Mammalian evidence of historical ecological change in the Caatinga semiarid vegetation of northeastern Brazil. J Comp Biol 2:65–73
del Hoyo J, Elliott A, Sargatal J, Christie D (1992–2013) Handbook of the birds of the world. Lynx Edicions, Barcelona

Denton GH, Prentice ML, Burckle LH (1991) Cenozoic history of the Antarctic ice sheet. In: Tingey RJ (ed) Geology of Antarctica. Oxford University Press, Oxford, pp 365–433

Derryberry EP, Claramunt S, Derryberry G, Chesser RT, Cracraft J, Aleixo A, Peérez-Emán J, Remsen JV Jr, Brumfield RT (2011) Lineage diversification and morphological evolution in a large-scale continental radiation: the Neotropical ovenbirds and woodcreepers (Aves: Furnariidae). Evolution 65:2973–2986

Dowsett HJ, Cronin TM, Poore RZ, Thompson RS, Whatley RC, Wood AM (1992) Micropaleontological evidence for increased meridional heat transport in the North Atlantic Ocean during the Pliocene. Science 258:1133–1135

Dowsett H, Barron J, Poore R (1996) Middle Pliocene sea surface temperatures: a global reconstruction. Mar Micropaleontol 27:13–25

Drewry DJ, Jordan SR, Jankowski E (1982) Measured properties of the Antarctic ice sheet: surface configurations, ice thickness, volume and bedrock characteristics. Ann Glaciol 3:83–91

Dutton A, Carlson AE, Long AJ, Milne GA, Clark PU, DeConto R (2015) Sea level rise due to polar ice-sheet mass loss during past warm periods. Science 349:4019

Dwyer GS, Chandler MA (2009) Mid-Pliocene sea level and continental ice volume based on coupled benthic Mg/Ca palaeotemperatures and oxygen isotopes. Phil Trans R Soc A 367:157–168

Ericson DB, Wolling G (1968) Pleistocene climates and chronology in deep-sea sediments. Science 162:1227–1234

Ericson DB, Ewing M, Wolling G (1964) The Pleistocene epoch in deep-sea sediments. Science 146:723–732

Erkens RHJ, Hoorn C (2016) The Panama Isthmus, ‘old’, ‘young’ or both? A reply to O’Dea et al. 2016. Sci Adv. Online e-letter

Fairbridge RW (1961) Eustatic changes in sea-level. In: Ahrens LH et al (eds) Physics and chemistry of the earth. Pergamon Press, London, pp 99–185

Fernandes AM, Wink M, Aleixo A (2012) Phylogeography of the Chestnut-tailed Antbird (*Myrmeciza hemimelaena*) clarifies the role of rivers in Amazonian biogeography. J Biogeogr 39:1524–1535

Fernandes AM, Wink M, Sardelli CH, Aleixo A (2014) Multiple speciation across the Andes and throughout Amazonia: the case of the spot-backed antbird species complex (*Hylophylax naevius/Hylophylax naevioides*). J Biogeogr 6:1094–1104

Ferreira M, Aleixo A, Ribas CC, Santos MPD (2017) Biogeography of the Neotropical genus *Malacoptila* (Aves: Bucconidae): the influence of the Andean orogeny, Amazonian drainage evolution and palaeoclimate. J Biogeogr 44:748–759

Figueiredo J, Hoorn C, van der Ven P (2009) Late Miocene onset of the Amazon River and the Amazon deep-sea fan: evidence from the Foz do Amazonas Basin. Geology 37:619–622

Figueiredo J, Hoorn C, van der Ven P, Soares E (2010) Late Miocene onset of the Amazon River and the Amazon deep-sea fan: evidence from the Foz do Amazonas Basin: reply. Geology e213:38

Finschii P, Pirani JR (2009) Review of plant biogeographic studies in Brazil. J Syst Evol 47:477–496

Fisher AG (1960) Latitudinal variation in organic diversity. Evolution 14:64–81

Fittkau EJ (1974) Zur ökologischen Gliederung Amazoniens. I: die erdgeschichtliche Entwicklung Amazoniens. Amazoniana 5:77–134

Flint RF (1957) Glacial and Pleistocene geology. Wiley, New York

Frailey CD (2002) Paleogeography of the Amazon basin. TER-QUA Symp Ser 3:71–97

Garzione CN, Hoke GD, Libarkin JC, Withers S, MacFadden B, Eiler J, Ghosh P, Mulch A (2008) Rise of the Andes. Science 5881:1304–1307

Gascon C, Lougheed SC, Bogart JP (1996) Genetic and morphological variation in *Vanzolinius discodactylus*: a test of the river hypothesis of speciation. Biotropica 28:376–387

Gascon C, Lougheed SC, Bogart JP (1998) Patterns of genetic population differentiation in four species of Amazonian frogs: a test of the Riverine Barrier Hypothesis. Biotropica 30:104–119

Gascon C, Malcon JR, da Silva MNF, Bogart JP, Lougheed SC, Peres CA, Neckel S, Boag PT (2000) Riverine barriers and the geographic distribution of Amazonian species. Proc Natl Acad Sci U S A 97:13672–13677
Gates WL (1976) Modeling the ice-age climate. Science 191:1138–1144
Gibbard P, van Kolfschoten T (2005) The Pleistocene and Holocene epochs. In: Gradstein FM, Ogg JG, Smithand A, Gilbert A (eds) Geologic time scale. Cambridge University Press, Cambridge, pp 441–452
Grau ET, Pereira SL, Silveira LF, Höfling E, Wajntal A (2005) Molecular phylogenetics and biogeography of Neotropical piping guans (Aves: Galliformes): Pipile Bonaparte, 1856 is synonym of *Aburria* Reichenbach, 1853. Mol Phylogenet Evol 35:637–645
Gregory-Wodzicki KM (2000) Uplift history of the central and northern Andes: a review. Geol Soc Am Bull 112:1091–1105
Haberle S (1997) Upper Quaternary vegetation and climate history of the Amazon basin: correlating marine and terrestrial pollen records. In: Flood RD, Piper DJW, Klaus A, Peterson LC (eds) Proceedings of the ocean program, scientific results. Ocean Drilling Program, College Station, TX, pp 381–396
Haffer J (1967) Speciation in Colombian forest birds west of the Andes. Am Mus Novit 2294:1–57
Haffer J (1969) Speciation in Amazonian forest birds. Science 165:131–137
Haffer J (1974) Avian speciation in tropical South America, No. 14. Publications of the Nuttall Ornithological Club, Cambridge, MA
Haffer J (1978) Distribution of Amazon forest birds. Bonner Zool Beirträge 29:38–78
Haffer J (1985) Avian zoogeography in the Neotropical lowlands. In: Buckley PA, Foster MS, Morton ES, Ridgely RS, Buckley FG (eds) Neotropical ornithology. AOU, Washington, DC, pp 113–146. (Orn Monogr 36)
Haffer J (1997) Alternative models of vertebrate speciation in Amazonia: an overview. Biodivers Conserv 6:451–476
Haffer J (2001) Hypotheses to explain the origin of species in Amazonia. In: Vieira ICG, Silva JMC, Oren DC, D'Incao MA (eds) Diversidade biológica e cultural de Amazonia. Museu Paraense Emílio Goeldi, Belém, pp 45–118
Haffer J, Prance GT (2001) Climatic forcing of evolution in Amazonia during the Cenozoic: on the refuge theory of biotic differentiation. Amazoniana 16:579–607
Hall JPW, Harvey DJ (2002) The phylogeography of Amazonia revisited: new evidence from Riodinid butterflies. Evolution 56:1489–1497
Hallam A (1971) Re-evaluation of the palaeogeographic argument for an expanding earth. Nature 232:180–182
Hannah L (2015) Past marine ecosystem changes. In: Hannah L (ed) Climate change biology. Academic Press, Cambridge, MA, pp 163–177
Haq BU, Hardenboland J, Vail PR (1987) Chronology of fluctuating sea levels since the Triassic. Science 235:1156–1167
Harrington HJ (1962) Paleogeographic development of South America. Bull Am Assoc Pet Geol 46:1773–1814
Harrison R (1991) Molecular changes at speciation. Annu Rev Ecol Syst 22:281–308
Hartley AJ (2003) Andean uplift and climate change. J Geol Soc Lond 160:7–10
Harvey MG, Brumfield RT (2015) Genomic variation in a widespread neotropical bird (*Xenops minutus*) reveals divergence, population expansion, and gene flow. Mol Phylogenet Evol 83:305–316
Hayes F, Sewlal J-AN (2004) The Amazon River as a dispersal barrier to passerine birds: effects of river width, habitat and taxonomy. J Biogeogr 31:1809–1818
Hays JD, Pitman WC (1973) Lithospheric plate motion, sea level changes and climatic and ecological consequences. Nature 246:18–22
Hewitt GM (2000) The genetic legacy of the Quaternary ice ages. Nature 405:907–913
Hillebrand H (2004) On the generality of the latitudinal diversity gradient. Am Nat 163:193–211
Hilty SL (2003) Birds of Venezuela. Princeton University Press, Princeton

Hilty SL, Brown WL (1986) A guide to the birds of Colombia. Princeton University Press, Princeton
Hooghiemstra H (1997) Tropical rain forest versus savannah: two sides of a precious medal? NWO/Huygenslezingen, The Hague
Hoorn C (1993) Marine incursions and the influence of Andean tectonics on the Miocene depositional history of northwestern Amazonia: results of a palynostratigraphic study. Palaeogeogr Palaeoclimatol Palaeoecol 105:267–309
Hoorn C (1997) Palynology of the Pleistocene glacial/interglacial cycles of the Amazon fan (holes 940A, 944A, and 946A). In: Flood RD, Piper DJW, Klaus A, Peterson LC (eds) Proceedings of the ocean drilling program, scientific results, vol 155. Ocean Drilling Program, College Station, TX, pp 397–409
Hoorn C, Wesselingh FP, ter Steege H, Bermudez MA, Mora A, Sevink J, Sanmartín I, Sanchez-Meseguer A, Anderson CL, Jaramillo JPFC, Riff D, Negri FR, Hooghiemstra H, Lundberg J, Stadler T, Särkinen T, Antonelli A (2010) Amazonia through time: Andean uplift, climate change, landscape evolution, and biodiversity. Science 330:927–931
Hueck H, Seibert P (1972) Vegetationskarte von Südamerika: Erläuterung zur Karte 1:8.000.000. Gustav Fischer, Stuttgart
Idnurm M, Cook PJ (1980) Palaeomagnetism of beach ridges in South Australia and the Milankovitch theory of ice ages. Nature 286:699–702
Ingólfsson Ó (2004) Quaternary glacial and climate history of Antarctica. Dev Quat Sci 2:3–43
Johnson KP, Weckstein JD (2011) The Central American land bridge as an engine of diversification in new world doves. J Biogeogr 38:1069–1076
Johnson MA, Saraiva PM, Coelho D (1999) The role of gallery forests in the distribution of Cerrado mammals. Rev Bras Biol 59:421–427
Kaufman DS, Brigham-Grette J (1993) Aminostratigraphic correlations and paleotemperature implications, Pliocene–Pleistocene high-sea-level deposits, northwestern Alaska. Quat Sci Rev 12:21–33
Keigwin LD (1982) Isotopic paleoceanography of the Caribbean and east Pacific: role of Panama uplift in late Neogene time. Science 217:350–353
Kennett JP, Shackleton NJ (1976) Oxygen isotopic evidence for the development of the psychrosphere 38 Myr ago. Nature 260:513–515
Kerr RA (1996) Ancient sea-level swings confirmed. Science 272:1097–1098
Klammer G (1971) Über plio-pleistozäne Terrassen und ihre Sedimente im unteren Amazonasgebiet. Z Geomorphol 15:62–106
Knapp S, Mallet J (2003) Refuting refugia? Science 300:71–72
Lambeck K, Esat TM, Potter EK (2002) Links between climate and sea levels for the past three million years. Nature 419:199–206
Latrubesse EM (2002) Evidence of quaternary paleohydrological changes in middle Amazonia: the Aripuana-Roosevelt and Jiparaná fans. Z Geomorphol Suppl 129:61–72
Latrubesse EM, Cozzuol M, da Silva-Caminha SAF, Rigsby CA, Absy ML, Jaramillo C (2010) The late Miocene paleogeography of the Amazon basin and the evolution of the Amazon river system. Earth Sci Rev 99:99–124
Leigh EG, O'Dea A, Vermeij GJ (2013) Historical biogeography of the Isthmus of Panama. Biol Rev Camb Philos Soc 89:148–172
Liu K, Colinvaux PA (1985) Forest changes in the Amazon Basin during the last glacial maximum. Nature 318:556–557
Lovejoy NR, Bermingham E, Martin AP (1998) Marine incursion into South America. Nature 396:421–422
Lovejoy NR, Albert JS, Crampton WGR (2006) Miocene marine incursions and marine/freshwater transitions: evidence from Neotropical fishes. J S Am Earth Sci 21:5–13
Maldonado-Coelho M, Blake JG, Silveira LF, Batalha-Filho H, Ricklefs RE (2013) Rivers, refuges and population divergence of fire-eye antbirds (*Pyriglena*) in the Amazon Basin. J Evol Biol 26:1090–1107

Malfait BT, Dinkelman MG (1972) Circum-Caribbean tectonic and igneous activity and the evolution of the Caribbean plate. Geol Soc Am Bull 83:251–271

Marchant DR, Denton GH (1996) Miocene and Pliocene paleoclimate of the Dry Valleys region, southern Victoria land: a geomorphological approach. Mar Micropaleontol 27:253–271

Markgraf V (1985) Paleoenvironmental history of the last 10,000 years in northwestern Argentina. In: Garleff K, Stingl H (eds) Sudamerika Geomorphologie und Paläoökologie de Jüngeren Quartars. Zentralblatt fur Geologie und Paläontologie, Stuttgart, pp 1739–1749

Marks BD, Hackett S, Capparella A (2002) Historical relationships among neotropical lowland forest areas of endemism as determined by mitochondrial DNA sequence variation within the Wedge-billed Woodcreeper (Aves: Dendrocolaptidae: *Glyphorhynchus spirurus*). Mol Phylogenet Evol 24:153–167

Marshall LG (1988) Land mammals and the Great American Interchange. Am Sci 76:380–388

Marshall LG, Webb SD, Sepkowski JJ, Raup DM (1982) Mammalian evolution and the Great American Interchange. Science 215:1351–1357

Matos MV, d'Horta F, Borges SH, Latrubesse E, Cohn-Haft CCM, Ribas CC (2016) Comparative phylogeography of two bird species, *Tachyphonus phoenicius* (Thraupidae) and *Polytmus theresiae* (Trochilidae), specialized in Amazonian white-sand vegetation. Biotropica 48:110–121

Mayle FE, Burbridge R, Killeen TJ (2000) Millennial-scale dynamics of southern Amazonian rain forests. Science 290:2291–2294

Mayle FE, Beerling DJ, Gosling WD, Bush MB (2004) Responses of Amazonian ecosystems to climatic and atmospheric carbon dioxide changes since the last glacial maximum. Philos Trans R Soc Lond B 359:499–514

Mégard F (1992) The evolution of the Pacific Ocean margin in South America north of Arica elbow (18° S). In: Ben-Avraham Z (ed) Evolution of the Pacific Ocean margins. Oxford University Press, New York, pp 208–230

Montes C, Bayona J, Cardona A, Buchs D, Silva CA, Morén S (2012) Arc-continent collision and orocline formation: closing of the Central American Seaway. J Geophys Res Atmos 117:4105–4129

Mora A, Baby P, Roddaz M, Parra M, Brusset S, Hermoza W, Espurt N (2010) Tectonic history of the Andes and sub-Andean zones: implications for the development of the Amazon drainage basin. In: Hoorn C, Wesselingh FP (eds) Amazonia, landscape and species evolution: a look into the past. Wiley, Oxford, pp 38–60

Moritz C, Patton JL, Schneider CJ, Smith TB (2000) Diversification of rainforest faunas: an integrated molecular approach. Annu Rev Ecol Syst 31:533–563

Müller P (1973) The dispersal centers of terrestrial vertebrates in the Neotropical realm. Biogeographica 2:1–250

Naish T (1997) Constraints of the amplitude of late Pliocene eustatic sea-level fluctuations: new evidence from the New Zealand shallow-marine sediment record. Geology 25:1139–1142

Naka LN, Bechtoldt CL, Pinto Henriques LM, Brumfield RT (2012) The role of physical barriers in the location of avian suture zones in the Guiana Shield, northern Amazonia. Am Nat 179:115–132

Nores M (1992) Bird speciation in subtropical South America in relation to forest expansion and retraction. Auk 109:346–357

Nores M (1999) An alternative hypothesis for the origin of Amazonian bird diversity. J Biogeogr 24:475–485

Nores M (2000) Species richness in the Amazonian bird fauna from an evolutionary perspective. EMU 100:419–430

Nores M (2004) The implications of Tertiary and Quaternary sea level rise events for avian distribution patterns in the lowlands of northern South America. Glob Ecol Biogeogr 13:149–162

Nores M (2009) Are bird populations in tropical and subtropical forests of South America affected by climate change? Clim Chang 97:543–551

Nores M (2011) The western Amazonian boundary for avifauna determined by species distribution patterns and geographical and ecological features. Int J Ecol 2011:1–7
Nores M (2013) Diversity, origin and status of Amazonian avifauna. In: Rodriguez Alcântara C (ed) Amazon: biodiversity conservation, economic development and human impact. Nova Science, New York, pp 115–154
Nores M, Cerana MM, Serra DA (2005) Dispersal of forest birds and trees along the Uruguay River in southern South America. Divers Distrib 11:205–217
O'Dea A et al (2016) Formation of the Isthmus of Panama. Sci Adv 2(8):e1600883. https://doi.org/10.1126/sciadv.1600883
Oliveira-Filho AT, Ratter JA (1995) A study of the origin of central Brazilian forests by the analysis of plant species distribution patterns. Edinb J Bot 52:141–194
Patel S, Weckstein JD, Patane JSL, Bates JM, Aleixo A (2011) Temporal and spatial diversification of *Pteroglossus* aracaris (Aves Ramphastidae) in the Neotropics: constant rate of diversification does not support an increase in radiation during the Pleistocene. Mol Phylogenet Evol 58:105–115
Patton JL, Silva MNF, Malcolm JR (1994) Gene genealogy and differentiation among arboreal spiny rats (Rodentia: Echmyidae) of the Amazon: a test of the riverine barrier hypothesis. Evolution 48:1314–1323
Patton JL, Silva MNF, Malcolm JR (1996) Hierarchical genetic structure and gene flow in three sympatric species of Amazonian rodents. Mol Ecol 5:229–238
Patton JL, Silva MNF, Malcolm JR (2000) Mammals of the Rio Juruá and the evolutionary and ecological diversification of Amazonia. Bull Am Mus Nat Hist 244:1–306
Pennington TR, Prado DE, Pendry CA (2000) Neotropical seasonally dry forest and Quaternary vegetation changes. J Biogeogr 27:261–273
Peterson AT (2006) Application of molecular clocks in ornithology revisited. J Avian Biol 37:541–544
Peterson AT (2009) Phylogeography is not enough: the need for multiple lines of evidence. Front Biogeogr 1:19–25
Peterson AT, Li X, Navarro-Sigüenza AG (2010) Joint effects of marine intrusion and climate change on the Mexican avifauna. Ann Assoc Am Geogr 100:908–916
Pielou EC (1979) Biogeography. Wiley, New York
Pinto OMO (1940) Aves de Pernambuco. Arq Zool 1:219–282
Pinto OMO (1978) Novo catálogo das aves do Brasil. Revista dos Tribunais, São Paulo, p 1
Piperno DR (1997) Phytoliths and microscopic charcoal from leg 155: a vegetational and fire history of the Amazon basin during the last 75 K.Y. In: Flood RD, Piper DJW, Klaus A, Peterson LC (eds) Proceedings of the ocean program, scientific results, vol 155. Ocean Drilling Program, College Station, TX, pp 411–418
Pomara LY, Ruokolainen K, Young KR (2014) Avian species composition across the Amazon River: the roles of dispersal limitation and environmental heterogeneity. J Biogeogr 41:784–796
Por FD (1992) Sooretama the Atlantic Rain Forest of Brazil. SPB Academic Publishing, The Hague
Potter PE, Szatmari P (2009) Global Miocene tectonics and the modern world. Earth Sci Rev 96:279–295
Prance GT (1973) Phytogeographic support for the theory of Pleistocene forest refuges in the Amazon basin, based on evidence from distribution patterns in Caryocaraceae, Chrysobalanaceae, Dichapetalaceae and Lecythidaceae. Acta Amazon 3:5–28
Ramos VA (2009) Anatomy and global context of the Andes: main geologic features and the Andean orogenic cycle. Geol Soc Am Mem 204:31–65
Räsänen ME, Linna AM, Santos JCR, Negri FR (1995) Late Miocene tidal deposits in the Amazonian foreland basin. Science 269:386–390
Raymo ME, Mitrovika JX, O'Leari MJ, DeConto RM, Earty PJ (2011) Departures from eustasy in Pliocene sea-level records. Nat Geosci 4:328–332
Raymo ME, Kozdon R, Evans D, Lisiecki L, Ford HL (2018) The accuracy of mid-Pliocene δ18O-based ice volume and sea level reconstructions. Earth-Sci Rev 177:291–302

Redford KE, da Fonseca GAB (1986) The role of gallery forests in the zoogeography of the Cerrado's non-volant mammalian fauna. Biotropica 18:126–135

Remsen JV Jr et al (2018) A classification of the bird species of South America. American Ornithologists' Union, Baton Rouge. www.museum.lsu.edu/~Remsen/SACCBaseline.htm

Ribas CC, Gaban-Lima R, Miyaki CY, Cracraft J (2005) Historical biogeography and diversification within the Neotropical parrot genus *Pionopsitta* (Aves: Psittacidae). J Biogeogr 32:1409–1427

Ribas CC, Aleixo A, Nogueira ACR, Miyaki CY, Cracraft J (2012) A palaeobiogeographic model for biotic diversification within Amazonia over the past three million years. Proc R Soc B 799:681–689

Ribas CC, Aleixo A, Gubili C, d'Horta FM, Brumfield RT, Cracraft J (2018) Biogeography and diversification of *Rhegmatorhina* (Aves: Thamnophilidae): implications for the evolution of Amazonian landscapes during the Quaternary. J Biogeogr 45:917–928

Richards PW (1952) The tropical rain forest. Cambridge University Press, Cambridge

Ridgely RS (1976) Birds of Panama. Princeton University Press, Princeton

Ridgely RS, Greenfield PJ (2001) The birds of Ecuador. Cornell University Press, Ithaca, NY

Ridgely RS, Tudor G (1989) The birds of South America, vol 1. Oxford University Press, Oxford

Ridgely RS, Tudor G (1994) The birds of South America, vol 2. Oxford University Press, Oxford

Rodríguez-Mata JR, Erize F, Rumboll M (2006) A field guide to the birds of South America. Non-passerines. Harper Collins, London

Rohling EJ, Grant K, Bolshaw M, Roberts AP, Siddall M, Hemleben C, Kucera M (2009) Antarctic temperature and global sea level closely coupled over the past five glacial cycles. Nat Geosci 2:500–504

Rull V (2006) Quaternary speciation in the Neotropics. Mol Ecol 15:4257–4259

Rull V (2011) Neotropical biodiversity: timing and potential drivers. Trends Ecol Evol 26:508–513

Rull V (2015) Pleistocene speciation is not refuge speciation. J Biogeogr 42:602–609

Sanders HL (1968) Marine benthic diversity: a comparative study. Am Nat 102:243–282

Sanders HL (1969) Benthic marine diversity and the stability-time hypothesis. Brookhaven Symp Biol 22:71–81

Schulenberg TS, Stotz DF, Lane DF, O'Neill JP, Parker TA III (2007) Birds of Peru. Princeton University Press, Princeton

Schultz ED, Burney CW, Brumfield RT, Polo EM, Cracraft J, Ribas CC (2017) Systematics and biogeography of the *Automolus infuscatus* complex (Aves; Furnariidae): cryptic diversity reveals western Amazonia as the origin of a transcontinental radiation. Mol Phylogenet Evol 107:503–515

Schwabe GH (1969) Towards an ecological characterisation of the South American continent. In: Fittkau EJ, Illies J, Klinge A, Schwabw GH, Sioli H (eds) Biogeography and ecology in South America. Dr. W. Junk Publishers, Dordrecht, pp 113–136

Shackleton NJ, Opdyke ND (1977) Oxygen isotope and palaeomagnetic evidence for early northern Hemisphere glaciation. Nature 270:216–219

Short LL (1975) A zoogeographic analysis of the South American Chaco avifauna. Bull Am Mus Nat Hist 154:163–352

Sick H (1967) Rios e enchentes na Amazonia como obstáculo para avifauna. In: Lent H (ed) Simpósio sobre a Biota Amazonica. CNPq, Rio de Janeiro, pp 495–520

Sick H (1993) Birds of Brazil. Princeton University Press, Princeton

Silva JMC (1995) Biogeographic analysis of the South American cerrado avifauna. Steenstrupia 21:49–67

Silva JMC (1996) Distribution of Amazonian and Atlantic birds in gallery forests of the Cerrado region, South America. Ornithol Neotrop 7:1–18

Silva JMC, De Sousa MC, Castelleti CHM (2004) Areas of endemism for passerine birds in Atlantic Forest, South America. Global Ecol Biogeogr 13:85–93

Silveira LF, Olmos F, Long AJ (2003) Birds in Atlantic Forest fragments in north-east Brazil. Cotinga 20:32–46

Simpson BB (1975) Pleistocene changes in the flora of the high tropical Andes. Paleobiology 1:273–294

Simpson BB, Haffer J (1978) Speciation patterns in the Amazonian forest biota. Annu Rev Ecol Syst 9:497–518

Smith LB (1962) Origins of the flora of southern Brazil. Bull US Natl Mus 32:215–249

Smith BT, Clicka J (2010) The profound influence of the Late Pliocene Panamanian uplift on the exchange, diversification, and distribution of New World birds. Ecography 33:333–342

Smith BT, McCormack JE, Cuervo AM, Hickerson MJ, Aleixo A, Cadena CD, Pérez-Emán J, Burney CW, Xie X, Harvey MG, Faircloth BC, Glenn TC, Derryberry EP, Prejean J, Fields S, Brumfield RT (2014) The drivers of tropical speciation. Nature 515:406–409

Sobral-Souza T, Lima-Ribeiro MS, Solferini VN (2015) Biogeography of Neotropical Rainforests: past connections between Amazon and Atlantic Forest detected by ecological niche modeling. Evol Ecol 29:643–655. https://doi.org/10.1007/s10682-015-9780-9

Sousa-Neves T, Aleixo A, Sequeira F (2013) Cryptic patterns of diversification of a widespread Amazonian woodcreeper species complex (Aves: Dendrocolaptidae) inferred from multilocus phylogenetic analysis: implications for historical biogeography and taxonomy. Mol Phylogenet Evol 68:410–424

Stattersfield AJ, Crosby MJ, Long AJ, Wege DC (1998) Endemic bird areas of the world. Priorities for biodiversity conservation. BirdLife International, Cambridge

Stotz DF, Fitzpatrick JW, Parker TA III, Moskovits DK (1996) Neotropical birds. Ecology and conservation. University of Chicago Press, Chicago

Stute M, Forster M, Frischkorn H, Serejo AJF, Clark PS, Broecker WS, Bonani G (1995) Cooling of tropical Brazil (5C) during the last glacial maximum. Science 269:379–383

Tanner WF (1968) Tertiary sea-level fluctuations. Palaeogeogr Palaeoclimatol Palaeoecol 5:7–171

Thistle D (1983) The role of biologically produced habitat heterogeneity in deep-sea diversity maintenance. Deep-Sea Res 30:1235–1245

Thom G, Aleixo A (2015) Cryptic speciation in the white-shouldered antshrike (*Thamnophilus aethiops*, Aves—Thamnophilidae): the tale of a transcontinental radiation across rivers in lowland Amazonia and the northeastern Atlantic Forest. Mol Phylogenet Evol 82:95–110

Thomé MTC, Sequeira F, Brusquetti F, Carstens B, Haddad CFB, Trefaut Rodrigues M, Alexandrino J (2016) Recurrent connections between Amazon and Atlantic forests shaped diversity in Caatinga four-eyed frogs. J Biogeogr 43:1045–1056

Trujillo-Arias N, Dantas GPM, Arbeláez-Cortés E, Naoki K, Gómez MI, Santos FR, Miyaki CY, Aleixo A, Tubaro PL, Cabanne GS (2017) The niche and phylogeography of a passerine reveal the history of biological diversification between the Andean and the Atlantic forests. Mol Phylogenet Evol 112:107–121

Van der Hammen T (1961) The Quaternary climatic changes of northern South America. Ann N Y Acad Sci 95:676–683

Van der Hammen T (1974) The Pleistocene changes of vegetation and climate in tropical South America. J Biogeogr 1:3–26

Van der Hammen T, Absy ML (1994) Amazonia during the last glacial. Palaeogeogr Palaeoclimatol Palaeoecol 109:247–261

Vanzolini PE (1968) Geography of the South American gekkonidae (Sauria). Arq Zool 17:85–111

Vanzolini PE (1981) A quasi-historical approach to the natural history of the differentiation of reptiles in tropical geographic isolates. Pap Avulsos Zool 34:189–204

Vanzolini PE, Williams EE (1970) South American anoles: the geographic differentiation and evolution of the *Anolis chrysolepis* species group (Sauria, Iguanidae). Arq Zool 19:1–298

Vaughan DG (2000) Bedmap subglacial and seabed topography for Antarctica. British Antarctic Survey, Cambridge. www.antarctica.ac.uk/BAS_Science/Highlights/2000/bedmap.html

Vergara J, Acosta LE, González-Ittig RE, Vaschetto LM, Gardenal CN (2017) The disjunct pattern of the Neotropical haverstman *Discocyrtus dilatatus* (Gonyleptidae) explained by climate-driven range shifts in the Quaternary: Paleodistributional and molecular evidence. PLoS One 12(11):e0187983. https://doi.org/10.1371/journal.pone.0187983

Vilela RV, Machado T, Ventura K, Fagundes V, Silva MJ, Yonenaga-Yassuda Y (2009) The taxonomic status of the endangered thin-spined porcupine, *Chaetomys subspinosus* (Olfers, 1818), based on molecular and karyologic data. BMC Evol Biol 9:29

Vonhof HB, Kaandorp RJG (2010) Climate variation in Amazonia during the Neogene and the Quaternary. In: Hoorn C, Wesselingh FP (eds) Amazonia: landscape and species evolution: a look into the past. Wiley Blackwell Publishing, New Jersey, pp 201–210

Vonhof HB, Wesselingh EP, Kaandorp RJG, Davies GR, van Hinte JE, Guerrero J, Räsänen M, Romero-Pitmann L, Ranzi A (2003) Paleogeography of Miocene Western Amazonia: isotopic composition of moluscan shells constrains the influence of marine incursions. Geol Soc Am Bull 115:983–993

Vuilleumier F (1985) Fossil and recent avifaunas and the interamerican interchange. In: Stehli FG, Webb SD (eds) The Great American Biotic Interchange. Springer, Boston, MA, pp 387–424

Wang X, Auler AS, Edwards RL, Cheng H, Cristalli PS, Smart PL, Richards DA, Shen CC (2004) Wet periods in northeastern Brazil over the past 210 kyr linked to distant climate anomalies. Nature 432:740–743

Wang X, Edwards RL, Auler AS, Cheng H, Kong X, Wang Y, Cruz FW, Dorale JA, Chiang HW (2017) Hydroclimate changes across the Amazon lowlands over the past 45,000 years. Nature 541:204–207

Ward WT, Jessup RW (1965) Changes of sea-levels in southern Australia. Nature 205:791–792

Webb SD (1991) Ecogeography and the Great American Interchange. Paleoebiology 17:266–280

Webb SD (1995) Biological implications of the middle Miocene Amazon seaway. Science 269:361–362

Webb RS, Rind D, Lehman SJ, Healy RJ, Sigman D (1997) Influence of ocean heat transport on the climate of the Last Glacial Maximum. Nature 385:695–699

Weir JT, Price M (2011) Andean uplift promotes lowland speciation through vicariance and dispersal in *Dendrocincla* woodcreepers. Mol Ecol 20:4550–4563

Weir JT, Bermingham E, Schluter D (2009) The Great American Biotic Interchange in birds. Proc Natl Acad Sci U S A 106:21737–21742

Weir JT, Faccio MS, Cruz PP, Barrera-Guzmán AO, Aleixo A (2015) Hybridization in headwater regions, and the role of rivers as drivers of speciation in Amazonian birds. Evolution 69:1823–1834

Wesselingh FP, Salo J (2006) A Miocene perspective of the evolution of Amazonian biota. Scri Geol 133:439–445

Wesselingh FP, Räsänen ME, Irion GE, Vonhof HB, Kaandorp R, Renema W, Romero Pitman L, Gingras M (2002) Lake Pebas: a palaeoecological reconstruction of a Miocene, long lived lake in western Amazonia. Cenozoic Res 1:35–81

Willis EO (1992) Zoogeographical origins of eastern Brazilian birds. Ornithol Neotrop 3:1–15

Willis KJ, Whittaker RJ (2000) The refugial debate. Science 287:1406–1407

Zeuner FE (1945) The Pleistocene period. In: Its climate, chronology and faunal successions. Ray Society, London

Zink RM, Barrowclough GF (2008) Mitochondrial DNA under siege in avian phylogeography. Mol Ecol 17:2107–2121

Chapter 9
Contrasting Patterns of Temporal Diversification in Neotropical Butterflies: An Overview

Andrew V. Z. Brower and Ivonne J. Garzón-Orduña

Abstract The exuberant diversity present in the Neotropics naturally leads to questions about its age of origin. The butterfly fossil record is so depauperate that most of what we know today about the age of butterflies is based on relaxed clock estimates. Here we compile relaxed clock-based crown age estimates for clades of Neotropical butterflies from 6 families, as well as age estimates for more than 400 pairs of sister species included in 102 genera. Alternative perspectives of the temporal diversification of the Neotropics based on these two sources of evidence are contrasted. While crown ages of more inclusive groups range back to the Miocene and Oligocene, most species pairs diverged within the Pleistocene.

Keywords Lepidoptera · Amazonia · Andes · South America · Divergence times · Crown ages · Sister species divergences · Molecular clock · Pleistocene refugia

1 Introduction: The Neotropics as a Region and as a Group of Related Regions

The Neotropical region, here considered to range from southern Mexico and the Antilles to the Chaco and Atlantic coastal forest (and excluding Patagonia) (Neotropical Region sensu lato of Morrone 2014) has a complex geological history over

---

Electronic supplementary material The online version of this chapter (https://doi.org/10.1007/978-3-030-31167-4_9) contains supplementary material, which is available to authorized users.

A. V. Z. Brower (✉)
Evolution and Ecology Group, Department of Biology, Middle Tennessee State University, Murfreesboro, TN, USA

USDA – APHIS National Identification Service, Riverdale, MD, USA
e-mail: andrew.brower@usda.gov

I. J. Garzón-Orduña
Colección Nacional de Insectos, Instituto de Biología, Universidad Nacional Autónoma de México, Ciudad de México, Mexico

V. Rull, A. C. Carnaval (eds.), *Neotropical Diversification: Patterns and Processes*, Fascinating Life Sciences, https://doi.org/10.1007/978-3-030-31167-4_9

hundreds of millions of years. However, some of the most important events that shaped the biotic landscape as we know it today happened rather recently in terms of geological time. The drainage of the Pebas wetland, which once occupied much of the western part of what is now the Amazon basin, the origin of the modern Amazon River and its tributaries, and the uplift of the northern Andes, each occurred since the Middle Miocene some 15 mya (Gregory-Wodzicki 2000; Hoorn et al. 2010). Today the Neotropical region is a composite landmass made up of formerly separate areas, currently featuring different biomes and a biota of composite origin as well (Amorim 2001).

Explaining the origin(s) of Neotropical biodiversity remains a puzzle, now over 200 years old. The problem at hand is a complex one by its historical nature, but it is also impaired by semantic disagreement. It is now evident that when authors speak of the "origins of diversity", two different kinds of data come to mind: the age of clades versus the age of extant species. In the context of crown ages of clades, authors are referring to the phylogenetic origin of an extant higher taxon, when the single ancestral species is inferred to have undergone its first speciation event. Thus, crown ages specify the age of the oldest lineage of a clade. In contrast, a focus on the ages of extant species provides a context for contemplating not only the "origins of diversity" in terms of the taxa that biologists count to compare species richness and other estimates of biodiversity, but also the historical origins of characters of phylogenetic and ecological interest, including current biogeographical distributions and biotic associations with other taxa.

Diversity, like maturity, is a condition reached gradually through time. The process of diversification for a clade in a region begins when a representative of that group migrates into the region from somewhere else, or diverges in situ from a sister taxon. The subsequent rate of increase in numbers of lineages through time can be estimated, or modeled, based on inferred ages of branching events (e.g., Matos-Maravi 2016), as long as we are willing to accept numerous assumptions about unobserved processes, such as extinction. Whether patterns so inferred have anything to do with what actually happened is speculative and to a great extent untestable, and diversification scenarios extrapolated from molecular clocks must be viewed as semi-fictional historical narratives, particularly when the clade in question is not well documented by fossils that provide estimates of minimum ages for the taxa to which they are assigned. Butterflies have a meager fossil record: de Jong (2017) indicated that the entirety of the world's known butterfly fossils may be used to calibrate 13 nodes upon the phylogenetic tree of 20,000+ extant species of Papilionoidea, including two in Papilionidae, two in Hesperiidae, five in Nymphalidae, two in Riodinidae and one each in Lycaenidae and Pieridae. These data points have been augmented to a limited degree with calibrations from biogeography and larval host plant associations (e.g., Wahlberg et al. 2009) at some empirical cost (Garzón-Orduña et al. 2015b).

Here we offer a descriptive account of the current understanding of diversification patterns of Neotropical butterflies based on published phylogenetic hypotheses for species-level groups. We touch specifically on the spatial and temporal origins of different clades, highlighting groups that have differentiated into a large number of species. We also expand our dataset on the age of divergence among pairs of sister species (Garzón-Orduña et al. 2014), compare these data with relaxed clock-based crown age estimates for the clades that include those species, and discuss patterns of diversification within individual clades.

2 Rationale, Data Collection and Data Analyses

In order to highlight the biogeographical origins of taxa that currently occupy the Neotropics, we have chosen to divide our discussion into two loosely defined groups: large, inclusive clades that evidently evolved in situ in the Neotropics (Endemics), and more restricted clades that diversified in the Neotropics but likely originated outside (called here Newcomers). For the temporal component, we report two complementary measures of taxon age: crown group ages of more inclusive Neotropical taxa, and inferred ages of divergence of extant sister species. Throughout the chapter we use either myr (million years) or mya (million years ago) as descriptors of these two metrics. The age of a crown group is defined as the inferred age of its most recent common ancestor (Jefferies 1979), indicated as the oldest branching point between extant members of the clade on a phylogenetic tree. Unless there were more specific published time trees for a particular group, crown age estimates (along with 95% credibility intervals) for major clades in Nymphalidae were derived from Wahlberg et al. (2009), for Hesperiidae from Sahoo et al. (2017), for Riodinidae from Espeland et al. (2015) and Seraphim et al. (2018), and for Papilionidae, Pieridae, Lycaenidae and Hedylidae ("moth-butterflies") from Espeland et al. (2018). In cases where 95% intervals were not reported as numbers but only as bars around the best estimated age in published figures, we estimated these values by comparing the bar length to the timescale in the figure.

The reader should keep in mind that evidence from molecular clock-based studies is affected by (1) the fact that taxonomic sampling in a given phylogenetic study is often incomplete, which may have a dramatic impact on the accuracy of date estimates, and (2) crown ages so estimated are inferred optima that come with wide confidence intervals—and even these may not take into account other sources of uncertainty, such as the accuracy and precision of ages of the calibration points themselves.

Given these diverse sources of uncertainty accompanying molecular clock estimation, we prefer a simple, fixed clock approach for estimates of divergence times between closely-related species, rather than the more complex and assumption-laden models employed in relaxed clock BEAST analyses (Garzón-Orduña et al. 2014), especially those that rely on secondary calibrations that may themselves be based on rather minimal fossil evidence (Garzón-Orduña et al. 2015). Brower (1994) published a widely cited clock estimate for insect mtDNA, providing a benchmark rate of 2.3% divergence per million years that is useful for calibrating ages for putatively young taxa, between which DNA sequence divergence is low and therefore unlikely to be prone to mutational saturation. Fortunately, almost all molecular phylogenetic work on Neotropical butterflies, including the mass-sampling of DNA barcodes from Guanacaste, Costa Rica (Janzen et al. 2011; Janzen and Hallwachs

2016), has employed mtDNA COI as a marker, rendering comparable sequences for age estimates between closely-related species in all groups.

We obtained data from putative sister species in one of three ways: if a species-level phylogenetic hypothesis for a group is available, we inferred pairwise sequence divergence from sister species as indicated on the published tree; in cases where large-scale neighbor-joining trees for barcodes implied a sister-taxon relationship between congeners, we used those pairs; and in cases where multiple species of a genus were represented in GenBank by sequences from different studies, we built COI matrices de novo and conducted heuristic parsimony searches using PAUP* (Swofford 2000) to determine putative sister species. Note that in the latter two protocols, there is no guarantee that the species compared are actually sister taxa. However, as discussed by Garzón-Orduña et al. (2014), such comparisons represent conservative estimates with regard to falsifying Pleistocene-age origins of extant species, since if they are not sister taxa the comparison would necessarily imply an older age of divergence than that of the true sister taxon pair. In all cases where multiple conspecific sequences were available for comparison with those of a putative sister species, we selected those that provided the longest sequence overlap, to maximize the empirical robustness of our age estimates.

Data mined from GenBank rely upon the assumption that the sources of the sequences are correctly identified. It is possible that in some instances, specimens identified as separate species are actually conspecific, which could lead to underestimates of species' ages. Another potential source of error in these estimates is differences in taxonomic circumscription from one group to another, depending on authors' proclivities to lump or split taxa. Thus, some genera may appear to have many, recent speciation events not so much because they are evolving rapidly, as because the taxonomist who revised the group separated geographical variants into separate species. Other taxa may have more broadly-circumscribed species that encompass tremendous geographical variability (e.g., *Heliconius*).

Of course, as we have noted previously (Garzón-Orduña et al. 2014), the notion of a "crown group" is a completely arbitrary higher taxonomic unit that can be selected to support whatever pattern a researcher may desire to emphasize. If we wish to say that group X diversified in the Paleogene, all we need to do is to circumscribe that group so that its crown age supports our claim. The crown ages we report here reflect "origins of neotropical diversification" to the extent that they are clades that appear to be Neotropical endemics, or to have originated in the Neotropics with some subsequent range expansion outside the region.

Taxonomic composition of clades is based on the most recent electronic update of the Lamas (2004) checklist of Neotropical butterflies, as well as ToLWEB pages (www.tolweb.org/Papilionoidea, accessed 31 July 2018). Numbers of genera and species indicated are those for the Neotropical fauna. Crown ages for each clade discussed are shown in Fig. 9.1, and ages of species pairs, by genus, are shown in Fig. 9.2 (Tables 9.1 and 9.2). Data supporting these graphical summaries is available as supplementary material.

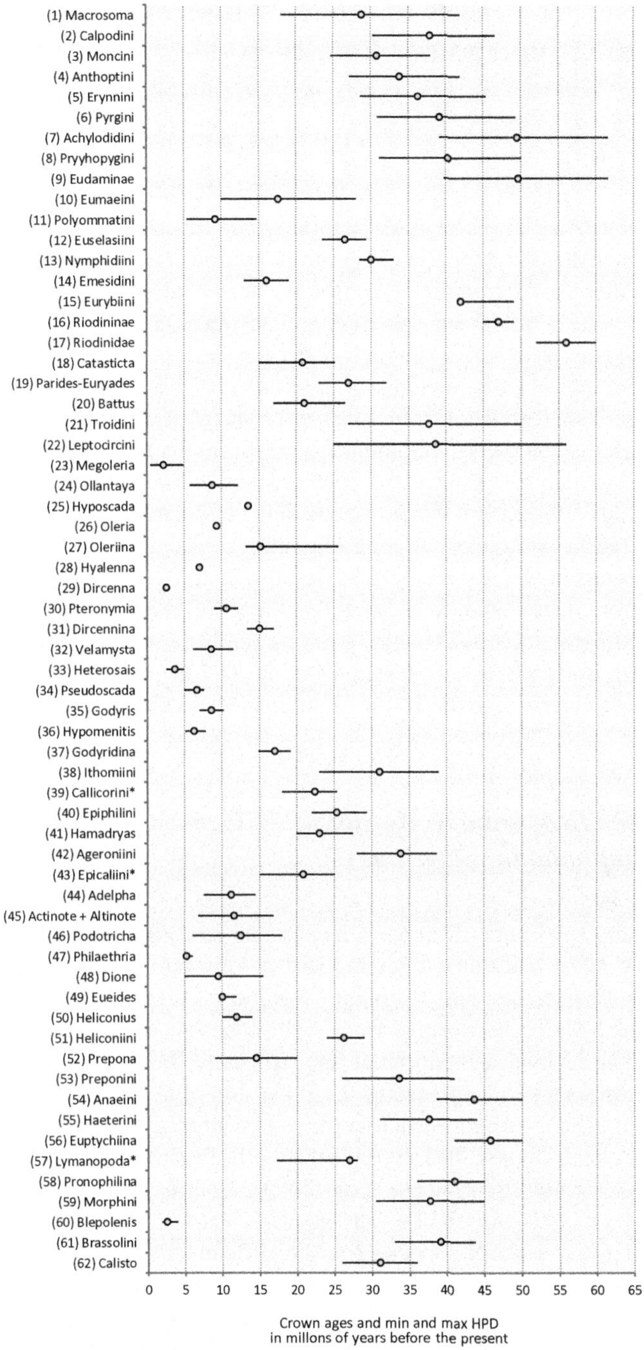

Fig. 9.1 Crown ages (and associated higher posterior density intervals) reported in the literature for 27 genera of neotropical butterflies (white circles) and 35 suprageneric groups (black circles). References to the studies used as source of these data can be found in Table 9.1. Colors in timescale represent the following geological periods in a present to past order: Holocene, Pleistocene, Pliocene, Miocene, Oligocene, Eocene and Paleocene

3 Endemic (or Largely Endemic) Butterfly Clades, Including Supergeneric Radiations

3.1 Tribe Ithomiini: 43 Genera, 354 spp. (Nymphalidae: Danainae)

Ithomiine butterflies, also known as clearwings, are endemic to the Neotropics, with a few representatives ranging as far north as central Mexico, and two species of *Greta* on Cuba and Hispaniola. The tribe includes clades associated more closely with lowlands (e.g. *Melinaea, Mechanitis*) and others associated with the Andes (e.g. *Napeogenes, Patricia*). Although noted for their diversity, mimicry, and patterns of geographical variation long ago by Bates (1862b), Brown (1977, 1979) was the first to provide detailed distributional maps and hypotheses about the origins of the group. In recent years, ithomiines have been flooded with time trees and diversification analyses (Elias et al. 2009; Chazot et al. 2016a, b 2017; De-Silva et al. 2016, 2017), making it the only clade of neotropical butterflies for which three alternative (and contradictory) dating hypotheses exist at the moment (Wahlberg et al. 2009; Garzón-Orduña et al. 2015; De-Silva et al. 2017). In general, crown age estimates from these studies suggest that most of the genera studied to date (12 included here) started diversifying during the Miocene, however according to strict clock estimations based on mitochondrial DNA, extant species pairs are on average of Pleistocene age (based on 54 sister species comparisons).

3.2 Tribes Brassolini + Morphini: 21 Genera, 137 spp. (Nymphalidae: Satyrinae)

The Brassolini (owl butterflies) are a crepuscular and mainly understory group of mostly very large butterflies, endemic to the lowland Neotropics, and sister to another endemic clade, the Morphini. Wahlberg et al. (2009) calculated that Brassolini and Morphini diverged from each other 48.5 mya, Brassolini began its diversification 39 mya and Morphini approximately 37 mya, in both cases their diversification seems to be related to the expansion of tropical forest (Blandin and Purser 2013). The most representative genus in Morphini is *Morpho* a charismatic and popular ambassador of the Neotropical butterfly fauna in butterfly houses around the world. According to Wahlberg et al. (2009), *Morpho* split from its sister group around 32 mya (with wide HPD). The genus contains 29 species. Based on Penz et al. (2012) it seems *Morpho* species began to differentiate during the early Miocene (~20 myr). A larval host shift from monocots to dicots seems to have been an important event in the diversification of the genus (Penz and DeVries 2002). According to our calculations and based on a phylogenetic hypothesis published by Chazot et al. (2016a, b), the average age of extant sister pairs in *Morpho* species is 1.67 myr. Brassolini contains 19 genera, the largest of which is *Caligo* with

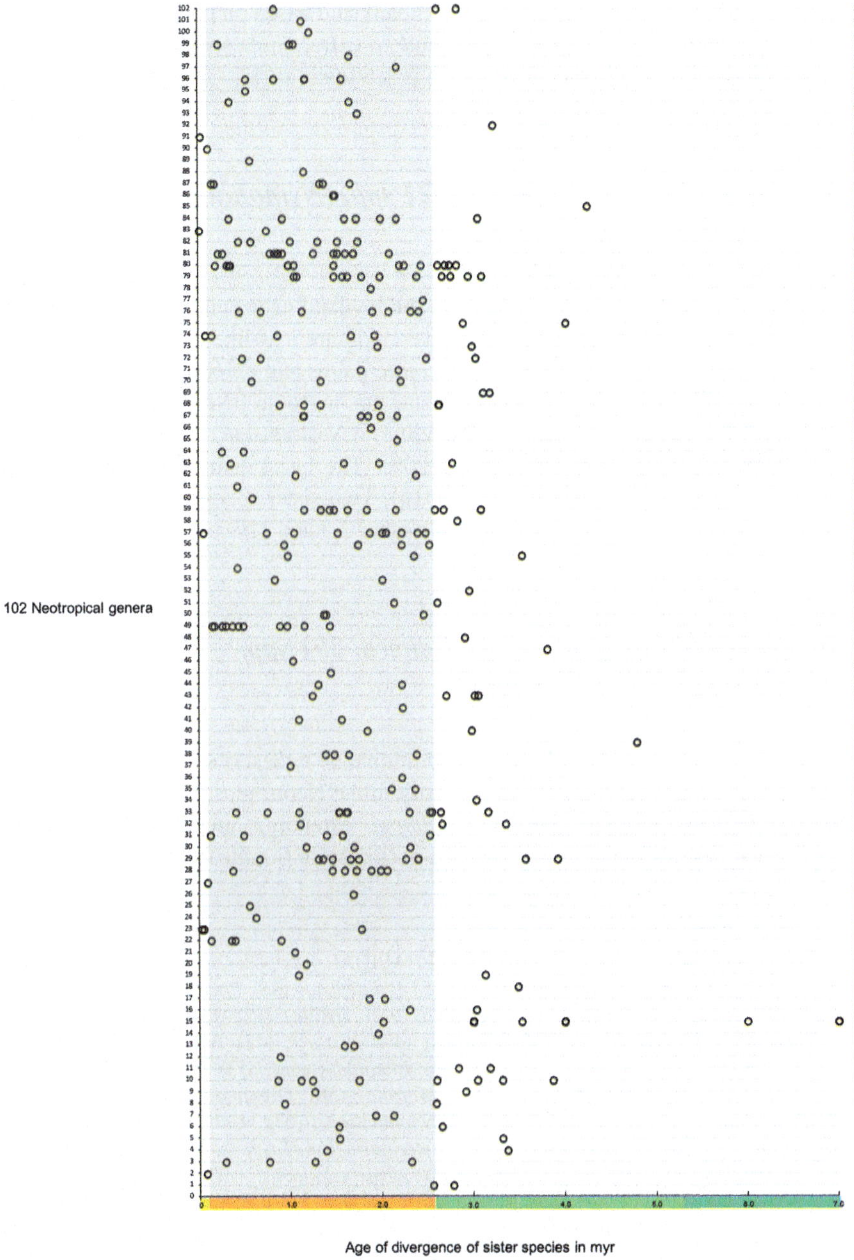

Fig. 9.2 Age of divergence between sister taxa of 102 neotropical genera of butterflies. From bottom to top: (1) HEDYLIDAE: *Macrosoma*, (2) HESPERIDAE: *Cobalus*, (3) *Astraptes*, (4) *Bungalotis*, (5) *Telemiades*, (6) *Urbanus*, (7) *Venada*, (8) *Carrhenes*, (9) *Synapte*, (10) PAPILIONIDAE: *Parides-Euryades*, (11) *Battus*, (12) *Protesilaus*, (13) *Mimiodes*, (14) *Neographium*, (15) PIERIDAE: *Catasticta*, (16) *Dismorphia*, (17) *Enantia*, (18) RIODINIDAE: *Theope*, (19) *Detritivora*, (20) *Sertania*, (21) *Riodina*, (22) LYCAENIDAE: *Pseudolucia*, (23) *Madeleinea*, (24) *Nabokovia*, (25) *Hemiargus*, (26) *Itylos*, (27) *Paralycaeides*,

21 species. Using COI sequences from Genbank and based on published cladograms (Penz 2009; Penz et al. 2011) we were able to extract nine species pairs from four genera, whose average age of divergence is 1.6 my.

3.3 Tribe Haeterini: 5 Genera, 21 spp. (Nymphalidae: Satyrinae)

The satyrine tribe Haeterini is endemic to the lowland neotropics from Amazonia to Central America. Adults in this distinctive clade are leaf-like or transparent forest floor dwellers whose larvae feed on *Heliconia*, palms and other non-Poaceae monocots (DeVries 1987). The group is sister taxon to a heterogeneous clade of Old-World tropical satyrines including Australasian Amathusiini and South African Dirini (Wahlberg et al. 2009). The latest estimated crown age for Haeterini is 37.5 myr (Céspedes et al. 2015). Two sister species of *Cithaerias* and two species of *Pierella* diverged from each other 2.5 and 1.89 myr ago, respectively.

3.4 Subtribe Euptychiina: 47 Genera, 391 spp. (Nymphalidae: Satyrinae)

The subtribe Euptychiina (including an additional 67 undescribed species, according to Marín et al. 2017) is distributed primarily in the Neotropics, but with a few taxa, such as *Megisto*, *Hermeuptychia* and *Cyllopsis* extending into the Nearctic realm, and one species (*Palaeonympha opalina*) occurring in Southeast China and Taiwan

---

Fig. 9.2 (continued) (28) *Oenomaus*, (29) NYMPHALIDAE: *Adelpha*, (30) *Altinote*, (31) *Actinote*, (32) *Eueides*, (33) *Heliconius*, (34) *Dione*, (35) *Philaethria*, (36) *Podotricha*, (37) *Janatella*, (38) *Eresia*, (39) *Gnathotriche*, (40) *Castilla*, (41) *Ortilia*, (42) *Telenassa*, (43) *Tegosa*, (44) *Anartia*, (45) *Hypanartia*, (46) *Junonia*, (47) *Historis*, (48) *Baeotus*, (49) *Perisama*, (50) *Hamadryas*, (51) *Paulogramma*, (52) *New Riodinid Genus*, (53) *Diaethria*, (54) *Consul*, (55) *Archaeoprepona*, (56) *Prepona*, (57) *Morpho*, (58) *Oressinoma*, (59) *Lymanopoda*, (60) *Harjesia*, (61) *Blepolenis*, (62) *Dasyophthalma*, (63) *Caligo*, (64) *Opsiphanes*, (65) *Parataygetis*, (66) *Taygetomorpha*, (67) *Forsterinaria*, (68) *Taygetis*, (69) *Taygetina* subclade, (70) *Altopedaliodes*, (71) *Corades*, (72) *Praepedaliodes*, (73) *Steromapedaliodes*, (74) *Caeruleuptychia*, (75) *Chloreuptychia*, (76) *Yphthimoides*, (77) *Cithaerias*, (78) *Pierella*, (79) *Calisto*, (80) *Oleria*, (81) *Pteronymia*, (82) *Episcada*, (83) *Dircenna*, (84) *Napeogenes*, (85) *Tithorea*, (86) *Hyposcada*, (87) *Ithomia*, (88) *Hypothyris*, (89) *Hyaliris*, (90) *Forbestra*, (91) *Megoleria*, (92) *Ollantaya*, (93) *Haenschia*, (94) *Hyalenna*, (95) *Pseudoscada*, (96) *Hypomenitis*, (97) *Greta*, (98) *Heterosais*, (99) *Godyris*, (100) *Veladyris*, (101) *Velamysta*, (102) *Hypoleria*. References to the studies used as source of these data can be found in Table 9.2. The area shaded on the figure represent the length of Pleistocene period. Colors in timescale represent the following geological periods in a present to past order: Holocene, Pleistocene, Pliocene and Miocene

Table 9.1 Source of crown ages reported in our study

Taxon	Family: Subfamily	References
Macrosoma	Hedylidae	Espeland et al. (2018)
Calpodini	Hesperidae: Hesperiinae	Sahoo et al. (2017)
Moncini	Hesperidae: Hesperiinae	Sahoo et al. (2017)
Anthoptini	Hesperidae: Hesperiinae	Sahoo et al. (2017)
Erynnini	Hesperidae: Pyrginae	Sahoo et al. (2017)
Pyrgini	Hesperidae: Pyrginae	Sahoo et al. (2017)
Achylodidini	Hesperid: Pyrginae	Sahoo et al. (2017)
Pryyhopygini	Hesperidae: Pyrginae	Sahoo et al. (2017)
Eudaminae	Hesperidae: Eudaminae	Sahoo et al. (2017)
Eumaeini	Lycaenidae: Theclinae	Espeland et al. (2018)
Polyommatini	Lycaenidae: Polyommatinae	Vila et al. (2011)
Euselasiini	Riodinidae: Euselasiinae	Seraphim et al. (2018)
Nymphidiini	Riodinidae: Riodininae	Seraphim et al. (2018)
Emesidini	Riodinidae: Riodininae	Seraphim et al. (2018)
Eurybiini	Riodinidae: Riodininae	Seraphim et al. (2018)
Riodininae	Riodinidae	Seraphim et al. (2018)
Riodinidae	Riodinidae	Seraphim et al. (2018)
Catasticta	Pieridae: Pierinae	Padron-Martinez (2014)
Parides-Euryades	Papilionidae: Papiloininae	Condamine et al. (2012)
Battus	Papilionidae: Papilioninae	Condamine et al. (2012)
Troidini	Papilionidae: Papilioninae	Espeland et al. (2018)
Leptocircini	Papilionidae: Papilioninae	Espeland et al. (2018)
Megoleria	Nymphalidae: Danainae	De-Silva et al. (2016)
Ollantaya	Nymphalidae: Danainae	De-Silva et al. (2016)
Hyposcada	Nymphalidae: Danainae	De-Silva et al. (2016)
Oleria	Nymphalidae: Danainae	De-Silva et al. (2016)
Oleriina	Nymphalidae: Danainae	De-Silva et al. (2016)
Hyalenna	Nymphalidae: Danainae	Chazot et al. (2017)
Dircenna	Nymphalidae: Danainae	Chazot et al. (2017)
Pteronymia	Nymphalidae: Danainae	De-Silva et al. (2017)
Dircennina	Nymphalidae: Danainae	Chazot et al. (2017)
Velamysta	Nymphalidae: Danainae	Chazot et al. (2016a, b)
Heterosais	Nymphalidae: Danainae	Chazot et al. (2016a, b)
Pseudoscada	Nymphalidae: Danainae	Chazot et al. (2016a, b)
Godyris	Nymphalidae: Danainae	Chazot et al. (2016a, b)
Hypomenitis	Nymphalidae: Danainae	Chazot et al. (2016a, b)
Godyridina	Nymphalidae: Danainae	Chazot et al. (2016a, b)
Ithomiini	Nymphalidae: Danainae	Garzón-Orduña et al. (2015a, b)
Callicorini*	Nymphalidae: Biblidinae	*includes *Lucinia*. Wahlberg et al. (2009)
Epiphilini	Nymphalidae: Biblidinae	Wahlberg et al. (2009)
Hamadryas	Nymphalidae: Biblidinae	Garzón-Orduña et al. (2013)
Ageroniini	Nymphalidae: Biblidinae	Wahlberg et al. (2009)

(continued)

Table 9.1 (continued)

Taxon	Family: Subfamily	References
Epicaliini*	Nymphalidae: Biblidinae	*excludes *Eunica* and *Sevenia*. Wahlberg et al. (2009)
Adelpha	Nymphalidae: Limenitidinae	Ebel et al. (2015)
Actinote + *Altinote*	Nymphalidae: Heliconiinae	Wahlberg et al. (2009)
Podotricha	Nymphalidae: Heliconiinae	Kozak et al. (2015)
Philaethria	Nymphalidae: Heliconiinae	Kozak et al. (2015)
Dione	Nymphalidae: Heliconiinae	Massardo et al. (2015)
Eueides	Nymphalidae: Heliconiinae	Kozak et al. (2015)
Heliconius	Nymphalidae: Heliconiinae	Kozak et al. (2015)
Heliconiini	Nymphalidae: Heliconiinae	Kozak et al. (2015)
Prepona	Nymphalidae: Charaxinae	Wahlberg et al. (2009)
Preponini	Nymphalidae: Charaxinae	Wahlberg et al. (2009)
Anaeini	Nymphalidae: Charaxinae	Wahlberg et al. (2009)
Haeterini	Nymphalidae: Satyrinae	Cespedes et al. (2015)
Euptychiina	Nymphalidae: Satyrinae	Wahlberg et al. (2009)
*Lymanopoda**	Nymphalidae: Satyrinae	* HPDs from Wahlberg et al. (2009), mean age from: Casner and Pyrcz (2010)
Pronophilina	Nymphalidae: Satyrinae	Wahlberg et al. (2009)
Morphini	Nymphalidae: Satyrinae	Wahlberg et al. (2009)
Blepolenis	Nymphalidae: Satyrinae	Penz et al. (2011a, b)
Brassolini	Nymphalidae: Satyrinae	Wahlberg et al. (2009)
Calisto	Nymphalidae: Satyrinae	Matos-Maravi et al. (2014)

(Yoshino 2001). Most of the euptychiines are lowland forest understory dwellers that as caterpillars feed on a variety of bamboos and grasses. According to the Wahlberg et al. (2009) comprehensive time tree of Nymphalidae, the crown age for the entire subtribe is 45.8 mya, and the age of the *Taygetis* clade (which only included five members in that study) is approximately 20.4 mya. More recent studies of this subgroup with increased taxon sampling (Peña et al. 2010; Matos-Maravi et al. 2013) have estimated the crown age of the *Taygetis* clade to be around 13 myr (though with large HPD intervals). We were able to compile ages of sister species pairs from numerous genera, including *Taygetis*, *Caeruleuptychia* and *Yphthimoides*. Species pairs in these genera were on average 1.7, 0.93 and 1.6 myr, respectively.

Table 9.2 Source of the data used to calculate the age of species pairs

Number in fig.	Taxon	Family	Source of DNA and sister species statements
(1)	*Macrosoma*	Hedylidae	Janzen et al. (unpublished); International Barcode of Life (iBOL)
(2)	*Cobalus*	Hesperidae	Hajibabaei et al. (2006)
(3)	*Astraptes*	Hesperidae	Hebert et al. (2004); Hajibabaei et al. (2006); Janzen et al. (2011)
(4)	*Bungalotis*	Hesperidae	Janzen et al. (2011); Sahoo et al. (2016); International Barcode of Life (iBOL)
(5)	*Telemiades*	Hesperidae	Janzen et al. (2011); International Barcode of Life (iBOL)
(6)	*Urbanus*	Hesperidae	Hajibabaei et al. (2006); Janzen et al. (2011)
(7)	*Venada*	Hesperidae	Janzen et al. (2011); International Barcode of Life (iBOL)
(8)	*Carrhenes*	Hesperidae	Janzen et al. (2011); International Barcode of Life (iBOL)
(9)	*Synapte*	Hesperidae	Janzen et al. (2011); Sahoo et al. (2016)
(10)	*Parides-Euryades*	Papilionidae	Condamine et al. (2012)
(11)	*Battus*	Papilionidae	Silva-Brandão et al. (2005); Condamine et al. (2012)
(12)	*Protesilaus*	Papilionidae	Dumar et al. (unpublished); Nabholz (unpublished)
(13)	*Mimiodes*	Papilionidae	Heikkila et al. (2011); Basset et al. (2015); International Barcode of Life (iBOL); Janzen and Hajibabaei (unpuplished)
(14)	*Neographium*	Papilionidae	Dumar et al. (unpublished); Nabholz (unpublished); International Barcode of Life (iBOL)
(15)	*Catasticta*	Pieridae	Padron-Martinez (2014)
(16)	*Dismorphia*	Pieridae	Chew and Watt (2006); Mitter et al. (2011); Basset et al. (2015); Janzen and Hajibabaei (unpuplished)
(17)	*Enantia*	Pieridae	Wahlberg et al. (2014); Jasso-Martinez and Castaneda-Sortibran (unpublished)
(18)	*Theope*	Riodinidae	Espeland et al. (2015)
(19)	*Detritivora*	Riodinidae	Espeland et al. (2015)
(20)	*Sertania*	Riodinidae	Kaminski et al. (2017)
(21)	*Riodina*	Riodinidae	Seraphim et al. (2018)
(22)	*Pseudolucia*	Lycaenidae	Vila et al. (2011)
(23)	*Madeleinea*	Lycaenidae	Vila et al. (2011)
(24)	*Nabokovia*	Lycaenidae	Vila et al. (2011)
(25)	*Hemiargus*	Lycaenidae	Vila et al. (2011)
(26)	*Itylos*	Lycaenidae	Vila et al. (2011)
(27)	*Paralycaeides*	Lycaenidae	Vila et al. (2011)
(28)	*Oenomaus*	Lycaenidae	Faynel et al. (2012); International Barcode of Life (iBOL)

(continued)

Table 9.2 (continued)

Number in fig.	Taxon	Family	Source of DNA and sister species statements
(29)	*Adelpha*	Nymphalidae	Ebel et al. (2015)
(30)	*Altinote*	Nymphalidae	Silva-Brandão et al. (2008)
(31)	*Actinote*	Nymphalidae	Silva-Brandão et al. (2008)
(32)	*Eueides*	Nymphalidae	Beltrán et al. (2007); Kozak et al. (2015)
(33)	*Heliconius*	Nymphalidae	Kozak et al. (2015)
(34)	*Dione*	Nymphalidae	Massardo et al. (2015)
(35)	*Philaethria*	Nymphalidae	Kozak et al. (2015)
(36)	*Podotricha*	Nymphalidae	Kozak et al. (2015)
(37)	*Janatella*	Nymphalidae	Wahlberg and Freitas (2007)
(38)	*Eresia*	Nymphalidae	Wahlberg and Freitas (2007)
(39)	*Gnathotriche*	Nymphalidae	Wahlberg and Freitas (2007)
(40)	*Castilla*	Nymphalidae	Wahlberg and Freitas (2007)
(41)	*Ortilia*	Nymphalidae	Wahlberg and Freitas (2007)
(42)	*Telenassa*	Nymphalidae	Wahlberg and Freitas (2007)
(43)	*Tegosa*	Nymphalidae	Wahlberg et al. (2005); Wahlberg and Freitas (2007)
(44)	*Anartia*	Nymphalidae	Blum et al. (2003)
(45)	*Hypanartia*	Nymphalidae	Willmott et al. (2001)
(46)	*Junonia*	Nymphalidae	Pfeiler et al. (2012)
(47)	*Historis*	Nymphalidae	Inferred from taxonomy
(48)	*Baeotus*	Nymphalidae	Inferred from taxonomy
(49)	*Perisama*	Nymphalidae	Anna Zubek unpublished data
(50)	*Hamadryas*	Nymphalidae	Garzón-Orduña et al. (2013)
(51)	*Paulogramma*	Nymphalidae	Freitas et al. (2014)
(52)	*New genus*	Nymphalidae	Freitas et al. (2014)
(53)	*Diaethria*	Nymphalidae	Freitas et al. (2014)
(54)	*Consul*	Nymphalidae	Inferred from taxonomy
(55)	*Archaeoprepona*	Nymphalidae	Ortiz-Acevedo and Willmott (2013); Ortiz-Acevedo et al. (2017)
(56)	*Prepona*	Nymphalidae	Ortiz-Acevedo and Willmott (2013); Ortiz-Acevedo et al. (2017)
(57)	*Morpho*	Nymphalidae	Chazot et al. (2016a, b)
(58)	*Oressinoma*	Nymphalidae	Kodandaramaiah et al. (2009)
(59)	*Lymanopoda*	Nymphalidae	Casner and Pyrcz (2010)
(60)	*Harjesia*	Nymphalidae	Matos-Maravi et al. (2013)
(61)	*Blepolenis*	Nymphalidae	Penz et al. (2011a, b)
(62)	*Dasyophthalma*	Nymphalidae	Penz (2009); Mielke and Rougerie (unpubliished)
(63)	*Caligo*	Nymphalidae	Wahlberg et al. (2003); Basset et al. (2015); Wahlberg (unpublished); International Barcode of Life (iBOL)
(64)	*Opsiphanes*	Nymphalidae	Penz et al. (2011a, b); Chacon et al. (2012)
(65)	*Parataygetis*	Nymphalidae	Matos-Maravi et al. (2013)

(continued)

Table 9.2 (continued)

Number in fig.	Taxon	Family	Source of DNA and sister species statements
(66)	*Taygetomorpha*	Nymphalidae	Matos-Maravi et al. (2013)
(67)	*Fosterinaria*	Nymphalidae	Matos-Maravi et al. (2013)
(68)	*Taygetis*	Nymphalidae	Matos-Maravi et al. (2013)
(69)	*Taygetina subclade*	Nymphalidae	Matos-Maravi et al. (2013)
(70)	*Altopedaliodes*	Nymphalidae	K. Willmott pers. comm.
(71)	*Corades*	Nymphalidae	Peña et al. (2006); A. Brower unpublished
(72)	*Praepedaliodes*	Nymphalidae	Pyrcz et al. (2018)
(73)	*Steromapedaliodes*	Nymphalidae	Matz and Brower (2016); Pyrcz et al. (2017)
(74)	*Caeruleuptychia*	Nymphalidae	Peña et al. (2006, 2010); Lavinia et al. (2017); Nakahara et al. (2018)
(75)	*Chloreuptychia*	Nymphalidae	Murray and Prowell (2005); Peña et al. (2010)
(76)	*Yphthimoides*	Nymphalidae	Murray and Prowell (2005); Peña et al. (2006, 2010); Freitas et al. (2012); Barbosa et al. (2015)
(77)	*Cithaerias*	Nymphalidae	Wahlberg et al. (2009)
(78)	*Pierella*	Nymphalidae	Wahlberg (unpublished)
(79)	*Calisto*	Nymphalidae	Matos-Maravi et al. (2014); Nuñez Aguila et al. (2017)
(80)	*Oleria*	Nymphalidae	De-Silva et al. (2010, 2016)
(81)	*Pteronymia*	Nymphalidae	De-Silva et al. (2017)
(82)	*Episcada*	Nymphalidae	Chazot et al. (2017)
(83)	*Dircenna*	Nymphalidae	Chazot et al. (2017)
(84)	*Napeogenes*	Nymphalidae	Elias et al. (2009)
(85)	*Tithorea*	Nymphalidae	Wahlberg et al. (2009)
(86)	*Hyposcada*	Nymphalidae	De-Silva et al. (2010, 2016); Garzón-Orduña et al. (2015a, b)
(87)	*Ithomia*	Nymphalidae	Mallarino et al. (2005)
(88)	*Hypothryris*	Nymphalidae	Arias-Mejia (2012)
(89)	*Hyaliris*	Nymphalidae	Arias-Mejia (2012)
(90)	*Forbestra*	Nymphalidae	Elias et al. (2007)
(91)	*Megoleria*	Nymphalidae	De-Silva et al. (2016)
(92)	*Ollantaya*	Nymphalidae	De-Silva et al. (2016)
(93)	*Haenschia*	Nymphalidae	Chazot et al. (2017)
(94)	*Hyalenna*	Nymphalidae	Chazot et al. (2017)
(95)	*Pseudoscada*	Nymphalidae	Chazot et al. (2016a, b)
(96)	*Hypomenitis*	Nymphalidae	Chazot et al. (2016a, b)
(97)	*Greta*	Nymphalidae	Chazot et al. (2016a, b)
(98)	*Heterosais*	Nymphalidae	Chazot et al. (2016a, b)
(99)	*Godyris*	Nymphalidae	Chazot et al. (2016a, b)
(100)	*Veladyris*	Nymphalidae	Chazot et al. (2016a, b)
(101)	*Velamysta*	Nymphalidae	Chazot et al. (2016a, b)
(102)	*Hypoleria*	Nymphalidae	Chazot et al. (2016a, b)

3.5 Subtribe Pronophilina: 64 Genera, 630 spp. (Nymphalidae: Satyrinae)

Unlike the satyrines in the subtribe Euptichiina, which mainly inhabit lowland tropical forests, most members of Pronophilina occur in the Andes, many occupying high elevations (e.g. 2000–3500 m). This extremely diverse Neotropical clade also includes 13 extralimital south temperate genera in Chile and Argentina (Matz and Brower 2016), and the recently-revised genus, *Praepedaliodes*, in the mountains of the Atlantic Coastal Forest of Brazil (Pyrcz et al. 2018). Evidence suggests that all feed on grasses, and many on the bamboo genus *Chusquea* in particular. Adams and Bernard (1977) proposed a biogeographical hypothesis for species generation based on alternating periods of isolation and range expansion, as habitat bands moved up and down the mountains due to sequential warming and cooling events in the Pleistocene (a sort of reverse refugium hypothesis, in which cooling promoted dispersal at lower elevations, and warming resulted in isolation at higher elevations). To test this hypothesis, Casner and Pyrcz (2010) studied the spatial and temporal patterns of diversification in the genus *Lymanopoda*, the second largest genus in the subtribe (64 species). Their results showed that most *Lymanopoda* species evolved at the same elevation, suggesting that latitude might play a slightly more important role in species differentiation than altitude. According to Casner and Pyrcz, the genus began diversifying during the Oligocene (~27 myr), and although many of its lineages split during the Miocene, our calculations suggest that extant *Lymanopoda* species pairs are on average 1.8 myr old. We were able to assess divergence between putative sister species for 18 additional pairs of pronophilines, and their average age of divergence was 1.9 my.

3.6 Tribe Preponini: 3 Genera, 21 spp. (Nymphalidae: Charaxinae)

Members of Preponini are fruit feeding butterflies characterized by their robust bodies and colorful wing patterns. At this moment the only phylogenetic hypothesis for relationships within the tribe is not dated. According to Wahlberg et al. (2009), the tribe began diversifying into the three extant genera approximately 33.6 (HPD 26–41) myr.

We extracted species pairs from Ortiz-Acevedo et al. (2017) and Ortiz-Acevedo and Willmott (2013) for *Prepona* and *Archeoprepona*. On average, extant species pairs of *Archeoprepona* are 2.2 myr and species pairs of *Prepona* 1.85 myr. Furthermore, since the study of Ortiz-Acevedo et al. (2017) synonymized the genus *Agrias* with *Prepona*, we therefore can hypothesize a previous stem age estimate for *Prepona* + *Agrias* (14.5, HPD 9–20 myr) published by Wahlberg et al. (2009) as the crown age for the genus *Prepona*.

3.7 *Tribe Anaeini: 9 Genera, 88 spp. (Nymphalidae: Charaxinae)*

Though there are a few other charaxines with ventral color patterns that resemble dead leaves, the best leaf-like phenotypes are found in species included in Anaeini. Like most Charaxinae, these butterflies predominantly feed on rotting fruits as adults. According to Espeland et al. (2018) the Anaenini (based on one exemplar) is the sister group to the remaining charaxines, splitting around 41.5 myr (29–59). This age largely agrees with the Wahlberg et al. (2009) crown age estimate (43.6 myr, HPD 38.6–50), although the nine genera of Anaeini sampled in the Wahlberg et al. topology form a derived clade that is sister to Preponini. *Memphis* is the largest genus in this tribe with 62 species (Lamas 2004). Because there is no published time tree for *Memphis,* we estimated the average age of the extant species pairs based on all the sequences available in Genbank. Species pairs of *Memphis* are on average 3.4 myr old.

3.8 *Subfamily Biblidinae: 32 Genera, 249 spp. (Nymphalidae)*

With the exception of a handful of genera which seem to have independently colonized tropical Africa and southeast Asia, the vast majority of biblidines currently inhabit the Neotropics. An ancestral area reconstruction on the Wahlberg et al. (2009, Supp. Mat. 2) topology implies that the subfamily originated and diversified within the Neotropics. Current age estimates suggest that stem biblidines split from their sister group (the Apaturinae) around 62 mya, and that diversification of modern clades began around 43 mya. It is not obvious what caused either of these events, but globally, the Paleogene is full of geological milestones: in South America, the expansion and diversity of tropical rain forest was underway but restricted to the west of the continent. In general, the caterpillars of the subfamily feed on different groups of Malpighiales [e.g., Euphorbioidea, crown age: 60–70 myr; Caryocaraceae (Oliveira and Freitas 1991)] and Sapindales (e.g.: Burseraceae, Rutaceae) (Robinson et al. 2010). As adults, they are exclusively fruit-feeding butterflies and currently show preference for undisturbed tropical forest (although many genera contain species that thrive in secondary growth forest or open areas). In terms of number of species, *Dynamine* and *Eunica* are the largest genera, each with over 40 species. According to the Wahlberg et al. (2009) time tree, stem *Dynamine* existed 39 myr ago. A fossil assigned to that genus is known from Dominican amber and estimated to be 15–20 my old (Peñalver and Grimaldi 2006), however, de Jong (2017) questioned the generic identity and suggested that the fossil should conservatively be assigned as a minimum age for the crown node of Biblidinae. A rough calculation of the average age of extant *Dynamine* species pairs for which COI sequences are available in Genbank resulted in 4.50 myr. We extracted sister species pairs of *Paulogramma*, and *Diaethria* from Freitas et al. (2014), which are 2.4 and 1.4 myr old, respectively. Based on three species pairs, the average age of species in *Hamadryas* is 1.7 myr (Garzón-Orduña et al. 2013) and species of *Perisama* are on average half a million years old, based on 10 species pairs (Anna Zubek pers. comm).

3.9 Subtribe Phyciodina: 21 Genera, 133 spp. (Nymphalidae: Nymphalinae)

Members of the New World subtribe Phyciodina are small to medium size butterflies that visit flowers for nectar. Some species are stunning Batesian mimics of Ithomiini and Heliconiina. According to Wahlberg and Freitas (2007), butterflies in the subtribe Phyciodina evolved from temperate Melitaeini ancestors and are hypothesized to have colonized the tropics via GAARlandia, the land bridge that once connected the Greater Antilles and South America. The majority of the species in the subtribe are neotropical (89 species) and diversified in situ, however others are thought to have dispersed back into the Neartic. Wahlberg et al. (2009) estimated the crown age of Phyciodina to be 26.8 myr. We were able to compile pairwise sequence divergence estimates from 15 species pairs in 7 genera, with an average age of 2.4 my.

3.10 Tribe Heliconiini 10 Genera, 72 spp. (Nymphalidae: Heliconiinae)

This is another widespread neotropical clade of unpalatable and aposematic butterflies. As caterpillars, Heliconiini feed on passion vines (Passifloraceae) from which they obtain cyanogenic chemical defenses against predators. As adults, all Heliconiini visit flowers for nectar, and some members of the genus *Heliconius* (considered to be the most derived in the tribe) are unique among butterflies for their ability to eat pollen. *Heliconius* is also the richest genus in the clade (with ~46 species and more than 300 described subspecies or geographical races; Lamas and Jiggins 2017). Although some species of *Heliconius* occur in montane forest, the diversity of the genus is at its highest in the western Amazon Basin. Kozak et al. (2015) estimated the crown age of the subtribe to be 26 my, and the crown age for the genus *Heliconius* to be 12 my. However, the ages of sister species pairs in *Dione*, *Philaethria*, *Eueides* and *Heliconius* each average less than 2.5 my. The genus *Heliconius* is most well-known for the occurrence of dramatic parallel geographical variation in wing color patterns among the species *H. erato* and *H. melpomene*, among others. Brower (1994, 1996) observed that much of this dramatic adaptation on the eastern and western sides of the Andes occurred within the past

200,000 years, entailing that the evolution of mimicry complexes in *Heliconius* is a Pleistocene phenomenon.

3.11 Eumaeini: 80 Genera, 1058 spp. (Lycaenidae: Theclinae)

The main representatives of Lycaenidae in the Neotropical Region are the eumaeine hairstreaks, which range from among the smallest butterflies to large, gaudy creatures with iridescent blue or green dorsal wing surfaces. They are primarily canopy-dwellers, who occasionally descend to perch in sun flecks on the forest floor. Many of them feed on Loranthaceae as larvae, the genus *Eumaeus* feeds on cycads, and others are detritovores. Although Tiago Quental is purported to have conducted an extensive molecular phylogenetic study of them for his dissertation work (cf. Duarte and Robbins 2010), this remains unpublished and the associated sequence data are not accessible in Genbank, causing Eumaeini to remain the least-studied major radiation of Neotropical butterflies, at least from a molecular phylogenetic perspective. We were able to infer a surprisingly young crown age for the clade of 17.6 my (10–28 HPD) from Espeland et al. (2018), and an average age of seven species pairs from the genus *Oenomaus* of 1.6 myr old (Faynel et al. 2012).

3.12 Family Riodinidae: 124 Genera, 1324 spp.

The family Riodinidae includes small butterflies that visit flowers, often resting on the underside of leaves with their wings open and shifting their bodies in circular movements. Recently, the phylogenetic relationships within the family have received a much awaited examination from two independent studies (Espeland et al. 2015; Seraphim et al. 2018). According to the ancestral area reconstruction and divergence times estimation conducted by Espeland et al., the family originated and began to diversify in the Neotropics approximately 81 ma (72–94), and by the end of the Cretaceous period (ca. 66 myr), the three Riodinid subfamilies were differentiated. However this hypothesis is challenged by the study of Seraphim et al. (2018), who produced much younger ages for all these clades, with riodinids' diversification dated at 56 myr (52–61), and none of the three major clades older than 50 myr. Outstanding diversifications within Riodinidae are exemplified by the genus *Euselasia*, a genus that includes more than 170 species, and is widespread across the tropical lowlands of South and Central America. According to Espeland et al., the crown age of the genus is ~50 myr (40–58 HPD), while in Seraphim et al. (2018), the genus is almost half that age (26.5 myr, HPD 23.5–29.4). Since most of the published phylogenetic work to date has examined higher-level relationships within the family, our calculations on the ages of extant riodinid species, are limited to a handful of species pairs from four genera. Species of *Detritivora* average 2.1 myr old, based on two comparisons between non-sister taxa; extant species pairs in the recently described genus *Sertania* (Kaminski et al. 2017) averaged 1.5 myr; one species pair obtained for the genus *Theope* diverged from one another 3.4 mya and three extant species of *Riodina* averaged 1.6 myr.

3.13 Family Hesperiidae

Approximately 2/3 of the world's diversity of skipper butterflies occurs in the Neotropics (some 2350 species). These mostly small, brown, fast-flying butterflies have only lately become the focus of phylogenetic research. Recent molecular systematic studies (Warren et al. 2009; Sahoo et al. 2016; Toussaint et al. 2018) have rearranged traditional classifications, indicating that Eudaminae is a separate subfamily from Pyrginae, and that the "fire-tailed skippers", Pyrrhopygini, is a tribe within Pyrginae. Both Eudaminae and Pyrrhopygini are neotropical clades. We present crown-group age estimates for eight monophyletic subfamilies or tribes that are endemic to, or at least have their probable origin in the Neotropics: Eudaminae, Erynnini, Moncini, Pyrgini, Pyrrhopygini (all four Pyrginae), Achylodini, Anthoptini, and Calpodini (all three Hesperiinae). Our pairwise comparisons of sister species are mostly based on COI barcode sequences from mass sampling of the fauna of Guanacaste, Costa Rica (e.g. Janzen et al. 2011; Janzen and Hallwachs 2016), and so may represent overestimates of sister-group ages, since some sister species could be allopatric. Among the putative sister species pairs we examined are several members of the eudamine genus *Astraptes*, which were the subject of the famous "ten species in one" hypothesis in the early days of DNA barcoding (Hebert et al. 2004; see also Brower 2006). One interesting genus for which insufficient data were available to estimate ages is the Andean *Dalla*. This group of 108 (and likely including more undescribed) species feed on *Chusquea* bamboo, like the diverse pronophiline satyrines, and are unlikely to be older than the Andean cloud forest habitats they now inhabit.

4 Endemic Generic Radiations Not Included in the Above

4.1 Genus **Catasticta**, *96 spp. (Pieridae: Pierinae: Aporiina)*

Although a detailed time tree is not yet available for Pieridae, Braby et al. (2007) estimated the crown age of Neotropical Aporiina as 32 myr (HPD 28–38). According to Braby et al. (2007) and Wahlberg et al. (2014), the *Catasticta* group and the genus *Catastica* in turn, represent the most derived Aporiina, which together with the other seven neotropical aporiine genera evolved from Palearctic ancestors (Braby et al. 2007). Reported larval host records of tropical Aporiina are on

Loranthaceaae (DeVries 1987; Larsen 1991; Parsons 1999), which may confer chemical defense. Liu et al. (2018) inferred the crown age of Neotropical members of Loranthaceae to be 47 myr, older than and therefore compatible with the inferred crown age of Aporiina. With 96 species, the genus *Catasticta* stands out among pierids as a major Neotropical radiation, particularly across the montane forests of Central and South America. The diversity of the genus peaks in the Andean forests of Ecuador, Peru and Bolivia. The data we document here on *Catasticta* are from Sebastian Padron Martinez's PhD dissertation on the genus (Padron-Martinez 2014). Padron estimated the crown age of the genus to be 20.8 myr (HPD 15–25.2) and the age of extant *Catasticta* species based on 19 pairs of sister taxa averaged 4 myr old. However, the genus contains many subspecies which originated during the Pleistocene (Padron-Martinez 2014).

4.2 Genus Calisto*, 47 spp. (Nymphalidae: Satyrinae)*

The enigmatic satyrine genus *Calisto* is endemic to the Greater Antilles and nearby smaller islands, representing the only large radiation of butterflies limited to the Caribbean (there are a few endemic skipper and blue genera as well; Smith et al. 1994). *Calisto* belongs neither to Pronophilina nor to Euptychiina, the two megadiverse neotropical clades of Satyrini, and appears to be most closely-related to Palearctic satyrines such as *Melanargia* (Wahlberg et al. 2009). The genus includes one species on Puerto Rico, one on Jamaica, two in the Bahamas, one on Anegada, 11 on Cuba, and 31 on Hispaniola. Matos-Maravi et al. (2014) conducted a dated phylogenetic analysis with secondary calibrations from Wahlberg et al. (2009) and suggested that the crown age of the group is consistent with the origin of the island chain in the late Oligocene (~28 mya), and that most species groups evolved in situ. The species from Anegada and Puerto Rico are the sister group to the remainder, but the diverse Hispaniolan fauna is paraphyletic with respect to both Jamaican and Cuban species. Pairwise comparison of mtDNA COI sequences indicates that the ages of divergence for 12 pairs of sister species ranges from 1.0 to 3.1 myr, with a mean of 2.0 myr, placing the origins of approximately half of the *Calisto* species within the Pleistocene.

5 Newcomers: Groups that Originated Outside the Neotropics

5.1 Tribe Leptocircini: 4 Genera, 42 spp. (Papilionidae)

The neotropical "kite swallowtails" Leptocircini, form an apparently monophyletic group, with one representative (*Neographium marcellus*) occurring in the Nearctic region. The bulk of the tribe occurs in the Old World tropics. Traditional taxonomy of the neotropical taxa suggests a trans-Pacific relationship with Australian *Protographium leothsenes*, but phylogenetic support is lacking, and the resemblance may be superficial. At 38.6 myr, the crown age of Leptocircini is too recent for the group's distribution to have resulted from Gondwanan vicariance (de Jong 2003), and the four pairs of sister taxa we document diverged from one another less than 2 mya.

5.2 *Tribe Troidini: 3 Genera, 48 spp. (Papilionidae)*

The extant Troidini are a diverse clade of mainly Old World swallowtails that are hypothesized to have originated in North America during the Eocene (33–42 myr). Members of Troidini feed exclusively on *Aristolochia* and are chemically defended by sequestering aristolochic acid. Most species exhibit aposematic coloration and are part of Müllerian rings, as well as serve as models for Batesian mimics in the genera *Mimoides*, *Heraclides* (Papilioninae), and *Archonias* (Pieridae) (Tyler et al. 1994). Condamine et al. (2012) hypothesized that the ancestors of the three neotropical troidine genera (*Battus*, *Parides* and *Euryades*) arrived in South America via the GAARlandia connection on two independent colonizations (*Parides* is more closely-related to the Australasian *Ornithoptera* than to *Battus* and *Euryades*). All three genera began diversifying during the Miocene: *Battus* and *Parides* around 22 mya and *Euryades* around 11 mya. Troidini is the only butterfly clade in which age and species richness has shown to be positively correlated (Condamine et al. 2012). Species pairs of *Parides* and *Euryades* were on average 2.2 myr old, and 3.0 myr old for *Battus*.

5.3 *Subfamily Dismorphiinae 6 Genera, 51 spp. (Pieridae)*

Dismorphiinae is the sister group to all other members of the family Pieridae (Wahlberg et al. 2014), and, aside from the Palearctic *Leptalis*, a subfamily endemic to the Neotropics (including two species of *Dismorphia* in the Greater Antilles). The neotropical representatives include some striking mimics of various ithomiines, and were the inspiration for the concept of Batesian mimicry (Bates 1862a, b). Sampling of the group in higher-level phylogenetic studies is insufficient to provide a crown age estimate, but the estimated age of divergence from other Pieridae is 87 myr (Espeland et al. 2018), suggesting that the group is quite ancient. Nevertheless, putative sister species divergences, of which we were able to estimate four pairs, average 2.3 my, in the same Pleistocene range as most other butterflies. Larvae of two genera have been recorded feeding on *Inga* species (DeVries 1987), which is a plant genus also noted for its recent radiation (Richardson et al. 2001).

5.4 *Genus* Adelpha, *Approx. 200 spp. (Nymphalidae: Limenitidinae)*

In terms of number of species, there are not many butterfly genera as large as *Adelpha*, which includes around 200 species, (inclusive of the Nearctic genus *Limenitis*, with respect to which *Adelpha* is paraphyletic) (Willmott 2003; Lamas 2004; Ebel et al. 2015). Currently, the initial diversification of the genus is hypothesized to be the result of ecological specialization, either through host plant shifts or by several instances of unrelated species converging on similar wing patterns (Mullen et al. 2011; Ebel et al. 2015). The crown age of *Adelpha* was estimated as 11 myr. Based on 11 species pairs, our calculations suggest that *Adelpha*'s extant diversity is fairly young, with species on average 2.0 myr old.

5.5 *Genera* Actinote, Altinote *and* Abananote, *Approx. 50 spp. (Heliconiinae: Acraeini)*

These three closely related genera form the Neotropical clade of Acraeini, a tribe of aposematic and unpalatable butterflies descended from African ancestors and which mainly inhabit montane forest. The three genera currently feed on Asteraceae, a preference that evolved after their split from a presumably polyphagous ancestor (Silva-Brandão et al. 2008). Of the three, the largest genus is *Actinote*, with 35 species, of which about half occur in Southern Brazil, the genus' center of diversity. The crown age of the Neotropical Acraeini (*Altinote* + *Actinote*) according to Wahlberg et al. (2009) is 11.6 myr. Our calculations suggest that extant species pairs in *Actinote* and *Altinote* are on average 1.1 and 1.8 myr old respectively, this is supported by ongoing molecular work (Freitas et al. 2018).

5.6 *Andean Polyommatini 7 Genera, 70 spp. (Lycaenidae)*

Vila et al. (2011) examined the phylogenetic relationships of a clade of high Andean blue butterflies that dwell in altiplano or puna habitats, corroborating a hypothesis proposed by Vladimir Nabokov regarding the butterflies' colonization of the Americas via Beringia some 10.7 my (stem age). Several successive sister taxa comprising a basal paraphyletic group related to this clade are distributed in the lowland Neotropics, and particularly in the Antilles. We estimated the age of sister species pairs in six genera, including *Pseudolucia* and *Madeleinea*. The mean age of Andean sister-species pairs is 0.6 my, suggesting that these are among the youngest species of neotropical butterflies.

6 Discussion

6.1 *Drivers of Diversification in Butterflies*

Few studies of Neotropical butterflies have put forward explicit hypotheses identifying potential (general) drivers of diversification. We mention here three of the most frequently discussed factors.

6.2 *Hostplants*

The evolutionary hypothesis that butterflies are closely tied to their larval host plants has existed for a long time (Brower and Brower 1964; Ehrlich and Raven 1965). Peña and Walhberg (2008) argued that global climate changes during the Tertiary allowed the expansion and radiation of grasses (Poaceae). Although clock-based crown ages imply that the Satyrini already existed before the spread of the grasses, the main satyrine lineages are hypothesized to have radiated (36–23 myr ago) simultaneously with the spread of grasses. Espeland et al. (2015) also suggested hostplant breadth to be positively correlated with diversity of clades within Riodinidae. Heliconiini are hypothesized to represent a radiation onto a chemically-defended host plant group. See also Brown et al. (1991) and Willmott and Mallet (2004) for Ithomiini, and Mullen et al. (2011) regarding *Adelpha*.

6.3 *Age*

Among butterfly groups so far examined, only the Condamine et al. (2012) study of troidine swallowtails supports an age-diversity positive correlation. Major components of the neotropical butterfly fauna are considered to have colonized South America from the Holarctic (e.g., Wahlberg and Freitas 2007; Mullen et al. 2011; Vila et al. 2011), which is inconsistent with the "museum hypothesis" for Neotropical diversity proposed by Stebbins (1974).

6.4 *Rivers*

Although suggested by Wallace (1853) as a mechanism responsible for separating not only monkeys, but also butterfly species of the genus *Callithea* (now *Asterope*), the riverine hypothesis has been little emphasized in butterflies, in comparison to the attention it has received in studies of vertebrates. Brown's (1979) synthesis of distributions of ithomiine and *Heliconius* butterflies suggested patterns of endemism in part bounded by major Amazonian rivers. More recently, the study of the riodinids in the *Charis cleonus* species complex (now the genus *Detritivora*) by Hall and Harvey (2002) suggested patterns of differentiation across rivers. Species in *Detritivora* (31 species) show preference for the understory of wet forest where they feed on dead leaves (hence the generic name). Described as "sedentary butterflies", the most important piece of evidence supporting the hypothesis is their parapatric distributional ranges. The relationships proposed by Hall and Harvey have not been corroborated by molecular data yet. Based on the very few COI sequences available from *Detritivora*, we obtained average species ages of 2.20 myr, from two comparisons between non-sister taxa.

6.5 Pleistocene Refugia

Haffer (1969) proposed that the most recent speciation events in Neotropical birds might have been driven by reduction of contiguous tropical rainforest during repeated cycles of Pleistocene cooling. Isolation in patches of forest in local, climate buffered refugia could have provided opportunities for vicariant divergence due to a combination of local adaptation and/or genetic drift of small populations. Brown (Brown et al. 1974; Brown 1979) extended this hypothesis to explain diversification of geographical variation among butterflies in the tribe Ithomiini and the genus *Heliconius*. Although the refugium hypothesis has been controversial (Colinvaux et al. 2001; Knapp and Mallet 2003), most criticisms set up Haffer's hypothesis as a straw man, focusing on the most recent period of glaciation 18–10,000 years ago, and ignoring that the glacial cycles extend at least 1.8 million years into the past. The current dataset (Fig. 9.2) reinforces the pattern that we have already shown (Garzón-Orduña et al. 2014): that some 70% of sister species pairs of Neotropical butterflies diverged from one another within the Pleistocene time frame.

For this contribution we have also compiled from the literature the crown ages of 27 genera and 35 suprageneric groups (Fig. 9.1). In doing so, we provide an illustration of how arbitrary crown ages are as temporal metrics of diversification. As we previously stated (Garzón-Orduña et al. 2014), there is no empirically-sound rationale to prefer the crown age of a subfamily, or a tribe, over the age of a subtribe or genus for a given taxon: one may simply select a convenient level of inclusiveness that provides evidence benevolent to one's preferred hypothesis about "ages of diversification." Furthermore, in most cases, these ages are orders of magnitude larger than the ages of extant species. If the crown ages are correct, the gap between them and the ages of extant species argues against hypotheses advocating that diversification in the Neotropics has been continuous throughout the tertiary and quaternary (Zink et al. 2004; Rull 2008), unless it has been accompanied by equally strong (and unobservable) bouts of extinction.

6.6 It Is Not About Rates

Matos-Maravi (2016: 8 pp.) characterized our 2014 study (Garzón-Orduña et al. 2014) as "gathering rough evidence to help uncover the period of time when most extant diversity originated". We largely agree with this statement, and to emphasize our point again: we were and remain agnostic regarding attempts to equate an effect of the Pleistocene cycles with increases in the rate of speciation. Neither in our previous papers on this subject (Garzón-Orduña et al. 2014, 2015), nor here are we suggesting that the rate of speciation of a given group was higher during the Pleistocene, rather, we simply document that a lot of speciation events took place during this period. Indeed, with virtually no pertinent fossil record, it is hard to imagine how estimates of evolutionary rates can be remotely accurate (of course, this remains an open question, since they are also untestable). One could assume a steady-state model, in which the number of speciation events and the number of extinctions are equal, but that would seem to beg the question regarding "diversification."

At least for the Neotropical component of the butterfly fauna, the one consistent finding from time-calibrated phylogenetic trees is that extant diversity of a group is not strictly correlated with its age. That is to say, species-rich groups have been found to be younger than expected if the accumulation of lineages followed time steadily. Thus, for butterflies, the so-called "museum hypothesis" (Stebbins 1974) has received little (Condamine et al. 2012) to no support (Mullen et al. 2011; Espeland et al. 2015).

6.7 Tempo and Mode of Speciation in the Neotropics-Substance or Semantics?

Hoorn et al. (2010: 927) argued that,

> Pleistocene forest remnants ('refugia') were long held to be responsible for Amazonian diversity. In the 1990s centers of diversity, postulated as prime evidence for the refuge theory, were shown to be sampling artifacts. Over time, the theory was abandoned and an older origin for the Amazonian diversity was proposed. Perhaps more important, regional diversification events, as inferred from the fossil record and molecular phylogenetic studies, mostly predate the Pleistocene.

Let's unpack these statements. The Pleistocene refugium hypothesis (PRH) was originally based upon the apparent congruent distributions of different groups of birds (Haffer 1969, 1974), which other authors noticed were corroborated by geographical patterns in lizards, butterflies, and plants (Vanzolini and Williams 1970; Prance 1973; Brown et al. 1974). The "sampling artifact" critique of these distributions (Nelson et al. 1990) provided an alternative explanation (a potentially confounding variable, not a refutation) of the pattern in the botanical data, and said nothing about the congruent zoogeographical patterns. The subsequent "abandonment" of the PRH was driven to great measure by aggressive extrapolation of fossil pollen core data that suggested that there was no evidence of aridification around three lakes during the past 40,000 years (Bush 1994; Colinvaux et al. 2001; Bush and De Oliveira 2006), using the most recent glacial cycle (LGM) as a temporal synecdoche for the entire Pleistocene epoch (see Garzón-Orduña et al. 2015 for further discussion). The paper cited regarding fossil evidence for pre-Pleistocene diversification is Jaramillo et al. (2006), a palynological study showing a diverse Eocene flora—an observation that, since the Eocene fossils are not conspecific with extant species, is irrelevant to hypotheses about speciation in the Pleistocene. The work cited for molecular phylogenetic evidence is Rull's (2008) compilation of time trees for various vertebrate and invertebrate groups.

Valenti Rull (2011) himself criticized Hoorn et al. (2010)'s invocation of his data to argue that "regional diversification dates ... mostly predate the Pleistocene", pointing out that he (Rull 2008) "concluded that about half of the dated extant neotropical species originated during the Pleistocene and the other half before it, and that speciation proceeded in continuous fashion with no evident bursts." As he articulated, the crown group age of a taxon hypothesizes its time of origin, which is necessarily older than the age of origin of most of the species that currently exist in that group. Countering Rull's observation, Hoorn et al. (2011) argued that "Rull's finding that about half of all extant species analyzed originated during Quaternary times is not surprising. Assuming the average species longevity is some 100,000 to a couple of million years, at any point we would expect to find that most species originated in the past few million years."

So is there a controversy, or is this simply a matter of different authors conflating different kinds of events? We doubt that anyone ever believed that the "diversity of the Neotropics" evolved de novo in the Quaternary (except maybe some young Earth creationists). Haffer (1969) said only that, "In view of the length of the Tertiary period (60 million years), during which the Amazonian fauna probably evolved rather slowly under quite uniform environmental conditions, the Quaternary faunal differentiation in tropical South America during the last 1–2 million years is, geologically speaking, very 'recent' and occurred rather 'rapidly'." Hoorn et al. (2011) and Rull (2011) would seem to agree with this, except for disputing the claim that speciation happened more slowly in the Tertiary. They evidently prefer a constant rate, a hypothesis as unsubstantiated by evidence as Haffer's. What can be said with empirical certainty is that extant Neotropical species are relatively young, and that biogeographically significant geological events in the Miocene only set the stage for their subsequent evolution.

It is our view that most of what is biologically interesting about the extant diversity of Neotropical butterflies is the result of fairly recent evolutionary events. For example, the geographical diversification of mimicry rings within and among species of *Heliconius* and ithomiine butterflies seems to be at least in part an intraspecific phenomenon (Brower 1996; Willmott et al. 2017), indicating that dramatic changes in wing patterns have taken place within the lifetimes of individual species, mostly within the Pleistocene. Likewise, the storied migration of the monarch butterfly appears to have originated during the Pleistocene, since *Danaus plexippus* diverged from its Neotropical sister species less than 2 mya (Brower and Jeansonne 2004). Thus, we consider the ages of divergence of species to be at least as relevant to our narrative about the history of diversification of the Neotropical region as the origins of the larger clades to which those species belong. We have previously documented a general pattern in which many pairs of sister taxa have Pleistocene-age divergence times (Garzón-Orduña et al. 2014). The data presented here augment our previous dataset, providing age estimates for pairs of sister taxa spanning across 102 genera in all families of Neotropical Lepidoptera, representing almost three times the amount of data we compiled on our first study. The resulting unambiguous pattern is the same as that which we documented before: most extant neotropical species of butterflies did not originate during the Tertiary (cf. Colinvaux et al. 2000), instead they originated in the Quaternary (Fig. 9.2).

Given this empirical fact, we are naturally predisposed as scientists to consider possible common mechanisms that might explain such a pattern. The Hoorn et al. (2011) claim that "at any point in time we would expect to find that most species originated in the past few million years", strikes us as explanatorily dismissive of the problem as saying "stuff happens." Scientists want to discover *why* stuff happens. The species that exist now are the species that have characteristics, behaviors and interactions that are accessible for study. Such species are "the units of evolution", not interchangeable tokens that anonymously populate some abstract, temporally homogeneous community.

Likewise, historical narratives addressing the vagaries of individual clades are of little value as general predictors of the histories of other taxa, unless patterns of diversification can be tied to extrinsic environmental processes that have the potential to impact those groups, as well. This is the reason that biogeographers prefer to invoke vicariance over dispersal, as the former process provides general explanatory hypotheses, while the latter is by its nature idiosyncratic and ad hoc (Nelson and Platnick 1981). At the beginning of the Quaternary, rivers and mountains in the Neotropics had largely achieved their current statures and positions. Thus, the most compelling extrinsic variable that could act as a general vicariant force driving temporally congruent speciation and geographically congruent patterns of distribution is climate-induced environmental change.

7 Concluding Remarks

We hope this contribution provides convincing evidence that the butterfly diversity in the neotropics is vast, that within each family there are interesting groups ripe for study, and that present critical gaps in our knowledge. While collecting data for this review, it was difficult to find genus-level phylogenetic hypotheses for clades in the Riodinidae, Lycaenidae and Hesperiidae. We note that, in comparison to the other butterfly families, and consistent with their conspicuousness in Neotropical forests, nymphalids are by far the best-studied group of butterflies.

As evolutionary biologists, we are in debt to traditional taxonomists whose dedication provides the raw materials of our research. Taxonomic progress can occur rapidly, as demonstrated by Faynel et al. (2012) who reported how in 10 years the hairstreak genus *Oenomaus* went from being a monotypic genus to including 28 species. DNA barcoding also has promise to suggest the existence of many cryptic taxa (e.g., Hebert et al. 2004). However, we are working against the clock: much of Neotropical biodiversity is in peril due to fragmentation and modification of the primary forest, and several species are deemed endangered or vulnerable, such as *Heliconiuis nattereri* (Nymphalidae) (Brown 1972), *Charonias theano* (Pieridae), (Freitas et al. 2011) and *Paulogramma hydarnis*, (Nymphalidae) (Freitas et al. 2014). For others, it is already too late—e.g., *Orobrassolis latusoris* (Nymphalidae) (Penz et al. 2011). The Anthropocene may prove to be a time of even more momentous biotic turnover than the Pleistocene. And that will be an extinction rate that we can measure.

Acknowledgements We thank editors Valentí Rull and Ana Carnaval for inviting us to write a contribution for this book. IJGO dedicates this chapter to her dear friend Kelly Jean Sherman whose inquisitiveness and love knows no boundaries. Many thanks also to Sebastián Padrón, Fabien Condamine, Nicolas Chazot, Marianne Espeland and Marianne Elias for responding to our requests and allowing us to use unpublished data. We are also grateful to Angel Viloria and Juan Jose Morrone for their observations on our manuscript.

Support for work in AVZB's lab at MTSU was provided through a collaborative grant, Dimensions US-Biota-São Paulo: Assembly and evolution of the Amazon biota and its environment: an integrated approach, supported by the US National Science Foundation (NSF DEB 1241056), National Aeronautics and Space Administration (NASA), and the Fundação de Amparo à Pesquisa do Estado de São Paulo (FAPESP Grant 2012/50260-6).

References

Adams MJ, Bernard GI (1977) Pronophiline butterflies (Satyridae) of the Sierra Nevada de Santa Marta, Colombia. Syst Entomol 2:263–281

Amorim D d S (2001) Dos Amazonias. In: Llorente-Bousquets JE, Morrone JJ (eds) Introducción a la biogeografia en Latinoamérica: teorías, conceptos, métodos y aplicaciones. Facultad de Ciencias. UNAM, Mexico D.F., pp 245–255

Arias-Mejia CF (2012) Patterns of diversification in Neotropical butterflies. PhD Dissertation, McGill University, Montreal, Quebec, Canada

Barbosa EP, Silva AK, Paluch M, Azeredo Espin AML, Freitas AVL (2015) Uncovering the hidden diversity of the Neotropical butterfly genus *Yphthimoides* Forster (Nymphalidae: Satyrinae): description of three new species based on morphological and molecular data. Org Divers Evol 15:577–589

Basset Y, Barrios H, Segar S, Srygley RB, Aiello A, Warren AD, Delgado F, Coronado J, Lezcano J, Arizala S, Rivera M, Perez F, Bobadilla R, Lopez Y, Ramirez JA (2015) The butterflies of Barro Colorado Island, Panama: local extinction since the 1930s. PLoS ONE 10:e0136623

Bates HW (1862a) Contributions to an insect fauna of the Amazon Valley. Lepidoptera—Papilionidae. J Entomol 1:218–245

Bates HW (1862b) Contributions to an insect fauna of the Amazon Valley. Lepidoptera: Heliconidae. Trans Linn Soc 23:495–566

Beltrán M, Jiggins CD, Brower AVZ, Bermingham E, Mallet J (2007) Do pollen feeding, pupal-mating and larval gregariousness have a single origin in *Heliconius* butterflies? Inferences from multilocus DNA sequence data. Biol J Linn Soc 92:221–239

Blandin P, Purser B (2013) Evolution and diversifcation of Neotropical butterflies: insights from biogeography and phylogeny of the genus *Morpho* Fabricius, 1807 (Nymphalidae: Morphinae), with a review of the geodynamics of South America. Trop Lepid Res 23:62–85

Blum MJ, Bermingham E, Dasmahapatra K (2003) A molecular phylogeny of the neotropical butterfly genus *Anartia* (Lepidoptera: Nymphalidae). Mol Phylogenet Evol 26:46–55

Braby MF, Pierce NE, Vila R (2007) Phylogeny and historical biogeography of the subtribe Aporiina (Lepidoptera: Pieridae): implications for the origin of Australian butterflies. Biol J Linn Soc 90:413–440

Brower AVZ (1994) Rapid morphological radiation and convergence among races of the butterfly *Heliconius erato* inferred from patterns of mitochondrial DNA evolution. Proc Nat Acad Sci USA 91:6491–6495

Brower AVZ (1996) Parallel race formation and the evolution of mimicry in *Heliconius* butterflies: a phylogenetic hypothesis from mitochondrial DNA sequences. Evolution 50:195–221

Brower AVZ (2006) Problems with DNA barcodes for species delimitation: "ten species" of *Astraptes fulgerator* reassessed (Lepidoptera: Hesperiidae). Syst Biodivers 4:127–132

Brower AVZ, Jeansonne MM (2004) Geographical populations and "subspecies" of New World monarch butterflies (Nymphalidae) share a recent origin and are not phylogenetically distinct. Ann Entomol Soc Amer 97:519–523

Brower LP, Brower JVZ (1964) Birds, butterflies, and plant poisons: a study in ecological chemistry. Zoologica NY 49:137–159

Brown KS Jr (1972) The heliconians of Brazil (Lepidoptera: Nymphalidae). Part III. Ecology and biology of *Heliconius nattereri*, a key primitive species near extinction, and comments on the evolutionary development of *Heliconius* and *Eueides*. Zoologica NY 57:41–69

Brown KS Jr (1977) Centros de evolução, refúgios quaternários e conservação de patrimônios genéticos na regiào neotropical: padrões de diferenciação em Ithomiinae (Lepidoptera: Nymphalidae). Acta Amazonica 7:75–137

Brown KS Jr (1979) Ecologia Geográfica e Evolução nas Florestas Neotropicais. Universidade Estadual de Campinas, Campinas, São Paulo, Brasil

Brown KS Jr, Sheppard PM, Turner JRG (1974) Quaternary refugia in tropical America: evidence from race formation in *Heliconius* butterflies. Proc R Soc Lond B 187:369–378

Brown KS Jr, Trigo JR, Francini RB, Barros de Morais AB, Motta PC (1991) Aposematic insects on toxic host plants: coevolution, colonization and chemical emancipation. In: Price PW, Lewinsohn TM, Fernandes GW, Benson WW (eds) Plant-animal interactions: evolutionary ecology in tropical and temperate regions. Wiley, New York, pp 375–402

Bush MB (1994) Amazonian speciation: a necessarily complex model. J Biogeog 21:5–17

Bush MB, De Oliveira PE (2006) The rise and fall of the refugial hypothesis of Amazonian speciation: a paleoecological perspective. Biota Neotropica 6:bn00106012006

Casner KL, Pyrcz TW (2010) Patterns and timing of diversification in a tropical montane butterfly genus, *Lymanopoda* (Nymphalidae, Satyrinae). Ecography 33:251–259

Cespedes A, Penz CM, DeVries PJ (2015) Cruising the rain forest floor: butterfly wing shape evolution and gliding in ground effect. J Animal Ecol 84:808–816

Chacon IA, Montero-Ramirez J, Janzen DH, Hallwachs W, Blandin P, Bristow CR, Hajibabaei M (2012) A new species of *Opsiphanes* Doubleday, [1849] from Costa Rica (Nymphalidae: Morphinae: Brassolini), as revealed by its DNA barcodes and habitus. Bull Allyn Museum 166:1–15

Chazot N, de-Silva DL, Willmott KR, Freitas AVL, Lamas G, Mallet J, Giraldo CE, Uribe S, Elias M (2017) Contrasting patterns of Andean diversification among three diverse clades of Neotropical clearwing butterflies. Ecol Evol 8:3965–3982

Chazot N, Panara S, Zilbermann N, Blandin P, Le Poul Y, Cornette R, Elias M, Debat V (2016a) *Morpho* morphometrics: shared ancestry and selection drive the evolution of wing size and shape in *Morpho* butterflies. Evolution 70:181–194

Chazot N, Willmott KR, Condamine FL, de-Silva DL, Freitas AVL, Lamas G, Morlon H, Giraldo CE, Jiggins CD, Joron M, Mallet J, Uribe S, Elias M (2016b) Into the Andes: multiple independent colonizations drive montane diversity in the Neotropical clearwing butterflies Godyridina. Mol Ecol 25:5765–5784

Chew FS, Watt WB (2006) The green-veined white (*Pieris napi* L.), its pierine relatives, and the systematic dilemmas of divergent character sets (Lepidoptera: Pieridae). Biol J Linn Soc 88:413–435

Colinvaux PA, De Oliveira PE, Bush MB (2000) Amazonian and neotropical plant communities on glacial time-scales: the failure of the aridity and refuge hypotheses. Quatern Sci Rev 19:141–169

Colinvaux PA, Irion G, Räsänen ME, Bush MB (2001) A paradigm to be discarded: Geological and paleoecological data falsify the HAFFER & PRANCE refuge hypothesis of Amazonian speciation. Amazoniana 16:609–646

Condamine FL, Silva Brandão KL, Kergoat GJ, Sperling FAH (2012) Biogeographic and diversification patterns of Neotropical Troidini butterflies (Papilionidae) support a museum model of diversity dynamics for Amazonia. BMC Evol Biol 12:82

de Jong R (2003) Are there butterflies with Gondwanan ancestry in the Australian region? Invert Syst 17:143–156

de Jong R (2017) Fossil butterflies, calibration points and the molecular clock (Lepidoptera: Papilionoidea). Zootaxa 4270:1–63

de-Silva DL, Day JJ, Elias M, Willmott KR, Whinnett A, Mallet J (2010) Molecular phylogenetics of the neotropical butterfly subtribe Oleriina (Nymphalidae: Danaini: Ithomiini). Mol Phylogenet Evol 55:1032–1041

de-Silva DL, Elias M, Willmott KR, Mallet J, Day JJ (2016) Diversification of clearwing butterflies with the rise of the Andes. J Biogeogr 43:44–58

de-Silva DL, Mota LL, Chazot N, Mallarino R, Silva Brandao KL, Gomez Pinerez LM, Freitas AVL, Lamas G, Joron M, Mallet J, Giraldo CE, Uribe S, Särkinen T, Knapp S, Jiggins CD, Willmott KR, Elias M (2017) North Andean origin and diversification of the largest ithomiine genus. Sci Rep 7:45966

DeVries PJ (1987) The butterflies of Costa Rica and their natural history. Papilionidae, Pieridae, Nymphalidae. Princeton University Press, Princeton

Duarte M, Robbins RK (2010) Description and phylogenetic analysis of the Calycopidina (Lepidoptera, Lycaenidae, Theclinae, Eumaeini): a subtribe of detritivores. Rev Bras Entomol 54:45–65

Ebel ER, Dacosta JM, Sorenson MD, Hill RI, Briscoe AD, Willmott KR, Mullen SP (2015) Rapid diversification associated with ecological specialization in Neotopical *Adelpha* butterflies. Mol Ecol 24:2392–2405

Ehrlich PR, Raven PH (1965) Butterflies and plants: a study in coevolution. Evolution 18:586–608

Elias M, Hill RI, Willmott KR, Dasmahaptra KK, Brower AVZ, Mallet J, Jiggins CD (2007) Limited performance of DNA barcoding in a diverse community of tropical butterflies. Proc R Soc London B 274:2881–2889

Elias M, Joron M, Willmott KR, Silva Brandão KL, Kaiser V, Arias CF, Gomez Piñerez LM, Uribe S, Brower AVZ, Freitas AVL, Jiggins CD (2009) Out of the Andes: patterns of diversification in clearwing butterflies. Mol Ecol 18:1716–1729

Espeland M, Breinholt JW, Willmott KR, Warren AD, Vila R, Toussaint EFA, Maunsell SC, Aduse-Poku K, Talavera G, Eastwood R, Jarzyna MA, Guralnick RP, Lohman DJ, Pierce NE, Kawahara AY (2018) A comprehensive and dated phylogenetic analysis of butterflies. Curr Biol 28:770–778

Espeland M, Hall JPW, DeVries PJ, Lees DC, Cornwall M, Hsu Y-F, Wu L-W, Campbell DL, Talavera G, Vila R, Salzman S, Ruehr S, Lohman DJ, Pierce NE (2015) Ancient Neotropical origin and recent colonisation: phylogeny, biogeography and diversification of the Riodinidae (Lepidoptera: Papilionoidea). Mol Phylogenet Evol 93:296–306

Faynel C, Busby RC, Robbins RK (2012) Review of the species level taxonomy of the neotropical butterfly genus *Oenomaus* (Lycaenidae, Theclinae, Eumaeini). ZooKeys 222:11–45

Freitas AVL, Bastos-Francini R, Paluch M, Barbosa EP (2018) A new species of *Actinote* Hübner (Nymphalidae: Heliconiinae: Acraeini) from southeast Brazil. Rev Bras Entomol 62:135–147

Freitas AVL, Kaminski LA, Iserhard CA, Magaldi LM, Wahlberg N, Silva-Brandão KL, Marini-Filho OJ (2014) *Paulogramma hydarnis* (n. comb.) (Nymphalidae: Biblidinae): distribution, systematic position, and conservation status of a rare and endangered butterfly. Neotrop Entomol 43:218–226

Freitas AVL, Kaminski LA, Mielke OHH, Barbosa EP, Silva-Brandão KL (2012) A new species of *Yphthimoides* (Lepidoptera: Nymphalidae: Satyrinae) from the southern Atlantic forest region. Zootaxa 3526:31–44

Garzón-Orduña IJ, Benetti-Longhini JE, Brower AVZ (2014) Timing the diversification of the Amazonian biota: butterfly divergences are consistent with Pleistocene refugia. J Biogeogr 41:1631–1638

Garzón-Orduña IJ, Benetti-Longhini JE, Brower AVZ (2015a) Competing paradigms of Amazonian diversification and the Pleistocene refugium hypothesis. J Biogeogr 42:1357–1360

Garzón-Orduña IJ, Marini-Filho OJ, Johnson SG, Penz CM (2013) Phylogenetic analysis of *Hamadryas* (Nymphalidae: Biblidinae) based on the combined analysis of morphological and molecular data. Cladistics 29:629–642

Garzón-Orduña IJ, Silva-Brandão KL, Willmott KR, Freitas AVL, Brower AVZ (2015b) Incompatible ages for clearwing butterflies based on alternative secondary calibrations. Syst Biol 64:752–767

Gregory-Wodzicki KM (2000) Uplift history of the Central and Northern Andes: a review. GSA Bull 112:1091–1105

Haffer J (1969) Speciation in Amazonian forest birds. Science 165:131–137

Haffer J (1974) Avian speciation in tropical South America. Nuttall Ornithological Club, Cambridge, MA

Hall JPW, Harvey DJ (2002) The phylogeography of Amazonia revisited: new evidence from riodinid butterflies. Evolution 56:1489–1497

Hajibabaei M, Janzen DH, Burns JM, Hallwachs W, Hebert PDN (2006) DNA barcodes distinguish species of tropical Lepidoptera. Proc Nat Acad Sci USA 103:968–971

Hebert PDN, Penton EH, Burns JM, Janzen DH, Hallwachs W (2004) Ten species in one: DNA barcoding reveals cryptic species in the Neotropical skipper butterfly *Astraptes fulgerator*. Proc Nat Acad Sci USA 101:14812–14817

Heikkila M, Kaila L, Mutanen M, Peña C, Wahlberg N (2011) Cretaceous origin and repeated Tertiary diversification of the redefined butterflies. Proc R Soc London B 279:1093–1099

Hoorn C, Wesselingh FP, Hovikoski J, Guerrero J (2010) The development of the Amazonian mega-wetland (Miocene; Brazil, Colombia, Peru, Bolivia). In: Hoorn C, Wesselingh FP (eds) Amazonia, Landscape and Species Evolution: A Look into the Past. Wiley-Blackwell, London, pp 123–142

Hoorn C, Wesselingh P, ter Steege H, Bermudez MA, Mora A, Sevink J, Sanmartin I, Sanchez-Meseguer A, Anderson CL, Figueredo JP, Jaramillo C, Riff D, Negri FR, Hooghiemstra H, Lundberg J, Stadler T, Sarkinen T, Antonelli A (2011) Response (to "Origins of biodiversity"). Science 331:399–400

Janzen DH, Hallwachs W (2016) DNA barcoding the Lepidoptera inventory of a large complex tropical conserved wildland, Area de Conservación Guanacaste, northwestern Costa Rica. Genome 59:641–660

Janzen DH, Hallwachs W, Burns JM, Hajibabaei M, Bertrand C, Hebert PDN (2011) Reading the complex skipper butterfly fauna of one tropical place. PLoS One 6:e19874

Jaramillo C, Rueda MJ, Mora G (2006) Cenozoic plant diversity in the neotropics. Science 311:1893–1896

Jefferies RPS (1979) The origin of the chordates – a methodological essay. In: House MR (ed) The origin of the major invertebrate groups. Academic Press (Systematics Association), London, pp 443–477

Kaminski LA, Callaghan CJ, Seraphim N, Magaldi LM, Volkmann L, Freitas AVL (2017) *Sertania* gen.nov., a new genus of butterflies (Lepidoptera: Riodinidae) from the South American dry diagonal. Zootaxa 4312:165–179

Knapp S, Mallet J (2003) Refuting refugia? Science 300:71–72

Kodandaramaiah U, Peña C, Braby MF, Grund R, Muller CJ, Nylin S, Wahlberg N (2009) Phylogenetics of Coenonymphina (Nymphalidae: Satyrinae) and the problem of rooting rapid radiations. Mol Phylogenet Evol 54:386–394

Kozak KM, Wahlberg N, Neild AFE, Dasmahapatra KK, Mallet J, Jiggins CD (2015) Multilocus species trees show the recent adaptive radiation of the mimetic *Heliconius* butterflies. Syst Biol 64:505–524

Lamas G (ed) (2004) Atlas of Neotropical Lepidoptera. Checklist: Part 4A Hesperioidea – Papiionoidea. Scientific/Association of Tropical Lepidoptera, Gainesville. (2014 electronic revision: http://www.ucl.ac.uk/taxome/gbn/)

Lamas G, Jiggins CD (2017) Taxonomic list. In: Jiggins CD (ed) The ecology and evoution of *Heliconius* butterflies. Oxford University Press, Oxford, pp 214–244

Larsen TB (1991) The butterflies of Kenya and their natural history. Oxford University Press, Oxford

Lavinia PD, Nuñez-Bustos EO, Kopuchian C, Lijtmaer DA, García NC, Hebert PDN, Tubaro PL (2017) Barcoding the butterflies of southern South America: species delimitation efficacy, cryptic diversityy and geographic patterns of divergence. PLoS One 12:e0186845

Liu B, Le CT, Barrett RL, Nickrent DL, Chen Z, Lu L, Vidal-Russell R (2018) Historical biogeography of Loranthaceae (Santalales): diversification agrees with emergence of tropical forests and radiation of songbirds. Mol Phylogenet Evol 124:199–212

Mallarino R, Bermingham E, Willmott KR, Whinnett A, Jiggins CD (2005) Molecular systematics of the butterfly genus *Ithomia* (Lepidoptera: Ithomiinae): a composite phylogenetic hypothesis based on seven genes. Mol Phylogenet Evol 34:625–644

Marín MA, Peña C, Uribe SI, Freitas AVL (2017) Morphology agrees with molecular data: phylogenetic affinities of Euptychiina butterflies. Syst Entomol 42:768–785

Massardo D, Fornel R, Kronforst MR, Goncalves GL, Moreira GRP (2015) Diversification of the silverspot butterflies (Nymphalidae) in the Neotropics inferred from multi-locus DNA sequences. Mol Phylogenet Evol 82:156–165

Matos-Maravi PF (2016) Investigating the timing of origin and evolutionary processes shaping regional species diversity: insights from simulated data and neotropical butterfly diversification rates. Evolution 70:1638–1650

Matos-Maravi PF, Nuñez-Aguila R, Peña C, Miller JY, Sourakov A, Wahlberg N (2014) Causes of endemic radiation in the Caribbean: evidence from historical biogeography and diversification of the butterfly genus *Calisto* (Nymphalidae: Satyrinae: Satyrini). BMC Evol Biol 14:199

Matos-Maravi PF, Peña C, Willmott KR, Freitas AVL, Wahlberg N (2013) Systematics and evolutionary history of butterflies in the "*Taygetis* clade" (Nymphalidae: Satyrinae: Euptychiina): towards a better understanding of Neotropical biogeography. Mol Phylogenet Evol 66:54–68

Matz J, Brower AVZ (2016) The South Temperate Pronophilina (Lepidoptera: Nymphalidae: Satyrinae): a phylogenetic hypothesis, redescriptions and revisionary notes. Zootaxa 4125:1–108

Mitter KT, Larsen TB, de Prins W, de Prins J, Collins S, Vande Weghe G, Safian S, Zakharov EV, Hawthorne DJ, Kawahara AY, Regier JC (2011) The butterfly subfamily Pseudopontiinae is not monobasic: marked genetic diversity and morphology reveal three new species of *Pseudopontia* (Lepidoptera: Pieridae). Syst Entomol 36:139–163

Morrone JJ (2014) Cladistic biogeography of the Neotropical Region: identifying the main events in the diversification of the terrestrial biota. Cladistics 30:202–214

Mullen SP, Savage WK, Wahlberg N, Willmott KR (2011) Rapid diversification and not clade age explains high diversity in neotropical *Adelpha* butterflies. Proc R Soc Lond B 278:1777–1785

Murray D, Prowell DP (2005) Molecular phylogenetics and evolutionary history of the neotropical satyrine subtribe Euptychiina (Nymphalidae: Satyrinae). Mol Phylogenet Evol 34:67–80

Nakahara S, Zacca T, Huertas B, Neild AFE, Hall JPW, Lamas G, Holian LA, Espeland M, Willmott KR (2018) Remarkable sexual dimorphism, rarity and cryptic species: revision of the '*aegrota* species group' of the Neotropical butterfly genus *Caeruleuptychia* Forster, 1964 with the description of three new species. Insect Syst Evol 49:130–182

Nelson BW, Ferreira CAC, da Silva MF, Kawasaki ML (1990) Endemism centres, refugia and botanical collection density in Brazilian Amazonia. Nature 345:714–716

Nelson GJ, Platnick NI (1981) Systematics and biogeography: cladistics and vicariance. Columbia University Press, New York

Nuñez Aguila R, Genaro JA, Peréz-Asso AR, Hausmann A (2017) A new species of the *hysius* species group of *Calisto* Hübner (Lepidoptera, Nymphalidae, Satyrinae) and insights into the status of different populations currently attributed to *C. grannus* Bates. Zootaxa 4317:1–44

Ortiz-Acevedo E, Bonfantti D, Casagrande MM, Mielke OHH, Espeland M, Willmott KR (2017) Using molecules and morphology to unravel systematics of Neotropical preponine butterflies (Lepidoptera: Charaxinae: Preponini). Insect Syst and Diversity 1:48–56

Ortiz-Acevedo E, Willmott KR (2013) Molecular systematics of the butterfly tribe Preponini (Nymphalidae: Charaxinae). Syst Entomol 38:440–449

Padron-Martinez SP (2014) Molecular phylogeny and biogeography of the genus *Catasticta* Butler, 1870. PhD Dissertation, University of Florida, Gainesville

Parsons M (1999) The butterflies of Papua New Guinea: their systematics and biology. Academic Press, San Diego

Peña C, Nylin S, Freitas AVL, Wahlberg N (2010) Biogeographical history of the butterfly subtribe Euptychiina (Lepidoptera, Nymphalidae, Satyrinae). Zool Scripta 39:243–258

Peña C, Wahlberg N (2008) Prehistorical climate change increased diversification of a group of butterflies. Biol Lett 4:274–278

Peña C, Wahlberg N, Weingartner E, Kodandaramaiah U, Nylin S, Freitas AVL, Brower AVZ (2006) Higher level phylogeny of Satyrinae butterflies (Lepidoptera: Nymphalidae) based on DNA sequence data. Mol Phylogenet Evol 40:29–49

Peñalver E, Grimaldi DA (2006) New data on Miocene butterflies in Dominican amber (Lepidoptera: Riodinidae and Nymphalidae) with the description of a new nymphalid. Amer Mus Novitates 3519(1):17

Penz CM (2009) Phylogeny of *Dasyophthalma* butterflies (Lepidoptera, Nymphalidae, Brassolini). Insecta Mundi 69:1–12

Penz CM, DeVries PJ (2002) Phylogenetic analysis of *Morpho* butterfles (Nymphalidae, Morphinae): implications for classification and natural history. Amer Mus Novitates 3374:1–33

Penz CM, DeVries PJ, Wahlberg N (2012) Diversification of *Morpho* butterflies (Lepidoptera, Nymphalidae): a re-evaluation of morphological characters and new insight from DNA sequence. Syst Entomol 37:670–685

Penz CM, Mohammadi N, Wahlberg N (2011a) Neotropical *Blepolenis* butterflies: wing pattern elements, phylogeny, and Pleistocene diversification (Lepidoptera: Nymphalidae). Zootaxa 2897(1):17

Penz CM, Simonsen TJ, DeVries PJ (2011b) A new *Orobrassolis* butterfly (Nymphalidae, Brassolini): a casualty of habitat destruction? Zootaxa 2740:35–43

Pfeiler E, Johnson S, Markow TA (2012) DNA barcodes and insights into the relationships and systematics of buckeye butterflies (Nymphalidae: Nymphalinae: *Junonia*) from the Americas. J Lep Soc 66:185–198

Prance GT (1973) Phytogeographic support for the theory of forest refuges in the Amazon Basin, based on evidence from distribution patterns in Caryocaraceae, Chrysobalanaceae, Dichapetalaceae and Lecythidaceae. Acta Amazonica 3:5–28

Pyrcz TW, Freitas AVL, Boyer P, Dias FMS, Dolibana DR, Barbosa EP, Magaldi LM, Mielke OHH, Casagrande MM, Lorenc-Brudecka J (2018) Uncovered diversity of a predominantly Andean butterfly clade in the Brazilian Atlantic forest: a revision of the genus *Praepedaliodes* Forster (Lepidoptera: Nymphalidae, Satyrinae, Satyrini). Neotrop Entomol 47:211–255

Pyrcz TW, Lorenc-Brudecka J, Zubek A, Boyer P, Gabaldon MC, Mavarez J (2017) Taxonomy, phylogeny and distribution of the genus *Steromapedaliodes sensu novo* in the Cordillera de Mérida, Venezuela (Lepidpotera: Nymphalidae: Satyrinae: Satyrini). Arthr Syst and Phylogen 75:195–243

Richardson JE, Pennington RT, Pennington TD, Hollingsworth PM (2001) Rapid diversification of a species-rich genus of neotropical rain forest trees. Science 293:2242–2245

Rull V (2008) Speciation timing and neotropical biodiversity: the Tertiary-Quaternary debate in the light of molecular phylogenetic evidence. Mol Ecol 17:2722–2729

Rull V (2011) Origins of biodiversity. Science 331:398–399

Sahoo RK, Warren AD, Collins SC, Kodandaramaiah U (2017) Hostplant change and paleoclimatic events explain diversification shifts in skipper butterflies (Family: Hesperiidae). BMC Evol Biol 17:174

Sahoo RK, Warren AD, Wahlberg N, Brower AVZ, Lukhtanov VA, Kodandaramaiah U (2016) Ten genes and two topologies: an exploration of higher relationships in skipper butterflies (Hesperiidae). PeerJ 4:e2653

Seraphim N, Kaminski LA, DeVries PJ, Penz C, Callaghan CJ, Wahlberg N, Silva-Brandão KL, Freitas AVL (2018) Molecular phylogeny and higher-level systematics of the metalmark butterfies (Lepidoptera: Riodinidae). Syst Entomol 43:407–425

Silva-Brandão KL, Freitas AVL, Brower AVZ, Solferini VN (2005) Phylogenetic relationships of the New World Troidini swallowtails (Lepidoptera: Papilionidae) based on COI, COII and Ef-1a genes. Mol Phylogenet Evol 36:468–483

Silva-Brandão KL, Wahlberg N, Francini RB, Azeredo-Espin AML, Brown KS Jr, Paluch M, Lees DC, Freitas AVL (2008) Phylogenetic relationships of butterflies of the tribe Acraeini (Lepidoptera, Nymphalidae, Heliconiinae) and the evolution of host plant use. Mol Phylogenet Evol 46:515–531

Smith DS, Miller LD, Miller JY (1994) The butterflies of the West Indies and South Florida. Oxford University Press, Oxford

Stebbins GL (1974) Flowering plants: evolution above the species level. Belknap Press, Cambridge, MA

Swofford DL (2000) PAUP* Phylogenetic analysis using parsimony (*and other methods). Sinauer Associates, Sunderland (Version 4.0a162)

Toussaint EFA, Breinholt JW, Earl C, Warren AD, Brower AVZ, Yago M, Dexter KM, Espeland M, Pierce NE, Lohman DJ, Kawahara AY (2018) Anchored phylogenomics illuminates the skipper tree of life. BMC Evol Biol 18:101

Tyler HA, Brown KS Jr, Wilson KA (1994) Swallowtail butterflies of the Americas. Scientific Publishers, Gainesville, FL

Vanzolini PE, Williams EE (1970) South American anoles: the grographic diffrentiation and evolution of the *Anolis chysolepis* species group (Sauria: Iguanidae). Arq Zool S Paulo 19:1–240

Vila R, Bell CD, Macniven R, Goldman-Huertas B, Ree RH, Marshall CR, Balint Z, Johnson K, Benyamini D, Pierce NE (2011) Phylogeny and paleoecology of *Polyommatus* blue butterflies show Beringia was a climate-regulated gateway to the New World. Proc R Soc Lond B 278:2737–2744

Wahlberg N, Brower AVZ, Nylin S (2005) Phylogenetic relationships of tribes and genera in the subfamily Nymphalinae (Lepidoptera: Nymphalidae) based on three gene sequences. Biol J Linn Soc 86:227–251

Wahlberg N, Freitas AVL (2007) Colonization of and radiation in South America by butterflies in the subtribe Phyciodina (Lepidoptera: Nymphalidae). Mol Phylogenet Evol 44:1257–1272

Wahlberg N, Leneveu J, Kodandaramaiah U, Peña C, Nylin S, Freitas AVL, Brower AVZ (2009) Nymphalid butterflies diversify following near demise at the cretaceous/tertiary boundary. Proc R Soc London B 276:4295–4302

Wahlberg N, Rota J, Braby MF, Pierce NE, Wheat CW (2014) Revised systematics and higher classification of pierid butterflies (Lepidoptera: Pieridae) based on molecular data. Zoologica Scripta 43:641–650

Wahlberg N, Weingartner E, Nylin S (2003) Towards a better understanding of the higher systematics of Nymphalidae (Lepidoptera: Papilionoidea). Mol Phylogenet Evol 28:473–484

Wallace AR (1853) A narrative of travels on the Amazon and Rio Negro. Reeve, London

Warren AD, Ogawa JR, Brower AVZ (2009) Revised classification of the family Hesperiidae (Lepidoptera: Hesperioidea) based on combined molecular and morphological data. Syst Entomol 34:467–523

Willmott KR (2003) The genus *Adelpha*: its systematics, biology and biogeography (Lepidoptera: Nymphalidae: Limenitidini). Scientific, Gainesville, FL

Willmott KR, Hall JPW, Lamas G (2001) Systematics of *Hypanartia* (Lepidoptera: Nymphalidae: Nymphalinae), with a test for geographic speciation mechanisms in the Andes. Syst Entomol 26:369–399

Willmott KR, Mallet J (2004) Correlations between adult mimicry and larval host plants in ithomiine butterflies. Proc R Soc Lond B 271:266–269

Willmott KR, Robinson Willmott JC, Elias M, Jiggins CD (2017) Maintaining mimicry diversity: optimal warning color patterns differ among microhabitats in Amazonian clearwing butterflies. Proc R Soc Lond B 284:20170744

Zink RM, Klicka J, Barber BR (2004) The tempo of avian diversification in the Quaternary. Phil Trans R Soc Lond B 359:215–220

Part II
Regional Biodiversity Patterns and Diversification Processes

Chapter 10
The Origin and Evolution of Amazonian Species Diversity

J. Cracraft, Camila C. Ribas, F. M. d'Horta, J. Bates, R. P. Almeida, A. Aleixo, J. P. Boubli, K. E. Campbell, F. W. Cruz, M. Ferreira, S. C. Fritz, C. H. Grohmann, E. M. Latrubesse, L. G. Lohmann, L. J. Musher, A. Nogueira, A. O. Sawakuchi, and P. Baker

J. Cracraft and Camila C. Ribas contributed equally to this work.

The original version of this chapter was revised. A correction to this chapter can be found at https://doi.org/10.1007/978-3-030-31167-4_31.

J. Cracraft (✉) · L. J. Musher
Department of Ornithology, American Museum of Natural History, New York, NY, USA
e-mail: jlc@amnh.org

C. C. Ribas (✉) · F. M. d'Horta · M. Ferreira
Coordenação de Biodiversidade, Instituto Nacional de Pesquisas da Amazônia, Manaus, AM, Brazil
e-mail: camila.ribas@inpa.gov.br; ribas@amnh.org

J. Bates
Department of Zoology, Field Museum of Natural History, Chicago, IL, USA

R. P. Almeida · F. W. Cruz · A. O. Sawakuchi
Instituto de Geociências, University of São Paulo, São Paulo, Brazil

A. Aleixo
Finnish Museum of Natural History, University of Helsinki, Helsinki, Finland

J. P. Boubli
School of Environment and Life Sciences, University of Salford, Salford, UK

K. E. Campbell
Vertebrate Zoology, Natural History Museum of Los Angeles County, Los Angeles, CA, USA

S. C. Fritz
Department of Earth and Atmospheric Sciences, University of Nebraska, Lincoln, NE, USA

C. H. Grohmann
Institute of Energy and Environment, University of São Paulo, São Paulo, Brazil

E. M. Latrubesse
Asian School of the Environment and Earth Observatory of Singapore, Nanyang Technological University, Singapore, Singapore

V. Rull, A. C. Carnaval (eds.), *Neotropical Diversification: Patterns and Processes*, Fascinating Life Sciences, https://doi.org/10.1007/978-3-030-31167-4_10

Abstract We identify key interdisciplinary questions and approaches to understanding the paleogeographic and biotic history of Amazonia. We discuss the importance of comparing evolutionary taxonomic units across groups and considering their particular environmental affinities as a framework for a mechanistic understanding of how the diversity of Amazonia was assembled across space and time. Although lineage diversity of the Amazonian biota is often quite old, current evidence indicates that much of Amazonia's observed "species-level" diversity, as exemplified by birds, mammals, and butterflies, is young, mostly dating to the Quaternary. Much of the modern Amazonian landscape, including in particular its hydrological and environmental systems, was shaped over the Pliocene and Quaternary (last 5 Ma), yet significant landscape change affecting the evolution of biotas is less than 2 million years old.

Keywords Amazonian landscape change · Diversification · Areas of endemism · Speciation · Extinction · Neotropical diversity

1 Introduction

No biome has generated more research addressing fundamental ecological and evolutionary questions about tropical species diversity than Amazonia. Yet, there still is little consensus regarding the causal underpinnings of Amazonian biotic history, partly as a result of several conceptual challenges (Boubli et al. 2015; Naka and Brumfield 2018; Prates et al. 2016; Smith et al. 2014; Antonelli et al. 2018a, b; Silva et al. 2019). First, understanding the history of the Amazonian biota and how it was structured over time requires integrating knowledge across multiple disparate disciplines often without a common conceptual or empirical language. Thus, geologists and biologists have found it difficult to reconcile how their data and interpretations might be relevant to each other. Second, investigators in different disciplines (e.g., systematists, ecologists, or palynologists) frequently frame questions from different conceptual worldviews or methodological norms. For example, researchers often use different concepts of taxa or phylogenetic and biogeographic methods to describe and understand patterns of diversity. In short, even the "data" are frequently in dispute. And, third, posing questions such as "what are the major drivers of Amazonian diversity?" is not straightforward when trying to understand one of the most species-rich and complex regions of the planet.

Here we address key questions about the origin and evolution of species diversity in Amazonia and attempt to frame current knowledge about them. Although our

L. G. Lohmann
Instituto de Biociências, University of São Paulo, São Paulo, Brazil

A. Nogueira
Instituto de Geociências, Federal University of Pará, Belém, Brazil

P. Baker
Division of Earth and Ocean Sciences, Duke University, Durham, NC, USA

answers to these questions may not be entirely new, we believe our integrative and broad approach leads to new insights. A primary goal is to address how investigators have often framed these questions differently, conceptually and empirically, and how this has led to disparate visions about Amazonian diversification and its causes. A key limitation is that detailed knowledge about Amazonian environmental history remains fragmentary and restricted primarily to the LGM or Holocene (Baker et al. 2014). This paucity of information can mislead biologists, who may over-interpret available data relevant for their biological problem. A simple, yet common, example is the debate over the age of formation of the present Amazonian drainage, which has been dated from 10 to 3.0 myr. Many biological papers adopt these differing dates rather than independently test them.

In this paper we will emphasize the importance of testing causal relationships between Amazonian diversification and Earth history, and advocate that consilience across multi-disciplinary data will be the only way to generate robust answers to these key questions. Our evidence is shaped predominately from the perspective of Amazonian vertebrates, particularly results from avian and mammalian diversification and biogeography. Therefore we recognize that not all our conclusions may be general across all biodiversity, and we incorporate relevant studies when appropriate.

2 How Much Amazonian Diversity Do We Really Know?

Investigating comparative questions about the evolutionary history of the Amazonian biota presupposes that investigators who work with different taxonomic groups and disciplines are employing comparable taxonomic units (Rull 2011). Indeed, quantifying diversity, comparing distributions, reconstructing phylogenetic or biogeographic patterns, unraveling speciation, studying community structure, and comparing patterns of diversification all depend crucially on having clearly defined comparable taxonomic units (Bortolus 2008). The proper delimitation of these "evolutionary units" has important implications for assessing the ages of Amazonian taxa, which is essential to link biotic diversification and Earth history events. None-the-less, biologists often compare dissimilar units in ways that confuse interpretations.

For most comparative biologists, species-level taxa are the logical units of diversity, but, as is widely recognized, even this unit has a long history of controversy. Current practices in taxonomy are largely based on partial knowledge about spatial variability of both molecular and phenotypic characters and, consequently, current taxonomy often underestimates diversity, confounds the discovery of biogeographic patterns, and misrepresents the spatial uniqueness of diversity at ecological scales (Chaves et al. 2015). This is especially important in Amazonia, as studies that have investigated taxonomic variation within widely distributed species have revealed many additional taxonomic units, even in supposedly well known groups like birds and primates (e.g., Lynch-Alfaro et al. 2015; Fernandes et al. 2013, 2014; Ribas et al. 2006, 2012; Thom and Aleixo 2015).

Diversification studies should focus on *evolutionary taxa*, which we define as the smallest population or set of populations that can be characterized as a taxon. These are entities produced largely by allopatric differentiation and distinguishable by evolutionary novelties from other such taxa. Gene lineages, regardless of whether populations are reciprocally monophyletic or not on a tree, do not necessarily constitute evolutionary taxa, because gene lineages might not yet have crossed the population-taxon boundary (i.e., they are not objectively diagnosable, even with genetic data; Sukumaran and Knowles 2017). Trying to “standardize” species-level units across groups based on the amount of genetic divergence, without a critical interpretation of their phenotype and geography, is at odds with systematic knowledge about geographic variation that goes back decades. No matter what species concept one adheres to, these approaches misrepresent evolutionary diversity and confound historical analyses. Moreover, counting supraspecific groups will always introduce error as, under standard classification schemes, “genera,” or “families” are neither evolutionary entities nor can they be circumscribed objectively.

The solution to circumscribing and comparing evolutionary diversity is conceptually simple, but it can be methodologically challenging, as good sampling and assessment of distributional and morphological data are often difficult in Amazonia. Taxa become comparable when they are singular and do not contain or include two or more other like-units. With a focus on “evolutionary taxa” we free ourselves from the debate about ranking (i.e., whether phylospecies, biospecies, or subspecies), which has often hampered dialogue and understanding of historical patterns. Yet, on a very practical level, systematists have to call these units something. We call these phylospecies, or simply species (basal units). It is important that many investigators understand this debate and incorporate this assessment as an essential step in biogeographic studies, so that taxonomic units are ontologically comparable.

An accurate understanding of time and age are at the heart of debates over the evolution of the Amazonian biota and are necessary for solving many key problems. Thus, linking biotic and Earth histories depends on knowing the ages of taxa and of the geologic and paleoclimatic events that have shaped biotic diversity. We emphasize that well sampled taxonomic revisions are not a step that should be overlooked when analyzing patterns of diversification, especially in Amazonia.

3 When Did the Modern Amazonian Biota Originate?

One of the most important, and debated, questions about Amazonian diversity is how old are most of its extant species? This is a fundamentally crucial question for investigating the abiotic and biotic factors that might have controlled observed patterns of diversification in space and time. Fortunately, methods used to build evolutionary time trees have become more sophisticated, and those calibrated by using well-dated fossils are the most rigorous. Time trees are important for testing geological hypotheses, and temporal data from the latter are key to testing many biological hypotheses (Baker et al. 2014). Importantly, this reciprocity assumes that age-calibrations for these data-sets are independent.

The broad question about when most modern Amazonian biota arose must be historically contextualized in terms of phylogeny and how those findings are reflected in taxonomy. Most authors have addressed this question referring to current species diversity, yet that is not always the case (e.g., Hoorn et al. 2010). Certainly, some elements of modern Amazonian biota have very old roots, and much of the remaining diversity has stem lineages going back into the Mesozoic. Thus, current diversity in Amazonia has been assembled and "layered" phylogenetically over time.

Many major Amazonian clades appear to have originated in the Late Cretaceous or Paleocene. Cretaceous floras from the New World tropics are poorly known (Burnham and Johnson 2004), but diverse "modern-aspect" rainforest, as indicated by modern plant families, was established in the Paleocene and was apparently widespread and abundant in South America by the early Eocene (Wilf et al. 2003; Burnham and Johnson 2004). This interpretation reflects a global Paleocene that was warm and wet (Zachos et al. 2001; Fine and Ree 2006), even at high latitudes (Jahren 2007). In addition to these Laurasian elements, the shared connections of southern Cretaceous and Paleocene floras across Gondwana are significant (Woodburne et al. 2014).

Paleontological data and inferences from time trees indicate that Late Cretaceous and especially early Paleocene tropical South American biomes contained an array of terrestrial vertebrate taxa having both Gondwanan and Laurasian histories, for example, hylid frogs (Wiens et al. 2006) and booid snakes (Noonan and Chippindale 2006). Non-therian mammals were present in South America by the latest Cretaceous (Woodburne et al. 2014), as were stem-marsupials (Westerman et al. 2016), and fossil eutherians were present in the Paleocene (Flynn and Wyss 1998; Woodburne et al. 2014). In South America, stem-didelphid marsupials apparently arose in the latest Cretaceous (Westerman et al. 2016), but most of their diversification was post-Oligocene (Jansa et al. 2014). Biogeographic and time tree analyses indicate that deep stem-lineages of modern birds arose in South America and/or other parts of Gondwana in the Paleocene, and some possibly extended back into the latest Cretaceous (Claramunt and Cracraft 2015). These reconstructions and fossils imply that the fauna of South American tropical forests was taxonomically and ecologically complex by the end of the Paleocene (Claramunt and Cracraft 2015).

Phylogenies and time-trees of South American bird clades show that many contemporary lineages (the stems of "higher" taxa) diversified throughout the Cenozoic. Indeed, by the Late Paleogene, a modern taxonomic aspect of the Amazonian avifauna was firmly established. All that was missing were Laurasian groups that entered during the Great American Biological Interchange, primarily oscines (Smith and Klicka 2010). However, large-scale phylogenetic trees provide only rough estimates of the temporal history of an evolving taxon or a biota. For example, Claramunt and Cracraft (2015) estimated the age of origin of stem-lineages, with whole families being typically represented by a single species. Similar approaches using genera instead of families (Hoorn et al. 2010) likewise provide little direct evidence for ages of contemporaneous species in the Amazonian biota (Rull 2011). As proposed above, it is most appropriate to individuate those *taxonomic* units that represent the end-points of differentiation, without bias as to how individual investigators might *rank* them.

Given this point of view, accumulating evidence suggests that the majority of Amazonian species-level (evolutionary) terrestrial vertebrate taxa found today originated in the last 5 million years (Ma) during the Pliocene (5.33–2.58 Ma) and Pleistocene (2.58–0.012 Ma) (Rull 2008, 2011; Silva et al. 2019). This is a frequency argument, similar to that made by Garzón-Orduña et al. (2014), who estimated that contemporary butterfly sister-species divergences occurred predominately (72%) less than 2.6 Ma. Multiple dated phylogenetic studies at fine taxonomic scale, and phylogeographic studies that reveal young and previously unrecognized microendemics across Amazonia for birds, primates and some anurans indicate that current species within these groups also differentiated within this period.

Most of the 19 primate sister-taxa separated across the Negro and Branco rivers diverged less than 2.5 Ma (Boubli et al. 2015). Crown-ages within the species themselves are much younger, mostly hundreds of thousands of years. Studies of other primate groups, including *Saimiri* squirrel monkeys (Alfaro et al. 2015), *Cebus* capuchins (Boubli et al. 2012, Lima et al. 2017), titi monkeys (Byrne et al. 2018), and callitrichid marmosets and tamarins (Buckner et al. 2015) found similar ages.

Perhaps the most cited example of recent speciation in Amazonia is the eight species of trumpeters (*Psophia*), each of which occupies one of the region's major areas of endemism (Ribas et al. 2012). The entire clade started diversifying about 3.0–2.5 Ma, and species ages range from 0.3 to 2 Ma. Remarkably, this is a clade of gruiform birds with a stem of about 40 million years, which implies significant turnover in the group. The time-frame for speciation in *Psophia* is not unusual, and in many other groups of birds differentiation may even be younger (e.g., Schultz et al. 2017; Silva et al. 2019; Ribas et al. 2018).

4 Pliocene and Pleistocene Landscape Evolution: Implications for Diversification

Many studies have investigated the drivers of Amazonian diversity, and multiple ecological correlates with Neotropical diversity have been proposed, including more energy, more niches, greater net primary productivity/resources, increased competition, ecological opportunity, and biotic age (e.g., Pianka 1966; Willig et al. 2003; Ricklefs 2004; Mittelbach et al. 2007). However, diversification occurs over time, and thus causation must be historically contextualized. Consequently, most literature on Amazonia's vast diversity does not adequately address mechanistic solutions to speciation and extinction rate-controls from a historical standpoint.

Allopatric speciation depends upon barriers to gene flow, either physical or ecological, and the main causes of barrier formation are events in Earth history. Two main drivers of Amazonian diversification have most frequently been implicated: (a) climate-mediated change to landscape dynamics influencing the distribution of forested/open ecosystems; and (b) the fragmentation of these ecosystems mediated by drainage evolution. Both landscape-level processes are jointly influenced by factors such as Andean uplift and global climate change, which means that parsing causation is difficult. Also, once barriers are created that hinder gene flow, other intrinsic processes may contribute to isolation and differentiation of populations, resulting in the origin of new taxa.

Importantly, extinction should also be linked to these processes and generally follows a pace similar to speciation. Paleontological analyses (Stanley 2007), as well as diversification analyses of the modern avifauna (Claramunt and Cracraft 2015; Musher et al. 2019) suggest that background rates of speciation and extinction are coupled mechanistically: if barriers are subdividing populations into smaller isolates, this should imply higher rates in both speciation and extinction in these smaller populations. In contrast, if barriers disappear and populations diffuse across landscapes, larger ranges and higher population sizes would be predicted to lower both rates. Of course, net-diversification rate can trend upward or downward over time, which means that deciphering Amazonian diversity will require keeping an integrated vision with these "first principles" in mind. Many "explanations" in the literature, however, are framed as single-variable correlations of diversity that speak to speciation, to the exclusion of extinction, or vice-versa. Thus, we endorse the importance of studying speciation and extinction by emphasizing the history of biotas within the context of Earth history. The key to developing an understanding of Amazonian biotic and environmental history lies in integration across disciplines as well as joint questions about the evolution of the environment and its biota (Baker et al. 2014).

Amazonian biogeographers have long used assumptions about the age of landscape features to interpret and explain the distributions of many upland terra firme species and their biogeographic patterns. One assumption, for example, is that the major rivers have been entrenched roughly in their current locations since initial development of the drainage system in the late Miocene (10–9 Ma) (Haffer 2008; Hoorn et al. 2010, 2017; Smith et al. 2014), which would imply these major barriers were in place well before the origin of much of the current species diversity, as discussed earlier. Other similar biogeographic assumptions could be made using the premise that the modern Amazon river drainage was primarily organized either around 6.5–5.0 Ma (Latrubesse et al. 2010) or near the Plio-Pleistocene boundary 3.0–2.5 Ma (Campbell et al. 2006). Each of these hypotheses has major implications for explaining the ages of species and patterns of endemism that are bounded by the major rivers. If river barriers were significantly involved in establishing these patterns, there might be an expectation of temporal congruence across them, but at this time the evidence for that is mixed (Boubli et al. 2015; Naka and Brumfield 2018; see below). An alternative is that the rivers, as barriers, had a more complex temporal history.

Beyond the debate about the age of its establishment, the Amazon river system has been dynamic, with even the largest rivers possibly shifting their channels over time. The evolution of each of the Amazonian rivers depends on its geological context: in western Amazonia, white water rivers with high sediment load like the

Solimões occupy extensive Andean-derived soft sedimentary terrains built by Plio-Pleistocene channel dynamics (Nogueira et al. 2013; Rossetti et al. 2015). These deposits formed by recurrent shifts in the position of the riverine channels due to their autogenic depositional dynamics as well as to precipitation changes in the watersheds. Eastern Amazonian rivers, partially draining the Brazilian and Guianan shields and running over old and more stable terrain, including part of the Madeira, Negro, and Tocantins, may also have shifted their channels over short time-spans through river capture and partial basin reorganizations (Almeida-Filho and Miranda 2007; Grohmann et al. 2011; Hayakawa and Rossetti 2015; Latrubesse 2002; Rossetti and Valeriano 2007). Other shield rivers like the Tapajós and Xingu have relatively stable valleys cutting into terrains under very low denudation rates (Wittmann et al. 2011). Variations in the efficiency of these rivers as biotic barriers could be related to variations in the water discharge modulated by rainfall changes in the watershed, in association with each species' dispersal capabilities. Independently of changes in river channel position, channel incision and aggradation can influence the effectiveness of rivers as barriers through time, especially in rivers flowing over sedimentary terrains in western Amazonia. This, in turn, controls floodplain width (Govin et al. 2014), and consequently, the degree of isolation between upland forests and their biotas on opposite margins. This environmental setting points to the fluvial system of Amazonia as being very dynamic during the Pliocene and Pleistocene, and biogeographers have likely significantly underestimated the importance of this dynamism in biotic evolution within both terrestrial and aquatic systems.

Landscape dynamics linked to climate changes have also been frequently linked to biogeography and diversification. Haffer's (1969) pioneering ideas on refuges and speciation are a prime example. But, criticism of these ideas has sometimes led to interpretations in the other extreme, namely that there has been relatively little change in forest cover in Amazonia during glacial cycles (e.g., Colinvaux et al. 2001; Bush 2017; Costa et al. 2018). Thus, based on that premise, several studies have relied on an interpretation of the sparse geological proxies suggesting that Amazonia was a relatively stable landscape during the Quaternary, following the Neogene rise of the Andes and the concurrent establishment of major contemporary landscape features (e.g., Smith et al. 2014). From this assumption, it has been proposed that Pleistocene diversification was driven mainly by stochastic dispersal over pre-existing barriers, and thus would mostly not be directly related to paleoenvironmental changes and landscape evolution. However, the idea of a stable Amazonia during the Quaternary is at variance with the repeated, large-amplitude, global and regional Pleistocene climate variations (Cheng et al. 2013; Wang et al. 2017) and evidence for consequent change in forest cover (Cowling et al. 2001; van der Hammen and Hooghiemstra 2000; D'Apolito et al. 2013; Arruda et al. 2018; Zular et al. 2018). Similarly, significantly younger dates have been proposed for development of the current drainage system (Rossetti et al. 2005; Nogueira et al. 2013) as well as for substantial fluxes in the drainage system reaching even Holocene times (Almeida-Filho and Miranda 2007; Hayakawa and Rossetti 2015). Moreover, the efficacy of dispersal across barriers as a mechanism for substantial differentiation seemingly conflicts with a basic premise of population genetics that even small amounts of gene flow can hinder differentiation. Yet, because rivers of all sizes seem to be geologically dynamic in relatively short time frames, thus altering the location and effectiveness of barriers through time, distinguishing between strict vicariance and dispersal scenarios may be difficult.

Haffer's vision of glacial-interglacial climate cycles as potential drivers of biome change leading to isolation in refugia and then speciation (Haffer 1969, 1974) has been challenged (e.g., Colinvaux et al. 2001), yet given the extremely sparse paleoclimatic and palynological sampling across lowland Amazonia (Bush et al. 2011), spatial and temporal changes in vegetation history at regional scales remain under-documented. There is no consensus that Amazonia has "remained forested even during the cooler and drier glacial periods" (Smith et al. 2014) for the past 2.6 Myr (van der Hammen and Absy 1994; Cheng et al. 2013; D'Apolito et al. 2013; Cohen et al. 2014). Some studies suggest that the magnitude, direction, and timing of climate change differed between eastern and western Amazonia, with more intense drying in the East (Cheng et al. 2013; Wang et al. 2017), but other proxies such as fluvial sediments and Late Pleistocene paleofauna suggest drier conditions during part of the late Pleistocene in the lowlands of southwestern Amazon (Latrubesse and Ramonell 1994). A remarkable decrease of precipitation in central-east Amazonia was postulated during the last glacial (Wang et al. 2017). This, in turn, presumably had effects on the river systems that mediated expansion and fragmentation of tropical forests, but also on the population dynamics of species within those habitats, mediated by the ecophysiological characteristics of each species (Papadopoulou and Knowles 2016). Although a strict interpretation of the Haffer (1969) refuge model is not supported, especially toward the west, paleoclimatic evidence suggests a dynamic history of large and sometimes abrupt changes in moisture (Cheng et al. 2013; Wang et al. 2017) and in vegetation (Häggi et al. 2017), both of which likely played a role in shaping Amazonian diversity (van der Hammen and Hooghiemstra 2000; D'Apolito et al. 2013, 2017; Baker and Fritz 2015; Silva et al. 2019). The extent of that role is still uncertain, but multiple studies of historical demography in Amazonian understory birds suggest recent changes in effective population sizes in several Amazonian regions (Ribas et al. 2012; d'Horta et al. 2013; Maldonado-Coelho et al. 2013; Harvey and Brumfield 2015), indicating habitat perturbations, possibly as a result of paleoclimate change. In addition, currently isolated specialist taxa of Amazonian open habitat patches show evidence of recent genetic connectivity, implying habitat connectivity during the Pleistocene (Wüster et al. 2005; Capurucho et al. 2013; Matos et al. 2016). And, climatic and niche models seem to suggest that some terra-firme bird lineages may have suffered local extirpation in some areas that has resulted in distributional gaps (Bonaccorso et al. 2006; Peterson and Nyári 2008; Silva et al. 2019). All of this suggests we still have much to learn about the vegetation dynamics across Amazonia and its effects on biotic history.

5 If the Amazonian Landscape Was Dynamic, How Did Biogeographic Patterns Arise?

Because vertebrate speciation is predominately allopatric (Coyne and Orr 2004), an initial approach to the above question targets the relationship between the ages of barriers leading to allopatry and the divergence time between sister taxa isolated by that barrier (Rull 2014). A rapidly changing paleogeographic setting, like Amazonia, predicts that most species-level divergences driven by landscape history would be relatively young, which seems to be the case (see also Haffer 1987). Yet, controversy about the importance of Earth history versus dispersal over pre-existing barriers followed by isolation and differentiation (decoupled from landscape change per se) has a long history within systematics and biogeography. The debate is not whether both mechanisms happen, but, rather, which one could be primarily responsible for the current spatial organization of diversity within Amazonia, including patterns of endemism recognized for some upland forest groups such as birds and primates (e.g., Zink et al. 2000; Crouch et al. 2018)?

The first assessments of the importance of landscape evolution in shaping Amazonian biota were based on empirical patterns of distributional congruence in specific taxonomic groups (e.g., Wallace 1852; Haffer 1974, 1982; Brown 1982; see papers in Duellman 1979; Prance 1982). It was observed that large rivers delimit the geographical distributions of many taxa, but many of these historical studies assumed that the Amazonian river system that we see today was not the primary cause of these biogeographic patterns; rather Pleistocene refugia were (Haffer 1982; Prance 1982). The assumption was that Pleistocene glacial cycles created large areas of dry/open habitat, thus driving vicariance of wet-forest and open-area biotas. This inference of vicariance was grounded on observed congruence of distributional patterns on a pre-existing drainage system. Haffer (1974, 1982) saw Amazonian history in mostly Pleistocene terms, including significant Andean uplift and the formation of the major drainage system, yet he emphasized the direct effect of climatic cycles and their moisture gradients on forest composition and extent as the primary driver of diversification.

Vicariance resulting from drainage evolution or climatic-ecological fragmentation of habitats, including shifts in terra firme versus floodplain forests, have become the standard models of diversification in well-studied groups, particularly within upland forest vertebrates (e.g., Haffer 1974; Patton and da Silva 1998; Ribas et al. 2012; Boubli et al. 2015; Silva et al. 2019), but also butterflies (Hall and Harvey 2002; Garzón-Orduña et al. 2014). Distributions change as populations expand or contract their ranges, but diversification is most clearly linked to the formation of barriers, either physical or ecological. This interpretation is built on considerable data spanning many decades, including congruence in biogeographic and phylogenetic pattern, co-distribution and endemism of close relatives linked to barriers, and dispersal abilities.

In a dynamic fluvial system like the Amazon basin, variation in the ages of species-splits across a riverine or ecological barrier might not falsify a role for landscape change in diversification; it may only suggest uncertainty in identifying a singular cause of allopatry inasmuch as barriers can vary in relative strength and position through time (Naka and Brumfield 2018). Thus, over the last 5 million years, and especially during the Pleistocene, major shifts of river channel position or discharge variation could account for multiple episodes of vicariance across the "same" river over time (e.g., Weeks et al. 2016). Adding to this complexity, the different ecological characteristics of individual species will probably result in their responding to barriers at different moments of their formation (Papadopoulou and Knowles 2016).

The complexity of Amazonian landscape change points to the possibility that taxon evolution involving what seem to be commonalities in distribution may have alternative explanations, because areas themselves may have had different histories. Thus, specifying *historically relevant* taxa for testing vicariance across a particular set of barriers becomes crucial. For example, many endemics occur within the "Guiana area" of endemism, but specifying a given barrier that isolated all or some of them is not so simple, as taxa on either side of that barrier may not be sister groups. A Guianan endemic, for example, may be sister to taxa across the Rio Negro; to taxa across the Amazon River; or even to a clade including taxa from multiple other South or Central American forests. The nuances of this scenario are found, for example, in *Schiffornis turdina* (Peterson and Nyári 2008), *Glyphorhynchus spirurus* (Fernandes et al. 2013), *Xenops minutus* (Harvey and Brumfield 2015), and *Dendrocincla fuliginosa* (Schultz et al. 2019), all of which could erroneously be regarded as potentially congruent, due to their current similar patterns of distribution (Smith et al. 2014). Yet, Naka and Brumfield (2018) recently showed that when species or subspecies replacements are compared across the Rio Negro, broad age differences were found. Because these comparisons involved both sister- and nonsister-taxa, they suggest different modes of speciation might be involved. It may well be, however, that barrier history in that region may have been more complicated than previously recognized.

6 Building a Picture of the Diversification and Assembly of the Amazonian Biota

A simple set of biotic processes, including biotic diffusion (geodispersal) across landscapes, dispersal across pre-existing barriers, isolation and differentiation (speciation) via vicariance, and extinction govern the origins and assembly of biotas. Yet, interpreting the spatial and temporal history of the Amazonian biota is still hampered by gaps in knowledge across multiple disciplines. We recognize three main empirical impediments. First, there remains a very incomplete knowledge about Amazonia's taxa and their distribution. This deficiency is acute within invertebrates, less so in plants (Hopkins 2007) and vertebrates. However, given our growing awareness of cryptic species (e.g., Aleixo et al. 2013; Portes et al. 2013) and microendemism (e.g., Ferreira et al. 2017; Jacobs et al. 1995), investigators cannot be complacent that their particular taxonomic group is "well known." Second, without phylogenetic knowledge at a fine taxonomic scale, the spatiotemporal history of a group is compromised. Given Amazonia's diversity, relatively few groups have been examined in the detail required for informative biogeographic and evolutionary analysis. Finally, the region's complex landscape and environmental history is only partially known, largely because of the lack of detailed correlative studies with a well-constrained timescale across Amazonia. Furthermore, the broader history of Amazonia cannot be completely decoupled from adjacent biogeographic regions, such as the Andes, Cerrado, Central America and the Atlantic forest (e.g., van der Hammen and Hooghiemstra 2000; Weir 2006; Lutz et al. 2013; Crouch et al. 2018; Antonelli et al. 2018a, b).

To achieve a better understanding of the relationship between landscape evolution and diversification in a system so complex and with so many uncertainties, many variables need to be considered. It is necessary to compare taxa whose distribution patterns, ecology, and evolutionary time frame are appropriate for testing the roles of the specific geological features and landscape events of interest. Additionally, the age of most geological features to be tested as barriers have large uncertainty and further geochronologic studies will be necessary to reconstruct a history of landscape change comparable to diversification events constrained by phylogenetic studies. Diversification of specific clades is dependent on their different habitat affinities, and also each may be differentially influenced by landscape evolution, depending on their ancestral distribution at the time the events were taking place. Eastern and western Amazonia had different histories both concerning drainage organization and climate, and clades with western or eastern ancestors would be influenced differently by landscape evolution during the same time frame (Silva et al. 2019).

In conclusion, deciphering Amazonian history is a shared interdisciplinary vision. Processes of speciation, extinction, land cover change, population history, and local ecological interactions commingle in space and time, and are influenced by tectonic history, drainage evolution, and climate change. We therefore advocate integration among these disciplines and their data. We have the capacity to go far beyond many current approaches to "explaining" Amazonian history through causally weak environmental correlations. Simplistic assumptions about the data of other disciplines ultimately do little to advance our understanding. Despite considerable attention over the past century to Amazonian biotic and environmental histories, no easy answers can be found, and many mysteries still remain. The last decade, however, has seen a resurgence of integrative studies and initiatives that point to new ways of understanding this common history (Baker et al. 2014). We believe this is the vision for the future.

For biogeographic thinking to achieve a more holistic picture of Amazonian history, we suggest that individual studies, such as the phylogeographic analysis of a small group or an analysis of assembly dynamics at a local scale, must be placed within the larger context of how biotas evolve across landscapes. Every local area of endemism is imbedded in a larger area. Areas of endemism are thus hierarchical. At the same time, it is crucial to recognize that biotas can diffuse across landscapes in relatively short periods of time, and the barriers that shape that diffusion can change as well and be of different ages. This is very relevant given that multiple abrupt wet-dry events occurred during the late Pleistocene, drastically impacting the tropical South American climate at large spatial scales (Cheng et al. 2013; Stríkis et al. 2015; Wang et al. 2017), inducing significant changes on Amazonian vegetation (Häggi et al. 2017; Zular et al. 2018). As barriers come and go, area-history becomes more complex. As a result, current methods of biogeography face analytical limitations in recovering landscape and biotic history.

In Amazonia, clear evidence of joint spatial history exists among some clades, but the patterns of other groups suggest conflict (e.g., Cracraft 1988; Bates et al. 1998; Silva et al. 2019). Understanding complex patterns requires thinking about both the characteristics of the lineages in question and how their temporal and spatial histories are responding to changes in forest structure and distribution induced by climate and drainage evolution. This suggests that we might want to think about the spatial history of groups and biotas as a whole, and ask where those biotas were distributed at different time-slices. The standard approach to biogeographic analysis has focused on individual clades, which remains important. What we suggest here, however, is the importance of recognizing an among-clade, or biotic, perspective. If we disentangle biotic history from Earth history, explanatory power disappears. Conversely, Earth history cannot be entirely disentangled from biotic history either. Each has the other's history imbedded inside it.

Acknowledgements These ideas have been developed with support from the grants: "Dimensions US-Biota-São Paulo: Assembly and evolution of the Amazon biota and its environment: an integrated approach", co-funded by the US National Science Foundation (NSF DEB 1241066), National Aeronautics and Space Administration (NASA), and the São Paulo Research Foundation (FAPESP, grant #2012/50260-6) to JC and LGL; PEER/USAID (AID-OAA-A-11-00012) grant to CCR, AA and JC; and NSF FESD (EAR-1338694) to PAB and SCF. CCR and CHG are supported by CNPq research fellowships (respectively #458311/2013-8 and 307647/2015-3). Support has also been provided by CNPq, CAPES and FAPEAM in the form of student fellowships and research grants.

References

Aleixo A, Portes CEB, Whittaker A, Weckstein JD, Gonzaga LP, Zimmer EA, Ribas C, Bates JM (2013) Molecular systematics and taxonomic revision of the curve-billed Scythebill complex (*Campylorhamphus procurvoides*: Dendrocolaptidae), with description of a new species from western Amazonian Brazil. In: del Hoyo J, Elliot A, Christie D (eds) Handbook of the birds of the world. Special volume: new species and global index. Lynx Ediciones, Barcelona, pp 253–257

Almeida-Filho R, Miranda FP (2007) Mega capture of the Rio Negro and formation of the Anavilhanas Archipelago, Central Amazônia, Brazil: evidences in an SRTM digital elevation model. Remote Sens Environ 110(3):387–392. https://doi.org/10.1016/j.rse.2007.03.005

Antonelli AM, Ariza M, Albert J, Andermann T, Azevedo J, Bacon C, Faurby S, Guedes T, Hoorn C, Lohmann LG, Matos-Maraví P, Ritter CD, Sanmartín I, Silvestro D, Tejedor M, ter

Steege H, Tuomisto H, Werneck FP, Zizka A, Edwards SV (2018a) Conceptual and empirical advances in neotropical biodiversity research. PeerJ 6:e5644. https://doi.org/10.7717/peerj.5644
Antonelli A, Zizka A, Carvalho FA, Scharn R, Bacon CD, Silvestro D, Condamine FL (2018b) Amazonia is the primary source of neotropical biodiversity. PNAS 115(23):6034–6039. https://doi.org/10.1073/pnas.1713819115
Arruda DM, Schaefer CEGR, Fonseca RS, Solar RRC, Fernandes-Filho EI (2018) Vegetation cover of Brazil in the last 21 ka: new insights into the Amazonian refugia and Pleistocenic arc hypotheses. Glob Ecol Biogeogr 27:47–56. https://doi.org/10.1111/geb.12646
Baker PA, Fritz SC (2015) Nature and causes of Quaternary climate variation of tropical South America. Quat Sci Rev 124:31–47. https://doi.org/10.1016/j.quascirev.2015.06.011
Baker PA, Fritz SC, Dick CW, Eckert AJ, Horton BK, Manzoni S, Ribas CC, Garzione CN, Battisti DS (2014) The emerging field of geogenomics: constraining geological problems with genetic data. Earth Sci Rev 135:38–47. https://doi.org/10.1016/j.earscirev.2014.04.001
Bates JM, Hackett SJ, Cracraft J (1998) Area-relationship in the Neotropical lowlands: an hypothesis based on raw distributions of Passerine birds. J Biogeogr 25(4):783–793. https://doi.org/10.1046/j.1365-2699.1998.2540783.x
Bonaccorso E, Koch I, Peterson AT (2006) Pleistocene fragmetation of Amazon species' ranges. Divers Distrib 12:157–164. https://doi.org/10.1111/j.1366-9516.2005.00212.x
Bortolus A (2008) Error cascades in the biological sciences: the unwanted consequences of using bad taxonomy in ecology. J Hum Environ 37(2):114–118. https://doi.org/10.1579/0044-7447(2008)37[114:ECITBS]2.0.CO;2
Boubli JP, Rylands AB, Farias IP, Alfaro ME, Alfaro JWL (2012) *Cebus* phylogenetic relationships: a preliminary reassessment of the diversity of the untufted capuchin monkeys. Am J Primatol 74:381–393. https://doi.org/10.1002/ajp.21998
Boubli JP, Ribas C, Lynch Alfaro JW, Alfaro ME, da Silva MN, Pinho GM, Farias IP (2015) Spatial and temporal patterns of diversification on the Amazon: a test of the riverine hypothesis for all diurnal primates of Rio Negro and Rio Branco in Brazil. Mol Phylogenet Evol 82 (Pt B):400–412. https://doi.org/10.1016/j.ympev.2014.09.005
Brown KS (1982) Paleoecology and regional patterns of evolution in Neotropical forest butterflies. In: Prance GT (ed) Biological diversification in the tropics. Columbia University Press, Columbia, pp 225–308
Buckner JC, Lynch Alfaro JW, Rylands AB, Alfaro ME (2015) Biogeography of the marmosets and tamarins (Callitrichidae). Mol Phylogenet Evol 82(Pt B):413–425. https://doi.org/10.1016/j.ympev.2014.04.031
Burnham RJ, Johnson KR (2004) South American palaeobotany and the origins of Neotropical rainforest. Philos Trans R Soc Lond Ser B Biol Sci 359(1450):1595–1610. https://doi.org/10.1098/rstb.2004.1531
Bush MB (2017) Climate science: the resilience of Amazonian forests. Nature 541(7636):167–168. https://doi.org/10.1038/541167a
Bush M, Flenley J, Gosling WD (2011) Tropical rainforest responses to climatic change, 2nd edn. Springer, Berlin. https://doi.org/10.1007/978-3-642-05383-2
Byrne H, Alfaro JWL, Sampaio I, Farias IP, Scheneider H, Hrbek T, Boubli JP (2018) Titi monkeys biogeography: parallel Pleistocene spread by *Plecturocebus* and *Cheracebus* into a post-Pebas Western Amazon. Zool Scr 47(5):499–517. https://doi.org/10.1111/zsc.12300
Campbell KE, Frailey CD, Romero-Pittman L (2006) The Pan-Amazonian Ucayali Peneplain, late Neogene sedimentation in Amazonia, and the birth of the modern Amazon River system. Palaeogeogr Palaeoclimatol Palaeoecol 239(1–2):166–219. https://doi.org/10.1016/j.palaeo.2006.01.020
Capurucho JMG, Cornelius C, Borges SH, Cohn-Haft M, Aleixo A, Metzger JP, Ribas CC (2013) Combining phylogeography and landscape genetics of *Xenopipo atronitens* (Aves: Pipridae), a white sand *campina* specialist, to understand Pleistocene landscape evolution in Amazonia. Biol J Linn Soc 110(1):60–76. https://doi.org/10.1111/bij.12102

Chaves BR, Chaves AV, Nascimento AC, Chevitarese J, Vasconcelos MF, Santos FR (2015) Barcoding Neotropical birds: assessing the impact of nonmonophyly in a highly diverse group. Mol Ecol Resour 15(4):921–931. https://doi.org/10.1111/1755-0998.12344

Cheng H, Sinha A, Cruz FW, Wang X, Edwards RL, d'Horta FM, Ribas CC, Vuille M, Stott LD, Auler AS (2013) Climate change patterns in Amazonia and biodiversity. Nat Commun 4:1411. https://doi.org/10.1038/ncomms2415

Claramunt S, Cracraft J (2015) A new time tree reveals Earth history's imprint on the evolution of modern birds. Sci Adv 1(11):e1501005. https://doi.org/10.1126/sciadv.1501005

Cohen MLC, Rossetti DF, Pessenda LCR, Friaes YS, Oliveira PE (2014) Late Pleistocene glacial forest of Humaitá – Western Amazonia. Palaeogeogr Palaeoclimatol Palaeoecol 415:37–47. https://doi.org/10.1016/j.palaeo.2013.12.025

Colinvaux PA, Irion G, Räsänen M, Bush MB, Mello JASN (2001) A paradigm to be discarded: geological and paleoecological data falsify the Haffer & Prance refuge hypothesis of Amazonian speciation. Amazoniana 16(3):609–646

Costa GC, Hampe A, Ledru M-P, Martinez PA, Mazzochini GG, Shepard DB, Werneck FP, Moritz C, Carnaval AC (2018) Biome stability in South America over the last 30 kyr: inferences from long-term vegetation dynamics and habitat modelling. Glob Ecol Biogeogr 27:285–297. https://doi.org/10.1111/geb.12694

Cowling SA, Maslin MA, Sykes MT (2001) Paleovegetation simulations of lowland Amazonia and implications for Neotropical allopatry and speciation. Quat Res 55(2):140–149. https://doi.org/10.1006/qres.2000.2197

Coyne JA, Orr HA (2004) Speciation. Sinauer Associates, Sunderland

Cracraft J (1988) Deep-history biogeography: retrieving the historical pattern of evolving continental biotas. Syst Zool 37(3):221–236. https://doi.org/10.1093/sysbio/37.3.221

Crouch NMA, Capurucho JMG, Hackett SJ, Bates JM (2018) Evaluating the contribution of dispersal to community structure in Neotropical passerine birds. Ecography 42(2):390–399. https://doi.org/10.1111/ecog.03927

D'Apolito C, Absy ML, Latrubesse EM (2013) The Hill of Six Lakes revisited: new data and re-evaluation of a key Pleistocene Amazon site. Quat Sci Rev 76:140–155. https://doi.org/10.1016/j.quascirev.2013.07.013

D'Apolito C, Absy ML, Latrubesse EM (2017) The movement of pre-adapted cool taxa in north-central Amazonia during the last glacial. Quat Sci Rev 169:1–12

d'Horta FM, Cuervo AM, Ribas CC, Brumfield RT, Miyaki CY (2013) Phylogeny and comparative phylogeography of *Sclerurus* (Aves: Furnariidae) reveal constant and cryptic diversification in an old radiation of rain forest understorey specialists. J Biogeogr 40(1):37–49. https://doi.org/10.1111/j.1365-2699.2012.02760.x

Duellman WE (1979) The South American herpetofauna: its origin, evolution, and dispersal. The Museum of Natural History, The University of Kansas Monograph n° 7. University of Kansas Printing Service, Lawrence

Fernandes AM, Gonzalez J, Wink M, Aleixo A (2013) Multilocus phylogeography of the Wedge-billed Woodcreeper *Glyphorynchus spirurus* (Aves, Furnariidae) in lowland Amazonia: widespread cryptic diversity and paraphyly reveal a complex diversification pattern. Mol Phylogenet Evol 66(1):270–282. https://doi.org/10.1016/j.ympev.2012.09.033

Fernandes AM, Cohn-Haft M, Hrbek T, Farias IP (2014) Rivers acting as barriers for bird dispersal in the Amazon. Rev Bras Ornitol 22(4):363–373

Ferreira M, Aleixo A, Ribas CC, Santos MPD (2017) Biogeography of the Neotropical genus *Malacoptila* (Aves: Bucconidae): the influence of the Andean orogeny, Amazonian drainage evolution and palaeoclimate. J Biogeogr 44(4):748–759. https://doi.org/10.1111/jbi.12888

Fine PVA, Ree RH (2006) Evidence for a time-integrated species-area effect on the latitudinal gradient in tree diversity. Am Nat 168(6):796–804. https://doi.org/10.1086/508635

Flynn JJ, Wyss AR (1998) Recent advances in South American mammalian paleontology. Trends Ecol Evol 13(11):449–454. https://doi.org/10.1016/S0169-5347(98)01457-8

Garzón-Orduña IJ, Benetti-Longhini JE, Brower AVZ, Riddle B (2014) Timing the diversification of the Amazonian biota: butterfly divergences are consistent with Pleistocene refugia. J Biogeogr 41(9):1631–1638. https://doi.org/10.1111/jbi.12330

Govin A, Chiessi CM, Zabel M, Sawakuchi AO, Heslop D, Hörner T, Zhang Y, Mulitza S (2014) Terrigenous input off northern South America driven by changes in Amazonian climate and the North Brazil Current retroflection during the last 250 ka. Clim Past 10(2):843–862. https://doi.org/10.5194/cp-10-843-2014

Grohmann CH, Riccomini C, Chamani MAC (2011) Regional scale analysis of landform configuration with base-level (isobase) maps. Hydrol Earth Syst Sci 15:1493–1504. https://doi.org/10.5194/hess-15-1493-2011

Haffer J (1969) Speciation in Amazonian forest birds. Science 165(3889):131–137. https://doi.org/10.1126/science.165.3889.131

Haffer J (1974) Avian speciation in tropical South America. Nuttall Ornithological Club, Harvard University, Cambridge

Haffer J (1982) General aspects of the refuge theory. In: Prance GT (ed) Biological diversification in the tropics. Columbia University Press, New York, pp 6–24

Haffer J (1987) Biogeography of Neotropical birds. In: Whitemore TC, Prance GT (eds) Biogeography and quaternary history in tropical America. Oxford University Press, Oxford, pp 105–150

Haffer J (2008) Hypotheses to explain the origin of species in Amazonia. Braz J Biol 68 (4):917–947. https://doi.org/10.1590/S1519-69842008000500003

Häggi C, Chiessi CM, Merkel U, Mulitza S, Prange M, Schulz M, Schefuß E (2017) Response of the Amazon rainforest to late Pleistocene climate variability. Earth Planet Sci Lett 479:50–59. https://doi.org/10.1016/j.epsl.2017.09.013

Hall JPW, Harvey DJ (2002) The phylogeography of Amazonia revisited: new evidence from Riodnid butterflies. Evolution 56(7):1489–1497. https://doi.org/10.1111/j.0014-3820.2002.tb01460.x

Harvey MG, Brumfield RT (2015) Genomic variation in a widespread Neotropical bird (*Xenops minutus*) reveals divergence, population expansion, and gene flow. Mol Phylogenet Evol 83:305–316. https://doi.org/10.1016/j.ympev.2014.10.023

Hayakawa EH, Rossetti DF (2015) Late quaternary dynamics in the Madeira River basin, southern Amazonia (Brazil), as revealed by paleomorphological analysis. An Acad Bras Cienc 87 (1):29–49. https://doi.org/10.1590/0001-3765201520130506

Hoorn C, Wesselingh FP, ter Steege H, Bermudez MA, Mora A, Sevink J, Sanmartin I, Sanchez-Meseguer A, Anderson CL, Figueiredo JP, Jaramillo C, Riff D, Negri FR, Hooghiemstra H, Lundberg J, Stadler T, Sarkinen T, Antonelli A (2010) Amazonia through time: Andean uplift, climate change, landscape evolution, and biodiversity. Science 330(6006):927–931. https://doi.org/10.1126/science.1194585

Hoorn C, Bogotá-A GR, Romero-Baez M, Lammertsma EI, Flantua SGA, Dantas EL, Dino R, do Carmo DA, Chemale F Jr (2017) The Amazon at sea: onset and stages of the Amazon River from a marine record, with special reference to Neogene plant turnover in the drainage basin. Global Planet Change 153:51–65. https://doi.org/10.1016/j.gloplacha.2017.02.005

Hopkins MJG (2007) Modelling the known and unkown plant biodiversity of the Amazon Basin. J Biogeogr 34:1400–1411. https://doi.org/10.1111/j.1365-2699.2007.01737.x

Jacobs SC, Larson A, Cheverud JM (1995) Phylogenetic relationships and orthogenetic evolution of coat color among Tamarins (Genus *Saguinus*). Syst Biol 44(4):515–532. https://doi.org/10.1093/sysbio/44.4.515

Jahren AH (2007) The Artic forest of the middle Eocene. Annu Rev Earth Planet Sci 35:509–540. https://doi.org/10.1146/annurev.earth.35.031306.140125

Jansa SA, Barker FK, Voss RS (2014) The early diversification history of didelphid marsupials: a window into South America's "Splendid Isolation". Evolution 68(3):684–695. https://doi.org/10.1111/evo.12290

Latrubesse EM (2002) Evidence of quaternary palaeohydrological changes in middle Amazonia: the Aripuanã-Roosevelt and Jiparaná "fans". Z Geomorph N F 129(Suppl):61–72

Latrubesse EM, Ramonell CG (1994) A climatic model for southwestern Amazonia in last glacial times. Quat Int 21:163–169

Latrubesse EM, Cozzuol M, Silva-Caminha SAF, Rigsby CA, Absy ML, Jaramillo C (2010) The Late Miocene paleogeography of the Amazon Basin and the evolution of the Amazon River system. Earth Sci Rev 99(3–4):99–124. https://doi.org/10.1016/j.earscirev.2010.02.005

Lima MGM, Buckner JC, Silva Junior JS, Aleixo A, Martins AB, Boubli JP, Link A, Farias IP, Silva MN, Röhe F, QUeiroz H, Chiou KL, Di Fiore A, Alfaro ME, Alfaro JWL (2017) Capuchin monkey biogeography: understanding *Sapajus* Pleistocene range expansion and the current sympatry between *Cebus* and *Sapajus*. J Biogeogr 44(4):810–820. https://doi.org/10.1111/jbi.12945

Lutz HL, Weckstein JD, Patane JS, Bates JM, Aleixo A (2013) Biogeography and spatio-temporal diversification of *Selenidera* and *Andigena* Toucans (Aves: Ramphastidae). Mol Phylogenet Evol 69(3):873–883. https://doi.org/10.1016/j.ympev.2013.06.017

Lynch-Alfaro JW, Boubli JP, Paim FP, Ribas CC, Silva MNF, Messias MR, Röhe F, Mercês MP, Júnior JSS, Silva CR, Pinho GM, Koshkarian G, Nguyen MTT, Harada ML, Rabelo RM, QUeiroz HL, Alfaro ME, Farias IP (2015) Biogeography of squirrel monkeys (genus *Saimiri*): south central Amazon origin and rapid pan-Amazonian diversification of a lowland primate. Mol Phylogenet Evol 82(B):436–454. https://doi.org/10.1016/j.ympev.2014.09.004

Maldonado-Coelho M, Blake JG, Silveira LF, Batalha-Filho H, Ricklefs RE (2013) Rivers, refuges and population divergence of fire-eye antbirds (*Pyriglena*) in the Amazon Basin. J Evol Biol 26 (5):1090–1107. https://doi.org/10.1111/jeb.12123

Matos MV, Borges SH, d'Horta FM, Cornelius C, Latrubesse E, Cohn-Haft M, Ribas CC (2016) Comparative Phylogeography of two bird species, *Tachyphonus phoenicius* (Thraupidae) and *Polytmus theresiae* (Trochilidae), specialized in Amazonian White-sand vegetation. Biotropica 48(1):110–120. https://doi.org/10.1111/btp.12292

Mittelbach GG, Schemske DW, Cornell HV, Allen AP, Brown JM, Bush MB, Harrison SP, Hurlbert AH, Knowlton N, Lessios HA, McCain CM, McCune AR, McDade LA, McPeek MA, Near TJ, Price TD, Ricklefs RE, Roy K, Sax DF, Schluter D, Sobel JM, Turelli M (2007) Evolution and latitudinal diversity gradient: speciation, extinction and biogeography. Ecol Lett 10(4):315–331. https://doi.org/10.1111/j.1461-0248.2007.01020.x

Musher LJ, Ferreira M, Auerbach A, McKay J, Cracraft J (2019) Why is Amazonia a 'source' of biodiversity? Climate-mediated dispersal and synchronous speciation across the Andes in an avian group (Tityrinae). Proc R Soc Lond 286:20182343

Naka LN, Brumfield RT (2018) The dual role of Amazonian rivers in the generation and maintenance of avian diversity. Sci Adv 4(8):eaar8575. https://doi.org/10.1126/sciadv.aar8575

Nogueira ACR, Silveira R, Guimarães JTF (2013) Neogene–Quaternary sedimentary and paleovegetation history of the eastern Solimões Basin, central Amazon region. J S Am Earth Sci 46:89–99. https://doi.org/10.1016/j.jsames.2013.05.004

Noonan BP, Chippindale PT (2006) Dispersal and vicariance: the complex evolutionary history of boid snakes. Mol Phylogenet Evol 40(2):347–358. https://doi.org/10.1016/j.ympev.2006.03.010

Papadopoulou A, Knowles LL (2016) Toward a paradigm shift in comparative phylogeography driven by trait-based hypotheses. PNAS 113(29):8018–8024. https://doi.org/10.1073/pnas.1601069113

Patton JL, da Silva MNF (1998) Rivers, refuges, and ridges: the geography of speciation of Amazonian mammals. In: Howard DJ, Berlocher SH (eds) Endless forms: species and speciation. Oxford University Press, New York, pp 202–213

Peterson AT, Nyári AS (2008) Ecological niche conservatism and Pleistocene refugia in the Thrush-like Mourner, *Schiffornis* sp., in the Neotropics. Evolution 62(1):173–184. https://doi.org/10.1111/j.1558-5646.2007.00258.x

Pianka ER (1966) Latitudinal gradients in species diversity: a review of concepts. Am Nat 100 (910):33–46

Portes CEB, Aleixo A, Zimmer KJ, Whittaker A, Weckstein JD, Gonzaga LP, Ribas C, Bates JM, Lees AC (2013) A new species of *Campylorhamphus* (Aves: Dedrocolaptidae) from the Tapajós-Xingu interfluve in Amazonian Brazil. In: del Hoyo J, Elliot A, Christie D (eds) Handbook of the birds of the world. Special volume, Lynx Ediciones, Barcelona, pp 258–262

Prance GT (1982) Forest refuges: evidence from woody angiosperms. In: Prance GT (ed) Biological diversification in the tropics. Columbia University Press, New York, pp 137–157

Prates I, Xue AT, Brown JL, Alvarado-Serrano DF, Rodrigues MT, Hickerson MJ, Carnaval AC (2016) Inferring responses to climate dynamics from historical demography in neotropical forest lizards. PNAS 113(29):7978–7985. https://doi.org/10.1073/pnas.1601063113

Ribas CC, Joseph L, Miyaki CY (2006) Molecular systematics and patterns of diversification of the Pyrrhura parakeets (Aves: Psittacidae) with special reference to the P. picta/P. leucotis complex. Auk 123:660–680

Ribas CC, Aleixo A, Nogueira AC, Miyaki CY, Cracraft J (2012) A palaeobiogeographic model for biotic diversification within Amazonia over the past three million years. Proc Biol Sci 279 (1729):681–689. https://doi.org/10.1098/rspb.2011.1120

Ribas CC, Aleixo A, Gubili C, d'Horta FM, Brumfield RT, Cracraft J (2018) Biogeography and diversification of *Rhegmatorhina* (Aves: Thamnophilidae): implications for the evolution of Amazonian landscapes during the Quaternary. J Biogeogr 45(4):917–928. https://doi.org/10.1111/jbi.13169

Ricklefs RE (2004) A comprehensive framework for global patterns in biodiversity. Ecol Lett 7 (1):1–15. https://doi.org/10.1046/j.1461-0248.2003.00554.x

Rossetti DF, Toledo PM, Goes AM (2005) New geological framework for Western Amazonia (Brazil) and implications for biogeography and evolution. Quat Res 63:78–89

Rossetti DF, Valeriano MM (2007) Evolution of the lowest amazon basin modeled from the integration of geological and SRTM topographic data. Catena 70(2):253–265. https://doi.org/10.1016/j.catena.2006.08.009

Rossetti DF, Cohen MCL, Tatumi SH, Sawakuchi AO, Cremon ÉH, Mittani JCR, Bertani TC, Munita CJAS, Tudela DRG, Yee M, Moya G (2015) Mid-Late Pleistocene OSL chronology in western Amazonia and implications for the transcontinental Amazon pathway. Sediment Geol 330:1–15. https://doi.org/10.1016/j.sedgeo.2015.10.001

Rull V (2008) Speciation timing and neotropical biodiversity: the Tertiary-Quaternary debate in the light of molecular phylogenetic evidence. Mol Ecol 17(11):2722–2729. https://doi.org/10.1111/j.1365-294X.2008.03789.x

Rull V (2011) Neotropical biodiversity: timing and potential drivers. Trends Ecol Evol 26 (10):508–513. https://doi.org/10.1016/j.tree.2011.05.011

Rull V (2014) Pleistocene speciation is not refuge speciation. J Biogeogr 42(3):602–604. https://doi.org/10.1111/jbi.12440

Schultz ED, Burney CW, Brumfield RT, Polo EM, Cracraft J, Ribas CC (2017) Systematics and biogeography of the *Automolus infuscatus* complex (Aves; Furnariidae): cryptic diversity reveals western Amazonia as the origin of a transcontinental radiation. Mol Phylogenet Evol 107:503–515. https://doi.org/10.1016/j.ympev.2016.12.023

Schultz ED, Pérez-Emán J, Aleixo A, Miyaki CY, Brumfield RT, Cracraft J, Ribas CC (2019) Diversification history in the Dendrocincla fuliginosa complex (Aves: Dendrocolaptidae): insights from broad geographic sampling. Mol Phylogenet Evol 140:106581

Silva SM, Peterson AT, Carneiro L, Burlamaqui TCT, Ribas CC, Sousa-Neves T, Miranda LS, Fernandes AM, d'Horta FM, Araújo-Silva LE, Batista R, Bandeira CHMM, Dantas SM, Ferreira M, Martins DM, Oliveira J, Rocha TC, Sardelli CH, Thom G, Rêgo PS, Santos MP, Sequeira F, Vallinoto M, Aleixo A (2019) A dynamic continental moisture gradient drove Amazonian bird diversification. Sci Adv 5:eaat5752

Smith BT, Klicka J (2010) The profound influence of the Late Pliocene Panamanian uplift on the exchange, diversification, and distribution of new world birds. Ecography 33(2):333–342. https://doi.org/10.1111/j.1600-0587.2009.06335.x

Smith BT, McCormack JE, Cuervo AM, Hickerson MJ, Aleixo A, Cadena CD, Perez-Eman J, Burney CW, Xie X, Harvey MG, Faircloth BC, Glenn TC, Derryberry EP, Prejean J, Fields S, Brumfield RT (2014) The drivers of tropical speciation. Nature 515(7527):406–409. https://doi.org/10.1038/nature13687

Stanley SM (2007) An analysis of the history of marine animal diversity. Paleobiology 33:1–56. https://doi.org/10.1666/06020.1

Stríkis NM, Chiessi CM, Cruz FW, Vuille M, Cheng H, Barreto EA, Mollenhauer G, Kasten S, Karmann I, Edwards RL, Bernal JP, Sales HR (2015) Timing and structure of Mega-SACZ events during the Heinrich Stadial 1. Geophys Res Lett 42(133):5477–5484A. https://doi.org/10.1002/2015GL064048

Sukumaran J, Knowles LL (2017) Multispecies coalescent delimits structure, not species. PNAS 114(7):1607–1612. https://doi.org/10.1073/pnas.1607921114

Thom G, Aleixo A (2015) Cryptic speciation in the white-shouldered antshrike (*Thamnophilus aethiops*, Aves – Thamnophilidae): the tale of a transcontinental radiation across rivers in lowland Amazonia and the northeastern Atlantic Forest. Mol Phylogenet Evol 82 (Pt A):95–110. https://doi.org/10.1016/j.ympev.2014.09.023

Van der Hammen T, Absy ML (1994) Amazonia during the last glacial. Palaeogeogr Palaeoclimatol Palaeoecol 109:247–261

van der Hammen T, Hooghiemstra H (2000) Neogene and quaternary history of vegetation, climate, and plant diversity in Amazonia. Quat Sci Rev 19:725–742. https://doi.org/10.1016/S0277-3791(99)00024-4

Wallace AR (1852) On the monkeys of the Amazon. Alfred Russel Wallace Writings 3

Wang X, Edwards RL, Auler AS, Cheng H, Kong X, Wang Y, Cruz FW, Dorale JA, Chiang H-W (2017) Hydroclimate changes across the Amazon lowlands over the past 45,000 years. Nature 541(7636):204–207. https://doi.org/10.1038/nature20787

Weeks BC, Claramunt S, Cracraft J (2016) Integrating systematics and biogeography to disentangle the roles of history and ecology in biotic assembly. J Biogeogr 43(8):1546–1559. https://doi.org/10.1111/jbi.12747

Weir JT (2006) Divergent timing and patterns of species accumulation in lowland and highland Neotropical birds. Evolution 60(4):842–855

Westerman M, Krajewski C, Kear BP, Meehan L, Meredith RW, Emerling CA, Springer MS (2016) Phylogenetic relationships of dasyuromorphian marsupials revisited. Zool J Linnean Soc 176 (3):686–701. https://doi.org/10.1111/zoj.12323

Wiens JJ, Graham CH, Moen DS, Smith SA, Reeder TW (2006) Evolutionary and ecological causes of the latitudinal diversity gradient in hylid frogs: treefog trees unearth the roots of high tropical diversity. Am Nat 168(5):579–596. https://doi.org/10.1086/507882

Wilf P, Cúneo NR, Johnson KR, Hicks JF, Wing SL, Obradovich JD (2003) High plant diversity in Eocene South America: evidence from Patagonia. Science 300(5616):122–125. https://doi.org/10.1126/science.1080475

Willig MR, Kaufman DM, Stevens RD (2003) Latitudinal gradients of biodiversity: pattern, process, scale, and synthesis. Ann Rev Ecol Evol Syst 34:273–309. https://doi.org/10.1146/annurev.ecolsys.34.012103.144032

Wittmann H, Blanckenburg F, Maurice L, Guyot J-L, Filizola N, Kubik PW (2011) Sediment production and delivery in the Amazon River basin quantified by in situ-produced cosmogenic nuclides and recent river loads. Geol Soc Am Bull 123(5/6):934–950. https://doi.org/10.1130/B30317.1

Woodburne MO, Goin FJ, Bond M, Carlini AA, Gelfo JN, López GM, Iglesias A, Zimicz AN (2014) Paleogene land mammal faunas of South America: a response to global climatic changes and indigenous floral diversity. J Mamm Evol 21(1):1–73. https://doi.org/10.1007/s10914-012-9222-1

Wüster W, Ferguson JE, Quijada-Mascareñas JA, Pook CE, Salomão MG, Thorpe RS (2005) Tracing an invasion: landbridges, refugia, and the phylogeography of the Neotropical

rattlesnake (Serpentes: Viperidae: *Crotalus durissus*). Mol Ecol 14(4):1095–1108. https://doi.org/10.1111/j.1365-294X.2005.02471.x
Zachos J, Pagani M, Sloan L, Thomas E, Billups K (2001) Trens, rhytms, and aberrations in global climate 65 Ma to present. Science 27:686–693. https://doi.org/10.1126/science.1059412
Zink RM, Blackwell RC, Ronquist F (2000) The shifting roles of dispesal and vicariance in biogeography. Proc R Soc Lond 267(1442):497–503. https://doi.org/10.1098/rspb.2000.1028
Zular A, Sawakuchi AO, Chiessi CM, d'Horta FM, Cruz FW, Demattê AM, Ribas CC, Hartmann GA, Giannini PCF, Soares EAA (2018) The role of abrupt climate change in the formation of an open vegetation enclave in northern Amazonia during the late Quaternary. Glob Planet Change 172:140–149. https://doi.org/10.1016/j.gloplacha.2018.09.006

Chapter 11
Patterns and Processes of Diversification in Amazonian White Sand Ecosystems: Insights from Birds and Plants

J. M. G. Capurucho, S. H. Borges, C. Cornelius, A. Vicentini, E. M. B. Prata, F. M. Costa, P. Campos, A. O. Sawakuchi, F. Rodrigues, A. Zular, A. Aleixo, J. M. Bates, and Camila C. Ribas

Abstract White sand ecosystems (WSE) occur in nutrient-poor sandy soils patchily distributed throughout the lowlands of Amazonia. The diversification and current patterns of diversity of birds and plants specialized to the WSE were likely affected by abiotic conditions in very different ways relative to those in the upland Terra Firme forest, which has been more frequently studied over the years. Here, we review information on the geological origin of WSE substrates in Amazonia, and analyze current patterns of distribution of species and genetic diversity of plants and birds associated with these environments. Specifically, we analyze data on community composition to assess patterns of species diversity and turnover, and review and summarize published genetic data to unravel phylogenetic origins and

The original version of this chapter was revised. A correction to this chapter can be found at https://doi.org/10.1007/978-3-030-31167-4_31.

Electronic supplementary material The online version of this chapter (https://doi.org/10.1007/978-3-030-31167-4_11) contains supplementary material, which is available to authorized users.

J. M. G. Capurucho (✉)
Department of Biological Sciences, University of Illinois at Chicago, Chicago, IL, USA

Life Sciences Section, Negaunee Integrative Research Center, The Field Museum of Natural History, Chicago, IL, USA

S. H. Borges · C. Cornelius
Universidade Federal do Amazonas (UFAM), Manaus, AM, Brazil

A. Vicentini · E. M. B. Prata · F. M. Costa · P. Campos · C. C. Ribas
Instituto Nacional de Pesquisas da Amazônia, Manaus, AM, Brazil

A. O. Sawakuchi · F. Rodrigues · A. Zular
Instituto de Geociências, Universidade de São Paulo, São Paulo, Brazil

A. Aleixo
Museu Paraense Emílio Goeldi, Coordenação de Zoologia, Belém, PA, Brazil

J. M. Bates
Life Sciences Section, Negaunee Integrative Research Center, The Field Museum of Natural History, Chicago, IL, USA

V. Rull, A. C. Carnaval (eds.), *Neotropical Diversification: Patterns and Processes*, Fascinating Life Sciences, https://doi.org/10.1007/978-3-030-31167-4_11

phylogeographic patterns. Despite having lower species richness relative to other Amazonian ecosystems, the WSE present unique assemblages of species. The WSE are older than the Quaternary, yet their Pleistocene history has been dynamic, and the phylogenetic and population genetic patterns of the WSE taxa are consistent with this dynamism. The WSE were sensitive to Pleistocene climatic changes, likely undergoing lineage diversification and local extinctions. With increasing threats to Amazonian ecosystems, the low resilience of the WSE to disturbance needs to be factored into conservation management—particularly in the face of future climate change.

Keywords Campinas · Campinaranas · Endemism · Geological history · Glacial cycles

1 Introduction

The environmental heterogeneity that exists in the Amazonian region influences patterns of species diversity, and the history of habitat-specialist taxa is linked to that of the habitats themselves (e.g. Ribas et al. 2012; Capurucho et al. 2013; Thom et al. 2018). This appears to be the case of a unique biota associated with oligotrophic soils, which evolved differently from the better studied Terra Firme forest species (TFF species; Table 11.1) in response to habitat and historical changes. These Amazonian White Sand Ecosystems (WSE; Table 11.1) occur in nutrient-poor sandy soils (Podzols or Spodosols) and are patchily distributed in the lowlands (<200 m; Adeney et al. 2016). This type of sandy substrate creates stressful conditions for vegetation growth due to oligotrophy, low water retention capacity, and flooding (Damasco et al. 2013). As a consequence, the WSE have distinct floristic composition and structure relative to other Amazonian ecosystems (Anderson 1981; Kubitzki 1990; Prance 1996; Fine et al. 2010; Stropp et al. 2011). Compared to the TFF, the WSE are characterized by lower plant species diversity and distinct vegetation structure, which ranges from open vegetation to forests of low stature.

Several lineages of birds and plants are specialized to the WSE and their diversification and patterns of diversity were likely affected by abiotic conditions in very different ways relative to the more commonly studied TFF taxa. This difference is often overlooked due to the comparatively small number of specialized WSE species. However, understanding the relationships between these taxa and their habitats can provide valuable insight into Amazonian biogeography, informing how different ecosystems respond to current and past climate and physical landscape heterogeneity (relief and substrate type) across the basin. Particularly relevant is to understand how this specialized biota is (and has been) connected throughout the

Table 11.1 Description of ecosystems discussed in this chapter (based on Anderson 1981; Junk et al. 2011; Werneck 2011; Adeney et al. 2016; Pennington et al. 2018)

Ecosystem	Description
White sand ecosystems (WSE)	All types of ecosystems associated with Amazonian white sand soils (Spodosol), occurring mainly in lowland areas. They have received several different names across the Amazon region such as Amazon caatinga, campinas, campinaranas, varillal, chamizal, muri bush, and wallaba forest
White sand shrublands and grasslands (WSC)	Vegetation physiognomies of the WSE in a gradient from closed and tall shrublands (up to around 3 m) to more open grasslands with scattered trees and palms
White sand forests (WSF)	Forested vegetation physiognomies of the WSE with trees up to 20 m, with lower biomass, relatively open canopy, and lower diversity compared to TFF
Terra Firme forest (TFF)	Lowland non-flooded Amazonian humid forest on clay soils
Igapó	Vegetation types growing on the floodplains of Amazonian black water and clear water rivers
Várzea	Vegetation types growing on the floodplains of Amazonian white water rivers
Savannas	South American savannas occur in areas of well-defined seasonality experiencing a long dry season with low rainfall. The soils are commonly Latosols with high acidity and low fertility. Fire is also an important feature. Enclaves of savanna occur in Amazonia and are referred to as *Amazonian savannas* herein
Restinga	South American vegetation types growing in the Atlantic coast sandy substrates. In this chapter we make a distinction between the *restingas* that occur within Amazonia's coastal region (Amazonian *restingas*) and those occurring in the Atlantic Forest region (Atlantic Forest *restingas*)

basin. Its distribution is currently fragmented, with vast expanses of tall, closed canopy, upland forest, as well as major rivers, isolating WSE patches.

WSE are mostly concentrated in northwestern and northern Amazonia, in the Branco and Negro River basins, but they also occur in other regions such as the Madeira River (Fig. 11.1). Amazonian WSE occur in areas of high precipitation, meaning that these habitats are not associated with modern dry conditions. However, soils of WSE have low water retention capacity and plants may suffer from water deficit in the dry season or experience flooding during the rainy season—both are important factors in determining vegetation structure and composition (Damasco et al. 2013; Mendonça et al. 2014). As such, the biota specialized in WSE may have been particularly affected by climatic changes. In a lower precipitation scenario, for instance, the WSE probably became hostile even to specialized plant lineages, given the low water retention capacity of its sandy substrates (Damasco et al. 2013). On the other hand, assuming that the TFF canopy cover became less dense during past drier phases (Cowling et al. 2001), as hypothesized for northern Amazonia during the Last Glacial Maximum (LGM) (Häggi et al. 2017) and the Heinrich Stadial 1 (HS1; 18.1–14.7 ka) (Zular et al. 2019), it is likely that those habitats became more

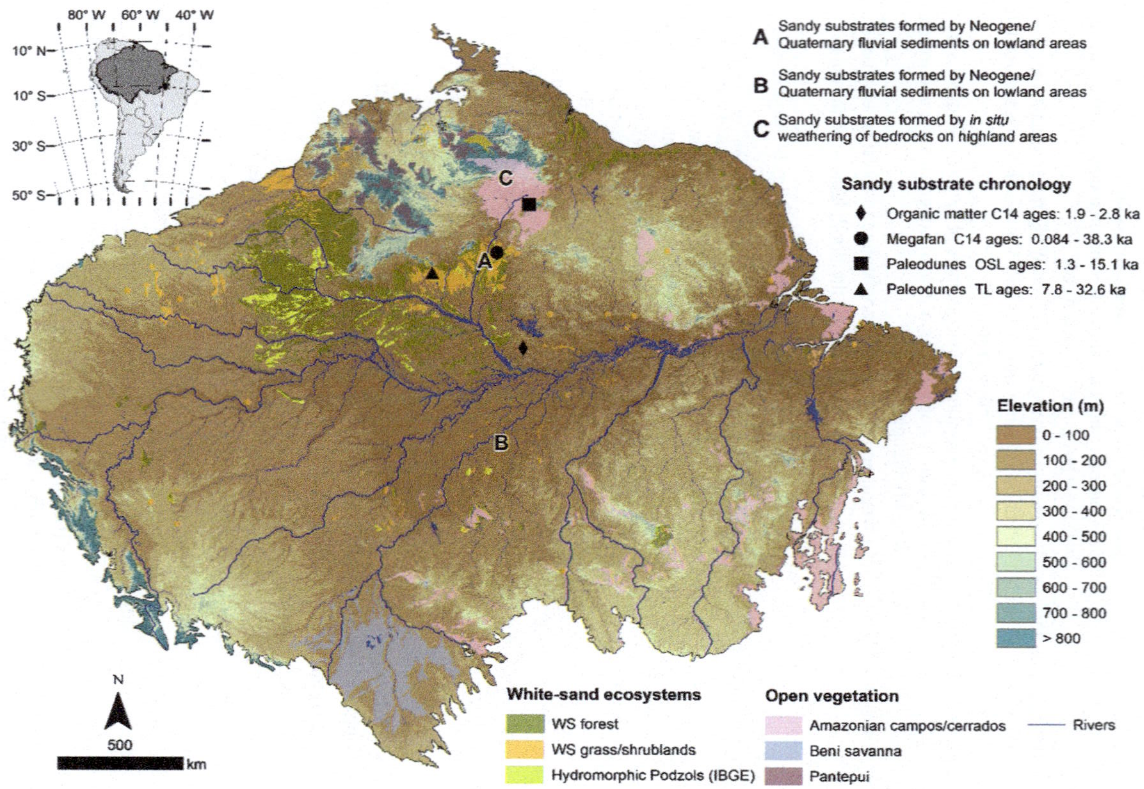

Fig. 11.1 Distribution of open vegetation and white sand ecosystems over Amazonia, in relation to elevation. A, B and C denote areas of white sand ecosystem or open vegetation developed over different geological substrates. Vegetation types are based on Adeney et al. (2016), and elevation follows Jarvis et al. (2008). Substrate ages were derived from Carneiro-Filho et al. (2002) (Paleodunes TL ages), Horbe et al. (2004) (Soil organic matter ^{14}C ages), Teeuw and Rhodes (2004) (Paleodunes OSL ages) and Rossetti et al. (2012a, b) (Megafan ^{14}C ages).

permeable to WSE organisms with ample dispersal capabilities (e.g. birds and animal-dispersed plants). Studying the current diversity and the evolution of WSE associated taxa may fill a significant knowledge gap in Amazonian biogeography, uncovering the magnitude and impact of past environmental changes.

We employ a multidisciplinary approach and review the evolution of the WSE, summarizing what is known about this ecosystem and incorporating new analyses of community composition and phylogenetic diversity. We also discuss questions that remain open about this distinctive environment and its biota. WSE have often been considered a "poor" environment, and as such have received little attention from a conservation perspective. However, these ecosystems have been strongly modified because of human activities, especially near urban areas. They are both burned or cleared for sand extraction, which is highly detrimental for their endemic biota. Characterizing the uniqueness of this fragile ecosystem is essential to direct needed conservation measures for WSE across Amazonia.

2 Origin and Evolution of White Sand Soils in Amazonia

The hypotheses about the origin of Amazonian white sand terrains are diverse and vary regionally. Few studies present absolute ages to constrain the timing of their development. They may have resulted from the formation of sandy plains related to Late Neogene (Miocene-Pliocene) and Quaternary fluvial and eolian depositional systems, as well as the weathering of ancient sandstones (pre-Cenozoic). The white sand areas near Iquitos, in western Amazonia (Peru), for example, are interpreted as sediments deposited by meandering rivers active between 8 and 6 Ma (Räsänen et al. 1998; Roddaz et al. 2005a, b, 2006). The white sand areas around Manaus, in central Amazonia, are considered to be the product of leaching of Quaternary fluvial sediments deposited on the edge of incised fluvial valleys (Klinge 1965). Likewise, the white sand patches along the Madeira River (Brazil) may have developed over Quaternary fluvial deposits, given their 'riverine shape' (Rossetti et al. 2012a, b). Similarly, large white sand areas in the Negro River basin may result from leaching of alluvial fan deposits. It has been suggested that these sediments were supplied by the weathering and erosion of Precambrian sandstones (Roraima Supergroup) of the Guiana Shield (Sombroek 1966; Janzen 1974). White sand patches distributed within the Branco River and the Negro River basins in northern Brazilian Amazonia are described as sediments deposited in large distributary fluvial systems (megafans) active during the Late Pleistocene and Holocene (Rossetti et al. 2012a, b, 2014; Zani and Rossetti 2012).

In addition to white sand patches over fluvial and alluvial sediments, some white sand areas in the middle Negro River basin display features characteristic of ancient eolian dune fields (Carneiro-Filho et al. 2002). The dune fields would have developed during drier phases of the Late Pleistocene and Holocene (32–7.8 ka). Iriondo and Latrubesse (1994) also report Quaternary eolian landforms associated with white sand areas around Parintins, Brazil, in central Amazonia.

Dubroeucq et al. (1991) and Mafra et al. (2002) propose that the white sand patches from the Negro River basin derive from in situ podzolization of Latosols, which may have occurred as early as 2.5 Ma ago (Bueno 2009). Richards (1941) and Bleackley and Khan (1963) suggest that the 'wallaba forest' and 'muri bush' WSE in Guyana may derive from the podzolization of sands and sandy clays of the Pleistocene Berbice Formation. Importantly, radiocarbon ages indicate the development of Spodosols over sandstones of the Cretaceous Alter do Chão Formation around Manaus in the last 3 ky (Horbe et al. 2004). Duivenvoorden and Lips (1995) also show the existence of Spodosols on fluvial terraces of the Caquetá River (Colombia), as well as over the Neogene sedimentary plain. These authors proposed the formation of white sands from in situ weathering of sandstone plateaus and from colluvial layers deposited on top and along the slopes of the plateaus. However, in situ weathering of bedrocks in Suriname (Heyligers 1963) and North of Manaus in Brazil (Horbe et al. 2004) also creates sandy soils without formation of Spodosols.

The above-mentioned studies suggest that Spodosols (white sands) can be formed at different times and over geological substrates of different ages. Additionally, it has been suggested that white sand substrates represent an ultimate phase of soil development under intense leaching conditions. This would argue for historical persistence of white sands in the landscape if they are protected from erosion or burial by younger sediments. WSE patches occur in areas covered by Neogene or Quaternary sandy sediments as well as in areas showing in situ weathering of presumed older bedrocks rich in quartz, like Cretaceous sandstones (Fig. 11.1). Thus, the formation of white sands and associated WSE are not exclusively related to specific geological units, and they may depend on specific conditions suitable for the development of soils with an upper heavily leached sand layer. These conditions include sandy substrates of diverse geological origin combined with low relief, favoring a shallow water table to promote leaching of the upper soil layer (Quesada et al. 2012). Neogene or Quaternary fluvial sediments are restricted to lowland areas (<200 m), as observed in the Viruá region in the Branco River basin (A in Fig. 11.1), or the white sand areas along the Madeira River (B in Fig. 11.1). Eolian sediments also prevail in lowland areas because they mainly represent eolian dunes formed by reworking of fluvial sands. White sands also can occur in highland areas (>200 m), where the development of open vegetation during drier periods can favor deflation (wind erosion) and the formation of eolian dunes. Sandy substrates formed by in situ weathering of bedrocks like sandstones are restricted to highland areas (>200 m, for example, in the Uiramutã area in Roraima, northern Amazonia; C in Fig. 11.1).

The effect of climate on the origin of sandy soils has been illustrated by the development of a sediment succession profile situated 2.5 km W of the Branco River, in Roraima, Brazil (Zular et al. 2019). At this site, the shift from fluvial (muddy) to eolian (sandy) sediment deposition occurred at the LGM (~23 ka), favoring high rates of sand accumulation that persisted until the HS1 (18–15 ka). There is a growing body of evidence from regional paleoclimate records suggesting that persistent and long-lasting southward shifts of the Intertropical Convergence Zone (ITCZ) during the LGM and HS1 promoted dry and windy conditions in Northern South America (e.g. Koutavas and Lynch-Stieglitz 2004; Baker et al.

2009; D'Apolito et al. 2017; Häggi et al. 2017). This may have made northern Amazonia more susceptible to the formation of sandy depositional systems like eolian dune fields and megafans. After HS1, climate became wetter in northern Amazonia, triggering the formation of Spodosols through weathering of the eolian sandy substrates. Accordingly, Zular et al. (2019) show that abrupt millennial climate events can lead to significant, long-lasting changes in substrates that may have favored the formation and expansion of WSE.

Initial geological and climate conditions required for the development of WSE likely precede the Cenozoic, assuming that uplifted sandstone units exposed to tropical weathering date to the late Cretaceous (>65 Ma) in the Amazon region. Significant landscape changes that may have favored an expansion of Amazonian WSE during the Late Cenozoic include, for example, the development of the transcontinental Amazon fluvial system variably dated as taking place during the Miocene (9.4–9.0 Ma, Hoorn et al. 2017), Miocene-Pliocene (6–4 Ma, Latrubesse et al. 2010) or Plio-Pleistocene (less than 3 Ma, Campbell et al. 2006). The emergence of the transcontinental Amazon River implies the formation of incised river valleys and the development of vast upland sedimentary terrains through the abandonment of fluvial floodplains. This landscape scenario has higher availability of sandy substrates for the formation of Spodosols in areas of adequate climate and relief conditions (e.g. shallow water table and low relief), favoring the expansion of vegetation adapted to stressing substrate environments.

3 Species Assemblages in Amazonian WSE: Low Richness But High Endemism

3.1 Plants

Amazonian WSE are species-poor compared to adjacent habitats on more fertile and less stressful edaphic conditions (e.g. Vicentini 2004). Lower plant species richness may be associated with multiple processes, such as the smaller area that WSE occupy in Amazonia, a strong filtering effect and selective pressure given by stressful abiotic conditions (Parolin and Wittmann 2010), and lower density dependence effects due to higher defense (concentration of secondary compounds) against herbivores (Janzen 1974; Fine et al. 2004). The oligotrophic conditions of WSE promote mutualistic associations between plant and fungi, and ectomycorrhizal associations are frequent on WSE in both flooded and non-flooded areas (Singer and Aguiar 1986; Roy et al. 2016). Furthermore, at any given site, micro-edaphic variation promotes high beta-diversity levels and structural variation from grasslands to 25 m tall forests. Edaphic effects on species distribution, vegetation structure, and species richness include flooding during the rainy season, drought during the dry season, and minor changes in soil fertility (Damasco et al. 2013; Demarchi et al.

2018). Fire- or drought-related mortality may also be important factors affecting vegetation structure and composition at local scales (Vicentini 2004).

Amazonian WSE also have a distinct floristic composition that may be characterized as composed of two levels of endemism. First, a *phylogenetic endemism*, given by the fact that adaptations to either flooding or drought are complex evolutionary transitions (Parolin and Wittmann 2010; Wittmann et al. 2013). Once a lineage adapts to the stressful abiotic conditions of the WSE, they have a greater chance to diversify along the ecological variation within WSE (e.g. *Pagamea*, Vicentini 2016). This phylogenetic endemism is supported by phylogenetic clustering and low phylogenetic dissimilarity among WSF plots in the Amazon (Guevara et al. 2016), and it is exemplified by many lineages that are endemic to oligotrophic systems on sandy soils in South America (Kubitzki 1990; Vicentini 2016). Second, although diversity is low and few lineages have adapted to WSE, a strong *spatial endemism* is observed at the species level (Costa et al. 2019; Guevara et al. 2016; Fine and Baraloto 2016). The fragmented distribution of WSE patches promotes such endemism by limiting species dispersal, and both area and connectivity among patches shape species occurrence at regional scales (Costa et al. 2019). Patchiness may also explain speciation in *Pagamea* (Vicentini et al. 2016). Hence, the WSE impose a strong filtering effect to the regional species pool, selecting the same lineages across Amazonia, while species distributions are strongly spatially structured.

Within the Amazon, the lowland WSE show greater floristic affinities with other oligotrophic systems at both low and high elevations than to the more fertile forests (Kubitzki 1990; Huber 1988; Steyermark 1986). The floristic relationship between WSE and forests flooded by blackwater river systems (*igapó*) is exemplified by the occurrence of plant lineages and morphologically similar species at adjacent habitats along the Negro River basin (e.g. in *Aldina*, *Pagamea*, *Ternstroemia*, *Aspidosperma*, *Macrolobium*, *Cybianthus*, *Endlicheria*, *Ouratea*, *Dimorphandra*) (Steyermark 1986; Kubitzki 1990; Givnish et al. 2000). These affinities may be related to evolutionary constraints, as flooding and drought, present in both these systems, trigger similar physiological responses (Wittmann et al. 2013). WSE also have many floristic links with high-altitude oligotrophic ecosystems, with specialized lineages being represented in both areas, particularly on the Guiana Shield (Steyermark 1986; Huber 1988; Kubitzki 1990; Vicentini 2016).

We explored the floristic affinity of the WSE with other oligotrophic systems by compiling a floristic dataset of 59 sampling sites characterized by oligotrophic (e.g. sand, rock, igapó) soils distributed over six main ecosystems (Fig. 11.2a): WSE, savanna/cerrado (Amazonian and extra-Amazonian), Amazonian *restinga*, Atlantic Forest restinga (AF-*restinga*), *igapó* forests and Tepui table-mountains of the Guiana Shield (detailed methods and the species list can be found online at Electronic Supplementary Material S1 and S2). We found 2060 taxa identified as species or subspecies belonging to 700 genera. The number of species reported per site varied from 12 (on a Tepui) to 278 species (WSE mosaic near Manaus). The total number of species and genera per ecosystem varied from 166 and 109 for the *igapó* to 925 and 344 for WSE, respectively. Between 57–84% of the species, and 17–42%

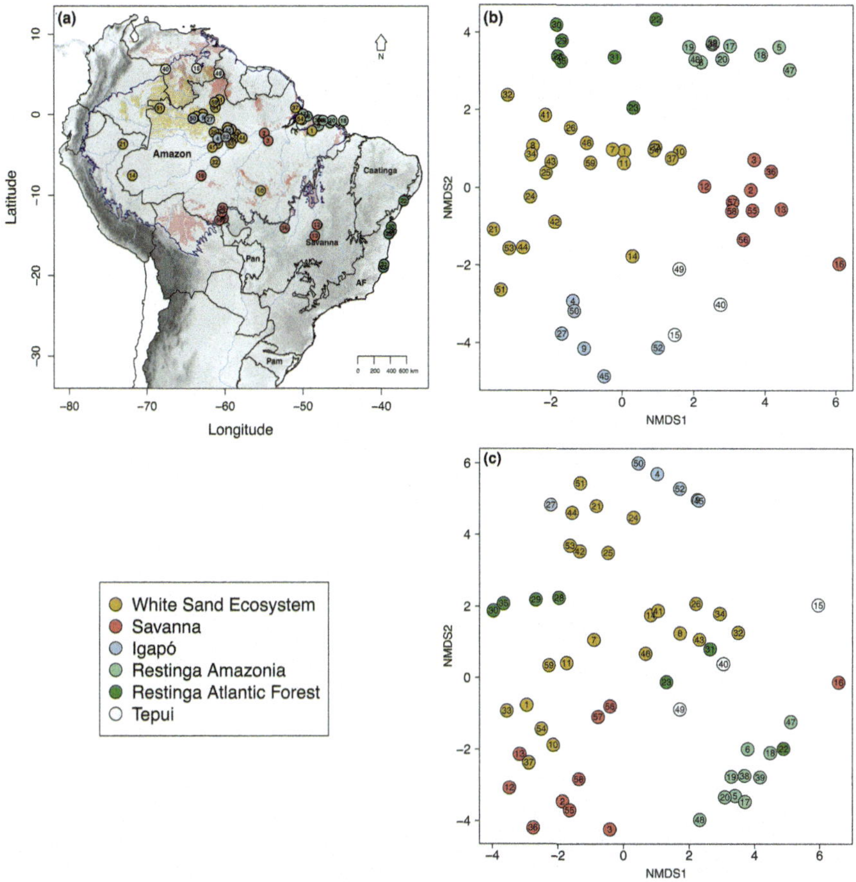

Fig. 11.2 (**a**) Map of the geographic distribution of the 59 sample sites (points are slightly displaced to avoid overlap). The Brazilian biomes Amazon, Savanna, Atlantic Forest (AF), Caatinga, Pantanal (Pan) and Pampa (Pam) are shown. The Amazon River basin is shown with a blue outline. The background includes a digital elevation model (SRTM) from low (white) to high (black) elevation; (**b**, **c**) Non-metric multidimensional ordination (NMDS) of the results of the Phylogenetic Community Dissimilarity (PCD) analysis. PCD is partitioned into a nonphylogenetic component (that reflects shared species between communities) and a phylogenetic component (that reflects the evolutionary relationships among nonshared species). The ordination of the nonphylogenetic component based on the distance matrix calculated (using Sørensen index) from the floristic dataset is shown in (**b**), and the ordination based in the distance matrix calculated as the product of the ecologic and the phylogenetic components is shown in (**c**)

of the genera, were restricted to a single ecosystem type, with the Tepui WSE having the highest endemism, and the *igapó* showing the lowest (Fig. 11.3). We found no species and only two genera shared among all ecosystems. These numbers increased to two species (*Hirtella racemosa* and *Tapirira guianensis*) and 10 genera shared among ecosystems when excluding the Tepuis. *Tapirira guianensis* is a widespread

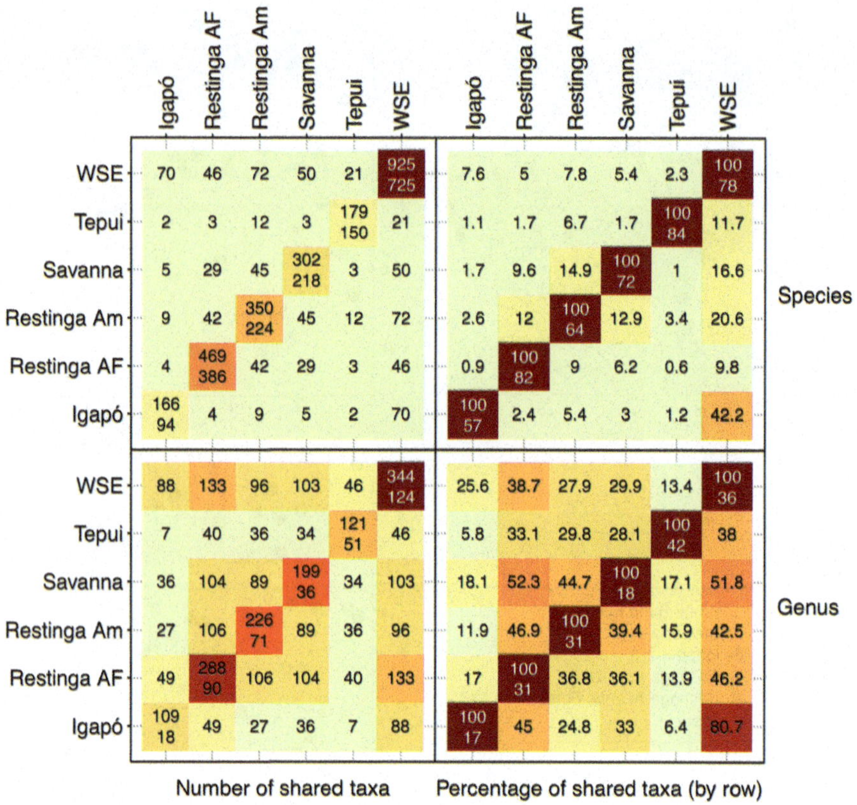

Fig. 11.3 Number of shared taxa among ecosystem categories based on our floristic dataset of 508 species and 59 sample sites. Left panels show the number of shared species (top) and genera (bottom). Diagonal show total number of taxa (above) and the number of endemics (below) in each category. Right panels show the proportions of the taxa in each category (rows) that are shared with other ecosystems. Colors are scaled to values

species in the Neotropics that is also very common in secondary vegetation on more fertile soils, while *Hirtella racemosa* is a WSE specialist.

Clustering the sites by the probability of non-identical species composition (using the Raup-Crick dissimilarity and a bootstrap procedure with 1000 re-sampling runs) allowed us to recognize groups that correspond to ecosystem categories, except for a few Tepui sites (Fig. 11.4). Tepui sites are the most distinct of all and share few species with the other categories, mostly with WSE (11.7%) and with Amazonian *restingas* (6.7%). AF-*restingas* also shares less than 10% of the species with any of the other categories. However, at the generic level, most AF-*restingas* are clustered within WSE, pointing to a strong phylogenetic link (Fig. 11.4). Savannas and Amazonian *restingas* form a group, and *igapó* and WSE form another. The WSE are further divided into three groups and a singleton (Cruzeiro do Sul, Acre), that correspond to different vegetation structures and geography: (1) WSF (see

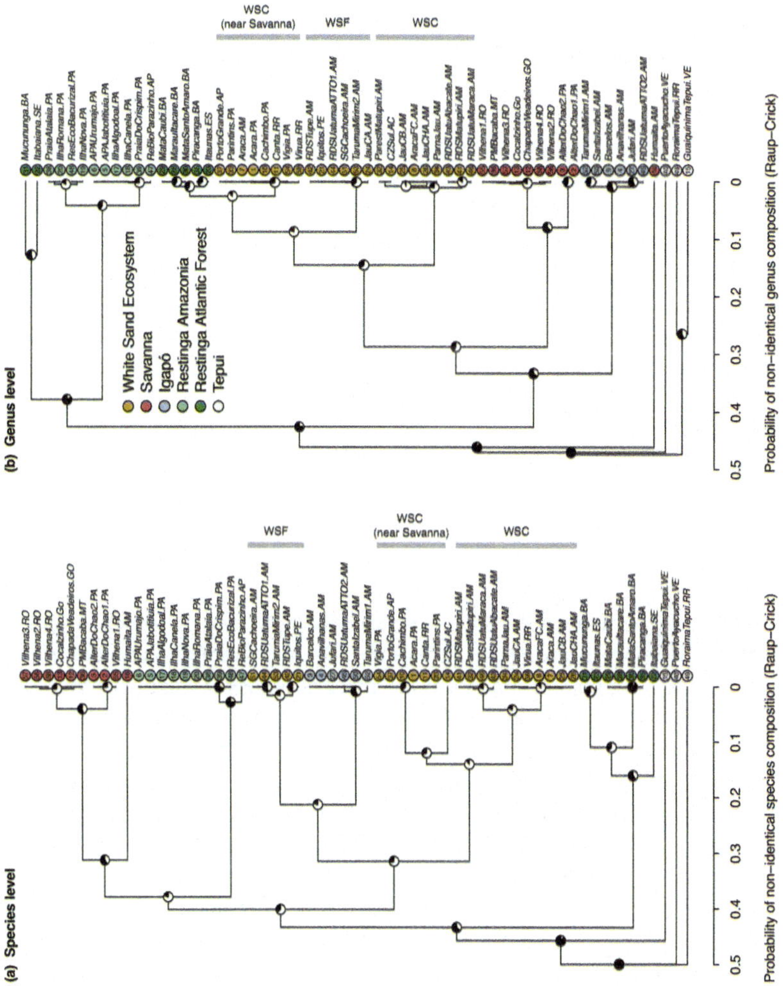

Fig. 11.4 Floristic affinities (based on our floristic dataset of 508 species and 59 sample sites) among sample sites from different oligotrophic ecosystems for both (**a**) plant species and (**b**) plant genera. Pies on nodes show the support for the floristic similarity among sites after bootstrapping the similarity matrix with the function *clusterboot* (Hennig 2018)

Table 11.1), (2) WSC; and, (3) WSE located nearby areas of Amazonian savannas (Figs. 11.2 and 11.4). By performing matrix swapping to calculate the probability of non-identical composition, the clustering of the *igapó* changed among the multiple runs of the analysis, as it was sometimes recovered within the WSE group (placed as "sister" to the WSF), or as "sister" to the WSE group. A large proportion of species (42%) and genera (80%) found in *igapó* are also found in WSE. Finally, the sandy savannas of eastern and southern Amazonia (sample sites 2, 3, 16, 55–58 in Fig. 11.2) are floristically more similar to the savannas of Central Brazil than to WSE, pointing to the climatic or soil effects that distinguish these classes.

The patterns described above show strong geographical and ecological structure of plant species composition. The high-level of endemism for all categories analyzed is particularly remarkable. Although physiognomically similar, the WSE flora was not strongly linked to the *restingas* and the sandy savannas, nor to the Tepui, as previously suggested (Huber 1988; Kubitzki 1990). Links are stronger with the other ecosystems, as many species are shared with the WSE. This is particularly strong for *igapó* at both specific and generic taxonomic levels, suggesting that the flora of the *igapós* is largely nested within and has a strong phylogenetic link with WSE (Kubitzki 1989, 1990). In addition to geography, this pattern may be in part explained by environmental filters selecting lineages adapted to the edaphic stressful conditions such as flooding and drought, present in both WSE and *igapós*.

3.2 Birds

The lower species richness of the WSE avifauna relative to other Amazonian ecosystems is striking. There also is substantial regional variation in species richness and community composition among WSE patches across Amazonia (Alonso and Whitney 2003; Alonso et al. 2013; Borges et al. 2016a). In the Jaú National Park (Central Amazon), WSF host around 150 bird species, while more than 250 were registered in adjacent TFF (Borges and Almeida 2011). This impoverishment is also noticed in comparisons between open vegetation types, such as between white sand grasslands and shrublands (WSC), and the Amazonian Savannas. Savanna relicts within Amazonia harbor 83 species on average, while 56 species occur in WSC. A sharper difference is obtained if only non-forest species (related to WSE or savanna habitats) are considered and the Amazonian Savannas host up to three times more species than WSE (Borges et al. 2016b).

The number of bird species found in WSE patches is influenced by the amount of habitat available in the landscape, but the regional context in which WSE patches are located is also important (Oren 1981; Borges et al. 2016a). Regions with larger and more connected WSE patches, like the Aracá and Viruá regions (Northern Brazilian Amazon), have from 70 to 100 species of non-forest birds. The Aracá and Viruá regions have a similar landscape structure, but the latter is notably more diverse, likely due to its proximity to other open habitat types such as the Northern

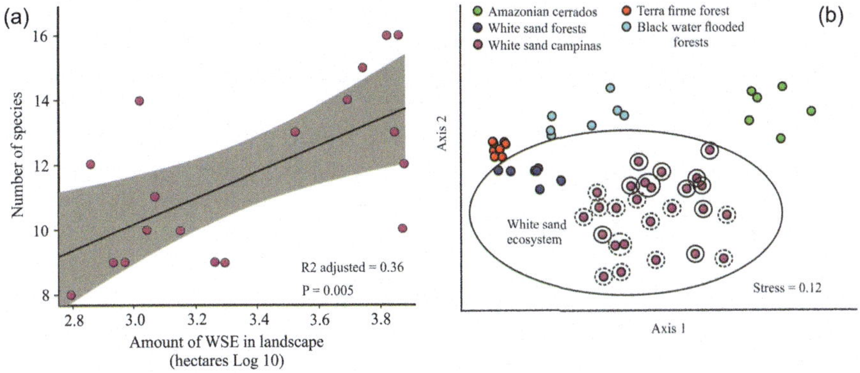

Fig. 11.5 Bird diversity patterns in WSE and Amazonian ecosystems. (**a**) The variation in the number of WSE specialist bird species is best explained by the amount of WSE habitats available in a 5 km circle around the sampling site (Borges et al. 2016a). (**b**) Bird species composition of WSE is clearly differentiated from other habitats in the Amazon. Note, however, that white sand forests tend to be more similar to Terra Firme Forest than white sand grasslands and shrublands (WSC; see text). Continuous circles over WSC sites represented data obtained in landscapes with large and connected patches, and dashed circles represent sites in landscapes with small and isolated patches. Data were analyzed through a non-metric multidimensional ordination (NMDS) of multiple sites in the Brazilian Amazon (data from Borges 2013; Borges et al. 2016a, b)

South America Savannas. However, smaller patches in highly fragmented landscapes of WSE in central Amazonia host, on average, less than 20 non-forest bird species. The number of WSE specialist birds (as defined in Borges et al. 2016b) is positively related to the proportion of WSE habitat available in the landscape, which explains 36% of the variation in species richness (Fig. 11.5a). Larger and more connected WSE patches sustain larger populations, likely reducing the probability of local extinctions. They also can present higher heterogeneity in vegetation structure providing more microhabitats and promoting higher local species richness.

Despite their lower species richness, WSE harbor unique avian assemblages within Amazonia with endemic species not shared with other Amazonian ecosystems (Fig. 11.5b). Even though common species from other Amazonian ecosystems consistently use WSE habitats (e.g. *Ceratopipra erythrocephalum*, *Willisornis poecilinotus*), the WSE bird community also has specialist species (e.g. *Elaenia ruficeps* and *Polioptila clementsi*) that are not known to breed outside of WSE. In general, WSF bird communities have a higher similarity to surrounding TFF than WSC do (Fig. 11.5b). This pattern is explained by the structural resemblance between these two types of forest and the sharper distinction between TFF and the open WSC. The importance of vegetation structure is further corroborated by the fact that WSC communities show higher similarity to *igapó* environments than to the WSF and TFF. Characteristic WSC species, such as *Aprositornis disjuncta*,

Xenopipo atronitens, and *Dolospingus fringilloides*, among others, often occur in low stature *igapó* forests (Borges et al. 2016b).

WSE bird communities show a disproportionate dominance of a few well-adapted bird species. The five most captured species with mist-nets in WSC and WSF represent 59 and 42% of all the captured individuals, respectively, while in TFF and *igapó* forests the top five species make up only 35% of the captures. In WSC bird communities, *Xenopipo atronitens*, *Elaenia ruficeps*, and *Tachyphonus phoenicius* are usually dominant species (Borges et al. 2016a). The lower diversity and higher dominance by a few species are probably related to the WSE oligotrophy, which imposes environmental limits to species adaptation, in addition to the comparatively reduced habitat complexity. This environmental filtering appears to lead to impoverished communities, dominated by few species highly adapted to the restrictive conditions of WSE.

The patchy distribution of WSE acts in combination with the degree of specialization and dispersal capabilities of WSE species to create a colonization/extinction dynamic driven by the size and isolation of habitat patches (Borges et al. 2016a). Regionally, the distribution of WSE specialist birds among patches is not random, as shown by a nested distribution pattern in the community (NODF = 50.73; $p < 0.05$; Fig. 11.6). Bird assemblages in sites with reduced amounts of WSE habitat cover are subsets of assemblages in sites with high amounts of WSE habitat cover. This pattern suggests a selective loss of WSE specialist birds, both locally and regionally, that is partially associated with a reduction of the amount of WSE habitat available in the landscape. However, the arrangement of sites along an isolation gradient (i.e. distance to the Aracá region as a potential species source) did not show significant nested structure (NODF = 42.6, $p > 0.05$). Nested patterns such as this one, that are better explained by area than by isolation, have been often interpreted as a distribution pattern that is extinction-driven rather than colonization-driven (Matthews 2015). This has important conservation implications for the ecologically restrictive WSE bird communities. Regions with a lower amount of WSE habitat encompass patches that are usually small and isolated, with reduced habitat heterogeneity, and supporting small population sizes that are more susceptible to local extinctions.

4 Phylogenetic Origin and Phylogeographic Patterns in Species Associated with WSE

4.1 Plants

The floristic composition and phylogenetic structure of WSE is the product of lineage diversification and dispersal events related to different processes: (1) habitat specialization of ancestral lineages from other non-WSE ecosystems; (2) habitat specialization within WSE; (3) allopatric speciation due to geographic isolation

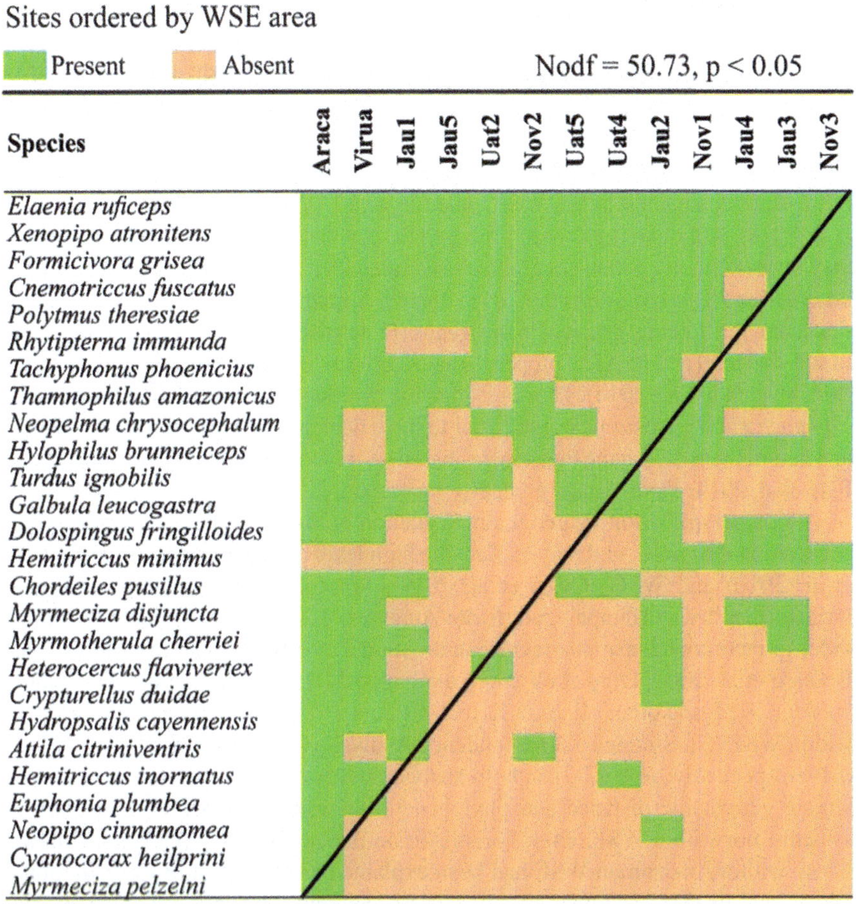

Fig. 11.6 The nested arrangement of WSE bird communities is affected by the amount of WSE habitat available in the landscape. The analysis was performed using records of 26 bird species obtained from our own fieldwork (Borges et al. 2016a, b; Aracá = 0°32′ N, 63°27′ W; Viruá = 1°36′ N, 61°13′ W; Jau = 1°50′ S, 61°37′ W; Uatumã (Uat) = 2°16′ S, 59°4′ W; Novo Airão (Nov) = 3°01′ S, 60°47′ W) and the literature (Laranjeiras et al. 2014). The nested metric applied was NOFD (Almeida-Neto et al. 2008), and the significance level was assessed through a null model (Strona et al. 2014)

among WSE patches or geographic barriers, and (4) historical dispersal events between WSE and other oligotrophic ecosystems within and between biomes. Although there is no comprehensive study to date addressing all these questions, we present some examples based on the literature and our floristic dataset to illustrate the role of these processes on the origin of the WSE lineages. To conduct the analyses, we generated a pruned version of the phylogenetic tree used by Zanne et al. (2014), resulting in a phylogenetic tree with 508 species and a reduced dataset of 508 species and 59 sample sites.

In the WSE flora, habitat specialization evolving from non-WSE stocks have been reported for Burseraceae, where 11 out of 14 WSE lineages have a sister species in the TFF, one in the Cerrado (Savanna), and only two in the WSE (Fine et al. 2014; Misiewicz and Fine 2014), and for *Pradosia* (Sapotaceae) where habitat transition from TFF to WSE occurred at least four times in ancestor lineages, with posterior diversification within WSE (Terra-Araujo et al. 2015). Diversification within WSE has been well-documented in the WSE endemic genus *Pagamea* (Vicentini 2016; Prata et al. 2018). Although it has been proposed that habitat specialization within the WSE may be less important than geographic isolation, given the level of niche conservatism observed in *Pagamea* (Vicentini 2016), there is strong evidence of habitat specialization along the flooding gradient in this genus (Prata et al. 2018). Evidence of adaptation to soil chemical and physical properties also has been reported in *Protium subserratum* (Misiewicz and Fine 2014). The diversification events of WSE lineages in Burseraceae, *Pagamea* and *Pradosia* were estimated to have occurred between 0.5 and 8 Ma ago, and half of them during the Pleistocene, based on molecular dating of these groups (see Fine et al. 2014; Terra-Araujo et al. 2015; Vicentini 2016).

Allopatric speciation by geographic isolation of the WSE patches may also have occurred in *Pagamea*, and may explain the high level of endemism in WSF (Guevara et al. 2016) and WSC (Costa et al. 2019). We found that geographic distance (transformed by a Principal Coordinate Analysis of Neighbor Matrices (PCNM) in order to incorporate spatial structure in the model; see Borcard and Legendre 2002; Borcard et al. 2004; Dray et al. 2006) explained 51.2% of the floristic dissimilarity between WSE and other oligotrophic ecosystems and 45.9% of the floristic variation within WSE (Constrained Correspondence Analysis (CCA); Constrained = 0.51 and 0.49, respectively; $P < 0.001$ in both analyses). However, no correlation was found among geographic distance and phylogenetic dissimilarity—neither between ecosystems, nor within WSE. High floristic dissimilarity and low phylogenetic dissimilarity within Amazonian WSE had been explained as a result of in situ radiation of white sand lineages followed by allopatric speciation (Guevara et al. 2016). Allopatric speciation promoted by vicariance events, such as the formation of rivers, has not yet been detected in Amazonian WSE plants. However, it has been proposed that an early Pleistocene diversification event may have isolated two lineages of *Pagamea* m. angustifolia (Prata et al. 2018) north and south of the Amazon River, possibly reflecting a vicariance event related to a Plio-Pleistocene (<3 Ma) origin of this river (Campbell et al. 2006; Ribas et al. 2012). Other patterns within *Pagamea* may also be related to the Amazon river formation (*Pagamea plicata* vs. *P. glabrescens*, *Pagamea coriacea* vs. *P. coriacea* var. *acuta*; Vicentini 2007). Additionally, despite the fragmented distribution and the geographic barriers, historical expansion and dispersal events may explain the occurrence of white sand lineages in different biomes, such as *Pagamea guianensis* and *Humiria balsamifera*, which are present in the WSE, the Amazonian *restingas* and the AF-*restingas*. For instance, it has been proposed that *P. guianensis* expanded from Eastern Amazonia to the AF-*restingas* during the Pleistocene (around 1 Ma ago), probably under favorable climatic conditions (Vicentini 2016; Prata 2016). The oldest divergences within *Pagamea* and *Pradosia* are detected between these biomes (Vicentini 2016).

To investigate the phylogenetic structure of WSE and other related oligotrophic ecosystems, we computed the Phylogenetic Diversity (PD), Mean Nearest Taxon Distance (MNTD), Mean Pairwise Distance (MPD), and the Phylogenetic Community Dissimilarity (PCD) followed by a Non-metric multidimensional ordination (NMDS) (see Webb 2000; Webb et al. 2008; Paradis 2012; Ives and Helmus 2010). We found a strong positive correlation between PD and species richness ($R^2 = 0.93$; $P < 0.001$). Most WSE presented phylogenetic clustering for both MPD and MNTD, suggesting that WSE communities are composed of lineages more closely related than expected by chance, given the species pool, which may be explained by the occurrence of lineages with conserved traits adapted to this stressful and selective ecosystem. The PCD revealed a strong floristic structure of the WSE (Fig. 11.2b) that decreases when the phylogenetic component (that reflects the evolutionary relationships among nonshared species) is included in the analysis (Fig. 11.2c), probably because of phylogenetic signals with other ecosystems. For example, WSE from the Negro River basin (sample sites 21, 24, 25, 42, 44, 51, 53) are phylogenetically more similar to *igapó* than to WSE from other regions in Amazonia, reinforcing the results of the previous analyses and showing that plant lineages from WSE are more likely to invade *igapó* because they already have the set of morphological and physiological traits required to survive in a semi-terrestrial environment. The WSE immersed in or close to Savannas (e.g. sample sites 1, 10, 33, 37 and 54 in Fig. 11.2) are phylogenetically closer to the Amazonian Savannas and the Savannas from Central Brazil than to other WSE sites, suggesting a phylogenetic similarity at deeper taxonomic levels between these communities. Most of the AF-*restinga* sites were phylogenetically grouped within WSE, signaling the historical connections between Amazonia and the Atlantic Forest (Costa 2003; Terra-Araujo et al. 2015; Vicentini 2016; Prata 2016).

4.2 Birds

The WSE harbor a diverse group of specialist bird species that can be widely distributed (e.g. *Xenopipo atronitens*, *Tachyphonus phoenicius*) or very restricted (e.g. *Polioptila clementsi*, *Ammonastes pelzelni*) across Amazonia. These specialists also present diverse phylogenetic affinities, as WSE birds' sister species or clades occur in different environments from within and outside Amazonia, for example, the Brazilian Cerrado (*Polytmus theresiae*, *Elaenia ruficeps*) and the Tepuis (*Xenopipo atronitens*). Few studies have focused on the diversification of WSE birds (Capurucho et al. 2013; Matos et al. 2016; Ferreira et al. 2018), but phylogenetic analyses of bird taxonomic groups often include one or a few WSE specialist species (e.g. Isler et al. 2013; Capurucho et al. 2018). These studies provide the opportunity to better understand the evolutionary origins of bird lineages that currently occupy WSE. We discuss published phylogenetic data and provide additional analyses to better understand the origin and recent population genetic patterns of WSE birds and their diversification within Amazonia.

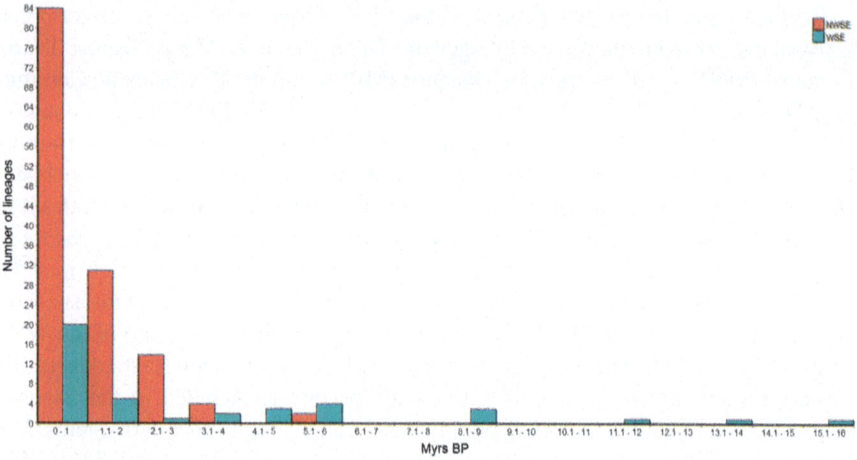

Fig. 11.7 Number of bird lineages from Amazonian white sand ecosystems (WSE, green bars) and non-white sand ecosystems (NWSE, red bars) that evolved every 1 million years. Data for WSE lineages was obtained from a literature compilation; for NWSE, we used Silva et al. (2019) data; most analyses in the literature were based on the molecular clock estimate for birds presented by Weir and Schluter (2008)

Based on molecular dating compiled from the literature, out of 41 lineages of birds near-restricted or restricted to WSE, 26 (63.4%) are estimated to have diverged from sister taxa during the Pleistocene (Fig. 11.7; the list of references used can be found at online Electronic Supplementary Material S1). Among the remaining 15 lineages, nine (21.9%) diverged in the Pliocene and six (14.6%) in the Miocene. This agrees with the view that the formation of white sand patches is diachronic, and that WSE of different ages occur across Amazonia. Several phylogenetic studies of species complexes (e.g. Ribas et al. 2012; Lutz et al. 2013; Ferreira et al. 2017), in addition to recent meta-analyses of Amazonian bird lineages (Smith et al. 2014; Naka and Brumfield 2018; Silva et al. 2019) suggest that most current TFF bird species or lineages originated in the Quaternary (94.8%; Fig. 11.7). A substantial portion of the WSE species are older than taxa found in other Amazonian ecosystems, particularly in the TFF. The reasons for this seemingly higher frequency of older lineages is not yet known, but the geological evidence suggests that soils supporting WSE have a long history in Amazonia. The distributions of these lineages may have been continuously changing according to changes in the WSE distribution across Amazonia. At the same time, as suggested for birds in other habitats and by geological data for some WSE regions, Quaternary climatic and landscape dynamics may have also shaped the diversification of WSE bird species, as more than half speciated in this period.

Using data from bird surveys from 54 sites (see above), the Phylogenetic Diversity (PD), Mean Nearest Taxon Distance (MNTD), and Mean Pairwise Distance (MPD) were calculated using a random sample of 100 bird phylogenies from Jetz et al. (2012; https://birdtree.org/). As expected, PD was positively correlated with species richness ($R^2 = 0.92$; $p < 0.01$), thus the impoverished WSE showed a lower

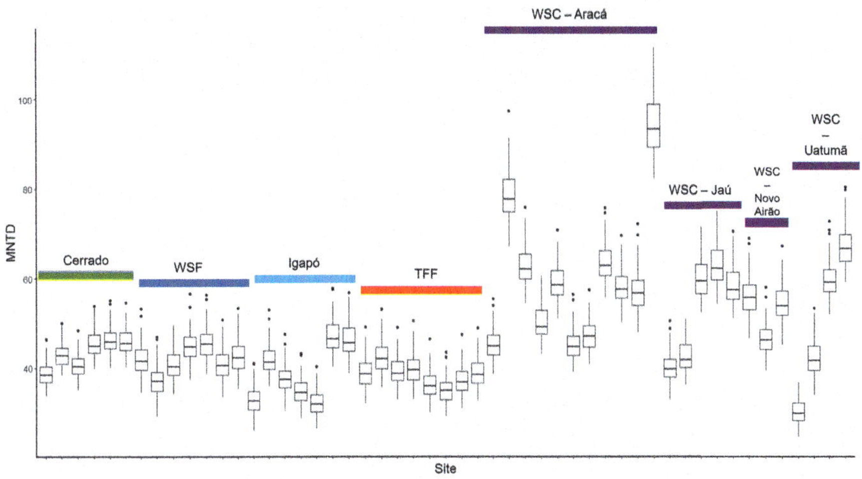

Fig. 11.8 Mean Nearest Taxon Distance (MNTD—the mean phylogenetic distance between each species and its closest taxon in a given site) values for sites within Amazonian Ecosystems (Cerrado, *igapó*, Terra Firme Forest (TFF), white sand forest (WSF), and white sand grasslands and shrublands (WSC)). Data from WSC were collected in different Amazonian regions, and are considered separately

PD than the other analyzed ecosystems (TFF, *igapó*, and cerrado). The values for MNTD were higher in several WSE sites, reflecting the older ages of some specialist birds and phylogenetic overdispersion (Fig. 11.8). The MNTD values are especially high in the Aracá region, where the highest number of specialist bird species is observed (Fig. 11.8). Aracá stands out as the only region with sites that are significantly different from the other Amazonian ecosystems, but not different from WSE sites in other regions (one-way ANOVA, $F(7, 43) = 5.38$, $p < 0.01$; Table 11.2). Higher values are also observed for MPD in several WSE sites, although the differences are subtle. Still, Aracá stands out as significantly different from TFF, but not different from the other ecosystems (one-way ANOVA, $F(7, 43) = 5.24$, $p < 0.01$; Table 11.2). These analyses show that a significant portion of WSE birds are older than species occupying other environments in Amazonia, and that species that make up WSE communities are more distantly related.

WSE bird species appear to be old and have geographic distributions comparable to the distribution of several related linages (species complexes) of TFF taxa. Still, shallow population level structuring is commonly observed in WSE birds (Capurucho et al. 2013, 2018; Matos et al. 2016), while upland forest species complexes occupying similar geographical extents have lineages delimited by the main Amazonian rivers (e.g. Ribas et al. 2012; Ferreira et al. 2017). Interestingly, for some WSE bird species (Capurucho et al. 2013; Matos et al. 2016; Ferreira et al. 2018), isolation between northern and southern Amazonian populations dates to about 1–2 Ma, which is similar to the age found for *Pagamea* (see above). Of the WSE species studied to date, only *Galbula leucogastra* does not present shared haplotypes between interfluvia (Ferreira et al. 2018), a pattern often found in TFF

Table 11.2 *P*-values for Bonferroni-corrected pairwise *t*-test comparing Mean Nearest Taxon Distance (MNTD) and Mean Pairwise Distance (MPD) values among different Amazonian ecosystems and WSE from different regions in Amazonia

MNTD		MPD	
	WSC—Aracá		*WSC—Aracá*
Cerrado	**0.03**	*Cerrado*	1.0
TFF	**0.0004**	*TFF*	**0.0008**
WSF	**0.009**	*WSF*	1.0
Igapó	**0.001**	*Igapó*	0.08
WSC—Jaú	1.0	*WSC—Jaú*	1.0
WSC—Uatumã	1.0	*WSC—Uatumã*	**0.049**
WSC—Novo Airão	1.0	*WSC—Novo Airão*	0.53

Only the main pairwise results are presented
Significant values, $P<0.05$, are shown in bold

birds. In other WSE birds, shallow population structure and recent gene flow is observed especially north of the Amazon River (Capurucho et al. 2013, 2018; Matos et al. 2016), a region spanning four bird areas of endemism for TFF taxa: Guiana, Negro, Napo, and Imeri (Cracraft 1985). When restriction of gene flow is observed, it is usually associated with white water rivers and their floodplains (*várzeas*), as in the Branco and Madeira Rivers (Capurucho et al. 2013; Matos et al. 2016). *Igapó* forests appear to be more permeable for WSE birds and recent gene flow is observed, for example, across the Negro River. This emphasizes an ongoing connection between the WSE and *igapó* environments, as shown floristically (see above).

In addition to recent gene flow, some WSE birds have experienced demographic fluctuations dating to the Late Pleistocene. Several specialist birds show a strong signal of demographic expansion after the LGM (~23 ka). Analyses of isotopic composition in speleothems and depositional chronology in sediment profiles indicate drier conditions in northern and eastern Amazonia during the LGM (Häggi et al. 2017; Wang et al. 2017; Zular et al. 2019). These climatic conditions may have affected WSE due to the low water retention capability associated with sandy soils, possibly leading to both habitat loss and local extinctions. Interestingly, drier LGM conditions have also been associated with the genesis and expansion of white sand areas in northern Amazonia (Zular et al. 2019). As wetter conditions returned after the LGM, populations adapted to WSE would face more favorable climatic conditions and at the same time have more available habitat, leading to the observed population expansions (Capurucho et al. 2013; Matos et al. 2016). In turn, changes in TFF and *igapó* vegetation structure during glacial cycles, such as more open canopy cover, may have facilitated dispersal between WSE patches, promoting gene flow (Capurucho et al. 2013; Matos et al. 2016). As barriers became more permeable, gene flow may have increased among areas that were able to sustain WSE (although local extinctions may also have occurred due to drier conditions).

The evidence from WSE avian studies supports both a deep evolutionary history for WSE birds and a dynamic recent history, consistent with a changing Amazonian landscape during the Pleistocene. This contradicts the views of stability in Amazonia during this period (Smith et al. 2014), but corroborates recent models of changes in vegetation cover since the LGM (Arruda et al. 2017) and the importance of Pleistocene climatic change in shaping current diversity patterns in Amazonia (Rangel et al. 2018).

5 Conclusions and Perspectives

Geological, botanical, and ornithological data presented and reviewed here document that the history of the WSE goes back millions of years, yet also involves dynamic changes during the Pleistocene and since the LGM. Despite their generally lower species richness relative to other Amazonian ecosystems, WSE present a unique assemblage of species. Regarding plant taxa composition, a closer relationship is observed between WSE and *igapó* and the Atlantic Forest *restingas*. Few WSE birds have sister groups in adjacent TFF; most are related to species that occur in habitats outside Amazonia, such as the Tepuis highlands and the Cerrado. It seems plausible that genetic connectivity of bird populations across Amazonia may be related to the use of *igapó* habitats, which also share botanical similarities with WSE.

Although a high proportion of the WSE and its associated species seem to be older than the Quaternary, the Pleistocene history of WSE has been very dynamic, with paleoclimatic changes having multiple consequences for these habitats, including the genesis of new WSE patches and local extinctions due to harsh climatic conditions in dry periods. The phylogenetic and population genetic patterns reported for WSE taxa are consistent with this dynamic and complex geological history. Phylogenetic structure across WSE plants is higher than that observed in WSE birds, likely due to differences in dispersal capabilities. Bird species may have been able to disperse between WSE patches in the recent past, and currently present low levels of intraspecific genetic structure.

WSE were sensitive to Pleistocene climatic changes which may have led to lineage diversification and local extinctions. The complex history of WSE provides valuable information about the biogeography of Amazonia and other South American ecosystems and deserves further attention in both scientific and conservation efforts. With increasing threats to the Amazonian ecosystems, the low resilience of WSE to disturbance needs to be considered for conservation planning, especially within a climate change scenario. Continuing work to catalog the diversity of the WSE, in addition to studies to better understand its historical and current ecological dynamics, will be pivotal to preserve this unique ecosystem and will improve our understanding of its historical and current contribution to Amazonian biodiversity.

Acknowledgements Grants: FAPESP-FAPEAM (Edital 006/2009); "Dimensions US-Biota-São Paulo", NSF (DEB 1241066)/NASA/FAPESP (grant #2012/50260-6); PEER/USAID (AID-OAA-A-11-00012); and Universal Amazonas (Edital FAPEAM 002/2018). Research fellowships: CNPq (308927/2016-8) to CCR. FAPEAM (Fixam Program, Edital 017/204) to SHB. SwB CAPES (11881-2013/5) to JMGC. Student fellowships from CNPq, CAPES and FAPEAM.

References

Adeney JM, Christensen NL, Vicentini A, Cohn-Haft M (2016) White-sand ecosystems in Amazonia. Biotropica 48:7–23. https://doi.org/10.1111/btp.12293

Almeida-Neto M, Guimarães P, Guimarães PR Jr, Loyola RD, Ulrich W (2008) A consistent metric for nestedness analysis in ecological systems: reconciling concept and measurement. Oikos 117:1227–1239

Alonso JA, Whitney B (2003) New distributional records of birds from white-sand forests of the northern Peruvian Amazon, with implications for biogeography of northern South America. Condor 105:552–566

Alonso JA, Metz MR, Fine PVA (2013) Habitat specialization by birds in western Amazonian white-sand forests. Biotropica 45:365–372

Anderson AB (1981) White-sand vegetation of Brazilian Amazonia. Biotropica 13:199–210

Arruda DM, Schaefer CEG, Fonseca RS, Solar RRC, Fernandes-Filho EI (2017) Vegetation cover of Brazil in the last 21 ka: new insights into the Amazonian refugia and Pleistocenic arc hypotheses. Glob Ecol Biogeogr 27:47–56

Baker PA, Fritz SC, Burns SJ, Ekdahl E, Rigsby CA (2009) The nature and origin of decadal to millennial scale climate variability in the southern tropics of South America: the Holocene record of Lago Umayo, Peru. In: Vimeux F, Sylvestre F, Khodri M (eds) Past climate variability in South America and surrounding regions: from the last glacial maximum to the Holocene. Springer, Amsterdam, pp 301–322. https://doi.org/10.1007/978-90-481-2672-9_13

Bleackley D, Khan EJA (1963) Observations on the white-sand areas of the Berbice Formation, British Guiana. J Soil Sci 14:44–51

Borcard D, Legendre P (2002) All-scale spatial analysis of ecological data by means of principal coordinates of neighbour matrices. Ecol Model 153:51–68

Borcard D, Legendre P, Avois-Jacquet C, Tuomisto H (2004) Dissecting the spatial structure of ecological data at multiple scales. Ecology 85:1826–1832

Borges SH (2013) Bird species distribution in a complex Amazonian landscape: species diversity, compositional variability and biotic-environmental relationships. Stud Neotropical Fauna Environ 48:106–118

Borges SH, Almeida RA (2011) Birds of the Jaú National Park and adjacent areas, Brazilian Amazon: new species records with reanalysis of a previous checklist. Rev Bras Ornitol 19:108–133

Borges SH, Cornelius C, Moreira M, Ribas C, Conh-Haft M, Capurucho JM, Vargas C, Almeida R (2016a) Bird communities in Amazonian white-sand vegetation patches: effects of landscape configuration and biogeographic context. Biotropica 48:121–131

Borges SH, Cornelius C, Ribas C, Almeida R, Guilherme E, Aleixo A, Dantas S, Santos MPD, Moreira M (2016b) What is the avifauna of Amazonian white-sand vegetation? Bird Conserv Int 26:192–204

Bueno GT (2009) Empobrecimento e Podzolização de Solos Lateríticos da Bacia do Rio Negro e Gênese dos Podzóis na Alta Bacia Amazônica. Tese de Doutorado, Instituto de Geociências e Ciências Exatas, Universidade Estadual Paulista

Campbell KE, Frailey CD, Romero-Pittman L (2006) The Pan-Amazonian Ucayali Peneplain, late Neogene sedimentation in Amazonia, and the birth of the modern Amazon River system. Palaeogeogr Palaeoclimatol Palaeoecol 239:166–219. https://doi.org/10.1016/j.palaeo.2006.01.020

Capurucho JMG, Cornelius C, Borges SH, Cohn-Haft M, Aleixo A, Metzger JP, Ribas CC (2013) Combining phylogeography and landscape genetics of *Xenopipo atronitens* (Aves: Pipridae), a white sand campina specialist, to understand Pleistocene landscape evolution in Amazonia. Biol J Linn Soc 110:60–76. https://doi.org/10.1111/bij.12102

Capurucho JMG, Ashley MV, Ribas CC, Bates JM (2018) Connecting Amazonian, Cerrado, and Atlantic forest histories: Paraphyly, old divergences, and modern population dynamics in tyrant-manakins (Neopelma/Tyranneutes, Aves: Pipridae). Mol Phylogenet Evol 127:696–705

Carneiro-Filho A, Schwartz D, Tatumi SH, Rosique T (2002) Amazonian paleodunes provide evidence for drier climate phases during the Late Pleistocene–Holocene. Quat Res 58:205–209
Costa LP (2003) The historical bridge between the Amazon and the Atlantic Forest of Brazil: a study of molecular phylogeography with small mammals. J Biogeogr 30:71–86
Costa FM, Terra-Araujo MH, Zartman CE, Frische CC, Carvalho FA, Hopkins MJG, Viana PL, Prata EMB, Vicentini A (2019) Islands in a green ocean: geographic endemism and biogeography of the lowland white-sand flora in Central Amazonia. Biotropica 00:1–12
Cowling SA, Maslin MA, Sykes MT (2001) Paleovegetation simulations of lowland Amazonia and implications for neotropical allopatry and speciation. Quat Res 55:140–149
Cracraft J (1985) Historical biogeography and patterns of differentiation within the South American avifauna: areas of endemism. Ornithol Monogr 36:49–84
D'Apolito C, Absy ML, Latrubesse EM (2017) The movement of pre-adapted cool taxa in north-central Amazonia during the last glacial. Quat Sci Rev 169:1–12. https://doi.org/10.1016/j.quascirev.2017.05.017
Damasco G, Vicentini A, Castilho CV, Pimentel TP, Nascimento HE (2013) Disentangling the role of edaphic variability, flooding regime and topography of Amazonian white-sand vegetation. J Veg Sci 24:384–394
Demarchi LO, Scudeller VV, Moura LC, Dias-Terceiro RG, Lopes A, Wittmann FK, Piedade MTF (2018) Floristic composition, structure and soil-vegetation relations in three white-sand soil patches in central Amazonia. Acta Amazon 48:46–56
Dray S, Legendre P, Peres-Neto PR (2006) Spatial modelling: a comprehensive framework for principal coordinate analysis of neighbour matrices (PCNM). Ecol Model 196:483–493
Dubroeucq D, Volkoff B, Pédro G (1991) La couverture pédologique du Bouclier du Nord de l'Amazonie (bassin du Haut Rio Negro). Séquence évolutive des sols et son rôle dans l'aplanissement généralisé des zones tropicales perhumides. C R Acad Sci Paris 312:663–671
Duivenvoorden JF, Lips JM (1995) A land-ecological study of soils, vegetation and plant diversity in Colombian Amazonia, Tropenbos edn. The Tropenbos Foundation, Wageningen, 438 p
Ferreira M, Aleixo A, Ribas CC, Santos MPD (2017) Biogeography of the Neotropical genus Malacoptila (Aves: Bucconidae): the influence of the Andean orogeny, Amazonian drainage evolution and palaeoclimate. J Biogeogr 44:748–759
Ferreira M, Fernandes AM, Aleixo A, Antonelli A, Olsson U, Bates JM, Cracraft J, Ribas CC (2018) Evidence for mtDNA capture in the jacamar Galbula leucogastra/chalcothorax species-complex and insights on the evolution of white-sand ecosystems in the Amazon basin. Mol Phylogenet Evol 129:149–157
Fine PV, García-Villacorta R, Pitman NC, Mesones I, Kembel SW (2010) A floristic study of the white-sand forests of Peru. Ann Mo Bot Gard 97:283–305
Fine PV, Baraloto C (2016) Habitat endemism in white-sand forests: insights into the mechanisms of lineage diversification and community assembly of the Neotropical flora. Biotropica 48:24–33
Fine PV, Mesones I, Coley PD (2004) Herbivores promote habitat specialization by trees in Amazonian forests. Science 305:663–665
Fine PV, Zapata F, Daly DC (2014) Investigating processes of neotropical rain forest tree diversification by examining the evolution and historical biogeography of the Protieae (Burseraceae). Evolution 68:1988–2004
Givnish TJ, Evans TM, Zjhra ML, Patterson TB, Berry PE, Sytsma KJ (2000) Molecular evolution, adaptive radiation, and geographic diversification in the amphiatlantic family Rapateaceae: evidence from ndhF sequences and morphology. Evolution 54:1915–1937
Guevara JE, Damasco G, Baraloto C et al (2016) Low phylogenetic beta diversity and geographic neo-endemism in Amazonian white-sand forests. Biotropica 48:34–46
Häggi C, Chiessi CM, Merkel U, Mulitza S, Prange M, Schulz M, Schefuß E (2017) Response of the Amazon rainforest to late Pleistocene climate variability. Earth Planet Sci Lett 479:50–59. https://doi.org/10.1016/J.EPSL.2017.09.013
Hennig C (2018) fpc: flexible Procedures for Clustering. R package version 2:2–3. https://CRAN.R-project.org/package=fpc.

Heyligers PC (1963) Vegetation and soil of a white-sand savanna in Suriname. N.V. Noord-Hollandsche Uitgevers Maatschappij, Amsterdam, p 148

Hoorn C, Bogotá-A GR, Romero-Baez M, Lammertsma EI, Flantua SGA, Dantas EL, Dino R, do Carmo DA, Chemale F (2017) The Amazon at sea: onset and stages of the Amazon River from a marine record, with special reference to Neogene plant turnover in the drainage basin. Glob Planet Change 153:51–65. https://doi.org/10.1016/j.gloplacha.2017.02.005

Horbe AMC, Horbe MA, Suguio K (2004) Tropical Spodosols in northeastern Amazonas State, Brazil. Geoderma 119:55–68. https://doi.org/10.1016/S0016-7061(03)00233-7

Huber O (1988) Guayana highlands versus Guayana lowlands, a reappraisal. Taxon 37:595–614

Iriondo M, Latrubesse EM (1994) A probable scenario for a dry climate in central Amazonia during the late Quaternary. Quat Int 21:121–128. https://doi.org/10.1016/1040-6182(94)90026-4

Isler ML, Bravo GA, Brumfield RT (2013) Taxonomic revision of Myrmeciza (Aves: Passeriformes: Thamnophilidae) into 12 genera based on phylogenetic, morphological, behavioral, and ecological data. Zootaxa 3717:469–497.

Ives AR, Helmus MR (2010) Phylogenetic metrics of community similarity. Am Nat 176:E128–E142

Janzen DH (1974) Tropical blackwater rivers, animals, and mast fruiting by the Dipterocarpaceae. Biotropica 6:69–103

Jarvis A, Reuter HI, Nelson A, Guevara E (2008) Hole-filled seamless SRTM data V4—SRTM Resampled 250m [Online]. Int Cent Trop Agric. http://gisweb.ciat.cgiar.org/TRMM/SRTM_Resampled_250m/. Accessed 17 June 2018

Jetz W et al (2012) The global diversity of birds in space and time. Nature 491:444–448

Junk WJ, Piedade MTF, Schöngart J, Cohn-Haft M, Adeney JM, Wittmann F (2011) A classification of major naturally-occurring amazonian lowland wetlands. Wetlands 31:623–640

Klinge H (1965) Podzol soils in the Amazon Basin. J Soil Sci 16:95–103

Koutavas A, Lynch-Stieglitz J (2004) Variability of the marine ITCZ over the eastern Pacific during the past 30,000 years. In: Diaz HF, Bradley RS (eds) The Hadley circulation: present, past and future, advances in global change research. Springer, Dordrecht, pp 347–369. https://doi.org/10.1007/978-1-4020-2944-8

Kubitzki K (1989) The ecogeographical differentiation of Amazonian inundation forests. Plant Syst Evol 162:285–304

Kubitzki K (1990) The psammophilous flora of northern South America. Mem N Y Bot Gard 64:248–253

Laranjeiras TO, Naka LN, Bechtoldt CL, Vieira TV, Andretti CB, Cerqueira MC, Torres MF (2014) The avifauna of Virua National Park, Roraima, reveals megadiversity in northern Amazonia. Rev Bras Ornitol 22:138–171

Latrubesse EM, Cozzuol M, Silva-Caminha SAF, Rigsby CA, Absy ML, Jaramillo C (2010) The Late Miocene paleogeography of the Amazon Basin and the evolution of the Amazon River system. Earth-Sci Rev 99:99–124

Lutz HL, Weckstein JD, Patane JSL, Bates JM, Aleixo A (2013) Biogeography and spatio-temporal diversification of Selenidera and Andigena Toucans (Aves: Ramphastidae). Mol Phylogenet Evol 69:873–883

Mafra AL, Miklós AAW, Volkoff B, Melfi AJ (2002) Pedogênese numa seqüência latossolo-espodossolo na região do alto rio Negro, Amazonas. Rev Bras Ciênc Solo 26:381–394. https://doi.org/10.1590/S0100-06832002000200012

Matos MV, Borges SH, d'Horta FM, Cornelius C, Latrubesse E, Cohn-Haft M, Ribas CC (2016) Comparative phylogeography of two bird species, *Tachyphonus phoenicius* (Thraupidae) and *Polytmus theresiae* (Trochilidae), specialized in Amazonian white-sand vegetation. Biotropica 48:110–120. https://doi.org/10.1111/btp.12292

Matthews T (2015) Analysing and modelling the impact of habitat fragmentation on species diversity: a macroecological perspective. Front Biogeogr 7. https://doi.org/10.21425/F57225942

Mendonça BAF, Simas FNB, Schafer CEGR, Fernandes Filho EI, Vale Júnior JF, Mendonça JGF (2014) Podzolized soils and paleoenvironmental implications of white-sand vegetation (Campinarana) in the Viruá National Park, Brazil. Geoderma Reg 2:9–20

Misiewicz TM, Fine PV (2014) Evidence for ecological divergence across a mosaic of soil types in an Amazonian tropical tree: *Protium subserratum* (Burseraceae). Mol Ecol 23:2543–2558

Naka LN, Brumfield RT (2018) The dual role of Amazonian rivers in the generation and maintenance of avian diversity. Sci Adv 4:eaar8575. https://doi.org/10.1126/sciadv.aar8575

Oren DC (1981) Zoogeographic analysis of the white sand campina avifauna of Amazonia. Ph.D. dissertation, Harvard University, Cambridge, MA

Paradis E (2012) Analysis of phylogenetics and evolution with R, 2nd edn. Springer, New York

Parolin P, Wittmann F (2010) Struggle in the flood: tree responses to flooding stress in four tropical floodplain systems. AoB Plants 2010:plq003

Pennington RT, Lehmann CER, Rowland LM (2018) Tropical savannas and dry forests. Curr Biol 28:R527–R548

Prance GT (1996) Islands in Amazonia. Philosophical Trans R Soc B: Biol Sci 351:823–833

Prata EMB (2016) Delimitação de espécies e história de diversificação do complexo Pagamea guianensis (Rubiaceae) na América do Sul Tropical. Unpublished D. Phil. Thesis, Insituto Nacional de Pesquisas da Amazônia

Prata EMB, Sass C, Rodrigues DP, Domingos FMCB, Specht CD, Damasco G, Ribas CC, Fine PVA, Vicentini A (2018) Towards integrative taxonomy in Neotropical botany: disentangling the *Pagamea guianensis* species complex (Rubiaceae). Bot J Linn Soc 188:213–231

Quesada CA, Phillips OL, Schwarz M, Czimczik CI, Baker TR, Patiño S et al (2012) Basin-wide variations in Amazon forest structure and function are mediated by both soils and climate. Biogeosciences 9:2203–2246

Rangel TF et al (2018) Modeling the ecology and evolution of biodiversity: biogeographical cradles, museums, and graves. Science 361:eaar5452. https://doi.org/10.1126/science.aar5452

Räsänen M, Linna A, Irion G, Hernani LR, Huaman RV, Wesselingh F (1998) Geologia y Geoformas de la Zona de Iquitos. In: Kalliola R, Flores Paitan S (eds) Geoecología y Desarrollo Amazónico: Estudio Integrado En La Zona de Iquitos, Perú. Turun Yliopisto, Turko, p 544

Ribas CC, Aleixo A, Nogueira ACR, Miyaki CY, Cracraft J (2012) A palaeobiogeographic model for biotic diversification within Amazonia over the past three million years. Proc R Soc B Biol Sci 279:681–689. https://doi.org/10.1098/rspb.2011.1120

Richards PW (1941) Lowland tropical podsols and their vegetation. Nature 148:129–131. https://doi.org/10.1038/148129a0

Roddaz M, Baby P, Brusset S, Hermoza W, Darrozes JM (2005a) Forebulge dynamics and environmental control in Western Amazonia: the case study of the Arch of Iquitos (Peru). Tectonophysics 399:87–108. https://doi.org/10.1016/j.tecto.2004.12.017

Roddaz M, Viers J, Brusset S, Baby P, Hérail G (2005b) Sediment provenances and drainage evolution of the Neogene Amazonian foreland basin. Earth Planet Sci Lett 239:57–78. https://doi.org/10.1016/j.epsl.2005.08.007

Roddaz M, Viers J, Brusset S, Baby P, Boucayrand C, Hérail G (2006) Controls on weathering and provenance in the Amazonian foreland basin: insights from major and trace element geochemistry of Neogene Amazonian sediments. Chem Geol 226:31–65. https://doi.org/10.1016/j.chemgeo.2005.08.010

Rossetti DF, Bertani TC, Zani H, Cremon EH, Hayakawa EH (2012a) Late Quaternary sedimentary dynamics in Western Amazonia: implications for the origin of open vegetation/forest contrasts. Geomorphology 177–178:74–92. https://doi.org/10.1016/j.geomorph.2012.07.015

Rossetti DF, Zani H, Cohen MCL, Cremon ÉH (2012b) A Late Pleistocene-Holocene wetland megafan in the Brazilian Amazonia. Sediment Geol 282:276–293. https://doi.org/10.1016/j.sedgeo.2012.09.015

Rossetti DF, Zani H, Cremon ÉH (2014) Fossil megafans evidenced by remote sensing in the Amazonian wetlands. Z Geomorphol 58:145–161. https://doi.org/10.1127/0372-8854/2013/0118

Roy M, Schimann H, Braga-Neto R et al (2016) Diversity and distribution of ectomycorrhizal fungi from Amazonian lowland white-sand forests in Brazil and French Guiana. Biotropica 48:90–100

Silva SM, Peterson TA, Carneiro L, Burlamaqui TCT, Ribas CC, Sousa-Neves T et al (2019) A dynamic continental moisture gradient drove Amazonian bird diversification. Sci Adv 5: eaat5752

Singer R, Aguiar IA (1986) Litter decomposing and ectomycorrhizal Basidiomycetes in an igapó forest. Plant Syst Evol 153:107–117

Smith BT et al (2014) The drivers of tropical speciation. Nature 515:406–409

Sombroek WG (1966) Amazon soils. Entre for Agricultural Publications and Documentation, Wageningen, p 292

Steyermark JA (1986) Speciation and endemism in the flora of the Venezuelan tepuis. In: Vuilleumier F, Monasterio M (eds) High altitude tropical biogeography. Oxford University Press, New York, pp 317–373

Strona G, Galli P, Seveso D, Montano S, Fattorini S (2014) Nestedness for Dummies (NeD): a user friendly web interface for exploratory nestedness analysis. J Stat Softw 59:1–9

Stropp J, Sleen PVD, Assunção PA, da Silva AL, ter Steege H (2011) Tree communities of white-sand and terra-firme forests of the upper Rio Negro. Acta Amazon 41:521–544

Teeuw RM, Rhodes EJ (2004) Aeolian activity in northern Amazonia: optical dating of Late Pleistocene and Holocene palaeodunes. J Quat Sci 19:49–54. https://doi.org/10.1002/jqs.815

Terra-Araujo MH, de Faria AD, Vicentini A, Nylinder S, Swenson U (2015) Species tree phylogeny and biogeography of the Neotropical genus *Pradosia* (Sapotaceae, Chrysophylloideae). Mol Phylogenet Evol 87:1–13

Thom G, Amaral FR, Hickerson MJ, Aleixo A, Araujo-Silva LE, Ribas CC, Choueri E, Miyaki CY (2018) Phenotypic and Genetic Structure Support Gene Flow Generating Gene Tree Discordances in an Amazonian floodplain endemic species. Syst Biol 67:700–718

Vicentini A (2004) A vegetação ao longo de um gradiente edáfico no Parque Nacional do Jaú. In: Borges SH, Iwanaga S, Durigan CC, Pinheiro MR (eds) Janelas para a biodiversidade no Parque Nacional do Jaú: uma estratégia par ao estudo da biodiversidade na Amazônia. Fundação Vitória Amazônica/WWF/IBAMA, Manaus, pp 117–143

Vicentini A (2007) *Pagamea* Aubl. (Rubiaceae), from species to processes, building the bridge. D. Phil. thesis, University of Missouri Saint Louis

Vicentini A (2016) The evolutionary history of Pagamea (Rubiaceae), a white-sand specialist lineage in tropical South America. Biotropica 48:58–69

Wang X, Edwards RL, Auler AS, Cheng H, Kong X, Wang Y, Cruz FW, Dorale JA, Chiang H-W (2017) Hydroclimate changes across the Amazon lowlands over the past 45,000 years. Nature 541:204–207

Webb CO (2000) Exploring the phylogenetic structure of ecological communities: an example for rain forest trees. Am Nat 156(2):145–155

Webb CO, Ackerly DD, Kembel SW (2008) Phylocom: software for the analysis of phylogenetic community structure and trait evolution. Version 4.0.1

Weir JT, Schluter D (2008) Calibrating the avian molecular clock. Mol Ecol 17:2321–2328.

Werneck FP (2011) The diversification of eastern South American open vegetation biomes: historical biogeography and perspectives. Quat Sci Rev 30:1630–1648

Wittmann F, Householder E, Piedade MT, de Assis RL, Schöngart J, Parolin P, Junk WJ (2013) Habitat specifity, endemism and the neotropical distribution of Amazonian white-water floodplain trees. Ecography 36:690–707

Zani H, Rossetti DF (2012) Multitemporal Landsat data applied for deciphering a megafan in northern Amazonia. Int J Remote Sens 33:6060–6075. https://doi.org/10.1080/01431161.2012.677865

Zanne AE, Tank DC, Cornwell WK et al (2014) Three keys to the radiation of angiosperms into freezing environments. Nature 506:89

Zular A, Sawakuchi AO, d'Horta FM, Cruz FW, Chiessi CM, Demattê JAR, Ribas CC, Hartmann GA, Giannini PCF, Soares EAA (2019) The role of abrupt climate change in the formation of an open vegetation enclave in northern Amazonia. Glob Planet Chang 172:140–149

Chapter 12
On the Young Savannas in the Land of Ancient Forests

Josué A. R. Azevedo, Rosane G. Collevatti, Carlos A. Jaramillo, Caroline A. E. Strömberg, Thaís B. Guedes, Pável Matos-Maraví, Christine D. Bacon, Juan David Carillo, Søren Faurby, and Alexandre Antonelli

Abstract Covering ancient geomorphological landscapes, and surrounded by some of the most diverse forests on Earth, the Neotropical savannas were once perceived by naturalists as ancient environments. However, current evidence suggests that tropical forests have existed in the Neotropics since the Paleocene, whereas most plant lineages present in South American savannas are recently derived from clades

Søren Faurby and Alexandre Antonelli are senior authors on this chapter.

J. A. R. Azevedo (✉) · C. D. Bacon · S. Faurby
Department of Biological and Environmental Sciences, University of Gothenburg, Gothenburg, Sweden

Gothenburg Global Biodiversity Centre, Gothenburg, Sweden
e-mail: soren.faurby@bioenv.gu.se

R. G. Collevatti
Laboratório de Genética and Biodiversidade, Universidade Federal de Goiás, Goiânia, Goiás, Brazil

C. A. Jaramillo
Smithsonian Tropical Research Institute, Balboa, Ancón, Panama

Institute of Evolution Sciences of Montpellier (ISEM), University of Montpellier, CNRS, EPHE, IRD, Montpellier, France
e-mail: JaramilloC@si.edu

C. A. E. Strömberg
Department of Biology and the Burke Museum of Natural History and Culture, University of Washington, Seattle, WA, USA
e-mail: caestrom@u.washington.edu

T. B. Guedes
Department of Biological and Environmental Sciences, University of Gothenburg, Gothenburg, Sweden

Gothenburg Global Biodiversity Centre, Gothenburg, Sweden

Center of Advanced Studies, State University of Maranhão, Caxias, Maranhão, Brazil

V. Rull, A. C. Carnaval (eds.), *Neotropical Diversification: Patterns and Processes*, Fascinating Life Sciences, https://doi.org/10.1007/978-3-030-31167-4_12

from the surrounding forested biomes. This chapter provides a multidisciplinary overview on the origin, assembly and expansion of Neotropical savannas, with focus on South America. For this, we consider available evidence from the fossil record, paleoenvironmental proxies (phytoliths), and phylogenetic information for both plants and animals. Paleoenvironmental reconstructions indicate suitable climates for central South American savannas since the middle Miocene, which is also when molecular phylogenies indicate the origin of some vertebrate groups typical of savannas. Fossil data indicate the ecological expansion of both C_3 and C_4 grasses in southern South America by the late Miocene. Fossil information also indicates the onset of savannas in northern South America during the Pliocene, a period in which most woody plants of the largest extension of Neotropical savannas (the Cerrado) are thought to have diversified, as inferred by dated phylogenies. Although the combined lines of evidence indicate that Neotropical savannas in South America are indeed younger than their surrounding forests, the precise timing and factors that influenced the origin, assembly and expansion of Neotropical savannas remain contentious. Future research should aim at (1) increasing and integrating knowledge about the diversification of important taxa characteristic to Neotropical savannas, (2) establishing continuous sequences of fossils, and (3) building accurate paleoenvironmental reconstructions for the entire Neogene.

Keywords Biome origin · Biome assembly · C_4 grasses · Cerrado · Molecular divergence times · Neotropical region · Phytoliths

P. Matos-Maraví
Department of Biological and Environmental Sciences, University of Gothenburg, Gothenburg, Sweden

Gothenburg Global Biodiversity Centre, Gothenburg, Sweden

Biology Centre CAS, Institute of Entomology, České Budějovice, Czech Republic
e-mail: pavel.matos@entu.cas.cz

J. D. Carillo
Department of Biological and Environmental Sciences, University of Gothenburg, Gothenburg, Sweden

Gothenburg Global Biodiversity Centre, Gothenburg, Sweden

Centre de Recherche en Paléontologie (CR2P), Muséum National d'Histoire Naturelle, Paris, France
e-mail: juan.carrillo@mnhn.fr

A. Antonelli
Department of Biological and Environmental Sciences, University of Gothenburg, Gothenburg, Sweden

Gothenburg Global Biodiversity Centre, Gothenburg, Sweden

Royal Botanic Gardens, Kew, Richmond, UK
e-mail: a.antonelli@kew.org

1 Introduction

The tropical savannas of South America were once considered to be among the oldest ecosystems on the continent—a notion mostly based on their geographical distribution, especially in the Cerrado domain (Fig. 12.1), which largely occurs on ancient geomorphological surfaces (Cole 1986; Ratter et al. 1997; Oliveira-Filho and Ratter 2002). The high species richness of the Neotropical savannas has also been interpreted as evidence of their antiquity, reflecting long periods of time for biological diversification over the Cenozoic (Cole 1986; Oliveira-Filho and Ratter 2002).

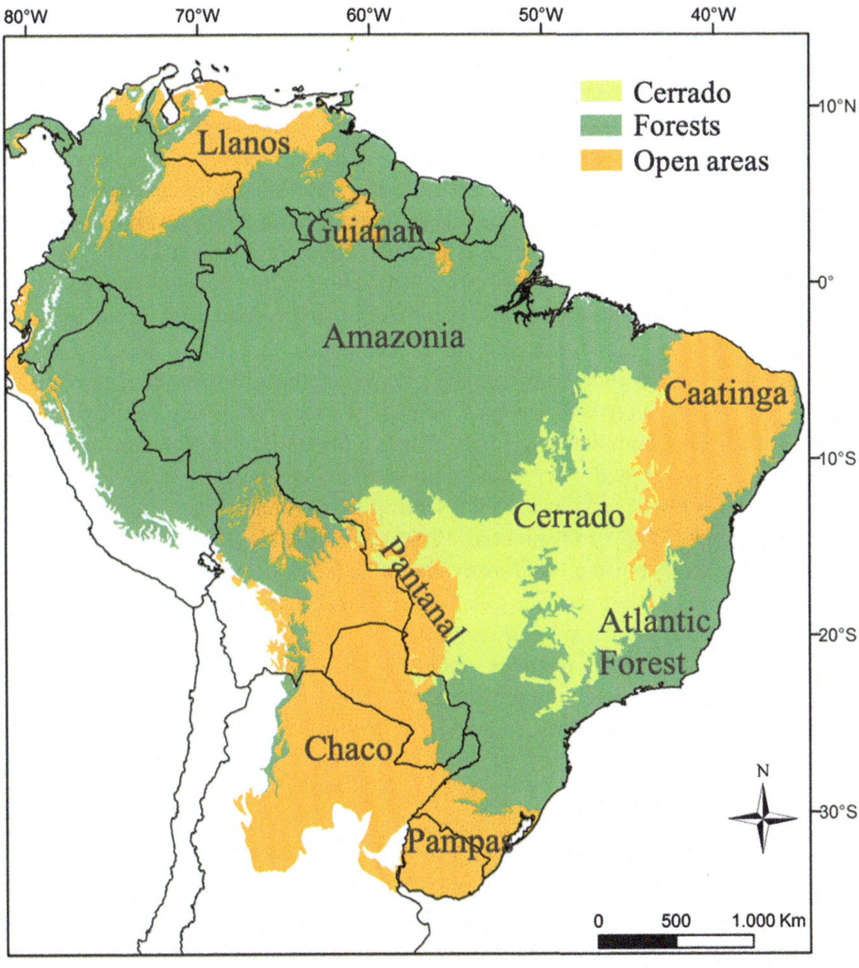

Fig. 12.1 Current distribution of the Cerrado domain and other open vegetation domains (open areas) discussed in this review in relation to broadly defined rain forests (Amazonia, Atlantic Forest). See text for details and definitions

However, this view has been challenged based on data from molecular phylogenies inferred for vascular plants, which indicate a more recent origin of the lineages that currently inhabit the Neotropical savannas. According to molecular results, the diversification of Neotropical savanna lineages started only in the late Miocene and early Pliocene, ca. 10–4 million of years ago, mya (Pennington et al. 2006a; Simon et al. 2009). In contrast, molecular evidence provides support for a much more ancient origin of Neotropical rainforest taxa, as exemplified by the radiation of major lineages of Malpighiales in the mid-Cretaceous, ca. 112–94 mya, and the origin of palms—characteristic of most current tropical rainforests—by ca. 100 mya (Couvreur et al. 2011). Although such lineages originated mainly in the Cretaceous, the establishment of rainforest ecosystems occurred only by the early Paleogene, as indicated by the fossil record (Davis et al. 2005; Wing et al. 2009). Consequently, South America may be considered a land of ancient forests co-occurring with relatively young savannas. However, the precise timing and factors that influenced the origin, assembly, and expansion of tropical savannas remain debated. For that, as exemplified by studies of the tropical forests, data should be sought from multiple sources (Antonelli et al. 2018b).

The term savanna has a wide and diverse use in the literature. Here we consider tropical savannas the ecosystems characterized by vegetation with continuous grassy ground cover, regardless of tree cover density, and with an ecological prominence of C_4 grasses as in the Cerrado *sensu lato* in central Brazil (the terms C_3 and C_4 refer to pathways that distinct groups of plants use to capture CO_2 for photosynthesis; the latter is often employed by plants found in warmer, drier, and more seasonal climates; Bourlière 1983; Sarmiento 1984; Lehmann et al. 2011). Globally, biomes in which tropical savannas predominate are also referred to as "savannas", with additional vegetation types occurring within the savanna biome (e.g., gallery forests; Olson et al. 2001; Moncrieff et al. 2016). Here we use the term "domain" to refer to each of the geographically separated core areas of the savanna biome (Ab'Sáber 2003). The savanna domains of the Neotropics are mainly coincident with the "terrestrial ecoregions" of Olson et al. (2001) and the "biogeographical provinces" of Morrone (2017), but the two former classifications treat parts of the Amazonian and the Atlantic forest domains as distinct ecoregions and provinces, which is less convenient for our discussion in this chapter.

In the Neotropics, there are three main domains of the savanna biome: the Llanos and the Guianan domains in northern South America, and the Cerrado domain in central South America—here we do not adtapt the classification of the Chaco and the Caatinga as tropical savannas, due to their more extreme climatic seasonality and their distinct vegetation structure (Olson et al. 2001; Ab'Sáber 2003; Fig. 12.1). There are also several smaller areas of savanna scattered across the Neotropics, such as within lowland Amazonia, in the Atlantic Forest, in Central America, and in the Caribbean islands (Pennington et al. 2006b). For simplicity, in this chapter we consider forests as any year-round vegetated ecosystem without a grassy ground cover, and with varying degrees of closed canopy (Lehmann et al. 2011). We distinguish between two main categories of forests: seasonally-dry tropical forest or SDTF (*sensu* Pennington et al. 2000) and tropical rain forests. We use these simplified definitions of forests and savannas in order to gather data across a wider geographical and temporal extent. Although in the dry season SDTFs have a similar physiognomy to savannas, a major factor distinguishing them is that savannas burn regularly and naturally, while SDTFs do not. This difference in fire regime is also reflected by their vegetation: fire-intolerant cacti, for instance, are particularly conspicuous in SDTFs (Banda et al. 2016). As these ecosystems and biomes are within the Neotropical region, they share approximately the same pool of lineages, making such simplifications less problematic than when comparing biomes across different continents (Corlett and Primack 2006; Moncrieff et al. 2016).

In terms of its geographical extent, its species richness, endemism, and floristic composition, the Cerrado domain can be considered the most cohesive of the Neotropical savannas (Silva and Bates 2002). It is mainly distributed across ancient well-drained plateaus (500–1500 m elevation), which are covered primarily by tropical savannas (Eiten 1972). It also extends across more recently dissected, adjacent peripheral depressions and valleys, where tropical savannas are less widespread. In some areas, they become replaced by SDTFs (Ratter et al. 1978). In contrast, the Llanos are not as geographically extensive and are located atop relatively recently formed geomorphological surfaces, which were subject to sporadic marine transgressions during the Neogene (Diaz de Gamero 1996; Jaramillo et al. 2017b). Additionally, the establishment of the climatic conditions associated with savannas in northern South America seems to be more recent. This recent formation is probably reflected in the relatively low species richness and endemism observed in the savanna of these northern domains (particularly for animals, but less so for plants) relative to the Cerrado savannas (Huber et al. 2006). Nevertheless, even young Neotropical systems, such as the high-elevation Páramos, may be extraordinarily diverse (Madriñán et al. 2013).

Most phylogenetic information regarding the ages of origination, assembly, and expansion of Neotropical savanna clades are derived from plant lineages of the Cerrado domain. However, phylogenetic information based on molecular data can at most only provide hypotheses for the ages of origination and diversification of extant lineages. Those dates may closely reflect the biomes they currently inhabit (e.g., Bytebier et al. 2011) but may also be of much more recent origin, and not necessarily reflect the ecological expansion of those biomes. The fossil record represents direct evidence, but, in the area currently occupied by the Cerrado domain, the number of available records is scarce; most of the information about the expansion of savannas or their fundamental components (e.g., C_4 grasses) relates to the northern and southern parts of South America. Given the pros and cons of each data source, to understand the evolution of tropical savannas in South America, it is necessary to integrate available data from all suitable sources and from across the whole continent.

Our goal in this chapter is to review and integrate different lines of evidence that can inform on the evolutionary history of Neotropical savannas in South America. Specifically, we combine information from dated molecular phylogenies, the fossil record, and paleoenvironmental reconstructions (Fig. 12.2) to better understand the origins, assembly and expansion of South American tropical savannas.

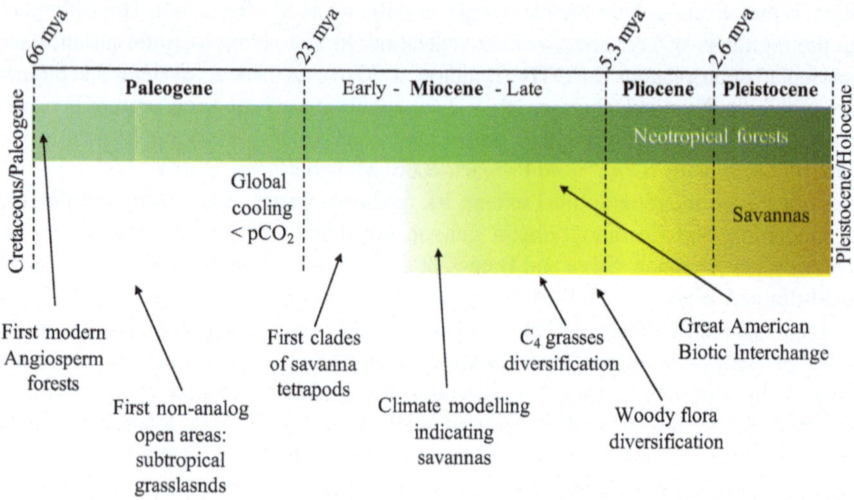

Fig. 12.2 Approximate timeline (not to scale) summarising the origin and expansion of South American tropical savannas. Arrows indicate earliest evidences. Rain forest ecosystems appeared in the early Paleocene (indicated by macrofossils and pollen). The first non-analog open biomes (shrublands) appeared in the middle Eocene (indicated by macrofossils, phytoliths, and pollen data). A decline in the concentration of CO_2 started already during the end of the Oligocene, preceding the Miocene global cooling. First divergences in clades of typical savanna vertebrates occurred during the middle Miocene (indicated by molecular phylogenies), the same time in which climatic modeling indicates the presence of appropriate conditions for the occurrence of savannas. C_4 grasses and the woody flora diversified from the late Miocene to the Pleistocene (indicated by molecular phylogenies). Savanna expansion occurred after the late Miocene (indicated by macrofossils, phytoliths, and pollen data). Finally, the Great American Biotic Interchange may have been increased by the expansion of savannas in South America (indicated by molecular phylogenies and fossil data)

2 Evolutionary Patterns of Extant Clades

Here we present an account of phylogenetic and biogeographical patterns of extant organisms, with a focus on the patterns of organisms in the savannas of the Cerrado domain, where such data are more complete. We do not aim at providing a comprehensive review, but rather highlight some of the earliest pieces of evidence of organisms typical of the Neotropical savannas available in the literature. Beyond the importance of phylogenetic and biogeographical information derived from plant clades, the presence of animals adapted to specific environmental conditions such as precipitation seasonality, fire regimes, and open vegetation structure can also give clues on the evolution of tropical savannas.

Plants Most phylogenetic evidence for the recent origin of tropical savannas is based on woody legumes (Fabaceae), which dominate most Neotropical biomes. In the Cerrado, the colonization of savannas by legume clades has been influenced by adaptations to fire and drought that emerged independently in several clades (Simon et al. 2009; Batalha et al. 2011; Simon and Pennington 2012). This suggests that habitat shifts may have been important in determining the high number of plant species and morphological diversity in savannas (Souza-Neto et al. 2016). For instance, species of the genera *Andira*, *Bauhinia*, *Lupinus*, *Stryphnodendron* and *Mimosa* have adapted to drought and fire, diversifying in savannas only recently from rainforest and SDTF lineages, ca. 10–4 mya (Simon et al. 2009, 2016; Souza-Neto et al. 2016).

An increasing amount of fossil and molecular data, particularly from woody legumes, supports the view that more ancient ecosystems surrounded the Neotropical savannas at the time of their assembly. For instance, woody legume clades that occur both in the Cerrado and in the Caatinga domains are generally older than in situ Cerrado radiations (e.g., *Cassia* clade, ca. 18–9 mya, Souza-Neto et al. 2016). The same is true for some clades of Neotropical SDTF legumes, such as *Coursetia*, *Poissonia*, and *Ruprechtia*, which are older than the savanna clades within these genera, ranging from ca. 20–8 mya (Pennington et al. 2004; de Queiroz and Lavin 2011). Estimations based on dated phylogenies are also congruent with the fossil record, which shows that SDTFs were already present around 13–12 mya (Burnham and Johnson 2004).

Other plant clades corroborate the results found within the Fabaceae, such as the Bignonieae lineages *Fridericia* and *Xylophragma*, in the Cerrado (Lohmann et al. 2012). Within *Pradosia* (Sapotaceae), savanna lineages evolved even more recently (less than ca. 1 mya), and the adaptation to savanna dry conditions and fire involved important morphological changes such as the enlargement of woody structures underneath the soil (geoxylic habits, Terra-Araujo et al. 2015). Moreover, colonization of the Cerrado and other seasonal habitats has involved adaptations to drought and fire-tolerance in palm species such as *Sabal palmetto* and *Serenoa repens* (McPherson and Williams 1998; Abrahamson 2007). Further, while the palm lineages *Allagoptera* and *Attalea* (Arecaceae) diversified ca. 10 mya, the savanna lineages of this clade diversified only after ca. 1 mya (Freitas et al. 2016; Bacon et al. 2017). In addition, some species of *Allagoptera* show progressive shifts towards increasingly dry habitats through time, with origins in tropical rain forests, followed by colonizations of SDTFs, the savannas of the Cerrado domain, and, finally, the sandy, coastal Restinga ecosystems (Bacon et al. 2017).

In summary, phylogenetic patterns of woody plants support the hypothesis of recent assembly of the savannas of the Cerrado domain, as proposed by Pennington et al. (2004) and Simon et al. (2009). However, it is important to mention that the major diversity in savanna ecosystems is composed of non-woody elements (more than 75% of all plant species; Filgueiras 2002). Therefore, we cannot reject the possibility that other plants in Neotropical savannas belong to lineages that are more ancient, or at least older than the woody elements of the Cerrado flora. For example, the typical open fields on rock outcrops in the higher elevation areas in the Cerrado domain ("*campos rupestres*") might include lineages potentially older and more diverse than the ones mentioned in this chapter (de Souza et al. 2013). However, we do not cover in detail the evolution of the *campos rupestres* here, as molecular data is still scarce, and in which evolution might have involved adaptations more specific to the poor soils and relatively colder climates where this physiognomy occurs (Neves et al. 2018).

Arthropods Most biogeographical studies of arthropods in South America examined diversification in the context of the set of eastern South American dry vegetation domains (Cerrado, Chaco and Caatinga), or the "South American Dry Diagonal" (Prado 1993; Werneck 2011). Most such studies rely on ancestral state reconstructions on molecular phylogenies, rather than the arthropod fossil record—which is scarce in the region. As such, they only provide indirect clues into the origins of the savannas in the region.

For example, a phylogenetic analysis indicates that turtle ants (Myrmicinae: *Cephalotes*) had a burst in speciation beginning ca. 12 mya, followed by a significant increase in dispersal out of the Cerrado and Chaco (Price et al. 2014). In addition, around 9 mya, the Cerrado and Chaco clades of *Zoniopoda* grasshoppers (Orthoptera: Romaleidae) diverged from each other (Pocco et al. 2018). Moreover, lineages adapted to wet environments of the eastern Andes and Atlantic Forest were apparently widespread until the appearance/expansion of Chaco and Cerrado vegetation. For instance, extant *Forsterinaria* satyr butterflies (Nymphalidae: Satyrinae) from the Andes and southeastern Brazil diverged at ca. 7–11 mya, and ancestral state reconstruction suggests that dispersal across dry domains has become less likely towards the present than before 11 mya (Matos-Maraví et al. 2013). Similarly, the divergence of southeastern and northwestern South American Cayenne ticks (Acari: Ixodidae: *Amblyomma cajennense*) at ca. 10 mya may have been driven by the onset of drier climatic regimes in the Cerrado domain (Beati et al. 2013).

On the other hand, some of the arthropod clades of the Cerrado are derived from ancestral clades of the surrounding forested domains, instead of the South American Dry Diagonal, as reported for butterflies and *Tropidopedia* bees (Brown and Gifford 2002; Aguiar and Melo 2007). Also, Cerrado spiders of the genus *Oligoxystre* (Theraphosidae) and *Araneus venatrix* (Araneidae) have their closest relatives in the Atlantic Forest, the former diverging from its sister clade in the late Miocene (Guadanucci 2011; Peres et al. 2017). In general, studies that focused on extant arthropods indicate a late Miocene origin of taxa associated to savannas, with open domains acting as a strong geographical barrier for taxa that occur in the surrounding rain forest domains (Morrone 2006, 2014; Ferrari et al. 2015). However, the absence of data on extinctions, together with several gaps in our understanding of the origins of important ecological and functional groups (e.g., termites), prevent more conclusive answers about the origins of the South American savanna arthropods.

Tetrapods Both mammals and birds show low levels of species endemism in the Cerrado (<10%; Macedo 2002; Marinho-Filho et al. 2002). In both groups, the oldest divergences of Cerrado lineages have been dated back to the late Miocene, with diversification inferred to the Pliocene and Pleistocene—thus temporally congruent with those reported for woody plants.

For birds, it has been hypothesized that the open ecosystems (both savannas and grasslands) of the Cerrado domain have been historically occupied by ancient endemic lineages, whereas younger endemic lineages have primarily occupied the forested areas (Silva 1997). The few published studies involving birds in the region indicate Pliocene (Passeriformes: *Neopelma*) or even Pleistocene (Passeriformes: *Paroaria capitate* and *P. gularis*) divergences between lineages from the Cerrado domain and lineages typical of Amazonia or the Atlantic Forest (Lopes and Gonzaga 2013; Capurucho et al. 2018). Divergences between lineages of the open ecosystems of the Cerrado domain and those of other open domains in South America (mostly grasslands) occurred from the Miocene to the Pleistocene (Chaves et al. 2015), suggesting a more complex assembly of birds in the Cerrado domain than proposed by Silva (1997).

For mammals, one of the most ancient divergences for a Cerrado endemic has been estimated for the monotypic rodent genus *Calassomys* (Muroidea), ca. 7 mya (Pardiñas et al. 2014). The divergence time between the rodent genus *Podoxymys* (Muroidea) of the tepuis and its closest relatives in the Cerrado has been used to suggest a connection between the central and northern savannas ca. 3 mya (Leite et al. 2015), although long-distance dispersal and extinctions should not be disregarded to explain the same pattern (Fine and Lohmann 2018). It has also been proposed that, still during the Pliocene, the Cerrado served as the center of diversification for the rodent genus *Calomys* (Muroidea; Almeida et al. 2007). For marsupials, a complex scenario has been inferred with groups colonizing the Cerrado domain from the Caatinga (e.g., Didelphidae: *Thylamys karimii*), between the Miocene and Pliocene whereas lineages derived from Amazonia colonized the Cerrado during the Pleistocene (e.g., Didelphidae: *Monodelphis domestica*; Pavan et al. 2016). In Neotropical primates, which are strongly associated with forests, there are cases of inferred diversification in savanna biomes during the Pleistocene, ca. 1–2 mya (Alfaro et al. 2015).

Amphibians and reptiles (collectively, the herpetofauna) are represented by more than 200 endemic species in the Cerrado domain, indicating their potential to unveil the history of South American savannas (Valdujo et al. 2012; Werneck 2011; Nogueira et al. 2011; Azevedo et al. 2016; Guedes et al. 2018). It can be argued that due to their relatively low mobility, herpetofaunal lineages preserve signals of the first evolutionary events in the savannas of the Cerrado domain. The oldest divergence times of herpetofaunal lineages studied so far in the Cerrado date back to the early and middle Miocene. Available studies suggest, for example, that the anole lizard *Norops meridionalis* diverged from its Amazonian sister clade ca. 21 mya, 15 mya lower bound of confidence interval (Guarnizo et al. 2016). Data indicate that the savanna clade of the lizard genus *Kentropyx* also diverged from its Amazonian sister lineage during the middle to late Miocene (Werneck et al. 2009), similar to some of the leptodactylid frogs of the genus *Adenomera* that colonized the Cerrado domain from Amazonia (Fouquet et al. 2014). The split of early diverging lineages in the region indicates a relatively early arrival of this group, before the origination and diversification of the woody plant species typical of today's savannas.

The topography of the region seems to have influenced the coincident patterns of geographical range restriction of herpetofaunal species in the highlands of central Brazil (Azevedo et al. 2016). This is also suggested by phylogenetic studies of the lizard lineages *Ameiva*, *Kentropyx*, *Phyllopezus*, *Polychrus acutirostris*, and

Gymnodactylus amaralli, as well as the frog genus *Rhinella* and snakes of the *Bothrops neuwedi* group (Werneck et al. 2009; Maciel et al. 2010; Thomé et al. 2010; Giugliano et al. 2013; Machado et al. 2013; Domingos et al. 2014; Fonseca et al. 2018). Although indicating relatively old initial divergences, geological and climatic events of the Pliocene and Pleistocene also appear to have affected the geographical distribution of lineages of the aforementioned clades, although we cannot disregard the influence of idiosyncratic colonization histories of individual clades (see discussion in Smith et al. 2014), nor the role of biotic interactions and other factors influencing the diversification of these taxa.

3 Inferences from the Fossil Record

While most information regarding extant groups of organisms of Neotropical savannas is derived from clades occurring in the Cerrado domain, the fossil record in this area is virtually absent for the proposed periods of origin and expansion of tropical savannas. For these reasons, we expand our focus to other regions of South America to find clues to the origin and expansion of environmental conditions and organisms associated with savannas.

Northern South America In this region, the plant fossil record indicates that most areas currently covered by savannas were occupied by rainforest during the Paleogene (Jaramillo and Cárdenas 2013). The Neogene witnessed a large expansion of several modern ecosystems, including savannas, which replaced areas formerly occupied by lowland forests. Several areas that nowadays are dominated by SDTFs and savannas were occupied by a different ecosystem during the early Neogene. For example, the upper Magdalena valley of Colombia was covered by humid forests ca. 13–11 mya (Kay et al. 1997), and northwestern Venezuela and northeastern Colombia were occupied by a humid forest up to at least the early Pliocene, ca. 3.5 mya (Hambalek 1993; Hambalek et al. 1994; Jaramillo et al. 2015; Carrillo et al. 2018). The extensive savannas of the Llanos of Colombia and Venezuela were occupied by a humid forest up to ca. 6 mya (Jaramillo et al. 2006, 2017b), which, during glacial times (last ca. 2.7 mya), developed extensive sand dunes free of vegetation (Morales 1979; Khobzi 1981; Carr et al. 2015; Tripaldi and Zarate 2016). However, the magnitude of local sand dune fields, and their effect on modern savanna distribution, are still poorly understood. Such northern South American sites indicate a recent expansion of dry climates and associated vegetation across a portion of the Neotropics.

Southern South America The functional morphology of endemic extinct groups of native ungulates (meridiungulates) was traditionally used to reconstruct the ecological shift of forest to grasslands during the Cenozoic. Specifically, the evolution of meridiungulates with high-crowned teeth, a trait that characterizes herbivorous ungulates in grassland ecosystems today, was long interpreted as an indication of the expansion of grasslands or savannas in the middle Eocene of Patagonia (e.g.,

Stebbins 1981; Jacobs et al. 1999). Analyses of middle and late Eocene fossil plant silica assemblages (phytoliths, see below), from fossilized soil horizons at Gran Barranca, Patagonia, Argentina, has supported this notion—pointing to the emergence of subtropical savannas already by ca. 40 mya (Mazzoni 1979; Zucol et al. 1999, 2007, 2010).

More recent and detailed studies have led to a very different interpretation of the paleosol biosilica record at Gran Barranca and nearby outcrops, namely that grasses were a relatively minor element in the landscape prior to the late Miocene (Strömberg et al. 2013, 2014; Selkin et al. 2015). This inference is consistent with the low relative abundance of grass phytoliths recorded in the outer shell of dung beetle ball trace fossils from Gran Barranca and other Patagonian sites (Strömberg and Stidham 2001; Sánchez et al. 2010; Strömberg 2011), as well as pollen data (Barreda and Palazzesi 2007; Palazzesi and Barreda 2012). However, these results contrast sharply with the previous interpretation of high abundance of grasses in the early-middle Eocene (e.g., Zucol et al. 2010). Dunn and co-authors (2015) further used epidermal phytolith shape to show that, despite being nearly grass-free, the vegetation in southern South America became increasingly open during the middle and late Eocene, culminating in non-analogue shrublands with abundant palms. This opening of landscapes may have been a consequence of the establishment of arid climates in southern South America by the middle Eocene, as inferred from stable isotopic, sedimentological, and paleosol climate proxy data (Bellosi 2010; Bellosi and Krause 2013; Kohn et al. 2015). Together, these lines of evidence are consistent with a hypothesis that the high-crowned cheek teeth of many meridiungulates may have evolved in response to eating plant tissues partially covered by dust or volcanic ash in the dry shrublands of Patagonia, rather than as an adaptation to feeding on grass (Dunn et al. 2015).

During the early Miocene, habitats became increasingly wet in Patagonia, a shift that seems to have preceded an expansion of more closed habitats by the middle Miocene, as shown by stable isotope data and biosilica records (Dunn et al. 2015; Kohn et al. 2015). A similar increase in forested areas in Patagonia can also be deducted from the fossil record of New World monkeys, which are commonly associated with closed canopies (Silvestro et al. 2018). Interestingly, palms decreased and grasses became more abundant, albeit still not dominant, during approximately the same time (Strömberg et al. 2013, 2014), lending support to the hypothesis that these open-habitat grasses were more mesic. The exact timing for an expansion of grass-dominated habitats in Patagonia is not known from phytoliths, but pollen data from southern South America indicate that typical grassland species became abundant only in the last ca. 10 myr (Barreda and Palazzesi 2007; Palazzesi and Barreda 2012). The spread of C_4 grasses appears to have occurred relatively soon thereafter (ca. 5 mya), based on stable carbon isotope ratios from mammal tooth enamel, paleosol carbonates, and leaf waxes (Latorre et al. 1997; Kleinert and Strecker 2001; Bywater-Reyes et al. 2010; Hynek et al. 2012; Rohrmann et al. 2016). A study using phytoliths, combined with stable isotopes, further showed that high-latitude C_3 environments existed alongside the lowland C_4-dominated vegetation (Cotton et al. 2014).

In summary, fossil evidence from southern South America suggest that the appearance of open vegetation predated that inferred for the northern parts of the continent by over 30 myr. Open, shrubland habitats that were not analogous to savannas appeared first in the south, as early as the middle Eocene, likely influencing the evolution of large, herbivorous mammals. Grass-dominated ecosystems, however, started to dominate the landscape only over the last 10 myr.

Savannas and the Great American Biotic Interchange After the opening of environments in both northern and southern South America in the late Miocene and early Pliocene, a large number of mammalian herbivores from North America arrived in South America in the Pliocene and Pleistocene following an expansion of savannas in Central America (Bacon et al. 2016). Prior to the Great American Biotic Interchange (GABI), the South American mid-Miocene large herbivore fauna was relatively disparate, at least at the ordinal level, consisting of giant sloths (Pilosa), armadillos/glyptodonts (Cingulata), a few large rodents, and three closely related clades of endemic south American ungulates [part of the Meridiungulata mentioned earlier (*sensu* McKenna and Bell 1997), i.e., Astrapotheria, Notoungulata, Litopterna; MacFadden 2006]. After the GABI, the record indicates the arrival of a diverse array of North American herbivores, such as ungulates (Artiodactyla and Perissodactyla) and relatives of elephants (Proboscidea) in South America. This resulted in an extremely diverse megaherbivore fauna in the late Pleistocene, which may have been the most diverse of any area of the world at that time in both species and phylogenetic diversity (Faurby and Svenning 2015). Because there is ample evidence from contemporary ecosystems that reductions in megafauna can sometimes lead to drastic increases in tree cover (Daskin et al. 2016), we postulate a coupling between the arrival of the North American mammals and an increase in the South American savannas, a hypothesis that could be potentially tested with the fossil record.

4 Climate Evolution and Savannas

The global distribution of savannas is largely constrained by annual precipitation and precipitation seasonality (Lehmann et al. 2011). Yet, extensive areas in the world in which tropical savannas currently occur are predicted to support forests instead (Bond 2005). Such areas are generally in more mesic environments and disturbances such as fires or herbivory help to keep the vegetation open (Bond and Midgley 2012). In less mesic environments, such as extensive areas of well-drained plateaus in central Brazil, precipitation seasonality can still be the main determinant of the predominance of savanna ecosystems, although soil composition is also an important predictor of the distribution of riparian forests, SDTFs and savannas (Ruggiero et al. 2002; Bueno et al. 2018).

Several factors interact to produce the precipitation seasonality in the area currently occupied by the Cerrado domain, including the geography of the whole continent and the latitudinal location, orientation and height of the Andean mountains—which according to global circulation models determine moisture transportation across the continent (Sepulchre et al. 2010). Seasonal and annual changes in the atmospheric circulation are in turn related to the thermal differences between the continental landmass and the surrounding oceans, characterized by a monsoonal system—the South American Convergence Zone (Gan et al. 2004; Liebmann and Mechoso 2011). This climatic system appeared first sometime during the Miocene, probably related to global scale changes in ocean circulation, tectonic movements, and variation in pCO_2 (Herbert et al. 2016). Due to its complexity, it is difficult to model more precisely when and where in South America a monsoonal system was first established, and how the climatic changes affected the origin and spread of savannas in central South America.

In contrast, in northern South America, a simple mechanism helps explains the origin of the climate seasonality determining the occurrence of savannas. In this region, precipitation is largely controlled by the amplitude and degree of migration of the Intertropical Convergence Zone (ITCZ). During the austral summer, the ITCZ migrates to the south and positions itself over southern Colombia, Ecuador, and the Amazon basin (Poveda et al. 2006), leaving large portions of northern South America under dry conditions. Precipitation increases over northern South America when the ITCZ migrates north during the boreal summer (Poveda et al. 2006). This shift of the ITCZ produces a long dry season over the region occupied by savannas and xerophytic forests in northern South America. We hence propose that the ITCZ has shifted at some point within the last 6 million years, yielding the modern climate configuration. Two events have been proposed to affect the ITCZ during the late Neogene. The first one relates to the onset of the thermohaline circulation ca. 10–12-mya, as a consequence of the closure of the Central American Seaway, which pushed the ITCZ southward to its modern position (Sepulchre et al. 2014; Bacon et al. 2015; Montes et al. 2015; Jaramillo et al. 2017a; Jaramillo 2018). The second relates to the onset of permanent extensive ice in the Northern Hemisphere ca. 2.7 mya, which would have pushed the ITZC south to its current position (Flohn 1981; Shackleton et al. 1984; Chiang and Bitz 2005), generating conditions for the expansion of the savannas in northern South America. However, no fossil record is yet available to document the period covering the late Pliocene to the early Pleistocene and hence test these alternatives.

Modeling Approaches Paleomodeling can provide insight on the origin and expansion of environmental conditions correlated to the occurrence of savannas. Beyond atmosphere-ocean circulation models, paleosols and paleotopography can also be modeled (Bragg et al. 2012; Dowsett et al. 2016), allowing the prediction of palaeovegetation (Henrot et al. 2017). The potential distribution of vegetation since the Last Glacial Maximum (LGM) inferred through such models suggests a surprising contraction of savannas and expansion of tropical forests (Costa et al. 2017), although former studies indicated precisely the contrary (Braconnot et al. 2007). Reconstruction of the forest/savanna cover for the Pliocene (mid-Piacenzian, ca. 3 mya) indicates the potential presence of savannas in the eastern part of current

Amazonia and central Brazil. During this time, forests are predicted to have occupied the southern part of the current Cerrado distribution, connecting the Atlantic Forest and Amazonia (as also predicted for the LGM; Dowsett et al. 2016). A reconstruction for the late Miocene biomes indicates the potential presence of savannas in wide areas of the eastern Amazon basin and the central and southern areas of the current Cerrado domain (Pound et al. 2011). Simulations for the predominance of trees, shrubs, or grasses (instead of modelling whole domains) for the late Miocene, predict that grasslands occurred in the eastern and northern Amazon basin but were not coincident with the current distribution of the Cerrado (Bradshaw et al. 2015). A model for the middle Miocene (17–15 mya) still indicates the potential occurrence of savannas, but with a predominance of forests for most areas of South America (Henrot et al. 2017). However, paleontological sites and data are rare in South America; in fact, middle Miocene fossil localities are almost all located in the temperate zone of the Northern Hemisphere (Henrot et al. 2017), preventing model validation. There is also great uncertainty in inferring modern vegetation analogues. For instance, Eocene-Oligocene (ca. 33.9–33.5 mya) biome classifications from fossilized plant communities do not point to the presence of any grass-dominated biomes in South America, but instead different types of forests and shrublands (Palazzesi and Barreda 2012; Strömberg et al. 2013; Pound and Salzmann 2017). In summary, despite the caveats and the difficulties involved in modeling biomes in deep geological time, the existing analyses suggest that environmental conditions favorable to savannas have existed since the middle Miocene, whereas environmental conditions suitable for forests and non-analog open ecosystems dominated the continent in earlier times.

5 Origin, Assembly, and Expansion of Neotropical Savannas

Origins The relatively recent emergence inferred from current evidence for Neotropical savannas provides a great opportunity to understand the origins, taxonomic assembly, and expansion of modern ecosystems and biomes.

Within the general framework of biome shifts and niche evolution (Donoghue and Edwards 2014), there are two main scenarios for the origin of savanna ecosystems. One envisions that multiple forest species gradually responded to a changing climate and independently developed adaptations to fire and seasonal drought. The competing alternative is that, once precipitation seasonality and a fire regime developed and created an open environment, species in surrounding environments and which were pre-adapted to those conditions succeeded in colonizing and diversifying the new environment. This latter model has been suggested for fire-prone habitats such as the South African *fynbos*, the *chaparral* in California and the *kwongan* in southwest Australia (see discussion in Bytebier et al. 2011).

In the Neotropics, information derived from molecular phylogenies and the biogeography of extant taxa appears to support the second scenario: early diversification of some vertebrate clades (e.g., herpetofauna) indicate the presence of open ecosystems in the early and middle Miocene, preceding the diversification of woody plants in the late Miocene and early Pliocene. In addition, the diversification of some lineages typical of the high elevation open grasslands in central South America (*campo rupestre*) pre-dates the diversification of the Cerrado woody-flora (Silveira et al. 2016), a pattern that may indicate that grassland ecosystems arose first in a region that remains under-sampled for fossil data.

Assembly Over evolutionary time, it has been suggested that lineages rarely cross the boundaries of major biomes across continents (Crisp et al. 2009). This does not seem to be the case of the savannas in Cerrado, since many congeners are found in Amazonia, the Atlantic Forest, and SDTFs. This indicates that habitat shifts have been an important process in the generation of savanna diversity (Souza-Neto et al. 2016). A recent study examining connectivity amongst all Neotropical biomes found that the Cerrado and Chaco have been colonized primarily by Amazonian species (Antonelli et al. 2018a). The telling cases revised earlier in this chapter suggest that the assembly of plants, arthropods and tetrapods typical of savannas in the Cerrado domain occurred mainly during the late Miocene to the Pleistocene, with lineages colonizing mainly from surrounding biomes, although, generally, very little is known on the degree of niche conservatism of such lineages.

Expansion of Savannas At a global scale, savanna expansion appears to differ in time among continents. Although the fossil record underlying this conclusion is still scarce (Edwards et al. 2010), most empirical data indicate that savannas had not yet expanded ca. 15 mya (Jacobs et al. 1999; Edwards et al. 2010; Strömberg 2011). What factors could have induced the expansion of Neotropical savannas in South America? Beyond changes in precipitation and seasonality (Lehmann et al. 2011), levels of CO_2 may also have played a critical role, as grasses cannot compete with trees under high levels of CO_2 or reduced water stress (e.g., Higgins and Scheiter 2012). Levels of CO_2 are thought to have been high (>500 parts per million, *ppm*) during most of the Paleogene (Royer et al. 2011), reaching relatively low levels by the mid-late Oligocene (ca. 34 mya). Although proxy data suggest that atmospheric CO_2 temporarily rose during the middle Miocene, to levels above 400 ppm (Kürschner et al. 2008), the issue is far from settled. It is believed that CO_2 levels decreased drastically to reach <200 ppm during glacial times in the beginning of the Pleistocene (Royer 2006, 2010; de Boer et al. 2010; Royer et al. 2011). During the glacial/interglacial cycles of the Pleistocene (starting at 2.6 mya), CO_2 oscillated in tandem with global temperature, ranging from ca. 280 ppm during interglacial periods to 180 ppm during glacial periods (Monnin et al. 2001; Siegenthaler et al. 2005; Lüthi et al. 2008; Tripati et al. 2009). Neogene global climate change leading to less precipitation and higher temperatures, coupled with lower atmosphere CO_2 concentrations favoring C_4 grasses has therefore been suggested to have driven vegetation changes in Neotropical savannas (Beerling and Osborne 2006; Osborne and Beerling 2006).

Although a coupling between CO_2 levels and the expansion of savannas is suggested, this is not coincident with the origins and diversification of C_4 grasses. Different groups of C_4 grasses have originated since at least ca. 30 mya, with most origins occurring since the early Miocene (ca. 20 mya; Spriggs et al. 2014). The diversification of C_4 grasses that dominate in South America (e.g., Andropogonae, *Paspalum*) occurred since the middle Miocene (ca. 15 mya; Spriggs et al. 2014). This predates, by a few million years, the commonly cited ages for when C_4 grasses become ecologically dominant on other continents (<10 mya). Given that CO_2 had reached relatively low levels already by the mid-late Oligocene, additional factors are also linked to the more recent rise and dominance of C_4 grasses (summarized in e.g., Edwards et al. 2010). Therefore, the causes of the late expansion of C_4-dominated ecosystems are still under debate. The fossil record revised in this chapter agrees with a late Miocene (Patagonia) to early Pliocene (north South America) increase in dominance and geographical extension of open ecosystems in South America.

6 Advancing Knowledge on the Origin and Evolution of Savannas

Phylogenies and the Age of Savannas Phylogenies of extant organisms can provide important information on the age of biomes, but the evidence should be corroborated with other data. Survival in savanna ecosystems likely requires drought adaptations for lineages originally from forested ecosystems and physiological and morphological adaptations are proposed to be linked to the transitions among biomes (Donoghue and Edwards 2014). However, a phylogeny cannot easily tell us if such adaptations originated after the biome transitions (Zanne et al. 2014). Therefore the lack of old lineages of plants or animals that are currently characteristic of the savannas in the Cerrado domain cannot be seen as definitive proof against an older origin of this ecosystem (Wang 1994; Wang et al. 1999). If dispersal between biomes is predominantly from forested to open biomes, as recent data for the Neotropics suggest (Antonelli et al. 2018a), we would not expect to observe more than a few old endemic lineages in the savannas of the Cerrado domain, especially if newly arising lineages regularly outcompete the existing ones. Distinct lineages of organisms can have an old and continuous history, but without continuous and reliable fossil data, all estimates of the ages of organisms in a certain area may be biased towards younger ages (Nagalingum et al. 2011; Matzke and Wright 2016). Clearly, phylogenetic estimation and the current distribution of taxa needs to be complemented by integrating information from the fossil record and about Earth's climatic conditions through time (Fritz et al. 2013), as reviewed in this chapter.

Modeling Savanna/Forest Transitions Tropical forests and savannas seem to represent alternative states in certain areas, and their transition may be associated with thresholds or tipping points in environmental variables, especially precipitation

(Archibald et al. 2011; Hirota et al. 2011; Lehmann et al. 2014). Savannas can potentially shift to SDTFs in drier areas with high soil fertility, or shift to semideciduous or evergreen forest in areas with higher water availability (Souza-Neto et al. 2016; Bueno et al. 2018). Therefore, modeling the past distribution of biomes without considering the different kinds of environmental thresholds from their different composing ecosystems can be imprecise. Modelling biomes or ecosystems can be also misleading due to the various assumptions made, such as the concordance between the distribution of linages and the distribution of a particular biome (Särkinen et al. 2011; Collevatti et al. 2013). Finally, although climatic conditions probably varied considerably in the past, recent evidence shows that stable biome states are possible for some combinations of climatic drivers, challenging the climatic determinism needed for biome reconstructions through time (Moncrieff et al. 2016). Despite these uncertainties, paleoclimatic reconstructions would benefit from the integration of specific knowledge on savanna thresholds and tipping points, as well as the validation and refinement of models based on fossil and phytolith data.

Filling Gaps in the Fossil Record

Fossil sites in northern South America could offer a poorly explored window to study the origination and evolution of tropical savannas, such as the fossil faunas of the Falcón basin in north-western Venezuela and in the Cocinetas basin in northern Colombia (Sánchez-Villagra et al. 2010; Jaramillo et al. 2015; Carrillo et al. 2018). The Pliocene-Pleistocene terrestrial mammal fauna of this region includes a high diversity of herbivores with a wide range of body sizes, which suggests there was enough vegetation cover to sustain a complex community of herbivores (Amson et al. 2016; Pérez et al. 2017; Carrillo et al. 2018). Quantifying the relationships between ecomorphological traits of the local fossil community (e.g., tooth structure, limb proportions) and the expected environmental properties (e.g., temperature, precipitation) may help characterize the replacement of forests by savannas in the region.

Fossil plant silica assemblages—phytoliths—have not been employed yet to study specifically the evolutionary history of the region currently occupied by the Cerrado biome. Such data could be a valuable addition to the scarce pollen and macrofossil information currently available. Studies described above using phytoliths from other parts of South America point to the potential of this record to uncover the non-analogue vegetation types and conditions that shaped the assembly of all grassland ecosystems on the continent. As amply demonstrated in recent work on modern grassy ecosystems (Hirota et al. 2011; Staver et al. 2011; Lehmann et al. 2014), such historical contingencies matter for predicting future vegetation responses to ongoing environmental change and therefore have direct conservation implications (Griffith et al. 2017).

7 Conclusions

Current knowledge indicates that the striking diversity found today in South American Neotropical savannas is most probably not the result of a long geological history or stability of this ecosystem, at least not relative to Neotropical rainforests. Environmental conditions for the occurrence of savannas have existed since the middle Miocene, and molecular phylogenies of some vertebrate groups indicate the presence of open ecosystems in central South America around then. This timing coincides broadly with the initial diversification of C_4 grasses inferred from phylogenies. However, the fossil data indicates a time lag for the expansion and ecological dominance of both C_3 and C_4 grasses a few million years later, in the late Miocene. The period of expansion and ecological dominance of C_4 grasses is coincident with the timing in which the woody flora of the savannas in the Cerrado started to diversify, as suggested by molecular phylogenies. This time lag between the origin and expansion of savannas in central South America could potentially suggest that lineages of trees that are adapted to the open fire-prone savannas diversified and became ecologically dominant after open habitats, presumably grasslands (e.g., "*campos rupestres*") and associated fauna, had emerged.

Further research on tropical savannas should focus on poorly known but potentially ecologically dominant or/and ancient taxa in the tropical savannas, which might include many herbs, small shrubs, termites, ants, and fossorial lizards (Filgueiras 2002; Constantino 2005; Costa et al. 2010; Colli et al. 2016). Unfortunately, areas in which the tropical savannas occur are also particularly favorable to agriculture and farming, undergoing an alarming rate of deforestation. Protecting those species-rich and fragile ecosystems will be crucial for our ability to answer many of the remaining questions about the origins and evolution of Neotropical savannas and their biotas.

Acknowledgments We thank Valentí Rull and Ana Carnaval for the invitation to contribute a chapter on this topic, an anonymous reviewer for the valuable comments on our manuscript, our research groups for valuable discussion, and S. Pineda-Muñoz for comments and discussion on ecometrics. Funding for this work was provided by the Swedish Research Council (B0569601), the Swedish Foundation for Strategic Research, the Biodiversity and Ecosystems in a Changing Climate (BECC) programme, and a Wallenberg Academy Fellowship to A.A.; continuous productivity grants from CNPq to R.G.C.; the Smithsonian Tropical Research Institute, the Anders Foundation, 1923 Fund and Gregory D. and Jennifer Walston Johnson to C.J.; FAPESP scholarship 2014/18837-7 and the State Univ. of Maranhão for the Senior Research fellowship to T.B.G.; the Swiss National Science Foundation fund P2ZHP3_174749 to J.D.C.; NSF EAR-1253713 and EAR-1349749 to C.A.E.S.; the Marie Skłodowska-Curie fellowship (project MARIPOSAS-704035) and the PPLZ programme of the Czech Academy of Sciences (grant L200961951) to P. M.M; Swedish Research Council (2017-04980) to C.D.B; Swedish Research Council (2017-03862) to S.F.; and the Coordenação de Aperfeiçoamento de Pessoal de Nível Superior (CAPES—99999.001292/2015-03) to J.A.R.A.

References

Ab'Sáber A (2003) Os domínios de natureza no Brasil: potencialidades paisagísticas. Ateliê Edi, São Paulo

Abrahamson WG (2007) Leaf traits and leaf life spans of two xeric-adapted palmettos. Am J Bot 94:1297–1308

Aguiar AJC, Melo GAR (2007) Taxonomic revision, phylogenetic analysis, and biogeography of the bee genus *Tropidopedia* (Hymenoptera, Apidae, Tapinotaspidini). Zool J Linnean Soc 151:511–554

Alfaro JWL, Cortés-Ortiz L, Di Fiore A, Boubli JP (2015) Comparative biogeography of Neotropical primates. Mol Phylogenet Evol 82:518–529

Almeida FC, Bonvicino CR, Cordeiro-Estrela P (2007) Phylogeny and temporal diversification of *Calomys* (Rodentia, Sigmodontinae): implications for the biogeography of an endemic genus of the open/dry biomes of South America. Mol Phylogenet Evol 42:449–466

Amson E, Carrillo JD, Jaramillo C (2016) Neogene sloth assemblages (Mammalia, Pilosa) of the Cocinetas basin (La Guajira, Colombia): implications for the Great American biotic interchange. Palaeontology 59:563–582

Antonelli A, Zizka A, Carvalho FA, Scharn R, Bacon CD, Silvestro D, Condamine FL (2018a) Amazonia is the primary source of Neotropical biodiversity. Proc Natl Acad Sci U S A 115 (23):6034–6039

Antonelli A, Ariza M, Albert J, Andermann T, Azevedo J, Bacon CD, Faurby S, Guedes T, Hoorn C, Lohmann LG, Matos-Maraví P, Ritter CD, Sanmartin I, Silvestro D, Tejedor M, ter Steege H, Tuomisto H, Werneck FP, Zizka A, Edwards SV (2018b) Conceptual and empirical advances in Neotropical biodiversity research. PeerJ 6:e5644

Archibald S, Levin S, Archibald SA, Hoffmann WA, Bond WJ, Hanan N, Ratnam J, Hawthorne W, Orgle T, Roy D, Wilgen B, Van Scholes R, Duff G, Bowman D, Cook G, Batalha MA, Pivello VR, Meirelles ST, Caylor KK, Kolle O, Lloyd J, Stocks B, Levine J, Jeugd HVD, Stock WD, Rensburg SJV, Waldram MS, Bond W, Stock W, Fairbanks D (2011) Global resilience of tropical forest. Science 334:232–235

Azevedo JAR, Valdujo PH, Nogueira C (2016) Biogeography of anurans and squamates in the Cerrado hotspot: coincident endemism patterns in the richest and most impacted savanna on the globe. J Biogeogr 43:2454–2464

Bacon CD, Silvestro D, Jaramillo C, Smith BT, Chakrabarty P, Antonelli A (2015) Biological evidence supports an early and complex emergence of the Isthmus of Panama. Proc Natl Acad Sci U S A 112:6110–6115

Bacon CD, Molnar P, Antonelli A, Crawford AJ, Montes C, Vallejo-Pareja MC (2016) Quaternary glaciation and the Great American biotic interchange. Geology 44:375–378

Bacon CD, Moraes RM, Jaramillo C, Antonelli A (2017) Endemic palm species shed light on habitat shifts and the assembly of the Cerrado and Restinga floras. Mol Phylogenet Evol 110:127–133

Banda K, Delgado-Salinas A, Dexter KG, Linares-Palomino R, Oliveira-Filho A, Prado D, Pullan M, Quintana C, Riina R, Rodriguez M GM, Weintritt J, Acevedo-Rodriguez P, Adarve J, Alvarez E, Aranguren BA, Arteaga JC, Aymard G, Castano A, Ceballos-Mago N et al (2016) Plant diversity patterns in neotropical dry forests and their conservation implications. Science 353:1383–1387

Barreda V, Palazzesi L (2007) Patagonian vegetation turnovers during the Paleogene-early Neogene: origin of arid-adapted floras. Bot Rev 73:31–50

Batalha MA, Silva IA, Cianciaruso MV, De Carvalho GH (2011) Trait diversity on the phylogeny of cerrado woody species. Oikos 120:1741–1751

Beati L, Nava S, Burkman EJ, Barros-Battesti DM, Labruna MB, Guglielmone AA, Cáceres AG, Guzmán-Cornejo CM, León R, Durden LA, Faccini JLH (2013) *Amblyomma cajennense* (Fabricius, 1787) (Acari: Ixodidae), the Cayenne tick: phylogeography and evidence for allopatric speciation. BMC Evol Biol 13:267

Beerling DJ, Osborne CP (2006) The origin of the savanna biome. Glob Chang Biol 12:2023–2031

Bellosi E (2010) Loessic and fluvial sedimentation in Sarmiento formation pyroclastics, middle Cenozoic of central Patagonia. In: Madden RH, Carlini AA, Vucetich MG, Kay RF (eds) The paleontology of Gran Barranca. Cambridge University Press, Cambridge, pp 278–292

Bellosi ES, Krause JM (2013) Onset of the Middle Eocene global cooling and expansion of open-vegetation habitats in central Patagonia. Andean Geol 1:29–48

Boer B, van de Wal RSW, Bintanja R, Lourens LJ, Tuenter E (2010) Cenozoic global ice-volume and temperature simulations with 1-D ice-sheet models forced by benthic δ18O records. Ann Glaciol 51:23–33

Bond WJ (2005) Large parts of the world are brown or black: a different view on the "Green World" hypothesis. J Veg Sci 16:261–266

Bond WJ, Midgley GF (2012) Carbon dioxide and the uneasy interactions of trees and savannah grasses. Philos Trans R Soc Lond B Biol Sci 367:601–612

Bourlière F (1983) Ecosystems of the world. In: Goodall DW (ed) Tropical savannas, vol 13. Elsevier, Amsterdam, pp 1–730

Braconnot P, Otto-Bliesner B, Harrison S, Joussaume S, Peterchmitt JY, Abe-Ouchi A, Crucifix M, Driesschaert E, Fichefet T, Hewitt CD (2007) Results of PMIP2 coupled simulations of the mid-Holocene and last glacial maximum—part 1: experiments and large-scale features. Clim Past 3:261–277

Bradshaw CD, Lunt DJ, Flecker R, Davies-Barnard T (2015) Disentangling the roles of late Miocene palaeogeography and vegetation—implications for climate sensitivity. Palaeogeogr Palaeoclimatol Palaeoecol 417:17–34

Bragg FJ, Lunt DJ, Haywood AM (2012) Mid-Pliocene climate modelled using the UK Hadley Centre Model: PlioMIP experiments 1 and 2. Geosci Model Dev 5:1109–1125

Brown KS, Gifford DC (2002) Lepidoptera in the Cerrado landscape and the conservation of vegetation, soil, and topographical mosaics. In: Oliveira PS, Marquis RJ (eds) The cerrados of Brazil: ecology and natural history of a neotropical savana. Columbia University Press, New York, pp 201–222

Bueno ML, Dexter KG, Pennington RT, Pontara V, Neves DM, Ratter JA, de Oliveira-Filho AT (2018) The environmental triangle of the Cerrado domain: ecological factors driving shifts in tree species composition between forests and savannas. J Ecol 106:2109–2120

Burnham RJ, Johnson KR (2004) South American palaeobotany and the origins of neotropical rainforests. Philos Trans R Soc Lond B Biol Sci 359:1595–1610

Bytebier B, Antonelli A, Bellstedt DU, Linder HP (2011) Estimating the age of fire in the Cape flora of South Africa from an orchid phylogeny. Proc R Soc B Biol Sci 278:188–195

Bywater-Reyes S, Carrapa B, Clementz M, Schoenbohm L (2010) Effect of late Cenozoic aridification on sedimentation in the Eastern Cordillera of northwest Argentina (Angastaco basin). Geology 38:235–238

Capurucho JMG, Ashley MV, Ribas CC, Bates JM (2018) Connecting Amazonian, Cerrado, and Atlantic Forest histories: Paraphyly, old divergences, and modern population dynamics in tyrant-manakins (*Neopelma/Tyranneutes*, Aves: Pipridae). Mol Phylogenet Evol 127:696–705

Carr AS, Armitage SJ, Berrio JC, Bilbao BA, Boom A (2015) An optical luminescence chronology for late Pleistocene aeolian activity in the Colombian and Venezuelan Llanos. Quat Res 85:299–312

Carrillo JD, Amson E, Jaramillo C, Sánchez R, Quiroz L, Cuartas C, Rincón AF, Sánchez-Villagra MR (2018) The Neogene record of northern South American native ungulates. Smithson Contrib Paleobiol 101:1–67

Chaves AV, Freitas GHS, Vasconcelos MF, Santos FR (2015) Biogeographic patterns, origin and speciation of the endemic birds from eastern Brazilian mountaintops: a review. Syst Biodivers 13:1–16

Chiang JCH, Bitz CM (2005) Influence of high latitude ice cover on the marine intertropical convergence zone. Clim Dyn 25:477–496

Cole MM (1986) The savannas: biogeography and geobotany. Academic, London, pp 1–438

Collevatti RG, Terribile LC, de Oliveira G, Lima-Ribeiro MS, Nabout JC, Rangel TF, Diniz-Filho JAF (2013) Drawbacks to palaeodistribution modelling: the case of South American seasonally dry forests. J Biogeogr 40:345–358
Colli GR, Fenker J, Tedeschi LG, Barreto-Lima AF, Mott T, Ribeiro SLB (2016) In the depths of obscurity: knowledge gaps and extinction risk of Brazilian worm lizards (Squamata, Amphisbaenidae). Biol Conserv 204:51–62
Constantino R (2005) Padrões de diversidade e endemismo de térmitas no bioma Cerrado. In: Scariot A, Silva JCS, Felfili JM (eds) Cerrado Ecologia biodiversidade e Conservação. Ministério do Meio Ambient, Brasília, pp 319–333
Corlett RT, Primack RB (2006) Tropical rainforests and the need for cross-continental comparisons. Trends Ecol Evol 21:104–110
Costa GC, Nogueira CC, Machado RB, Colli GR (2010) Sampling bias and the use of ecological niche modeling in conservation planning: a field evaluation in a biodiversity hotspot. Biodivers Conserv 19:883–899
Costa GC, Hampe A, Ledru MP, Martinez PA, Mazzochini GG, Shepard DB, Werneck FP, Moritz C, Carnaval AC (2017) Biome stability in South America over the last 30 kyr: inferences from long-term vegetation dynamics and habitat modelling. Glob Ecol Biogeogr 27:285–297
Cotton JM, Hyland EG, Sheldon ND (2014) Multi-proxy evidence for tectonic control on the expansion of C_4 grasses in northwest Argentina. Earth Planet Sci Lett 395:41–50
Couvreur TLP, Forest F, Baker WJ (2011) Origin and global diversification patterns of tropical rain forests: inferences from a complete genus-level phylogeny of palms. BMC Biol 1:1–144
Crisp MD, Arroyo MT, Cook LG, Gandolfo MA, Jordan GJ, McGlone MS, Weston PH, Westoby M, Wilf P, Linder HP (2009) Phylogenetic biome conservatism on a global scale. Nature 458:754–756
Daskin JH, Stalmans M, Pringle RM (2016) Ecological legacies of civil war: 35-year increase in savanna tree cover following wholesale large-mammal declines. J Ecol 104:79–89
Davis CC, Webb CO, Wurdack KJ, Jaramillo CA, Donoghue MJ (2005) Explosive radiation of malpighiales supports a mid-cretaceous origin of modern tropical rain forests. Am Nat 165:E36–E65
de Souza ER, Lewis GP, Forest F, Schnadelbach AS, Berg CVD, de Queiroz LP (2013) Phylogeny of *Calliandra* (Leguminosae: Mimosoideae) based on nuclear and plastid molecular markers. Taxon 62:1200–1219
Domingos FMCB, Bosque RJ, Cassimiro J, Colli GR, Rodrigues MT, Santos MG, Beheregaray LB (2014) Out of the deep: cryptic speciation in a Neotropical gecko (Squamata, Phyllodactylidae) revealed by species delimitation methods. Mol Phylogenet Evol 80:113–124
Donoghue MJ, Edwards EJ (2014) Biome shifts and niche evolution in plants. Annu Rev Ecol Evol Syst 45:547–572
Dowsett H, Dolan A, Rowley D, Pound M, Salzmann U, Robinson M, Chandler M, Foley K, Haywood A (2016) The PRISM4 (mid-Piacenzian) palaeoenvironmental reconstruction. Clim Past Discuss 4:1–39
Dunn RE, Strömberg CAE, Madden RH, Kohn MJ, Carlini AA (2015) Linked canopy, climate and faunal evolution in the Cenozoic of Patagonia. Science 347:258–261
Edwards EJ, Osborne CP, Strömberg CAE, Smith SA, Bond WJ, Christin PA, Cousins AB, Duvall MR, Fox DL, Freckleton RP, Ghannoum O, Hartwell J, Huang Y, Janis CM, Keeley JE, Kellogg EA, Knapp AK, Leakey ADB, Nelson DM, Saarela JM, Sage RF, Sala OE, Salamin N, Still CJ, Tipple B (2010) The origins of C_4 grasslands: integrating evolutionary and ecosystem science. Science 328:587–591
Eiten G (1972) The cerrado vegetation of Brazil. Bot Rev 38:201–341
Faurby S, Svenning JC (2015) Historic and prehistoric human-driven extinctions have reshaped global mammal diversity patterns. Divers Distrib 21:1155–1166
Ferrari A, Barão KR, Bianchi FM, Campos LA, Grazia J (2015) Classification and biogeography of neotropical true bugs. In: Panizzi Antônio R, Jocélia G (eds) True bugs (Heteroptera) of the Neotropics. Springer, Dordrecht, pp 57–87

Filgueiras T (2002) Herbaceous plant communities. In: Oliveira PS, Marquis RJ (eds) The cerrados of Brazil: ecology and natural history of a neotropical savana. Columbia University Press, New York, pp 121–139

Fine PVA, Lohmann LG (2018) Importance of dispersal in the assembly of the Neotropical biota. Proc Natl Acad Sci U S A 115:201807012

Flohn H (1981) A hemispheric circulation asymmetry during late Tertiary. Geol Rundsch 70:725–736

Fonseca EM, Gehara M, Werneck FP, Lanna FM, Colli GR, Sites JW, Rodrigues MT, Garda AA (2018) Diversification with gene flow and niche divergence in a lizard species along the South American "diagonal of open formations". J Biogeogr 45:1688–1700

Fouquet A, Cassini SC, Haddad CFB, Pech N, Rodrigues MT (2014) Species delimitation, patterns of diversification and historical biogeography of the Neotropical frog genus Adenomera (Anura, Leptodactylidae). J Biogeogr 41:855–870

Freitas C, Meerow AW, Pintaud JC, Henderson A, Noblick L, Costa FRC, Barbosa CE, Barrington D (2016) Phylogenetic analysis of *Attalea* (Arecaceae): insights into the historical biogeography of a recently diversified Neotropical plant group. Bot J Linn Soc 182:287–302

Fritz SA, Schnitzler J, Eronen JT, Hof C, Böhning-Gaese K, Graham CH (2013) Diversity in time and space: wanted dead and alive. Trends Ecol Evol 28:509–516

Gamero MLD (1996) The changing course of the Orinoco river during the Neogene: a review. Palaeogeogr Palaeoclimatol Palaeoecol 123:385–402

Gan MA, Kousky VE, Ropelewski CF (2004) The South America monsoon circulation and its relationship to rainfall over west-central Brazil. J Clim 17:47–66

Giugliano LG, Nogueira C, Valdujo PH, Collevatti RG, Colli GR (2013) Cryptic diversity in South American Teiinae (Squamata, Teiidae) lizards. Zool Scr 42:473–487

Griffith DM, Lehmann CER, Strömberg CAE, Parr CL, Pennington RT, Sankaran M, Ratnam J, Still CJ, Powell RL, Hanan NP (2017) Comment on "The extent of forest in dryland biomes". Science 358:27–30

Guadanucci JPL (2011) Cladistic analysis and biogeography of the genus *Oligoxystre* Vellard 1924 (Araneae: Mygalomorphae: Theraphosidae). J Arachnol 39:320–326

Guarnizo CE, Werneck FP, Giugliano LG, Santos MG, Fenker J, Sousa L, D'Angiolella AB, dos Santos AR, Strüssmann C, Rodrigues MT, Dorado-Rodrigues TF, Gamble T, Colli GR (2016) Cryptic lineages and diversification of an endemic anole lizard (Squamata, Dactyloidae) of the Cerrado hotspot. Mol Phylogenet Evol 94:279–289

Guedes TB, Sawaya RJ, Zizka A, Laffan S, Faurby S, Pyron RA, Bernils RS, Jansen M, Passos P, Prudente ALC et al (2018) Patterns, biases and prospects in the distribution and diversity of Neotropical snakes. Glob Ecol Biogeogr 27:14–21

Hambalek N (1993) Palinoestratigrafia del Mioceno-Pliocene de la región de Urumaco, Falcon Noroccidental. Universidad Central de Venezuela, Caracas

Hambalek N, Rull V, Digiacomo E, Gamero MLD (1994) Evolución paleoecológica y paleoambiental de la secuencia del Neógeno en el surco de Urumaco. Estudio palinológico y litológico. Bol Soc Venez Geol 191:7–19

Henrot AJ, Utescher T, Erdei B, Dury M, Hamon N, Ramstein G, Krapp M, Herold N, Goldner A, Favre E, Munhoven G, François L (2017) Middle Miocene climate and vegetation models and their validation with proxy data. Palaeogeogr Palaeoclimatol Palaeoecol 467:95–119

Herbert TD, Lawrence KT, Tzanova A, Peterson LC, Caballero-Gill R, Kelly CS (2016) Late Miocene global cooling and the rise of modern ecosystems. Nat Geosci 9:843–847

Higgins SI, Scheiter S (2012) Atmospheric CO2 forces abrupt vegetation shifts locally, but not globally. Nature 488:209–212

Hirota M, Holmgren M, Van Nes EH, Scheffer M (2011) Global resilience of tropical forest and savanna to critical transitions. Science 334:232–235

Huber O, de Stefano RD, Aymard G, Riina R (2006) Flora and vegetation of the Venezuelan Llanos: a review. In: Ratter JA, Pennington RT (eds) Neotropical savannas and seasonally dry forests: plant diversity, biogeography, and conservation. CRC, Boca Raton, FL, pp 95–120

Hynek SA, Passey BH, Prado JL, Brown FH, Cerling TE, Quade J (2012) Small mammal carbon isotope ecology across the Miocene–Pliocene boundary, northwestern Argentina. Earth Planet Sci Lett 321:177–188
Jacobs BF, Kingston JD, Jacobs LL (1999) The origin of grass-dominated ecosystems. Ann Mo Bot Gard 86:590–643
Jaramillo C (2018) Evolution of the Isthmus of Panama: biological, paleoceanographic, and paleoclimatological implications. In: Hoorn C, Perrigo A, Antonelli A (eds) Mountains, climate and biodiversity. Wiley, Oxford, pp 323–338
Jaramillo C, Cárdenas A (2013) Global warming and neotropical rainforests: a historical perspective. Annu Rev Earth Planet Sci 41:741–766
Jaramillo C, Rueda M, Mora G (2006) Cenozoic plant diversity in the Neotropics. Science 311:1893–1896
Jaramillo C, Moreno F, Hendy AJW, Sánchez-Villagra MR, Marty D (2015) Preface: La Guajira, Colombia: a new window into the Cenozoic neotropical biodiversity and the Great American biotic interchange. Swiss J Palaeontol 134:1–4
Jaramillo C, Montes C, Cardona A, Silvestro D, Antonelli A, Bacon C (2017a) Comment (1) on "Formation of the Isthmus of Panama" by O'Dea et al. Sci Adv 3:e1602321
Jaramillo C, Romero I, D'Apolito C, Bayona G, Duarte E, Louwye S, Escobar J, Luque J, Carrillo-Briceno J, Zapata V, Mora A, Schouten S, Zavada M, Harrington G, Ortiz J, Wesselingh F (2017b) Miocene flooding events of Western Amazonia. Sci Adv 3:e1601693
Kay RF, Madden RH, Cifelli RL, Flynn JJ (1997) Vertebrate paleontology in the Neotropics. The Miocene Fauna of La Venta, Colombia. Smithsonian Institution Press, Washington, DC
Khobzi J (1981) Los campos de dunas del norte de Colombia y de los Llanos del Orinoco (Colombia y Venezuela). Rev CIAF 6:257–292
Kleinert K, Strecker MR (2001) Climate change in response to orographic barrier uplift: paleosol and stable isotope evidence from the late Neogene Santa Maria basin, northwestern Argentina. Geol Soc Am Bull 113:728–742
Kohn MJ, Strömberg CAE, Madden RH, Dunn RE, Evans S, Palacios A, Carlini AA (2015) Quasi-static Eocene-Oligocene climate in Patagonia promotes slow faunal evolution and mid-Cenozoic global cooling. Palaeogeogr Palaeoclimatol Palaeoecol 435:24–37
Kürschner WM, Kvaček Z, Dilcher DL (2008) The impact of Miocene atmospheric carbon dioxide fluctuations on climate and the evolution of terrestrial ecosystems. Proc Natl Acad Sci 105:449–453
Latorre C, Quade J, McIntosh WC (1997) The expansion of C_4 grasses and global change in the late Miocene: stable isotope evidence from the Americas. Earth Planet Sci Lett 146:83–96
Lehmann CE, Archibald SA, Hoffmann WA, Bond WJ (2011) Deciphering the distribution of the savanna biome. New Phytol 191:197–209
Lehmann CER, Anderson TM, Sankaran M, Higgins SI, Archibald S, Hoffmann WA, Hanan NP, Williams RJ, Fensham RJ, Felfili J, Hutley LB, Ratnam J, San Jose J, Montes R, Franklin D, Russell-Smith J, Ryan CM, Durigan G, Hiernaux P, Haidar R, Bowman D, Bond WJ (2014) Savanna vegetation-fire-climate relationships differ among continents. Science 343:548–552
Leite YLR, Kok PJR, Weksler M (2015) Evolutionary affinities of the "Lost World" mouse suggest a late Pliocene connection between the Guiana and Brazilian shields. J Biogeogr 42:706–715
Liebmann B, Mechoso CR (2011) The South American monsoon system. In: Chang CP, Ding Y, Lau NC, Johnson RH, Wang B, Yasunari T (eds) The global monsoon system: research and forecast. World Scientific Publishing, New Jersey, pp 137–157
Lohmann LG, Bell CD, Calió MF, Winkworth RC (2012) Pattern and timing of biogeographical history in the Neotropical tribe Bignonieae (Bignoniaceae). Bot J Linn Soc 171:154–170
Lopes LE, Gonzaga LP (2013) Taxonomy, natural history, and conservation of *Paroaria baeri* (Aves: Thraupidae). Trop Zool 26:87–103
Lüthi D, Le Floch M, Bereiter B, Blunier T, Barnola JM, Siegenthaler U, Raynaud D, Jouzel J, Fischer H, Kawamura K, Stocker TF (2008) High-resolution carbon dioxide concentration record 650,000–800,000 years before present. Nature 453:379–382

Macedo RHF (2002) The avifauna: ecology, biogeography, and behavior. In: Oliveira PS, Marquis RJ (eds) The cerrados of Brazil: ecology and natural history of a neotropical savana. Columbia University Press, New York, pp 242–265

MacFadden BJ (2006) Extinct mammalian biodiversity of the ancient New World tropics. Trends Ecol Evol 21:157–165

Machado T, Silva VX, Silva MJJ (2013) Phylogenetic relationships within *Bothrops neuwiedi* group (Serpentes, Squamata): geographically highly-structured lineages, evidence of introgressive hybridization and Neogene/Quaternary diversification. Mol Phylogenet Evol 71:1–14

Maciel NM, Collevatti RG, Colli GR, Schwartz EF (2010) Late Miocene diversification and phylogenetic relationships of the huge toads in the *Rhinella marina* (Linnaeus, 1758) species group (Anura: Bufonidae). Mol Phylogenet Evol 57:787–797

Madriñán S, Cortés AJ, Richardson JE (2013) Páramo is the world's fastest evolving and coolest biodiversity hotspot. Front Genet 4:1–7

Marinho-Filho J, Rodrigues FHG, Juarez KM (2002) The Cerrado mammals: diversity, ecology, and natural history. In: Oliveira PS, Marquis RJ (eds) The cerrados of Brazil: ecology and natural history of a neotropical savana. Columbia University Press, New York, pp 266–284

Matos-Maraví P, Peña C, Willmott KR, Freitas AVL, Wahlberg N (2013) Systematics and evolutionary history of butterflies in the "*Taygetis* clade" (Nymphalidae: Satyrinae: Euptychiina): towards a better understanding of Neotropical biogeography. Mol Phylogenet Evol 66:54–68

Matzke NJ, Wright A (2016) Inferring node dates from tip dates in fossil Canidae: the importance of tree priors. Biol Lett 12:1–4

Mazzoni MM (1979) Contribución al conocimiento petrográfico de la Formación Sarmiento, barranca sur del lago Colhue Huapi, provincia de Chubut. Rev Asoc Argent Mineral Petrol Sedimentol 10:33–54

McKenna MC, Bell SK (1997) Classification of mammals above the species level. Columbia University Press, New York

McPherson K, Williams K (1998) Fire resistance of cabbage palms (*Sabal palmetto*) in the southeastern USA. For Ecol Manag 109:197–207

Moncrieff GR, Bond WJ, Higgins SI (2016) Revising the biome concept for understanding and predicting global change impacts. J Biogeogr 43:863–873

Monnin E, Indermühle A, Dällenbach A, Flückiger J, Stauffer B, Stocker TF, Raynaud D, Barnola J-M (2001) Atmospheric CO_2 concentrations over the last glacial termination. Science 291:112–114

Montes C, Cardona A, Jaramillo C, Pardo A, Silva JC, Valencia V, Ayala C, Pérez-Angel LC, Rodriguez-Parra LA, Ramirez V, Niño H (2015) Middle Miocene closure of the Central American Seaway. Science 348:226–229

Morales PR (1979) Estudio de los médanos de los Llanos Centrales de Venezuela: evidencias de un clima desértico. Acta Biol Venez 10:19–49

Morrone JJ (2006) Biogeographic areas and transition zones of Latin America and the Caribbean islands based on panbiogeographic and cladistic analyses of the entomofauna. Annu Rev Entomol 51:467–494

Morrone JJ (2014) Biogeographical regionalisation of the Neotropical region. Zootaxa 3782:1–110

Morrone JJ (2017) Neotropical biogeography: regionalization and evolution. CRC/Taylor & Francis Group, Boca Raton, FL

Nagalingum NS, Marshall CR, Quental TB, Rai HS, Little DP, Mathews S (2011) Recent synchronous radiation of a living fossil. Science 334:796–799

Neves DM et al (2018) Lack of floristic identity in campos rupestres—a hyperdiverse mosaic of rocky montane savannas in South America. Flora 238:24–31

Nogueira C, Ribeiro S, Costa GC, Colli GR (2011) Vicariance and endemism in a Neotropical savanna hotspot: distribution patterns of Cerrado squamate reptiles. J Biogeogr 38:1907–1922

Oliveira-Filho AT, Ratter JA (2002) Vegetation physiognomies and wood flora of the Cerrado biome. In: The cerrados of Brazil: ecology and natural history of a Neotropical savanna. Columbia University Press, New York, pp 91–120

Olson DM, Dinerstein E, Wikramanayake ED, Burgess ND, Powell GVN, Underwood EC, D'Amico JA, Itoua I, Strand HE, Morrison JC et al (2001) Terrestrial ecoregions of the world: a new map of life on earth. A new global map of terrestrial ecoregions provides an innovative tool for conserving biodiversity. Bioscience 51:933–938

Osborne CP, Beerling DJ (2006) Nature's green revolution: the remarkable evolutionary rise of C4 plants. Philos Trans R Soc Lond B Biol Sci 361:173–194

Palazzesi L, Barreda V (2012) Fossil pollen records reveal a late rise of open-habitat ecosystems in Patagonia. Nat Commun 3:1294

Pardiñas UFJ, Lessa G, Teta P, Salazar-Bravo J, Câmara EMVC (2014) A new genus of sigmodontine rodent from eastern Brazil and the origin of the tribe Phyllotini. J Mammal 95:201–215

Pavan SE, Jansa SA, Voss RS (2016) Spatiotemporal diversification of a low-vagility Neotropical vertebrate clade (short-tailed opossums, Didelphidae: *Monodelphis*). J Biogeogr 43:1299–1309

Pennington RT, Prado DE, Pendry CA (2000) Neotropical seasonally dry forests and Quaternary vegetation changes. J Biogeogr 27:261–273

Pennington RT, Lavin M, Prado DE, Pendry CA, Pell SK, Butterworth CA (2004) Historical climate change and speciation: neotropical seasonally dry forest plants show patterns of both tertiary and quaternary diversification. Philos Trans R Soc Lond B Biol Sci 1443:515–537

Pennington RT, Richardson JE, Lavin M (2006a) Insights into the historical construction of species-rich biomes from dated plant phylogenies, neutral ecological theory and phylogenetic community structure. New Phytol 172:605–616

Pennington TR, Lewis GP, Ratter JA (2006b) Neotropical savannas and seasonally dry forests: plant diversity, biogeography, and conservation. CRC, London

Peres EA, Silva MJ, Solferini VN (2017) Phylogeography of the spider *Araneus venatrix* (Araneidae) suggests past connections between Amazon and Atlantic rainforests. Biol J Linn Soc 121:771–785

Pérez ME, Vallejo-Pareja MC, Carrillo JD, Jaramillo C (2017) A new Pliocene *Capybara* (Rodentia, Caviidae) from northern South America (Guajira, Colombia), and its implications for the Great American biotic interchange. J Mamm Evol 24:111–125

Pocco ME, Guzmán N, Plischuk S, Confalonieri V, Lange CE, Cigliano MM (2018) Diversification patterns of the grasshopper genus *Zoniopoda* Stål (Romaleidae, Acridoidea, Orthoptera) in open vegetation biomes of South America. Syst Entomol 43:290–307

Pound MJ, Salzmann U (2017) Heterogeneity in global vegetation and terrestrial climate change during the late Eocene to early Oligocene transition. Sci Rep 7:1–12

Pound MJ, Haywood AM, Salzmann U, Riding JB, Lunt DJ, Hunter SJ (2011) A Tortonian (Late Miocene, 11.61–7.25Ma) global vegetation reconstruction. Palaeogeogr Palaeoclimatol Palaeoecol 300:29–45

Poveda G, Waylen PR, Pulwarty RS (2006) Annual and inter-annual variability of the present climate in northern South America and southern Mesoamerica. Palaeogeogr Palaeoclimatol Palaeoecol 234:3–27

Prado DE (1993) What is the Gran Chaco vegetation in South America? Candollea 145:29

Price SL, Powell S, Kronauer DJC, Tran LAP, Pierce NE, Wayne RK (2014) Renewed diversification is associated with new ecological opportunity in the Neotropical turtle ants. J Evol Biol 27:242–258

Queiroz LP, Lavin M (2011) *Coursetia* (Leguminosae) from eastern Brazil: nuclear ribosomal and chloroplast DNA sequence analysis reveal the monophyly of three Caatinga-inhabiting species. Syst Bot 36:69–79

Ratter JA, Askew GP, Montgomery RF, Gifford DR (1978) Observations on forests of some mesotrophic soils in Central Brazil (Observações sobre florestas de alguns solos mesotroficos no Brasil Central). Rev Bras Bot 1:47–58

Ratter JA, Ribeiro JF, Bridgewater S (1997) The Brazilian cerrado vegetation and threats to its biodiversity. Ann Bot 80:223–230

Rohrmann A, Sachse D, Mulch A, Pingel H, Tofelde S, Alonso RN, Strecker MR (2016) Miocene orographic uplift forces rapid hydrological change in the southern central Andes. Sci Rep 6:35678

Royer DL (2006) CO_2-forced climate thresholds during the Phanerozoic. Geochim Cosmochim Acta 70:5665–5675

Royer DL (2010) Fossil soils constrain ancient climate sensitivity. Proc Natl Acad Sci U S A 107:517–518

Royer DL, Pagani M, Beerling DJ (2011) Geologic constraints on earth system sensitivity to CO_2 during the Cretaceous and early Paleogene. Earth Syst Dyn Discuss 2:1–30

Ruggiero PGC, Batalha MA, Pivello VR, Meirelles ST (2002) Soil-vegetation relationships in cerrado (Brazilian savanna) and semideciduous forest, Southeastern Brazil. Plant Ecol 160:1–16

Sánchez MV, Laza JH, Bellosi ES, Genise JF (2010) Ichnostratigraphy of middle Cenozoic Coprinisphaera from central Patagonia: insights into the evolution of dung beetles, herbivores and grass-dominated habitats. Palaeogeogr Palaeoclimatol Palaeoecol 297:633–648

Sánchez-Villagra MR, Aguilera OA, Carlini AA (2010) Urumaco and Venezuelan Paleontology. In: Orangel A et al (eds) The fossil record of the northern Neotropics The fossil record of the northern Neotropics. Indiana University Press, Bloomington, IN, pp 192–213

Särkinen T, Iganci JRV, Linares-Palomino R, Simon MF, Prado DE (2011) Forgotten forests—issues and prospects in biome mapping using seasonally dry tropical forests as a case study. BMC Ecol 11:1–15

Sarmiento G (1984) The ecology of neotropical savannas. Harvard University Press, Cambridge

Selkin PA, Strömberg CAE, Dunn RE, Kohn MJ, Carlini AA, Davies-Vollum KS, Madden RH (2015) Climate, dust, and fire across the Eocene-Oligocene transition, Patagonia. Geology G36664.1:1–4

Sepulchre P, Sloan LC, Fluteau F (2010) Modelling the response of Amazonian climate to the uplift of the Andean mountain range. In: Hoorn MC, Wesselingh FP (eds) Amazonia, landscape and species evolution: a look into the past. Wiley-Blackwell, Oxford, pp 211–222

Sepulchre P, Arsouze T, Donnadieu Y, Dutay JC, Jaramillo C, Le Bras J, Martin E, Montes C, Waite AJ (2014) Consequences of shoaling of the Central American Seaway determined from modeling Nd isotope. Paleoceanography 29:176–189

Shackleton NJ, Backman J, Zimmerman H, Kent DV, Hall MA, Roberts DG, Schnitker D, Baldauf JG, Desprairies A, Homrighausen R, Huddlestun P, Keene JB, Kaltenback AJ, Krumsiek KAO, Morton AC, Murray JW, Westberg-Smith J (1984) Oxygen isotope calibration of the onset of ice-rafting and history of glaciation in the North Atlantic region. Nature 307:620–623

Siegenthaler U, Stocker TF, Monnin E, Lüthi D, Schwander J, Stauffer B, Raynaud D, Barnola JM, Fischer H, Masson-Delmotte V, Jouzel J (2005) Stable carbon cycle-climate relationship during the late Pleistocene. Science 310:1313–1317

Silva JMC (1997) Endemic bird species and conservation in the Cerrado Region, South America. Biodivers Conserv 6:435–450

Da Silva JC, Bates J (2002) Biogeographic patterns and conservation in the South American Cerrado: a tropical savanna hotspot. Bioscience 52:225–233

Silveira FAO, Negreiros D, Barbosa NPU, Buisson E, Carmo FF, Carstensen DW, Conceição AA, Cornelissen TG, Echternacht L, Fernandes GW, Garcia QS, Guerra TJ, Jacobi CM, Lemos-Filho JP, Le Stradic S, Morellato LPC, Neves FS, Oliveira RS, Schaefer CE, Viana PL, Lambers H (2016) Ecology and evolution of plant diversity in the endangered campo rupestre: a neglected conservation priority. Plant Soil 403:129–152

Silvestro D, Tejedor MF, Serrano-Serrano ML, Loiseau O, Rossier V, Rolland J, Zizka A, Höhna S, Antonelli A, Salamin N (2018) Early arrival and climatically-linked geographic expansion of New World monkeys from tiny African ancestors. Syst Biol 68:78–92

Simon MF, Pennington T (2012) Evidence for adaptation to fire regimes in the tropical savannas of the Brazilian Cerrado. Int J Plant Sci 173:711–723

Simon MF, Grether R, Queiroz LP, Skema C, Pennington RT, Hughes CE (2009) Recent assembly of the Cerrado, a neotropical plant diversity hotspot, by in situ evolution of adaptations to fire. Proc Natl Acad Sci U S A 106:20359–20364

Simon MF, Pastore JFB, Souza AF, Borges LM, Scalon VR, Ribeiro PG, Santos-Silva J, Souza VC, Queiroz LP (2016) Molecular phylogeny of *Stryphnodendron* (Mimosoideae, Leguminosae) and generic delimitations in the *Piptadenia* group. Int J Plant Sci 177:41–59

Smith BT, McCormack JE, Cuervo AM, Hickerson MJ, Aleixo A, Cadena CD, Pérez-Emán J, Burney CW, Xie X, Harvey MG, Faircloth BC, Glenn TC, Derryberry EP, Prejean J, Fields S, Brumfield RT (2014) The drivers of tropical speciation. Nature 515:406–409

Souza-Neto AC, Cianciaruso MV, Collevatti RG (2016) Habitat shifts shaping the diversity of a biodiversity hotspot through time: insights from the phylogenetic structure of Caesalpinioideae in the Brazilian Cerrado. J Biogeogr 43:340–350

Spriggs EL, Christinb PA, Edwards EJ (2014) C_4 photosynthesis promoted species diversification during the Miocene grassland expansion. PLoS One 9:e97722

Staver AC, Archibald S, Levin SA (2011) The global extent and determinants of savanna and forest as alternative biome states. Science 334:230–232

Stebbins GL (1981) Coevolution of grasses and herbivores. Ann Mo Bot Gard 68:75–86

Strömberg CAE (2011) Evolution of grasses and grassland ecosystems. Annu Rev Earth Planet Sci 39:517–544

Strömberg CAE, Stidham TA (2001) Dung beetle brood balls and notoungulate diet. J Vertebr Paleontol 21:105A

Strömberg CAE, Dunn RE, Madden RH, Kohn MJ, Carlini AA (2013) Decoupling the spread of grasslands from the evolution of grazer-type herbivores in South America. Nat Commun 4:1478

Strömberg CAE, Dunn RE, Madden RH, Kohn MJ, Carlini AA (2014) Evolution of grazer morphologies in the absence of grasslands in southern South America. In: North American paleontological convention. The Paleontological Society Special Publications, Gainesville, FL, p 113

Terra-Araujo MH, de Faria AD, Vicentini A, Nylinder S, Swenson U (2015) Species tree phylogeny and biogeography of the Neotropical genus *Pradosia* (Sapotaceae, Chrysophylloideae). Mol Phylogenet Evol 87:1–13

Thomé MTC, Zamudio KR, Giovanelli JGR, Haddad CFB, Baldissera FA Jr, Alexandrino J (2010) Phylogeography of endemic toads and post-Pliocene persistence of the Brazilian Atlantic Forest. Mol Phylogenet Evol 55:1018–1031

Tripaldi A, Zarate M (2016) A review of Late Quaternary inland dune systems of South America east of the Andes. Quat Int 410:96–110

Tripati AK, Roberts CD, Eagle RA (2009) Coupling of CO2 and ice sheet stability over major climate transitions of the last 20 million years. Science 326:1394

Valdujo PH, Silvano DL, Colli G, Martins M (2012) Anuran species composition and distribution patterns in Brazilian Cerrado, a Neotropical hotspot. S Am J Herpetol 7:63–78

Wang X (1994) Phylogenetic systematics of the Hesperocyoninae (Carnivora, Canidae). Bull Am Mus Nat Hist 221:1–27

Wang X, Tedford RH, Taylor BE (1999) Phylogenetic systematics of the Borophaginae (Carnivora, Canidae). Bull Am Mus Nat Hist 243:1–391

Werneck FP (2011) The diversification of eastern South American open vegetation biomes: historical biogeography and perspectives. Quat Sci Rev 30:1630–1648

Werneck FP, Giugliano LG, Collevatti RG, Colli GR (2009) Phylogeny, biogeography and evolution of clutch size in South American lizards of the genus *Kentropyx* (Squamata: Teiidae). Mol Ecol 18:262–278

Wing SL, Herrera F, Jaramillo CA, Gomez-Navarro C, Wilf P, Labandeira CC (2009) Late Paleocene fossils from the Cerrejon formation, Colombia, are the earliest record of Neotropical rainforest. Proc Natl Acad Sci U S A 106:18627–18632

Zanne AE, Tank DC, Cornwell WK, Eastman JM, Smith SA, Fitzjohn RG, McGlinn DJ, O'Meara BC, Moles AT, Reich PB, Royer DL, Soltis DE, Stevens PF, Westoby M, Wright IJ, Aarssen L,

Bertin RI, Calaminus A, Govaerts R, Hemmings F, Leishman MR, Oleksyn J, Soltis PS, Swenson NG, Warman L, Beaulieu JM (2014) Three keys to the radiation of angiosperms into freezing environments. Nature 506:89–92

Zucol AF, Mazzoni MM, Madden RH (1999) Análisis fitolíticos en la secuencia sedimentaria de Gran Barranca, Chubut. In: Frenguelli J (ed) Primer Encuentro Argentino de Investigaciones Fitolíticas. Asociación de Ciencias Naturales del Litoral, Diamante, pp 11–12

Zucol AF, Brea M, Bellosi E, Carlini AA, Vucetich MG (2007) Preliminary phytolith analysis of Sarmiento formation in the Gran Barranca (central Patagonia, Argentina). In: Madella M, Zurro D (eds) Plants, people and places. Recent studies in phytolith analysis. Oxbow Books, Oxford, pp 189–195

Zucol AF, Brea M, Bellosi E (2010) Phytolith studies in Gran Barranca (central Patagonia, Argentina): the middle-late Eocene. In: Madden RH, Carlini AA, Vucetich MG, Kay RF (eds) The paleontology of Gran Barranca. Cambridge University Press, Cambridge, pp 317–340

Chapter 13
Diversity, Endemism, and Evolutionary History of Montane Biotas Outside the Andean Region

Thaís B. Guedes, Josué A. R. Azevedo, Christine D. Bacon, Diogo B. Provete, and Alexandre Antonelli

Abstract Mountain ranges are important centers of biodiversity around the world. This high diversity is the result of the presence of different soil types and underlying bedrock, a variety of micro-climatic regimes, high topographic heterogeneity, a heterogeneous and complex vegetation cline, and a dynamic geo-climatic history. Neotropical research on mountains has focused on the Andes, while other mountain ranges are lacking in biodiversity and biogeographic studies. However, the non-Andean mountains comprise important elements of the South American relief, are home to a substantial proportion of Neotropical species, and exhibit a complex and reticulate history of diversification of their biota. Here, we provide a brief review

T. B. Guedes (✉)
Center of Advanced Studies, State University of Maranhão, Caxias, Maranhão, Brazil

Department of Biological and Environmental Sciences, University of Gothenburg, Göteborg, Sweden

Gothenburg Global Biodiversity Centre, Göteborg, Sweden

J. A. R. Azevedo · C. D. Bacon
Department of Biological and Environmental Sciences, University of Gothenburg, Göteborg, Sweden

Gothenburg Global Biodiversity Centre, Göteborg, Sweden

D. B. Provete
Gothenburg Global Biodiversity Centre, Göteborg, Sweden

Institute of Biosciences, Federal University of Mato Grosso do Sul, Cidade Universitária, Campo Grande, Mato Grosso do Sul, Brazil
e-mail: diogo.provete@ufms.br

A. Antonelli
Department of Biological and Environmental Sciences, University of Gothenburg, Göteborg, Sweden

Gothenburg Global Biodiversity Centre, Göteborg, Sweden

Royal Botanic Gardens Kew, Richmond, UK
e-mail: a.antonelli@kew.org

V. Rull, A. C. Carnaval (eds.), *Neotropical Diversification: Patterns and Processes*, Fascinating Life Sciences, https://doi.org/10.1007/978-3-030-31167-4_13

of the biological and biogeographical importance of the major non-Andean South American mountain ranges, discussing their role for diversification and maintenance of Neotropical biodiversity. We focus on six regions: the Serra do Mar Range, the Mantiqueira Mountains, the Espinhaço Mountains, the Northeastern Highlands, the Central Brazilian Highlands, and the Pantepui region. We summarize the main geophysical and biotic characteristics of each mountain range, as well as key results from phylogenetic studies, the fossil record, and studies tackling biogeographical patterns of diversity, richness, and endemism. Moreover, mountain biodiversity studies can incorporate not only environmental data, but also information on more recent man-made landscape shifts. Here, we provide an example of how human population density interacts with climate and species traits to explain richness patterns in one group of montane organisms particularly vulnerable to environmental changes: anuran amphibians. Our results and the evidence published to date indicate that the Neogene and Quaternary were pivotal periods of Neotropical diversification for many terrestrial taxa, promoting endemism in non-Andean mountains. In general, all non-Andean mountain ranges have high levels of species richness and endemism as compared to their surrounding lowlands. Biotic interchange among them, the Andes, and their surrounding biotas has been intensive over tens of millions of years, greatly contributing to the outstanding levels of Neotropical biodiversity observed today. Despite their vast and understudied biodiversity, mountain ecosystems are fragile, facing severe challenges in the face of climate change, habitat loss, and extinctions. Efforts to better understand and protect South American mountain ecosystems are urgently needed.

Keywords Atlantic Forest · Caatinga · *Campos de altitude* · *Campos rupestres* · Cerrado · Diversification · Montane habitats · Tepuis

1 Introduction

There is a strong connection between mountains and biodiversity. Topographic variation, heterogeneity of soil types, and altitudinal gradients are important factors that, together with climatic variability, generate habitat diversity in mountains and increase species richness and endemism (Tuomisto et al. 2003; Körner 2004; Fischer et al. 2011; Fjeldså et al. 2012; Luebert and Muller 2015; Badgley et al. 2017; Antonelli et al. 2018a). Furthermore, the composition and spatial distribution of biodiversity in mountain ranges may also reflect environmental tolerances, habitat fragmentation, and distinct life histories of species (Marquet et al. 2004; Leibold et al. 2004; Ricklefs 2004; Schipper et al. 2008). With their often-isolated positions on continents, like islands in a surrounding ocean, mountain ranges are key to understanding evolutionary processes since they generate, receive, and maintain biodiversity (Hughes and Atchison 2015; Antonelli et al. 2009, 2018a)—thereby influencing spatial patterns of biological diversity (Homeier et al. 2010; Bonaccorso and Guayasamin 2013; Hoorn et al. 2013, 2018a; Guedes et al. 2014; Moura et al.

2016; Bacon et al. 2018a). Mountain ranges around the world are known to hold about one-third of all terrestrial species and are recognized as important centers of biological diversity (Körner et al. 2017). These areas are crucial for the maintenance of biodiversity and have been of interest since the days of early naturalists, including Alexander von Humboldt and Charles Darwin (Wulf 2016).

Owing to their remoteness and, sometimes, extremely difficult access, the biodiversity of mountains remains poorly explored. In the Neotropics, much attention has been paid to the patterns and processes of diversity of the rich Andean biota, the longest mountain range on Earth (e.g., Rundel et al. 1994; Kessler 2001; Cadena 2007; Castroviejo-Fisher et al. 2014; Luebert and Weigend 2014; Bacon et al. 2016, 2018a, b; Sanín et al. 2016; Hoorn et al. 2018b). However, besides the Andes, the Neotropics harbor several other important mountain ranges.

In the Neotropics, outside South America, there is pronounced relief (<3100 m) on several Caribbean islands such as Cuba, Hispaniola, and Jamaica. Mexican and Central American mountains and plateaus are also extensive and highly complex, including volcanos that reach above 4200 m of altitude in Guatemala, and several volcanos in Mexico that exceed 5000 m.

In South America, non-Andean mountain ranges are generally lower than 3000 m and include the Serra do Mar Range, the Mantiqueira Mountains, the Espinhaço Mountains, the Diamantina Plateau, the Central Brazilian Highlands (e.g., the Guimarães Plateau, the Serra Geral Plateau, the Caiaponia Mountain, the Veadeiros Plateau, the Central Brazilian Plateau, the Canastra Range, the Mesas Plateau region, the Parecis Plateau, and the Bodoquena Range), the Northeastern Highlands and the Pantepui Region (Fig. 13.1). These non-Andean mountains act as refugia and centers of endemism and diversification for Neotropical organisms (Rull 2005; Carnaval et al. 2009; Désamoré et al. 2010; Bonacorso and Guayasamin 2013; Chaves et al. 2014; Azevedo et al. 2016; Huber et al. 2018).

The Atlantic Forest is a hotspot of biodiversity (Myers et al. 2000) that comprises three non-Andean mountain ranges: the Serra do Mar Range, the Mantiqueira Mountains, and Espinhaço Mountains. The biodiversity of this complex rainforest system is one of the best documented in South America (e.g., Bello et al. 2017; Bovendorp et al. 2017; Culot et al. 2019; Santos et al. 2018; Vancine et al. 2018). However, most of its original vegetation has been degraded, and a large portion of it has been completely lost as a result of human disturbances (Ribeiro et al. 2009), leaving only 9%–16% of its original extent. The effects of degradation are seen at a broad range of spatial scales, including changes in microclimatic conditions (Didham and Lawton 1999), species abundances and community composition (Ewers and Didham 2005), geographic distributions of species (Ewers and Didham 2005), and effects on global climate change (Travis 2003). There is also evidence that deforestation has changed the distributional patterns of some species in the Atlantic Forest (Sancha et al. 2014).

The scarcity of species inventories for most biological groups inhabiting South American mountains, combined with an even more severe lack of phylogenetic and fossil information, are major obstacles to understanding the origin and maintenance of the huge biological diversity on mountains (Zizka and Antonelli 2018). In this

Fig. 13.1 The non-Andean Mountains of South America. Map of the Neotropical region showing the major non-Andean Mountains of continental South America reviewed here, according to the indication and naming of the Global Mountains Biodiversity Assessment (GMBA 2018). The map was produced on QGis 2.14.10 using a relief mask provided by Natural Earth Data (https://www.naturalearthdata.com)

chapter, we (1) provide an overview of non-Andean mountains in South America from an environmental and a biological perspective, and (2) summarize what is known about the mechanisms potentially underlying local diversification. To address these goals, we combine evidence from phylogenies, the fossil record, and biogeographical patterns. As an empirical example of how biodiversity data can be studied and interpreted in the light of anthropogenic pressures, we (3) select one specific group (amphibians), and one mountain range (the Serra do Mar, in the Brazilian Atlantic Forest), to analyze the influence of bioclimatic variables and human population density on species richness.

2 Non-Andean South American Mountains: What Are They?

Central and Eastern South America have several mountain ranges that often support relatively high levels of biodiversity. In this section, we focus on the most prominent non-Andean mountains (Fig. 13.1) based on a recent shapefile provided by The Global Mountain Biodiversity Assessment (GMBA 2018). Our aim is to discuss the origin and diversification of their biota, as well their patterns of species richness, endemism, and distribution. Below, we summarize the main geophysical and biotic characteristics of each mountain range.

Serra do Mar Range, or Serra do Mar. This is a continuous mountain range extending about 1500 km along the east coast of Brazil between the states of Rio de Janeiro in the north to northern Rio Grande do Sul in the south (Fig. 13.1). The Serra do Mar Range forms a narrow strip of cliffs and eroded escarpments on granite-gneiss bedrock (Gontijo-Pascutti et al. 2012). In the region that faces to the coast, these cliffs drop 1000–1300 m (all altitudes are provided as meters above sea level), while the interior (continental) face has small peaks where elevation reaches between 500 and 1100 m. The range's highest peaks are in the Serra dos Órgãos (e.g., 2366 m in Maior Peak, 2257 m in Caledônia Peak, and 2255 m in Pedra do Sino) in the state of Rio de Janeiro, and in the Serra da Bocaina (1550 m) in the state of São Paulo (Almeida 1964; Ab'Saber 1971; Gontijo-Pascutti et al. 2012). The Serra do Mar Range is mostly covered by the Atlantic Forest (with the exception of barren granite outcrops), with a complex set of physiognomies dominated by ombrophilous montane forest (Fig. 13.2a, b) (Veloso et al. 1991; Morellato and Haddad 2000; Medeiros et al. 2012), and patches of highland grasslands above 1000 m (Garey and Provete 2016). The region has outstanding levels of species richness and endemism and is therefore recognized as a global hotspot of biodiversity (Fig. 13.2i–l; Myers et al. 2000). Currently, less than 10% of the original area of the Atlantic Forest remains, of which most is concentrated in the Serra do Mar region (Galindo-Leal and Câmara 2003; Ribeiro et al. 2009).

Mantiqueira Mountains, or Serra da Mantiqueira. These mountains derive from the same tectonic events that formed the Serra do Mar Range, making their exact

Fig. 13.2 Landscapes and biodiversity of non-Andean South American Mountains. Examples of the diversity of habitats in mountains along with some characteristic taxa: (**a**) Serra do Mar Range, dominated by ombrophilous Atlantic Forest in the state of São Paulo, Brazil; (**b**) View from the top of Serra da Bocaina, one of the highest peaks of the Serra do Mar in the state of São Paulo, Brazil, reaching 1500 m above sea level; (**c**) View of the mountainous relief of the Itatiaia massif, in the Mantiqueira Mountains at the border between the states of Minas Gerais and Rio de Janeiro, Brazil; (**d**) The highland grassland *campos de altitude* of the Serra da Bocaina; (**e**) Forested areas in the Northeastern Highlands in Areia, Paraíba, Brazil; (**f**) View of the Pantepui Region, Guiana Highland, in French Guiana; (**g**) General view of the Diamantina Plateau showing a complex physiognomy including savanna (Cerrado), semi-arid (Caatinga), and forested vegetation; (**h**) Blue

delimitation difficult (Modenesi-Gauttieri et al. 2002; Gontijo-Pascutti et al. 2012). The Mantiqueira region stretches for about 900 km along the borders of three Brazilian states (Minas Gerais, Rio de Janeiro, and São Paulo; Figs. 13.1 and 13.2c). It is divided into two distinct geomorphological units, Campos do Jordão (in the states of São Paulo and Minas Gerais) and Itatiaia (reaching the states of Minas Gerais and Rio de Janeiro) massifs, formed by crystalline rocks at altitudes between 1700 and 2000 m (Ab'Saber and Bernardes 1958; Almeida 1964; Ab'Saber 1970; Gontijo-Pascutti et al. 2012). The highest portions of the Mantiqueira Mountains are the Agulhas Negras Peak (2792 m), Três Estados Peak (2665 m), Mina Peak (2798 m), and the isolated Bandeira Peak (inside Serra do Caparó, 2891 m). The region is also part of the Atlantic Forest bioregion and besides the typical rainforests of the region, it also includes highland grasslands or *campos de altitude*. The *campos de altitude* are a series of cool-humid, grass-dominated formations found exclusively on the uplifted blocks of igneous or high-grade metamorphic rocks above the treeline (up to 1000 m; Fig. 13.2d). The vegetation consists mainly of grasses and herbaceous daises and allies (Asteraceae) and Melastomataceae, with many endemic species (Safford 1999a, b; Almeida et al. 2004; Garey and Provete 2016; Silva et al. 2018).

Espinhaço Mountains, or Cadeia do Espinhaço. This is the second largest South American mountain range after the Andes, extending for about 1000 km and up to 75 km wide (Fig. 13.1). The Espinhaço range is formed by sets of smaller ranges, with many local names, and the altitude generally varies between 800 and 1000 m. Some higher elevations can be found along its extension, the highest ones being the Sol (2072 m) and Itambé Peaks (2002 m), both in the state of Minas Gerais. The Espinhaço Mountains separate two large river basins: the São Francisco to the west and the Doce to the east; and they also form a border between the Atlantic Forest and the open, savanna landscapes of the Cerrado and Caatinga bioregions. The Espinhaço Mountains are an important geographical and ecological barrier that shaped the distribution and diversification of many Neotropical organisms occurring across the region (Ab'Saber 1977; Alkmin 2012). The special soil conditions (acidic, nutrient-impoverished and low in phosphorus), climate, and relief harbor a unique physiognomy, the rocky grasslands (*campos rupestres*) being the most peculiar one. The *campos rupestres* are not a single homogeneous vegetation type, but instead are a mosaic of related communities composed mainly of grasses (Poaceae) and

Fig. 13.2 (continued) lily *Vellozia* sp., common in the Serra da Canastra after the fire season; (**i**) *Dendropsophus elegans*, an endemic frog species of the Atlantic Forest, shown here in the Serra do Mar Range; (**j**) The Black-cheeked Gnateater *Conopophaga melanops*, an endemic species of the Atlantic Forest photographed in the limits of the Serra do Mar Range, São Paulo, Brazil; (**k**) *Echinanthera amoena*, an endemic snake of the Atlantic Forest, whose distribution coincides with the limits of the Serra do Mar Range; (**l**) *Mazama guazoubira*, a species of mammal found on Serra do Mar Range, state of São Paulo, Brazil. Photo credits: **a**, **e**, **h**, **i**, **k**: TBG; **b**–**d**: DBP; **f**: AA; **g**: Daniela Coelho; **j**: Giulia B. D'Angelo; **l**: Marcela Nascimento

Asteraceae, also containing many narrowly distributed and threatened taxa, such as orchids growing directly on rocky outcrops (Antonelli et al. 2010). This landscape is interspersed with savanna vegetation patches, including denser shrubs and sparse treelets, which cover the higher slopes and mountaintops (between 700 and 2000 m of elevation), especially in the Espinhaço Mountains (Giulietti and Pirani 1988; Almeida et al. 2004; Chaves et al. 2014). The *campos rupestres*, also present over other non-Andean highlands, cover less than 1% of the Brazilian territory, yet they harbor a disproportionate 17% of the country's estimated plant diversity, and almost half of the diversity of the Cerrado (Fernandes et al. 2018).

Diamantina Plateau, or Chapada Diamantina. This is the northernmost portion of the Espinhaço Mountains. The Diamantina Plateau comprises an area of about 35,000 km^2 in the state of Bahia, northeastern Brazil (Fig. 13.1). The highest altitudes can be found in Almas, Itobira, and Barbados Peaks, reaching from 1958 to 2033 m, respectively (Alkmin 2012). The Diamantina Plateau is part of the Caatinga bioregion (Ab'Saber 1977; Morrone 2014), but it contains a diverse set of landscapes and physiognomies, including typical elements of the Caatinga, open grasslands typical of the Cerrado, *campos rupestres*, and semi-deciduous Atlantic Forest (Fig. 13.2g). The altitudinal variation along the relief and the complex mosaic of vegetation provide a large variety of habitats, supporting a high diversity of species and an endemic biota (Funch et al. 2005; Harley et al. 2005; Queiroz et al. 2005; Rocha et al. 2005; Conceição et al. 2005; Conceição and Pirani 2007; Echternacht et al. 2011; Fernandes and Hamdan 2014).

Central Brazilian Highlands, or Planalto Brasileiro. This is set of distinct plateaus or hills covered mostly by the Cerrado vegetation, including pre-Cambrian formations such as the Central Brazilian Plateau, Veadeiros Plateau and the Canastra Range as well as multiple other plateaus or hills of diverse geological origins, generally uplifted Phanerozoic sedimentary basins (e.g., Guimarães Plateau, Serra Geral Plateau, Mesas Plateau region, Parecis Plateau, Bodoquena Range) (Fig. 13.1; Bartorelli 2012; Petri and Sanches 2012; Carvalho Junior et al. 2015). The altitude across the Central Brazilian Highlands ranges from 400 to 1650 m. The Veadeiros Plateau is one of the highest areas (600–1600 m). The plateau's topographic heterogeneity determines many features of the Cerrado bioregion, such as soil composition, local climate, vegetation mosaics, and distribution of many groups of organisms simultaneously (Brown and Gifford 2002; Azevedo et al. 2016). The Central Brazilian Highlands is characterized by the most diverse physiognomies among the regions surveyed here, including several types of savannas (from open grasslands—*campinas, cerrado limpo*—to more densely forested areas or *campo sujo*), *campos rupestres* or rocky grasslands, and seasonal forests, all regulated by seasonality of precipitation and fire (Henriques 2005). The Cerrado harbors a relatively high species richness and endemism level as compared to other savannas around the world, which could be related to the presence of such diverse plateau systems, and is a global biodiversity hotspot (Myers et al. 2000; DRYFLOR 2016; Bacon et al. 2017; Guedes et al. 2018).

Northeastern Highlands, or Brejos Nordestinos. The region comprises about 43 mountains and plateaus between 600 and 1200 m altitude, situated in northeastern

Brazil in the states of Bahia, Ceará, Pernambuco, and Paraíba (Fig. 13.1; Ab'Sáber 2003; Tabarelli and Santos 2004). The vegetation on each of these mountains (Caatinga moist forest enclaves, according to WWF's ecoregion delimitation) is recognized as an archipelago of "exception landscapes" with higher humidity than the semi-arid surrounding Caatinga (Andrade-Lima 1966; Ab'Sáber 2003; Tabarelli and Santos 2004). In general, the vegetation in these Northeastern Highlands is complex and includes caatinga vegetation, elements of the Cerrado such as *campos rupestres*, *carrasco* (a kind of xerophilous vegetation in sandy soils present on some plateaus contiguous to the Caatinga bioregion; Andrade-Lima 1978; Araújo et al. 1999), and dense and humid relictual forested areas (Fig. 13.2e; Ab'Saber 1967, 1974).

The Pantepui Region, or Pantepui. This is the third major mountain system in South America by area (Figs. 13.1 and 13.2f). It is a biogeographic province composed of extensive table mountains—"tepuis"—derived from the sandstone of the Roraima Group, which in turn is part of the Guiana Shield that covers a large portion of northern South America (Fig. 13.1) (Rull 2004, 2005; Rull and Nogué 2007). The summits of the tepuis range from ~1500 to 3000 m, are sometimes flat and in those cases termed "table mountains" (Gibbs and Barron 1993; McDiarmid and Donnelly 2005; Rull 2005; Rull and Nogué 2007). The Pantepui Region has a varied and specialized diversity of flowering plants comprising about 2500 species (Huber 1988). Since the surface of the tepui summits represents only about 0.5% of the total area of the Pantepui Region, its density of endemic species is one of the highest on Earth, making the Pantepui one of the most important centers of species endemism in the Neotropics (Huber 1988, 1995; Berry and Riina 2005; Rull 2004, 2005; Rull and Nogué 2007; Désamoré et al. 2010).

3 Non-Andean South American Mountains: What Lives There?

Due to its large area and high topographic and habitat heterogeneity, the **Serra do Mar Range** is a center of endemism for many animal and plant taxa (Cardoso-da-Silva et al. 2004; Loyola et al. 2009; Villalobos et al. 2013). A recent spatial prioritization study (Loyola et al. 2009) identified the Serra do Mar Range as an important area for conservation of terrestrial vertebrates, including both endemic and threatened species. Villalobos et al. (2013) found similar results and pointed to the uniqueness of the central portion of the Atlantic Forest bioregion as having high richness of small-ranged anuran species. The concentration of small-ranged species in this region suggests that the complex topography of the region, coupled with the dispersal limitation of anurans, may have contributed to the speciation process. A biogeographic regionalization scheme based on taxonomic species composition (Vasconcelos et al. 2014) also found a similar result, further highlighting the distinctive anuran species composition between the southeastern and northern portion of the Atlantic Forest. Those authors also found that temperature, topographic variation, and precipitation seasonality best predicted the formation of the regionalization pattern. Overall, these results agree with the division of the Atlantic Forest bioregion into two distinct regions: one north and one south of the Doce River, as distinguished by many lineages of vertebrates and a major shift in climatic space through time (Carnaval et al. 2014).

A particular habitat in the montaintops of the Serra do Mar Range and **Mantiqueira Mountains** known to comprise several endemic taxa is the *campos de altitude* (Garey and Provete 2016; Silva et al. 2018). Open vegetation physiognomies of the Andes and *campos de altitude* of the Serra do Mar Range and Mantiqueira Mountains share about one third of plant genera (Safford 1999b, 2007). Among animals, an example of disjunct distribution is the sister relationship between the bufonid frog genera *Amazophrynella* and *Dendrophryniscus*, in which the former occurs in the Amazonia and Andes, whereas the latter occurs in the Atlantic rainforest, including the Serra do Mar Range; their split dates to the Eocene (Fouquet et al. 2012).

Several phylogeographic studies found evidence that birds, squamate reptiles (Cabanne et al. 2007; Batalha-Filho and Miyaki 2011), anuran amphibians (Thomé et al. 2014), and small mammals (Dantas et al. 2011) have only recently expanded their ranges towards the southern portion of the Atlantic Forest, as a result of the warmer climate compared to the Last Glacial Maximum. Many of these species appear to have diverged recently from their ancestors (Gaston 2003). The speciation mode of most anuran species is allopatric (Lynch 1989; Hua and Wiens 2010; Skeels and Cardillo 2019), which points to the role of mountains as vicariant barriers to lowland species, promoting endemism. Taken together, these results help explain the high number of small-ranged species in the mountainous areas of southeastern Brazil in general, and the Serra do Mar Range in particular.

The **Espinhaço Mountains** are geologically more stable than the Andes, with a Precambrian origin followed by comparatively fewer tectonic events in the Cenozoic (Saadi 1995). Similar to the Andes, the Espinhaço Mountains are an orographic barrier (Derby 1906), important in maintaining a minimum level of precipitation and humidity in the windward side of the range through geological time (Magalhães et al. 2015). In general, such geologic and climatic stability are related to high levels of endemism of both recent and relatively unique and ancient lineages (Loarie et al. 2009). Endemism levels are also high in different portions of the mountain range, which from a biogeographic viewpoint may be considered as a series of sky islands with various levels of differentiation among different lineages (Bonatelli et al. 2014; Ramos et al. 2018) or distinct species of the same genera occurring on different mountain tops (Echternacht et al. 2011). The Espinhaço Mountains harbor more than 4000 species of vascular plants, with endemism levels around 30% (Giulietti et al. 1997). Most of the endemic species of plants and animals across the region are associated with high altitude open fields on rock outcrops. These include one endemic monotypic genus of rodent (*Calassomys apicalis*), four bird species, and 10 *Mimosa* plant species (Silva 1997; Simon and Proença 2000; Almeida et al. 2007). The herpetofauna is narrowly distributed, with at least 18 species of frogs and five endemic reptiles only found in the southern portion of the Espinhaço (Azevedo et al. 2016). Although the *campos rupestres* harbor the majority of the endemics of the Espinhaço Mountains, some species are endemic to other environments, such as gallery forests on mountaintops, as is the case of some *Bokermannohyla* frogs and the lizard *Enyalius erythroceneus* (Napoli and Juncá 2006; Rodrigues et al. 2006; Leite et al. 2008).

The **Central Brazilian Highlands** have a higher diversity of plants than lower elevation areas within the Cerrado (Munhoz and Felfili 2006). The plant genus *Mimosa* (Fabaceae) alone contains more than 20 species restricted to the Veadeiros Plateau (Simon and Proença 2000). The same pattern of high richness and endemism is known for amphibians of the Veadeiros Plateau, with the occurrence of five endemic frogs from a total of 54 species, one of the highest levels of amphibian richness in South America (Santoro and Brandão 2014). Some other species are more widespread within the Central Brazilian Highlands, occurring also in other highlands to the south of the Veadeiros Plateau (500–1300 m). The Central Brazilian Plateau harbors several small-ranged species of reptiles, amphibians, plants, and rodents, but no endemic bird species (Silva 1997; Simon and Proença 2000; Marinho-Filho et al. 2002; Azevedo et al. 2016). However, comparisons of the whole Central Brazilian Plateau with other mountain ranges are difficult, given the many differences in area and climate, and the fact that the plateau has a more gentle and eroded relief and deeper and well drained soils than other mountain systems discussed here (Ab'Sáber 2003). The Central Brazilian Plateau does not serve as an orogenic barrier (Derby 1906; Bookhagen and Strecker 2008) as the Espinhaço Mountains do, the latter being more subject to the extreme climatic seasonality of central South America. To the west of the Cerrado domain, there are several small isolated plateaus and table mountains (e.g., Bodoquena, Guimarães, Huanchaca, and Parecis Ranges), generally below 900 m. Although lower in elevation and with relatively more gentle relief, these mountains and plateaus also harbor endemic species of reptiles, amphibians, and plants (Simon and Proença 2000; Azevedo et al. 2016), but most of them do not have any endemic mammal or bird species (with exception of the rodent *Jucelinomis huanchaca* from the Huanchaca Plateau; Emmons 1999).

The summits of the **Pantepui Region** harbor vegetation more similar to the Páramos and the *campos de altitude* of the Atlantic Forest than to the surrounding lowland Amazonian Forest. However, sometimes these table mountains can also be surrounded by tropical savanna (Huber 1995). The entire region harbors 2100 species of vascular plants, of which around 1300 are endemic (Huber et al. 2018). The difference in endemism levels for the Pantepui Region compared to other Brazilian ranges is remarkable, with 43 species of birds compared to only four for the (southern) Espinhaço (Borges et al. 2018), and nine in the **Diamantina Plateau** (Guedes et al. 2014). Regional endemism for the herpetofauna is even more extreme, reaching 87.6% of the amphibians and 74.2% of the reptiles (McDiarmid and Donnelly 2005). In contrast to plants and birds, there are fewer/lower affinities between the Pantepui Region and the Andean herpetofauna (Kok 2013). This is an interesting pattern that contrasts with most biogeographical hypotheses proposed based on the distribution of organisms with good dispersal abilities (many vascular plants and birds; Antonelli et al. 2009; Borges et al. 2018; Huber et al. 2018).

Campos rupestres from different mountains share more vascular plant species with the surrounding Cerrado savannas than to any other mountains or bioregions (Neves et al. 2018a, b), which contrasts with the situation for the tepuis and their distant biotic affinities. High mountains in the Atlantic Forest domain also harbor *campos de altitude* (Safford 1999a, b; Ribeiro et al. 2007). Besides being primarily related to central Brazilian *campos rupestres*, the vascular flora of the *campos de altitude* share a strong floristic similarity to the Páramos in the Andean mountains, and to southern temperate grasslands (DeForest Safford 2007). This can be exemplified by the species radiation of the fern genus *Jamesonia* (Pteridaceae) in the Páramos, which is the sister taxon of *Eriosorus myriophyllus*, a fern typical of the highland grasslands in the Atlantic Forest (Sánchez-Baracaldo 2004). The birds *Asthenes luizae* and *Cinclodes espinhacensis* from the *campos rupestres* of the Espinhaço Mountains also show biogeographical connections among southern Brazilian mountains and the Andes, probably connected by montane regions across the Patagonian-Chacoan region c. 8 million years ago (Ma) (Derryberry et al. 2011).

Biotic connections, or interchange, across montane habitats have taken place across different routes, including the Southern route (southern Brazilian mountains—Patagonia—southern Andes). The hummingbird *Colibri delphinae* has a disjunct distribution across the northern Espinhaço Mountains, Pantepui Region, northern Andes, Central America, and the island of Trinidad (Schuchmann 1999). Similarly, Chaves et al. (2014) provide an extensive list of widespread plant species across the Pantepui Region, and south-eastern and southern Brazil, including ferns, grasses, bromeliads, and plants of the families Cyperaceae, Eriocaulaceae, Velloziaceae, and Xyridaceae (Costa et al. 2008; Rapini et al. 2008; Salino and Almeida 2008; Versieux et al. 2008; Viana and Filgueiras 2008; Mello-Silva 2010). Finally, the rodent species *Podoxymys roraimae* also suggests an old biogeographical connection between the Cerrado of the Central Brazilian Plateau and the Pantepui Region during the Pliocene (Leite et al. 2015).

Alternatively, at least some disjunct distributions may not reflect bioregion connectivity, but rather represent sporadic dispersal events. Indeed, intensive dispersal among Neotropical mountain ranges and with surrounding lowlands may explain the reticulate evolutionary history of many Neotropical bioregions, as recently shown by a large cross-comparative analysis of dated phylogenies of plants and animals (Antonelli et al. 2018b; Fine and Lohmann 2018).

4 Evolutionary Origins and Species Diversification in Non-Andean Mountains: Insights from Phylogenetic Studies

One of the main goals of evolutionary biology is to understand the mechanisms that drive spatial variation in biodiversity (Quintero et al. 2015). One way to investigate such mechanisms is by using molecular phylogenies, which provide insights into the origin and diversification of extant lineages. When integrated with fossil information, phylogenies are particularly powerful tools for historical inference by enabling time calibration through node or tip constraints (Ronquist et al. 2012) or through diversification rates and biogeographic analyses (Silvestro et al. 2016). When lineages are endemic to an area—or are highly characteristic of a particular bioregion—their evolutionary history can be used to interpret the history of the region itself (Eiserhardt et al. 2017). Further, the inference of ancestral areas enables the investigation of how entire biotas and local communities are assembled through time (Bacon 2013; Antonelli et al. 2018b). The similarity of divergence times for diverse taxa can inform on the formation of a bioregion or geological feature (e.g., Bacon et al. 2015), although a conceptual and methodological reliance on concordance may be unnecessary (Papadopoulou and Knowles 2016; Zamudio et al. 2016).

Molecular data indicate that the diversification of many Neotropical terrestrial organisms took place during the Neogene, with lineages accruing their present diversity and distribution in the Quaternary (since 2.6 Ma; Rull 2011). For example, Machado et al. (2014) found a Neogene origin for snakes of *Bothrops neuwiedi* species group, with population differentiation likely driven by Quaternary climate change. Specifically, the species from the *campos de altitude* of the **Mantiqueira Mountains** diverged around 1.2–0.58 Ma, similar to the even younger ages of the **Espinhaço Mountains** (less than 1 Ma; Machado et al. 2014). Neogene origins were also found in the radiation of *Minaria* plants (Apocynaceae), comprising 21 species that are primarily endemic to the *campos rupestres* (Ribeiro et al. 2014). *Minaria* arose in the **Diamantina Plateau** during the Late Miocene and Early Pliocene in *campos rupestres*, and during the Pliocene the genus diversified across its entire range, particularly in the northern Diamantina Plateau of Bahia. These plant divergence times are older than those of the endemic birds in *Cinclodes* (Late Pleistocene; Freitas et al. 2012) and younger than those inferred for another characteristic *campos rupestres* plant group, *Hoffmannseggella* (syn. *Cattleya*) orchids (Orchidaceae; 14–11 Ma; Antonelli et al. 2010).

Historical biogeography uses phylogenies to infer the ancestral areas of lineages through time, which can in turn be used to suggest geological and/or climatic connectivity among regions. Yet, evidence about the origins and diversification along the non-Andean mountains seems far from settled, and more data are needed to understand temporal patterns. Sometimes, the reticulate and complex history of clades may not always reflect the geological history of the area. For instance, although the final upheaval of the **Central Brazilian Highlands** took place c. 4–2 Ma, Beerling and Osborne (2006) and Werneck et al. (2012) inferred that widespread lizard populations (*Phyllopezus pollicaris*) of the Caatinga and Cerrado diverged during the Miocene. In contrast, Carnaval and Bates (2007) found the divergence of two frog species from the **Northeastern Highlands** of Ceará to be much more recent (ca. 0.1 Ma). These mountain enclaves of the Northeastern Highlands are some of the least sampled among all non-Andean mountains, with few phylogenetic studies published to date. In contrast, the mountains of Central Brazil are fairly well known, and increased interest in the patterns and causes of high Cerrado diversity has shown divergence of endemic clades primarily occurring since 4 Ma (e.g., Almeida et al. 2007; Giugliano et al. 2007; Simon et al. 2009).

Climatic changes through time also have an important role in shaping the evolutionary trajectory of species, which may affect both geographic distribution patterns and species composition, as well as genetic variability across landscapes. The evolutionary history of the titi monkey genus *Callicebus* suggests a separation between the Central Brazilian Highlands and Northeastern Highlands lineages from those of the **Serra do Mar Range** in the Pleistocene (crown ages of ca. 2 and 6 Ma, respectively; Carneiro et al. 2018). Similarly, the manakin bird *Neopelma pallescens* originated in the Serra do Mar Range and dispersed to areas including Central Brazilian Highlands in the Late Miocene (Capurucho et al. 2018). In recent geological times, besides glacial cycling during the Quaternary the central Serra do Mar Range remained climatically more stable in comparison to southern parts of the range (Carnaval et al. 2009), leading to complex patterns of genetic diversity within frog species across their distribution. These types of climatic changes have also led to different responses among communities, creating different evolutionary trajectories for lizard species (Prates et al. 2016).

Usually, ancient geomorphological landscapes also have ancient environments occupied by ancient lineages. However, the **Pantepui mountain system** is one of the oldest in South America, but its biota contains both ancient and more recent lineages. Among the earliest diverging lineages are tepuian bromeliads (Bromeliaceae; Givnish et al. 2011), dating to c. 9.1 Ma. More recent endemic lineages include a species of opossum (*Podoxymys roraimae*; Pavan et al. 2016) and a toucanet (*Aulacorhynchus whitelianus*; Bonaccorso and Guayasamin 2013), which diverged in the Pleistocene. A review of the diversification of the Pantepui biota is provided by Huber et al. (2018).

5 Connecting Studies of Biodiversity and Human Impacts on Non-Andean Montane Biotas: A Case Study from the Serra Do Mar Range

Among the non-Andean montane regions, the Serra do Mar Range is one of the most heavily altered by humans, with a long history of deforestation dating back to colonial times between 1500 and 1815 (Dean 1997). This is one of the most densely populated areas in Brazil, comprising São Paulo and Rio de Janeiro, the two largest Brazilian cities (ca. 19 million inhabitants). Because evidence suggests that deforestation has changed the distributional patterns of small mammals in the Atlantic

Forest (Sancha et al. 2014), possibly driven by extinction, we expect that more densely occupied regions within this range will generally have lower species richness.

Anurans are especially diverse in the Serra do Mar Range (Villalobos et al. 2013). However, it is unknown whether the anthropogenic impacts detected in small mammals (Sancha et al. 2014) are reflected in these taxa, especially because the effect of deforestation in natural communities varies depending on life history and species traits. In general, species exhibiting narrow environmental requirements should be more heavily affected than ecological generalists.

To further understand whether and how human population density interacts with climate (reflecting climatic niche space; Wiens et al. 2006) and the evolutionary history of the local biota, and hence influences species richness patterns along the Serra do Mar Range, we deconstructed the total species richness into groups based on three life-history traits correlated with resource use: habitat use (terrestrial pond dwellers, burrowing, stream dwellers, and arboreal species), reproductive mode (species with free-swimming larvae vs. direct developers), and body size (50.99 mm $\leq$ SVL $\geq$ 51 mm, based on breaks of the snout-vent length data; Fig. 13.3), following Marquet et al. (2004). With this, we sought to test how important life-history traits are in determining species richness patterns at a broad spatial scale, given the human footprint. Direct-developing anuran species (i.e., without larval phase) require high levels of humidity (da Silva et al. 2012). Thus, we expect them to show a nested distributional pattern along a rainfall seasonality gradient. We expect the highest species richness of direct-developing species to be in areas with high temperature and precipitation and low human influence. Since streams are more common in regions with high altitude, we expect the richness of stream-dwellers to be higher in high altitude areas. Yet it is unknown if these areas have been preferentially altered by humans.

For our analysis, we built a grid with 57 cells of 0.5° covering the Serra do Mar Range in the SAM software (Rangel et al. 2010), using the limits of this ecoregion as provided by Olson et al. (2001). To calculate species richness for each cell, we overlaid polygons depicting the extent of occurrence ('range maps') of all anuran species available in IUCN (2009). Species were considered present if at least 50% of the range polygon covered the cell. We then used six climatic variables (following Wiens et al. 2006; Qian and Ricklefs 2007) obtained from Worldclim v. 1.4 (Hijmans et al. 2005). These variables (see Table 13.1) are related to physiological limits of amphibian species and influence their broad-scale distribution (e.g., Wiens et al. 2006). We built a global stationary Generalized Least Squares (GLS) model to predict the total richness and the richness of each group of species separately (response variables) as a function of macroclimatic variables and human population density (predictor variables; Fig. 13.4), while also accounting for the spatial autocorrelation in the data. Finally, we used a semi-variogram to explicitly model the residuals and to build their variance-covariance matrix (Dormann et al. 2013) into the model. Moran's I of residuals was used to diagnose if the GLS model successfully accommodated spatial autocorrelation.

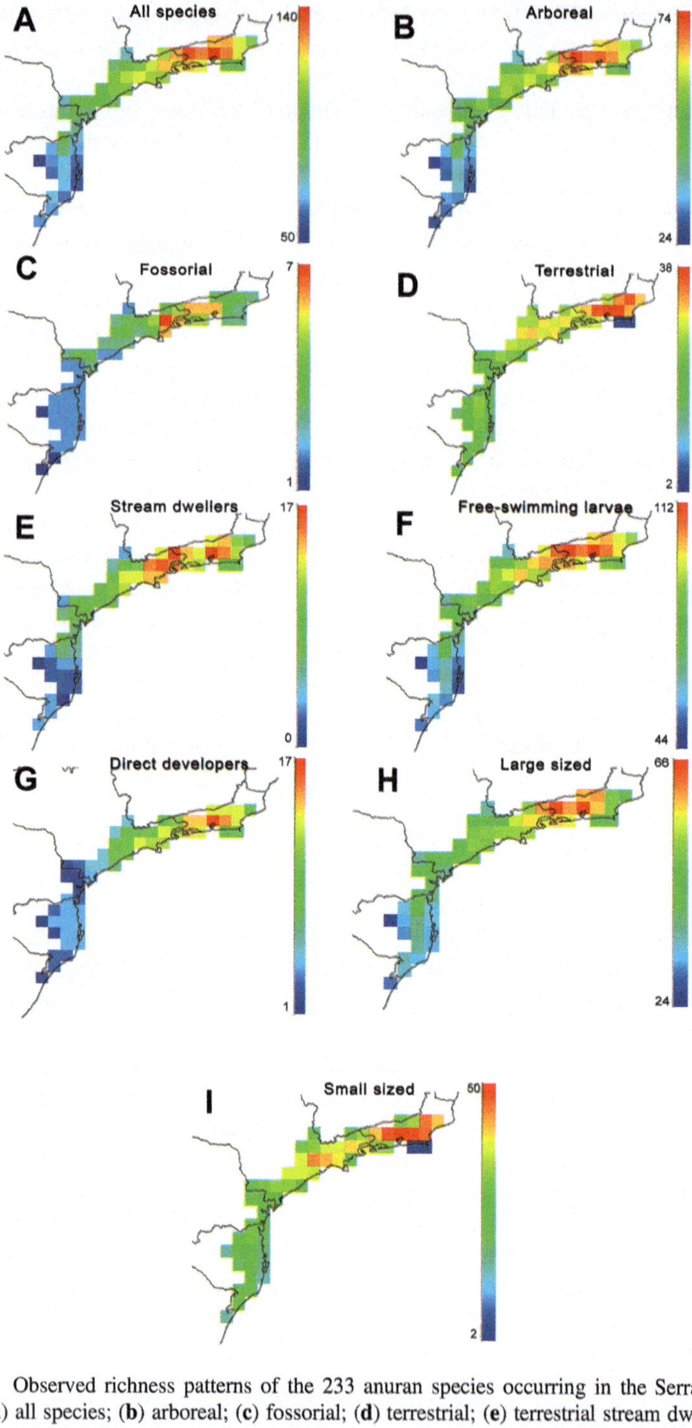

Fig. 13.3 Observed richness patterns of the 233 anuran species occurring in the Serra do Mar Range. (**a**) all species; (**b**) arboreal; (**c**) fossorial; (**d**) terrestrial; (**e**) terrestrial stream dwellers; (**f**) terrestrial with free-swimming larvae; (**g**) direct developers; (**h**) large-bodied species; (**i**) small-bodied species. The figure shows that the richness of most species peaks in the northern portion of

We found that the variable that most influenced species richness was temperature seasonality (Table 13.1). This supported our expectations, since the Serra do Mar Range has generally low rainfall seasonality and high species richness. However, the relative importance of each individual variable analyzed differed for each group of species, depending on their life-history traits. Surprisingly, altitude and human population size were negatively related to the richness of small-bodied and pond-dweller species, yet positively related to the richness of other species groups (Fig. 13.4; Table 13.1). Ponds are much more common in lowlands, where flat terrain necessary for their formation is found, while streams are more common in the escarpments and high-altitude areas, due to the sloped terrain. Data available for ectotherms demonstrate that larger species occur in cooler places as predicted by Bergmann's rule (Ashton 2002; Ashton and Feldman 2003; Morrison and Hero 2003; Vinarski 2014; Zamora-Camacho et al. 2014; Amado et al. 2018; but see Adams and Church 2008; Romano and Ficetola 2010 for contrasting results). Together, these factors may explain the negative relationship between altitude and richness of small-bodied species we found.

Overall, our empirical results show that life-history traits are important for assessing and explaining species–climate relationships in mountains. Such traits are part of the functional diversity of ecosystems and should be included alongside total richness in similar analyses whenever possible. Also, some predictor variables showed shifts in coefficient, such as temperature seasonality, positively affecting the richness of fossorial species, but negatively affecting that of stream-dwellers and large species, besides total richness.

Ecological theory predicts that highly seasonal environments present harsh conditions for most species, acting like a strong environmental filter (Ricklefs 2004; Wiens et al. 2006; Qian and Ricklefs 2007). We therefore expected temperature seasonality to negatively affect species richness. However, surprisingly, fossorial species were positively correlated with it, probably because belowground habitats would buffer them against variations in ambient temperature.

Contrary to our initial hypothesis, we found human population density to be positively correlated with richness of direct-developing species, and also free-swimming larvae, fossorial species, stream dwellers, and arboreal species (Table 13.1), with standardized slope varying from 0.047 to 0.21. This unexpected result may be an artefact of sampling bias, since urban centers also concentrate more researchers and research institutions (Oliveira et al. 2016).

Fig. 13.3 (continued) the Serra do Mar Range, in the state of Rio de Janeiro, decreasing towards the south. The region with the highest richness is also the one with the highest peaks (see main text)

Table 13.1 Results from the analysis of anuran species richness along the Serra do Mar Range

Predictors	Total richness	Direct developers	Free-swimming larvae	Terrestrial pond dwellers	Fossorials	Stream dwellers	Arboreals	Large-bodied (≥51 mm SVL)	Small-bodied (≤50.99 mm SVL)
ALT	–	0.134*	0.232**	−0.053	0.197*	0.213**	0.245**	0.179**	−0.017
POP	–	0.185*	0.122*	0.193	0.21*	0.047**	0.166**	0.103	0.173
BIO15	–	–	–	−0.287	–	–	–	–	−0.238
BIO12	–	–	–	–	–	–	−0.006	–	–
BIO16	–	–	−0.384**	–	–	–	–	–	–
BIO18	–	0.141	–	–	–	0.029	–	–	–
BIO2	0.093	0.011	0.014	–	0.232**	–	0.062	0.129	–
BIO4	−0.025	–	–	–	0.025	−0.256	–	−0.041	–
R^2	0.767	0.579	0.843	0.204	0.851	0.78	0.843	0.826	0.358
Moran's *I* of the residuals in the first distance class	0.25	0.286	0.11	0.071	0.248	0.213	0.222	0.068	−0.064

Standardized partial coefficients and partial R^2 of variables included in the GLS model used to portray the total and deconstructed anuran species richness patterns observed; *indicates $P < 0.05$; **indicates $P < 0.001$. A dash indicates that a variable was not included in the model because its Variance Inflation Factor was higher than 3, in order to avoid collinearity. These results show that life-history traits drive distribution patterns and must be considered when investigating how climate influences biodiversity. The variables used were mean diurnal temperature range (BIO2); annual precipitation (BIO12); temperature seasonality (BIO4); precipitation seasonality (BIO15); precipitation of wettest quarter (BIO16); precipitation of warmest quarter (BIO18). We also included the current human population size (POP), obtained from http://sedac.ciesin.columbia.edu/gpw and altitude (ALT) as additional predictor variables

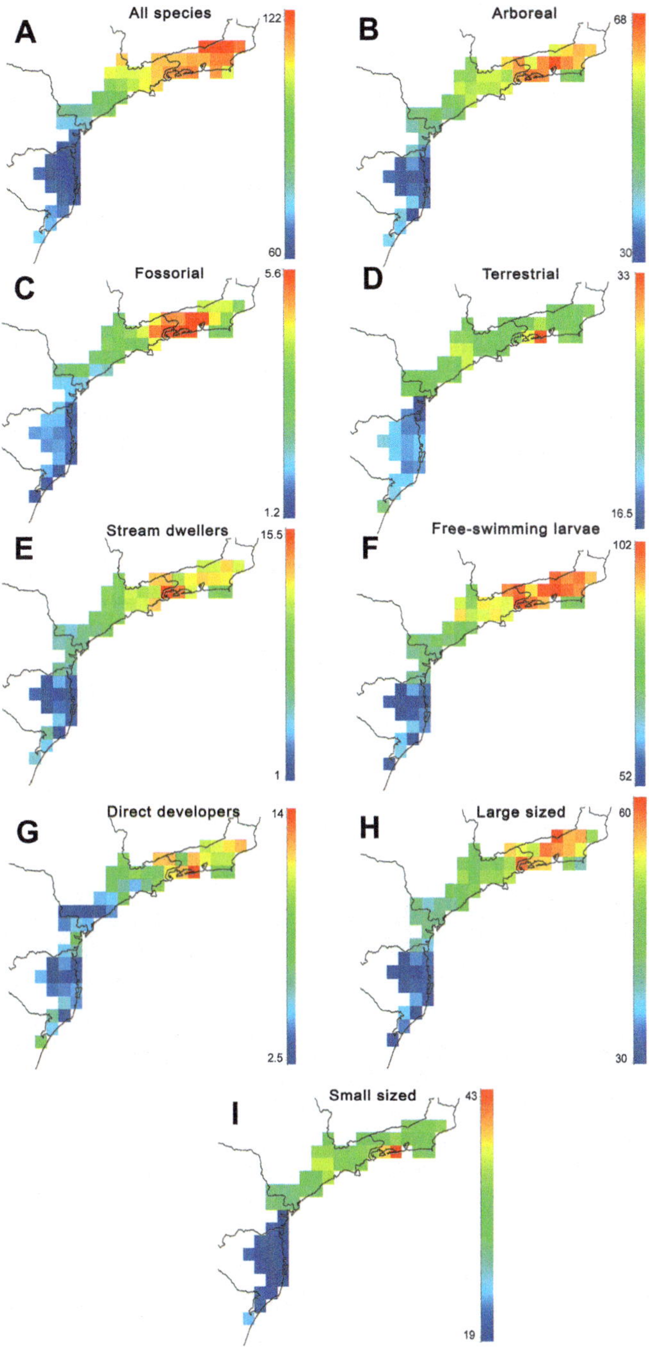

Fig. 13.4 Predicted richness patterns of the 233 anuran species occurring in the Serra do Mar Range, using a generalized least square (GLS) approach. (**a**) all species; (**b**) arboreal; (**c**) fossorial; (**d**) terrestrial; (**e**) terrestrial stream dwellers; (**f**) terrestrial with free-swimming larvae; (**g**) direct developers; (**h**) large-bodied species; (**i**) small-bodied species. A visual comparison between this

6 Future Directions

The generally low level of biological sampling carried out in most mountain ranges surveyed to date precludes robust evolutionary inferences. Only fragmentary evolutionary stories have been told, showing how particular species are related and when they derived from common ancestors in the surrounding landscapes or from distant regions. *How general are the patterns described?* We still do not know. Our review motivates several additional questions, which could be addressed by a more thorough genetic, and if possible paleontological, sampling in each mountain range. For instance, *are the rainforest species in mountain enclaves in the Caatinga remnants of a once much larger Atlantic rainforest, or the result of dispersals from the Andean slopes, Amazonia, or eastern Brazil? How did surface uplift and associated landscape changes influence the rates of speciation and extinction in Neotropical mountains, and how were those related to biotic and abiotic changes? How often have biotic corridors been created in deep history that linked the montane biotas across South America, and what biological impact did such connections have on the distribution and diversification of Neotropical diversity* (*e.g.*, Costa et al. 2017)? *What was the importance of* in situ *adaptation of lowland ancestors* versus *long-distance establishment of pre-adapted species for the high levels of montane endemicity* (Antonelli 2015; Merckx et al. 2015)? *What role did Neotropical mountains play in protecting species from episodes of past climate changes, due to their lower climatic velocity in comparison with lowlands* (Sandel et al. 2011)? *Will mountains provide refugia for biodiversity from escalating human pressures in the future?* Addressing these important questions will require concerted efforts from researchers across disciplines.

Across the world, mountains play key roles as cradles and reservoirs of biodiversity. Although the contribution of the Andes to Neotropical diversity cannot be understated (Antonelli et al. 2018c; Rangel et al. 2018), it is now time to increase efforts to better understand the evolution and distribution patterns of the rich and highly endemic biota found on other Neotropical mountain ranges. On an even more urgent level, we need to investigate the effects of deforestation and other changes in land use on distributional patterns among communities, in order to facilitate the prioritization of areas for conservation. Through a case study of the amphibians of the Serra do Mar Range, we provide one example of how human impacts can be assessed alongside current climatic conditions and species traits, which reflect the evolutionary history of local lineages, in correlative studies of observed biodiversity patterns. Although we were unable to detect a strong human imprint in the system of

Fig. 13.4 (continued) and the previous figure shows that patterns predicted by the model closely resemble the observed ones, with small extensions towards the borders of the Serra do Mar Range in the predicted richness. The results suggest that including human population size in the statistical niche model, as well as analyzing both total species richness and smaller groups of species, can improve the prediction of species richness in a global biodiversity hotspot focus, we hope that our review will inspire work on the various aspects of research outlined here, and most importantly lead to a recognition of the complexity, challenges, and amazing prospects ahead of us.

Acknowledgments We are grateful to Valentí Rull and Ana Carnaval for inviting us to contribute this chapter. One anonymous reviewer and Ana Carnaval provided excellent feedback that improved the manuscript. We thank various colleagues in our research groups and scientific networks who have discussed the ideas presented here. We also thank Giulia B. D'Angelo and Marcela Nascimento for providing the bird and mammal photos from the Serra do Mar Range in Fig. 13.2. Funding for this work was provided by Fundação de Amparo à Pesquisa do Estado de São Paulo (FAPESP 2014/18837-7) and State University of Maranhão, "Coordenação de Aperfeiçoamento de Pessoal de Nível Superior" (CAPES 99999.001292/2015-03), the Swedish Research Council (B0569601), the European Research Council under the European Union's Seventh Framework Programme (FP/2007-2013, ERC Grant Agreement n. 331024), the Swedish Foundation for Strategic Research, the Biodiversity and Ecosystems in a Changing Climate (BECC) programme, and a Wallenberg Academy Fellowship to A.A.

References

Ab'Saber AN (1967) Domínios morfoclimáticos e províncias fitogeográficas do Brasil. Orient Dep Geografia USP São Paul 3:45–48

Ab'Saber AN (1970) Províncias geológicas e domínios morfoclimáticos no Brasil. Geomorfologia 20:1–26

Ab'Saber AN (1971) A organização natural das paisagens inter e subtropicais brasileiras In: Modenesi-Gauttieri A, Bartorelli V, Mantesso Neto CDR, Carneiro MA, Lisboa MA (Orgs) 2010 A obra de Aziz Nacib Ab'Saber. Beca, São Paulo, pp 404–414

Ab'Saber AN (1974) O domínio morfoclimático semi-árido das Caatinga brasileiras. Geomorfologia 43:1–39

Ab'Saber AN (1977) Os domínios morfoclimáticos da América do Sul. Primeira Aproximação. Geomorfologia 52:1–21

Ab'Sáber AN (2003) Os domínios de natureza no Brasil. Potencialidades paisagísticas. Ateliê, Cotia

Ab'Saber AN, Bernardes N (1958) Vale do Paraíba, Serra da Mantiqueira e arredores de São Paulo: guia da excursão no 4, realizada por ocasião do XVIII Congresso Internacional de Geografia. Conselho Nacional de Geografia, Rio de Janeiro

Adams DC, Church JO (2008) Amphibians do not follow Bergmann's rule. Evolution 62:413–420

Alkmin FF (2012) Serra do Espinhaço e Chapada Diamantina. In: Hasui Y, Carneiro CDR, Almeida FFM, Bartorelli A (Orgs) Geologia do Brasil. Beca, São Paulo, pp 236–244

Almeida FFM (1964) Fundamentos geológicos do relevo paulista. In: Inst Geol Geogr Geologia do Estado de São Paulo. IGG, São Paulo, pp 167–262

Almeida AM, Prado PI, Lewinsohn TM (2004) Geographical distribution of Eupatorieae (Asteraceae) in the south-eastern and South Brazilian mountain ranges. Plant Ecol 174:163–181

Almeida FC, Bonvicino CR, Cordeiro-Estrela P (2007) Phylogeny and temporal diversification of *Calomys* (Rodentia, Sigmodontinae): implications for the biogeography of an endemic genus of the open/dry biomes of South America. Mol Phylogenet Evol 42:449–466

Amado TF, Bidau CJ, Olalla-Tárraga MÁ (2018) Geographic variation of body size in New World anurans: energy and water in a balance. Ecography 41:1–11

Andrade-Lima D (1966) Esboço fitoecológico de alguns "brejos" de Pernambuco. Bol Técnico Inst Pesq Agron Pernamb 8:3–9
Andrade-Lima D (1978) Vegetação. In: Lins RC (ed) Bacia do Parnaíba: aspectos fisiográficos. Instituto Joaquim Nabuco de Pesquisas Sociais, Recife, pp 131–135
Antonelli A (2015) Multiple origins of mountain life. Nature 524:300–301
Antonelli A, Nylander JAA, Persson C, Sanmartín I (2009) Tracing the impact of the Andean uplift on Neotropical plant evolution. Proc Natl Acad Sci U S A 106:9749–9754
Antonelli A, Verola CF, Parisod C, Gustafsson ALS (2010) Climate cooling promoted the expansion and radiation of a threatened group of South American orchids (Epidendroideae: Laeliinae). Biol J Linn Soc 100:597–607
Antonelli A, Kissling WD, Flantua SGA, Bermúdez MA, Mulch AM, Muellner-Riehl AN, Kreft H, Linder HP, Badgley C, Fjeldså J, Fritz SA, Rahbek C, Herman F, Hooghiemstra H, Hoorn C (2018a) Geological and climatic influences on mountain biodiversity. Nat Geosci 11:718–725
Antonelli A, Zizka A, Carvalho FA, Scharn R, Bacon CD, Silvestro D, Condamine FL (2018b) Amazonia is the primary source of Neotropical biodiversity. Proc Natl Acad Sci U S A 115:6034–6039
Antonelli A, Ariza M, Albert J, Andermann T, Azevedo JAR, Bacon CD, Faurby S, Guedes TB, Hoorn C, Lohmann LG, Matos-Maraví P, Ritter CD, Sanmartín I, Silvestro D, Tejedor M, ter Steege H, Tuomisto H, Werneck FP, Zizka A, Edwards S (2018c) Conceptual and empirical advances in Neotropical biodiversity research. PeerJ 6:e5644. 1–53
Araújo FS, Martins FR, Shephers GJ (1999) Variações estruturais e florísticas do carrasco no Planalto da Ibiapaba, estado do Ceará. Rev Bras Biol 59:663–678
Ashton KG (2002) Do amphibians follow Bergmann's rule? Can J Zool 80:708–716
Ashton KG, Feldman CR (2003) Bergmann's rule in nonavian reptiles: turtles follow it, lizards and snakes reverse it. Evolution 57:1151–1163
Azevedo JAR, Valdujo PH, Nogueira C (2016) Biogeography of anurans and squamates in the Cerrado hotspot: coincident endemism patterns in the richest and most impacted savanna on the globe. J Biogeogr 43:2454–2464
Bacon CD (2013) Biome evolution and biogeographical change through time. Front Biogeogr 5:227–231
Bacon CD, Silvestro D, Jaramillo CA, Tilston Smith B, Chakrabarty P, Antonelli A (2015) Biological evidence shows earlier emergence of the Isthmus of Panama. Proc Natl Acad Sci U S A 112:6110–6115
Bacon CD, Velásquez-Puentes F, Flórez-Rodríguez A, Balslev H, Galeano G, Bernal R, Antonelli A (2016) Phylogenetics of Iriarteeae (Arecaceae), cross-Andean disjunctions and convergence of clustered infructescence morphology in *Wettinia*. Bot J 182:272–286
Bacon CD, Moraes M, Jaramillo C, Antonelli A (2017) Endemic palm species shed light on habitat shifts and the assembly of the Cerrado and Restinga Floras. Mol Phylogenet Evol 110:127–133
Bacon CD, Velásquez-Puentes F, Hoorn C, Antonelli A (2018a) Iriarteeae palms tracked the uplift of Andean Cordilleras. J Biogeogr 45:1653–1663
Bacon CD, Velásquez-Puentes FJ, Hinojosa LF, Schwartz T, Oxelman B, Pfeil B, Arroyo MTK, Wanntorp L, Antonelli A (2018b) Evolutionary persistence in Gunnera and the contribution of southern plant groups to the tropical Andes biodiversity hotspot. PeerJ 6:e4388. 1–21
Badgley C, Smiley TM, Terry R, Davis EB, DeSantis LRG, Fox DL, Hopkins SSB, Jezkova T, Matocq MD, Matzke N, McGuire JL, Mulch A, Roddle BR, Roth VL, Samuels JX, Strômberg CAE, Yanites BJ (2017) Biodiversity and topographic complexity: modern and geohistorical perspectives. Trends Ecol Evol 32:211–226
Bartorelli A (2012) Chapada dos Veadeiros. In: Hasui Y, Carneiro CDR, Almeida FFM, Bartorelli A (Orgs) Geologia do Brasil. Beca, São Paulo, pp 326–330
Batalha-Filho H, Miyaki CY (2011) Filogeografia da Mata Atlântica. Revista da Biologia, USP Volume Especial Biogeografia, pp 31–34
Beerling DJ, Osborne CP (2006) The origin of the savanna biome. Glob Chang Biol 12:2023–2031

Bello C, Galetti M, Montan D, Pizo MA et al (2017) Atlantic frugivory: a plant-frugivory interaction data set for the Atlantic Forest. Ecology 98:1729–1729

Berry PE, Riina R (2005) Insights into the diversity of the Pantepui Flora and the biogeographic complexity of the Guayana Shield. Biol Skr 55:145–167

Bonaccorso E, Guayasamin JM (2013) On the origin of Pantepui montane biotas. A perspective based on the phylogeny of *Aulacorhynchus toucanets*. PLoS One 8:e67321

Bonatelli IAS, Perez MF, Peterson AT, Taylor NP, Zappi DC, Machado MC, Koch I, Pires AHC, Moraes EM (2014) Interglacial microrefugia and diversification of a cactus species complex: phylogeography and palaeodistributional reconstructions for *Pilosocereus aurisetus* and allies. Mol Ecol 23:3044–3063

Bookhagen B, Strecker MR (2008) Orographic barriers, high-resolution TRMM rainfall, and relief variations along the eastern Andes. Geophys Res Lett 35:1–6. https://doi.org/10.1029/2007GL032011

Borges SH, Santos MPD, Moreira M, Baccaro F, Capurucho JMG, Ribas C (2018) Dissecting bird diversity in the Pantepui area of endemism, northern South America. J Ornithol 159:1–14

Bovendorp RS, Villar N, Abreu-Junior EF, Bello C, Regolin AL, Percequillo A, Galetti M (2017) Atlantic small-mammal: a dataset of communities of rodents and marsupials of the Atlantic forests of South America. Ecology 99:2226–2226

Brown KS Jr, Gifford DR (2002) Lepidoptera in the Cerrado landscape and conservation of vegetation, soil and topographical mosaics. In: Oliveira PS, Marquis RJ (eds) The Cerrados of Brazil: ecology and natural history of a Neotropical savanna. Columbia University Press, New York, NY, pp 201–222

Cabanne GS, Santos FR, Miyaki CY (2007) Phylogeography of *Xiphorhynchus fuscus* (Passeriformes, Dendrocolaptidae): vicariance and recent demographic expansion in southern Atlantic forest. Biol J Linn Soc 91:73–84

Cadena CD (2007) Testing the role of interspecific competition in the evolutionary origin of elevational zonation: an example with *Buarremon* brush-finches (Aves, Emberizidae) in the Neotropical Mountains. Evolution 61:1120–1136

Capurucho JMG, Ashley MV, Ribas CC, Bates JM (2018) Connecting Amazonian, Cerrado, and Atlantic forest histories: paraphyly, old divergences, and modern population dynamics in tyrant-manakins (*Neopelma/Tyranneutes*, Aves: Pipridae). Mol Phylogenet Evol 127:696–705

Cardoso-da-Silva JM, Cardoso-de-Sousa M, Castelletti CHM (2004) Areas of endemism for passerine birds in the Atlantic forest, South America. Glob Ecol Biogeogr 13:85–92

Carnaval AC, Bates JM (2007) Amphibian DNA shows marked genetic structure and tracks Pleistocene climate change in northeastern Brazil. Evolution 61:2942–2957

Carnaval AC, Moritz C, Hickerson M, Haddad C, Rodrigues M (2009) Stability predicts diversity in the Brazilian Atlantic forest hotspot. Science 323:785–789

Carnaval AC, Waltari E, Rodrigues MT, Rosauer D, VanDerWal J, Damasceno R, Prates I, Strangas M, Spanos Z, Rivera D, Pie MR, Firkowski CR, Bornschein MR, Ribeiro LF, Moritz C (2014) Prediction of phylogeographic endemism in an environmentally complex biome. Proc R Soc B Biol Sci 281:1–8

Carneiro J, Sampaio I, de Sousa e Silva-Júnior J, Farias I, Hrbek T, Pissinatti A, Silva R, Martins-Junior A, Boubli J, Ferrari SF, Schneider H (2018) Phylogeny, molecular dating and zoogeographic history of the titi monkeys (Callicebus, Pitheciidae) of eastern Brazil. Mol Phylogenet Evol 124:10–15

Carvalho Junior OA, Guimarães RF, Martins ES, Gomes RAT (2015) Chapada dos Veadeiros: The highest landscapes in the Brazilian Central Plateau. In: Vieira B, Salgado A, Santos L (eds) Landscapes and landforms of Brazil. world geomorphological landscapes. Springer, Dordrecht, pp 221–230

Castroviejo-Fisher S, Guayasamin JM, Gonzales-Voyer A, Vilà C (2014) Neotropical diversification seen through glassfrogs. J Biogeogr 41:66–80

Chaves AV, Freitas GHS, Vasconcelos MF, Santos FR (2014) Biogeographic patterns, origin and speciation of the endemic birds from eastern Brazilian mountaintops: a review. Syst Biodivers 13:1–16

Conceição AA, Pirani JR (2007) Diversidade em quatro áreas de campos rupestres na Chapada Diamantina, Bahia, Brazil: Espécies distintas, mas riquezas similares. Rodriguesia 58:193–206

Conceição AA, Rapini A, Pirani JR, Giulietti AM, Harley RM, Silva TRS, Santos AKA, Correia C, Andrade IM, Costa JAS, Souza LRS, Andrade MJG, Funch RR, Freitas AMM, Oliveira AA (2005) Campos Rupestres. In: Juncá FA, Funch L, Rocha W (Orgs) Biodiversidade e Conservação da Chapada Diamantina. Ministério do Meio Ambiente, Brasília, pp 153–168

Costa FN, Trovó M, Sano PT (2008) Eriocaulaceae na Cadeia do Espinhaço: riqueza, endemismo e ameaças. Megadiversidade 4:91–99

Costa GC, Hampe A, Ledru M, Martinez PA, Mazzochini GG, Shepard DB, Werneck FP, Moritz C, Carnaval AC (2017) Biome stability in South America over the last 30 kyr: inferences from long-term vegetation dynamics and habitat modelling. Glob Ecol Biogeogr 27:285–297

Culot L, Pereira LA, Agostini I, Almeida MAB et al (2019) Atlantic-primates: a dataset of communities and occurrences of primates in the Atlantic forest of South America. Ecology 100:2525–2525

da Silva FR, Almeida-Neto M, do Prado VHM, Haddad CFB, Rossa-Feres DC (2012) Humidity levels drive reproductive modes and phylogenetic diversity of amphibians in the Brazilian Atlantic Forest. J Biogeogr 39:1720–1732

Dantas GPMA, Cabanne GSN, Santos FCR (2011) How past vicariant events can explain the Atlantic Forest biodiversity? In: Grillo O, Venora G (eds) Ecosystems biodiversity. InTech, Rijeka, pp 429–442

de Mello-Silva R (2010) Circumscribing *Vellozia hirsuta* and *V. tubiflora* (Velloziaceae). Hoehnea 37:617–646

Dean W (1997) With broadax and firebrand: the destruction of the Brazilian Atlantic Forest. University of California Press, Berkeley, CA

Derby OA (1906) The Serra do Espinhaço, Brazil. J Geol 14:374–401

Derryberry EP, Claramunt S, Derryberry G, Chesser RT, Cracraft J, Aleixo A, Pérez-Emán J, Remsen JV, Brumfield RT (2011) Lineage diversification and morphological evolution in a large-scale continental radiation: the Neotropical ovenbirds and woodcreepers (Aves: Furnariidae). Evolution 65:2973–2986

Désamoré A, Vanderpoorten A, Laenen B, Gradstein SR, Kok PJR (2010) Biogeography of the lost world (Pantepui region, northeastern South America): insights from bryophytes. Phytotaxa 9:254–265

Didham RK, Lawton JH (1999) Edge structure determines the magnitude of changes in microclimate and vegetation structure in tropical forest fragments. Biotropica 31:17–30

Dormann CF, Elith J, Bacher S, Buchmann C, Carl G, Carré G, García Marquéz JR, Gruber B, Lafourcade B, Leitão PJ, Münkemüller T, McClean C, Osborne PE, Reineking B, Schröder B, Skidmore AK, Kurell D, Lautenbach S (2013) Collinearity: a review of methods to deal with it and a simulation study evaluating their performance. Ecography 36:27–46

DRYFLOR (2016) Plant diversity patterns and their conservation implications in Neotropical dry Forests. Science 353:1383–1387

Echternacht L, Trovó M, Oliveira CT, Pirani JR (2011) Areas of endemism in the Espinhaço Range in Minas Gerais, Brazil. Flora 206:782–791

Eiserhardt WL, Couvreur TLP, Baker WJ (2017) Plant phylogeny as a window on the evolution of hyperdiversity in the tropical rainforest biome. New Phytol 214:1408–1422

Emmons LH (1999) Two new species of *Juscelinomys* (Rodentia: Muridae) from Bolivia. Am Mus Novit 3280:1–15

Ewers RM, Didham RK (2005) Confounding factors in the detection of species responses to habitat fragmentation. Biol Rev 81:11–142

Fernandes D, Hamdan B (2014) A new species of *Chironius* Fitzinger, 1826 from the state of Bahia, Northeastern Brazil (Serpentes: Colubridae). Zootaxa 3881:563–575

Fernandes GW, Barbosa NPU, Alberton B, Barbieri A, Dirzo R, Goulart F, Guerra TJ, Morellato LPC, Solar RRC (2018) The deadly route to collapse and the uncertain fate of Brazilian rupestrian grasslands. Biodivers Conserv 27:1–17

Fine PVA, Lohmann LG (2018) Importance of dispersal in the assembly of the Neotropical biota. Proc Natl Acad Sci U S A 115:5829–5831

Fischer A, Blaschke M, Bässler C (2011) Altitudinal gradients in biodiversity research: the state of the art and future perspectives under climate change aspects. Waldökologie, Landschaftsforschung und Naturschutz 11:35–47

Fjeldså J, Bowie RCK, Rahbek C (2012) The role of mountain ranges in the diversification of birds. Annu Rev Ecol Evol Syst 43:249–265

Fouquet A, Recoder R, Teixeira M Jr, Cassimiro J, Amaro RC, Camacho A, Damasceno R, Carnaval AC, Moritz C, Rodrigues MT (2012) Molecular phylogeny and morphometric analyses reveal deep divergence between Amazonia and Atlantic Forest species of *Dendrophryniscus*. Mol Phylogenet Evol 62:823–838

Freitas GHS, Chaves AV, Costa LM, Santos FR, Rodrigues M (2012) A new species of *Cinclodes* from the Espinhaço Range, southeastern Brazil: insights into the biogeographical history of the South American highlands. Int J Avian Sci 154:738–755

Funch LS, Funch RR, Harley R, Giulietti AM, Queiroz LP, França F, de Melo E, Gonçalves CN, Santos T (2005) Florestas Estacionais Semideciduais. In: Juncá FA, Funch L, Rocha W (Orgs) Biodiversidade e Conservação da Chapada Diamantina. Ministério do Meio Ambiente, Brasília, pp 180–193

Galindo-Leal C, Câmara IG (2003) Atlantic Forest hotspots status: an Overview. In: Galindo-Leal C, Câmara IG (eds) The Atlantic Forest of South America: biodiversity status, threats, and outlook. Island Press, Washington, DC, pp 3–11

Garey MV, Provete DB (2016) Conservation status, threat sources, and endemism of anurans in highland grasslands in Southern and Southeastern Brazil. Oecol Aust 20:94–108

Gaston KJ (2003) The structure and dynamics of geographic ranges. Oxford University Press, New York

Gibbs AK, Barron CN (1993) The geology of the Guiana Shield. Oxford University Press, New York

Giugliano LG, Collevatti RG, Colli GR (2007) Molecular dating and phylogenetic relationships among Teiidae (Squamata) inferred by molecular and morphological data. Mol Phylogenet Evol 45:168–179

Giulietti AM, Pirani JR (1988) Patterns of geographic distribution of some plant species from the Espinhaço Range, Minas Gerais and Bahia, Brazil. In: Vanzolini PE, Heyer WR (eds) Proceedings of a workshop on Neotropical distribution patterns. Academia Brasileira de Ciências, Rio de Janeiro, pp 39–69

Giulietti AM, Pirani JR, Harley RM (1997) Espinhaço range region, eastern Brazil. In: Davis SD, Heywood VH, Herrera-MacBryde O, Villa-Lobos J, Hamilton AC (eds) Centres of plant diversity: a guide strategy for their conservation, vol 3. IUCN Publication Unity, Cambridge, pp 397–404

Givnish TJ, Barfuss MH, Van Ee B, Riina R, Schulte K, Horres R, Gonsiska PA, Jabailey RS, Crayn DM, Smith JA, Winter K, Brown GK, Evans TM, Holst BK, Luther H, Till W, Zizka G, Berry PE, Systsma KJ (2011) Phylogeny, adaptive radiation, and historical biogeography in Bromeliaceae: insights from an eight-locus plastid phylogeny. Am J Bot 98:872–895

GMBA (2018) Global Mountain Biodiversity Assessment. http://www.gmba.unibe.ch/about_us/background/. Accessed 23 July 2018

Gontijo-Pascutti AHF, Hasui Y, Santos M, Júnior AVS, Souza IA (2012) As Serras do Mar e Mantiqueira. In: Hasui Y, Carneiro CDR, Almeida FFM, Bartorelli A (Orgs) Geologia do Brasil. Beca, São Paulo, pp 549–571

Guedes TB, Sawaya RJ, Nogueira CC (2014) Biogeography, vicariance and conservation of snakes of the neglected and endangered Caatinga region, northeastern Brazil. J Biogeogr 41:19–931
Guedes TB, Sawaya R, Zizka A, Laffan S, Faurby S, Pyron RA, Bérnils RS, Jansen M, Passos P, Prudente ALC, Cisneros-Heredia DF, Braz HB, Nogueira CC, Antonelli A (2018) Patterns, biases and prospects in the distribution and diversity of Neotropical snakes. Glob Ecol Biogeogr 27:14–21
Harley RM, Giulietti AM, Grilo AS, Silva TRS, Funch L, Funch RR, Queiroz LP, França F, Melo E, Gonçalves CN, Nascimento FHF (2005) Cerrado. In: Juncá FA, Funch L, Rocha W (Orgs) Biodiversidade e Conservação da Chapada Diamantina. Ministério do Meio Ambiente, Brasília, pp 121–152
Henriques RPB (2005) Influência da história, solo e fogo na distribuição e dinâmica das fitofisionomias no bioma Cerrado. In: Souza Silva JC, Felfili JM (eds) Cerrado: ecologia, biodiversidade e conservação. Ministério do Meio Ambiente, Brasília, pp 73–92
Hijmans RJ, Cameron SE, Parra JL, Jones PG, Jarvis A (2005) Very high resolution interpolated climate surfaces for global land areas. Int J Climatol 25:1965–1978
Homeier J, Breckle SW, Günter S, Rollenbeck RT, Leuschner C (2010) Tree diversity, forest structure and productivity along altitudinal and topographical gradients in a species-rich Ecuadorian montane rain forest. Biotropica 42:140–148
Hoorn C, Mosbrugger V, Mulch A, Antonelli A (2013) Mountain building and biodiversity. Nat Geosci 6:154
Hoorn C, Perrigo A, Antonelli A (2018a) Mountains, climate and biodiversity: an introduction. In: Hoorn C, Perrigo A, Antonelli A (eds) Mountains, climate and biodiversity. Wiley-Blackwell, Hoboken, NJ, pp 1–15
Hoorn C, Perrigo A, Antonelli A (2018b) Mountains, climate and biodiversity. Wiley-Blackwell, Hoboken, NJ
Hua X, Wiens JJ (2010) Latitudinal variation in speciation mechanisms in frogs. Evolution 64:429–443
Huber O (1988) Vegetation y Flora de Panteui, Region Guayana. Acta Bot Bras 1:41–52
Huber O (1995) Geographical and physical features. In: Berry PE, Holst BK, Yatskievych K (eds) Flora of the venezuelan Guayana. Voli. 1. Introduction. Missouri Botanical Garden, St Louis, pp 1–61
Huber O, Prance GT, Kroonenberg SB, Antonelli A (2018) The Tepuis of the Guiana Highlands. In: Hoorn C, Perrigo A, Antonelli A (eds) Mountains, climate and biodiversity. Wiley-Blackwell, Hoboken, NJ, pp 339–353
Hughes CE, Atchison GW (2015) The ubiquity of alpine plant radiations: from the Andes to the Hengduan Mountains. New Phytol 207:275–282
IUCN (International Union for Conservation of Nature), Conservation International, and NatureServe (2009) Global amphibian assessment. IUCN, Gland
Kessler M (2001) Patterns of diversity and range size of selected plant groups along an elevation transect in the Bolivia Andes. Biodivers Conserv 10:1897–1921
Kok PJR (2013) Islands in the sky: species diversity, evolutionary history, and patterns of endemism of the Pantepui Herpetofauna. Institute of Biology (IBL)/National Herbarium Nederland (NHN), Faculty of Science, Leiden University
Körner C (2004) Mountain biodiversity, its causes and function. Ambio 13:11–17
Körner C, Jetz W, Paulsen J, Payne D, Rudmann-Maurer K, Spehn EM (2017) A global inventory of mountains for bio-geographical applications. Alp Bot 127:1–15
Leibold MA, Holyoak M, Mouquet N, Amarasekare P, Chase JM, Hoopes MF, Holt RD, Shurin JB, Law R, Tilman D, Loreau M, Gonzalez A (2004) The metacomunity concept: a framework for multi-scale community ecology. Ecol Lett 7:601–613
Leite FSF, Juncá FA, Eterovick PC (2008) Status do conhecimento, endemismo e conservação de anfíbios anuros da Cadeia do Espinhaço, Brasil. Megadiversidade 4:158–176
Leite YLR, Kok PJR, Weksler M (2015) Evolutionary affinities of the "Lost World" mouse suggest a late Pliocene connection between the Guiana and Brazilian shields. J Biogeogr 42:706–715

Loarie SR, Duffy PB, Hamilton H, Asner GP, Field CB, Ackerly DD (2009) The velocity of climate change. Nature 462:1052

Loyola RD, Kubota U, Fonseca GAB, Lewinsohn TM (2009) Key Neotropical ecoregions for conservation of terrestrial vertebrates. Biodivers Conserv 18:2017–2031

Luebert F, Muller LAH (2015) Effects of mountain formation and uplift on biological diversity. Front Genet 6:1–2

Luebert F, Weigend M (2014) Phylogenetic insights into Andean plant diversification. Front Ecol Evol 2:1–17

Lynch JD (1989) The gauge of speciation: on the frequencies of modes of speciation. In: Otte D, Endler J (eds) Speciation and its consequences. Sinauer Associates, Sunderland, MA, pp 527–553

Machado T, Silva VX, de Silva MJ (2014) Phylogenetic relationships within *Bothrops neuwiedi* group (Serpentes, Squamata): geographically high-structured lineages, evidence of introgressive hybridization and Neogene/Quaternary diversification. Mol Phylogenet Evol 71:1–14

Magalhães AP, Barros LF, Felippe MF (2015) Southern Serra do Espinhaço: the impressive plateau of quartzite ridges. In: Vieira B, Salgado A, Santos L (eds) Landscapes and landforms of Brazil. World geomorphological landscapes. Springer, Dordrecht, pp 359–370

Marinho-Filho J, Rodrigues FHG, Juarez KM (2002) The Cerrado mammals: diversity, ecology, and natural history. In: Oliveira PS, Marquis RJ (eds) The Cerrados of Brazil: ecology and natural history of a Neotropical savana. Columbia University Press, New York, pp 266–284

Marquet PA, Fernández M, Navarrete SA, Valdovinos C (2004) Species richness emerging: toward a deconstruction of species richness patterns. In: Lomolino M, Lawrence R (eds) Frontiers in biogeography: new directions in the geography of nature. Sinauer Associates, Sunderland, MA, pp 191–209

McDiarmid RW, Donnelly MA (2005) The herpetofauna of the Guayana Highlands: amphibians and reptiles of the lost world. In: Donnelly MA, Crother BI, Guyer C, Wake MH, White ME (eds) Ecology and evolution in the tropics: a herpetological perspective. University of Chicago Press, Chicago, IL, pp 461–560

Medeiros MCMP, Mattos IFA, Kanashiro MM, Tamashiro JY, Aidar MPM (2012) Vegetation mapping in an area of Ombrophilous Dense Forest at Parque Estadual da Serra do Mar, São Paulo State, Brazil, and floristic composition of the tree component of some physiognomies. Hoehnea 39:219–233

Merckx VSFT, Hendriks KP, Beentjes KK et al (2015) Evolution of endemism on a young tropical mountain. Nature 524:347–350

Modenesi-Gauttieri MC, Hiruma ST, Riccomini C (2002) Morphotectonics of a high plateau on the northwestern flank of the Continental Rift of southeastern Brazil. Geomorphology 43:257–271

Morellato LPC, Haddad CFB (2000) Introduction: the Brazilian Atlantic Forest. Biotropica 32:786–792

Morrison C, Hero JM (2003) Geographic variation in life-history characteristics of amphibians: a review. J Anim Ecol 72:270–279

Morrone J (2014) Biogeographical regionalisation of the Neotropical region. Zootaxa 3782:1–110

Moura MR, Villalobos F, Costa GC, Garcia PCA (2016) Disentangling the role of climatic, topography and vegetation in species richness gradients. PLoS One 3:1–16

Munhoz CBR, Felfili JM (2006) Floristics of the herbaceous and subshrub layer of a moist grassland in the Cerrado biosphere reserve (Alto Paraíso de Goiás), Brazil. Edinb J Bot 63:343–354

Myers N, Mittermeier RA, Mittermeier CG, Fonseca GAB, Kent J (2000) Biodiversity hotspots for conservation priorities. Nature 403:853–858

Napoli MF, Juncá FA (2006) A new species of the *Bokermannohyla circumdata* group (Amphibia: Anura: Hylidae) from Chapada Diamantina, State of Bahia, Brazil. Zootaxa 1244:57–68

Neves DM, Dexter KG, Pennington RT, Bueno ML, de Miranda PLS, Oliveira-Filho AT (2018a) Lack of floristic identity in campos rupestres: a hyperdiverse mosaic of rocky montane savannas in South America. Flora 238:24–31

Neves MO, Pereira EA, Sugai JLMM, Rocha SB, Da Feio RN, Santana DJ (2018b) Distribution pattern of anurans from three mountain complexes in southeastern Brazil and their conservation implications. An Acad Bras Cienc 90:1611–1623

Oliveira U, Paglia AP, Brescovit AD, Carvalho CJB, Silva DP, Rezende DT, Leite FS, Batista JA, Barbosa JPP, Stehmann JR, Ascher JS, Vasconcelos MF, De Marco P Jr, Löwenberg-Neto P, Dias PG, Ferro VG, Santos AJ (2016) The strong influence of collection bias on biodiversity kwnoledge shortfalls of Brazilian terrestrial biodiversity. Divers Distrib 22:1232–1244

Olson DM, Dinerstein E, Wikramanayake ED, Burgess ND, Powell GVN, Underwood EC, D'Amico JA, Itoua I, Strand HE, Morrison JC, Loucks CJ, Allnutt TF, Ricketts TH, Kura HE, Lamoreux JF, Wettngel WW, Hedao P, Kassem KR (2001) Terrestrial ecoregions of the world: a new map of life on earth. BioSciences 51:933–938

Papadopoulou A, Knowles LL (2016) Toward a paradigm shift in comparative phylogeography driven by trait-based hypothesis. Proc Natl Acad Sci U S A 29:8018–8024

Pavan SE, Jansa S, Voss RS (2016) Spatiotemporal diversification of a low-vagility Neotropical vertebrate clade (short-tailed opossums, Didelphidae: *Monodelphis*). J Biogeogr 43:1299–1309

Petri S, Sanches EAM (2012) Chapada dos Guimarães. In: Hasui Y, Carneiro CDR, Almeida FFM, Bartorelli A (Orgs) Geologia do Brasil. Beca, São Paulo, pp 413–418

Prates I, Xue AT, Brown JL, Alvarado-Serrano DF, MTU R, Hickerson MJ, Carnaval AC (2016) Inferring responses to climate dynamics from historical demography in neotropical forest lizards. Proc Natl Acad Sci U S A 113:7978–7985

Qian H, Ricklefs RE (2007) A latitudinal gradient in large-scale beta diversity for vascular plants in North America. Ecol Lett 10:737–744

Queiroz LP, França F, Giulietti AM Melo E, Gonçalves CN, Funch LS, Harley RM, Funch RR, Silva TS (2005) Caatinga. In: Juncá FA, Funch L, Rocha W (Orgs) Biodiversidade e Conservação da Chapada Diamantina. Ministério do Meio Ambiente, Brasília, pp 95–120

Quintero I, Keil P, Jetz W, Crawford FW (2015) Historical biogeography using species grographical ranges. Syst Biol 64:1059–1073

Ramos EKS, de Magalhães RF, Sari EHR, Rosa AHB, Garcia PCA, Santos FR (2018) Population genetics and distribution data reveal conservation concerns to the sky island endemic *Pithecopus megacephalus* (Anura, Phyllomedusidae). Conserv Genet 19:99–110

Rangel TF, Diniz-Filho JAF, Bini LM (2010) SAM: a comprehensive application for spatial analysis in macroecology. Ecography 33:46–50

Rangel TF, Edwards NR, Holden PB, Diniz-Filho JAF, Gosling WD, Coelho MTP, Cassemiro FAS, Rahbek C, Colwell RK (2018) Modeling the ecology and evolution of biodiversity: biogeographical cradles, museums, and graves. Science 361:1–15

Rapini A, Ribeiro PL, Lambert S, Pirani JR (2008) A flora dos campos rupestres da Cadeia do Espinhaço. Megadiversidade 4:16–24

Ribeiro KT, Medina BMO, Scarano FR (2007) Species composition and biogeographic relations of the rock outcrop flora on the high plateau of Itatiaia, SE-Brazil. Rev Bras Bot 30:623–639. https://doi.org/10.1590/S0100-84042007000400008

Ribeiro MC, Metzger JP, Martensen AC, Ponzoni FJ, Hirota MM (2009) The Brazilian Atlantic Forest: how much is left, and how is the remaining forest distributed? Implications for conservation. Biol Conserv 142:1141–1153

Ribeiro PL, Rapini A, Damascena LS, van den Berg C (2014) Plant diversification in the Espinhaço Range: insights from the biogeography of *Minaria* (Apocynaceae). Taxon 63:1253–1264

Ricklefs RE (2004) A comprehensive framework for global patterns in biodiversity. Ecol Lett 7:1–15

Rocha WJSF, Chaves JM, Rocha CC, Lobão JB (2005) Unidades de paisagem da Chapada Diamantina-BA. In: Juncá FA, Funch L, Rocha W (Orgs) Biodiversidade e Conservação da Chapada Diamantina. Ministério do Meio Ambiente, Brasília, pp 47–64

Rodrigues MT, de Freitas MA, Silva TFS, Bertolotto CEV (2006) A new species of lizard genus *Enyalius* (Squamata, Leiosauridae) from the highlands of Chapada Diamantina, state of Bahia, Brazil, with a key to species. Phyllomedusa 5:11–24

Romano A, Ficetola GF (2010) Ecogeographic variation of body sixe in the spectacled salamanders (Salamandrina): influence of genetic structure and local factors. J Biogeogr 27:2358–2370
Ronquist F, Klopfstein S, Vilhelmsen L, Schulmeister S, Murray DL, Rasnityn AP (2012) A total-evidence approach to dating with fossils applied to the early radiation of the hymenoptera. Syst Biol 61:973–999
Rull V (2004) Biogeography of the 'lost world': a palaeoecological perspective. Earth Sci Rev 67:125–137
Rull V (2005) Biotic diversification in the Guayana Highlands: a proposal. J Biogeogr 32:921–927
Rull V (2011) Neotropical biodiversity: timing and potential drivers. Trends Ecol Evol 26:508–513
Rull V, Nogué S (2007) Potential migration routes and barriers for vascular plants of the Neotropical Guyana Highlands during the Quaternary. J Biogeogr 34:1327–1341
Rundel PW, Smith AP, Meinzer FC (1994) Tropical alpine environments: plant form and function. Cambridge University Press, Cambridge
Saadi A (1995) A Geomorfologia da Serra Do Espinhaço em Minas Gerais e de suas Margens. Geonomos 3:41–63
Safford HD (1999a) Brazilian Páramos I. An introduction to the physical environment and vegetation of the campos de altitude. J Biogeogr 26:693–712
Safford HD (1999b) Brazilian Páramos II. Macro- and mesoclimate of the Campos de Altitude and affinities with high mountain climates of the tropical Andes and Costa Rica. J Biogeogr 26:713–737
Safford HD (2007) Brazilian Páramos IV. Phytogeography of the Campos de Altitude. J Biogeogr 34:1701–1722
Salino A, Almeida TE (2008) Diversidade e conservação das pteridófitas na Cadeia do Espinhaço, Brasil. Megadiversidade 4(1–2):78–98
Sancha NU, Higgins CL, Presley SJ, Strauss RE (2014) Metacommunity structure in a highly fragmented forest: has deforestation in the Atlantic Forest altered historic biogeographic patterns? Divers Distrib 20:1058–1070
Sánchez-Baracaldo P (2004) Phylogenetics and biogeography of the Neotropical fern genera *Jamesonia* and *Eriosorus* (Pteridaceae). Am J Bot 91:274–284
Sandel B, Arge L, Dalsgaard B, Davies RG, Gaston KJ, Sutherland WJ, Svenning JC (2011) The influence of Late Quaternary climate-change velocity on species endemism. Science 334:660–664
Sanín MJ, Kissling WD, Bacon CD, Borchsenius F, Galeano G, Svenning JC, Olivera J, Ramírez R, Trénel P, Pintaud JC (2016) The Neogene rise of the tropical Andes facilitated diversification of wax palms (*Ceroxylon*: Arecaceae) through geographical colonization and climatic niche separation. Bot J 182:303–317
Santoro GRCC, Brandão RA (2014) Reproductive modes, habitat use, and richness of anurans from Chapada dos Veadeiros, central Brazil. North West J Zool 10:365–373
Santos JP, Freitas AVL, Brown KS Jr, Carreira JYO et al (2018) Atlantic butterflies: a data set of fruit-feeding butterfly communities from the Atlantic Forests. Ecology 99:2875–2875
Schipper J, Chanson JS, Chiozza F et al (2008) The status of the world's land and marine mammals: diversity, threat, and knowledge. Science 322:225–230
Schuchmann KL (1999) Family Trochilidae (Hummingbirds). In: Del Hoyo J, Elliott A, Sargatal J (eds) Handbook of the birds of the world. Lynx Edicions, Barcelona, pp 468–680
Silva JMC (1997) Endemic bird species and conservation in the Cerrado Region, South America. Biodivers Conserv 6:435–450
Silva ET, Peixoto MAA, Leite FSF, Feio RN, Garcia PCA (2018) Anuran distribution in a highly diverse region of the Atlantic Forest: the Mantiqueira Mountain Range in Southeastern Brazil. Herpetologica 74:294–305
Silvestro D, Zizka A, Bacon CD, Cascales-Miñna B, Salamin N, Antonelli A (2016) Fossil biogeography: a new model to infer dispersal, extinction and sampling from palaeontological data. Philos Trans R Soc B 371:20150225

Simon MF, Proença C (2000) Phytogeographic patterns of *Mimosa* (Mimosoideae, Leguminosae) in the Cerrado biome of Brazil: an indicator genus of high-altitude centers of endemism? Biol Conserv 96:279–296

Simon MF, Grether R, Quieroz LP, Skema C, Pennington RT, Hughes CE (2009) Recent assembly of the Cerrado, a Neotropical plant diversity hotspot. Proc Natl Acad Sci U S A 106:20359–20364

Skeels A, Cardillo M (2019) Reconstructing the geography of speciation from contemporary biodiversity data. Am Nat 193:000–000

Tabarelli M, Santos AMM (2004) Uma breve descrição sobre a história natural dos brejos nordestinos. In: Pôrto KC, Cabral JJP, Tabarelli M (Orgs) Brejos de altitude em Pernambuco e Paraíba. História natural, ecologia e conservação. Ministério do Meio Ambiente, Brasília, pp 17–24

Thomé MTC, Zamudio KR, Haddad CFBF, Alexandrino JO (2014) Barriers, rather than refugia, underlie the origin of diversity in toads endemic to the Brazilian Atlantic Forest. Mol Ecol 23:6152–6164

Travis JMJ (2003) Climate change and habitat destruction: a deadly anthropogenic cocktail. Proc R Soc Lond Ser B Biol Sci 270:467–473

Tuomisto H, Ruokolainen K, Yli-Halla M (2003) Dispersal, environment, and floristic variation of Wester Amazonian Forests. Science 299:241–244

Vancine MH, Duarte KS, Souza YS, Giovanelli JGR et al (2018) Atlantic Amphibians: a data set of amphibian communities from the Atlantic Forests of South America. Ecology 99:1692–1692

Vasconcelos TS, Prado VHM, da Silva FR, Haddad CFB (2014) Biogeographic distribution patterns and their correlates in the diverse frog fauna of the Atlantic Forest hotspot. PLoS One 9:e104130

Veloso HP, Rangel-Filho ALR, Lima JC (1991) Classificação da vegetação brasileira adaptada a um sistema universal. Instituto Brasileiro de Geografia e Estatística, Rio de Janeiro

Versieux LM, Wendt T, Louzada RB, Wanderleu MGL (2008) Bromeliaceae da Cadeia do Espinhaço. Megadiversidade 4:99–110

Viana PL, Filgueiras TS (2008) Inverntário e distribuicão geográfica das gramíneas (Poaceae) na Cadeia do Espinhaço, Brasil. Megadiversidade 4:71–88

Villalobos F, Dobrovolski R, Provete DB, Gouvea SF (2013) Is rich and rare the common share? Describing biodiversity and informing conservation practices for South American amphibians. PLoS One 8:e56073

Vinarski MV (2014) On the applicability of Bergmann's rule to ectotherms: the state of the art. Biol Bull Rev 4:232–242

Werneck FP, Gamble T, Colli GR, Rodrigues MT, Sites JW Jr (2012) Deep diversification and long-term persistence in the South American "dry diagonal": integrating continent-wide phylogeography and distribution modeling of Gekos. Evolution 66:3014–3034

Wiens JJ, Graham CH, Moen DS, Smith SA, Reeder TW (2006) Evolutionary and ecological causes of the latitudinal diversity gradient in hylid frogs: treefrog trees unearth the roots of high tropical diversity. Am Nat 168:579–596

Wulf A (2016) The invention of nature: Alexander von Humboldt's New World. Vintage Publisher, London

Zamora-Camacho FJ, Reguera S, Moreno-Rueda G (2014) Bergmann's rule rules body size in an ectotherm: heat conservation in a lizard along a 2200-metre elevational gradient. J Evol Biol 27:2820–2828

Zamudio KR, Bell RC, Mason NA (2016) Phenotypes in phylogeography: species' traits, environmental variation, and vertebrate diversification. Proc Natl Acad Sci U S A 113:8041–8048

Zizka A, Antonelli A (2018) Mountains of diversity. Nature 555:173–174

Chapter 14
Diversification in Ancient and Nutrient-Poor Neotropical Ecosystems: How Geological and Climatic Buffering Shaped Plant Diversity in Some of the World's Neglected Hotspots

Fernando A. O. Silveira, Roberta L. C. Dayrell, Cecilia F. Fiorini, Daniel Negreiros, and Eduardo L. Borba

Abstract South America harbors the highest plant diversity on Earth. The causes of such exceptionally high diversity remain poorly understood, despite great attention devoted to the ecology and evolution of biota in productive and geologically recent ecosystems such as the Amazon forest and the Andes. Evidence suggests ancient and extremely nutrient-poor landscapes are major centers of plant diversity and endemism, and acted as interglacial refugia, but singularities of their evolutionary history have been overlooked. Here, we examine to what extent Ocbil theory (old, climatically-buffered, infertile landscapes) may prove useful in explaining diversification patterns in some of the most diverse Neotropical ecosystems. We propose a theoretical framework that encompasses a mechanistic explanation for the predictions of Ocbil theory, and links ecological and evolutionary processes to vegetation patterns and functional traits. We review diversification patterns and population genetics in *campos rupestres* vegetation in light of Ocbil theory. We propose areas of future research that will accelerate and improve our understanding on the ecology and evolution of Neotropical biota on ancient, nutrient-poor vegetation. This knowledge is expected to shed light on the complex history of Neotropical plant diversification and, ultimately, provide tools for their sustainable use and conservation.

F. A. O. Silveira (✉) · C. F. Fiorini · E. L. Borba
Departamento de Botânica, Universidade Federal de Minas Gerais, Belo Horizonte, MG, Brazil
e-mail: faosilveira@icb.ufmg.br

R. L. C. Dayrell
Departamento de Botânica, Universidade Federal de Minas Gerais, Belo Horizonte, MG, Brazil

School of Biological Sciences, University of Western Australia (UWA), Perth, WA, Australia

D. Negreiros
Instituto de Ciências Biológicas e Saúde, Centro Universitário UNA, Belo Horizonte, MG, Brazil

V. Rull, A. C. Carnaval (eds.), *Neotropical Diversification: Patterns and Processes*, Fascinating Life Sciences, https://doi.org/10.1007/978-3-030-31167-4_14

Keywords *Campos rupestres* · Diversification · Molecular markers · Ocbil · Population genetics · Sky islands

1 Introduction

The Neotropical region holds the highest plant diversity in the world (Gentry 1982; Kier et al. 2005; Antonelli and Sanmartín 2011). The astounding Neotropical diversity has fascinated Europeans explorers and naturalists during early scientific exploration of the South American continent, and continues to wonder and puzzle present-day biologists (Antonelli and Sanmartín 2011; Antonelli et al. 2015). Despite such incredible high levels of diversity and endemism (Hughes et al. 2013), the mechanisms generating these patterns are not yet fully understood.

Much of the discussion of biogeographical gradients in species diversity has stemmed from the marked contrast between tropical versus temperate ecosystems (Pianka 1966). Pianka (1966) summarized six major theories (time, spatial heterogeneity, competition, predation, climatic stability, productivity) aiming to explain the latitudinal species gradient, but surprisingly, after for more than 50 years, there is yet no consensus on the factors driving species latitudinal gradients (Schemske and Mittelbach 2017). The fact that some temperate areas harbor higher diversity and endemism compared to some tropical ones (Cowling et al. 1996) suggests that additional overlooked factors may play a role in driving large-scale species diversity, or that the main six drivers may operate on other organizational scales, not only on a global one. In addition, the fact that adjacent areas at the same latitude may strongly differ in diversity and endemism patterns (Neves et al. 2018) reinforce the need of understanding the role of additional drivers of plant diversity and endemism (Fig. 14.1). With particular reference to the Neotropical regions, topography, soil adaptation, niche conservatism, and dispersal ability have been proposed as additional key drivers of continental-scale assembly of the biota (Antonelli and Sanmartín 2011), but the relative contribution of each driver remains elusive (but see Rangel et al. 2018).

Hopper (2009) developed Ocbil theory, an integrated set of principles to explain plant ecology and evolution in old, climatically-buffered, infertile landscapes (OCBILs). Ocbil theory proposes a series of testable hypotheses that span several levels of biological organization (from individuals to landscape), and helps explain why these ancient landscapes challenge patterns of latitudinal gradients with regard to species diversity and endemism. OCBILs are in the extreme end of a multivariate *continuum* of variation in landscape age, climatic fluctuation and soil fertility, whereas YODFELs (young, often-disturbed, fertile landscapes) stand at the other end of this eco-evolutionary *continuum* (Hopper et al. 2016).

Most of the theories underpinning modern Biology were generated in YODFELs (Hopper 2009), biasing the understanding of Neotropical vegetation and leading to detrimental consequences for the long-term conservation of these ancient, nutrient-poor vegetation types. A realistic example is the misunderstanding by some European and North American ecologists that grassy biomes are deforested land

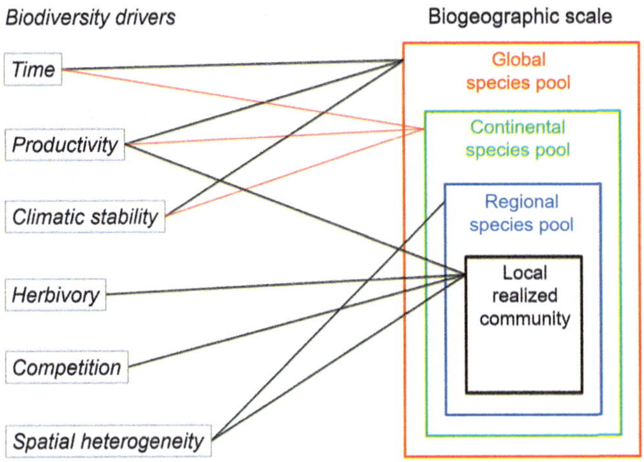

Fig. 14.1 Theoretical model explaining biogeographic assembly of Neotropical flora across scales (inspired by Götzenberger et al. 2012). Each box depicts one of the six classical drivers of diversity summarized by Pianka (1966). Lines indicate the role of each driver across each ecological scale. Red lines indicate the three dimensions of Ocbil theory, with time reflecting landscape age, productivity reflecting soil fertility and climatic stability reflecting climatic buffering (*sensu* Hopper 2009). The continental scale added here is included to provide complimentary explanations on why latitudinal species gradient do not always hold true and why different sites at the same latitude have contrasting diversities and endemism

cleared by humans (Veldman et al. 2017). As a result, many ill-conceived policies have been advocated such as afforestation, despite clear evidence that planting trees in biodiverse grassy biomes will erode biodiversity and associated ecosystem services (Veldman et al. 2015). Gaining an evolutionary perspective into the assembly of OCBILs can help to clarify this misconception and provide strong arguments to improve conservation and restoration of extremely impoverished biodiversity hotspots (Dayrell et al. 2016).

Ocbil theory has gained recent traction by Neotropical plant biologists, plant ecologists, vegetation scientists and restoration ecologists, especially those studying vegetation types that fulfill the criteria for a classic OCBIL. Until July 2018, Hopper's original paper (2009) has been cited by 21 studies in *campos rupestres*, 7 studies in inselbergs vegetation, 2 studies in *tepuis* and *cangas* each, and 1 in *campos de altitude* and southern grasslands each. Evidence supporting Ocbil theory is reviewed and discussed below for each of these vegetation types. Papers inspired by Ocbil theory addressed a broad of topics, ranging from population genetics (e.g. Hmeljevski et al. 2017b) to restoration ecology (e.g. Le Stradic et al. 2018), attesting the value of Ocbil theory to both theoretical and applied sciences (Morellato and Silveira 2018).

There are two main reasons why Ocbil theory is invoked to help explaining the community-scale assembly of Neotropical communities. First, it addresses factors that operate on a different scale than the classic Pianka's drivers (Schemske and Mittelbach 2017; Fig. 14.1), thus providing complementary explanations for global patterns of species diversity and endemism. Second, significant fractions of the diversification of the Neotropical biota are related to the relatively recent Andean uplift in western South America (Antonelli et al. 2009; Rangel et al. 2018), but most Neotropical OCBILs are concentrated in eastern South America and are unlikely to have been strongly affected by the Andean uplift. Present-day floristic similarities between OCBILs and the Andes are relatively uncommon, suggesting limited biota interchange (Alves and Kolbek 2010).

In this chapter we examine to what extent Ocbil theory may prove useful in explaining diversification patterns in some of the most biodiverse Neotropical vegetation types. We developed a mechanistic explanation for the Ocbil theory by proposing a theoretical framework linking ecological and evolutionary processes to vegetation patterns. Next, we address the ecology and evolution of vegetation types that fulfill the criteria for a classic OCBIL. These vegetation types are amongst the least studied Neotropical ecosystems, despite emerging evidence suggesting they may be the most ancient (Hughes et al. 2013) and the most species-rich ones (BFG 2015) in South America. We scrutinize diversification predictions of Ocbil theory by critically reviewing the literature on diversification and population genetics in *campos rupestres* vegetation. Finally, we propose areas of future research that will accelerate and improve our understanding on the ecology and evolution of Neotropical biota on ancient, nutrient-poor vegetation. We expect this knowledge shed light on the complex history of Neotropical plant diversification.

2 Towards a Mechanistic Understanding of Ocbil Theory

Ocbil theory was conceived to better understand the origins, ecology and devise conservation strategies tailored for biodiversity on OCBILs. These ecosystems are particularly common in the Southern Hemisphere (Hopper 2009, 2018; Hopper et al. 2016) and remain largely unknown by most Northern Hemisphere plant scientists, ecologists and evolutionary biologists, who were born and educated in YODFELs. Two exceptions are the Southwestern Australia Floristic Region (SWAFR) and the Greater Cape Floristic Region (GCFR) in South Africa, which have received considerable attention, probably due to their Mediterranean-climate, enabling their comparisons with climatically similar ecosystems of the Northern Hemisphere (e.g. Cowling et al. 2015).

Ocbil theory proposes a series of testable hypotheses (see Table 14.1) that have inspired research and resonated with the ecology of some Neotropical OCBILs. However, since its original inception, Hopper (2009) recognized the need for more theoretical development of Ocbil theory, and the need for more quantitative assessments of these predictions (Hopper et al. 2016). Here, we provide a conceptual framework attempting to improve the mechanistic understanding of Ocbil theory by connecting the three drivers of diversity (old landscapes, infertile soils and buffered climates) with the expectations regarding diversification, vegetation patterns, and species functional traits (Fig. 14.2). Two original predictions, the Semiarid Cradle

Table 14.1 Mechanistic explanations for nine predictions of Ocbil theory and examples of ways of testing them through direct and indirect evidence

Prediction	Mechanistic explanations	Direct evidence	Indirect evidence
Reduced dispersability	Species are adapted to impoverished, patchily-distributed soils. High risk of seeds landing on sites unsuitable for seedling establishment	Reduced seed dispersal distance determined under field conditions	Predominance of species with no obvious mechanisms of seed dispersal, few species with fleshy fruits, high genetic structure from cpDNA
High local endemism and rarity	Reduced dispersability results in local genetic divergence and allopatric speciation. Prolonged interpopulational divergence should lead to the evolution of suites of local endemic species	Quantitative determination of species geographic areas and accurate estimates of population sizes	High proportion of geographically restricted species and occurrence of species with small populations in patches
Old lineages	Lack of strong climatic changes since at least the Permian has allowed the persistence of lineages that arose during Gondwanan times	Presence of ancient lineages determined through time-calibrated phylogenies and fossil record	Significant representation of ancient lineages in floristic lists
Old individuals	Selective pressures on strategies for persistence. Resource limitation leads to a resource-conservative strategy that implies in low growth rates	Determining reliable estimates of plant age under field conditions	Combining estimates of plant relative growth rate and plant size
The James effect	Small population sizes should select for strategies avoiding deleterious effects of inbreeding	Obtaining reliable estimates of pollen dispersal distance, translocation, heterozygosity, dysploidy, polyploidy, and the evolution of B chromosome systems and large genome sizes	High proportions of species pollinated by long-distance pollinators, larger genomes
Nutritional specialization	Long-term landscape weathering should select for adaptations to cope with extremely-impoverished soils	Experimental data showing effective functioning of root specializations	High proportions of carnivore and parasitic species and species with root specializations including cluster roots, dauciform roots and sand-binding roots

(continued)

Table 14.1 (continued)

Prediction	Mechanistic explanations	Direct evidence	Indirect evidence
Vulnerability to soil removal	Lack of large-scale disturbances have not selected for traits favoring habitat recolonization. Intrinsic low growth rates and limited dispersal prevent natural regeneration	Experimental assessments of natural regeneration following dated disturbances	Assessing natural regeneration in sites with different disturbance ages along chronosequences
Resilience to fragmentation	Evolution for millions of years in fragmented populations have selected for unexpected persistence and resilience to human-derived habitat fragmentation	Low decrease in species richness in artificially fragmented sites, low or absent decrease in fitness (seed output) in plants occurring in fragmented sites	Persistence of small populations on isolated patches
Slow individual growth	Pervasive resource limitation has selected for traits associated with resource conservation	Assessment of individual growth under field and experimental conditions	Data on functional traits for plant communities

Hypothesis and adaptation to saline soils, can be expected only for the SWAFR, and were excluded from our framework. The recent natural hybridization hypothesis (Hopper 2018) was also excluded due to lack of available data in Neotropical vegetation. To expand the predictive power of Ocbil theory, we also included an additional prediction, the evolution of resource-conservation strategies, which we discuss below (Table 14.1). The dynamic and flexible nature of our conceptual framework accommodates the fact that some of the predicted patterns (see black boxes in Fig. 14.2) are not only determined by the three drivers of diversity, but can also indirectly provide feedback to other vegetation and trait expectations, thus playing dual roles in the framework (see gray arrows in Fig. 14.2).

2.1 *Driver-Pattern Feedbacks*

Old landscapes, extremely-impoverished soils and buffered climate have been proposed to be the main drivers of diversification, vegetation patterns and species traits in a complex way (Hopper 2009; Table 14.1). Since the Cenozoic some areas of the world have remained relatively free from extreme geological and climatic events such as glaciation, mountain-building, volcanism, inundations, which has resulted in continuous physical and chemical weathering, causing marked decreases in nutrient availability such as Phosphorus (Lambers et al. 2008). In contrast, other areas had been recurrently exposed to these large-scale climatic and geologic

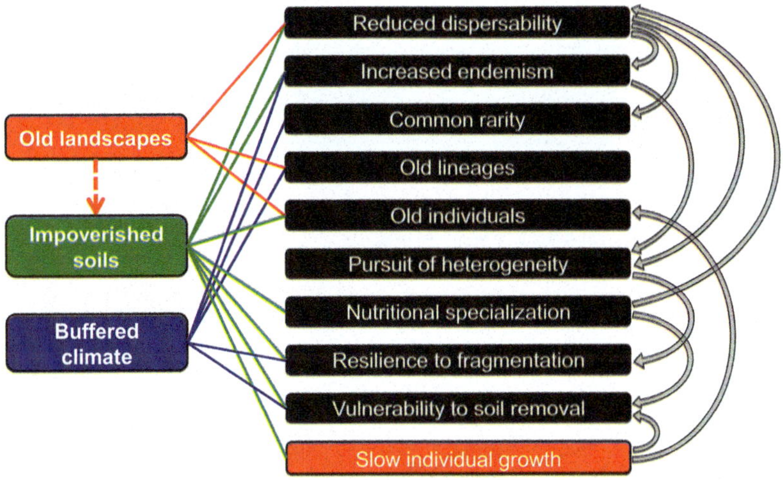

Fig. 14.2 A mechanistic framework showing the connections between three major edapho-climatic drivers of diversity (left panel boxes) and the predictions derived from Ocbil theory. These predictions result in a set of plant traits and vegetation patterns shown in the right panel boxes. The dashed arrow indicates that impoverished soils are the result of long-term physical and chemical weathering in some cases. Here, we expand the predictions of Ocbil theory by including an additional prediction that OCBIL plants have evolved a resource-conservation strategy in the red box (explained in the text). Our framework also allows positive feedbacks among plant traits and vegetation patterns (shown by gray arrows)

disturbances, resulting in rejuvenated soils with increased levels of soil nutrients (Fig. 14.3).

Mucina and Wardell-Johnson (2011) challenged the central tenets of Ocbil theory by identifying soil-impoverishment as a function of landscape age. It is undisputed that old soils are of low fertility, thus landscape age drives soil fertility to some extent (see the dashed arrow in Fig. 14.2). However, Schaefer et al. (2016) argue that some quarzitic bedrocks at the Espinhaço Range are intrinsically nutrient-poor. Therefore, soil fertility may be driven by long-term weathering, but this is not always the case (Walker and Syers 1976). As soil progressively becomes infertile, nutritional adaptations and biological specializations arise only in clades from OCBILs, whereas some traits likely evolve only in clades from YODFELs (e.g. traits selecting for long-distance seed dispersal) (Krüger et al. 2015; Oliveira et al. 2015; Zemunik et al. 2015; Turner et al. 2018).

Old lineages experienced prolonged speciation in ancient sites, where lower climatic fluctuations did not prompt diversification, and at the same time had decreased rates of species extinctions. In contrast, higher climatic fluctuations in YODFELs not only resulted in species extinction, but also created opportunities for diversification (Fig. 14.3). As a consequence of contrasting diversification dynamics in OCBILs versus YODFELs, ancient lineages (museums) occur only in the former, whereas recent lineages (cradles) occur in both sites (Hopper 2009).

More recent events, the Andean uplift in the Neogene and geomorphological changes during the Pliocene at the Espinhaço Range (Eastern Brazil; Saadi 1995),

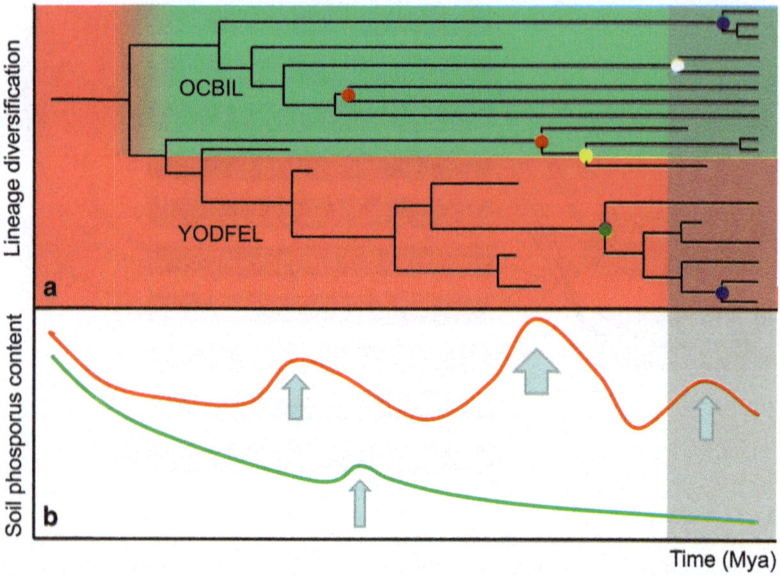

Fig. 14.3 Theoretical predictions of lineage diversification in Neotropical OCBILs (green panel and line) and YODFELs (red panel and line) (**a**) as a function of climatic and geological stability (**b**). The model begins with a YODFEL transitioning to an OCBIL as climatic stability increases and soil fertility decreases (overlapping area). Note that OCBILs are both museums (old lineages) and cradles (recent lineages), whereas YODFELs contain only recent lineages. Rates of extinction (graves) are higher in YODFELs and coincide with periods or disturbance. OCBILs are also biodiversity pumps generating lineages to adjacent lowland YODFELs. Note a high speciation rate during the Pleistocene indicated by a shaded area. Large-scale disturbance (indicated by arrows with different width) results in soil rejuvenation, increasing soil Phosphorus content. Extensive chemical and physical weathering in OCBILs results in loss of soil phosphorus generating extremely-impoverished soils. Throughout time species evolve different traits indicated by circles. Some traits are exclusive from each ecosystem (red, white and green circles), whereas others are common in both ecosystems, arising from a common ancestor with shifted ecosystem (yellow circle) or by convergent evolution (blue circles)

played a major role in diversification dynamics of Neotropical plants, which extended throughout South America (Antonelli et al. 2009; Potter and Szatmari 2009; Antonelli and Sanmartín 2011; Armijo et al. 2015; Rangel et al. 2018). It is expected that such pattern of rapid diversification may resulted in the evolution of similar functional traits in both OCBILs and YODFELs during the Quaternary (Fig. 14.3). This is supported by time-calibrated phylogenies showing recent and rapid diversification of lineages in both ecosystems (Richardson et al. 2001; Hughes and Eastwood 2006; Loeuille et al. 2015; Rando et al. 2016; Inglis and Cavalcanti 2018), but with exceptionally high diversification in the Andes (Madriñán et al. 2013). Persistence of old lineages, in turn, was only reported for OCBILs (Zappi et al. 2017; Alcântara et al. 2018).

2.2 The Interplay Between Patterns and Mechanisms

The combination of old landscapes, infertile soils and buffered climates leads to specific vegetation and trait patterns (see the black boxes in Fig. 14.2), but these can also influence and feedback one another. For instance, dispersal from parental habitat has high risks on OCBILs because plants are highly specialized to the rocky or sandy soils where they occur and are thus susceptible to physiological constraints that could limit growth in other habitat types (Porembski and Barthlott 2000; Jacobi et al. 2007; Poot et al. 2012; Silveira et al. 2016). Reduced dispersability is therefore a common strategy among OCBIL plants (Hopper et al. 2016; Fig. 14.2) that promotes divergence of local populations and allopatric speciation, and consequently contributes to increased endemism and common rarity (Hopper 2009; Echternacht et al. 2011; Silveira et al. 2016). The constraints in seed dispersal also limit population sizes and push towards mechanisms to conserve heterozygosity and escape from the deleterious effects of inbreeding.

Soil infertility, especially severe P-impoverishment, is a strong environmental filter that leads to a clear predominance of nutrient-conserving, slow-growing strategy (de Paula et al. 2015; Oliveira et al. 2015; Pierce et al. 2017; Fig. 14.5c). OCBILs are, therefore, mainly dominated by long-lived perennials that are able to survive and resprout after fire (Bond and Midgley 2001; Le Stradic et al. 2015). The great investment in persistent tissues in the nutrient conserving strategy may imply in relatively less investment in sexual reproduction (Bazzaz et al. 1987; Ehrlén and van Groenendael 1998; Gomes et al. 2018), often resulting either in a low seed set per individual (Stock et al. 1989), or in high proportion of seeds that lack embryos or are otherwise nonviable (Dayrell et al. 2016). The trade-off between nutrient-conservation and investment in sexual reproduction reinforces the vulnerability to soil removal, since OCBIL plants are not expected to have strategies for effective habitat recolonization. Old individuals appear to be consequences of the slow growth rate of plants (Negreiros et al. 2014; de Oliveira and Dickman 2017). The nutrient-poor soils are also associated with well-known nutritional specializations to efficiently capture and use soil nutrients (Lambers et al. 2010; Oliveira et al. 2015). It leads to further vulnerability to soil removal (Fig. 14.5f, g) because: (1) disturbances increases nutrients in the soil and these can be toxic for the plant (Barbosa et al. 2010); (2) disassociates species from the microhabitat they are extremely specialized in; (3) potentially alters the soil microbiota (Lambers et al. 2018).

Finally, natural selection in buffered climates and impoverished soils during millions of years favoured disjunct populations restricted to very specific soils in fine-scale mosaics with other soil types, resulting in fragmented population systems of many OCBIL plants. For this reason, these plants are also naturally resilient to fragmentation caused by human activities (Hopper 2009). The mechanisms OCBIL plants evolved to pursuit and conserve heterozygosity, including long-distance pollination (Fig. 14.5e), should also help population persistence in the face of fragmentation by maintaining the flow of nuclear genes in these taxa (Fig. 14.2).

3 Plant Diversity and Endemism in Neotropical Ocbils: The Untold History of Ancient and Nutrient-Poor South American Vegetation

Hopper (2009) originally included three sites as classic OCBILs, but anticipated the existence of many other candidates across the southern hemisphere. The three original OCBILs were the SWAFR, GCFR, and the *tepuis* in northern South America. More recently, Hopper et al. (2016) identified significant areas covered by OCBILs in at least 12 biodiversity hotspots around the globe, concentrated, but not limited to the southern hemisphere (Fig. 14.4). Altogether the *tepuis*, *campos rupestres*, SWAFR and GCFR cover only 0.27% of Earth's land surface (Hopper 2009; Silveira et al. 2016), yet are home to 7.7% of the known vascular plant diversity (Christenhusz and Byng 2016).

In the Neotropics, classic OCBILs are represented by the *tepuis*, *campos rupestres*, *cangas*, *campos de altitude* and inselbergs (Fig. 14.4). The *tepuis* are famous worldwide even outside the scientific literature, with the publication of "The Lost World" by Sir Arthur Conan Doyle (Conan Doyle 1912). However, the other four ecosystems have not enjoyed the same level of recognition. All five ecosystems share and fulfill to varying extents the criteria for being included as OCBILs, occurring at the extreme of old geological age, climatic stability, and soil infertility. Below, we describe features of these vegetation types, although we recognize that many other vegetation types (e.g. subtropical highland grasslands; Iganci et al. 2011) and white-sand ecosystems in throughout Amazonia (Adeney et al. 2016) may also fit the criteria for inclusion in Ocbil theory to some extent. Our focus is on diversification patterns, but we also discuss vegetation patterns and plant traits predicted by Ocbil theory (Table 14.2).

Two additional key geographic and geomorphologic characteristics are shared by all Neotropical OCBILs. First, all five Neotropical OCBILs are associated with rocky outcrops. Rocky outcrops are often long-lasting landscape features with stable micro-climates and constitute ecological refuges (Fitzsimons and Michael 2017). Also, populations growing on rocky outcrops with different geology and mineralogy may be genetically differentiated (Borba et al. 2001a; Lousada et al. 2013; Leles et al. 2015). At the community-level, there are clear differences in species composition and distribution of plant traits among different rock outcrop types (Carmo and Jacobi 2016; Carmo et al. 2016). The *tepuis* and the *campos rupestres* are of quartzite and/or sandstone geology, *cangas* occur on ironstone outcrops (mostly banded iron formations), and inselbergs and *campos de altitude* on granitoids.

The second characteristic is that all of them can fit the definition of sky islands (isolated mountaintops surrounded by drastically different lowland environments; McCormack et al. 2009). This raises the question on whether Ocbil theory should be invoked to explain the high species diversity and endemism in these ecosystems, since mountains *per se* outside polar regions and deserts are home to an exceptional biodiversity and high levels of endemism regardless of landscape age and soil fertility (Colwell et al. 2008; Steinbauer et al. 2016; Hoorn et al. 2018). There are

Fig. 14.4 Typical landscape of Neotropical Ocbils include the *campos rupestres* (**a**, **b**), vegetation on ironstone outcrops (*cangas*) (**c**, **d**), Inselbergs (**e**, **f**), and *campos de altitude* (**g**, **h**). Note the sky islands distribution across all vegetation types in Bahia (**b**), Espírito Santo (**e**) and Rio de Janeiro states (**h**). Individuals of *Vellozia* are shown in a dehydrated state (**c**) and on flowering (**d**). Pictures **a**, **b**, **d**, **g**, **h**—Augusto Gomes, **c**—Luiza C. Martins, **e**, **f**—Luiza de Paula

Table 14.2 Summary of empirical support for the predictions of Ocbil theory across four Neotropical Ocbils

Prediction	*Campos rupestres*	*Cangas*	Inselbergs	*Campos de altitude*
Reduced dispersability	Inferred	Inferred	Inferred	NC
High local endemism and rarity	Confirmed	Confirmed	Confirmed	Confirmed
Old lineages	Confirmed	Confirmed	Confirmed	Confirmed
Old individuals	Confirmed	NA	Inferred	NA
The James effect	Confirmed	NC	Confirmed	NC
Nutritional specialization	Confirmed	NA	NA	NA
Vulnerability to soil removal	Confirmed	Confirmed	Inferred	Inferred
Resilience to fragmentation	Inferred	Inferred	Inferred	Inferred
Slow individual growth	Confirmed	Inferred	Confirmed	NC

Confirmed indicates support from direct evidence, whereas *inferred* indicates support from indirect evidence, *NA* indicates non-available data and *NC* indicates not-conclusive data

marked genetic differentiations in populations growing on different sky islands (Pinheiro et al. 2011), so these ecosystems are excellent models to study species diversification, which is discussed in the next section.

Neotropical OCBILs are represented by grassy-shrub, open vegetation usually associated with outcrops of different origin. The *campos rupestres* and *campos de altitude* are fire-prone ecosystems, although human-caused fires may occur on *cangas* and inselbergs occasionally. Neotropical OCBILs often occur on some of the most spectacular landforms and landscapes of eastern South America, often on mountains that are spatially distributed as terrestrial islands (Fig. 14.4). Such landscape is better described as a mosaic of islands of ancient climatically-buffered, infertile highlands immersed in a matrix of relatively recent and fertile landscapes. Typical Neotropical OCBILs discussed include the *tepuis* (scattered throughout the Amazon forest), *campos rupestres* (immersed in the Atlantic Rain Forest, *Cerrado* and *Caatinga* biomes), on ironstone outcrops such as the *cangas* (in Central Amazon, *Cerrado* and Atlantic Forest) and on granitic-gneissic outcrops such as inselbergs and the *campos de altitude* (embedded in *Caatinga* and the Atlantic Rain Forest biomes). They all cover diminutive geographic areas, but host disproportionately high levels of diversity and endemism (Safford 1999a; Porembski 2000; Jacobi et al. 2007; Silveira et al. 2016).

The geological origin of banded iron formations, quartzite and granitic outcrops in Eastern South America dates back to the Precambrian with dates ranging from 2.5 Ga to a few hundred million years (Twidale 1982; Klein 2005; Gradim et al. 2014; Vieira et al. 2015). Throughout late Proterozoic and the Phanerozoic, long-term chemical and physical weathering denudated such landscapes (Barreto et al. 2013), creating snow-free mountains ranging in altitudes from 900 to 2020 m above sea level. This suggests that their present-day altitudes are only a fraction of what they once were (Safford 1999b). This ancient geological origin has set the scene for the establishment of the some of the oldest open vegetation types in Eastern South America (Hughes et al. 2013; Zappi et al. 2017).

The ancient origin of OCBILs markedly contrasts with the recent Andean uplift during the Neogene and diversification of lineages associated with the geomorphological and hydrological changes in Western South America. These recent changes have been extensive documented and explored (Antonelli et al. 2009), but the evolutionary and ecological history of Neotropical OCBILs is not well understood. Below, we discuss these issues for four typical vegetation types that are centers of diversity and endemism (Fig. 14.4). For the ecology and evolution of the *tepuis*, a fifth typical OCBIL, refer to Rull (2005, 2009).

3.1 Campos Rupestres

Campos rupestres (or rupestrian grasslands) is defined as a montane, grassy-shrubby, fire-prone vegetation mosaic with outcrops of quartzite or sandstone, along with sandy, stony, and waterlogged grasslands. Patches of transitional vegetation such as *cerrado*, gallery forests, and relictual hilltop forests are also within the *sensu lato* definition of *campos rupestres* (Silveira et al. 2016; Morellato and Silveira 2018). *Campos rupestres* vegetation dominates the highest elevation sites at the Espinhaço Range, the third largest mountain chain in South America. The Espinhaço Range is bordered on the eastern slopes by the Atlantic Rain Forest, on the western slopes by the *Cerrado*, and on the northern slopes by the *Caatinga* (see map in Morellato and Silveira 2018). However, isolated sites also occur in central and southern Brazil, central Amazonia, northeastern Brazil, and eastern Bolivia. Despite occupying an area smaller than 0.8% of Brazil, *campos rupestres* host nearly 15% of plant diversity in the country (Silveira et al. 2016). Endemism in *campos rupestres* is nearly 40%, the highest among Brazilian vegetation types (BFG 2015).

The acknowledgment of *campos rupestres'* exceptionally high diversity, endemism, and typical harsh environment has emerged only recently (Fernandes 2016; but see Giulietti et al. 1997), and the causes of such diversity are still being debated. Nevertheless, studies on population genetics in *campos rupestres* species are relatively common (see Sect. 4). While the prediction of persistence of old lineages (the Gondwanan heritage hypothesis) has been supported by literature (see reviews in Silveira et al. 2016; Zappi et al. 2017; Alcântara et al. 2018), recent diversification of some lineages also indicates the *campos rupestres* as cradles of endemic lineages (e.g. Ribeiro et al. 2013; Loeuille et al. 2015; Rando et al. 2016; Inglis and Cavalcanti 2018). Unfortunately, there are few time-calibrated trees for *campos rupestres* lineages, and those available do not focus on the most diverse and dominant clades.

3.2 *Vegetation on* Cangas

Vegetation growing on ironstone outcrops (locally known as *cangas*), are geologically, pedologically, structurally, floristically and functionally different from previously identified quartzitic and sandstone *campos rupestres* (Mucina 2018). Therefore, vegetation establishing on ironstone outcrops is treated here separately from the *campos rupestres* (*sensu* Silveira et al. 2016). Ironstone outcrops originate from Precambrian deposits (Klein 2005) and constitute some of the world's most important sites of iron ore mining. In Brazil they are best represented by sites at the Iron Quadrangle in southeastern Brazil (Jacobi et al. 2007) and the Carajás Range in eastern Amazon (Viana et al. 2016), but small patches are found elsewhere (Carmo and Kamino 2015).

Ironstone outcrops share most of the characteristics of other outcrops, such as isolation and edapho-climatic harshness, but differ in that they metal-rich substrates targeted for extensive and rapidly increasing opencast mining, and thus subjected to irrecoverable degradation (Jacobi et al. 2007). Iron caves are singular habitats that are extremely sensitive to human impacts (Jaffé et al. 2018). Vegetation on *cangas* is recognized as centers of diversity and endemism, with southeastern Brazil hosting nearly 3000 species (Carmo et al. 2018). Species establishing on *cangas* show adaptations to heavy metals in the soils (Jacobi et al. 2007), but their ecology is poorly understood (Giannini et al. 2017; Lanes et al. 2018).

Population genetic studies on a *cangas* endemic and threatened bromeliad suggest that heterozygosity is lower than expected due to selfing or biparental inbreeding and that low genetic differentiation probably results from long distance pollination by hummingbirds (Lavor et al. 2014). However, knowledge on diversification patterns in *cangas* is virtually unknown.

3.3 Campos de Altitude

Campos de altitude (or high-altitude grasslands) are a series of cool-humid, grass-dominated formations restricted to the highest summits of the Serra da Mantiqueira and Serra do Mar Range, the second largest mountain range in South America, only to the Andes (Safford 1999a). The Serra do Mar Range stretches along the Atlantic Coast fully immersed within the Atlantic Rain Forest biome, and *campos de altitude* are found exclusively on uplifted blocks of igneous or high-grade metamorphic rocks, ranging from Archean gneisses to Late Proterozoic granites and granitoid gneisses (Safford 1999a). The *campos de altitude* has been present on highest summits (from 1800 to 2000 m upwards) from at least since the Late Pleistocene and occupies an area of only 350 km^2 in the present-day (Safford 1999a).

The flora of the *campos de altitude* is highly diverse and characterized by a high degree of endemism, and has stronger floristic similarities (at least at genus-level) with the flora established on the equatorial alpine formations of the Andean and

Central American Cordillera; these similarities also extend to climate, soils, and landscapes (Safford 1999a, b; Alves and Kolbek 2010). Macroclimatic similarities between the Andes and the Serra do Mar Range may form the basis for the strong biogeographic connections, and the context within which evolutionary and ecological parallelisms shaped the biota of these two widely separated Neotropical mountains (Safford 1999b). Ancient elements of the flora are represented by Gondwanan-heritage lineages such as Velloziaceae, Eriocaulaceae and Xyridaceae. Pollination by long-distance pollinators such as hummingbirds is not high as predicted by the James effect hypotheses (Freitas and Sazima 2006).

3.4 Inselbergs

Inselbergs, or granitic and gneissic monolithic rock outcrops are emblematic examples of ancient, nutrient-poor ecosystems that are scattered across the Neotropics (Porembski et al. 1997; Porembski 2000; Scarano 2002; Neves et al. 2017). Inselbergs rise abruptly above the surrounding lowland landscape as sky islands. These outcrops are found embedded in a matrix of Amazon forest (Sarthou et al. 2017; Villa et al. 2018), Atlantic Rain Forest (de Paula et al. 2016), *Caatinga* (Silva et al. 2018), and the southern grasslands (Carlucci et al. 2015). They fulfill many criteria of Ocbil theory, with their geological dating back to the Precambrian (Twidale 1982), and their prevailing stressful conditions including shallow (or absent), nutrient-poor soils, water stress, high temperatures and constant winds (Porembski 2000; Scarano 2002; de Paula et al. 2015).

Inselbergs harbor a highly diverse, endemic and threatened flora (Porembski 2000, 2007; Porembski et al. 2016). The spatial configuration of terrestrial islands is likely the cause of high beta diversity among inselbergs, with species turnover being the major driver of changes in species composition (Martinelli 1989; Sarthou et al. 2017). Inselbergs are centers of endemism for several plant clades, but especially for Bromeliaceae (de Paula et al. 2016). The origin and diversification of this Neotropical dominant family is associated with rocky outcrop formation in the *tepuis* and southeastern Brazil (Givnish et al. 2014; Gomes-da-Silva et al. 2017).

To date several studies have addressed population genetics, reproductive ecology, phylogeography and radiation in inselbergs endemics (e.g. Barbará et al. 2008, 2009; Palma-Silva et al. 2011; Pinheiro et al. 2014). Pollination by long-distance pollinators (large bees, bats and hummingbirds), short seed dispersal distances, high genetic differentiation and structure, low genetic connectivity and long-term persistence of populations emerge as ubiquitous patterns (e.g. Paggi et al. 2010; Hmeljevski et al. 2017a, b). The sky islands of inselbergs also isolate populations on inselbergs summits. Some of these studies have explicitly tested the idea of dispersal limitation by comparing population genetic structure from nuclear genes (mediated by pollen dispersal) and chloroplast markers (mediated by seed dispersal). Some studies have found higher genetic structure from cpDNA (plastid DNA) compared to nuclear DNA (Hmeljevski et al. 2017a, b), therefore suggesting seed dispersal is much more limited than pollen dispersal (Sarthou et al. 2001) and providing indirect support for Ocbil theory in inselbergs endemics. Nevertheless, most studies on inselbergs plants are restricted to bromeliads and orchids (but see Duputié et al. 2009), limiting our ability to draw general conclusions for inselbergs vegetation.

4 Diversification and Population Genetics in *Campos Rupestres*

Island-like environments are recognized as cradles for endemic plants, due to (1) restricted gene flow between populations, followed by speciation by genetic drift, or (2) distinct selective pressures, leading to speciation by local adaptation (Kier et al. 2009; Stuessy et al. 2014; Crawford and Archibald 2017). Besides geographic isolation, elevation is also positively correlated to an increase in endemism, due to topography-driven isolation (Steinbauer et al. 2016). In a seminal work, Giulietti and Pirani (1988) suggested that the disjunct distribution of the *campos rupestres* leads to disjunct population distribution of its plant species, especially the rupicolous ones, potentially constituting one of the main engines to the great plant diversity and high endemism observed in this ecosystem. Here we summarize some of the findings of population genetics studies investigating this hypothesis.

In the *campos rupestres* and other Neotropical island-like OCBILs, the barrier to gene flow imposed by the matrix is strong, because plants would be physiologically constrained by their high specialization to the rock or sandy soils (Porembski and Barthlott 2000; Jacobi et al. 2007; Poot et al. 2012; Silveira et al. 2016). It has also been proposed that in this environment the lineages persist *in loco* for long periods of time, since it is expected that the climate is buffered and that physical characteristics of the outcrops allow the maintenance of stable microhabitats during climatic oscillation (Main 1997; Hopper 2009; Silveira et al. 2016). Indeed, simulations indicate that outcrop flora remains in local refugia during drier climate periods (Schut et al. 2014). All these factors are expected to lead to population systems with prolonged independent evolution and higher coalescence times than the observed in naturally less fragmented landscapes, as observed in OCBILs in Australia (Byrne and Hopper 2008; Tapper et al. 2014a, b). Differently from the vegetation of lowland Neotropical forests and savannas, which are alternative stable states maintained mainly by vegetation-fire feedbacks (Murphy and Bowman 2012), the soil-specialized vegetation of *campos rupestres* is not expected to expand significantly out of the geographical limits of the outcrops and sandy soils during climatic oscillations, sporadically becoming more continuous. However, it is expected that the populations could suffer demographic changes during climatic oscillations, as population sizes are not strictly linked to geographical range sizes and migration could be affected by matrix characteristics (Barbosa et al. 2015).

The first genetic studies testing the hypothesis by Giulietti and Pirani (1988) began only in the twenty-first century, which found very divergent results regarding genetic diversity and structuring between species of orchids (Borba et al. 2001a) and Asteraceae (Jesus et al. 2001). Since then, genetic diversity of populations of more than 80 *campos rupestres* plants species have been evaluated (Table 14.3). Allozymes were the more frequent marker employed, representing nearly 60% of all studies. Sequencing of nuclear and plastid regions, RAPD, SSR, and ISSR were also a source of information. The geographical distribution of these species covered all the latitudinal distribution of the *campos rupestres*, but there was a strong focus in some groups such as Orchidaceae (24 species, Fig. 14.5a, b), Fabaceae and Cactaceae (14 each) and Asteraceae (10 species). Bromeliaceae and Apocynaceae (6 species each), Eriocaulaceae and Velloziaceae (5), Melastomataceae (1) and Polygonaceae (1) were also studied (Table 14.3).

Campos rupestres species showed either low or high intra-population genetic diversity, with only a few species showing regular levels (meaning close to the average observed for different markers in plants; refer to Hamrick and Godt 1990, 1996; Nybom 2004) (Table 14.3). Narrowly distributed taxa show lower levels of variation in some comparisons with congenerics, but definitive conclusions cannot be drawn since studies on rare endemics or endangered species did not concentrate on taxa with lower levels of genetic diversity (Franceschinelli et al. 2006; da Silva et al. 2007). Some endangered species also present low intrapopulation genetic diversity (e.g. Lambert et al. 2006a, b; Pereira et al. 2007; Jesus et al. 2009), what would be expected for species with small populations threatened by human activities. In addition, some studies that showed lower levels of genetic diversity point out to limitations inherent to the markers as one possible masking factor for this pattern. Generally, species of Orchidaceae showed high values of genetic diversity, while species in Asteraceae, Eriocaulaceae, and Fabaceae generally presented lower values. These trends have been tentatively explained by life-history traits, mating systems, demography, and ecological characteristics, but unfortunately, data on more species are still not available to test these hypotheses.

The population genetics of *campos rupestres* species corroborates that gene flow between populations is limited. Twenty-four out of 60 studies where population differentiation was evaluated (Table 14.3) showed high fixation index (F_{ST}), giving support to the hypothesis that the huge diversity of species, especially of endemics, observed in the *campos rupestres* is generated due to limited gene flow among populations. From the 23 studies that showed low levels of differentiation, 14 encompassed endemics with narrow geographic distributions. In addition, low genetic differentiation between populations may be related to ancestral polymorphisms due to limited marker variability, mainly in allozymes. For example, Borba et al. (2001a) pointed out that some close areas possess exclusive alleles, despite the overall similarity between populations, indicating that this similarity could be an artifact of the marker resolution.

Campos rupestres plants from distinct families exhibited greater genetic structuring in plastid markers compared to nuclear markers, as a result of lower gene flow by seed dispersal than by pollen dispersal (Barbosa 2011; Palma-Silva et al. 2011;

Table 14.3 Summary of results of plant population genetics studies on *campos rupestres*

Species	Family (-ceae)	Rare[a]	Dist	Pop	Markers	Intra-pop diversity measures	Intra-pop diversity	Mean F_{IS}	F_{ST}	Pop-dif	IBD	Refs.[b]
Baccharis concinna	Astera-	Yes	Medium	6	RAPD (113 fragments)		High			Low		1
Lychnophora ericoides	Astera-	Yes	Broad	12	cpDNA (2 region), nrDNA[c] (1 locus)	Hp = 1–6, H = 0–0.783, Pi = 0–0.0042	High			High	Yes/No	2
Minasia alpestris	Astera-	Yes	Medium	5	Isozyme (9 loci)	He = 0.022–0.067	Low	0.374	0.321	High	No	3
Minasia cabralensis	Astera-	Yes	Narrow	6	Isozyme (9 loci)	He = 0.011–0.141	Low	0.105	0.621	High	No	3
Minasia lewinsohnii	Astera-	Yes	Narrow	1	Isozyme (9 loci)	He = 0.053	Low	–0.150			No	3
Minasia pereirae	Astera-	Yes	Medium	5	Isozyme (9 loci)	He = 0.007–0.062	Low	0.051	0.041	Low	No	3
Minasia scapigera	Astera-	Yes	Medium	4	Isozyme (9 loci)	He = 0.023–0.101	Low	–0.094	0.189	Regular	No	3
Minasia splettiae	Astera-		Narrow	1	Isozyme (9 loci)	He = 0.150	Low	0.050			No	3
Proteopsis argentea	Astera-		Medium	11	Isozyme (9 loci)	He = 0.015–0.119	Low	0.127	0.300	High	Yes/No	4
Wunderlichia mirabilis	Astera-		Broad	9	cpDNA (1 region), nrDNA[c] (1 locus)				0.182		No	5
Encholirium biflorum	Bromelia-	Yes	Narrow	1	RAPD (58 fragments)							6
Encholirium pedicellatum	Bromelia-	Yes	Narrow	1	RAPD (59 fragments)							6
Encholirium spectabile	Bromelia-		Broad	20	nrSSR[c] (8 loci), cpSSR (4 regions)	He = 0.565–0.805		0.257	0.334	Regular	No	7

Encholirium subsecundum	Bromelia-	Yes	Medium	4	RAPD (60 fragments)				0.150	Regular		6
Vriesea cacuminis	Bromelia-	Yes	Narrow	2	ISSR (86 fragments)	I = 0.22–0.31	High		0.160	Low		8
Vriesea minarum	Bromelia-	Yes	Narrow	12	SSR (10 loci)	He = 0.529–0.620	Regular	0.341	0.088	Low	No	9
Facheiroa squamosa	Cacta-		Broad	1	Isozyme (14 loci)	He = 0.401	High	0.518				10
Melocactus × *albicephalus*	Cacta-		Narrow	1	Isozyme (12 loci)	He = 0.017	Low[d]					11
Melocactus concinnus	Cacta-		Medium	3	Isozyme (12 loci)	He = 0.000–0.104	Low	0.901	0.349	High		12
Melocactus concinnus	Cacta-		Medium	2	Isozyme (12 loci)	He = 0.086–0.096	Low[d]	0.901	0.022	Low		11
Melocactus ernestii	Cacta-		Broad	1	Isozyme (12 loci)	He = 0.009	Low					11
Melocactus glaucescens	Cacta-	Yes	Narrow	4	Isozyme (12 loci)	He = 0.030–0.081	Low	0.579	0.045	Low		11
Melocactus paucispinus	Cacta-	Yes	Medium	10	Isozyme (12 loci)	He = 0.000–0.123	Low[d]	0.732	0.504	High		12
Pilosocereus aureispinus	Cacta-	Yes	Narrow	1	SSR[c] (8 loci), nrDNA (1 locus), cpDNA (2 regions)	He = 0.61						13
Pilosocereus aureispinus	Cacta-	Yes	Narrow	5	SSR[c] (8 loci), cpDNA (2 regions)	He = 0.390–0.525	High	−0.067	0.071	Low	No	14
Pilosocereus aureispinus	Cacta-		Narrow	1	Isozyme (17 loci)	He = 0.284	High	0.623				10

(continued)

Table 14.3 (continued)

Species	Family (-ceae)	Rare[a]	Dist	Pop	Markers	Intra-pop diversity measures	Intra-pop diversity	Mean F_{IS}	F_{ST}	Pop-dif	IBD	Refs.[b]
Pilosocereus aurisetus	Cacta-		Medium	11	SSR[c] (8 loci), nrDNA (1 locus), cpDNA (2 regions)	He = 0.63–0.69				High	Yes	13
Pilosocereus aurisetus	Cacta-		Medium	2	Anonymous nuclear markers (25 loci)							15
Pilosocereus aurisetus	Cacta-		Medium	11	SSR (10 loci), cpDNA (2 regions)							16
Pilosocereus bohlei	Cacta-		Narrow	1	nrDNA (1 locus), cpDNA (2 regions)							13
Pilosocereus jauruensis	Cacta-		Medium	4	SSR[c] (8 loci), nrDNA (1 locus), cpDNA (2 regions)	He = 0.62				High	Yes	13
Pilosocereus jauruensis	Cacta-		Medium	1	Anonymous nuclear markers (25 loci)							15
Pilosocereus machrisii	Cacta-		Broad	13	SSR[c] (8 loci), nrDNA (1 locus), cpDNA (2 regions)	He = 0.70–0.50				High	Yes	13
Pilosocereus machrisii	Cacta-		Broad	6	Isozyme (19 loci)	He = 0.179–0.329	High	0.427	0.281	Regular[d]	No	10

Pilosocereus machrisii	Cacta-		Broad	4	Anonymous nuclear markers (25 loci)							15
Pilosocereus machrisii	Cacta-		Broad	13	SSR (11 loci), cpDNA (2 regions)							16
Pilosocereus parvus	Cacta-		Narrow	1	SSR[c] (8 loci), nrDNA (1 locus), cpDNA (2 region)	He = 0.55						13
Pilosocereus vilaboensis	Cacta-		Medium	2	SSR[c] (8 loci), nrDNA (1 locus), cpDNA (2 region)	He = 0.59–0.60				High	Yes	13
Pilosocereus vilaboensis	Cacta-		Medium	1	Isozyme (17 loci)	He = 0.292	High	0.296				10
Pilosocereus vilaboensis	Cacta-		Medium	1	Anonymous nuclear markers (25 loci)							15
Praecereus euchlorus	Cacta-		Broad	6	Isozyme (17 loci)	He = 0.181–0.330	High	0.242	0.484	High	No	10
Comanthera borbae	Eriocaula-	Yes	Broad	1	Isozyme (8 loci)	He = 0.167	High					17
Comanthera curralensis	Eriocaula-	Yes	Narrow	6	Isozyme (8 loci)	He = 0.017–0.160	Low	0.612	0.427	High	No	17
Comanthera harleyi	Eriocaula-	Yes	Narrow	1	Isozyme (8 loci)	He = 0.078	Low					17
Comanthera hatschbachii	Eriocaula-	Yes	Medium	5	Isozyme (8 loci)	He = 0.052–0.091	Low	0.585	0.473	High	No	17
Syngonanthus mucugensis	Eriocaula-	Yes	Narrow	10	Isozyme (14 loci)	He = 0.026–0.164	Low		0.512	High	Yes	18

(continued)

Table 14.3 (continued)

Species	Family (-ceae)	Rare[a]	Dist	Pop	Markers	Intra-pop diversity measures	Intra-pop diversity	Mean F_{IS}	F_{ST}	Pop-dif	IBD	Refs.[b]
Chamaecrista coriacea	Faba-		Broad	1	Isozyme (15 loci)	He = 0.010	Low					19
Chamaecrista cytisoides "f. *prostrata*"	Faba-		Narrow	3	Isozyme (15 loci)	He = 0.033–0.036	Low	−0.111	0.035	Low		19
Chamaecrista cytisoides var. *blanchetii*	Faba-		Broad	7	Isozyme (15 loci)	He = 0.011–0.037	Low	−0.122	0.539	High		19
Chamaecrista cytisoides var. *brachystachya*	Faba-		Broad	5	Isozyme (15 loci)	He = 0.010–0.058	Low	−0.105	0.648	High		19
Chamaecrista cytisoides var. *confertiformis*	Faba-		Medium	5	Isozyme (15 loci)	He = 0.029–0.052	Low	0.476	0.109	Regular		19
Chamaecrista cytisoides var. *cytisoides*	Faba-		Broad	1	Isozyme (15 loci)	He = 0.066	Low					19
Chamaecrista cytisoides var. *decora*	Faba-		Narrow	3	Isozyme (15 loci)	He = 0.045–0.058	Low	−0.177	0.005	Low		19
Chamaecrista cytisoides var. *micrantha*	Faba-		Narrow	4	Isozyme (15 loci)	He = 0.012–0.034	Low	0.084	0.049	Low		19
Chamaecrista cytisoides var. *unijuga*	Faba-		Narrow	2	Isozyme (15 loci)	He = 0.046–0.054	Low	−0.141	0.007	Low		19

Chamaecrista depauperata	Faba-		Narrow	2	Isozyme (15 loci)	He = 0.038–0.067	Low	−0.141	0.016	Low		19
Chamaecrista mucronata	Faba-		Broad	2	RAPD (75 fragments)	I = 0.299	High		0.378	High[d]		20
Chamaecrista semaphora	Faba-	Yes	Medium	2	RAPD (65 fragments)	I = 0.124	Low		0.198	Regular[d]		20
Tibouchina papyrus	Melastomata-	Yes	Medium	3	SSR[c] (10 loci), cpDNA (3 regions)	He = 0.205–0.409	High[d]	0.127	0.712	High		21
Bulbophyllum adiamantinum	Orchida-		Narrow	2	Isozyme (14 loci)	He = 0.391–0.49	High	0.684	0.030	Low		22
Bulbophyllum bidentata	Orchida-		Narrow	1	Isozyme (14 loci)	He = 0.612	High	0.505				22
Bulbophyllum epiphytum	Orchida-		Broad	2	Isozyme (14 loci)	He = 0.413–0.518	High	0.450	0.110			22
Bulbophyllum exaltatum	Orchida-		Broad	20	Isozyme (9 loci)	He = 0.165–0.404	High	0.676	0.230	Regular[d]	No	23
Bulbophyllum insectiferum	Orchida-		Broad	1	Isozyme (14 loci)	He = 0.439	High	0.654				22
Bulbophyllum involutum	Orchida-		Broad	1	Isozyme (8 loci)	He = 0.351	High					24
Bulbophyllum involutum	Orchida-		Broad	7	Isozyme (9 loci)	He = 0.149–0.372	High	0.658	0.232	Regular[d]		23
Bulbophyllum plumosum	Orchida-		Broad	4	Isozyme (14 loci)	He = 0.419–0.466	High	0.604	0.020	Low		22
Bulbophyllum regnellii	Orchida-		Broad	1	Isozyme (14 loci)	He = 0.481	High	0.399				22
Bulbophyllum rupicolum	Orchida-		Broad	1	Isozyme (14 loci)	He = 0.490	High	0.592				22
Bulbophyllum sanderianum	Orchida-		Medium	2	Isozyme (9 loci)	He = 0.086–0.234	High	0.693	0.145	Regular[d]		23

(continued)

Table 14.3 (continued)

Species	Family (-ceae)	Rare[a]	Dist	Pop	Markers	Intra-pop diversity measures	Intra-pop diversity	Mean F_{IS}	F_{ST}	Pop-dif	IBD	Refs.[b]
Bulbophyllum weddellii	Orchida-		Broad	1	Isozyme (8 loci)	He = 0.207	High					24
Bulbophyllum weddellii	Orchida-		Broad	4	Isozyme (9 loci)	He = 0.098–0.225	High	0.517	0.269	Regular[d]		23
Cattleya × *tenuata*	Orchida-	Yes	Narrow	1	Isozyme (14 loci)	He = 0.223	High	0.909				25
Cattleya brevi-pedunculata	Orchida-		Medium	5	SSR (7 loci)	He = 0.684–0.757	Regular	0.137				26
Cattleya coccinea	Orchida-		Broad	1	SSR (7 loci)	He = 0.800	Regular	0.176				26
Cattleya elongata	Orchida-		Medium	1	Isozyme (14 loci)	He = 0.285	High	0.884				25
Cattleya elongata	Orchida-		Medium	9	Isozyme[c] (10 loci), ISSR (119 fragments)	He = 0.123–0.251	High		0.140	Regular[d]	No	27
Cattleya liliputana	Orchida-		Narrow	8	SSR (7 loci)	He = 0.712–0.793	High		0.138	High	No	28
Cattleya mantiqueirae	Orchida-		Medium	1	SSR (7 loci)	He = 0.805	Regular	0.222				26
Cattleya tenuis	Orchida-	Yes	Medium	3	Isozyme (14 loci)	He = 0.108–0.286	High	0.884	0.240	High		25
Pleurothallis adamantinensis	Orchida-	Yes	Narrow	2	Isozyme (12 loci)	He = 0.262–0.281	High	0.077	0.049	Low		29
Pleurothallis fabiobarrosii	Orchida-		Medium	3	Isozyme (1 locus)							30

Pleurothallis fabiobarrosii	Orchida-		Medium	2	Isozyme (12 loci)	He = 0.366–0.423	High	0.128	0.081	Low		29
Pleurothallis fabiobarrosii	Orchida-		Medium	2	Pyrrolizidine alkaloids (2 markers)		Low[d]			Low[d]		31
Pleurothallis johannensis	Orchida-		Narrow	7	Isozyme (1 locus)							30
Pleurothallis johannensis	Orchida-		Narrow	7	Isozyme (12 loci)	He = 0.306–0.371	High	0.190	0.046	Low		29
Pleurothallis johannensis	Orchida-		Narrow	5	Pyrrolizidine alkaloids (2 markers)		Low[d]			Low[d]		31
Pleurothallis ochreata	Orchida-		Broad	4	Isozyme (1 locus)							30
Pleurothallis ochreata	Orchida-		Broad	4	Isozyme (12 loci)	He = 0.209–0.318	High	0.190	0.175	Low		29
Pleurothallis ochreata	Orchida-		Broad	4	Pyrrolizidine alkaloids (2 markers)		Low[d]			Low[d]		31
Pleurothallis teres	Orchida-		Medium	7	Isozyme (12 loci)	He = 0.128–0.417	High	0.091	0.205	High		29
Pleurothallis teres	Orchida-		Medium	7	Pyrrolizidine alkaloids (2 markers)		Low[d]			Low[d]		31
Sophronitis sincorana	Orchida-	Yes	Medium	5	Isozyme (6 loci)	He = 0.33–0.48	High	0.491	0.053	Low	No	32
Coccoloba cereifera	Polygona-	Yes	Narrow	9	SSR (13 loci)	He = 0.324–0.566	Low	−0.061	0.123	Low	Yes	33
Vellozia compacta	Vellozia-		Medium	10	ISSR (141 fragments)	I = 0.068–0.235	Low		0.559	High	No	34
Vellozia epidendroides	Vellozia-		Medium	5	Isozyme (5 loci)	He = 0.27–0.476	High		0.266	Regular	No	35

(continued)

Table 14.3 (continued)

Species	Family (-ceae)	Rare[a]	Dist	Pop	Markers	Intra-pop diversity measures	Intra-pop diversity	Mean F_{IS}	F_{ST}	Pop-dif	IBD	Refs.[b]
Vellozia gigantea	Vellozia-	Yes	Narrow	9	ISSR (89 fragments)	I = 0.216–0.339	High		0.280	High	Yes	36
Vellozia hirsuta	Vellozia-		Broad	23	cpDNA (1 region)	Hp = 1–4, H = 0–0.788, Pi = 0–0.0077	High		0.818	High		37
Vellozia leptopetala	Vellozia-	Yes	Narrow	5	Isozyme (3 loci)	He = 0.184–0.259	Low		0.439	High	No	35

Dist geographic distribution, *Pop* number of populations evaluated in the study, *Mean* F_{IS} mean inbreeding coefficient, F_{ST} fixation index, *Pop-dif* population differentiation, *IBD* evidence of isolation by distance, *Ref.* reference, *He* expected mean heterozygosity, *I* Shannon index, *H* haplotype diversity, *Pi* nucleotide diversity, *Hp* number of haplotypes

Please, refer to the original papers to accesses the variance and significance of the summary statistics

[a]According to paper's authors

[b]1—Gomes et al. (2004), 2—Collevatti et al. (2009), 3—Jesus et al. (2009), 4—Jesus et al. (2001), 5—Feres et al. (2009), 6—Cavallari et al. (2006), 7—Gonçalves-Oliveira et al. (2017), 8—Ribeiro et al. (2013), 9—Lavor et al. (2014), 10—Moraes et al. (2005), 11—Lambert et al. (2006b), 12—Lambert et al. (2006a), 13—Bonatelli et al. (2014), 14—Khan et al. (2018), 15—Perez et al. (2016a), 16—Perez et al. (2016b), 17—Ribeiro et al. (2018), 18—Pereira et al. (2007), 19—Conceição et al. (2008), 20—da Silva et al. (2007), 21—Collevatti et al. (2012), 22—Azevedo et al. (2007), 23—Ribeiro et al. (2008), 24—Azevedo et al. (2006), 25—Borba et al. (2007a), 26—Leal et al. (2016), 27—da Cruz et al. (2011), 28—Leles et al. (2015), 29—Borba et al. (2001a), 30—Borba et al. (2000), 31—Borba et al. (2001b), 32—Borba et al. (2007b), 33—Moreira et al. (2010), 34—Lousada et al. (2013), 35—Franceschinelli et al. (2006), 36—Lousada et al. (2011), and 37—Barbosa et al. (2012)

[c]Values refer to this marker

[d]Qualitative level inferred by the authors of this chapter from the quantitative measures

Fig. 14.5 Flowers of fly-pollinated orchids (*Bulbophyllum perii*, **a** and *Bulbophyllum exaltatum*, **b**) from *campos rupestres*. Young seedling of *Pleroma* (Melastomataceae) established directly on a granitic outcrop showing a developed aquiferous pith related to water conservation (**c**). Dry capsules of *Lavoisiera subulata* (Melastomataceae) showing fruits with no obvious mechanisms for seed dispersal (**d**). Hummingbird-pollinated flowers of the mistletoe *Psittacanthus robustus* (Loranthaceae) (**e**). Lack of spontaneous natural regeneration following soil removal due to road building in *campos rupestres* (**f**) and surface iron ore mining in *cangas* (**g**). Pictures **a**, **b**—Cecília Fiorini, **c**—Luiza de Paula, **d**, **f**—Fernando Silveira, **e**—Tadeu Guerra, **g**—Lucas Perillo

Pinheiro et al. 2014). Even in the presence of pollen flow, isolation due to low seed dispersal may lead to conflicts between nuclear and plastid genes, and consequent reproductive isolation and speciation (Greiner et al. 2011; Greiner and Bock 2013;

Barnard-Kubow et al. 2016). Studies investigating speciation by gene conflict (Crespi and Nosil 2013) should be performed to examine the extension of their commonality in OCBILs.

Isolation by distance (IBD) was found in 18 of the 42 taxa tested (Table 14.3). IBD was more common in species with smaller geographic distributions, suggesting that long distances restrict plant species gene flow in *campos rupestres*. For some species, even on short distances, there were no signs of IBD, indicating stochastic colonization of the patches. However, some cacti showed IBD, despite their broad or medium distributions. This could be related to seed and pollen dispersion strategies (Bonatelli et al. 2014). One Asteraceae species with intermediate geographic distribution also showed evidence of IBD, but data may have been influenced by two populations that are highly isolated from the species core area (Jesus et al. 2001).

The population genetics studies in *campos rupestres* show that the disjunct distribution of this environment is an important factor in the diversification of its species. This pattern could be expected for species with traits leading to low dispersability, such as seeds of dry fruits without adaptations for anemochory (e.g. Fig. 14.5d) and flowers pollinated by small insects, or animals presenting optimal foraging behavior (e.g. social bees and territorialist hummingbirds). The presence of such traits has been invoked as justifying the high genetic structuring in species of some plant groups, such as Velloziaceae (Franceschinelli et al. 2006; Lousada et al. 2011, 2013; Barbosa et al. 2012), Eriocaulaceae (Pereira et al. 2007; Ribeiro et al. 2018), and some Asteraceae (Jesus et al. 2001, 2009; Collevatti et al. 2009). The data presented by these studies support the hypothesis of low dispersability as a characteristic of the plant species in OCBILs, particularly expected in environments such as *campos rupestres*, with the occurrence of populations clearly delimited in small areas, due to local peculiarities of characteristics of sandy soils and small, isolated outcrops.

The moderate to high genetic structuring observed in some species of Orchidaceae would not be expected, at least theoretically (e.g. Azevedo et al. 2007; Borba et al. 2001a, b, 2007a, b; Ribeiro et al. 2008; da Cruz et al. 2011; Leal et al. 2016). Orchid seeds are among the smallest and lightest in plants, and can travel hundreds of kilometers by the action of the wind (Arditti and Ghani 2000). However, the absence of individuals in rocky outcrops nearby established populations (a few tens or hundreds of meters), and apparently very similar to outcrops with large populations of the same species is remarkable. We suggest that in these cases the potential of physical dispersal of these seeds is not a good indicator of effective dispersal (considering seed germinability and seedling establishment; Schupp et al. 2010), probably due to small variations in physical, chemical and biological characteristics (in the case of orchids, occurrence of symbiotic mycorrhizae) of the substrate. Unfortunately, no study has been conducted so far to test this hypothesis.

Future studies might make an effort to achieve better resolution about the gene flow dynamics by using more powerful markers and analysis (Bertorelle et al. 2010; Ellegren 2014; Andrews et al. 2016), bearing in mind the need for careful sampling to avoid spurious patterns of genetic structure. We also emphasize the need to encourage the development of studies determining the effective dispersal and its restrictions in the different groups of plants, especially in those presenting both significant genetic structuring and characteristics that would not favor such structuring, such as pollination by birds and bats (e.g. Bromeliaceae) and long-distance seed dispersal by wind (e.g. Orchidaceae).

As generally *campos rupestres* plant species show a pattern of high population differentiation, a better understanding of *campos rupestres* species life-history and demography is fundamental for their conservation. In population systems where the genetic variability is structured, as occurs in the *campos rupestres*, the loss of a single population could have a great impact on total species diversity and may compromise its conservation. Besides, as an island-like environment, *campos rupestres* also offer opportunities for the study of the evolution of species and a better comprehension of differentiation processes.

5 Conclusions

Here, we argued that Ocbil theory is useful for explaining diversification, vegetation patterns and functional traits in old, climatically-buffered and infertile landscapes. Notably, all Neotropical OCBILs overlap with montane areas and their geographic distribution can be described as sky islands. Sky island theory plays a role in structuring species and genetic diversity in these four ecosystems, but it does not aim to explain the evolution of functional traits related to resource acquisition and conservation. Therefore, we contend that Ocbil theory and sky island theory both are useful for explaining plant ecology and evolution in South America's most ancient and infertile soils.

The four examples presented here—*campos rupestres*, *cangas*, inselbergs and *campos de altitude*—illustrate ecosystems which occupy a diminutive area, yet harbor exceptionally high diversity and endemism. Compared to Australian, European and North American ecosystems, the study of ecology and evolution of Neotropical OCBILs is still in the first development steps. However, the results that we already have show an astonishing opportunity for a better understanding of the drivers of biological diversification. These ecosystems are excellent models to study speciation, diversification and evolution, and should be given special conservation priority. Bringing ancient, nutrient-poor open vegetation to the forefront of Neotropical plant biologists is critical to increase awareness of their conservation and restoration (Fiedler 2015; Overbeck et al. 2015; Veldman et al. 2015, 2017; Morellato and Silveira 2018). Some of these ecosystems are among the most threatened by human-impact (Fernandes et al. 2018) and we run the risk of rapidly losing an irreplaceable evolutionary history. Such understanding is not only important to reconstruct the complex biogeography of the Neotropics, but also is vital for sustaining ecosystem services (Fernandes et al. 2018; Pontara et al. 2018).

5.1 Areas of Future Research

Despite recent developments and progress towards the understanding of the ecology and evolution of biota in OCBILs, opportunities for future investigation remain vast. Moving from qualitative to quantitative assessments of ecosystem properties would provide robust evidence to test the predictions of Ocbil theory. Available evidence supports the *campos rupestres* and inselbergs as classic OCBILs, but more research is needed on the diversification patterns and species traits in *cangas* vegetation and *campos de altitude* in order to determine their position in the OCBIL-YODFEL multivariate *continuum* (Table 14.2).

Particularly, we need to move from indirect to direct indicators of diversification, vegetation patterns and functional traits. We call for further measurements of both pollen and seed deposition patterns under field conditions (Schupp et al. 2010), quantitative meta-analyses of diversification patterns (e.g. Madriñán et al. 2013), controlled experiments determining the effectiveness of root specializations (Güsewell and Schroth 2017), and long-term assessments of the impacts of soil removal and habitat fragmentation. Such data will be important to support the development of empirical quantitative modeling. We should also have a better appreciation on the role played by habitat heterogeneity in determining species diversity (Schemske and Mittelbach 2017), which is not addressed extensively by Ocbil theory.

The recent developments in sequencing technologies and the progressive increase in computer power are also offering us an unseen capability of evaluating past biological dynamics on non-model organisms with more resolution and accuracy (Ellegren 2014). Many of the 50 fundamental questions about island biology proposed by Patiño et al. (2017) could be answered in the context of the Neotropical OCBILs with molecular tools, boosting the understanding of this field in a context still little explored. Researchers studying Neotropical OCBILs should take advantage of these to better explore evolutionary hypotheses and shed light on the exciting life on OCBILs.

The pervasive bias in the ecological literature towards young and fertile environments (Martin et al. 2012) has prevented the evolution of a theoretical framework for Gondwanan vegetation. We contend that theory of plant ecology and evolution in the Neotropics needs to challenge pre-established paradigms and offer fresh perspectives that can be derived by Ocbil theory.

Acknowledgements FAOS thanks financial support from FAPEMIG and CNPq and support from Luiza C. Martins. RLCD received a scholarship from CAPES and a Scholarship for International Research Fees at UWA. CFF thanks financial support from CAPES and CNPq. ELB receives a scholarship from CNPq. Prof. S. D. Hopper provided criticism that improved the quality of this manuscript.

References

Adeney JM, Christensen NL, Vicentini A, Cohn-Haft M (2016) White-sand ecosystems in Amazonia. Biotropica 48:7–23

Alcântara S, Ree RH, Mello-Silva R (2018) Accelerated diversification and functional trait evolution in Velloziaceae reveal new insights into the origins of the *campos rupestres*' exceptional floristic richness. Ann Bot 122:165–180

Alves RVJ, Kolbek J (2010) Can *campo rupestre* vegetation be floristically delimited based on vascular plant genera? Plant Ecol 207:67–79

Andrews KR, Good JM, Miller MR, Luikart G, Hohenlohe PA (2016) Harnessing the power of RADseq for ecological and evolutionary genomics. Nat Rev Genet 17:81–93

Antonelli A, Sanmartín I (2011) Why are there so many plant species in the Neotropics? Taxon 60:403–414

Antonelli A, Nylander JA, Persson C, Sanmartín I (2009) Tracing the impact of the Andean uplift on Neotropical plant evolution. Proc Natl Acad Sci USA 106:9749–9754

Antonelli A, Zizka A, Silvestro D, Scharn R, Cascales-Miñana B, Bacon CD (2015) An engine for global plant diversity: highest evolutionary turnover and emigration in the American tropics. Front Genet 6(130). https://doi.org/10.3389/fgene.2015.00130

Arditti J, Ghani AKA (2000) Numerical and physical properties of orchid seeds and their biological implications. New Phytol 145:367–421

Armijo R, Lacassin R, Coudurier-Curveur A, Carrizo D (2015) Coupled tectonic evolution of Andean orogeny and global climate. Earth-Sci Rev 143:1–35

Azevedo CO, Borba EL, van den Berg C (2006) Evidence of natural hybridization and introgression in *Bulbophyllum involutum* Borba, Semir & F. Barros and *B. weddellii* (Lindl.) Rchb. f. (Orchidaceae) in the Chapada Diamantina, Brazil, by using allozyme markers. Braz J Bot 29:415–421

Azevedo MTA, Borba EL, Semir J, Solferini VN (2007) High genetic variability in Neotropical myophilous orchids. Bot J Linn Soc 153:33–40

Barbará T, Lexer C, Martinelli G, Mayo S, Fay MF, Heuertz M (2008) Within-population spatial genetic structure in four naturally fragmented species of a neotropical inselberg radiation, *Alcantarea imperialis*, *A. geniculata*, *A. glaziouana* and *A. regina* (Bromeliaceae). Heredity 101:285–296

Barbará T, Martinelli G, Palma-Silva C, Fay MF, Mayo S, Lexer C (2009) Genetic relationships and variation in reproductive strategies in four closely related bromeliads adapted to neotropical 'inselbergs': *Alcantarea glaziouana*, *A. regina*, *A. geniculata* and *A. imperialis* (Bromeliaceae). Ann Bot 103:65–77

Barbosa AR (2011) Biossistemática do complexo *Vellozia hirsuta* (Velloziaceae) baseada em análise filogeográfica e genética de populações. MSc Dissertation, Universidade Federal de Minas Gerais

Barbosa NPU, Fernandes GW, Carneiro MAA, Júnior LA (2010) Distribution of non-native invasive species and soil properties in proximity to paved roads and unpaved roads in a quartzitic mountainous grassland of southeastern Brazil (rupestrian fields). Biol Invasions 12:3745–3755

Barbosa AR, Fiorini CF, Silva-Pereira V, Mello-Silva R, Borba EL (2012) Geographical genetic structuring and phenotypic variation in the *Vellozia hirsuta* (Velloziaceae) ochlospecies complex. Am J Bot 99:1477–1488

Barbosa NPU, Fernandes GW, Sanchez-Azofeifa A (2015) A relict species restricted to a quartzitic mountain in tropical America: an example of microrefugium? Acta Bot Bras 29:299–309

Barnard-Kubow KB, So N, Galloway LF (2016) Cytonuclear incompatibility contributes to the early stages of speciation. Evolution 70:2752–2766

Barreto HN, Varajão CA, Braucher R, Bourlès DL, Salgado AA, Varajão AF (2013) Denudation rates of the Southern Espinhaço Range, Minas Gerais, Brazil, determined by in situ-produced cosmogenic beryllium-10. Geomorphology 191:1–13

Lambers H, Brundrett MC, Raven JA, Hopper SD (2010) Plant mineral nutrition in ancient landscapes: high plant species diversity on infertile soils is linked to functional diversity for nutritional strategies. Plant Soil 348:7–27

Lambers H, Albornoz F, Kotula L, Laliberté E, Ranathunge K, Teste FP, Zemunik G (2018) How belowground interactions contribute to the coexistence of mycorrhizal and non-mycorrhizal species in severely phosphorus-impoverished hyperdiverse ecosystems. Plant Soil 424:11–33

Lambert SM, Borba EL, Machado MC (2006a) Allozyme diversity and morphometrics of the endangered *Melocactus glaucescens* (Cactaceae), and investigation of the putative hybrid origin of *Melocactus* × *albicephalus* (*Melocactus ernestii* × *M. glaucescens*) in north-eastern Brazil. Plant Species Biol 21:93–108

Lambert SM, Borba EL, Machado MC, Andrade SCDS (2006b) Allozyme diversity and morphometrics of *Melocactus paucispinus* (Cactaceae) and evidence for hybridization with *M. concinnus* in the Chapada Diamantina, north-eastern Brazil. Ann Bot 97:389–403

Lanes ÉC, Pope NS, Alves R, Carvalho Filho NM, Giannini TC, Giulietti AM et al (2018) Landscape genomic conservation assessment of a narrow-endemic and a widespread morning glory from Amazonian savannas. Front Plant Sci 9(532). https://doi.org/10.3389/fpls.2018.00532

Lavor P, van den Berg C, Jacobi CM, Carmo FF, Versieux LM (2014) Population genetics of the endemic and endangered *Vriesea minarum* (Bromeliaceae) in the Iron Quadrangle, Espinhaço Range, Brazil. Am J Bot 101:1167–1175

Le Stradic S, Buisson E, Fernandes GW (2015) Vegetation composition and structure of some Neotropical mountain grasslands in Brazil. J Mt Sci 12:864–877

Le Stradic S, Fernandes GW, Buisson E (2018) No recovery of *campo rupestre* grasslands after gravel extraction: implications for conservation and restoration. Restor Ecol 26(S2):S151–S159

Leal BS, Chaves CJ, Koehler S, Borba EL (2016) When hybrids are not hybrids: a case study of a putative hybrid zone between *Cattleya coccinea* and *C. brevipedunculata* (Orchidaceae). Bot J Linn Soc 181:621–639

Leles B, Chaves AV, Russo P, Batista JA, Lovato MB (2015) Genetic structure is associated with phenotypic divergence in floral traits and reproductive investment in a high-altitude orchid from the Iron Quadrangle, southeastern Brazil. PLoS One 10(3):e0120645. https://doi.org/10.1371/journal.pone.0120645

Loeuille B, Semir J, Lohmann LG, Pirani JR (2015) A phylogenetic analysis of Lychnophorinae (Asteraceae: Vernonieae) based on molecular and morphological data. Syst Bot 40:299–315

Lousada JM, Borba EL, Ribeiro KT, Ribeiro LC, Lovato MB (2011) Genetic structure and variability of the endemic and vulnerable *Vellozia gigantea* (Velloziaceae) associated with the landscape in the Espinhaço Range, in southeastern Brazil: implications for conservation. Genetica 139:431–440

Lousada JM, Lovato MB, Borba EL (2013) High genetic divergence and low genetic variability in disjunct populations of the endemic *Vellozia compacta* (Velloziaceae) occurring in two edaphic environments of Brazilian *campos rupestres*. Braz J Bot 36:45–53

Madriñán S, Cortés AJ, Richardson JE (2013) Páramo is the world's fastest evolving and coolest biodiversity hotspot. Front Genet 4(192). https://doi.org/10.3389/fgene.2013.00192

Main BY (1997) Granite outcrops: a collective ecosystem. J R Soc West Aust 80:113–122

Martin LJ, Blossey B, Ellis E (2012) Mapping where ecologists work: biases in the global distribution of terrestrial ecological observations. Front Ecol Environ 10:195–201

Martinelli G (1989) Campos de altitude. Editora Index, Rio de Janeiro

McCormack JE, Huang H, Knowles LL (2009) Sky islands. In: Gillespie R, Clague D (eds) Encyclopedia of islands. University of California Press, Berkeley, CA, pp 841–843

Moraes EM, Abreu AG, Andrade SC, Sene FM, Solferini VN (2005) Population genetic structure of two columnar cacti with a patchy distribution in eastern Brazil. Genetica 125:311–323

Moreira RG, McCauley RA, Cortés-Palomec AC, Fernandes GW, Oyama K (2010) Spatial genetic structure of *Coccoloba cereifera* (Polygonaceae), a critically endangered microendemic species of Brazilian rupestrian fields. Conserv Genet 11:1247–1255

Morellato LPC, Silveira FAO (2018) Plant life in *campo rupestre*: new lessons from an ancient biodiversity hotspot. Flora 238:1–10

Mucina L (2018) Vegetation of Brazilian campos rupestres on siliceous substrates and their global analogues. Flora 238:11–23

Mucina L, Wardell-Johnson GW (2011) Landscape age and soil fertility, climatic stability, and fire regime predictability: beyond the OCBIL framework. Plant Soil 341:1–23

Murphy BP, Bowman DM (2012) What controls the distribution of tropical forest and savanna? Ecol Lett 15:748–758

Negreiros D, Le Stradic S, Fernandes GW, Rennó HC (2014) CSR analysis of plant functional types in highly diverse tropical grasslands of harsh environments. Plant Ecol 215:379–388

Neves DM, Dexter KG, Pennington RT, Valente AS, Bueno ML, Eisenlohr PV et al (2017) Dissecting a biodiversity hotspot: the importance of environmentally marginal habitats in the Atlantic Forest Domain of South America. Divers Distrib 23:898–909

Neves DM, Dexter KG, Pennington RT, Bueno ML, de Miranda PL, Oliveira-Filho AT (2018) Lack of floristic identity in campos rupestres—a hyperdiverse mosaic of rocky montane savannas in South America. Flora 238:24–31

Nybom H (2004) Comparison of different nuclear DNA markers for estimating intraspecific genetic diversity in plants. Mol Ecol 13:1143–1155

Oliveira RS, Galvão HC, de Campos MC, Eller CB, Pearse SJ, Lambers H (2015) Mineral nutrition of campos rupestres plant species on contrasting nutrient-impoverished soil types. New Phytol 205:1183–1194

Overbeck GE, Vélez-Martin E, Scarano FR, Lewinsohn TM, Fonseca CR, Meyer ST, Müller SC, Ceotto P, Dadalt L, Durigan G, Ganade G, Gossner MM, Guadagnin DL, Lorenzen K, Jacobi CM, Weisser WW, Pillar VD (2015) Conservation in Brazil needs to include non-forest ecosystems. Divers Distrib 21:1455–1460

Paggi GM, Sampaio JAT, Bruxel M, Zanella CM, Goeetze M, Buettow MV, Palma-Silva C, Bered F (2010) Seed dispersal and population structure in *Vriesea gigantea*, a bromeliad from the Brazilian Atlantic Rainforest. Bot J Linn Soc 164:317–325

Palma-Silva C, Wendt T, Pinheiro F, Barbará T, Fay MF, Cozzolino S, Lexer C (2011) Sympatric bromeliad species (*Pitcairnia* spp.) facilitate tests of mechanisms involved in species cohesion and reproductive isolation in Neotropical inselbergs. Mol Ecol 20:3185–3201

Patiño J, Whittaker RJ, Borges PA, Fernández-Palacios JM, Ah-Peng C, Araújo MB et al (2017) A roadmap for island biology: 50 fundamental questions after 50 years of The Theory of Island Biogeography. J Biogeogr 44:963–983

Pereira ACS, Borba EL, Giulietti AM (2007) Genetic and morphological variability of the endangered *Syngonanthus mucugensis* Giul. (Eriocaulaceae) from the Chapada Diamantina, Brazil: implications for conservation and taxonomy. Bot J Linn Soc 153:401–416

Perez MF, Bonatelli IAS, Moraes EM, Carstens BC (2016a) Model-based analysis supports interglacial refugia over long-dispersal events in the diversification of two South American cactus species. Heredity 116:550–557

Perez MF, Carstens BC, Rodrigues GL, Moraes EM (2016b) Anonymous nuclear markers reveal taxonomic incongruence and long-term disjunction in a cactus species complex with continental-island distribution in South America. Mol Phylogenet Evol 95:11–19

Pianka ER (1966) Latitudinal gradients in species diversity: a review of concepts. Am Nat 100 (910):33–46

Pierce S, Negreiros D, Cerabolini BE, Kattge J, Díaz S, Kleyer M et al (2017) A global method for calculating plant CSR ecological strategies applied across biomes world-wide. Funct Ecol 31:444–457

Pinheiro F, de Barros F, Palma-Silva C, Fay MF, Lexer C, Cozzolino S (2011) Phylogeography and genetic differentiation along the distributional range of the orchid *Epidendrum fulgens*: a Neotropical coastal species not restricted to glacial refugia. J Biogeogr 38:1923–1935

Pinheiro F, Cozzolino S, Draper D, de Barros F, Félix LP, Fay MF, Palma-Silva C (2014) Rock outcrop orchids reveal the genetic connectivity and diversity of inselbergs of northeastern Brazil. BMC Evol Biol 14:49. https://doi.org/10.1186/1471-2148-14-49

Pontara V, Bueno ML, Rezende VL, de Oliveira-Filho AT, Gastauer M, Meira-Neto JAA (2018) Evolutionary history of *campo rupestre*: an approach for conservation of woody plant communities. Biodivers Conserv 27:2877–2896

Poot P, Hopper SD, van Diggelen JM (2012) Exploring rock fissures: does a specialized root morphology explain endemism on granite outcrops? Ann Bot 110:291–300

Porembski S (2000) The invasibility of tropical granite outcrops ('inselbergs') by exotic weeds. J R Soc West Aust 83:131–134

Porembski S (2007) Tropical inselbergs: habitat types, adaptive strategies and diversity patterns. Braz J Bot 30:579–586

Porembski S, Barthlott W (2000) Granitic and gneissic outcrops (inselbergs) as centers of diversity for desiccation-tolerant vascular plants. Plant Ecol 151:19–28

Porembski S, Seine R, Barthlott W (1997) Inselberg vegetation and the biodiversity of granite outcrops. J R Soc West Aust 80:193–199

Porembski S, Silveira FAO, Fiedler PL, Watve A, Rabarimanarivo M, Kouame F, Hopper SD (2016) Worldwide destruction of inselbergs and related rock outcrops threatens a unique ecosystem. Biodivers Conserv 25:2827–2830

Potter PE, Szatmari P (2009) Global Miocene tectonics and the modern world. Earth-Sci Rev 96:279–295

Rando JG, Zuntini AR, Conceição AS, van den Berg C, Pirani JR, de Queiroz LP (2016) Phylogeny of *Chamaecrista* ser. Coriaceae (Leguminosae) unveils a lineage recently diversified in Brazilian campo rupestre vegetation. Int J Plant Sci 177:3–17

Rangel TF, Edwards NR, Holden PB, Diniz-Filho JAF, Gosling WD, Coelho MTP, Cassemiro FAS, Rahbek C, Colwell RK (2018) Modeling the ecology and evolution of biodiversity: biogeographical cradles, museums, and graves. Science 361(6399):eaar5452. https://doi.org/10.1126/science.aar5452

Ribeiro PL, Borba EL, Smidt EC, Lambert SM, Selbach-Schnadelbach AS, van den Berg C (2008) Genetic and morphological variation in the *Bulbophyllum exaltatum* (Orchidaceae) complex occurring in the Brazilian "campos rupestres": implications for taxonomy and biogeography. Plant Syst Evol 270:109–137

Ribeiro PCC, Pinheiro LC, Domingues R, Forzza RC, Machado MA, Viccini LF (2013) Genetic diversity of *Vriesea cacuminis* (Bromeliaceae): an endangered and endemic Brazilian species. Genet Mol Res 12:1934–1943

Ribeiro PL, Pereira ACS, Borba EL, Giulietti AM (2018) Genetic and morphological diversity and evidence of hybridization in the "sempre-vivas" (*Comanthera*, Eriocaulaceae) endemic to the Chapada Diamantina, Bahia, Brazil. Flora 238:60–71

Richardson JE, Pennington RT, Pennington TD, Hollingsworth PM (2001) Rapid diversification of a species-rich genus of neotropical rain forest trees. Science 293(5538):2242–2245

Rull V (2005) Biotic diversification in the Guayana highlands: a proposal. J Biogeogr 32:921–927

Rull V (2009) Pantepui. In: Gillespie R, Clague D (eds) Encyclopedia of islands. University of California Press, Berkeley, CA, pp 717–720

Saadi A (1995) A geomorfologia da Serra do Espinhaço em Minas Gerais e de suas margens. Rev Geonomos 3:41–63

Safford HD (1999a) Brazilian Páramos I. An introduction to the physical environment and vegetation of the campos de altitude. J Biogeogr 26:693–712

Safford HD (1999b) Brazilian Páramos II. Macro-and mesoclimate of the campos de altitude and affinities with high mountain climates of the tropical Andes and Costa Rica. J Biogeogr 26:713–737

Sarthou C, Samadi S, Boisselier-Dubayle MC (2001) Genetic structure of the saxicole *Pitcairnia geyskesii* (Bromeliaceae) on inselbergs in French Guiana. Am J Bot 88:861–868

Sarthou C, Pavoine S, Gasc JP, De Massary JC, Ponge JF (2017) From inselberg to inselberg: floristic patterns across scales in French Guiana (South America). Flora 229:147–158

Scarano FR (2002) Structure, function and floristic relationships of plant communities in stressful habitats marginal to the Brazilian Atlantic rainforest. Ann Bot 90:517–524

Schaefer CE, Corrêa GR, Candido HG, Arruda DM, Nunes JA, Araujo RW, Rodrigues PMS, Fernandes Filho EI, Pereira AFS, Brandão PC, Neri AV (2016) The physical environment of rupestrian grasslands (Campos Rupestres) in Brazil: geological, geomorphological and pedological characteristics, and interplays. In: Fernandes GW (ed) Ecology and conservation of mountaintop grasslands in Brazil. Springer, Cham, pp 15–53

Schemske DW, Mittelbach GG (2017) "Latitudinal gradients in species diversity": reflections on Pianka's 1966 article and a look forward. Am Nat 189:599–603

Schupp EW, Jordano P, Gómez JM (2010) Seed dispersal effectiveness revisited: a conceptual review. New Phytol 188:333–353

Schut AG, Wardell-Johnson GW, Yates CJ, Keppel G, Baran I, Franklin SE, Hopper SD, Van Niel KP, Mucina L, Byrne M (2014) Rapid characterisation of vegetation structure to predict refugia and climate change impacts across a global biodiversity hotspot. PLoS One 9(1):e82778. https://doi.org/10.1371/journal.pone.0082778

Silva JB, Sfair JC, dos Santos ND, Pôrto KC (2018) Different trait arrangements can blur the significance of ecological drivers of community assembly of mosses from rocky outcrops. Flora 238:43–50

Silveira FAO, Negreiros D, Barbosa NPU, Buisson E, Carmo FF, Carstensen DW, Conceição AA, Cornelissen TG, Echternacht L, Fernandes GW, Garcia QS, Guerra TJ, Jacobi CM, Lemos-Filho JP, Le Stradic S, Morellato LPC, Neves FS, Oliveira RS, Schaefer CE, Viana PL, Lambers H (2016) Ecology and evolution of plant diversity in the endangered campo rupestre: a neglected conservation priority. Plant Soil 403:129–152

Steinbauer MJ, Field R, Grytnes JA, Trigas P, Ah-Peng C, Attorre F et al (2016) Topography-driven isolation, speciation and a global increase of endemism with elevation. Glob Ecol Biogeogr 25:1097–1107

Stock WD, Pate JS, Kuo J, Hansen AP (1989) Resource control of seed set in *Banksia laricina* C. Gardner (Proteaceae). Funct Ecol 3:453–460

Stuessy TF, Takayama K, López-Sepúlveda P, Crawford DJ (2014) Interpretation of patterns of genetic variation in endemic plant species of oceanic islands. Bot J Linn Soc 174:276–288

Tapper SL, Byrne M, Yates CJ, Keppel G, Hopper SD, Van Niel K, Schut AGT, Mucina L, Wardell-Johnson GW (2014a) Isolated with persistence or dynamically connected? Genetic patterns in a common granite outcrop endemic. Divers Distrib 20:987–1001

Tapper SL, Byrne M, Yates CJ, Keppel G, Hopper SD, Van Niel K, Schut AGT, Mucina L, Wardell-Johnson GW (2014b) Prolonged isolation and persistence of a common endemic on granite outcrops in both mesic and semi-arid environments in south-western Australia. J Biogeogr 41:2032–2044

Turner BL, Hayes PE, Laliberté E (2018) A climosequence of chronosequences in southwestern Australia. Eur J Soil Sci 69:69–85

Twidale CR (1982) Granite landforms. Elsevier Scientific Publishing, Amsterdam

Veldman JW, Overbeck GE, Negreiros D, Mahy G, Le Stradic S, Fernandes GW, Durigan G, Buisson E, Putz FE, Bond WJ (2015) Where tree planting and forest expansion are bad for biodiversity and ecosystem services. Bioscience 65:1011–1018

Veldman JW, Silveira FAO, Fleischman FD, Ascarrunz NL, Durigan G (2017) Grassy biomes: an inconvenient reality for large-scale forest restoration? A comment on the essay by Chazdon and Laestadius. Am J Bot 104:649–651

Viana PL, Mota NFDO, Gil ADSB, Salino A, Zappi DC, Harley RM, Ilkiu-Borges AL, Secco RS, Almeida TE, Watanabe MTC, Santos JUM, Trovó M, Maurity C, Giulietti AM (2016) Flora of the cangas of the Serra dos Carajás, Pará, Brazil: history, study area and methodology. Rodriguésia 67:1107–1124

Vieira BC, Salgado AAR, Santos LJC (eds) (2015) Landscapes and landforms of Brazil. Springer, London

Villa PM, Gastauer M, Martins SV, Carrión JF, Campos PV, Rodrigues AC, Heringer G, Meira-Neto JAA (2018) Phylogenetic structure is determined by patch size in rock outcrop vegetation on an inselberg in the northern Amazon region. Acta Amazon 48:248–256

Walker TW, Syers JK (1976) The fate of phosphorus during pedogenesis. Geoderma 15:1–19

Zappi DC, Moro MF, Meagher TR, Lughadha EN (2017) Plant biodiversity drivers in Brazilian Campos Rupestres: insights from phylogenetic structure. Front Plant Sci 8:2141. https://doi.org/10.3389/fpls.2017.02141

Zemunik G, Turner BL, Lambers H, Laliberté E (2015) Diversity of plant nutrient-acquisition strategies increases during long-term ecosystem development. Nat Plants 1(5):15050. https://doi.org/10.1038/NPLANTS.2015.50

Chapter 15
The Pantepui "Lost World": Towards a Biogeographical, Ecological and Evolutionary Synthesis of a Pristine Neotropical Sky-Island Archipelago

Valentí Rull and Teresa Vegas-Vilarrúbia

Abstract Pantepui is an archipelago of sky islands formed by the flat summits of the Neotropical Guiana table mountains (tepuis) situated between the Orinoco and Amazon basins. Pantepui is a virtually pristine land and a natural laboratory to study the origin and evolution of Neotropical biodiversity. This review aims to synthesize the existing biological knowledge of Pantepui, with an emphasis on the latest developments in biogeographical, ecological and evolutionary studies. Biogeographically, Pantepui is a province of the Guiana region, within the Neotropical realm, but the precise definition of this province varies according to the taxonomic group studied. Here we adopt a definition based on elevation, with a diffuse lower boundary at 1200–1500 m and an upper boundary at the uppermost elevations of the Guiana Highlands (ca. 3000 m). The biodiversity and endemism patterns of Pantepui are outstanding. With almost 2600 known species (>5000 species/10,000 km^2), plants are the most diverse organisms and situate Pantepui among the most diverse regions of the world. Endemism usually ranges from 30 to 40% but may reach 55% in amphibians. Ecology is poorly known. Autecological studies are lacking, and community studies are available only for vegetation and solely in descriptive terms. Paleoecological studies have shown that plant communities have changed through time under the action of Holocene climatic changes and fire. Glacial-interglacial alternation has deeply modified the Pantepui biota and this biogeographical unit has been recurrently disassembled during glaciations and reassembled during interglacials. The origin and evolution of the Pantepui biota has been explained by diverse evolutionary processes involving a variety of environmental drivers and diversification mechanisms. Most of these hypotheses

V. Rull (✉)
Institute of Earth Sciences Jaume Almera (ICTJA-CSIC), Barcelona, Spain
e-mail: vrull@ictja.csic.es

T. Vegas-Vilarrúbia
Department of Evolutionary Biology, Ecology and Environmental Sciences, University of Barcelona, Barcelona, Spain
e-mail: tvegas@ub.es

V. Rull, A. C. Carnaval (eds.), *Neotropical Diversification: Patterns and Processes*, Fascinating Life Sciences, https://doi.org/10.1007/978-3-030-31167-4_15

emerged from the study of extant biogeographical patterns and geological-paleoecological reconstructions. The inception of molecular phylogenetics, albeit still incipient in Pantepui, has provided evidence useful for testing these proposals. Taken individually, none of the proposed hypotheses can explain the evolution of the whole Pantepui biota, whose proper understanding requires complex thinking and the consideration of multiple drivers and a diversity of ecological and evolutionary processes and mechanisms acting together across spatiotemporal scales. Pantepui pristinity could be threatened by direct human disturbance and global warming. Preliminary estimates suggest that, under the worst warming scenario, >80% (including >50% of endemics) of the unique vascular flora could lose their habitat by the end of this century. *In situ* conservation actions are difficult to implement, and *ex situ* strategies (germplasm banks, botanical gardens, managed relocation) should thus be considered. More systematic and target-focused, rather than exploratory, approaches are needed for future research on Pantepui. International cooperation and the improvement of bureaucratic facilities are required to preserve the still-pristine Pantepui biota and ecosystems.

Keywords Biogeography · Climate change · Conservation · Ecology · Evolution · Glacial cycles · Global warming · Guiana highlands · Neogene · Paleoecology · Quaternary

1 Introduction

Pantepui is one of the few pristine areas remaining in the world and constitutes a unique laboratory for studying the origin and evolution of tropical biodiversity (Rull 2010). Pantepui is located on the Guiana Highlands, one of the most spectacular Neotropical landscapes, characterized by remote table mountains (the tepuis) with flat and largely inaccessible summits (Fig. 15.1) that form an archipelago of sky islands (McCormack et al. 2009) of approximately 1500–3000 m in elevation and from <1 to approximately 1000 km^2 in surface area, separated from the surrounding lowlands and uplands by vertical cliffs up to 1000 m high (Huber 1995a) (Fig. 15.1). The assemblage of these tepui summits is known as Pantepui (from the Greek *pan*, meaning "all", and the local Pemon word *tepui*, meaning "table mountain"). Pantepui pristinity is guaranteed by its remoteness, difficulty of access due to the complex topography and the lack of natural resources to exploit, as well as the fact that indigenous peoples inhabiting the surrounding uplands and lowlands do not climb to these summits because of religious constraints (Huber 1995d). The unique features of the Pantepui biota and the communities it forms, together with the amazing levels of biodiversity and endemism, are the basis for the definition of the Pantepui biogeographical province (Berry et al. 1995a).

The scientific exploration of the tepuis began in the mid-nineteenth century, when the German brothers Robert and Richard Schomburgk collected plant and animal specimens on the southern slopes of the Roraima-tepui, between 1838 and 1842.

Fig. 15.1 The Guiana Highlands landscape and examples of tepuis (see Fig. 15.2 for location). (1) Roraima-Kukenán massif; from left to right: Kukenán-tepui (summit area of 20 km^2 and 2650 m of maximum elevation), Roraima-tepui (35 km^2 and 2720 m) and Wei-Assipu-tepui (3 km^2 and 2260 m). (2) Vertical cliffs of the Chimantá massif. (3) General view of the NE sector of the Chimantá massif. (4) Tirepón-tepui (2600 m) in the Chimantá massif. (5) Gran Sabana uplands with

Approximately 40 years later (1881–1883), British ornithologist Henry Whitely explored the upper slopes of the Roraima-Kukenán massif but did not reach its summits. Only a few years later (1884 and 1898), British botanist Everard im Thurn and his colleagues climbed to the summit of the Roraima-tepui (Fig. 15.1) and collected the first rare plants and animals from this hitherto new and strange life zone (Huber 1995b). The oddity of the biological specimens collected during these first expeditions suggested that the tepui summits were a separate world different from what was known at that time and inspired the famous Arthur Conan Doyle's fantastic novel entitled "The Lost World" (Doyle 1912). The first scientific explorers managed to access some tepui summits by foot after long and difficult trips but this was only possible—and still is today—on a few of these table mountains. The exploration of Pantepui underwent a decisive bourgeoning after the Second World War, with the use of helicopters, which are still the preferred—and in many cases, the only—means to reach the tepui summits. A detailed account of the history of Pantepui scientific exploration can be found in Huber (1995b).

With time and the intensification of scientific exploration of the tepui summits, the perception of the rarity of the Pantepui biota has decreased, but its uniqueness still captivates the most experienced researchers and has contributed to the progress of general disciplines, notably biogeography and evolution (Rull 2010). Unfortunately, the ecological study of Pantepui is still embryonic, likely because of the difficulty in conducting regular and extended field studies atop the tepuis. This is due in part to the remoteness and inaccessibility of the tepui summits in addition to the permanent problems in regard to obtaining official fieldtrip permits for scientific collection, especially for genetic studies (Rull and Vegas-Vilarrúbia 2008). This not only hinders the progress of ecological and evolutionary study but also delays the adoption of conservation measures to protect the Pantepui biota from eventual extinction by habitat loss under the action of ongoing global warming (Rull et al. 2016). Another handicap is that until recently, knowledge of Pantepui biota was dispersed among hundreds of publications, including a significant amount of grey literature, and a synthesis of the state of the art was lacking. Recently, Rull et al. (2019a) summarized the most relevant information available for biodiversity and endemism patterns of the better-known taxonomic groups in Pantepui. Although the available information is largely descriptive, it constitutes the first empirical basis for a still-inexistent integrative biogeographical, ecological and evolutionary perspective of the unique Pantepui region.

Fig. 15.1 (continued) Angasima-tepui (2 km^2 and 2250 m) (left) and the Upuigma-tepui (>1 km^2 and 2100 m) (right) in the background. (6) Landing at the Eruoda-tepui summit in the Chimantá massif. (7) Angasima-tepui. (8) Ilú-Tramen massif (5.6 km^2 and 2700 m): Tramen-tepui (left), Ilú-tepui (right) and the Karaurín-tepui (2 km^2 and 2500 m) in the front. (9) Upuigma-tepui. (10) Los Testigos massif. From left to right: Aparamán-tepui (1 km^2 and 2100 m), Murisipán-tepui (5 km^2 and 2350 m), Tereké-yurén-tepui (<1 km^2 and 1900 m) and Kamarkawarai-tepui (5 km^2 and 2400 m). Photos: V. Rull

This review is a first trial towards such integrative view aimed at developing a preliminary holistic framework encompassing biogeographical patterns, ecological and evolutionary processes and mechanisms, and environmental drivers such as past climatic changes and paleogeographical reorganizations. The paper is subdivided into three main sections. The first section briefly describes the main physical and environmental features of Pantepui. The second section is a synthesis of the existing knowledge of its biogeography, ecology, evolution and conservation. The third part suggests future research directions in the same fields. It is hoped that this review will help set the stage for further studies, hopefully leading to an integrative understanding of the biological features of the unique Pantepui region and, eventually, the general progress of tropical ecology, evolution and biogeography, especially in relation to the origin of Neotropical biodiversity and latitudinal biodiversity gradients (Rull 2019a). It is also hoped that the ensuing research will contribute to the establishment of a more suitable framework for the preservation of Pantepui biota and ecosystems in the face of ongoing threats derived from human actions.

2 Pantepui

2.1 *Physical Setting*

Physiographically, the Guiana region has been subdivided into three main altitudinal levels, characterized by different physical, environmental and biotic features to define six main life zones (Fig. 15.3). The lowlands (<500 m elevation) are characterized by premontane and the basimontane life zones, the uplands (500–1500 m) are characterized by lower montane and montane life zones, and the highlands, where the upper montane and high-tepui life zones develop (Fig. 15.2). The tepui summits are part of the highlands. Approximately 70 tepuis and high

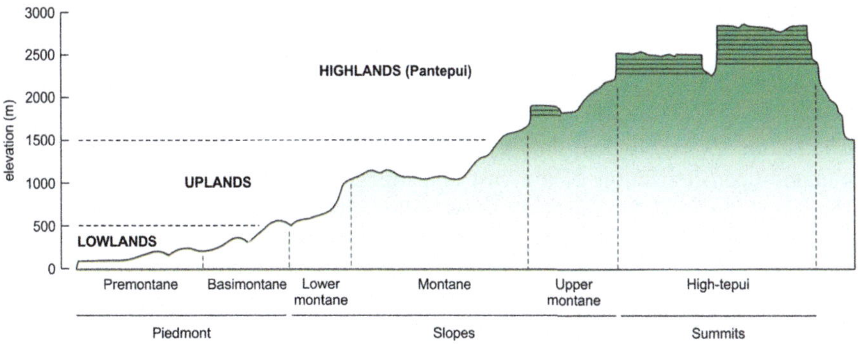

Fig. 15.2 Physiographic sketch of the Guiana region around Pantepui indicating the elevational belts (lowlands, uplands, highlands) and the altitudinal sequence of life zones (piedmont, slopes, summits). Redrawn from Huber (1995c)

mountains have been recognized, attaining a total surface area of >5000 km^2; most of them are located in Venezuela, with a few representatives in Brazil and Guyana (Table 15.1, Fig. 15.3). In Venezuela, the term "Guayana" is frequently used instead of "Guiana" but the meaning is exactly the same (Berry et al. 1995b). Care should be taken to not confound these terms with the country name "Guyana".

Geologically, the Guiana region lies on the igneous-metamorphic Guiana Shield, one of the oldest South American rock complexes formed by Archaeo-Proterozoic granites and gneisses (Huber 1995a). This is the basement on which the quartzites/sandstones of the Roraima Group sedimented during the Precambrian. Further erosion of the Roraima Group caused the development of several planation surfaces from which the typical tabular physiography of the Guiana Highlands originated (Fig. 15.4). The tepui summits that constitute the Pantepui archipelago of sky islands are the isolated remnants of the highest (and oldest) of these erosion surfaces (Briceño and Schubert 1990). Details on the geochemical weathering of the Roraima sandstones can be found in Mecchia et al. (2019) and the literature therein.

The tepui summits are a mixture of bare rock, extensive peat blankets approximately 3–4 m deep and occasional diabase outcrops. Peatland soils (histosols) are highly acidic and poor in nutrients, which largely constrains the type of plant communities growing on them and has fostered particular morphological and ecophysiological adaptions to such substrates (Cuevas 1992; Zink and Huber 2011). Soils developed on diabase outcrops (entisols) are slightly more fertile, which promotes the establishment of biotic communities that cannot grow on other substrates (Huber 1995c). The tepui summits have been included in the so called "old, climatically buffered, infertile landscapes" (OCBILs), to differentiate them from the "young, often disturbed, fertile landscapes" (YODFELs), on which conventional ecological and evolutionary theories have developed. OCBILs are rare—other examples are the Southwest Australian Floristic Region and the South Africa Greater Cape—but it is believed that their extreme environmental conditions and the peculiar ecological traits and adaptions of their biotas will be able to provide new and interesting ecological and evolutionary insights (Hopper 2009).

2.2 Climate

The scarce meteorological data available for Pantepui is insufficient for a sound climatic characterization of the vast and topographically complex Pantepui area. However, the available data—12 years (1997–2009) of measurements from three weather stations situated between 1750 m and 2600 m elevation—enable some preliminary insights (Huber and García 2011). The annual average temperature ranges between 16.5 °C and 11.4 °C with a general elevational decrease (adiabatic lapse rate) of approximately −0.6 °C/100 m. This is in agreement with the general topographic temperature decrease for the whole Guiana region, which largely determines the altitudinal distribution of the above-mentioned life zones (Huber 1995c). Minimum air temperatures are always above 0 °C and frost has never been

Table 15.1 High mountains and tepuis of Pantepui (see Fig. 15.2 for location)

Mountain complex	Tepui/mountain unit	Abbreviation (Fig. 15.1)	Country	Maximum elevation (m)	Summit area (km^2)	Summit-slope connection	Slope area (km^2)
Eastern tepui chain	Uei-tepui	Ui	GU	2150	2.5	3	20
	Roraima-tepui	Ro	VE-GU-BR	2723	34.38	0	300
	Kukenán (Matauí)-tepui	Ku	VE	2650	20.63	0	
	Yuruaní-tepui	Yu	VE	2400	4.38	0	
	Wadakapiapué-tepui	Wd	VE	2000	<0.01	0	
	Karaurín-tepui	Kr	VE	2500	1.88	3	
	Ilú-tepui	Iu	VE	2700	5.63	0	
	Tramen-tepui	Tr	VE				
	Wei-Assipu-tepui	Wa	GU-BR	ca. 2260	3	NA	NA
	Mount Ayanganna	An	GU	2080	NA	0	NA
	Mount Wokomong	Wo	GU	1680	300	3	NA
	Mount Maringma	Mm	GU	2134	1.7	NA	NA
	Monte Caburaí	Cb	GU-BR	1465	NA	NA	NA
Gran Sabana uplands	Cerro Venamo (Waukauyengtipu)	Vn	VE-GU	1600	NA	NA	NA
	Sierra de Lema	Le	VE	800–1650	NA	3	NA
Ptari massif	Ptari-tepui	Pt	VE	2400	1.25	0	28
	Carrao-tepui	Ca	VE	2200	1.25	NA	NA
	Sororopán-tepui	Sp	VE	2050	NA	NA	30
Los Testigos massif	Kamarkawarai-tepui	Ts	VE	2400	5	0	NA
	Tereke-yurén-tepui	Ts	VE	1900	0.63	0	NA
	Murisipán-tepui	Ts	VE	2350	5	1	88
	Aparamán-tepui	Ts	VE	2100	1.25	0	28

(continued)

Table 15.1 (continued)

Mountain complex	Tepui/mountain unit	Abbreviation (Fig. 15.1)	Country	Maximum elevation (m)	Summit area (km^2)	Summit-slope connection	Slope area (km^2)
Auyán massif	Auyán-tepui	Ay	VE	2450	666.9	2	715
	Cerro La Luna	Lu	VE	1650	0.2	1	NA
	Cerro El Sol	So	VE	1750	0.6	1	NA
	Uaipán-tepui	Ua	VE	1950	2.5	3	60
Canaima uplands	Cerro Venado	Ve	VE	1320	0.6	1	17
	Kurún-tepui	Kn	VE	1100	0.6	1	NA
Aprada massif	Aprada-tepui	Ap	VE	2500	4.37	1	210
	Araopán-tepui	Ar	VE	2450	1.25	1	
Chimantá massif	Sumits of 10 undifferentiated tepuis: Abapaká, Akopán, Agparamán, Amurí, Apakará, Eruoda (Murey), Chimantá, Churí, Tirepón and Toronó.	Ch	VE	2200–2650	615	2–3	915
	Angasima-tepui	Ag	VE	2250	2	0	32
	Upuigma-tepui	Ug	VE	2100	0.63	0	13
Paragua uplands	Sierra Marutaní (Pia-Zoi)	Mt	VE	1500	740	3	NA
	Cerro Ichún	Ic	VE	1400	2460	1	798
	Cerro Guanacoco	Gc	VE	1500	526.25	2	400
	Cerro Guaiquinima	Gq	VE	700–1650	1096.26	2	410
Jaua massif	Cerro Sarisariñama	Sr	VE	2350	546.88	2	286
	Cerro Jaua	Ja	VE	2250	625.62	2	482
Maigualida massif	Sierra de Maigualida	Mg	VE	2400	440	3	NA
	Serranía de Uasadi	Us	VE	1300–1800	N.A	3	NA
Yaví massif	Cerro Ualipano	Up	VE	1800	N.A	NA	NA
	Cerro Yaví	Yv	VE	2300	5.62	1	70
Yutajé massif	Serranía Yutajé	Yt	VE	2140	95.63	1	NA
	Coro Coro	Co	VE	2400	179.38	2	143
	Cerro Guanay	Gy	VE	2080	165	1	113
	Cerro Camani	Cm	VE	1800	1.88	3	NA

Sipapo uplands	Cerro Ovaña (Ouana)	Ov	VE	1800	NA	NA	NA
	Cerro Moriche	Mo	VE	1250	0.2	3	31
Parú massif	Cerro Parú (Asisa)	Pr	VE	2200	724.38	2	580
	Cerro Euaja	Eu	VE	2000	205.62	2	NA
Cuao-Sipapo massif	Cerro Autana	Au	VE	1300	1.88	0	NA
	Cerro Cuao	Cu	VE	2000	80	2	282
	Cerro Sipapo	Si	VE	1800	56	3	NA
Yapacana upland	Cerro Yapacana	Yp	VE	1300	10.5	1	38
Duida-Marahuaka massif	Cerro Huachamakari	Hu	VE	1900	8.75	1	60
	Cerro Marahuaka	Mk	VE	2800	121	0	325
	Cerro Duida	Du	VE	2358	1089	2	715
Parima uplands	Sierra de Parima	Pm	VE	800–1600	NA	3	NA
Unturán uplands	Cerro Aratitiyope	At	VE	1700	<0.01	0	NA
	Sierra Unturán	Un	VE	1600	NA	3	NA
Tapirapecó massif	Cerro Tamacuari	Tc	VE	2340	<0.01	NA	NA
	Serranía Tapirapecó	Tp	VE-BR	2000	NA	1?	NA
Aracamuni-Avispa uplands	Cerro Arakamuni	Am	VE	1600	NA	2	NA
	Cerro Avispa	Av	VE	1600	238	1	658
Imeri massif	Sierra de la Neblina	Nb	VE-BR	2994	235	2	857
Mocidade massif	Serra da Mocidade	Mc	BR	1900	NA	NA	NA
Aracá massif	Serra do Aracá	Ac	BR	1700	NA	NA	NA

Countries: *BR* Brazil, *GU* Guyana, *SU* Surinam, *VE* Venezuela; summit/slope connection: 0—no connection, 1—deep rocky canyons, 2—valleys, 3—slopes without major interruptions; *NA* not ascertained/not applicable. Reproduced from Rull et al. (2019b)

Fig. 15.3 Topographic map of the Guiana Highlands, with the locations of its tepuis and tepuian massifs (Rull et al. 2019b). Abbreviations in Table 15.1. Base map: radar image courtesy of NASA/JPL/SRTM, Feb 2000 (freely available at https://photojournal.jpl.nasa.gov/catalog/pia03388)

reported (Huber 1995a). Total annual precipitation increases with elevation from 2800 to 5300 mm per year, at a rate of approximately 30 mm/100 m. Additional moisture is provided by frequent dense mists. Seasonality in temperature is negligible and rainfall may vary throughout the year (rainfalls may be less common between December and March) but there is not a true dry season (<60 mm/month), especially on higher tepuis, where rainfall seasonality is minor. Winds and thunderstorms are frequent, especially during the rainier season (March to November). These data agree with previous inferences regarding mesothermic (lower tepui summits) to microthermic (higher tepui summits) ombrophilous climates, based on altitudinal inferences (Huber 1995a).

3 Existing Knowledge

The term "Pantepui" was coined by Mayr and Phelps (1967) to designate the assemblage of "…sandstone tabletop mountains in the Venezuelan Territorio Amazonas and Estado Bolivar and in the adjacent border regions of Brazil and Guyana", whose avifauna is highly endemic and strikingly different from that in the

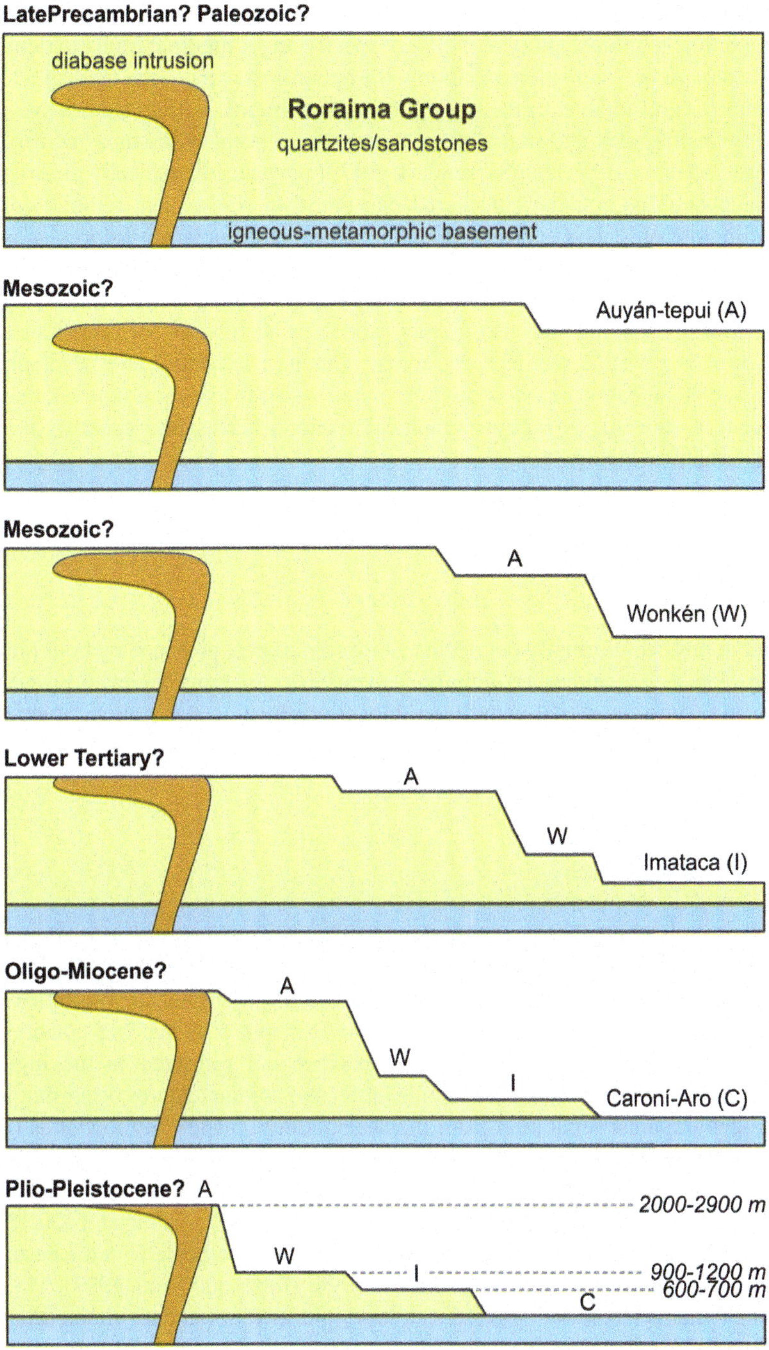

Fig. 15.4 Formation of the different erosion surfaces (Auyán-tepui, Wonkén and Imataca) on the Precambrian Roraima Group and on the igneous-metamorphic basement (Caroní-Aro erosion surface) from the Precambrian to the Pleistocene. Redrawn from Briceño and Schubert (1990)

surrounding lowlands. These authors considered Pantepui as an artificial unit rather than a biogeographical entity and did not specify any altitudinal limits to define this unit, although they considered that only the uppermost parts of the tepuis (>1500 m elevation) were suitable for the typical Pantepui avifauna. Further studies on insects, amphibians, reptiles and mammals adopted this concept with some modifications (review in Huber 1987), but Steyermark (1979) introduced a radically different idea of Pantepui. This botanist considered that Pantepui was a plant refugial complex including not only the Guiana Highlands but also some uplands and lowlands from the Central and Western Guiana provinces. This expanded Pantepui would have been a refugial area for rainforests and other tropical plant communities to persist and diversify during the assumedly arid glacial phases of the Pleistocene (Steyermark 1979). Sometimes, the terms "Guiana Highlands" and "Pantepui" are used interchangeably but it should be noted that the concepts behind them are different, as the first is a physiographical setting whereas the second is a formal biogeographical subdivision.

3.1 Biogeography

Pantepui has been formally defined as a biogeographical province by both botanists and zoologists but neither the definitions themselves, in terms of elevation and biotic characterization, nor the underlying biogeographical classification coincide. A recent update of the main plant and animal groups from Pantepui has provided a more comprehensive and integrative view.

3.1.1 Phytogeography

The Pantepui province was first defined in botanical terms, as part of the Guiana region, within the Neotropical realm (Fig. 15.5). The Guiana region has been subdivided into four provinces: Central Guiana, Eastern Guiana, Western Guiana and Pantepui (Fig. 15.3). Geographically, the Pantepui province was considered to be located within the Central Guiana province but restricted to the highlands (1500–3000 m elevation), where unique biotic and abiotic features occur that justify its biogeographical separation from the surrounding uplands and lowlands (Huber 1994). The higher limit of the Pantepui phytogeographical province is the maximum elevation of the Guiana Highlands (Sierra de la Neblina; 2994 m) but the lower limit has been modified over time. Huber (1987) set this lower limit at 1200/1500 m elevation. Later, the lower boundary was considered to coincide with the base of the Guiana Highlands, that is, 1500 m (Huber 1994; Huber and Riina 1997). The latest botanical characterization returned to a less rigid lower boundary situated between 1300 and 1500 m elevation (Huber et al. 2018).

According to Huber et al. (2018), the most characteristic features of the Pantepui province are the typical broadleaved meadows and hardleaved (sclerophyllous) woodlands growing on peat, the pioneering plant communities living on bare

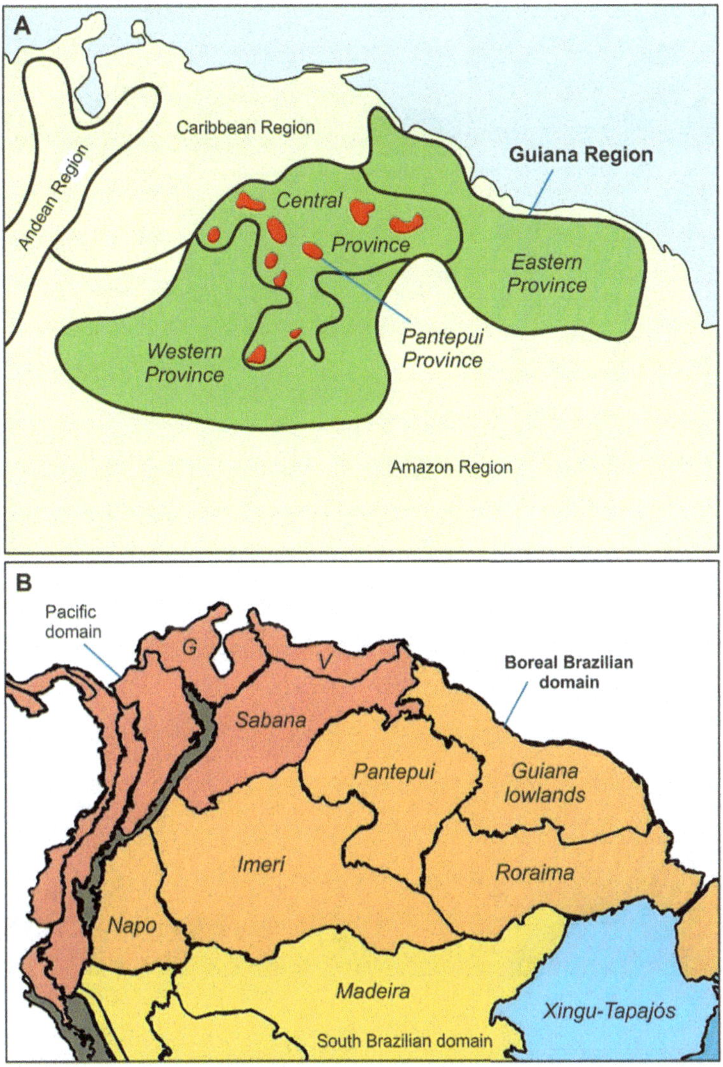

Fig. 15.5 Graphical representation of the phytogeographical (**a**) and zoogeographical (**b**) definitions of the Pantepui province. (**a**) Phytogeographical Pantepui. The Guiana region (green) and its four biogeographical provinces. The Pantepui province is represented as a discontinuous surface of red spots representing the major tepuis and tepuian massifs. Redrawn from Berry et al. (1995a). (**b**) Zoogeographical Pantepui. Zoogeographical provinces of northern South America. Colors represent the domains (pink—Pacific, orange—Boreal Brazilian, yellow—South Brazilian, blue—South-eastern Amazonian). G—Guajira, V—Venezuela, P—Pará. Modified from Morrone (2014)

rocks (saxicolous) and the comparatively small patches of low forests growing in depressions and crevasses and along water courses (Fig. 15.6). The most characteristic plant families of Pantepui are the herbaceous Rapateaceae, the woody Bonnetiaceae and the pitcher plant family Sarraceniaceae, with their Pantepui

Fig. 15.6 Vegetation types representative of Pantepui (see Fig. 15.6 for representative species). (1) *Bonnetia roraimae* forest stands on protected sites of the Eruoda-tepui (Chimantá massif). (2) Aerial view of the Chimantá summit showing the typical gallery forests of *Bonnetia roraimae* (Bonnetiaceae) along a water course surrounded by broad-lived meadows dominated by *Stegolepis ligulata* (Rapateaceae). (3) General view of a paramoid shrubland on the Apakará summit of the

endemic genus *Heliamphora* (Fig. 15.7). In general, montane shrublands, meadows and pioneer communities dominate the landscape, whereas forests play a subordinate role and are typically low and relatively species-poor. Shrublands are highly varied, ranging from dense high-tepui scrub to paramoid (for their similarity with the Andean paramo biome) and ericoid (i.e., dominated by plants of the family Ericaceae) scrub. Meadows are well-differentiated into broad-leaved, tubuliform and rosette meadows with several areas of grasslands (Berry et al. 1995b). Using the compilations of Berry and Riina (2005) and Riina et al. (2019), it has been estimated that vascular plant richness of Pantepui is approximately 5000 species per 10,000 km^2, making it one of the most biodiverse areas worldwide (Mutke and Bathlott 2005), and endemism is 35–40%, which is also very high and comparable to that of many oceanic archipelagos.

3.1.2 Zoogeography

Some zoologists have used the phytogeographical concept as a working framework (e.g., McDiarmid and Donnelly 2005), whereas others have included both lowlands and uplands when defining Pantepui (e.g., Kok et al. 2016; Borges et al. 2018). A number of zoologists believe that the concept of Pantepui as a biogeographical province should be reconsidered and redefined (e.g., Costa et al. 2013). A recent revision of the Neotropical biogeography by Morrone (2014) used primarily zoological criteria to formally define the Pantepui province (Fig. 15.5). This definition was explicitly considered by the author to be analogous to the informal Pantepui concept of Mayr and Phelps (1967) and the formal Pantepui phytogeographical province definition of Huber (1994). In this case, however, no distinction was made among lowlands, uplands and highlands; therefore, elevation was not used as a criterion to differentiate Pantepui from the surrounding terrains that, in the phytogeographical classification, define the Central Guiana province (Fig. 15.1). Another significant difference is that, whereas the phytogeographical definition places Pantepui within the Guiana region of the Neotropical realm, the zoogeographical conception does not define a Guiana unit and Pantepui is placed within the Boreal Brazilian dominion. In addition, the Neotropics is considered to be a biogeographical region, rather than a realm (Morrone 2014). The zoogeographical Pantepui

Fig. 15.6 (continued) Chimantá massif, dominated by *Chimantaea mirabilis* (Asteraceae) with an herbaceous layer dominated by *Myriocladus steyermarkii* (Poaceae, Bambusoideae). (4) Broad-leaved meadows of *Stegolepis guianensis* (Rapateaceae) with *Brocchinia tatei* (Bromeliaceae) rosettes on the summit of the Uei-tepui. (5) Peat bog on the summit of the Eruoda tepui (Chimantá massif) dominated by tubular rosettes of *Brocchinia hechtioides* (Bromeliaceae), with emergent caulirossulae of *Chimantaea lanocaulis* (Asteraceae). (6) Aerial view of the Chimantá summit showing wide extensions of cracked and bare rock with scattered stands of pioneer vegetation. (7) Closer view of the initial stages of rock colonization by pioneer plant communities (Eruoda summit, Chimantá massif). Photos: V. Rull

Fig. 15.7 Representative vascular plants of Pantepui. (1) *Bonnetia roraimae* (Bonnetiaceae), Eruoda-tepui (Chimantá massif). (2) *Chimantaea mirabilis* (Asteraceae), Apakará-tepui (Chimantá massif). (3 and 4) *Stegolepis ligulata* (Rapateaceae), Apakará-tepui (Chimantá massif). (5) *Chimantaea lanocaulis* (Asteraceae), Eruoda-tepui (Chimantá massif). (6) *Brocchinia tatei* (Bromeliaceae), Uei-tepui. (7) *Brocchinia hechtioides* (Bromeliaceae), Eruoda-tepui (Chimantá massif). (8 and 9) *Heliamphora minor* (Sarraceniaceae), Eruoda-tepui (Chimantá massif). Photos: V. Rull

was based on an extensive list of endemics, including a single plant genus, and 100 genera/species of arthropods and vertebrates.

3.1.3 An Integrative View

The latest comprehensive review based on the main plant and animal groups studied to date in Pantepui (Rull et al. 2019a) is used here as the basis for a more holistic definition of the Pantepui province in biotic terms. Table 15.2 summarizes the documented richness and endemism of each taxonomic group studied to date in Pantepui, as well as the lower boundary of this biogeographical province considered for each of these groups. It should be highlighted that this information is still far from

Table 15.2 Lower Pantepui limit considered for each taxonomic group, total richness and endemism patterns

Taxonomic groups	Lower Pantepui limit (m)	Species richness	Pantepui endemics	Single-tepui endemics	References
Algae	1300	>300	~10 (~33%)	NA	Kaštovský et al. (2019)
Bryophytes	1500	~320	30 (~9%)	9 (~3%)	Désamoré et al. (2010), Riina et al. (2019)
Vascular plants	**1500**	**2579**	**885 (34%)**	**640 (25%)**	Riina et al. (2019)
Aquatic insects	1500	~140	~80 (57%)	~60 (~43%)	Derka et al. (2019)
Butterflies	~1000[a]	NA	94 (24–29%)[b]	NA	Viloria and Costa (2019)
Scorpions	1400	11	8 (73%)	8 (73%)	Ochoa and Rojas-Runjaic (2019)
Land snails	1500	24	20 (83%)	12 (50%)	Breure (2019)
Parasites (insects, acari, worms)	NA	67	1 (<2%)	0 (0%)	Guerrero (2019)
Amphibians	**1200**	**109**	**60 (55%)**	**46 (42%)**	Señaris and Rojas-Runjaic (2019)
Reptiles	**1200**	**63**	**28 (44%)**	**22 (35%)**	Señaris and Rojas-Runjaic (2019)
Birds	**1300**[c]	**141**	**41 (29%)**	**2 (<1%)**	Pérez-Emán et al. (2019)
Mammals	**1500**	**63**	**9 (14%)**	**0 (0%)**	Lew and Lim (2019)

The better known groups are in bold. *NA* not ascertained
[a]Not explicitly stated but deduced from the fact that Pantepui endemics appear at this elevation and increase upwards
[b]Percentage based on estimations from the Eastern District, which is the best known for this group
[c]Not explicitly stated but deduced form the maximum elevation of Cerro Yapacana, which is mentioned by the authors as the lower limit for the region considered in this paper

being complete and that some groups are better known than others. The most studied groups are vascular plants, amphibians, reptiles, birds and mammals and the poorly known groups are algae, bryophytes, insects, arachnids and snails. Fishes, a group that is especially diverse in the Neotropics, is absent on Pantepui. No typically highland fish species have been found to date and lowland/upland species rarely cross the lower elevational boundary of Pantepui, with very few exceptions (Lasso et al. 1989). Some examples of emblematic and representative species of groups displayed in Table 15.2 are depicted in Figs. 15.8 and 15.9.

From approximately half of the cases (bryophytes, vascular plants, aquatic insects, land snails and mammals), the lower Pantepui phytogeographical boundary of 1500 m is explicitly adopted, whereas for others, lower limits of 1200–1400 m are preferred. No explanation is provided in the case of algae (1300 m) and scorpions (1400 m) but in the case of butterflies, amphibians, reptiles and birds (1200–1300 m), it is explicitly argued that the 1500 m boundary is not appropriate, as tepui summits below this elevation hold faunas characteristic of Pantepui. Butterflies could be considered as an outlier, as the proposed boundary at ~1000 m includes most of the Gran Sabana uplands, which are biogeographically very different for all other organisms. However, an artifact due to flying ability could not be disregarded (Viloria and Costa 2019). The same could hold true for birds. According to these numbers, the former phytogeographical definition of Huber (1987), who situated the lower boundary between 1200 and 1500 m, would be adopted, and the specific elevation for each taxonomic group may be defined on the basis of its particular biogeographical patterns. Following these criteria, based on the widest range of taxonomic groups available to date, it is recommended that the zoogeographical definition of the Pantepui province (Morrone 2014), which does not explicitly consider altitudinal boundaries and hence implicitly includes lowlands, uplands and highlands, is revised on the basis of the most recent updates.

Regarding richness and endemism, it is necessary to distinguish between the better-known Pantepui organisms, whose numbers may be considered to be more stable, and the less-known groups, for which studies are still incipient and reliable estimates require further evaluation. Considering only the first group, plants are by far the more diverse group with almost 2600 known species, whereas the number of animal species ranges from approximately 60 to 100 (Table 15.2). Endemism ranges from 14% in mammals to 55% in amphibians, with intermediate values for birds (29%), plants (34%) and reptiles (44%). The percentage of single-tepui endemics is also high for amphibians and reptiles (42% and 35%, respectively), intermediate for plants (25%) and negligible for birds and mammals. In general, further studies are expected to increase richness and decrease endemism estimates, which is common in biogeographical studies.

Fig. 15.8 Examples of Pantepui bryophytes, algae and animals (aquatic insects, butterflies and scorpions). (1) Cushion-like *Sphagnum* colonies widespread in wetlands on the Apakará-tepui (Chimantá massif) (Photo: V. Rull). (2) Algal mats colonizing bare rock surfaces on the Eruoda-tepui (Chimantá massif) (Photo: V. Rull). (3) *Enderleina preclara*, found on Akopán-tepui (Chimantá massif) and Kukenán-tepui (Photo: T. Derka). (4) *Parakari churiensis*, endemic to

3.2 Ecology

Ecology is the Cinderella of Pantepui research. Autecological studies of plant and animal species are almost nonexistent except for scattered observations of apparent preferences of certain species for specific substrates (bare rocks, peat bogs, sandstones, diabases, caves) or assumed biotic associations such as symbiotic or pollination relationships. However, these are sporadic field annotations that eventually appear in the description of certain species, and no systematic autecological studies exist for any Pantepui species. The most complete information regarding habitat preferences is the record of the elevational ranges of occurrence of all known vascular plant species available in the outstanding *Flora of the Venezuelan Guayana* (Steyermark et al. 1995–2005), which represents a milestone of Guianan biological research. Further updates are available from Berry and Riina (2005) and Riina et al. (2019). Similar information is available for amphibians, reptiles, birds and mammals (Señaris and Rojas-Runjaic 2019; Pérez-Emán et al. 2019; Lew and Lim 2019). Ecological studies at community level are better developed for plants but they are rather descriptive (review in Huber and Rull 2019), investigations of internal ecosystem dynamics and their relationships with external drivers are absent. Information regarding long-term ecological processes, such as community assembly and the potential role of Quaternary climatic changes in the shaping of present-day ecological communities, has also been provided by paleoecological studies.

3.2.1 Present-Day Patterns

The more systematic ecological studies have involved the identification, description and characterization of Pantepui vegetation types. Whereas most classical botanists have been primarily interested in its floristic aspects, Huber (1992, 1995c) developed an intensive and extensive record of plant associations that culminated in the definition of five main vegetation types: forests, shrublands, grasslands, meadows and pioneer communities. Within these general types, 40 plant communities have been defined, including 12 types of forests, 10 types of shrublands, 2 types of grasslands and 16 types of meadows (Huber and Rull 2019). Pioneer communities have not yet been classified. Based on spatial vegetation patterns, Vareschi (1992)

Fig. 15.8 (continued) Churí-tepui (Chimantá massif) (Photo: T. Derka). (5) *Entophysalis arboriformis* endemic to the Eastern District (Photo: J. Kaštovský). (6) *Ekerewekia churicola*, endemic to Churí-tepui (Photo: J. Kaštovský). (7) *Hydrolutos breweri*, endemic to Pantepui (Photo: T. Derka). (8) *Heliconius elevatus roraima*, endemic to Pantepui (Photo: M. Costa). (9) *Mesotaenia vaninka delafuentei*, endemic to Pantepui (Photo: M. Costa). (10) *Callithomia lenea bella*, endemic to Pantepui (Photo: M. Costa). (11) *Oxeoschistus romeo*, endemic to Pantepui (Photo: M. Costa). (12) *Taurepania porosa* from Roraima-tepui, endemic to Pantepui (Photo: F. Rojas-Runjaic). (13) *Vachoniochactas amazonicus* from Sierra de la Neblina, endemic to Pantepui (Photo: F. Rojas-Runjaic)

Fig. 15.9 Examples of Pantepui animals (vertebrate parasites, snails, amphibians reptiles, birds and mammals). (1) *Laelaps conula* found on the mouse *Rhipidomys macconnelli*, from Auyán-tepui. The only parasite considered to be endemic to Pantepui (Photo: R. Guerrero). (2) *Plekocheilus (Eurytus) juliani* from the Chimantá massif, endemic to Pantepui (Photo: A. Breure). (3) *Plekocheilus (Eurytus) sophiae* from Yuruaní-tepui, endemic to Pantepui (Photo: A. Breure).

speculated about the potential successional trends leading to the extant vegetation and considered that forests were the climax communities toward which natural succession was directed. However, no empirical evidence supporting or dismissing this interpretation has been found thus far.

3.2.2 Long-Term Processes

Paleoecological records have shown that present-day Pantepui communities originated following the complex succession of climatic shifts that fostered up-and-down migrations of sensitive species from the Last Glacial Maximum (LGM; ~21,000 years ago) to the present (see Rull et al. 2019c for a detailed review). Overall, this process can be viewed as a millennial-scale upward migration trend, punctuated with minor downward displacements driven by smaller temperature and moisture oscillations that occurred during the Holocene (the last 11,700 years). Species that migrated downwards during the LGM could have spread across the surrounding lowlands and uplands thus having the possibility of climbing to several tepuis during the Holocene warming (Fig. 15.10). However, not all tepui summits could have had the possibility of sharing species in this way, as a number of them are higher than the magnitude of the estimated downward biotic displacement, are totally isolated by vertical cliffs preventing elevational migrations, or both. The irregularity of climatic oscillations, combined with the complicated topography of the Guiana Highlands and the fact that species' responses to climatic shifts are not homogeneous but idiosyncratic—i.e., dependent on the particular climatic requirements and tolerances of each involved species—makes the general upward migration since the LGM—and, therefore, the assembly of extant Pantepui communities—a very complex and unpredictable process. A geographical information systems (GIS) modelling approach has been used to reconstruct the potential routes and barriers for upward migration, and their changes over time and could be considered as a hypothetical framework to be tested in future studies (Rull and Nogué 2007).

Fire seems to have played some role in vegetation shifts, as major community replacements coincided with Holocene fires, especially those that occurred between 6000 and 5000 years ago and approximately 3500 years ago (Rull and Montoya

Fig. 15.9 (continued) (4) *Oreophrynella vazquezi*, endemic to Ilú and Tramen tepuis (Photo: J. Mesa). (5) *Anomaloglossus rufulus*, endemic to Eruoda-tepui (Chimantá massif) (Photo: J. M. Rojas-Runjaic). (6) *Tepuiphyla edelcae*, endemic to Auyán-tepui and Los Testigos massif (Photo: J. M. Rojas-Runjaic). (7) *Oreosaurus mcdiarmidi*, endemic to the Chimantá massif (Photo: J. M. Rojas-Runjaic). (8) *Anolis carlostoddi*, endemic to Abakapá-tepui (Chimantá massif) (Photo: J. Mesa). (9) *Thamnodynastes chimanta*, endemic to the Chimantá massif (Photo: J. M. Rojas-Runjaic). (10) *Campylopterus hyperythrus*, endemic to the Eastern District (Photo: D. Southall). (11) *Diglossa major*, endemic to the Eastern District (Photo: D. Southall). (12) *Myioborus castaneocapilla*, endemic to Pantepui (Photo: D. Southall). (13) *Marmosops pakaraimae*, endemic to the Eastern tepui chain (Photo: B. Lim). (14) *Platyrrhinus aritus*, endemic to Pantepui (Photo: B. Lim)

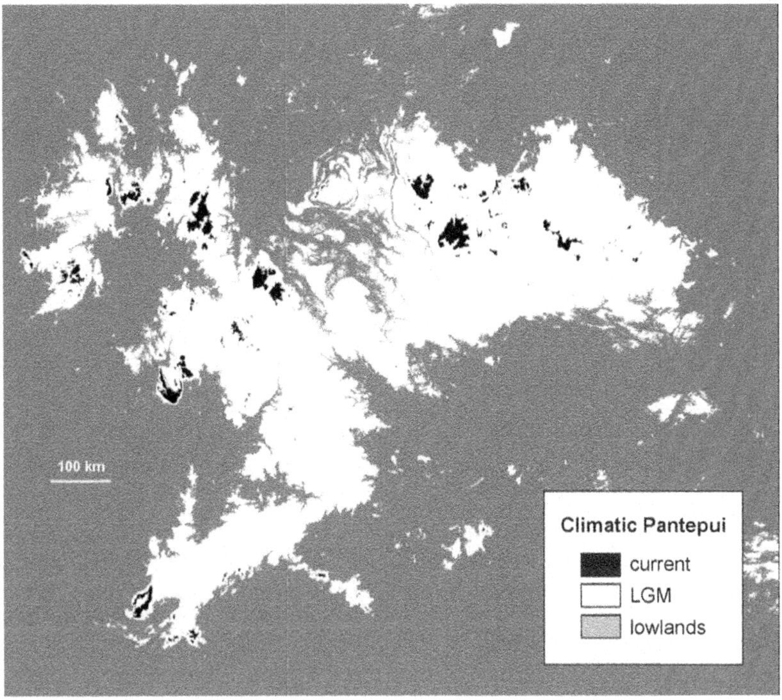

Fig. 15.10 Present (black) and simulated Last Glacial Maximum (LGM) extent of typical Pantepui climates (white). Lowlands below 400 m elevation are in grey. LGM simulation performed assuming a downward migration of climatic conditions up to 1100 m below the present lower boundary of Pantepui (1500 m) on a digital elevation model from the Shuttle Radar Topography Mission (USGS/NGA/NASA) of 5 arcsec (90 m) precision. Modified from Rull and Nogué (2007)

2017). Thus far, it is not possible to know whether these fires were natural or anthropogenic, but the lack of evidence regarding human presence and the coincidence of charcoal peaks with climate shifts, especially in terms of moisture, suggest that humans were not involved. During the last millennium, fire, likely of human origin, has deeply modified the ecosystems of a tepui summit (Uei-tepui), which is connected to the adjacent Gran Sabana uplands by gradual and fully vegetated slopes (Safont et al. 2016). No similar fire events have been recorded on other tepui summits, during historical times.

3.2.3 Biogeographical Implications

In addition to their relevance for community assembly and succession, Quaternary climate changes also affected the integrity of the Pantepui biogeographical province itself. The Pantepui concept has been subdivided into three main components: the orographic Pantepui, the climatic Pantepui and the biotic Pantepui. The first

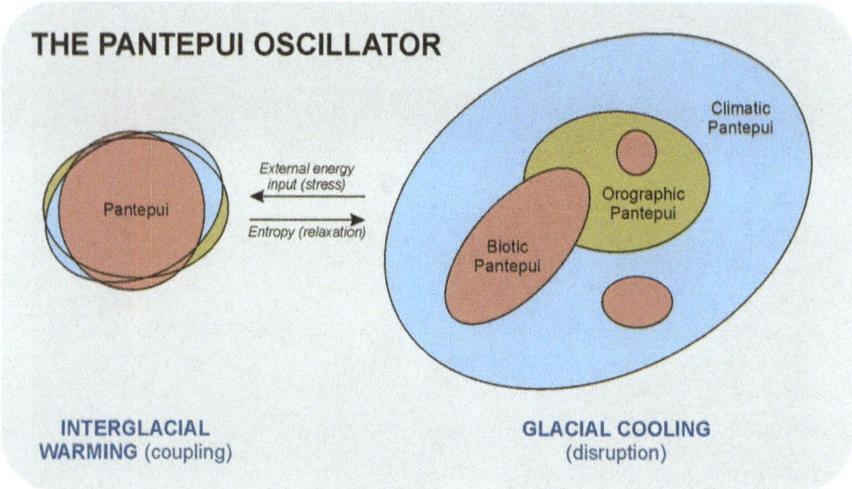

Fig. 15.11 The Pantepui oscillator. The orographic (solid lines), the climatic (dashed lines) and the biotic (dotted lines) Pantepui assemble during the interglacials and disassemble during glaciations, via the downslope migration of the latter two Pantepui components. The interglacial state is less stable, as it is maintained by external energy inputs (i.e., temperature maxima), whereas the glacial state is more stable, as entropy maximizes. Redrawn from Rull and Vegas-Vilarrúbia (2019a)

remained constant during the Quaternary but the other two underwent significant changes. The climatic Pantepui fluctuated approximately 50 times from glacial to interglacial conditions following a cyclicity of 41,000–100,000-year periods. During glacials, the biotic Pantepui disassembled as a result of the differential downward migration of sensitive species, according to their particular climatic requirements and tolerances. During interglacials, the biotic Pantepui reassembled as a result of upward migration but the species composition was not necessarily the same due to the species' climatic idiosyncrasies and eventual extinctions of summit elements. This recurrence has been called the Pantepui oscillator (Fig. 15.11). Quaternary evolution could also have played a role in the modification of the Pantepui biota. This introduces a new dynamic concept of Pantepui, as a biogeographical province, allowing it to be viewed as a typical interglacial feature that could have adopted approximately 50 different states during the Quaternary (Rull and Vegas-Vilarrúbia 2019a).

Glacial conditions have been the norm during the Quaternary, as they have persisted 80% of the time, whereas interglacials consisted of short warming peaks (Willis and Whittaker 2000; Bush et al. 2001). Therefore, glacial Pantepui disassembly could be considered the normal state, whereas the currently observed interglacial coincidence of the orographic, the climatic and the biotic pantepuis may be viewed as the exception. In addition, glacial-interglacial recurrence is typically asymmetric, with abrupt interglacial warming and gradual glacial cooling episodes. In thermodynamic terms, interglacials could be viewed as short unstable states maintained by external energy inputs that, once terminated, enable system relaxation and entropy maximization, which is characteristic of glaciations (Ellis and Palmer 2016). Therefore, glacial Pantepui disassembly could be viewed as the more stable state of Pantepui, with maximum entropy (disorder), whereas interglacial assembly (order) would be a transient condition maintained by the incoming of external energy inputs (Fig. 15.11).

3.3 *Evolution*

As has occurred in the Neotropics, in general, knowledge on the origin and evolution of Pantepui biodiversity can be subdivided into three main historical phases (Chap. 2). During the first phase, evolutionary inferences were based on biogeographical evidence, the second phase started with the advent of paleoecology and the third stage has been characterized by the recent remarkable development of molecular phylogenetic and phylogeographical studies (Rull 2019a).

3.3.1 The Biogeographical Phase

Mayr and Phelps (1967), the creators of the term "Pantepui", summarized the first hypotheses that emerged from the study of birds and mammals, into five main categories (Table 15.3). The Plateau Theory (PT) proposes that the faunal differences among the tepui summits are due to vicariance by physical isolation, after a long-standing process of the maintained erosion of a former plateau formed by the Roraima quartzites/sandstones. The Cool Climate Theory (CCT) contends that the Pantepui fauna colonized the tepui summits from the surrounding lowlands during the Quaternary glaciations, which could have facilitated connectivity between the lowlands and summits thus allowing biotic interchange. According to the Habitat Shift Theory (HST), the Pantepui fauna is derived from long-ranging species (i.e., able to live from lowlands to highlands), after lowland extinction and highland survival and further differentiation (parapatric speciation). The Distance Dispersal Theory (DDT) maintains that the Pantepui fauna was derived from other montane regions (the Andes, Caribbean coastal ranges, the Brazilian Highlands) by island hopping, or jump dispersal using their flying capacity to cross inhospitable terrains. The Specialized Habitat Theory (SHT) proposes that Pantepui species are ecological specialists with very specific habitat requirements available only in Pantepui.

Other authors did not explicitly refer to the above hypotheses although they used similar or identical proposals. For example, Maguire (1970) implicitly favored the PT, although he did not mention it, for the origin of the Pantepui flora. In contrast, Steyermark and Dunsterville (1980) and Huber (1988) favored the idea that Pantepui plant species could have migrated up and down following the Quaternary glacial-interglacial cycles. These contrasting views were later called the Lost World Hypothesis (LWH) and the Vertical Displacement Hypothesis (VDH), respectively (Rull 2004). The VDH is sometimes confused with the CCT, likely due to the

Table 15.3 Hypotheses proposed to explain the origin and evolution of the Pantepui biota

Hypotheses	Geographical extent	Abbreviation	Proposed mechanism	References
Plateau theory	Pantepui	PT	Vicariance on isolated tepui summits after erosional dissection of a former continuous high plateau	Chapman (1931), Tate (1938)
Cool climate theory	Pantepui	CCT	Upward migration from lowlands during glacial phases and further speciation	Chapman (1931), Tate (1938)
Habitat shift theory	Pantepui	HST	Parapatric speciation in former long-ranging species after lowland extinction and highland survival	Mayr and Phelps (1967)
Distance dispersal theory	Pantepui	DDT	Active dispersal from the Andes and other Neotropical mountain ranges and further differentiation	Mayr and Phelps (1967)
Specialized habitat theory	Pantepui	SHT	Spatial restriction due to specific autecological requirements present only on Pantepui	Mayr and Phelps (1967)
Lost world hypothesis	Pantepui	LWH	Adaptive radiation on the tepui summits isolated by Miocene uplift and progressive erosion (largely coincides with the PT)	Maguire (1970), Rull (2004)
Vertical displacement hypothesis	Pantepui	VDH	Downward (glacials) and upward (interglacials) migration by available routes connecting lowland/uplads and highlands	Steyermark and Dunsterville (1980), Huber (1988), Rull (2004)
Vicariance-migration hypothesis	Pantepui	VMH	Combination of the LWH and the VDH	Rull (2004, 2005)
Isolation-cooling hypothesis	Pantepui	ICH	Long-term Cenozoic vicariance (isolation), downward migration (cooling) and jump dispersal from other mountains	Rull et al. (2019b)
Disturbance-vicariance hypothesis	Neotropics	DVH	Similar to the VMH but including moderate moisture reduction and atmospheric CO_2 depletion, in addition to cooling, during glacials	Bush (1994), Colinvaux (1998)

(continued)

Table 15.3 (continued)

Hypotheses	Geographical extent	Abbreviation	Proposed mechanism	References
Continuum multifactor hypothesis	Neotropics	CMH	Coupled action of diversification drivers, processes and mechanisms, with different impacts on diverse taxonomic groups and geographical areas, at a variety of spatial and temporal scales during the Neogene and the Quaternary	Rull (2008, 2011)

See Rull et al. (2019b) for a more detailed review. Some hypotheses have been specifically developed for Panteui, whereas others have been proposed for the Neotropics, in general

possibility of vertical migration, but they are actually opposite views. The CCT proposes upward migration of the lowland fauna to the tepui summits during glacials, with no reference to the interglacials and is based on theoretical considerations. On the contrary, the VDH proposes downward, rather than upward, glacial migrations and upward migrations during interglacials and is based on worldwide empirical evidence, including the Neotropics.

3.3.2 The Paleoecological Phase

Paleoecological studies have provided evidence for testing the LWH and VDH, and for suggesting new proposals. Past climatic and ecological records suggest that a combination of these two hypotheses is needed to explain the Pantepui biotic patterns. This is called the Vicariance-Migration Hypothesis (VMH) (Table 15.3). According to this proposal, glaciations are characterized by the biotic spread of Pantepui species across uplands and lowlands thus promoting hybridization and adaptive radiation (Fig. 15.12). Interglacial upward migration would have favored vicariance and extinction by habitat loss (Rull 2005). The classic debate between vicariance vs. dispersalism is not compatible with the VMH. This hypothesis was useful to evaluate Quaternary diversification but meta-analyses including a wide range of organisms have demonstrated that Neotropical biodiversity originated in a continuous manner during Neogene and Quaternary times with no diversification bursts at any particular time period (Rull 2008). Therefore, the analysis of pre-Quaternary diversification requires a different approach.

The apparent absence of pre-Quaternary sediments in Pantepui and surrounding areas prevented empirical evidence from illustrating biotic diversification and the drivers involved during that time. In contrast to other Neotropical areas (as for example the Andes or the Panama Isthmus), which were subjected to intense tectonically-driven orographic and paleogeographical changes, the Guiana region remained relatively stable during the Neogene, the major geological process being the continued erosion of the Roraima quartzites and sandstones leading to the

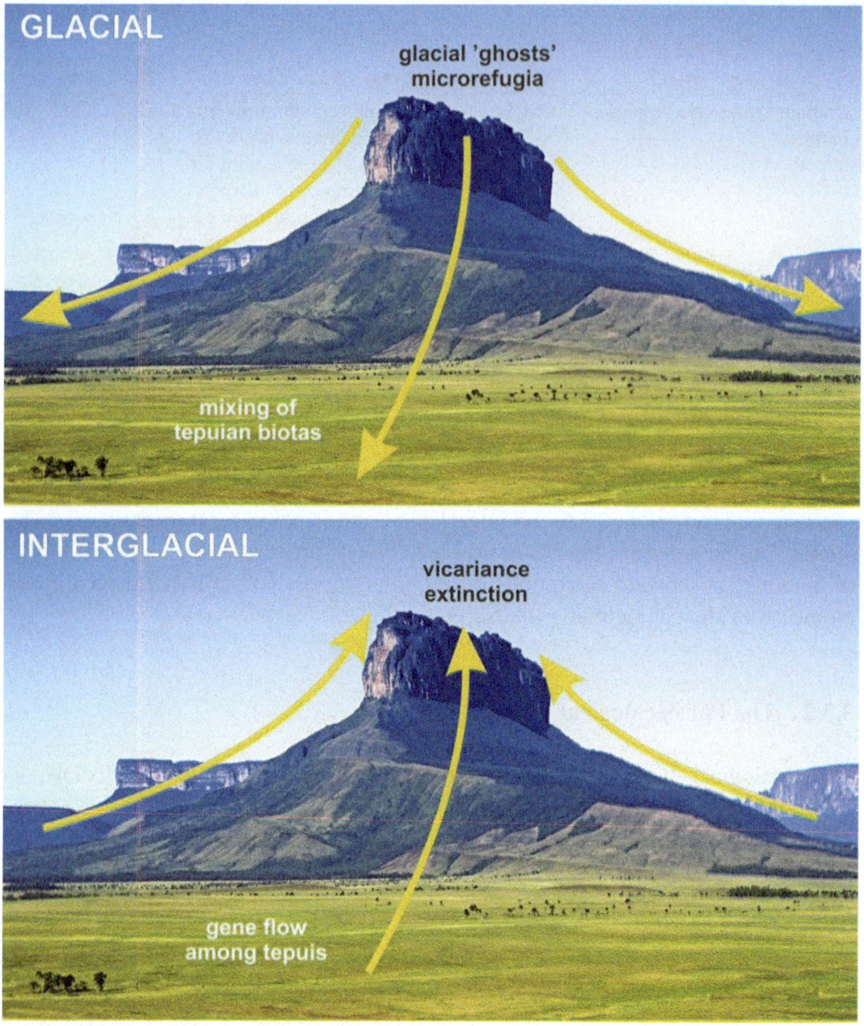

Fig. 15.12 Graphical representation of the Vicariance-Migration Hypothesis (VMH) (Rull 2005) using the Upuigma-tepui. Photo: V. Rull

formation of the tepuis (Fig. 15.4). However, the climate did not remain constant. During the Cenozoic, our biosphere has experienced a maintained million-scale global cooling since the Paleocene/Eocene boundary, when temperatures were approximately 14 °C higher than they are today on average (Hansen et al. 2013). Under these conditions, the early Cenozoic Pantepui biota was likely more similar, in bioclimatic terms, to the extant lowland biota. Therefore, the Cenozoic evolution of this biota took place on summits that were progressively smaller, lower, more isolated and cooler. Topographical isolation would have favored vicariance and extinction, whereas cooling would have fostered downward migration and species "pumping" to the surrounding lower terrains. This hypothesis has been called the Isolation-Cooling Hypothesis (ICH) (Table 15.3).

3.3.3 Molecular Phylogenetics and Phylogeography

Molecular phylogenetic studies and their geographical expression (phylogeography) are still scarce in Pantepui due to its remoteness and the difficulty of obtaining fieldwork permits for collecting genetic material (Rull and Vegas-Vilarrúbia 2008; Rull et al. 2009). However, the results obtained to date have been used to evaluate the existing hypotheses of Pantepui biotic evolution. The divergence time of the different taxa studied, inferred from time-calibrated phylogenetic trees, is a key parameter that has been used to support the different evolutionary hypotheses and theories, mainly for vascular plants, amphibians, birds and mammals. The main results obtained are as follows.

In plants, most studies of this type correspond to the families Rapateaceae and Bromeliaceae. The genus *Stegolepis* (Rapateaceae) would have invaded Pantepui at ~12 Ma and started to diversify at ~6 Ma by a combination of vicariance and bird-mediated dispersal (Givnish et al. 2000, 2004). In the Bromeliaceae, the *Brocchinia* species studied would have originated between 12 and 17 Ma, whereas the *Lindmannia* species would have started to diverge by 2.5 Ma, at the beginning of the Pleistocene (Givnish et al. 2007). Although the ICH was not yet proposed when these taxa were studied, the above results seem to be consistent with this proposal.

Among amphibians, the species of the genus *Tepuiphyla* were proposed to have originated between ~5 and ~2 Ma (Pliocene to Pleistocene), which was considered to support a combination of the HST and the VDH, thus dismissing the DDT (Salerno et al. 2012, 2014). Using the same genus and two other amphibian genera (*Oreophrynella* and *Stefania*), Kok et al. (2012) favored active/passive dispersal mechanisms, rather than ancient endemism, for the explanation of the lower than expected genetic variability. The genus *Stephania*, whose diversification occurred during the Miocene, was used by Kok et al. (2017) to support the PT, rather than the Disturbance-Vicariance Hypothesis (DVH) (Table 15.3), according to which current Neotropical biogeographical patterns could be explained by a combination of LGM cooling, a moderate precipitation reduction and atmospheric CO_2 depletion (Bush 1994; Colinvaux 1998). A combination of the DDT and PT was proposed by Kok et al. (2018) after studying the genera *Oreophrynella* and *Atelopus*, whose divergence occurred ~40 Ma (Eocene) in the proto-Andes, after which they reached Pantepui by jump dispersal and diversified there by vicariance.

Regarding birds, a study of the genus *Myoborus* (redstars) was used to support the DVH, after initial dispersal from the Andes and/or the Venezuelan Coastal Range and further *in situ* diversification on the tepui summits (Pérez-Emán 2005). The Pantepui species of the genus *Aulacorhynchus* (toucanets) emerged during the Pliocene, which was used to support the DDT followed by Pleistocene population differentiation (Bonaccorso and Guayasamin 2013). Studies on other Neotropical bird genera with representatives on Pantepui also favored dispersal from either the Andes or the surrounding Guiana lowlands (Smith et al. 2014; Berv and Prum 2014).

The representatives of Pantepui mammal fauna studied to date using molecular methods seem to favor the DDT from a variety of Neotropical areas, including the Andes, the Amazon lowlands and the Brazilian Shield. For example, the ancestor of the opossum *Monodelphis reigi* was proposed to have dispersed to Pantepui from the Andes during the Miocene (Velazco and Patterson 2008). A further study of the whole genus situated the origin of *M. reigi* in the Amazon lowlands and its establishment on the tepui summits in the Quaternary, between 2.4 and 1.2 Ma (Pavan et al. 2016). A similar situation was proposed for the opossum *Marmosops pakaraimae* (Voss et al. 2013). A mouse endemic to the Roraima-tepui, *Podoxymys roraimae*, diverged from its closest relative at the end of the Pliocene, between 3.7 and 2.5 Ma, and dispersed to Pantepui from the Brazilian Shield. In this case, the CCT and the HST were dismissed, and the DDT was favored (Leite et al. 2015).

A preliminary conclusion is that most hypotheses proposed to date (Table 15.3), as well as combinations of them, have been supported by one or more case studies; therefore, based on the available evidence, none of the hypotheses can be dismissed. Different taxonomic groups seem to have acquired their current biodiversity and endemism patterns by diverse processes and mechanisms under the action of different drivers in different Pantepui areas and at different times. Therefore, single case studies based on particular taxonomic groups and/or specific geographical areas cannot be generalized to all of Pantepui, which is also true for the entire Neotropics (Rull 2013). As occurs in the Neotropics, in general, a proper understanding of the origin of the Pantepui biota requires complex thinking and the consideration of multiple drivers (tectonics, paleogeographical reorganizations, marine incursions, climatic changes) as well as a diversity of ecological and biogeographical processes (migration, long-distance dispersal, *in situ* evolution, microrefugia, extinction) and evolutionary mechanisms (vicariance, gene flow, hybridization, adaptive radiation) acting together across spatial and temporal scales (Rull 2011, 2015). This has been called the Continuum Multifactor Hypothesis (CMH) (Rull 2019b and Table 15.3).

3.4 Conservation

Due to its remoteness and pristinity, it could be expected that the biota and ecosystems of Pantepui are not threatened by human activities. On the tepui summits, activities such as hydrocarbon exploitation, mining, hydroelectricity production, forestry and farming are not possible due to the special geological, edaphic and vegetation features (Huber 1995d). The main activities atop the tepuis are tourism and scientific exploration, although permanent facilities for such tasks (hotels, scientific stations, etc.) are nonexistent and these activities are carried out by camping. Only a few tepui summits, notably the Roraima-tepui and Auyán-tepui, are accessible by foot, while others can be reached by helicopter. The tepuis have been declared natural monuments and some Pantepui areas are under other

Fig. 15.13 Examples of Gran Sabana fires and their effects on tepuis in the Eastern District. (1 and 2) Active Gran Sabana fires near Santa Elena de Uairén. (3) Gallery forest reduced by recent fires near Wonkén. (4) General view of the southern Gran Sabana showing a recent fire (brow patch in the front) and the mostly deforested landscape. In the background, the Kukenán-tepui (left) and the Roraima-tepui (right). (5 and 6) SE flanks of the Chimantá massif near Yunek, showing the slopes extensively deforested by fire. The brown area corresponds to a major recent fire that reached the cliffs. (7 and 8) The summit of the Uei-tepui, showing standing charred trunks of the former *Bonnetia* forests removed by fire and the secondary broad-leaved meadows of *Stegolepis* and *Brocchinia* with the fire-tolerant shrub *Cyrilla*. Photos: V. Rull

designations, such as national parks, biosphere reserves and a human heritage site (Huber 1995d; Bevilacqua et al. 2019). However, the available resources for effective control and surveillance of such a huge and remote area are insufficient (Rull et al. 2016; Rull and Vegas-Vilarrúbia 2017). In 1989, official permits to visit the tepui summits were suspended. Since then, a long, complex, erratic and uncertain process was established to obtain permits for visiting the tepuis that makes scientific research virtually impossible (Rull and Vegas-Vilarrúbia 2008; Rull et al. 2009). In spite of this, illegal touristic and scientific activities have not stopped and are difficult to control. Other threats for Pantepui are human activities that occur in the surrounding uplands and lowlands (notably fire), which can eventually affect the tepui summits and their slopes. This is especially true for the Eastern Tepui District, where the tepuis are surrounded by the Gran Sabana uplands, the most populated and active area in the region. In addition to direct human impacts, the ongoing anthropogenic global warming may also affect the Pantepui biota, as occurs in other Neotropical and worldwide mountain areas. This section briefly discusses the potential direct and indirect consequences of local and global human activities on Pantepui.

3.4.1 Direct Impacts

Humans visit the tepui summits for a variety of purposes including scientific collection, recreation, adventure tourism, extreme sports (climbing, paragliding, skydiving), photography, documentary filming and biopiracy (illegal biodiversity trade and commercial exploitation of biochemical and genetic resources), among others. Presently, most of these activities are carried out without the corresponding official permits, likely due to the difficulty of obtaining them. It should be noted that indigenous people do not choose to reach these summits on their own because they consider them to be sacred places, although they are often hired by tourists as guides and/or porters. The main impacts of these activities are vegetation trampling, plant/animal extraction and garbage accumulation. These impacts are most significant on the only two tepuis that are reachable by foot and open to touristic visits, the Auyán-tepui and the Roraima-tepui, especially on the second, which is the most visited by far. The summit of the Roraima-tepui (~34 km^2) receives 3000–4000 tourists each year and vegetation trampling/damage is a common problem, together with the presence of garbage, food scraps, toilet paper, human excrement and graffiti (Safont et al. 2014). Recently, two novel threats have been identified: invasive plants and water contamination.

The latest botanical survey of the Roraima summit, conducted with the corresponding official permits, documented the presence of 13 plant species introduced by humans (Safont et al. 2014). Most of these species are cultivated or occur as small and localized populations and, therefore, do not represent an immediate menace. However, two of these introduced species, the grasses *Polypogon elongatus* and *Poa annua*, may be more dangerous. *P. elongatus* is a well-known invasive elsewhere and its abundance and patterns of spatial occurrence on the Roraima summit fulfil the criteria to be considered naturalized and invasive (Richardson et al. 2000). *P. annua* is of cosmopolitan distribution and is considered to be one of the most aggressive weeds (Holm et al. 1997). The eventual invasion of the Roraima and other summits by these plants could lead to large-scale degradation of the vegetation and the eventual extinctions of the autochthonous biota (Rull et al. 2016). Another recent study documented the presence of *Helicobacter pylori*—a well-known dweller of the human gastrointestinal tract that causes gastritis, ulcers and cancer—in the Roraima freshwaters near the usual camp sites (Fernández-Delgado et al. 2016). This bacterium may be easily transported by water to the uplands and lowlands and spread across the Guiana region through the fluvial network (Rull et al. 2016).

Human fires also constitute a direct threat, especially in the Gran Sabana tepuis (Eastern District). Currently, summit fires do not occur, but it is well known that Gran Sabana fires are active deforestation agents in gallery forests and on tepui slopes (Fig. 15.13). These fires are frequent and recurrent, and it has been estimated that more than 10,000 fires are lit by humans each year and that a given Gran Sabana area is recurrently burnt every 2–3 years (Huber 1995d). The danger of these fires reaching a tepui summit is latent. This fact has already been documented on the Uei-tepui summit, which is not totally isolated by vertical cliffs from the surrounding uplands (Safont et al. 2016) (Fig. 15.13). As we have seen above, many tepui summits are connected to uplands and lowlands by ridges and extended valleys (Huber 1988) through which fires could propagate to the highlands.

3.4.2 Global Warming

Global warming is considered to be a significant threat to mountain biota worldwide, as species may respond by shifting their altitudinal distributions, leading to changes in the diversity and composition of their communities and reductions, fragmentation or loss of their habitat. Tropical mountains are of particular concern because of their high degree of biodiversity and endemism (Safont et al. 2014 and literature therein). In Pantepui, several studies have been conducted to quantitatively estimate the potential extinction of vascular plant species by GW-induced habitat loss and to evaluate the possible conservation actions (Rull and Vegas-Vilarrúbia 2006, 2017; Nogué et al. 2009; Safont et al. 2012, 2014; Vegas-Vilarrúbia et al. 2012). These estimates have been based on three complementary methods, namely the altitudinal range shift (ARS) method, the species-area relationship (SAR) method and the climate envelope distribution model (CEDM) method (see Rull and Vegas-Vilarrúbia 2019b for more details). The climatic scenarios used as a reference correspond to the IPCC (Intergovernmental Panel on Climate Change) projections of a 2–4 °C increase for northern South America, by the end of the twenty-first century (Houghton et al. 2001; Solomon et al. 2007; Stocker et al. 2013).

The first projections estimate a plant diversity reduction of 75% (28% Pantepui endemics) for the more optimistic scenario (2 °C) and of 83% (54% endemics) for the more pessimistic scenario (4 °C) (Nogué et al. 2009). Further research did not improve expectations as the estimates for endemic extinction by habitat loss ranged from approximately 30% to 50% for the best and the worst-case scenarios, respectively (Safont et al. 2012). These preliminary studies were conducted on the basis of the available data, which consisted of only presence/absence records for the studied tepui summits and altitudinal ranges of occurrence for the known species. The lack of autecological knowledge on Pantepui species prevented the consideration of idiosyncratic features such as differential phenotypic plasticity among species and the resulting acclimation capacity, as well as the potential for rapid genetic change and the ensuing evolutionary adaption to novel environments. Therefore, the first estimates should be considered preliminary and likely inflated, and should be revised when more autecological knowledge is available. Rull and Vegas-Vilarrúbia (2017) suggested future studies that should be developed for more realistic estimates of the potential effect of global warming on Pantepui species.

Several *in situ* and *ex situ* actions have been discussed to minimize extinction by habitat loss due to global warming. *In situ* actions are problematic because warming is a global phenomenon and, hence, difficult to palliate at the local and regional levels. Under the current estimates, most tepui summits are set to totally lose their climatic Pantepui envelope by 2100 except the Chimantá massif and Cerro Marahuaka, which would retain the current Pantepui climates across approximately 40% of their surface, and the Roraima-Kukenán massif, with a typical Pantepui climate remaining across approximately 20% of its surface by 2100 (Vegas-Vilarrúbia et al. 2012) (Fig. 15.14). Therefore, eventual *in situ* conservation efforts should be focused on these summits. Regarding *ex situ* possibilities, the creation of germplasm banks and botanical gardens, as well as the viability of managed relocation practices, have been discussed (Safont et al. 2012). The first two possibilities are common and would be relatively easy to implement. However, managed relocation is a highly controversial conservation option, especially in the case of Pantepui, where ecological knowledge on its biota is lacking. Much more ecological and evolutionary research is still needed to determine whether the managed relocation of Pantepui species is a feasible and convenient option.

4 Future Research

The potential of Pantepui to provide relevant insights on the origin and evolution of the Neotropical biota has not been fully exploited, due to the still incipient and biased research conducted in this biogeographical province. Most studies have been aimed at reporting and documenting the biodiversity and endemism patterns of different taxonomic groups and have been carried out by specialists on these groups. General ecological, evolutionary and conservation studies are still lacking. This is likely due to three main factors. First, the uniqueness of the species and life forms has fostered the interest of specialists avid in identifying new rare species and documenting high levels of endemism. This has largely led research regarding Pantepui since its discovery. Second, the remoteness and inaccessibility of the

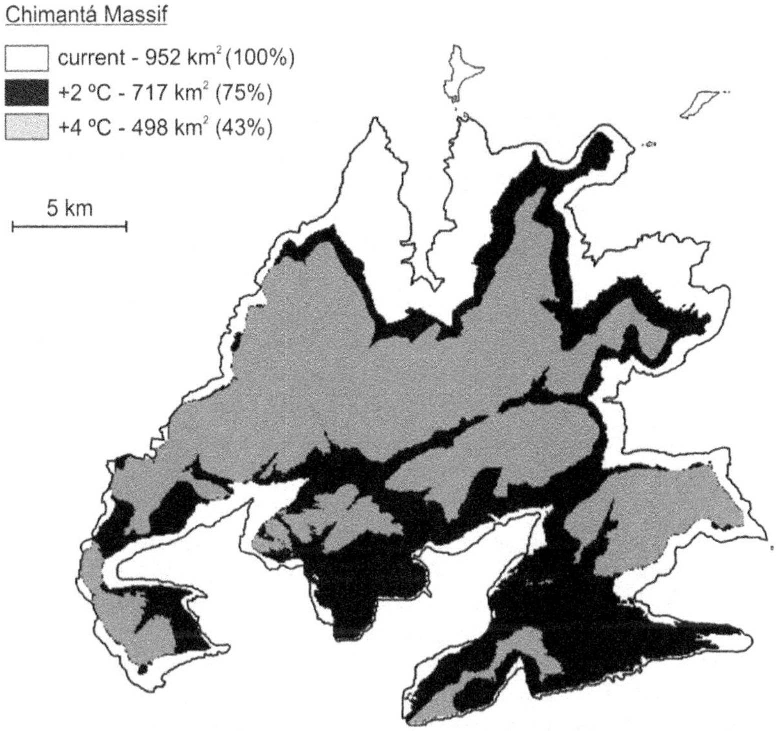

Fig. 15.14 ARS-GIS simulation (see text) of the Pantepui area at the top of the Chimantá massif for the end of the present century under the more optimistic (2 °C increase) and the more pessimistic (4 °C increase) IPCC scenarios (Solomon et al. 2007). Redrawn from Nogué et al. (2009) and Vegas-Vilarrúbia et al. (2012)

Guiana Highlands, together with the lack of *in situ* facilities, make any study based on periodic and/or systematic campaigns using complex logistics and equipment difficult and expensive. Therefore, research on these mountains has largely been based on sporadic studies consisting of short and occasional visits. It seems that, more than a century and a half after its scientific discovery, Pantepui is still in an exploratory phase. Third, the difficulty—and sometimes the impossibility—of obtaining official permits has aggravated this situation, as it not only makes biological sampling difficult but also prevents the establishment of permanent or semipermanent recording instruments. These handicaps should be circumvented to exploit the full ecological and evolutionary potential still hidden in Pantepui. After the review presented here, some major knowledge gaps have been identified that should be urgently addressed. Suggestions for future studies can be classified into four broad categories: (1) biodiversity and endemism, (2) ecology and paleoecology, (3) genomics and evolution and (4) conservation.

4.1 Biodiversity and Endemism

Despite the numerous visits to the tepui summits during the last century and a half, knowledge on the diversity and endemism patterns of Pantepui organisms is only fragmentary. On the one hand, research on some organism groups is still incipient, and no general insights on their diversity or geographical patterns are yet possible. This is the case, for example, for bryophytes, algae, aquatic insects, butterflies, arachnids and snails when considering only to refer only the organisms included in this review. Other groups are even less known or completely unknown. Even within the better known organisms (vascular plants, amphibians, reptiles, birds and mammals) there are important gaps that prevent robust generalizations. Therefore, taxonomic research should continue on Pantepui for a sound understanding of its biodiversity and endemism patterns. Such research, however, should leave behind the exploratory mindset and open a new research era based on more specific aims and more systematic studies according to taxonomic group, geographical area, or both. Pioneers of Pantepui scientific exploration were fundamental for acquiring current knowledge on this unique biogeographical province, which is the seed from which all research has emerged and will continue to develop. However, the existing biodiversity background for major groups seems already sufficient to advance towards a more systematic and target-oriented research framework, including not only biodiversity but also ecological, evolutionary and conservation studies. For example, the theory of island biogeography (Whittaker and Fernández-Palacios 2007) could be a suitable framework for conducting Pantepui research. Some attempts have been made using species-area relationships and the distance to possible sources of Pantepui species, notably the Andes (Michelangeli 2000; Berry and Riina 2005; Nogué et al. 2009; Riina et al. 2019; Pérez-Emán et al. 2019), but the potential of this approach has not been fully exploited.

4.2 Ecology and Paleoecology

Ecology is perhaps the most overlooked discipline in Pantepui research. Incipient ecological studies are available for plants after the definition of the main vegetation types and plant communities, with indication of their preferred substrates (bare rock, peat, sandstone, gneiss). However, the ecological preferences and tolerances of individual species are completely unknown due to the lack of autecological studies and the scarcity of environmental data to define species niches. This is true not only for plants but also for all taxonomic groups reported here. Sometimes, the elevational ranges of species have been used as a proxy for temperature tolerance, but this is not enough. Autecological and ecophysiological research on Pantepui species, with emphasis on the most relevant elements, is mandatory and urgent. From a synecological point of view, knowledge on community assembly and the main internal and external ecological drivers is still insufficient. The available paleoecological studies useful for this purpose are restricted to the Holocene and should be extended backwards. To date, it has not been possible to find LGM and earlier sediments but this is fundamental to advance the ecological knowledge. Functional ecology is also completely unknown and multidisciplinary studies on the composition and ecological dynamics (biotic and abiotic interactions) should be prioritized to obtain sound ecological knowledge.

4.3 Genomics and Evolution

Molecular phylogenetic and phylogeographical research on Pantepui organisms has already begun but it should be continued and focused on specific problems. Genetic analyses not only resolve group-specific taxonomies but also contribute to inferring evolutionary relationships. As we have seen above, evolutionary inference based solely on biogeographical patterns and geological/paleoecological data are necessary but insufficient, as they provide circumstantial evidence. Molecular phylogenies render direct evolutionary evidence and their geographical expression contributes to causal inferences regarding diversification processes and their mechanisms. It is noteworthy, however, that molecular studies conducted to date have been aimed at testing the existing hypotheses on the origin and evolution of Pantepui biota, which are still largely based on pioneer biogeographical studies. It could be expected that genomic research—with the corresponding biogeographical, geological and paleoecological inputs—may provide new possible explanations but this has not yet occurred. For this, it is necessary that molecular phylogenetic and phylogeographical research is more target-focused and based on fully sampled taxonomic and/or geographical evolutionary units, rather than on partial evidence. This could help circumvent generalizations from single case studies thus favoring complex thinking, which seems mandatory to properly understand Neotropical biotic patterns and processes.

4.4 Conservation

Robust biogeographical, ecological and evolutionary knowledge is mandatory for estimating the potential effects of direct and indirect threats to the Pantepui biota and ecosystems, as well as to inform conservation strategies and, eventually, restoration options. Autecological knowledge, notably regarding species tolerances to environmental parameters, is fundamental for forecasting the potential responses—i.e., in situ survival, migration, and extinction—of each species to future environmental change, which affects not only biodiversity but also the composition of future communities and therefore intra-community ecological dynamics. In addition, surviving climate change depends not only on the phenotypic plasticity of species but also on their capacity to undergo short-term evolutionary adaption, which is largely dependent on their intra- and interpopulation genetic variability. Therefore, genomic studies can also contribute to estimating the capacity of species to resist environmental change. In summary, ecological and evolutionary research can be useful for both understanding the origin and evolution of the Pantepui biota and conserving it in the face of future global warming. The existing studies on the subject have already provided a list of the most endangered plant species sorted by risk, which could be useful for establishing priorities for more detailed autecological and evolutionary studies. Similar studies may be developed for other organisms. International worldwide initiatives, such as GLORIA (GLobal Observation Research Initiative in Alpine environments; www.gloria.ac.at), which has a network of stations to record changes in mountain vegetation driven by global warming, should be considered for Pantepui. Regarding restoration practices, paleoecology may be useful to identify better communities for present-day environmental conditions thus avoiding eventual unrealistic or nonviable combinations, such as LGM or Late Glacial communities.

4.5 Summary

A general message from the above considerations is that future Pantepui research should be more systematic and target focused. The available knowledge seems sufficient to consider the pioneering and exploratory times to have already finished, and a new mentality should be adopted based on more scientific procedures. For this to occur, it is essential that the three main handicaps mentioned above are circumvented. Only one of these handicaps—i.e., the new/rare species approach as the only target—depends on scientists' ability to shift the goal of research, whereas the other two—i.e., the difficulty of access and of obtaining official permits—are beyond the control of researchers. However, nothing on Earth seems more remote and inaccessible than the poles, which are now a global priority for research. Perhaps it would be possible to convince the countries sharing the Guiana Highlands of the importance of knowing and preserving Pantepui as one of the few remaining pristine areas on Earth. The Guiana Shield Facility (www.guianashield.org) could be a suitable platform, provided it is able to secure funds from international agencies and provide a suitable political environment for conducting systematic and target-focused scientific research on Pantepui.

5 Conclusions

- Pantepui is a highly fragmented biogeographical province with a >5000 km^2 total surface area, formed by the assemblage of remote and pristine flat summits (up to approximately 3000 m elevation) of tabular Guiana mountains (tepuis) situated between the Orinoco and Amazon basins. The lower elevational boundary of the Pantepui province varies according to the taxonomic group considered. After the analysis of the better-known Pantepui floral and faunal groups and their respective distribution patterns, a lower boundary between 1200 m and 1500 m elevations seems more appropriate than a rigid boundary situated at 1500 m, as was usual in former classifications.

- The Pantepui biodiversity and endemism patterns are outstanding and vary across taxonomic groups. The most diverse organisms are plants, with almost 2600 known species (>5000 species/10,000 km^2), which situates Pantepui among the most diverse regions of the world. Endemism usually ranges from 30 to 40% in birds, plants and reptiles and may reach 55% in amphibians, which are similar to the percentages on many oceanic islands. The frequency of local endemics—i.e., species restricted to a single tepui summit—is high (25%) for plants and very high (35–40%) for amphibians and reptiles.
- Ecological knowledge of Pantepui biota and the communities it forms is almost inexistent. There are no autecological studies useful for defining the niche preferences of the different species and their tolerance ranges to biotic and abiotic variables. Only preliminary elevational ranges are available for the species of some taxonomic groups. Regarding communities, only descriptive studies of vegetation types are available, and almost nothing is known about community assembly and ecosystem dynamics. The situation is even worse for animals, where synecological studies are absent.
- Long-term ecological studies using paleoecological methods have shown that Pantepui plant communities have changed through time under the influence of Holocene climatic changes (notably temperature and moisture balance) and fire events. The biotic responses, notably in the form of altitudinal range displacements, to these environmental shifts have been idiosyncratic—i.e., dependent on the particular tolerances of each species—rather than at the community level. This has led to continuous changes in community composition that have shaped the present communities.
- The recurrent glacial-interglacial cycles and the idiosyncratic responses of the different species would have affected the taxonomic composition of the tepui summits, turning Pantepui into a dynamic biogeographical unit. In this framework, Pantepui seems to have been a typically interglacial feature that has adopted different states during the different interglacials and has been disassembled during glacials, when highland species migrated to the surrounding lowlands and uplands. Pantepui assembly is viewed as a less stable state in comparison with glacial disassembly, as the latter has been much more frequent during the Quaternary and does not need external energy inputs to persist.
- The origin and evolution of the Pantepui biota has been explained by diverse evolutionary processes involving a variety of environmental drivers (notably climatic changes and paleogeographical rearrangements) and diversification mechanisms (vicariance, jump dispersal and allopatry, parapatry, adaptive radiation). Most of these hypotheses emerged from the study of extant biogeographical patterns and geological-paleoecological surveys. The inception of molecular phylogenetics and phylogeography, albeit still incipient in Pantepui, has provided evidence useful for testing these hypotheses. Taken individually, none of the

proposed explanations can explain the evolution of the whole Pantepui biota, whose proper understanding requires complex thinking and the consideration of multiple drivers and a diversity of ecological and evolutionary processes and mechanisms acting together across spatiotemporal scales.
- In spite of their current pristinity, the Pantepui biota and ecosystems are threatened by direct (mechanical damage, introduction of invasive species, garbage accumulation, water contamination, fire) and indirect (global warming) consequences of human activities. Direct threats might be addressed by *in situ* actions (access control, surveillance), but unfortunately, the resources available are insufficient considering the huge size, the topographical complexity and the remoteness of the Guiana Highlands. The main consequence of global warming may be habitat loss for a large number of Pantepui species (up to 85% of vascular plant species, including 55% of endemics), which can be combatted only by *ex situ* actions such as the creation of germplasm banks and botanical gardens and, eventually, managed relocation.
- At present, Pantepui research is at an impasse that slows advancement towards a biological synthesis of this singular Neotropical "lost world" and prevents gathering the information necessary to inform the conservation of its unique biota. The main drawback seems to be the current permitting policy, which directly affects research tasks, especially by blocking fieldwork. However, there is another handicap that is under the control of researchers. After more than a century and a half of research on Pantepui, the pioneering and exploratory times should be declared over, and a different mentality should be adopted based on more scientific procedures, including more systematic and target-focused activities.

Acknowledgements This paper is a homage to the pioneers of Pantepui scientific exploration, without whom nothing reported here would have been possible ("we stand on the shoulders of giants"). The first author is especially grateful to Otto Huber, who introduced him to Pantepui research in 1984 and led many of the expeditions in which he has participated. The authors are grateful to Abrahram Breure, Mauro Costa, Tomas Derka, Ricardo Guerrero, Jan Kaštovský, Burton Lim, Javier Mesa, José Ochoa, Jorge-Pérez-Emán, Fernando Rojas-Runjaic, Celsa Señaris, David Southall and Ángel Viloria for providing photographs and information for Figs. 15.8 and 15.9.

References

Berry PE, Riina R (2005) Insights into the diversity of the Pantepui flora and the biogeographical complexity of the Guayana Shield. Biol Skrift 55:145–167

Berry PE, Huber O, Holst BK (1995a) Floristic analysis and phytogeography. In: Berry P, Holst BK, Yatskievych K (eds) Flora of the Venezuelan Guayana, Introduction, vol 1. Missouri Bot Gard Press, St. Louis, pp 161–191

Berry PE, Holst BK, Yatskievych K (1995b) Introduction. In: Berry P, Holst BK, Yatskievych K (eds) Flora of the Venezuelan Guayana, Introduction, vol 1. Missouri Botanical Garden Press, St. Louis, pp xv–xx

Berv JS, Prum RO (2014) A comprehensive multilocus phylogeny of the Neotropical cotingas (Cotingidae, Aves) with a comparative evolutionary analysis of breeding system and plumage dimorphism and a revised phyologenetic classification. Mol Phylogenet Evol 81:120–136
Bevilacqua M, Señaris JC, Huber O (2019) Conservation of Pantepui. In: Rull V, Vegas-Vilarrúbia T, Huber O, Señaris C (eds) Biodiversity of Pantepui, the pristine “lost world” of the neotropical Guiana highlands. Elsevier, London, pp 391–402
Bonaccorso E, Guayasamin JM (2013) On the origin of Pantepui montane biotas: a perspective based on the phylogeny of Aulancorhynchus tocanets. PLoS One 8:6e67321
Borges SH, Santos MPD, Moreira M et al (2018) Dissecting bird diversity in the Pantepui area of endemism, northern South America. J Ornithol 159:1073–1086
Breure ASH (2019) Land snalis. In: Rull V, Vegas-Vilarrúbia T, Huber O, Señaris C (eds) Biodiversity of Pantepui, the pristine “lost world” of the neotropical Guiana highlands. Elsevier, London, pp 247–261
Briceño HO, Schubert C (1990) Geomorphology of the Gran Sabana, Guayana Shield, southeastern Venezuela. Geomorphology 3:125–141
Bush MB (1994) Amazonian speciation: a necessarily complex model. J Biogeogr 21:5–17
Bush M, Stute M, Ledru M-P et al (2001) Palaeotemperature estimates for the lowland Americas between 30°S and 30°N at the Last Glacial Maximum. In: Margkgraf V (ed) Interhemispheric climate linkages. Academic, San Diego, pp 293–306
Chapman J (1931) The upper zonal bird-life of Mts. Roraima and Duida. Bull Am Mus Nat Hist 63:1–135
Colinvaux PA (1998) A new model for Amazon endemics. Glob Ecol Biogeogr Lett 7:95–96
Costa M, Viloria AL, Huber O et al (2013) Lepidoptera del Pantepui. Parte I: Endemismo y caracterización biogeográfica. Entomotropica 28:193–217
Cuevas E (1992) Relaciones nutricionales de la vegetación de turberas alto-tepuyanas. In: Huber O (ed) El Macizo del Chimantá. Oscar Todtmann, Caracas, pp 203–218
Derka T, Zamora-Muñoz C, Tierno de Figueroa JM (2019) Aquatic insects. In: Rull V, Vegas-Vilarrúbia T, Huber O, Señaris C (eds) Biodiversity of Pantepui, the pristine “lost world” of the neotropical Guiana highlands. Elsevier, London, pp 167–192
Désamoré A, Vanderpoorten A, Laenen B, Gradstein SR, Kok PJR (2010) Biogeography of the lost world (Pantepui region, northeastern South America): insights from bryophytes. Phytotaxa 9:254–265
Doyle AC (1912) The lost world. Hodder & Stoughton, London
Ellis R, Palmer M (2016) Modulation of ice ages via precession and dust-albedo feedbacks. Geosci Front 7:891–909
Fernández-Delgado M, Giarrizzo JG, García-Amado MA et al (2016) Evidence of *Helicobacter* spp. in freshwaters from Roraima Tepui, Guayana Shield, South America. Antoine van Leeuwenhoeck 109:529–542
Givnish TJ, Evans TM, Zjhra ML et al (2000) Molecular evolution, adaptive radiation, and geographic diversification in the amphiatlantic family Rapateaceae: edidence from ndhF sequences and morphology. Evolution 54:1815–1937
Givnish TJ, Millam KC, Evans TM et al (2004) Ancient vicariance or recent long-distance dispersal? Inferences about phylogeny and South American-African disjunctions in Rapateaceae and Bromeliaceae based on ndhF sequence data. Int J Plant Sci 165:S35–S54
Givnish TJ, Millam KC, Berry PE, Sytsma KJ (2007) Phylogeny, adpative radiation, and historical biogeography of Bromeliaceae inferred from sdhF sequence data. Aliso 23:3–26
Guerrero R (2019) Vertebrate parasites. In: Rull V, Vegas-Vilarrúbia T, Huber O, Señaris C (eds) Biodiversity of Pantepui, the pristine “lost world” of the neotropical Guiana highlands. Elsevier, London, pp 373–385
Hansen J, Sato M, Russell G, Kharecha P (2013) Climate sensitivity, sea level and atmospheric carbon dioxide. Philos Trans R Soc A 371:20120294
Holm LG, Holm E, Pancho JV, Herberger JP (1997) World weeds: natural histories and distribution. Wiley, New York

Hopper SD (2009) OCBIL theory: towards an integrated understanding of the evolution, ecology and conservation of biodiversity on old, climatically buffered, infertile landscapes. Plant Soil 322:49–86

Houghton JT, Ding Y, Griggs DJ et al (2001) Climate change 2001: the scientific basis. Cambridge University Press, Cambridge

Huber O (1987) Consideraciones sobre el concepto de Pantepui. Pantepui 1:2–10

Huber O (1988) Guayana highlands versus Guayan lowlands, a reappraisal. Taxon 37:595–614

Huber O (1992) La vegetación. In: Huber O (ed) El Macizo del Chimantá. Oscar Todtmann, Caracas, pp 161–178

Huber O (1994) Recent advances in the phytogeography of the Guayana region, South America. Mém Soc Biogéogr 4(3ème sèr):53–63

Huber O (1995a) Geographical and physical features. In: Berry PE, Holst BK, Yatskievych K (eds) Flora of the Venezuelan Guayana, Introduction, vol 1. Missouri Bot Gard Press, St. Louis, pp 1–61

Huber O (1995b) History of botanical exploration. In: Berry PE, Holst BK, Yatskievych K (eds) Flora of the Venezuelan Guayana, Introduction, vol 1. Missouri Bot Gard Press, St. Louis, pp 63–95

Huber O (1995c) Vegetation. In: Berry PE, Holst BK, Yatskievych K (eds) Flora of the Venezuelan Guayana, Introduction, vol 1. Missouri Bot Gard Press, St. Louis, pp 97–160

Huber O (1995d) Conservation of the Venezuelan Guayana. In: Berry PE, Holst BK, Yatskievych K (eds) Flora of the Venezuelan Guayana, Introduction, vol 1. Missouri Bot Gard Press, St. Louis, pp 193–208

Huber O, García P (2011) The Venezuelan Guayana region and the study areas: geo-ecological characteristics. In: Zink A, Huber O (eds) Peatlands of the Western Guayana highlands, Venezuela. Springer, Berlin, pp 29–89

Huber O, Riina R (1997) Glosario fitoecológico de las Américas. In: América del Sur, países hispanoparlantes, vol 1. Tamandúa, Caracas

Huber O, Rull V (2019) Plant communities. In: Rull V, Vegas-Vilarrúbia T, Huber O, Señaris C (eds) Biodiversity of Pantepui, the pristine "lost world" of the neotropical Guiana highlands. Elsevier, London, pp 149–164

Huber O, Prance GT, Kroonenberg S, Antonelli A (2018) The tepuis of the Guiana highlands. In: Hoorn C, Perrigo A, Antonelli A (eds) Mountains, climate and biodiversity. Wiley, Hoboken, NJ, pp 339–353

Kaštovský J, Fučiková K, Veselá J, Brewer Carías C, Vegas-Vilarrúbia T (2019) Algae. In: Rull V, Vegas-Vilarrúbia T, Huber O, Señaris C (eds) Biodiversity of Pantepui, the pristine "lost world" of the neotropical Guiana highlands. Elsevier, London, pp 95–120

Kok PJR, MacCulloch RD, Means DB et al (2012) Low genetic diversity in tepui summit vertebrates. Curr Biol 22:R589–R590

Kok PJR, Russo VG, Ratz S, Aubret F (2016) On the distribution and conservation of two "lost world" tepui summit endemic frogs, *Stefania ginesi* Rivero, 1968 and *S. satelles* Señaris, Ayarzagüena, and Gorzula, 1997. Amphibian Reptile Conserv 10:5–12

Kok PJR, Russo VG, Ratz S et al (2017) Evolution in the South American 'lost world': insights from multilocus phylogeography of stefanias (Anura, Hemiphractidae, *Stefania*). J Biogeogr 44:170–181

Kok PJR, Ratz S, MacCulloch RD et al (2018) Historical biogeography of the palaeoendemic toad genus *Oreophrynella* (Amphibia: Bufonidae) sheds a new light on the origin of the Pantepui endemic terrestrial biota. J Biogeogr 45:26–36

Lasso C, Machado-Allison C, Pérez R (1989) Consideraciones zoogeográficas del los peces de la Gran Sabana (Alto Caroní) Venezuela, y sus relaciones con las cuencas vecinas. Mem Soc Cien Nat La Salle 49:109–129

Leite YLR, Kok PJR, Weksler M (2015) Evolutionary affinities of the 'lost world' mouse suggest a late Pliocene connection between the Guiana and Brzailian shields. J Biogeogr 42:706–715

Lew D, Lim B (2019) Mammals. In: Rull V, Vegas-Vilarrúbia T, Huber O, Señaris C (eds) Biodiversity of Pantepui, the pristine "lost world" of the neotropical Guiana highlands. Elsevier, London, pp 333–356
Maguire B (1970) On the flora of the Guayana highland. Biotropica 2:85–100
Mayr E, Phelps WH (1967) The origin of the bird fauna of the South Venezuela highlands. Bull Am Mus Nat Hist 136:273–327
McCormack JE, Huang H, Knowles LL (2009) Sky islands. In: Gillespie R, Clague D (eds) Encyclopedia of islands. University of California Press, Berkeley, pp 839–843
McDiarmid RW, Donnelly MA (2005) The herpetofauna of the Guayana highlands: amphibiands and reptiles of the lost world. In: Donnelly MA, Crother BI, Guyer C, Marvalee H, White ME (eds) Ecology and evolution in the tropics. University of Chicago Press, Chicago, pp 461–560
Mecchia M, Sauro F, Piccini L, Columbu A, De Waele J (2019) A hybrid model to evaluate subsurface chemical weathering and fractures kartsification in quartz sandstone. J Hydrogeol 572:745–760
Michelangeli F (2000) Species composition and species-area relationships in vegetation isolates on a summit of a sandstone mountain in southern Venezuela. J Trop Ecol 16:69–82
Morrone JJ (2014) Biogeographical regionalisation of the Neotropical region. Zootaxa 3782:1–110
Mutke J, Bathlott W (2005) Patterns of vascular plant diversity at continental to global scales. Biol Skr 55:521–531
Nogué S, Rull V, Vegas-Vilarrúbia T (2009) Modelling biodiversity loss by global warming on Pantepui, northern South America: projected upward migration and potential habitat loss. Clim Chang 96:77–85
Ochoa JA, Rojas-Runjaic F (2019) Scorpions. In: Rull V, Vegas-Vilarrúbia T, Huber O, Señaris C (eds) Biodiversity of Pantepui, the pristine "lost world" of the neotropical Guiana highlands. Elsevier, London, pp 223–241
Pavan SE, Jansa SA, Voss RS (2016) Spatiotemporal diversification of a low-vagility Neotropical vertebrate clade (short-tailed opossums, Didelphidae: Monodelphis). J Biogeogr 43:1299–1309
Pérez-Emán JL (2005) Molecular phylogenetics and biogeography of the Neotropical redstars (*Myoborus miniatus*). Ornitol Monogr 67:511–528
Pérez-Emán JL, Lentino M, Bonaccorso E (2019) Birds. In: Rull V, Vegas-Vilarrúbia T, Huber O, Señaris C (eds) Biodiversity of Pantepui, the pristine "lost world" of the neotropical Guiana highlands. Elsevier, London, pp 299–332
Richardson DM, Pyšek P, Remánek M et al (2000) Naturalization and invasion of alien plants: concepts and definitions. Divers Distrib 6:93–107
Riina R, Berry PE, Huber O, Michelangeli F (2019) Vascular plants and bryophytes. In: Rull V, Vegas-Vilarrúbia T, Huber O, Señaris C (eds) Biodiversity of Pantepui, the pristine "lost world" of the neotropical Guiana highlands. Elsevier, London, pp 121–147
Rull V (2004) Biogeography of the "lost world": a paleoecological perspective. Earth-Sci Rev 67:125–137
Rull V (2005) Biotic diversification in the Guayana highlands: a proposal. J Biogeogr 32:921–927
Rull V (2008) Speciation timing and neotropical biodiversity: the Tertary-Quaternary debate in the light of molecular phylogenetic evidence. Mol Ecol 17:2722–2729
Rull V (2009) Pantepui. In: Gillespie R, Clague D (eds) Encyclopedia of islands. University of California Press, Berkeley, pp 717–720
Rull V (2010) The Guayana highlands: a natural laboratory for the biogeographical and evolutionary study of the neotropical flora. In: Sanchez-Villagra MM, Aguilera O, Carlini AA (eds) Urumaco and Venezuelan Paleontology. The fossil record of the northern neotropics. Indiana University Press, Bloomington, pp 84–102
Rull V (2011) Neotropical biodiversity: timing and potential drivers. Trends Ecol Evol 26:508–513
Rull V (2013) Some problems in the study of the origin of neotropical biodiversity using palaeoecological and molecular phylogenetic evidence. Syst Biodivers 11:415–423
Rull V (2015) Pleistocene speciation is not refuge speciation. J Biogeogr 42:602–609

Rull V (2019a) Neotropical diversification: historical overview and conceptual insights. In: Rull V, Carnaval A (eds) Neotropical diversification: patterns and processes. Springer, Berlin
Rull V (2019b) Origin and evolution of the Pantepui biota. In: Rull V, Vegas-Vilarrúbia T, Huber O, Señaris C (eds) Biodiversity of Pantepui, the pristine "lost world" of the neotropical Guiana highlands. Elsevier, London, pp 69–91
Rull V, Montoya E (2017) Holocene vegetation dynamics on the Apakará summit of the neotropical Guayana highlands and potential environmental drivers. Rev Palaeobot Palynol 240:22–32
Rull V, Nogué S (2007) Potential migration routes and barriers for vascular plants of the neotropical Guyana highlands during the quaternary. J Biogeogr 34:1327–1341
Rull V, Vegas-Vilarrúbia T (2006) Unexpected biodiversity loss under global warming in the neotropical Guayana highlands: a preliminary appraisal. Glob Chang Biol 12:1–9
Rull V, Vegas-Vilarrúbia T (2008) Biopiracy rules hinder conservation efforts. Nature 453:26
Rull V, Vegas-Vilarrúbia T (2017) Potential responses of vascular plants from the pristine 'lost world' of the neotropical Guayana highlands to global warming: review and perspectives. Front Plant Sci 8:81
Rull V, Vegas-Vilarrúbia T (2019a) Pantepui as a dynamic biogeographical concept. In: Rull V, Vegas-Vilarrúbia T, Huber O, Señaris C (eds) Biodiversity of Pantepui, the pristine "lost world" of the neotropical Guiana highlands. Elsevier, London, pp 55–67
Rull V, Vegas-Vilarrúbia T (2019b) Pantepui and global warming. In: Rull V, Vegas-Vilarrúbia T, Huber O, Señaris C (eds) Biodiversity of Pantepui, the pristine "lost world" of the neotropical Guiana highlands. Elsevier, London, pp 403–417
Rull V, Vegas-Vilarrúbia T, Nogué S, Montoya E (2009) Bureaucratic obstruction of conservation science in the Guayana highlands. Conserv Biol 22:508–509
Rull V, Vegas-Vilarrúbia T, Safont E (2016) The lost world's pristinity at risk. Divers Distrib 22:995–999
Rull V, Vegas-Vilarrúbia T, Huber O, Señaris C (2019a) Biodiversity of Pantepui, the pristine "lost world" of the neotropical Guiana highlands. Elsevier, London
Rull V, Vegas-Vilarrúbia T, Huber O, Señaris C (2019b) Definition and characterization of the Pantepui biogeographical province. In: Rull V, Vegas-Vilarrúbia T, Huber O, Señaris C (eds) Biodiversity of Pantepui, the pristine "lost world" of the neotropical Guiana highlands. Elsevier, London, pp 3–32
Rull V, Montoya E, Nogué S, Safont E, Vegas-Vilarrúbia T (2019c) Climatic and ecological history of Pantepui and surrounding areas. In: Rull V, Vegas-Vilarrúbia T, Huber O, Señaris C (eds) Biodiversity of Pantepui, the pristine "lost world" of the neotropical Guiana highlands. Elsevier, London, pp 33–54
Safont E, Vegas-Vilarrúbia T, Rull V (2012) Use of environmental impact assessment (EIA) tools to set priorities and optimize strategies in biodiversity conservation. Biol Conserv 149:113–121
Safont E, Rull V, Vegas-Vilarrúbia T et al (2014) Establishing a baseline of plant diversity and endemism on a neotropical mountain summit for future comparative studies assessing upward migration: an approach from biogeography and nature conservation. Syst Biodivers 12:292–314
Safont E, Rull V, Vegas-Vilarrúbia T et al (2016) Late-Holocene vegetation and fire dynamics on the summits of the Guayana highlands: the Uei-tepui palynological record. Palaeogeogr Palaeoclimatol Palaeoecol 455:33–43
Salerno PE, Ron SR, Señaris JC et al (2012) Ancient tepui summits harbor young rather than old lineages of endemic frogs. Evolution 66:3000–3013
Salerno PE, Señaris JC, Rojas-Runjaic FJM, Cannatella DC (2014) Recent evolutionary history of lost world endemics: population genetics, species delimitation, and phylogeography of sky-island treefrogs. Mol Phylogenet Evol 82:314–323
Señaris JC, Rojas-Runjaic F (2019) Amphibians and reptiles. In: Rull V, Vegas-Vilarrúbia T, Huber O, Señaris C (eds) Biodiversity of Pantepui, the pristine "lost world" of the neotropical Guiana highlands. Elsevier, London, pp 263–297
Smith BT, McCormack JE, Cuervo AM et al (2014) The drivers of tropical speciation. Nature 515:406–409

Solomon S, Qin D, Manning M (2007) Climate change 2007: the physical science basis. Cambridge University Press, Cambridge

Steyermark JA (1979) Plant refuge and dispersal centres in Venezuela: their relict and endemic element. In: Larsen K, Holm-Nielsen L (eds) Tropical botany. Academic, New York, pp 185–221

Steyermark JA, Dunsterville GCK (1980) The lowland floral element of the summit of Cerro Guaiquinima and other cerros of the Guayana Highlands of Venezuela. J Biogeogr 7:285–303

Steyermark JA, Berry PE, Holst BK (1995–2005) Flora of the Venezuelan Guayana, vol 1–9. Missouri Bot Gard Press, St. Louis

Stocker TF, Qin D, Plattner G-K et al (2013) Climate change 2013: the physical science basis. Cambridge University Press, Cambridge

Tate GHH (1938) A new 'lost world'. Nat Hist 2:107–120

Vareschi V (1992) Observaciones sobre la dinámica vegetal en el Macizo del Chimantá. In: Huber O (ed) El Macizo del Chimantá. Oscar Todtmann, Caracas, pp 179–188

Vegas-Vilarrúbia T, Nogué S, Rull V (2012) Global warming, habitat shifts and potential refugia for biodiversity conservation in the neotropical Guayana highlands. Biol Conserv 152:159–168

Velazco PM, Patterson BD (2008) Phylogenetics and biogeography of the broad-nosed bats, genus *Platyrrhinus* (Chiroptera: Phyllostomidae). Mol Philogenet Evol 49:749–759

Viloria AL, Costa M (2019) Butterflies. In: Rull V, Vegas-Vilarrúbia T, Huber O, Señaris C (eds) Biodiversity of Pantepui, the pristine "lost world" of the neotropical Guiana highlands. Elsevier, London, pp 193–222

Voss RS, Lim BK, Díaz-Nieto JF, Jansa SA (2013) A new species of Marmosops (Marsupialia: Didelphidae) from the Pakaraima Highlands of Guyana, with remarks on the origin of the endemic Pantepui Mammal fauna. Am Mus Novit 3778:27

Whittaker RJ, Fernández-Palacios JM (2007) Island biogeography: ecology, evolution and conservation. Oxford University Press, Oxford

Willis KJ, Whittaker RJ (2000) The refugial debate. Science 287:1406–1407

Zink JA, Huber O (2011) Peatlands of the Western Guayana Highlands, Venezuela. Springer, Berlin

Chapter 16
Patterns of Species and Lineage Diversity in the Atlantic Rainforest of Brazil

Elen A. Peres, Ricardo Pinto-da-Rocha, Lúcia G. Lohmann, Fabián A. Michelangeli, Cristina Y. Miyaki, and Ana Carolina Carnaval

Abstract The Atlantic rainforest of Brazil harbors outstanding species richness and levels of endemism, representing one of the most biodiverse regions on Earth. Yet, a lot is still unknown about the spatial and temporal evolution of its biota. While the history of the region is complex, with taxa often showing distinct richness patterns and evolutionary histories, some trends are common to a variety of lineages. Higher species richness is often found in the topographically complex coast of Rio de Janeiro and São Paulo, but the variety of environmental spaces available within this domain, tied to the different niches explored by its species, also support the existence of other patterns of species accumulation. The biological communities that inhabit the southern portions of the Atlantic forest are clearly distinct from those that occupy the northern range of the domain: several taxa undergo turnover in the Doce river valley (state of Espírito Santo), while others shift further south (near the Rio de Janeiro/São Paulo state boundary), or further north (state of Bahia). Areas of endemism have been recognized within the forest; while their boundaries do not perfectly match across all taxonomic groups, a nested pattern is detected, with contiguous areas of endemism defined by low-dispersal organisms (e.g., harvestmen) often fitting within more inclusive endemism centers defined by high-dispersal groups (e.g., birds). To date, the fauna and flora has been compartmentalized into five main areas of endemism: (1) Pernambuco, (2) Coastal Bahia, (3) Central Bahia, (4) Serra do Mar (often with further subregions), and (5) Paraná/Araucaria (Fig. 16.2). Here, we summarize main spatial patterns of diversity, flagging areas of higher species accumulation, turnover, and endemism. We also review main patterns of lineage divergence and population structure recovered from molecular phylogenies and phylogeographic studies, and explore some of the most common diversification hypotheses proposed for the Atlantic forest biota to date. We

E. A. Peres (✉) · R. Pinto-da-Rocha · L. G. Lohmann · C. Y. Miyaki
Instituto de Biociências, Universidade de São Paulo, São Paulo, SP, Brazil

F. A. Michelangeli
Institute of Systematic Botany, The New York Botanical Garden, Bronx, NY, USA

A. C. Carnaval
Department of Biology, City College of New York, New York, NY, USA

Biology Program, Graduate Center of CUNY, New York, NY, USA

V. Rull, A. C. Carnaval (eds.), *Neotropical Diversification: Patterns and Processes*, Fascinating Life Sciences, https://doi.org/10.1007/978-3-030-31167-4_16

demonstrate that the spatial turnover of lineages within species largely mirrors that of species within communities; both climate change and geographic barriers appear to have acted in combination to produce patterns of diversification. However, while geographic patterns of species and genetic diversity are similar across Atlantic forest organisms, their underlying processes are not.

Keywords Biodiversity · Beta diversity · Biogeography · Community ecology · Neotropics · Phylogeography

1 Introduction

Known as one of the world's biological hotspots (Myers et al. 2000), the Atlantic rainforest of Brazil stretches over extensive latitudinal (3°S–30°S), longitudinal (35°W–60°W), altitudinal (sea level to 2900 m a.s.l.), and climatic gradients (1000–4200 mm annual rainfall), ranging from the northeastern coast of Brazil into northern Paraguay and Argentina (Ribeiro et al. 2011a). In Pre-Columbian times, its vegetation covered approximately 1,300,000 km^2 along the eastern coast of South America (Morellato 2000), including a variety of forest physiognomies such as coastal evergreen ombrophilous forests, semi-deciduous and deciduous forests, mangroves, swamps, restingas, high-altitude grasslands (known as "campos rupestres" and "campos de altitude"), and subtropical mixed forests ("matas de Araucárias," Figs. 16.1 and 16.2; Morellato 2000; Olson et al. 2001; Oliveira-Filho and Fontes 2000; Ribeiro et al. 2011a). Since the sixteenth century, however, the forest has been under intense human pressure; today, it is reduced to less than 12% of its Pre-Columbian coverage (Oliveira-Filho and Fontes 2000).

Inventory work in the Brazilian Atlantic rainforest dates back to the eighteenth Century, with A. Saint-Hilaire since 1816, J.B. von Spix & C.F.P. von Martius since 1817, and A.F. Glaziou since 1858. However, it was only in the last 30–40 years that substantial effort was devoted to studying the processes driving local biodiversity patterns. Despite the urgent need for in-depth studies in the region, its high beta diversity and topographic complexity complicate the understanding of species richness patterns (Moura et al. 2017; Leão-Pires et al. 2018).

We present a comprehensive multi-taxon review of diversity patterns along the Atlantic forest, and summarize main spatial patterns of plant and animal diversity, flagging areas of higher species accumulation, turnover, and endemism. We also review main patterns of lineage divergence and population structure recovered from molecular phylogenies and phylogeographic studies, and explore some of the hypotheses commonly invoked to explain the history of the Atlantic rainforest biota. We hope the information summarized here can provide a basis for further conservation planning in the region.

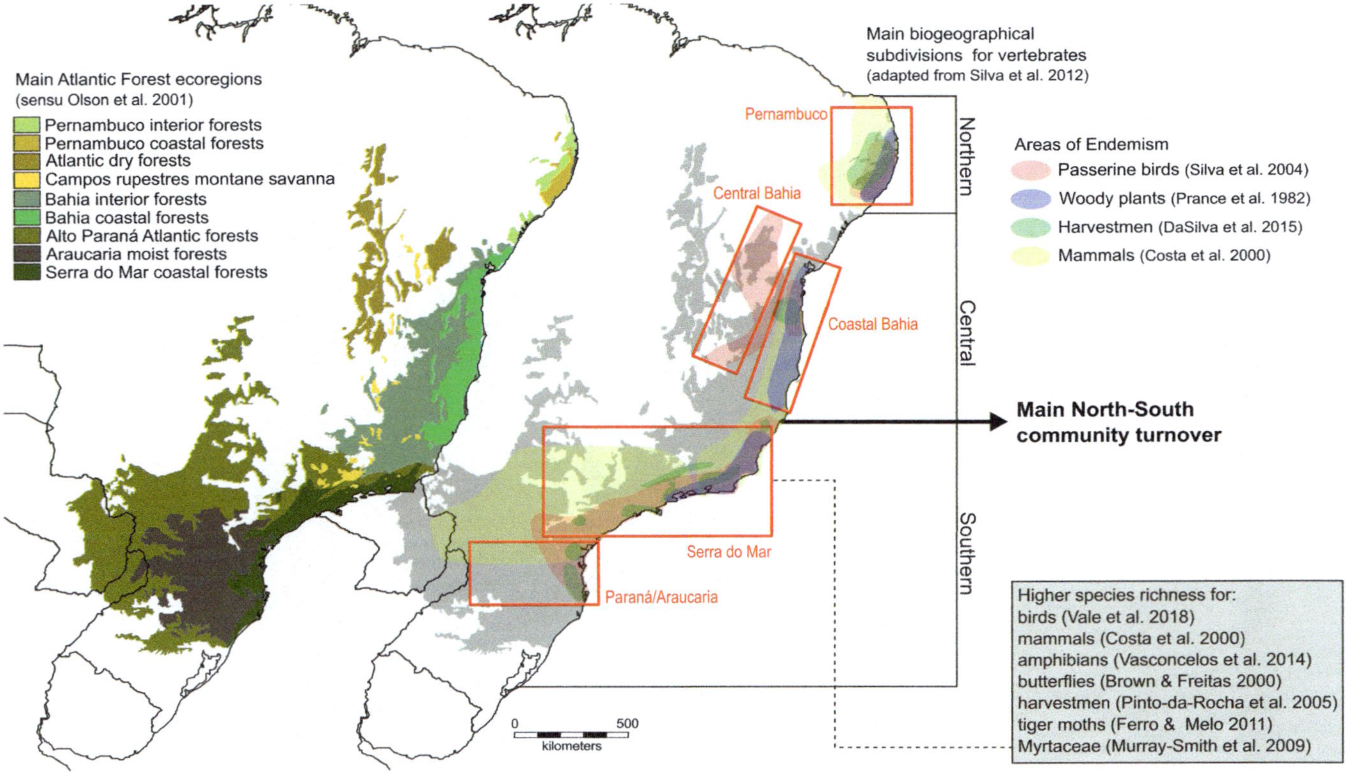

Fig. 16.1 Atlantic Forest ecoregions (left) and spatial patterns of diversity and endemism based on multi-taxon data (right). Red polygons indicate areas of endemism mentioned throughout the text. Colored ranges exemplify taxon-specific endemism regions or regionalization schemes. Main point of community turnover (north/south) is indicated to the right of the map

Fig. 16.2 Multiple forests in one: landscapes of the Atlantic Forest in Paraná (Araucaria forest, top left), Rio de Janeiro (montane forest, top right), Espírito Santo (lowland forest, center left), São Paulo (with the capital, São Paulo, in the back; center right), Central Bahia (Chapada Diamantina, bottom left), and Pernambuco ("brejo de altitude" in the township of Jaqueira, bottom right)

2 Spatial Patterns of Species Richness

The Atlantic rainforest of Brazil stands out as one of the most biologically rich environments in the Neotropics, accumulating a number of species that is only comparable to those found in Amazonia and the Andes (Myers et al. 2000; Fiaschi and Pirani 2009; Jenkins et al. 2015). Within its domain, congruent richness patterns emerge among butterflies (Brown and Freitas 2000), harvestmen (Pinto-da-Rocha et al. 2005; DaSilva et al. 2015), non-volant mammals (Costa et al. 2000), birds (Vale et al. 2018), amphibians (Vasconcelos et al. 2014), tiger moths (Ferro and Melo 2011), antbirds (Durães and Loiselle 2004), and members of the plant family Myrtaceae (Murray-Smith et al. 2009): all these taxa show elevated number of species in the coastal and montane regions of Rio de Janeiro and São Paulo, in southeastern Brazil (Fig. 16.1). In butterflies, this is observed in the Morphinae, Brassolinae, Acraeini, Papilionidae, and Pieridae groups (Brown and Freitas 2000). The latitudinal bands of 22–25°S, which include the coast and a portion of the Serra da Mantiqueira mountain range in the states of São Paulo and Rio de Janeiro, also host the highest number of marsupials, rodents, and primates within the Atlantic forest (Costa et al. 2000). In harvestmen, the coastal mountain ranges from Paraná to southern Bahia harbor an impressive diversity, with the Serra dos Órgãos National Park (Rio de Janeiro) housing the highest species richness in the whole world; inland regions and northern Atlantic forest sites, characterized by lower altitudes, lower and more seasonal rainfall, house much lower diversity (Nogueira et al. 2019). Data from tiger moths also identify the mountains and coastal regions of Rio de Janeiro and São Paulo as the most species rich (Ferro and Melo 2011).

Other groups show high diversity in this region, but also elsewhere. Endemic antbirds (Formicariidae), for instance, are highly diverse in the coastal areas and mountain ranges of Rio de Janeiro and São Paulo, but show equally high species richness in the neighboring states of Paraná and Espírito Santo (Durães and Loiselle 2004). In the plant genus *Myrcia*, high species richness is observed in the mountain ranges of Rio de Janeiro and São Paulo, but also in Paraná, northern Espírito Santo, and southern Bahia (Murray-Smith et al. 2009). The same pattern is recovered in an analysis of all vascular plants (Reginato and Michelangeli 2020).

There are, however, exceptions to this pattern. Butterflies of the Ithomiinae, Satyrinae, and Heliconini clades peak in richness further north of Rio de Janeiro and São Paulo (20°S), in the state of Espírito Santo, whereas groups such as the Eurytelinae and Charaxinae are more diverse in interior seasonal forests (Brown and Freitas 2000). Non-endemic antbirds have a distinct pattern of diversity relative to endemic species, with sites further inland accumulating more taxa (Durães and Loiselle 2004). In termites (Isoptera), diversity is highest in the northern portion of the Atlantic Forest and decreases towards the south, regardless of taxonomic groups (Cancello et al. 2014). In contrast, patterns in groups that heavily rely on humidity (e.g., amphibians) seem to be directly associated with water availability, regardless of latitude (Silva et al. 2012), illustrating how taxon-specific life-history traits influence regional spatial richness patterns.

The variation observed across taxonomic groups suggests that generalizations of regional spatial patterns of species richness within the Atlantic rainforest domain must be made with care. Differences in scope, grain, and degree of biological sampling across studies add complexity to cross-taxonomic comparisons based on the literature only. Moreover, several taxon-centric analyses performed to date have been built on somewhat biased spatial sampling, as species records and inventory sites have historically been more frequent in the central corridor of the Atlantic rainforest (e.g., São Paulo, Rio de Janeiro, Espírito Santo) relative to the northern and southern extremes (Ferro and Melo 2011). While it appears that high levels of diversity tend to accumulate in the coastal and montane regions of Rio de Janeiro and São Paulo, accurate comparisons of the patterns of diversity shown by multiple forest taxa will require those occurrence datasets to be completed and analyzed under the same framework, and at comparable spatial scales.

3 Large-Scale Community Turnover

Community composition has been reported to differ strikingly between the southern and the northern portions of the Atlantic forest, in groups as diverse as plants (Oliveira-Filho and Fontes 2000; Oliveira-Filho et al. 2006; Fiaschi and Pirani 2009), reptiles (Vanzolini 1988), birds (Bates et al. 1998), mammals (Costa et al. 2000), termites (Cancello et al. 2014), and harvestmen (Nogueira et al. 2019; Fig. 16.1), regardless of methodology and beta-diversity metrics. Although the exact point of turnover varies across taxa, or often goes unreported, most studies have detected a compositional shift along the central corridor of the forest, somewhere between the states of Rio de Janeiro and Bahia (Fig. 16.1). Within this region, most studies indicate a break around the Doce river valley, in Espírito Santo (e.g., Cracraft and Prum 1988; Amorim and Pires 1996; Costa 2003; Silva et al. 2004; Pinto-da-Rocha et al. 2005; Perret et al. 2006; Fiaschi and Pirani 2009; Costa and Leite 2013).

In plants, for instance, a northern Atlantic forest has been recognized, extending from Rio Grande do Norte (ca. 5°S) to northern Espírito Santo (ca. 19°S; Fiaschi and Pirani 2009). This area is mostly formed by a narrow strip of forest standing between the semi-arid Caatinga and the Atlantic Ocean, as well as small inland forest enclaves (i.e., "Brejos de Altitude"; Rodal and Sales 2008), and the inland forests of the Chapada Diamantina (Funch et al. 2008; Fig. 16.2). In contrast, a southern portion of the Atlantic forest has been described to extend from Espírito Santo (ca. 19°S) to southern Santa Catarina (ca. 29°S) (Fiaschi and Pirani 2009). This region encompasses not only evergreen forests along the coast (both at sea level and montane and sub-montane regions), but also the southern Araucaria forest province (Morrone 2006), and the seasonally dry forests of Southern Brazil, Paraguay, and Missiones in Argentina (Oliveira-Filho and Fontes 2000; Oliveira-Filho et al. 2006). The latter two, however, have also been interpreted as distinct phytogeographic units (Prado 2000; Pennington et al. 2006; Cabrera and Willink 1973).

Straightforward superimpositions of range maps of animal groups support this north/south distinction (Fig. 16.1). Differences in the composition of species of non-volant mammals exist between the northern and the southern Atlantic forest (Costa 2003; Costa and Leite 2013)—but the latitudinal placement of the turnover varies across genera, ranging from the southern boundary of Espírito Santo (e.g., *Rhipidomys*), to its northern limit (e.g., *Micoureus*; Costa 2003; Costa and Leite 2013). Superimposed maps of suboscine birds also show major distinctions between northern and southern bird communities, with 21 species limited to the southern range of the forest, and 37 taxa restricted to the northern region; turnover occurs along the coast of Espírito Santo (Batalha-Filho et al. 2013). Similarly, biogeographical studies identified north/south breaks in the distribution of amphibians (Lynch 1979; Müller 1973), and lizards (Vanzolini 1988). While spatial patterns of Atlantic forest invertebrates are less known, harvestmen show a very distinctive pattern with the southern portion being predominantly occupied by the Gonyleptidae family, whereas the northern forests are occupied by the Cosmetidae and the Stygnidae; in the state of Bahia, species composition is mixed, with co-occurring members of all three families (Nogueira et al. 2019). Lastly, community turnover has also been documented through large-scale macroecological analyses. In line with the findings of Costa (2003) and Batalha-Filho et al. (2013), data from 3836 bird and 1641 mammal species from throughout the New World demonstrate spatial differences across biological groups within the Atlantic forest (Melo et al. 2009): while mammals showed areas of high turnover in Espírito Santo (especially near the Doce river), higher within-quadrat turnover was observed further north, in coastal Bahia (south of the San Francisco river), for birds (Melo et al. 2009).

4 Areas of Endemism and Regionalization

Several analyses of community composition within the Atlantic rainforest's southern and northern regions indicate fractal patterns of biodiversity, with further differentiation within these two large units. One such attempt, which used published databases of Atlantic forest vertebrates in a panbiogeographical analysis, broke down the Atlantic forest into three major regions of unique community composition along the latitudinal gradient (Silva et al. 2012; Fig. 16.1): a southern region, a central area (extending from the Doce river in Espírito Santo, to the São Francisco river in Bahia), and a northern coastal region. The later two biogeographic units are recognized within what was originally described as the northern half of the forest, and largely agree with two areas with unique floristic compositions (Thomas et al. 1998): the Pernambuco region (ca. 8°S), and the Bahia ("central corridor") unit extending from ca. 13°S to 19°S.

A more recent, comprehensive Parsimony Analysis of Endemism (i.e., PAE; Morrone 1994; DaSilva et al. 2015), based on a presence/absence matrix of 140 endemic passerine birds of the Atlantic forest (Silva et al. 2004), identified four areas of endemism that are largely consistent with subsequent regionalization studies: Pernambuco, Central Bahia (the interior of the Diamantina region), Coastal Bahia (southern coast of Bahia), and Serra do Mar (Fig. 16.1). By clustering spatial regions based on their shared species, this and other PAEs seek to identify centers of endemism or regions with common biotic differentiation (Cracraft 1985). While the first three areas identified by Silva et al. (2004) are subsets of the northern Atlantic forest component previously discussed in the literature, those passerine areas of endemism have been shown to have significant predictive power, matching patterns reported for several other organisms. Because these biogeographic regions are supported by multiple taxonomic groups, they represent useful regionalization schemes for biodiversity studies and conservation. Below we describe each of these four areas of endemism, plus the Paraná/Araucaria area (Fig. 16.1).

4.1 The Pernambuco Area of Endemism

The Pernambuco area of endemism was originally described from passerine data (Silva et al. 2004). It encompasses the strip of evergreen, semideciduous and deciduous forests that follows the Atlantic Ocean coast north of the São Francisco River, including the states of Alagoas, Pernambuco, and Paraíba. Support for the recognition of this region as biologically unique has been widespread among animals and plants. For instance, it matches the biologically unique region of "the Atlantic slope of Alagoas and Pernambuco" described earlier for birds (Stattersfield et al. 1998), and plants (Prance 1982). This pattern is consistent with invertebrate turnover (see DaSilva et al. 2015, Nogueira et al. 2019 for harvestmen; Tyler et al. 1994, Brown and Freitas 2000 for butterflies).

Some of the analyses of species ranges and centers of endemism that identified the uniqueness of the Pernambuco region have extended this area. For instance, a study of pooled vertebrate datasets suggested that the southern bank of the São Francisco river should be included within the Pernambuco center of endemism (Müller 1973). A Parsimony Analysis of Endemism (PAE) of lowland forest mammals (i.e., marsupials, rodents, and primates) also identified a slightly broader Pernambuco area of endemism that extended to the north bank of the São Francisco river along the coast and further south to the inland portion of northeastern Brazil (Costa et al. 2000). More recently, a southward extension of a Pernambuco-centered area of endemism was proposed based on snake (Moura et al. 2017), and harvestmen (DaSilva et al. 2015) data. An unconstrained ordination and clustering analysis of 198 snake species indicates the existence of a unique biogeographical subregion (Pernambuco Coastal/Interior Forests; Moura et al. 2017) that includes the coastal region of northeastern Brazil, from the northern range of the Atlantic forest to the coast of Northern Bahia, including the "Brejos de Altitude" located within the dry Caatinga.

4.2 The Central Bahia Area of Endemism

Passerine data support the recognition of a center of endemism in the topographically complex interior region of northeastern Brazil encompassing the slopes of the Chapada Diamantina, as well as the plateaus of central Bahia and northern Minas Gerais, including patches of evergreen, semideciduous and deciduous forests (Silva et al. 2004). A previous analysis of bird ranges also identified a unique region of community composition centered in the "deciduous forests of Bahia" (Stattersfield et al. 1998), which are contained within the area recovered based on the passerine data. Butterfly communities (Brown and Freitas 2000) and vascular plants (Reginato and Michelangeli 2020) are also unique in the Brazilian central plateau, particularly in the interior of Bahia (Chapada Diamantina region).

4.3 The Coastal Bahia Area of Endemism

The narrow coastal forests of Bahia, dominated by evergreen elements, have been consistently recognized as biologically unique. Little agreement exists, however, relative to the full latitudinal extent of this area. An area of endemism was recognized to occupy the coast of Bahia based on data from woody plants (Bahia endemism region; Prance 1982) and bamboos (Bahia region; Soderstrom et al. 1988). In both harvestmen (DaSilva et al. 2015) and passerines (Silva et al. 2004), however, this area of endemism fails to occupy the entire coast of the state—in the latter, for instance, it is limited to the central region of costal Bahia, between the Jiquiriça and Jequitinhonha rivers (Coastal Bahia Endemism Center; Silva et al. 2004). In butterflies, the spatial pattern is even more different, with unique communities being found form southern Bahia to northern Espírito Santo (Brown and Freitas 2000). An analysis of 15 invertebrate genera, including flies, harvestmen, butterflies, beetles, spiders, hemipterans, and heteropteran, also recognized a center of insect endemism that extends from the southern bank of the São Francisco river to the northern portions of Espírito Santo (Sigrist and Carvalho 2008).

4.4 The Serra do Mar Area of Endemism

The Serra do Mar area of endemism encompasses a gradient of habitats that range from montane (ca. 1700 m a.s.l.) to sub-montane and lowland forests, as well as coastal restingas; this system is associated with the mountain chain that runs parallel to the coast of southeastern Brazil. Multiple organisms show unique community composition in this region, although, again, the exact geographical limits of the areas of endemism reported to date vary across taxonomic groups. Data from woody plants (Prance 1982), bamboos (Soderstrom et al. 1988), amphibians (Vasconcelos et al. 2014), and passerine birds (Silva et al. 2004) suggest an area of endemism that extends from Santa Catarina to central Espírito Santo (Prance 1982). Analyses of a pooled set of vertebrates (Müller 1973), butterflies (Tyler et al. 1994), and snake species (Moura et al. 2017), however, flag a more restricted area of endemism, between Santa Catarina and Rio de Janeiro (and not reaching Espírito Santo). In lowland mammals (Costa et al. 2000), an area of endemism is thought to reach its northern limit in the state of São Paulo. In insects (Sigrist and Carvalho 2008), amphibians (Vasconcelos et al. 2014), and snakes (Moura et al. 2017), unique community composition has been detected only north of northern São Paulo (i.e., excluding the southern state of Santa Catarina).

Further community differentiation, and smaller sub-centers of endemism, have been proposed to occur within the Serra do Mar region (e.g., Cracraft 1985; Silva et al. 2004). This pattern is particularly common in taxa with reduced dispersal ability and narrower environmental tolerances. For instance, differences between lowland and highland avifauna communities were noted along the Serra do Mar, leading to the proposition of an 'Atlantic forest mountains' area of endemism (Stattersfield et al. 1998). On the other hand, butterfly community differentiation has been observed between the montane areas of southeastern Brazil (with localities in the states of Rio de Janeiro, São Paulo, Minas Gerais), and the inland, central plateau (including semideciduous forests in São Paulo; Brown and Freitas 2000). Further differentiation of biological communities within the Serra do Mar region, separating communities in Espírito Santo, interior of Minas Gerais, and the Northern São Paulo/Southern Rio de Janeiro, was proposed in an analysis of ranges of 15 invertebrate genera, one species of plant, and one amphibian species (Sigrist and Carvalho 2008). Tiger moth (Arctiidae) data also support a sharp distinction between communities in lowland, coastal localities (within 62 km from the ocean, and mostly below 200 m a.s.l.) vs. montane sites in southeastern Brazil, usually near the ocean (Ferro and Melo 2011). Finally, harvestmen data support the recognition of eight smaller centers (i.e., Espırito Santo, Serra dos Orgãos, Serra da Bocaina, coastal southern Rio de Janeiro, Serra do Mar in São Paulo, Southern São Paulo, Paraná, Santa Catarina) within the generally reported "Serra do Mar" area of endemism (DaSilva et al. 2015). It is expected that other biological groups similarly impacted by water availability, or with reduced dispersal ability (e.g., amphibians, crickets), may share high turnover rates in geographic space (DaSilva et al. 2015).

4.5 The Paraná/Araucaria Area of Endemism

A region comprising a portion of the southern states of Paraná, Santa Catarina, and Rio Grande do Sul, which is characterized by the presence of the Araucaria forest, has been once recognized as a center of bird endemism (Cracraft 1985). This region, which is bordered by São Paulo (in the north), the Paraná river (in the west), and the Jacui river (in the south, Müller 1973), is also known to host distinct communities of tiger moths (Ferro and Melo 2011), butterflies (particularly the frost-prone areas south of 24°S; Brown and Freitas 2000), amphibians (Vasconcelos et al. 2014), harvestmen (DaSilva et al. 2015), and snakes (Moura et al. 2017).

5 Phylogenetic Relationships Among Regional Biotas

Analyses of spatial patterns of taxonomic diversity, species richness, and areas of endemism allow us to formulate hypotheses about the origin and evolution of biotas (Cracraft 1985). Testing the latter, however, requires knowledge of phylogenetic relationships of individual lineages: historical inference is crucial to this effort by shedding light on the timing and processes guiding biotic assembly (Weeks et al. 2016). As such, understanding the evolutionary relationships of lineages with spatially restricted distributions along the Atlantic rainforest can provide critical information about the historical processes that may have shaped diversification in this region.

Differences in community composition between the northern and southern Atlantic forest are well supported by evolutionary studies that recover entire clades that occur exclusively within each region (Fiaschi and Pirani 2009). Yet, the relationships among northern and southern lineages seem to vary considerably among taxonomic groups. While some phylogenetic studies have reconstructed Atlantic forest clades (e.g., Amorim and Pires 1996; Costa et al. 2000; Sigrist and Carvalho 2008; Pires and Marinoni 2010), others have recovered Atlantic forest taxa that are more closely related to members of other Neotropical biomes (e.g., Amazonia, Cerrado, Andes) than to other Atlantic forest taxa (Cracraft and Prum 1988; Costa 2003; Perret et al. 2006; Santos et al. 2007; Nogueira et al. 2019). While the former scenario supports *in situ* diversification, the latter supports an *ex situ* origin for the Atlantic forest biota, followed by subsequent migration into this biome. As of now, evidence suggest that both processes have helped shape diversity patterns in this region. Fine-scale analyses, based on comprehensive population sampling within individual species, may provide further understanding of the origin and diversification history of the Atlantic forest biota.

5.1 Intra-specific Patterns of Lineage Diversity

Studies of the patterns of lineage divergence within species (phylogeography; Avise et al. 1987) have been increasingly applied to Atlantic forest organisms. These molecular datasets have been integrated with the pollen record, ecological niche models, and paleoclimatic reconstructions, and used to propose and test hypotheses about the historical processes that shaped lineage divergence, as well as to estimate population structure and genetic breaks along the domain.

Most Atlantic forest species exhibit genetic breaks along their ranges (Table 16.1, Fig. 16.3). Understanding the processes that influenced these lineage distribution

Table 16.1 Phylogeographic studies of species completely or mostly restricted to the Atlantic forest, showing inferred lineages divergence times, latitudinal breaks, and approximate timing of demographic events as detected in one or more lineages

Taxon (order/family)	Genetic marker	Lineage divergence period (CI)	Location of main spatial break(s)	Demographic events	References
Mammals					
Alouatta guariba (Primates, Atelidae)	mtDNA	1.2 Mya (0.38–2 Mya)	CRSB	No	Martins et al. (2011)
Akodon cursor (Rodentia, Cricetidae)	mtDNA	NA	Central corridor	NA	Colombi et al. (2010)
Bradypus torquatus (Pilosa, Bradypodidae)	mtDNA, nuDNA	0.39–5.36 Mya (0.08–8 Mya)	Central corridor	No	Schetino et al. (2017)
Cerdocyon thous (Carnivora, Canidae)	mtDNA, nuDNA	0.58 Mya (0.18–1.64 Mya)	Central corridor	Expansion of southern clade ca. 0.1 Mya	Tchaicka et al. (2007)
Desmodus rotundus (Chiroptera, Phyllostomidae)	mtDNA, nuDNA	0.35–0.7 Mya	CRSB	Expansion of southern clade	Martins et al. (2009)
Euryoryzomys russatus (Rodentia, Cricetidae)	mtDNA, nuDNA	NA	Central corridor	No	Miranda et al. (2007)
Nectomys squamipes (Rodentia, Cricetidae)	mtDNA, microsatellites	0.48–1.01 (0.15–1.64)	CRSB	Expansion between LIG and LGM, decline after LGM	Dalapicolla and Leite (2018)
Small rodents and marsupials	mtDNA	Miocene, Pliocene, Pleistocene	Central corridor	NA	Costa (2003)
Trinomys (Rodentia, Echimyidae)	mtDNA	1.64–6.22 Mya	Interior x Coastal	NA	Lara and Patton (2000)
Birds					
Basileuterus leucoblepharus (Passeriformes, Parulidae)	mtDNA, nuDNA	NA	No subdivision	Expansion ca. 0.3 Mya	Batalha-Filho et al. (2012)
Conopophaga lineata (Passeriformes, Conopophagidae)	mtDNA, nuDNA	NA	Central corridor, CRSB	Expansion ca. 0.25 Mya	Dantas et al. (2015)

(continued)

Table 16.1 (continued)

Taxon (order/family)	Genetic marker	Lineage divergence period (CI)	Location of main spatial break(s)	Demographic events	References
Myrmeciza spp. (Passeriformes, Thamnophilidae)	mtDNA, nuDNA	0.13–3.8 Mya (0.04–5.4 Mya)	Central corridor, CRSB	Expansion of northern clade ca. 2 Mya; expansion of southern clade after LGM (0.02 Mya)	Amaral et al. (2013)
Pyriglena (Passeriformes, Thamnophilidae)	mtDNA	1.83–4.54 Mya (0–8.25 Mya)	Central corridor	Expansion of southern clade in LGM (0.03 Mya)	Maldonado-Coelho (2012)
Rhopias gularis (Passeriformes, Thamnophilidae)	mtDNA, nuDNA	0.3 Mya (0.09–0.93 Mya)	Central corridor	Expansion of southern clade after LGM (0.02 Mya)	Batalha-Filho and Miyaki (2016)
Schiffornis virescens (Passeriformes, Tityridae)	mtDNA	0.43 Mya (0.23–0.68 Mya)	No subdivision	Expansion ca. 0.2 Mya	Cabanne et al. (2013)
Sclerurus scansor (Passeriformes, Furnariidae)	mtDNA, nuDNA	0.06–0.54 Mya (0.04–0.76 Mya)	Central corridor, CRSB, SFRV	Expansion of southern clade	D'Horta et al. (2011)
Synallaxis ruficapilla (Passeriformes, Furnariidae)	mtDNA, nuDNA	0.53–2.24 Mya (0.16–2.75 Mya)	Central corridor, SFRV	NA	Batalha-Filho et al. (2013)
Xiphorhynchus fuscus (Passeriformes, Furnariidae)	mtDNA, nuDNA	0.39–0.79 Mya (0.08–1.66 Mya)	Central corridor, CRSB, SFRV	Expansion of most clades in Late Pleistocene	Cabanne et al. (2007, 2008)
Amphibians					
Hypsiboas spp. (Anura, Hylidae)	mtDNA, nuDNA	Late Pleistocene	Central corridor, CRSB, SFRV	Expansion of southern clade after LGM	Carnaval et al. (2009)
Thoropa miliaris (Anura, Cycloramphidae)	mtDNA, nuDNA	0.2–12 Mya (0.06–13.2 Mya)	Central corridor	Expansions ca. 0.15–0.2 Mya	Fitzpatrick et al. (2009)
Phyllomedusa burmeister (Anura, Hylidae)	mtDNA, nuDNA	0.4–5 Mya (0–6.63 Mya)	Central corridor, CRSB	NA	Brunes et al. (2010)
Rhynella crucifer (Anura, Bufonidae)	mtDNA, nuDNA	0.6–2.7 Mya (0.27–10.68 Mya)	Central corridor, CRSB	Moderate expansion of northern and central clades	Thomé et al. (2014)

(continued)

Table 16.1 (continued)

Taxon (order/family)	Genetic marker	Lineage divergence period (CI)	Location of main spatial break(s)	Demographic events	References
Proceratophrys boiei (Anura, Odontophrynidae)	mtDNA, nuDNA	1.56–7.86 Mya (0.74–11.7 Mya)	Central corridor, CRSB	No	Amaro et al. (2012)
Dendropsophus elegans (Anura, Hylidae), *Chiasmocleis carvalhoi* (Anura, Microhylidae)	mtDNA	0.9–6.56 Mya (0.53–8.4 Mya)	Central corridor, CRSB	Expansions in several clades	Tonini et al. (2013)
Phyllomedusa distincta (Anura, Hylidae)	mtDNA, nuDNA	~0.6 Mya	CRSB	Expansion of southern clade in Holocene	Brunes et al. (2015)
Scinax eurydice (Anura, Hylidae)	mtDNA, nuDNA	0.8–2.1 Mya (0.4–3.9 Mya)	Central corridor, CRSB, SFRV	Decline and expansion of southern clades ca. 0.13 and 0.5, respectively	Menezes et al. (2016)
Reptiles					
Bothrops jararaca (Squamata, Viperidae)	mtDNA	1.54–6 Mya (0.38–8.9 Mya)	CRSB	Expansion ca. 0.1 Mya	Grazziotin et al. (2006)
Enyalius (Squamata, Leiosauridae)	mtDNA, nuDNA	2.21–27.83 Mya (0.99–34.53 Mya)	Central corridor, CRSB, SFRV	NA	Rodrigues et al. (2014)
Gymnodactylus darwinii (Squamata, Phyllodactylidae)	mtDNA	0.9–23.3 Mya	Central corridor, CRSB, SFRV	NA	Pellegrino et al. (2005)
Arthropods					
Acutisoma longipes (Opiliones, Gonyleptidae)	mtDNA, nuDNA	7.9–14.5 Mya (4.2–16.4 Mya)	CRSB	Decline of some clades ca. 0.3 Mya	Peres et al. (2018)
Araneus omnicolor (Araneae, Araneidae)	mtDNA, nuDNA	0.8 Mya (0.32–1.67 Mya)	No subdivision	Expansion ca. 0.1 Mya	Peres et al. (2015)
Araneus venatrix (Araneae, Araneidae)	mtDNA, nuDNA	0.4–8.35 Mya (0.23–15.54 Mya)	Central corridor	Expansion of southern clade ca. 0.25 Mya	Peres et al. (2017)

(continued)

Table 16.1 (continued)

Taxon (order/family)	Genetic marker	Lineage divergence period (CI)	Location of main spatial break(s)	Demographic events	References
Bombus spp. (Hymenoptera, Apidae)	mtDNA, microsatellites	0.1–0.2 Mya	CRSB	Expansions during LGM	Françoso et al. (2016)
Dinoponera lucida (Hymenoptera, Formicidae)	mtDNA	0.08–3.34 Mya (0.02–4.38 Mya)	Central corridor	Decline ca. 0.2 Mya, expansion ca. 0.025 Mya	Resende et al. (2010)
Euglossa iopoecila (Hymenoptera, Apidae)	mtDNA	NA	Central corridor, SFRV	Expansion of southern clade ca. 0.2 Mya	Frantine-Silva et al. (2017)
Melipona quadrifasciata (Hymenoptera, Apidae)	mtDNA	0.39–0.49 Mya (0.23–0.84 Mya)	CRSB	Expansions in both clades	Batalha-Filho et al. (2010)
Neoleucinodes elegantalis (Lepidoptera, Crambidae)	mtDNA	0.25–0.6 Mya (0.15–1 Mya)	Central corridor	No	Maia et al. (2016)
Promitobates (Opiliones, Gonyleptidae)	mtDNA, nuDNA	5.3–58.5 Mya (5.3–62.4 Mya)	CRSB	Expansions in several clades	Bragagnolo et al. (2015)
Synoeca spp. (Hymenoptera, Vespidae)	mtDNA, nuDNA	0.09–2.6 Mya	Central corridor, CRSB, SFRV	Expansion of central and southern clades ca. 0.1 Mya	Menezes et al. (2017)
Platyhelminthes					
Cephaloflexa bergi (Tricladida, Geoplanidae)	mtDNA, nuDNA	7.02 Mya (5.8–8.5 Mya)	Central corridor, CRSB	Expansion of southern clade	Álvarez-Presas et al. (2014)
Plants					
Dalbergia nigra (Fabales, Fabaceae)	cpDNA	0.35–0.78 Mya (0.32–1.18 Mya)	Central corridor, CRSB	Expansions in both clades	Ribeiro et al. (2011b)
Epidendrum fulgens (Asparagales, Orchidaceae)	cpDNA, microsatellites	NA	Extreme south	Bottleneck in northern clade	Pinheiro et al. (2011)
Eugenia uniflora (Myrtales, Myrtaceae)	cpDNA, nuDNA	~4.9 Mya	CRSB	Expansion of northern clade ca. 0.1 Mya	Turchetto-Zolet et al. (2016)
Passiflora contracta (Malpighiales, Passifloraceae)	cpDNA, nuDNA	0.6–4.5 Mya (0.1–6.7 Mya)	Central corridor	No	Cazé et al. (2016)

(continued)

Table 16.1 (continued)

Taxon (order/ family)	Genetic marker	Lineage divergence period (CI)	Location of main spatial break(s)	Demographic events	References
Vriesea gigantea (Poales, Bromeliaceae)	cpDNA, microsatellites	NA	Central corridor	NA	Palma-Silva et al. (2009)

CI = confidence interval, when available
NA = data not available; Central corridor = genetic breaks around Doce or Jequitinhonha rivers; CRSB = genetic break in the region of the Continental Rift of Southeastern Brazil (see Sect. 6.3 for more details); SFRV = genetic breaks in the São Francisco river valley

patterns (intra-specific structure) is a common question among most phylogeographic studies conducted in the region. In general, these data are consistent with the patterns of community turnover detected via species occurrence information. Despite differences in molecular markers and sampling ranges, lineage turnover is frequently observed in the central corridor of the Atlantic forest (i.e., between the northern and southern forest regions), as well as between documented areas of endemism (e.g., breaks around the Paraíba do Sul river valley, and the São Francisco river valley). This pattern has been documented in mammals (e.g., Costa 2003; Moraes-Barros et al. 2006; Tchaicka et al. 2007; Martins et al. 2009), birds (Cabanne et al. 2007, 2008; D'Horta et al. 2011; Maldonado-Coelho 2012; Amaral et al. 2013; Batalha-Filho et al. 2013; Batalha-Filho and Miyaki 2016), reptiles (Pellegrino et al. 2005; Grazziotin et al. 2006), amphibians (Carnaval et al. 2009; Fitzpatrick et al. 2009; Brunes et al. 2010; Thomé et al. 2010, 2014; Amaro et al. 2012; Tonini et al. 2013; Menezes et al. 2015), invertebrates (Batalha-Filho et al. 2010; Resende et al. 2010; Álvarez-Presas et al. 2014; Bragagnolo et al. 2015; Françoso et al. 2016; Peres et al. 2017, 2018, 2019), and plants (Palma-Silva et al. 2009; Ribeiro et al. 2011b; Table 16.1, Fig. 16.3).

While the spatial distribution of lineages is broadly consistent among groups, the estimated divergence times of co-distributed lineages vary widely across species. Several genetic breaks around the Doce river have been estimated to be recent, e.g., 0.4 Mya for passerine birds (Cabanne et al. 2008), 0.4–0.6 Mya for rosewood trees (Ribeiro et al. 2011b), 1.6–2.1 Mya for treefrogs (Brunes et al. 2010; Menezes et al. 2016), 2.5 Mya for ants (Resende et al. 2010). Others, in contrast, go back to the Plio-Pleistocene: 5.4 Mya for sloths (Schetino et al. 2017), 5.7–6.6 Mya for treefrogs (Tonini et al. 2013), 8–23 Mya for lizards (Pellegrino et al. 2005). In the Ribeira de Iguape river region (Fig. 16.3), Plio-Pleistocene phylogeographic breaks were observed for bees (~0.4 Mya, Batalha-Filho et al. 2010), wasps (2.6 Mya, Menezes et al. 2017), and amphibians (0.6 Mya, Brunes et al. 2015; Thomé et al. 2010), while older divergences were reported for snakes (3.9 Mya, Grazziotin et al. 2006), frogs (4.9 Mya, Brunes et al. 2010), and harvestmen (31.5 Mya, Peres et al. 2019).

It has been observed that the general timing of divergences within species also varies according to the target group. For instance, most reported divergence times of

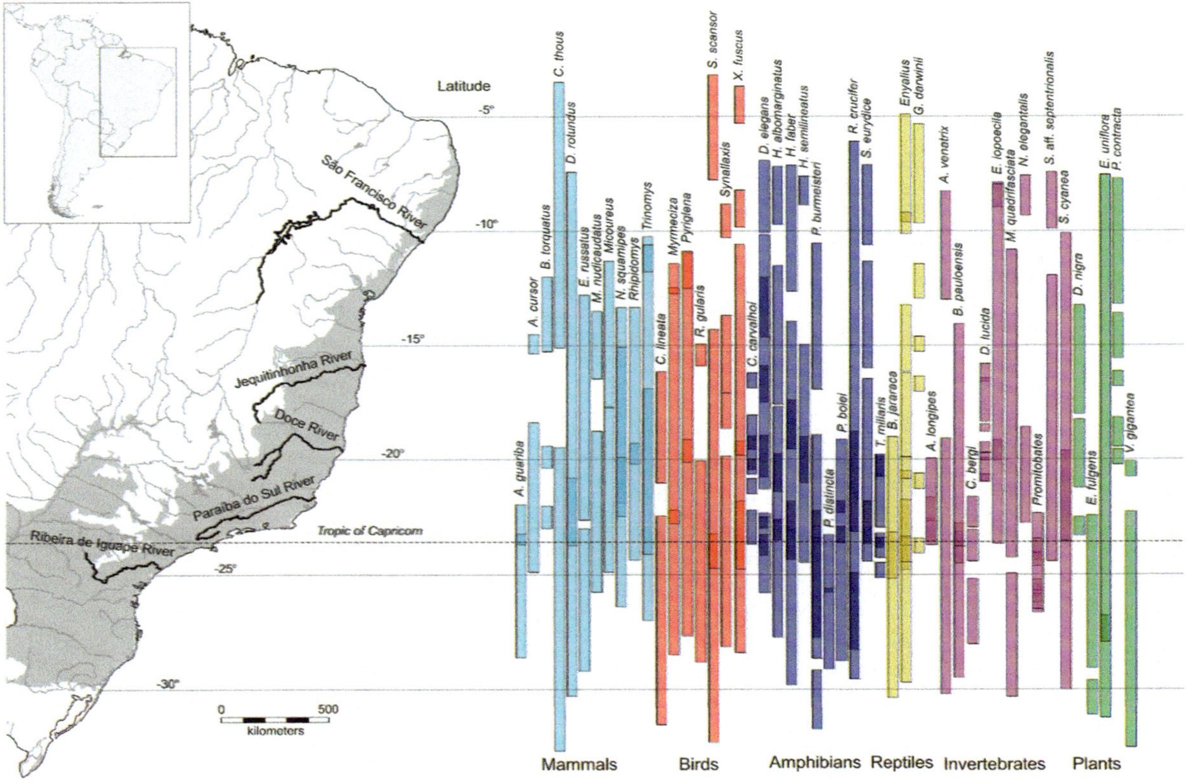

Fig. 16.3 Latitudinal breaks reported by phylogeographic studies of Atlantic forest groups. Map shows the distribution of the domain and the location of rivers most commonly associated genetic breaks within the forest. Bars (right) represent the approximate latitudinal distribution of the main lineages detected across taxa (see Table 16.1)

bird lineages in the Atlantic forest fall within the Pleistocene (e.g., Batalha-Filho and Miyaki 2016; Cabanne et al. 2007, 2008; Dantas et al. 2015; D'Horta et al. 2011). On the other hand, mammals, amphibians and invertebrates show higher variability in terms of the age of lineage divergence, with some lineages going back to the Miocene, or even earlier (Table 16.1). Some lineages of birds (e.g., Batalha-Filho et al. 2012), and invertebrates (e.g., Cabanne et al. 2013; Françoso et al. 2016; Peres et al. 2015) do not seem to respond to historical barriers to gene flow; no clear pattern of genetic structure is detected.

Several Atlantic forest species seem to have undergone demographic change, such as population expansion or retraction, in the Late Pleistocene—likely associated to climatic oscillations (e.g., Batalha-Filho and Myiaki 2016; Cabanne et al. 2013; Carnaval et al. 2009). Yet, the degree of congruence in the timing of the inferred demographic events is relatively low. We argue that this spatial and temporal incongruence may be explained by differential species responses to environmental changes mediated by taxon-specific traits, dispersal ability, population dynamics, physiological constraints, and/or ecological interactions (Cabanne et al. 2016). It is important to keep in mind, also, that the molecular markers, geographical scale, and taxon sampling might impact the demographic inferences being compared. While terrestrial vertebrates have been more intensely sampled from a population genetic standpoint, plants, invertebrates, and aquatic species have received less attention (Turchetto-Zolet et al. 2013). Large-scale genomic studies (e.g., Pie et al. 2018; Prates et al. 2016), especially cross-taxonomic comparative studies, can provide a rigorous test of cross-species synchrony in their responses to landscape or climate shifts. Already, they suggest that the similar spatial patterns of biodiversity observed along the Atlantic forest may result from distinct historical and ecological processes (Prates et al. 2016).

6 Biogeographical Hypotheses

Lineage and species diversification are commonly attributed to vicariance or dispersal events; when spatial patterns are shared by numerous co-occurring taxa in a region, they are often interpreted as the result of common historical processes. However, the relative contribution of topographic, climatic, and ecological factors toward the biodiversity patterns observed in the Atlantic rainforest remain controversial; as of now, it seems that multiple asynchronic drivers may have led to similar spatial patterns of community, species, and lineage distribution (Prates et al. 2016). Although many hypothetical scenarios have been proposed for the origin and evolution of this biome, the historical causes of biotic diversification in this region remain poorly understood. Below we review some of the main hypotheses available to date, and their associated evidence.

6.1 Historical Connections Between the Atlantic Forest and Other Neotropical Biomes

It is possible that the marked north-south species (and lineage) turnover observed in the Atlantic forest results from a distinct biogeographic origin of the northern vs. southern biota, reflecting past connections between the Atlantic, Amazonian, and Andean forests. Until the Late Paleocene and Early Eocene, the Earth's climate was predominantly warmer and more humid than today, suggesting that South America was predominantly covered by continuous rainforests (Morley 2000; Zachos et al. 2001). During the Late Eocene and Oligocene, global episodes of cooling and dryness favored the expansion of grasslands in the southern and central regions of the continent (Flower and Kennett 1994; Morley 2000), which culminated in the formation of a diagonal belt of more open and drier biomes (known as "dry diagonal"; Prado and Gibbs 1993; Werneck 2011; Zanella 2011). The formation of the dry diagonal marked the formation of the Atlantic forest in the east and Amazonia in the west. However, floristic and faunistic inventories, fossil record, and paleoclimate models suggest that those rainforests were re-connected multiple times in the past (Por 1992; Costa 2003; Batalha-Filho et al. 2013; Fiaschi and Pirani 2009; Sobral-Souza et al. 2015; Ledo and Colli 2017).

Different historical forest connections are thought to have impacted the biological composition of the Atlantic forest (see Ledo and Colli 2017). Past forest corridors are thought to have connected the Atlantic rainforest and the Amazon through northeastern Brazil, either through the coastal region (possibly through areas now occupied by the "brejos de altitude") or through the interior of the present-day Caatinga (Por 1992). Other corridors have been suggested to exist through gallery forests in the central southern portion of Cerrado savannah or through forests associated to the Paraná and Paraguay Rivers, possibly linking the southern Atlantic forest to Western Amazonia (Por 1992; Batalha-Filho et al. 2013). Based on a phylogenetic study of subsoscine birds, Batalha-Filho et al. (2013) proposed that these distinct connections may have occurred in different periods, being the western Amazonia—southern Atlantic forest bridges older (Middle/Late Miocene) than the connections through the Brazilian northeast (Pliocene/Pleistocene). These ideas are supported by phylogenetic and phylogeographic studies of mammals (e.g., Costa 2003; Martins et al. 2009; Pavan et al. 2011), amphibians (Fouquet et al. 2012a, b, 2014; Rodrigues et al. 2014; Prates et al. 2016; Thomé et al. 2016; de Sá et al. 2019), reptiles (Pellegrino et al. 2011; Prates et al. 2016, 2018; Rodrigues et al. 2014), snakes (Dal Vechio et al. 2018), and arthropods (Bartoleti et al. 2018; Menezes et al. 2015; Peres et al. 2017). Phylogenetic data also support a hypothesis of past contact between the southern Atlantic forest and the Andean rainforests. These connections are believed to have contributed to the diversification of lineages in the Atlantic forest (see Costa and Leite 2013; Batalha-Filho et al. 2013; Trujillo-Arias et al. 2018), reinforcing the notion that historical shifts in forest range, tied to dispersal events, likely played a key role for the current patterns of Neotropical biodiversity.

Neotropical climate dynamics are recognized as one potential driver of these ancient forest connections, and are thought to have prompted marked differentiation between the northern and southern blocks of the Atlantic forest. Based on information from speleothem d18O records, Cheng et al. (2013) detected an east-west dipole-like pattern of precipitation changes in South America in the past: while northern Atlantic forests underwent climate changes more similar to those observed in eastern Amazonia, southern Atlantic forests climate dynamics have been in-sync with those in western Amazonia and the Andes. Cheng et al. (2013) proposed this climate dynamics to have enabled the formation of contact bridges between different regions of the Atlantic, Amazonian, and Andean forests, favoring the interchange of species during wetter periods (Por 1992; Costa 2003; Batalha-Filho et al. 2013). However, genomic-scale data from lizards distributed both in Amazonia and the Atlantic forest (Prates et al. 2016) were unable to detect demographic responses of the magnitude of those inferred by Cheng et al. (2013), suggesting in-synch (instead of out-of-synch) responses between populations in eastern and western Amazonia, and the Atlantic forest.

6.2 Regionalization Hypotheses

Several hypotheses proposed to explain Amazonian patterns of biodiversity have been applied to the study of Atlantic forest taxa. One example is the Pleistocene refugia hypothesis (Haffer 1969), which suggested that Pleistocene climatic fluctuations led to rainforest fragmentation and promoted divergence of lineages or species in isolated forest fragments or refugia. This hypothesis received early support as a model to explain the distribution of endemism centers and diversity patterns in the Atlantic forest (Vanzolini and Williams 1981). Yet, this hypothesis was heavily criticized due to the lack of concordance with empirical phylogenetic data, as well as by the evidence that shifts in forest species distribution, rather than fragmentation, have been the main consequences of global glaciations in the Neotropics (Antonelli et al. 2018; Colinvaux and De Oliveira 2001). An alternative and more comprehensive version of this model, the vanishing refuge model (Vanzolini and Williams 1981; Damasceno et al. 2014), combined forest retractions and expansions, habitat shifts associated with climate change, vicariance, and adaptation, to explain biological diversification; nevertheless, this issue is still debated (Antonelli et al. 2018).

Still, ecological and phylogeographic studies support the notion that the Atlantic forest biota tracked climatic shifts experienced over the past 21 kyrs. Correlative (climate-based) models of the paleodistribution of the forest and forest animals, tied to population genetic studies, suggest regional differences in the climatic stability of the northern and southern regions of the Atlantic forest (Carnaval and Moritz 2008; Carnaval et al. 2014; Leite et al. 2016). Together, these data suggest that the southern, montane species (and forests) of coastal Brazil may have expanded their distributions during cool phases of the Late Quaternary, being today restricted to small refugia. Conversely, lowland species and forests show signals consistent with a scenario of range contraction during cooler phases. They also suggest that the geographic extent of climatically stable areas of the forest was larger in the central corridor of the forest—particularly in Bahia, Espirito Santo, and northern Rio de Janeiro—relative to mountaintop refuges in both southern and northeastern Brazil. These proposed dynamics appear to match endemism patterns in the Atlantic forest and were validated with genetic data from lowland frogs (Carnaval et al. 2009), birds (Cabanne et al. 2007, 2008; D'Horta et al. 2011; Maldonado-Coelho 2012), mammals (Costa 2003; Moraes-Barros et al. 2006), bees (Batalha-Filho et al. 2010), and ants (Resende et al. 2010). Data from all of these organisms show lower genetic variation in southern populations, possibly as a result of genetic bottlenecks and recent colonization events. On the other hand, the distribution of montane or cool-associated groups during the LGM suggests that those taxa responded differently to forest shifts according to their ecological requirements and behavior (Porto et al. 2013; Cabanne et al. 2016).

Molecular studies have also suggested that pre-Pleistocene tectonic and climatic events shaped current patterns of genetic diversity of many Atlantic forest taxa. Several lineages are much older and spatially coincident with orogenic landmarks (Lara and Patton 2000; Grazziotin et al. 2006; Pellegrino et al. 2005; Thomé et al. 2010, 2014; Bragagnolo et al. 2015; Peres et al. 2018, 2019). The location of rivers and mountain ranges, for instance, have been frequently correlated with spatial turnover in species distribution limits (e.g., the Doce river, in the central region of major turnover within the Atlantic forest; the São Francisco River, separating the Pernambuco and Bahia areas of endemism; Costa et al. 2000; Costa 2003; Silva et al. 2004; Pinto-da-Rocha et al. 2005; Sigrist and Carvalho 2008), as well as lineage turnover within species (e.g., Brunes et al. 2010; Grazziotin et al. 2006; Paz et al. 2018; Pellegrino et al. 2005; Peres et al. 2018; Thomé et al. 2010). Given the spatial congruence between the limits of most areas of endemism and the location of important rivers crossing the Atlantic forest, several authors have discussed whether Wallace's (1852) "Riverine barrier hypothesis," proposed in the context of the Amazon biota, should also apply to this region. This theory states that rivers could represent barriers to species distribution, thus promoting genetic differentiation by preventing gene flow between populations on opposite banks. Atlantic forest rivers, however, are not as wide as those in the Amazonian region, putting into question the actual role of rivers as physical barriers for many taxa (Colombi et al. 2010). On the other hand, the evolution of the landscape for the formation of river basins involves regional climate and vegetation changes, which might have acted as vicariant barriers in some period(s) in the past. Turnover in plant communities in the central corridor of the Atlantic forest, for example, seems to be more closely linked to climatic differences north and south of the Doce river basin, rather than to the river itself (Saiter et al. 2016). Different characteristics of the relief, north and south of river banks, may also be involved in such biotic breaks—especially in the southern Atlantic forest, which encompasses higher mountains and a more complex topography.

6.3 *Narrow-Scale Diversification and Microrefugia in Southern Atlantic Forest*

Molecular and geological studies in the southern portion of the Atlantic forest suggest that both Quaternary climatic oscillations and the evolution of a complex geological mosaic may have affected the regional biota. During the Late Cretaceous, the region was affected by intense uplift and denudation processes (Gallagher et al. 1994; Hiruma et al. 2010). Tectonic reactivations in the Paleogene led to the Continental rift of Southeastern Brazil, a depression filled by Cenozoic deposits, and tied to important rivers in the region, as the Ribeira do Iguape and Paraíba do Sul (Riccomini et al. 2004). Further geomorphologic processes along the rift, dated back to the Neogene–Quaternary, were likely responsible for deforming the original depression, for increasing topographical differences between the mountain chain and nearby valleys, and for rearranging the river network (Almeida 1976; Riccomini et al. 2004). The rift currently encompasses some of the most important relief structures found in the Atlantic forest domain, such as the Serra do Mar and the Serra da Mantiqueira mountain ranges, reaching more than 2000 m a.s.l.

In addition, the southern Atlantic forest exhibits great variation of vegetation types (Oliveira-Filho and Fontes 2000), allowing a greater range of different microhabitats in comparison to the central and northern portions of the forest. Paleoclimate models and fossil vegetation evidence indicate that the southern Atlantic forest was subjected to drastic climatic oscillations during the Late Pleistocene, which caused multiple forest sites to be replaced by grasslands in short periods of time (Behling 2002; Costa et al. 2018; Sobral-Souza et al. 2015). Still, predicted periods of higher or less humidity cannot be generalized, as there is evidence of wet conditions in certain forest regions even during the Last Glacial Maximum (Cruz et al. 2009; Ledru et al. 2005; Pessenda et al. 2009), making the hypothesis of restricted humid refuges in the Atlantic forest, during glacial periods, the focus of intense debate. Fossil vegetation records indicate high rates of biome shifts and rapid forest recovery events in this region (Costa et al. 2018), suggesting that the biome configuration was not homogeneous and that multiple forest microrefugia (Carnaval et al. 2014; Rull 2009) may have helped to maintain diversity over time. This dynamic history matches phylogeographic data reporting high lineage divergence, high genetic variability, and putative refugia in the south (e.g., Amaral et al. 2013; Amaro et al. 2012; Bragagnolo et al. 2015; Brunes et al. 2015; Carnaval et al. 2014; Peres et al. 2018; Porto et al. 2013; Thomé et al. 2010). This higher complexity relative to other portions of the forest agrees with the narrow endemism observed in several biological groups, especially those with movement or physiological restrictions as harvestmen and amphibians.

Several authors attempt to distinguish between historical events that occurred during the Quaternary, mainly the cyclic climatic oscillations characteristic from this period, from those that affected the Neotropics in the Tertiary, commonly referring to geomorphological events. In fact, this distinction is a more functional classification, as lineage divergences often occur continuously through time. As such, it is not possible to completely refute one explanation in detriment to others. Both forest refugia and geographic barriers acted in combination to produce taxon-specific idiosyncratic patterns of diversification in the Atlantic forest (Paz et al. 2018).

7 Implications for Conservation

Knowing the distributional range of Atlantic forest taxa and the general patterns of diversity and endemism in this biome is essential for conservation purposes, particularly considering the vulnerability of most endemic species. Analyzing the spatial patterns of vertebrate diversity in Brazil, Jenkins et al. (2015) showed that, although the Amazonian and Atlantic forests have equivalent levels of species richness for birds, mammals, and amphibians, endemic and small-ranged species are strongly concentrated in the southern portion of the latter.

Biodiversity metrics that incorporate the evolutionary distinctiveness of biotas have been increasingly used in conservation studies. While the effects of fragmentation on species richness, abundance and composition is better understood, it is also important to evaluate the impact of environmental changes on phylogenetic diversity. Phenotypic traits associated with species vulnerability to habitat loss might be conserved along evolutionary lineages; thus, forest reduction could impact more drastically the highly susceptible lineages, leading to a reduction in the community's phylogenetic diversity (Andrade et al. 2015; Webb et al. 2002). Conserving the phylogenetic diversity of a regional community not only reduces the chance of losing unique phenotypic and ecological traits, but is also crucial to preserve ecosystem stability (Jetz et al. 2014; Matos et al. 2017).

Studies on the fragmentation effects on phylogenetic diversity and structure in the Atlantic forest are more common in arboreal plant communities. Santos et al. (2010) detected a significant loss of tree phylogenetic diversity in forest remnant edges (higher than in the core areas), highlighting the importance of conservation efforts against edge effects in a biome with highly fragmented spatial configuration. Andrade et al. (2015) reported that forest cover reduction negatively affects the local phylogenetic diversity in Rubiaceae, suggesting that deforestation results in the extinction of lineages and phylogenetic clustering. In addition, these authors detected high phylogenetic beta-diversity among different forest remnants, reinforcing the importance of Atlantic forest patches for the maintenance of regional diversity. Using combined metrics, Matos et al. (2017) showed that edge density (the length of all forest edges divided by the total area of the landscape) and number of forest patches affect the phylogenetic diversity and structure of remaining tree assemblages. This data highlight that increasing landscape fragmentation can bring negative effects to the genetic diversity of Atlantic forest lineages. However, this study did not recover significant edge effects on the extirpation of lineages, and suggested that the loss of tree species in recently fragmented areas (<100 years) occurs randomly across the phylogenetic tree, supporting that losses are not clustered within specific lineages (contradicting hypotheses of phylogenetic trait conservatism, although the conclusions should be discussed with caution; Matos et al. 2017).

8 Conclusions and Future Prospects

Although much information has been accumulated in the last century, the mechanisms involved in the origin and spatial diversification of Atlantic forest lineages are still not completely understood. The marked turnover regions detected along the biome, and the distinctiveness among endemism centers, support that vicariant events played an important role, isolating regional biotas and favoring independent evolutionary processes. However, how these regional communities were affected by orogenic events and climate changes in the past is still not clear. Furthremore, the extent of historical connectivity between the Atlantic forest and other Neotropical regions also remains to be understood. Answering these questions will require linking information from field surveys and molecular inferences with paleoecological and paleoclimatic data. Adding further complexity, the patterns exhibited by each taxon are strongly affected by specific traits, highlighting the need of more ecological studies to inform biogeographical reconstructions. Moreover, expanding and publicizing Brazil's georeferenced biological databases represents a key step towards improving our knowledge about the geographical patterns of diversity and divergence of regional biotas. Surveys focused on neglected groups and undersampled regions are specially important. The Atlantic forest is a megadiverse biome and might still hide numerous unknown taxa, as indicated by the growing number of recognized species. More refined examinations of putative cryptic species, tied to intensified inventories in poorly sampled areas and in transition zones are crucial for a better comprehension of lineage distribution.

Analytical advances will also contribute to clarify the evolutionary history of the Atlantic forest. Even though its first molecular-based biogeographic studies were published in the late 1990s (e.g., Mustrangi and Patton 1997), the scenario emerging from molecular phylogenetic and phylogeographic studies is still puzzling. This is partially due to methodological and sampling gaps, given that most groups have only been genetically studied recently. The incongruent inferences obtained to date seem to reflect distinct geomorphological or climatic events that affect Atlantic forest organisms in different ways. Alternatively, these incongruences might have also resulted from unequal sampling effort and/or biases associated with the genetic markers chosen. Lastly, most available phylogeographic studies in the Atlantic forest were based on one or few molecular markers, especially mitochondrial loci, limiting the assessment of demographic events. Synchrony in lineage divergence, or shifts in population size, for instance, may be wrongfully inferred in datasets composed by few loci (Amaral et al. 2018; McCormack et al. 2013). The employment of more statistically robust analyses and the increasing availability of large amounts of molecular data through high-throughput sequencing (e.g., Prates et al. 2016) are among the most promising advances in Neotropical comparative phylogeography.

The newly available data and tools already suggest that geographic patterns of genetic diversity are similar across forest species yet, their underlying processes are not.

References

Almeida FFM (1976) The system of continental rifts bordering the Santos basin Brazil. An Acad Bras Cienc 48:15–26

Álvarez-Presas M, Sánchez-Garcia A, Carbayo F, Rozas J, Riutort M (2014) Insights into the origin and distribution of biodiversity in the Brazilian Atlantic forest hot spot: a statistical phylogeographic study using a low-dispersal organism. Heredity 112:656–665

Amaral FR, Albers PK, Edwards SV, Miyaki CY (2013) Multilocus tests of Pleistocene refugia and ancient divergence in a pair of Atlantic forest antbirds (Myrmeciza). Mol Ecol 22:3996–4013

Amaral FR, Maldonado-Coelho M, Aleixo A et al (2018) Recent chapters of neotropical history overlooked in phylogeography: shallow divergence explains phenotype and genotype uncoupling in Antilophia manakins. Mol Ecol 27:4108–4120. https://doi.org/10.1111/mec.14843

Amaro RC, Rodrigues MT, Yonenaga-Yassuda Y, Carnaval AC (2012) Demographic processes in the montane Atlantic rainforest: molecular and cytogenetic evidence from the endemic frog *Proceratophrys boiei*. Mol Phylogenet Evol 62:880–888

Amorim DS, Pires MRS (1996) Neotropical biogeography and a method for a maximum biodiversity estimation. In: Bicudo CEM, Menezes NA (eds) Biodiversity in Brazil: a first approach. Conselho Nacional de Desenvolvimento Científico e Tecnológico, São Paulo, pp 183–219

Andrade ER, Jardim JG, Santos BA, Melo FPL, Talora DC, Faria D, Cazetta E (2015) Effects of habitat loss on taxonomic and phylogenetic diversity of understory Rubiaceae in Atlantic forest landscapes. For Ecol Manag 349:73–84

Antonelli A, Ariza M, Albert J et al (2018) Conceptual and empirical advances in neotropical biodiversity research. PeerJ 6:e5644

Avise JC, Arnold J, Ball RM, Bermingham E, Lamb T, Neigel JE, Reeb CA, Saunders NC (1987) Intraspecific phylogeography: the mitochondrial DNA bridge between population genetics and systematics. Annu Rev Ecol Syst 18(1):489–522

Bartoleti LFM, Peres EA, Fontes FHM, Silva MJ, Solferini VN (2018) Phylogeography of the widespread spider *Nephila clavipes* (Araneae: Araneidae) in South America indicates geologically and climatically driven lineage diversification. J Biogeogr 45:1246–1260

Batalha-Filho H, Miyaki CY (2016) Late Pleistocene divergence and postglacial expansion in the Brazilian Atlantic forest: multilocus phylogeography of *Rhopias gularis* (Aves: Passeriformes). J Zool Syst Evol Res 54:137–147

Batalha-Filho H, Waldschmidt AM, Campos LAO, Tavares MG, Fernandes-Salomão TM (2010) Phylogeography and historical demography of the neotropical stingless bee *Melipona quadrifasciata* (Hymenoptera, Apidae): incongruence between morphology and mitochondrial DNA. Apidologie 41:534–547

Batalha-Filho H, Cabanne GS, Miyaki CY (2012) Phylogeography of an Atlantic forest passerine reveals demographic stability through the last glacial maximum. Mol Phylogenet Evol 65:892–902

Batalha-Filho H, Fjeldså J, Fabre P-H, Miyaki CY (2013) Connections between the Atlantic and the Amazonian forest avifaunas represent distinct historical events. J Ornithol 154:41–50

Bates JM, Hackett SJ, Cracraft J (1998) Area-relationships in the neotropical lowlands: an hypothesis based on raw distributions of Passerine birds. J Biogeogr 25:783–793

Behling H (2002) South and southeast Brazilian grasslands during late quaternary times: a synthesis. Palaeogeogr Palaeoclimatol Palaeoecol 177:19–27

Bragagnolo C, Pinto-da-Rocha R, Antunes M, Clouse RM (2015) Phylogenetics and phylogeography of a long-legged harvestman (Arachnida: Opiliones) in the Brazilian Atlantic Rain Forest reveals poor dispersal, low diversity and extensive introgression. Invertebr Syst 29 (4):386–404

Brown KS, Freitas AVL (2000) Atlantic forest butterflies: indicators for landscape conservation 1. Biotropica 32:934–956

Brunes TO, Sequeira F, Haddad CFB, Alexandrino J (2010) Gene and species trees of a neotropical group of treefrogs: genetic diversification in the Brazilian Atlantic forest and the origin of a polyploid species. Mol Phylogenet Evol 57:1120–1133

Brunes TO, Thomé MTC, Alexandrino J, Haddad CFB, Sequeira F (2015) Ancient divergence and recent population expansion in a leaf frog endemic to the southern Brazilian Atlantic forest. Org Divers Evol 15:695–710

Cabanne GS, Santos FR, Miyaki CY (2007) Phylogeography of *Xiphorhynchus fuscus* (Passeriformes, Dendrocolaptidae): vicariance and recent demographic expansion in southern Atlantic forest. Biol J Linn Soc 91:73–84

Cabanne GS, D'Horta FM, Sari EHR, Santos FR, Miyaki CY (2008) Nuclear and mitochondrial phylogeography of the Atlantic forest endemic *Xiphorhynchus fuscus* (Aves: Dendrocolaptidae): biogeography and systematics implications. Mol Phylogenet Evol 49:760–773

Cabanne GS, Sari HER, Meyer D, Santos FR, Miyaki CY (2013) Matrilineal evidence for demographic expansion, low diversity and lack of phylogeographic structure in the Atlantic forest endemic Greenish Schiffornis *Schiffornis virescens* (Aves: Tityridae). J Ornithol 154:371–384

Cabanne GS, Calderón L, Trujillo Arias N et al (2016) Effects of Pleistocene climate changes on species ranges and evolutionary processes in the neotropical Atlantic forest. Biol J Linn Soc 119:856–872

Cabrera AL, Willink A (1973) Biogeografia de America Latina. Monografia 13, série de biología. Organización de los Estados Americanos (OEA), Washington, DC, p 120

Cancello EM, Silva RR, Vasconcellos A, Reis YT, Oliveira LM (2014) Latitudinal variation in termite species richness and abundance along the Brazilian Atlantic forest hotspot. Biotropica 46:441–450

Carnaval AC, Moritz C (2008) Historical climate modelling predicts patterns of current biodiversity in the Brazilian Atlantic forest. J Biogeogr 35:1187–1201

Carnaval AC, Hickerson MJ, Haddad CFB, Rodrigues MT, Moritz C (2009) Stability predicts genetic diversity in the Brazilian Atlantic forest hotspot. Science 323:785–789

Carnaval AC, Waltari E, Rodrigues MT et al (2014) Prediction of phylogeographic endemism in an environmentally complex biome. Proc R Soc B Biol Sci 281(1792):20141461

Cazé ALR, Mäder G, Nunes TS, Queiroz LP, Oliveira G, Diniz-Filho JA, Bonatto SL, Freitas LB (2016) Could refuge theory and rivers acting as barriers explain the genetic variability distribution in the Atlantic forest? Mol Phylogenet Evol 101:242–251

Cheng H, Sinha A, Cruz FW et al (2013) Climate change patterns in Amazonia and biodiversity. Nat Commun 4:1411

Colinvaux PA, De Oliveira PE (2001) Amazon plant diversity and climate through the Cenozoic. Palaeogeogr Palaeoclimatol Palaeoecol 166:51–63

Colombi VH, Lopes SR, Fagundes V (2010) Testing the Rio Doce as a riverine barrier in shaping the Atlantic rainforest population divergence in the rodent *Akodon cursor*. Genet Mol Biol 33:785–789

Costa LP (2003) The historical bridge between the Amazon and the Atlantic forest of Brazil: a study of molecular phylogeography with small mammals. J Biogeogr 30:71–86

Costa LP, Leite YLR (2013) Historical fragmentation shaping vertebrate diversification in the Atlantic forest biodiversity hotspot. In: Patterson BD, Costa LP (eds) Bones, clones, and biomes: the history and geography of recent neotropical mammals. Chicago University Press, Chicago, pp 283–306

Costa LP, Leite YLR, da Fonseca GAB, da Fonseca MT (2000) Biogeography of South American forest mammals: endemism and diversity in the Atlantic forest. Biotropica 32:872–881

Costa GC, Hampe A, Ledru M-P et al (2018) Biome stability in South America over the last 30 kyr: inferences from long-term vegetation dynamics and habitat modelling. Glob Ecol Biogeogr 27:285–297

Cracraft J (1985) Historical biogeography and patterns of differentiation within the South American avifauna: areas of endemisms. Ornithol Monogr 36:49–84

Cracraft J, Prum RO (1988) Patterns and processes of diversification: speciation and historical congruence in some Neotropical birds. Evolution 42(3):603–620

Cruz FW, Wang X, Auler A, Vuille M, Burns SJ, Edwards LR, Karmann I, Cheng H (2009) Orbital and millennial-scale precipitation changes in Brazil from speleothem records. In: Vimeux F, Sylvestre F, Khodri M (eds) Past Climate Variability in South America and Surrounding Regions: from the Last Glacial Maximum to the Holocene. Springer, Dordrecht, pp 29–60

Dal Vechio F, Prates I, Grazziotin FG, Zaher H, Rodrigues MT (2018) Phylogeography and historical demography of the arboreal pit viper *Bothrops bilineatus* (Serpentes, Crotalinae) reveal multiple connections between Amazonian and Atlantic rainforests. J Biogeogr 45:2415–2426

Dalapicolla J, Leite YLR (2018) Historical connections among river basins and climatic changes explain the biogeographic history of a water rat. PeerJ 6:e5333

Damasceno R, Strangas ML, Carnaval AC, Rodrigues MT, Moritz C (2014) Revisiting the vanishing refuge model of diversification. Front Genet 5:353

Dantas GPM, Sari HER, Cabanne GS, Pessoa RO, Marini MA, Miyaki CY, Santos FR (2015) Population genetic structure of the Atlantic forest endemic *Conopophaga lineata* (Passeriformes: Conopophagidae) reveals a contact zone in the Atlantic forest. J Ornithol 156:85–99

DaSilva MB, Pinto-da-Rocha R, DeSouza AM (2015) A protocol for the delimitation of areas of endemism and the historical regionalization of the Brazilian Atlantic Rain Forest using harvestmen distribution data. Cladistics 31(6):692–705. https://doi.org/10.1111/cla.12121

de Sá RO, Tonini JFR, van Huss H, Long A, Cuddy T, Forlani MC, Peloso PLV, Zaher H, Haddad CFB (2019) Multiple connections between Amazonia and Atlantic forest shaped the phylogenetic and morphological diversity of *Chiasmocleis* Mehely, 1904 (Anura: Microhylidae: Gastrophryninae). Mol Phylogenet Evol 130:198–210

D'Horta FM, Cabanne GS, Meyer D, Miyaki CY (2011) The genetic effects of late quaternary climatic changes over a tropical latitudinal gradient: diversification of an Atlantic forest passerine. Mol Ecol 20:1923–1935

Durães R, Loiselle BA (2004) Inter-scale relationship between species richness and environmental heterogeneity: a study case with antbirds in the Brazilian Atlantic forest. Ornitol Neotrop 15:127–135

Ferro VG, Melo AS (2011) Diversity of tiger moths in a neotropical hotspot: determinants of species composition and identification of biogeographic units. J Insect Conserv 15:643–651

Fiaschi P, Pirani JR (2009) Review of plant biogeographic studies in Brazil. J Syst Evol 47:477–496

Fitzpatrick SW, Brasileiro CA, Haddad CFB, Zamudio KR (2009) Geographical variation in genetic structure of an Atlantic Coastal Forest frog reveals regional differences in habitat stability. Mol Ecol 18:2877–2896

Flower BP, Kennett JP (1994) The middle Miocene climate transition: East Antarctic ice sheet development, deep ocean circulation and global carbon cycling. Palaeogeogr Palaeoclimatol Palaeoecol 108:537–555

Fouquet A, Recoder R, Teixeira M Jr, Cassimiro J, Amaro RC, Camacho A, Damasceno R, Carnaval AC, Moritz C, Rodrigues MT (2012a) Molecular phylogeny and morphometric analyses reveal deep divergence between Amazonia and Atlantic forest species of *Dendrophryniscus*. Mol Phylogenet Evol 62:826–838

Fouquet A, Loebmann D, Castroviejo-Fisher S, Padial JM, Orrico VGD, Lyra ML, Roberto IJ, Kok PJR, Haddad CFB, Rodrigues MT (2012b) From Amazonia to the Atlantic forest: molecular

phylogeny of Phyzelaphryninae frogs reveals unexpected diversity and a striking biogeographic pattern emphasizing conservation challenges. Mol Phylogenet Evol 65:547–561
Fouquet A, Cassini CS, Haddad CFB, Pech N, Rodrigues MT (2014) Species delimitation, patterns of diversification and historical biogeography of the neotropical frog genus *Adenomera* (Anura, Leptodactylidae). J Biogeogr 41:855–870
Françoso E, Zuntini AR, Carnaval AC, Arias MC (2016) Comparative phylogeography in the Atlantic forest and Brazilian savannas: pleistocene fluctuations and dispersal shape spatial patterns in two bumblebees. BMC Evol Biol 16:1–16
Frantine-Silva W, Giangarelli DC, Penha RES, Suzuki KM, Dec E, Gaglianone MC, Alves-dos-Santos I, Sofia SH (2017) Phylogeography and historical demography of the orchid bee *Euglossa iopoecila*: signs of vicariant events associated to quaternary climatic changes. Conserv Genet 18:539–552
Funch LS, Rodal MJN, Funch RR (2008) Floristic aspects of the forests of the Chapada Diamantina, Bahia, Brazil. In: Thomas W, Briton EG (org.) The Atlantic coastal forest of Northeastern Brazil, vol 100. Mem. of the New York Botanical Garden Press, pp 193–220
Gallagher K, Hawkesworth CJ, Mantovani MSM (1994) The denudation history of the onshore continental margin of SE Brazil inferred from apatite fission track data. J. Geophys. Res. Solid Earth 99:18117–18145. https://doi.org/10.1029/94JB00661
Grazziotin FG, Monzel M, Echeverrigaray S, Bonatto SL (2006) Phylogeography of the *Bothrops jararaca* complex (Serpentes: Viperidae): past fragmentation and island colonization in the Brazilian Atlantic forest. Mol Ecol 15:3969–3982
Haffer J (1969) Speciation in Amazonian forest birds. Science 165:131–137
Hiruma ST, Riccomini C, Modenesi-Gauttieri MC, Hackspacher PC, Hadler Neto JC, Franco-Magalhães AOB (2010) Denudation history of the Bocaina Plateau, Serra do Mar, southeastern Brazil: relationships to Gondwana breakup and passive margin development. Gondwana Res 18:674–687
Jenkins CN, Alves MAS, Uezu A, Vale MM (2015) Patterns of vertebrate diversity and protection in Brazil. PLoS One 10(12):e0145064. https://doi.org/10.1371/journal.pone.0145064
Jetz W, Thomas GH, Joy JB, Redding DW, Hartmann K, Mooers AO (2014) Global distribution and conservation of evolutionary distinctness in birds. Curr Biol 24:919–930
Lara MC, Patton JL (2000) Evolutionary diversification of spiny rats (genus Trinomys, Rodentia: Echimyidae) in the Atlantic forest of Brazil. Zool J Linnean Soc 130:661–686
Leão-Pires TA, Luiz AM, Sawaya RJ (2018) The complex roles of space and environment in structuring functional, taxonomic and phylogenetic beta diversity of frogs in the Atlantic forest. PLoS One 13:e0196066
Ledo R, Colli GR (2017) The historical connections between the Amazon and the Atlantic forest revisited. J Biogeogr 44:2551–2563
Ledru M-P, Rousseau DD, Cruz FW, Riccomini C, Karmann I, Martin L (2005) Paleoclimate changes during the last 100,000 yr from a record in the Brazilian Atlantic rainforest region and interhemispheric comparison. Quat Res 64:444–450
Leite YLR, Costa LP, Loss AC, Rocha RG, Batalha-Filho H, Bastos AC, Quaresma VS, Fagundes V, Paresque R, Passamani M, Pardini R (2016) Neotropical forest expansion during the last glacial period challenges refuge hypothesis. Proc Natl Acad Sci 113:1008–1013
Lynch JD (1979) The amphibians of the lowland tropical forests. In: Duellman WE (ed) The South American Herpetofauna: its origin, evolution, and dispersal. Monogr Mus Nat Hist, Univ Kans, vol 7, pp 189–215
Maia AVP, Almeida C, Santoro KR, Melo JLA, Oliveira JV, Guedes RNC, Badji CA (2016) High-level phylogeographic structuring of *Neoleucinodes elegantalis* Guenée (Lepidoptera, Crambridae) in Brazil: an important tomato pest. Revista Brasileira de Entomologia 60:206–210
Maldonado-Coelho M (2012) Climatic oscillations shape the phylogeographical structure of Atlantic forest fire-eye antbirds (Aves: Thamnophilidae). Biol J Linn Soc 105:900–924
Martins FM, Templeton AR, Pavan AC, Kohlbach BC, Morgante JS (2009) Phylogeography of the common vampire bat (*Desmodus rotundus*): marked population structure, neotropical

Pleistocene vicariance and incongruence between nuclear and mtDNA markers. BMC Evol Biol 9:294. https://doi.org/10.1186/1471-2148-9-294
Martins FM, Gifali-Iughetti C, Koiffman CP, Harris EE (2011) Coalescent analysis of mtDNA indicates Pleistocene divergence among three species of howler monkey (*Alouatta* spp.) and population subdivision within the Atlantic Coastal Forest species, *A. guariba*. Primates 52:77–87
Matos FA, Magnago LF, Gastauer M, Carreiras JM, Simonelli M, Meira-Neto JA, Edwards DP (2017) Effects of landscape configuration and composition on phylogenetic diversity of trees in a highly fragmented tropical forest. J Ecol 105:265–276
McCormack JE, Hird SM, Zellmer AJ, Carstens BC, Brumfield RT (2013) Applications of next-generation sequencing to phylogeography and phylogenetics. Mol Phylogenet Evol 66:526–538
Melo AS, Rangel TFLVB, Diniz-Filho JAF (2009) Environmental drivers of beta-diversity patterns in New-World birds and mammals. Ecography 32:226–236
Menezes RST, Brady SG, Carvalho AF, Del Lama MA, Costa MA (2015) Molecular phylogeny and historical biogeography of the neotropical swarm-founding social Wasp Genus *Synoeca* (Hymenoptera: Vespidae). PLoS One 10(3):e0119151. https://doi.org/10.1371/journal.pone.0119151
Menezes L, Canedo C, Batalha-Filho H, Garda AA, Gehara M, Napoli MF (2016) Multilocus phylogeography of the Treefrog *Scinax eurydice* (Anura, Hylidae) reveals a Plio-Pleistocene diversification in the Atlantic forest. PLoS One 11(6):e0154626
Menezes RST, Brady SG, Carvalho AF, Del Lama MA, Costa MA (2017) The roles of barriers, refugia, and chromosomal clines underlying diversification in Atlantic forest social wasps. Sci Rep 7:7689
Miranda GB, Andrades-Miranda J, Oliveira LFB, Langguth A, Mattevi MS (2007) Geographic patterns of genetic variation and conservation consequences in three South American rodents. Biochem Genet 45:839–856
Moraes-Barros N, Silva JAB, Miyaki CY, Morgante JS (2006) Comparative phylogeography of the Atlantic forest endemic sloth (*Bradypus torquatus*) and the widespread three-toed sloth (*Bradypus variegatus*) (Bradypodidae, Xenarthra). Genetica 126:189–198
Morellato LPC (2000) Introduction: the Brazilian Atlantic forest. Biotropica 32:786–792
Morley RJ (2000) Origin and evolution of tropical rain forests. Wiley, Chichester, p 362
Morrone JJ (1994) On the identification of areas of endemism. Syst Biol 43:438–441
Morrone JJ (2006) Biogeographic areas and transition zones of Latin America and the Caribbean Islands based on panbiogeographic and cladistic analyses of the entomofauna. Annu Rev Entomol 51:467–494
Moura MR, Argôlo AJ, Costa HC (2017) Historical and contemporary correlates of snake biogeographical subregions in the Atlantic forest hotspot. J Biogeogr 44:640–650
Müller P (1973) The dispersal centers of terrestrial vertebrates in the neotropical realm: a study in the evolution of the neotropical biota and its native landscape. Biogeographica 2:1–250
Murray-Smith C, Brummitt NA, Oliveira-Filho AT et al (2009) Plant diversity hotspots in the Atlantic coastal forests of Brazil. Conserv Biol 23:151–163
Mustrangi MA, Patton J (1997) Phylogeography and systematics of the slender mouse opossum Marmosops (Marsupialia, Didelphidae). Univ Calif Publ Zool 130:1–86
Myers N, Mittermeier RA, Mittermeier CG, da Fonseca GA, Kent J (2000) Biodiversity hotspots for conservation priorities. Nature 403:853–858
Nogueira AA, Bragagnolo C, DaSilva MB, Martins TK, Lorenzo EP, Perbiche-Neves G, Pinto-da-Rocha R (2019) Historical signatures in the alpha and beta diversity patterns of Atlantic forest harvestman communities (Opiliones-Arachnida). Can J Zool 97(7):631. https://doi.org/10.1139/cjz-2018-0032
Oliveira-Filho AT, Jarenkow JA, Rodal MJN (2006) Floristic relationships of seasonally dry forests of eastern South America based on tree species distribution patterns. In: Pennington RT, Ratter JA, Lewis JP (eds) Neotropical savannas and dry forests: plant diversity, biogeography and

conservation, the systematics association special volume series, vol 69. CRC Press, Boca Raton, pp 159–192

Oliveira-Filho AT, Fontes MAL (2000) Patterns of floristic differentiation among Atlantic forests in Southeastern Brazil and the influence of climate. Biotropica 32:793–810

Olson DM, Dinerstein E, Wikramanayake ED, Burgess ND, Powell GVN, Underwood EC, D'amico JA, Itoua I, Strand HE, Morrison JC, Loucks CJ, Allnutt TF, Ricketts TH, Kura T, Lamoreux JF, Wettengel W, Hedao P, Kassem KR (2001) Terrestrial ecoregions of the world: a new map of life on earth: a new global map of terrestrial ecoregions provides an innovative tool for conserving biodiversity. Bioscience 51:933–938

Palma-Silva C, Lexer C, Paggi GM, Barbará T, Bered F, Bodanese-Zanettini MH (2009) Range-wide patterns of nuclear and chloroplast DNA diversity in *Vriesea gigantea* (Bromeliaceae), a neotropical forest species. Heredity 103:503–512

Pavan AC, Martins FM, Santos FR, Ditchfield A, Redondo RAF (2011) Patterns of diversification in two species of short-tailed bats (Carollia Gray, 1838): the effects of historical fragmentation of Brazilian rainforests. Biol J Linn Soc 102:527–539

Paz A, Spanos Z, Brown JL, Lyra M, Haddad C, Rodrigues M, Carnaval A (2018) Phylogeography of Atlantic forest glassfrogs (*Vitreorana*): when geography, climate dynamics and rivers matter. Heredity 122(5):545. https://doi.org/10.1038/s41437-018-0155-1

Pellegrino KCM, Rodrigues MT, Waite AN et al (2005) Phylogeography and species limits in the *Gymnodactylus darwinii* complex (Gekkonidae, Squamata): genetic structure coincides with river systems in the Brazilian Atlantic forest. Biol J Linn Soc 85:13–26

Pellegrino KC, Rodrigues MT, Harris DJ, Yonenaga-Yassuda Y, Sites JW Jr (2011) Molecular phylogeny, biogeography and insights into the origin of parthenogenesis in the neotropical genus *Leposoma* (Squamata: Gymnophthalmidae): ancient links between the Atlantic forest and Amazonia. Mol Phylogenet Evol 61:446–459

Pennington RT, Lewis GP, Ratter JA (2006) An overview of the plant diversity, biogeography and conservation of neotropical savannas and seasonally dry forests. In: Pennington RT, Lewis GP, Ratter JA (eds) Neotropical savannas and dry forests: plant diversity, biogeography, and conservation. CRC Press, Oxford, pp 1–29

Peres EA, Sobral-Souza T, Perez MF et al (2015) Pleistocene niche stability and lineage diversification in the subtropical spider *Araneus omnicolor* (Araneidae). PLoS One 10:e0121543

Peres EA, Silva MJ, Solferini VN (2017) Phylogeography of the spider *Araneus venatrix* (Araneidae) suggests past connections between Amazon and Atlantic rainforests. Biol J Linn Soc 121:771–785

Peres EA, DaSilva MB, Antunes M Jr, Pinto-da-Rocha R (2018) A short-range endemic species from south-eastern Atlantic Rain Forest shows deep signature of historical events: phylogeography of harvestmen *Acutisoma longipes* (Arachnida: Opiliones). Syst Biodivers 16:171–187

Peres EA, Benedetti AR, Hiruma ST, Sobral-Souza T, Pinto-da-Rocha R (2019) Phylogeography of Sodreaninae harvestmen (Arachnida: Opiliones: Gonyleptidae): insights into the biogeography of the southern Brazilian Atlantic Forest. Mol Phylogenet Evol 138:1–16

Perret M, Chautems A, Spichiger R (2006) Dispersal-Vicariance analyses in the tribe Sinningieae (Gesneriaceae): a clue to understanding biogeographical history of the Brazilian Atlantic forest. Ann Mo Bot Gard 93:340–358

Pessenda LCR, Oliveira PE, Mofatto M, Medeiros VB, Garcia RJF, Aravena R, Bendassoli JA, Leite AZ, Saad AR, Etchebehere ML (2009) The evolution of a tropical rainforest/grassland mosaic in southeastern Brazil since 28,000 14C yr BP based on carbon isotopes and pollen records. Quat Res 71:437–452

Pie MR, Faircloth BC, Ribeiro LF, Bornschein R, McCormack JE (2018) Phylogenomics of montane frogs of the Brazilian Atlantic forest is consistent with isolation in sky islands followed by climatic stability. Biol J Linn Soc 125:72–82

Pinheiro F, Barros F, Palma-Silva C, Fay MF, Lexer C, Cozzolino S (2011) Phylogeography and genetic differentiation along the distributional range of the orchid *Epidendrum fulgens*: a neotropical coastal species not restricted to glacial refugia. J Biogeogr 38:1923–1935

Pinto-da-Rocha R, DaSilva MB, Bragagnolo C (2005) Faunistic similarity and historic biogeography of the harvestmen of southern and southeastern Atlantic rain forest of Brazil. J Arachnol 33:290–299

Pires AC, Marinoni L (2010) DNA barcoding and traditional taxonomy unified through integrative taxonomy: a view that challenges the debate questioning both methodologies. Biota Neotropica 10(2):339–346. https://doi.org/10.1590/S1676-06032010000200035

Por FD (1992) Sooretama: the Atlantic rain forest of Brazil. SPB Academic, The Hague

Porto TJ, Carnaval AC, da Rocha PLB (2013) Evaluating forest refugial models using species distribution models, model filling and inclusion: a case study with 14 Brazilian species. Divers Distrib 19:330–340

Prado DE (2000) Seasonal dry forest of tropical South America: from forgotten ecosystems to a new phytogeographic unit. Edinb J Bot 57(3):437–461

Prado DE, Gibbs PE (1993) Patterns of species distributions in the dry seasonal forests of South America. Ann Mo Bot Gard 80:902–927

Prance GT (1982) Forest refuges: evidence from woody angiosperms. In: Prance GT (ed) Biological diversification in the tropics. Columbia University Press, New York, pp 137–158

Prates I, Rivera D, Rodrigues MT, Carnaval AC (2016) A mid-Pleistocene rainforest corridor enabled synchronous invasions of the Atlantic forest by Amazonian anole lizards. Mol Ecol 25:5174–5186

Prates I, Penna A, Rodrigues MT, Carnaval AC (2018) Local adaptation in mainland anole lizards: integrating population history and genome–environment associations. Ecol Evol 8:11932–11944

Reginato M, Michelangeli FA (2020) Bioregions of eastern Brazil based on vascular plant occurrence data. In: Rull V, Carnaval AC (eds) Neotropical diversification. Springer

Resende HC, Yotoko KSC, Delabie JHC, Costa MA, Campiolo S, Tavares MG, Campos LAO, Fernandes-Salomão TM (2010) Pliocene and Pleistocene events shaping the genetic diversity within the central corridor of the Brazilian Atlantic forest. Biol J Linn Soc 101:949–960

Ribeiro MC, Martensen AC, Metzger JP, Tabarelli M, Scarano F, Fortin M-J (2011a) The Brazilian Atlantic forest: a shrinking biodiversity hotspot. In: Zachos F, Habel J (eds) Biodiversity hotspots. Springer, Berlin

Ribeiro RA, Lemos-Filho JP, Ramos ACS, Lovato MB (2011b) Phylogeography of the endangered rosewood *Dalbergia nigra* (Fabaceae): insights into the evolutionary history and conservation of the Brazilian Atlantic forest. Heredity 106:46–57

Riccomini C, Santanna L, Ferrari A (2004) Evolução geológica do Rift Continental do Sudeste do Brasil. In: Mantesso Neto V, Bartorelli A, Carneiro C, Brito Neves B (eds) Geologia Do Continente Sul-Americano: Evolução Da Obra de Fernando Flávio Marques de Almeida. Beca, São Paulo, pp 383–405

Rodal MJN, Sales MF (2008) Composição da flora vascular em um remanescente de floresta montana no semi-árido do nordeste do Brasil. Hoehnea 34(4):433–446

Rodrigues MT, Bertolotto CEV, Amaro RC, Yonenaga-Yassuda Y, Freire EMX, Pellegrino KCM (2014) Molecular phylogeny, species limits, and biogeography of the Brazilian endemic lizard genus Enyalius (Squamata: Leiosauridae): an example of the historical relationship between Atlantic forests and Amazonia. Mol Phylogenet Evol 81:137–146

Rull V (2009) Microrefugia. J Biogeogr 36:481–484

Saiter FZ, Brown JL, Thomas WW, Oliveira-Filho AT, Carnaval AC (2016) Environmental correlates of floristic regions and plant turnover in the Atlantic forest hotspot. J Biogeogr 43:2322–2331

Santos AMM, Cavalcanti DR, Silva JMCD, Tabarelli M (2007) Biogeographical relationships in north-eastern Brazil. J Biogeogr 34:437–446

Santos BA, Arroyo-Rodrıguez V, Moreno CE, Tabarelli M (2010) Edge-related loss of tree phylogenetic diversity in the severely fragmented Brazilian Atlantic forest. PLoS One 5:e12625
Schetino MAA, Coimbra RTF, Santos FR (2017) Time scaled phylogeography and demography of *Bradypus torquatus* (Pilosa: Bradypodidae). Glob Ecol Conserv 11:224–235
Sigrist MS, Carvalho CJBD (2008) Historical relationships among areas of endemism in the tropical South America using Brooks Parsimony Analysis (BPA). Biota Neotropica 9:79–90
Silva JMC, Sousa MC, Castelletti CHM (2004) Areas of endemism for passerine birds in the Atlantic forest, South America. Glob Ecol Biogeogr 13:85–92
Silva SM, Moraes-Barros N, Ribas CC, Ferrand N, Morgante JS (2012) Divide to conquer: a complex pattern of biodiversity depicted by vertebrate components in the Brazilian Atlantic forest. Biol J Linn Soc 107:39–55
Sobral-Souza T, LIma-Ribeiro MS, Solferini VN (2015) Biogeography of neotropical rainforests: past connections between Amazon and Atlantic forest detected by ecological niche modeling. Evol Ecol 29(5):643–655
Soderstrom T, Judziewicz E, Clark L (1988) Distribution patterns of neotropical bamboos. In: Vanzolini P, Heyer W (eds) Proceedings of a workshop on neotropical distribution patterns. Academia Brasileira de Ciências, Rio de Janeiro, pp 121–157
Stattersfield AJ, Crosby MJ, Long AJ, Wege DC (1998) Endemic bird areas of the world: priorities for biodiversity conservation. BirdLife International, Cambridge, UK
Tchaicka L, Eizirik E, Oliveira TG, Cândido JF Jr, Freitas TO (2007) Phylogeography and population history of the crab-eating fox (*Cerdocyon thous*). Mol Ecol 16:819–838
Thomas WW, Carvalho AMV, Amorim AM, Garrison J, Arbelaez AL (1998) Plant endemism in two forests in southern Bahia, Brazil. Biodivers Conserv 7:311–322
Thomé MTC, Zamudio KR, Giovanelli JGR et al (2010) Phylogeography of endemic toads and post-Pliocene persistence of the Brazilian Atlantic forest. Mol Phylogenet Evol 55:1018–1031
Thomé MTC, Zamudio KR, Haddad CFB, Alexandrino J (2014) Barriers, rather than refugia, underlie the origin of diversity in toads endemic to the Brazilian Atlantic forest. Mol Ecol 23:6152–6164
Thomé MTC, Sequeira F, Brusquetti F, Carstens B, Haddad CF, Rodrigues MT, Alexandrino J (2016) Recurrent connections between Amazon and Atlantic forests shaped diversity in Caatinga four-eyed frogs. J Biogeogr 43:1045–1056
Tonini JFR, Costa LP, Carnaval AC (2013) Phylogeographic structure is strong in the Atlantic forest; predictive power of correlative paleodistribution models, not always. J Zool Syst Evol Res 51:114–121
Trujillo-Arias N, Calderón L, Santos FR, Miyaki CY, Aleixo A, Witt CC, Tubaro PL, Cabanne GS (2018) Forest corridors between the central Andes and the southern Atlantic forest enabled dispersal and peripatric diversification without niche divergence in a passerine. Mol Phylogenet Evol 128:221–232
Turchetto-Zolet AC, Pinheiro F, Salgueiro F, Palma-Silva C (2013) Phylogeographical patterns shed light on evolutionary process in South America. Mol Ecol 22:1193–1213
Turchetto-Zolet AC, Salgueiro F, Turchetto C, Cruz F, Veto NM, Barros MJ, Segatto ANA, Freitas LB, Margis R (2016) Phylogeography and ecological niche modelling in *Eugenia uniflora* (Myrtaceae) suggest distinct vegetational responses to climate change between the southern and the northern Atlantic forest. Bot J Linn Soc 182:670–688
Tyler HA, Brown KS Jr, Wilson KH (1994) Swallowtail butterflies of the Americas: a study in biological dynamics, ecological diversity, biosystematics, and conservation. Scientific Publishers, Gainesville
Vale MM, Tourinho L, Lorini ML, Rajão H, Figueiredo MSL (2018) Endemic birds of the Atlantic forest: traits, conservation status, and patterns of biodiversity. J Field Ornithol 89:193–206
Vanzolini PE (1988) Distributional patterns of South American lizards. In: Proceedings of a workshop on neotropical distribution. Academia Brasileira de Ciências, Rio de Janeiro
Vanzolini PE, Williams EF (1981) The vanishing refuge: a mechanism for ecogeographic speciation. Papéis Avulsos de Zoologia 34:251–255

Vasconcelos TS, Prado VHM, da Silva FR, Haddad CFB (2014) Biogeographic distribution patterns and their correlates in the diverse Frog Fauna of the Atlantic forest hotspot. PLoS One 9:e104130

Wallace AR (1852) A narrative of travels on the Amazon and Rio Negro. Dover, New York

Webb CO, Ackerly DD, McPeek MA, Donoghue MJ (2002) Phylogenies and community ecology. Annu Rev Ecol Syst 33(1):475–505

Weeks B, Claramunt S, Cracraft J (2016) Integrating systematics and biogeography to disentangle the roles of history and ecology in biotic assembly. J Biogeogr 43(8):1546–1559

Werneck FP (2011) The diversification of eastern South American open vegetation biomes: historical biogeography and perspectives. Quat Sci Rev 30:1630–1648

Zachos J, Pagani M, Sloan L, Thomas E, Billups K (2001) Trends, rhythms, and aberrations in global climate 65 Ma to present. Science 292:686–693

Zanella FCV (2011) Evolução da Biota da Diagonal de Formações Abertas Secas da América do Sul. In: Carvalho CB, Almeida E (eds) Biogeografia da América do Sul: padrões e processos. Roca, São Paulo, pp 198–220

Chapter 17
Forests Diversity in the Mexican Neotropics: A Paleoecological View

Blanca L. Figueroa-Rangel, Miguel Olvera-Vargas, Socorro Lozano-García, Gerald Islebe, Nuria Torrescano, Susana Sosa-Najera, and Ana P. Del Castillo-Batista

Abstract The forests of Mexico have conclusive environmental characteristics mainly shaped by geology and climate change. The western region is geologically linked to volcanic forcing, which in turn was responsible for mountain uplift while the east was formed by tectonic movement in a steady and gradual process allowing the development of rare taxa. These natural processes have produced dissimilarities in forest composition along the different regions of the Mexican territory. For the Holocene, palaeoclimatic and palaeoecological records have revealed that climate change is an essential factor involved in this diversification. Therefore, in order to discern differences in taxon diversity in tropical and temperate ecosystems during the Holocene, 15 sites with palaeoecological data retrieved from several sources (lakes, lagoons, ponds, a moraine depression and forest hollows) were studied. They comprised highland and lowland vegetation from 0 to 3860 m asl representing local and regional signals in mangrove forest, tropical evergreen forest, tropical subdeciduous forest, pine forest, pine-oak forest and cloud forests in different regions of Mexico. We used seven diversity metrics along with abundance estimation (Rarefaction, *Pielou* evenness, *Shannon* diversity, *Hill* numbers: N0, N1 and N2 and Rate of Change). Our results indicated that forests with a tropical component were more diverse than forests with Holarctic affinities. Nonetheless, it was not clear if local signals were more diverse than regional signals or that lowland signals were more diverse than highland signals.

Keywords Highlands · Hill numbers · Local diversity · Lowlands · Palaeoecology · Regional diversity · Z-values

---

B. L. Figueroa-Rangel (✉) · M. Olvera-Vargas · A. P. Del Castillo-Batista
Departamento de Ecología y Recursos Naturales, Centro Universitario de la Costa Sur, Universidad de Guadalajara, Autlán de Navarro, Jalisco, Mexico
e-mail: bfrangel@cucsur.udg.mx

S. Lozano-García · S. Sosa-Najera
Instituto de Geología, Universidad Nacional Autónoma de México, México DF, Mexico

G. Islebe · N. Torrescano
Departamento Conservación de la Biodiversidad, El Colegio de la Frontera Sur, Chetumal, Quintana Roo, Mexico

V. Rull, A. C. Carnaval (eds.), *Neotropical Diversification: Patterns and Processes*, Fascinating Life Sciences, https://doi.org/10.1007/978-3-030-31167-4_17

1 Introduction

1.1 The Mexican Neotropics

The geographical position of Mexico, located in the intersection of the Nearctic and the Neotropic regions, has significantly contributed to the diversity and expansion of the present-day biota (Halffter et al. 2008). However, the connection between the Nearctic and the Neotropics is not a sharp defined geographical area and many descriptions depend upon the taxonomic group under study; e.g. for plants (Rzedowski 2006); for insects (Morrone and Márquez 2001) and for vertebrates (Savage 1982).

In general, the northern part, together with the Central Mexican Plateau, Sierra Madre Occidental y Oriental belongs to the Nearctic, while the highlands in Sierra Madre del Sur and Sierra Madre de Chiapas, the Yucatan Peninsula and the tropical lowlands are part of the Neotropics. The overlapping region between the Nearctic and the Neotropics is named Transition Zone [after (Halffter 1974, 1976)], an area highly influenced by the orography and present climate, but also by the geological and the biological history of the Cenozoic era. These conditions created, together with Pleistocene glaciations, biogeographical archipelagos with abundant opportunities for speciation, and hence diversification (Halffter et al. 2008). This transition zone is formed by the Trans-Mexican Volcanic Belt (TMVB), which is the largest volcanic arc in North America built upon Cretaceous and Cenozoic magmatic provinces. Its western end lies near the southern Gulf of California oblique rift; to the east, it ends at the Gulf of Mexico ocean basin (Ferrari et al. 2012). The TMVB is a geological province crossing Mexico from west to east around 20° N. Its volcanic activity started in the Neogene (~19 Ma) creating around 8000 volcanic structures corresponding to the highest volcanoes chain in the country (Gómez-Tuena et al. 2005; Caballero et al. 2010). In this province, many intermountain water basins are found (Lozano-Garcia et al. 2013), which allow the preservation of sedimentary sequences for palaeoenvironmental reconstruction.

1.2 Biodiversity in Mexico

Mexico is one of the 17-megadiverse countries of the world. With 94,412 species, this figure corresponds to approximately 8.59% of the global diversity according to Martínez-Meyer et al. (2014). The most diverse group of plants is Magnoliophyta (flowering plants) with 21,841 species (Villaseñor and Ortiz 2014), followed by Lepidoptera (Butterflies) with 14,507 species and fungi with 6500 species (Aguirre-Acosta et al. 2014). Other important taxa are less diverse in terms of number of species; e.g. lichens with 2833; Pteridophyta with 1014 (Martínez-Salas and Ramos 2014); Bryophyta with 984 (Delgadillo-Moya 2014) and Pinophyta with 94 species (Gernandt and Pérez-De La Rosa 2014). Plant biodiversity in Mexico is vast;

therefore different vegetation classifications have been developed. Some authors have described 10 vegetation types (Rzedowski 1981); Miranda and Hernández (Miranda and Hernández 1963) reported 32 types and the most recent and widely used is the one established by INEGI with 58 types (INEGI 2017). This classification is based on the two previous outlined classifications. The main difference among those classifications is how they separate main vegetation types: tropical forests (deciduous and evergreen forests), temperate forests (conifers and broadleaved forests), cloud forests, xerophytic forests, grasslands and wetlands. Modern plant taxa of these vegetation types originated as a result of evolutionary and geological processes, which generated physiographical (mainly altitude) and climatic differences influencing taxa radiation and diversification. This favored plant species to establish in sites with heterogeneous environmental conditions.

The palaeobotanical perspective of Mexico outlines biodiversity that began ca. 132 million years ago. Main divergences occurred during the Cretaceous with communities developing under hot-humid conditions. Under such scenarios, vegetation types diverged at two different times in two distinctive regions. During the Paleogene it occurred in northern Mexico while for central and southern Mexico, it occurred in the Neogene (Cevallos-Ferriz et al. 2012). Especially for the Mexican highlands, diversification was reported as a result of climate fluctuations and recent volcanism under a sky-island dynamic occurred throughout the last 1 Myr of the Pleistocene (Mastretta-Yanes et al. 2015).

1.3 The Palaeoecological View in Long-Term Studies

The study of processes and mechanisms involved in the origin of extant biodiversity gradients exceed neoecological studies (Rull 2013, 2014), as those processes act between the ecology and the evolutionary interface. For this purpose, the use of long-term research ranging from centuries to tens of millennia is the most appropriate method to study ecological processes such as succession, migration, community stability, niche conservatism and extinction (Rull and Vegas-Vilarrúbia 2011). Hence, palaeoecology is a discipline included in long-term studies; it can be defined as the ecology of the past using proxies (Birks and Birks 1980; Rull 2010). Palaeoecology is concerned with time scales from decades to millennia where different processes and mechanisms allow the study of diversification. Regarding plants, the main proxies are plant macrofossils and fossil pollen. In Mexico, the foremost source of past biodiversity information derives from palaeoecological studies using sediment retrieved from lakes, lagoons, pits and forest hollows (Fig. 17.1). A number of synthetizing efforts using fossil pollen have shown changes in plant taxonomic composition at different regions of the country, in an arrangement of vegetation types, mainly during the Holocene. Main drivers for potential diversification involve palaeogeographic and palaeoclimatic events, together with anthropogenic activities (Lozano-Garcia et al. 2015; Islebe et al. 2016; Caballero-Rodríguez et al. 2018).

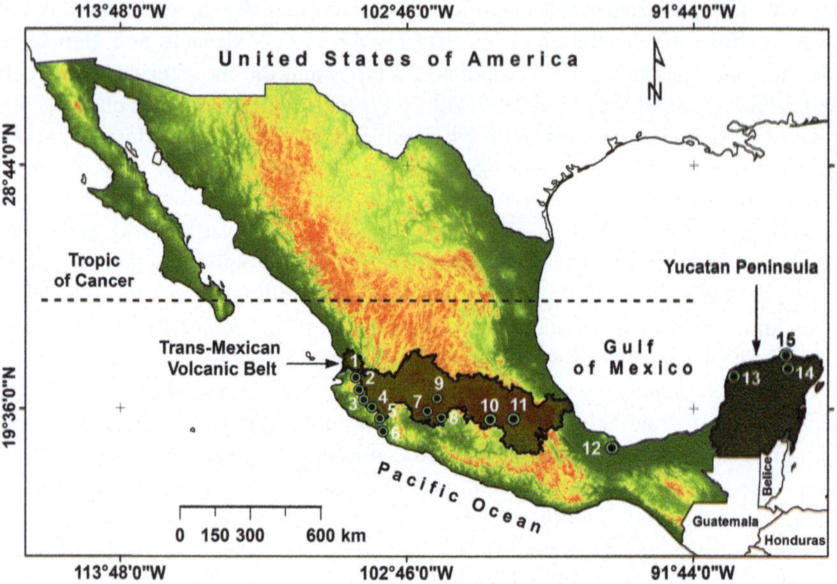

Fig. 17.1 Palaeoecological sites in the Mexican Neotropics. See Table 17.1 for the explanation of the site numbers

1.4 *Records of Palaeodiversity in the Mexican Neotropics*

The most recurrent ecosystems with palaeoecological studies in Mexico are composed by temperate elements such as *Pinus* forests, *Quercus* forests and *Pinus-Quercus* forest (Lozano-Garcia and Vazquez-Selem 2005; Lozano-Garcia et al. 2013; Caballero-Rodríguez et al. 2017). For the lowlands of the Yucatan Peninsula, there is also an ample number of studies, mainly developed in mangrove vegetation (Whitmore et al. 1996; Carrillo-Bastos et al. 2012, 2013; Cook et al. 2012; Torrescano-Valle and Islebe 2015; Vela-Pelaez et al. 2018). Nevertheless, there are few palaeoecological studies undertaken in ecosystems with the highest distribution per unit area. For instance, the *Pinus-Quercus* forest only represents 5.2% of the total vegetation cover in Mexico; the *Pinus* forest, 3.2% and mangrove forest only 0.6%. (Challenger and Soberón 2008). The highest extension, 28.8%, corresponds to xerophytic scrub forest (Villaseñor 2016); nonetheless, there are no reports on palaeoecological studies undertaken in this forest, except evidence on its past existence: e.g. Pérez-Crespo et al. (Pérez-Crespo et al. 2013), using a bioclimatic model based on mammal species, marked the presence of xerophytic scrub forest in Oaxaca between 9000 and 4500 years BP.

1.5 *Palaeodiversity Measures*

Biodiversity estimation of the past based on palynological data is a challenging task due to the imprecision in translating pollen or spore taxa into specific plant taxa. After all, taxonomic identity of fossil pollen and spores barely reaches the genus level. The source of pollen and/or spore source (e.g. lakes, forest hollows, etc.), and its effect on the spatial scale of the reconstruction, is also a significant topic to be contemplated in diversity estimation. In general, estimates of species diversity are dependent on the spatial scale and the size of the sample (Gotelli and Colwell 2001; McGill et al. 2015).

Plant diversity in a particular location can be evaluated using species richness (the number of different species found in a community). This is the simplest way to describe community patterns and regional diversity (Magurran 1988). In palynological studies, the number of fossil pollen grains in a sample of sediment is used as a surrogate to estimate the number of plant individuals. Consequently, estimates of biodiversity in palaeoecology set emphasis on palynological richness in order to represent pollen diversity (Weng et al. 2006). However, the different indices used to measure biodiversity should consider, not only richness but also evenness (the relative abundance of the different species). The main difficulty arise with the differential pollen each taxon produce, generating bias in the estimation of plant abundances (Birks et al. 2016). More perplexing is the need to compare species diversity of different sites along time. Therefore, in the present chapter we used seven metrics to compare diversity of 15 sites with palaeorecords of the Mexican Neotropics in the Holocene combining estimates of richness and evenness.

2 Methods

We used 15 sites with palaeorecords previously published (Table 17.1). These palaeorecords are distributed from west to east in the Trans-Mexican Volcanic Belt and in the Yucatan Peninsula. Sites include sediments retrieved from different sources: five forest hollows, four lakes, one lagoon, a moraine depression, two ponds and two crater lakes. They cover an assortment of vegetation types; from lowland locations (e.g. mangrove forest and tropical evergreen forest) to the highlands (e.g. *Pinus* forest, *Pinus-Quercus* forest and Cloud forest).

Number of taxa, as well as taxon abundances, were determined for each site. To compare taxon abundances among sites we estimated pollen percentages selecting the most abundant tree taxa (four in the highlands and five in the lowlands) and three of the most abundant herbaceous and epiphytic taxa (Figs. 17.2 and 17.3).

In view of the heterogeneous composition of the 15 sites (sources, forest types, sampling resolutions and chronologies) and to allow for a robust comparison among sites by controlling the scale of the variable used for analysis (in this case, taxa percentages used as a proxy for plant abundances), we standardized pollen

Table 17.1 Paleoecological sites from the Mexican Neotropics

Site number	Site name	Acronym	Location	Source	Signal	Altitude (m asl)	Vegetation classification	Vegetation type	Time span (years)	References
1	Talpa	TAL	20°23′ N, 104°49′ W	Forest hollow	Local	1800	Highland	Cloud forest	720	Del Castillo-Batista et al. (2016)
2	La Cumbre	CUM	20°10′20″ N, 104°42′45″ W	Forest hollow	Local	2120	Highland	*Pinus-Quercus-Abies* forest	1580	Del Castillo-Batista et al. (2018)
3	Xilosuchitlán	XIL	19°35′32″ N, 104°16′55″ W	Forest hollow	Local	1850	Highland	Cloud forest	1388	Figueroa-Rangel et al. (2010)
4	Neverías	NEV	19°33′95″ N, 104°10′53″ W	Forest hollow	Local	2550	Highland	*Pinus* forest	1313	Figueroa-Rangel et al. (2008)
5	Las Cabañas	CAB	19°35′34″ N, 104°16′46″ W	Forest hollow	Local	2100	Highland	*Pinus-Quercus-Carpinus* forest	1230	Figueroa-Rangel et al. (2012)
6	Cuyutlán	CUY	18°51′ N, 103°53′ W	Lagoon	Regional	0	Lowland	Mangrove forest	1300	Figueroa-Rangel et al. (2016)
7	Zirahuén	ZIR	19°26′11″ N, 101°44′20″ W	Lake	Regional	2075	Highland	*Pinus-Quercus* forest	11,500	Lozano-García et al. (2013)
8	Tacámbaro	TAC	19°12′38″ N, 101°27′29″ W	Crater lake	Regional	1475	Highland	*Pinus-Quercus* and Tropical deciduous forests	9220	Caballero-Rodríguez et al. (2018)
9	Zacapu	ZAC	19°55′ N, 101°40′ W	Lake	Regional	1970	Highland	*Pinus-Quercus* forest	9000	Correa-Metrio et al. (2012)

10	Santa Cruz Atizapán	SCA	19°10′67″ N, 99°32′56″ W	Lake	Regional	2570	Highland	*Pinus-Quercus* forest	9900	Lozano-García and Vazquez-Selem (2005)
11	Valle Agua El Marrano	VAM	19°12′35″ N, 98°39′57″ W	Moraine depression	Regional	3860	Highland	*Pinus* forest	10,900	Lozano-García and Vazquez-Selem (2005), Caballero-Rodríguez et al. (2018)
12	Verde	VER	18°36′53″ N, 95°20′55″ W	Crater lake	Regional	100	Lowland	Tropical Ever-green forest	2800	Lozano-García et al. (2010)
13	Silvituc	SIL	18°37′2″ N, 90°17′41″ W	Lake	Regional	59	Lowland	tropical Subdeciduous forest	7900	Torrescano-Valle and Islebe (2015)
14	Ría Lagartos 1	RIA1	21°56′61″ N, 88°08′65″ W	Pond	Regional	0	Lowland	Mangrove forest	3800	Carrillo-Bastos et al. (2013)
15	Ría Lagartos 2	RIA2	21°34′ N, 88° 04′ W	Pond	Regional	0	Lowland	Mangrove forest	3800	Aragon-Moreno et al. (2012)

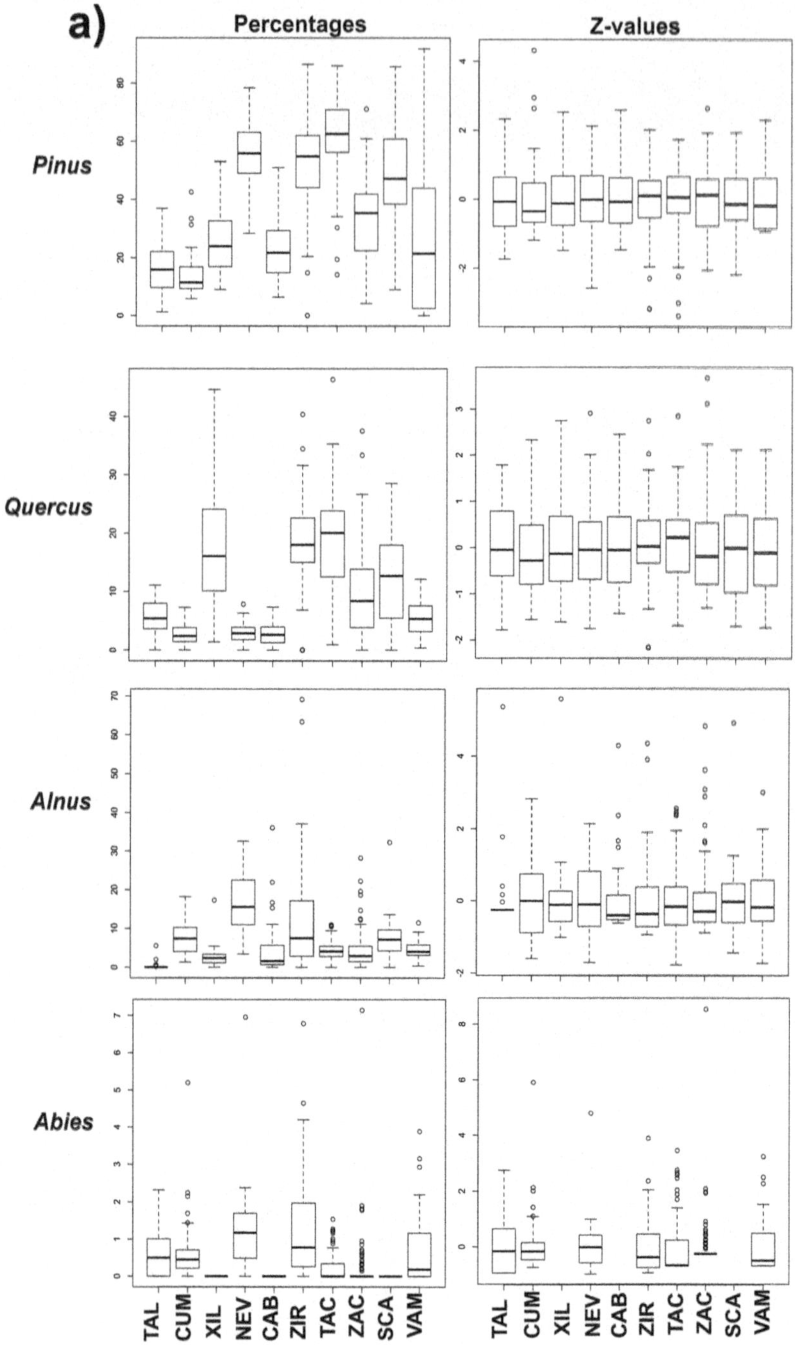

Fig. 17.2 (**a**) Abundant tree taxa from highland sites estimated in percentages (to the left) and Z-values (to the right). (**b**) Abundant herbaceous taxa from highland sites estimated in percentages (to the left) and Z-values (to the right). (**c**) Abundant epiphytic taxa from highland sites estimated in percentages (to the left) and Z-values (to the right)

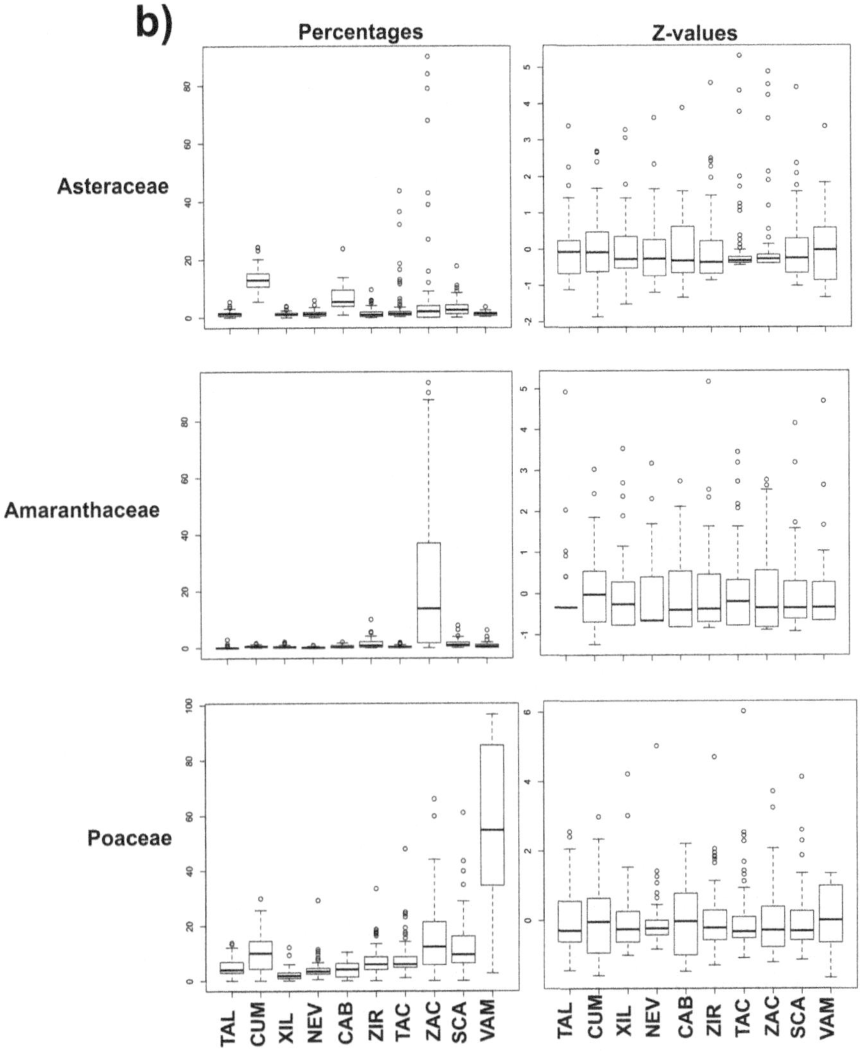

Fig. 17.2 (continued)

percentages using Z-values (Figs. 17.2 and 17.3) with the *scale* function in the *standardize* package available in R-v3.4.3 (R Development Core Team 2017). Standardization was undertaken to account for the high variation in ranges of pollen percentages among taxa. Some taxa ranges between 0 and 100 while other ranges between 0 and 1. Using variables without standardization can give variables with larger ranges greater importance in the comparison. Transforming the data to comparable scales prevented this problem. The equation to transform pollen percentages to Z-values follows:

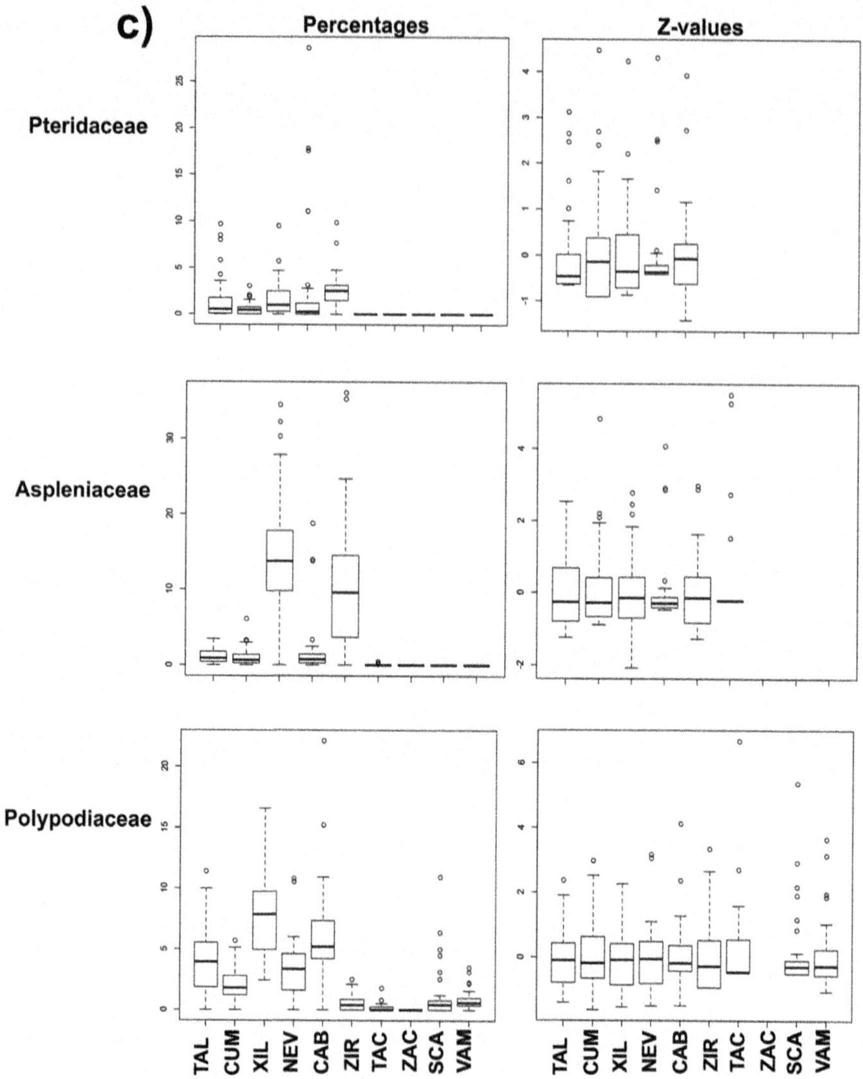

Fig. 17.2 (continued)

$$Z = \frac{x_i - \mu}{\sigma}$$

where x_i is the original pollen percentage value for taxa i in each age of the sequence; μ is the mean of the pollen percentage of taxon i for all ages in the sequence; σ is the standard deviation of the mean (Zar 1999). Z-values represent the distance between the original pollen percentage value and the mean in units of the standard deviation.

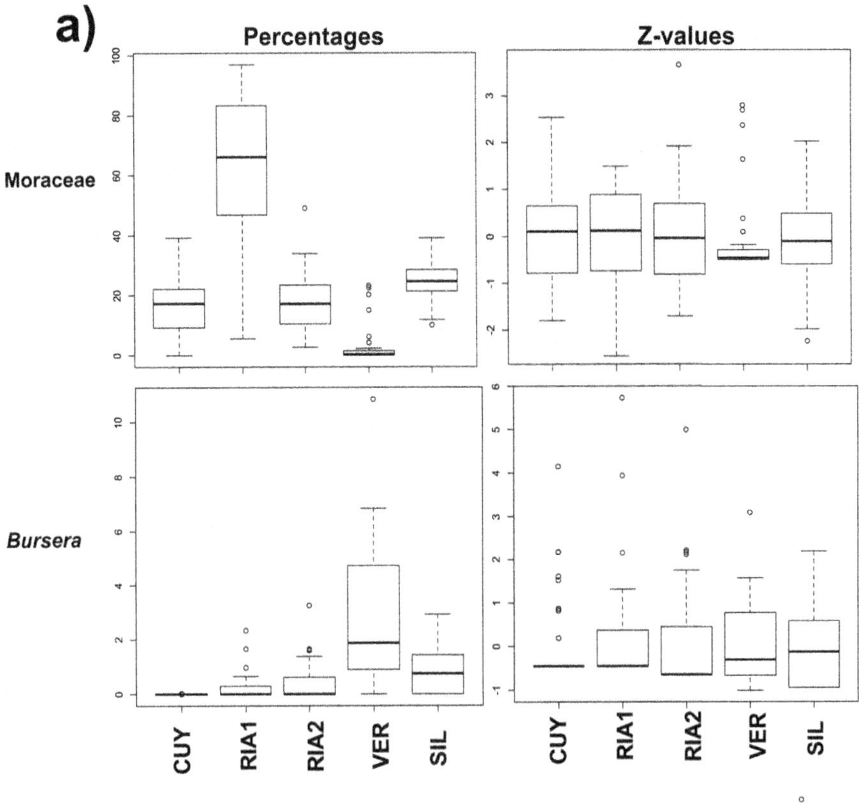

Fig. 17.3 (**a**) Abundant tree taxa from lowland sites estimated in percentages (to the left) and Z-values (to the right). (**b**) Abundant tree taxa from mangrove vegetation in lowland sites estimated in percentages (to the left) and Z-values (to the right). (**c**) Abundant herbaceous taxa from lowland sites estimated in percentages (to the left) and Z-values (to the right). (**d**) Abundant aquatic taxa from lowland sites estimated in percentages (to the left) and Z-values (to the right)

Z-value is negative when the original value is below the mean and positive when it is above.

Pollen percentages and their Z-values were graphically displayed using box and whiskers plots to compare the mean of pollen percentages distribution values among sequences, as well as their variances along time.

For diversity comparison among sites, we used metrics that included species (taxon) richness and species (taxon) relative abundances (evenness). Every metric was calculated by sample (age) for each of the 15 sites; subsequently confidence intervals (95%) were estimated (Table 17.2). The following metrics were used:

Rarefaction This is the most robust estimate of richness for palynological data (Birks and Line 1992). Rarefaction was estimated as the expected number of fossil pollen taxa ($E(S_n)$) in samples of equal size (n). This process was achieved counting

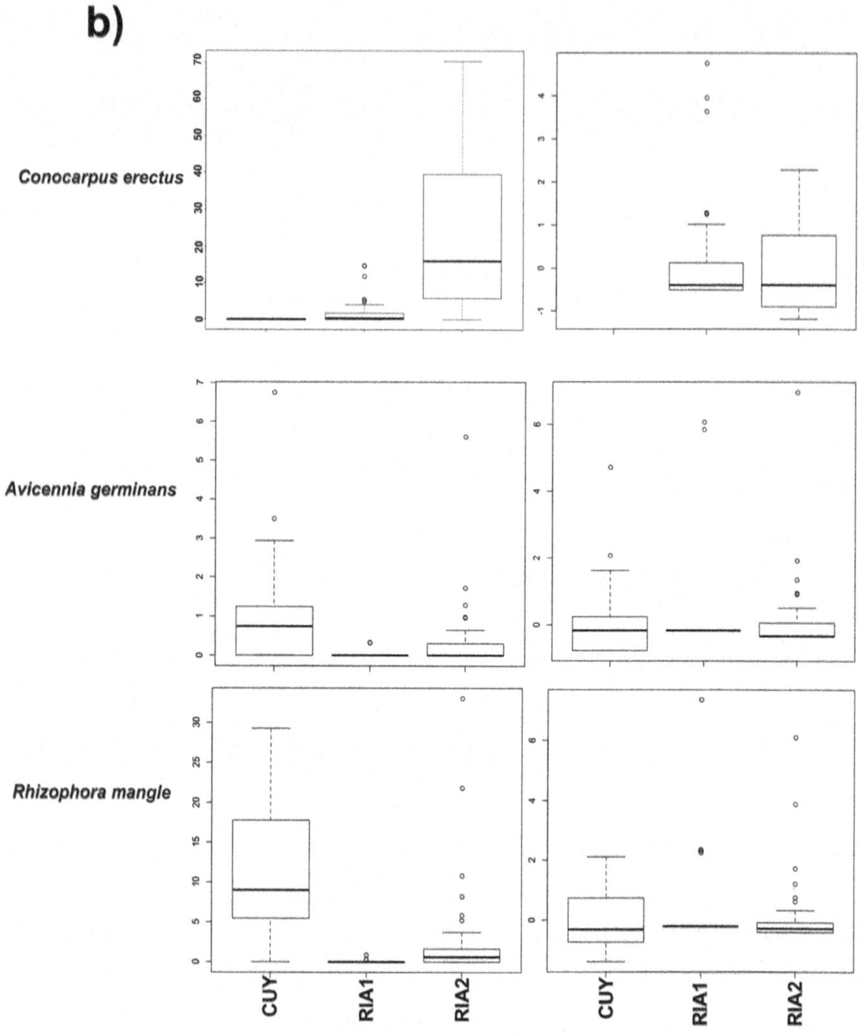

Fig. 17.3 (continued)

a standard number of pollen counts in each sediment sample (Birks et al. 2016), which varied depending on the site (Table 17.2).

Shannon-Weaver index This is an index based on the proportional abundances of species; it assumes that individuals are randomly sampled from an infinitely large community and that all species are represented in the sample (Magurran 1988). It was calculated using:

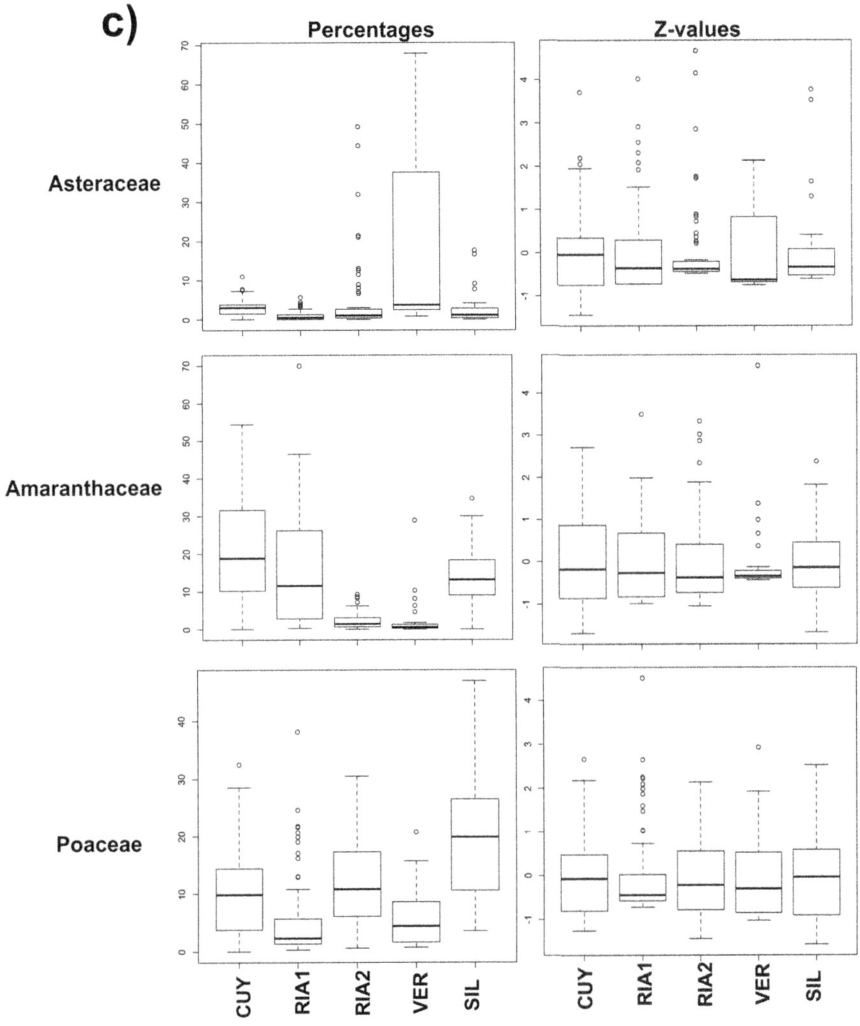

Fig. 17.3 (continued)

$$H = -\sum_{i=1}^{S} p_i \log_b p_i$$

where p_i is the proportion of species i, and S is the number of species (in this case the number of fossil pollen taxa), $b = 2$ (Oksanen 2017).

Pielou's Evenness Index It calculates a separate evenness measure based on the Shannon-Weaver index (H), which accounts for the degree of evenness in species abundances (Magurran 2004). It was calculated according to (Oksanen 2017) as:

(a) *Hill N0*: This is the number of species (*S*) found in a sample notwithstanding their abundances.

$$N0 = S$$

(b) *Hill N1*: This is the number of abundant taxa at each site, weighting each taxon by its relative abundance. It was estimated as the exponential of the Shannon-Weaver index (*H*):

$$N1 = \exp(H)$$

(c) *Hill N2*: This metric represents the number of very abundant fossil pollen taxa (Gotelli and Ellison 2013; Gosling et al. 2018). It was calculated as the inverse of Simpson's index (*D*), which is commonly referred as an evenness measure (Magurran 2004):

$$N2 = \frac{1}{D}$$

where: $D = 1 - \sum_{i=1}^{S} p_i^2$ Simpson's index.

Rate of Change This metric in generally used to indicate how fast changes taxon composition changes at a site. We used the dissimilarity approach (Birks et al. 2012) estimating, for each paleorecord, the chord distance dissimilarity between adjacent pairs of fossil pollen samples and then divide it by the age interval between the pair of samples (Bennett et al. 1992).

Rarefaction and rate of change were estimated using PSIMPOLL 4.25 and PSCOMB 1.03 software (Bennett 2005). *Hill Numbers*, *Shannon-Weaver* and *Pielou's* indexes were estimated using the *vegan* package (Oksanen et al. 2018) available in R-v3.4.3 (R Development Core Team 2017).

3 Results and Discussion

3.1 Fossil Pollen Abundances in the Highlands

If values were compared in percentages, *Pinus* in NEV (a site with local signal) showed high values, similarly to regional signals (e.g. TAC). The same occurred for *Quercus* in XIL (Xilosuchitlán) site, which showed very high percentages. In turn, *Abies* and *Alnus* showed a highly variable pattern in both percentages and Z-values (Fig. 17.2a). The highest abundance for *Abies* (ZIR and TAL) did not correspond to sites of higher altitudinal ranges (from 2000 to 3600 m asl), where this genus commonly developed (Lara-González et al. 2009).

We anticipated that sites with regional signals would contain high pollen abundances coming from genera such as *Pinus*, *Quercus* and *Alnus*, due to the fact their pollen is associated to false richness. False-richness, also termed false-presence, is given to pollen which is dispersed over long-distances such as *Pinus*, but also to re-deposited pollen such as *Quercus* and *Alnus* (Birks et al. 2016). However, our results showed that pollen associated to false-richness did not produce higher abundances in sites with regional signals.

For herbaceous taxa, percentage was not a convenient metric for comparison due to the elevated variability among sites. The use of Z-values exposed many outliers at all sites but not a sharp distinction between regional and local signals. One consistent configuration for the three herbaceous taxa was represented by medians <0 (Fig. 17.2b) implying that 50% of the observations presented pollen abundances below the mean [Z-value $= 0$ corresponds to the mean in a normal probability distribution (Zar 1999)]. If we take into account patterns of Z-values for specific sites, Asteraceae showed the highest and the most variable scores in the *Pinus* forest of VAM (Valle Agua El Marrano). This result agrees with the current report for Asteraceae diversity as temperate forest is the most diverse ecosystem in terms of number of species of Asteraceae family (1906) and the second regarding genera (266) (Villaseñor 2018).

With exception of the *Pinus-Quercus* forest in ZAC (Zacapu), which produced high percentages of Amaranthaceae, the rest of the sites produced low values for this family ($<5\%$). Z-values revealed that most of the sites, except CUM, had Amaranthaceae percentages below the mean along the sequences. Species that belong to this family prefer coastal and xeric habitats of temperate and subtropical climates. They are pioneers occurring in bare soils of open sites, and commonly associated to anthropogenic disturbance (Kuhn 1993; Ralska-Jasiewiczowa et al. 2004; Sandoval-Ortega et al. 2017). Their high abundance in ZAC site is related to the occurrence of dry conditions along the Holocene (Correa-Metrio et al. 2012).

For Poaceae, a strikingly elevated percentage was present in VAM, followed by ZAC. Using Z-values, sites CUM and CAB (Las Cabañas) resembled that of VAM distribution with similar values above and below the mean. In Mexico, Poaceae, distributed over 197 genera, represents the third family with the highest number of species (1127). Poaceae is a well-adapted family located in many aquatic and terrestrial ecosystems, from temperate to tropical affinities and from humid to arid zones. It is also a family of high economic importance used as food, forage, ornament and construction material (Valdez Reyna 1995; Sánchez-Ken 2011). Their high abundance in *Pinus* forest and *Pinus-Quercus* forest is similar to the pattern observed at present times where Poaceae species thrive well in the understory (Rzedowski 2006).

Pteridaceae and Aspleniaceae families were the most distinctive epiphytes in sources with local signals, nonetheless with low percentages (Fig. 17.2c). On the contrary, they were absent in sites of regional signals. This result is coincident with the present-day epiphyte diversity in Mexico: Pteridaceae is the richest epiphytic family with 218 species while Aspleniaceae is in fourth place with 89 species (Martínez-Salas and Ramos 2014). The highest percentage corresponded to the cloud forest in XIL and the lowest to the *Pinus* forest in CUM. This is an expected result as cloud forests in Mexico are characterized by a high epiphyte diversity with around 526 species classified in 104 genera and 25 families (Villaseñor 2010). On the contrary, epiphytes are scarce in *Pinus* forests; probably the resin secreted by pine trees do not provide an appropriate substrate for their development (Rzedowski 2006).

In terms of Z-values, the highest variation was present in CUM for Pteridaceae while that of Aspleniaceae was higher in TAL (Fig. 17.2c). For Pteridaceae, a right-skewed distribution was observed in XIL, while for Aspleniaceae a well-defined normal distribution emerged in the same site. Polypodiaceae was present in nine out of ten sites, with the highest percentages in sites with a local signal. A more homogeneous distribution arose for NEV (Fig. 17.2c).

Polypodiaceae is also a very diverse family in Mexico with 92 species. It represents the third epiphytic family with the highest number of species (Martínez-Salas and Ramos 2014). *Polypodium* is one of the main genera of the Polypodiaceae family at several sites. This genus is represented by around 120 species world-wide. Mexico is considered as centre of diversification for *Polypodium* (Vázquez-Pérez et al. 2012). Biogeographical studies for Mexico and Central America reported that *Polypodium* has a wide habitat specificity, present in different vegetation types including cloud forest, pine forest, oak forest, pine-oak forest, tropical evergreen forest, tropical sub-deciduous forest and gallery forest. The genus is also widely distributed in terms of altitude, from 100 to 3000 m asl (Luna-Vega et al. 2012).

3.2 Fossil Pollen Abundances in the Lowlands

Moraceae percentages were highly variable in the mangrove forest of RIA1 (Ría Lagartos 1). The tropical evergreen forest in VER (Verde) had very low percentages. A reverse pattern was observed for *Bursera* (Fig. 17.3a). Z-values for the three mangrove forests: CUY (Cuyutlan), RIA1 and RIA2 (Ría Lagartos 2) revealed a rather similar variation above and below the mean [Z-value = 0 correspond to the mean in a normal probability distribution; Z = 1 represents 1σ above the mean and -1σ below the mean (Zar 1999)] for Moraceae (Fig. 17.3a). This results means that 50% of the observations (median or first quartile Q_1) in the sequence were above the pollen percentage mean for Moraceae and 50% below the pollen percentage mean.

For *Bursera*, the highest variation corresponded to the tropical subdeciduous forest in SIL (Silvituc), while CUY presented low values with many outliers above the mean.

For mangrove vegetation, *Conocarpus erectus* was abundant in RIA2 (up to 70%) with no presence in CUY and very low percentages in RIA1. *Avicennia germinans* presented low values (no more than 3%) in CUY, minor values in RIA2 and around 2% in RIA1. *Rhizophora mangle* reached 30% in CUY with negligible values in RIA1. Z-values performed the same pattern as the percentages among the three sites (Fig. 17.3b).

Disparities in mangrove species abundances among lowland sites could be related to their regional distribution at present; *Rhizophora mangle* is reported as the most common species from Baja California Sur to Chiapas and from Tamaulipas to Quintana Roo (Agraz-Hernández et al. 2007). In the Yucatan Peninsula, mangrove communities include combinations of *Conocarpus erectus L.*, *Rhizophora mangle*, *Avicennia germinans* and *Laguncularia racemosa* (López-Portillo and Excurra 2012). *Conocarpus erectus var. sericeous* is only present in the Yucatan Peninsula (Agraz-Hernández et al. 2007).

The herbaceous taxa that were abundant in the highlands were also very abundant in the lowlands. Using percentage data, VER reached up to 70% of Asteraceae whereas for Amaranthaceae it reached values closes to 0. SIL reported the highest percentages in Poaceae (Fig. 17.3c). In terms of Z-values, Asteraceae had more variation with most of the values above the mean in VER. However, for Amaranthaceae CUY was the most variable of the sites.

For Poaceae, most sites showed a more or less similar variation except for RIA1, which showed a low variation in Z-values (Fig. 17.3c). The high abundance of Asteraceae for VER may be related to anthropogenic disturbance by agriculture and/or cattle ranching in this ecosystem during the last decades, which promoted the establishment of plant communities dominated by herbaceous taxa (Lozano-Garcia et al. 2010).

For aquatic taxa, *Nymphaea*, using percentages, high numbers corresponded to SIL, followed by CUY. Scarce values emerged for RIA1 and RIA2. *Typha* median values were around 8% in CUY, approximately 2% in SIL and close to 0 in RIA1 and RIA2. Concerning Z-values, medians at all sites were below zero. For *Nymphaea* CUY and SIL were very similar; for *Typha* SIL was more variable than CUY (Fig. 17.3d).

Cyperaceae showed percentages as high as 40% in RIA2 but less than 10% in RIA1. Nevertheless, considering Z-values, RIA1 presented higher and variable values than the rest of the sites while CUY showed an opposite pattern with low values (Fig. 17.3d). The scarce number of aquatic taxa, as well as their low abundance, is probably related with the present-day richness of aquatic taxa in Mexico. Permanent wetlands such as RIA1, RIA2, CUY and SIL with saline water bodies contain, in general, less aquatic taxa than lakes, dams, rivers and streams (Mora-Olivo et al. 2013). From the three aquatic taxa reported for the lowland sites, Cyperaceae is the aquatic family with the highest number of species (125) in Mexico; *Typha* only contains 2 species (Lot 2012) and both (*Typha domingensis* and *Typha latifolia*) are reported as invasive species in Mexico (Bonilla-Barbosa and Santamaría-Araúz 2013). Finally, Nymphaea has 11 species (Bonilla-Barbosa et al. 2000), two reported as invasive as well: *Nymphaea ampla* and *N. mexicana* (Bonilla-Barbosa and Santamaría-Araúz 2013).

3.3 Holocene Diversity

Palynological richness using rarefaction reported that the tropical evergreen forest in VER was the most diverse of the 15 sites under study while the least diverse was the tropical subdeciduous forest in SIL. The same arrangement arose using Hill numbers. For number of taxa, VER accounted for the highest N0; the lowest to the *Pinus-Quercus* forest in ZAC. Nonetheless, this metrics did not match with the actual number of taxa (S) counted in those sequences (Table 17.2).

Hill number N1, which truly expressed the effective number of taxa in the sequence, showed that around 17 equally common taxa were present in VER and TAL. The site with fewer equally common taxa was the *Pinus-Quercus* and the tropical deciduous forests reconstructed in Tacambaro (TAC) followed by the *Pinus* forest in VAM. For Hill number N2, where rare taxa had practically no contribution to the summation, the most diverse site was the cloud forest in TAL; the less diverse was VAM and TAC (Table 17.2).

Overall, the results indicated that forest with a tropical component such as tropical evergreen forest in VER and cloud forest in TAL were more diverse than the forest with Holarctic affinities such as *Pinus* forest in VAM and *Pinus-Quercus* in ZAC and TAC. This finding is not consistent with the rest of the forests with tropical affinities such as the cloud forest in XIL, the tropical subdeciduous forest in SIL and the three sites with mangrove forests (CUY, RIA1 and RIA2) (Table 17.2).

Palaeoecological results from this study are consistent with the present-day descriptions for plant diversity in Mexico. For instance, humid tropical forest is the most species-rich terrestrial ecosystem. In particular the tropical evergreen forest of VER, contains hundreds plant species per hectare with 75–99% of the genera with Neotropical affinity (Challenger and Soberón 2008). Cloud forest is notorious for its high number of species by area unit; it is extremely rich in taxonomic composition enclosing 82% of the families, 52% of the genera and 10% of the species reported for the whole flora of Mexico (Villaseñor 2003, 2004).

The highest rate of change or taxa turnover was present in the *Pinus-Quercus-Carpinus* forests in Las Cabañas (CAB), followed by the cloud forests in XIL, *Pinus*-forests in NEV and the mangrove forests of RIA1 and CUY. Minor values were present in VAM, Santa Cruz Atizapán (SCA), ZIR, CUM and TAL (Table 17.2). These results suggest there is not an obvious pattern of higher taxa turnover for lowlands versus highlands, for local versus regional signals or for Tropical versus Holarctic forests. Probably this is an effect of the different time intervals compared among sites, the time resolution of the samples and mostly the chronologies developed in the 15 sites; these conditions had an important effect on the computation of rate of change which divide dissimilarity by the temporal difference among samples.

4 Conclusions

Although palaeoecological techniques based on palynological data are contributing to the study of past forest diversity in Mexico, they have focused on few forest ecosystems: *Pinus* and *Pinus-Quercus* forests are the most commonly studied. There are some efforts to analyze mangrove vegetation and cloud forests, few efforts to study tropical assemblages (tropical evergreen and subdeciduous) but none to study xerophytic forests. The present analysis, involving several diversity metrics, revealed the difficulty and complexity of the task to compare punctual disparities on biodiversity in highly diverse environments. Additionally, fossil pollen identification is based on family and genus level. Depending on the metrics used to measure diversity, the tropical evergreen forest and the cloud forest seemed the most diverse ecosystems in the Mexican Neotropics. Metric values differed mainly depending on the use of rare or abundant taxa on the diversity estimation. Therefore, we recommend the use of both richness and evenness approaches to compute and to compare diversity among temporal studies. For abundance estimation, standardization is essential to account for differential pollen production and dispersion among species. Additionally, it is compelling to standardize the time resolution in the chronologies and to homogenize taxa identification, in order to decrease the uncertainties of the age-depth models developed in each site, which hinders the accurate comparison of rate of change among sites.

Acknowledgements Research undertaken in forest hollows was supported by Conacyt (Mexican National Council for Science and Technology, project 106435) and CoecytJal (State Council for Science and Technology Jalisco, project 5-2010-875).

References

Agraz-Hernández CM, Noriega-Trejo R, López-Portillo J et al (2007) Guía de Campo. Identificación de los Manglares en México

Aguirre-Acosta E, Ulloa M, Aguilar S et al (2014) Biodiversidad de hongos en México. Rev Mex Biodivers 85:76–81

Aragón-Moreno AA, Islebe GA, Torrescano-Valle N (2012) A ~3800-yr, high-resolution record of vegetation and climate change on the north coast of the Yucatan Peninsula. Rev Palaeobot Palynol 178:35–42. https://doi.org/10.1016/j.revpalbo.2012.04.002

Bennett K (2005) Psimpoll 4.25 and pscomb 1.03

Bennett KD, Boreham S, Sharp MJ, Switsur VR (1992) Holocene history of environment, vegetation and human settlement on Catta Ness, Lunnasting, Shetland. J Ecol 80:241–273. https://doi.org/10.3109/10826084.2012.705685

Birks HJB, Birks HH (1980) Quaternary palaeoecology. Edward Arnold, London

Birks HJB, Line JM (1992) The use of rarefaction analysis for estimating palynological richness from quaternary pollen-analytical data. Holocene 2:1–10. https://doi.org/10.1177/095968369200200101

Birks HJB, Lotter AF, Juggins S, Smol JP (2012) Tracking environmental change using lake sediments. Springer, London

Birks HJB, Felde VA, Bjune AE et al (2016) Does pollen-assemblage richness reflect floristic richness? A review of recent developments and future challenges. Rev Palaeobot Palynol 228:1–25

Bonilla-Barbosa JR, Santamaría-Araúz B (2013) Plantas acuáticas exóticas y traslocadas invasoras. In: Mendoza R, Koleff P (eds) Especies acuáticas invasoras en México. Comisión Nacional para el Conocimiento y Uso de la Biodiversidad, pp 223–247

Bonilla-Barbosa J, Novelo A, Hornelas Orozco Y, Márquez-Guzmán J (2000) Comparative seed morphology of Mexican Nymphaea species. Aquat Bot 68:189–204. https://doi.org/10.1016/S0304-3770(00)00125-X

Caballero M, Lozano-Garcia S, Vazquez-Selem L, Ortega B (2010) Evidencias de cambio climático y ambiental en registros glaciales y en cuencas lacustres del centro de México durante el último maximo glacial. Bol Soc Geol Mex. https://doi.org/10.1108/13683041011027445

Caballero-Rodríguez D, Lozano-García S, Correa-Metrio A (2017) Vegetation assemblages of central Mexico through the late Quaternary: modern analogs and compositional turnover. J Veg Sci 28:504–514. https://doi.org/10.1111/jvs.12515

Caballero-Rodríguez D, Correa-Metrio A, Lozano-García S et al (2018) Late-quaternary spatio-temporal dynamics of vegetation in Central Mexico. Rev Palaeobot Palynol 250:44–52. https://doi.org/10.1016/j.revpalbo.2017.12.004

Carrillo-Bastos A, Islebe GA, Torrescano-Valle N (2012) Geospatial analysis of pollen records from the Yucatan peninsula, Mexico. Veg Hist Archaeobot 21:429–437. https://doi.org/10.1007/s00334-012-0355-1

Carrillo-Bastos A, Islebe GA, Torrescano-Valle N (2013) 3800 years of quantitative precipitation reconstruction from the northwest Yucatan peninsula. PLoS One 8:e84333. https://doi.org/10.1371/journal.pone.0084333

Cevallos-Ferriz S, González-Torres EA, Calvillo-Canadell L (2012) Perspectiva paleobotánica y geológica de la biodiversidad en méxico. Acta Bot Mex (100):317–350

Challenger A, Soberón J (2008) 3. Los ecosistemas terrestres. In: Capital natural de México, vol. I: Conocimiento actual de la biodiversidad, pp 87–108

Cook BI, Anchukaitis KJ, Kaplan JO et al (2012) Pre-Columbian deforestation as an amplifier of drought in Mesoamerica. Geophys Res Lett 39. https://doi.org/10.1029/2012GL052565

Correa-Metrio A, Lozano-García S, Xelhuantzi-López S et al (2012) Vegetation in western Central Mexico during the last 50 000 years: modern analogs and climate in the Zacapu Basin. J Quat Sci 27:509–518. https://doi.org/10.1002/jqs.2540

Del Castillo-Batista AP, Figueroa-Rangel BL, Lozano-García S et al (2016) Historia florística y ambiental del bosque mesófilo de montaña en el centro-occidente de México durante la pequeña edad de hielo. Rev Mex Biodivers 87:216–229. https://doi.org/10.1016/j.rmb.2016.01.021

Del Castillo-Batista AP, Figueroa-Rangel BL, Lozano-García S et al (2018) 1580 years of human impact and climate change on the dynamics of a Pinus-Quercus-Abies forest in west-central Mexico. Rev Mex Biodiv 89:208–225. https://doi.org/10.22201/ib.20078706e.2018.1.2117

Delgadillo-Moya C (2014) Biodiversidad de Bryophyta (musgos) en México. Rev Mex Biodivers 85. https://doi.org/10.7550/rmb.30953

Ferrari L, Orozco-Esquivel T, Manea V, Manea M (2012) The dynamic history of the Trans-Mexican Volcanic Belt and the Mexico subduction zone. Tectonophysics 522–523:122–149

Figueroa-Rangel BL, Willis KJ, Olvera-Vargas M (2008) 4200 years of pine-dominated upland forest dynamics in west-central Mexico: human or natural legacy? Ecology 89:1893–1907. https://doi.org/10.1890/07-0830.1

Figueroa-Rangel BL, WIllis KJ, Olvera-Vargas M (2010) Cloud forest dynamics in the mexican neotropics during the last 1300 years. Glob Chang Biol 16:1689–1704. https://doi.org/10.1111/j.1365-2486.2009.02024.x

Figueroa-Rangel BL, Willis KJ, Olvera-Vargas M (2012) Late-Holocene successional dynamics in a transitional forest of west-central Mexico. Holocene 22:143–153. https://doi.org/10.1177/0959683611414929

Figueroa-Rangel B, Olvera-Vargas M, Vázquez-López JM et al (2016) Modern and fossil pollen assemblages revealing forest taxonomic changes in the Mexican subtropics during the last 1300 years. Rev Palaeobot Palynol 231:1–13. https://doi.org/10.1016/j.revpalbo.2016.04.007

Gernandt DS, Pérez-De La Rosa JA (2014) Biodiversidad de Pinophyta (coníferas) en México. Rev Mex Biodivers 85:S126–S133. https://doi.org/10.7550/rmb.32195

Gómez-Tuena A, Orozco-Esquivel M, Ferrari L (2005) Petrogénesis ígnea de la faja Volcánica Transmexicana. Bol Soc Geol Mex. https://doi.org/10.18268/BSGM2005v57n3a2

Gosling WD, Julier ACM, Adu-Bredu S et al (2018) Pollen-vegetation richness and diversity relationships in the tropics. Veg Hist Archaeobot 27:411–418. https://doi.org/10.1007/s00334-017-0642-y

Gotelli NJ, Colwell RK (2001) Quantifying biodiversity: procedures and pitfalls in the measurement and comparison of species richness. Ecol Lett 4:379–391

Gotelli NJ, Ellison AM (2013) A primer of ecological statistics. Sinauer Associates, Sunderland

Halffter G (1974) Elements anciens de l'entomofaune neo-tropicale: ses implications biogeographiques. Quaest Entomol 10:223–262

Halffter G (1976) Distribución de los insectos en la zona de transición mexicana: Relaciones con la entomofauna de Norteamérica. Folia Entomol Mex 35:1–64

Halffter G, Llorente-Bousquets J, Morrone JJ (2008) La perspectiva biogeográfica histórica. In: Soberón J, Halfter G, Llorente-Bousquets J (eds) Capital Natural de México. Conabio, México, p 627

Hill MO (1973) Diversity and evenness: a unifying notation and its consequences. Ecology 54:427–432. https://doi.org/10.2307/1934352

INEGI (2017) Guía para la interpretación de cartografía: Uso del suelo y la vegetación: escala 1:250:000. México

Islebe GA, Domínguez Vázquez G, Espadas Manrique C et al (2016) Cambio climático: contexto histórico, paleoecológico y paleoclimático. Tendencias actuales y perspectivas. In: Balvanera P, Arias E, Ricardo Rodríguez Estrella R et al (eds) Ecosistemas de México: una mirada a su conocimiento. Consejo Nacional de Ciencia y Tecnología y Universidad Nacional Autonoma de México, México, p 412

Kuhn U (1993) Chenopodiaceae. In: Kubitzki K, Rohwer JG, Bittrich V (eds) Flowering plants. Dicotyledons. Springer, Berlin, pp 253–281

Lara-González R, Sánchez-Velásquez LR, Corral-Aguirre J (2009) Regeneration of Abies religiosa in canopy gaps. Agrociencia 43:739–747

López-Portillo J, Excurra E (2012) Los manglares de México: Una revisión. Madera y Bosques Número esp:27–51

Lot A (2012) Las monocotiledóneas acuáticas y subacuáticas de méxico. Acta Bot Mex 100:135–148

Lozano-García S, Vazquez-Selem L (2005) A high-elevation Holocene pollen record from Iztaccihuatl volcano, central Mexico. Holocene 15:329–338. https://doi.org/10.1191/0959683605hl814rp

Lozano-García S, Caballero M, Ortega-Guerrero B et al (2010) Late Holocene palaeoecology of Lago Verde: evidence of human impact and climate change in the northern limit of the neotropics during the late formative and classic periods. Veg Hist Archaeobot 19:177–190

Lozano-García S, Torres-Rodríguez E, Ortega-Guerrero B et al (2013) Ecosystem responses to climate and disturbances in western central Mexico during the late Pleistocene and Holocene. Palaeogoegr Palaeoclimatol Paleoecol 370:184–195. https://doi.org/10.1016/j.palaeo.2012.12.006

Lozano-Garcia S, Roy P, Correa-Metrio A et al (2015) Registros paleoclimáticos. In: Gay y García CRA (ed) Reporte Mexicano de Cambio Climático. Universidad Nacional Autonoma de México y Programa de Investigación en Cambio Climático, México, DF, p 294

Luna-Vega I, Tejero-Díez JD, Contreras-Medina R et al (2012) Biogeographical analysis of two Polypodium species complexes (Polypodiaceae) in Mexico and Central America. Biol J Linn Soc 106:940–955. https://doi.org/10.1111/j.1095-8312.2012.01921.x

Magurran AE (1988) Ecological diversity and its measurement. Princeton University Press, Princeton

Magurran AE (2004) Measuring of biological diversity. Blackwell, Oxford

Martínez-Meyer E, Sosa-Escalante JE, Alvarez F (2014) El estudio de la biodiversidad en México: ¿una ruta con dirección? The study of the biodiversity in Mexico : a route with a course? Rev Mex Biodivers 85:S1–S9. https://doi.org/10.7550/rmb.43248

Martínez-Salas E, Ramos CH (2014) Biodiversidad de Pteridophyta en México. Rev Mex Biodivers 85:S110–S113. https://doi.org/10.7550/rmb.31827

Mastretta-Yanes A, Moreno-Letelier A, Piñero D et al (2015) Biodiversity in the Mexican highlands and the interaction of geology, geography and climate within the Trans-Mexican Volcanic Belt. J Biogeogr 42:1586–1600. https://doi.org/10.1111/jbi.12546

McGill BJ, Dornelas M, Gotelli NJ, Magurran AE (2015) Fifteen forms of biodiversity trend in the anthropocene. Trends Ecol Evol

Miranda F, Hernández X (1963) Los tipos de vegetación en México y su clasificación. Bol Soc Bot Méx 28:29–179

Mora-Olivo A, Villaseñor JL, Martínez M (2013) Las plantas vasculares acuáticas estrictas y su conservación en México. Acta Bot Mex:27. https://doi.org/10.21829/abm103.2013.50

Morrone JJ, Márquez J (2001) Halffter's Mexican Transition Zone, beetle generalized tracks, and geographical homology. J Biogeogr 28:635–650. https://doi.org/10.1046/j.1365-2699.2001.00571.x

Oksanen J (2017) Vegan: ecological diversity. R Packag. Version 2.4-4 12

Oksanen J, Guillaume Blanchet F, Friendly M et al (2018) Vegan: community ecology package. R package. Version 2.5-3

Pérez-Crespo VA, Rodríguez J, Arroyo-Cabrales J, Alva-Valdivia LM (2013) Variación ambiental durante el pleistoceno tardío y holoceno temprano en Guilá Naquitz (Oaxaca, México). Rev Bras Paleontol 16:487–494. https://doi.org/10.4072/rbp.2013.3.08

R Development Core Team R (2017) R: a language and environment for statistical computing

Ralska-Jasiewiczowa M, Latalowa M, Wasylikowa K, Tobolski E, Madeyska E, Wright H Jr, Turner C (2004) History of vegetation in Poland based on isopollen maps. W. Szafer Institute of Botany, Polish Academy of Science, Kraków

Rull V (2010) Ecology and palaeoecology: two approaches, one objective. Open Ecol J 3:1–5. https://doi.org/10.2174/1874213001003020001

Rull V (2013) Some problems in the study of the origin of neotropical biodiversity using palaeoecological and molecular phylogenetic evidence. Syst Biodivers 11:415–423. https://doi.org/10.1080/14772000.2013.865682

Rull V (2014) Ecological palaeoecology: a missing link between ecology and evolution. Callectanea Bot 33:65–73

Rull V, Vegas-Vilarrúbia T (2011) What is long-term in ecology? Trends Ecol Evol 26:3–4

Rzedowski J (1981) Vegetacion de Mexico. Limusa, Mexico City

Rzedowski J (2006) Vegetación de México, Edición di. Comisión Nacional para el Conocimiento y Uso de la Biodiversidad

Sánchez-Ken JG (2011) Flora del Valle de Tehuacán-Cuicatlán. Fascículo 81. Poaceae, Primera. Universidad Nacional Autonoma de México, DF, México

Sandoval-Ortega H, Siqueiros-Delgado E, Sosa-Ramírez J, Cerros Tlatilpa R (2017) Riqueza y Distribución de la familia Amaranthaceae en el Estado de Aguascalientes, México. Bot Sci 95:203–220. https://doi.org/10.17129/botsci.909

Savage JM (1982) The enigma of the Central American herpetofauna: dispersals or vicariance? Ann Missouri Bot Gard 69:464–547. https://doi.org/10.2307/2399082

Torrescano-Valle N, Islebe GA (2015) Holocene paleoecology, climate history and human influence in the southwestern Yucatan Peninsula. Rev Palaeobot Palynol 217:1–8. https://doi.org/10.1016/j.revpalbo.2015.03.003

Valdez Reyna J (1995) Clasificación de los géneros de Gramíneas (Poaceae) mexicanas. Acta Bot Mex 33:37–50

Vázquez-Pérez N, Mendoza-Ruiz A, Pérez-García B (2012) Morfogénesis de la fase sexual de siete taxa epífitos de Polypodium (Polypodiaceae) de México. Acta Bot Mex 98:5–21
Vela-Pelaez AA, Torrescano-Valle N, Islebe GA et al (2018) Holocene precipitation changes in the Maya forest, Yucatán peninsula, Mexico. Palaeogeogr Palaeoclimatol Palaeoecol 505:42–52. https://doi.org/10.1016/j.palaeo.2018.05.024
Villaseñor JL (2003) Diversidad y distribución de las Magnoliophyta de Mexico. Interciencia 28:160–167+184. https://doi.org/10.4414/smw.2017.14531
Villaseñor JL (2004) Los géneros de plantas vasculares de la flora de México. Bol Soc Bot Méx 75:105–135. https://doi.org/10.17129/botsci.1694
Villaseñor JL (2010) El bosque húmedo de montaña en México y sus plantas vasculares: catálogo florístico-taxonómico. Universidad Nacional Autonoma de México, México, DF
Villaseñor JL (2016) Catálogo de las plantas vasculares nativas de México. Rev Mex Biodivers 87:559–902. https://doi.org/10.1016/j.rmb.2016.06.017
Villaseñor JL (2018) Diversidad y distribución de la familia Asteraceae en México. Bot Sci 96:332–358
Villaseñor JL, Ortiz E (2014) Biodiversidad de las plantas con flores (División Magnoliophyta) en México. Rev Mex Biodivers 85:134–142. https://doi.org/10.7550/rmb.31987
Weng C, Hooghiemstra H, Duivenvoorden JF (2006) Challenges in estimating past plant diversity from fossil pollen data: statistical assessment, problems, and possible solutions. Divers Distrib 12:310–318. https://doi.org/10.1111/j.1366-9516.2006.00230.x
Whitmore TJ, Brenner M, Curtis JH et al (1996) Holocene climatic and human influences on lakes of the Yucatan Peninsula, Mexico: an interdisciplinary, palaeolimnological approach. Holocene 6:273–287. https://doi.org/10.1177/095968369600600303
Zar J (1999) Biostatistical analysis, 4th edn. Prentice Hall, Upper Saddle River

Chapter 18
Bioregions of Eastern Brazil, Based on Vascular Plant Occurrence Data

Marcelo Reginato and Fabián A. Michelangeli

Abstract The geographical division of the Earth into meaningful biodiversity units (e.g., biomes, areas of endemism, ecoregions or bioregions) is a necessary step for the study of biodiversity and its conservation. Eastern Brazil harbors a significant proportion of the Earth's terrestrial biodiversity in a geographically complex area. However, the delimitation of biogeographical areas in eastern Brazil has relied on the simultaneous use of biotic (animal and plant distributions and vegetation physiognomy) and abiotic factors, often without an explicit methodology. Here we take advantage of the availability of large numbers of vascular plant specimens and their digitized data, the existence of a well curated taxonomy for the plants that occur in the area, and the emergence of new biogeographic tools in order to identify bioregions (geographic areas that contain similar taxa) of eastern Brazil. To provide a classification scheme suitable to studies of lineages that differ in dispersion, species richness, and endemism, we provide three levels of grouping. The dataset analyzed here had a comparable number of species across eastern Brazil relative to the recent taxonomic synthesis of the Brazilian flora. Maps of richness and endemism are provided for the region, and confirm regions of eastern Brazil recognized for both highly diversity and endemism across both coastal (Serra do Mar and Mantiqueira) and inland mountain ranges (Campos Rupestres), as well as in southern Bahia. The first network analysis divided eastern Brazil into 10 bioregions, which were clustered in five super-bioregions and divided in 23 sub-bioregions in the two additional network analyses. The super-bioregions recovered correspond to the Southern Atlantic Forest/Paraná Forests, the Northern Atlantic Forest, the Espinhaço/Mantiqueira, the Cerrado, and the Caatinga/Diamantina. To some extent, these areas present some congruence with domains, where a major incongruence is distinctiveness of the Espinhaço/Mantiqueira super-bioregion, while the Caatinga/Diamantina super-bioregion presented the highest congruence. Comparisons of species richness, endemism and overlapping of the bioregions with three other

M. Reginato (✉)
Departamento de Botânica, Universidade Federal do Rio Grande do Sul, Porto Alegre, Rio Grande do Sul, Brazil

F. A. Michelangeli
Institute of Systematic Botany, The New York Botanical Garden, Bronx, NY, USA

V. Rull, A. C. Carnaval (eds.), *Neotropical Diversification: Patterns and Processes*, Fascinating Life Sciences, https://doi.org/10.1007/978-3-030-31167-4_18

classifications (domains, ecoregions and biogeographical provinces) are presented. The shapefiles of the recovered bioregions are available for public use.

Keywords Angiosperms · Araucaria · Atlantic forest · Biogeography · Bioregion · Caatinga · Campo rupestre · Cerrado · Mata Atlântica

1 Introduction

The geographical division of the Earth's surface into meaningful biodiversity units (e.g., biomes, areas of endemism, ecoregions, or bioregions) is a necessary step for the study of biogeography, macroevolution, macroecology, and community ecology, as well as for policy-making and conservation (Gaston 2000; Morrone 2009). The fact that life on Earth is not distributed randomly, and that there are recurring patterns of plant and animal distribution, has been the subject of many studies since the early European naturalists and taxonomists started to catalog the South American flora and fauna (von Humboldt and Bonpland 1805; von Humboldt 1808; de Candolle 1820; Sclater 1858; Wallace 1876; Buffon 1761).

However, how biogeographical units are defined and delimited can vary considerably, due to different data, methods, and objectives. Early attempts relied on a combination of geography, climate and qualitative assessments of the distribution of selected organisms, but how each of these factors was specifically used to delineate areas was rarely explicit, and not reproducible (Kreft and Jetz 2010). More recently, the identification of biogeographical areas has also relied on data that do not directly involve species distributions, such as climatic records (Holdridge 1947; Woodward 1987), vegetation structure, and physiognomy (Takhtajan 1986), sometimes implicitly assuming that species composition is directly determined only by one or both of these factors. Cladistic biogeography developed a more explicit methodology, which is focused especially on the definition of areas of endemism (Platnick and Nelson 1978; Nelson and Platnick 1981; Crisci et al. 1991; Morrone 1994). This led to the development of several methods that identify biogeographical areas, and subsequent variations that use spatial organismal data (Morrone 1994, 2014b; Linder 2001; Porzecanski and Cracraft 2005; Cracraft 1991; Szumik et al. 2012; Szumik and Goloboff 2004, 2015). Even though these approaches are more reproducible than those of the nineteenth and early twentieth centuries, there is still confusion on which methods perform better under different conditions, or with different taxa (see Ferrari 2017).

Proper delimitation of biogeographic units is relevant to the study of the ecology and evolution of the Neotropical biota. For example, they directly inform the estimation of ancestral distributional ranges, as well as the comparison of diversification rates of a given lineage across discrete biogeographical units, which became frequent exercises in the past decade, particularly within the fields of systematics, macroevolution and biogeography (Goldberg et al. 2011; Ronquist and Sanmartin 2011; Matzke 2014). In such analyses, all species of a given clade are coded as present in or absent from discrete units in geographic space, and the resulting matrix, along with a phylogeny, are used as input. In discrete-model analysis the processes correspond to events that change the biogeographic state, that is, the distribution of organisms (Ronquist and Sanmartin 2011). The influence of area delimitation in such analyses, however, is still neglected in the literature. Suitable areas for discrete-model analysis are usually established on the basis of geological features, habitat data, or the question under study (Ronquist and Sanmartin 2011).

For the Neotropical region, two widely used classifications (or user-modified versions of them) are the ecoregions of Olson et al. (2001) and the biogeographical provinces of Morrone (2014a). Nonetheless, depending on the scale of the lineage under study, these classifications may not be informative for historical analyses. In some cases, for instance, the entire distribution of a lineage is included in a single biogeographical province or ecoregion. In addition, because most biogeographical studies of eastern Brazilian domains have targeted the identification of areas of endemism (Costa et al. 2000; Silva et al. 2004; DaSilva et al. 2015), their outcomes cannot be directly applied in discrete-model analyses given that they tend to identify only a few hotspot areas and leave out several uncovered regions.

Clustering analyses using similarity indices have been commonly used to identify biogeographical regions (Gonzalez-Orozco et al. 2014; Bloomfield et al. 2018), and, more recently, network methods have been applied to detect bioregions as an alternative to clustering methods (Vilhena and Antonelli 2015). Network methods adapted to biogeography have been implemented with user-friendly tools (Edler et al. 2016; Topel et al. 2017), and have rapidly been applied to a variety of regions, specific environments, or taxa (e. g. Costello et al. 2017; Droissart et al. 2018; Hazzi et al. 2018; Hipp et al. 2018; Perera et al. 2018; Rojas et al. 2017; Segatto et al. 2017; Spalink et al. 2018). Overall, these network approaches result in less scale-dependent outcomes, allowing endemic taxa to contribute more strongly to the identification of bioregions (Vilhena and Antonelli 2015). These methods also appear to be less affected by sampling biases, and might provide new insights when applied to the identification and delimitation of bioregions (Bloomfield et al. 2018; Vilhena and Antonelli 2015).

Here, we use a network approach to propose a new bioregionalization of a particularly biodiverse Neotropical region: Eastern Brazil. It includes three distinct domains (or biomes *sensu* IBGE 2004): the Atlantic Forest, the Cerrado, and the Caatinga, which all have been subjected to subdivisions in some biogeographical studies (reviewed in Fiaschi and Pirani 2009). The first two of these domains are among the top ten biodiversity hotspots in the world, each with high levels of species richness, endemism, and growing pressure from humans (Myers et al. 2000). The Atlantic Forest runs along the Brazilian coast, ranging over 25 degrees of latitude, and covers over 2000 m of elevation. It is characterized by rain forests, subtropical forests and seasonal forests, as well as several smaller vegetation types (Morellato and Haddad 2000; Stehmann et al. 2009). It harbors over 15,000 species of vascular plants, more than 45% of them endemic (Stehmann et al. 2009). In contrast, the Caatinga is characterized by a hot and xeric environment with shrubby, deciduous, and thorny vegetation (Prado 2003; Sampaio 1995; Da Costa et al. 2007). Over 4000 species of vascular plants have been reported for this region (Prado et al. 2015). The

Cerrado is characterized by its seasonal climate, and the woody savanna environments with small trees, shrubs and dominant C4 grasses (Oliveira-Filho and Ratter 2002). Over 12,000 species of vascular plants, ca. 35% endemic, have been reported for the Cerrado (de Mendonça et al. 2008). Even though each of these domains is well characterized, reality on the ground is more complex due to a variety of reasons. First, many areas are so degraded, especially in the Atlantic Forest and the Cerrado, that it is hard to infer their original vegetation. Secondly, each environment can show a high degree of heterogeneity in soils, elevation, and climate, all of which may vary across longitude. Finally, areas of contact of between domains are not clear-cut due to the presence of rivers or ridges where the different physiognomies interdigitate, creating a variety of transitional zones.

In the last 15 years, our understanding of how organisms are distributed over the landscape has seen much improvement. Biome-focused studies in eastern Brazil have provided baseline data on plant species richness and taxonomy (Stehmann et al. 2009; de Mendonça et al. 2008). This allowed for the Catalogo de Plantas and Fungos do Brasil to be completed (Forzza et al. 2010), and this in turn has spawned updates to our understanding of the Brazilian Flora (BFG 2015, 2018). As a result of multiple efforts in Brazil and elsewhere, most botanical collections are now digitized, with their data freely available through individual collection websites and through data aggregators such as the Global Biodiversity Information Facility (https://www.gbif.org/) and the SpeciesLink Network (http://www.splink.org.br/).

In this chapter, we take advantage of the availability of large datasets on vascular plant diversity, the existence of a well-curated taxonomy for the plants that occur in Eastern Brazil (e.g., Flora do Brasil 2020, under construction), and the emergence of new biogeographic tools, to identify bioregions across eastern Brazil. For the purpose of this analysis, we define bioregions as geographic areas that contain similar taxa. Although eastern Brazil is a reasonably well-sampled relative to the Amazon, collection gaps still exist in this region; if a species has not yet been found in a given area, it is difficult to know whether it is truly absent there, or if the perceived absence is in fact an artifact of incomplete sampling. This is further complicated by the amount of habitat degradation and transformation in Eastern Brazil, which often times prevents verification via field work. To overcome this challenge, we derived distributional ranges based on climatic niche models, and used the latter to conduct our analyses. To provide a classification scheme suitable to studies of lineages that differ in dispersal ability, species richness, and endemism, we propose three levels of grouping. Comparisons of species richness, endemism and overlapping of the bioregions with other three classifications available in the literature (of domains, ecoregions and biogeographical provinces) are also presented.

2 Methods

Study Area Eastern Brazil was defined here to include the domains Cerrado, Mata Atlântica and Caatinga, for which the boundaries were extracted from the Terrestrial Ecoregions of the World (Olson et al. 2001).

Distribution Data Records of all vascular plants from Brazil and neighboring countries (Argentina, Bolivia, Paraguay and Uruguay) were downloaded from the data portals GBIF (GBIF.org; GBIF Occurrence Download https://doi.org/10.15468/dl.yt1wzc) and speciesLink (spLink.cria.org). The two datasets were merged and the records were filtered in several ways in order to flag both coordinate misplacements and putative taxonomic misidentifications. All filtering steps were performed in R (R_Core_Team 2018) with custom scripts (available under request from the first author), using several functions of the packages maptools (Bivand and Lewin-Koh 2013), raster (Hijmans 2016), rgeos (Bivand and Rundel 2013), sp. (Pebesma and Bivand 2005) and flora (Carvalho 2017). Briefly, coordinate filtering consisted of removing records that were considered suspicious or faulty, such as those with invalid coordinates, located in non-terrestrial areas, older than 1960 (where legacy georeferencing may lead to more problems), with less than four decimals in the georefenced data, with coordinates intersecting countries centroids, with inconsistency between the coordinates and reported country, and with inconsistency between the coordinates and the recorded Brazilian state. The taxonomy of each record was updated following the database of Flora do Brasil 2020 (under construction) through the R package flora (Carvalho 2017). In order to filter out putative misidentifications, the records of any given species were removed when their coordinates intersected a Brazilian state or a domain not reported in the Flora do Brasil 2020 database. Species reported as non-native were also removed from the data set (Flora do Brasil 2020, under construction). The cleaned records were then intersected with the study area region (eastern Brazil mask), and the species were classified as endemic or widespread (i.e., also presenting records outside eastern Brazil). Endemics were defined by having all points falling within the polygon of eastern Brazil.

Geographical Ranges The geographical ranges of all species were estimated with correlative climatic niche models. Models were built and evaluated using the filtered records with Maxent 3.3.3 (Phillips and Dudík 2008) through the R package dismo (Hijmans et al. 2017). Model predictors included 19 bioclimatic variables from the WorldClim 2 data set (30″ spatial resolution; Fick and Hijmans 2017) that are interpolated climate surfaces describing precipitation and temperature variables. In order to reduce the effect of biased species occurrence collections on spatial model outcomes, the data set was spatially thinned with the R package spThin 0.1.0 (Aiello-Lammens et al. 2014). Only points with a minimum distance of 10 km were kept for downstream analyses. For each species, a mask was generated with a circular buffer of 1000 km of diameter around its known distribution. Additionally, in order to exclude the immediate area around the known localities from the background, a buffer of 100 km in diameter was generated for each known point and subtracted from the main mask. Climatic layers were cropped and masked using the individual species masks, and 10,000 random pseudo-absence points were sampled in the masked area for modeling. Default parameters were applied, and models were trained based on the presence-only records (75% training and 25% testing). Models were evaluated with area under the receiver operating characteristic curve (AUC), a widely used statistic for assessing the discriminatory capacity of species distribution models. Models with AUC greater than 0.85 were projected in geographic space and their thresholds (kappa, sum of the sensitivity and specificity, equal sensitivity and specificity, and sensitivity) estimated. The threshold is needed to transform the continuous results into a binary product. Often require binary models of distributions, and a threshold is needed to transform the continuous results into a binary product.

The projected model was binarized using the mean of all thresholds, and transformed into polygons. The resulting polygons of each species were then intersected with its known occurrences; polygons that did not intersect any known point of occurrence were excluded from its inferred distribution. For species with models with AUC lower than 0.85, a different procedure was applied. We created a buffer of 30 km surrounding the known point localities, cropped the buffered range using the occurrences' elevational range (plus a buffer of 200 m), and intersected the cropped range to known localities, again removing non-intersecting polygons. Both procedures were performed in R with several functions from the packages maptools (Bivand and Lewin-Koh 2013), raster (Hijmans 2016), rgeos (Bivand and Rundel 2013) and sp. (Pebesma and Bivand 2005). Species distributions were derived from the resulting polygons and used to build richness (via stacked polygons) and weighted endemism maps. In the latter, species richness is weighted by the inverse of the range size of each species: this allows the identification of areas where highly endemic species are concentrated. These maps were generated using the R package raster (Hijmans 2016) and the R function weighted.endemism (Guerin et al. 2015).

Bioregions Eastern Brazilian bioregions were identified with a network approach (Vilhena and Antonelli 2015) implemented in the Infomap Bioregions web application (Edler et al. 2016). This method uses the species distributions mapped over grid cells, generates a bipartite network between species and the grid cells, and produces bioregions by clustering the network with the Infomap algorithm (Rosvall and Bergstrom 2008). This approach has been suggested as more efficient than turnover measures and cluster analyses frequently employed in biogeography (Bloomfield et al. 2018). Network analyses were performed in the application with grid cell size ranging between 0.125° and 0.25°, with 10 trials, and the remaining parameters were left as the default settings (maximum cell capacity = 100; minimum cell capacity = 10; cluster cost = 1.0).

In order to provide both coarser and finer scale classifications, the clustering resulting from the first analysis (bioregions) was used on two additional network analyses to identify super and sub-bioregions. For the coarser analysis (i.e., to identify super-bioregions), the bioregions were treated as grid cells and a new presence/absence data set with all species was created and analyzed. In contrast, to further divide the recovered bioregions into sub-bioregions, new presence/absence data sets were scored for each recovered bioregion, based only on species that are endemic to Eastern Brazil, and new network analyses were performed. The settings of all secondary network analyses were similar to those of the primary analysis. Our resulting super-bioregions, bioregions and sub-bioregions were compared with the eastern Brazilian domains (IBGE 2004), biogeographical provinces (Löwenberg-Neto 2014; Morrone 2014a), and ecoregions (Olson et al. 2001), respectively, given that these are the most widely used classifications, and that they have a comparable number of areas to those that we derive. The shapefile of the recovered bioregions is available for public use (https://github.com/mreginato/EBR_Bioregions) and could be modified as users see fit.

3 Results

Our pre-filtering data set included a total of 1,930,852 records. Coordinate misplacement filtering removed 566,254 records (ca. 29%), and putative taxonomic misidentifications filters based on "Flora do Brasil" crosschecking removed an additional of 237,099 records (ca. 12%). After excluding records outside eastern Brazil (Fig. 18.1a), the final clean dataset included 875,770 records (45%), from which 232,955 (ca. 27%) belong to species classified as eastern Brazilian endemics (Fig. 18.1b–c). The final data set included 272 families of vascular plants, 3816 genera and 20,642 species, where 12,460 species (ca. 60%) were classified as eastern Brazilian endemics. Following the domains classification (IBGE 2004), our dataset included 14,847 species in the Atlantic Forest (5227 endemics, ca. 35%), 11,075 species in the Cerrado (2130 endemics, ca. 19%) and 5192 species in the Caatinga (922 endemics, ca. 18%).

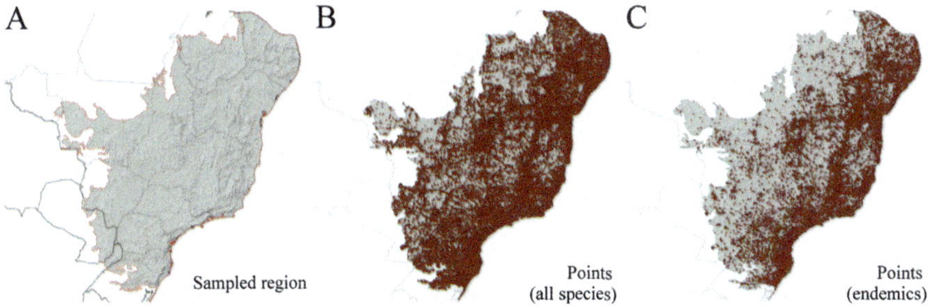

Fig. 18.1 (**a**) Study area, including the eastern Brazilian domains of Atlantic Forest, Cerrado and Caatinga. (**b**) Filtered dataset (occurrence points) including all species (875,770 records of 20,754 species). (**c**) Filtered dataset (occurrence points) of eastern Brazilian endemics (232,955 records of 12,460 species)

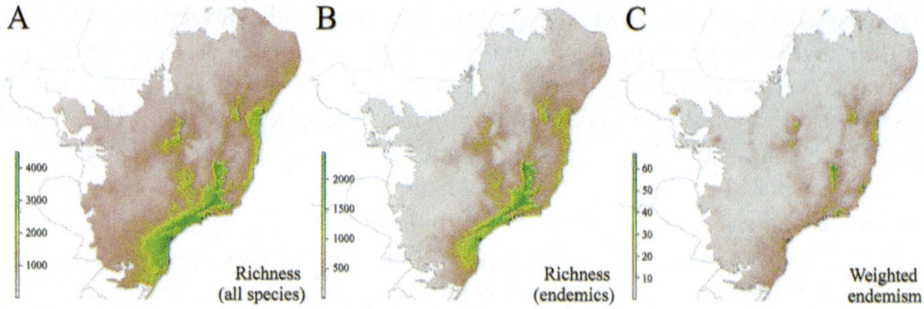

Fig. 18.2 Richness and endemism heat maps based on stacked ranges distributions. (**a**) Richness of all species. (**b**) Richness of eastern Brazilian endemics. (**c**) Weighted endemism of all species

Distribution models presented a median AUC of 0.96 (s.d. of 0.19), from which ca. 24% presented AUC lower than 0.85 and were not used to derive species ranges. Richness heat maps of stacked ranges including all species (Fig. 18.2a) and only eastern Brazilian endemics (Fig. 18.2b) present a similar richness pattern where areas of "Serra do Mar", "Serra da Mantiqueira" and "Serra do Espinhaço" higher species accumulation, whereas intermediate richness levels scattered throughout coastal regions and other areas of the Brazilian shield like "Chapada Diamantina" (around latitudes 41° W, 13° S), "Chapada dos Veadeiros" (47° W, 14° S), and "Serra da Canastra" (26° W, 20° S). Weighted endemism generally followed the richness heat maps (Fig. 18.2c). However, major differences include the relatively lower endemism detected in the southern portion of "Serra do Mar", and the higher endemism observed in the mountains of the state of Espírito Santo. Additional highly endemic areas are found in the "Serra do Espinhaço" (43° W, 19° S), "Serra dos Órgãos" (43° W, 22° S), "Chapada dos Veadeiros" (47° W, 14° S), "Chapada Diamantina" (41° W, 13° S), and southern Bahia.

The first network analysis divided eastern Brazil into 10 bioregions, which were subsequently clustered into five super-bioregions and divided in 23 sub-bioregions (Fig. 18.3). Across the five super-bioregions, ca. 49% of the species are endemic to a single area, 22% are endemic of two areas, 15% of three areas, 8% of four areas, and 6% are found in all areas. The Espinhaço/Mantiqueira super-bioregion presents the highest richness (10,369 species, 21% endemics), followed by Southern Atlantic Forest/Paraná Forest (9331 species, 14% endemics), Northern Atlantic Forest (8711 species, 21% endemics), Cerrado (7628 species, 15% endemics) and Caatinga (5476 species, 16% endemics). A graphical comparison of richness and endemism across the recovered bioregions in this study against the previously defined "domains" (IBGE 2004), biogeographical "provinces" (Morrone 2014a) and "ecoregions" (Olson et al. 2001) are presented in Fig. 18.4. To describe the components of each super-bioregion uncovered by our analysis (below), we use the biogeographic provinces named by Morrone (2014a), and the ecoregions described by Olson et al. (2001). All overlap proportions between the super-bioregions and biogeographical provinces (sensu Morrone 2014a) are provided in Table 18.1, while the overlap between the bioregions with the ecoregions (sensu Olson et al. 2001) is

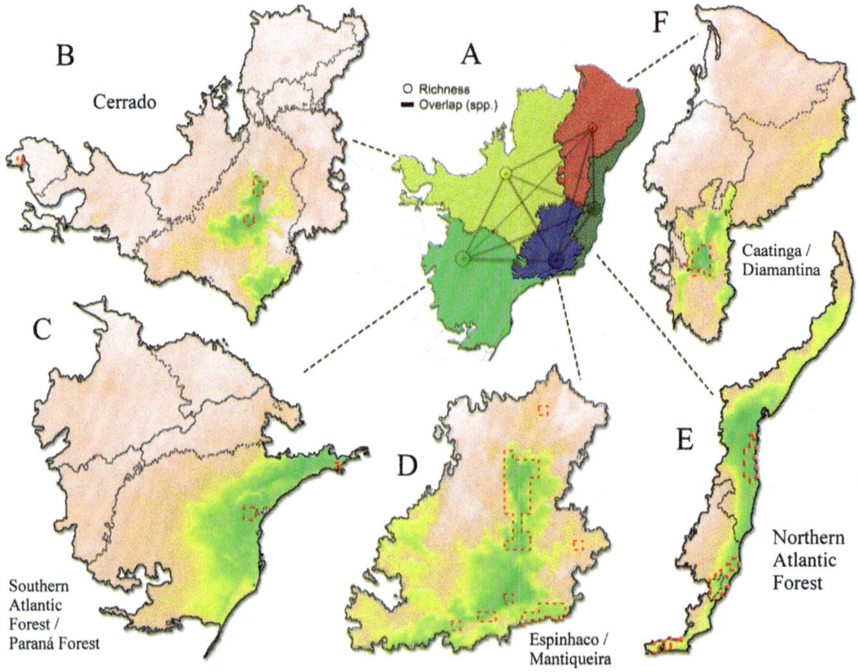

Fig. 18.3 (**a**) Eastern Brazilian super-bioregions. (**b–f**) Detailed bioregions and sub-bioregions within each super-bioregion (black lines). Richness, including all species (endemics and non-endemics), is depicted in the colored gradient following Fig. 18.2a. Highly endemic regions are highlighted in red (dashed lines)

presented in Table 18.2. A characterization of the bioregions, using as base the well-known ecoregions and provinces, are presented in the following paragraphs.

Southern Atlantic Forest/Paraná Forest This super-bioregion clustered two bioregions (Fig. 18.3c; areas 8 and 9 in Fig. 18.4c) and includes regions of southern Brazil. Its western portion includes areas of the Paraná River basin and its northern coastal limit is in Angra dos Reis (southern extreme of the state of Rio de Janeiro, at a latitude of 23° S). The area is composed by the Araucaria province as described by Morrone (2014a; area 1 in Fig. 18.4d), a subtropical portion of the Atlantic province (area 3 in Fig. 18.4d), a southwestern portion of the Paraná Forest province (area 4 in Fig. 18.4d) and a small fraction of the southwestern Cerrado province (area 5 in Fig. 18.4d). The southern Atlantic Forest bioregion (area 9 in Fig. 18.4c) is predominantly composed by areas of the Alto Paraná (area 1 in Fig. 18.4f), Araucaria (area 2 in Fig. 18.4f) and Serra do Mar (area 7 in Fig. 18.4f) ecoregions, as defined by Olson et al. (2001). The Paraná Forest bioregion (area 8 in Fig. 18.4c) also includes areas of the Alto Paraná ecoregion plus areas of southwestern Cerrado ecoregion (area 9 in Fig. 18.4f). The southern Atlantic Forest bioregion was not further divided (area 5 in Fig. 18.4e), while the Paraná Forest bioregion was subdivided into four areas (areas 1, 2, 3, and 4 in Fig. 18.4e).

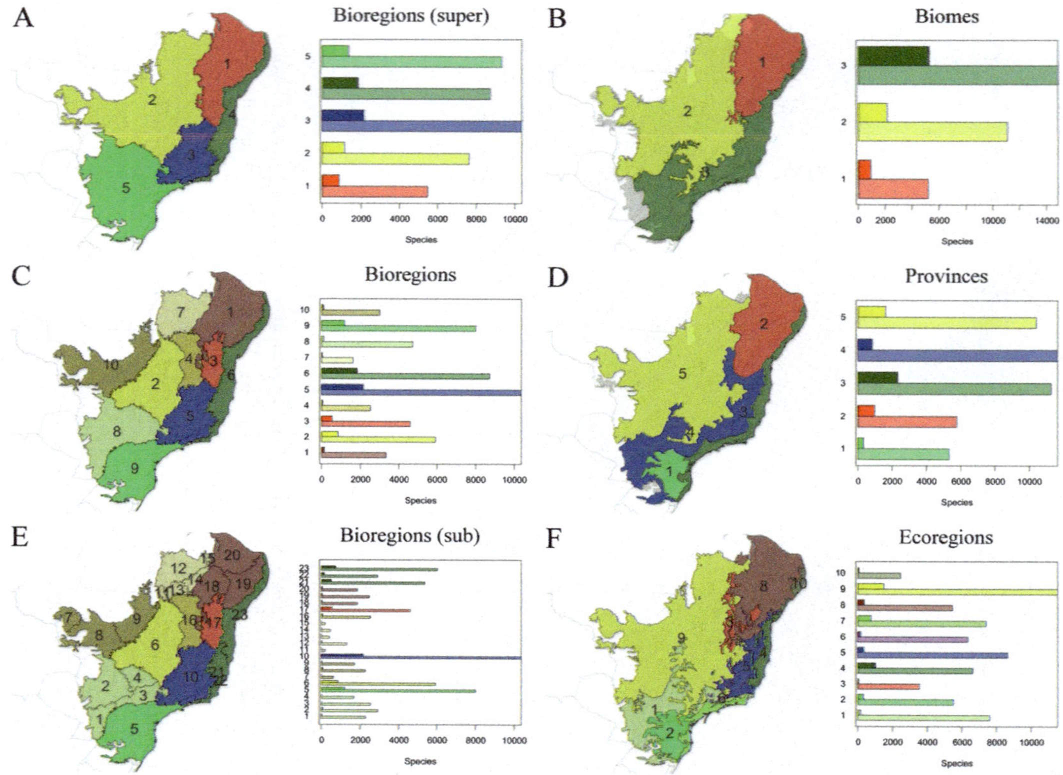

Fig. 18.4 Richness (lower bar) and number of endemic species (upper bar) across bioregions and other classifications. (**a**) Super-bioregions. (**b**) Domains (biomes *sensu* IBGE 2004). (**c**) Bioregions. (**d**) Biogeographical provinces (Morrone 2014a, b). (**e**) Sub-bioregions. (**f**) Ecoregions (Olson et al. 2001)

Table 18.1 A comparison between the super-bioregions identified by this study with the biogeographical provinces described by Morrone (2014a)

		Biogeographical provinces (%)				
		Araucaria Forest	Caatinga	Atlantic	Parana Forest	Cerrado
Super-bioregions	Caatinga/Diamantina	0	92	0	5	3
	Cerrado	0	1	0	7	92
	Espinhaço/Mantiqueira	0	0	11	50	39
	Northern Atlantic Forest	0	20	58	22	0
	Southern Atlantic Forest/ Paraná Forest	20	0	10	42	28

Numbers indicate what percent of the super-bioregions overlaps with each biogeographical provinces

Northern Atlantic Forest This super-bioregion includes only one bioregion (Fig. 18.3e; area 6 in Fig. 18.4c), corresponding to the northern portion of the Brazilian coast (from Rio de Janeiro to Rio Grande do Norte states). The area is composed by the tropical portion of the Atlantic province as per Morrone (2014a; area 3 in Fig. 18.4d), the northeastern portion of the Paraná Forest province (area 4 in Fig. 18.4d), and eastern areas of the Caatinga province (area 2 in Fig. 18.4d). If compared to the ecoregions, it predominantly includes areas from Bahia Coastal and Interior forests (areas 4 and 5 in Fig. 18.4f), Caatinga (area 8 in Fig. 18.4f) and Pernambuco (area 10 in Fig. 18.4f) ecoregions of Olson et al. (2001). The northern Atlantic Forest bioregion was further divided into three sub-areas (areas 21, 22 and 23 in Fig. 18.4e). The first sub-area includes its southern portion, up north to the Rio Doce region, the second area by the northern region, and the third is composed by the mountains of Espírito Santo state, in the region of Santa Lúcia municipality.

Espinhaço/Mantiqueira This super-bioregion includes only one bioregion (Fig. 18.3d; area 5 in Fig. 18.4c), corresponding to the interior areas of southeastern Brazil, including "Serra do Espinhaço", "Quadrilátero Ferrífero", "Serra da Canastra" and "Serra da Mantiqueira". The area is composed by southeastern portions of the Paraná Forest (area 4 in Fig. 18.4d) and Cerrado provinces of Morrone (2014a; area 5 in Fig. 18.4d), and a small fraction of the Atlantic province (area 3 in Fig. 18.4d). If compared to Olson et al.'s (2001) ecoregions, it predominantly includes areas from Cerrado (area 9 in Fig. 18.4f), Bahia Interior forests (area 5 in Fig. 18.4f), and Alto Paraná (area 1 in Fig. 18.4f) ecoregions, but also includes portions of the Campos Rupestres and Serra do Mar ecoregions (areas 6 and 7 in Fig. 18.4f). The Espinhaço/Mantiqueira bioregion was not further divided into sub-areas (area 10 in Fig. 18.4e).

Cerrado This super-bioregion clustered four bioregions (Fig. 18.3b; areas 2, 4, 7 and 10 in Fig. 18.4c), including most of the Cerrado domain in central Brazil. The area is predominantly composed by the Cerrado province of Morrone (2014a; area 5 in Fig. 18.4d), but also includes a small fraction of the Paraná Forest province (area 4 in Fig. 18.4d). The Central (area 2 in Fig. 18.4c), Northern (area 7 in Fig. 18.4c),

Table 18.2 A comparison between the eastern Brazilian bioregions (this study) with the ecoregions identified by Olson et al. (2001)

		Ecoregions (%)									
		Alto Paraná	Araucaria	Atlantic dry	Bahia coastal	Bahia interior	Campos Rupestres	Serra do Mar	Caatinga	Cerrado	Pernambuco
Bioregions	Caatinga	0	0	3	0	0	0	0	94	2	1
	Diamantina	0	0	16	1	16	5	0	60	3	0
	Central Cerrado	7	0	0	0	0	0	0	0	93	0
	Eastern Cerrado	0	0	31	0	0	0	0	9	60	0
	Northern Cerrado	0	0	5	0	0	0	0	7	88	0
	Western Cerrado	0	0	0	0	0	0	0	0	100	0
	Espinhaço/ Mantiqueira	13	0	1	1	30	5	6	0	44	0
	Northern Atlantic Forest	1	0	1	34	29	0	4	19	0	12
	Paraná Forest	45	0	0	0	0	0	0	0	55	0
	Southern Atlantic Forest	32	48	0	0	0	0	16	0	4	0

Numbers indicate what percent of the bioregions overlaps with each ecoregion and Western Cerrado (area 10 in Fig. 18.4c) bioregions are predominantly composed by areas of the Cerrado ecoregion as per Olson et al. (2001; area 9 in Fig. 18.4f), while the Eastern Cerrado (area 4 in Fig. 18.4c) also includes areas of the Atlantic Dry forests ecoregion (area 3 in Fig. 18.4f). The Central and Eastern Cerrado bioregions were not further divided (areas 6 and 16 in Fig. 18.4e), while the Western Cerrado was subdivided into three areas (areas 7, 8 and 9 in Fig. 18.4e) and the Northern Cerrado into five areas (areas 11, 12, 13, 14 and 15 in Fig. 18.4e).

Caatinga/Diamantina This super-bioregion clustered two bioregions (Fig. 18.3f; areas 1 and 3 in Fig. 18.4c), including most of the Caatinga domain in northeastern Brazil. The area is predominantly composed by the Caatinga province as defined by Morrone (2014a; area 2 in Fig. 18.4d), and small fractions of the Cerrado and Paraná provinces (areas 5 and 2 in Fig. 18.4d). The Caatinga bioregion (area 1 in Fig. 18.4c) is predominantly composed by the Caatinga ecoregion as per Olson et al. (2001; area 8 in Fig. 18.4f), while the Diamantina bioregion (area 3 in Fig. 18.4c) also includes areas of the Caatinga ecoregion plus areas from the Atlantic Dry Forests, Bahia Interior Forests, Campos Rupestres and Cerrado ecoregions (areas 3, 4, 6 and 9 in Fig. 18.4f). The Diamantina bioregion was not further divided (area 5 in Fig. 18.4e), while the Caatinga bioregion was subdivided into three areas (areas 18, 19 and 20 in Fig. 18.4e).

4 Discussion

The extensive specimen-based dataset analyzed in this study included similar numbers of species richness across the domains to the synthesis of Brazilian angiosperms by the BFG (2015) and for ferns and lycophytes by Prado et al. (2015). Here, we found 14,847 species in the Atlantic Forest [in contrast to the 15,884 species listed by BFG (2015) and Prado et al. (2015)], 11,075 versus 12,364 species in the Cerrado, and 5192 versus 4416 species in the Caatinga. The lower numbers included in our analyses may reflect the fact that rare species with few specimens were frequently filtered out during the data set cleaning procedures. Interestingly, we observed more species in the Caatinga than reported by those taxonomic synthesis (BFG 2015; Prado et al. 2015). This difference might be due to the continuous update efforts in the Flora do Brasil project since 2015, to the presence of under-reported taxa for this domain (e.g., taxa occurring within the domain boundaries, but on a different vegetation type), or occurrence errors in our dataset. Our estimates of endemism are lower than the values reported in BFG (2015), except for the Caatinga. Our analysis found *ca.* 35% of the species are endemic to the Atlantic Forest domain versus 49% of endemics in BFG (2015; also Prado et al. 2015). In the Cerrado, we found 15% endemism (vs. 35% in BFG 2015), and in the Caatinga *ca.* 18% (vs. 20%). These differences may have been caused by our use of a strict classification of endemics. Thus, minor erroneous points, that were still not filtered out, may have contributed to the lower numbers. Additionally, most species that were removed from the data set (with few points), or not represented at all in online collections, are likely to be narrow endemics.

Here we created maps of both richness and weighted endemism for eastern Brazilian vascular plants. We are not aware of similar maps and comparable sampling. Nonetheless, the regions recovered as highly diverse in our analysis are well-known in the literature for their richness and endemism (Martini et al. 2007; Fiaschi and Pirani 2009; Murray-smith et al. 2009; Silveira et al. 2016). Overall, locations with high richness were observed in the coastal mountain ranges of Serra do Mar and Serra da Mantiqueira, and in the inland mountain ranges of the "Campos Rupestres" region (Serra do Espinhaço, Chapada Diamantina, Chapada dos Veadeiros and Serra da Canastra). Mountain ranges are widely recognized by their exceptionally high biodiversity worldwide (Quintero and Jetz 2018; Spehn et al. 2011). In such areas, rapid turnover in communities and environmental conditions are observed across elevations, and different geomorphological and climatic attributes seem to be pivotal determinants of the remarkable richness gradients observed today (Quintero and Jetz 2018). Isolation and micro habitats may also be responsible for high species richness in mountain areas, as these are also the areas with the highest endemism and weighted endemism. One exception to the "mountain range/highly diverse" pattern is the southern Bahia region, which has some topography complexity yet does not have the elevational differences found in the other mentioned regions. In southern Bahia, it has been suggested that variation in soils and topography are still relevant to form a patchwork of diverse microhabitats within the superficially uniform forest (Thomas et al. 2008). Additionally, a stable climate during the Pleistocene could have contributed to its high levels of richness and endemism (Carnaval et al. 2009). In general, areas with the highest levels of weighted endemism correspond to the richest regions (e.g., Serra do Espinhaço). The mountains of the state of Espírito Santo show the greatest deviation from this pattern: high endemism is detected in an area of moderate levels of richness. Such pattern has been noticed by taxonomists working on the flora of this state (Fiaschi and Pirani 2005; Reginato and Goldenberg 2013; Dutra et al. 2015), but the processes involved are still neglected in the literature. To better understand the processes that have led to these patterns of richness and endemism in the AF (and all other regions), we need better dated and species-dense phylogenies to ascertain the dynamics of diversification and explore how with observation correlates with geological, climatological and other explanatory variables.

The five super-bioregions recovered in our analyses present some degree of congruence with the three major domains of eastern Brazil (IBGE 2004). A major deviation is the Espinhaço/Mantiqueira super-bioregion, which includes areas of the Atlantic Forest and the Cerrado domains. On the other hand, the Caatinga super-bioregion presented the highest congruence to the domain classification. The Southern + Northern Atlantic Forest super-bioregions are also highly overlapping, but some areas included in the Atlantic Forest domain were classified into the Espinhaço/Mantiqueira super-bioregion. The Cerrado bioregion did not include some southwestern regions of the Cerrado domain, which were included in the Southern Atlantic Forest super-bioregion, and some southeastern areas, which were included in the Espinhaço/Mantiqueira super-bioregion. Among the eastern Brazilian domains, the most analyzed in a biogeographical framework is the Atlantic Forest. Several studies have suggested an historical separation of the Atlantic Forest in two blocks (northern and southern), with limits more or less coincident with the Rio Doce valley (reviewed in Fiaschi and Pirani 2009). Recently, it has been suggested that instead of being caused by a riverine barrier, the biogeographical break near the Rio Doce may have a climatic basis (Carnaval et al. 2014; Saiter et al. 2016). Although such break was recovered in our analyses within the bioregion of Northern Atlantic Forest, our results indicated a stronger separation of the Atlantic Forest in two blocks located further south. Our analyses recovered a division of southern and northern Atlantic Forest that closely follows the Tropic of Capricorn, therefore dividing the Atlantic Forest into tropical and subtropical bioregions. A north-south divergence in composition was found for the arboreal component of both rain (coastal) and semi-deciduous (interior) forests that correlated with climatic variables (temperature and rainfall regime) and have been reported previously (Oliveira-Filho and Fontes 2000), and it is likely to explain the patterns we found here as well. Floristic comparisons among several sites have already indicated heterogeneity across the Cerrado domain (reviewed in Fiaschi and Pirani 2009). Those studies found that most differences seem to be associated with soil type and geographic location (Ratter et al. 2006; Fiaschi and Pirani 2009).

Despite issues associated with online collections databases (Zizka et al. 2018), the automated compiled dataset for this study recovered, with great overlap, most of the well-established biogeographical units of eastern Brazil (domains, ecoregions, or biogeographical provinces). Our results highlight the value of efforts to digitize herbarium collections, and demonstrate how their usefulness is enhanced when coupled with online accessible taxonomic information based on those collections (Flora do Brasil 2020, under construction). This combination permitted assembly of data and facilitated a data filtering process. Although the proposed bioregions are primarily based on species distributions, it is important to note that some influence of climatic conditions might be expected, since we derived species ranges based on climatic niche models. Unfortunately, working without the niche models would results in areas also being defined on the lack of data due to habitat degradation, which is significant in Eastern Brazil.

Our results also have provided new insights into the biogeography of eastern Brazil, particularly the Espinhaço/Mantiqueira bioregion, and the sub-divisions of other bioregions. The script-based automated workflow for both data gathering and analysis applied in this study can certainly be improved as both the methods and data sets are updated. Such approaches can be useful to detect meaningful biogeographical units for historical analyses in macroevolution, complementing classifications based on geomorphology, climate, vegetation and endemic species. In addition to the high species richness and endemism observed in eastern Brazil, the region also harbors several endemic or nearly endemic lineages of plants, some of them quite species-rich (Loeuille et al. 2015; Staggemeier et al. 2015; Reginato and Michelangeli 2016; Rocha et al. 2016). We propose that the bioregions identified in this study may be used to test further hypotheses and elucidate macroevolutionary processes responsible for this remarkable diversity.

References

Aiello-Lammens ME, Boria RA, Radosavljevic A, Vilela B, Anderson RP (2014) spThin: functions for spatial thinning of species occurrence records for use in ecological models. R package version 0.1.0 [online]. Available from https://CRAN.R-project.org/package=spThin

BFG (2015) Growing knowledge: an overview of seed plant diversity in Brazil. Rodriguésia 66:1085–1113. https://doi.org/10.1590/2175-7860201566411

BFG (2018) Brazilian Flora 2020: innovation and collaboration to meet Target 1 of the global strategy for plant conservation (GSPC). Rodriguésia 69:1513–1527. https://doi.org/10.1590/2175-7860201869402

Bivand R, Lewin-Koh N (2013) maptools: tools for reading and handling spatial objects. R package version 0.8-27. http://CRAN.R-project.org/package=maptools

Bivand R, Rundel C (2013) Rgeos: interface to geometry engine – open source (GEOS). R package version 0.3-2. http://CRAN.R-project.org/package=rgeos

Bloomfield NJ, Knerr N, Encinas-Viso F (2018) A comparison of network and clustering methods to detect biogeographical regions. Ecography 41:1–10

Buffon GLLC (1761) Histoire Naturelle Générale. Imprimerie Royale, Paris

Carnaval AC, Hickerson MJ, Haddad CF, Rodrigues MT, Moritz C (2009) Stability predicts genetic diversity in the Brazilian Atlantic forest hotspot. Science 323:785–789

Carnaval AC, Waltari E, Rodrigues MT, Rosauer D, VanDerWal J, Damasceno R, Prates I, Strangas M, Spanos Z, Rivera D, Pie MR, Firkowski CR, Bornschein MR, Ribeiro LF, Moritz C (2014) Prediction of phylogeographic endemism in an environmentally complex biome. Proc R Soc B-Biol Sci 281:20141461

Carvalho G (2017) Flora: tools for Interacting with the Brazilian Flora 2020. R package version 0.3.0. https://CRAN.R-project.org/package=flora

Costa LP, Leite YL, da Fonseca GA, da Fonseca MT (2000) Biogeography of South American forest mammals: endemism and diversity in the Atlantic Forest 1. Biotropica 32 (4b):872–881

Costello MJ, Tsai P, Wong PS, Cheung AKL, Basher Z, Chaudhary C (2017) Marine biogeographic realms and species endemicity. Nat Commun 8. https://doi.org/10.1038/s41467-017-01121-2

Cracraft J (1991) Patterns of diversification within continetal biotas: hierarchical congruence among the areas of endemism of Australia vertebrates. Aust Syst Bot 4:211–227

Crisci JV, Cigliano MM, Morrone JJ, Roig Junent S (1991) A comparative review of cladistic approaches to historical biogeography of Southern South America. Aust Syst Bot 4:117–126

Da Costa RC, de Araujo FS, Lima-Verde LW (2007) Flora and life-form spectrum of deciduous thorn woodland (caatinga) in northeastern, Brazil. J Arid Environ 68:237–247

DaSilva MB, Pinto-da-Rocha R, DeSouza AM (2015) A protocol for the delimitation of areas of endemism and the historical regionalization of the Brazilian Atlantic Rain Forest using harvestmen distribution data. Cladistics 31:692–705

de Candolle AP (1820) Essai elementaire de geographie botanique, volume Dictionnaire des sciences naturelles. F. Levrault, Strasbourg

de Mendonça RC, Felfili JM, Walter BMT, Junior MCdS, Rezende AV, Filgueiras TdS, Nogueira PE, Fagg CW (2008) Flora vascular do Bioma Cerrado. Checklist com 12.356 especies. In: Sano SM, Almeida SP, Ribeiro JF (eds) Cerrado: ecology and flora, vol 2. Embrapa Informaçâo Tecnológica, Brasilia, pp 10–1279

Droissart V, Dauby G, Hardy OJ, Deblauwe V, Harris DJ, Janssens S, Mackinder B, Blach-Overgaard A, Sonke B, Sosef MSM, Stevart T, Svenning JC, Wieringa JJ, Couvreur TLP

(2018) Beyond trees: biogeographical regionalization of tropical Africa. J Biogeogr 45:1153–1167. https://doi.org/10.1111/jbi.13190
Dutra VF, Alves-Araújo A, Carrijo TT (2015) Angiosperm checklist of Espírito Santo: using electronic tools to improve the knowledge of an Atlantic Forest biodiversity hotspot. Rodriguésia 66:1145–1152
Edler D, Guedes T, Zizka A, Rosvall M, Antonelli A (2016) Infomap bioregions: interactive mapping of biogeographical regions from species distributions. Syst Biol 66:197–204
Ferrari A (2017) Biogeographical units matter. Aust Syst Bot 30:391–402. https://doi.org/10.1071/SB16054
Fiaschi P, Pirani JR (2005) Four new species of *Schefflera* (Araliaceae) from Espírito Santo State, Brazil. Kew Bull 60:77–85
Fiaschi P, Pirani JR (2009) Review of plant biogeographic studies in Brazil. J Syst Evol 47:477–496
Fick SE, Hijmans RJ (2017) WorldClim 2: new 1-km spatial resolution climate surfaces for global land areas. Int J Climatol 37:4302–4315
Flora do Brasil 2020 (under construction). Jardim Botânico do Rio de Janeiro. http://floradobrasil.jbrj.gov.br/. Accessed 29 Mar 2019
Flora do Brasil 2020, under construction. Jardim Botânico do Rio de Janeiro. http://floradobrasil.jbrj.gov.br/. Accessed 16 Feb 2019
Forzza RC, Baumgratz JFA, Bicudo CEM, Carvalho AA Jr, Costa A, Costa DP, Hopkins M, Leitman PM, Lohmann LG, Maia LC, Martinelli G, Menezes M, Morim MP, Coelho MAN, Peixoto AL, Pirani JR, Prado J, Queiroz LP, Souza VC, Stehmann JR, Sylvestre LS, Walter BMT, Zappi D (eds) (2010) Catálogo de plantas e fungos do Brasil. Vol. 1, 2. Andrea Jakobsson Estúdio Editorial, Jardim Botânico do Rio de Janeiro, Rio de Janeiro
Gaston KJ (2000) Global patterns in biodiversity. Nature 405:220–227
Goldberg EE, Lancaster LT, Ree RH (2011) Phylogenetic inference of reciprocal effects between geographic range evolution and diversification. Syst Biol 60:451–465
Gonzalez-Orozco CE, Ebach MC, Laffan S, Thornhill AH, Knerr NJ, Schmidt-Lebuhn AN, Cargill CC, Clements M, Nagalingum NS, Mishler BD, Miller JT (2014) Quantifying phytogeographical regions of Australia using geospatial turnover in species composition. PLoS One 9(3): e92558
Guerin GR, Ruokolainen L, Lowe AJ (2015) A georeferenced implementation of weighted endemism. Methods Ecol Evol 6:845–852
Hazzi NA, Moreno JS, Ortiz-Movliav C, Palacio RD (2018) Biogeographic regions and events of isolation and diversification of the endemic biota of the tropical Andes. Proc Natl Acad Sci USA 115:7985–7990. https://doi.org/10.1073/pnas.1803908115
Hijmans RJ (2016) Raster: geographic data analysis and modeling. R package version 2.5-8. https://CRAN.R-project.org/package=raster
Hijmans RJ, Phillips S, Leathwick J, Elith J (2017) Dismo: species distribution modeling. R package version 1.1-4. https://CRAN.R-project.org/package=dismo
Hipp AL, Manos PS, Gonzalez-Rodriguez A, Hahn M, Kaproth M, McVay JD, Avalos SV, Cavender-Bares J (2018) Sympatric parallel diversification of major oak clades in the Americas and the origins of Mexican species diversity. New Phytol 217:439–452. https://doi.org/10.1111/nph.14773
Holdridge LR (1947) Determination of world formations from simple climatic data. Science 105:367–368
IBGE (2004) Mapa de Biomas do Brasil, primeira aproximação. Available at http://www.ibge.gov.br
Kreft H, Jetz W (2010) A framework for delineating biogeographical regions based on species distributions. J Biogeogr 37:2029–2053
Linder HP (2001) On areas of endemism, with an example from the African Restionaceae. Syst Biol 50:892–912
Loeuille B, Semir J, Lohmann LG, Pirani JR (2015) A phylogenetic analysis of Lychnophorinae (Asteraceae: Vernonieae) based on molecular and morphological data. Syst Bot 40:299–315

Löwenberg-Neto P (2014) Neotropical region: a shapefile of Morrone's (2014) biogeographical regionalisation. Zootaxa 3802:300

Martini AMZ, Fiaschi P, Amorim AM, Paixao JM (2007) A hot-point within a hot-spot: a high diversity site in Brazil's Atlantic Forest. Biodivers Conserv 16:3111–3128. https://doi.org/10.1007/s10531-007-9166-6

Matzke NJ (2014) Model selection in historical biogeography reveals that founder-event speciation is a crucial process in island clades. Syst Biol 63:951–970

Morellato LPC, Haddad CFB (2000) Introduction: the Brazilian Atlantic forest. Biotropica 32 (4b):786–792

Morrone JJ (1994) On the identification of areas of endemism. Syst Biol 43:438–441. https://doi.org/10.1093/sysbio/43.3.438

Morrone JJ (2009) Evolutionary biogeography: an integrative approach with case studies. Columbia University Press, New York

Morrone JJ (2014a) Biogeographical regionalisation of the Neotropical region. Zootaxa 3782:1–110

Morrone JJ (2014b) Parsimony analysis of endemicity (PAE) revisited. J Biogeogr 41:842–854. https://doi.org/10.1111/jbi.12251

Murray-smith C, Brummitt NA, Oliveira-Filho AT, Bachman S, Moat J, Lughadha EMN, Lucas EJ (2009) Plant diversity hotspots in the Atlantic coastal forests of Brazil. Conserv Biol 23:151–163

Myers N, Mittermeier RA, Mittermeier CG, da Fonseca GA, Kent J (2000) Biodiversity hotspots for conservation priorities. Nature 403:853–858. https://doi.org/10.1038/35002501

Nelson GJ, Platnick NI (1981) Systematics and biogeography, cladistics and vicariance. Columbia University Press, New York

Oliveira-Filho AT, Fontes MAL (2000) Patterns of floristic differentiation among Atlantic Forests in Southeastern Brazil and the influence of climate. Biotropica 32(4b):793–810

Oliveira-Filho AT, Ratter JA (2002) Vegetation physiognomies and woody flora of the Cerrado Biome. In: Oliveira PS, Marquis RJ (eds) The Cerrados of Brazil. Ecology and natural history of a Neotropical Savanna. Columbia University Press, New York, pp 91–120

Olson DM, Dinerstein E, Wikramanayake ED, Burgess ND, Powell GVN, Underwood EC, D'Amico JA, Itoua I, Strand HE, Morrison JC, Loucks CJ, Allnutt TF, Ricketts TH, Kura Y, Lamoreux JF, Wettengel WW, Hedao P, Kassem KR (2001) Terrestrial ecoregions of the world: a new map of life on Earth. Bioscience 51:933–938. https://doi.org/10.1641/0006-3568(2001)051[0933:TEOTWA]2.0.C

Pebesma EJ, Bivand R (2005) Classes and methods for spatial data in R. R News 5(2). http://cran.r-project.org/doc/Rnews/

Perera SJ, Proches S, Ratnayake-Perera D, Ramdhani S (2018) Vertebrate endemism in south-eastern Africa numerically redefines a biodiversity hotspot. Zootaxa 4382:56–92. https://doi.org/10.11646/zootaxa.4382.1.2

Phillips SJ, Dudík M (2008) Modeling of species distributions with Maxent: new extensions and a comprehensive evaluation. Ecography 31:161–175

Platnick NI, Nelson GJ (1978) A method of analysis for historical biogeography. Syst Zool 27 (1–16). https://doi.org/10.2307/2412808

Porzecanski AL, Cracraft J (2005) Cladistic analysis of distributions and endemism (CADE): using raw distributions of birds to unravel the biogeography of the South American aridlands. J Biogeogr 32:261–275

Prado DE (2003) As Caatingas da America do Sul. In: Leal IR, Tabarelli M, da Silva JMC (eds) Ecologia e conservacao da caatinga. Univwersitaria da UFPE, Recife, pp 3–74

Prado J, Sylvestre LS, Labiak PH, Windisch PG, Salino A, Barros ICL, Hirai RY, Almeida TE, Santiago ACP, Kieling-Rubio MA, Pereira AFN, Øllgaard B, Ramos CGV, Mickel JT, Dittrich VAO, Mynssen CM, Schwartsburd PB, Condack JPS, Pereira JBS, Matos FB (2015) Diversity of ferns and lycophytes in Brazil. Rodriguésia 66:1073–1085. https://doi.org/10.1590/2175-7860201566410

Quintero I, Jetz W (2018) Global elevational diversity and diversification of birds. Nature 555:246

R_Core_Team (2018) R: a language and environment for statistical computing. R Foundation for Statistical Computing, Vienna

Ratter JA, Bridgewater S, Ribeiro JF (2006) Biodiversity patterns of the woody vegetation of the Brazilian Cerrado. In: Pennington RT, Lewis GP, Ratter JA (eds) Neotropical savannas and seasonally dry forests: plant diversity, biogeography and conservation. The Systematics Association Special Volume. Vol Series 69. CRC, Boca Raton, pp 31–66

Reginato M, Goldenberg R (2013) Two new species of Leandra s.str. (Melastomataceae) from the Atlantic Forest in Espirito Santo, Brazil. Blumea 57:210–214

Reginato M, Michelangeli FA (2016) Untangling the phylogeny of *Leandra* s.str. (Melastomataceae, Miconieae). Mol Phylogenet Evol 96:17–32

Rocha MJR, Batista JAN, Guimaraes PJF, Michelangeli FA (2016) Phylogenetic relationships in the Marcetia alliance (Melastomeae, Melastomataceae) and implications for generic circumscription. Bot J Linn Soc 181:585–609. https://doi.org/10.1111/boj.12429

Rojas A, Patarroyo P, Mao L, Bengtson P, Kowalewski M (2017) Global biogeography of Albian ammonoids: a network-based approach. Geology 45:659–662. https://doi.org/10.1130/g38944.1

Ronquist F, Sanmartin I (2011) Phylogenetic methods in biogeography. Annu Rev Ecol Evol Syst 42:441–464. https://doi.org/10.1146/annurev-ecolsys-102209-144710

Rosvall M, Bergstrom C (2008) Maps of information flow reveal community structure in complex networks. Proc Natl Acad Sci 105:1118

Saiter FZ, Brown JL, Thomas WW, de Oliveira-Filho AT, Carnaval AC (2016) Environmental correlates of floristic regions and plant turnover in the Atlantic Forest hotspot. J Biogeogr 43:2322–2331

Sampaio EVSB (1995) Overview of the Brazilian Caatinga. In: Bullock SH, Mooney HA, Medina E (eds) Seasonally tropical dry forests. Cambridge University Press, Cambridge, pp 35–63

Sclater PL (1858) On the general geographical distribution of the members of the class aves. J Proc Linn Soc Zool 2:130–136

Segatto ALA, Reck-Kortmann M, Turchetto C, Freitas LB (2017) Multiple markers, niche modelling, and bioregions analyses to evaluate the genetic diversity of a plant species complex. BMC Evol Biol 17. https://doi.org/10.1186/s12862-017-1084-y

Silva JMC, de Sousa MC, Castelletti CH (2004) Areas of endemism for passerine birds in the Atlantic forest, South America. Glob Ecol Biogeogr 13(1):85–92

Silveira FA, Negreiros D, Barbosa NP, Buisson E, Carmo FF, Carstensen DW, Conceição AA, Cornelissen TG, Echternacht L, Fernandes GW, Garcia QS (2016) Ecology and evolution of plant diversity in the endangered campo rupestre: a neglected conservation priority. Plant Soil 403:129–152

Spalink D, Kriebel R, Li P, Pace MC, Drew BT, Zaborsky JG, Rose J, Drummond CP, Feist MA, Alverson WS, Waller DM, Cameron KM, Givnish TJ, Sytsma KJ (2018) Spatial phylogenetics reveals evolutionary constraints on the assembly of a large regional flora. Am J Bot 105:1938–1950. https://doi.org/10.1002/ajb2.1191

Spehn EM, Rudmann-Maurer K, Körner C (2011) Moutain biodiversity. Plant Ecol Divers 4:301–302

Staggemeier VG, Diniz-Filho JAF, Forest F, Lucas E (2015) Phylogenetic analysis in *Myrcia* section *Aulomyrcia* and inferences on plant diversity in the Atlantic rainforest. Ann Bot 115:747–761

Stehmann JR, Forzza RC, Alexandre S, Sobral M, DPd C, Kamino LHY (2009) Diversidade taxonômica na Floresta Atlântica. In: Stehmann JR, Forzza RC, Alexandre S, Sobral M, DPd C, Kamino LHY (eds) Plantas da Floresta Atlântica. Instituto de Pesquisas Jardim Botânico do Rio de Janeiro, Rio de Janeiro, pp 3–40

Szumik CA, Goloboff PA (2004) Areas of endemism: an improved optimality criterion. Syst Biol 53:968–977. https://doi.org/10.1080/10635150490888859

Szumik CA, Goloboff PA (2015) Higher taxa and the identification of areas of endemism. Cladistics 31:568–572. https://doi.org/10.1111/cla.12112

Szumik C, Aagesen L, Casagranda D, Arzamendia V, Baldo D, Claps LE, Cuezzo F, Gomez JMD, Di Giacomo A, Giraudo A, Goloboff P, Gramajo C, Kopuchian C, Kretzschmar S, Lizarralde M, Molina A, Mollerach M, Navarro F, Nomdedeu S, Panizza A, Pereyra VV, Sandoval M, Scrocchi G, Zuloaga FO (2012) Detecting areas of endemism with a taxonomically diverse data set: plants, mammals, reptiles, amphibians, birds, and insects from Argentina. Cladistics 28:317–329. https://doi.org/10.1111/j.1096-0031.2011.00385.x

Takhtajan A (1986) Floristic regions of the world. University of California Press, Berkeley

Thomas WW, Carvalho AD, Amorim AM, Garrison J, Santos TD (2008) Diversity of woody plants in the Atlantic coastal forest of southern Bahia, Brazil. Mem NY Bot Gard 100:21–66

Topel M, Zizka A, Calio MF, Scharn R, Silvestro D, Antonelli A (2017) SpeciesGeoCoder: fast categorization of species occurrences for analyses of biodiversity, biogeography, ecology, and evolution. Syst Biol 66:145–151. https://doi.org/10.1093/sysbio/syw064

Vilhena DA, Antonelli A (2015) A network approach for identifying and delimiting biogeographical regions. Nat Commun 6. https://doi.org/10.1038/ncomms7848

von Humboldt A (1808) Ansichten der Natur mit wissenschaftlichen Erlauterungen. J.G. Cotta, Tubingen

von Humboldt A, Bonpland A (1805) Essai sur la geographie des plantes. Paris

Wallace AR (1876) The geographical distribution of animals, vol 1. Harper & Brothers, New York

Woodward FI (1987) Climate and plant distribution. Cambridge University Press, Cambridge

Zizka A, Silvestro D, Andermann T, Azevedo J, Duarte RItter C, Edler D, Farooq H, Herdean A, Ariza M, Scharn R, Svanteson S, Wengtrom N, Zizka V, Antonelli A (2018) CoordinateCleaner: standardized cleaning of occurrence records from biological collection databases. https://github.com/ropensci/CoordinateCleaner

Chapter 19
Human Contribution to Amazonian Plant Diversity: Legacy of Pre-Columbian Land Use in Modern Plant Communities

Encarni Montoya, Umberto Lombardo, Carolina Levis, Gerardo A. Aymard, and Francis E. Mayle

Abstract Amazonia is the world's largest tropical forest and is globally important in terms of its ecosystem services and extraordinarily high levels of biodiversity. The origin of this biodiversity has long been attributed to purely natural drivers, with little consideration given to the legacy of millennia of human land use. Here, the potential contribution of pre-Columbian human activity (prior 1492 CE) to current patterns of plant diversity in Amazonia is explored via long-term (palaeoecology, archaeology) and short-term (botany, plant ecology) studies. The aim of the chapter is to examine the information available to date, and discuss recent advances and persisting shortcomings relevant to the extent to which pre-Columbian human societies influenced patterns of Amazonian plant diversity. This topic has been the subject of long-standing scientific debate over several decades, and among diverse disciplines. In recent years, this debate has intensified following the development of new techniques and data. The findings indicate that humans have had an impact upon

E. Montoya (✉)
Institute of Earth Sciences Jaume Almera (ICTJA-CSIC), Barcelona, Spain
e-mail: emontoya@liverpool.ac.uk

U. Lombardo
Institute of Geography, University of Bern, Bern, Switzerland
e-mail: umberto.lombardo@giub.unibe.ch

C. Levis
Instituto Nacional de Pesquisas da Amazônia – INPA, Manaus, Brazil

Universidade Federal de Santa Catarina – UFSC, Florianópolis, Brazil
e-mail: carolina.levis@pq.cnpq.br

G. A. Aymard
UNELLEZ-Guanare, Programa de Ciencias del Agro y el Mar, Herbario Universitario (PORT), Mesa de Cavacas, Estado Portuguesa, Venezuela

Compensation International Progress S. A. – Ciprogress Greenlife, Bogotá, DC, Colombia
e-mail: gaymard@ciprogress.com

F. E. Mayle
University of Reading, Reading, UK
e-mail: f.mayle@reading.ac.uk

V. Rull, A. C. Carnaval (eds.), *Neotropical Diversification: Patterns and Processes*, Fascinating Life Sciences, https://doi.org/10.1007/978-3-030-31167-4_19

Amazonian plant diversity for over 13,000 years. Late Pleistocene/early Holocene humans domesticated numerous plant species and may have inadvertently caused long-lasting ecosystem changes by contributing to Pleistocene megafauna extinction. Based on our literature review, we identify four key types of pre-Columbian anthropogenic impact, leaving a clear legacy upon current patterns of plant diversity: (1) construction of vast earthworks, which has altered forest and savannah cover through changes in micro-topography, fire use and hydrology, (2) widespread distribution and dispersal of domesticated plants, (3) the creation of exceptionally fertile, anthropogenic soils, which enabled continuous, intensive agro-forestry, and (4) the enrichment of plant communities with edible and useful species. We argue that knowledge of the degree to which humans have shaped plant diversity over the past several millennia has relevance for developing sustainable land use and improving our understanding of the likely responses of Amazonian ecosystems to environmental and anthropogenic disturbance.

Keywords Anthropogenic soils · Archaeology · Biogeography · Botany · Edible plants · Indigenous peoples · Land use · Landscape transformation · Palaeoecology · Plant composition

1 Introduction

Amazonia contains the largest area of tropical rainforest on Earth (6 million km^2), encompassing Amazonia *sensu stricto* (Olson et al. 2001; Keller et al. 2004) and three peripheral regions: the Guayana Shield and Plateau, the Andean foothills, and the Gurupi basin (*sensu* ter Steege et al. 2013; Antonelli et al. 2018; Alvez-Valles et al. 2018). This vast area is characterised by high levels of biodiversity, which have long been the focus of scientific interest across disciplines (Antonelli et al. 2018). Yet, the origin and patterns of Neotropical diversity have largely been explored in the context of natural processes and drivers, with little consideration given to the potential role of humans, despite the ancient human occupancy of the Americas (Goulding et al. 2003; Lyons et al. 2016).

Humans have interacted with tropical forests for at least 100 millennia (Roberts et al. 2016, 2017). While the type and scale of these activities has varied across the world, sometimes leading to significant landscape impact (van der Kaars et al. 2017), the oldest evidence of human occupation in the tropical Americas dates back to 13,000 cal year BP (calibrated years before 1950 CE) (Roosevelt et al. 2002). We hypothesize that early scattered human populations may have been responsible for some of the current patterns of forest biodiversity and structure in Amazonia, even when they attained low population densities (Goldberg et al. 2016).

The disappearance of Pleistocene megafauna in South America, for instance, was roughly coeval with the arrival of humans (Barnosky and Lindsey 2010). As ecological engineers, megafauna had profound impacts on biodiversity, plant distribution, forest structure, nutrient cycling and carbon storage in Amazonia (Barnosky and Lindsey 2010; Doughty et al. 2013, 2016). While it could be argued that their extinction entailed a shift in the natural dynamics of Amazonian ecosystems and led to long-lasting effects on present-day ecosystems, the role of humans in this process has yet to be fully resolved (Barnosky and Lindsey 2010), and therefore will not be discussed in this chapter. Besides megafauna depletion, it also has been suggested that early hunter-gatherers may also have modified the distribution and density of edible and useful tree species, including examples from Amazonia (Balée 1994), and the South American Atlantic forest (Lauterjung et al. 2018).

A clear example of direct anthropogenic influence on current Amazonian plant diversity emerges from studies of domestication of globally important crops (Clement et al. 2010; Doughty 2010; Piperno 2011; Watling et al. 2018). Humans are known to have managed Amazonian landscapes since the early-mid Holocene (Levis et al. 2018), through practices that involved (1) the transformation of soil properties, through the addition of nutrients or hydrological changes in watersheds (Glaser and Birk 2012; Lombardo et al. 2011a; Rostain 2010), (2) the domestication and cultivation of edible species (Bush et al. 1989; Clement 1999; Perry et al. 2007; Hilbert et al. 2017), (3) the contribution to the dispersal success of selected species (Shepard and Ramirez 2011), and (4) the elimination of competitors or predators of selected species (Bruno et al. 2003; Posey 1985). Beyond the global importance of Amazonian crop staples such as manioc (*Manihot esculenta*, also known as *cassava* or *yuca*), for instance, special attention has been paid to the potential impact of humans on the abundance and distribution of several species of palm (Arecaceae), which have been widely used by indigenous peoples, both in the past (Morcote-Ríos and Bernal 2001) and present (Smith 2015). As we explain below, Arecaceae contains a large number of Amazonian hyperdominant species (ter Steege et al. 2013), whose unusually high abundance has been attributed, in some cases, to their relationship with humans (Levis et al. 2017). It is therefore not surprising that several authors have explored the potential anthropogenic influence on the current distribution, or domestication routes, of some of the most useful palm species (Rull and Montoya 2014; Bush and McMichael 2016; Galluzzi et al. 2015; Clement et al. 2017).

We attribute the ongoing controversy over the relationship between past human activities and Amazonian plant diversity to multiple factors: (1) a scarcity of data and differing land-use practices compared to extra-tropical regions (Whitehouse and Kirleis 2014); (2) the massive depopulation of Amazonia over the last 500 years, which gives the impression of an 'empty' pristine wilderness (Denevan 1992a); and (3) a disconnect between scientific fields, and lack of interdisciplinary research, which led to an unbalanced understanding of human-environment interactions through time (Barlow et al. 2012; Maezumi et al. 2018). To fill this knowledge gap, we here review key lines of evidence from multiple scientific fields—archaeology, biogeography, botany, ecology, and palaeoecology—to examine the potential legacy of past human land use upon Amazonian plant diversity.

2 Historical Scientific Background and Hypotheses

The extraordinarily biodiverse rainforests of Amazonia have long been considered by Europeans to be evidence of a largely untouched pristine wilderness, ever since the Columbian Encounter (Jameson 1858; Denevan 1992a). Despite early European reports of numerous settlements of high population density (Medina 1934), this view of Amazonia as a pristine wilderness was at least partially due to the absence of the elaborate and ornate stone architecture, urban plazas, or population centres that characterised Peru and Mexico at the time of the Columbian Encounter. Later, the romantic idea of the 'noble savage' living in harmony with the surrounding environment, akin to a 'Garden of Eden', was embraced by European societies that spread the 'pristine myth' during the following centuries (Hemming 2006; Redford 1991).

In this sense, historical records from the nineteenth century described the Amazonian peoples as sparse, scattered populations, largely confined to the banks of major rivers (Jameson 1858; Medina 1934; Denevan 1996; Livi-Bacci 2016). After the Columbian Encounter, there was widespread depopulation of the most accessible sites, due to a combination of population collapse (arising from the introduction of European diseases and slavery) and abandonment to avoid further interactions with Europeans (Dobyns 1966; Crosby 2004; Chambouleyron et al. 2011). The net effect of site abandonment was that lands formerly cleared and/or cultivated underwent secondary succession (Chambouleyron et al. 2011). Consequently, the 'wilderness' described by European naturalists in the nineteenth century as pristine, was actually, in many cases, secondary forest regrowth (Loughlin et al. 2018). It was in this framework that research programmes on pre-Columbian environmental-human interactions began to develop during the second half of the twentieth century.

Observations of lifestyles and land-use practices of modern Amazonian indigenous groups formed the basis of theories, hypotheses, and assumptions developed by anthropologists and cultural ecologists. They observed that Amazonia was populated by small and mobile indigenous groups, mostly hunter-gatherers, and linked the low level of social complexity of these populations with limited agricultural potential due to low soil fertility and high rainfall (Meggers 1954), and scarcity of protein (Gross 1975). According to Meggers, the archaeological evidence of pre-Columbian complex societies in Amazonia (e.g. Marajó Island, Meggers 2001) was best explained by the migration of complex societies from the Andes, rather than in situ cultural development. Meggers thought that, once these groups from the Andes settled in Amazonia, the harsh, 'unproductive' Amazonian environment led to cultural regression towards simpler societies such as the small indigenous groups we see in modern Amazonia. On the other hand, Lathrap (1970) disagreed with Meggers' view, instead arguing that the *várzea* (the white-water river floodplains) was relatively fertile and nutrient-rich, enabling social complexity to emerge along the large rivers of Amazonia. Both of these scenarios of social complexity are consistent with 'environmental determinism', by which human development and activities are controlled by the physical environment (Meggers 1954).

The role of 'environmental determinism' in Amazonia, however, has been increasingly challenged thanks to the emergence of evidence for complex pre-Columbian urban systems in riverine settings, *terra firme* (upland interfluve), and wetland areas. Such evidence includes anthropogenic soils (*terra pretas*) and a diverse range of artificial earthworks, including fishery ponds, ring ditches, large habitation mounds and raised fields (Heckenberger and Neves 2009; Blatrix et al. 2018; Carson et al. 2016; Iriarte et al. 2012; Lombardo et al. 2011b; Prestes-Caneiro et al. 2016; Rostain 2010; de Souza et al. 2018)—the most impressive of which are the geometric geoglyphs of Acre and the urban road and plaza network of the Upper Xingu (Heckenberger et al. 2003, 2008; Watling et al. 2017). The best evidence for social complexity comes from Marajó Island and Santarém (Roosevelt 1991; Stenborg et al. 2012) in Brazil, and the Llanos de Moxos in Bolivia (Lombardo and Prümers 2010). The existence of archaeological evidence of large cultural diversity has promoted a shift in how the Amazonian forests are traditionally perceived. Thus, the general view of pre-Columbian Amazonia has been shifting from the traditional paradigm of virgin wilderness toward a radically different view of Amazonia as a human-created, domesticated landscape or cultural parkland (Denevan 1992a; Heckenberger et al. 2003; Balée and Erickson 2006; Erickson 2006)—especially among the archaeologists and anthropologists.

3 The Debate Reactivated

This progressive accumulation of archaeological and ethnobotanical data has led to the development of a new research program for Amazonia within the natural and social sciences called Historical Ecology, which attempts to understand the cumulative effects of centuries and millennia of interactions between past societies and environments (Balée 2006, 2013). Currently, a scientific debate has been reactivated and focused on critical questions related to the scale of human activities and their impact on the vegetation that remain unanswered. The debate is directly related to whether or not there is enough evidence accumulated to infer the main drivers of the ecological dynamics for the entire region, which results in open questions in and outside academia. The uncertainty about the spatial extent of human legacy on Amazonian plant diversity can be exemplified with the following questions: Can we assume that a particular spot of Amazonia is pristine in the absence of evidence of pre-Columbian human impact? Or should we assume that people significantly impacted all areas of Amazonia, in which case no place in Amazonia should be considered pristine? Clearly, these hypotheses have profound consequences in the way Amazonian vegetation diversity should be perceived nowadays and in the future.

Besides the contribution of archaeology and anthropology to the subject, additional input has come from the study of long-term (palaeoecology) and short-term (botany, ecology) dynamics of Amazonian vegetation. Even though Neotropical palaeoecology began in the late 1960s, most studies focused on unravelling

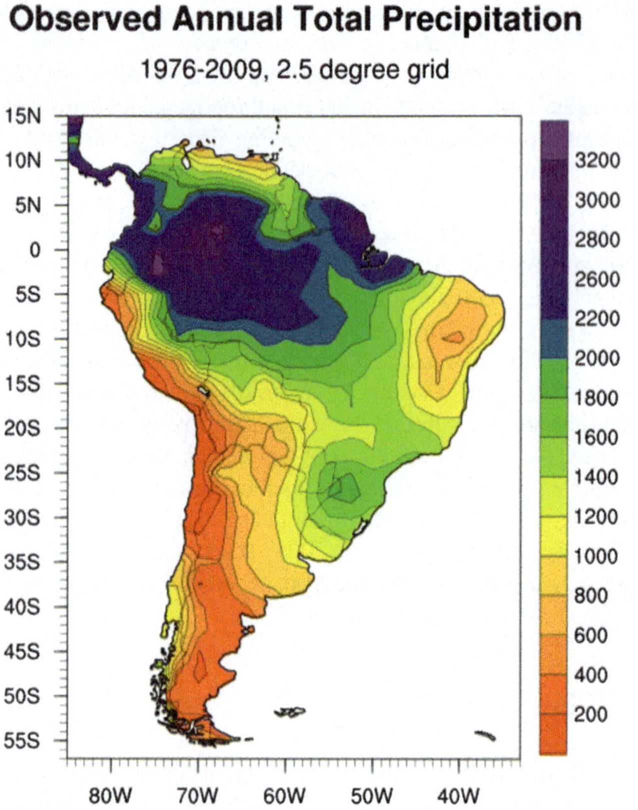

Fig. 19.1 Map of observed annual total precipitation in South America during the time interval between 1976 and 2009 at 2.5° grids resolution. Source: NOAA/OAR/ESRL PSD, Boulder, Colorado, USA, from their Web site at https://www.esrl.noaa.gov/psd, based on Liebmann and allured (2005)

vegetation responses to climate change, especially during the last Ice Age (Colinvaux 2007). It was not until the 1990s that most palaeoecologists routinely included charcoal analysis to infer past fire regimes as disturbance drivers of the vegetation. Recent studies indicate that natural fires are rare in the wettest parts of Amazonia under today's climatic conditions (e.g. eastern Ecuador), and are normally associated with human-caused ignition (Fig. 19.1; Bush et al. 2007a; de Toledo and Bush 2007; Mayle and Power 2008). In this sense, Charles Lyell's principle of uniformitarianism (summarised as the present is the key to the past; Tomkeieff 1962) is often used to interpret the occurrence of high fire incidence in Amazonian palaeoecological records as indication of human presence (Bush et al. 2016). Two hypothetical scenarios are provided below to exemplify potential interpretations of human presence based on the charcoal record. First, an abrupt increase in past fire regime in study sites nowadays inhabited and with no further major changes in the record (thus, maintaining a similar signal since the former and abrupt increase), can be indicative of the presence of indigenous peoples (Bush et al. 2007a; de Toledo and Bush 2007; Montoya and Rull 2011). Second, past high fire regimes in todays' uninhabited locations (presenting low charcoal values nowadays) might point to abandoned settlements (Carson et al. 2014). However, assuming human presence as the exclusive driver of fire occurrence may result in wrong interpretations, as the (past and present) natural flammability potential of the ecosystems must also be considered. For instance, SW Amazonia was characterised by a climate much drier in the mid Holocene relative to now, supporting more flammable dry forests and savannahs than today's rainforests. Thus, peaks in mid Holocene fire frequency in this region could have been caused by natural processes, with or without any human intervention (Mayle and Power 2008). Such differences in the interpretations of the same proxy highlight the importance of developing a truly multi-factor scenario for palaeoecological reconstructions, considering the role of several drivers and their different importance as forcing factors through time (Rull 2018). Also, it manifests the shortcomings of using only one proxy to infer human presence. Besides charcoal particles, the occurrence of pollen grains from cultivated plants in sedimentary archives, even in low abundances as in the case of maize, is considered direct evidence of human occupation (Liu and Colinvaux 1988).

The increase of multiproxy investigations, as well as the development of new techniques including phytoliths (microscopic biogenic silica formed in plants, which enables reconstructions of past vegetation), non-pollen palynomorphs (microscopic remains of biological origin other than pollen and pteridophyte spores preserved in palynological slides), and remote sensing and GIS, have provided clear evidence of pre-Columbian occupation in several areas previously thought to be uninhabited (Piperno et al. 2002; Lombardo et al. 2013a; Palace et al. 2017). Site selection is crucial, as searching for palaeoecological sites near to archaeological sites would enable direct comparison and integration of different lines of evidence. In this sense, the creation of truly inter-disciplinary studies and networks has enabled the development of a more holistic, accurate picture of the human dimension to Amazonian biodiversity and ecology than has previously been possible (Mayle and Iriarte 2014; Lombardo et al. 2011b; Clement et al. 2015a). Such an improvement is essential for deciphering past land use, especially for those archaeological sites where long-term human impacts upon the environment are subtle or difficult to determine in sedimentary profiles (Kelly et al. 2018). Nevertheless, given the heterogeneity and spatial magnitude of Amazonia, the current debate does not deal with the pre-Columbian population density per se, but whether or not the indigenous peoples had a long-lasting effect on the Amazonian vegetation that could be recognised or inferred throughout the basin. Specifically, the controversy centres on the magnitude, spatial extent, and type of pre-Columbian anthropogenic impacts and the legacy of those impacts observable within the floristic and structural patterns of vegetation across Amazonia today (Tollefson 2013; Bush et al. 2015; Clement et al. 2015a, b; McMichael et al. 2015a; Piperno et al. 2015).

3.1 Broad-Scale Patterns

Amazonian landscapes have been managed by humans over at least the past 13,000 years of occupation, although the intensity and type of management likely varied through the Holocene (Roosevelt 2013). Through time, some ecosystems and plants were modified by humans to make them more productive and more suitable for the resource needs of human societies (Clement et al. 2015a). As a general pattern, Amazonian people today often cut and burn the vegetation, plant useful species, and alter soils to improve their drainage and fertility, resulting in considerable transformation of forests near their homes (Levis et al. 2018). Recent archaeological and anthropological studies have shown that small-scale clearing and burning activities by indigenous communities have not been practised with the same magnitude over the last several millennia (Siren 2014; Riris 2018). Based on modelling data, interviews with current indigenous settlements, and the review of historical records literature, these studies suggest more extensive slash-and-burn forest clearing activities after the European Encounter compared to the pre-Columbian period than previously thought (Siren 2014; Riris 2018). Anthropogenic soils (*terra preta* and *terra marrom*, also known as Amazonian Dark Earth—ADE—, and *terra mulata* respectively), indigenous pottery, mounds, raised fields and charcoal integrated with pollen and phytolith records of cultivated plants (Neves et al. 2003; McKey et al. 2010; Iriarte et al. 2012; Bush et al. 2007a; McMichael et al. 2012a, b; Morcote-Ríos and León-Sicard 2012; Morcote-Ríos et al. 2013; Piperno et al. 2015) have been used to detect past human activities. Reconstructing past forest management practices, however, may require other techniques, such as an assessment of the distribution and abundance of useful (in the sense of socio-economically important or used by peoples for different reasons) tree species in modern vegetation, on the assumption that such distributions are not explainable by natural ecological processes alone but are instead a legacy of historic forest management practices (Balée 1993; Levis et al. 2012; Clement 2014).

Although Amazonian forests have been shown to be very diverse, some tree species are consistently more abundant than would be expected from chance alone (ter Steege et al. 2013). An estimated 10,071 tree species, belonging to 1225 genera and 140 families, make up Amazonian forests (ter Steege et al. 2019), yet over half the stems belong to just 227 species. This subset of disproportionately common trees has been dubbed the "hyperdominants" (ter Steege et al. 2013). The contribution of rare species to this diversity has been recognized (Wills et al. 2006), but their spatial distribution remains poorly understood (Zizka et al. 2018). Likewise, the distribution range of some Amazonian tree species extends across the entire Amazon basin, but most are restricted to much smaller areas (Kristiansen et al. 2009). A similar imbalance is observed in species to genus ratios. Over half of all Amazonian tree species belong to genera with 100 or more species, while the majority of genera (52%) have ten or fewer species (Gentry 1993; Dexter and Chave 2016). Some of these hyperdominants are tree species with some evidence of domestication that dominate large forest areas, especially in those locations near archaeological sites

(Levis et al. 2017). Ancient forest management may at least partly explain the hyperdominance of these species, corroborated by pollen evidence for the cultivation and domestication of several tree species before European conquest (Maezumi et al. 2018). However, some ecologists (Barlow et al. 2012) argue that large monodominant stands of a given species do not necessarily imply human intervention, but instead a natural phenomenon whereby particular edaphic and/or hydrological conditions favour only that species. In addition to studies dealing with the dominance of certain species within vegetation plots, it has been argued that useful tree species with some evidence of domestication (e.g., *Bertholletia excelsa, Theobroma cacao, Bactris gasipaes, Inga* (*edulis*) *ynga* and *Crescentia cujete*) were likely dispersed across Amazonia by past societies (Morcote-Ríos and Bernal 2001; Guix 2005; Shepard and Ramirez 2011; Thomas et al. 2012, 2015; Levis et al. 2017; Moreira et al. 2017; Clement et al. 2017). Moreover, genetic and morphological variation of fruit shapes and types between Amazonian regions illustrates how humans have been important drivers of fruit shape diversity and distribution over several millennia (Moreira et al. 2017; Clement et al. 2017). In this sense, management and domestication of Amazonian ecosystems and their constituent flora by past peoples would have contributed to explain current ecological patterns.

Predictions of where humans transformed Amazonian ecosystems also remain controversial because vast portions of Amazonia have not been studied to date (McMichael et al. 2017a). Knowledge of Amazonia's environmental and human history has increased significantly in recent decades, but unfortunately remains geographically patchy. Particular regions have received far less scientific attention than others due, for example, to political turmoil (i.e. Colombia and Venezuela), bureaucracy (Rull et al. 2008), or lack of roads/trails preventing access. Even for those areas which are accessible and have been studied, a lack of evidence of landscape domestication may arise from (1) the use of inappropriate proxies or lack of an inter-disciplinary research strategy, (2) the absence of long-lasting effects clearly visible today (e.g., forest recovery after land abandonment), or (3) a poor visibility of past land use in the palaeo record (e.g. lack of charcoal due to poor flammability, use of cultigens/domesticates which are poorly preserved/represented in the palaeoecological record). Nevertheless, some insights about the human contribution to biodiversity, understood hereafter in terms of species abundance and plant community assemblages, can be obtained. For instance, humans controlled local-scale forest expansion around past occupation sites in some forest-savannah ecotones (Carson et al. 2014) and expanded the distribution and abundance of socio-economically important tree species across the region (Balée 2013). Legacies of past human activities, such as landscapes and plant populations with different degrees of domestication, are commonly found near archaeological sites with and without *terra preta* (Levis et al. 2017, 2018). Moreover, remote-sensing data have revealed past human influence on forest structure (Palace et al. 2017; Stenborg et al. 2018). In *terra preta* sites, forest biomass and tree height were found to be lower compared to random sites (Palace et al. 2017). Large-scale studies have described a widespread distribution, although heterogeneous, of archaeological sites (Clement et al. 2015a; McMichael et al. 2017a, see also Fig. 19.2) and socio-economically important plants

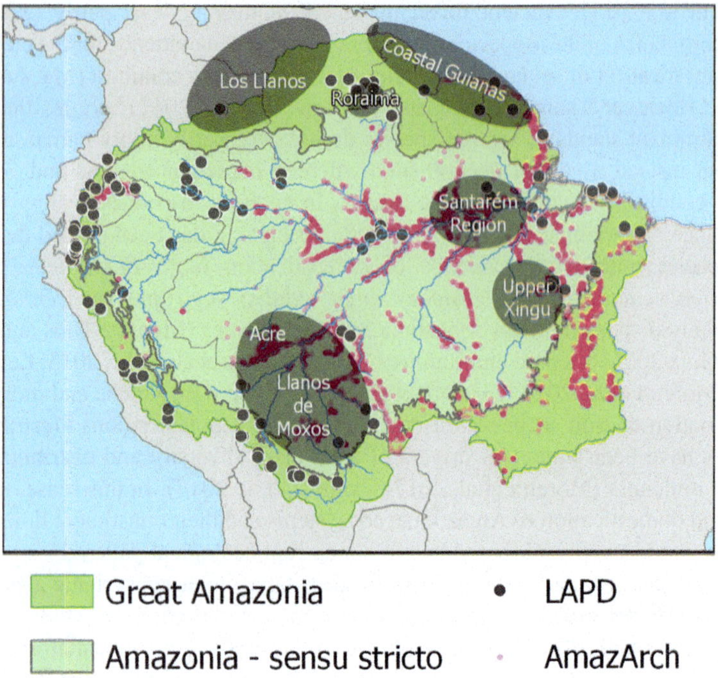

Fig. 19.2 Map of Amazonia (following Eva et al. 2005) showing the archaeological sites contained in the database AmazonArch and the palaeoecological records uploaded in LAPDB (Latin American Pollen Database; Flantua et al. 2015). The main regions mentioned in the text are marked with dark shades. Modified from Lombardo et al. (2018b)

across Amazonia (Levis et al. 2017). Balée (1989, 2013) estimated that around 11% of forest has been transformed by the long-term process of plant cultivation and forest management among indigenous peoples. However, caution must be taken when considering this rough estimate, as it is based on a list of potential useful species that encompasses species with and without known active management, and extrapolates sub-regional data across the large extent of the entire Amazon basin. In addition, modern assemblages with useful, hyperdominant trees may not necessarily derive from pre-Columbian activities, but could be historically more recent, for instance from the last few centuries (McMichael et al. 2017b).

To prevent unwarranted extrapolations and assumptions, the spatial extent and heterogeneity of the entire Amazonia must first be established and understood, as the evidence obtained and interpretations made in a given region are not necessarily representative of other regions in the basin. Furthermore, palaeoecological evidence of past human land use can be highly localised: clearly evident at some sites, but absent from others only a few kilometres (km) away (Bush et al. 2007a). In addition, the magnitude of human impact upon vegetation is not necessarily correlated with human population density, and some types of land use, especially those not requiring fire, may not be recognisable with current palaeoecological techniques. For instance,

Bush and Silman (2007) reconstructed the regional fire and vegetation history to infer associated human presence and activity using 22 palaeoecological records of lakes of western and eastern Amazonia. The study perfectly shows the danger of extrapolations to supra-local scale data from "dot-data maps", and highlights the need for (1) incorporating climate seasonality information to infer fire origins, and (2) sampling away from human-inhabited areas to prevent over-estimates regarding human occupation and impact (Bush and Silman 2007). In another study, Bush et al. (2007b) employed a multi-site approach with different lacustrine sequences near an archaeological site in Peruvian Amazonia (total study area of 50 km radius), which again demonstrated highly localised land use. Similarly, McMichael et al. (2012a), using phytolith and charcoal data from 55 soil cores, found low frequencies of charcoal and lack of phytoliths of maize and cultivated plants in most soil samples of western and interfluvial parts of the basin. These findings were used to support the hypothesis that wetter and remoter forests were occupied by smaller human groups that created only sporadic and highly localised impacts on the vegetation (Bush et al. 2015; McMichael et al. 2012a, b; Piperno et al. 2015). However, care must be taken again, highlighting the danger of extrapolations from point data, and also avoiding the implicit unwarranted assumption that absence of charcoal signifies absence of human occupation or land use (Watling et al. 2017)—especially in wet areas where forests are not easily burned.

Below we review evidence of past land use in those areas of Amazonia where most archaeological studies have been undertaken: southern (including southwestern), central, and eastern (including coastal and Guianas) Amazonia (Fig. 19.2).

3.2 *Regional Patterns*

3.2.1 Eastern Amazonia: Coastal Regions of Amazonia and the Guiana Shield

Coastal Amazonia includes raised fields, which are artificially elevated berms used for agriculture, in both French Guiana (Rostain 2008) and the island of Marajó, in the mouth of the Amazon River (Schaan 2008). Data indicate that a mixed diet with maize, fruits and legumes (*Acronomia aculeata*, *Euterpe oleracea*, *Inga* spp., *Spondias mombin* and *Byrsonima sp.*) and fish sustained pre-Columbian societies on Marajó Island, and likely other complex societies in coastal regions as well (Roosevelt 2013). Today, these Marajó mounds are covered by secondary forests enriched with edible species (Roosevelt 2013). The coasts of the Guianas and areas north of the mouth of the Amazon river underwent major earthwork engineering, including mounds and raised fields to avoid seasonal flooding (Rostain 1991; Iriarte et al. 2012). Inter-disciplinary studies of the French Guiana raised fields have shown that they were cultivated in pre-Columbian times without the use of fire (Iriarte et al. 2012), demonstrating the danger of relying on charcoal alone to infer the scale of pre-Columbian land use or population density. The modified landscape and soil left a long-term legacy upon the composition of the vegetation community (Rostain 2010). These earthworks not only altered the vegetation composition, but also had long-lasting effects on ecosystem structure and functioning (which are still evident today) due to the introduction of ecosystem engineers such as ants, termites, earthworms, and woody plants in raised fields that persisted well after humans abandoned them (McKey et al. 2010).

Further inland, more evidence of human contribution to current plant diversity has been proposed in the Roraima savannahs, near the northeastern margin of Amazonia—specifically the Gran Sabana of the Venezuelan Guayana Shied (Fig. 19.2). In this unusual treeless area located between the Amazon and the Orinoco basins and rainforests, an increase in fire frequency 2000 years ago was coeval with the arrival and establishment of a new type of vegetation in the Gran Sabana, the monospecific palm stands of *Mauritia flexuosa*, locally referred to as *morichales* (Montoya and Rull 2011; da Silva Meneses et al. 2013). It has been suggested that the vast expanses of *morichales* in Gran Sabana could be anthropogenic, whereby selective burning by humans facilitated their expansion at the expense of other species (Rull et al. 2013; Rull and Montoya 2014).

The large extent of "Los Llanos del Orinoco", or the Orinoquia savannah, has also been tentatively linked to past human activities (Berrío et al. 2012). Los Llanos is the largest savannah area in northwestern South America—a broad lowland zone that extends from northeastern Colombia to western Venezuela, occupying an area of ca. 532,000 km^2 (Huber et al. 2006). The Llanos vegetation is composed of a mosaic of savannahs, gallery forests, and dry semi-deciduous to evergreen forests (Aymard 2015). It has been proposed that humans played a key role in this region due to the occurrence of fire and humans in the adjacent Río Negro area, since the mid Holocene (Sanford et al. 1985; Sánchez et al. 2017). However, while humans may have modified these savannahs through use of fire and selection of economically important palms, this savannah biome is likely natural in origin, resulting from the combination of edaphic, hydrological, and climatic conditions unsuitable for forest growth (Huber et al. 2006). This is supported by drier late-glacial and early-mid Holocene conditions leading to savannah expansion and development of extensive dune fields (Sánchez et al. 2017). Because no palaeoecological record and charcoal analysis are available for a time period prior to human occupation, any inference of a fire regime shift (anthropogenically driven or not) remains elusive in this area. That being said, the hypothesis of humans driving savannah expansion has also been suggested for other treeless locations, such as the Amapá savannahs in eastern Brazil and the Venezuelan Gran Sabana (de Toledo and Bush 2007; Montoya et al. 2011)—although the presence of savannahs has been reported since the late Glacial (Rull et al. 2015). Today, savannahs might be expanding in floodplain ecosystems due to the close relationship of these ecosystems to fire (Flores et al. 2017) and their proximity to long-term human settlements along major Amazonian rivers (Denevan 1996). While humans may have been involved in the savannah expansion and maintenance in certain locations where forest-savannah transitions zones exist today, this assumption cannot be generalised to all Amazonian savannahs.

3.2.2 Central Amazonia

Central Amazonia hosts most of the anthropogenic soils, i.e. *terra preta* sites, mainly along the major water courses (Neves et al. 2004). Patches of *terra preta* are relatively large (1–2 ha, sometimes reaching 90 ha) and comprise very dark, fertile soil, and coincide with locations that were densely occupied by humans since at least 2500 cal year BP (Neves et al. 2003). Prolonged occupation resulted in the enrichment of soil with high amounts of pottery sherds, calcium, phosphorous, charcoal and organic matter. The latter accounts for the soil's dark colour and high fertility (Glaser and Birk 2012). *Terra preta* sites are often surrounded by *terra marrom*, which is also an anthropogenic soil, similar to *terra preta*, but less dark and without pottery and less nutrient-rich. *Terra marrom* is believed to be formed in areas cultivated in the past. Past occupations and cultivation activities left very clear signals in terms of soil properties, abundance of charcoal (anthropogenic originated) and phytoliths. These proxies are increasingly used to infer past human presence and land use (Woods and Glaser 2004; Neves et al. 2003). However, the absence of charcoal and cultigen phytoliths does not necessarily imply the absence of people, as shown by Levis et al. (2012). Diverse plant communities, rich in socio-economically important plants, have been found near archaeological sites in central Amazonian old-growth forests (Levis et al. 2012; Junqueira et al. 2017), secondary forests (Junqueira et al. 2010), and home gardens (Lins et al. 2015). Many native crops cultivated today predominate on *terra preta* soils, implying that these soils may be considered reservoirs of agrobiodiversity (Clement et al. 2003; Junqueira et al. 2010, 2016). Moreover, multiple cultures contributed to the diversification of plant communities in *terra preta* sites by creating 'home gardens'. The latter were successively occupied by different pre-Columbian cultures and hold a greater variety of socio-economically important species than those occupied by a single cultural group (Lins et al. 2015). Some forests, therefore, may be perceived as mosaics of numerous and diverse vegetation patches with different degrees of cultural intervention (Levis et al. 2018).

Santarém (including Tapajós, Belterra Plateau and Monte Alegre areas), in Pará state, NE Brazil, is a region that deserves special attention. Since the 1920s, numerous pre-Columbian settlements have been discovered in this area, predominantly in *terra firme* (not in floodplains or *várzeas*) locations, and occasionally near the river banks (Stenborg 2009; Stenborg et al. 2018). These settlements are associated with *terra preta* and *terra marrom* soils, cavities or depressions (*poços de água*), and a system of roads that connected the settlements further inland. It has been suggested that these settlements, near the junction of major rivers, occupied a highly strategic location and facilitated efficient communication between several cultures along the north-south and east-west riverine axes (Stenborg et al. 2012, 2018). Besides the concentration of *terra pretas* and *terra marrom*, this area is also renowned for the longest human occupation history in Amazonia, extending to the beginning of the Holocene at the cave Pedra Pintada (Roosevelt et al. 1996). Recently published palaeoecological evidence (Maezumi et al. 2018) shows that, prior to the construction of *terra preta* soils (ca. 2000 cal year BP), the peoples of Santarém had developed several agroforestry practices for subsistence since 4500 cal year BP, which, by 2500 cal year BP, had deeply altered the floristic composition of the forests via selective enrichment with edible plant species. Although this human intervention began 2500 cal yearr BP, it is clearly evident in the composition of today's forests.

3.2.3 South and Southwestern Amazonia

Southwestern Amazonia is one of the best studied regions of the basin. It has, by far, the clearest evidence of past human occupation and modification of the landscape, as pre-Columbian peoples built thousands of earthworks in this area—something easily visible in satellite imagery (Lombardo et al. 2011a). Acre State in western Brazil, for instance, is covered by more than 450 geometric structures called geoglyphs (Pärssinen et al. 2003). The forest surrounding these sites has been managed for millennia (Watling et al. 2017). Archaeobotanical studies detected changes in plant community composition, especially in palm abundance, when human activities intensified during the Holocene (McMichael et al. 2015b; Watling et al. 2017). In the Bolivian Amazon, much of which is covered by a seasonally flooded savannah called the *Llanos de Moxos*, raised fields (Rodrigues et al. 2018), monumental mounds (Lombardo and Prümers 2010), canals and causeways (Erickson 2001), fish weirs (Blatrix et al. 2018) and ring ditches (Carson et al. 2014) form a 100,000 km^2 anthropogenic landscape (Erickson 2008; Prümers and Jaimes Betancourt 2014). Within this region, studies reveal interactions between pre-Columbian peoples and river networks over several millennia. People settled in this region more than 10,000 cal year BP (Lombardo et al. 2013b), but abandoned the area around 4000 cal year BP due to catastrophic flooding from the Río Grande (Lombardo et al. 2018a). These floods transformed the landscape in the southern *Llanos de Moxos* by building a sedimentary lobe of fertile and relatively well drained land that made possible the establishment of the populous, agriculture-based, monumental mounds culture in this region 2000 years later. Different cultures transformed the eastern *Llanos de Moxos* by building other types of earthworks—hundreds of canals, ring ditches and causeways (Lombardo et al. 2012). The *Llanos de Moxos* can therefore be considered to be probably the most spectacular example of pre-Colombian human-environment interactions anywhere in Amazonia.

Interestingly, centuries after the abandonment of anthropogenic mounds and soils, plant communities still hold legacies of past transformations (Erickson and Balée 2006; Quintero-Vallejo et al. 2015; Levis et al. 2017). For instance, southwestern Amazonia is the region where the highest abundance and diversity of native domesticated plants has been found (Levis et al. 2017). Important crops were first domesticated in this region, such as manioc (*Manihot esculenta*), peach palm (*Bactris gasipaes*) and hot peppers (*Capsicum baccatum, C. anuum*) (Clement et al. 2010, 2016), making this one of the most important domestication hotspots of South America (Piperno 2011; Watling et al. 2018). Genetic analysis also confirms that peach palm was first domesticated in southwestern Amazonia, followed by two dispersal events into western and eastern Amazonia (Clement et al. 2017). Today, domesticated plant diversity (Levis et al. 2017) and cultural diversity (Crevels and Van der Voort 2008) is striking in this region.

Further east, in the Upper Xingu River basin, ethnoarchaeological studies have mapped territorial polities, urbanism and extensive landscape transformations (e.g., roads, raised causeways and *terra preta*) covering an area of approximately 50,000 km^2 (Heckenberger et al. 2003, 2008). In addition, a recent study in the Upper Tapajós River basin using remote sensing, archaeological survey, and excavations has detected more than a hundred earthworks with ceramics and *terra preta*, confirming that the southern rim of the Amazon basin was densely settled between 750 and 500 cal year BP (de Souza et al. 2018). These authors also estimated that approximately 400,000 km^2 of southern Amazonia was occupied by pre-Columbian earth-builders.

4 Towards a Common View

Determining the extent of the human footprint on current Amazonian biodiversity is essential for the development of appropriate conservation and management strategies, as this information is directly relevant for ascertaining the degree of resilience of plant communities to drivers of change (Maslin et al. 2005). First, studying vegetation response to past climatic conditions as a rough analogue for predicted responses to future projected climate change (e.g. with respect to drought) may inform the development of appropriate mitigation or adaptation strategies (Mayle and Power 2008; IPCC 2013; Nolan et al. 2018). This palaeoecological information has been proposed to establish adequate restoration strategies (Bush et al. 2014), and has also been inadequately used by companies for exploitation activities (Bush and Silman 2007). Second, some of the pre-Columbian practices have been proposed for a future sustainable economy such as the creation of new *terra preta* soils for agriculture (Glaser et al. 2001). However, this initiative is in its infancy and further information and deeper knowledge are required before its viability can be ascertained (Renard et al. 2012; Bezerra et al. 2016).

However, this determination of the past human footprint in Amazonia is not always an easy task. As argued above, not all the aforementioned human activities left an impact upon the vegetation that remains preserved in the sedimentary record (Kelly et al. 2018). Neither the proxy preserved in the sedimentary records has to be indicative about the same environmental driver in different settings, preventing extrapolations. The examples discussed in this chapter show, for instance, that the presence of charcoal particles is indicative of fire occurrence (environmental process)—but does not imply an anthropogenic origin (environmental driver) per se (Carson et al. 2014; Maezumi et al. 2017). This calls for alternative approaches to investigate past human presence and activities, as well as other drivers of landscape change in sedimentary archives (Rull 2018). With respect to palaeoecological proxies, the search for pollen grains and phytoliths from either edible or cultivated plants, and disturbance indicators such as *Cecropia* sp. (Marchant et al. 2002), has yielded important information concerning pre-Columbian land use. Yet, it still has important limitations which need to be overcome. A particular challenge, for example, is differentiating between disturbances which are anthropogenic *versus* those which are natural, especially in geomorphologically dynamic environments such as southwestern Amazonia (Lombardo 2014, 2016, 2017). In most cases, the information obtained in palaeoecological records can only provide coarse evidence of human land use (Bush and Colinvaux 1988), which results in general (not detailed) interpretations of human-driven modification of the landscape (Bush et al. 2007b, 2016). Accurate evaluation of the legacy of past indigenous peoples within current ecosystems—particularly regarding species richness—requires interdisciplinary studies that combine long- and short- term data such as archaeological, palaeoecological, and modern vegetation inventories (e.g., Maezumi et al. 2018).

Climatic conditions also differ largely within Amazonia (Fig. 19.1), preventing supra-local extrapolations in most cases (i.e., to make basin-wide extrapolations from local-scale or region-specific data). For instance, equatorial western Amazonia is one of the wettest regions of the basin, containing forests which experience no dry season (i.e. months with <100 mm precipitation). In such areas, it has been argued that slash-and-burn deforestation was difficult without readily available stones (for the production of axes) and given the poor flammability of vegetation under wet conditions (Bush et al. 2015). Clearly, absence of charcoal in these wet aseasonal regions would not necessarily signify absence of human occupation or land use, in which case alternative lines of evidence of human occupation would need to be found (Kelly et al. 2018). Although Charles Lyell's analogue model is certainly useful for inferring the past, it does not necessarily hold true for all regions or time periods, especially with respect to slash-and-burn agriculture, which, prior to the introduction of metal by Europeans, was likely far less common in pre-Columbian times than today (Lathrap et al. 1985; Denevan 1992b). Instead, agroforestry systems which did not require forest clear-cutting and shifting cultivation was likely a much more common form of land use in pre-Columbian times than today (Denevan et al. 1984; Peters 2000; Levis et al. 2018; Maezumi et al. 2018). Therefore, forest enrichment of socio-economically important and edible plants could have been a much more common form of land use than burning in pre-Columbian times (Politis 1996; Maezumi et al. 2018). To deal with this issue, a major challenge for palaeoecologists is to increase the taxonomic resolution of their vegetation proxies to species level, in order to identify socio-economically useful plant species (Rull 2014). Another limitation that requires attention is the poor understanding of the spatial scale of human settlements and its signal in sedimentary archives. Whereas archaeology generally focuses on the local area of habitation, palaeoecological records are collected from water bodies that might be capturing a regional signal depending on the basin size. This feature of the palaeoecological records result in the difficulty of quantifying the spatial scale represented by a pollen or charcoal record. In addition, archaeological and palaeoecological archives differ in the chronological resolution that can be achieved. Given these differences, it is often challenging to effectively integrate data from different disciplines. So far, what it is clear is that more interdisciplinary collaboration is needed (Mayle and Iriarte 2014).

Following the accumulation of relevant data over the last decades, the dichotomy between the extreme and exclusive views of pristine or manufactured parklands has shifted to whether the heterogeneous past human impact was or was not strong throughout the basin (e.g., McMichael et al. 2012b). An additional conclusion from these studies relates to the spatial scale considered. Taking into account the size of Amazonia, the number of both archaeological and palaeoecological studies published so far is very limited. This is shown in Fig. 19.2, where vast areas lack sampling points. Even within well-studied regions it is clear how little overlap exists between palaeoecological studies (concentrated toward the margin of Amazonia) and archaeological sites (distributed along the main rivers) (Lombardo et al. 2018b). In this sense, some authors argue that small sample sizes, lack of interdisciplinary (multi-proxies) overlap, and the proxies used to date are insufficient to (1) detect subtle human activities across vast areas, (2) reveal the diversity of ancient domestication and management processes, and finally (3) extrapolate results from one region to the entire heterogeneous Amazonia (Stahl 2015; Clement et al. 2015a; Montoya 2018; Lombardo et al. 2018b). Caution must be taken especially regarding this last argument: whereas in archaeological studies there is a *priori* assumption of human presence, this is not necessarily the case in palaeoecology. However, given the water resources supply that lakes offer, it is not rare to expect a higher likelihood of human presence near the locations where the sedimentary records are retrieved.

Although this chapter has focused on the human contribution to diversity prior to the arrival of Europeans (pre-1492 CE), we must remember that the anthropogenic legacy upon Amazonian plant communities did not stop at the contact eve. The colonial period and, more recently, globalization, have produced important biotic interchanges that has contributed to Amazonian plant diversity by the introduction of exotic cultigens (e.g., sugar, banana) and animals from elsewhere. On the one hand, the deforestation of vast areas of the Amazonia involving fire– for industrialised agriculture and cattle ranching (Carmenta et al. 2018)—is rapidly altering the composition, structure and functioning of rainforest ecosystems in recent decades (Longo et al. 2018). On the other hand, increased indigenous political activism, the establishment of protective laws, and the recognition of indigenous territories (indigenous lands), may lead to the expansion of indigenous forest management (Muehlebach 2001). We will need to await the consequences of these changes for the future of Amazonia's biodiversity.

5 Final Remarks

Pre-Columbian human impact on Amazonian plant diversity has been occurring since humans first arrived on the continent, and a clear legacy of this disturbance remains in today's forests. Late Pleistocene/early Holocene human practices such as the onset of plant domestication and cultivation, and perhaps the anthropogenic influence on the terminal Pleistocene megafaunal extinctions, had long-lasting impacts still evident in today's ecosystems. The scientific debate about the presence, population density, and impact of pre-Columbian peoples upon their rainforest environment prior to the arrival of Europeans has now extended to studies of the human legacy on current Amazonian plant diversity. Some of the interpretations obtained are the result of studies developed in specific locations and cannot be extrapolated to other areas or the entire region given the spatial extent and heterogeneity of Amazonia. Nevertheless, based on the available evidence, a potential human contribution to current diversity recognisable in an Amazonian context (supra-regionally, in a wider scale of more than a single site), would be suggested by the following general processes: (1) construction of vast earthworks, which has altered forest versus savannah cover via changes to micro-topography, fire use and hydrology, (2) widespread distribution and dispersal of domesticated plants, (3) change in soil properties, and (4) enrichment of plant communities with edible and useful species.

To obtain a better understanding of the degree to which pre-Columbian peoples have shaped Amazonia's plant diversity (and where), a greater range of complementary approaches and techniques are needed from different disciplines. First, greater sampling density is needed across Amazonia. New study sites should prioritise random locations as well as sites deliberately located close to and far from archaeological sites. In order to improve our understanding of the historical human role in Amazonian biodiversity, one must consider the characteristics of the study area (climate, landscape and cultures) to determine whether broader scale extrapolations are warranted (e.g. with respect to fire regime). Thus, independent proxies are needed to determine the degree to which disturbances, such as fire, are anthropogenic or natural (climate driven). These include a combination of climatic proxies (speleothems, biomarkers), soil properties (isotopic composition), and evidence of direct human presence (non-pollen palynomorphs, biomarkers). Increased understanding of the degree to which current patterns of Amazonian biodiversity were shaped by pre-Columbian indigenous peoples, and the process by which this was achieved, may provide important insights into elucidating rainforest resilience to future land-use and climate change, as well as the development of long-term, sustainable land-use strategies.

Acknowledgments The authors would like to thank Valentí Rull for the invitation to participate in the book. Encarni Montoya was funded by the Spanish Ministry of Economy and Competitivity (Juan de la Cierva Incorporación contract, ref. IJCI-2015-24273) and the Catalan Agency for Universities and Research AGAUR and the EU programme Marie Curie COFUND (Beatriu de Pinos—Marie Curie COFUND fellowship, ref.: 2014 BP-B 00094). Umberto Lombardo was funded by the European Union's Horizon 2020 research and innovation programme (Marie Skłodowska-Curie actions, EU project 703045). Carolina Levis thanks CNPq for a doctoral and postdoctoral fellowships (CNPq 141652/2014-4 and 159440/2018-1). Special thanks to Hans ter Steege and Ana Carvajal for comments of an earlier version of the manuscript.

References

Alvez-Valles CM, Balslev H, García-Villacorta R et al (2018) Palm species richness, latitudinal gradients, sampling effort, and deforestation in the Amazon region. Acta Bot Bras 32(4). https://doi.org/10.1590/010233062017abb0400

Antonelli A, Zizka A, Carvalho FA et al (2018) Amazonia is the primary source of Neotropical biodiversity. Proc Natl Acad Sci U S A 115(23):6034–6039. https://doi.org/10.1073/pnas.1713819115

Aymard G (2015) Bosques de los Llanos de Venezuela: estructura, composición florística, diversidad y estado actual de conservación. In: López-Falcón R, Hétier JM, López HD et al (eds) Tierras Llaneras de Venezuela. IRD-CIDIAT, Mérida, pp 241–268

Balée W (1989) The culture of Amazonian forests. In: Posey DA, Balée W (eds) Resource management in Amazonia: indigenous and folk strategies. NYGB Press, New York, pp 1–21

Balée W (1993) Indigenous transformation of Amazonian forests: an example from Maranhão, Brazil. L'Homme 33:231–254

Balée W (1994) Footprints of the forest: Ka'apar ethnobotany-the historical ecology of plant utilization by an Amazonian people. Columbia University Press, Columbia

Balée W (2006) The research program of historical ecology. Annu Rev Anthropol 35:75–98

Balée W (2013) Cultural forests of the Amazon: a historical ecology of people and their landscapes. University of Alabama Press, Tuscaloosa

Balée W, Erickson CL (2006) Time, complexity, and historical ecology. Columbia University Press, New York

Barlow J, Gardner TA, Lees AC et al (2012) How pristine are tropical forests? An ecological perspective on the pre-Columbian human footprint in Amazonia and implications for contemporary conservation. Biol Conserv 151:45–49

Barnosky AD, Lindsey EL (2010) Timing of Quaternary megafaunal extinction in South America in relation to human arrival and climate change. Quat Int 217:10–29

Berrío JC, Wouters H, Hooghiemstra H et al (2012) Using paleoecological data to define main vegetation dynamics along the savannah-forest ecotone in Colombia: impliactions for accurate assessment of human impacts. In: Myster RW (ed) Ecotones between forest and grassland. Springer, New York, pp 209–225

Bezerra J, Turhout E, Melo Vasquez I et al (2016) The promises of the Amazonian soil: shifts on discourses of terra preta and biochar. J Environ Policy Plan 21(5):623–635. https://doi.org/10.1080/1523908X.2016.1269644

Blatrix R, Roux B, Béarez P et al (2018) The unique functioning of a pre-Columbian Amazonian floodplain fishery. Sci Rep 8:5998

Bruno JF, Stachowicz JL, Bertness MD (2003) Inclusion of facilitation into ecological theory. Trends Ecol Evol 18:119–125

Bush MB, Colinvaux PA (1988) A 7000-year pollen record from the Amazon lowlands, Ecuador. Vegetatio 76:141–154

Bush MB, McMichael CHN (2016) Holocene variability of an Amazonian hyperdominant. J Ecol 104:1370–1378

Bush MB, Silman MR (2007) Amazonian exploitation revisited: ecological asymmetry and the policy pendulum. Front Ecol Environ 5:457–465

Bush MB, Piperno DR, Colinvaux PA (1989) A 6,000 year history of Amazonian maize cultivation. Nature 340:303–305

Bush MB, Silman MR, de Toledo MB et al (2007a) Holocene fire and occupation in Amazonia: records from two lake districts. Philos Trans R Soc Lond B Biol Sci 362:209–218

Bush MB, Silman MR, Listopad CMCS (2007b) A regional study of Holocene climate change and human occupation in Peruvian Amazonia. J Biogeogr 34:1342–1356

Bush MB, Restrepo A, Collins AF (2014) Galápagos history, restoration and a shifted baseline. Restor Ecol 22:296–298

Bush MB, McMichael CH, Piperno DR et al (2015) Anthropogenic influence on Amazonian forest in pre-history: an ecological perspective. J Biogeogr 42:2277–2288
Bush MB, Correa-Metrio A, McMichael CH et al (2016) A 6900-year history of landscape modification by humans in lowland Amazonia. Quat Sci Rev 141:52–64
Carmenta R, Coudel E, Stewards AM (2018) Forbidden fire: does criminalising fire hinder conservation efforts in swidden landscapes of the Brazilian Amazon? Geogr J 184:1–15
Carson JF, Whitney BS, Mayle FE et al (2014) Environmental impact of geometric earthwork construction in pre-Columbian Amazonia. Proc Natl Acad Sci U S A 111:10497–10502
Carson JF, Mayle FE, Whitney BS et al (2016) Pre-Columbian ring ditch construction and land use on a "chocolate forest island" in the Bolivian Amazon. J Quat Sci 31:337–347
Chambouleyron R, Barbosa BC, Bombardi FA et al (2011) 'Formidable contagion': epidemics, work and recruitment in Colonial Amazonia (1660–1750). Hist Cienc Saúde Manguinhos 18:987–1004
Clement CR (1999) 1492 and the loss of Amazonian crop genetic resources. I. The relation between domestication and human population decline. Econ Bot 53:188–202
Clement CR (2014) Landscape domestication and archaeology. In: Smith C (ed) Encyclopedia of global archaeology. Springer, New York, pp 4388–4394
Clement CR, McCann JM, Smith NJ (2003) Agrobiodiversity in Amazonia and its relationship with dark earths. In: Lehmann J, Kern DC, Glaser B, Woods WI (eds) Amazonian dark earths: origin, properties, management. Springer, Berlin, pp 159–178
Clement CR, de Cristo-Araújo M, Coppens d'Eeckenbrugge G et al (2010) Origin and domestication of native Amazonian crops. Diversity 2:72–106
Clement CR, Denevan WM, Heckenberger MJ et al (2015a) The domestication of Amazonia before European conquest. Proc R Soc B Biol Sci 282:20150813
Clement CR, Denevan WM, Heckenberger MJ et al (2015b) Response to comment by McMichael, Piperno and Bush. Proc R Soc B Biol Sci 282:20152459
Clement CR, Rodrigues DP, Alves-Pereira A et al (2016) Crop domestication in the upper Madeira River basin. Bol Mus Para Emflio Goeldi Cienc Hum 11:193–205
Clement CR, Cristo-Araújo M, Coppens d'Eeckenbrugge G et al (2017) Origin and dispersal of domesticated peach palm. Front Ecol Evol 5:148
Colinvaux P (2007) Amazon expeditions: my quest for the ice-age equator. Yale University Press, New Haven
Crevels M, Van der Voort H (2008) The Guaporé-Mamoré region as a linguistic area. In: Muysken P (ed) From linguistic areas to areal linguistics. John Benjamins Press, Amsterdam, pp 151–179
Crosby AW (2004) Ecological imperialism: the biological expansion of Europe, 900–1900. Cambridge University Press, Cambridge
da Silva Meneses MEN, Lima da Costa M, Behling H (2013) Late Holocene vegetation and fire dynamics from a savanna-forest ecotone in Roraima state, northern Brazilian Amazon. J S Am Earth Sci 42:17–26
de Souza JG, Schaan DP, Robinson M et al (2018) Pre-Columbian earth-builders settled along the entire southern rim of the Amazon. Nat Commun 9:1125
de Toledo MB, Bush MB (2007) A Mid-Holocene environmental change in Amazonian savannas. J Biogeogr 34:1313–1326
Denevan WM (1992a) The pristine myth: the landscapes of the Americas in 1492. Ann Assoc Am Geogr 82:369–385
Denevan WM (1992b) Stone vs metal axes: the ambiguity of shifting cultivation in prehistoric Amazonia. J Steward Anthropol Soc 20:153–165
Denevan WM (1996) A bluff model of riverine settlement in prehistoric Amazonia. Ann Assoc Am Geogr 86:654–681
Denevan WM, Treacy JM, Alcorn JB et al (1984) Indigenous agroforestry in the Bora Indian management of swidden fallows. Interciencia 9:346–357
Dexter K, Chave J (2016) Evolutionary patterns of range size, abundance and species richness in Amazonian angiosperm trees. PeerJ 4:e2402

Dobyns HF (1966) Estimating aboriginal American population: an appraisal of techniques with a new hemispheric estimate. Curr Anthropol 7:395–416
Doughty CE (2010) The development of agriculture in the Americas: an ecological perspective. Ecosphere 1:art21
Doughty CE, Wolf A, Malhi Y (2013) The legacy of the Pleistocene megafauna extinctions on nutrient availability in Amazonia. Nat Geosci 6:761–764
Doughty CE, Wolf A, Morueta-Holme N et al (2016) Megafauna extinction, tree species range reduction, and carbon storage in Amazonian forests. Ecography 39:194–203
Erickson CL (2001) Pre-Columbian roads of the Amazon. Expedition 43:21–30
Erickson CL (2006) The domesticated landscapes of the Bolivian Amazon. In: Balée W, Erickson CL (eds) Time and complexity in historical ecology: studies in the Neotropical Lowlands. Columbia University Press, New York, pp 235–278
Erickson CL (2008) Amazonia: the historical ecology of a domesticated landscape. In: Silverman H, Isbell WH (eds) Handbook of South American archaeology. Springer, Berlin, pp 157–183
Erickson CL, Balée WL (2006) The historical ecology of a complex landscape in Bolivia. In: Balée W, Erickson CL (eds) Time and complexity in historical ecology: studies in the Neotropical Lowlands. Columbia University Press, New York, pp 187–233
Eva HD, Huber O, Achard F et al (2005) Synthesis of the results from an expert consultation workshop organized by the European Commission in collaboration with the Amazon Cooperation Treaty Organization – JRC Ispra, 7–8 June 2005, EC, Luxembourg, p 40
Flantua SGA, Hooghiemstra H, Grimm EC et al (2015) Updated site compilation of the Latin American Pollen Database. Rev Palaeobot Palynol 223:104–115
Flores BM, Holmgren M, Xu C et al (2017) Floodplains as an Achilles' heel of Amazonian forest resilience. Proc Natl Acad Sci U S A 114:4442–4446
Galluzzi G, Dufour D, Thomas E et al (2015) An integrated hypothesis on the domestication of *Bactris gasipaes*. PLoS One 10:e0144644
Gentry AH (1993) A field guide to the families and genera of woody plants of northwest South America. Conservation International, Washington, DC
Glaser B, Birk JJ (2012) State of the scientific knowledge on properties and genesis of anthropogenic dark earths in Central Amazonia (terra preta de Índio). Geochim Cosmochim Acta 82:39–51
Glaser B, Haumaier L, Guggenberger G et al (2001) The "terra preta" phenomenon: a model for sustainable agriculture in the humid tropics. Naturwissenschaften 88:37–41
Goldberg A, Mychajliw AM, Hadly EA (2016) Post-invasion demography of prehistoric humans in South America. Nature 532:232–235
Goulding M, Barthem R, Ferreira EJG (2003) The Smithsonian atlas of the Amazon. Smithsonian Books, Washington, DC
Gross DR (1975) Protein capture and cultural development in the Amazon Basin. Am Anthropol 77:526–549
Guix JC (2005) Evidence of old anthropic effects in forests at the confluence of the Caurés and Negro Rivers, NW Amazonia: the role of Indians and Caboclos. Grup Estud Ecol 8:1–27
Heckenberger M, Neves EG (2009) Amazonian archaeology. Annu Rev Anthropol 38:251–266
Heckenberger MJ, Kuiluro A, Kuikuro UT et al (2003) Amazonia 1492: pristine forest or cultural parkland? Science 301:1710–1714
Heckenberger MJ, Russell JC, Fausto C et al (2008) Pre-Columbian urbanism, anthropogenic landscapes, and the future of the Amazon. Science 321:1214–1217
Hemming J (2006) Romance and reality: the first European vision of Brazilian Indians. In: Posey DA, Balick MJ (eds) Human Impacts on Amazonia: the role of traditional ecological knowledge in conservation and development. Columbia University Press, New York, pp 5–16
Hilbert L, Góes Neves E, Pugliese F et al (2017) Evidence for mid-Holocene rice domestication in the Americas. Nat Ecol Evol 1:1693–1698

Huber O, Duno de Stefano R, Aymard G et al (2006) Flora and vegetation of the Venezuelan Llanos: a review. In: Pennington RT, Lewis GP, Ratter JA (eds) Neotropical savannas and dry forests: plant diversity, biogeography and conservation. Taylor & Francis Group, LLC de Boca Ratón, FL, pp 95–120

IPCC (2013) Climate change 2013: the physical science basis. In: Stocker TF, D. Qin, G-K Plattner, M Tignor, SK Allen, J Boschung, A Nauels, Y Xia, V Bex, PM Midgley (eds) Contribution of working group I to the fifth assessment report of the intergovernmental panel on climate change. Cambridge University Press, Cambridge, 1535 pp

Iriarte J, Power MJ, Rostain S et al (2012) Fire-free land use in pre-1492 Amazonian savannas. Proc Natl Acad Sci U S A 109:6473–6478

Jameson W (1858) Excursion made from Quito to the River Napo, January to May, 1857. J Roy Geogr Soc Lond 28:337–349

Junqueira AB, Shepard GH, Clement CR (2010) Secondary forests on anthropogenic soils in Brazilian Amazonia conserve agrobiodiversity. Biodivers Conserv 19:1933–1961

Junqueira AB, Souza NB, Stomph TJ et al (2016) Soil fertility gradients shape the agrobiodiversity of Amazonian homegardens. Agric Ecosyst Environ 221:270–281

Junqueira AB, Levis C, Bongers F et al (2017) Response to comment on "Persistent effects of pre-Columbian plant domestication on Amazonian forest composition". Science 358:aan8837

Keller MA, Alencar A, Asner GP et al (2004) Ecological research in the large-scale biosphere atmosphere experiment in Amazonia (LBA): early results. Ecol Appl 14:S3–S16

Kelly TJ, Lawson IT, Roucoux KH et al (2018) Continuous human presence without extensive reductions in forest cover over the past 2500 years in an aseasonal Amazonian rainforest. J Quat Sci 33:369–379

Kristiansen T, Svenning JC, Grandez C et al (2009) Commonness of Amazonian palm (Arecaceae) species: cross-scale links and potential determinants. Acta Oecol Int J Ecol 35:554–562

Lathrap DW (1970) The upper Amazon. Thames and Hudson, Southampton

Lathrap DW, Gebhart-Sayer A, Mester AM (1985) The roots of the Shipibo art style: three waves on Imiríacocha, or there were "Incas" before Incas. J Lat Am Lore 11:31–119

Lauterjung MB, Bernardi AP, Montagna T et al (2018) Phylogeography of Brazilian pine (*Araucaria angustifolia*): integrative evidence for pre-Columbian anthropogenic dispersal. Tree Genet Genomes 14:36

Levis C, de Souza PF, Schietti J et al (2012) Historical human footprint on modern tree species composition in the Purus-Madeira interfluve, Central Amazonia. PLoS One 7:e48559

Levis C, Costa FRC, Bongers F et al (2017) Persistent effects of pre-Columbian plant domestication on Amazonian forest composition. Science 355:925–931

Levis C, Flores BM, Moreira PA et al (2018) How people domesticated Amazonian forests. Front Ecol Evol 5:171

Liebmann B, Allured D (2005) Daily precipitation grids for South America. Bull Am Meteorol Soc 86:1567–1570

Lins J, Lima HP, Baccaro FB et al (2015) Pre-Columbian floristic legacies in modern homegardens of Central Amazonia. PLoS One 10:e0127067

Liu KB, Colinvaux PA (1988) A 5200-year history of Amazonian rain forest. J Biogeogr 15:231–248

Livi-Bacci M (2016) The depopulation of upper Amazonia in colonial times. Rev Indias 76:419–448

Lombardo U (2014) Neotectonics, flooding patterns and landscape evolution in southern Amazonia. Earth Surf Dyn 2:493–511

Lombardo U (2016) Alluvial plain dynamics in the southern Amazonian foreland basin. Earth Surf Dyn 7:453–467

Lombardo U (2017) River logjams cause frequent large-scale forest die-off events in southwestern Amazonia. Earth Surf Dyn 8:565–575

Lombardo U, Prümers H (2010) Pre-Columbian human occupation patterns in the eastern plains of the Llanos de Moxos, Bolivian Amazonia. J Archaeol Sci 37:1875–1885

Lombardo U, Canal-Beeby E, Fehr S et al (2011a) Raised fields in the Bolivian Amazonia: a prehistoric green revolution or a flood risk mitigation strategy? J Archaeol Sci 38:502–512
Lombardo U, Canal-Beeby E, Veit H (2011b) Eco-archaeological regions in the Bolivian Amazon. Geogr Helv 66:173–182
Lombardo U, May JH, Veit H (2012) Mid- to late-Holocene fluvial activity behind pre-Columbian social complexity in the southwestern Amazon basin. The Holocene 22:1035–1045
Lombardo U, Denier S, May J-H et al (2013a) Human-environment interactions in pre-Columbian Amazonia: the case of the Llanos de Moxos, Bolivia. Quat Int 312:109–119
Lombardo U, Szabo K, Capriles JM et al (2013b) Early and middle Holocene hunter-gatherer occupations in Western Amazonia: the hidden shell middens. PLoS One 8:e72746
Lombardo U, Rodrigues L, Veit H (2018a) Alluvial plain dynamics and human occupation in SW Amazonia during the Holocene: a paleosol-based reconstruction. Quat Sci Rev 180:30–41
Lombardo U, McMichael C, Tamanaha EK et al (2018b) Mapping pre-Columbian land use in Amazonia. PAGES Mag 26:14–15
Longo M, Knox RG, Levine NM et al (2018) Ecosystem heterogeneity and diversity mitigate Amazon forest resilience to frequent extreme droughts. New Phytol 219(3):914–931. https://doi.org/10.1111/nph.15185
Loughlin NJD, Gosling WD, Mothes P et al (2018) Ecological consequences of post-Columbian depopulation in the Andean-Amazonian corridor. Nat Ecol Evol 2:1233–1236
Lyons SK, Amatangelo KL, Behrensmeyer AK et al (2016) Holocene shifts in the assembly of plant and animal communities implicate human impacts. Nature 529:80–83
Maezumi SY, Whitney B, Mayle F et al (2017) Reassessing climate and pre-Columbian drivers of paleofire activity in the Bolivian Amazon. Quat Int 3:1–14. https://doi.org/10.1016/j.quaint.2017.11.053
Maezumi SY, Alves D, Robinson M et al (2018) The legacy of 4500 years of polyculture agroforesty in the Eastern Amazon. Nat Plants 4(8):540–547. https://doi.org/10.1038/s41477-018-0205-y
Marchant R, Almeida L, Behling H et al (2002) Distribution and ecology of parent taxa of pollen lodged within the Latin American Pollen Database. Rev Palaeobot Palynol 121:1–75
Maslin M, Malhi Y, Phillips O et al (2005) New views on an old forest: assessing the longevity, resilience and future of the Amazon rainforest. Trans Inst Br Geogr 30:477–499
Mayle FE, Iriarte J (2014) Integrated palaeoecology and archaeology – a powerful approach for understanding pre-Columbian Amazonia. J Archaeol Sci 51:54–64
Mayle FE, Power MJ (2008) Impact of a drier Early-Mid-Holocene climate upon Amazonian forests. Philos Trans R Soc B 363:1829–1838
McKey D, Rostain S, Iriarte J et al (2010) Pre-Columbian agricultural landscapes, and self-organized patchiness in Amazonia. Proc Natl Acad Sci U S A 107:7823–7828
McMichael CH, Piperno DR, Bush MB et al (2012a) Sparse pre-Columbian human habitation in western Amazonia. Science 336:1429–1431
McMichael CH, Bush MB, Piperno DR et al (2012b) Spatial and temporal scales of pre-Columbian disturbance associated with western Amazonian lakes. The Holocene 22:131–141
McMichael CH, Piperno DR, Bush MB (2015a) Comment on Clement et al. 2015 "The domestication of Amazonia before European conquest". Proc R Soc B 282:20151837
McMichael CH, Piperno DR, Neves EG et al (2015b) Phytolith assemblages along a gradient of ancient human disturbance in western Amazonia. Front Ecol Evol 3:1–15
McMichael CH, Matthews-Bird F, Farfan-Rios W et al (2017a) Ancient human disturbances may be skewing our understanding of Amazonian forests. Proc Natl Acad Sci U S A 114:522–527
McMichael CH, Feeley KJ, Dick CW, Piperno DR, Bush MB (2017b) Comment on "Persistent effects of pre-Columbian plant domestication on Amazonian forest composition". Science 358: eaan8347
Medina JT (1934) The discovery of the Amazon according to the account of Friar Gaspar de Carvajal and other documents. The American Geographical Society, New York, NY

Meggers BJ (1954) Environmental limitation on the development of culture. Am Anthropol 56:801–824
Meggers BJ (2001) The mystery of the Marajoara: an ecological solution. Amazoniana 16:421–440
Montoya E (2018) Historia de la Amazonía: contribución de la paleoecología al debate de ocupación precolombina y sus efectos en el ecosistema. Ecosistemas 27:18–25
Montoya E, Rull V (2011) Gran Sabana fires (SE Venezuela): a paleoecological perspective. Quat Sci Rev 30:3430–3444
Montoya E, Rull V, Nogué S (2011) Early human occupation and land use changes near the boundary of the Orinoco and the Amazon basins (SE Venezuela): palynological evidence from El Paují record. Palaeogeogr Palaeoclimatol Palaeoecol 310:413–426
Morcote-Ríos G, Bernal R (2001) Remains of palms (Palmae) at archaeological sites in the New World: a review. Bot Rev 67:309–350
Morcote-Ríos G, León-Sicard T (2012) Las *Terras Pretas* del Igarape Takana: un sistema de cultivo precolombino en Leticia Amazonas –Colombia. Universidad Nacional de Colombia, Facultad de Ciencias, Bogotá
Morcote-Ríos G, Raz L, Giraldo-Cañas D et al (2013) Terras Pretas de Índio of the Caquetá-Japurá River (Colombian Amazonia). Tipití 11:30–39
Moreira PA, Aguirre-Dugua X, Mariac C et al (2017) Diversity of treegourd (Crescentia cujete) suggests introduction and prehistoric dispersal routes into Amazonia. Front Ecol Evol 5:150
Muehlebach A (2001) "Making place" at the United Nations: indigenous cultural politics at the UN working group on indigenous populations. Cult Anthropol 16:415–448
Neves EG, Petersen JB, Bartone RN et al (2003) Historical and socio-cultural origins of Amazonian dark earths. In: Lehmann J, Kern DC, Glaser B, Woods WI (eds) Amazonian dark earths: origin, properties, management. Springer, Berlin, pp 29–50
Neves EG, Petersen JB, Bartone RN et al (2004) The timing of terra preta formation in the central Amazon: archaeological data from three sites. In: Glaser B, Woods WI (eds) Amazonian dark earths: explorations in space and time. Springer, Berlin, pp 125–134
Nolan C, Overpeck JT, Allen JRM et al (2018) Past and future global transformation of terrestrial ecosystems under climate change. Science 361:920–923
Olson DM, Dinerstein E, Wikramanayake ED et al (2001) Terrestrial ecoregions of the world: a new map of life on Earth. Bioscience 51:933–938
Palace MW, McMichael CNH, Braswell BH et al (2017) Ancient Amazonian populations left lasting impacts on forest structure. Ecosphere 8:e02035
Pärssinen M, Ranzi A, Saunaluoma S et al (2003) Geometrically patterned ancient earthworks in the Rio Branco Region of Acre, Brazil: new evidence of ancient chiefdom formations in Amazonian interfluvial *terra firme* environment. In: Pärssinin M, Korpisaari A (eds) Western Amazonia – Amazônia Ocidental. Multidisciplinary studies on ancient expansionistic movements, fortifications and sedentary life. Helsinki, Renvall Institute for Area and Cultural Studies, University of Helsinki, pp 97–133
Perry L, Dickau R, Zarrillo S et al (2007) Starch fossils and the domestication and dispersal of chili peppers (*Capsicum* spp. L.) in the Americas. Science 315:986–988
Peters CM (2000) Precolumbian silviculture and indigenous management of neotropical forests. In: Lentz DL (ed) Imperfect balance: landscape transformations in the Precolumbian Americas. Columbia University Press, New York, NY, pp 203–223
Piperno DR (2011) The origins of plant cultivation and domestication in the New World tropics. Curr Anthropol 52:S453–S470
Piperno DR, Holst I, Wessel-Beaver L et al (2002) Evidence for the control of phytolith formation in Cucurbita fruits by the hard rind (Hr) genetic locus: archaeological and ecological implications. Proc Natl Acad Sci U S A 99:10923–10928
Piperno DR, McMichael C, Bush MB (2015) Amazonia and the Anthropocene: what was the spatial extent and intensity of human landscape modification in the Amazon basin at the end of prehistory? The Holocene 25:1588–1597

Politis GG (1996) Moving to produce: Nukaka mobility and settlement patterns in Amazonia. World Archaeol 27:492–511
Posey DA (1985) Indigenous management of tropical forest ecosystems: the case of the Kayapó Indians of the Brazielian Amazon. Agrofor Syst 3:139–158
Prestes-Caneiro G, Béarez P, Bailon S et al (2016) Subsistence fishery at Hatahara (750–1230 CE), a pre-Columbian central Amazonian village. J Archaeol Sci Rep 8:454–462
Prümers H, Jaimes Betancourt C (2014) 100 años de investigación arqueológica en los Llanos de Mojos. Arqueoantropológicas 4:11–53
Quintero-Vallejo E, Klomberg Y, Bongers F et al (2015) Amazonian dark earth shapes the understory plant community in a Bolivian forest. Biotropica 47:152–161
Redford KH (1991) The ecologically noble savage. Cult Surviv Q 1:46–48
Renard D, Iriarte J, Birks JJ et al (2012) Ecological engineers ahead of their time: the functioning of pre-Columbian raised-field agriculture and its potential contributions to sustainability today. Ecol Eng 45:30–44
Riris P (2018) Assessing the impact and legacy of swidden farming in neotropical interfluvial environments through exploratory modelling of post-contact Piaroa land use (Upper Orinoco, Venezuela). The Holocene 28:945–954
Roberts P, Boivin N, Lee-Thorp J, Petraglia M, Stock J (2016) Tropical forests and the genus *Homo*. Evol Anthropol 25:306–317
Roberts P, Hunt C, Arroyo-Kalin M et al (2017) The deep human prehistory of global tropical forest and its relevance for modern conservation. Nat Plant 3:17093
Rodrigues L, Lombardo U, Veit H (2018) Design of pre-Columbian raised fields in the Llanos de Moxos, Bolivian Amazon: differential adaptations to the local environment? J Archaeol Sci Rep 17:366–378
Roosevelt AC (1991) Moundbuilders of the Amazon: geophysical archaeology on Marajo Island, Brazil. Academic, San Diego, CA
Roosevelt AC (2013) The Amazon and the Anthropocene: 13,000 years of human influence in a tropical rainforest. Anthropocene 4:69–87
Roosevelt AC, Lima da Costa M, lopes Machado C et al (1996) Paleoindian cave dwellers in the Amazon: the peopling of the Americas. Science 272:373–384
Roosevelt AC, Douglas J, Brown L (2002) Migrations and adaptations of the first Americans: Clovis and pre-Clovis viewed from South America. In: Jablonski N (ed) The first Americans: the Pleistocene colonization of the New World, vol 27. University of California, Berkeley, pp 159–236
Rostain S (1991) Les champs surélevés amérindiens de la Guyane, Coll° La Nature et l'Homme. Centre ORSTOM de Cayenne/Institut Géographique National, Cayenne
Rostain S (2008) Agricultural earthworks on the French Guiana Coast. In: Silverman H, Isbell WH (eds) Handbook of South American archaeology. Springer, Berlin, pp 217–233
Rostain S (2010) Pre-Columbian earthworks in coastal Amazonia. Diversity 2:331–352
Rull V (2014) Time continuum and true long-term ecology: from theory to practice. Front Ecol Evol 2:75
Rull V (2018) Strong fuzzy EHLFS: a general conceptual framework to address past records of environmental, ecological and cultural change. Quaternary 1:10
Rull V, Montoya E (2014) *Mauritia flexuosa* palm swamp communities: natural or human-made? A palynological study of the Gran Sabana region (northern South America) within a neotropical context. Quat Sci Rev 99:17–33
Rull V, Vegas-Villarrúbia T, Nogué S et al (2008) Bureaucratic obstruction of conservation science in the Guayana highlands. Conserv Biol 22:508–509
Rull V, Montoya E, Nogué S et al (2013) Ecological palaeoecology in the neotropical Gran Sabana region: long-term records of vegetation dynamics as a basis for ecological hypothesis testing. Perspect Plant Ecol Evol Syst 15:338–359
Rull V, Montoya E, Vegas-Vilarrúbia T et al (2015) New insights on paleofires and savannisation in northern South America. Quat Sci Rev 122:158–165

Sánchez F, Fernández J, Gassón R et al (2017) Paleoecología y ocupación humana durante el Holoceno en los Llanos del Orinoco: una revisión y nuevos datos. Biollania 15:297–333
Sanford RL, Saldarriaga J, Clark KE et al (1985) Amazon rain-forest fires. Science 227:53–55
Schaan D (2008) The nonagricultural chiefdoms of Marajó Island. In: Silverman H, Isbell WH (eds) Handbook of South American archaeology. Springer, Berlin, pp 339–357
Shepard GH Jr, Ramirez H (2011) "Made in Brazil": human dispersal of the Brazil Nut (*Bertholletia excelsa*, Lecythidaceae) in ancient Amazonia. Econ Bot 65:44–65
Siren A (2014) History of natural resource use and environmental impacts in an interfluvial upland forest area in western Amazonia. Fennia 192:36–53
Smith N (2015) Palms and people in the Amazon. Geobotany studies: basics, methods and case studies. Springer, Berlin
Stahl PW (2015) Interpreting interfluvial landscape transformations in the pre-Columbian Amazon. The Holocene 25:1598–1603
Stenborg P (2009) Points of convergence-routes of divergence: some considerations based on Curt Nimuendajú's archaeological work in the Satarém-Trombetas Area and Amapá. In: Whitehead NL, Alemán SW (eds) Anthropologies of Guayana: cultural spaces in northeastern Amazonia. University of Arizona Press, Tucson, pp 55–73
Stenborg P, Schaan DP, Amaral-Lima M (2012) Precolumbian land use and settlement pattern in the Santarém region, lower Amazon. Amazônica 4:222–250
Stenborg P, Schaan DP, Figueiredo CG (2018) Contours of the past: LiDAR data expands the limits of late pre-Columbian human settlement in the Santarém region, lower Amazon. J Field Archaeol 43:1–14
ter Steege H, Pitman NCA, Sabatier D et al (2013) Hyperdominance in the Amazonian tree flora. Science 342:1243092
ter Steege H, Mota de Oliveira S, Pitman NCA, Sabatier D, Antonelli A, Guevara Andino JE, Aymard GA, Salomão RP (2019) Towards a dynamic list of Amazonian tree species. Sci Rep 9 (1):3501
Thomas E, van Zonneveld M, Loo J et al (2012) Present spatial diversity patterns of Theobroma cacao L. in the neotropics reflect genetic differentiation in Pleistocene refugia followed by human-influenced dispersal. PLoS One 7:e47676
Thomas E, Alcázar Caicedo C, McMichael CH et al (2015) Uncovering spatial patterns in the natural and human history of Brazil nut (Bertholletia excelsa) across the Amazon Basin. J Biogeogr 42:1367–1382
Tollefson J (2013) Footprints in the forest. Nature 502:160–162
Tomkeieff SI (1962) Unconformity – an historical study. Proc Geol Assoc 73:383–417
van der Kaars S, Miller GH, Turney CSM et al (2017) Humans rather than climate the primary cause of Pleistocene megafaunal extinction in Australia. Nat Commun 8:14142
Watling J, Iriarte J, Mayle FE et al (2017) Impact of pre-Columbian "geogliph" builders on Amazonian forests. Proc Natl Acad Sci U S A 114:1868–1873
Watling J, Shock MP, Mongeló GZ, Almeida FO, Kater T, De Oliveira PE et al (2018) Direct archaeological evidence for southwestern Amazonia as an early plant domestication and food production centre. PLoS One 13:e0199868
Whitehouse NJ, Kirleis W (2014) The world reshaped: practices and impacts of early agrarian societies. J Archaeol Sci 51:1–11
Wills C, Harms KE, Condit R et al (2006) Nonrandon processes maintain diversity in tropical forests. Science 311:527–531
Woods WI, Glaser B (2004) Towards an understanding of Amazonian dark earths. In: Glaser B, Woods WI (eds) Amazonian dark earths: explorations in space and time. Springer, Berlin, pp 1–8
Zizka A, ter Steege H, Do Céo M et al (2018) Finding needles in the haystack: where to look for rare species in the American tropics. Ecography 41:321–330

Chapter 20
Historical Biogeography of Caribbean Plants Revises Regional Paleogeography

Julissa Roncal, María Esther Nieto-Blázquez, Agustín Cardona, and Christine D. Bacon

Abstract The Caribbean archipelago is the largest and most species rich insular system in the Neotropics, yet the origins of this biodiversity hotspot are poorly understood. Four main hypotheses have been proposed to explain Caribbean diversity: vicariance, GAARlandia landspan, long-distance dispersal, and in situ speciation. Here, we use phylogenetic data and historical biogeographical inferences of Caribbean endemic plants to test amongst these four hypotheses. Together with a revision of Caribbean paleogeography, we compile the colonization times, in situ speciation, and ancestral areas of Caribbean endemic lineages. We model species richness differences among endemic lineages as a function of their colonization time in the Caribbean, resulting in an estimation of the time-for-speciation effect. Our results show that plants repeatedly colonized the West Indies over the last 60 million years from continental America, especially from central and south America. Plant colonization times do not cluster during the GAARlandia timeframe and most colonization events occurred after its putative subsidence. We did not find a relationship between the number of species in each endemic clade and the estimated time when each colonized the Caribbean (no time-for-speciation effect). Published geological constraints from the Lesser Antilles and Aves Ridge fail to support the emergent land between the Antilles and south America.

Keywords Colonization · Divergence time · Endemic flora · Evolutionary assembly · GAARlandia · In situ speciation · Paleogeography · Time-for-speciation · Vicariance · West Indies

J. Roncal (✉) · M. E. Nieto-Blázquez
Department of Biology, Memorial University of Newfoundland, St. John's, NL, Canada
e-mail: jroncal@mun.ca; menb13@mun.ca

A. Cardona
Department of Processes and Energy, National University of Colombia, Medellín, Colombia
e-mail: agcardonamo@unal.edu.co

C. D. Bacon
Department of Biological and Environmental Sciences, University of Göteborg, Göteborg, Sweden

Gothenburg Global Biodiversity Centre, Göteborg, Sweden

V. Rull, A. C. Carnaval (eds.), *Neotropical Diversification: Patterns and Processes*, Fascinating Life Sciences, https://doi.org/10.1007/978-3-030-31167-4_20

1 Introduction

Considered as laboratories of evolution and biogeography, islands have inspired scientists for two centuries since the time of Darwin (1859) and Wallace (1869). The study of island biota has improved our understanding of important concepts in ecology and evolution such as speciation, adaptive radiation, extinction, colonization, and community assembly (MacArthur and Wilson 1967; Whittaker 1998; Schluter 2000; Gillespie 2004; Savolainen et al. 2006).

1.1 The Caribbean Islands or West Indies

The Caribbean archipelago, comprising the Bahamas, Greater, Lesser and Leeward Antilles, with its crescent shape that extends for ca. 3200 km in longitude, is the most important insular system in the Neotropics in terms of land area and species richness (Maunder et al. 2008). It is also one of the world's 34 biodiversity hotspots sensu Mittermeier et al. (2004). This archipelago (229,549 km^2 in 900 islands, Bellard et al. 2014) hosts nearly 13,000 seed plant species, up to 8000 of which are endemic (Mittermeier et al. 2004; Acevedo-Rodriguez and Strong 2008). The Caribbean archipelago houses several examples of adaptive radiations, like the *Anolis* lizards, which exhibit convergent evolution of morphological traits under similar environments (Losos et al. 2006) and the greater Antillean boas, which show microhabitat specialization linked to body size evolution and cranial morphological diversification (Reynolds et al. 2016). Unfortunately, the Caribbean is also the most threatened insular hotspot by high levels of habitat reduction as a result of sea level rise (Bellard et al. 2014) and has the highest deforestation rates among 13 tropical forest hotspots (Brooks et al. 2002).

1.2 A Paleogeographical Review of the Antilles

The Antilles and the submerged Aves Ridge (Fig. 20.1a) have a ca. 130 million year history during which their geographic position and size, including their exposure below sea level, have changed due to the northeastern displacement of the Caribbean oceanic plate and the changing nature of volcanic activity. The geological knowledge of the Greater and Leeward Antilles has been strong since the beginning of the Caribbean biogeographic discussions (Iturralde-Vinent and MacPhee 1999; Iturralde-Vinent 2006), becoming more sophisticated in the last years (Pindell and Kennan 2009; Lidiak and Anderson 2015). However, our understanding of the mostly submerged Aves Ridge (Neill et al. 2011) and the long term volcanic record of the Lesser Antilles (MacDonald et al. 2000) have not been critically considered in any paleogeographic model.

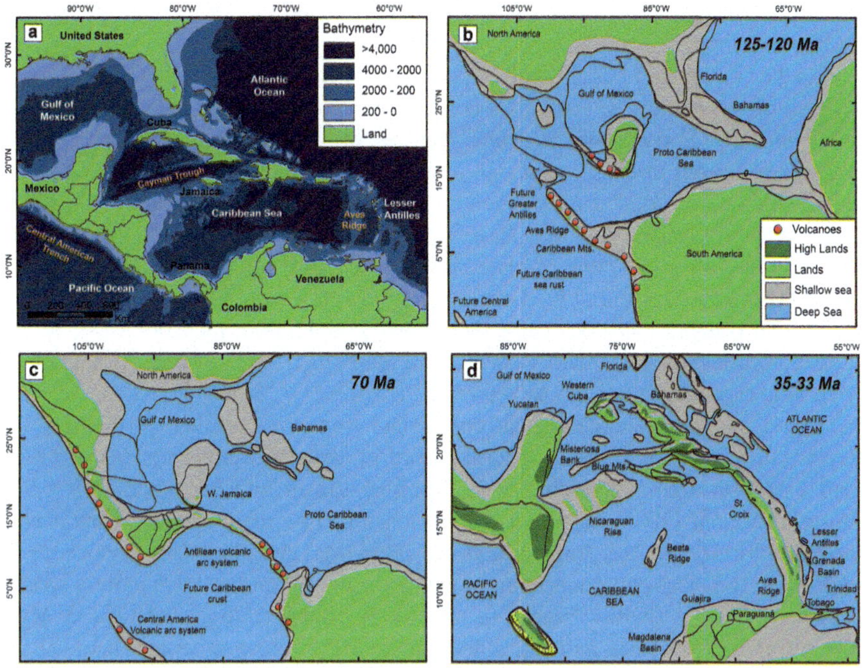

Fig. 20.1 Modern and paleogeographic configuration of the Antilles during selected time spans. Bathymetry and elevation in the Caribbean and the Antilles are after Sandwell et al. (2014). Cretaceous and Oligocene reconstruction with the distribution of sedimentary environments after Iturralde-Vinent (2006) who presented the GAARlandia hypothesis

After considerable paleogeographic debates which reached a climax by the end of last decade (James et al. 2009), it is currently accepted that the Greater Antilles and Aves Ridge (Fig. 20.1b) originated as a submerged to subaerially exposed chain of volcanic edifices located in a southwestern position within the ocean that linked North and South America during the lower Cretaceous (130–110 Mya, Pindell and Kennan 2009). This chain of volcanic edifices, known as proto-Antilles, subsequently drifted northeastwards as the ocean that separated the two Americas was subducted, colliding with the Yucatan Peninsula and northwestern Colombia and Ecuador 70–84 Mya as the volcanic edifices and associated oceanic plate were inserted between the Americas (Fig. 20.1c, Iturralde-Vinent 2006; Vallejo et al. 2006; Pindell and Kennan 2009; Villagómez et al. 2011). The continuous advance of this plate between the Americas created two major deformational belts since the early Eocene (c. 57 Mya). Cuba, Hispaniola, Puerto Rico and the Virgin Islands in the north, and Tobago and the Leeward Antilles in the south which collided with Florida and the Bahamas Platform, and with northern Colombia and Venezuela, respectively (Graham 2003). The landmasses of the Greater Antilles are thought to have been permanently subaerial only since the mid-Eocene when the collision with the Bahamas took place (<45 Mya, Iturralde-Vinent and MacPhee 1999; Iturralde-

Vinent 2006) due to recurrent transgression and subsidence of the Greater Antilles before that time.

As the Greater and Lesser Antilles collided and were uplifted with the continental margin of the Americas, volcanism continued in the Aves Ridge between 75 Mya and the early Eocene (48–55 Mya, Neill et al. 2011). After this time period, volcanism ceased in the Aves Ridge and a new volcanic chain was formed to the east, which corresponded to the modern Lesser Antilles (MacDonald et al. 2000). This latter volcanism that may have started 37.4 Mya (White et al. 2017) was characterized by slow rates of plate convergence and low generation of volcanic products since its beginning (2–4 cma^{-1} and 3–5 $\text{km}^3\ \text{Mya}^{-1}\ \text{km}^{-1}$) when compared with other oceanic volcanic chains (MacDonald et al. 2000; Jackson 2013). As the Lesser Antilles volcanic chain was formed, the extinct Aves Ridge volcanic chain was submerged and thinned (Aitken et al. 2011).

Based on published geological constraints from the Antilles, particularly the review of stratigraphic and structural record of terrestrial environments or uplift, and based on sea level global drop, Iturralde-Vinent and MacPhee (1999) and Iturralde-Vinent (2006) proposed the existence of two main intercontinental connections formed by the Antilles and the Americas. An older late Cretaceous to Paleocene (65–75 Mya) connection in which the proto-Antilles were closely spaced and separated by shallow seas (Fig. 20.1c). A younger Oligocene (35–33 Mya) GAARlandia (MacPhee and Iturralde-Vinent 1995), where emergent parts of the present-day Greater Antilles were in a close-packed array until the commencement of the Neogene, and the Aves Ridge and the Lesser and Leeward Antilles were completely emerged forming a landspan (not land bridge because it did not completely connect north with south America). Studies have confirmed that the Greater and Lesser Antilles were deformed and uplifted as they interacted with the Americas (Pindell and Kennan 2009; Escalona and Mann 2011; Lidiak and Anderson 2015). The Aves Ridge and the Lesser Antilles, however, which may represent a major element of GAARlandia were far from the continental margins that promote uplift (Fig. 20.1), and their low rates of volcanic activity do not promote the formation of big islands (MacDonald et al. 2000; Jackson 2013). Ocean-floor drilling data on the Aves Ridge to demonstrate the magnitude of land exposure are needed (Ali 2012).

1.3 Hypotheses that Explain the Origin of Caribbean Biota

Four leading hypotheses explain the historical assembly of Caribbean biota: late Cretaceous vicariance, colonization through the GAARlandia landspan, transoceanic dispersal, and in situ speciation (Fig. 20.2). These hypotheses are not necessarily mutually exclusive and it is more likely that all four have contributed to assembly. With the advent of more sophisticated historical biogeography methods added to the molecular-clock dating and increasingly inexpensive DNA sequencing, many recent molecular phylogenetic studies have sampled Caribbean taxa and inferred their

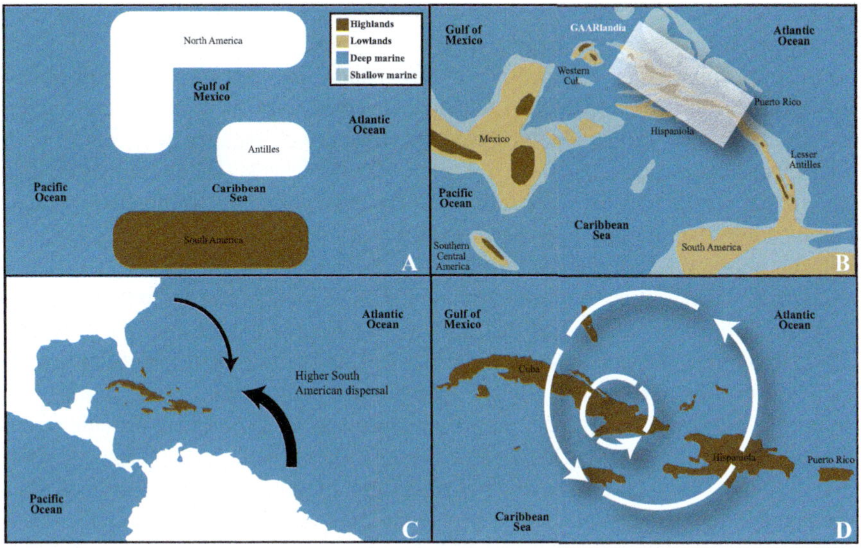

Fig. 20.2 Hypotheses that explain the origin of biota in the Caribbean: (**a**) Vicariance through the eastward movement of the proto-Greater Antilles, adapted from Rosen (1975), (**b**) the GAARlandia landspan provided a means of land migration 33–35 Mya, adapted from Iturralde-Vinent and MacPhee (1999), (**c**) long-distance dispersal from continental America, where dispersal differs between North and South America, and (**d**) in situ speciation, both within islands and between islands. The different source areas of Caribbean lineages are identified in white

ancestral areas. It is thus timely to evaluate these hypotheses across a wide spectrum of taxa and synthesize emergent biogeographic patterns.

The late Cretaceous vicariance hypothesis states that between 80 and 70 Mya, the proto-Antilles land mass, or masses, which occupied a location similar to today's Central America, experienced a vast overland rupture as it migrated eastward on the Caribbean plate and carrying ancient biota with it (Rosen 1975; Iturralde-Vinent 2006). Subsequent fragmentation of the archipelago resulted in allopatric speciation as evidenced by phylogenetic splitting events (Matos-Maraví et al. 2014). Hedges (2006) revised molecular clock analyses for several ancient, relictual groups (e.g. xantusiid lizards and solenodontoid shrews) providing support for the proto-Antillean vicariance. Other studies have used this hypothesis as the best explanation for current phylogenetic relationships, species distribution patterns of endemic lineages and their Central American sister taxa, the fossil evidence, and age of immunological data (Crother and Guyer 1996; Felix and Mejdalani 2011). Opponents of the vicariance model, however, argue that no proto-Antillean land masses existed that could have supported biota from before the middle Eocene (ca. 40 Mya) due to the recurrent transgression, subsidence, and the Cretaceous bolide impact (Iturralde-Vinent and MacPhee 1999).

In 1999, Iturralde-Vinent and MacPhee proposed that colonization of the Antilles was possible from northeast South America through the GAARlandia landspan or island chain described above. At the Eocene-Oligocene boundary, a major drop in temperature caused a rapid ice-sheet growth in Antarctica and a drop in sea level, which might have exposed large portions of the Aves Ridge (Hedges 2001, 2006) facilitating migrations especially for non-volant animals. This hypothesis of GAARlandia as a colonization route has received support from several studies using dated molecular phylogenies in a wide diversity of lineages including amphibians (Alonso et al. 2012), mammals (Davalos 2004), insects (Deler-Hernandez et al. 2018), and also plants (Bacon et al. 2012). The corroboration of this landspan as a colonization facilitator, however, requires more geological, biological and palaeoceanographical data (Ali 2012).

Long distance rafting and over-water dispersal from continental landmasses have been the preferred mechanism of Caribbean biotic occupation for many decades (e.g. Darlington 1938; Hedges et al. 1992; Regalado et al. 2018). Species vary in their ability to colonize islands due to their different dispersal capabilities, and contrary to expectations, plants with multiple dispersal syndromes provide only a marginal advantage for long-distance dispersal (Vargas et al. 2015). Discussions on the possible colonization routes that ancestors of Antillean biota utilized have dominated the biogeographic studies of this region (e.g. Tucker et al. 2017; Nieto-Blázquez et al. 2017; Cano et al. 2018). South America is the preferred hypothesized source area for lineages that arrived in the Caribbean through water dispersal or rafting since ocean–atmosphere models for various points in the Cenozoic show currents flowing from northeast South America to the Greater Antilles (e.g. Huber and Caballero 2003; Sarnthein et al. 2009). However, the complex history of migrations and divergence amongst continental and insular lineages revealed in some studies (e.g. Roncal et al. 2008; Cacho and Baum 2012; Matos-Maraví et al. 2014; Antonelli et al. 2018) cannot discard the effects of the complex Caribbean palaeogeographical history, and stochasticity.

In situ speciation has been proposed as a mechanism to explain the maintenance of high species richness and endemism (Losos and Schluter 2000). Within an island or archipelago, strong evidence for in situ speciation is evidenced by the monophyly of endemic species, supporting the hypothesis that in situ speciation contributes significantly to island biota assembly (Warren et al. 2015). A remarkable example is the 1000 species of picture-winged *Drosophila* all arising from a single common ancestor in the last 25 million years within the Hawaiian archipelago (O'Grady et al. 2011). In the classic Caribbean *Anolis* case of adaptive radiation, parallel radiation with trait convergence of entire faunas was observed in four islands (Mahler et al. 2013). The relative importance of in situ speciation and long distance dispersal on the evolutionary assembly of island biota is subject to characteristics specific to each island. For example, anagenetic or cladogenetic speciation is hypothesized to be an important driver in older and larger islands, while recurrent immigration affects more the composition of younger and smaller islands (Borregaard et al. 2017).

Other forces hypothesized to drive the ecological and evolutionary assembly of insular biotas are extinctions and biotic interactions (Lomolino et al. 2017). Higher local extinction rates on smaller islands have been reported (e.g. Palmeirim et al. 2018) conforming to predictions of island biogeography theory. As biotic density and diversity increase on an island, species interactions become more important in determining the success of colonization and in situ diversification, and the governing selective forces for adaptive radiations (Thompson 2013). In this chapter, however, we do not focus on extinctions or biotic interactions because this information is less readily available from the molecular phylogenetic studies published in the last 15 years that sampled Caribbean taxa.

1.4 Time-for-Speciation Effect in the Caribbean

The hypothesis that species richness in an area is constrained by the time since the area was colonized is referred to in the literature as the "evolutionary time hypothesis" (Stebbins 1974) or the "time-for-speciation effect" (Stephens and Wiens 2003). This idea predicts a positive correlation between evolutionary time of occupation within a region and its species richness, assuming no diversity-dependent control of speciation and extinction rates (Rabosky 2012). Evidence supporting this hypothesis comes from well-dated molecular phylogenies (e.g Wiens et al. 2007; Li et al. 2009a; Mota Rodrigues et al. 2017), and few studies have analyzed this effect at different spatial scales (Stephens and Wiens 2003; Roncal et al. 2011). Further, time of colonization is a good predictor of species richness distributions along elevation gradients (Wiens et al. 2007; Li et al. 2009a). Since lineage diversification follows an exponential pattern, it is expected that early immigrants will have a proportionally large effect on species diversity in a region (Eiserhardt et al. 2017). The time-for-speciation effect has not been evaluated in the context of Caribbean biota, and the relative contribution of old clades to the total Caribbean diversity remains to be quantified. Here we use a modified version of the time-for-speciation effect of Stephens and Wiens (2003) by modeling species richness differences among endemic lineages (not areas or regions) as a function of their colonization time in the Caribbean. Under this modified time-for-speciation effect, early Caribbean colonizers are expected to hold higher biodiversity because they have had more time for in situ speciation.

In this study we compiled published phylogenetic dated trees that contain Caribbean endemic plant lineages to test the following hypotheses: (1) stem divergence and minimum colonization times of endemic lineages are too young to support the Cretaceous vicariance and GAARlandia landspan hypotheses; (2) South America was a more important colonization source than North or Central America given the hypothesized spatial location of GAARlandia, ocean currents, and hurricane wind direction, and (3) early immigrant lineages contain more species because they have had more time to speciate. We use our review on the paleogeography of the Antilles (Sect. 1.2) to contrast past geological events in the Caribbean with the colonization patterns that emerged from the phylogenies.

2 Methods

We searched the Web of Science database with the keywords: "Caribbean" AND "Clade" and "Caribbean" AND "Biogeogra*", and "Caribbean" AND "phylogen*" to obtain phylogenetic studies that sampled Caribbean endemic plant species until

May 2018. We complemented this dataset with studies that were referenced in these publications and personal communications. We compiled phylogenetic trees that were calibrated with fossils or known rates of nucleotide substitution, and excluded those with calibrations based on geological events, such as the GAARlandia land bridge. When two studies sampled the same endemic lineage, we chose the study with the most comprehensive sampling within the lineage or taxonomic rank for which it belongs.

For each endemic plant lineage, we extracted the crown and stem node ages including their confidence intervals when available from the dated phylogenetic tree. Crown node ages provided a time estimate for the origin of the in situ speciation process for each endemic lineage. Stem node ages provided a time estimate for the split between the endemic lineage and its continental sister and thus useful insight on the robustness of the late Cretaceous vicariance hypothesis. Furthermore, we obtained minimum and maximum colonization times from studies that conducted a formal biogeographic range evolution analysis using any method (e.g. Lagrange, BioGeoBears, DIVA, RASP). In some instances, it was necessary to go back one or two nodes before the stem of the endemic lineage if the colonization was inferred to occur before the stem node. When no formal biogeographic analysis was used to infer the colonization event, we used the crown and stem node ages to represent the minimum and maximum colonization times, respectively.

We also harvested ancestral areas for the lineages that colonized the Antilles (i.e. source areas) from the formal biogeographic analysis or as hypothesized by the authors in the study. We coded ancestors' occurrence in one of the following four biogeographic regions following Nieto-Blázquez et al. (2017): South America, Central America, North America, and the rest of the world. We considered the Trans-Mexican Volcanic Belt as the limit between North and Central America separating Nearctic and Neotropical regions, and where biotas from both regions overlap (Mastretta-Yanes et al. 2015). The Caribbean archipelago was treated as a single area to facilitate understanding of biotic exchanges amongst insular and continental landmasses.

To provide insight on the time-for-speciation effect we elaborated a set of ordinary least square linear and nonlinear models (lm function in R, R Core Team 2018) that could explain the variation of species number in the endemic lineage. Regression models included minimum colonization time, its quadratic form, and ancestral area as explanatory variables. We excluded lineage crown age because it was highly correlated (Pearson $R^2 = 0.61$) with the minimum colonization time, and used the latter since it is more biologically relevant. We followed the recommendations of Rabosky (2012) that colonization time should not be log-transformed because it can lead to substantial residual error when modelling age-diversity relationships. The total number of species in the endemic lineage, and the phylogenetic sampling intensity were used as the response variables. We obtained the total number of endemic species for each lineage from the literature or from the Smithsonian National Museum of Natural History database: Flora of the West Indies (available at http://botany.si.edu/antilles/Westindies/query.cfm). For some non-endemic genera with endemic species it was not possible to obtain the total number of species in the endemic lineage because we could not assume that all endemic species form one clade. Using Akaike's Information Criterion (AIC), we selected the best model among the candidate set of five models for each of the two response variables. Because our sample size was small (n = 87) relative to K (the total number of estimable parameters in the model), we calculated AIC adjusted for small sample sizes as: $AICc = -2 \log likelihood + 2K + 2K(K + 1)/(n - K - 1)$. We derived the −2 log likelihoods from linear regressions of the models and calculated differences in AICc values as: $\Delta AICci = AICci - \text{minimum } AICc$. The best model had lowest AICc value (Burnham and Anderson 2002).

3 Results

Our compiled dataset consisted of 87 Caribbean endemic plant lineages with a phylogenetic sampling of 1–17 species each, including a total of 330 endemic species sampled. The complete list of lineages, the plant families to which they belong, and the 42 studies from which we obtained the data are listed in Table 20.1. Each of these lineages represents a unique dispersal event to the Caribbean. We found a minimum colonization time of 0.3–1.2 Mya in *Harrisia* series *Harrisia* (Cactaceae), and a maximum colonization time of 106–122 Mya in *Neobracea* (Apocynaceae) (Fig. 20.3). Seven lineages had minimum and maximum colonization ages that predate the final subaerial nature of the Antillean masses in the mid-Eocene (40 Mya, Iturralde-Vinent 2006). These lineages were *Neobracea*, *Microcycas* (Zamiaceae, 67–60 Mya), *Cubanola* (Rubiaceae, 63–49 Mya), *Dilomilis-Neocogniauxia* (Orchidaceae, 61–47 Mya), *Arcoa* (Leguminosae, 58–55 Mya), *Colleteria* (Rubiaceae, 40–51 Mya) and *Hedyosmum arborescens* (Chloranthaceae, 45 Mya). Most (81.4%) colonization events occurred after the hypothesized GAARlandia landspan, eight (9.3%) lineages had minimum and maximum colonization timeframes that crossed the GAARlandia timespan, and only seven (8.1%) dispersal events were inferred to have occurred well before this landspan (Fig. 20.3). The palm genus *Pseudophoenix* had a very wide colonization timeframe (4–60 Mya) and thus was impossible to interpret whether this genus migrated before or after the hypothesized GAARlandia.

All endemic lineages examined had lower estimates of stem divergence times than would be predicted by proto-Antillean vicariance (80–70 Mya, Fig. 20.4). Most (65.5%) lineages initiated in situ speciation in the Caribbean less than 10 Mya. 20.7% had crown ages between 10 and 20 Mya, and only 8% between 20 and 30 Mya (Fig. 20.4). The most species-rich plant lineages (≥30 species) have produced conspicuous radiations with in situ divergence times that range from 3 to 27 Mya. The youngest of these radiations corresponds to a clade in the Lobelioideae (Campanulaceae), and the oldest to the genus *Calycogonium* (Melastomataceae).

Overall, South America was the most important ancestral area for Caribbean lineages, closely followed by Central America. 38 and 36% of the dispersal events had South and Central America, as source areas, respectively (Fig. 20.5). When

Table 20.1 List of studies from which areas of origin, colonization, and in situ speciation times for Caribbean endemic seed plant lineages were compiled

Plant family	Endemic lineage	Study references
Acanthaceae	*Ruellia blechioides* (1/7)	Tripp and McDade (2013)
Apocynaceae	*Neobracea* (3/7)	Nieto-Blázquez et al. (2017)
Arecaceae	*Coccothrinax-Zombia* (11/51), *Thrinax-Hemithrinax-Leucothrinax* (7/7), *Sabal maritima-S. dominguensis-S. causiarum* (3/3)	Cano et al. (2018)
	Copernicia (11/19)	Bacon et al. (2012)
	Pseudophoenix (4/4)	Trénel et al. (2007)
	Calyptronoma (3/3)	Roncal et al. (2010)
	Attalea crassispatha (1/1)	Freitas et al. (2016)
	Gaussia (3/3)	Cuenca et al. (2008)
	Prestoea acuminata var. montana (1/1), *Euterpe broadwayi* (1/1)	Pichardo-Marcano et al. (2019)
Asteraceae	*Anastraphia* (17/33)	Funk et al. (2014)
Bromeliaceae	*Wittmackia* (16/17)	Aguirre-Santoro (2015)
Burseraceae	*Protium clade of 5 spp.* (5/5), *Protium fragrans* (1/1), *Protium attenuatum* (1/1), *Tetragastris balsamifera* (1/1)	Fine et al. (2014)
Cactaceae	*Leptocereus-Dendrocereus* (3/17), *Melocactus* (2/7)	Hernandez-Hernandez et al. (2014)
	Harrisia series Harrisia (6/13)	Franck et al. (2013)
Campanulaceae	Caribbean clade of Lobelioideae (14/54)	Lagomarsino et al. (2016)
Chloranthaceae	*Hedyosmum arborescens* (1/5)	Zhang and Renner (2003)
Cucurbitaceae	*Penelopeia* (2/2)	Schaefer et al. (2009)
	Linnaeosicyos (1/1)	de Boer et al. (2012)
	Cayaponia americana (1/1)	Duchen and Renner (2010)
Dioscoreaceae	*Dioscorea sect. Rajania* (4/18)	Viruel et al. (2016)
Gesneraceae	*Gesneria-Rhytidophyllum* (15/75)	Roalson et al. (2008)
Ericaceae	*Pieris cubensis* (1/1)	Li et al. (2009b)
Euphorbiaceae	*Bonania-Grimmeodendron* (2/9), *Ditta* (1/2), *Broughtonia-Psychilis-Quisqueya-Tetramicra* (8/38)	Nieto-Blázquez et al. (2017)
	Lasiocroton-Leucocroton-Garciadelia (11/37), *Acidoton-Platygyna* (8/12), *Bernardia dichotoma-B.*	Cervantes et al. (2016)

(continued)

Table 20.1 (continued)

Plant family	Endemic lineage	Study references
	tenuifolia (2/5), *Acalypha chamaedrifolia-A. glechomifolia* (2/29)	
	Croton ekmanii-C. trigonocarpus-C. maestrensis (3/8)	Van Ee et al. (2008)
Leguminosae	*Arcoa* (1/1)	Bruneau et al. (2008)
	Brya (1/7), *Rhodopis* (1/2), *Stahlia* (1/1),	Nieto-Blázquez et al. (2017)
	Hebestigma (1/1),	Lavin et al. (2018)
	Pictetia (7/8), *Poitea* (11/12)	Lavin and Beyra Matos (2008)
Loranthaceae	*Dendropemon* (1/33)	Nieto-Blázquez et al. (2017)
Melastomataceae	*Calycogonium* (1/33)	Nieto-Blázquez et al. (2017)
	Charianthus (6/6)	Abrahamczyk et al. (2015)
Myrtaceae	*Mitranthes* (2/7)	Vasconcelos et al. (2017)
	Calyptranthes garciae-C. thomasiana-C. eriocephala (3/?), *Myrcia maricaensis-M. abbotiana* (2/?), *Calyptranthes pallens-C. barkeri-C. laevigata* (3/?)	Santos et al. (2017)
Onagraceae	*Fuchsia triphylla* (1/2)	Berry et al. (2004)
Orchidaceae	*Dilomilis-Neocogniauxia* (2/7)	Sosa et al. (2016)
	Domingoa (2/3)	Nieto-Blázquez et al. (2017)
	Encyclia (4/29)	Leopardi-Verde et al. (2017)
Passifloraceae	*Passiflora subg. Decaloba* (9/?)	Abrahamczyk et al. (2015)
Phyllanthaceae	*Chascotheca* (1/2)	Nieto-Blázquez et al. (2017)
Picrodendraceae	*Picrodendron* (1/1)	Nieto-Blázquez et al. (2017)
Pinaceae	*Pinus caribaea* (1/1), *Pinus occidentalis-P. cubensis* (2/2)	Hernandez-Leon et al. (2013)
Podocarpaceae	*Podocarpus* Greater Antilles clade (6/8), *Podocarpus* Lesser Antilles clade (2/2)	Quiroga et al. (2016)
Rubiaceae	*Cubanola* (1/2)	Antonelli et al. (2009)
	Suberanthus-Acuneanthus (2/10), *Scolosanthus* (1/27), *Colleteria* (1/2), *Siemensia* (1/1), *Ottoschmidtia* (1/1), *Ceratopyxis-Eosanthe-Phialanthus-Schmidtottia* (4/39), *Catesbaea-Phyllacanthus* (3/18), *Isidorea-Portlandia* (2/24), *Acrosynanthus-Mazaea-Phyllomelia-Roigella-Rondeletia* (11/140),	Manns et al. (2012)

(continued)

Table 20.1 (continued)

Plant family	Endemic lineage	Study references
	Stenostomum acutatum-S. resinosum (2/42), *Exostema clade of 3 spp.* (3/29), *Hamelia papilosa-H. cuprea* (2/4), *Machaonia portoricensis* (1/19), *Erithalis harrisii* (1/9)	
Rutaceae	*Spathelia* (9/9)	Appelhans et al. (2012)
Scrophulariaceae	*Scrophularia clade of 5 spp.* (5/7)	Scheunert and Heubl (2011)
Solanaceae	*Espadaea-Goetzea* (3/3), *Cestrum* Caribbean clade (7/43)	Abrahamczyk et al. (2015)
	Brunfelsia Caribbean clade (12/23)	Filipowicz and Renner (2012)
Symplocaceae	*Symplocos ser Urbaniocharis* (3/7), *Symplocos clade Symplocastrum* (3/6), *Symplocos salicifolia* (1/?), *Symplocos guadeloupensis* (1/?)	Fritsch et al. (2015)
Zamiaceae	*Microcycas* (1/1)	Salas-Leiva et al. (2013)

Genera or species separated by a dash form one monophyletic group in the phylogenetic tree. A comma separates lineages within a plant family. Numbers in parenthesis are species sampled in phylogenetic tree/total number of species in endemic clade. ? = unknown total number of species in endemic clade

assessing the ancestral areas for each of the four best-sampled plant families, we found that Central America was the most important source area for the Arecaceae and Rubiaceae. For the Leguminosae, Central and South America shared importance, and for the Euphorbiaceae all three continental areas were equally important (Fig. 20.5).

All regression models were of poor fit to explain the variation of species number within endemic lineages, with non-significant p-values (Table 20.2, Fig. 20.6). The adjusted R^2 values were all less than 0.1. This result suggests that the timing since colonization in the Caribbean has not influenced the number of species currently present in this archipelago (i.e. older lineages do not have more species), and therefore does not support the time-for-speciation effect in the Caribbean. This result awaits confirmation using extinction rates.

4 Discussion

The argument of over-water versus land bridge dispersal to explain the origin of Caribbean biota began as early as 1916 with the published disagreements between Thomas Barbour and William Mathew (Barbour 1914, 1916) and continues to date. Barbour (referring to amphibians and reptiles) thought that over-water dispersal was untenable and that dispersal through flotsam and jetsam never occurred. Mathew on

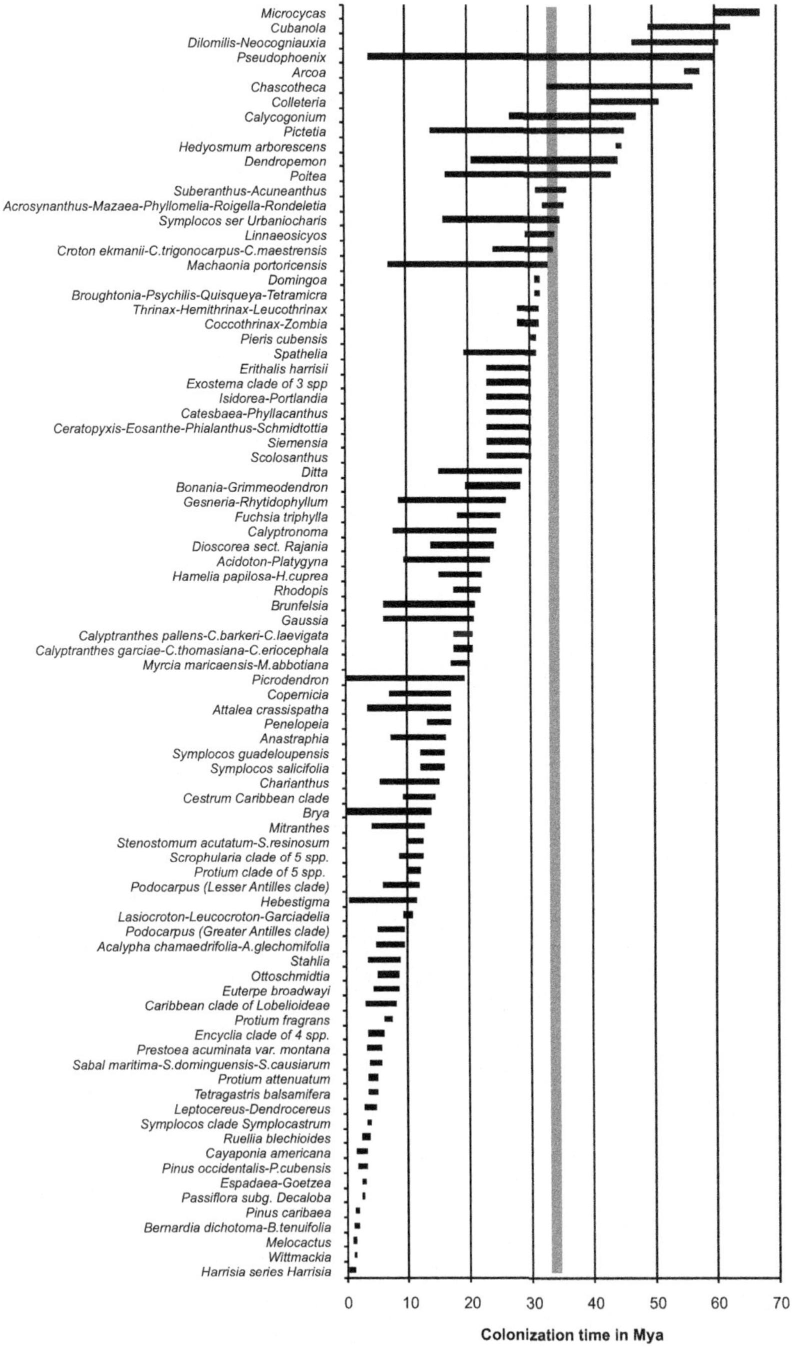

Fig. 20.3 Minimum and maximum colonization times of Caribbean ancestral lineages compiled from molecular phylogenetic trees (see Table 20.1)

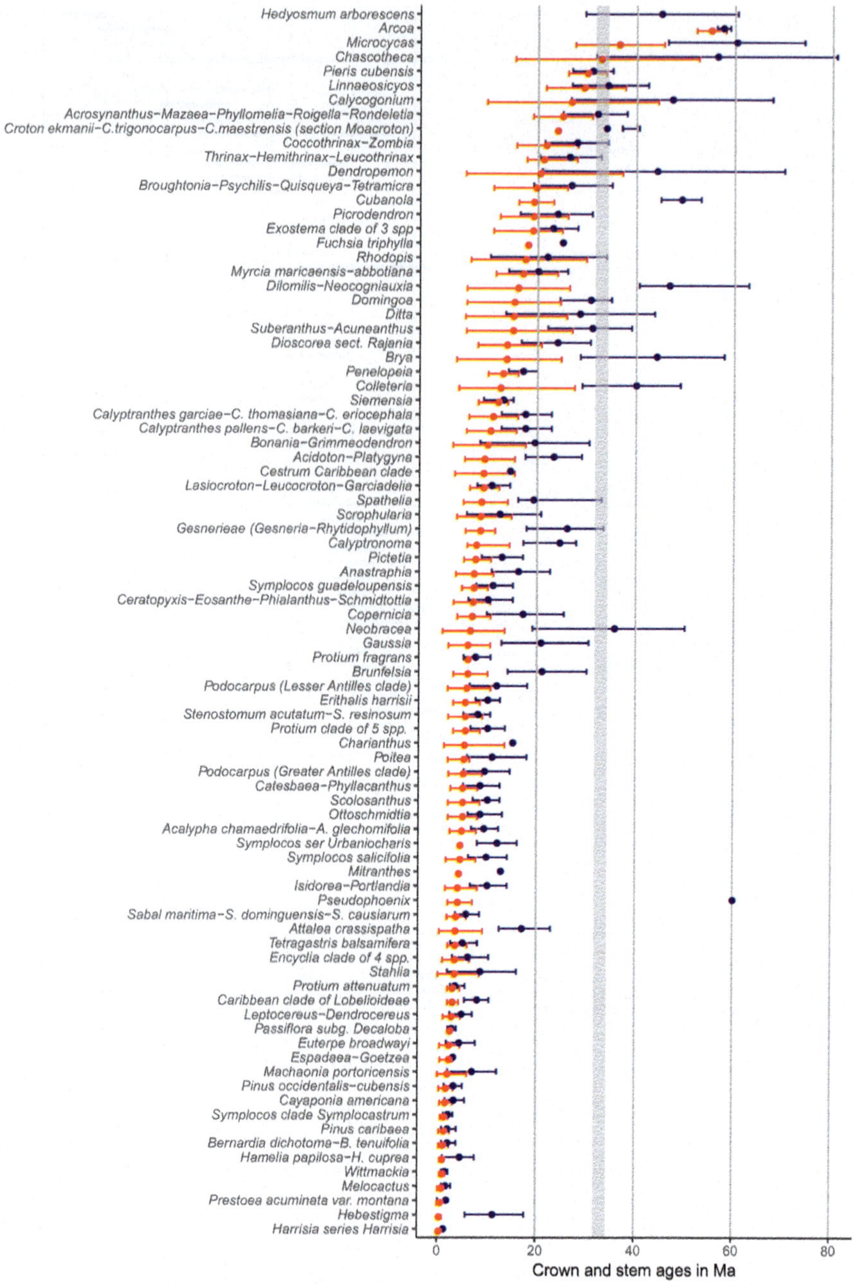

Fig. 20.4 Crown and stem node ages of endemic clades compiled from molecular phylogenetic trees to show the age of in situ speciation in the Caribbean and to evaluate the robustness of the late Cretaceous vicariance hypothesis, respectively. Circles represent mean ages and bars are confidence intervals

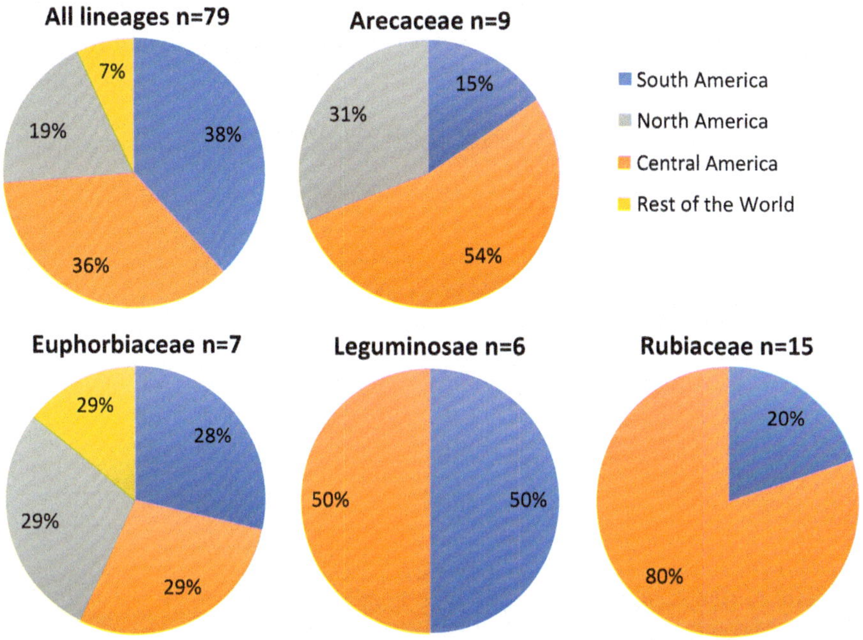

Fig. 20.5 Proposed ancestral area from phylogenetic or biogeographic analyses for Caribbean endemic plant lineages. Pie charts indicate the percentage of each continent as a source area for all lineages available and for four plant families with the highest number of studied lineages

the other hand, influenced by Alfred Wallace's work, considered over-water dispersal as the main explanation for animal distributions. Our review highlights that plants have repeatedly colonized the West Indies for the last 60 million years from continental America, especially from Central and South America. The fact that colonization times do not cluster during the GAARlandia timeframe, and that most colonization events (81.4%) occurred after its putative subsidence, diminishes the importance of land bridge dispersal in the debate and highlights the predominance of over-water dispersal, at least for seed plants. In agreement with our review, numerous studies have shown that plants don't need continuous land to colonize new environments (Andrus et al. 2009; Duchen and Renner 2010; Bacon et al. 2013; Vargas et al. 2015).

The general dynamic model (GDM) of island biogeography of Whittaker and his colleagues incorporates the geological ontogeny of an oceanic island (volcanic activity, erosion, downcutting, and subsidence) into estimates of fundamental biogeographic processes such as immigration, speciation, extinction, which ultimately drive species richness of the island's biota (Whittaker et al. 2008, 2010). This model assumes an initial high immigration rate during island emergence that is insufficient to fill out the potential carrying capacity of the new island followed by high speciation rates of early colonizers (Lomolino et al. 2017; Borregaard et al. 2017). Our present review did not estimate relative colonization rates to the Caribbean

Table 20.2 Selection of regression models using the Akaike Information Criterion (AICc) hypothesized to explain the total number of species in an endemic clade, and the phylogenetic sampling intensity (i.e. number of species sampled in that clade in a phylogenetic tree)

	K[a]	Log lik[b]	AICc[c]	ΔAICc[d]	w_i[e]	Adj R^2[f]	p-value
Total number of species in an endemic clade							
mct	2	−355.54	715.24	63.1	0.00	−0.01	0.64
mct + ancestral area	7	−318.18	652.11	0	0.67	−1.3 e-3	0.44
mct * ancestral area	12	−315.80	660.89	8.78	8 e-3	−0.02	0.54
mct + mct^2	3	−354.29	714.90	62.8	0.00	8.2 e-3	0.27
mct + mct^2 + ancestral area	8	−317.66	653.61	1.50	0.32	2.4 e-3	0.46
Phylogenetic sampling							
mct	2	−238.93	482.00	42.4	0.00	−3.8 e-3	0.25
mct + ancestral area	7	−212.01	439.62	0	0.75	0.04	0.17
mct * ancestral area	12	−210.31	449.42	9.8	5 e-3	0.01	0.37
mct + mct^2	3	−238.91	484.11	44.5	0.00	−7.6 e-3	0.51
mct + mct^2 + ancestral area	8	−211.91	441.91	2.29	0.24	0.03	0.24

mct minimum colonization time
[a]Number of estimable parameters in the model including the intercept
[b]Logarithm of the likelihood of each model given the data
[c]Akaike Information Criterion corrected for small sample size
[d]Relative AICc for each model compared to the best-supported model
[e]Akaike weight indicating the degree of support for each model relative to the entire model set (values range from 0 to 1)
[f]Adjusted R^2 representing the proportion of the variability in the data explained by the model.
*Denotes models where the simple effects are tested first before the interactions

through time, but it revealed that half of the lineages (51.7%) had minimum colonization times during the last 10 My. This pattern is consistent with a recent study of biotic interchanges among Neotropical biogeographic regions and biomes, where Antonelli et al. (2018) found a general increase in the number of relative dispersal events toward the present, especially during the last 10 My. A formal test of colonization rates in the Lesser Antilles through time across a broad spectrum of plant and animal taxa awaits investigation to test the GDM prediction of a declining immigration rate after island formation. Since the GDM was developed for short-lived volcanic oceanic islands, predictions of this model on the Greater Antilles are not possible until further theoretical improvements on the model are made.

Different explanations could be proposed for the seven lineages that showed colonization times predating the final subaerial exposure of the Antillean masses in the mid-Eocene. The origin of the ancestor of the Cuban endemic genus *Microcycas* was suggested to result from a vicariant event between Africa and America 60 Mya, with subsequent extinctions of several cycad taxa in the South American continent prior to the most recent cycad diversification events (Salas-Leiva et al. 2013). This was the only study in our literature search that proposed vicariance as a possible mechanism to explain the origin of Caribbean lineages. The estimated colonization time of the ancestor of *Neobracea* (Apocynaceae) as inferred in Nieto-Blázquez

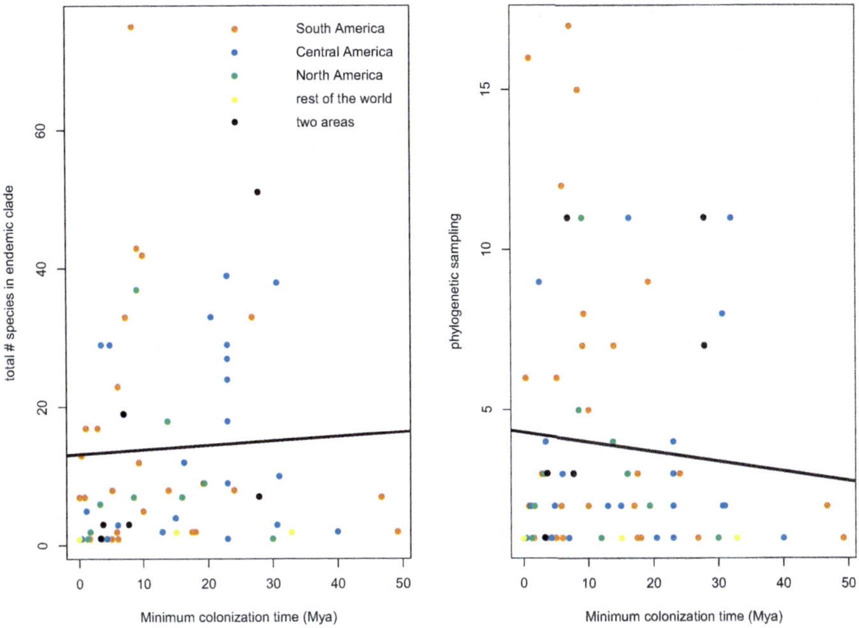

Fig. 20.6 Scatterplot of endemic lineage minimum colonization time (mct) and (**a**) total number of endemic species in clade, and (**b**) phylogenetic sampling effort in clade. Lines show the fit of the regression model of the two variables plotted. Ancestral area from which the endemic lineage colonized the Antilles is color-coded

et al. (2017, 106–122 Mya) may be an artifact of the small Apocynaceae sampling in this dated phylogenetic tree (7 spp.). This colonization time should, therefore, be taken with caution.

Hedges et al. (1992) suggested that the catastrophic bolide impact in the Caribbean region at the end of the Cretaceous is another potential explanation for the rarity of ancient West Indian biota (see Crother and Guyer (1996) for evidence that some organisms probably survived this impact). Results from Hedges (2006) supported the late Cretaceous vicariance hypothesis, arguing that the biota of an archipelago can be maintained over time even while individual islands rise and fall because of the high resilience of biota in changing landscapes through dispersal among ephemeral islands. Furthermore, the age of the split of the proto-Antilles from the continent is contentious (Table 20.1 in Crother and Guyer 1996) and may have occurred as late as 48 Mya (Sykes et al. 1982). The proto-Antillean vicariance hypothesis could thus explain the evolutionary origin of some of these early Caribbean lineages, especially those with Central American ancestors such as *Colleteria* (Rubiaceae). The Caribbean fossil record has unfortunately been of limited value for testing this vicariance hypothesis, mainly because it is almost exclusively Quaternary (MacPhee and Wyss 1990).

South and Central America were the most common source areas occupied by the ancestors of the Caribbean lineages analyzed. This result is concordant with the observation of Acevedo-Rodriguez and Strong (2008) that floristic affinities between the West Indies and the surrounding continents are stronger to Central and South America as a whole than with either one of them separately or with North America. This result also diminishes the importance of South America as the prime ancestral area of Caribbean lineages as hypothesized by the GAARlandia hypothesis and the direction of ocean currents. Iturralde-Vinent and MacPhee (1999) remarked on the highly unpredictable fate of passively floating objects despite the general tendency of Caribbean surface currents to flow northward with respect to South America. These authors also suggest that until the late Miocene, Caribbean surface-water currents were unsuited for over-water dispersal between South America and the Greater Antilles. Instead, alternative regional paleoceanographic models (Mullins et al. 1987; Duque-Caro 1990; Droxler et al. 1998) showed that current-flow patterns from major rivers would have delivered South American waifs to the Central American coast, not to the Greater or Lesser Antilles. The importance of Central America as a source area of Caribbean lineages in our review (36%) is concordant with this alternative water flow pattern.

We did not find a time-for-speciation effect for the endemic Caribbean lineages analyzed—older lineages are not more species rich. Minimum colonization time and ancestral area were poor predictors of clade species richness and phylogenetic sampling. Early colonizers that landed as soon as the Antilles were subaerial (minimum colonization time greater than 30 Mya) accounted for just ca. 10% of the species diversity sampled in phylogenies, or ca. 19% of the total number of species these clades represent. The majority of these species belong to two clades that include several genera in the Orchidaceae and Rubiaceae. Comparatively, recent colonizers with a maximum colonization time of less than 10 Mya accounted for ca. 9% of the species diversity sampled in phylogenies, or 22% of the total number of species these clades represent. Thus, early and recent colonizers contribute almost equally to species richness in the studies we have reviewed. Changes in speciation rates and extinctions throughout the history of a clade could potentially counteract the time-for-speciation effect (Eiserhardt et al. 2017). For example, the Menispermaceae (Wang et al. 2012) and the tropical American Protieae (Fine et al. 2014) showed initially high then decreasing diversification rates masking this effect. Recent lineages such as the Lobeliads (Campanulaceae) attained high species diversity over a short period of time (Lagomarsino et al. 2016). To more comprehensively understand the contribution of old immigrants to Caribbean biodiversity, lineage diversification rates (speciation and extinction) should be assessed comparatively across several groups.

Current geological knowledge suggests that the Greater Antilles were uplifted since the late Cretaceous with final emergence by 45 Mya, and the existence of a major Oligocene uplift event that took place as they advanced between the Americas. The Aves Ridge and the Lesser Antilles volcanic chain putatively represent the connecting segment of the emergent GAARlandia landspan. The position of this volcanic chain far from the continental margins that promote uplift, together with the volcanic waning and beginning of submergence of the Aves Ridge after ca. 48 Mya, and the low production rate of volcanic products since 40 Mya (MacDonald et al.

2000; Jackson 2013) precludes the formation of big islands and may maintain significant distance among islands as seen today (Ali 2012), casting doubts on the real existence of GAARlandia in the Oligocene. Moreover, the potential amount of global sea level drop during the Oligocene glaciation (Houben et al. 2012) has been presented as a fundamental process for the existence of GAARlandia (Iturralde-Vinent 2006). This 100 m sea level drop was probably not enough for promoting big land extensions in the Lesser Antilles or the Aves Ridge as discussed by Ali (2012) and seen in the Caribbean bathymetry (Fig. 20.1). Studies on the evolution of the Lesser Antilles volcanic activity and uplift history have shown that volcanic waning and deformation on the island can cause submergence or emergence of the islands in a 5 Mya scale (Jackson 2013), suggesting that a very dynamic geological activity underlies the biogeography of Caribbean biota.

Acknowledgements We are grateful to Valenti Rull and Ana Carnaval for the invitation to participate in this book. This work was funded by an NSERC-Discovery grant (RGPIN-2014-03976) to JR, the Knut and Alice Wallenberg Foundation to A. Antonelli supported CDB. J.S. Jaramillo and S. Leon are acknowledged for their help with Fig. 20.1, and Jose Maria Fernandez-Palacios for providing useful comments on the text.

References

Abrahamczyk S, Souto-Vilaros D, McGuire JA, Renner SS (2015) Diversity and clade ages of West Indian hummingbirds and the largest plant clades dependent on them: a 5–9 Myr young mutualistic system. Biol J Linn Soc 114(4):848–859. https://doi.org/10.1111/bij.12476

Acevedo-Rodriguez P, Strong MT (2008) Floristic richness and affinities in the West Indies. Bot Rev 74(1):5–36. https://doi.org/10.1007/s12229-008-9000-1

Aguirre-Santoro JA (2015) Systematics and evolution of the *Ronnbergia* alliance (Bromeliaceae): history of disjunct diversification in three biodiversity hotspots of the Neotropics. Ph.D. City University of New York, New York

Aitken T, Mann P, Escalona A et al (2011) Evolution of the Grenada and Tobago basins and implications for arc migration. Mar Pet Geol 28(1):235–258. https://doi.org/10.1016/j.marpetgeo.2009.10.003

Ali JR (2012) Colonizing the Caribbean: is the GAARlandia land-bridge hypothesis gaining a foothold? J Biogeogr 39(3):431–433. https://doi.org/10.1111/j.1365-2699.2011.02674.x

Alonso R, Crawford AJ, Bermingham E (2012) Molecular phylogeny of an endemic radiation of Cuban toads (Bufonidae: Peltophryne) based on mitochondrial and nuclear genes. J Biogeogr 39 (3):434–451. https://doi.org/10.1111/j.1365-2699.2011.02594.x

Andrus N, Tye A, Nesom G et al (2009) Phylogenetics of *Darwiniothamnus* (Asteraceae: Astereae)—molecular evidence for multiple origins in the endemic flora of the Galápagos Islands. J Biogeogr 36(6):1055–1069. https://doi.org/10.1111/j.1365-2699.2008.02064.x

Antonelli A, Nylander JAA, Persson C et al (2009) Tracing the impact of the Andean uplift on Neotropical plant evolution. Proc Natl Acad Sci U S A 106(24):9749–9754. https://doi.org/10.1073/pnas.0811421106

Antonelli A, Zizka A, Carvalho FA et al (2018) Amazonia is the primary source of Neotropical biodiversity. Proc Natl Acad Sci U S A 115(23):6034–6039. https://doi.org/10.1073/pnas.1713819115

Appelhans MS, Kessler PJA, Smets E et al (2012) Age and historical biogeography of the pantropically distributed Spathelioideae (Rutaceae, Sapindales). J Biogeogr 39(7):1235–1250. https://doi.org/10.1111/j.1365-2699.2012.02686.x

Bacon CD, Baker WJ, Simmons MP (2012) Miocene dispersal drives island radiations in the palm tribe Trachycarpeae (Arecaceae). Syst Biol 61(3):426–442. https://doi.org/10.1093/sysbio/syr123

Bacon CD, Mora A, Wagner WL et al (2013) Testing geological models of evolution of the Isthmus of Panama in a phylogenetic framework. Bot J Linn Soc 171(1):287–300. https://doi.org/10.1111/j.1095-8339.2012.01281.x

Barbour T (1914) A contribution to the zoogeography of the West Indies, with especial reference to amphibians and reptiles. Mem Mus Comp Zool 44:209–359

Barbour T (1916) Some remarks upon Matthew's climate and evolution, with supplementary note by W. D. Matthew. Ann N Y Acad Sci 27:1–15

Bellard C, Leclerc C, Courchamp F (2014) Impact of sea level rise on the 10 insular biodiversity hotspots. Glob Ecol Biogeogr 23(2):203–212. https://doi.org/10.1111/geb.12093

Berry PE, Hahn WJ, Sytsma KJ et al (2004) Phylogenetic relationships and biogeography of *Fuchsia* (Onagraceae) based on noncoding nuclear and chloroplast DNA data. Am J Bot 91 (4):601–614. https://doi.org/10.3732/ajb.91.4.601

Borregaard MK, Amorim IR, Borges PAV et al (2017) Oceanic island biogeography through the lens of the general dynamic model: assessment and prospect. Biol Rev 92(2):830–853. https://doi.org/10.1111/brv.12256

Brooks TM, Mittermeier RA, Mittermeier CG et al (2002) Habitat loss and extinction in the hotspots of biodiversity. Conserv Biol 16(4):909–923. https://doi.org/10.1046/j.1523-1739.2002.00530.x

Bruneau A, Mercure M, Lewis GP, Herendeen PS (2008) Phylogenetic patterns and diversification in the caesalpinioid legumes. Botany-Botanique 86(7):697–718. https://doi.org/10.1139/B08-058

Burnham KP, Anderson DR (2002) Model selection and multimodel inference: a practical information-theoretical approach. Springer, New York

Cacho NI, Baum DA (2012) The Caribbean slipper spurge *Euphorbia tithymaloides*: the first example of a ring species in plants. Proc Biol Sci 279(1742):3377–3383. https://doi.org/10.1098/rspb.2012.0498

Cano A, Bacon CD, Stauffer FW et al (2018) The roles of dispersal and mass extinction in shaping palm diversity across the Caribbean. J Biogeogr 45(6):1432–1443. https://doi.org/10.1111/jbi.13225

Cervantes A, Fuentes S, Gutierrez J et al (2016) Successive arrivals since the Miocene shaped the diversity of the Caribbean Acalyphoideae (Euphorbiaceae). J Biogeogr 43(9):1773–1785. https://doi.org/10.1111/jbi.12790

Crother BI, Guyer C (1996) Caribbean historical biogeography: was the dispersal-vicariance debate eliminated by an extraterrestrial bolide? Herpetologica 52(3):440–465. https://doi.org/10.2307/3892664

Cuenca A, Asmussen-Lange CB, Borchsenius F (2008) A dated phylogeny of the palm tribe Chamaedoreeae supports Eocene dispersal between Africa, North and South America. Mol Phylogenet Evol 46(2):760–775. https://doi.org/10.1016/j.ympev.2007.10.010

Darlington PJJ (1938) The origin of the fauna of the Greater Antilles, with discussion of dispersal of animals over water and through the air. Q Rev Biol 13(3):274–300. https://doi.org/10.1086/394561

Darwin CR (1859) The origin of species by means of natural selection, or the preservation of favoured races in the struggle for life. John Murray, London

Davalos L (2004) Phylogeny and biogeography of Caribbean mammals. Biol J Linn Soc 81 (3):373–394. https://doi.org/10.1111/j.1095-8312.2003.00302.x

de Boer HJ, Schaefer H, Thulin M, Renner SS (2012) Evolution and loss of long-fringed petals: a case study using a dated phylogeny of the snake gourds, *Trichosanthes* (Cucurbitaceae). BMC Evol Biol 12:108. https://doi.org/10.1186/1471-2148-12-108

Deler-Hernandez A, Sykora V, Seidel M et al (2018) Multiple origins of the *Phaenonotum* beetles in the Greater Antilles (Coleoptera: Hydrophilidae): phylogeny, biogeography and systematics. Zool J Linnean Soc 183(1):97–120. https://doi.org/10.1093/zoolinnean/zlx071

Droxler AW, Burke K, Cunningham AD et al (1998) Caribbean constraints on circulation between Atlantic and Pacific oceans over the past 40 million years. In: Crowley T, Burke K (eds) Tectonic boundary conditions for climate reconstruction. Oxford University Press, Oxford, pp 160–191

Duchen P, Renner SS (2010) The evolution of *Cayaponia* (Cucurbitaceae): repeated shifts from bat to bee pollination and long-distance dispersal to Africa 2–5 million years ago. Am J Bot 97 (7):1129–1141. https://doi.org/10.3732/ajb.0900385

Duque-Caro H (1990) Major Neogene events in Panamic South America. In: Tsuchi R (ed) Pacific Neogene events, their timing, nature and interrelationships. University Tokyo Press, Tokyo, pp 101–114

Eiserhardt WL, Couvreur TLP, Baker WJ (2017) Plant phylogeny as a window on the evolution of hyperdiversity in the tropical rainforest biome. New Phytol 214(4):1408–1422. https://doi.org/10.1111/nph.14516

Escalona A, Mann P (2011) Tectonics, basin subsidence mechanisms, and paleogeography of the Caribbean-South American plate boundary zone. Mar Pet Geol 28(1):8–39. https://doi.org/10.1016/j.marpetgeo.2010.01.016

Felix M, Mejdalani G (2011) Phylogenetic analysis of the leafhopper genus *Apogonalia* (Insecta: Hemiptera: Cicadellidae) and comments on the biogeography of the Caribbean islands. Zool J Linnean Soc 163(2):548–570. https://doi.org/10.1111/j.1096-3642.2011.00724.x

Filipowicz N, Renner SS (2012) *Brunfelsia* (Solanaceae): a genus evenly divided between South America and radiations on Cuba and other Antillean islands. Mol Phylogenet Evol 64(1):1–11. https://doi.org/10.1016/j.ympev.2012.02.026

Fine PVA, Zapata F, Daly DC (2014) Investigating processes of Neotropical rain forest tree diversification by examining the evolution and historical biogeography of the Protieae (Burseraceae). Evolution 68(7):1988–2004. https://doi.org/10.1111/evo.12414

Franck AR, Cochrane BJ, Garey JR (2013) Phylogeny, biogeography, and infrageneric classification of *Harrisia* (Cactaceae). Syst Bot 38(1):210–223. https://doi.org/10.1600/036364413X662105

Freitas C, Meerow AW, Pintaud J, Henderson A, Noblick L, Costa FRC, Barbosa CE, Barrington D (2016) Phylogenetic analysis of *Attalea* (Arecaceae): insights into the historical biogeography of a recently diversified Neotropical plant group. Bot J Linn Soc 182(2):287–302. https://doi.org/10.1111/boj.12466

Fritsch PW, Manchester SR, Stone RD et al (2015) Northern hemisphere origins of the amphi-Pacific tropical plant family Symplocaceae. J Biogeogr 42(5):891–901. https://doi.org/10.1111/jbi.12442

Funk VA, Sancho G, Roque N et al (2014) A phylogeny of the Gochnatieae: understanding a critically placed tribe in the Compositae. Taxon 63(4):859–882. https://doi.org/10.12705/634.27

Gillespie R (2004) Community assembly through adaptive radiation in Hawaiian spiders. Science 303(5656):356–359. https://doi.org/10.1126/science.1091875

Graham A (2003) Geohistory models and cenozoic paleoenvironments of the Caribbean region. Syst Bot 28(2):378–386. https://doi.org/10.1043/0363-6445-28.2.378

Hedges SB (2001) Biogeography of the West Indies: an overview. In: Woods CA, Sergile FE (eds) Biogeography of the West Indies: patterns and perspectives. CRC Press, Boca Raton, FL, pp 15–33

Hedges SB (2006) Paleogeography of the Antilles and origin of West Indian terrestrial vertebrates. Ann Mo Bot Gard 93(2):231–244. https://doi.org/10.3417/0026-6493(2006)93[231:POTAAO]2.0.CO;2

Hedges SB, Hass CA, Maxson L (1992) Caribbean biogeography—molecular evidence for dispersal in West Indian terrestrial vertebrates. Proc Natl Acad Sci U S A 89(5):1909–1913. https://doi.org/10.1073/pnas.89.5.1909

Hernandez-Hernandez T, Brown JW, Schlumpberger BO et al (2014) Beyond aridification: multiple explanations for the elevated diversification of cacti in the New World Succulent Biome. New Phytol 202(4):1382–1397. https://doi.org/10.1111/nph.12752

Hernandez-Leon S, Gernandt DS, Perez de la Rosa JA, Jardon-Barbolla L (2013) Phylogenetic relationships and species delimitation in *Pinus* Section *Trifoliae* inferrred from plastid DNA. PLoS One 8(7):e70501. https://doi.org/10.1371/journal.pone.0070501

Houben AJP, van Mourik CA, Montanari A et al (2012) The Eocene-Oligocene transition: changes in sea level, temperature or both? Paleogeogr Paleoclimatol Paleoecol 335:75–83. https://doi.org/10.1016/j.palaeo.2011.04.008

Huber M, Caballero R (2003) Eocene El Nino: evidence for robust tropical dynamics in the "hothouse". Science 299(5608):877–881. https://doi.org/10.1126/science.1078766

Iturralde-Vinent MA (2006) Meso-Cenozoic Caribbean paleogeography: implications for the historical biogeography of the region. Int Geol Rev 48(9):791–827. https://doi.org/10.2747/0020-6814.48.9.791

Iturralde-Vinent MA, MacPhee RDE (1999) Paleogeography of the Caribbean region: implications for cenozoic biogeography. Bull Am Mus Nat Hist 238:1–95

Jackson TA (2013) A review of volcanic island evolution and magma production rate: an example from a Cenozoic island arc in the Caribbean. J Geol Soc 170(3):547–556. https://doi.org/10.1144/jgs2011-166

James KH, Lorente MA, Pindell JL (eds) (2009) The origin and evolution of the Caribbean plate—special publication 328, New ed edn. Geological Society of London, London

Lagomarsino LP, Condamine FL, Antonelli A et al (2016) The abiotic and biotic drivers of rapid diversification in Andean bellflowers (Campanulaceae). New Phytol 210(4):1430–1442. https://doi.org/10.1111/nph.13920

Lavin M, Beyra Matos A (2008) The impact of ecology and biogeography on legume diversity, endemism, and phylogeny in the Caribbean region: a new direction in historical biogeography. Bot Rev 74(1):178–196. https://doi.org/10.1007/s12229-008-9006-8

Lavin M, Pennington T, Hughes CE, Lewis GP, Delgado-Salinas A, Duno de Stefano R, de Queiroz LP, Cardoso D, Wojciechowski MF (2018) DNA sequence variation among conspecific accessions of the legume *Coursetia caribaea* reveals geographically localized clades here ranked as species. Syst Bot 43(3):664–675. https://doi.org/10.1600/036364418X697382

Leopardi-Verde CL, Carnevali G, Romero-Gonzalez GA (2017) A phylogeny of the genus *Encyclia* (Orchidaceae: Laeliinae), with emphasis on the species of the northern hemisphere. J Syst Evol 55(2):110–123. https://doi.org/10.1111/jse.12225

Li J, He Q, Hua X et al (2009a) Climate and history explain the species richness peak at mid-elevation for *Schizothorax* fishes (Cypriniformes: Cyprinidae) distributed in the Tibetan plateau and its adjacent regions. Glob Ecol Biogeogr 18(2):264–272. https://doi.org/10.1111/j.1466-8238.2008.00430.x

Li M, Li J, Jiang J et al (2009b) Phylogenetics and biogeography of *Pieris* (Lyonieae, Ericaceae) inferred from sequences of nuclear and chloroplast genomes. Syst Bot 34(3):553–560. https://doi.org/10.1600/036364409789271272

Lidiak EG, Anderson TH (2015) Evolution of the Caribbean plate and origin of the Gulf of Mexico in light of plate motions accommodated by strike-slip faulting. In: Anderson TH, Didenko AN, Johnson CL et al (eds) Late Jurassic margin of Laurasia—a record of faulting accommodating plate rotation, vol 513. Geological Society of America, Boulder, CO, pp 1–88

Lomolino MV, Riddle BR, Whittaker RJ (2017) Biogeography. Biological diversity across space and time, 5th edn. Sinauer Associates, Sunderland, MA

Losos JB, Schluter D (2000) Analysis of an evolutionary species-area relationship. Nature 408 (6814):847–850. https://doi.org/10.1038/35048558

Losos JB, Glor RE, Kolbe JJ et al (2006) Adaptation, speciation, and convergence: a hierarchical analysis of adaptive radiation in Caribbean Anolis lizards. Ann Mo Bot Gard 93(1):24–33. https://doi.org/10.3417/0026-6493(2006)93[24:ASACAH]2.0.CO;2

MacArthur RH, Wilson EO (1967) The theory of island biogeography. Princeton University Press, Princeton, NJ

MacDonald R, Hawkesworth CJ, Heath E (2000) The Lesser Antilles volcanic chain: a study in arc magmatism. Earth-Sci Rev 49(1–4):1–76. https://doi.org/10.1016/S0012-8252(99)00069-0

MacPhee RDE, Iturralde-Vinent M (1995) Origin of the Greater Antillean land mammal fauna. 1, New Tertiary fossils from Cuba and Puerto Rico. Am Mus Novit 3141:1–31

MacPhee RDE, Wyss AR (1990) Oligo-Miocene vertebrates from Puerto Rico, with a catalog of localities. Am Mus Novit 2965:45 pp

Mahler DL, Ingram T, Revell LJ et al (2013) Exceptional convergence on the macroevolutionary landscape in island lizard radiations. Science 341(6143):292–295. https://doi.org/10.1126/science.1232392

Manns U, Wikstrom N, Taylor CM et al (2012) Historical biogeography of the predominantly Neotropical subfamily Cinchonoideae (Rubiaceae): into or out of America? Int J Plant Sci 173 (3):261–289. https://doi.org/10.1086/663971

Mastretta-Yanes A, Moreno-Letelier A, Pinero D et al (2015) Biodiversity in the Mexican highlands and the interaction of geology, geography and climate within the trans-Mexican volcanic belt. J Biogeogr 42(9):1586–1600. https://doi.org/10.1111/jbi.12546

Matos-Maraví P, Núñez Águila R, Pena C et al (2014) Causes of endemic radiation in the Caribbean: evidence from the historical biogeography and diversification of the butterfly genus *Calisto* (Nymphalidae: Satyrinae: Satyrini). BMC Evol Biol 14:199. https://doi.org/10.1186/s12862-014-0199-7

Maunder M, Leiva A, Santiago-Valentin E et al (2008) Plant conservation in the Caribbean island biodiversity hotspot. Bot Rev 74(1):197–207. https://doi.org/10.1007/s12229-008-9007-7

Mittermeier RA, Gil RR, Hoffman M et al (2004) Hotspots revisited: earth's biologically richest and most threatened terrestrial ecoregions. CEMEX, Mexico, DF

Mota Rodrigues JF, Olalla-Tarraga MA, Iverson JB et al (2017) Time and environment explain the current richness distribution of non-marine turtles worldwide. Ecography 40(12):1402–1411. https://doi.org/10.1111/ecog.02649

Mullins HT, Gardulski AF, Wise SW et al (1987) Middle Miocene oceanographic event in the eastern gulf of Mexico—implications for seismic stratigraphic succession and loop current gulf-stream circulation. Geol Soc Am Bull 98(6):702–713. https://doi.org/10.1130/0016-7606(1987)982.0.CO;2

Neill I, Kerr AC, Hastie AR et al (2011) Origin of the aves ridge and Dutch-Venezuelan Antilles: interaction of the Cretaceous 'great arc' and Caribbean-Colombian oceanic plateau? J Geol Soc 168(2):333–347. https://doi.org/10.1144/0016-76492010-067

Nieto-Blázquez ME, Antonelli A, Roncal J (2017) Historical biogeography of endemic seed plant genera in the Caribbean: did GAARlandia play a role? Ecol Evol 7(23):10158–10174. https://doi.org/10.1002/ece3.3521

O'Grady PM, Lapoint RT, Bonacum J et al (2011) Phylogenetic and ecological relationships of the Hawaiian *Drosophila* inferred by mitochondrial DNA analysis. Mol Phylogenet Evol 58 (2):244–256. https://doi.org/10.1016/j.ympev.2010.11.022

Palmeirim AF, Benchimol M, Vieira MV et al (2018) Small mammal responses to Amazonian forest islands are modulated by their forest dependence. Oecologia 187(1):191–204. https://doi.org/10.1007/s00442-018-4114-6

Pichardo-Marcano FJ, Nieto-Blázquez ME, MacDonald AN, Galeano G, Roncal J (2019) Phylogeny, historical biogeography and diversification rates in an economically important group of Neotropical palms: tribe Euterpeae. Mol Phylogenet Evol 133:67–81

Pindell JL, Kennan L (2009) Tectonic evolution of the Gulf of Mexico, Caribbean and northern South America in the mantle reference frame: an update. Geol Soc Spec Publ 328:1–55. https://doi.org/10.1144/SP328.1

Quiroga PM, Mathiasen P, Iglesias A et al (2016) Molecular and fossil evidence disentangle the biogeographical history of *Podocarpus*, a key genus in plant geography. J Biogeogr 43 (2):372–383. https://doi.org/10.1111/jbi.12630

R Core Team (2018) R: a language and environment for statistical computing. R Foundation for Statistical Computing, Vienna. https://www.r-project.org/

Rabosky DL (2012) Testing the time-for-speciation effect in the assembly of regional biotas. Methods Ecol Evol 3(2):224–233. https://doi.org/10.1111/j.2041-210X.2011.00166.x

Regalado L, Loriga J, Bechteler J et al (2018) Phylogenetic biogeography reveals the timing and source areas of the *Adiantum* species (Pteridaceae) in the West Indies, with a special focus on Cuba. J Biogeogr 45(3):541–551. https://doi.org/10.1111/jbi.13119

Reynolds RG, Collar DC, Pasachnik SA et al (2016) Ecological specialization and morphological diversification in Greater Antillean boas. Evolution 70(8):1882–1895. https://doi.org/10.1111/evo.12987

Roalson EH, Skog LE, Zimmer EA (2008) Untangling Gloxinieae (Gesneriaceae). II Reconstructing biogeographic patterns and estimating divergence times among new world continental and island lineages. Syst Bot 33(1):159–175. https://doi.org/10.1600/036364408783887429

Roncal J, Zona S, Lewis CE (2008) Molecular phylogenetic studies of Caribbean palms (Arecaceae) and their relationships to biogeography and conservation. Bot Rev 74(1):78–102. https://doi.org/10.1007/s12229-008-9005-9

Roncal J, Borchsenius F, Asmussen-Lange CB, Balslev H (2010) Divergence times in the tribe Geonomateae (Arecaceae) coincide with Tertiary geological events. In: Seberg O, Petersen G, Barfod AS, Davis JI (eds) Diversity, phylogeny, and evolution in the monocotyledons. Aarhus University Press, Aarhus, pp 245–265

Roncal J, Blach-Overgaard A, Borchsenius F et al (2011) A dated phylogeny complements macroecological analysis to explain the diversity patterns in *Geonoma* (Arecaceae). Biotropica 43(3):324–334. https://doi.org/10.1111/j.1744-7429.2010.00696.x

Rosen DE (1975) A vicariance model of Caribbean biogeography. Syst Zool 24(4):431–464. https://doi.org/10.2307/2412905

Salas-Leiva DE, Meerow AW, Calonje M et al (2013) Phylogeny of the cycads based on multiple single-copy nuclear genes: congruence of concatenated parsimony, likelihood and species tree inference methods. Ann Bot 112(7):1263–1278. https://doi.org/10.1093/aob/mct192

Sandwell DT, Müller RD, Smith WHF, Garcia E, Francis R (2014) New global marine gravity model from Cryo-Sat-2 and jason-1 reveals buried tectonic structure. Science 346(6205):65–67. https://doi.org/10.1126/science.1258213

Santos MF, Lucas E, Sano PT et al (2017) Biogeographical patterns of *Myrcia* s.l. (Myrtaceae) and their correlation with geological and climatic history in the Neotropics. Mol Phylogenet Evol 108:34–48. https://doi.org/10.1016/j.ympev.2017.01.012

Sarnthein M, Bartoli G, Prange M et al (2009) Mid-Pliocene shifts in ocean overturning circulation and the onset of Quaternary-style climates. Clim Past 5(2):269–283. https://doi.org/10.5194/cp-5-269-2009

Savolainen V, Anstett MC, Lexer C et al (2006) Sympatric speciation in palms on an oceanic island. Nature 441(7090):210–213. https://doi.org/10.1038/nature04566

Schaefer H, Heibl C, Renner SS (2009) Gourds afloat: a dated phylogeny reveals an Asian origin of the gourd family (Cucurbitaceae) and numerous oversea dispersal events. Proc R Soc B Biol Sci 276(1658):843–851. https://doi.org/10.1098/rspb.2008.1447

Scheunert A, Heubl G (2011) Phylogenetic relationships among new world *Scrophularia* L. (Scrophulariaceae): new insights inferred from DNA sequence data. Plant Syst Evol 291 (1–2):69–89. https://doi.org/10.1007/s00606-010-0369-z

Schluter D (2000) The ecology of adaptive radiation. Oxford University Press, Oxford

Sosa V, Cameron KM, Angulo DF, Hernandez-Hernandez T (2016) Life form evolution in epidendroid orchids: ecological consequences of the shift from epiphytism to terrestrial habit in *Hexalectris*. Taxon 65(2):235–248. https://doi.org/10.12705/652.2
Stebbins GL (1974) Flowering plants: evolution above the species level. Harvard University Press, Cambridge, MA
Stephens PR, Wiens JJ (2003) Explaining species richness from continents to communities: the time-for-speciation effect in emydid turtles. Am Nat 161(1):112–128. https://doi.org/10.1086/345091
Sykes LR, McCann WR, Kafka AL (1982) Motion of Caribbean plate during last 7 million years and implications for earlier Cenozoic movements. J Geophys Res 87(NB13):656–676. https://doi.org/10.1029/JB087iB13p10656
Thompson JN (2013) Relentless evolution. Chicago University Press, Chicago
Trénel P, Gustafsson MHG, Baker WJ, Asmussen-Lange CB, Dransfield J, Borchsenius F (2007) Mid-Tertiary dispersal, not Gondwanan vicariance explains distribution patterns in the wax palm subfamily (Ceroxyloideae: Arecaceae). Mol Phylogenet Evol 45:272–288. https://doi.org/10.1016/j.ympev.2007.03.018
Tripp EA, McDade LA (2013) Time-calibrated phylogenies of hummingbirds and hummingbird-pollinated plants reject a hypothesis of diffuse co-evolution. Aliso 31(2):89–103. https://doi.org/10.5642/aliso.20133102.05
Tucker DB, Hedges SB, Colli GR et al (2017) Genomic timetree and historical biogeography of Caribbean island ameiva lizards (*Pholidoscelis*: Teiidae). Ecol Evol 7(17):7080–7090. https://doi.org/10.1002/ece3.3157
Vallejo C, Spikings RA, Luzieux L et al (2006) The early interaction between the Caribbean plateau and the NW South American plate. Terr Nova 18(4):264–269. https://doi.org/10.1111/j.1365-3121.2006.00688.x
Van Ee BW, Berry PE, Riina R, Amaro JEG (2008) Molecular phylogenetics and biogeography of the Caribbean-centered *Croton* subgenus *Moacroton* (Euphorbiaceae s.s.). Bot Rev 74(1):132–165. https://doi.org/10.1007/s12229-008-9003-y
Vargas P, Arjona Y, Nogales M et al (2015) Long-distance dispersal to oceanic islands: success of plants with multiple diaspore specializations. AoB Plants 7:plv073. https://doi.org/10.1093/aobpla/plv073
Vasconcelos TNC, Proenca CEB, Ahmad B et al (2017) Myrteae phylogeny, calibration, biogeography and diversification patterns: increased understanding in the most species rich tribe of Myrtaceae. Mol Phylogenet Evol 109:113–137. https://doi.org/10.1016/j.ympev.2017.01.002
Villagómez D, Spikings R, Magna T et al (2011) Geochronology, geochemistry and tectonic evolution of the western and central cordilleras of Colombia. Lithos 125(3–4):875–896. https://doi.org/10.1016/j.lithos.2011.05.003
Viruel J, Segarra-Moragues GJ, Raz L et al (2016) Late Cretaceous-Early Eocene origin of yams (*Dioscorea*, Dioscoreaceae) in the Laurasian Palaearctic and their subsequent Oligocene-Miocene diversification. J Biogeogr 43(4):750–762. https://doi.org/10.1111/jbi.12678
Wallace AR (1869) The Malay archipielago. Macmillan, London
Wang W, Ortiz RDC, Jacques FMB et al (2012) Menispermaceae and the diversification of tropical rainforests near the Cretaceous-Paleogene boundary. New Phytol 195(2):470–478. https://doi.org/10.1111/j.1469-8137.2012.04158.x
Warren BH, Simberloff D, Ricklefs RE et al (2015) Islands as model systems in ecology and evolution: prospects fifty years after MacArthur-Wilson. Ecol Lett 18(2):200–217. https://doi.org/10.1111/ele.12398
White W, Copeland P, Gravatt DR et al (2017) Geochemistry and geochronology of Grenada and Union islands, Lesser Antilles: the case for mixing between two magma series generated from distinct sources. Geosphere 13(5):1359–1391. https://doi.org/10.1130/GES01414.1
Whittaker RJ (1998) Island biogeography. Oxford University Press, Oxford
Whittaker RJ, Triantis KA, Ladle RJ (2008) A general dynamic theory of oceanic island biogeography. J Biogeogr 35(6):977–994. https://doi.org/10.1111/j.1365-2699.2008.01892.x

Whittaker RJ, Triantis KA, Ladle RJ (2010) A general dynamic theory of oceanic island biogeography: extending the MacArthur-Wilson theory to accommodate the rise and fall of volcanic islands. Princeton University Press, Princeton, NJ

Wiens JJ, Parra-Olea G, Garcia-Paris M et al (2007) Phylogenetic history underlies elevational biodiversity patterns in tropical salamanders. Proc R Soc B Biol Sci 274(1612):919–928. https://doi.org/10.1098/rspb.2006.0301

Zhang L, Renner S (2003) The deepest splits in Chloranthaceae as resolved by chloroplast sequences. Int J Plant Sci 164(5):S383–S392. https://doi.org/10.1086/376882

Chapter 21
The Diversification of Extant Angiosperms in the South America Dry Diagonal

Rosane G. Collevatti, Natácia E. Lima, and Luciana C. Vitorino

Abstract The evolutionary processes leading to lineage diversification in Neotropical plants are still poorly understood. Here, we provide a synthesis of phylogeographic patterns and unravel whether the Neogene geological events or the Quaternary climatic changes drove lineage diversification of Angiosperms in the South America dry diagonal. Despite the high number of plant species in the dry diagonal (~19,000) only few species (30) were studied. Major lineage divergences occurred in the Pliocene but most lineage diversifications occurred at the Early and Middle Pleistocene. The Last Glacial Maximum (LGM) may have had a more local and regional effect in differentiation among populations and patterns of genetic diversity distribution. Species responded differently to the Quaternary climate changes leading to high variation in spatial patterns in genetic diversity and phylogeographic patterns. Finally, our findings challenge the hypothesis of glacial refugia and the importance of the last glacial maximum (LGM) in the diversification of Angiosperms in the South America dry diagonal.

Keywords Caatinga · Cerrado · Chaco · Pleistocene refugia · Quaternary climatic changes · Savanna · Seasonally dry forests · Neogene

1 Introduction

Here, we discuss the diversification and biogeographic patterns of extant Angiosperms in the dry diagonal (Prado and Gibbs 1993; Vanzolini 1963, 1974), the open vegetation that extends from the Northeastern Brazil to the Northern Argentina and includes Caatinga, Cerrado and Chaco biomes (Fig. 21.1), based on a review of phylogeography. Our goal is to provide a synthesis of phylogeographic patterns and

R. G. Collevatti (✉) · N. E. Lima
Laboratório de Genética & Biodiversidade, Instituto de Ciências Biológicas, Universidade Federal de Goiás, Goiânia, GO, Brasil

L. C. Vitorino
Instituto Federal Goiano Campus Rio Verde, Rio Verde, GO, Brasil

V. Rull, A. C. Carnaval (eds.), *Neotropical Diversification: Patterns and Processes*, Fascinating Life Sciences, https://doi.org/10.1007/978-3-030-31167-4_21

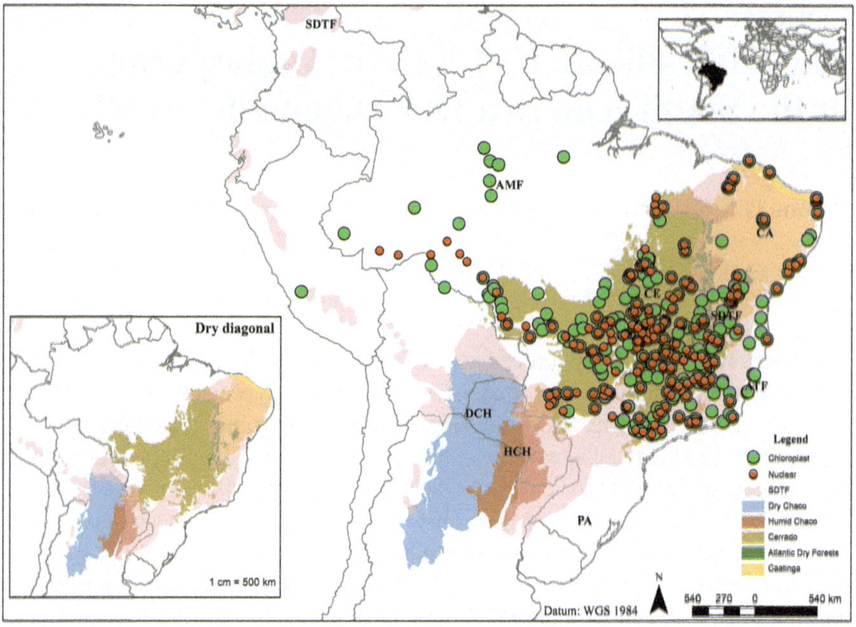

Fig. 21.1 Distribution of the dry diagonal in South America and phylogeographic data for Angiosperms. The distribution of seasonally dry tropical forests (SDTFs) was modified from Collevatti et al. (2012a, b). Biome distribution was obtained from IBGE (2019) and The Nature Conservancy (2009)

unravel whether the Neogene geological events or the Quaternary climate changes drove lineage diversification in the dry diagonal. The dry diagonal is not necessarily xeromorphic, but composed by diverse seasonally dry vegetation, from seasonally dry tropical forests (SDTFs) to more open xeric shrublands and savannas (see Pennington et al. 2006; Prado and Gibbs 1993). The number of plant species in the dry diagonal is astonishingly high, nearly 19,114, estimated from the number of species in Cerrado (11,392) and Caatinga (4322) (Forzza et al. 2012), and in the Gran Chaco (~3400).

This seasonally dry vegetation separates two tropical rain forests, the Amazon and the Atlantic rainforests, and thus has been considered a barrier to the migration of species between these two rainforests, explaining some floristic differences between them (Oliveira-Filho and Ratter 1995). On the other hand, the disjunction distribution for some taxa suggests important routes of species migration between the two rainforests in the past, mainly via riparian forests (Oliveira-Filho and Ratter 1995; Costa 2003; Ratter et al. 1997; Sobral-Souza et al. 2015; Souza-Neto et al. 2016). Moreover, savannas and SDTFs have a more recent origin than the rain forests and thus species immigration from rainforests into savannas and SDTFs may have shaped species assembly and the evolution of high diversity in both biomes (Pennington et al. 2009; Souza-Neto et al. 2016).

SDTFs are deciduous and semi-deciduous forests distributed in regions with mean annual rainfall below 1600 mm with pronounced seasonality (Mooney et al. 1995). They are usually found in eutrophic and oligotrophic soils with moderate pH and low levels of aluminum (Furley and Ratter 1988; Oliveira-Filho and Ratter 2002). In the dry diagonal, the SDTFs are usually scattered throughout other vegetation types such as riparian forests, savannas and xeric shrublands with high heterogeneity between these areas (Oliveira-Filho and Ratter 2002; Pennington et al. 2006; Miranda et al. 2018). In a broad definition, xeric shrublands are considered as SDTFs following Murphy and Lugo (1995) and Pennington et al. (2000). They define SDTFs to include from tall forest on moister sites to cactus scrubs on the driest, and tropical and subtropical dry forests and xeric shrublands. Under this view, Caatinga biome is currently considered the largest remnant of SDTF in South America (Pennington et al. 2000). Although species composition differs among separate SDTF areas (DRYFLOR 2016), there are common features such as the high Fabaceae species richness in all areas, except the Caribbean (Lugo et al. 2006), where Myrtaceae are predominant. Cactaceae is also one of the most species-rich families in SDTFs (Pennington et al. 2006).

Savanna is a global biodiversity "hotspot", with high levels of endemism (Myers et al. 2000). Occupying more than 20° of latitude in Central Brazil, savannas occur in strongly seasonal environments with a dry season during the winter and a mean annual precipitation of 1500 mm. They usually occur over ancient, dystrophic, acid, and aluminum-rich soils, with low concentrations of calcium and magnesium and high density of C4 grasses, and are subject to recurrent fires every 3–4 years (Ratter et al. 2006). The vegetation composition varies from savanna grasslands with dense grasses masses with shrubs and small sparse trees to savanna woodlands with higher density of trees, and forested savannas, comprising a near-closed forest (Oliveira-Filho and Ratter 1995).

The Chaco is included in the Gran Chaco biome, comprising semi-arid forests and woodlands covering the plains of Northeastern Argentina, Western Paraguay, South-eastern Bolivia and the southwestern Brazil with dry season in winter-spring and precipitation ranging from ~1000 to 350 mm (see Prado and Gibbs (1993) for details). The dry season is generally negligible at the Chaco's eastern edge, and increases in duration from east to west. Thus, the vegetation of the Chaco is subjected to low soil moisture and freezing in the dry season and waterlogging and extremely high temperatures during part of the rainy season.

2 Paleovegetation Reconstruction in the Dry Diagonal

The traditional model of paleovegetation dynamics in South America during the Quaternary states that the irregular contraction of humid forests formed isolated refuges during the dry and cooler periods of glaciation, whereas dry forests and savannas expanded (Brown and Ab'Saber 1979; Haffer 1969; Prance 1973; van der Hammen 1974). However, this model has been challenged for Amazonia based on recent evidence of a more complex history and variable patterns of vegetation dynamics, mainly due to a non-uniform reduction in precipitation during the glaciations (e.g., Colinvaux et al. 1996; Haberle and Maslin 1999; Mayle 2004; van der Hammen and Absy 1994).

The South American climate was spatially and temporally complex in both open and forest biomes during the Quaternary (Ab'Saber 2000). Paleoclimatic and palynological evidence show that the climate of western Amazonia has been continuously humid enough to support a permanent rain forest (e.g., Colinvaux et al. 2000; Mayle et al. 2007). Southeastern Brazil was cooler and drier during the glacial periods than it is in the present day and the grasslands and subtropical Araucaria forest spread northward, replacing the Atlantic Forest and savannas to latitudes up to 20°S, and retreating to the south only after the glaciation (Behling 2002, 2003; Salgado-Labouriau 1997). Modern, wetter climatic conditions in Central Brazil, with seasonal dry periods became established during the late Holocene (Behling 2003). In Central Brazil, the palynological evidence supports the presence of either stable grasslands where forest exists today or repeated alternations between forest and savanna (Behling and Hooghiemstra 2000; Salgado-Labouriau 1997).

Studies of the distribution patterns of a suite of plant species widespread in SDTFs lead to the hypothesis that during the dry-cool period of the last glaciation, SDTFs expanded throughout the dry diagonal, towards the east Andes slopes and the Mesoamerica (the 'Pleistocene Arc Hypothesis', Prado and Gibbs 1993). However, palynological data suggest that SDTFs expanded towards the Amazon basin either, but contracted in some areas of the dry diagonal, such as in southeastern Brazil, causing the disjunction of the Caatinga and Misiones nuclei (Mayle et al. 2004). The paleovegetation evidence raises questions regarding the long-term stability of the dry diagonal vegetation structure and whether the Quaternary climate changes promoted species diversification through range fragmentation. On one hand, the geological events in the Neogene and Paleogene changed the climate and the landscape of South America affecting species geographical range and population connectivity. The Andean orogeny in the Palaeogene increased the latitudinal temperature gradient in South America (Hoorn et al. 2010; Rull 2011), and the uplift of the Central Brazilian Plateau in the Miocene–Pliocene transition caused the geographical reorganization of the Central Brazil ecosystems that may have favored species diversification by vicariance (e.g. Luebert and Muller 2015; Luebert and Weigend 2014), and by the formation of new landscape features with compartmentalization between plateaus and depressions (Rull 2011). On the other hand, Quaternary climate oscillations also affected species distributions through space and time and thus might have affected species diversity and diversification (Rull 2008, 2018). However, due to differences in climatic tolerances and niche characteristics, species responded individualistically to the common drier and colder conditions of glacial periods (Cheng et al. 2013; Collevatti et al. 2013a, b; Summerhayes and Charman 2014), and thus, we might expect different phylogeographic patterns in dry diagonal species. In addition, if the Neogene geological events were more important in lineage diversification in the dry diagonal species, we expect more ancient lineage

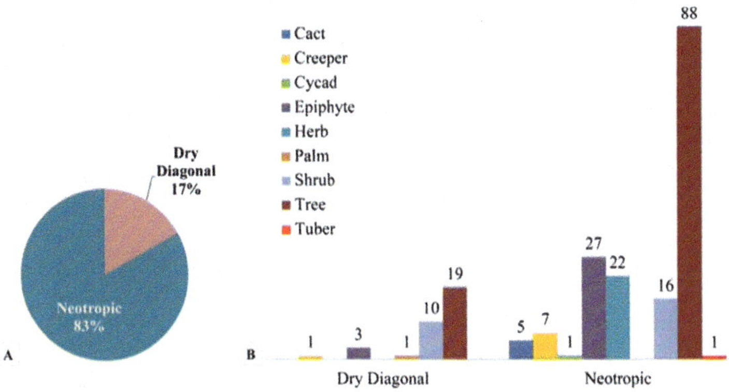

Fig. 21.2 Phylogeographic studies performed with Angiosperms in the Neotropics and in the South America dry diagonal. (**a**) Percentage of phylogeographic studies. (**b**) Number of studies per life forms

diversification than the expected for a major effect of the Quaternary climate changes.

3 The State of the Art: A Scientometric View of Phylogeography in the Dry Diagonal

We performed an exhaustive search for phylogeography publications in the Neotropics using databases such as ISI Web of Science platform (www.webofknowledge.com), Scopus (www.elsevier.com/online-tools/scopus) and Scielo (www.scielo.br) and including other grey literature (mainly thesis), from the first register in 1945 until January 2018. From this search, we obtained the articles from dry diagonal for the review of phylogeography. The first article for the Neotropics was published in 1998 (Aide and Rivera 1998) and for the dry diagonal in 2002 (Olsen 2002). Only a small number of studies have been performed with Angiosperm species in the dry diagonal (Fig. 21.2). From 201 studies available for Angiosperm species from the Neotropics (164 plant species analyzed), only 34 (17%) included species from the dry diagonal encompassing 30 species (Fig. 21.2a). Most studied plant species in the Neotropics and the dry diagonal were trees and shrubs (Fig. 21.2b). In the dry diagonal, most studies were performed with trees (60%) and shrubs (23%, Fig. 21.3a). The studies included 20 genera (Fig. 21.3b) from 16 families (Fig. 21.3c). Most species are endemic to the Cerrado or have more populations studied in this biome (Fig. 21.4). Only one species endemic to Caatinga was studied (*Mandevilla bahiensis*, Fig. 21.4). No phylogeographic study with Angiosperm species has been carried out in the Chaco. The number of studies using chloroplast sequence and both nuclear (mainly ITS rDNA) and chloroplast sequences were the same (13), but only five used

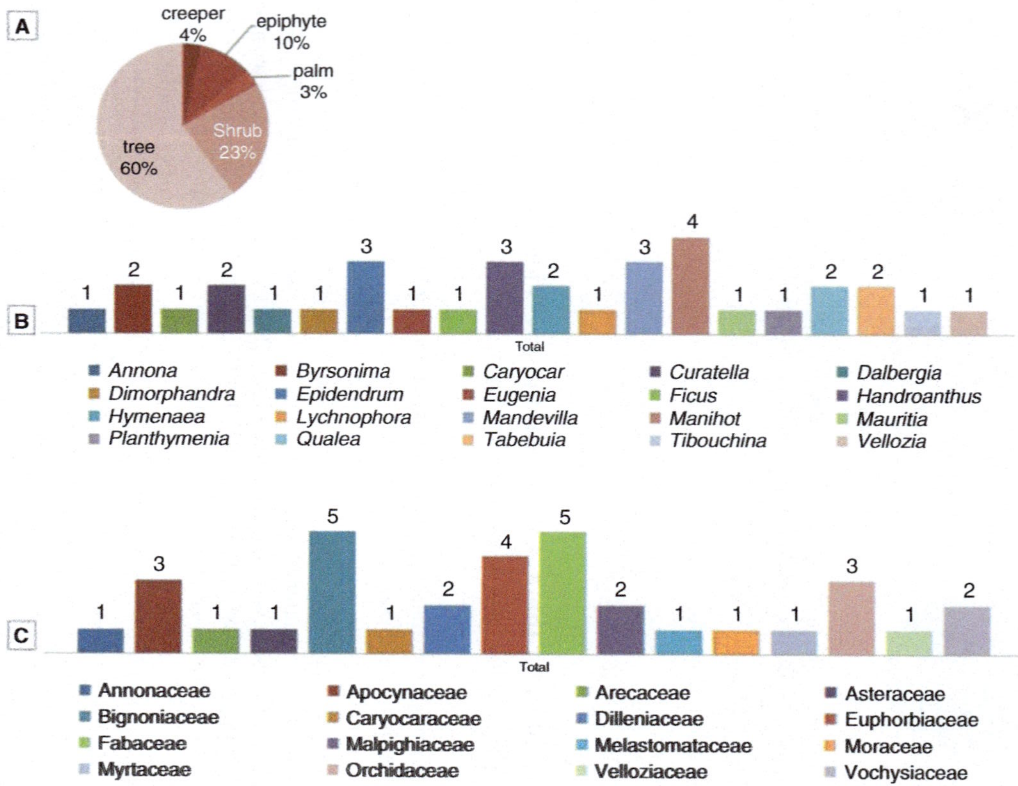

Fig. 21.3 Phylogeographic studies performed with Angiosperms in the South America dry diagonal. **(a)** Percentage of phylogeographic studies per life form. **(b)** Number of phylogeographic studies per genera. **(c)** Number of phylogeographic studies per family

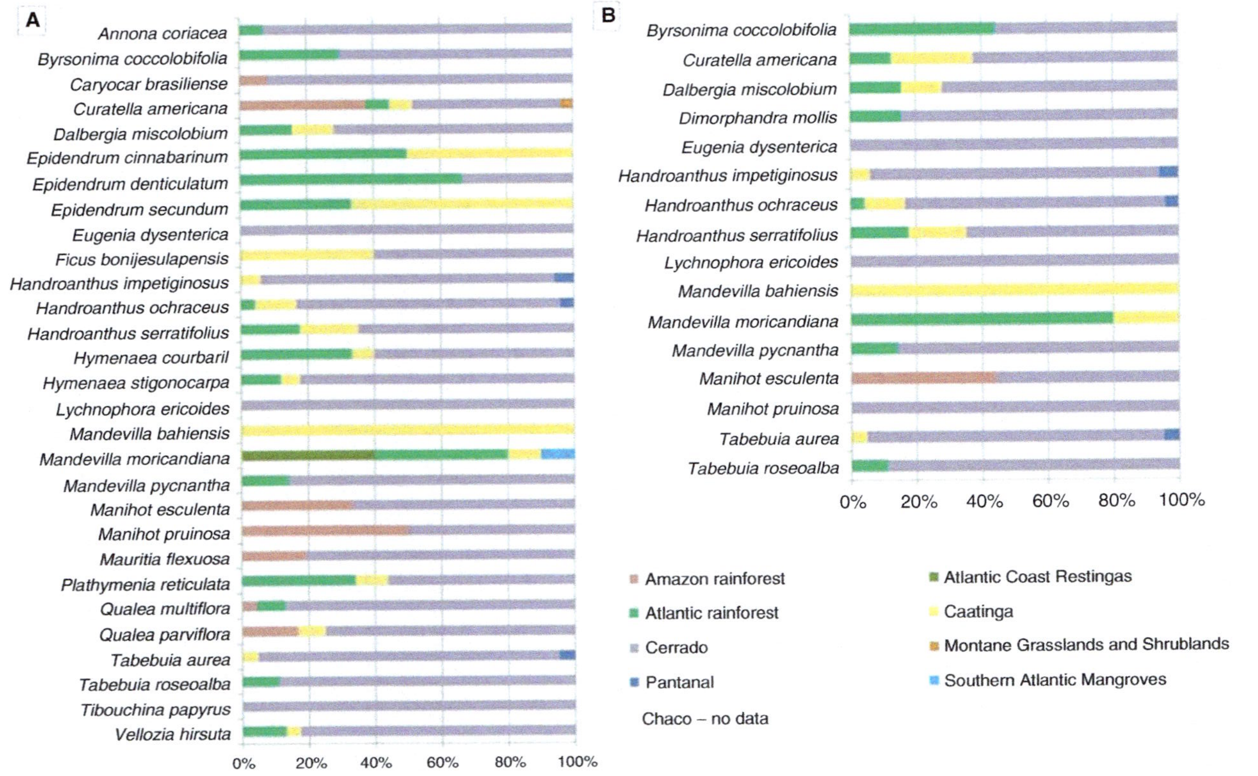

Fig. 21.4 Angiosperm species studied in each biome of the south America dry diagonal, per molecular marker. The bars represent the percentage of populations studied in each biome in the "dry diagonal". (**a**) Chloroplast and (**b**) nuclear data

exclusively nuclear sequences. The number of populations and individuals studied for chloroplast genome was higher (496 and 5507 respectively) than for nuclear genome (275 and 3449) (Fig. 21.1).

Despite the high number of plant species in the dry diagonal (19,114), we currently have phylogeographic information for only 0.15%. Considering the current extinction rate of ~0.01%/year (http://wwf.panda.org/about_our_earth/biodiversity/biodiversity/), ~2 plant species would be expected to go extinct in the dry diagonal each year. Using the regression for the number of articles on 'population genetics of Neotropical plants' per year (Tinoco et al. 2015; $Y = -12.07 + 0.006086^{*}X$) and assuming that each newly published article addresses a different species in dry diagonal, in 2052 (34 years), 1.0% (191 species) of the dry diagonal will have been studied. During this same period, 0.36% of the species will go extinct (~68 species). We believe this is an important indicator of the paucity of scientific production in the dry diagonal and of the need to accelerate and increase the funding for research. In fact, the conservation of dry biomes has been neglected worldwide (Sunderland et al. 2015; DRYFLOR 2016), and most remaining areas of SDFs and savannas in South America and particularly in Brazil are threatened mainly by agricultural expansion, harvesting for wood products and the increase of fire frequency due to agricultural practices (see Collevatti et al. 2013b).

4 The Role of Neogene Geological Events and Quaternary Climate Changes in Lineage Diversification in the Dry Diagonal

The response of plant species to the Neogene geological events or the Quaternary climate changes remains poorly understood in the dry diagonal due to the low number of studies explicitly testing it with coalescent simulation (e.g. Collevatti et al. 2012a, 2015; Lima et al. 2014, 2017) or using niche modeling and coalescent analyses (e.g. Melo et al. 2016; Souza et al. 2017; Vitorino et al. 2016, 2018). Many studies show only indirect evidence based on patterns of haplotype network or neutrality tests (e.g. Caetano et al. 2008; Novaes et al. 2010; Ramos et al. 2007; Ribeiro et al. 2016) and thus have no dated divergence or simulations to test hypotheses of demographical history. Only a few studies (13 for 17 species) analyzed lineage diversification through time, using a dated phylogenetic or coalescent tree and most of them (12 studies encompassing 16 species) provided credibility internals of time to most recent common ancestors (TMRCA) for chloroplast, nuclear or concatenated data (chloroplast and nuclear markers). Regardless the molecular marker used, studies show that most dry diagonal lineages originated in the Neogene, especially in the Pliocene, whereas diversification of lineages within species occurred mainly during the Early and Middle Pleistocene (Fig. 21.5). Moreover, time of diversification is not related to the biome, habitat or life form. For instance, most lineage divergences of *Tabebuia aurea*, *Handroanthus ochraceus*,

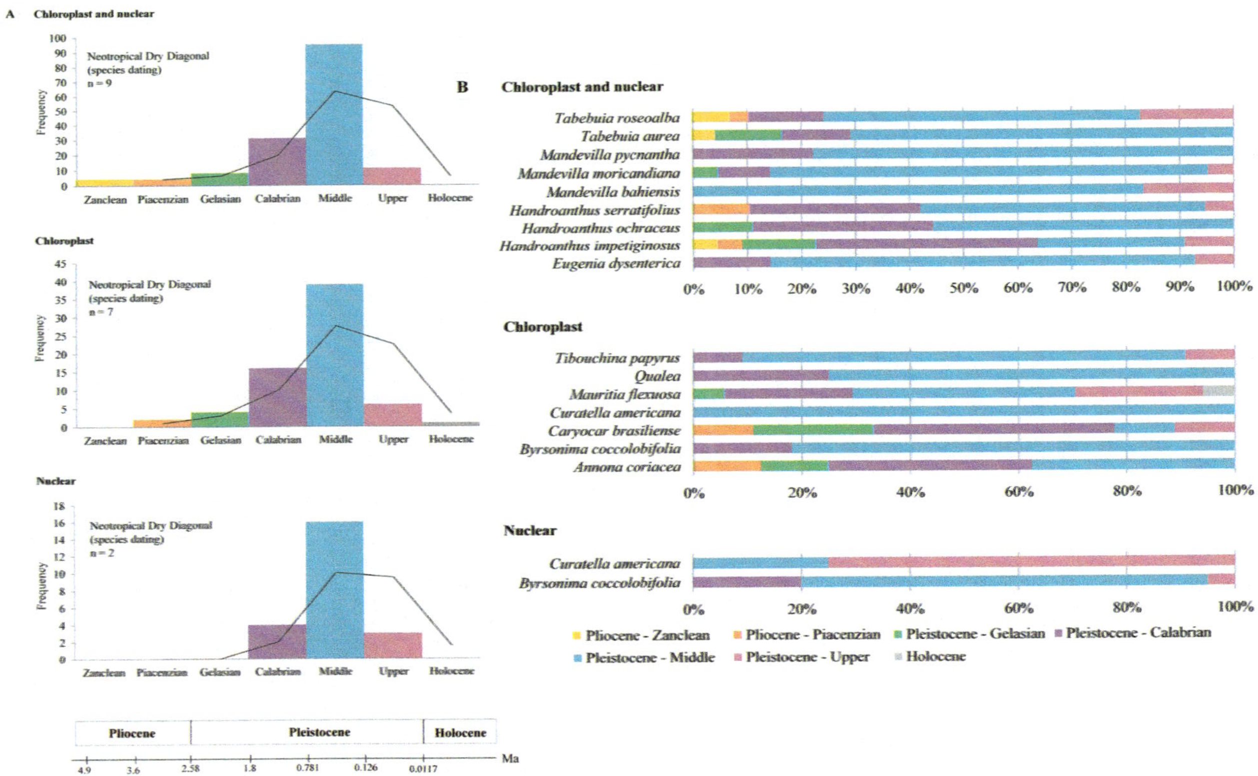

Fig. 21.5 Timing of Angiosperms lineages diversification in the South America dry diagonal. (**a**) Frequency distribution and normalized density function of the ages of diversification overall species using coalescent approach for concatenated nuclear and chloroplast, chloroplast, and nuclear data. (**b**) Percentage of divergences in time dated per species. The timescale is in millions of years ago (Ma) before present. Each color represents the Epoch and Stage in the geologic time scale

Qualea spp. and *Caryocar brasiliense* from savanna occurred during the Early and Middle Pleistocene (Fig. 21.6), but for the other savanna species, such as *Eugenia dysenterica*, *Curatella americana* and *Byrsonima coccolobifolia* lineages diverged mostly at the Middle and Late Pleistocene (Fig. 21.6). Similarly, species from SDTFs showed variation in the frequency distribution of intraspecific lineage diversification time. *Handroanthus impetiginosus* lineages diversified predominantly in the Early and Middle Pleistocene (Fig. 21.6), but *H. serratifolius* and *T. roseoalba* lineages diversified mostly in the Middle and Late Pleistocene. Thus, although glaciation cycles might have been important in the diversification and differentiation of lineages, the data show a minor importance of the last glaciation, especially the last glacial maximum (LGM). This result challenges the hypothesis of the glacial refugia during the LGM and the importance of the LGM in the diversification of Neotropical flora, specifically to the dry diagonal (see Hooghiemstra and van der Hammen (1998) and Mayle (2004), for reviews). In addition, the more ancient origin of the MRCA of most species from the dry diagonal studied so far emphasizes the potential importance of tectonic and paleogeographical events of the Pliocene establishing barriers and different connections among biomes and habitats, leading to lineage origin and diversification (see Pennington et al. (2004) and Rull (2011) for a review).

Population genetic differentiation was high for most species (Table 21.1) and we found no pattern in genetic differentiation and life form (see also Ballesteros-Mejia et al. 2016). Species also showed different patterns of diversity distribution and lineage differentiation. Some species showed an east-west pattern in lineage differentiation and some showed a northeast-southeast differentiation (Figs. 21.7 and 21.8, Table 21.1). For instance, the savanna species *Caryocar brasiliense* (Collevatti et al. 2003, 2012a), *Eugenia dysenterica* (Lima et al. 2017), *Dimorphandra mollis* (Souza et al. 2017), *Dalbergia miscolobium* (Novaes et al. 2010) and *Hymenaea stigonocarpa* (Ramos et al. 2007) show evidence of recent colonization of the southern region of the Cerrado biome in contrast to the northern or central region. On the other hand, the SDTF species *Handroanthus impetiginosus* (Collevatti et al. 2012b) and *Tabebuia roseoalba* (Melo et al. 2016) have no phylogeographic pattern in lineage differentiation (Table 21.1). This is most likely due to differences in species historical demography leading to different patterns in genetic diversity distribution. The cycles of range retraction during the Quaternary glacial periods and the subsequent expansion in the interglacial may have contributed to the high genetic differentiation among populations from the dry diagonal (Table 21.1). In addition, species responded differently to the Quaternary climate changes (Table 21.1), and the spatial patterns in genetic diversity may be the outcome of different patterns of range shifts. For instance, *Dimorphandra mollis* (Souza et al. 2017), *Handroanthus ochraceus* (Vitorino et al. 2018) and *Eugenia dysenterica* (Lima et al. 2017) showed a quasi stability in geographical range through the last glacial cycle, while *Caryocar brasiliense* (Collevatti et al. 2012a), *Tabebuia aurea* (Collevatti et al. 2015) and *Mauritia flexuosa* (Lima et al. 2014) showed a retraction during the glaciation with multiple refugia. Species from SDTFs also showed contrasting response to climate changes during the Quaternary and thus in patterns

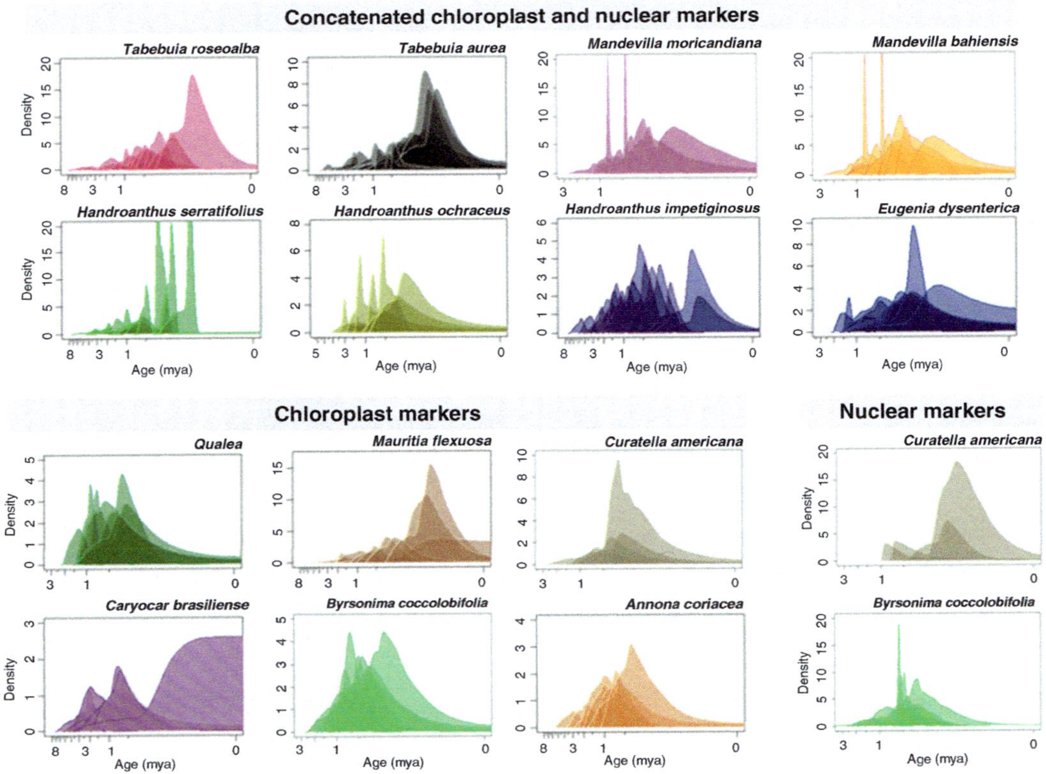

Fig. 21.6 Density function of the ages of lineages diversification for each Angiosepem species in the South America dry diagonal, using coalescent approach for concatenated nuclear and chloroplast, chloroplast, and nuclear data. The timescale is in millions of years ago (Ma) before present

Table 21.1 Phylogeographic patterns in plant species from the dry diagonal

Species	Pop	Ind	Marker	F_{ST}	Demographic pattern—LMG	Method	Phylogeographic pattern
Annona coriacea	28	151	Chloroplast	0.75	Expansion	Neutrality tests	East–West
Byrsonima coccolobifolia	10	49	Chloroplast	0.57	–	–	East–West
Byrsonima coccolobifolia	10	47	Chloroplast	0.31	Constant	EBSP, Simulation	East–West
Byrsonima coccolobifolia	10	152	Nuclear	0.17	Constant	EBSP, Simulation	East–West
Caryocar brasiliense	12	147	Chloroplast	0.79	Multiple refugia	Simulation, Coalescence, Niche modeling	West–East
Epidendrum cinnabarinum	11	177	Chloroplast	0.644	Expansion	Niche modelling	No pattern
Epidendrum denticulatum	13	201	Chloroplast	0.7514	Multiple refugia	Network	Southwest–Northeast
Epidendrum secundum	9	174	Chloroplast	0.636	Retraction	Niche modelling	No pattern
Curatella americana	22	112	Chloroplast	0.28	Expansion	Neutrality tests	Central–North
Curatella americana	10	49	Chloroplast	0.64	Constant	EBSP, Simulation	Southwest–Northeast
Curatella americana	10	96	Nuclear	0.24	Constant	EBSP, Simulation	Southwest–Northeast
Dalbergia miscolobium	32	287	Chloroplast	0.95	Expansion	Neutrality tests	Eastern–Western
Dalbergia miscolobium	32	287	Nuclear	0.77	Expansion	Neutrality tests	Eastern–Western
Dimorphandra mollis	32	151	Nuclear	0.30	Constant	BSP, Coalescence, Niche modeling	No pattern

Eugenia dysenterica	23	333	Chloroplast	0.59	Constant	EBSP, Simulation, Coalescence, Niche modeling	Southeast–Central
Eugenia dysenterica	23	333	Nuclear	0.43	Constant	EBSP, Simulation, Coalescence, Niche modeling	No pattern
Ficus bonijesulapensis	15	126	Chloroplast	0.84	Expansion	Neutrality tests	Central-West, Central-East and scattered
Handroanthus impetiginosus	17	258	Chloroplast	0.89	Expansion	BSP, Coalescence, Niche modeling	No pattern
Handroanthus impetiginosus	17	258	Nuclear	0.81	Expansion	BSP, Coalescence, Niche modeling	No pattern
Handroanthus ochraceus	24	468	Chloroplast	0.74	Multiple refugia	EBSP, Simulation, Coalescence, Niche modeling	No pattern
Handroanthus ochraceus	24	468	Nuclear	0.54	Multiple refugia	EBSP, Simulation, Coalescence, Niche modelling	No pattern
Handroanthus serratifolius	17	257	Chloroplast	0.53	Expansion	EBSP, Simulation, Coalescence, Niche modelling	Northeast and Central-West, South-east and Central-East
Handroanthus serratifolius	17	257	Nuclear	0.74	Expansion	EBSP, Simulation, Coalescence, Niche modelling	North–Northeast and Central-West, Central-West
Hymenaea courbaril	15	149	Chloroplast	0.60	Expansion	Neutrality tests	Western–Central–Northeastern
Hymenaea stigonocarpa	17	174	Chloroplast	0.69	–	–	Western–Central–Northeastern
Lychnophora ericoides	12	192	Chloroplast	0.84	Expansion	Coalescence	Southeast and Central
Lychnophora ericoides	12	192	Nuclear	0.70	Expansion	Coalescence	Southeast and Central
Mandevilla bahiensis	9	138	Chloroplast	0.71	Expansion	EBSP, Niche modelling	No pattern
Mandevilla bahiensis	9	138	Nuclear	0.06	Expansion	EBSP, Niche modelling	No pattern

(continued)

Table 21.1 (continued)

Species	Pop	Ind	Marker	F_{ST}	Demographic pattern—LMG	Method	Phylogeographic pattern
Mandevilla moricandiana	10	152	Chloroplast	0.59	Expansion	EBSP, Niche modelling	No pattern
Mandevilla moricandiana	10	152	Nuclear	0.71	Expansion	EBSP, Niche modelling	No pattern
Mandevilla pycnantha	7	83	Chloroplast	0.09	Expansion	EBSP, Coalescence, Niche modelling	No pattern
Mandevilla pycnantha	7	61	Nuclear	0.15	Expansion	EBSP, Coalescence, Niche modelling	No pattern
Manihot esculenta	27	314	Nuclear	–	–	–	Northeast–West
Manihot esculenta	4	157	Chloroplast	–	–	–	No pattern
Manihot pruinosa	6	70	Nuclear	–	–	–	Northeast–West
Manihot pruinosa	2	35	Chloroplast	–	–	–	No pattern
Mauritia flexuosa	26	257	Chloroplast	0.88	Multiple refugia	Simulation, Coalescence, Niche modelling	West, Central-South, Central-North and East
Plathymenia reticulata	41	220	Chloroplast	0.93	Expansion	Neutrality tests	Central-East, Central-North and Central-South
Qualea multiflora	23	179	Chloroplast	0.71	Multiple refugia	BSP, Niche modelling	West, North-Western, North-Eastern and South-Western
Qualea parviflora	12	75	Chloroplast	0.71	Multiple refugia	BSP, Niche modelling	West, North-Western and North-Eastern
Tabebuia aurea	21	285	Chloroplast	0.77	Multiple refugia	EBSP, Simulation, Coalescence, Niche modelling	No pattern
Tabebuia aurea	21	285	Nuclear	0.97	Multiple refugia	EBSP, Simulation, Coalescence, Niche modelling	No pattern
Tabebuia roseoalba	18	235	Chloroplast	0.62	Retraction	EBSP, Simulation, Coalescence, Niche modelling	No pattern

Tabebuia roseoalba	18	235	Nuclear	0.75	Retraction	EBSP, Simulation, Coalescence, Niche modeling	No pattern
Tibouchina papyrus	3	96	Chloroplast	0.68	Expansion	EBSP, Coalescence	No pattern
Vellozia hirsuta	23	244	Chloroplast	0.82	Expansion	Neutrality tests	Southeast–Northeast

Demographic pattern is the demographic history during the Last Glacial Maximum. Phylogeographic pattern is the general spatial pattern of lineage differentiation. Method referrers to the main method used to infer the demographic patterns. Pop, number of populations studied for each species; Ind, number of individuals sampled; F_{ST}, genetic differentiation among populations

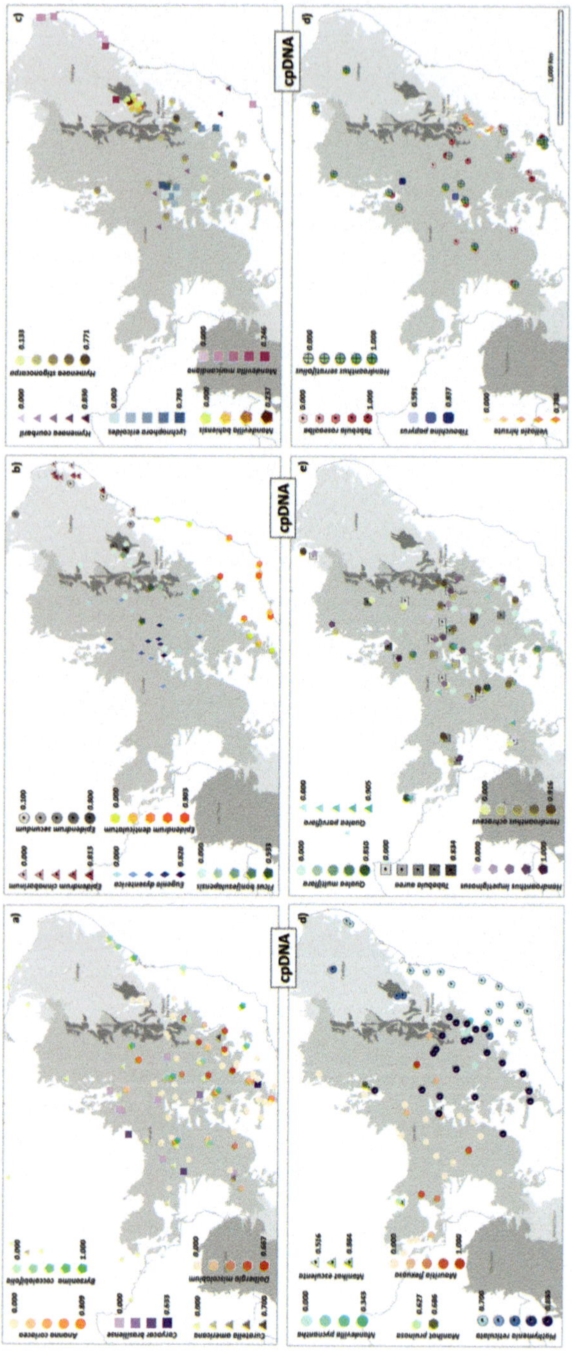

Fig. 21.7 Geographic distribution of haplotype diversity for each Angiosperm species of the South America dry diagonal, based on chloroplast genome. The symbols represent different species and the colors the haplotype diversity in each population, following the legend in the figure

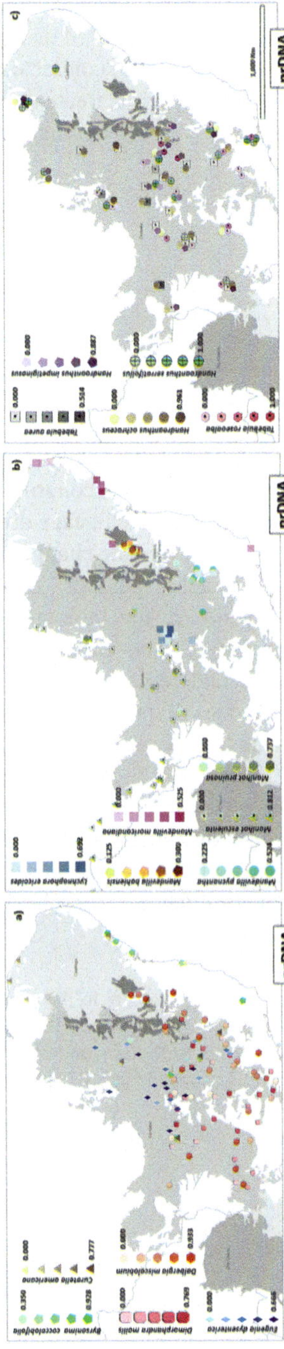

Fig. 21.8 Geographic distribution of haplotype diversity for each Angiosperm species of the South America dry diagonal, based on nuclear genome. The symbols represent different species and the colors the haplotype diversity in each population, following the legend in the figure

of genetic diversity. While *Handroanthus impetiginosus* (Collevatti et al. 2012b) and *Ficus bonijesulapensis* (Vieira et al. 2015) expanded their range during glacial periods, *H. serratifolius* (Vitorino et al. 2016) shifted its range towards the Amazon basin and *Tabebuia roseoalba* (Melo et al. 2016) contracted its range. It is important to note that most studies were performed with trees, so lineage divergences and patterns in genetic differentiation and diversity are highly biased to this plant life form.

5 Conclusion

The low number of Angiosperm species studied and the wide variation in response among species still hinders a synthesis of the biogeography of the South America dry diagonal and the effects of Neogene and Quaternary climate changes in species distribution dynamics. However, despite these caveats, our review unravels some phylogeographic patterns. Major lineage origin is congruent in timing with geological events in the Pliocene, but most intraspecific lineage divergences occurred in the Early and Middle Pleistocene. Quaternary climate changes have differently affected the genetic diversity and phylogeographic patterns of plant species in the dry diagonal because their effects depend on the biological characteristics of each species, as well as geographical features and climate changes at each site. The LGM may have had a more local and regional effect in differentiation among populations and patterns of genetic diversity distribution (Collevatti et al. 2012a, b, 2015). While most savanna species show a range retraction during the glaciation of the Quaternary, SDTF species showed an expansion, although species response was highly variable. The high variation in species response to Quaternary climate changes may be responsible for the variation in phylogeographic patterns. Finally, our study highlights the lack of phylogeographic studies in the Chaco and the overall paucity of studies in the dry diagonal, despite the high level of threat to these unique environments.

Acknowledgments This work was supported by grants to the research network Rede Cerrado CNPq/PPBio (project no. 457406/2012-7) and CAPES/PROCAD (project no. 88881.068425/2014-01). NEL received a CAPES/PNPD fellowship. RGC has continuously been supported by productivity grant from CNPq, which we gratefully acknowledge. We thank Toby Pennington for helpful comments in the manuscript.

References

Ab'Saber AN (2000) Spaces occupied by the expansion of dry climates in South America during the Quaternary ice ages. Rev Inst Geol 21:71–78

Aide TM, Rivera E (1998) Geographic patterns of genetic diversity in *Poulsenia armata* (Moraceae): implications for the theory of Pleistocene refugia and the importance of riparian forest. J Biogeogr 25:695–705

Ballesteros-Mejia L, Lima NE, Lima-Ribeiro MS, Collevatti RG (2016) Pollination mode and mating system explain patterns in genetic differentiation in Neotropical plants. PLoS One 11: e0158660

Behling H (2002) South and southeast Brazilian grasslands during Late Quaternary times: a synthesis. Palaeogeogr Palaeoclimatol Palaeoecol 177:19–27

Behling H (2003) Late glacial and Holocene vegetation, climate and fire history inferred from Lagoa Nova in the southeastern Brazilian lowland. Veg Hist Archaeobotany 12:263–270

Behling H, Hooghiemstra H (2000) Holocene Amazon rainforest-savanna dynamics and climatic implications: high-resolution pollen record from Laguna Loma Linda in eastern Colombia. J Quat Sci 15:687–695

Brown KS, Ab'Saber AN (1979) Ice-age refuges and evolution in the neotropics: correlation of paleoclimatological, geomorphological and pedological data with modern biological endemism. Paleoclimas 5:1–30

Caetano S, Prado D, Pennington RT, Beck S, Oliveira-Filho A, Spichiger R et al (2008) The history of seasonally dry tropical forests in eastern South America: inferences from the genetic structure of the tree *Astronium urundeuva* (Anacardiaceae). Mol Ecol 17:3147–3159

Cheng H, Sinha A, Cruz FW, Wang X, Edwards RL, d'Horta FM et al (2013) Climate change patterns in Amazonia and biodiversity. Nat Commun 4:1411

Colinvaux PA, de Oliveira PE, Bush MB (2000) Amazonian and Neotropical plant communities on glacial time-scales: the failure of the aridity and refuge hypotheses. Quat Sci Rev 19:141–169

Colinvaux PA, DeOliveira PE, Moreno JE, Miller MC, Bush MB (1996) A long pollen record from lowland Amazonia: forest and cooling in glacial times. Science 274:85–88

Collevatti RG, Grattapaglia D, Hay JD (2003) Evidences for multiple maternal lineages of Caryocar brasiliense populations in the Brazilian Cerrado based on the analysis of chloroplast DNA sequences and microsatellite haplotype variation. Mol Ecol 12:105–115

Collevatti RG, Terribile LV, Oliveira G, Lima-Ribeiro MS, Nabout JC, Rangel TF, Diniz-Filho JAF (2013a) Drawbacks to palaeodistribution modelling: the case of South American seasonally dry forests. J Biogeogr 40:345–358

Collevatti RG, Lima-Ribeiro MS, Diniz-Filho JAF, De Oliveira G, Dobrovolski R, Terribile LC (2013b) Stability of Brazilian seasonally dry forests under climate change: inferences for long-term conservation. Am J Plant Sci 4:792–805

Collevatti RG, Lima-Ribeiro MS, Souza-Neto AC, Franco AA, de Oliveira G, Terribile LC (2012a) Recovering the demographical history of a Brazilian Cerrado tree species *Caryocar brasiliense*: coupling ecological niche modeling and coalescent analyses. Nat Conserv 10:169–176

Collevatti RG, Terribile LC, Lima-Ribeiro MS, Nabout JC, de Oliveira G, Rangel TF et al (2012b) A coupled phylogeographical and species distribution modelling approach recovers the demographical history of a Neotropical seasonally dry forest tree species. Mol Ecol 21:5845–5863

Collevatti RG, Terribile LC, Rabelo SG, Lima-Ribeiro MS (2015) Relaxed random walk model coupled with ecological niche modeling unravel the dispersal dynamics of a Neotropical savanna tree species in the deeper Quaternary. Front Plant Sci 6:1–15

Costa LP (2003) The historical bridge between the Amazon and the Atlantic Forest of Brazil: a study of molecular phylogeography with small mammals. J Biogeogr 30:71–86

DRYFLOR (2016) Plant diversity patterns in Neotropical dry forests and their conservation implications. Science 353:1383–1387

Forzza RC, Baumgratz JFA, Bicudo CEM, Canhos DAL, Carvalho JAA, Coelho MAN et al (2012) New Brazilian floristic list highlights conservation Challenges. Bioscience 62:39–45

Furley P, Ratter J (1988) Soil resources and plant communities of the central Brazilian cerrado and their development. J Biogeogr 15:97–108

Haberle SG, Maslin MA (1999) Late Quaternary vegetation and climate change in the Amazon basin based on a 50,000 year pollen record from the Amazon fan, ODP site 932. Quat Res 51:27–38

Haffer J (1969) Speciation in Amazonian forest birds. Science 165:131–137

Hoorn C, Wesselingh FP, ter Steege H, Bermudez MA, Mora A, Sevink J et al (2010) Amazonia through time: Andean uplift, climate change, landscape evolution, and biodiversity. Science 330:927–931

Hooghiemstra H, van der Hammen T (1998) Neogene and Quaternary development of the Neotropical rain forest: the forest refugia hypothesis, and a literature overview. Earth Sci Rev 44:147–183

IBGE (2019) Diomas e Sistemas Marinhos Costeiros do Brasil 1:250.000. Instituto Brasileiro de Geografia e Estatística. Available at https://www.ibge.gov.br/geociencias/informacoes-ambientais/15842-biomas.html?=&t=downloads

Lima JS, Telles MPC, Chaves LJ, Lima-Ribeiro MS, Collevatti RG (2017) Demographic stability and high historical connectivity explain the diversity of a savanna tree species in the Quaternary. Ann Bot 119:645–657

Lima NE, Lima-Ribeiro MS, Tinoco CF, Terribile LC, Collevatti RG (2014) Phylogeography and ecological niche modelling, coupled with the fossil pollen record, unravel the demographic history of a Neotropical swamp palm through the Quaternary. J Biogeogr 41:673–686

Luebert F, Muller LAH (2015) Effects of mountain formation and uplift on biological diversity. Front Genet 6:1–2

Luebert F, Weigend M (2014) Phylogenetic insights into Andean plant diversification. Front Ecol Evol 2:1–17

Lugo AE, Medina E, Trejo-Torres CJ, Helmer E (2006) Botanical and ecological basis for the resilience of antillean dry forests. In: Pennington RT, Ratter JA (eds) Neotropical savannas and seasonally dry forests plant diversity, biogeography, and conservation. Taylor & Francis, Boca Raton, FL, pp 359–381

Mayle FE (2004) Assessment of the Neotropical dry forest refugia hypothesis in the light of palaeoecological data and vegetation model simulations. J Quat Sci 19:713–720

Mayle FE, Beerling DJ, Gosling WD, Bush MB (2004) Responses of Amazonian ecosystems to climatic and atmospheric carbon dioxide changes since the last glacial maximum. Philos Trans R Soc Lond B 359:499–514

Mayle FE, Langstroth RP, Fisher RA, Meir P (2007) Long-term forest–savannah dynamics in the Bolivian Amazon: implications for conservation. Philos Trans R Soc Lond B 362:291–307

Melo WA, Lima-Ribeiro MS, Terribile LC, Collevatti RG (2016) Coalescent simulation and paleodistribution modeling for *Tabebuia rosealba* do not support South American dry forest refugia hypothesis. PLoS One 11:e0159314

Miranda PLS, Oliveira-Filho A, Pennington RT, Neves D, Baker T, Dexter KG (2018) Using tree species inventories to map biomes and assess their climatic overlaps in lowland tropical South America. Global Ecol Biogeogr 27:899–912

Mooney HA, Bullock SH, Medina E (1995) Introduction. In: Mooney HA, Bullock SH, Medina E (eds) Dry tropical forests. Cambridge University Press, Cambridge

Murphy PG, Lugo AE (1995) Dry forests of Central America and the Caribbean. In: Bullock SH, Mooney HA, Medina E (eds) Seasonally dry tropical forests. Cambridge University Press, Cambridge, pp 9–34

Myers N, Mittermeier RA, Mittermeier CG, da Fonseca GAB, Kent J (2000) Biodiversity hotspots for conservation priorities. Nature 403:853–858

Novaes RML, De Lemos JP, Ribeiro RA, Lovato MB (2010) Phylogeography of *Plathymenia reticulata* (Leguminosae) reveals patterns of recent range expansion towards northeastern Brazil and southern Cerrados in Eastern Tropical South America. Mol Ecol 19:985–998

Oliveira-Filho AT, Ratter JA (1995) A study of the origin of central Brazilian forests by the analysis of plant species distribution patterns. Edinb J Bot 52:141–194

Oliveira-Filho AT, Ratter JA (2002) Vegetation physiognomies and wood flora of the bioma Cerrado. In: Oliveira PS, Marquis RJ (eds) The Cerrados of Brazil: ecology and natural history of a neotropical Savanna. Columbia University Press, New York, pp 91–120

Olsen KM (2002) Population history of *Manihot esculenta* (Euphorbiaceae) inferred from nuclear DNA sequences. Mol Ecol 11:901–911

Pennington RT, Lavin M, Prado DE, Pendry CA, Pell S, Butterworth CA (2004) Historical climate change and speciation: Neotropical seasonally dry forest plants show patterns of both Tertiary and Quaternary diversification. Philos Trans R Soc Lond Ser B Biol Sci 359:515–538

Pennington RT, Lavin M, Oliveira A (2009) Woody plant diversity, evolution, and ecology in the tropics: perspectives from seasonally dry tropical forests. Annu Rev Ecol Evol Syst 40:437–457
Pennington RT, Lewis GP, Ratter JA (2006) An overview of the plant diversity, biogeography and conservation of Neotropical savannas and seasonally dry Forests. In: Pennington RT, Lewis GP, Ratter JA (eds) Neotropical savannas and seasonally dry forests: plant diversity, biogeography and conservation. Taylor & Francis, Boca Raton, FL, pp 1–29
Pennington RT, Prado DE, Pendry CA (2000) Neotropical seasonally dry forests and Quaternary vegetation changes. J Biogeogr 27:261–273
Prado DE, Gibbs PE (1993) Patterns of Species Distributions in the Dry Seasonal Forests of South-America. Ann Mo Bot Gard 80:902–927
Prance GT (1973) Phytogeographic support tor the theory of Pleistocene forest refuges in the Amazon Basin, based on evidente from distribution patterns in Caryocaraceae, Chrysobalanaceae, Dichapetalaceae and Lecythidaceae. Acta Amaz 3:5–28
Ramos ACS, Lemos-Filho JP, Ribeiro RA, Santos FR, Lovato MB (2007) Phylogeography of the tree *Hymenaea stigonocarpa* (Fabaceae: Caesalpinioideae) and the influence of Quaternary climate changes in the Brazilian Cerrado. Ann Bot 100:1219–1228
Ratter JA, Bridgewater S, Ribeiro F (2006) Biodiversity patterns of the woody vegetation of the Brazilian Cerrado. In: Pennington RT, Lewis GP, Ratter JA (eds) Neotropical savannas and seasonally dry forests: plant diversity, biogeography and conservation. Taylor & Francis, Boca Raton, FL, pp 31–66
Ratter JA, Ribeiro JF, Bridgewater S (1997) The Brazilian cerrado vegetation and threats to its biodiversity. Ann Bot 80:223–230
Ribeiro PC, Lemos JP, Buzatti RSD, Lovato MB, Heuertz M (2016) Species-specific phylogeographical patterns and Pleistocene east-west divergence in *Annona* (Annonaceae) in the Brazilian Cerrado. Bot J Linn Soc 181:21–36
Rull V (2008) Speciation timing and neotropical biodiversity: the Tertiary-Quaternary debate in the light of molecular phylogenetic evidence. Mol Ecol 17:2722–2729
Rull V (2011) Neotropical biodiversity: timing and potential drivers. Trends Ecol Evol 26:508–513
Rull V (2018) Neotropical diversification: historical overview and conceptual insights. In: Rull V, Carnaval A (eds) Neotropical Diversification. Springer, Berlin
Salgado-Labouriau ML (1997) Late Quaternary palaeoclimate in the savannas of South America. J Quat Sci 12:371–379
Sobral-Souza T, Lima-Ribeiro MS, Solferini VN (2015) Biogeography of Neotropical Rainforests: past connections between Amazon and Atlantic Forest detected by ecological niche modeling. Evol Ecol Res 29:643–655
Souza-Neto AC, Cianciaruso MV, Collevatti RG (2016) Habitat shifts shaping the diversity of a biodiversity hotspot through time: insights from the phylogenetic structure of Caesalpinioideae in the Brazilian Cerrado. J Biogeogr 43:340–350
Souza HA, Collevatti RG, Lima-Ribeiro MS, Lemos-Filho JP, Lovato MB (2017) A large historical refugium explains spatial patterns of genetic diversity in a Neotropical savanna tree species. Ann Bot 119:239–252
Summerhayes C, Charman D (2014) Introduction to Holocene climate change: new perspectives. J Geol Soc Lond 172:251–253
Sunderland T, Apgaua D, Baldauf C, Blackie R, Colfer C, Cunningham AB, Dexter K, Djoudi H, Gautier D, Gumbo D, Ickowitz A, Kassa H, Parthasarathy N, Pennington RT, Paumgarten F, Pulla S, Sola P, TNG D, Waeber P, Wilmé L (2015) Global dry forests: a prologue. Int For Rev 17:1–9
Tinoco CF, Lima NE, Lima-Ribeiro MS, Collevatti RG (2015) Research and partnerships in studies on population genetics of Neotropical plants: a scientometric evaluation. Biochem Syst Ecol 61:357–365
The Nature Conservancy (2009) Terrestrial ecoregions of the world. Available at http://maps.tnc.org/files/metadata/TerrEcos.xml. Accessed 15 Dec 2017

van der Hammen T (1974) The pleistocene changes of vegetation and climate in tropical South America. J Biogeogr 1:3–26

van der Hammen T, Absy ML (1994) Amazonia during the last glacial. Palaeogeogr Palaeoclimatol Palaeoecol 109:247–261

Vanzolini PE (1963) Problemas faunisticos do Cerrado, p. 307–320. Ferri MG (ed) Simpósio sôbre o Cerrado. Universidade de São Paulo, São Paulo, 469 p

Vanzolini PE (1974) Ecological and geographical distribution of lizards in Pernambuco, northeastern Brazil (Sauria). Pap Avulsos Zool 28:61–90

Vieira FA, Novaes RML, Fajardo CG, Santos RM, Almeida HS, Carvalho D et al (2015) Holocene southward expansion in seasonally dry tropical forests in South America: phylogeography of *Ficus bonijesulapensis* (Moraceae). Bot J Linn Soc 177:189–201

Vitorino LC, Lima-Ribeiro MS, Terribile LC, Collevatti RG (2016) Demographical history and palaeodistribution modelling show range shift towards Amazon Basin for a Neotropical tree species in the LGM. BMC Evol Biol 16:1–15

Vitorino LC, Lima-Ribeiro MS, Terribile LC, Collevatti RG (2018) Demographical expansion of *Handroanthus ochraceus* in the Cerrado during the Quaternary: implications for the genetic diversity of Neotropical trees. Biol J Linn Soc Lond 123:561–577

Part III
Taxon-Based Local and Regional Studies

Chapter 22
Amphibians and Reptiles of Venezuelan Guayana: Diversity, Biogeography and Conservation

Celsa Señaris and Fernando J. M. Rojas-Runjaic

Abstract Venezuelan Guayana covers a fifth of the Precambrian Guiana Shield in South America, one of the largest wilderness areas on the planet. Scientific knowledge of its herpetofauna has increased in a remarkable and sustained way in recent decades, but is still incomplete, with enormous geographical sampling gaps and almost total ignorance about species ecology. Currently, records exist of 195 species of amphibians (189 anurans and six caecilians) and 221 of reptiles (202 lizards and snakes, 15 turtles, and four crocodilians). Together, they represent nearly the 14% of the megadiverse South America herpetofauna. Such diversity and taxonomic composition of amphibians and reptiles are unevenly distributed in the Guayanan region: richness decreases with elevation, while endemism increases. Thirty-four percent of the amphibians and 22% of the reptiles are exclusive to the uplands and the highlands of Central Guayana province, including four genera. In the peripheral lowlands, herpetofaunal communities are dominated by taxa that are widespread or that have an Amazonian-Guianan distribution. Presently, Venezuelan Guayana biodiversity is seriously threatened by increasing deforestation, uncontrolled illegal mining, and uncontrolled mega-developments. Few highlands are affected by uncontrolled tourism, but climate change has emerged recently as a significant threat for the endemic Pantepui diversity. It is imperative and urgent to increase national and international efforts to study and monitor the most vulnerable populations of amphibians and reptiles in Venezuelan Guayana, before many of them disappear.

Keywords Herpetofauna · Community assemblage · Elevational diversity gradient · Endemism · Diversification · Biodiversity threats

C. Señaris (✉)
Laboratorio de Ecología y Genética de Poblaciones, Centro de Ecología, Instituto Venezolano de Investigaciones Científicas, Caracas, Venezuela

F. J. M. Rojas-Runjaic
Fundación La Salle de Ciencias Naturales, Museo de Historia Natural La Salle, Caracas, Venezuela

Laboratório de Sistemática de Vertebrados, Pontifícia Universidade Católica do Rio Grande do Sul (PUCRS), Porto Alegre, Brazil

V. Rull, A. C. Carnaval (eds.), *Neotropical Diversification: Patterns and Processes*, Fascinating Life Sciences, https://doi.org/10.1007/978-3-030-31167-4_22

1 Introduction

The Guiana Shield is a ca. 2.3 million km^2 Neotropical region of tremendous environmental, hydrological, and topographical complexity. It includes savannas and white-sand forests—the largest patches known in Amazonia—, humid and flooded pristine forests, dome-shaped rocky outcrops (i.e., *inselbergs* or *lajas*), and the spectacular flat-topped sandstone mountains, known as *tepuis,* that rise abruptly above the surrounding lowlands. Its diversity is remarkable, but even more striking is the degree of endemism of its flora and fauna (Hoogmoed 1979; Duellman 1999; Berry and Riina 2005; Hollowell and Reynolds 2005; Naka 2011). Such high number of lineages and clades, found nowhere else on Earth, seems to be the result of multiple and intricate evolutionary and ecological processes.

About 20% of the total surface of the Guiana Shield is found in southern Venezuela (Hammond 2005), largely covered by a variety of tropical and montane forests (Huber and García 2011). This region, the Venezuelan Guayana, is one of the most pristine areas of the planet. Although significant, our understanding of its diversity has been accumulating at a pace lower than ideal, especially considering that the threats to this region are severe and increasing rapidly (RAISG 2012, 2015; Bovolo et al. 2018).

Based on more than 15 years of field work in 410 locations, as well as museum records, and scientific literature search, Gorzula and Señaris (1999) offered the first compendium of the herpetofauna of the Venezuelan Guayana. Later, within the framework of Guiana Shield Priority-Setting Workshop held in Paramaribo 2002 (Conservation International 2003), the taxonomic lists of amphibians (Señaris and MacCulloch 2005) and reptiles (Ávila-Pires 2005) for the region—including the Venezuelan Guayana —were updated and used as a basis for the establishment of conservation areas. In that same year, McDiarmid and Donnelly (2005) published a detailed review of the Guayana highland herpetofauna ($\geq$1500 m elevation), resulting in an updated species list, the identification of distribution patterns, and a discussion of possible drivers, guided by data analysis. Since then, new data and analyses have resulted in relevant changes to our understanding of the biodiversity of the region thanks to the discovery and description of new species, novel phylogenetic relationships, and new geographic records. The amount of new information, and its implications to our knowledge of the Guayana biodiversity, call for a review that synthetizes the most relevant advances since the revisionary studies of 2005.

In this chapter, we update and summarize what is known about the diversity and distribution of amphibians and reptiles of the Venezuelan Guayana until mid 2018, highlighting relevant biogeographical aspects that we consider a priority to define and characterize the region. We also identify information gaps, report on species conservation status, and discuss current and potential threats. Our hope is that this information will promote further and much needed study of the Guayanan biodiversity, and, specially, the development of effective conservation strategies. Along the

XVII century, explorations towards the mysterious territory of Guiana were greatly stimulated by the myth of "El Dorado" (Caballero 2014)—a lost city of gold on the shores of the mythical Lake Parima, waiting for discovery by an adventurous conqueror. The legendary "El Dorado" may exist, not as fabulous amounts of gold stones, but as an extraordinary and unique legacy of organismal richness, the product of the intricate and continuous mechanisms of millions of years of diversification. Unfortunately, we seem too eager to destroy this treasure, even without grasping its diversity and potential.

2 The Venezuelan Guayana

Guayana (=Guiana) is a physiographic region of northern South America centered on one of the two Precambrian crystalline cratons in the continent (Berry et al. 1995a). The region includes whole Guyana, Suriname, and French Guiana, as well as southern Venezuela and minor areas in southeastern Colombia and northern Brazil. In Venezuela, Guayana is a political unit and an administrative region comprised of three states: Amazonas, Bolívar, and Delta Amacuro. Geographically, the Venezuela Guayana is bordered by the Orinoco river and its delta to the north, by the Sierra de la Neblina, Sierra Parima, and Sierra Pakaraima mountains to the south, by the Venamo and Cuyuní rivers to the east, and, finally, by the Orinoco, Atabapo, Guainía, and Negro rivers to the west (Huber 1995) (Fig. 22.1). The Venezuelan Guayana (VG) covers half of Venezuela's national continental territory (453,950 km^2).

Venezuela is known to host most of the tepuis of the Guiana Shield. Extending from sea level to 3014 m, the characteristics tepuis (also known as Guiana Highlands) emerge abruptly as high-elevation 'islands' in a matrix of lowland forests (Fig. 22.2), and are recognized by their outstanding plant diversity and high levels of endemism (Berry and Riina 2005; Hollowell and Reynolds 2005). Most of the ~60 tepuis are in Venezuelan territory, with only a few in Guyana and Brazil (Huber 1995). A few other lower sandstone massifs are found in central Surinam (Wilhelmina Mountains ca. 1280 m a.s.l., and Tafelberg 1090 m a.s.l.) and in central-eastern Colombia (Serranía de Chiribiquete, ca. 840 m a.s.l.) (Huber and García 2011).

Based on species composition and distribution patterns, early herpetological studies have attempted to delimit the Guiana biogeographical region (Lescure 1976; Descamps et al. 1978; Hoogmoed 1979), as well as the Venezuelan Guayana (Barrio-Amorós 1999; Gorzula and Señaris 1999; Péfaur and Rivero 2000). However, they failed to arrive at a consensus. To make matters more complex, the "Guayana" has been occasionally treated as part of Amazonia (e.g., Ávila-Pires 1995; Godinho and da Silva 2018). To synthesize the discrepancies between these two regions, Pérez-Hernández and Lew (2001) documented the different phytogeographic and zoogeographic classifications of this area in Venezuela.

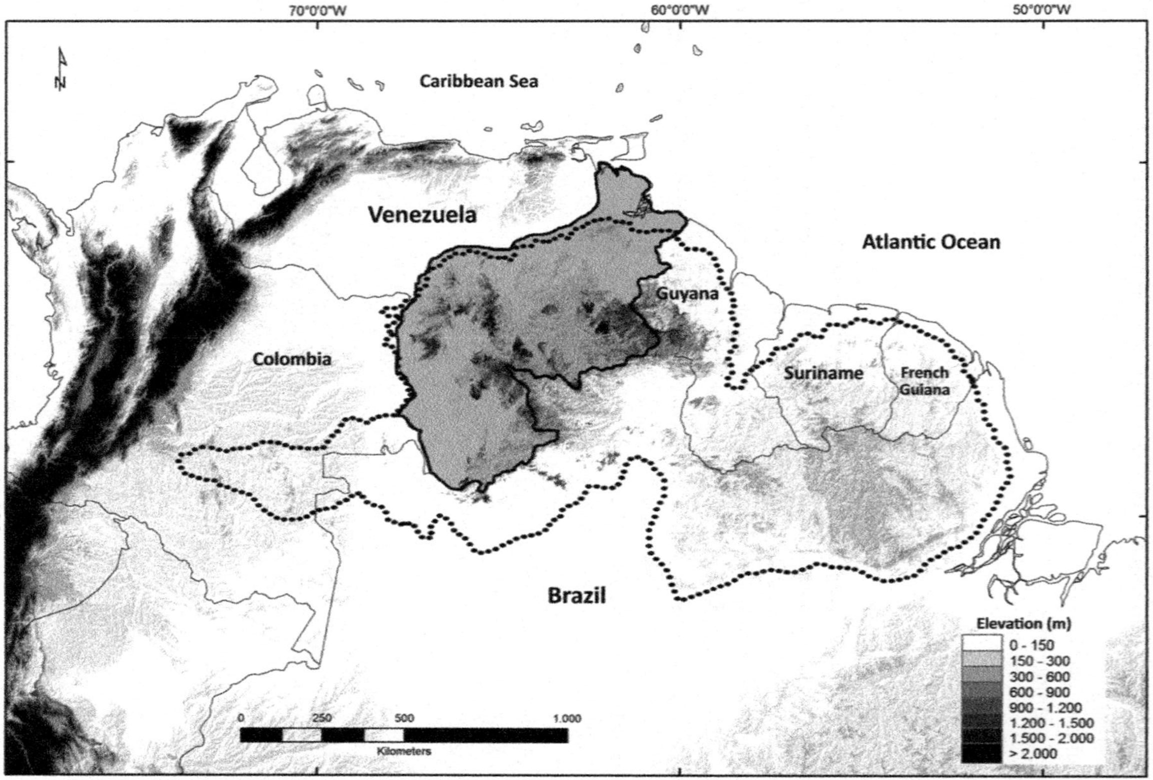

Fig. 22.1 Map of the Guiana Shield (dotted line, following Gibbs and Baron 1993) and Venezuelan Guayana as defined in the present study (gray area)

Fig. 22.2 Mount Roraima on the Venezuela-Guyana border, (a) and Cerro Autana seen from the Sipapo River, Amazonas State (b) Photos: Javier Mesa

For this chapter, we define Venezuelan Guayana in the broadest sense: it is the territory of the Venezuelan states of Amazonas, Bolívar and Delta Amacuro. This follows the definition of Huber (1995), Berry et al. (1995b), and Huber and García (2011), which are based on geographical, climatic, ecological and floristic criteria.

The inclusion of the main course of the Orinoco River and its delta in the Guayana region follows Péfaur and Rivero (2000), Molina et al. (2004), Señaris (2004), and Señaris and Ayarzagüena (2004). Likewise, the southern lowlands of the Venezuelan state of Amazonas are included in our analysis, although some authors consider them to be part of Amazonia (e.g., Barrio-Amorós 1999; Péfaur and Rivero 2000).

3 Herpetofauna of the Venezuelan Guayana

3.1 State of Knowledge

Knowledge of the amphibians and reptiles of the Venezuelan Guayana (VG) has greatly improved over the last 60 years, thanks to increased scientific explorations and helicopter-enabled collections to reach remote locations and highlands (Huber 1995; Gorzula and Señaris 1999; McDiarmid and Donnelly 2005). Those explorations resulted in new taxon descriptions and new records for lowland and uplands species previously known solely from neighboring countries such as Guyana, Brazil, and Colombia. Of the known species and distribution records of VG amphibians today, 40.6% represent contributions published after 1980 (Fig. 22.3a). The pace of species discovery through time is also an increasing curve for reptiles, although the study of these vertebrates had already begun by the nineteenth century (Fig. 22.3b). Only 29 reptile species were known in the VG as of 1950, most of them large-bodied turtles and crocodilians. After 1950, 56 species of reptiles have been described and another 134 were newly reported, many of them snakes (Roze 1958a, b, 1966). Since ~2010, however, the number of scientific expeditions to the Venezuelan Guayana has notably decreased, partly because of the difficulty of obtaining permits for fieldwork, sampling, and molecular studies (Rull and Vegas-Vilarrúbia 2008; Rull et al. 2016), but also because of insufficient scientific and technical personal, supplies, and food—along with increased security risks as a result of the political and economic crisis afflicting Venezuela (Bifano 2016; Requeña and Caputo 2016).

Undoubtedly, however, the last three decades have been the most productive period for amphibian and reptile species discovery in the Guiana Shield, in spite of the fact that only part of the herpetological collections obtained decades ago has been studied in detail. Recent taxonomic efforts, including the use of molecular methods, have revealed high levels of cryptic species diversity in the region. Because several studies have uncovered numerous yet unnamed lineages (e.g., Jungfer et al. 2013; Gehara et al. 2014; Passos et al. 2013; Kok et al. 2016, 2018a; Jablonski et al. 2017; Vacher et al. 2017), we further expect a steady increase of species in the next decades. These recent phylogenetic studies also have provided important information about the origin and drivers of diversification of the highland amphibians and reptiles (Heinicke et al. 2009; Kok et al. 2012, 2017, 2018b; Salerno et al. 2012, 2014; Duellman et al. 2016).

From a geographical point of view, the areas explored herpetologically in the Venezuelan Guayana represent an extremely small portion within the total expanse

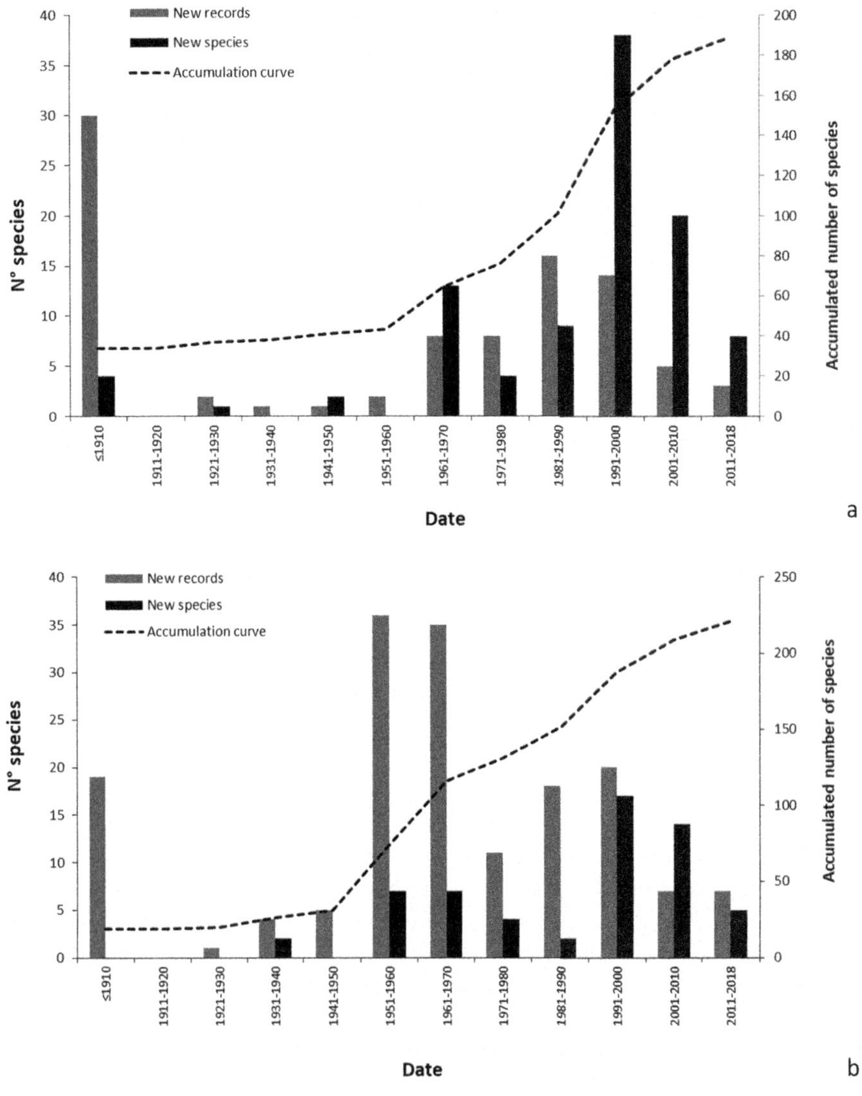

Fig. 22.3 Number of species described, new records, and the species accumulation curve in Venezuelan Guayana since 1910 for (**a**) amphibians, and (**b**) reptiles

of the region. Señaris and Ávila-Pires (2003) and McDiarmid and Donnelly (2005) considered that sampling efforts have been relatively high in La Escalera (part of the Sierra de Lema), Canaima National Park, and in the summits of Auyán-tepui, Chimantá Massif and the Roraima-Kukenán-Yuruaní chain. However, in a detailed study of the vertebrate fauna of Canaima—the best studied and most emblematic National Park in the VG—Rojas and Lew (2009) show that only 23% of the cells in a grid of 100 km^2 have at least one museum record. This demonstrates the sparse zoological exploration, even in areas considered relatively well-sampled. Lowland floral and faunal collections come mostly from areas adjacent to navigable rivers, highways, pathways and human settlements, leaving the rest of the territory virtually unexplored (Huber 1995; Gorzula and Señaris 1999; Molina et al. 2004; Rojas and Lew 2009). Due to the difficulty of access, the uplands and the upper basins of the majority of Guayana rivers are practically unexplored, as it has been illustrated in the Paragua River basin (Señaris and Rivas 2008), and Cuyuní River (Lasso et al. 2009). It is clear that our knowledge of the richness of amphibians and reptiles of the Venezuelan Guayana is still incomplete. For this reason, the analyses presented below should be understood as preliminary.

3.2 Richness and Endemism

Until mid-2018, the herpetofauna of the Venezuelan Guayana includes 416 described species; 26.9% of them are endemic to this region (Appendices 1 and 2). Amphibians are represented by 189 anuran species and six caecilians, and reptiles include 202 species of lizards and snakes, 15 turtles, and four crocodilians (Table 22.1). This richness represents more than half (56.2%) of the total number of amphibians and reptiles in Venezuela, one of the world's top 10 megadiverse countries. Even more remarkably, the herpetofauna of the VG comprises 6.6% (amphibians) and 22% (reptiles) of the megadiverse fauna of South America, and 2.2% of the world total herpetofauna.

The most species-rich anuran families in the VG are Hylidae (10 genera and 55 species), Leptodactylidae (7 genera and 26 species), Strabomantidae (2 genera and 21 species), Aromobatidae (2 genera and 20 species), Bufonidae (5 genera and 16 species), and Hemiphactidae (1 genus with 12 species). These six families account for 77% of the amphibian fauna of the Venezuelan Guayana, in more or less the same taxonomic proportion as they occur in whole Guiana Shield (Señaris and MacCulloch 2005)—with the exception of the Hemiphractidae (genus *Stefania*), which are mostly restricted to the uplands and highlands of the VG and Guyana (Señaris et al. 1996; MacCulloch and Lathrop 2002; MacCulloch et al. 2006; Kok et al. 2017). *Pristimantis* (Strabomantidae) and *Boana* (Hylidae) are the most species-rich genera, followed by *Anomaloglossus* (Aromobatidae), *Leptodactylus* (Leptodactylidae), *Scinax* (Hylidae), and *Stefania* (Hemiphractidae) (Table 22.2).

Table 22.1 Number of amphibians and reptiles by major taxonomic group in Venezuelan Guayana

Class	Order	Families	Genera	Species
Amphibia	Anura	16	50	189
	Gymnophiona	4	6	6
Reptilia	Crocodilia	2	3	4
	Testudines	5	9	15
	Squamata	20	82	202
Total		47	150	416

Table 22.2 Taxonomic composition of the Order Anura in Venezuelan Guayana

Family	N° species/family	Genera	N° species	N° endemics
Allophrynidae	1	*Allophryne*	1	–
Aromobatidae	20	*Allobates*	4	2
		Anomaloglossus	16	12
Bufonidae	16	*Amazophrynella*	1	–
		Metaphryniscus	1	1
		Oreophrynella	6	4
		Rhaebo	2	–
		Rhinella	6	–
Centrolenidae	9	*"Cochranella"*	2	2
		Vitreorana	2	–
		Hyalinobatrachium	5	1
Ceratophryidae	1	*Ceratophrys*	1	–
Ceuthomantidae	3	*Ceuthomantis*	3	2
Dendrobatidae	4	*Ameerega*	2	–
		Dendrobates	1	–
		Minyobates	1	1
Eleutherodactylidae	2	*Adelophryne*	1	–
		Eleutherodactylus	1	–
Hemiphractidae	12	*Stefania*	12	11
Hylidae	55	*Boana*	18	2
		Myersiohyla	5	5
		Dendropsophus	6	–
		Aparasphenodon	1	–
		Osteocephalus	3	–
		Tepuihyla	6	4
		Trachycephalus	2	–
		Pseudis	1	–
		Scinax	12	2
		Sphaenorhynchus	1	–
Leptodactylidae	26	*Engystomops*	1	–
		Physalaemus	3	–
		Pleurodema	1	–
		Pseudopaludicola	2	–
		Adenomera	2	–
		Leptodactylus	16	–
		Lithodytes	1	–
Microhylidae	11	*Adelastes*	1	–
		Chiasmocleis	1	–
		Ctenophryne	1	–
		Elachistocleis	2	–
		Hamptophryne	1	–
		Otophryne	3	–
		Synapturanus	2	–

(continued)

Table 22.2 (continued)

Family	N° species/family	Genera	N° species	N° endemics
Phyllomedusidae	5	*Callimedusa*	1	–
		Phyllomedusa	3	–
		Pithecopus	1	–
Pipidae	2	*Pipa*	2	–
Ranidae	1	*Lithobates*	1	–
Strabomantidae	21	*Dischidodactylus*	2	2
		Pristimantis	19	14
		Total	189	65

Of the 195 amphibian species known in the Venezuela Guayana, more than a third is endemic (Table 22.2). Ceuthomantidae, recently reestablished at the family level and restricted to the genus *Ceuthomantis* (Heinicke et al. 2018), only occurs in the uplands of the Guiana Shield; two of the four described *Ceuthomantis* species are endemic to the VG. Three genera—*Metaphryniscus* (Bufonidae), *Minyobates* (Dendrobatidae), and *Dischidodactylus* (Strabomantidae)—, which inhabit the summits and slopes of tepuis in Amazonas state, are entirely endemic to the VG. After the recent description of the genus *Ectopoglossus*, which includes the trans-Andean aromobatids having a median lingual process (Grant et al. 2017), the genus *Anomaloglossus*—with 29 species—became restricted to the Guiana Shield (Fouquet et al. 2015, 2018; Vacher et al. 2017) and 41% of its members are VG endemics.

Recent amphibian and reptiles records from neighboring countries have expanded the geographic range of several taxa originally know only from the Venezuelan Guayana. The Microhylidae monotypic genus *Adelastes*, described from the lowlands of La Neblina Massif in Venezuela (Zweifel 1986), was recently recorded in a second locality in Brazil (Almeida et al. 2014). This is also the case of four genera closely associated with the uplands and highlands of the Guayana region (the Bufonidea genus *Oreophrynella,* and the Hylidae genera *Myersiohyla, Stefania* and *Tepuihyla*) which have the greatest species richness and endemism in the VG but some species had been found in neighboring territories of Guyana and Brazil (MacCulloch and Lathrop 2002, MacCulloch et al. 2006; Lathrop and MacCulloch 2007; Carvalho et al. 2010; Cole et al. 2013; Faivovich et al. 2013; Kok et al. 2012, 2015). Other ca. 20 additional species have small binational ranges between Venezuela and Guyana or Brazil (Heyer 1994; Caramaschi and de Niemeyer 2005a, b; Barrio-Amorós et al. 2011; Cole et al. 2013; Almeida et al. 2014; de Sá et al. 2014; Señaris et al. 2014; Araujo-Vieira 2017; Jablonski et al. 2017). In this sense, we expect that some taxa considered so far as exclusive to Venezuelan Guayana will be found in those neighboring countries (and vice versa), especially for those species inhabiting geographically continuous areas (e.g., binational massifs and mountain ranges).

Snake species numerically dominate the reptiles of the VG (54% of the total), as in the whole Guiana Shield (Ávila-Pires 2005). Colubridae (31 genera and

88 species) is by far the most speciose family. The genera *Erythrolamprus, Atractus*, and *Thamnodynastes* contain the greatest richness, and comprise a quarter of the Guayanan snake fauna. Lizards are also a very diverse group in the region, especially the family Gymnophthalmidae. The most species rich lizard genera, in order of numerical importance, are *Anolis* (Dactyloidae), *Gonatodes* (Sphaerodactylidae), *Arthrosaura* and *Neusticurus* (Gymnophthalmidae).

With 21.3% of the reptile taxa being endemic, Venezuelan Guayana ranks second in reptile endemism across the Venezuelan biogeographic units, standing just after the Cordillera de Mérida mountains (Rivas et al. 2012). Except for one turtle (*Rhinoclemmys flammigera*) and two amphisbaena (*Amphisbaena gracilis*, and *A. rozei*), all VG endemics are lizards (28 species) and snakes (15 species). The subfamily Riolaminae (Gymnophthalmidae) is closely restricted to the VG (Kok 2015; Goicoechea et al. 2016), although one of its four members—*Riolama leucosticta*— has been found on Wei-Assipu-tepui and Maringma-tepui in Guyana (Kok 2015). Furthermore, two undescribed *Riolama* species from La Neblina Massif in Venezuela (McDiarmid and Paolillo 1988; McDiarmid and Donnelly 2005) will most likely eventually be found on the Brazilian side of this massif. So far, only the monospecific lizard genus *Adercosaurus* from Cerro Yutajé in Amazonas state (Myers and Donnelly 2001) is actually endemic to VG. The genus *Arthrosaura*, although widespread in the Guiana Shield, has five of its seven species distributed in Venezuela, with four of them exclusive to the highlands of the region (Table 22.3) and another undescribed species from Chimantá Massif (Gorzula 1992). Among snakes, endemism is concentrated in Colubridae, with endemic species in the genera *Thamnodynastes* (five species), *Atractus* (four species), *Erythrolamprus* (three species) and *Philodryas* (one species), in addition to a recently-described blind snake in the genus *Epictia*, from Cerro Guaiquinima (Esqueda et al. 2015) (Table 22.4). As with amphibians, is likely that reptile endemism is presently inflated, and that some species currently considered endemics to the VG actually have a wider distribution (e.g., Moraes et al. 2017; Fraga et al. 2017; Kok et al. 2018a).

3.2.1 Altitudinal Shifts in Richness and Composition

Venezuelan Guayana encompasses one of the most diverse landscapes in South America, from nearly sea level to the highest peaks in the Sierra de la Neblina mountains, at 3014 m (Huber 1995). Along this elevational gradient, the herpetofauna diversity is unevenly distributed and has marked differences in taxonomic composition. At a large (regional) scale, amphibian and reptile species richness decrease monotonically as one moves from low to high elevations, while, conversely, the number of endemic species increases markedly with elevation (Fig. 22.4)—a pattern commonly observed in other herpetological communities (McCain 2010). This general trend is also observed at a small (local) scale as have been documented in Roraima-tepui (MacCulloch et al. 2007), Cerro Sarisariñama

Table 22.3 Taxonomic composition of Amphisbaenia and Sauria (Order Squamata) in Venezuelan Guayana

Family	N° species/ family	Genera	N° species	N° endemics
Amphisbaenidae	5	*Amphisbaena*	5	2
		Mesobaena	1	–
Iguanidae	1	*Iguana*	1	–
Dactyloidae	9	*Anolis*	9	4
Polychrotidae	1	*Polychrus*	1	–
Tropiduridae	9	*Plica*	5	3
		Tropidurus	2	1
		Uracentron	1	–
		Uranoscodon	1	–
Gekkonidae	1	*Hemidactylus*	1	–
Phyllodactylidae	2	*Phyllodactylus*	1	1
		Thecadactylus	1	–
Sphaerodactylidae	12	*Chatogekko*	1	–
		Coleodactylus	1	–
		Gonatodes	8	3
		Pseudogonatodes	1	–
		Sphaerodactylus	1	–
Scincidae	2	*Copeoglossum*	1	–
		Panopa	1	–
Gymnophthalmidae	32	*Adercosaurus*	1	1
		Anadia	1	1
		Arthrosaura	5	4
		Bachia	4	–
		Cercosaura	3	2
		Euspondylus	1	1
		Loxopholis	2	–
		Neusticurus	6	1
		Gymnophthalmus	2	1
		Oreosaurus	1	1
		Riolama	4	4
		Tretioscincus	1	–
Teiidae	9	*Ameiva*	2	–
		Cnemidophorus	2	–
		Crocodilurus	1	–
		Kentropix	3	–
		Tupinambis	1	–

(Barrio-Amorós and Brewer-Carías 2008), Sierra de Lema (Barrio-Amorós and Duellman 2009) and in Canaima National Park (Señaris et al. 2009).

In the lowlands (≤500 m) all amphibian families are present, except for Ceuthomantidae, whose lowest record (for *Ceuthomantis cavernibardus*) is at 930 m, in the Brazilian side of the Sierra Tapirapecó (Caramaschi and de Niemeyer

Table 22.4 Taxonomic composition of Serpentes (Order Squamata) in Venezuelan Guayana

Family	N° species/ family	Genera	N° species	N° endemics
Anillidae	1	*Anilius*	1	–
Anomalepididae	1	*Typhlophis*	1	–
Boidae	7	*Boa*	1	–
		Corallus	3	–
		Epicrates	2	–
		Eunectes	1	–
Colubridae	88	*Atractus*	8	3
		Chironius	7	–
		Clelia	1	–
		Dendrophidion	1	–
		Dipsas	5	–
		Drymarchon	1	–
		Drymobius	1	–
		Drymoluber	1	–
		Erythrolamprus	13	3
		Helicops	3	–
		Hydrodynastes	1	–
		Hydrops	2	–
		Imantodes	2	–
		Leptodeira	1	–
		Leptophis	3	–
		Lygophis	1	–
		Masticophis	1	–
		Mastigodryas	3	–
		Oxybelis	2	–
		Oxyrhopus	3	–
		Philodryas	4	1
		Phimophis	1	–
		Phrynonax	1	–
		Pseudoboa	2	–
		Pseudoeryx	1	–
		Rhinobothrium	1	–
		Sibon	1	–
		Siphlophis	2	–
		Spilotes	2	–
		Tantilla	1	–
		Thamnodynastes	8	5
		Xenodon	3	–
		Xenopholis	1	–
Elapidae	8	*Leptomicrurus*	1	–
		Micrurus	7	–
Leptotyphlopidae	4	*Epictia*	3	1
		Siagonodon	1	–

(continued)

Table 22.4 (continued)

Family	N° species/ family	Genera	N° species	N° endemics
Typhlopidae	3	*Amerotyphlops*	3	–
Viperidae	6	*Bothrops*	4	–
		Crotalus	1	–
		Lachesis	1	–

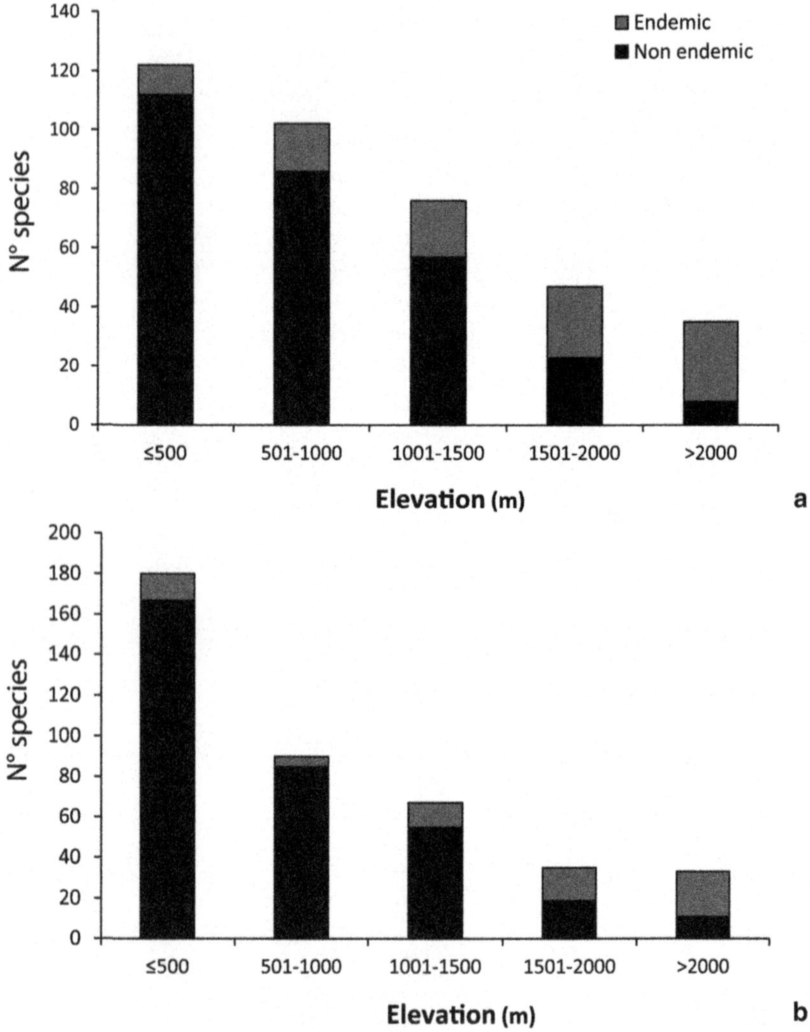

Fig. 22.4 Number of endemic and non-endemic species along the elevational gradient of Venezuelan Guayana for (**a**) amphibians, and (**b**) reptiles

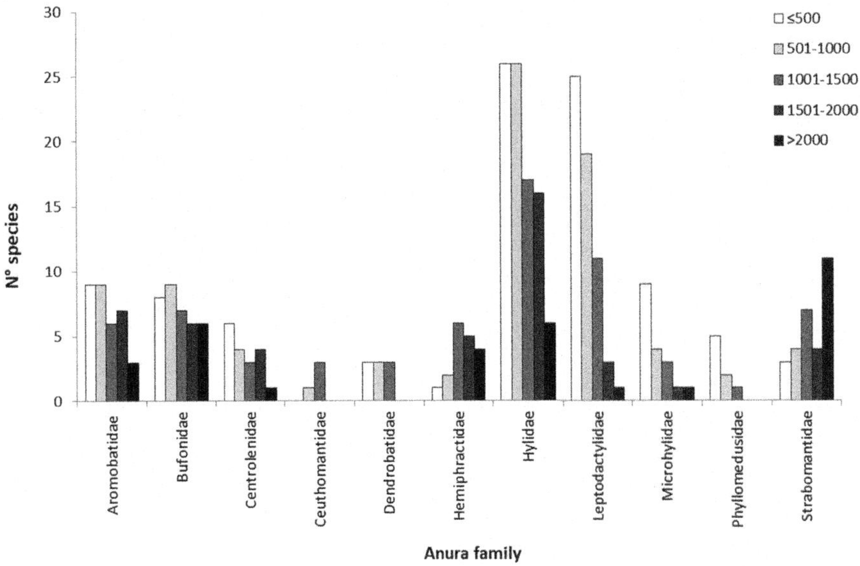

Fig. 22.5 Species richness of the most specious Anuran families at different elevations in Venezuelan Guayana

2005a). From lower elevations up to 1000 m, the anuran families Allophrynidae, Ceratophryidae, Pipidae, and Ranidae are absent. Dendrobatidae, Phyllomedusidae, and all caecilians drop out at 1500 m. In contrast, at the highest elevations (≥2000 m) the numerical importance of Strabomantidae markedly increases, dominating amphibian communities, along with Bufonidae, Hylidae, Hemiphractidae and Aromobatidae, in similar proportions as the anurofauna of the Andes (Armesto and Señaris 2017).

In the Venezuelan Guayana some anuran families show a mid-elevation diversity peak and thereafter a drastic reduction, in a pattern also observed in the Tropical Andes (e.g. Hutter et al. 2013, 2017; Armesto and Señaris 2017). Hemiphractidae have mid-elevation peak richness at 1000–1500 m—a range to which Ceuthomantidae is also restricted, and in which Strabomantidae frogs diversity increases (Fig. 22.5). All of these families have direct embryonic development, without a free-living tadpole stage (Padial et al. 2014; Castroviejo-Fisher et al. 2015; Heinicke et al. 2018), an attribute also present in the highland genus *Oreophrynella* (Bufonidae) (McDiarmid and Gorzula 1989; Señaris et al. 1994). This reproductive attribute has allowed these species to become independent of aquatic environments—permanent or semi-permanent water bodies—and to utilize terrestrial habitats with sufficient moisture for the survival of eggs, hatchlings, and adults. Consequently, these taxa can occupies a large variety of microhabitats, which facilitates their wide distribution. For example, the most diverse direct development genus *Pristimantis* occupy a huge variety of microhabitats from lowlands to

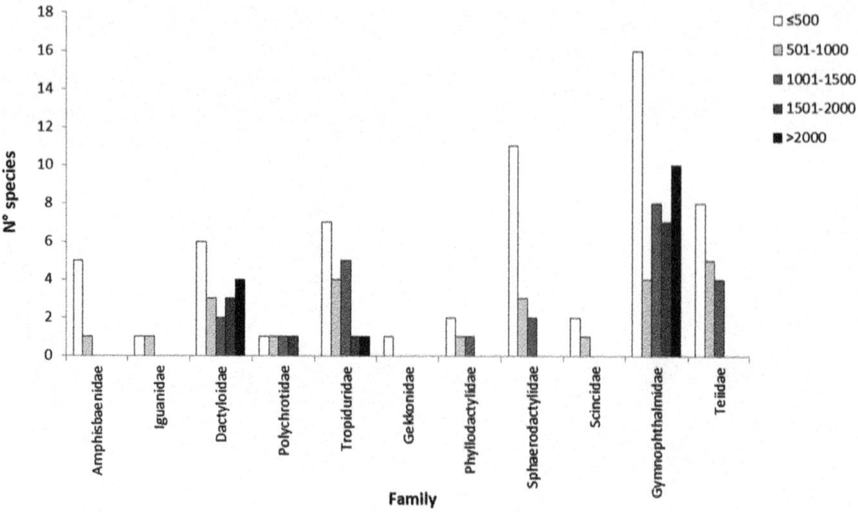

Fig. 22.6 Species richness of amphisbaenian and lizard families at different elevational ranges in Venezuelan Guayana

highlands (Heinicke et al. 2007), and also survives in the extreme climatic conditions on tepui summits (high winds, wide temperature variation, high solar radiation).

Turtles and crocodilians in the VG are distributed mostly at elevations below 500 m, with only the Smooth-fronted Caiman *Paleosuchus trigonatus* having been recorded above 1000 m (Gorzula and Señaris 1999). Other groups absent over 1000 m, include Amphisbaenids, the lizard families Iguanidae, Gekkonidae, and Scincidae, and the snake families Aniliidae and Anomalepidae. Other Squamata that are highly diverse in the lowlands are poorly represented or absent above 1500 m, including geckos (Phyllodactylidae, Sphaerodactylidae), whiptails and tegus (Teiidae), boids (Boidae), coral snakes (Elapidae), and slender blind snakes (Leptotyphlopidae and Typhlopidae). In the highlands, reptiles consist almost exclusively of lizards—Gymnophthalmidae, Dactyloidae, Tropiduridae—with a few colubrid snakes, and very occasionally a viperid snake.

At a regional scale, most reptile groups in VG have a striking pattern of richness decreasing with elevation, but two lizard families are the exception: Dactyloidae and Gymnophthalmidae. *Anolis* and gymnophthalmids lizards have their lowest species richness at middle elevations, and a high proportion of endemic species close to summits (≥2000 m; Fig. 22.6). However, based on our limited knowledge of the diversity of lizards in the Guiana Shield highlands, this atypical richness pattern in Dactyloidae and Gymnophthalmidae lizards may be influenced by taxonomic and sampling biases.

Although amphibian and reptile species richness decreases with elevation, 21% of the total species in the Venezuelan Guayana are found in highlands (≥1500 m a.s. l.). This means that one fifth of the species are concentrated in only ~1.1% (ca. 5000 km^2) of the total area of the VG region (ca. 453,950 km^2). This high density of species on tepuis summits has also been observed in plants (Berry and Riina 2005), and may be due to the accumulation of microendemic species of different linages that persist on evolutionary timescales. Several hypotheses proposed to explain global-scale patterns in species richness can also be applied to elevational patterns. Proposed drivers fall into four main categories: climate, space, evolutionary history and biotic processes (McCain 2009; McCain and Grytnes 2010; Cadena et al. 2011). We argue that the observed variation in species richness and endemism along elevational gradients are ultimately the result of differences between processes of speciation, extinction and dispersal, modulated continually by the shifting of abiotic factors and ecological process over time (Rull 2011; Andresen et al. 2018; Quintero and Jetz 2018).

3.3 Distribution Patterns

The Venezuelan Guayana herpetofauna is a combination of taxa with different geographical distribution patterns, including species that are widely distributed across northern South America, others that are limited to the Guiana Shield, and taxa that are highly restricted to massifs and tepui summits—i.e. microendemics. Only four species have been introduced into VG: the Antillean coqui *Eleutherodactylus johnstonei*, the geckos *Gonatodes albogularis* and *G. vittatus*, and Cope's lizard *Ameiva bifrontata* (Hoogmoed 1979; Gorzula and Señaris 1999; Molina et al. 2004).

In general, about 59% of the amphibians that inhabit the Venezuelan Guayana have geographic distributions limited to the Guiana Shield or part of it, including those species endemic to the country. Another 23% of the species extended their distributions throughout Amazonia, while the remaining 18% are widespread across South America, or are shared with the flooded grasslands (Llanos) or other regions in northern South America. In contrast, reptiles have much broader geographical distributions (Ávila-Pires 2005): 68% of the reptile species found in the VG are widespread in northern South America or extend into the Llanos and Amazonian regions. Only one third of reptile species are restricted to the Guiana Shield, and most of them are also VG endemics.

Based on geographic, floristic, and ecological criteria, Berry et al. (1995b) defined four phytogeographical provinces in the Guayana Region: Eastern Guayana, Central Guayana, Western Guayana, and Pantepui. An analysis of the presence/absence of amphibian species in these provinces shows that the Eastern and Western Guayana provinces are the most similar to each other (Jaccard's coefficient of similarity: 0.63), followed by Central Guayana, which is joined to the previous group (0.44). The Pantepui ($\geq$1500 m) is clearly unique in its amphibian fauna, with very few taxa shared with other provinces; its greatest similarity is with the Central Guayana province, but even then the coefficient of similarity is only 0.12. In the case of reptile fauna, as a whole or considering lizards and snakes separately, the greatest similarity occurs between the Central and the Western Guayana provinces (0.6), with

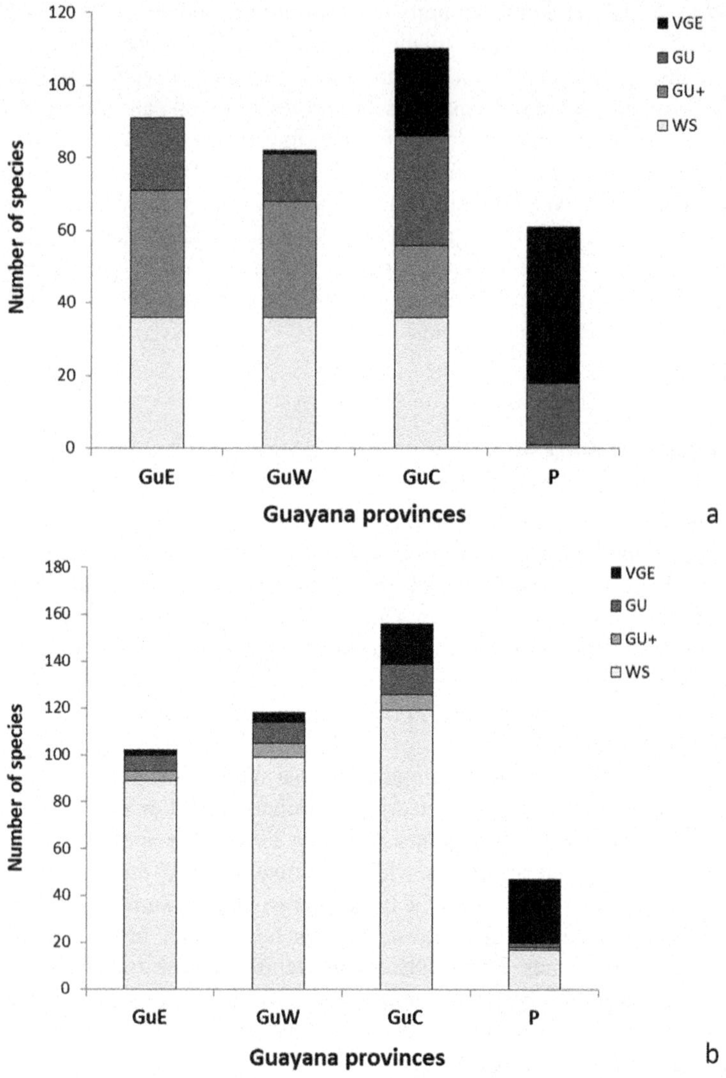

Fig. 22.7 Composition of distributional patterns for (**a**) amphibians, and (**b**) reptiles in the phytogeographical provinces (sensu Berry et al. 1995a, b). WS: widespread in South America; GU+: Guayana Shield and areas of adjacent regions; GU: restricted to the Guayana Shield; VGE: restricted to Venezuelan Guayana (endemics)

something more than 100 shared taxa, followed by Eastern Guayana with a slightly lower similarity coefficient (0.5). As with amphibians, Pantepui reptile communities have the most exclusive taxa, with a similarity lower than 20% with respect to the other Guayana provinces.

Figure 22.7 illustrates herpetological assemblages in each phytogeographical province according to species distribution patterns. In general, amphibian and reptile communities in the Guayana lowlands—Eastern and Western provinces—are diverse, but are relatively similar and homogenous in taxonomic composition and distributional patterns because of the numerical dominance of widespread or Amazonian-Guianan taxa, and given the very rare endemic species. In the irregular submontane and montane landscape of Central Guayana the richness is higher, because restricted-range species increase in frequency, with an endemism of 21.2% for amphibians and 11% in reptiles. Finally, the Pantepui province is unique in taxonomic composition. It has reduced richness (Myers and Donnelly 1997, 2001; McDiarmid and Donnelly 2005; MacCulloch et al. 2007), but a species-area relationship more than five times higher than the uplands (Table 22.5) due to the presence of taxa with extremely small distributions. Endemism reaches 57.4% in reptiles, and 70.5% in amphibians.

The northern and northwestern border of the VG (i.e. middle Orinoco River) is not included in Berry's provinces, and it is considered a phytogeographical and ecological transition area (Berry et al. 1995b). Amphibian and reptile communities in these peripheral zones of the Guiana Shield, including the Orinoco Delta, are a mixture of species shared principally with the Llanos region, but also with the Amazon, or across South America. An analysis of the aquatic reptile assemblages (turtles and crocodiles) of Venezuelan Guayana reveals three large groupings (Fig. 22.8, based on a similarity analyses of species presence/absence), suggesting a division of the Orinoco basin into three main sections. This indicates that the biogeographical patterns of extant aquatic reptiles closely follow the physiographic, limnological and ecological criteria that define the subdivision of this basin (Rosales et al. 2010). Based on new and more precise data, future studies will be able to rigorously evaluate the different hypotheses that have been proposed to explain patterns of Guiana species distribution—including the Contemporary Climate hypothesis, the Pleistocene Climate Variation hypothesis, the Topography hypothesis, the Vegetation Structure hypothesis, and the Riverine Barrier hypothesis—as recently done by Godinho and da Silva (2018) for the Amazonian anurans.

At first glance, herpetological communities in Venezuelan Guayana seem to be associated to spatial heterogeneity, and in particular with altitudinal variation. This variation is gradual and continuous in some areas (e.g., Sierra de Maigualida, Sierra Parima, Sipapo Massif, Sierra Tapirapecó), but is abrupt in others, with gaps due to the sheer cliffs characteristic of some tepuis (e.g., Ptari-tepui, Angasima-tepui, Upuigma-tepui, Cerro Marahuaca, Cerro Huachamacari) (Huber 1995). Additionally, the altitudinal gradient ends abruptly in a flat-topped summit on many tepuis. Thus, summits may be completely isolated from their slopes—topographically and ecologically—with drastically different environmental conditions in tightly adjacent areas, creating "sky islands" (Aubrecht et al. 2012; Kok 2013). As documented on Roraima-tepui, the composition of the herpetofauna is very different between the slopes and the summit (MacCulloch et al. 2007). By contrast, on lower tepuis (those with a maximum elevation $\leq$1600 m), or those with a series of consecutive escarpments and no abrupt division between slopes and summit, the herpetofauna show no

Table 22.5 Total number of species, endemism, and species–area relationships at different elevational ranges in Venezuelan Guayana

		Amphibians				Reptiles			
Elevation range (m)	Area (10^3 km^2)	Total species	N° endemics	Total species/area	N° endemics/area	Total species	N° endemics	Total species/area	N° endemics/area
≤500	316.3471	102	10	0.322	0.032	162	13	0.512	0.041
501–1000	85.2533	87	16	1.020	0.188	89	5	1.044	0.059
1001–1500	33.96365	68	19	2.002	0.559	66	12	1.943	0.353
1501–2000	4.84576	47	24	9.699	4.953	34	16	7.016	3.302
>2000	1.33857	33	27	24.653	20.171	29	22	21.665	16.435

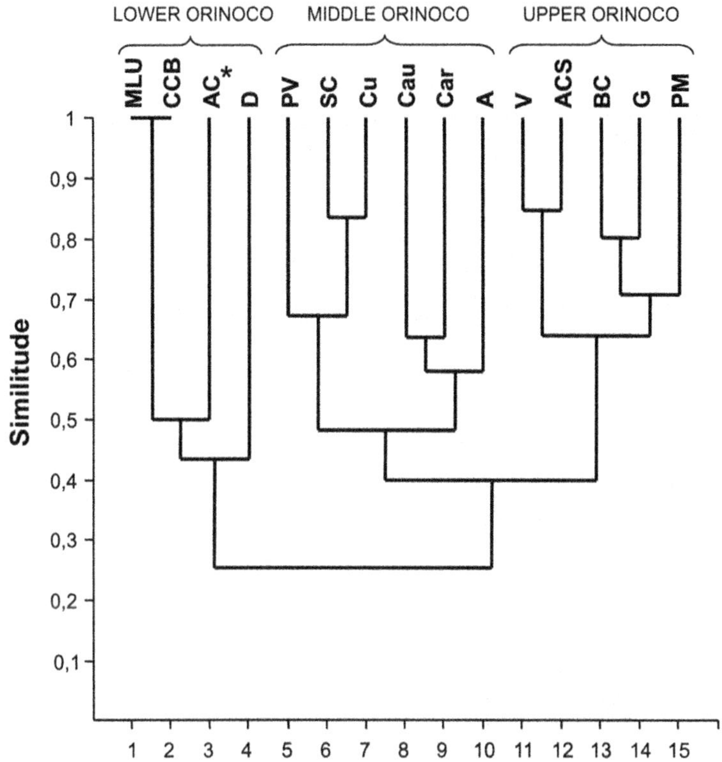

Fig. 22.8 Similarity dendrogram (Jaccard's coefficient) for the sub-basins of the Orinoco River based on the composition of their river turtles and crocodile communities. *MLU* Morichal Largo—Uracoa, *CCB* Casacoima—Caño Basama, *AC* Alto Cuyuní, *D* Orinoco Delta, *PV* Parguaza—Villacoa, *SC* Suapure—Chiviripa, *Cu* Cuchivero, *Cau* Caura, *Car* Caroní, *A* Aro, *V* Ventuari, *ACS* Atabapo—Guayapo—Cuao—Sipapo—Autana, *BC* Brazo Casiquiare, *G* Headwaters of the Guainía, *PM* Headwaters of the Orinoco, Padamo—Matacuni. Asterisk, Cuyuní River basin

sharp differences along the elevational gradient, and widespread taxa are found also on summits (e.g., Guaiquinima: Donnelly and Myers 1991; McDiarmid and Donnelly 2005; Sierra de Lema: Duellman 1997; Barrio-Amorós and Duellman 2009).

In addition to the elevational effect, the distribution of amphibians and reptiles across the Guiana Shield is influenced by the physiological and ecological capacity of each group or taxa, just as Navas (2003) pointed out for Andean herpetofauna. Most of the amphibians and reptiles in VG have narrow elevational ranges (the difference between the maximum and minimum recorded elevation for the species) of less than 200 m, including practically all highland endemic taxa (Fig. 22.9). This can suggests that most of these ectothermic vertebrates are thermally specialized, particularly those inhabiting the upper ranges of the mountains. At a broad-scale in the Neotropics, Laurance et al. (2011) showed that a relatively high proportion of plants and ectothermic vertebrates (amphibians and reptiles) are upper-zone

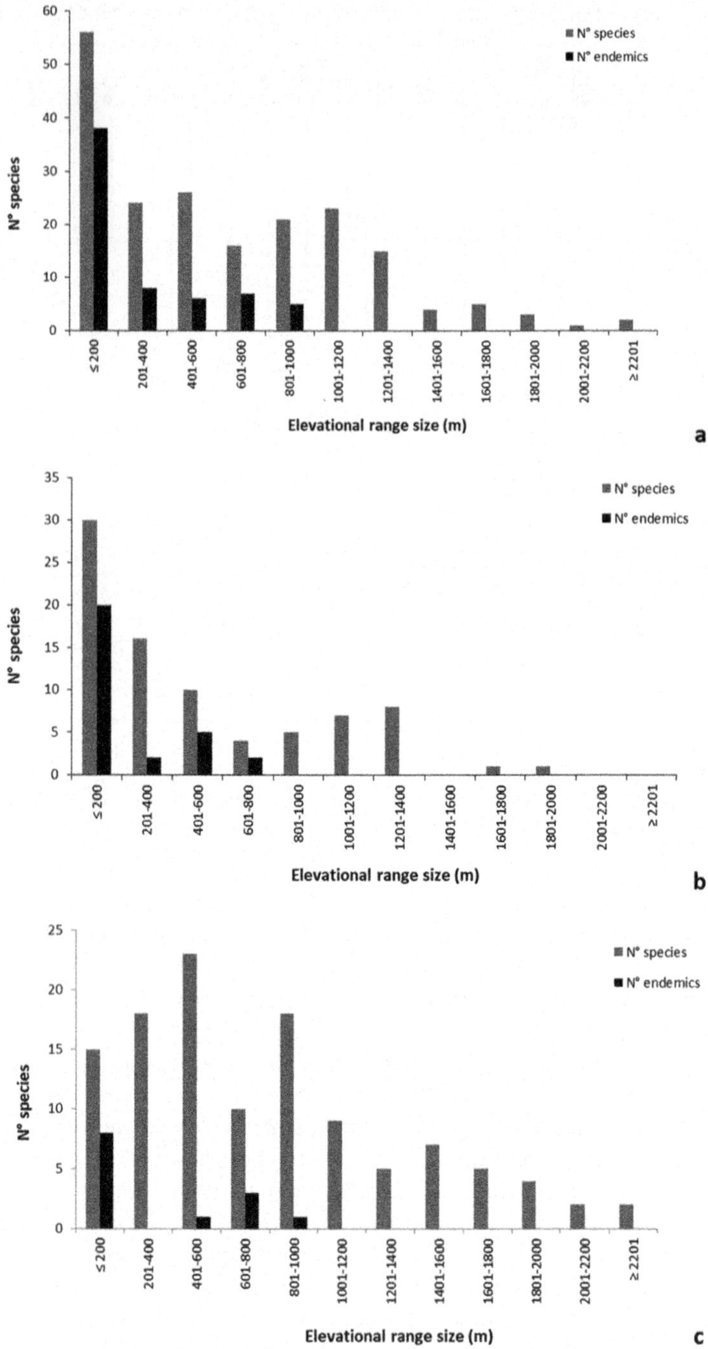

Fig. 22.9 Number of species by elevational range size in (**a**) amphibians, (**b**) lizards, and (**c**) snakes

specialists, in sharp contrast to relatively few endothermic vertebrates (birds and mammals). If the amphibians and reptiles with small and highland ranges in the VG indeed show limited physiological capacity, as reflected in their narrow altitudinal ranges, they may be especially vulnerable to future global climate change and so deserve research priorities.

3.3.1 The Tepui Amphibians and Reptiles

Early, Myers and Donnelly (2001) synthesized important generalities about the tepui amphibians and reptiles, which they also found on a smaller scale on Roraima-tepui (MacCulloch et al. 2007) and Auyán-tepui (Myers and Donnelly 2008). McDiarmid and Donnelly (2005) examined the Pantepui herpetofauna assemblage, taking into account abiotic factors (maximum elevation, summit and slope areas, and distance to nearest neighbor tepui), along with biotic factors (flora and vegetation attributes, including number of forest types, richness and endemism of vascular and primitive plants), as well as the elevational ranges of amphibian and reptile species, and an index of sampling efforts. In general, they found that floristic diversity, the slope and summit area, and the connectivity ("accessibility" *sensu* McDiarmid and Donnelly 2005) positively influence amphibian and reptile richness atop a tepui. For example, the herpetofauna in the small and lower Wei-Assipu-tepui (also called "Little Roraima") forested summit (3 km^2 of area; 2260 m a.s.l.) is twice as diverse as in the 10 times greater rocky summit of the Roraima-tepui (34 km^2; 2700 m a.s.l.) (Villareal et al. 2003; MacCulloch et al. 2007). These two sandstone table mountains are very close each other, with a ca. 2 km airline distance between summit to summit.

On the other hand, McDiarmid and Donnelly (2005) countered that the number of endemic taxa on a massif is mainly influenced by the number of discrete tepuis, their maximum elevation, and their relative degree of isolation. The complexity of a massif with more or fewer discrete tepuis appears to explain the great diversification in the clade including *Stefania ginesi* on the Chimantá Massif and peripheral tepuis (Kok et al. 2017), but does not explain the presence of a single species of *Tepuihyla* (*T. obscura*), the lizard *Oreosaurus mcdiarmidi* or the snake *Thamnodynastes chimanta* on most of the tepui summits that comprise the massif (Gorzula 1992; Gorzula and Señaris 1999; McDiarmid and Donnelly 2005; Kok and Rivas 2011; Kok et al. 2015). Likewise, the absence of *Oreophrynella* toads or glass frogs (Centrolenidae) in Chimantá —highly diverse anurans groups in nearby eastern tepuis—and the presence of other species on only one summit of this complex massif remains a mystery (e.g., *Pristimantis muchimuk* is found only on Churí-tepui, and *P. abakapa* is exclusive to Abakapá-tepui). Patterns must therefore also involve local extinctions or other factors (e.g., ecological process) that define the survival of species atop tepuis over time (Rull 2005; Kok et al. 2018b). If so, and if there is a general pattern of richness and endemics, it remains difficult to interpret with current data and knowledge.

Several hypotheses have been proposed to explain the origin and evolution of the Pantepui biota, mostly differing in the relative roles of vicariance vs. dispersal processes (Rull 2004). The name Plateau Theory, for instance, was coined by Mayr and Phleps (1967) in reference to a process envisioned by Chapman (1931) and Tate (1938) to explain distribution patterns of birds and mammals. It proposed the extant tepuian summit biota to have resulted from vicariance processes due to erosion of a former Pantepui plateau and the subsequent isolation of previously continuous highland ancestors. Chapman (1931) and Tate (1938) had also proposed an alternative mechanism to explain those similarity patterns (named the Cool Climate Theory by Mayr and Phleps in 1967): a decrease in temperatures in the Amazon and the Orinoco lowlands, during the cool phases of the Pleistocene, would have enabled connectivity (i.e. ecological corridors) and hence facilitated biotic exchange between the Andes and the tepuis—two areas previously isolated by warmer intervening areas. In slight contrast, the Distance Dispersal Theory (Chapman, 1931; Mayr and Phelps 1967) maintained that some tepuian summit organisms with flying capacity (i.e. birds) derived from other mountain regions by jump dispersal, with most colonists coming from the Andes and the northern Venezuelan coastal ranges. Haffer (1974), in turn, proposed the Modified Cool Climate Theory—suggesting that glacial cooling could have promoted range expansions (in Andean bird species), allowing the organisms to use low elevation mountains or hills (sub-optimal habitats) as resources that aided in a stepping-stone dispersal process to the Pantepui. He also formulated the Mountain Bridge Theory to explain affinities between the avifauna of the Andean and the tepuian summits, assuming past physical connection (i.e. a "bridge") between the Andes and the tepuis that would have allowed bird dispersal in their respective life zone (Haffer 1974). Alternatively, Mayr and Phelps (1967) presented the Habitat Shift Theory, which stated that some of the Pantepuian fauna is derived from lowland ancestors that shifted in habitat preference. Those authors also extended the last hypothesis in the Specialized Habitat Theory, for some specialist birds found only in the Pantepui (and with particular habitat requirements such as tepuian cliffs and crevices) are missing in adjacent lowlands or uplands.

These hypotheses were reviewed from a herpetological perspective by Hoogmoed (1979), Gorzula (1992), and McDiarmid and Donnelly (2005) using the available distributional data of amphibians and reptile. Early, the Plateau theory (i.e. vicariance) has been suggested as the explanation of the distribution pattern of the species of the genera *Stefania* (Duellman and Hoogmoed 1984), *Oreophrynella* and *Metaphryniscus*, and the *Riolama* lizards (McDiarmid and Donnelly 2005). Other theories—as the Distance Dispersal, the Specialized Habitat, and the Mountain Bridge hypotheses—were discarded as possible explanations for the origin of the herpetofauna, due the impossibility of aerial dispersion and the lack of geological evidence of a physical connection between the Andes and the Pantepui. Recent molecular phylogenetic and phylogeographic studies of Guiana amphibians and reptiles—despite a notable absence of samples from Western Guayana—suggest more complex historical scenarios involving not only different origins for the Pantepui herpetofauna, but also multiple diversification mechanisms (long distance dispersals, vicariance, and habitat shifts; Salerno et al. 2012, 2014; Kok et al. 2012, 2017, 2018a, b), as once proposed by Rull (2005, 2011) based in palaeoecological evidence.

4 Conservation of the Herpetofauna of Venezuelan Guayana

Amphibians are the most threatened vertebrate group in the world, with 41% of assessed species classified as Vulnerable, Endangered, or Critically Endangered (IUCN 2018). According to the International Union for Conservation of Nature (IUCN) Red List, 8.6% assessed species of the Venezuelan Guayana are globally threatened. Of these, only one frog is categorized as Critically Endangered, and another 14 species are Vulnerable (Appendix 1). Moreover, 46 species (26.3% of the amphibian species in the VG) are categorized as Data Deficient. The remaining 64.6% of amphibian species are considered of Least Concern (IUCN 2018). Yet, 20 anurans recently discovered in the Venezuelan Guayana have not been evaluated by the IUCN, demonstrating the difficulty of keeping global assessments updated for taxonomic groups with high rates of species discovery (Tapley et al. 2018). The most recent national assessment contains fewer amphibian species in threatened categories —one as Critically Endangered, and 10 as Vulnerable— but a substantially greater number of Near Threatened species (Table 22.6; Rodríguez et al. 2015). Although taxa in the latter category do not qualify as threatened based on IUCN criteria, the status of these taxa is subject to change pending reassessment in the near future (e.g., Kok et al. 2016). This is also the case of amphibians presently categorized as Data Deficient, which are often later found to be threatened (Morais et al. 2013; Howard and Bickford 2014). Additional research on the distribution, population ecology, and threats of these species should be conducted as a matter of urgency.

Most of the threatened Venezuelan Guayana amphibians are microendemics. The Critically Endangered Yapacana poison arrow-frog *Minyobates steyermarki* has an area of occupancy less than 10 km^2, with all individuals in a single location; the extent of its forest habitat on the Cerro Yapacana has dramatically declined because intensive illegal open gold mining in the area has increased greatly in the last 25 years (Señaris et al. 2015). All anurans considered Vulnerable or Near Threatened

Table 22.6 Number of amphibian and reptile species in each category of extinction risk (IUCN Red list: global assessment; http://animalesamenazados.provita.org.ve/: National assessment)

	Amphibia		Crocodilia		Testudines		Squamata	
Category	Global	National	Global	National	Global	National	Global	National
CR	1	1	1	–	–	1	1	–
EN	–	–	–	1	–	1	–	–
VU	14	10	–	–	4	3	1	–
NT	1	22	–	–	–	2	–	–
LC	113	127	3	1	1	7	71	189
DD	46	35	–	2	–	1	2	2
NE	20	–	–	–	10	–	128	12

CR Critical Endangered, *EN* Endangered, *VU* Vulnerable, *NT* Near Threatened, *LC* Least Concern, *DD* Data Deficient, *NE* Not Evaluated are also tepui endemics, so their extinction risk is based fundamentally on their very small and restricted ranges, even though nothing is known about their demographic parameters and trends over the time. Most of these highland taxa have among the smallest known geographic distributions for any vertebrate on the planet (Gorzula and Señaris 1999), and they are potentially very sensitive to the effects of disease and global warming (Pounds et al. 2006; Hof et al. 2011; Laurance et al. 2011; Foden et al. 2013).

We know less about the conservation status of the VG reptiles. Of the 13,000 species and subspecies of reptiles known in the world (Uetz and Stylianou 2018) only 58% have been evaluated by the IUCN, in most cases in a non-systematic manner, and with taxonomic and geographical biases (Böhm et al. 2013). Furthermore, many of the species assessments that have been conducted need updating. This is especially noticeable among reptiles of Venezuelan Guayana, of which 62.2% of species have not been assessed by the IUCN Red List, including most squamates but also turtles (Table 22.6). In a Venezuelan national assessment, 5.4% of the VG reptiles were not evaluated because of their recent description or because of taxonomic issues.

Threatened reptiles basically correspond to the large crocodilians (the Orinoco Crocodile, *Crocodylus intermedius*) and turtles associated with freshwater habitats (*Podocnemis*spp). These species are at risk due to historical commercial overexploitation, subsistence hunting, and habitat loss across most of their geographical distributions (Seijas et al. 2015; Hernández and Ferrer-Pérez 2015; Hernández et al. 2015)—not only in Venezuela, but along their entire ranges (Balaguera-Reina et al. 2017). A total number of 22 species of reptiles from Venezuelan Guayana—mostly crocodiles, turtles, iguanas and boas—are included in an Appendix of the Convention on International Trade in Endangered Species of Wild Fauna and Flora—CITES (www.cites.org). Ninety-one percent are in Appendix II (e.g. species not necessarily threatened now, but potentially becoming so unless trade is controlled), while the rattlesnake *Crotalus durissus* is in Appendix III (e.g. a species subject to regulation that prevents or restricts exploitation, and which demands cooperation to control trade). The Orinoco Crocodile is the only species listed in Appendix I, due to its critically endangered status (Appendix 2).

The other threatened reptiles of VG are the lowland geckos *Gonatodes infernalis* and *G. astralis* (Rivas and Schargel 2008; Schargel et al. 2010), which have small distributions associated with isolated granitic outcrops (inselbergs), and are threatened by habitat destruction (fires for agricultural purposes; Rivas and Schargel 2017a, b). Although other lizards and snakes in Venezuelan Guayana also have restricted distributions (e.g., the Pantepui species), none are considered threatened, demonstrating an unequal application of criteria and a lag between national and global assessments—the former of which are more up to date.

Venezuela has an extensive protected area system (Category II and III *sensu* IUCN 1994), which includes 43 national parks and 36 natural monuments comprising 17% of the continental territory (Naveda and Yerena 2010). A representative part of Venezuelan Guayana ecosystems is protected by seven national parks—Canaima National Park, also named a World Heritage Site by UNESCO in 1994—and 19 natural monuments covering 97,118.15 km^2 (22.5%) in the region, encompassing the entire elevational gradient, but especially protecting areas above 1000 m (Naveda and Yerena 2010). Additionally, two biosphere reserves were created—Delta del Orinoco (11,250 km^2) and Alto Orinoco-Casiquiare (83,830 km^2, the second largest biosphere reserve in the world)—that join to previous areas, and cover about half of VG. An important protected area is the refuge of the Arrau turtle (*Podocnemis expansa*), which preserves 24 km of the main channel of the Orinoco River as well as nesting sites utilized by this species. Four forest reserves complete the network of special administration areas system in Venezuelan Guayana, but these reserves are subject to legal logging and mining, making them less effective for biodiversity conservation. Recently, the government established the Caura National Park, with 75,339.52 km^2 (Official Gazette N° 41,118, March 21, 2017; Inparques 2018 https://www.inparques.gob.ve/parque-nacional-caura-2/), and containing four natural monuments, the National Park Jaua-Sarisariñama, and the El Caura Forestal Reserve. This declaration was made without consultation and other legal requirements, and exemplifies contradictions observed between the conservation and the extractive use of natural and abiotic resources. The new national park partly overlaps the "Zone of the Orinoco Mining Arc for National Strategic Development" (ZDEN-AMO for its initials in Spanish), which is intensively developed for mining and lacks legislative approval as well as support from the local indigenous communities.

The ZDEN-AMO was created on February 24, 2016 (Official Gazette N° 40,855) to promote the mining of gold, diamond, coltan, bauxite, iron and rare-earth elements in 11,844 km^2 of the northern part of the Guayana region—including the state of Bolívar, and part of Amazonas and Delta Amacuro. This mega-development was initiated without environmental controls required by law—despite social rejection and technical considerations—and it is expected to lead to adverse environmental consequences in the Venezuelan Guayana and the Caribbean region at large (Cavada-Blanco and Aguado-Adriani 2018), in addition to having deeply negative impacts on the indigenous communities within ZDEN-AMO (Vitti 2018; Rodríguez and Aguilar-Castro 2018). Alvarez-Berríos and Aide (2015) demonstrate than 684 km^2 of forest in the Guiana ecoregion (principally in Guyana, Suriname, and French Guiana) was lost to gold mining between 2001 and 2013, associated with increases in global demand for gold after the international financial crisis. Venezuela is in a growing political and economic crisis, and deforestation emerging from the development in the ZDEN-AMO has the potential to be similar to, or worse than, the environmental degradation documented in the others areas of the Guiana Shield.

In addition to both legal and illegal mining, Venezuelan Guayana biodiversity is threatened by deforestation for crops and livestock, logging, hydroelectric and infrastructure projects, oil extraction, and forest fires (Conservation International 2003; RAISG 2015). The lowlands of northern VG are the most threatened due to the synergy among these anthropogenic activities, especially along the Orinoco River, its floodplain and the delta, the Caroní River basin, the Sierra de Imataca, and the lower basin of Caura and Paragua Rivers. RAISG (2015) estimated that Venezuelan Guayana had lost approximately 8900 km^2 (2.2%) of its original forests by 2000, and that deforestation covered an additional 4150 km^2 between 2000 and 2013, representing a marked acceleration in deforestation. Other areas in the states of Bolívar and Amazonas are at risk, even within national parks. For example, the Cerro Yapacana National Park is threatened by historical illegal mining, which has resulted in habitat destruction, mercury contamination, and overexploitation of natural resources (Lasso et al. 2009). Moreover, armed groups are associated with illegal mining in different basins of the state of Amazonas (RAISG 2015). In the uplands and highlands of Canaima National Park, especially the Gran Sabana, further impacts are generated by uncontrolled tourism, the development of infrastructure and roads, hunting, intentional burning, and more recently an accelerated increase in artisanal mining. The Roraima and Auyán-tepui summits have tourist routes, and the most important threats in these highlands are the introduction of exotic species, the contamination of waters by human feces, forest fires, and the illegal extraction of local flora and fauna (Sanfont et al. 2014; Rull et al. 2016).

Finally, climate change is becoming a significant new threat to biodiversity, and the herpetofauna of Venezuelan Guayana is no exception. A global assessment predicted that amphibians and birds are the vertebrates with greatest relative vulnerability to climate change, especially in the Amazon basin (Foden et al. 2013). The unique and isolated Pantepui ecosystems appear to be particularly sensitive to climate change: the potential consequences of global warming have been estimated to include the loss of >80% of highland habitats (Rödder et al. 2010, Vegas-Vilarrúbia et al. 2012), along with the extinction of 30–50% of its endemic vascular flora (Nogué et al. 2009; Safont et al. 2012). Many dead adult frogs of *Tepuihyla edelcae* and *T. obscura* were found in pools at Auyán-tepui and Chimantá Massif during expeditions in 1984 (Ayarzagüena et al. 1992; Gorzula and Señaris 1999). Because no evidence of the deadly fungal pathogen *Batrachochytrium dendrobatidis* was found in those dead frogs found in 1984, Lampo and Señaris (2006) concluded that climatic changes may be the most reasonable cause of those mortalities. Amphibians have biological and ecological characteristics that make them particularly sensitive to climate change. Pantepui amphibians are even more vulnerable than common and widespread species, due to their specialized habitat requirements, narrow physiological tolerance, high dependence on environmental triggers for breeding, poor dispersal ability, and extremely restricted distributions (Foden et al. 2013).

5 Final Remarks

After decades of exploration and research, our knowledge of the amphibians and reptiles of Venezuelan Guayana still has significant gaps. Since ~2008, socio-political conditions have practically paralyzed fieldwork in the south of Venezuela. At the same time, changing conditions have increased threats alarmingly. Only by joining national and international efforts we can continue to advance in our understanding of the herpetofauna diversity in Venezuelan Guayana, while reducing threats and developing effective conservation action. National taxonomic efforts are insufficient, and conservations programs are in decline. In the short term, we suggest that one feasible and useful action would be the development of an international program to revise, update and elaborate an online database of the museum records from Venezuelan Guayana. There are still numerous samples waiting to be studied in museums, including the interesting amphibian and reptile collection obtained in explorations of the La Neblina Massif, 30 years ago (McDiarmid and Paolillo 1988). Such efforts will undoubtedly provide new information at much lower cost than those implied by new expeditions under current adverse conditions. However, multidisciplinary biological expeditions to completely unexplored areas are essential, especially the uplands and highlands of the southern Venezuelan Guayana as Sierra de Maigualida, Sierra Parima, Parú massif, and Unturán uplands (Table 22.7). We hope these will take place in the near future; without a doubt they will result in the discovery of new species, new records, and a better understanding of distribution patterns of the Guayana herpetofauna. Biological samples resulting from these new collections may be used in molecular studies to improve the resolution of previous phylogenetic analyses.

Present knowledge indicates that the spatial heterogeneity of the uplands and highlands of Venezuelan Guayana explains the distribution of the majority of the unique taxa of the region, and appears to have been crucial for different processes of diversification. Regarding the herpetological fauna, the fragmented Guayana mountains represent a relatively small biogeographical region —in contrast to the adjacent and transitional lowlands— but they contribute notably to overall Neotropical diversity. Some of these VG taxa have high evolutionary distinctiveness (Jetz and Pyron 2018) and should be given priority for conservation. Threats to the herpetofauna of Venezuelan Guayana vary across taxonomic groups and areas, and each must be confronted through a specific set of approaches. Nonetheless, in all cases, the obligatory incorporation of research, conservation and management programs into national and regional public policies remains a fundamental and key requirement to reduce diversity loss, and to face the challenges that future climate change will bring.

Table 22.7 High mountains and tepuis of Venezuelan Guayana, each ranked according to available herpetological data

Drainage	Mountain massif	Tepui/mountain unit	Country	Maximum elevation (m a.s.l.)	Summit area (km^2)	Relative degree of herpetological knowledge[a]
Caroní basin						
Eastern tepui chain		Roraima-tepui	VE-GU-BR	2723	34.38	5*[b,c,d]
		Kukenán (Matauí)- tepui	VE	2650	20.63	3*[b,c]
		Yuruaní-tepui	VE	2400	4.38	3*[b,e,c,f]
		Wadakapiapué-tepui	VE	2000	<0.01	1*[b,g]
		Karaurín-tepui	VE	2500	1.88	1
		Ilú-tepui	VE	2700	5.63	3*[b,c]
		Tramen-tepui	VE			
Gran Sabana uplands		Gran Sabana	VE	800–1450	NA	4*[b,h,i,d]
		Cerro Venamo	VE	1600	NA	1
		Sierra de Lema	VE	800–1650	NA	4*[b,h,e,i]
Ptari massif		Ptari-tepui	VE	2400	1.25	3*[g]
		Carrao-tepui	VE	2200	1.25	0
		Sororopán-tepui	VE	2050	NA	1
Los Testigos massif		(undifferentiated)	VE	1900–2400	12	0
		Kamarkawarai-tepui	VE	2400	5	1
		Tereke-yurén-tepui	VE	1900	0.63	1
		Murisipán-tepui	VE	2350	5	3*[e,f]
		Aparamán-tepui	VE	2100	1.25	2
Auyán massif		Auyán-tepui	VE	2450	666.9	5*[b,g,h,e,c,i]
		Cerro La Luna	VE	1650	0.2	1
		Cerro El Sol	VE	1750	0.6	1*[c]
		Uaipán-tepui	VE	1950	2.5	0
Canaima uplands		Cerro Venado	VE	1320	0.6	1
		Kurún-tepui	VE	1100	0.6	0
Aprada massif		Aprada-tepui	VE	2500	4.37	2*[e]
		Araopán-tepui	VE	2450	1.25	1

Chimantá massif	Chimantá (undifferentiated, incl. 11 tepuis)	VE	2200–2650	615	3*[b,g,h,e,i,d]
	Angasima-tepui	VE	2250	2	
	Upuigma-tepui	VE	2100	0.63	
Paragua basin					
Paragua uplands	Sierra Marutaní (Pia-Zoi)	VE	1500	740	0
	Cerro Ichún	VE	1400	2460	1
	Cerro Guanacoco	VE	1500	526.25	0
	Cerro Guaiquinima	VE	1650	1096.26	4
Caura basin					
Jaua massif	Cerro Sarisariñama	VE	2350	546.88	2*[b,e]
	Cerro Jaua	VE	2250	625.62	2
Maigualida massif	Sierra de Maigualida	VE	2400	440	0
	Serranía de Uasadi	VE	1300–1800	NA	0
Ventuari basin					
Yaví massif	Cerro Ualipano	VE	1800	NA	0
	Cerro Yaví	VE	2300	5.62	2
Yutajé massif	Serranía Yutajé	VE	2140	95.63	2
	Coro Coro	VE	2400	179.38	2
	Cerro Guanay	VE	2080	165	2
	Cerro Camani	VE	1800	1.88	0
Sipapo uplands	Cerro Ovaña (Ouana)	VE	1800	NA	1*[d]
	Cerro Moriche	VE	1250	0.2	0
Parú massif	Cerro Parú (A'roko)	VE	2200	724.38	0
	Cerro Asisa	VE	1700		0
	Cerro Euaja	VE	2000	205.62	0
Middle Orinoco basin					
Cuao-Sipapo massif	Cerro Autana	VE	1300	1.88	2
	Cerro Cuao	VE	2000	80	1
	Cerro Sipapo	VE	1800	56	1*[b]

(continued)

Table 22.7 (continued)

Drainage	Mountain massif	Tepui/mountain unit	Country	Maximum elevation (m a.s.l.)	Summit area (km^2)	Relative degree of herpetological knowledge[a]
Upper Orinoco basin						
Yapacana upland		Cerro Yapacana	VE	1300	10.5	3*[d]
Duida-Marahuaka massif		Cerro Huachamacari	VE	1900	8.75	2
		Cerro Marahuaca	VE	2800	121	3
		Cerro Duida	VE	2358	1089	3*[g]
Parima uplands		Sierra Parima	VE	800–1600	NA	1
Casiquiare-Rio Negro basin						
Unturán uplands		Cerro Vinilla	VE	815	NA	0
		Cerro Aratitiyope	VE	1700	<0.01	0
		Sierra Unturán	VE	1600	NA	0
Tapirapecó massif		Cerro Tamacuari	VE	2340	<0.01	3*[d]
		Serranía Tapirapecó	VE-BR	2000	NA	1
Arakamuni-Avispa uplands		Cerro Aracamuni	VE	1600	238	1
		Cerro Avispa	VE	1600		0
Imeri massif		Sierra de la Neblina	VE-BR	2994	235	3*[j]

List and geographical features follow Huber (1995); *NA* not available.

[a]Relative degree of herpetological knowledge (N° of samples and explorations) based on McDiarmid and Donnelly (2005), recent literature, and the review of herpetological collections in national museums (0 = none, 5 = max); *at least one or more species/samples included in molecular studies

[b]Kok et al. (2012)

[c]Kok et al. (2018b)

[d]Grant et al. (2017)

[e]Kok et al. (2016)

[f]Kok (2015)

[g]Salerno et al. (2012)

[h]Salerno et al. (2014)

[i]Kok et al. (2018a)

[j]Faivovich et al. (2013)

Acknowledgements We are very grateful to Santiago Castroviejo-Fisher, Kathryn M. Rodríguez-Clark, Lourdes Echevarria, anonymous reviews, and Ana C. Carnaval for their critical review of the first draft of this manuscript. They provided valuable comments and suggestions that improved and enriched the original text. Grisel Velázquez (UniSIG, IVIC) prepared the base map for Fig. 22.1, and estimated the area of the different elevational ranges for Table 22.5. We thank Javier Mesa for the extraordinary photos in Fig. 22.2.

Appendix 1. Amphibians of the Venezuelan Guayana

TAXA	Elevation range (m)	Category of extinction risk		CITES (2017)	Distribution pattern
		National LRFV 2015	Global (UICN 2018)		
CLASS AMPHIBIA					
ORDER ANURA					
Family Allophrynidae					
Allophryne ruthveni Gaige, 1926	100–130	LC	LC	–	AM-GU
Family Bufonidae					
Amazophrynella minuta (Melin, 1941)	0–300	LC	LC	–	AM-GUo
***Metaphryniscus sosae* Señaris, Ayarzagüena *et* Gorzula, 1994**	2480–2600	NT	VU	–	P
***Oreophrynella cryptica* Señaris, 1995 "1993"**	1750–2400	NT	VU	–	P
***Oreophrynella huberi* Diego-Aransay *et* Gorzula, 1990 "1987"**	1700–1755	VU	VU	–	P
Oreophrynella macconnelli Boulenger, 1900	700–1830	NT	VU	–	GUc
***Oreophrynella nigra* Señaris, Ayarzagüena *et* Gorzula, 1994**	2300–2700	VU	VU	–	P
Oreophrynella quelchii (Boulenger, 1895)	1700–2750	VU	VU	–	P
***Oreophrynella vasquezi* Señaris, Ayarzagüena *et* Gorzula, 1994**	2371–2680	VU	VU	–	P
Rhaebo guttatus (Schneider, 1799)	50–860	LC	LC	–	AM-GU
Rhaebo nasicus (Werner, 1903)	100–1410	LC	LC	–	GUc
Rhinella ceratophrys (Boulenger, 1882)	412–2713	LC	LC	–	AM-GUo
Rhinella margaritifera complex (Laurenti, 1768)	100–1500	LC	LC	–	WS
Rhinella marina (Linnaeus, 1758)	0–2000	LC	LC	–	WS
Rhinella beebei (Gallardo, 1965)	0–1015	LC	LC	–	WS

(continued)

TAXA	Elevation range (m)	Category of extinction risk		CITES (2017)	Distribution pattern
		National LRFV 2015	Global (UICN 2018)		
Rhinella merianae (Gallardo, 1965)	20–1000	LC	NE	–	GUe
Rhinella nattereri (Bokermman *in* Lent, 1967)	600–1300	DD	LC	–	GUe
Family Centrolenidae					
Subfamily Centroleninae					
***"Cochranella" duidaeana* (Ayarzagüena, 1992)**	2140	NT	DD	–	P
***"Cochranella" riveroi* (Ayarzagüena, 1992)**	1600	NT	VU	–	P
Vitreorana gorzulae (Ayarzagüena, 1992)	430–1850	LC	DD	–	GUc
Vitreorana helenae (Ayarzagüena, 1992)	850–1000	LC	DD	–	GUc
Subfamily Hyalinobatrachinae					
Hyalinobatrachium cappellei van Lidth de Jeude, 1904	45–2000	LC	LC	–	AM-GUc, GUe
Hyalinobatrachium iaspidiense (Ayarzagüena, 1992)	25–1000	LC	DD	–	AM-GUc, GUe
***Hyalinobatrachium mesai* Barrio-Amorós *et* Brewer-Carías, 2008**	420	DD	NE	–	GUc
Hyalinobatrachium mondolfii Ayarzagüena *et* Señaris, 2001	0–200	LC	LC	–	AM-GUe
Hyalinobatrachium taylori (Goin, 1968 "1967")	30–1850	LC	LC	–	GUc, GUe
Family Ceratophryidae					
Ceratophrys calcarata Boulenger, 1890	0–500	LC	LC	–	GUo
Family Ceuthomantidae					
***Ceuthomantis aracamuni* (Barrio-Amorós *et* Molina, 2006)**	1493	DD	VU	–	P
Ceuthomantis cavernibardus (Myers *et* Donnelly, 1997)	930–1270	DD	DD	–	GUc
***Ceuthomantis duellmani* Barrio-Amorós, 2010**	1100-1375	DD	LC	–	GUc
Superfamily Dendrobatoidea					
Family Aromobatidae					
Subfamily Allobatinae					
Allobates femoralis (Boulenger, 1884)	185–1000	LC	LC	II	AM-GU
Allobates aff. *marchesianus* (Melin, 1941)	0–800	LC	LC	–	AM-Guo
***Allobates sanmartini* (Rivero, Langone *et* Prigioni, 1986)**	±70	DD	DD	–	GUc

(continued)

TAXA	Elevation range (m)	Category of extinction risk		CITES (2017)	Distribution pattern
		National LRFV 2015	Global (UICN 2018)		
***Allobates undulatus* (Myers *et* Donelly, 2001)**	1750	DD	DD	–	P
Subfamily Anomaloglossinae					
***Anomaloglossus ayarzaguenai* (La Marca, 1997 "1996")**	1600	DD	DD	–	P
***Anomaloglossus breweri* (Barrio-Amorós, 2006)**	660	DD	VU	–	GUc
***Anomaloglossus guanayensis* (La Marca, 1997 "1996")**	1650–1800	DD	DD	–	P
***Anomaloglossus moffetti* Barrio-Amorós *et* Brewer-Carías, 2008**	1108	DD	NE	–	GUc
***Anomaloglossus murisipanensis* (La Marca, 1997 "1996")**	2350	VU	VU	–	P
***Anomaloglossus parimae* (La Marca, 1997 "1996")**	670	DD	DD	–	GUc
Anomaloglossus parkerae (Meinhardt *et* Parmelee, 1996)	860–1300	NT	DD	–	GUc
Anomaloglossus praderioi (La Marca, 1997 "1996")	1310–1950	LC	DD	–	P
***Anomaloglossus roraima* (La Marca, 1997 "1996")**	1860–2700	NT	DD	–	P
***Anomaloglossus rufulus* (Gorzula, 1990 "1988")**	2100–2600	LC	DD	II	P
***Anomaloglossus shrevei* (Rivero, 1961)**	350–1829	DD	DD	–	GUc
Anomaloglossus tamacuarensis (Myers *et* Donelly, 1997)	350–1200	DD	DD	–	GUc
Anomaloglossus tepuyensis (La Marca, 1997 "1996")	390–1850	LC	DD	–	GUc
***Anomaloglossus triunfo* (Barrio-Amorós, Fuentes *et* Rivas, 2004)**	350–680	DD	DD	–	GUc
***Anomaloglossus verbeeksnyderorum* Barrio-Amorós, Santos et Jovanovic, 2010**	56–300	DD	NE	–	GUc
***Anomaloglossus wothuja* (Barrio-Amorós, Fuentes *et* Rivas, 2004)**	150	DD	DD	–	GUc
Family Dendrobatidae					
Subfamily Colostethinae					
Ameerega picta (Bibron *in* Tshudi, 1838)	200–1200	LC	LC	II	AM-GUe
Ameerega trivittata (Spix, 1824)	200–400	LC	LC	II	AM-GUe

(continued)

TAXA	Elevation range (m)	Category of extinction risk		CITES (2017)	Distribution pattern
		National LRFV 2015	Global (UICN 2018)		
Subfamily Dendrobatinae					
Dendrobates leucomelas Steindachner, 1864	0–1300	LC	LC	II	GU
***Minyobates steyermarki* (Rivero, 1971)**	600–1300	CR	CR	II	GUc
Family Eleutherodactylidae					
Subfamily Eleutherodactylinae					
Eleutherodactylus johnstonei Barbour, 1914	0–1300	LC	LC	–	I
Subfamily Phyzelaphryninae					
Adelophryne gutturosa Hoogmoed *et* Lescure, 1984	40–2200	LC	LC	–	GUc, GUe
Family Hemiphractidae					
Subfamily Hemiphractinae					
***Stefania breweri* Barrio-Amorós *et* Fuentes-Ramos, 2003**	1250	DD	DD	–	GUc
***Stefania ginesi* Rivero, 1968 "1966"**	1850–2600	LC	LC	–	P
***Stefania goini* Rivero, 1968 "1966"**	1400–1700	NT	DD	–	P
***Stefania marahuaquensis* (Rivero, 1961)**	340–1200	DD	DD	–	GUc
***Stefania oculosa* Señaris, Ayarzagüena *et* Gorzula, 1997 "1996"**	1600	DD	DD	–	P
***Stefania percristata* Señaris, Ayarzagüena *et* Gorzula, 1997 "1996"**	1600	DD	DD	–	P
***Stefania riae* Duellman *et* Hoogmoed, 1984**	1108–1400	DD	DD	–	GUc
***Stefania riveroi* Señaris, Ayarzagüena *et* Gorzula, 1997 "1996"**	2300	VU	VU	–	P
***Stefania satelles* Señaris, Ayarzagüena *et* Gorzula, 1997 "1996"**	2000–2500	VU	NT	–	P
***Stefania scalae* Rivero, 1970**	600–1360	LC	LC	–	GUc
***Stefania schuberti* Señaris, Ayarzagüena *et* Gorzula, 1997 "1996"**	1750–2400	NT	VU	–	P
Stefania tamacuarina Myers *et* Donnelly, 1997	1160–1270	DD	DD	–	GUc
Family Hylidae					
Subfamily Cophomantinae					
Boana benitezi (Rivero, 1961)	400–1600	LC	NE	–	GUc

(continued)

TAXA	Elevation range (m)	Category of extinction risk		CITES (2017)	Distribution pattern
		National LRFV 2015	Global (UICN 2018)		
Boana boans (Linnaeus, 1758)	0–1000	LC	LC	–	WS
Boana calcarata (Troschel, 1848)	0–650	LC	LC	–	AM-GUe
Boana cinerascens (Spix, 1824)	±0–1000	LC	LC	–	AM-GU
Boana geographica (Spix, 1824)	±0–1200	LC	LC	–	AM-GU
Boana hobbsi (Cochran *et* Goin, 1970)	90–200	LC	LC	–	GUo
***Boana jimenezi* (Señaris *et* Ayarzagüena, 2006)**	970–2100	LC	LC	–	GUc
Boana lanciformis (Cope, 1871 "1870")	100–1500	LC	LC	–	AM-GU
Boana lemai (Rivero, 1972 "1971")	600–1400	LC	LC	–	GUc
Boana multifasciata (Günther, 1859 "1858")	150–1400	LC	LC	–	AM-GUe
Boana ornatissima (Noble, 1923)	100–900	LC	LC	–	GUe, GUo
Boana punctata (Schneider, 1799)	0–900	LC	LC	–	WS
***Boana rhythmica* (Señaris *et* Ayarzagüena, 2002)**	1600	DD	DD	–	P
Boana roraima (Duellman *et* Hoogmoed, 1992)	490–1600	LC	DD	–	GUc
Boana sibleszi (Rivero, 1972 "1971")	500–1850	LC	LC	–	GUc
Boana tepuiana (Barrio-Amorós *et* Brewer-Carías, 2008)	420–1800	LC	LC	–	GUc
Boana wavrini (Parker, 1936)	0–400	LC	LC	–	AM-GUo
Boana xerophylla (Duméril *et* Bibron, 1841)	0–1700	LC	LC	–	WS
***Myersiohyla aromatica* (Ayarzagüena *et* Señaris, 1994 "1993")**	1700	VU	DD	–	P
***Myersiohyla chamaeleo* Faivovich, McDiarmid *et* Myers, 2013**	1450–2100	DD	NE	–	P
***Myersiohyla inparquesi* (Ayarzagüena *et* Señaris, 1994 "1993")**	2600	NT	DD	–	P
***Myersiohyla loveridgei* (Rivero, 1961)**	614–910	DD	DD	–	GUc
***Myersiohyla neblinaria* Faivovich, McDiarmid *et* Myers, 2013**	1250–2100	DD	NE	–	P

(continued)

TAXA	Elevation range (m)	Category of extinction risk		CITES (2017)	Distribution pattern
		National LRFV 2015	Global (UICN 2018)		
Subfamily Dendropsophinae					
Dendropsophus marmoratus (Laurenti, 1768)	±100–1000	LC	LC	–	AM-GU
Dendropsophus microcephalus (Cope, 1886)	0–1300	LC	LC	–	WS
Dendropsophus minusculus (Rivero, 1971)	0–600	LC	LC	–	WS
Dendropsophus minutus (Peters, 1872)	0–2000	LC	LC	–	WS
Dendropsophus parviceps (Boulenger, 1882)	186–1600	LC	LC	–	AM-GUo
Dendropsophus sarayacuensis (Shreve, 1935)	90–1200	LC	LC	–	AM-GUo
Subfamily Hylinae					
Aparasphenodon venezolanus (Mertens, 1950)	±100	LC	LC	–	GUo
Osteocephalus helenae (Ruthven, 1919)	±0–700	LC	NE	–	AM-GU
Osteocephalus leprieurii (Duméril *et* Bibron, 1841)	±0–1000	LC	LC	–	AM-GU
Osteocephalus aff. *taurinus* Steindachner, 1862	0–1365	LC	LC	–	AM-GU
***Tepuihyla aecii* (Ayarzagüena, Señaris *et* Gorzula, 1993 "1992")**	2150	NT	DD	–	P
***Tepuihyla edelcae* (Ayarzagüena, Señaris *et* Gorzula, 1993 "1992")**	1700–2100	LC	LC	–	P
Tepuihyla exophthalma (Smith *et* Noonan, 2001)	585–1550	LC	DD	–	GUc
***Tepuihyla luteolabris* (Ayarzagüena, Señaris *et* Gorzula, 1993 "1992")**	2550	NT	DD	–	P
***Tepuihyla obscura* Kok, Ratz, Tagelaar, Aubret *et* Means, 2015**	1800–2600	LC	NE	–	P
Tepuihyla rodriguezi (Rivero, 1968)	366–2440	LC	DD	–	GUc
Trachycephalus resinifictrix (Goeldi, 1907)	90–140	LC	LC	–	AM-GUe
Trachycephalus typhonius (Linneus, 1758)	±0–500	LC	LC	–	WS

(continued)

TAXA	Elevation range (m)	Category of extinction risk		CITES (2017)	Distribution pattern
		National LRFV 2015	Global (UICN 2018)		
Subfamily Pseudinae					
Pseudis paradoxa (Linnaeus, 1758)	0–300	LC	LC	–	WS
Subfamily Scinaxinae					
Scinax baumgardneri **(Rivero, 1961)**	±50–609	DD	DD	–	GUo
Scinax boesemani (Goin, 1966)	0–650	LC	LC	–	AM-GU
Scinax danae **(Duellman, 1986)**	180–1250	LC	DD	–	Guc
Scinax exiguus (Duellman, 1986)	650–1230	LC	LC	–	GUc
Scinax garbei (Miranda-Ribeiro, 1926)	100–1000	LC	LC	–	AM-GUo
Scinax kennedyi (Pyburn, 1973)	±100–400	LC	LC	–	LL-GUo
Scinax nebulosus (Spix, 1824)	0–400	LC	LC	–	AM-GUe
Scinax rostratus (Peters, 1863)	±0–1300	LC	LC	–	WS
Scinax ruber (Laurenti, 1768)	±0–2600	LC	LC	–	WS
Scinax fuscomarginatus (Lutz, 1925)	0–2000	LC	LC	–	WS
Scinax wandae (Pyburn *et* Fouquette, 1971)	±100–880	LC	LC	–	LL-GUo
Scinax x-signatus (Spix, 1824)	±0–900	LC	LC	–	WS
Sphaenorhynchus lacteus (Daudin, 1800)	0–500	LC	LC	–	WS
Family Leptodactylidae					
Subfamily Leiuperinae					
Engystomops pustulosus (Cope, 1864)	±0–1590	LC	LC	–	WS
Physalaemus cf. *cuvieri* Fitzinger, 1826	0–2000	LC	NE	–	GUe
Physalaemus ephippifer (Steindachner, 1864)	0–350	LC	LC	–	GUe
Physalaemus fischeri (Boulenger, 1890)	±0–800	LC	LC	–	LL-GU
Pleurodema brachyops (Cope, 1869 "1868")	±0–500	LC	LC	–	WS
Pseudopaludicola boliviana Parker, 1927	±50–700	LC	LC	–	WS
Pseudopaludicola llanera Lynch, 1989	100–1220	LC	LC	–	WS

(continued)

TAXA	Elevation range (m)	Category of extinction risk		CITES (2017)	Distribution pattern
		National LRFV 2015	Global (UICN 2018)		
Subfamily Leptodactylinae					
Adenomera andreae (Müller, 1923)	±50–500	LC	LC	–	WS
Adenomera hylaedactyla (Cope, 1868)	0–1000	LC	LC	–	WS
Leptodactylus bolivianus Boulenger, 1898	±100–800	LC	LC	–	WS
Leptodactylus diedrus Heyer, 1994	90–400	LC	LC	–	AM-GUo
Leptodactylus fuscus (Schneider, 1799)	±0–1200	LC	LC	–	WS
Leptodactylus guianensis Heyer *et* De Sa, 2011	80–1300	LC	NE	–	GUe
Leptodactylus knudseni Heyer, 1972	0–1800	LC	LC	–	WS
Leptodactylus leptodactyloides (Andersson, 1945)	15–400	LC	LC	–	AM-GUe
Leptodactylus lithonaetes Heyer, 1995	100–1250	LC	LC	–	GUo
Leptodactylus longirostris Boulenger, 1882	±100–1300	LC	LC	–	GU
Leptodactylus mystaceus (Spix, 1824)	±0–1000	LC	LC	–	AM-GU
Leptodactylus pentadactylus (Laurenti, 1768)	±0–1000	LC	LC	–	AM-GUe
Leptodactylus petersii (Steindachner, 1864)	<600	LC	LC	–	AM-GU
Leptodactylus rhodomystax Boulenger, 1884	50–520	LC	LC	–	AM-GUe, GUo
Leptodactylus riveroi Heyer *et* Pyburn, 1983	90–450	LC	LC	–	AM-GUo
Leptodactylus rugosus Noble, 1923	230–2100	LC	LC	–	Guc, Gue, P
Leptodactylus sabanensis Heyer, 1994	800–1250	LC	LC	–	GUc
Leptodactylus validus Garman, 1888 "1887"	0–360	LC	LC	–	WS
Lithodytes lineatus (Schneider, 1799)	100–1800	LC	LC	–	WS
Family Microhylidae					
Subfamily Adelastinae					
Adelastes hylonomos Zweifel, 1986	100–140	DD	DD	–	GUc

(continued)

TAXA	Elevation range (m)	Category of extinction risk		CITES (2017)	Distribution pattern
		National LRFV 2015	Global (UICN 2018)		
Subfamily Gastrophryninae					
Chiasmocleis hudsoni (Parker, 1940)	80–140	LC	LC	–	AM-GU
Ctenophryne geayi Mocquard, 1904	0–600	LC	LC	–	AM-GU
Elaschistocleis "ovalis" (Schneider, 1799)	±0–500	LC	LC	–	WS
Elaschistocleis surinamensis (Daudin, 1802)	±0–500	LC	LC	–	GUe
Hamptophryne boliviana (Parker, 1927)	0–400	LC	LC	–	AM-GUe, GUo
Subfamily Otophryninae					
Otophryne pyburni Campbell *et* Clarke, 1998	100–1100	LC	LC	–	GUe, GUo
Otophryne robusta Boulenger *in* Lankester, 1900	600–1411	LC	LC	–	GUe
Otophryne steyermarki Rivero, 1968 "1967"	1234–2140	LC	LC	–	GUc
Synapturanus mirandaribeiroi Nelson *et* Lescure, 1975	100–700	LC	LC	–	GUe, GUo
Synapturanus salseri Pyburn, 1975	90–140	LC	LC	–	GUe, GUo
Family Phyllomedusidae					
Callimedusa tomopterna (Cope, 1868)	0–500	LC	LC	–	AM-GUe, GUo
Phyllomedusa tarsius (Cope, 1868)	0–450	LC	LC		WS
Phyllomedusa bicolor (Boddaert, 1772)	0–800	LC	LC	–	AM-GUe, GUo
Phyllomedusa vaillantii Boulenger, 1882	0–450	LC	LC	–	AM-GUe, GUo
Pithecopus hypochondrialis (Daudin, 1800)	50–1500	LC	LC	–	WS
Family Pipidae					
Pipa arrabali Izecksohn, 1976	350–860	LC	LC	–	AM-GUe
Pipa pipa (Linnaeus, 1758)	0–400	LC	LC	–	WS
Family Ranidae					
Lithobates palmipes (Spix, 1824)	±100–1100	LC	LC	–	WS
Family Strabomantidae					
***Dischidodactylus colonnelloi* Ayarzagüena, 1985**	2550	NT	DD	–	P
***Dischidodactylus duidensis* (Rivero, 1968)**	990–1530	NT	DD	–	GUc

(continued)

TAXA	Elevation range (m)	Category of extinction risk		CITES (2017)	Distribution pattern
		National LRFV 2015	Global (UICN 2018)		
***Pristimantis abakapa* Rojas-Runjaic, Salerno, Señaris *et* Pauly, 2013**	2245	DD	NE	–	P
***Pristimantis auricarens* (Myers *et* Donnelly, 2008)**	1600–1750	NT	NE	–	P
Pristimantis aureoventris Kok, Means *et* Bossuyt, 2011	2210–2260	NT	NE	–	P
***Pristimantis avius* (Myers *et* Donelly, 1997)**	1160–1460	DD	DD	–	GUc
***Pristimantis cantitans* (Myers *et* Donnelly, 1996)**	1700–2150	NT	DD	–	P
***Pristimantis guaiquinimensis* (Schlüter *et* Rödder, 2007 "2006")**	780–980	DD	NE	–	GUc
***Pristimantis imthurni* Kok,** 2013	2471	NT	NE	–	P
***Pristimantis jamescameroni* Kok,** 2013	2557–2571	NT	NE	–	P
***Pristimantis marahuaka* (Fuentes-Ramos *et* Barrio-Amorós, 2004)**	2450	NT	VU	–	P
Pristimantis marmoratus (Boulenger, 1900)	600–1800	LC	LC	–	GUc, GUe
Pristimantis memorans (Myers *et* Donnelly, 1997)	1160–1270	NT	DD	–	GUc
***Pristimantis muchimuk* Barrio-Amorós, Mesa, Brewer-Carías *et* McDiarmid, 2010**	2325	NT	NE	–	P
***Pristimantis pruinatus* (Myers *et* Donnelly, 1996)**	2150	VU	DD	–	P
Pristimantis pulvinatus (Rivero, 1968)	950–1700	LC	LC	–	GUc, GUe
***Pristimantis sarisarinama* Barrio-Amorós *et* Brewer-Carías, 2008**	1100–1400	DD	NE	–	GUc
Pristimantis vilarsi (Melin, 1941)	100–1230	LC	LC	–	AM-GU
***Pristimantis yaviensis* (Myers *et* Donnelly, 1996)**	1700–2150	DD	DD	–	P
***Pristimantis yuruaniensis* Rödder *et* Jungfer, 2008**	2300	VU	NE	–	P
Pristimantis zeuctotylus (Lynch *et* Hoogmoed, 1977)	10–300	LC	LC	–	GUe, GUo

(continued)

TAXA	Elevation range (m)	Category of extinction risk		CITES (2017)	Distribution pattern
		National LRFV 2015	Global (UICN 2018)		
ORDER GYMNOPHIONA					
Family Caecilidae					
Caecilia tentaculata Linnaeus, 1758	±100–1000	LC	LC	–	AM-GU
Family Rhinatrematidae					
***Rhinatrema nigrum* Dunn, 1942**	100–1700	LC	LC	–	GUe
Family Siphonopidae					
Microcaecilia rabei (Roze *et* Solano, 1963)	±100–300	DD	DD	–	GUe
Family Typhlonectidae					
Nectocaecilia petersii (Boulenger, 1882)	±100	LC	LC	–	GUo
Potomotyphlus kaupii (Berthold, 1859)	0–100	LC	LC	–	WS
Typhlonectes compressicauda (Duméril *et* Bibron, 1841)	0–200	LC	LC	–	AM-GU

Species list, geographical and elevational distribution based on references (Gorzula and Señaris 1999, McDiarmid and Donnelly 2005; Señaris and McCulloch 2005; Señaris et al. 2009), recent taxa descriptions, and review of herpetological collections in national museums [Museo de Historia Natural La Salle, Caracas (MHNLS), Museo de Biología de la Universidad Central de Venezuela, Caracas (MBUCV) and Museo de la Estación Biológica Rancho Grande, Maracay (EBRG)]. Taxonomic arrangement follows Frost (2018). Distribution patterns: WS: widespread, taxa known from areas of South America outside the Guiana Shield; LL: Llanos Province; GU: all Guiana Shield; GUo: taxa know only in the Western Guayana, GUc: taxa know only in the Central Guayana; GUe: taxa know only in the Eastern Guayana, P: taxa know only in Pantepui (Guiana Highlands ≥1500 m a.s.l.). Endemic Venezuelan Guayana species in bold

Appendix 2. Reptiles of the Venezuelan Guayana

TAXA	Elevation range (m)	Category of extinction risk		CITES (2017)	Distribution pattern
		National LRFV 2015	Global (UICN 2017)		
CLASS REPTILIA					
ORDER CROCODYLIA					
Family Alligatoridae					
Caiman crocodilus Linnaeus, 1758	±0–350	LC	LC	II	WS
Paleosuchus palpebrosus (Cuvier, 1807)	±0–500	DD	LC	II	WS
Paleosuchus trigonatus (Schneider, 1801)	±0–1340	DD	LC	II	WS
Family Crocodylidae					
Crocodylus intermedius Graves, 1819	±0–300	EN C1 + 2a (i)	CR A1c, C2a	I	LL, GU
ORDEN TESTUDINES					
Family Chelidae					
Chelus fimbriata (Schneider, 1783)	±0–200	LC	NE	–	WS
Mesoclemmys gibba (Schweigger, 1812)	±0–400	LC	NE	–	WS
Mesoclemmys heliostemma (McCord, Ouni *et* Lamar, 2001)	±0–400	DD	NE	–	WS
Phrynops tuberosus (Peters, 1870)	±0–400	LC	NE	–	WS
Platemys platycephala (Schneider, 1792)	±0–500	LC	NE	–	WS
Family Geoemydidae					
***Rhinoclemmys flammigera* (Paolillo, 1985)**	±90–100	EN B1ab (iii)	NE	–	GUw
Rhinoclemmys punctularia (Daudin, 1802)	±0–300	LC	NE	–	WS
Family Kinosternidae					
Kinosternon scorpioides (Linnaeus, 1766)	±0–300	LC	NE	–	WS
Family Podocnemididae					
Peltocephalus dumerilianus (Schweigger, 1812)	±50–250	VU A2d	VU A1acd	II	WS
Podocnemis erythrocephala (Spix, 1824)	±100–250	VU A2cde	VU A1bd	II	GUw, AM

(continued)

TAXA	Elevation range (m)	Category of extinction risk		CITES (2017)	Distribution pattern
		National LRFV 2015	Global (UICN 2017)		
Podocnemis expansa (Schweigger, 1812)	±0–250	CR A2acde	LC	II	WS
Podocnemis unifilis Troschel, 1848	±0–500	VU A2d	VU A1acd	II	WS
Podocnemys vogli Muller, 1935	±0–500	LC	NE	II	WS
Family Testudinidae					
Chelonoidis carbonaria (Spix, 1824)	±0–400	NT	NE	II	WS
Chelonoidis denticulata (Linnaeus, 1766)	±0–600	NT	VU A1cd + 2 cd	II	WS
ORDEN SQUAMATA					
Family Amphisbaenidae					
Amphisbaena alba Linnaeus, 1758	±0–300	LC	LC	–	WS
Amphisbaena fuliginosa Linnaeus, 1758	±60–1000	LC	NE	–	WS
***Amphisbaena gracilis* Strauch, 1881**	±0–100	LC	LC	–	GUe
***Amphisbaena rozei* Lancini, 1963**	±350	LC	NE	–	GUc
Mesobaena huebneri Mertens, 1925	±200	LC	LC	–	GUw
Family Iguanidae					
Iguana iguana Linnaeus, 1758	0–900	LC	NE	II	WS
Family Dactyloidae					
Anolis auratus Daudin, 1802	±50–300	LC	NE	–	WS
***Anolis bellipeniculus* (Myers *et* Donnelly, 1995)**	2150	LC	NE	–	P
***Anolis carlostoddi* (Williams, Praderio *et* Gorzula, 1996)**	2200	LC	NE	–	P
***Anolis deltae* Williams, 1974**	±0–50	LC	NE	–	GUe
Anolis fuscoauratus D'Orbigny en Duméril *et* Bibron, 1837	±100–1500	LC	NE	–	WS
***Anolis neblininus* (Myers, Williams *et* McDiarmid, 1993)**	1690–2100	LC	NE	–	P
Anolis ortonii Cope, 1868	±100–1200	LC	NE	–	WS

(continued)

TAXA	Elevation range (m)	Category of extinction risk		CITES (2017)	Distribution pattern
		National LRFV 2015	Global (UICN 2017)		
Anolis planiceps Troeschel, 1848	±0–1800	LC	NE	–	WS
Anolis punctatus Daudin, 1802	±100–320	LC	NE	–	WS
Family Polychrotidae					
Polychrus marmoratus (Linnaeus, 1758)	±0–1800	LC	NE	–	WS
Family Tropiduridae					
***Plica lumaria* Donnelly *et* Myers, 1991**	1060–1090	NE	LC	–	GUc
***Plica pansticta* Myers *et* Donnelly, 2001**	180–1220	LC	LC	–	GUc
Plica plica (Linnaeus, 1758)	±30–1500	LC	NE	–	WS
Plica rayi Murphy *et* Jowers, 2013	±40–500	NE	NE	–	GUw
Plica umbra (Linnaeus, 1758)	±0–1500	LC	NE	–	GU-AM
***Tropidurus bogerti* Roze, 1958**	1600–2080	LC	LC	–	P
Tropidurus hispidus (Spix 1825)	20–1420	LC	NE	–	WS
Uracentron azureum (Linnaeus, 1758)	±100–200	LC	NE	–	WS
Uranoscodon superciliosus (Linnaeus, 1758)	±0–250	LC	NE	–	WS
Family Gekkonidae					
Hemidactylus palaichthus Kluge, 1969	±0–250	LC	LC	–	WS
Family Phyllodactylidae					
***Phyllodactylus dixoni* Rivero-Blanco *et* Lancini, 1968**	±30–50	LC	LC	–	GUc
Thecadactylus rapicauda (Houttuyn, 1782)	±0–1500	LC	NE	–	WS
Family Sphaerodactylidae					
Chatogekko amazonicus (Anderson, 1918)	±0–1500	LC	NE	–	WS
Coleodactylus septentrionalis Vanzolini, 1980	±100–300	LC	LC	–	GU
Gonatodes albogularis (Duméril *et* Bibron, 1836)	0–50	LC	NE	–	IL

(continued)

TAXA	Elevation range (m)	Category of extinction risk		CITES (2017)	Distribution pattern
		National LRFV 2015	Global (UICN 2017)		
Gonatodes alexandermendesi Cole *et* Kok, 2006	±317–900	LC	LC	–	GU
Gonatodes annularis Boulenger, 1887	0–300	LC	NE	–	GU
***Gonatodes astralis* Schargel et al.,** 2010	91–600	LC	NE	–	GU
Gonatodes humeralis (Guichenot, 1855)	0–300	LC	NE	–	WS
***Gonatodes infernalis* Rivas *et* Schargel, 2008**	±100	LC	NE	–	GUw
***Gonatodes superciliaris* Barrio-Amorós *et* Brewer-Carías, 2008**	1100	LC	LC	–	GUc
Gonatodes vittatus (Lichtenstein, 1856)	0–1100	LC	LC	–	IL
Pseudogonatodes guianensis (Parker, 1935)	±100–300	LC	NE	–	WS
Sphaerodactylus molei Boettger, 1894	±0–100	LC	LC	–	WS
Family Scincidae					
Copeoglossum nigropunctatum Spix, 1825	±0–1800	LC	NE	–	WS
Panopa carvalhoi (Reboucas-Spieker *et* Vanzolini, 1990)	0–500	LC	LC	–	GU
Family Gymnophthalmidae					
***Adercosaurus vixadnexus* Myers *et* Donnelly, 2001**	1700	LC	NE	–	P
***Anadia escalerae* Myers, Rivas *et* Jadin, 2009**	1385–1430	LC	LC	–	GUc
***Arthrosaura montigena* Myers *et* Donnelly, 2008**	940–1700	LC	LC	–	P
Arthrosaura reticulata (O'Shaughnessy, 1881)	±50–250	LC	NE	–	WS
***Arthrosaura synaptolepis* Donelly, McDiarmid *et* Myers, 1992**	1160–1450	LC	LC	–	P
***Arthrosaura testigensis* Gorzula *et* Señaris, 1998**	1800–2350	LC	LC	–	P
***Arthrosaura tyleri* (Burt *et* Burt, 1931)**	1402–2164	LC	LC	–	P
Bachia flavescens (Bonnaterre, 1789)	±84–940	LC	LC	–	WS

(continued)

TAXA	Elevation range (m)	Category of extinction risk		CITES (2017)	Distribution pattern
		National LRFV 2015	Global (UICN 2017)		
Bachia guianensis Hoogmoed *et* Dixon, 1977	230–300	LC	NE	–	GU
Bachia heteropa (Lichhtenstein, 1856)	±0–500	LC	NE	–	WS
Bachia pyburni Kizirian *et* McDiarmid, 1998	±50–300	LC	NE	–	GU, AM
***Cercosaura nigroventris* (Gorzula *et* Señaris, 1998)**	1650	LC	LC	–	P
Cercosaura ocellata Wagler, 1830	±250	LC	NE	–	WS
***Cercosaura phelpsorum* (Lancini, 1968)**	1917–2150	LC	LC	–	P
***Euspondylus auyanensis* Myers, Rivas *et* Jadin, 2009**	2325	LC	LC	–	P
Loxopholis hexalepis (Ayala *et* Harris, 1982)	35–700	LC	LC	–	GU
Loxopholis percarinatum (Müller, 1923)	100–500	LC	LC	–	WS
Neusticurus arekuna Kok et al. 2018	900–2203	NE	NE	–	GU
Neusticurus bicarinatus (Linnaeus, 1758)	±1000–1300	LC	NE	–	WS
Neusticurus medemi Dixon *et* Lamar, 1981	±50–300	LC	LC	–	GU
Neusticurus racenisi Roze, 1958	100–1215	LC	NE	–	GU
Neusticurus rudis Boulenger, 1900	313–1200	LC	NE	–	GU
***Neusticurus tatei* (Burt *et* Burt, 1931)**	±400–1402	LC	LC	–	GUc
***Gymnophthalmus cryptus* Hoogmoed, Cole *et* Ayarzaguena, 1992**	±40–100	LC	LC	–	GUc
Gymnophthalmus speciosus (Hallowell, 1861)	100–300	LC	NE	–	WS
***Oreosaurus mcdiarmidi* Kok *et* Rivas, 2011**	2100–2600	LC	LC	–	P
***Riolama inopinata* Kok,** 2015	2400–2413	NE	NE	–	P
Riolama leucosticta (Boulenger, 1900)	1940–2600	LC	NE	–	P

(continued)

TAXA	Elevation range (m)	Category of extinction risk		CITES (2017)	Distribution pattern
		National LRFV 2015	Global (UICN 2017)		
***Riolama luridiventris* Esqueda, La Marca *et* Praderio, 2004**	2470–2480	DD	LC	–	P
***Riolama uzzelli* Molina *et* Señaris, 2003**	1850–2600	DD	LC	–	P
Tretioscincus oriximinensis Avila-Pires, 1995	±100–150	LC	NE	–	GU, AM
Family Teiidae					
Ameiva ameiva (Linnaeus, 1758)	±0–1180	LC	NE	–	WS
Ameiva bifrontata Cope, 1862	±0–300	LC	LC	–	IL
Cnemidophorus cryptus Cole *et* Dessauer, 1993	±500–1000	LC	NE	–	GU
Cnemidophorus lemniscatus (Linnaeus, 1758)	±0–1100	LC	NE	–	WS
Crocodilurus amazonicus (Spix, 1825)	±100–200	LC	LC	II	WS
Kentropyx altamazonica Cope, 1876	±100–300	LC	NE	–	WS
Kentropyx calcarata Spix, 1825	±0–1500	LC	NE	–	WS
Kentropyx striata (Daudin, 1802)	±15–1100	LC	NE	–	WS
Tupinambis cryptus Murphy, Jowers, Lehtinen, Charles, Colli, Peres, Hendry *et* Pyron, 2016	0–800	NE	NE	–	WS
Family Aniliidae					
Anilius scytale (Linnaeus, 1758)	±50–600	LC	NE	–	WS
Family Anomalepididae					
Typhlophis squamosus (Schlegel, 1839)	±0–300	LC	NE	–	WS
Family Boidae					
Boa constrictor Linnaeus 1758	±0–1500	LC	NE	II	WS
Corallus caninus (Linnaeus, 1758)	±0–1000	LC	LC	II	WS
Corallus hortulanus (Linnaeus, 1758)	±0–500	LC	LC	II	WS
Corallus ruschenbergerii (Cope, 1876)	±0–500	LC	LC	II	WS

(continued)

TAXA	Elevation range (m)	Category of extinction risk		CITES (2017)	Distribution pattern
		National LRFV 2015	Global (UICN 2017)		
Epicrates cenchria (Linnaeus, 1758)	±0–900	LC	NE	II	WS
Epicrates maurus (Gray, 1849)	±0–1500	LC	NE	II	WS
Eunectes murinus Linnaeus, 1758	±0–850	LC	NE	II	WS
Family Colubridae					
***Atractus duidensis* Roze, 1961**	2050–2150	LC	LC	–	P
Atractus elaps (Gunther, 1858)	±50–200	LC	NE	–	WS
***Atractus insipidus* Roze, 1961**	952	LC	NE	–	Guw
Atractus riveroi Roze, 1961	944–1800	LC	NE	–	Guw, P
***Atractus steyermarki* Roze, 1958**	1430–2160	LC	LC	–	P
Atractus tamessari Kok, 2006	±500–2200	LC	LC	–	G
Atractus torquatus (Duméril, Bibron *et* Duméril, 1854)	0–600	LC	NE	–	WS
Atractus trilineatus Wagler, 1828	±0–200	LC	NE	–	GU
Chironius carinatus (Linnaeus, 1758)	±0–750	LC	NE	–	WS
Chironius challenger Kok, 2010	1400–2088	LC	LC	–	GU
Chironius exoletus (Linnaeus, 1758)	0–1553	LC	NE	–	WS
Chironius fuscus (Linnaeus, 1758)	±50–1520	LC	NE	–	WS
Chironius multiventris Schmidt *et* Walker, 1943	±10–1000	LC	NE	–	WS
Chironius scurrulus (Wagler, 1824)	±50–1000	LC	NE	–	WS
Chironius spixi (Hallowell, 1845)	±0–550	LC	NE	–	WS
Clelia clelia (Daudin 1803)	±15–1500	LC	NE	II	WS
Dendrophidion dendrophis (Schlegel, 1837)	±150–1000	LC	NE	–	WS
Dipsas catesbyi (Sentzen, 1796)	±150–900	LC	LC	–	WS

(continued)

TAXA	Elevation range (m)	Category of extinction risk		CITES (2017)	Distribution pattern
		National LRFV 2015	Global (UICN 2017)		
Dipsas copei (Gunther, 1872)	±700–1180	LC	LC	–	GU
Dipsas indica Laurenti, 1768	80–1000	LC	NE	–	WS
Dipsas pavonina Schlegel, 1837	±30–100	LC	LC	–	WS
Dipsas variegata (Duméril, Bibron *et* Duméril, 1854)	300–1300	LC	NE	–	WS
Drymarchon corais (Boie, 1827)	±30–250	LC	NE	–	WS
Drymobius rhombifer (Gunther, 1860)	±200–1400	LC	NE	–	WS
Drymoluber dichrous (Peteres, 1863)	±100–800	LC	LC	–	WS
Erythrolamprus aesculapii (Linnaeus, 1766)	±10–1200	LC	NE	–	WS
Erythrolamprus breviceps (Cope, 1861 "1860")	±15–1950	LC	NE	–	WS
Erythrolamprus cobellus (Linnaeus, 1758)	±0–2100	LC	NE	–	WS
***Erythrolamprus ingeri* (Roze, 1958)**	1900–2100	LC	LC	–	P
Erythrolamprus melanotus (Shaw, 1802)	0–500	LC	LC	–	WS
Erythrolamprus miliaris (Linnaeus, 1758)	±150	LC	NE	–	WS
Erythrolamprus oligolepis (Wied, 1825)	±0–500	NE	NE	–	WS
Erythrolamprus poecilogyrus (Wied-Neuwied, 1825)	±50–300	LC	NE	–	WS
Erythrolamprus pygmaeus (Cope, 1868)	±10–500	NE	NE	–	WS
Erythrolamprus reginae (Linnaeus, 1758)	±0–1400	LC	NE	–	WS
***Erythrolamprus torrenicola* (Donnelly *et* Myers, 1991)**	300–1180	LC	LC	–	GUc
***Erythrolamprus trebbaui* (Roze, 1958)**	1000–1750	LC	NE	–	GUc, P
Erythrolamprus typhlus (Linnaeus, 1758)	±20–600	LC	NE	–	WS
Helicops angulatus (Linnaeus, 1758)	±0–500	LC	NE	–	WS

(continued)

TAXA	Elevation range (m)	Category of extinction risk		CITES (2017)	Distribution pattern
		National LRFV 2015	Global (UICN 2017)		
Helicops hagmanni Roux, 1910	±0–500	LC	NE	–	WS
Helicops hogei Lancini, 1979	±0–300	LC	NE	–	GU
Hydrodynastes bicinctus (Herrmann, 1804)	±0–500	LC	NE	–	WS
Hydrops martii (Wagler, 1824)	±0–300	LC	LC	–	GU, AM
Hydrops triangularis (Wagler, 1824)	±0–300	LC	NE	–	WS
Imantodes cenchoa (Linnaeus, 1758)	±230–1340	LC	NE	–	WS
Imantodes lentiferus (Cope, 1894)	115–1030	LC	NE	–	WS
Leptodeira annulata (Linnaeus, 1758)	0–2150	LC	NE	–	WS
Leptophis ahaetulla (Linnaeus, 1758)	±0–1800	LC	NE	–	WS
Leptophis coeruleodorsus Oliver, 1942	±0–300	NE	NE	–	WS
Leptophis cupreus (Cope, 1868)	±100–1850	LC	LC	–	GU, AM
Lygophis lineatus (Linnaeus, 1758)	±0–300	LC	NE	–	WS
Masticophis mentovarius (Duméril, Bibron *et* Duméril, 1854)	0–250	LC	LC	–	WS
Mastigodryas bifossatus (Raddi, 1820)	200–1300	LC	NE	–	WS
Mastigodryas boddaerti (Sentzen, 1796)	±500–2300	LC	NE	–	WS
Mastigodryas pleei (Duméril, Bibron *et* Duméril, 1854)	±0–300	LC	NE	–	WS
Oxybelis aeneus (Wagler, 1824)	0–2500	LC	NE	–	WS
Oxybelis fulgidus (Daudin, 1803)	±0–400	LC	NE	–	WS
Oxyrhopus occipitalis (Wagler *in* Spix, 1824)	±50–140	LC	LC	–	GU, AM
Oxyrhopus melanogenys (Tschudi, 1854)	±50–1100	LC	LC	–	WS
Oxyrhopus petolarius (Linnaeus, 1758)	±0–1700	LC	NE	–	WS

(continued)

TAXA	Elevation range (m)	Category of extinction risk		CITES (2017)	Distribution pattern
		National LRFV 2015	Global (UICN 2017)		
Philodryas argentea (Daudin, 1803)	±90–1250	LC	LC	–	WS
***Philodryas cordata* Donnelly *et* Myers, 1991**	1030–1520	LC	LC	–	GUc
Philodryas olfersii (Lichenstein, 1823)	±0–500	LC	NE	–	WS
Philodryas viridissimus (Linnaeus, 1758)	±0–1360	LC	NE	–	WS
Phimophis guianensis (Troschel *in* Schomburgk, 1848)	±0–800	LC	NE	–	WS
Phrynonax polylepis (Peters, 1867)	±0–2000	LC	NE	–	WS
Pseudoboa coronata Schneider, 1801	±0–400	LC	NE	–	WS
Pseudoboa neuwieddii (Duméril, Bibron *et* Duméril, 1854)	±0–1200	LC	NE	–	WS
Pseudoeryx plicatilis (Linnaeus, 1758)	0–300	LC	LC	–	WS
Rhinobothrium lentiginosum (Scopoli, 1785)	±0–500	LC	NE	–	WS
Sibon nebulatus (Linnaeus, 1758)	0–1000	LC	NE	–	WS
Siphlophis cervinus (Laurenti 1768)	±0–700	LC	NE	–	WS
Siphlophis compressus (Daudin 1803)	±0–1000	LC	LC	–	WS
Spilotes pullatus (Linnaeus, 1758)	0–1500	LC	NE	–	WS
Spilotes sulphureus (Wagler *in* Spix, 1824)	±0–600	LC	NE	–	WS
Tantilla melanocephala (Linnaeus, 1758)	0–3000	LC	NE	–	WS
***Thamnodynastes chimanta* Roze, 1958**	1900–2600	LC	LC	–	P
***Thamnodynastes corocorensis* Gorzula *et* Ayarzagüena, 1995**	2150	LC	LC	–	P
Thamnodynastes dixoni Bailey *et* Thomas, 2007 “2006”	±50–500	LC	LC	–	WS
***Thamnodynastes duida* Myers *et* Donnelly, 1995**	2015	LC	LC	–	P

(continued)

TAXA	Elevation range (m)	Category of extinction risk		CITES (2017)	Distribution pattern
		National LRFV 2015	Global (UICN 2017)		
***Thamnodynastes marahuaquensis* Gorzula *et* Aryarzagüena, 1995**	2500	LC	LC	–	P
Thamnodynastes pallidus (Linnaeus, 1758)	±50–1270	LC	LC	–	WS
Thamnodynastes ramonriveroi Manzanilla *et* Sánchez, 2005	±0–300	LC	NE	–	WS
***Thamnodynastes yavi* Myers *et* Donnelly, 1995**	2150	LC	LC	–	P
Xenodon merremii (Wagler, 1824)	±10–500	LC	NE	–	WS
Xenodon rhabdocephalus (Wierd, 1824)	0–500	LC	NE	–	WS
Xenodon severus (Linnaeus, 1758)	±10–900	LC	NE	–	WS
Xenopholis scalaris (Wucherer, 1861)	±0–900	NE	LC	–	WS
Family Elapidae					
Leptomicrurus collaris (Schlegel, 1837)	±100–900	LC	LC	–	GU
Micrurus hemprichii (Jan, 1858)	±110–1000	LC	NE	–	WS
Micrurus isozonus (Cope, 1860)	±0–600	LC	LC	–	WS
Micrurus lemniscatus (Linnaeus, 1758)	±0–360	LC	NE	–	WS
Micrurus nattereri (Schmidt, 1952)	50–250	LC	NE	–	GU
Micrurus psyches (Daudin, 1803)	±100–1400	LC	NE	–	GU
Micrurus remotus Roze, 1987	90–1700	LC	NE	–	GU
Micrurus obscurus (Jan, 1872)	±100–600	LC	NE	–	GU, AM
Family Leptotyphlopidae					
Epictia albifrons (Wagler, 1824)	±40–1200	LC	NE	–	WS
***Epictia hobartsmithi* Esqueda, Schlüter, Machado, Castelaín *et* Natera, 2015**	1180	NE	NE	–	GUc

(continued)

TAXA	Elevation range (m)	Category of extinction risk National LRFV 2015	Global (UICN 2017)	CITES (2017)	Distribution pattern
Epictia signata (Jan, 1861)	±150	LC	NE	–	GU
Siagonodon septemstriatus (Schneider, 1801)	±100–1000	LC	NE	–	WS
Family Typhlopidae					
Amerotyphlops brongersmianus (Vanzolini, 1972)	±0–1300	LC	NE	–	WS
Amerotyphlops minuisquamus (Dixon *et* Hendricks, 1979)	±0–250	LC	NE	–	WS
Amerotyphlops reticulatus (Linnaeus, 1758)	±0–350	LC	LC	–	WS
Family Viperidae					
Bothrops bilineatus (Wied-Neuwied, 1825)	±150–530	LC	NE	–	WS
Bothrops taeniatus Wagler *in* Spix, 1824	0–2000	LC	NE	–	WS
Bothrops atrox (Linnaeus, 1758)	±50–1584	LC	NE	–	WS
Bothrops brazili Hoge, 1953	±10–500	LC	NE	–	WS
Crotalus durissus (Linnaeus, 1758)	±0–1920	LC	LC	III	WS
Lachesis muta (Linnaeus, 1766)	±0–900	LC	NE	–	WS

Species list, geographical and elevational distribution based on references (Gorzula and Señaris 1999; Ávila-Pires 2005; McDiarmid and Donnelly 2005; Señaris et al. 2009; Rivas et al. 2012; Natera et al. 2015), recent taxa descriptions, and review of herpetological collections in national museums [Museo de Historia Natural La Salle, Caracas (MHNLS), Museo de Biología de la Universidad Central de Venezuela, Caracas (MBUCV) and Museo de la Estación Biológica Rancho Grande, Maracay (EBRG)]. Taxonomic arrangement follows Uetz et al. (2018). Distribution patterns: WS: widespread, taxa known from areas of South America outside the Guiana Shield; LL: Llanos Province; GU: all Guiana Shield; GUo: taxa know only in the Western Guayana, GUc: taxa know only in the Central Guayana; GUe: taxa know only in the Eastern Guayana, P: taxa know only in Pantepui (Guiana Highlands ≥1500 m a.s.l.); IL: Introduced locally. Endemic Venezuelan Guayana species in bold

References

Almeida AP, de Carvalho VT, Gordo M et al (2014) First record *of Adelastes hylonomos* Zweifel, 1986 (Amphibia, Anura, Microhylidae) outside the type locality, with notes on the advertisement call. Check List 10:1226–1228

Alvarez-Berríos NL, Aide TM (2015) Global demand for gold is another threat for tropical forests. Environ Res Lett 10:014006. https://doi.org/10.1088/1748-9326/10/1/014006
Andresen E, Arroyo-Rodríguez V, Escobar F (2018) Tropical biodiversity: the importance of biotic interactions for its origin, maintenance, function, and conservation. In: Dáttilo W, Rico-Gray V (eds) Ecological networks in the tropics. Springer, Cham, pp 1–13
Araujo-Vieira K (2017) Geographic distribution: *Scinax exiguus*. Herpetol Rev 47:624
Armesto O, Señaris JC (2017) Anuros del Norte de los Andes: riqueza de especies y estado de conservación. Pap Avulsos Zool 57:491–526
Aubrecht R, Barrio-Amorós CL, Breure ASH et al. (2012) Venezuelan tepuis: their caves and biota, Comenius University Bratislava
Ávila-Pires TCS (1995) Lizards of Brazilian Amazonia (Reptilia: Squamata). Zool Verhandel 299:1–706
Ávila-Pires TCS (2005) Reptiles. In: Hollowell T, Reynolds RP (eds) Checklist of the terrestrial vertebrates of the Guiana shield. Bull Biol Soc Wash, Washington, pp 25–40
Ayarzagüena J, Señaris JC, Gorzula S (1992) El Grupo *Osteocephalus rodriguezi* de las tierras altas de la Guayana venezolana: descripción de cinco nuevas especies. Mem Soc Ci Nat La Salle 137:113–142
Balaguera-Reina SA, Espinosa-Blanco AS, Morales-Betancourt MA et al (2017) Conservation status and regional habitat priorities for the Orinoco crocodile: past, present, and future. PLoS One 12(2):e0172439. https://doi.org/10.1371/journal.pone.0172439
Barrio-Amorós CL (1999) Sistemática y Biogeografía de los anfibios (Amphibia) de Venezuela. Acta Biol Venez 18:1–93
Barrio-Amorós CL, Brewer-Carías C (2008) Herpetological results of the 2002 expedition to Sarisariñama, a tepui in Venezuelan Guayana, with the description of five new species. Zootaxa 1942:1–68
Barrio-Amorós CL, Duellman WE (2009) Herpetofauna de la Sierra de Lema, Estado Bolívar, Venezuela. In: Lasso CA, Señaris JC, Flores AL et al (eds) Evaluación Rápida de la Biodiversidad de los Ecosistemas Acuáticos del Alto Río Cuyuní, Guayana Venezolana. RAP Bulletin of Biological Assessment 55. Conservation Internacional, Arligton, pp 137–155
Barrio-Amorós CL, Señaris JC, MacCulloch RD et al (2011) Distribution, vocalization and taxonomic status of *Hypsiboas roraima* and *H. angelicus* (Amphibia: Anura: Hylidae). Pap Avulsos Zool 51:21–28
Berry PE, Riina R (2005) Insights into the diversity of the Pantepui Flora and the biogeographic complexity of the Guayana Shield. Biol Skr 55:145–167
Berry P, Holst BK, Yatskievych K (1995a) Introduction. In: Steyermark J, Berry E, Holst BK (eds) Flora of the Venezuelan Guayana, Introduction, vol 1. Timber Press Inc, Oregon, pp xv–xx
Berry P, Huber O, Holst BK (1995b) Floristic analysis and phytogeography. In: Steyermark J, Berry E, Holst BK (eds) Flora of the Venezuelan Guayana, Introduction, vol Vol 1. Timber Press Inc, Oregon, pp 161–191
Bifano C (2016) Venezuelan science under siege. Nature 535:336–337
Böhm M, Collen B, Baillie JEM et al (2013) The conservation status of the world's reptiles. Biol Conserv 157:372–385. https://doi.org/10.1016/j.biocon.2012.07.015
Bovolo CI, Wagner T, Parkin G et al (2018) The Guiana Shield rainforests—overlooked guardians of South American climate. Environ Res Lett 13:074029. https://doi.org/10.1088/1748-9326/aacf60
Caballero H (2014) Desencuentros y encuentros en el Alto Orinoco: incursiones en territorio Yanomami, siglos XVIII-XIX. Ediciones Ivic, Caracas
Cadena CD, Kozak KH, Gómez JP et al (2011) Latitude, elevational climatic zonation and speciation in New World vertebrates. Proc R Soc B 279:194–201. https://doi.org/10.1098/rspb.2011.0720
Caramaschi U, de Niemeyer H (2005a) Geographic distribution: *Colostethus tamacuarensis*. Herpetol Rev 36:73

Caramaschi U, de Niemeyer H (2005b) Geographic distribution: *Stefania tamacuarina*. Herpetol Rev 36:77

Carvalho VT, MacCulloch RD, Bonora L et al (2010) New species of *Stefania* (Anura: Cryptobatrachidae) from Northern Amazonas. Brazil J Herpetol 44:229–235

Castroviejo-Fisher S, Padial JM, De la Riva I et al (2015) Phylogenetic systematics of egg-brooding frogs (Anura: Hemiphractidae) and the evolution of direct development. Zootaxa 4004:1–75

Cavada-Blanco F, Aguado-Adriani E (2018) El Orinoco y el Caribe. Exp Dermatol 2018:119–134

Chapman RM (1931) The upper zonal bird-life of Mts. Roraima and Duida. Bull Am Mus Nat Hist 63:1–135

Cole CJ, Townsend CR, Reynolds RP et al (2013) Amphibians and reptiles of Guyana, South America: illustrated keys, annotated species accounts, and a biogeographic synopsis. Proc Biol Soc Wash 125:317–578

Conservation Internacional (2003) Conservation priorities for the Guayana Shield: 2002 Consensus. Center for Applied Biodiversity Science, CI, Washington, DC

de Sá RO, Grant T, Camargo A et al (2014) Systematics of the Neotropical genus *Leptodactylus* Fitzinger, 1826 (Anura: Leptodactylidae): phylogeny, the relevance of non-molecular evidence, and species accounts. South Am J Herpetol 9(Spec. Issue 1):1–128

Descamps M, Gasc J, Lescure J et al (1978) Etude des écosystemes Guyanais. II Données biogéographiques sur la partie orientale des Guyanes. CR Soc Biogéogr 467:55–82

Donnelly MA, Myers CW (1991) Herpetological results of the 1990 Venezuelan expedition to the summit of Cerro Guaiquinima, with new tepui reptiles. Am Mus Novit 3017:1–54

Duellman WE (1997) Amphibians of La Escalera region, southeastern Venezuela: taxonomy, ecology, and biogeography. Sci Pap Nat Hist Mus Univ Kans 2:1–52

Duellman WE (1999) Distribution patterns of amphibians in South America. In: Duellman WE (ed) Patterns of distribution of amphibians. The Johns Hopkins University Press, Baltimore and London, pp 255–327

Duellman WE, Hoogmoed MS (1984) The taxonomy and phylogenetic relationships of the hylid frog genus *Stefania*. Misc Publ Univ Kans Mus Nat Hist 75:1–39

Duellman WE, Marion AB, Hedges SB (2016) Phylogenetics, classification, and biogeography of the treefrogs (Amphibia: Anura: Arboranae). Zootaxa 4104:1–109

Esqueda LF, Schlüter A, Machado C et al. (2015) Una nueva especie de cieguita o serpiente de gusano (Serpentes: Leptotyphlopidae: *Epictia*) nativa del tepui Guaiquinima, Provincia Pantepui en el Escudo de Guayana, Venezuela. In: Natera M, Esqueda LF, Castelaín M (eds) Atlas Serpientes de Venezuela: Appendix II. Santiago de Chile, pp 414–433

Faivovich J, McDiarmid RW, Myers CW (2013) Two new species of *Myersiohyla* (Anura: Hylidae) from Cerro de la Neblina, Venezuela, with comments on other species of the genus. Am Mus Novit 3792:1–63

Foden WB, Butchart SHM, Stuart SN et al (2013) Identifying the World's Most climate change vulnerable species: a systematic trait-based assessment of all birds, amphibians and corals. PLoS One 8:e65427. https://doi.org/10.1371/journal.pone.0065427

Fouquet A, Souza SM, Nunes PMS et al (2015) Two new endangered species of *Anomaloglossus* (Anura: Aromobatidae) from Roraima State, northern Brazil. Zootaxa 3926:191–210

Fouquet A, Vacher JP, Courtois EA et al (2018) On the brink of extinction: two new species of *Anomaloglossus* from French Guiana and amended definitions of *Anomaloglossus degranvillei* and *A. surinamensis* (Anura: Aromobatidae). Zootaxa 4379:1–23

Fraga R, Almeida AP, Moraes LJCL et al (2017) Narrow endemism or insufficient sampling? Geographical range extension and morphological variation of the poorly known *Atractus riveroi* Roze, 1961 (Serpentes: Dipsadidae). Herpetol Rev 48:281–284

Frost DR (2018) Amphibian Species of the World: an Online Reference. Version 6.0. http://research.amnh.org/herpetology/amphibia/index.html

Gehara MCM, Crawford AJ, Orrico VGD et al (2014) High levels of diversity uncovered in a widespread nominal taxon: continental phylogeography of the Neotropical tree frog *Dendropsophus minutus*. PLoS One 9:e103958. https://doi.org/10.1371/journal.pone.0103958

Gibbs AK, Barron CN (1993) The geology of the guiana shield. Oxford monographs on geology and geophysics, vol 22. Oxford University Press, New York, NY
Godinho MB, da Silva FR (2018) The influence of riverine barriers, climate, and topography on the biogeographic regionalization of Amazonian anurans. Sci Rep 8:3427. https://doi.org/10.1038/s41598-018-21879-9
Goicoechea N, Frost D, De la Riva I et al (2016) Molecular systematics of teioid lizards (Teioidea/Gymnophthalmoidea: Squamata) based on the analysis of 48 loci under tree-alignment and similarity alignment. Cladistics 32:1–48
Gorzula S (1992) La herpetofauna del macizo del Chimantá. In: Huber O (ed) El macizo del Chimantá Escudo de Guayana, Venezuela. Un ensayo ecológico tepuyano. Oscar Todtmann Editores, Caracas. p 267-280 + 304-310, photos 152-171
Gorzula S, Señaris JC (1999) Contribution to the herpetofauna of the Venezuelan Guayana I. A data base. Scientia Guaianae 8:xviii+270+32
Grant T, Rada M, Anganoy-Criollo MA et al (2017) Phylogenetic systematics of dart-poison frogs and their relatives revisited (Anura: Dendrobatoidea). South Am J Herpetol 12(Special Issue):1–90
Haffer J (1974) Avian speciation in tropical South America. Publ Nuttall Ornith Club 14:390
Hammond DS (2005) Ancient land in a modern world. In: Hammond DS (ed) Tropical forests of the Guiana Shield: ancient forests of the modern world. CABI, Cambridge, MA, pp 1–14
Heinicke MP, Duellman WE, Hedges SB (2007) Major Caribbean and Central American frog faunas originated by ancient oceanic dispersal. Proc Natl Acad Sci U S A 104:10092–10097
Heinicke MP, Duellman WE, Trueb L et al (2009) A new frog family (Anura: Terrarana) from South America and an expanded direct developing clade revealed by molecular phylogeny. Zootaxa 2211:1–35
Heinicke MP, Lemmon AR, Lemmon EM et al (2018) Phylogenomic support for evolutionary relationships of New World Direct-developing frogs (Anura: Terraranae). Mol Phylogenet Evol 118:145–155
Hernández O, Ferrer-Pérez A (2015) Chipiro, *Podocnemis erythrocephala*. In: Rodríguez JP, García-Rawlins A, Rojas-Suárez F (eds) Libro Rojo de la Fauna Venezolana. Cuarta edición. Provita y Fundación Empresas Polar, Caracas
Hernández O, Marín E, Ferrer-Pérez A (2015) Tortuga arrau, *Podocnemis expansa*. In: Rodríguez JP, García-Rawlins A, Rojas-Suárez F (eds) Libro Rojo de la Fauna Venezolana. Cuarta edición. Provita y Fundación Empresas Polar, Caracas
Heyer WR (1994) *Hyla benitezi* (Amphibia: Anura: Hylidae): first record from Brazil and its biogeographical significance. J Herpetol 28:497–499
Hof C, Araújo MB, Jetz W et al (2011) Additive threats from pathogens, climate and land-use change for global amphibian diversity. Nature 480:516–519
Hollowell T, Reynolds RP (eds) (2005) Checklist of the terrestrial vertebrates of the Guiana Shield, Bulletin of the Biological Society of Washington No. 13. Patuxent Wildlife Research Center, Washington, pp 1–98
Hoogmoed MS (1979) The Herpetofauna of the Guianan Region. In: Duellman WE (ed) The South American Herpetofauna: its origin, evolution and dispersal. Monograph of the Museum of Natural History. University of Kansas, Kansas, pp 241–279
Howard SD, Bickford DP (2014) Amphibians over the edge: silent extinction risk of data deficient species. Diversity Distrib 20:837–846
Huber O (1995) Geographical and physical features. In: Steyermark JA, Berry PE, Holst BK (eds) Flora of the Venezuelan Guayana, Introduction, vol Vol 1. Missouri Botanical Garden and Timber Press, Portland, pp 1–61
Huber O, García P (2011) The Venezuelan Guayana Region and the study areas: geo-ecological characteristics. In: Zinck JA, Huber O (eds) Peatlands of the Western Guayana highlands, Venezuela. Springer, Berlin Heidelberg, pp 29–89
Hutter CR, Guayasamin JM, Wiens JJ (2013) Explaining Andean megadiversity: the evolutionary and ecological causes of glassfrog elevational richness patterns. Ecol Lett 16:1135–1144

Hutter CR, Lambert SM, Wiens JJ (2017) Rapid diversification and time explain amphibian richness at different scales in the tropical Andes, Earth's Most biodiverse hotspot. Am Nat 190:828–843

IUCN (1994) Guidelines for protected areas management categories. IUCN, Gland

IUCN (2018) The IUCN Red List of Threatened Species. Version 2018-1. http://www.iucnredlist.org

Jablonski D, Grul'a D, Barrio-Amorós CL et al (2017) Molecular phylogenetic relationships among *Pristimantis* summit populations in the eastern tepui chain: insights from *P. aureoventris* (Anura: Craugastoridae). Salamandra 53:473–478

Jetz W, Pyron RA (2018) The interplay of past diversification and evolutionary isolation with present imperilment across the amphibian tree of life. Nat Ecol Evol 2:850–858. https://doi.org/10.1038/s41559-018-0515-5

Jungfer KH, Faivovich J, Padial JM et al (2013) Systematics of spiny-backed treefrogs (Hylidae: *Osteocephalus*): an Amazonian puzzle. Zool Scr 42:351–380

Kok PJR (2013) Islands in the sky: species diversity, evolutionary history, and patterns of endemism of the Pantepui Herpetofauna. PhD Thesis, Leiden University, The Netherlands

Kok PJR (2015) A new species of the Pantepui endemic genus *Riolama* (Squamata: Gymnophthalmidae) from the summit of Murisipán-tepui, with the erection of a new gymnophthalmid subfamily. Zool J Linnean Soc 174:500–518

Kok PJR, Rivas GA (2011) A new species of *Anadia* (Reptilia, Squamata) from the Venezuelan 'lost world', northern South America. Eur J Taxon 3:1–18

Kok PJR, MacCulloch RD, Means DB et al (2012) Low genetic diversity in tepui summit vertebrates. Curr Biol 22:R589–R590. https://doi.org/10.1016/j.cub.2012.06.034

Kok PJR, Ratz S, Tegelaar M et al (2015) Out of taxonomic limbo: a name for the species of *Tepuihyla* (Anura: Hylidae) from the Chimantá Massif, Pantepui region, northern South America. Salamandra 51:283–314

Kok PJR, Russo VG, Ratz S et al (2016) On the distribution and conservation of two "lost world" tepui summit endemic frogs, *Stefania ginesi* Rivero, 1968 and *S. satelles* Señaris, Ayarzagüena and Gorzula, 1997. Amphib Reptile Conserv 10:5–12

Kok PJR, Russo VG, Ratz S et al (2017) Evolution in the South American 'lost world': insights from multilocus phylogeography of *Stefanias* (Anura, Hemiphractidae, *Stefania*). J Biogeogr 44:170–181

Kok PJR, Bittenbinder MA, van den Berg JK et al (2018a) Integrative taxonomy of the gymnophthalmid lizard *Neusticurus rudis* Boulenger, 1900 identifies a new species in the eastern Pantepui region, north-eastern South America. J Nat Hist 52:1029–1066. https://doi.org/10.1080/00222933.2018.1439541

Kok PJR, Ratz S, MacCulloch RD et al (2018b) Historical biogeography of the palaeoendemic toad genus *Oreophrynella* (Amphibia: Bufonidae) sheds a new light on the origin of the Pantepui endemic terrestrial biota. J Biogeogr 45:26–36

Lampo M, Señaris JC (2006) Unexplained amphibian mortalities at the secluded mountains of the Venezuelan Guayana: is there any evidence of chitridiomycosis? Herpetol Rev 37:47–49

Lasso CA, Señaris JC, Flores AL et al (eds) (2009) Evaluación Rápida de la Biodiversidad de los Ecosistemas Acuáticos del Alto Río Cuyuní, Guayana Venezolana. RAP Bulletin of Biological Assessment N° 55. Conservation Internacional, Arligton

Lathrop A, MacCulloch RD (2007) A new species of *Oreophrynella* (Anura: Bufonidae) from Mount Ayanganna, Guyana. Herpetologica 63:87–93

Laurance WF, Useche C, Shoo LP et al (2011) Global warming, elevational ranges and the vulnerability of tropical biota. Biol Conserv 144:548–557

Lescure J (1976) Contribution à l'étude des Amphibiens de Guyane française. VI. Liste préliminaire des Anoures. Bull Mus Natl Hist Nat (Zoologie) 265:475–525

MacCulloch RD, Lathrop A (2002) Exceptional diversity of *Stefania* (Anura: Hylidae) on Mount Ayanganna, Guyana: three new species and new distribution records. Herpetologica 58:327–346

MacCulloch RD, Lathrop A, Khan SZ (2006) Exceptional diversity of *Stefania* (Cryptobatrachidae) II: six species from Mount Wokomung, Guyana. Phyllomedusa 5:31–41
MacCulloch R, Lathrop A, Reynolds B et al (2007) Herpetofauna of Mount Roraima, Guiana Shield Region, northeastern South America. Herpetol Rev 38:24–30
Mayr E, Phelps WH Jr (1967) The origin of the bird fauna of the South Venezuela highlands. Bull Am Mus Nat Hist 136:273–327
McCain CM (2009) Global analysis of bird elevational diversity. Glob Ecol Biogeogr 18:346–360
McCain CM (2010) Global analysis of reptile elevational diversity. Glob Ecol Biogeogr 19:541–553
McCain CM, Grytnes JA (2010) Elevational gradients in species richness. In: Encyclopedia of life sciences (ELS). John Wiley & Sons, Ltd, Chichester. https://doi.org/10.1002/9780470015902.a0022548
McDiarmid RW, Donnelly MA (2005) The Herpetofauna of the Guayana highlands: amphibians and reptiles of the lost world. In: Donnely MA, Crother BI, Guyer C et al (eds) Ecology and evolution in the tropics: a herpetological perspective. University of Chicago Press, Chicago, pp 461–560
McDiarmid RW, Gorzula S (1989) Aspects on the reproductive ecology of the tepui toads, genus *Oreophrynella* (Anura: Bufonidae). Copeia 1989:445–451
McDiarmid RW, Paolillo A (1988) Herpetological collections – Cerro de la Neblina, updated January 1988. In: Brewer-Carías C (ed) Cerro de la Neblina. Resultados de la expedición 1983-1987. Fundación para el Desarrollo de las Ciencias Físicas. Matemáticas y Naturales, Caracas, pp 667–670
Molina C, Señaris JC, Rivas G (2004) Los reptiles del Delta del Orinoco, Venezuela. Mem Fund La Salle Cien Nat 159-160:235–264
Moraes LJCL, Almeida AP, Fraga R et al (2017) Integrative overview of the herpetofauna from Serra da Mocidade, a granitic mountain range in northern Brazil. ZooKeys 715:103–159
Morais AR, Siqueira MN, Lemes PM et al (2013) Unraveling the conservation status of data deficient species. Biol Conserv 166:98–102
Myers CW, Donnelly MA (1997) A tepui herpetofauna on a granitic mountain (Tamacuri) in the borderland between Venezuela and Brazil: report from the Phipps Tapirapecó expedition. Am Mus Novit 3213:1–71
Myers CW, Donnelly MA (2001) Herpetofauna of the Yutajé-Corocoro Massif, Venezuela: second report from the Robert G. Goelet American Museum-Terramar expedition to the northwestern tepuis. Bull Am Mus Nat Hist 261:1–85
Myers CW, Donnelly MA (2008) The summit herpetofauna of Auyantepui, Venezuela: report from the Robert G. Goelet American Museum-Terramar expedition. Bull Am Mus Nat Hist 308:1–147
Naka LN (2011) Avian distribution patterns in the Guiana Shield: implications for the delimitation of Amazonian areas of endemism. J Biogeogr 38:681–696
Natera MM, Esqueda LFG, Castelaín MF (2015) Atlas serpientes de Venezuela. Una visión actual de su diversidad, Santiago de Chile
Navas CA (2003) Herpetological diversity along Andean elevational gradients: links with physiological ecology and evolutionary physiology. Comp Biochem Physiol 133:469–485
Naveda J, Yerena E (2010) Sistema de Parques Nacionales de Venezuela: una evaluación aproximada de su situación actual. In: Machado-Allison A (ed) Simposio: Investigación y Manejo de Fauna Silvestre en Venezuela en Homenaje al Dr. Juhani Ojasti. Academia de Ciencias Físicas, Matemáticas y Naturales, Caracas, pp 225–246
Nogué S, Rull V, Vegas-Vilarrúbia T (2009) Modeling biodiversity loss by global warming on Pantepui, northern South America. Clim Chang 99:77–85
Padial JM, Grant T, Frost DR (2014) Molecular systematics of terraranas (Anura: Brachycephaloidea) with an assessment of the effects of alignment and optimality criteria. Zootaxa 3825:1–132

Passos P, Kok PJR, de Albuquerque NR et al (2013) Groundsnakes of the lost world: a review of *Atractus* (Serpentes: Dipsadidae) from the Pantepui region, northern South America. Herpetol Monogr 27:52–86
Péfaur J, Rivero J (2000) Distribution, species-richness, endemism, and conservation of Venezuelan amphibians and reptiles. Amphib Reptile Conserv 2:42–70
Pérez-Hernández R, Lew D (2001) Las clasificaciones e hipótesis biogeográficas para la Guayana Venezolana. Interciencia 26:373–382
Pounds A, Bustamante MR, Coloma LA et al (2006) Widespread amphibian extinctions from epidemic disease driven by global warming. Nature 439:161–167
Quintero I, Jetz W (2018) Global elevational diversity and diversification of birds. Nature 555:246–250. https://doi.org/10.1038/nature25794
RAISG (2012) Amazonian network of georeferenced socio-environmental information: Amazonia under pressure. (www.raisg.socioambiental.org)
RAISG (2015) Deforestación en la Amazonía (1970–2013). Lima, Perú. www.raisg.socioambiental.org
Requena J, Caputo C (2016) Pérdida de talento en Venezuela: migración de sus investigadores. Interciencia 41:444–453
Rivas GA, Schargel WE (2008) Gecko on the rocks: an enigmatic new species of *Gonatodes* (Sphaerodactylidae) from Inselbergs of the Venezuelan Guayana. Zootaxa 1925:39–50
Rivas G, Schargel W (2017a) *Gonatodes astralis*. The IUCN Red List of Threatened Species 2017: e.T66083022A66083024, doi IUCN.UK.2017-2.RLTS.T66083022A66083024.en
Rivas G, Schargel W (2017b) *Gonatodes infernalis*. The IUCN Red List of Threatened Species 2017: e.T66083295A66083297, doi IUCN.UK.2017-2.RLTS.T66083295A66083297.en
Rivas GA, Molina CR, Ugueto GN et al (2012) Reptiles of Venezuela: an updated and commented checklist. Zootaxa 3211:1–64
Rödder D, Schlüter A, Lötters S (2010) Is the 'lost world' lost? High endemism of amphibians and reptiles on South American tepuis in a changing climate. In: Habel JC, Assmann T (eds) Relict species: phylogeography and conservation biology. Springer, Berlin, pp 401–416
Rodríguez I, Aguilar-Castro V (2018) Derechos indígenas y los proyectos de desarrollo al sur del Orinoco. Exp Dermatol 2018:157–166
Rodríguez JP, García-Rawlins A, Rojas-Suárez F (2015) Libro Rojo de la Fauna Venezolana, 4ta edición edn. Provita y Fundación Empresas Polar, Caracas
Rojas H, Lew D (2009) Análisis espacial de los registros de vertebrados del Parque Nacional Canaima. In: Señaris JC, Lew D, Lasso C (eds) Biodiversidad del Parque Nacional Canaima: Bases Científicas para la Conservación de la Guayana Venezolana. Fundación La Salle de Ciencias Naturales and The Nature Conservancy, Caracas, pp 183–205
Rosales J, Suárez CF, Lasso CA (2010) Descripción del medio natural de la cuenca del Orinoco. In: Lasso CA, Usma JS, Trujillo F et al (eds) Biodiversidad de la cuenca del Orinoco: bases científicas para la identificación de áreas prioritarias para la conservación y uso sostenible de la biodiversidad. Instituto de Investigación de Recursos Biológicos Alexander von Humboldt, WWF Colombia, Fundación Omacha, Fundación La Salle e Instituto de Estudios de la Orinoquia Universidad Nacional de Colombia, Bogotá, DC, pp 51–73
Roze JA (1958a) Resultados zoológicos de la expedición de la Universidad Central de Venezuela a la región del Auyan-tepui, Venezuela, basándose en las colecciones de las expediciones de Phelps-Tate, del American Museum of Natural History, 1937-1938, y de la Universidad Central de Venezuela, 1956. Acta Biol Venez 2:243–270
Roze JA (1958b) Los reptiles del Chimantá Tepui (Estado Bolívar, Venezuela) colectados por la expedición botánica del Chicago Natural History Museum. Acta Biol Venez 2:299–314
Roze JA (1966) La taxonomía y zoogeografía de los ofidios en Venezuela. Universidad Central de Venezuela, Caracas
Rull V (2004) Biogeography of the 'lost world': a palaeoecological perspective. Earth-Sci Rev 67:125–137
Rull V (2005) Biotic diversification in the Guayana highlands: a proposal. J Biogeogr 32:921–927

Rull V (2011) Neotropical biodiversity: timing and potential drivers. Trends Ecol Evol 26:508–513

Rull V, Vegas-Vilarrúbia T (2008) Biopiracy rules hinder conservation efforts. Nature 453:26

Rull V, Vegas-Vilarrúbia T, Safont E (2016) The lost World's pristinity at risk. Divers Distrib 22:995–999

Safont E, Vegas-Vilarrúbia T, Rull V (2012) Use of environmental impact assessment (EIA) tools to set priorities and optimize strategies in biodiversity conservation. Biol Conserv 149:113–121

Safont E, Rull V, Vegas-Vilarrúbia T et al (2014) Establishing a baseline of plant diversity and endemism on a neotropical Mountain summit for future comparative studies assessing up ward migration: an approach from biogeography and nature conservation. Syst Biodivers 12:292–314

Salerno PE, Ron SR, Señaris JC et al (2012) Ancient Tepui summits harbor young rather than old lineages of endemic frogs. Evolution 66:3000–3013

Salerno PE, Señaris JC, Rojas-Runjaic FJM et al (2014) Recent evolutionary history of lost world endemics: population genetics, species delimitation, and phylogeography of sky island treefrogs. Mol Phylogenet Evol 82:314–323

Schargel WE, Rivas GA, Makowsky R et al (2010) Phylogenetic systematics of the genus *Gonatodes* (Squamata: Sphaerodactylidae) in the Guayana region, with description of a new species from Venezuela. Syst Biodivers 8:321–339

Seijas AE, Antelo R, Hernández O (2015) Caimán del Orinoco, *Crocodylus intermedius*. In: Rodríguez JP, García-Rawlins A, Rojas-Suárez F (eds) Libro Rojo de la Fauna Venezolana, 4ta edición edn. Provita y Fundación Empresas Polar, Caracas

Señaris JC (2004) Herpetofauna of the Paria Gulf and Orinoco Delta, Venezuela. In: Lasso CA, Alonso LE, Flores AL et al (eds) Evaluación rápida de la biodiversidad y aspectos sociales de los ecosistemas acuáticos del delta del río Orinoco y golfo de Paria, Venezuela. RAP Bulletin of Biological Assessment, Washington DC, pp 103–113

Señaris JC, Ávila-Pires T (2003) Anfibios y reptiles. In: Conservation priorities for the Guayana Shield: 2002 consensus. Conservation International, Washington DC, pp 11–13

Señaris JC, Ayarzagüena J (2004) Contribución al conocimiento de la Anurofauna del Delta del Orinoco, Venezuela: Diversidad, Ecología y Biogeografía. Mem Fund La Salle Cien Nat 157:129–152

Señaris JC, MacCulloch R (2005) Amphibians. In: Hollowell T, Reynolds RP (eds) Checklist of the terrestrial vertebrates of the Guiana Shield, Bull Biol Soc Wash 13. Smithsonian Institution, Washington, pp 9–23

Señaris JC, Rivas G (2008) Anfibios y reptiles de la cuenca alta del río Paragua, Estado Bolívar, Venezuela. In: Señaris JC, Lasso CA, Flores AL, Alonso LE (eds) Evaluación rápida de la Biodiversidad de los Ecosistemas Acuáticos en el alto río Paragua, Estado Bolívar, Venezuela. RAP Bulletin of Biological Assessment, Washington DC, pp 123–131

Señaris JC, Ayarzagüena J, Gorzula S (1994) Los sapos de la Familia Bufonidae (Amphibia; Anura) de las tierras altas de la Guayana Venezolana: Descripción de un nuevo género y tres especies. Publ Aso Am Doñana 3:1–37

Señaris JC, Ayarzagüena J, Gorzula S (1996) Revisión taxonómica del género *Stefania* (Anura; Hylidae) en Venezuela con la descripción de cinco nuevas especies. Publ Aso Am Doñana 7:1–56

Señaris JC, Rivas G, Molina C (2009) Anfibios y reptiles del Parque Nacional Canaima. In: Señaris JC, Lew D, Lasso CA (eds) Biodiversidad del Parque Nacional Canaima: Bases Científicas para la Conservación de la Guayana Venezolana. Fundación La Salle de Ciencias Naturales y The Nature Conservancy, Caracas, pp 70–92

Señaris JC, Lampo M, Rojas-Runjaic FJM et al (2014) Guía ilustrada de los anfibios del Parque Nacional Canaima, Venezuela. Ediciones IVIC, Caracas

Señaris JC, Rojas-Runjaic FJM, La Marca E (2015) Sapito rojo del Yapacana, *Minyobates steyermarki*. In: Rodríguez JP, García-Rawlins A, Rojas-Suárez F (eds) Libro Rojo de la Fauna Venezolana. Cuarta edición Provita y Fundación Empresas Polar, Caracas

Tapley B, Michaels CJ, Gumbs R et al (2018) The disparity between species description and conservation assessment: a case study in taxa with high rates of species discovery. Biol Conserv 220:209–214

Tate GHH (1938) Notes on the Phelps Venezuelan expedition. Geogr Rev 28:452–474

Uetz P, Stylianou A (2018) The original descriptions of reptiles and their subspecies. Zootaxa 4375:257–264

Uetz P, Freed P, Hošek J (eds) (2018) The Reptile Database. www.reptile-database.org

Vacher JP, Kok PJR, Rodrigues MT et al (2017) Cryptic diversity in Amazonian frogs: integrative taxonomy of the genus *Anomaloglossus* (Amphibia: Anura: Aromobatidae) reveals a unique case of diversification within the Guiana Shield. Mol Phylogenet Evol 112:158–173

Vegas-Vilarrúbia T, Nogué S, Rull V (2012) Global warming, habitat shifts and potential refugia for biodiversity conservation in the neotropical Guayana Highlands. Biol Conserv 152:159–168

Villareal O, Señaris JC, DoNascimiento C (2003) Contribución al conocimiento de la fauna del Wei-Assipu-tepui, Macizo del Roraima, con énfasis en la anurofauna y opiliofauna. Bol Soc Venez Espeleol 36:46–50

Vitti N (2018) Pueblos indígenas, los grandes perdedores del ZDEN-AMO. Explora 2018:141–156

Zweifel RG (1986) A new genus and species of microhylid frog from the Cerro de la Neblina Region of Venezuela and a discussion of relationships among New World microhylid genera. Am Mus Novit 2863:1–24

Chapter 23
Origin of Elevational Replacements in a Clade of Nearly Flightless Birds: Most Diversity in Tropical Mountains Accumulates via Secondary Contact Following Allopatric Speciation

Carlos Daniel Cadena and Laura N. Céspedes

Abstract Tropical mountains are biodiversity hotspots. In particular, mountains in the Neotropics exhibit remarkable beta diversity reflecting species turnover along elevational gradients. Elevational replacements of species have been known since early surveys of the tropics, but data on how such replacements arise are scarce, limiting our understanding of mechanisms underlying patterns of diversity. We employed a phylogenetic framework to evaluate hypotheses accounting for the origin of elevational replacements in the genus *Scytalopus* (Rhinocryptidae), a speciose clade of passerine birds with limited dispersal abilities occurring broadly in the Neotropical montane region. We found that species of *Scytalopus* have relatively narrow elevational ranges, that closely related species resemble each other in elevational distributions, and that most species replacing each other along elevational gradients are distantly related to each other. Although we cannot reject the hypothesis that a few elevational replacements may reflect parapatric speciation along mountain slopes, we conclude that speciation in *Scytalopus* occurs predominantly in allopatry within elevational zones, with most elevational replacements resulting from secondary contact of formerly allopatric lineages. Our study suggests that accumulation of species diversity in montane environments reflects colonization processes as opposed to *in situ* divergence even in dispersal-limited animals.

Keywords Andes · Diversification · Elevational gradient · Species turnover

---

C. D. Cadena (✉) · L. N. Céspedes
Departamento de Ciencias Biológicas, Universidad de los Andes, Bogotá, Colombia
e-mail: ccadena@uniandes.edu.co

V. Rull, A. C. Carnaval (eds.), *Neotropical Diversification: Patterns and Processes*, Fascinating Life Sciences, https://doi.org/10.1007/978-3-030-31167-4_23

1 Introduction

Species turnover along elevational gradients is a salient pattern in tropical biogeography. Ever since pioneering work by Francisco José de Caldas and Alexander von Humboldt on plant geography, naturalists have noticed that many species occur over narrow ranges of elevation and replace each other along mountain slopes (Nieto 2006; von Humboldt and Bonpland 2009). Elevational replacements of closely related species are prevalent in the tropics (Terborgh 1971; Diamond 1973; Wake and Lynch 1976), where organisms likely have narrower physiological tolerances than in temperate zones (Janzen 1967; McCain 2009). Marked changes in species assemblages with elevation (e.g. of plants, invertebrates and vertebrates; Patterson et al. 1998; Kessler 2001; Jankowski et al. 2013b; García-Robledo et al. 2016; Gill et al. 2016; Badgley et al. 2018) render tropical mountains hotspots of beta diversity (Melo et al. 2009; Fjeldså et al. 2012). Therefore, knowledge about the evolutionary and ecological mechanisms involved in the origin of elevational replacements is central to understanding major patterns in the distribution of life (Janzen 1967; Huey 1978).

Abutting species distributions along elevational gradients may reflect either (1) parapatric ecological speciation leading to divergence of a formerly widespread species into two or more daughter species with restricted ranges, or (2) secondary contact following range expansions of species originating in allopatry (Endler 1982; Hua 2016). Although one should exercise caution when making inferences about the geographic context of speciation based on current geographic distributions (Losos and Glor 2003), these alternative hypotheses are, in principle, amenable to testing by means of phylogenetic analyses: parapatric divergence predicts that species replacing each other with elevation are sister to each other, whereas secondary contact predicts they are not (Patton and Smith 1992; Moritz et al. 2000). The few studies testing these predictions on animals indicate that most speciation in the montane Neotropics occurs in allopatry and that species replacing each other along elevational gradients are not each other's closest relatives (Patton and Smith 1992; Cadena et al. 2012; Caro et al. 2013). Therefore, elevational replacements may primarily reflect secondary contact, but more studies are necessary to confirm this pattern.

As shown by some of the first large-scale surveys of the geographic and ecological distributions of species (Chapman 1917, 1926; Todd and Carriker 1922), most birds living in montane areas of the Neotropics have narrow elevational ranges (Jankowski et al. 2013a; but see Gadek et al. 2018 for exceptions). For example, median ranges of species across their geographic distributions span only ca. 1100 m in three families of Neotropical birds (Fig. 23.1; see also Graves 1988). Narrow elevational ranges of individual species are often coupled with segregation with ecologically similar species along mountain slopes. For instance, a series of landmark studies in the Peruvian Andes documented multiple cases of pairs of congeneric species of birds replacing each other with elevation, as well as scenarios where up to 4–5 congeners occur successively along mountain slopes with minimum

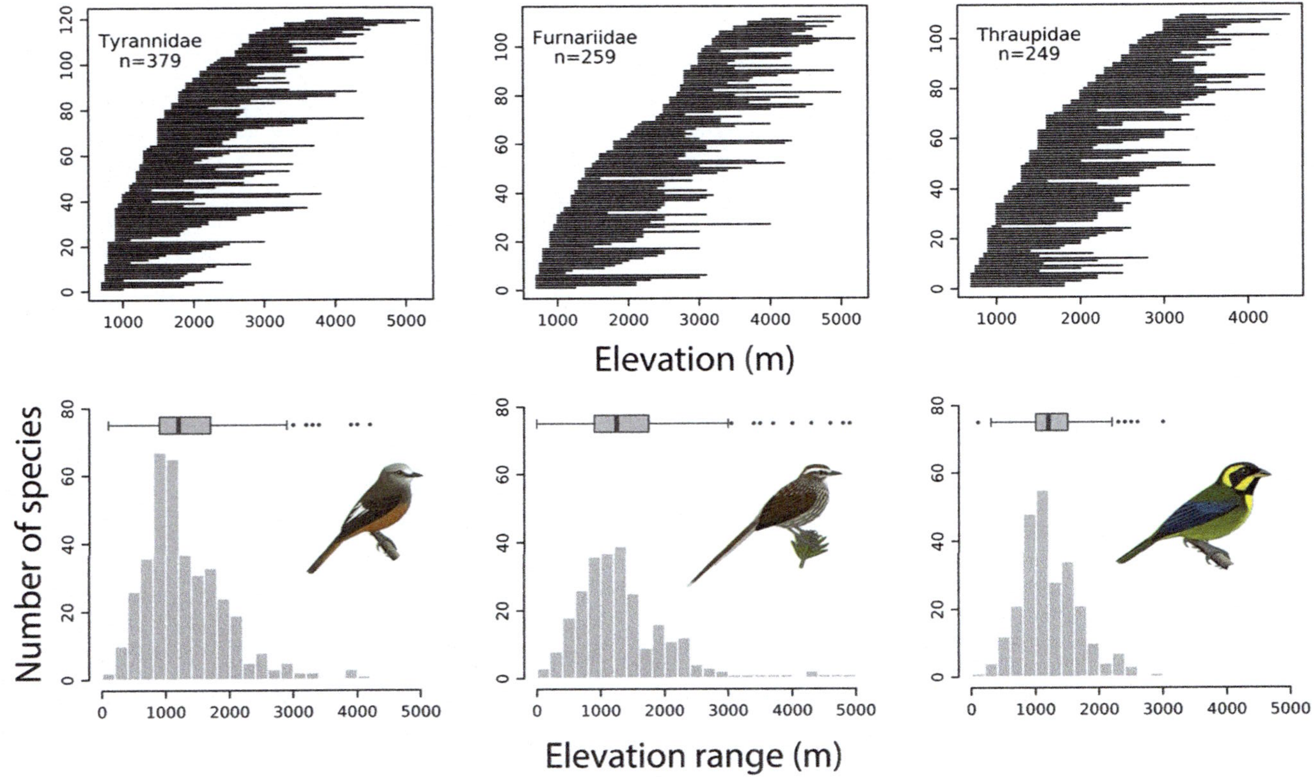

Fig. 23.1 Restricted elevational distributions of species in three diverse families of Neotropical birds: ovenbirds (Furnariidae), tyrant flycatchers (Tyrannidae), and tanagers (Thraupidae). The panels in the top row depict the range of elevations occupied by species with lower elevational limits at or above 700 m ordered

overlap (Terborgh 1971, 1977; Terborgh and Weske 1975). Understanding the ecological and physiological mechanisms that maintain patterns of elevational segregation in tropical birds has been the focus of multiple studies (reviewed by Jankowski et al. 2013a), yet analyses of the origins of elevational replacements remain scarce (Cadena 2007; Freeman 2015; Cadena et al. 2019a). In particular, we are unaware of studies attempting to test alternative hypotheses accounting for the origin of elevational replacements in a diverse clade of widespread organisms, and in the context of a robust phylogeny.

Tapaculos within the genus *Scytalopus* (Rhinocryptidae), are small passerine birds with poor dispersal abilities and ranging broadly in the Neotropical montane region. Except for the Pantepui, species in the genus occupy all major mountains in the Neotropics, from Costa Rica to Patagonia, reaching lowland areas and foothills in southern South America and eastern Brazil. *Scytalopus* are mouse-like birds which forage by walking or hopping on or near the ground in dense vegetation; they are unable to engage in long, powered flights because they have small and rounded wings and unfused clavicles (Fig. 23.2; Krabbe and Schulenberg 2003; Maurício et al. 2008). *Scytalopus* avoid highly lit open areas and are rarely found far from vegetation cover except in barren high-elevation environments. Local diversity of *Scytalopus* is typically low, yet multiple species may be found in different habitats in a given landscape, being prime examples of patterns of elevational replacement (Krabbe and Schulenberg 1997, 2003). For example, on a single morning walking trails upslope in forests along the Cerro de Montezuma on the Pacific-facing slope of the Andes of Colombia, birdwatchers may successively encounter *S. chocoensis*, *S. alvarezlopezi*, *S. vicinior*, *S. spillmanni*, and *S. latrans*; if they visit paramos at higher elevations in the region, they may also find *S. canus* (Stiles et al. 2017; see below). Elevational replacements in the genus are typically sharp. Species seldom co-occur at the same elevations, and when they do so, they often segregate by habitat. Current taxonomy recognizes 44 species of *Scytalopus*, but this is most certainly an underestimate given marked genetic structure within species, geographic variation in vocalizations, and the potential for discovery of new taxa in unexplored regions (Cadena et al. 2019b).

Elevational replacements in *Scytalopus* are somewhat paradoxical given the two main hypotheses posed to account for such patterns. First, because morphological and life-history traits make these birds poor dispersers, one would expect that long-distance dispersal or range expansions would hardly bring multiple species together in a given slope, especially because mountain regions are separated by geographic barriers associated with population isolation and diversification in other birds (Weir 2009; Hazzi et al. 2018). Alternatively, given what we know about the geographic

Fig. 23.1 (continued) from lowest to highest along the vertical axis, showing turnover of species with elevation. Panels in the bottom row summarize the data above with histograms and boxplots, indicating that most species inhabit only a fraction of the elevational ranges existing on mountains like the Andes; across the three families, median elevational ranges span only ca. 1100 m and very few species have ranges greater than 2500 m. Data are from Parker et al. (1996) and illustrations from Ayerbe-Quiñones (2018)

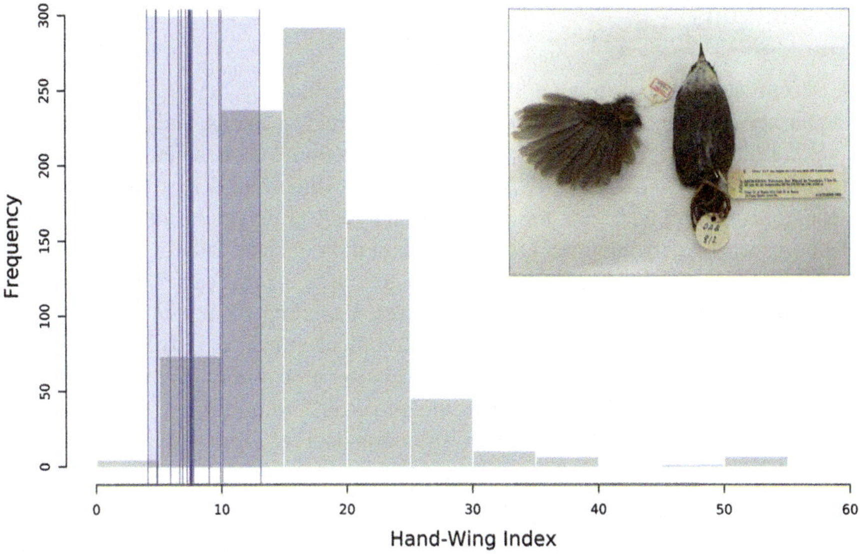

Fig. 23.2 The frequency distribution of the hand-wing index, a proxy for dispersal abilities in birds, shows that *Scytalopus* tapaculos exhibit reduced potential for flighted dispersal relative to other birds. The dimensionless hand-wing index describes wing size and shape as a function of wing length (measured from the carpal joint to the longest primary feather) and secondary length (measured from the carpal joint to the tip of the first secondary feather), with larger values indicating greater dispersal abilities (Claramunt et al. 2012). The histogram in grey depicts the distribution of the hand-wing index across a sample of 851 species of New World passerine birds (measurements from Claramunt et al. 2012; P. Montoya, G. Bravo, and E. Tenorio, unpubl. data). Data for *Scytalopus* are in purple, showing the median (vertical thick bar) and the range (shaded area) of the hand-wing index across 12 species (thin bars are mean values per species). The inset illustrates the small and rounded wing of a specimen of *S. superciliaris* from Argentina housed at the Burke Museum of Natural History and Culture at the University of Washington (photograph by Cooper French)

context of speciation in birds in general (Phillimore et al. 2008; Price 2008) and in the Andes in particular (García-Moreno and Fjeldså 2000; Caro et al. 2013), parapatric divergence of multiple species along elevational gradients in several geographic regions would also appear unlikely. A preliminary analysis of diversification in *Scytalopus* focusing on few taxa from eastern Ecuador and Peru suggested secondary contact is a more likely explanation for elevational replacements (Arctander and Fjeldså 1994; Roy et al. 1997). However, revisiting these questions is important because the phylogeny employed for analyses may have been problematic (Arctander 1995), taxonomic and geographic coverage were limited, and our understanding of geographic ranges, species limits and diversity in *Scytalopus* has improved considerably (Krabbe and Schulenberg 1997; Cadena et al. 2019b). Here, we use a comprehensive molecular phylogeny and data on geographic and elevational distributions of species to describe the mode of speciation and thereby examine the origin of elevational replacements in *Scytalopus*.

2 Methods

A recent study reconstructed molecular phylogenies of *Scytalopus* employing sequences of mitochondrial DNA (mtDNA), 80 nuclear exons, and >1800 regions flanking ultraconserved elements in the nuclear genome (Cadena et al. 2019b). We used the mtDNA ND2 data set from that study, which had denser taxonomic and geographic sampling, as the basis for our analyses; phylogenies inferred with this data set were largely congruent with those based on nuclear data for nodes relevant to analyses we present here. We inferred a gene tree using Beast 2.4.8 (Bouckaert et al. 2014) for an alignment of 90 sequences, including 83 *Scytalopus* and seven outgroups. We applied a relaxed uncorrelated clock (mean = 0.0125, SD = 0.15; Smith and Klicka 2010) and a Yule speciation tree prior. We ran chains for 100 million generations and discarded the initial 50% as burn-in.

We were able to gather elevational distribution data for 57 taxa of *Scytalopus* (data available at https://github.com/cdanielcadena/Scytalopus); these represent all named species as well as several distinct populations, which may be undescribed species given genetic divergence, vocal variation, or geographic distributions (see Cadena et al. 2019b). We obtained information on elevational ranges from taxonomic descriptions and other papers (Whitney 1994; Krabbe and Schulenberg 1997; Cuervo et al. 2005; Krabbe et al. 2005; Maurício 2005; Bornschein et al. 2007; Donegan and Avendaño 2008; Vasconcelos et al. 2008; Krabbe and Cadena 2010; Whitney et al. 2010; Donegan et al. 2013; Hosner et al. 2013; Maurício et al. 2014; Avendaño et al. 2015; Avendaño and Donegan 2015; Stiles et al. 2017), regional handbooks (Fjeldså and Krabbe 1990; Ridgely and Greenfield 2001; Hilty 2003; Restall et al. 2007; Schulenberg et al. 2007; Herzog et al. 2016; Ayerbe-Quiñones 2018), and the Handbook of the Birds of the World (del Hoyo et al. 2018). Additionally, fo several taxa we defined elevational distributions based on expert knowledge (A M. Cuervo, N. K. Krabbe, D. F. Lane, T. S. Schulenberg, and V. Piacentini, unpubl. data); this was particularly useful for unnamed populations differing phenotypically or genetically. Our compiled data base is available online at https://github.com/cdanielcadena/Scytalopus. We performed all analyses described below using all 57 taxa and, separately, we also used a data set that included only the 47 samples representing all currently recognized species (American Ornithologists' Union 1998; Stiles et al. 2017; Remsen Jr. et al. 2018), plus three unnamed species from the Andes of Peru (an unnamed form referred to *S. altirostris* mentioned by Cadena et al. (2019b) was not considered). To conduct the analyses, we trimmed the phylogeny constructed with the complete data set (90 terminals) to include only the 57 or 47 taxa considered. The elevation ranges of those species represented once in the 47-taxon data set, yet represented by more than one taxon in the 57-taxon data set, correspond to the combined ranges of these taxa (e.g. the range of *S. rodriguezi* in the 47-taxon data set is 1700–2300 m, i.e. the combined range of forms *rodriguezi and yariguiorum*).

To describe changes in elevational distributions over the history of *Scytalopus,* we mapped the midpoint of the elevation range of each taxon in the phylogeny using the *contMap* function of *phytools*, which estimates character states at nodes and along branches using a maximum-likelihood approach (Revell 2012). Additionally, we calculated Pagel's λ (Pagel 1999) as a measure of phylogenetic signal of midpoint elevation and accounted for phylogenetic uncertainty by calculating this statistic across the final 1001 trees in the posterior distribution using *phytools*. Using likelihood-ratio tests, we also tested whether Pagel's λ in each tree was significantly different from 0 (i.e. no phylogenetic signal) and 1 (i.e. the value expected under pure Brownian motion). Phylogenetic signal is often interpreted in terms of the degree of conservatism or lability of a trait (e.g. Blomberg et al. 2003), but inferences about factors underlying patterns should be cautious because different evolutionary processes may produce similar values of phylogenetic signal (Revell et al. 2008). Nonetheless, high phylogenetic signal can be interpreted as a strong tendency of closely related species to resemble each other in a given trait (Revell et al. 2008). Therefore, high phylogenetic signal in elevational ranges could indicate close resemblance between close relatives in such ranges. In addition to examining the midpoint of elevational ranges of species, we also conducted analyses based on minimum and maximum elevation, obtaining qualitatively similar results. We comment on the assumptions involved in treating elevational ranges as if they were phenotypic attributes amenable to evolution in the discussion.

To evaluate whether sister taxa have similar or contrasting elevational distributions as predicted by allopatric and parapatric speciation, respectively (Patton and Smith 1992), we identified sister taxa across the 1001 final trees in the posterior distribution. We then calculated the elevational overlap of pairs of sister taxa by dividing the amount of overlap by the elevational range of the taxon with narrower range (Kozak and Wiens 2007; Cadena et al. 2012). A value of 1 indicates that either ranges are exactly the same or that the narrower range is entirely contained in the wider range; a value of 0 indicates that elevational distributions do not overlap. The number of pairs of sister taxa employed for analyses varied across trees in the posterior from 18 to 23 (median = 20.3, total across trees = 51 pairs) in the 57-tip data set, and from 12 to 18 (median = 15.1, total across trees = 52 pairs) in the 47-tip data set. These analyses were restricted to sister taxa representing terminal branches in the phylogeny (i.e., we did not employ ancestral state reconstructions to compare elevational ranges in cases when a species representing a long terminal branch was sister to a clade formed by $\geq$2 species).

Finally, we graphically examined the phylogenetic affinities of species of *Scytalopus* replacing each other along elevational gradients in four regions of South America to examine whether such replacements more likely reflect secondary contact or parapatric speciation along mountain slopes. These regions were: (A) the Sierra Nevada de Santa Marta in northern Colombia, (B) the Pacific slope of the Western Cordillera of Colombia, (C) Zamora-Chinchipe Province on the Amazonian slope of the Andes of Ecuador, and (D) the Río Satipo Valley in Junín Department, eastern Andean slope of Peru. All analyses were conducted and figures plotted in the R programming environment (R Core Team 2018).

3 Results

Scytalopus tapaculos jointly occupy a wide range of elevations in the Neotropics, from sea level up to 4600 m in the high Andes (Table 23.1, Fig. 23.3). Elevational ranges vary substantially among species, from very narrow (200 m) to quite broad (3500 m), yet most species occupy only a relatively small fraction of the elevational gradients in which they occur: mean ranges were 1079 m (SD = 625 m) in the 57-taxa data set and 1166 m (SD = 641 m) in the 47-taxa data set. Given the geographic setting where most species of *Scytalopus* occur, where mountains reach very high altitudes and habitats for birds may extend over several thousand meters (e.g., Fig. 23.1), the elevational ranges of species are generally rather narrow. Overall, taxa within main clades of *Scytalopus* have roughly similar elevational distributions. For example, most species in the Southern Andean clade (Fig. 23.3, clade B) occur exclusively at high elevations; exceptions include *S. fuscus*, found in lowlands, and *S. magellanicus*, a temperate-zone species with the widest range in the genus (0 to 3500 m). Species from the tropical Andes and Central America (Fig. 23.3, clade C) show wide variation in elevational distributions, but species within subclades in the region tend to have similar ranges. For example, species in clades D, G and I all occur at high elevations in the tropical Andes except for *S. femoralis*, *S. micropterus* and *S. caracae,* which inhabit lower elevations than their close relatives (Fig. 23.3). All taxa in clades E, F and H occur at mid elevations, with some ranging to lower elevations (Fig. 23.3). All species from Brazil (Fig. 23.3, clade A) occur at low to mid elevations, an unsurprising pattern given that mountains reach much lower elevations in that region than in the Andes.

Phylogenetic signal for the midpoint of the elevational range of species measured by Pagel's λ, which differed significantly from 0 across all trees and for both data sets ($p < 0.001$), was relatively high (mean = 0.89 for the 57-taxon data set and 0.86 for the 47-taxon data set; Fig. 23.3). However, phylogenetic signal was also significantly different from 1 ($p < 0.05$) in most trees (77% in the 57-taxon data set and 64% in the 47-taxon data set), implying that elevational ranges are more divergent than expected given evolution under pure Brownian motion. These results indicate that closely related *Scytalopus* tend to strongly resemble each other in the midpoint of their elevation ranges, but differences among species cannot be fully accounted for by time since their divergence. The minimum and maximum elevations in the ranges of species also had significant phylogenetic signal; estimates of Pagel's λ for maximum elevation were very similar to those we obtained for midpoint elevation, while those for minimum elevation were slightly lower yet qualitatively similar (data not shown).

Most sister taxa in *Scytalopus* have similar elevational distributions (Fig. 23.4). Mean overlap in elevational ranges between sister taxa was slightly higher in the 57-taxon data set (pooled mean across the 1001 trees = 0.76, SD = 0.33) than in the 47-taxon data set (mean = 0.70, SD = 0.38). Regardless of the data set employed for analyses, half or more pairs of sister taxa overlapped substantially in elevational ranges (overlap >0.8) while less than a quarter of pairs of sister taxa had distinct

Table 23.1 Elevational distributions of *Scytalopus* tapaculos considered in analyses

Taxa	Min.	Max.	Mid.	Source
Scytalopus magellanicus	0	3500	1750	del Hoyo et al. (2018)
Scytalopus altirostris	2900	3700	3300	Schulenberg et al. (2007)
Scytalopus affinis	3000	4600	3800	Schulenberg et al. (2007)
Scytalopus urubambae	3500	4200	3850	Schulenberg et al. (2007)
Scytalopus simosi	3000	4150	3575	Herzog et al. (2016)
Scytalopus zimmeri	1700	3200	2450	del Hoyo et al. (2018)
Scytalopus superciliaris	1500	3350	2425	del Hoyo et al. (2018)
Scytalopus fuscus	0	800	400	del Hoyo et al. (2018)
Scytalopus canus	3300	3800	3550	A. Cuervo
Scytalopus opacus opacus	3000	3800	3400	Ayerbe-Quiñones (2018)
Scytalopus opacus androstictus∗	3000	3650	3325	Krabbe and Cadena (2010)
Scytalopus schulenbergi	2975	3400	3187.5	Whitney (1994)
Scytalopus sp. Ampay	3450	4200	3825	N. Krabbe
Scytalopus sp. Millpo	3450	4220	3835	N. Krabbe
Scytalopus speluncae	600	2870	1735	del Hoyo et al. (2018)/V. Piacentini
Scytalopus gonzagai	660	1140	900	Maurício et al. (2014)
Scytalopus petrophilus	900	2100	1500	Whitney et al. (2010)
Scytalopus diamantinensis	850	1600	1225	Bornschein et al. (2007)
Scytalopus pachecoi	10	1500	755	Maurício (2005)
Scytalopus iraiensis	730	1850	1290	Vasconcelos et al. (2008)
Scytalopus novacapitalis	800	1000	900	del Hoyo et al. (2018)
Scytalopus parvirostris (La Paz)	2000	3250	2625	N. Krabbe
Scytalopus parvirostris (Junín)∗	2050	3350	2700	N. Krabbe
Scytalopus panamensis	1050	1500	1275	del Hoyo et al. (2018)
Scytalopus chocoensis	250	1465	857.5	Krabbe and Schulenberg (1997)
Scytalopus rodriguezi rodriguezi	2000	2300	2150	Krabbe et al. (2005)
Scytalopus rodriguezi yariguiorum∗	1700	2200	1950	Donegan et al. (2013)
Scytalopus stilesi	1420	2130	1775	Cuervo et al. (2005)
Scytalopus robbinsi	700	1250	975	Krabbe and Schulenberg (1997)
Scytalopus vicinior	1250	2000	1625	Ridgely and Greenfield (2001)
Scytalopus latebricola	2000	3660	2830	del Hoyo et al. (2018)
Scytalopus meridanus meridanus	1600	3600	2600	Restall et al. (2007)
Scytalopus meridanus fuscicauda∗	2500	3200	2850	Fjeldså and Krabbe (1990), Restall et al. (2007)
Scytalopus argentifrons	1000	3000	2000	del Hoyo et al. (2018)
Scytalopus caracae	1200	2400	1800	del Hoyo et al. (2018)
Scytalopus spillmanni	1900	3700	2800	del Hoyo et al. (2018)

(continued)

Table 23.1 (continued)

Taxa	Min.	Max.	Mid.	Source
Scytalopus parkeri	2250	3350	2800	Krabbe and Schulenberg (1997)
Scytalopus griseicollis griseicollis	2600	3900	3250	Krabbe and Schulenberg (1997)
Scytalopus griseicollis gilesi∗	2500	3200	2850	Donegan and Avendaño (2008)
Scytalopus griseicollis morenoi∗	2000	3900	2950	Avendaño and Donegan (2015)
Scytalopus alvarezlopezi	1300	1750	1525	Stiles et al. (2017)
Scytalopus perijanus	1600	3225	2412.5	Avendaño et al. (2015)
Scytalopus sanctaemartae	900	1700	1300	del Hoyo et al. (2018)
Scytalopus atratus nigricans (Tamá)∗	1150	1900	1525	Hilty (2003)
Scytalopus atratus (Loreto)	1300	1500	1400	T. Schulenberg/D. Lane
Scytalopus atratus (SanMartín)∗	1125	1800	1462.5	T. Schulenberg/D. Lane
Scytalopus atratus (Cusco & Junín)∗	1100	1900	1500	T. Schulenberg/D. Lane
Scytalopus bolivianus	1000	2300	1650	del Hoyo et al. (2018)
Scytalopus latrans	1500	4000	2750	del Hoyo et al. (2018)
Scytalopus latrans intermedius∗	2620	3200	2910	N. Krabbe
Scytalopus unicolor	2400	3400	2900	Schulenberg et al. (2007)
Scytalopus macropus	2400	3500	2950	Schulenberg et al. (2007), Fjeldså and Krabbe (1990)
Scytalopus micropterus	1250	2300	1775	del Hoyo et al. (2018)
Scytalopus femoralis	1000	2600	1800	Schulenberg et al. (2007)
Scytalopus acutirostris	2500	3700	3100	Schulenberg et al. (2007)
Scytalopus gettyae	2400	3200	2800	Hosner et al. (2013)
Scytalopus sp. Lambayeque	3000	3400	3200	D. Lane

For each taxon, we provide the minimum, maximum and midpoint of the elevation range (in m), as well as sources for these data. Taxa with asterisks were those excluded from analyses involving only the 47 species recognized (or soon to be recognized) by taxonomists (see text). Files with the compiled 47-taxa and 57-taxa data sets are available online at https://github.com/cdanielcadena/Scytalopus

elevational ranges (overlap <0.2; Fig. 23.4). Furthermore, the majority of sister taxa identified across the 1001 trees (10 pairs out of 11, or 12 out of 13 depending on the data set) showing overlap <0.2 do not occur on the same gradient, i.e. they are allopatric. These results suggest speciation in *Scytalopus* occurs predominantly within elevational zones and not in parapatry along mountain slopes. The only possible exception to this pattern is divergence between *S. acutirostris* and *S. femoralis*, which may or may not be sister species, but are close relatives replacing each other along the Amazonian slope of the Central Andes of Peru (see below). We note that our analyses included data for at most 57 taxa yet the true number of species

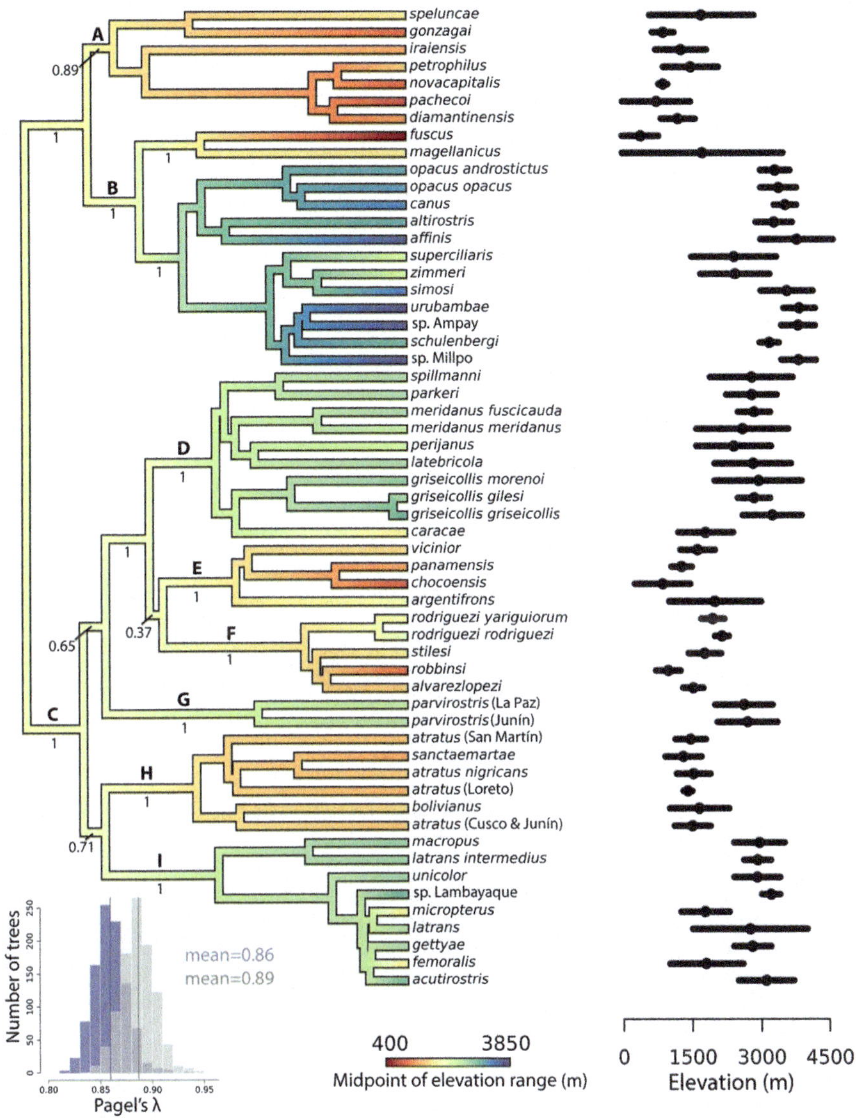

Fig. 23.3 Elevational ranges of species are broadly similar within main clades of *Scytalopus* tapaculos and there is relatively high phylogenetic signal in midpoint of the elevational range. The phylogeny is the maximum clade credibility tree with midpoint of elevation mapped on branches using maximum-likelihood; posterior probabilities are shown only for major clades. The elevation range and midpoint elevation of each taxon are shown in front of each tip. The histogram in the bottom left depicts the distribution of phylogenetic signal in midpoint elevation measured by Pagel's λ estimated across 1001 posterior trees (purple for the 47-taxa data set, gray for the 57-taxa data set)

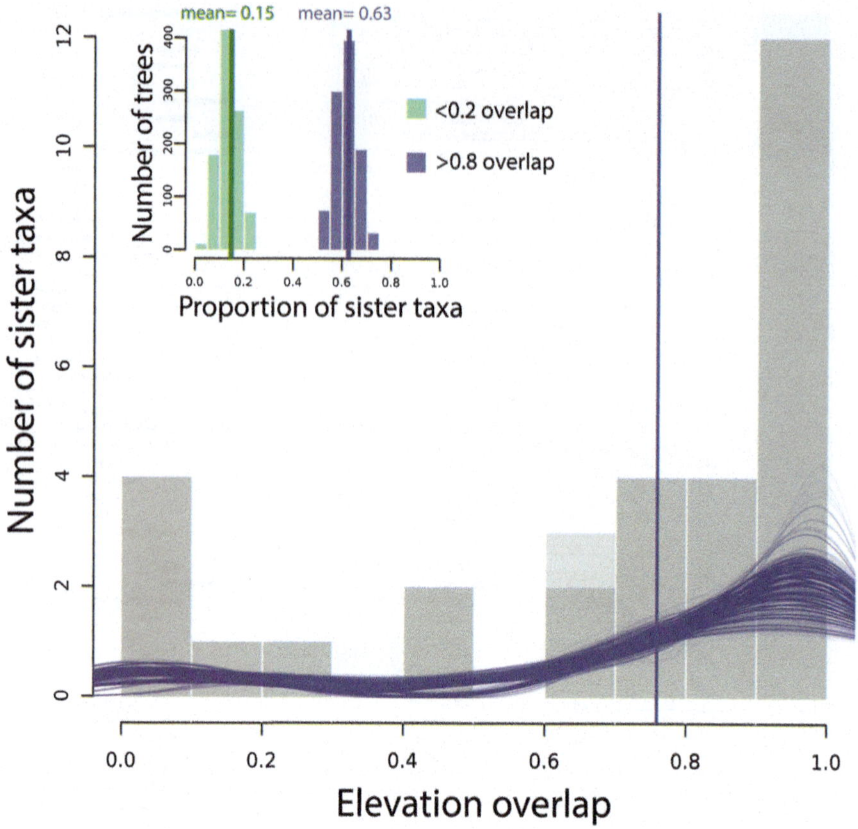

Fig. 23.4 Most sister species of *Scytalopus* have similar elevational distributions. The main figure shows an overlay of 1001 histograms, each corresponding to a phylogenetic tree in the posterior, showing the distribution of values of elevational overlap between sister taxa (0 indicates no overlap, 1 indicates complete overlap of elevational ranges); purple lines are density plots and the vertical line signals the pooled mean across trees. Histograms in the inset show the distribution of the proportion of sister taxa in the 57-taxon data set with little (<0.2) and high overlap (>0.8) in elevational distributions across the 1001 trees. We obtained qualitatively similar results with the 47-taxon data set (see text)

in *Scytalopus* is likely higher. Because most of the lineages which we did not consider are closely allied to other lineages with similar elevational ranges (e.g. groups within *S. atratus* or *S. parvirostris*; Cadena et al. 2019b) we believe that, if there is any bias in our approach, it tends to underestimate the true overlap in elevational ranges of close relatives. In other words, we predict that greater taxonomic coverage will likely reinforce our conclusion that sister taxa have similar elevational distributions.

The majority of *Scytalopus* replacing each other along the elevation gradients we examined are distant relatives (Fig. 23.5). The two species endemic to the Sierra Nevada de Santa Marta belong to distinct clades last sharing a common ancestor

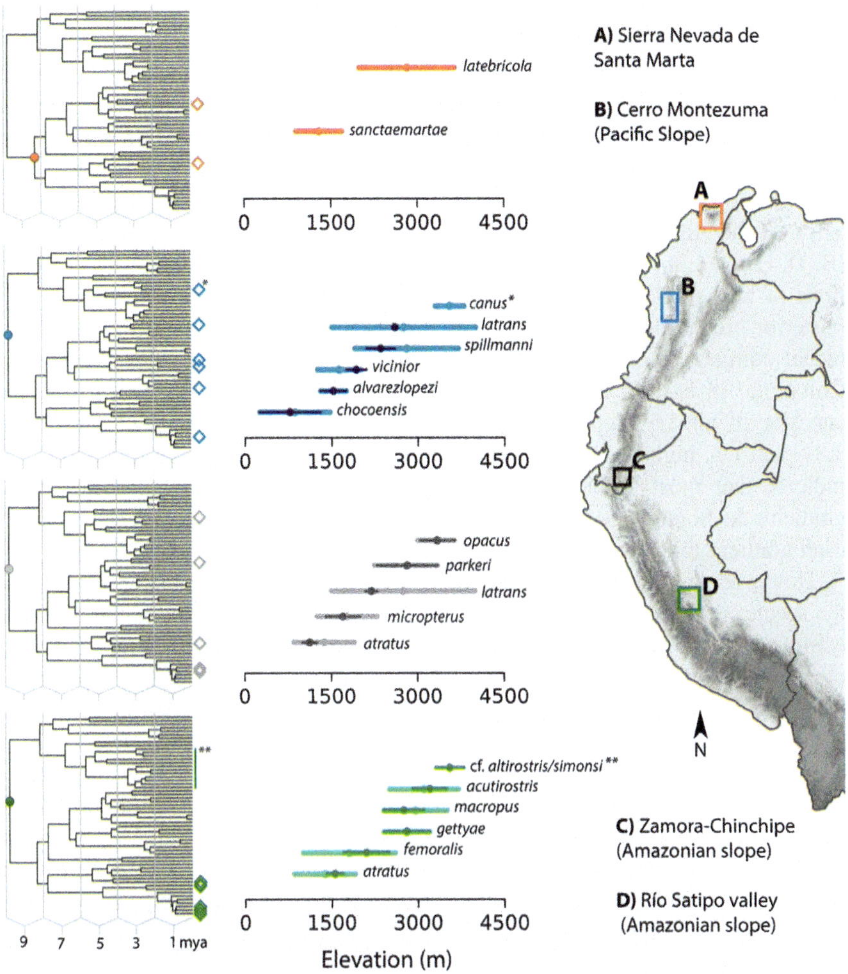

Fig. 23.5 *Scytalopus* species replacing each other with elevation in four regions of South America are often, but not always, distantly related. Rhombuses at the tips of the phylogeny highlight species that replace each other along each elevational gradient, and the dot denotes the most recent common ancestor of these species. Horizontal bars depict the elevational range of species, where the lighter color represents the elevation range for the species across its distribution and the darker color the elevational range reported for each specific gradient (**b**: Stiles et al. 2017; **c**: Krabbe and Schulenberg 1997; **d**: Hosner et al. 2013); a single bar is shown for the species from the Sierra Nevada de Santa Marta because both are endemic and for *S. canus* because the species does not occur in the specific gradient we highlight (i.e., Cerro Montezuma). The range of *S. latrans* in the Cerro de Montezuma is shown as a dot because it is only known from the very highest elevations in the area. Although the taxonomic identity of the species occurring at the highest elevations in the Satipo Valley gradient is uncertain, it very likely belongs to the clade depicted with the vertical line on the tree (Hosner et al. 2013)

ca. 8 million years ago (clades D and H in Figs. 23.2 and 23.5). Likewise, the four species occurring in forests on the Pacific slope of the Western Andes of Colombia belong to four different clades (clades D, E, F and I in Fig. 23.3), whereas a fifth species (*S. canus*) occurring in páramo habitats above treeline belongs to yet another clade (clade B). The species found above and below *S. alvarezlopezi* in the region (*S. vicinior* and *S. chocoensis*, respectively; Fig. 23.5) belong to clade E, but they are not sister to each other. The five species replacing each other along the Amazonian slope of the Andes in Zamora-Chinchipe, Ecuador, belong to four distinct clades (B, D, H and I in Figs. 23.3 and 23.5). Two of the species with parapatric distributions in this gradient (*S. latrans* and *S. micropterus*) are closely related, but the posterior probability of the hypothesis that they are sisters is only 0.43. The most recent common ancestor of species occurring in elevational gradients in western Colombia, in eastern Ecuador, and in eastern Peru is the most recent common ancestor of all *Scytalopus*, which existed ca. 9.8 million years ago (crown age 7.8–12.3 m.a. highest posterior density; Fig. 23.5). Taken together, the above data indicate that most species in the genus replacing each other along elevational gradients in South America did not evolve in parapatry *in situ*, but rather met in each gradient following divergence in allopatry.

However, the possibility of parapatric speciation remains plausible for some of the species replacing each other with elevation on the eastern slope of the Andes. In contrast to patterns observed in other regions, four of the six species replacing each other with elevation the Río Satipo Valley in Junín, Peru, belong to clade I (Figs. 23.3 and 23.5). Moreover, three of these species (*S. femoralis*, *S. gettyae* and *S. acutirostris*) belong to a group nested within clade I including several closely allied taxa with shallow divergence in mtDNA (Figs. 23.3 and 23.5). The remaining two species found in this gradient belong to clade B (an unidentified taxon; Hosner et al. 2013) and clade H (*S. atratus*). Unpublished evidence indicates a seventh species (*S.* aff. *parvirostris*) belonging to clade G also occurs in the Río Satipo Valley (N. Krabbe, pers. comm.) but we did not include it in Fig. 23.5 because information on its elevational range in the area is lacking.

4 Discussion

Many species of birds and other organisms have restricted elevational distributions, particularly in the tropics. This results in biodiversity patterns observable globally (e.g., tropical mountains are hotspots of species turnover in space because species replace each other with elevation; Fjeldså et al. 2012) and regionally (e.g., diversity may peak at mid elevations or decline with elevation in a given mountain; Quintero and Jetz 2018). We probed into evolutionary processes resulting in replacements of species along mountain slopes by examining the elevational ranges of species in the context of a phylogeny of *Scytalopus* tapaculos, a speciose clade of Neotropical montane birds in which the elevational replacement of species is commonplace. We found that (1) elevational ranges of species of *Scytalopus* are relatively narrow given the broad elevational gradients of the mountains where they live, (2) closely related species in the genus usually resemble each other in elevational distributions, (3) most pairs of sister taxa have largely overlapping elevational ranges, and (4) species coexisting regionally with elevational segregation on mountains are very often—but not always—distantly related to each other. Assuming that current distributional ranges are informative about the geographic context of speciation (Barraclough and Vogler 2000), our study thus suggests that speciation in *Scytalopus* occurs predominantly in allopatry within elevational zones, and that elevational replacements typically result from secondary contact of formerly allopatric species rather than from primary divergence in parapatry (see also Arctander and Fjeldså 1994). However, the latter possibility cannot be entirely excluded for some species and regions.

Our analyses of phylogenetic signal in elevational ranges and of overlap in elevational ranges between sister taxa treat such ranges as if they were phenotypic attributes amenable to evolution. Although elevational ranges are not organismal traits, we consider them emergent properties of populations which do reflect heritable phenotypes allowing organisms to persist over a given range of environments, justifying our study of the evolution of such ranges in a phylogenetic framework (Cadena 2007). We acknowledge, however, that because elevation *per se* is not a factor directly influencing organisms, one needs to exercise caution when comparing elevational distributions among species living in regions differing in the way in which elevation covaries with conditions and resources to which organisms directly respond (Cadena and Loiselle 2007). In this regard, we point out that some factors likely limiting organisms such as partial O_2 pressure do vary consistently with elevation regardless of geographic context, but others like temperature do not. However, because closely related species of *Scytalopus* typically occur in the same geographic region (i.e. at similar latitudes; Cadena et al. 2019b), we believe it is generally sensible to assume that similar elevational distributions reflect similar ecology and underlying phenotypic traits.

In agreement with our results, previous work on birds (García-Moreno and Fjeldså 2000; Caro et al. 2013) and other animals (Patton and Smith 1992; Lynch 1999) indicates that species replacing each other with elevation in the Neotropics are often not sister to each other. Furthermore, sister species in several vertebrate clades overlap considerably in their elevational ranges in Neotropical mountains (Cadena et al. 2012), suggesting that speciation occurs most often in allopatry within elevational zones and thus that elevational replacements result predominantly from secondary contact (but see Kozak and Wiens 2007). Work on this topic in other tropical regions has been more limited, yet evidence from Africa (Fuchs et al. 2011) and southeast Asia (Moyle et al. 2017) also indicates secondary contact is the most likely explanation for elevational replacements (but see exceptions in Bryja et al. 2018; Eldridge et al. 2018). Likewise, assembly of biotas in other mountain systems often results more from colonization by lineages from other regions than from diversification within mountains (Johansson et al. 2007; Merckx et al. 2015). Therefore, understanding the processes influencing the dynamics of geographic ranges which lead to secondary sympatry following divergence in allopatry is central to establishing how diversity accumulates in montane regions.

Those unfamiliar with *Scytalopus* tapaculos might be unsurprised by our finding that species originating in distinct mountains may come together into regional sympatry with elevational segregation in a given mountain. After all, tapaculos are birds, and birds fly around. We, however, find this result quite striking because, unlike many birds, *Scytalopus* are notably poor dispersers (Krabbe and Schulenberg 2003). Birds in the genus walk and hop much more than they fly, have tiny and rounded wings which preclude long-distance powered flight, and have even lost fused clavicles, one of the most exquisite putative adaptations of birds in general to their flighted life style. The behavior of *Scytalopus* also makes them highly reluctant to disperse: most species very rarely venture far from forest cover, having been described as agoraphobic (Krabbe and Schulenberg 1997) or photophobic (Sick 1993). Even during storms in the high Andes, tapaculos tend to stay put: rather than moving downslope to avoid inclement weather, individuals maintain their territories and forage in tunnel systems under the snow (Fjeldså 1991).

How did such undispersive birds manage to get around, colonizing an isolated mountain system like the Sierra Nevada de Santa Marta twice or the two slopes of the northern Andes multiple times? Tapaculos are not alone in achieving such feats. Phylogeographic analyses of *Henicorhina* wood-wrens (Troglodytidae), another group of poorly dispersive birds, also revealed that elevational replacements result from secondary contact of formerly allopatric lineages (Caro et al. 2013; Cadena et al. 2019a). An explanation for the apparent paradox of poor dispersers repeatedly coming into contact from disjunct areas, even in highly isolated mountains, is that individual birds did not disperse over large distances crossing barriers now appearing unsurmountable. Rather, populations likely tracked the dynamics of their favored environments, gradually expanding their geographic distributions in concert with climatic change. During cool periods in Earth history, montane environments in tropical mountains were displaced downslope, which increased opportunities for formerly isolated areas to become connected by vegetation; in turn, isolation among such areas likely increased during warmer periods when vegetation zones retreated upslope (Hooghiemstra and Van der Hammen 2004; Bush et al. 2011). Repeated cycles of disconnection and connection of montane areas (Ramírez-Barahona and Eguiarte 2013) may thus have spurred cycles of allopatric speciation and subsequent secondary contact, thereby enabling the regional accumulation of diversity (Roy et al. 1997). Owing to the expected narrow thermal tolerance of tropical montane organisms (Janzen 1967), one would expect this mechanism of divergence and accumulation of diversity in mountains to be especially prevalent in the tropics (Ghalambor et al. 2006; Kozak and Wiens 2007; Cadena et al. 2012). An alternative, nonexclusive explanation for cycles of allopatric speciation followed by establishment of secondary sympatry, is that species may go through phases of expansion and contraction of their geographic ranges even in the absence of marked changes in the physical environment (Cadena et al. 2019a). This may occur owing to evolution of phenotypic traits influencing dispersal (Hosner et al. 2017), or to changes in ecological specialization and interactions with other species (Ricklefs 2010).

Although adaptation to divergent selective pressures along gradients of elevation may seem like a prime precursor to the origin of new species (e.g. Funk et al. 2016; Hua 2016), parapatric speciation along mountain slopes appears to be rare. In contrast to data discussed above, several studies do suggest that species replacing each other along elevational gradients may be closely related (Bates and Zink 1994; Hall 2005; DuBay and Witt 2012), yet evidence that these replacements do not reflect separate colonization events of elevation belts or lowland-highland vicariance resulting from uplift processes (Brumfield and Edwards 2007; Ribas et al. 2007; Santos et al. 2009) is lacking. To our knowledge, the clearest example of parapatric speciation in mountains involves sister species in the plant genus *Senecio* occurring on Mount Etna, Italy, which differ strikingly in ecology and phenotype despite experiencing extensive gene flow (Chapman et al. 2013, 2016; Osborne et al. 2013). A promising additional case is that of *Syma* kingfishers in New Guinea, in which two distinct species co-occuring with elevational segregation have experienced gene flow yet maintain divergence in regions of the genome likely involved in adaptation and, presumably, mate choice (Linck et al. 2019). In contrast to the *Senecio* and *Syma* examples, avian taxa replacing each other with elevation in Neotropical mountains seldom show evidence of gene flow, with the only documented cases of interbreeding between elevational replacements in the region we are aware of being those of *Anairetes* tit-tyrants in Peru (Dubay and Witt 2014), *Myiarchus* flycatchers in Bolivia (Lanyon 1978), *Henicorhina* wood-wrens in Ecuador (Halfwerk et al. 2016), and *Ramphocelus* tanagers in Colombia (Sibley 1958; Morales-Rozo et al. 2017). The apparent paucity of hybridization between birds replacing each other with elevation in the Neotropics further suggests that elevational replacements did not originate through primary divergence in parapatry in the absence of barriers to gene flow, but instead via secondary contact of reproductively isolated populations. However, our inference of lack of hybridization between elevational replacements in the Andes is largely based on patterns of phenotypic variation; in most cases it remains to be seen whether genetic data reveal cryptic gene flow (Weir et al. 2015).

While we cannot reject the hypothesis that closely allied species of *Scytalopus* replacing each other along the eastern slope of the Andes colonized such regions independently, our analyses indicate that parapatric speciation along the elevational gradient may have occurred there. Several of the species replacing each other with elevation in the Satipo Valley of Peru are closely related to each other, belonging to a clade of relatively recent origin in which mtDNA divergence is shallow and rates of speciation appear faster than in the rest of the genus (Cadena et al. 2019b). Two members of this clade, *S. latrans* and *S. micropterus*, also replace each other with elevation in eastern Ecuador and Colombia. Because shallow divergence in putatively neutral loci and high rates of speciation may reflect rapid divergence mediated by adaptation in the face of gene flow, future studies should explicitly test predictions of ecological speciation (Smith et al. 2005) to determine whether elevational replacements on the eastern Andean slope may indeed be uniquely explained by parapatric divergence. The same is true for the western slope of the northern Andes, where phylogeographic patterns suggest parapatric speciation may have occurred in amphibians and reptiles (Arteaga et al. 2016; Guayasamin et al. 2017).

Beyond examining patterns of relationships among species and populations, studies of the mechanisms underlying adaptation and of how adaptive evolution in the face of gene flow may lead to speciation along elevational gradients are needed (see Hua 2016 for a theoretical perspective). In birds, for example, putatively adaptive variation with elevation has been documented in various traits influencing functions such as respiratory physiology (Scott 2011; Dawson et al. 2016; York et al. 2017), thermoregulation (Scott et al. 2008; Symonds and Tattersall 2010), foraging (Kleindorfer et al. 2006; McCormack and Smith 2008), locomotion (Altshuler et al. 2004; Milá et al. 2009), and vocal signalling (Dingle et al. 2008; Kirschel et al. 2009). Whether any of such selective pressures may account for speciation by directly or pleiotropically influencing mating patterns in *Scytalopus* in the eastern Andean slope and in other groups is essentially unknown.

Given that elevational replacements more often reflect secondary contact than parapatric divergence, a central question involving the origin of non-overlapping ranges characterizing many species assemblages from tropical mountains remains unanswered. Do contrasting elevational distributions of species originate in allopatry or upon secondary contact? Diamond (1973) reasoned that elevational parapatry reflects the outcome of competitive interactions, whereby interspecific competition between formerly isolated species favors divergence of elevational ranges when they come into contact. Alternatively, contrasting elevational ranges may arise in allopatry, with the non-overlapping ranges of species one observes reflecting sorting processes, such that only species differing in elevational ranges *a priori* may successfully attain regional sympatry with segregation along mountain slopes (Cadena 2007; McEntee et al. 2018). A recent analysis revealed that sympatric sister species of birds in the tropics have more different elevational ranges than allopatric sister species regardless of their age, which was interpreted as evidence in favor of the hypothesis that elevational divergence is driven by competition upon secondary contact (Freeman 2015). While abutting elevational ranges may indeed be maintained by competition in some cases (Cadena and Loiselle 2007; Jankowski et al. 2010; Freeman and Montgomery 2016), other biotic and abiotic forces may also mediate species turnover with elevation (Elsen et al. 2017). Moreover, because most cases of elevational replacements do not involve sister species, more work is necessary to determine the geographic context in which contrasting elevational ranges arise. Analyses incorporating species interactions into models of trait evolution (Nuismer and Harmon 2015) while jointly considering the potential for such interactions to occur given geographic distributions of species (Drury et al. 2016; Clarke et al. 2017) would be a fruitful avenue for future studies on the topic (Quintero and Landis 2019). Other mechanisms through which elevational distributions of species may change including tectonic processes of uplift or subsidence which may displace organisms vertically in passive fashion also merit consideration (Heads 1989, 2005; Ribas et al. 2007). *Scytalopus* tapaculos are well suited for additional studies on the dynamics of elevational ranges integrating ecology, evolution, and Earth history.

Acknowledgments We thank Ana Carnaval and Valentí Rull for inviting us to contribute this chapter. Andrés Cuervo, Niels Krabbe, Daniel Lane, Thomas Schulenberg, and Vitor Piacentini shared valuable information on the distributions of tapaculos. We are grateful to Ana Carnaval, Santiago Herrera, Ethan Linck, Ignacio Quintero, Glenn Seeholzer, Felipe Zapata, an anonymous reviewer, and CDC's laboratory group for helpful discussion and comments on the manuscript. Fernando Ayerbe kindly allowed us to use his illustrations of birds in our figures. Paola Montoya, Gustavo Bravo and Elkin Tenorio provided measurement data, and Cooper French and Ethan Linck provided photographs for Fig. 23.2. We received financial support from the Facultad de Ciencias at Universidad de los Andes (Programa de Investigación to CDC).

References

Altshuler DL, Dudley R, McGuire JA (2004) Resolution of a paradox: hummingbird flight at high elevation does not come without a cost. Proc Natl Acad Sci U S A 101:17731–17736

American Ornithologists' Union (1998) Checklist of North American birds, 7th edn. American Ornithologists' Union, Washington, DC

Arctander P (1995) Comparison of a mitochondrial gene and a corresponding nuclear pseudogene. Proc R Soc Lond B 262:13–19

Arctander P, Fjeldså J (1994) Andean tapaculos of the genus *Scytalopus* (Aves, Rhinocryptidae): a study of speciation using DNA sequence data. In: Loeschcke V, Tomiuk J, Jain SK (eds) Conservation genetics. Birkhauser Verlag, Basel, pp 205–225

Arteaga A, Pyron RA, Peñafiel N, Romero-Barreto P, Culebras J, Bustamante L, Yánez-Muñoz MH, Guayasamin JM (2016) Comparative phylogeography reveals cryptic diversity and repeated patterns of cladogenesis for amphibians and reptiles in northwestern Ecuador. PLoS One 11:e0151746

Avendaño JE, Donegan TM (2015) A distinctive new subspecies of *Scytalopus griseicollis* (Aves, Passeriformes, Rhinocryptidae) from the northern Eastern Cordillera of Colombia and Venezuela. ZooKeys 506:137–153

Avendaño JE, Cuervo AM, López-Ordóñez JP, Gutiérrez-Pinto N, Cortés-Diago A, Cadena CD (2015) A new species of tapaculo (Rhinocryptidae: *Scytalopus*) from the Serranía de Perijá of Colombia and Venezuela. Auk 132:450–466

Ayerbe-Quiñones F (2018) Guía ilustrada de la avifauna colombiana. Wildlife Conservation Society, Bogotá, Colombia

Badgley C, Smiley TM, Cable R (2018) Mountains, climate and mammals. In: Hoorn C, Perrigo A, Antonelli A (eds) Mountains, climate and biodiversity. Wiley, Chichester

Barraclough TG, Vogler AP (2000) Detecting the geographical pattern of speciation from species-level phylogenies. Am Nat 155:419–434

Bates JM, Zink RM (1994) Evolution into the Andes: molecular evidence for species relationships in the genus *Leptopogon*. Auk 111:507–515

Blomberg SP, Garland T, Ives AR (2003) Testing for phylogenetic signal in comparative data: behavioral traits are more labile. Evolution 57:717–745

Bornschein MR, Maurício GN, Belmonte-Lopes R, Mata H, Bonatto SL (2007) Diamantina Tapaculo, a new *Scytalopus* endemic to the Chapada Diamantina, northeastern Brazil (Passeriformes: Rhinocryptidae). Revista Brasileira de Ornitologia 15:151–174

Bouckaert R, Heled J, Kühnert D, Vaughan T, Wu C-H, Xie D, Suchard MA, Rambaut A, Drummond AJ (2014) BEAST 2: a software platform for Bayesian evolutionary analysis. PLoS Comput Biol 10:e1003537

Brumfield RT, Edwards SV (2007) Evolution into and out of the Andes: a Bayesian analysis of historical diversification in *Thamnophilus* antshrikes. Evolution 61:346–367

Bryja J, Kostin D, Meheretu Y, Šumbera R, Bryjová A, Kasso M, Mikula O, Lavrenchenko LA (2018) Reticulate Pleistocene evolution of Ethiopian rodent genus along remarkable altitudinal gradient. Mol Phylogenet Evol 118:75–87

Bush MB, Hanselman JA, Hooghiemstra H (2011) Andean montane forests and climate change. In: Bush MB, Flenley JR, Gosling WD (eds) Tropical rainforest responses to climatic change, 2nd edn. Springer, Berlin, pp 35–60

Cadena CD (2007) Testing the role of interspecific competition in the evolutionary origin of elevational zonation: an example with *Buarremon* brush-finches (Aves, Emberizidae) in the neotropical mountains. Evolution 61:1120–1136

Cadena CD, Loiselle BA (2007) Limits to elevational distributions in two species of emberizine finches: disentangling the role of interspecific competition, autoecology, and geographic variation in the environment. Ecography 30:491–504

Cadena CD, Kozak KH, Gómez JP, Parra JL, McCain CM, Bowie RCK, Carnaval AC, Moritz C, Rahbek C, Roberts TE, Sanders NJ, Schneider CJ, VanDerWal J, Zamudio KR, Graham CH (2012) Latitude, elevational climatic zonation, and speciation in New World vertebrates. Proc R Soc Lond B 279:194–201

Cadena CD, Pérez-Emán JL, Cuervo AM, Céspedes LN, Epperly KL, Klicka JT (2019a) Extreme genetic structure and dynamic range evolution in a montane passerine bird: implications for tropical diversification. Biol J Linn Soc 126:487–506

Cadena CD, Cuervo AM, Céspedes LN, Bravo GA, Krabbe N, Schulenberg TS, Derryberry GE, Silveira LF, Derryberry EP, Brumfield RT, Fjeldså J (2019b) Systematics, biogeography and diversification of *Scytalopus* tapaculos (Rhinocryptidae), an enigmatic radiation of Neotropical montane birds. bioRxiv. https://doi.org/10.1101/600775

Caro LM, Caycedo-Rosales PC, Bowie RCK, Slabbekoorn H, Cadena CD (2013) Ecological speciation along an elevational gradient in a tropical passerine bird? J Evol Biol 26:357–374

Chapman FM (1917) The distribution of bird-life in Colombia; a contribution to a biological survey of South America. Bull Am Mus Nat Hist 36:1–169

Chapman FM (1926) The distribution of bird-life in Ecuador: a contribution to a study of the origin of Andean bird-life. Bull Am Mus Nat Hist 55:1–784

Chapman MA, Hiscock SJ, Filatov DA (2013) Genomic divergence during speciation driven by adaptation to altitude. Mol Biol Evol 30:2553–2567

Chapman MA, Hiscock SJ, Filatov DA (2016) The genomic bases of morphological divergence and reproductive isolation driven by ecological speciation in *Senecio* (Asteraceae). J Evol Biol 29:98–113

Claramunt S, Derryberry EP, Remsen JV Jr, Brumfield RT (2012) High dispersal ability inhibits speciation in a continental radiation of passerine birds. Proc R Soc Lond B 279:1567–1574

Clarke M, Thomas GH, Freckleton RP (2017) Trait evolution in adaptive radiations: modeling and measuring interspecific competition on phylogenies. Am Nat 189:121–137

Cuervo AM, Cadena CD, Krabbe N, Renjifo LM (2005) *Scytalopus stilesi*, a new species of tapaculo (Rhinocryptidae) from the Cordillera Central of Colombia. Auk 122:445–463

Dawson NJ, Ivy CM, Alza L, Cheek R, York JM, Chua B, Milsom WK, McCracken KG, Scott GR (2016) Mitochondrial physiology in the skeletal and cardiac muscles is altered in torrent ducks, *Merganetta armata*, from high altitudes in the Andes. J Exp Biol 219:3719–3728

del Hoyo J, Elliott A, Sargatal J, Christie DA, de Juana E (2018) Handbook of the birds of the world alive. Lynx Edicions. Retrieved from http://www.hbw.com/ on (18 Sep 2018). Barcelona

Diamond JM (1973) Distributional ecology of New Guinea birds. Science 179:759–769

Dingle C, Halfwer W, Slabbekoorn H (2008) Habitat-dependent song divergence at subspecies level in the grey-breasted wood-wren. J Evol Biol 21:1079–1089

Donegan TM, Avendaño CJE (2008) Notes on tapaculos (Passeriformes: Rhinocryptidae) of the Eastern Andes of Colombia and the Venezuelan Andes, with a new subspecies of *Scytalopus griseicollis* from Colombia. Ornitología Colombiana 6:24–65

Donegan TM, Avendaño JE, Lambert F (2013) A new tapaculo related to *Scytalopus rodriguezi* from Serranía de los Yariguíes, Colombia. Bull Br Ornithol Club 133:256–271

Drury J, Clavel J, Manceau M, Morlon H (2016) Estimating the effect of competition on trait evolution using maximum likelihood inference. Syst Biol 65:700–710

DuBay SG, Witt CC (2012) An improved phylogeny of the Andean tit-tyrants (Aves, Tyrannidae): more characters trump sophisticated analyses. Mol Phylogenet Evol 64:285–296

Dubay SG, Witt CC (2014) Differential high-altitude adaptation and restricted gene flow across a mid-elevation hybrid zone in Andean tit-tyrant flycatchers. Mol Ecol 23:3551–3565

Eldridge RA, Achmadi AS, Giarla TC, Rowe KC, Esselstyn JA (2018) Geographic isolation and elevational gradients promote diversification in an endemic shrew on Sulawesi. Mol Phylogenet Evol 118:306–317

Elsen PR, Tingley MW, Kalyanaraman R, Ramesh K, Wilcove DS (2017) The role of competition, ecotones, and temperature in the elevational distribution of Himalayan birds. Ecology 98:337–348

Endler JA (1982) Problems in distinguishing historical from ecological factors in biogeography. Am Zool 22:441–452

Fjeldså J (1991) The activity of birds during snow-storms in high-level woodlands in Peru. Bull Br Ornithol Club 111:4–11

Fjeldså J, Krabbe N (1990) Birds of the high Andes. Zoological Museum, University of Copenhagen and Apollo Books, Svendborg, Denmark

Fjeldså J, Bowie RCK, Rahbek C (2012) The role of mountain ranges in the diversification of birds. Annu Rev Ecol Evol Syst 43:249–265

Freeman B (2015) Competitive interactions upon secondary contact drive elevational divergence in tropical birds. Am Nat 186:470–479

Freeman BG, Montgomery G (2016) Interspecific aggression by the Swainson's Thrush (*Catharus ustulatus*) may limit the distribution of the threatened Bicknell's Thrush (*Catharus bicknelli*) in the Adirondack Mountains. Condor 118:168–178

Fuchs J, Fjeldså J, Bowie RCK (2011) Diversification across an altitudinal gradient in the Tiny Greenbul (*Phyllastrephus debilis*) from the Eastern Arc Mountains of Africa. BMC Evol Biol 11:117

Funk WC, Murphy MA, Hoke KL, Muths E, Amburgey SM, Lemmon EM, Lemmon AR (2016) Elevational speciation in action? Restricted gene flow associated with adaptive divergence across an altitudinal gradient. J Evol Biol 29:241–252

Gadek CR, Newsome SD, Beckman EJ, Chavez AN, Galen SC, Bautista E, Witt CC (2018) Why are tropical mountain passes "low" for some species? Genetic and stable-isotope tests for differentiation, migration and expansion in elevational generalist songbirds. J Anim Ecol 87:741–753

García-Moreno J, Fjeldså J (2000) Chronology and mode of speciation in the Andean avifauna. In: Rheinwald G (ed) Isolated vertebrate communities in the tropics, Proceedings of the 4th international symposium, Bonn. Bonn zoological monographs, vol 46, pp 25–46

García-Robledo C, Kuprewicz EK, Staines CL, Erwin TL, Kress WJ (2016) Limited tolerance by insects to high temperatures across tropical elevational gradients and the implications of global warming for extinction. Proc Natl Acad Sci U S A 113:680–685

Ghalambor C, Huey RB, Martin PR, Tewksbury JJ, Wang G (2006) Are mountain passes higher in the tropics? Janzen's hypothesis revisited. Integr Comp Biol 46:5–17

Gill BA, Kondratieff BC, Casner KL, Encalada AC, Flecker AS, Gannon DG, Ghalambor CK, Guayasamin JM, Poff NL, Simmons MP, Thomas SA, Zamudio KR, Funk WC (2016) Cryptic species diversity reveals biogeographic support for the 'mountain passes are higher in the tropics' hypothesis. Proc R Soc B Biol Sci 283:20160553

Graves GR (1988) Linearity of geographic range and its possible effect on the population structure of Andean birds. Auk 105:47–52

Guayasamin JM, Hutter CR, Tapia EE, Culebras J, Peñafiel N, Pyron RA, Morochz C, Funk WC, Arteaga A (2017) Diversification of the rainfrog *Pristimantis ornatissimus* in the lowlands and Andean foothills of Ecuador. PLoS One 12:e0172615

Halfwerk W, Dingle C, Brinkhuizen DM, Poelstra JW, Komdeur J, Slabbekoorn H (2016) Sharp acoustic boundaries across an altitudinal avian hybrid zone despite asymmetric introgression. J Evol Biol 29:1356–1367
Hall JPW (2005) Montane speciation patterns in *Ithomiola* butterflies (Lepidoptera: Riodinidae): are they consistently moving up in the world? Proc R Soc Lond B 272:2457–2466
Hazzi NA, Moreno JS, Ortiz-Movliav C, Palacio RD (2018) Biogeographic regions and events of isolation and diversification of the endemic biota of the tropical Andes. Proc Natl Acad Sci U S A 115:7985–7990
Heads M (1989) Integrating earth and life sciences in New Zealand natural history: the parallel arcs model. New Zeal J Zool 16:549–585
Heads M (2005) Towards a panbiogeography of the seas. Biol J Linn Soc 84:675–723
Herzog SK, Terril RS, Jahn AE, Remsen JV Jr, Maillard ZO, García-Solíz VH, MacLeod R, Maccormick A, Vidoz JQ (2016) Birds of Bolivia: field guide. Asociación Armonía, Santa Cru
Hilty SL (2003) Birds of Venezuela, 2nd edn. Princeton University Press, Princeton, NJ
Hooghiemstra H, Van der Hammen T (2004) Quaternary ice-age dynamics in the Colombian Andes: developing an understanding of our legacy. Philos Trans R Soc Lond B 359:173–181
Hosner PA, Robbins MB, Valqui T, Peterson AT (2013) A new species of *Scytalopus* tapaculo (Aves: Passeriformes: Rhinocryptidae) from the Andes of Central Peru. Wilson J Ornithol 125:233–242
Hosner PA, Tobias JA, Braun EL, Kimball RT (2017) How do seemingly non-vagile clades accomplish trans-marine dispersal? Trait and dispersal evolution in the landfowl (Aves: Galliformes). Proc R Soc Lond B 284:20170210
Hua X (2016) The impact of seasonality on niche breadth, distribution range and species richness: a theoretical exploration of Janzen's hypothesis. Proc R Soc Lond B 283:20160349
Huey RB (1978) Latitudinal pattern of between-altitude faunal similarity: mountains might be higher in the tropics. Am Nat 112:225–229
Jankowski JE, Robinson SK, Levey DJ (2010) Squeezed at the top: interspecific aggression may constrain elevational ranges in tropical birds. Ecology 91:1877–1884
Jankowski JE, Londoño GA, Robinson SK, Chapell MA (2013a) Exploring the role of physiology and biotic interactions in determining elevational ranges of tropical animals. Ecography 36:1–12
Jankowski JE, Merkrod CL, Farfan Rios W, García Cabrera K, Salinas Revilla N, Silman MR (2013b) The relationship of tropical bird communities to tree species composition and vegetation structure along an Andean elevational gradient. J Biogeogr 40:950–962
Janzen DH (1967) Why mountain passes are higher in the tropics. Am Nat 101:233–249
Johansson US, Alstrom P, Olsson U, Ericson PGP, Sundberg P, Price TD (2007) Build-up of the Himalayan avifauna through immigration: a biogeographical analysis of the *Phylloscopus* and *Seicercus* warblers. Evolution 61:324–333
Kessler M (2001) Patterns of diversity and range size of selected plant groups along an elevational transect in the Bolivian Andes. Biodivers Conserv 10:1897–1921
Kirschel ANG, Blumstein DT, Cohen RE, Buermann W, Smith TB, Slabbekoorn H (2009) Birdsong tuned to the environment: green hylia song varies with elevation, tree cover, and noise. Behav Ecol 20:1089–1095
Kleindorfer S, Chapman TW, Winkler H, Sulloway FJ (2006) Adaptive divergence in contiguous populations of Darwin's Small Ground Finch (*Geospiza fuliginosa*). Evol Ecol Res 8:357–372
Kozak KH, Wiens JJ (2007) Climatic zonation drives latitudinal variation in speciation mechanisms. Proc R Soc Lond B 274:2995–3003
Krabbe N, Cadena CD (2010) A taxonomic revision of the Paramo Tapaculo *Scytalopus canus* Chapman (Aves: Rhinocryptidae), with description of a new subspecies from Ecuador and Peru. Zootaxa 2354:56–66
Krabbe N, Schulenberg TS (1997) Species limits and natural history of *Scytalopus* tapaculos (Rhinocryptidae), with descriptions of Ecuadorian taxa, including three new species. Ornithol Monogr 48:47–88

Krabbe N, Schulenberg TS (2003) Family Rhinocryptidae (Tapaculos). In: Del Hoyo J, Elliott A, Christie D (eds) Handbook of the birds of the world, *Broadbills to tapaculos*, vol 8. Lynx Edicions, Barcelona, pp 748–787

Krabbe N, Salaman P, Cortés A, Quevedo A, Ortega LA, Cadena CD (2005) A new species of *Scytalopus* tapaculo from the upper Magdalena Valley, Colombia. Bull Br Ornithol Club 125:93–108

Lanyon WE (1978) Revision of the *Myiarchus* flycatchers of South America. Bull Am Mus Nat Hist 161:427–628

Linck E, Freeman BG, Dumbacher JP (2019) Speciation with gene flow across an elevational gradient in New Guinea kingfishers. bioRxi*v*. https://doi.org/10.1101/589044

Losos JB, Glor RE (2003) Phylogenetic comparative methods and the geography of speciation. Trends Ecol Evol 18:220–227

Lynch JD (1999) Ranas pequeñas, la geometría de evolución, y la especiación en los Andes colombianos. Revista de la Academia Colombiana de Ciencias Exactas, Físicas y Naturales 23:143–159

Maurício GN (2005) Taxonomy of southern populations in the *Scytalopus speluncae* group, with description of a new species and remarks on the systematics and biogeography of the complex (Passeriformes: Rhinocryptidae). Ararajuba 13:7–28

Maurício GN, Mata H, Bornschein MR, Cadena CD, Alvarenga H, Bonatto SL (2008) Hidden generic diversity in Neotropical birds: molecular and anatomical data support a new genus for the "*Scytalopus*" *indigoticus* species-group (Aves: Rhinocryptidae). Mol Phylogenet Evol 49:125–135

Maurício GN, Belmonte-Lopes R, Pacheco JF, Silveira LF, Whitney BM, Bornschein MR (2014) Taxonomy of "Mouse-colored Tapaculos" (II): an endangered new species from the montane Atlantic Forest of southern Bahia, Brazil (Passeriformes: Rhinocryptidae: *Scytalopus*). Auk 131:643–659

McCain CM (2009) Vertebrate range sizes indicate that mountains may be 'higher' in the tropics. Ecol Lett 12:550–560

McCormack JE, Smith TB (2008) Niche expansion leads to small-scale adaptive divergence along an elevation gradient in a medium-sized passerine bird. Proc R Soc Lond B 275:2155–2164

McEntee JP, Tobias JA, Sheard C, Burleigh JG (2018) Tempo and timing of ecological trait divergence in bird speciation. Nat Ecol Evol 2:1120–1127

Melo AS, Rangel TFLVB, Diniz-Filho JAF (2009) Environmental drivers of beta-diversity patterns in New-World birds and mammals. Ecography 32:226–236

Merckx VSFT, Hendriks KP, Beentjes KK, Mennes CB, Becking LE, Peijnenburg KTCA, Afendy A, Arumugam N, de Boer H, Biun A, Buang MM, Chen P-P, Chung AYC, Dow R, Feijen FAA, Feijen H, Soest CF-v, Geml J, Geurts R, Gravendeel B, Hovenkamp P, Imbun P, Ipor I, Janssens SB, Jocqué M, Kappes H, Khoo E, Koomen P, Lens F, Majapun RJ, Morgado LN, Neupane S, Nieser N, Pereira JT, Rahman H, Sabran S, Sawang A, Schwallier RM, Shim P-S, Smit H, Sol N, Spait M, Stech M, Stokvis F, Sugau JB, Suleiman M, Sumail S, Thomas DC, van Tol J, Tuh FYY, Yahya BE, Nais J, Repin R, Lakim M, Schilthuizen M (2015) Evolution of endemism on a young tropical mountain. Nature 524:347

Milá B, Wayne RK, Fitze P, Smith TB (2009) Divergence with gene flow and fine-scale phylogeographical structure in the wedge-billed woodcreeper, *Glyphorynchus spirurus*, a Neotropical rainforest bird. Mol Ecol 18:2979–2995

Morales-Rozo A, Tenorio EA, Carling MD, Cadena CD (2017) Origin and cross-century dynamics of an avian hybrid zone. BMC Evol Biol 17:257

Moritz C, Patton JL, Schneider CJ, Smith TB (2000) Diversification of rainforest faunas: an integrated molecular approach. Annu Rev Ecol Syst 31:533–563

Moyle RG, Manthey JD, Hosner PA, Rahman M, Lakim M, Sheldon FH (2017) A genome-wide assessment of stages of elevational parapatry in Bornean passerine birds reveals no introgression: implications for processes and patterns of speciation. PeerJ 5:e3335

Nieto M (2006) La obra cartográfica de Francisco José de Caldas. Universidad de los Andes, Bogotá, Colombia

Nuismer SL, Harmon LJ (2015) Predicting rates of interspecific interaction from phylogenetic trees. Ecol Lett 18:17–27

Osborne OG, Batstone TE, Hiscock SJ, Filatov DA (2013) Rapid speciation with gene flow following the formation of Mt. Etna. Genome Biol Evol 5:1704–1715

Pagel M (1999) Inferring the historical patterns of biological evolution. Nature 401:877–884

Parker TAI, Stotz DF, Fitzpatrick JW (1996) Ecological and distributional databases. In: Stotz DF, Fitzpatrick JW, Parker TAI, Moskovits DK (eds) Neotropical birds: ecology and conservation. The University of Chicago Press, Chicago

Patterson BD, Stotz DF, Solari S, Fitzpatrick JW, Pacheco V (1998) Contrasting patterns of elevational zonation for birds and mammals in the Andes of southeastern Peru. J Biogeogr 25:593–607

Patton JL, Smith MF (1992) mtDNA phylogeny of Andean mice: a test of diversification across ecological gradients. Evolution 46:174–183

Phillimore AB, Orme CDL, Thomas GH, Blackburn TM, Bennett PM, Gaston KJ, Owens IPF (2008) Sympatric speciation in birds is rare: insights from range data and simulations. Am Nat 171:646–657

Price T (2008) Speciation in birds. Roberts and Company, Greenwood Village, Colorado

Quintero I, Jetz W (2018) Global elevational diversity and diversification of birds. Nature 555:246–250

Quintero I, Landis MJ (2019) Interdependent phenotypic and biogeographic evolution driven by biotic interactions. bioRxiv, https://doi.org/10.1101/560912

R Core Team (2018) R: A language and environment for statistical computing. R Foundation for Statistical Computing, Vienna, Austria. http://www.r-project.org/

Ramírez-Barahona S, Eguiarte LE (2013) The role of glacial cycles in promoting genetic diversity in the Neotropics: the case of cloud forests during the Last Glacial Maximum. Ecol Evol 3:725–738

Remsen JV Jr, Areta JI, Cadena CD, Claramunt S, Jaramillo A, Pacheco JF, Robbins MB, Stiles FG, Stotz DF, Zimmer KJ (2018) A classification of the bird species of South America, version July 7 2018. American Ornithologists' Union, Baton Rouge. http://www.museum.lsu.edu/~Remsen/SACCBaseline.htm

Restall R, Rodner C, Lentino M (2007) Birds of northern South America: an identification guide. Volume 1: species accounts. Yale University Press, New Haven, CT

Revell LJ (2012) Phytools: an R package for phylogenetic comparative biology (and other things). Methods Ecol Evol 3:217–223

Revell LJ, Harmon LJ, Collar DC (2008) Phylogenetic signal, evolutionary process, and rate. Syst Biol 57:591–601

Ribas CC, Moyle RG, Miyaki CY, Cracraft J (2007) The assembly of montane biotas: linking Andean tectonics and climatic oscillations to independent regimes of diversification in *Pionus* parrots. Proc R Soc Lond B 274:2399–2408

Ricklefs RE (2010) Host-pathogen coevolution, secondary sympatry and species diversification. Philos Trans R Soc Lond B 365:1139–1147

Ridgely RS, Greenfield PJ (2001) The birds of Ecuador: status, distribution, and taxonomy. Cornell University Press, Ithaca, NY

Roy MS, Silva JMC, Arctander P, García-Moreno J, Fjeldså J (1997) The speciation of south American and African birds in montane regions. In: Mindell DP (ed) Avian molecular evolution and systematics. Academic, San Diego, pp 325–343

Santos JC, Coloma LA, Summers K, Caldwell JP, Ree R, Cannatella DC (2009) Amazonian amphibian diversity is primarily derived from late Miocene Andean lineages. PLoS Biol 7:e56

Schulenberg TS, Stotz DF, Lane DF, O'Neill JP, Parker TAI (2007) Birds of Peru. Princeton University Press, Princeton, NJ

Scott GR (2011) Elevated performance: the unique physiology of birds that fly at high altitudes. J Exp Biol 214:2455–2462
Scott GR, Cadena V, Tattersall GJ, Milsom WK (2008) Body temperature depression and peripheral heat loss accompany the metabolic and ventilatory responses to hypoxia in low and high altitude birds. J Exp Biol 211:1326–1335
Sibley CG (1958) Hybridization in some Colombian tanagers, avian genus "*Ramphocelus*". Proc Am Philos Soc 102:448–453
Sick H (1993) Birds in Brazil: a natural history. Princeton University Press, Princeton, NJ
Smith BT, Klicka J (2010) The profound influence of the late Pliocene Panamanian uplift on the exchange, diversification, and distribution of New World birds. Ecography 33:333–342
Smith TB, Wayne RK, Girman D, Bruford MW (2005) Evaluating the divergence-with-gene-flow model in natural populations: the importance of ecotones in rainforest speciation. In: Bermingham E, Dick CW, Moritz C (eds) Tropical rainforests: past, present, and future. The University of Chicago Press, Chicago, pp 148–165
Stiles FG, Laverde-R O, Cadena CD (2017) A new species of tapaculo (Rhinocryptidae: *Scytalopus*) from the Western Andes of Colombia. Auk 134:377–392
Symonds MRE, Tattersall GJ (2010) Geographical variation in bill size across bird species provides evidence for Allen's rule. Am Nat 176:188–197
Terborgh J (1971) Distribution on environmental gradients: theory and a preliminary interpretation of distributional patterns in the avifauna of the Cordillera Vilcabamba, Peru. Ecology 52:23–40
Terborgh J (1977) Bird species diversity on an Andean elevational gradient. Ecology 58:1007–1019
Terborgh J, Weske JS (1975) The role of competition in the distribution of Andean birds. Ecology 56:562–576
Todd WEC, Carriker MA Jr (1922) The birds of the Santa Marta region of Colombia: a study in altitudinal distribution. Ann Carnegie Mus 14:1–611
Vasconcelos MF, Maurício GN, Kirwan GM, Silveira LF (2008) Range extension for Marsh Tapaculo *Scytalopus iraiensis* to the highlands of Minas Gerais, Brazil, with an overview of the species'distribution. Bull Br Ornithol Club 128:101–106
von Humboldt A, Bonpland A (2009) Essay on the geography of plants—with a physical tableau of the equinoctial regions (1807). In: Jackson ST (ed) Essay on the geography of plants. University of Chicago Press, Chicago, pp 61–155
Wake DB, Lynch JF (1976) The distribution, ecology, and evolutionary history of plethodontid salamanders in tropical America. Sci Bull Nat Hist Mus Los Angeles County 25:1–65
Weir JT (2009) Implications of genetic differentiation in Neotropical montane forest birds. Ann Mo Bot Gard 96:410–433
Weir JT, Faccio MS, Pulido-Santacruz P, Barrera-Guzmán AO, Aleixo A (2015) Hybridization in headwater regions, and the role of rivers as drivers of speciation in Amazonian birds. Evolution 69:1823–1834
Whitney BM (1994) A new *Scytalopus* tapaculo (Rhinocryptidae) from Bolivia, with notes on other Bolivian members of the genus and the *magellanicus* complex. Wilson Bull 106:585–614
Whitney BM, Vasconcelos MF, Silveira LF, Pacheco JF (2010) *Scytalopus petrophilus* (Rock Tapaculo): a new species from Minas Gerais, Brazil. Rev Bras Biol 18:73–88
York JM, Chua BA, Ivy CM, Alza L, Cheek R, Scott GR, McCracken KG, Frappell PB, Dawson NJ, Laguë SL, Milsom WK (2017) Respiratory mechanics of eleven avian species resident at high and low altitude. J Exp Biol 220:1079–1089

Chapter 24
Clade-Specific Biogeographic History and Climatic Niche Shifts of the Southern Andean-Southern Brazilian Disjunction in Plants

Federico Luebert, Maximilian Lörch, Rafael Acuña, Renato Mello-Silva, Maximilian Weigend, and Jens Mutke

Abstract Several plant clades show a characteristic disjunction between the southern Andes (SA) and southern Brazil (SB). This disjunction has been explained by Neogene geologic and climatic changes, especially due to the Andean uplift, the Mid-Miocene marine transgressions, and the development of Patagonian aridity. We compared phylogeny-derived divergence times and realized climatic niches for ten clades disjunct between the SA and the SB, and compared them with the timing proposed for the above-mentioned geologic and climatic events. Divergence times obtained for the node corresponding to the SA-SB split ranged from 27 to 1 Ma. Climatic niche overlap between SA and SB is generally low and niche equivalency (but not similarity) between SA and SB is significantly lower than expected by chance. Precipitation tends to be higher in SA and SB than in the intervening areas, where no species of the disjunct clades occur, but this trend does not apply to all clades. Clades ages correlate to differences in minimum temperature of coldest month and mean temperature of coldest quarter. Our results indicate that while different Neogene climatic and geological events can explain the disjunction of most clades, there is no single explanation for all of them: SA-SB disjunctions go

F. Luebert (✉)
Nees-Institut für Biodiversität der Pflanzen, Universität Bonn, Bonn, Germany

Departmento de Silvicultura y Conservación de la Naturaleza, Universidad de Chile, Santiago, Chile
e-mail: fluebert@u.uchile.cl

M. Lörch · M. Weigend · J. Mutke
Nees-Institut für Biodiversität der Pflanzen, Universität Bonn, Bonn, Germany

R. Acuña
Nees-Institut für Biodiversität der Pflanzen, Universität Bonn, Bonn, Germany

Escuela de Biología, Universidad de Costa Rica, San Pedro de Montes de Oca, Costa Rica

R. Mello-Silva
Department of Botany, University of São Paulo, São Paulo, SP, Brazil

V. Rull, A. C. Carnaval (eds.), *Neotropical Diversification: Patterns and Processes*, Fascinating Life Sciences, https://doi.org/10.1007/978-3-030-31167-4_24

back to different clade histories. Climatic niche shifts could explain the generally low climatic overlap and lack of niche equivalency between SA and SB, and might have played a central role in the biogeographic history of the SA-SB disjunction. A proper test of this hypothesis will require measuring differences in the fundamental niches of the species—which is beyond the scope of this study.

Keywords Andes · Arid diagonal · Marine transgressions · Paranean Sea · Patagonia · South America

1 Introduction

The origins of disjunct distribution patterns, i.e., the occurrence of closely related species in geographically separated areas, has long interested biogeographers. Disjunctions provide an ideal model for studying the interplay between past climatic and geological events and biogeographical processes such as dispersal, vicariance and extinction. A range of studies have addressed intercontinental disjunctions in order to examine the relative roles of long distance dispersal and continental drift in shaping those patterns (e.g., Wen 1999; Morley 2003; Givnish and Renner 2004; Wen and Ickert-Bond 2009; Simpson et al. 2018). Disjunctions within continents are increasingly attracting attention because they provide an opportunity to study the biological effect of more recent climatic and geological changes. Similar disjunct patterns in different lineages might imply parallel processes in historical biogeography, i.e., a common cause for the disjunctions. In Africa, the disjunct patterns of the so-called Rand flora have been associated with the development of aridity (Pokorny et al. 2015), but each lineage appears to show a different biogeographic history (Mairal et al. 2017). Similarly, Andean uplift in South America is believed to have created a dispersal barrier and caused East-West vicariance, but again each lineage appears to have had a different history (Luebert and Weigend 2014).

Several plant taxa are disjunct between the southern Andes, southern Brazil, and the tropical Andes (Villagrán and Hinojosa 1997). One typical disjunct pattern is the southern Andean-southern Brazilian (SA-SB) disjunction, found, for instance, in *Araucaria* sect. *Araucaria* (Golte 1993), *Fuchsia* sect. *Quelusia* (Berry 1989) and *Myrceugenia* (Landrum 1981). Schmithüsen (1956) suggested that these elements are vicariant relics of a formerly continuous tropical Cenozoic vegetation that covered a broad latitudinal band across South America. Landrum (1981) similarly postulated that the disjunction between the southern Andes (SA) and Southern Brazil (SB) in *Myrceugenia* (Myrtaceae) corresponds to the vicariant remnants of a more continuous Oligocene and early Miocene subtropical flora. More recent paleobotanical data provide evidence for the presence of a more or less continuous subtropical flora in South America during the Oligocene (Axelrod et al. 1991), with many of the nowadays disjunct taxa represented in the fossil record (see Hinojosa and Villagrán 1997 and references therein).

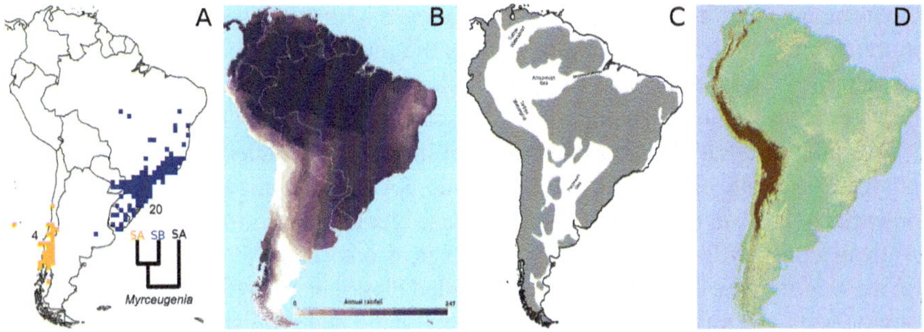

Fig. 24.1 (**a**) Geographical distribution of the disjunct clade of *Myrceugenia* (based on Landrum 1981; Murillo-A et al. 2016). To build the map, occurrence data were projected onto a raster of 1 × 1 degree cells. Numbers next to the map indicate how many species of the corresponding clade have phylogenetic data available. Color codes of the grid cells correspond to the southern Andean (SA; orange) and southern Brazilian (SB; blue) clades. The sister relationship of the disjunct clade is shown in the inset. (**b**) Distribution of annual rainfall in South America showing the most arid zones in white and increasing precipitation with darker colors (from SAGE 2018). (**c**) Mid-Miocene marine transgressions in South America (white areas, from Ortiz-Jaureguizar and Cladera 2006). (**d**) Distribution of the South American highlands (brown tones, after Luebert and Weigend 2014). Dark brown color represents areas with elevation higher than 3000 m

If the hypothesis of a vicariant origin of the disjunctions is accepted, then three major events can be invoked as causes. Schmithüsen (1956), for instance, argued that the fragmentation of this (sub-)tropical flora was caused by Andean uplift during the late Miocene. Some studies indeed found evidence of an increased Andean uplift rate during the late Miocene (e.g., Blisniuk et al. 2005; Garzione et al. 2008), although this remains controversial (Barnes and Ehlers 2009). In addition, Mid-Miocene marine transgressions in southern South America are well documented (Aceñolaza and Sprechmann 2002; Hernández et al. 2005) and may have interrupted any previously existing continuous range in that area. Lastly, fragmentation of a continuous mesic, subtropical flora may have taken place with the onset of aridity in Patagonia and the Chaco since the late Miocene. This aridisation is clearly documented in the fossil record (e.g., Latorre et al. 1997; Palazzesi and Barreda 2012; Palazzesi et al. 2014) and the origin of arid lands in southern South America has been used as an explanation for similar disjunct patterns of other taxa, e.g., *Alstroemeria* (Chacón et al. 2012). Figure 24.1 contrasts the present range of *Myrceugenia* with the distribution of arid lands in South America, the extension of the Andes, and the Mid-Miocene marine transgressions.

Assuming a vicariant origin of the disjunctions as argued by Schmithüsen (1956) and Landrum (1981), one should expect divergence times of the different clades to coincide between lineages and with the ages of the corresponding climatic or geological events. Only few studies have so far estimated divergence times for the SA-SB disjunction. One of them was carried out in *Myrceugenia* by Murillo-A et al. (2016), who estimated the median age for the disjunction to be ca. 20 Ma, long before the onset of aridity in Patagonia. Furthermore, if past climatic events were responsible for the vicariance, niche conservatism would imply that the environments within current geographical ranges of disjunct species are climatically more similar to each other than they are to the intervening areas, where the species are absent. That being said, climatic niches tend to be less conserved between lineages separated over evolutionary timescales (Peterson 2011).

We analyzed ten plant taxa sharing a similar disjunct pattern in order to provide phylogenetic and climatic insights into the SA-SB disjunction, investigating the potential effect of the three major Neogene geological and climatic events, i.e., Andean uplift, Mid-Miocene marine transgressions, and Patagonian aridification, on the origin of the disjunctions. We compiled divergence time estimates from dated phylogenies generated from our own original data, and published sources. We also characterized realized climatic niches and compared the climatic conditions within the current ranges of the target species with those of the intervening areas, over a background comprising the whole South American southern cone, to evaluate whether they are consistent with the hypothesis that past climatic changes resulted in the disjunct pattern, assuming climatic niche conservatism. If niche shifts have occurred, one could expect that clades in SA occupy climatic conditions different from those occupied by clades in SB, after controlling for available climates in SA and SB. In this case, the degree of climatic difference in SA-SB niches is expected to be correlated to clade age.

2 Material and Methods

We compiled a list of eleven disjunct clades between the southern Andes and southern Brazil, for which information is available (Table 24.1). These clades are well distributed across the phylogeny of seed plants (see plant families in Table 24.1), meaning that these disjunctions likely represent multiple independent events. Life forms range from annual herbs (*Blumenbachia*, *Micropsis*) to perennial herbs (*Bipinnula*, *Blumenbachia*, *Viviania*), shrubs (*Colliguaja*, *Fuchsia*, *Griselinia*, *Myrceugenia*), or trees (e.g., *Araucaria*, *Butia* + *Jubaea*, *Myrceugenia*, *Quillaja*), and occupy a wide variety of habitats, including montane forests, dry scrubs and wetlands. In SA, the distribution of some groups extends into southern Patagonia, while the SB groups usually also range into northeastern Argentina, Uruguay and southeastern Paraguay.

We compiled divergence time estimates for the SA-SB disjunct clades. For *Araucaria* sect. *Araucaria*, divergence time estimates were obtained from Hedges et al. (2015); for *Blumenbachia*, estimates were obtained from Acuña et al. (2019); for *Butia-Jubaea*, estimates were obtained from Meerow et al. (2015); for *Myrceugenia*, estimates were obtained from Murillo-A et al. (2016). For the remaining groups we either re-analyzed already published data or used newly generated molecular data (Genbank accession numbers MK807007–MK807020) to estimate divergence times. All dating analyses were conducted with a Bayesian uncorrelated relaxed-clock approach implemented in BEAST v.1.8.4 (Drummond and Rambaut 2007). Convergence was checked in Tracer v.1.6 (Rambaut et al.

Table 24.1 Plant taxa disjunctly distributed between the southern Andes and southern Brazil indicating total number of species (Nsp total), number of Neotropical species (Nsp Neotr), number of species known to be part of the disjunct clade as shown in phylogenetic studies indicated in the last column (Nsp disjunct); number of species of the disjunct clade present in the southern Andes (Nsp SA) and in southern Brazil (Nsp SB)

Family	Taxon	Nsp total	Nsp Neotr	Nsp disjunct	Nsp SA	Nsp SB	Distribution data	Reference phylogeny
Araucariaceae	*Araucaria* Juss. sect. *Araucaria*[a]	2	2	2	1	1	1, 2	18
Orchidaceae	*Bipinnula* Comm. ex A.Juss.[b,c]	11	11	4	2	2	1, 2, 3, 4, 5	19
Loasaceae	*Blumenbachia* Schrad.[d]	11	11	9	3	6	2, 6, 7, 8	20
Arecaceae	*Butia* (Becc.) Becc.-*Jubaea* Kunth[e]	21	21	8	1	7	1, 9	21
Euphorbiaceae	*Colliguaja* Molina[f]	5	5	5	4	1	1, 2, 5	22
Onagraceae	*Fuchsia* sect. *Quelusia* (Vand.) DC.	9	9	6	1	5	10	23
Griseliniaceae	*Griselinia* J.R.Forst. & G.Forst.	7	5	1	1	1	11	22
Asteraceae	*Micropsis* DC.[g]	5	5	3	1	2	2, 5, 12	24
Myrtaceae	*Myrceugenia* O. Berg	37	37	24	4	20	1, 13, 14	25
Quillajaceae	*Quillaja* Molina	2	2	2	1	1	15	26
Vivianaceae	*Viviania* Cav.	6	6	3	1	2	1, 16, 17	27

Numbers of species as given in the sources of distribution data unless otherwise indicated. Sources of distribution data as follows: 1. GBIF; 2. Reflora; 3. Buzatto et al. (2014); 4. Novoa et al. (2015); 5. Scherson et al. (2017); 6. Sleumer (1955); 7. Pérez-Moreau and Crespo (1988); 8. Acuña (unpublished data); 9. González et al. (2009); 10. Berry (1989); 11. Dillon and Munoz-Schick (1993); 12. Cabrera (1932); 13. Landrum (1981); 14. Landrum (1984); 15. Luebert (2014); 16. Lefor (1975); 17. Ferreira et al. (2016). Reference Phylogeny as follows: 18. Kranitz et al. (2014); 19. Cisternas et al. (2012); 20. Acuña et al. (2017); 21. Meerow et al. (2015); 22. This work; 23. Berry et al. (2004); 24. Luebert et al. (2017); 25. Murillo-A et al. (2016); 26. Luebert (unpublished); 27. Palazzesi et al. (2012)

[a]Number of species based on Golte (1993)

[b]Number of species based on Pridgeon et al. (2003)

[c]Incl. *Geoblasta* Barb. Rodr

[d]Number of species based on Henning et al. (2015)

[e]Number of species of *Butia* based on Hoffmann et al. (2014)

[f]Number of species based on Pax and Hoffmann (1912)

[g]Number of species based on Ward et al. (2009)

2013) with the criterion ESS > 200 for all parameters. Burn-in was set to 10% except for *Viviania*, where 20% ensured convergence of the posterior distribution. From the taxa listed in Table 24.1, we were able to obtain divergence time estimates for all except *Quillaja*.

Bipinnula We re-analyzed sequences of four plastid markers (*rbcL*, *matK*, *trnL-trnF*, *rpoB-trnC*) published by Cisternas et al. (2012). A secondary calibration point was placed at the root of the tree (the MRCA of *Chloraea* Lindl. and *Codonorchis* Lindl.), following Hoffmann et al. (2015) apud Hedges et al. (2015). Therefore, we set a uniform prior at the root, with bounds 40 and 62 Ma. The analysis was conducted with 300,000,000 generations, sampling every 10,000 generations.

Colliguaja We downloaded one *ndhF* sequence of each genus of Euphorbiaceae available in GenBank as of 16 January 2018 and complemented them with newly generated sequences of all species of *Colliguaja*, and a sequence of *Pera* Mutis (Peraceae), as an outgroup for Euphorbiaceae (see Appendix). Sequences were aligned with mafft v.7.271 (Katoh et al. 2002). After a preliminary RaxML analysis (Stamatakis et al. 2008), two fossil calibration points were placed following Cervantes et al. (2016) and using the relationships presented by Wurdack et al. (2005). We constrained the crown node of the clade including *Excoecaria* L., *Colliguaja* and *Bonania* A.Rich. with an offset of 40.5 Ma, lognormal mean = 2 and SD = 1. We further placed a lognormal calibration on the crown node of the clade containing *Acalypha* L., *Conceveiba* Aubl., *Dalechampia* L. and *Ricinus* L. with an offset of 60 Ma, lognormal mean = 1.5 and SD = 1. The analysis lasted for 10,000,000 generations sampling every 1000 generations.

Fuchsia* Sect. *Quelusia We re-analyzed the data of Berry et al. (2004). Based on Berry et al. (2004), we set a calibration point at the split of *Fuchsia* L. and *Circaea* L. as normal distribution with mean = 41.2 Ma and standard deviation = 0.5. The analysis lasted for 100,000,000 generations sampling every 10,000 generations.

Griselinia We analyzed a dataset of two plastid markers (*trnH-psbA* and *trnL-trnF*), based on published sequences plus two new samples including the Brazilian species of the genus (see Appendix). Outgroup taxa were selected within Apialaes including representatives of Araliaceae and Apiaceae, and the tree was rooted with *Torricellia* DC. (Beaulieu et al. 2013). Following Beaulieu et al. (2013), we placed a fossil calibration at the node including *Dendropanax* Decne. & Planch., *Hedera* L. and *Schefflera* J.R.Forst. & G.Forst. using the fossil of *Dendropanax* reported by Dilcher and Dolph (1970) for the middle Eocene. Therefore, an offset was set to 41.2 Ma with a lognormal mean of 1.2 and standard deviation 1.0. The analysis lasted for 100,000,000 generations sampling every 10,000 generations.

Micropsis We used the dataset of ETS and ITS published by Luebert et al. (2017). Based on Nie et al. (2016), we placed two secondary calibration points with a normal distribution, one at the MRCA of *Leontopodium* R.Br. ex Cass. and the *Lucilia* Cass.-group and one at the root of Gnaphalieae. For the former we assigned a mean of 10.3 Ma and standard deviation 1.7. For the latter, we assigned a mean of 29.4 Ma and standard deviation 3.5. The analysis lasted for 66,410,000 generations sampling every 10,000 generations.

Viviania We based our analysis on the ITS+*trnLF* alignment provided by Sytsma et al. (2014), which is in turn based on Palazzesi et al. (2012). To this alignment we added two newly sequenced Brazilian species of *Viviania* (Genbank accession numbers MK807018–MK807020). Calibration points and topological constraints were set following Sytsma et al. (2014). The analysis lasted for 50,000,000 generations, sampling every 5000 generations.

We built a geographical dataset of all clades listed in Table 24.1. Georeferenced data points were obtained from taxonomic revisions, regional floras and online databases, especially GBIF (https://www.gbif.org/) and Reflora (http://reflora.jbrj.gov.br/). In cases where collection locality information was available and detailed enough, but no coordinates were provided, we added the approximate coordinates using Google Earth or Geonames (http://www.geonames.org/). Since GBIF data have sometimes problems associated with taxonomic uncertainty and geographic inaccuracy (Maldonado et al. 2015), we filtered out those georeferenced records without locality information as well as duplicate entries and those without information about determination by a specialist. Extreme outliers and georeferenced records at country/county centroids were checked manually for plausibility and excluded when appropriate. A total of 4623 data points were compiled, of which 2832 correspond to species sampled in the phylogenies. Georeferenced data points used in the analyses are available at https://doi.org/10.5281/zenodo.2647591.

From these data points, we excluded from further analysis those species not included in phylogenetic analyses or not retrieved as part of a disjunct clade in the phylogenetic studies (see references in Table 24.1). The disjunct clades were defined at the shallowest possible node connecting all the species occurring either in SA or in SB and their sister group. Geographical distributions of all ten clades included in this study are depicted in Fig. 24.2, except for *Myrceugenia*, which is presented in Fig. 24.1. To reduce possible collection bias, data points closer to each other than 0.1 degree were filtered out using the desaggregation function of the R-package ecospat (Di Cola et al. 2017).

For all georeferenced data points, we extracted 19 standard bioclimatic variables (see http://chelsa-climate.org/bioclim/) from the CHELSA database v.1.2 (Karger et al. 2017). We did the same for a random sampling of 1000 points within the intervening area where no species of the disjunct clades are present. The latter was done for each disjunct clade separately. The intervening areas were delimited on the basis of the ecoregions (Olson et al. 2011) located between the disjunct areas (Dry Chaco, Low Monte, Humid Pampas and Espinal), which were clipped with the longitudinal range defined by the easternmost and westernmost points of the occurrences in SA and SB, respectively.

We conducted a principal component analysis (PCA) of the 19 bioclimatic variables matching the occurrences of the species, the random points of the intervening area and a selection of 10,000 random points for the background area. The background area was defined as a rectangle, using the maximum and minimum x and y

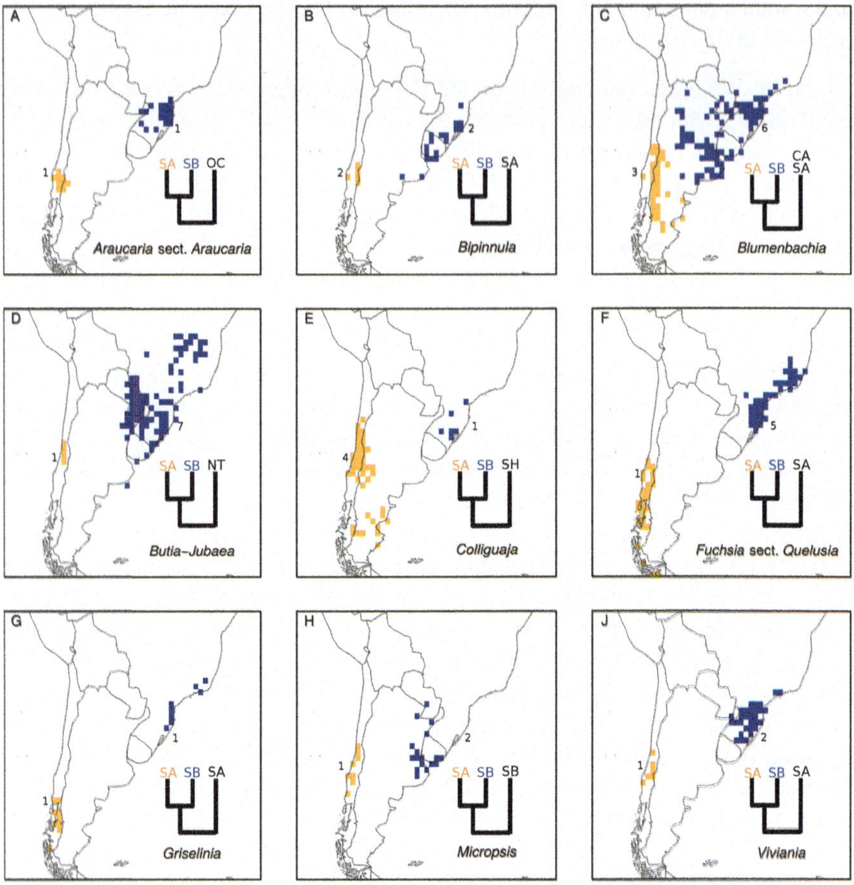

Fig. 24.2 Distribution maps of nine taxa showing a SA-SB disjunction. To build the maps, occurrence data were projected onto a raster of 1 × 1 degree cells. Numbers next to the map indicate how many species of the corresponding clade have phylogenetic data available. Color codes of the grid cells correspond to the SA (orange) and SB (blue) clades/species. The sister relationships of the disjunct clades are shown in the inset of each map. *SA* Southern Andes, *SB* Southern Brazil, *OC* Oceania, *CA* Central Andes, *NT* Neotropics, *SH* Southern Hemisphere

coordinates from the data points across all disjunct clades, roughly corresponding to the Southern Cone of South America. PCA scores of species data points were used to evaluate, for each clade, whether the climatic space of species in SA are available in the surrounding area of the species in SB and vice-versa. To this end, we employed the method for identifying analogous and non-analogous environments proposed by Mesgaran et al. (2014), which uses Mahalanobis distances to define the surrounding areas. It then measures the environmental similarity between these two areas, thus controlling for the environments actually accessible to each clade. PCA scores were also used to generate density grids charts where the occupancy of the environmental space of the disjunct clades and the intervening areas were compared over the whole background area (Broennimann et al. 2012). We used the density grids to calculate Schoener's D and Hellinger's I overlap metrics (Warren et al. 2008) between SA, SB and the intervening areas of each disjunct clade. D is a measure based on arithmetic differences, while I is based on normalized distances (see Warren et al. 2008 for details). We further conducted one-tailed tests of niche similarity and niche equivalency for both overlap metrics D and I (Warren et al. 2008; Broennimann et al. 2012) between SA, SB and the intervening areas of each disjunct clade. While the niche similarity test evaluates the differences in overlap between the observed niches of two ranges and the overlap of randomly selected niches from the one range and the observed niche of the other range, the niche equivalency test does it by randomly reallocating the occurrences between the two ranges, keeping constant the number of occurrences of each range, and comparing the overlap between these random reallocations (Broennimann et al. 2012). All comparisons were conducted in R with the packages ecospat (Di Cola et al. 2017), raster (Hijmans 2016), maptools (Bivand et al. 2017) and rgdal (Bivand et al. 2018).

Using the PCA scores, we also calculated the average pairwise climatic distance between the points in SA and SB of each disjunct clade, for which we have molecular data. These distances were compared to the median ages at the split of each disjunct clade. The same was done for each bioclimatic variable separately. Ages were also compared to the absolute mean of each bioclimatic variable from the occurrence points of each disjunct clade. Linear models were fitted for all above comparisons between climate and clade ages.

Finally, we compared the range of the bioclimatic variables between SA, SB and the intervening area for each clade.

3 Results

All studied groups show a clade/species in one of the disjunct areas that is sister to a clade/species in the other disjunct area, i.e., the disjunct clade does not include species beyond the SA-SB disjunction. In most clades, species richness is higher in SB (*Blumenbachia*, *Butia-Jubaea*, *Fuchsia* sect. *Quelusia*, *Micropsis Myrceugenia*, *Viviania*, see Figs. 24.1 and 24.2, Table 24.1), but in some the richness is equal in both SA and SB (*Araucaria* sect. *Araucaria*, *Bipinnula*, *Griselinia*) or higher in the SA (*Colliguaja*). Conversely, the sister group of most disjunct clades is distributed in the SA (*Bipinnula*, *Blumenbachia*, *Fuchsia* sect. *Quelusia*, *Griselinia*, *Myrceugenia*, *Viviania*) or broadly distributed in the Neotropics (*Butia-Jubaea*), the southern hemisphere (*Colliguaja*) or in Oceania (*Araucaria* sect. *Araucaria*), with only one sister group restricted to eastern South America (*Micropsis*).

Divergence times range from ~27 Ma in *Blumenbachia* to ~1 Ma in *Micropsis* (Table 24.2, Fig. 24.3), with most disjunct clades showing an age for their separation older than 5 Ma.

The two first components of the climatic PCA analysis performed on the occurrence points of the species in SA and SB, given the Southern Cone background,

Table 24.2 Divergence time estimates of ten SA-SB (southern Andes-southern Brazil) disjunct taxa

Taxon	Age (Ma)	Markers	Source
Araucaria sect. *Araucaria*	10.19 (6.16–14.22)	*rbcL*, *matK*, *atpH-atpI*, *rps*12-*rps*20	Hedges et al. (2015)
Bipinnula	4.09 (2.73–5.82)	*trnL-trnF*, *rbcL*, *matK-trnK*, *rpoB-trnC*	Re-analysis of Cisternas et al. (2012)
Blumenbachia	27.48 (20.72–34.22)	*matK*, *rps16*, *trnL-trnF*, *trnS-trnG*	Acuña et al. (2019)
Butia-Jubaea	14.71 (8.87–21.39)	Six combined WRKY loci	Meerow et al. (2015)
Colliguaja	15.05 (7.23–25.25)	*ndhF*	This work
Fuchsia sect. *Quelusia*	11.19 (7.22–15.91)	ITS, *trnL-trnF*, *rpl*16	Re-analysis of Berry et al. (2004)
Griselinia	4.38 (0.01–25.58)	*trnH-psbA*, *trnL-trnF*	This work
Micropsis	0.84 (0.34–1.51)	ITS, ETS	Re-analysis of Luebert et al. (2017)
Myrceugenia	20.28 (18.86–21.67)	*matK-trnK*, *rpl*32-*trnL*, *trnQ*- *rps*16, *rpl*16, ITS, ETS	Murillo-A et al. (2016)
Viviania	8.11 (5.21–11.12)	ITS, *trnL-trnF*	Re-analysis of Palazzesi et al. (2012) and Sytsma et al. (2014)

Indicated are median ages and age ranges for the node of the split between SA and SB clades/ species as provided in the respective source

cxplain 68.41% of the total variance (PC1: 46,59%, PC2: 21,82%). PC1 is a combination of most temperature and precipitation variables. Temperature ranges (mean diurnal range and annual temperature range) and temperature seasonality influence PC1 negatively, while the rest do it positively. PC2 is mostly negatively influenced by the precipitation of the driest period (precipitation of driest month, precipitation of driest quarter) and positively so by precipitation seasonality. Climatic spaces of the species in SA are at least partially available in SB for six clades (*Bipinnula*, *Blumenbachia*, *Fuchsia*, *Griselinia*, *Micropsis*, *Myrceugenia*) and are completely novel in four of them (*Araucaria*, *Butia-Jubaea*, *Colliguaja* and *Viviania*), which also true in the opposite direction (Fig. 24.4). Climatic spaces of SA and SB tend to be segregated from each other as observed in the density grids charts (Fig. 24.5), but that varies across clades. This is also revealed in the climatic niche overlap values, which in general tend to be low (<0.5, Table 24.3). Tests of niche equivalency rejected the null hypothesis that niches are identical between SA and SB in all clades, with significant results towards lower equivalency of niches than expected by chance (Table 24.3). However, the test of niche similarity gives non significant results: similarity of niches in SA and SB is neither greater nor lower than expected by chance (Table 24.3).

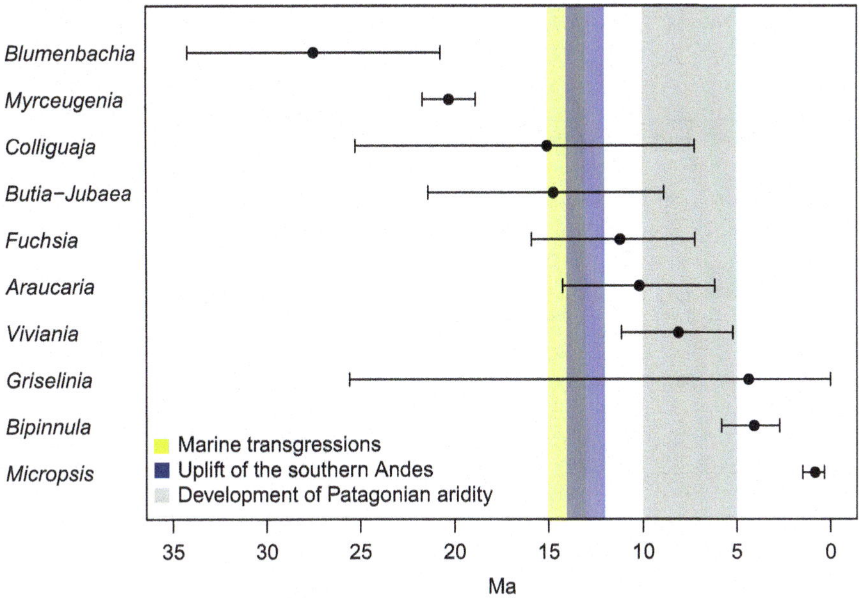

Fig. 24.3 Median divergence time estimates and 95% high density posterior intervals of ten SA-SB disjunct taxa compared to the age ranges proposed for the Mid-Miocene marine transgressions in South America (after Hernández et al. 2005), the uplift of the southern Andes (after Blisniuk et al. 2005) and the development of Patagonian aridity (after Palazzesi et al. 2014). Note that the age ranges of the marine transgressions and the southern Andean uplift overlap (dark blue color)

No trends were observed when comparing the SA-SB PCA distances with clade ages (data not shown), and our data are thus unable to identify a trend towards overall climatic niche divergence. However, when climatic SA-SB average differences of individual bioclimatic variables (see Methods) were compared to clade ages, significant trends ($p < 0.05$) were observed for winter temperatures (minimum temperature of coldest month and mean temperature of coldest quarter) (Fig. 24.6). When the means of single bioclimatic variables (by pooling values in SA and SB) were compared against clade ages, no significant trends were revealed (data not shown).

Comparison of the climatic conditions of the disjunct areas (SA and SB) of each clade reveals a clear pattern for winter precipitation (precipitation of the coldest quarter, PCQ), where both disjunct areas show rainfall amounts higher than, and more similar to each other than to the intervening area (Fig. 24.7, left panel). Annual precipitation (P) tends to be lower in the intervening area than in SB in all disjunct clades (not shown), but lower than in SA only in those clades with SA species located in the temperate zone, i.e., the southern portion of the southern Andes (*Araucaria*, *Bipinnula*, *Fuchsia*, *Griselinia*, *Myrceugenia*). For the remaining clades, which inhabit the Mediterranean climate zone of Central Chile (*Blumenbachia*, *Jubaea*, *Colliguaja*, *Micropsis*, *Viviania*), annual precipitation in the intervening area is either similar to or higher than in SA. Temperature variables

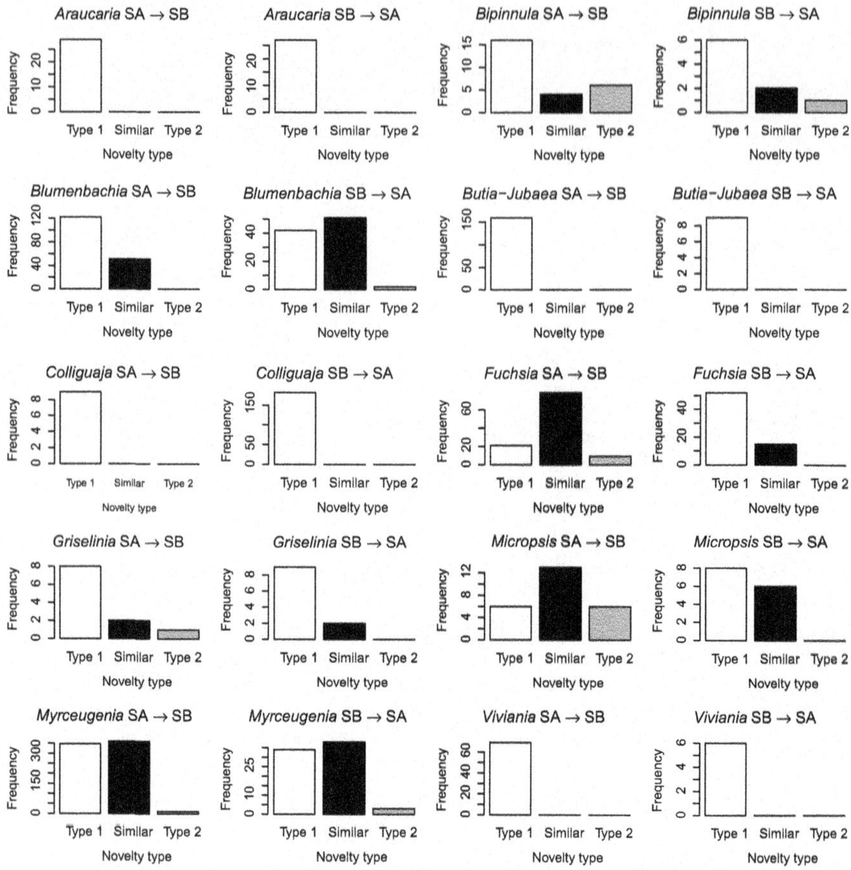

Fig. 24.4 Novelty types between SA and SB based on the first two axes of the PCA. Bars represent frequencies of the points in the projected area (right of arrow) based on the climatic space of the reference area (left of arrow). Type 1 are points novel in the projected area because their environmental space is outside the range of all individual variables. Type 2 are points within the univariate range but constitute novel combinations between covariates in the projected area. Analog points, where the climatic space of the reference area is available in the projected area, are designated as "Similar"

show a different trend, with SA species generally occurring at colder places than both SB species and the intervening area (Fig. 24.7, right panel). The opposite is true for the temperature annual range, with higher values in both SA and the intervening area than in SB (not shown).

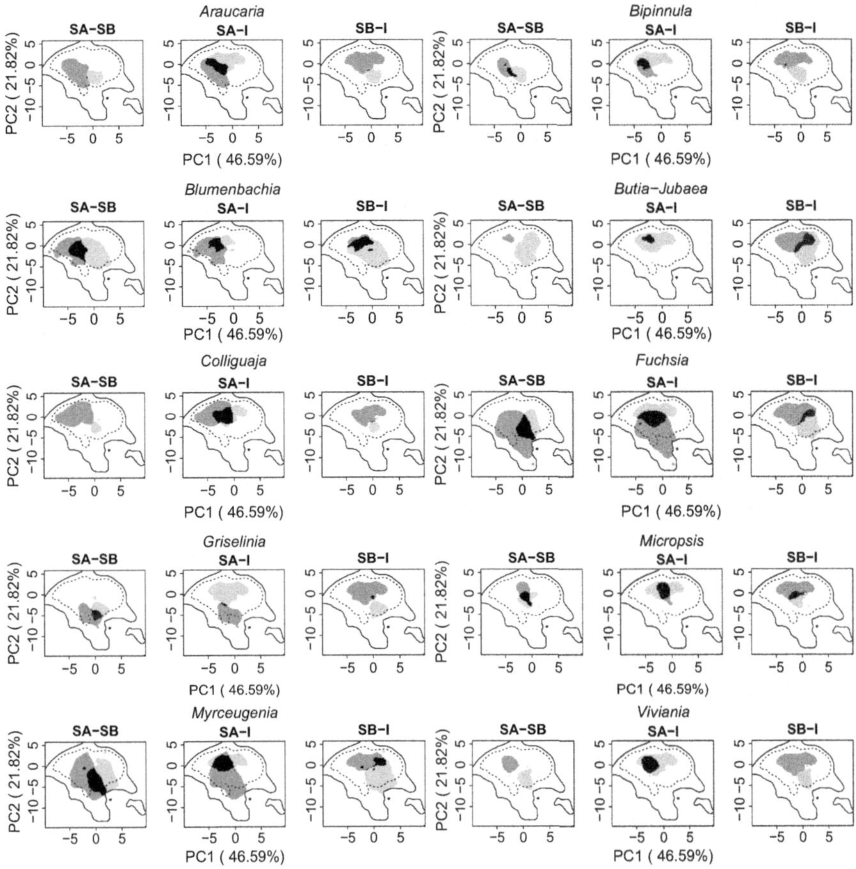

Fig. 24.5 Density grid charts based on PCA scores for the species of each clade in SA, SB and the intervening areas (I) showing their overlap in climatic space (black areas). Solid lines show the boundaries of the background area and the dotted line show the core climatic space within it (threshold = 50%)

4 Discussion

Our analyses show divergence times of the disjunctions ranging from ~27 Ma to ~1 Ma—indicating widely divergent evolutionary and biogeographical histories of the studied taxa. This is at odds with the general assumption of a common biogeographic history of clades with similar disjunct patterns (e.g., Schmithüsen 1956; Smith 1962; Villagrán and Hinojosa 1997), but agrees with the analysis of Rull (2008), who showed Neotropical lineages to have originated in a rather continuous way since the Oligocene. It also fits in the picture of most extant Neotropical lineages being of Neogene origin rather than representing ancient Paleogene or Cretaceous lineages of Gondwana origin (Hughes et al. 2013). Comparing the timing of the disjunctions with the three events that have been proposed as causes in the literature,

Table 24.3 Climatic niche overlap between the species occurring in the southern Andes (SA), southern Brazil (SB) and the respective intervening area (I) for each disjunct clade

	Schoener's D			Hellinger's I		
Clade	SA-SB	SA-I	SB-I	SA-SB	SA-I	SB-I
Araucaria sect. *Araucaria*	0.001 n.s. **(l)	0.088 n.s. **(l)	0.001 n.s. **(l)	0.002 n.s. **(l)	0.172 n.s. **(l)	0.001 n.s. **(l)
Bipinnula	0.117 n.s. **(l)	0.107 n.s. **(l)	0.002 n.s. **(l)	0.167 n.s. **(l)	0.192 n.s. **(l)	0.005 n.s. **(l)
Blumenbachia	0.133 n.s. **(l)	0.067 n.s. **(l)	0.080 n.s. **(l)	0.209 n.s. **(l)	0.166 n.s. **(l)	0.207 n.s. **(l)
Butia-Jubaea	0.000 n.s. **(l)	0.195 *(g) **(l)	0.057 n.s. **(l)	0.000 n.s. **(l)	0.421 **(g) **(l)	0.099 n.s. **(l)
Colliguaja	0.000 n.s. **(l)	0.200 n.s. **(l)	0.001 n.s. **(l)	0.000 n.s. **(l)	0.358 n.s. **(l)	0.003 n.s. **(l)
Fuchsia sect. *Quelusia*	0.116 n.s. **(l)	0.032 n.s. **(l)	0.014 n.s. **(l)	0.238 n.s. **(l)	0.104 n.s. **(l)	0.024 n.s. **(l)
Griselinia	0.214 n.s. **(l)	0.004 n.s. **(l)	0.005 n.s. **(l)	0.286 n.s. **(l)	0.005 n.s. **(l)	0.006 n.s. **(l)
Micropsis	0.255 n.s. *(l)	0.415 *(g) *(l)	0.084 n.s. **(l)	0.353 n.s. **(l)	0.584 *(g) **(l)	0.146 n.s. **(l)
Myrceugenia	0.237 n.s. **(l)	0.033 n.s. **(l)	0.004 n.s. **(l)	0.375 n.s. **(l)	0.133 n.s. **(l)	0.013 n.s. **(l)
Viviania	0.000 n.s. **(l)	0.495 *(g) n.s.	0.000 n.s. **(l)	0.000 n.s. **(l)	0.667 *(g) n.s.	0.000 n.s. **(l)

P-values of the niche similarity and equivalency one-tailed tests are shown in that order under D and I metrics. Niche similarity and equivalency might be greater (g) or lower (l) than expected by chance. *n.s.* non significant; $*P < 0.05$; $**P = 0.01$

a rather mixed pattern emerges: The ages for *Griselinia* are not informative due to the large uncertainty in the divergence time estimates. The splits into sub-clades in *Blumenbachia* and *Myrceugenia* seem to have been related to other factors, because their divergence times predate all the events that have been proposed as causal. Based on the dated phylogenies, the southern Andean uplift and the Mid-Miocene marine transgressions seem to be plausible factors for the SA-SB disjunction in *Colliguaja* and the *Butia-Jubaea* clade, because their disjunctions fall into the corresponding time frame (see Fig. 24.3). The disjunctions in *Araucaria*, *Fuchsia*, and *Viviania* coincide with the development of the arid corridor in South America, likely causing extinctions in the intervening area. In *Bipinnula* and *Micropsis*, mean divergence time estimates postdate the onset of aridity. However, the estimated

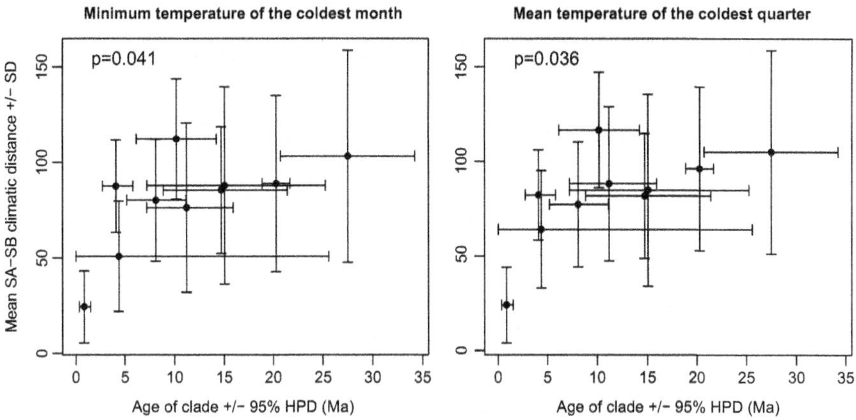

Fig. 24.6 Relation between clade age and minimum temperature of coldest month (left) and mean temperature of coldest quarter (right). P-values of the linear models are also shown

divergence time for *Bipinnula* coincides with a probable middle Pliocene (3.6–3.3 Ma) aridification event in the southern central Andes in N Argentina (Amidon et al. 2017) and a major mid-Pliocene faunal turnover in E Argentina dated at ca. 3.3 Ma (Vizcaíno et al. 2004).

Our analyses cannot resolve the question whether the SA-SB disjunction is the product of the fragmentation of a formerly continuous area, or is the result of colonization events from one area of the disjunction into the other. Most authors have favored the vicariance scenario (Schmithüsen 1956; Landrum 1981; Villagrán and Hinojosa 1997), which seems to be supported by the age and ancestral area reconstructed for *Myrceugenia* (Murillo-A et al. 2016) and paleobotanical records, suggesting a more or less continuous tropical-subtropical climatic belt in southern South America in the Eocene and Oligocene (Axelrod et al. 1991; Gayó et al. 2005). The second scenario, proposed by Smith (1962), assumes a colonization of SB from SA ancestors across the currently arid barrier. In this case, dispersal events and thus the divergence times may well postdate the arisal of the biogeographical barriers. The most recent disjunctions correspond to *Bipinnula* and *Micropsis*, both belonging to hyperdispersed plant families (Orchidaceae—dust fliers resp. Asteraceae—parachute fliers); we find long-distance dispersal in these two plant groups to be more plausible, and vicariance inherently less likely. If the geographical distribution of the clades sister to the disjunct ones provides any indication of geographical origin, then a SA origin should be postulated for most clades (see Fig. 24.2). The exception here is *Micropsis*, the only clade whose sister group is distributed in eastern South America, with one species introduced in North America (Morefield 2006).

Palazzesi et al. (2012) proposed an alternative explanation for *Viviania* based on the very good fossil record of that genus in SE Patagonia, where the group is now absent: They proposed a primary diversification of Vivianiaceae and *Viviania* in the—then subtropical—southern section of South America, with a subsequent independent northwards displacement due to the global cooling trend. Our dated phylogeny retrieves the split between SA- and SB-*Viviania* in the late Miocene, at a

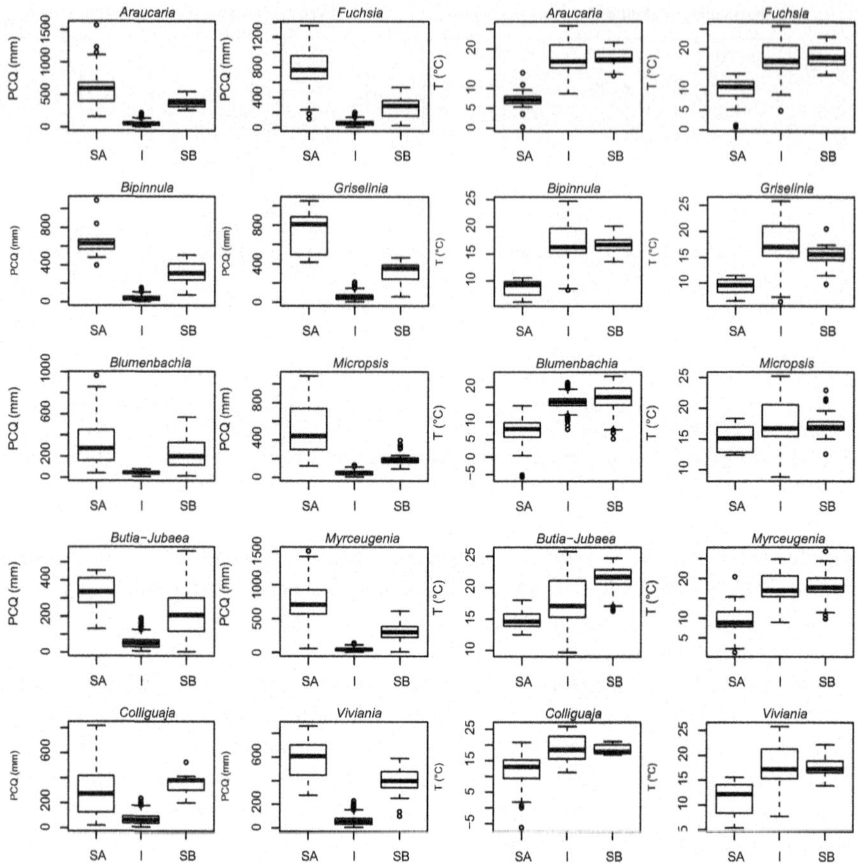

Fig. 24.7 Boxplots for precipitation of the coldest quarter (PCQ, left panel) and annual mean temperature (T, right panel) of each clade separated for the species in SA, SB and the respective intervening areas (I)

stage when the main global cooling trend had already taken place. We can thus clearly infer an extinction in the southern part of the range, but cannot provide conclusive evidence for an actual northward displacement.

The climatic analyses detect changes in the realized niches of the clades involved in the SA-SB disjunction, as shown by the generally low climatic overlap values between SA and SB and the rejection of niche equivalency in all clades. A suggestion of niche shifts should however be seen with caution at least for two reasons. First, tests of niche identity are prone to Type I errors (Peterson 2011). Second, rather than shifts, climatic differences between SA and SB clades might also be the result of the geographical fragmentation of their fundamental niche. The latter is underscored by the lack of analog climates in SA and SB in four of the clades (Fig. 24.4), which is fully coincident with zero overlap values. However, for more than half of the clades, the climatic niche overlap between SA and the intervening areas is higher than between SA and SB—indicating that geographical truncation of the climatic niche might not be the main reason for the observed disjunctions or that the biologically relevant ecological variables are not captured by the bioclimatic dataset. Interestingly, the climatic space occupied by the different clades in SB is relatively small (except for *Blumenbachia*) and is similar between the different clades (see Fig. 24.5). The differences between the climatic spaces occupied by the different clades in SA are larger, since some groups occur in the Mediterranean climate zone whereas others are restricted to the temperate zone of Chile and Argentina. The former groups are those whose climatic spaces are not analog in SA and SB. Differences in minimum temperatures between SA and SB appear to be correlated to divergence time, suggesting that niche shifts may have gradually progressed during the history of the lineages in coincidence with the global cooling that has taken place since the mid-Miocene (Zachos et al. 2001), a trend that appears to be mirrored in southern South America (Hinojosa and Villagrán 2005). Since SB clades tend to occupy warmer niches than SA ones, these shifts appear to be more pronounced in SA clades (Fig. 24.7), which is at odds with the conclusion recently drawn by Rezende et al. (2017) supporting the so-called austral conservatism hypothesis. A correlation between clade age and temperature was also reported for the disjunct lineages of the African rand flora (Mairal et al. 2017). However, the case in Africa seems to be one of niche conservatism, because older clades tend to retain a warmer niche than younger ones. In other words, in our case the magnitude of the variables per se does not relate to clade age, but the magnitude the difference in temperature between the two areas. Temperature shifts in the SA-SB disjunction may be associated with the adaptation of SA species to cooler conditions, since these species tend to occupy areas with lower temperatures (see Fig. 24.7). This relationship is also coincident with the general trend towards less niche conservatism in clades with older divergence times (Peterson 2011).

Aridity, especially low winter rainfall (see Fig. 24.7), is the main parameter that clearly differentiates the intervening area from SA and SB and thus might constitute a climatic barrier for all disjunct clades. However, for both *Colliguaja* and *Butia-Jubaea* clades, where the SA representatives occur in the Mediterranean zone of central Chile, this trend does not hold regarding overall annual precipitation. Coincidentally, these two clades have median divergence times that match the mid Miocene marine transgressions and predate the more recent uplift pulse of the Patagonian Andes (see Fig. 24.3). The interplay between temperature and precipitation, not captured in the bioclimatic variables used in this study, could also explain this trend, with evapotranspiration being too high in the warmer parts and therefore displacing SA taxa into cooler regions, not available in the intervening area.

This cross-taxonomic comparison of the SA-SB disjunction confirms the idea that every lineage has its own idiosyncratic biogeographic history. The disjunct clades analyzed here show mean divergence times that range from the late Oligocene to the Pleistocene. Our results suggest that climatic niches of the disjunct clades are only moderately conserved, and that differentiation in minimum temperature is correlated to clade age. Yet, because high niche differentiation might be the result of lack of analog climates in SA and SB in four clades, truncation of niches (rather than niche shifts) cannot be ruled out as a possibility to explain these differences. Andean uplift, Marine transgressions, and Patagonian aridification - the three major events that have taken place in southern South America since the Mid-Miocene -may have played a role in the fragmentation of the distribution of disjunct SA-SB lineages. However, the relative contribution of these events is likely clade-specific.

Acknowledgements FL thanks the DAAD for travel support. RA is supported by the ALECOSTA joint program with funds from both the Universidad de Costa Rica and the DAAD. RMS is supported by CNPq research fund and CAPES. We thank Ana C. Carnaval and Patricio Pliscoff for insightful discussions on niche shift analysis, Paul Berry and Corinna Treydte for help with the geographical dataset of *Fuchsia* and Rafael Trevisan for material of Brazilian *Viviania*.

Appendix

Genbank accession numbers of *ndhF* sequences used for the analysis of *Colliguaja*: JX662766; AY425027; EF416920; JN249078; EF416921; JN249079; EF416922; AF538255; FJ670079; JN249081; FJ670080; EF416947; FJ670081; JN249082; AY374313; AY374314; JN249083; FJ670082; AY374315; FJ670083; AF538261; AF538264; JN249086; JN249088; JN249091; JN249242; KY628420; JF937588; FJ695500; EU117376; MK807007; MK807008; MK807009; MK807010; MK807011; MK807012; MK807012.

Genbank accession numbers of *trnH-psbA* sequences used for the analysis of *Griselinia*: AJ536845; AJ536846; AJ536847; AJ536848; AJ536849; AJ536850; AJ536851; KX648359; KP271241; AY582139; KC456163; KC456166; NC_015113; NC_008325; NC_016430; AJ537465; MK807014; MK807015.

Genbank accession numbers of *trnL-trnF* sequences used for the analysis of *Griselinia*: AJ536854; AJ537453; AJ537455; AJ537456; AJ537458; AJ537460; AJ537461; KX648359; KP271241; AY582139; KC456163; KC456166; NC_015113; NC_008325; NC_016430; AJ537464; MK807016; MK807017.

References

Aceñolaza F, Sprechmann P (2002) The Miocene marine transgression in the meridional Atlantic of South America. N Jb Geol Paläont (Abh) 225:75–84

Acuña R, Fließwasser S, Ackermann M, Henning T, Luebert F, Weigend M (2017) Phylogenetic relationships and generic re-arrangements in “South Andean Loasas” (Loasaceae). Taxon 66:365–378

Acuña R, Luebert F, Henning T, Weigend M (2019) Major lineages of Loasaceae subfam. Loasoideae diversified during the Andean uplift. Mol Phylogenet Evol 141:106616

Amidon WH, Fisher GB, Burbank DW, Ciccioli PL, Alonso RN, Gorin AL, Silverhart PH, Kylander-Clark ARC, Christoffersen MS (2017) Mio-Pliocene aridity in the south-central Andes associated with Southern Hemisphere cold periods. Proc Natl Acad Sci U S A 114:6474

Axelrod DI, Arroyo MTK, Raven PH (1991) Historical development of temperate vegetation in the Americas. Rev Chil Hist Nat 64:413–446

Barnes JB, Ehlers TA (2009) End member models for Andean Plateau uplift. Earth Sci Rev 97:105–132

Beaulieu JM, Tank DC, Donoghue MJ (2013) A Southern Hemisphere origin for campanulid angiosperms, with traces of the break-up of Gondwana. BMC Evol Biol 13:80

Berry PE (1989) A systematic revision of *Fuchsia* sect. *Quelusia* (Onagraceae). Ann Mo Bot Gard 76:532–584

Berry PE, Hahn WJ, Sytsma KJ, Hall JC, Mast A (2004) Phylogenetic relationships and biogeography of *Fuchsia* (Onagraceae) based on noncoding nuclear and chloroplast DNA data. Am J Bot 91:601–614

Bivand R, Lewin-Koh N, Pebesma E, Archer E, Baddeley A, Bearman N, Bibiko H-J, Brey S, Callahan J, Carrillo G, Dray S, Forrest D, Friendly M, Giraudoux P, Golicher D, Rubio VG, Hausmann P, Hufthammer KO, Jagger T, Johnson K, Luque S, MacQueen D, Niccolai A, Pebesma E, Lamigueiro OP, Short T, Snow G, Stabler B, Stokely M, Turner R (2017) Maptools: tools for reading and handling spatial objects. Available from https://CRAN.R-project.org/package=maptools. Accessed 12 July 2018

Bivand R, Keitt T, Rowlingson B, Pebesma E, Sumner M, Hijmans R, Rouault E, Warmerdam F, Ooms J, Rundel C (2018) rgdal: bindings for the "geospatial" data abstraction library. Available from https://CRAN.R-project.org/package=rgdal. Accessed 12 July 2018

Blisniuk PM, Stern LA, Chamberlain CP, Idleman B, Zeitler PK (2005) Climatic and ecologic changes during Miocene surface uplift in the Southern Patagonian Andes. Earth Planet Sci Lett 230:125–142

Broennimann O, Fitzpatrick MC, Pearman PB, Petitpierre B, Pellissier L, Yoccoz NG, Thuiller W, Fortin M-J, Randin C, Zimmermann NE, Graham CH, Guisan A (2012) Measuring ecological niche overlap from occurrence and spatial environmental data. Glob Ecol Biogeogr 21:481–497

Buzatto CR, Sanguinetti A, Romero-González GA, van den Berg C, Singer RB (2014) A taxonomic synopsis of Brazilian Chloraeinae (Orchidaceae: Orchidoideae). Phytotaxa 158:1–22

Cabrera AL (1932) La distribución geográfica del género *Micropsis* (Compositae). Boletín de la Sociedad Española de Historia Natural 32:427–434

Cervantes A, Fuentes S, Gutiérrez J, Magallón S, Borsch T (2016) Successive arrivals since the Miocene shaped the diversity of the Caribbean Acalyphoideae (Euphorbiaceae). J Biogeogr 43:1773–1785

Chacón J, Assis MC, Meerow AW, Renner SS (2012) From East Gondwana to Central America: historical biogeography of the Alstroemeriaceae. J Biogeogr 39:1806–1818

Cisternas MA, Salazar GA, Verdugo G, Novoa P, Calderón X, Negritto MA (2012) Phylogenetic analysis of Chloraeinae (Orchidaceae) based on plastid and nuclear DNA sequences. Bot J Linn Soc 168:258–277

Di Cola V, Broennimann O, Petitpierre B, Breiner FT, D'Amen M, Randin C, Engler R, Pottier J, Pio D, Dubuis A, Pellissier L, Mateo RG, Hordijk W, Salamin N, Guisan A (2017) ecospat: an R package to support spatial analyses and modeling of species niches and distributions. Ecography 40:774–787

Dilcher DL, Dolph GE (1970) Fossil leaves of *Dendropanax* from Eocene sediments of southeastern North America. Am J Bot 57:153–160

Dillon MO, Munoz-Schick M (1993) A revision of the dioecious genus *Griselinia* (Griseliniaceae), including a new species from the coastal Atacama Desert of northern Chile. Brittonia 45:261–274

Drummond AJ, Rambaut A (2007) BEAST: Bayesian evolutionary analysis by sampling trees. BMC Evol Biol 7:214

Ferreira JPR, Hassemer G, Campestrini S, Weigend M, Trevisan R (2016) A revision of the extra-Andean Vivianiaceae. Phytotaxa 246:23–36

Garzione CN, Hoke GD, Libarkin JC, Withers S, MacFadden B, Eiler J, Ghosh P, Mulch A (2008) Rise of the Andes. Science 320:1304–1307

Gayó E, Hinojosa LF, Villagrán C (2005) On the persistence of tropical Paleofloras in Central Chile during the early eocene. Rev Palaeobot Palynol 137:41–50

Givnish TJ, Renner SS (2004) Tropical intercontinental disjunctions: Gondwana breakup, immigration from the Boreotropics, and transoceanic dispersal. Int J Plant Sci 165:S1–S6

Golte W (1993) Araucaria*: Verbreitung und Standortansprüche einer Coniferengattung in vergleichender Sicht*. Steiner Franz Verlag, Stuttgart

González LA, Bustamante RO, Navarro RM, Herrera MA, Toral M (2009) Ecology and management of the Chilean Palm (*Jubaea chilensis*): history, current situation and perspectives. Palms 53:68–74

Hedges SB, Marin J, Suleski M, Paymer M, Kumar S (2015) Tree of life reveals clock-like speciation and diversification. Mol Biol Evol 32:835–845. Reported in http://www.timetree.org/. Accessed 3 July 2018

Henning T, Oliveira SS, Schlindwein C, Weigend M (2015) A new, narrowly endemic species of *Blumenbachia* (Loasaceae subfam. Loasoideae) from Brazil. Phytotaxa 236:196–200

Hernández RM, Jordan TE, Dalenz Farjat A, Echavarría L, Idleman BD, Reynolds JH (2005) Age, distribution, tectonics, and eustatic controls of the Paranense and Caribbean marine transgressions in southern Bolivia and Argentina. J S Am Earth Sci 19:495–512

Hijmans RJ (2016) raster: geographic data analysis and modeling. Available from https://CRAN.R-project.org/package=raster

Hinojosa LF, Villagrán C (1997) Historia de los bosques del sur de Sudamérica, I: Antecedentes paleobotánicos, geológicos y climáticos del Terciario del cono sur de América. Rev Chil Hist Nat 70:225–239

Hinojosa LF, Villagrán C (2005) Did South American mixed Paleofloras evolve under thermal equability or in the absence of an effective Andean barrier during the Cenozoic? Palaeogeogr Palaeoclimatol Palaeoecol 217:1–23

Hoffmann JF, Barbieri RL, Rombaldi CV, Chaves FC (2014) *Butia* spp. (Arecaceae): an overview. Sci Hortic 179:122–131

Hoffmann V, Verboom GA, Cotterill FPD (2015) Dated plant phylogenies resolve neogene climate and landscape evolution in the Cape Floristic Region. PLoS One 10:e0137847

Hughes CE, Pennington RT, Antonelli A (2013) Neotropical plant evolution: assembling the big picture. Bot J Linn Soc 171:1–18

Karger DN, Conrad O, Böhner J, Kawohl T, Kreft H, Soria-Auza RW, Zimmermann NE, Linder HP, Kessler M (2017) Climatologies at high resolution for the earth's land surface areas. Sci Data 4:170122

Katoh K, Misawa K, Kuma K, Miyata T (2002) MAFFT: a novel method for rapid multiple sequence alignment based on fast Fourier transform. Nucleic Acids Res 30:3059–3066

Kranitz ML, Biffin E, Clark A, Hollingsworth ML, Ruhsam M, Gardner MF, Thomas P, Mill RR, Ennos RA, Gaudeul M, Lowe AJ, Hollingsworth PM (2014) Evolutionary diversification of new Caledonian *Araucaria*. PLoS One 9:e110308

Landrum LR (1981) A monograph of the genus *Myrceugenia* (Myrtaceae). Flora Neotropica 29:1–135

Landrum LR (1984) Taxonomic implications of the discovery of calyptrate species of *Myrceugenia* (Myrtaceae). Brittonia 36:161–166

Latorre C, Quade J, McIntosh WC (1997) The expansion of C4 grasses and global change in the late Miocene: stable isotope evidence from the Americas. Earth Planet Sci Lett 146:83–96

Lefor MW (1975) A taxonomic revision of the vivianiaceae. Univ Conn Occas Pap Biol Sci Ser 2:225–255

Luebert F (2014) Taxonomy and distribution of the genus *Quillaja* Molina (Quillajaceae). Feddes Repertorium 124:157–162

Luebert F, Weigend M (2014) Phylogenetic insights into Andean plant diversification. Front Ecol Evol 2:27

Luebert F, Moreira-Muñoz A, Wilke K, Dillon MO (2017) Phylogeny and evolution of achenial trichomes in the *Lucilia*-group (Asteraceae: Gnaphalieae) and their systematic significance. Taxon 66:1184–1199

Mairal M, Sanmartín I, Pellissier L (2017) Lineage-specific climatic niche drives the tempo of vicariance in the Rand Flora. J Biogeogr 44:911–923

Maldonado C, Molina CI, Zizka A, Persson C, Taylor CM, Albán J, Chilquillo E, Rønsted N, Antonelli A (2015) Estimating species diversity and distribution in the era of big data: to what extent can we trust public databases? Glob Ecol Biogeogr 24:973–984

Meerow AW, Noblick L, Salas-Leiva DE, Sanchez V, Francisco-Ortega J, Jestrow B, Nakamura K (2015) Phylogeny and historical biogeography of the cocosoid palms (Arecaceae, Arecoideae, Cocoseae) inferred from sequences of six WRKY gene family loci. Cladistics 31:509–534

Mesgaran MB, Cousens RD, Webber BL (2014) Here be dragons: a tool for quantifying novelty due to covariate range and correlation change when projecting species distribution models. Divers Distrib 20:1147–1159

Morefield JD (2006) 103. *Micropsis*. In: FNA Editorial Committee (ed) Flora of North America North of Mexico. Oxford University Press, New York, pp 463–465

Morley RJ (2003) Interplate dispersal paths for megathermal angiosperms. Perspect Plant Ecol Evol Syst 6:5–20

Murillo-A JC, Stuessy TF, Ruiz E (2016) Explaining disjunct distributions in the flora of southern South America: evolutionary history and biogeography of *Myrceugenia* (Myrtaceae). J Biogeogr 43:979–990

Nie Z-L, Funk VA, Meng Y, Deng T, Sun H, Wen J (2016) Recent assembly of the global herbaceous flora: evidence from the paper daisies (Asteraceae: Gnaphalieae). New Phytol 209:1795–1806

Novoa P, Espejo J, Alarcón D, Cisternas M, Domínguez E (2015) Guía de campo de las orquídeas chilenas, 2nd edn. Corporacion Chilena de la Madera, Concepción

Olson DM, Dinerstein E, Wikramanayake ED, Burgess ND, Powell GVN, Underwood EC, D'Amico JA, Itoua I, Strand HE, Morrison JC, Loucks CJ, Allnutt TF, Ricketts TH, Kura Y, Lamoreux JF, Wettengel WW, Hedao P, Kassem KR (2011) Terrestrial ecoregions of the world: a new map of life on Earth. Bioscience 51:933–938

Ortiz-Jaureguizar E, Cladera GA (2006) Paleoenvironmental evolution of southern South America during the Cenozoic. J Arid Environ 66:498–532

Palazzesi L, Barreda V (2012) Fossil pollen records reveal a late rise of open-habitat ecosystems in Patagonia. Nat Commun 3:1294

Palazzesi L, Gottschling M, Barreda V, Weigend M (2012) First Miocene fossils of Vivianiaceae shed new light on phylogeny, divergence times, and historical biogeography of Geraniales. Biol J Linn Soc 107:67–85

Palazzesi L, Barreda VD, Cuitiño JI, Guler MV, Tellería MC, Ventura Santos R (2014) Fossil pollen records indicate that Patagonian desertification was not solely a consequence of Andean uplift. Nat Commun 5:3558

Pax F, Hoffmann K (1912) Euphorbiaceae-Hippomaneae. In: Engler A (ed) Das Pflanzenreich. Wilhelm Engelmann, Leipzig, pp 1–319

Pérez-Moreau RL, Crespo S (1988) Loasaceae. In: Correa MN (ed) Flora Patagónica. Colección Científica del INTA, Buenos Aires, pp 199–217

Peterson AT (2011) Ecological niche conservatism: a time-structured review of evidence. J Biogeogr 38:817–827

Pokorny L, Riina R, Mairal M, Meseguer AS, Culshaw V, Cendoya J, Serrano M, Carbajal R, Ortiz S, Heuertz M, Sanmartín I (2015) Living on the edge: timing of Rand Flora disjunctions congruent with ongoing aridification in Africa. Front Genet 6:154

Pridgeon AM, Cribb PJ, Chase MW, Rasmussen FN (eds) (2003) Genera Orchidacearum, Orchidoideae (Part 2), Vanilloideae, vol 3. Oxford University Press, Oxford

Rambaut A, Suchard MA, Drummond AJ (2013) Tracer v. 1.6. University of Edinburgh, Edinburgh. Available from http://tree.bio.ed.ac.uk/software/tracer/. Accessed 3 July 2018

Rezende VL, Dexter KG, Pennington RT, Oliveira-Filho AT (2017) Geographical variation in the evolutionary diversity of tree communities across southern South America. J Biogeogr 44:2365–2375

Rull V (2008) Speciation timing and Neotropical biodiversity: the tertiary–quaternary debate in the light of molecular phylogenetic evidence. Mol Ecol 17:2722–2729

SAGE (2018) *Atlas of the Biosphere*. Center for Sustainability and the Global Environment, Nelson Institute for Environmental Studies (SAGE), University of Wisconsin-Madison, USA. Available from https://nelson.wisc.edu/sage/data-and-models/atlas/

Scherson RA, Thornhill AH, Urbina-Casanova R, Freyman WA, Pliscoff PA, Mishler BD (2017) Spatial phylogenetics of the vascular flora of Chile. Mol Phylogenet Evol 112:88–95

Schmithüsen J (1956) Die raumliche Ordnung der chilenischen Vegetation. Bonner Geographische Abhandlungen 17:1–86

Simpson MG, Johnson LA, Villaverde T, Guilliams CM (2018) American amphitropical disjuncts: perspectives from vascular plant analyses and prospects for future research. Am J Bot 104:1600–1650

Sleumer H (1955) Die Loasaceen Argentiniens. Bot Jahrb Syst Pflanzengesch Pflanzengeogr 76:411–462

Smith LB (1962) Origins of the flora of southern Brazil. Contr US Natl Herb 35:215–249

Stamatakis A, Hoover P, Rougemont J (2008) A rapid bootstrap algorithm for the RAxML web servers. Syst Biol 57:758–771

Sytsma KJ, Spalink D, Berger B (2014) Calibrated chronograms, fossils, outgroup relationships, and root priors: re-examining the historical biogeography of Geraniales. Biol J Linn Soc 113:29–49

Villagrán C, Hinojosa LF (1997) Historia de los bosques del sur de Sudamérica, II: Análisis fitogeográfico. Rev Chil Hist Nat 70:241–267

Vizcaíno SF, Fariña RA, Zárate MA, Bargo MS, Schultz P (2004) Palaeoecological implications of the mid-Pliocene faunal turnover in the Pampean Region (Argentina). Palaeogeogr Palaeoclimatol Palaeoecol 213:101–113

Ward J, Bayer R, Breitwieser I, Smissen R, Galbany-Casals M, Unwin M (2009) Gnaphalieae. In: Funk VA, Susanna A, Stuessy TF, Bayer RJ (eds) Systematics, evolution, and biogeography of Compositae. International Association for Plant Taxonomy, University of Vienna, Vienna, pp 537–585

Warren DL, Glor RE, Turelli M (2008) Environmental niche equivalency versus conservatism: quantitative approaches to niche evolution. Evolution 62:2868–2883

Wen J (1999) Evolution of Eastern Asian and Eastern North American disjunct distributions in flowering plants. Annu Rev Ecol Syst 30:421–455

Wen J, Ickert-Bond SM (2009) Evolution of the Madrean–Tethyan disjunctions and the North and South American amphitropical disjunctions in plants. J Syst Evol 47:331–348

Wurdack KJ, Hoffmann P, Chase MW (2005) Molecular phylogenetic analysis of uniovulate Euphorbiaceae (Euphorbiaceae sensu stricto) using plastid rbcL and trnL-F DNA sequences. Am J Bot 92:1397–1420

Zachos J, Pagani H, Sloan L, Thomas E, Billups K (2001) Trends, rhythms, and aberrations in global climate 65 ma to present. Science 292:686–693

Chapter 25
Biotic and Landscape Evolution in an Amazonian Contact Zone: Insights from the Herpetofauna of the Tapajós River Basin, Brazil

Leandro J. C. L. Moraes, Camila C. Ribas, Dante Pavan, and Fernanda P. Werneck

Abstract The Amazon basin region, which harbors the largest river system in the world, and which is drained by the Amazon River, has experienced several geomorphological and climatic changes over time. These shifts, as we know, led to distinct signatures in the abiotic variables and the diversification patterns of its native biota. One excellent and yet still understudied model system for biogeographic studies relating biotic diversification to landscape evolution in this region is the Tapajós River basin. Located in an ecotonal zone marked by the presence of geomorphological, climatic and biotic gradients, the basin is crossed by the geographical barrier determined by the Tapajós River, one of the large southern tributaries of the Amazon River. The basin is also marked by high diversity of amphibians and reptiles, with distinct assemblages across its extent and riverbanks. We compare the landscape dynamics within the middle Tapajós River region and the whole basin with a

---

Electronic supplementary material The online version of this chapter (https://doi.org/10.1007/978-3-030-31167-4_25) contains supplementary material, which is available to authorized users.

L. J. C. L. Moraes
Coordenação de Biodiversidade, Instituto Nacional de Pesquisas da Amazônia (INPA), Manaus, AM, Brazil

Ecosfera Consultoria e Pesquisa em Meio Ambiente Ltda., São Paulo, SP, Brazil

C. C. Ribas
Programa de Coleções Científicas Biológicas, Coordenação de Biodiversidade, Instituto Nacional de Pesquisas da Amazônia, Manaus, AM, Brazil

D. Pavan
Ecosfera Consultoria e Pesquisa em Meio Ambiente Ltda., São Paulo, SP, Brazil

F. P. Werneck (✉)
Programa de Coleções Científicas Biológicas, Coordenação de Biodiversidade, Instituto Nacional de Pesquisas da Amazônia, Manaus, AM, Brazil

Department of Organismic and Evolutionary Biology, Museum of Comparative Biology, Harvard University, Cambridge, MA, USA

V. Rull, A. C. Carnaval (eds.), *Neotropical Diversification: Patterns and Processes*, Fascinating Life Sciences, https://doi.org/10.1007/978-3-030-31167-4_25

comparison of molecular diversification patterns across 16 taxa of amphibians and lizards. For that, we analyze spatial and temporal congruence in diversification through Bayesian gene trees, inter-bank genetic p-distances, and molecular dating based on novel mtDNA data. We find that most groups present spatial diversification patterns concordant with the current position of the Tapajós River, showing high genetic distance between banks. We recovered a continuous range of riverine-associated divergence times, which we group in three main time periods for discussion purposes: (1) ancient divergences, dating from the middle Miocene (ca. 10 Ma); (2) divergences in the late Miocene-Pliocene (ca. 5–3.5 Ma), and (3) more recent divergences, dating to Pleistocene (ca. 2–1 Ma). These results are consistent with a hypothesis of high landscape dynamism in the Tapajós River basin over time, leading to multiple vicariance or colonization events, but cannot be used to discard alternative hypotheses, such as that the river or basin represent areas of secondary contact only. Based on literature and new molecular data, we discuss the evolution of the landscape of the Tapajós River basin and its impact on the diversification of different taxonomic and functional groups.

Keywords Comparative phylogeography · Amazon · Amphibians · Reptiles · Ecotone

1 Landscape Evolution in Amazonia and Its Signatures

There is consensus that the megadiverse Amazon region has been subject to dramatic landscape and biotic changes throughout its history (Leite and Rogers 2013). Since the segregation of the western Gondwana portion and the rising of the South American Plate (Almeida and Hasui 1984), at approximately 140–120 million year ago (Ma; Clark 2018), the region was subject to cycles of erosion and deposition of sediments (Caputo and Soares 2016), uplift and subsidence of terrain (Hoorn et al. 2010a; Latrubesse et al. 2010), mountain orogeny (Horton et al. 2010; Garzione et al. 2017), tectonic adjustments (Rossetti 2014; Caputo and Soares 2016), marine incursions (Jaramillo et al. 2017), climate change (Haffer 1969; Cheng et al. 2013) and vegetational connections to adjacent biomes (Duellman 1979; Silva and Bates 2002; Ledo and Colli 2017). In this historically dynamic region currently lies the largest hydrographic basin of the world, covering much of northern South America and draining through the monumental Amazonas River (Sioli 1968). The emergence and dynamics of this basin over time are intrinsically linked to the spatio-temporal evolution of the South American landscape (Hoorn et al. 2010a; Ribas et al. 2012).

A logical sequence of key events contributed to the formation of the present Amazonian fluvial system (Hoorn et al. 2010b, 2017; Latrubesse et al. 2010). Before the Andean uplift, drainage in the region ran mainly from the Brazilian and Guianan Shields towards the west, flowing into the Pacific Ocean (Hoorn et al. 2010a; Latrubesse et al. 2010). The Andean uplift, with final stages dating from 5–3 Ma, dammed the western flow of the fluvial system, and, allied to a western depression of the plate, generated a system dominated by lacustrine or swampy environments in the region (the Pebas System; Wesselingh and Salo 2006; Hoorn et al. 2010a, b; Shephard et al. 2010). The subsequent erosion of the Andean mountains led to increasing and intense sediment deposition, which reversed the river system flow eastward, giving birth to the transcontinental Amazonas River (Hoorn et al. 2010b, 2017; Latrubesse et al. 2010; Mora et al. 2010; Shephard et al. 2010; Albert et al. 2018). The river was supplied by the tributaries originated at the recent uplifted Andes and at the ancient terrains of Amazonian Cráton (Hoorn et al. 2010b; Albert et al. 2018).

In addition to geomorphological changes, climatic variation also contributed to the high evolutionary dynamism of the Amazon basin (Cheng et al. 2013; Wang et al. 2017). It cyclically altered the hydrological characteristics of the rivers (e.g. flow, water level, sediment load, and sedimentation rates; Irion et al. 2006), causing marine incursions (Jaramillo et al. 2017), changing the types and extent of habitats present in the interfluves (Haffer 1969; Vanzolini and Williams 1970; Wang et al. 2017; Arruda et al. 2018) and episodically connecting these habitats with adjacent biomes (Batalha-Filho et al. 2012; Ledo and Colli 2017; Arruda et al. 2018). However, scientists still disagree about the precise time scale in which processes such as the establishment of the transcontinental Amazon River (Hoorn et al. 2010b; Latrubesse et al. 2010; Albert et al. 2018), and possible connections between biomes (Batalha-Filho et al. 2012; Ledo and Colli 2017) may have occurred. Such discordance may be caused by the fact that the evolution of the Amazonian landscape and its biota is the result of multiple spatial and temporal processes that triggered different speciation mechanisms (Tuomisto et al. 1995; Tuomisto and Ruokolainen 1997). Multiple sources of evidence are needed to understand these processes and their consequences to the current organization of Amazonian diversity.

Signatures of evolutionary process in the Amazon region may be accessed through currently observed patterns based on two main lines of evidence (Baker et al. 2014; Antonelli et al. 2018): (1) *abiotic*, which investigates, for example, ages and sequences of sediment deposition, location of ancient hydrographic basins, geological breaks, remote sensing, riverine paleochannels, and climatic variation (e.g., Mora et al. 2010; Soares et al. 2010; Rossetti 2014; Rossetti et al. 2014a, b; Latrubesse et al. 2007; Cheng et al. 2013) and (2) *biotic*, investigating the presence and age of fossils, ancient pollens, vegetational dynamics, and levels of intra- and interspecific diversification at distinct biological scales, such as molecular, population and community levels (e.g., Latrubesse et al. 2007; Ferreira et al. 2017; Maia et al. 2017; Soares et al. 2017; Rossetti et al. 2018; Ribas et al. 2018; Ortiz et al. 2018; Godinho and da Silva 2018). Such biotic approach is relevant as the native biota of the Amazon region suffered over time the effects of the intense environmental dynamism, becoming the main players and outcomes of landscape evolution (Antonelli et al. 2009; Baker et al. 2014). Here we integrate abiotic and biotic lines of evidence to understand the evolutionary processes acting in a regional Amazonian set up (the Tapajós River basin). For that, we provide a review of the landscape history combined with novel molecular data obtained for amphibians and reptiles distributed in this region.

Recognized through sub-disciplines such as "geo-genomics" (Baker et al. 2014) and "trans-disciplinary biogeography" (Antonelli et al. 2018), the integration of different conceptual and methodological approaches based on abiotic and biotic data is becoming more common in studies investigating the evolution of Amazonian landscapes. In many cases, consideration of a single evidence line does not provide substantial advance (Baker et al. 2014), and an integrative framework becomes fundamental to clarify the historical and ecological mechanisms driving current patterns of diversity. For environments with a historical depositional record, such as the Amazonian sedimentary basins (Latrubesse et al. 2005; Hoorn et al. 2010a, b), abiotic information is accessible through sediment layers deposited across time, which act as pages of a book, reporting the sequence of geological and climatic events, even though erosion may periodically erase parts of this sedimentary sequence. Sediment data has led to a substantial accumulation of scientific knowledge regarding the landscape evolution of western Amazon basin, particularly in the Solimões, Madeira, Purus, and Juruá river basins (e.g., Mertes et al. 1996; Latrubesse and Rancy 1998; Latrubesse and Kalicki 2002; Latrubesse and Franzinelli 2005; Latrubesse et al. 2007; Nogueira et al. 2013; Hayakawa and Rossetti 2015; Rossetti et al. 2018). However, in environments with an erosive character, such as the regions of the ancient Brazilian and Guiana Shields (Latrubesse et al. 2005; Hoorn et al. 2010a), the abiotic history is more difficult to reconstruct due to the absence of sedimentological information, making this region a book with several missing pages. This is the case of the eastern Amazonian basins of the Tapajós, Xingu, Tocantins, and Trombetas rivers. In these environments, neotectonic events, mapped through abandoned paleochannels and fault alignments, appear extremely relevant for landscape shaping (Rossetti and Valeriano 2007; Valente and Latrubesse 2012; Leite and Rogers 2013; Rossetti 2014). However, integration with biotic data is fundamental to build a more complete evolutionary history within these regions.

2 The Tapajós River Basin and Its Dual Role as a Barrier for the Biota

The Tapajós River basin is located in northern Brazil, south of the Amazon River (Fig. 25.1), covering portions of the Brazilian states of Pará, Amazonas, and Mato Grosso. This basin has the Juruena, Teles Pires, Jamanxim, and Tapajós rivers as the main tributaries, and its spatial range corresponds to about 10% of the Amazon River basin (Sioli 1968), being the fifth largest tributary sub-basin of the Amazonas River, and supplying about 6% of the water in this river (Latrubesse et al. 2005).

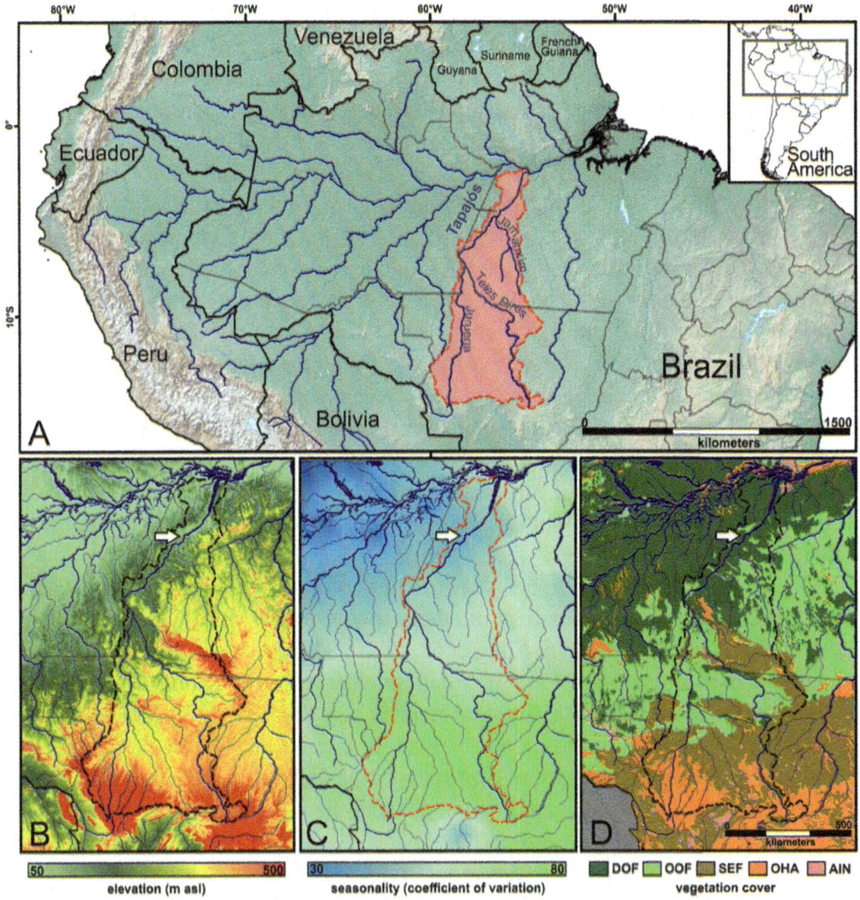

Fig. 25.1 Location of the Tapajós River basin (dashed lines) in relation to the Northern South America and the Amazon basin (**a**) and upon elevation (**b**), seasonality (**c**) and vegetation (**d**) backgrounds, showing the ecotonal character in which the basin is located based on geomorphological, climatic and vegetational variation, in both North-South and west-east directions. The white arrow represents the zone of geological transition between the Brazilian Shield and Amazonas sedimentary basin. Vegetation cover: (DOF) Dense Ombrophylous Forest; (OOF) Open Ombrophylous Forest; (SEF) Seasonal forests; (OHA) Open habitats; (AIN) Anthropic influence

Geologically, the Tapajós River basin runs through two main compartments: its upper and middle courses are located in the Brazilian Shield, lying on ancient Pre-Cambrian (>541 Ma) rocks (Santos et al. 1975), presenting 2–3 km of horizontal extension between the banks, and draining with a predominantly erosive character. Its lower course, in contrast, is located in the more recent Cenozoic (66 Ma to present) sedimentary basin of the Amazon River, extending to ca. 10–20 km of extension between the banks, and favoring the depositional character and the formation of alluvial plains (Santos et al. 1975; Araujo et al. 1976; Silva et al.

1980). These two geological compartments are segregated by a short region of Paleozoic sediments (455–300 Ma) in their contact zone, resulting from the historical sedimentation process of the Brazilian Shield (Santos et al. 1975). Due to this origin and geomorphological setting, the current spatial conformation of this basin is asymmetric, especially in its upper and middle course, where the main trunk river runs through the border of the Brazilian Shield (in contact with the Solimões sedimentary basin), and its main tributaries drain from the east (Santos et al. 1975; Araujo et al. 1976; Silva et al. 1980). This conformation generates a scenario in which the larger tributaries in the left bank of the Tapajós River are influenced by the Solimões sedimentary basin, while the larger tributaries on the right bank drain older terrain in the Brazilian Shield (Santos et al. 1975; Araujo et al. 1976; Silva et al. 1980; Moraes et al. 2016). Because of this erosive character over shield rock, with low sediment input, the rivers of this basin transport little amounts of dissolved sediments (Latrubesse et al. 2005), leading to their classification as "clear-water" rivers, in contrast to other Amazonian "black water" (with low amount of sediments and high amount of organic material dissolved in the water column) and "white water" rivers (with high amount of sediments dissolved in the water column; Sioli 1968; Junk et al. 2011).

This geological difference between the upper/middle and lower courses of the Tapajós River basin is also evident in other aspects of the current landscape (Santos et al. 1975; Araujo et al. 1976; Silva et al. 1980). The upstream courses, which run embedded in the rocky bed of the Brazilian Shield, present numerous rocky outcrops in the river channels, especially evident in the annual dry period (Fig. 25.2d). For this reason, the water flow in these courses is more intense, and rocky rapids, unevenness, and waterfalls are common, becoming especially abrupt in the contact zone between the main geological compartments, near the village of São Luis do Tapajós (Fig. 25.1). Given this hydromorphology, these courses of the Tapajós River basin are frequent targets of projects for generation of hydroelectric energy (Latrubesse et al. 2017). As the course of the Tapajós River reaches its lower portion, rocky outcrops and rapids become scarce and the river acquires a slower water flow, becoming a huge lentic "ria-lake", draining to the channel of the Amazon River (Irion et al. 2006) (Fig. 25.1).

Climatically, the Tapajós River basin presents a mean annual temperature range from 21 °C to 27 °C (Alvares et al. 2013). Its precipitation regime is intermediate relative to the two distinct dynamics observed in western vs. eastern Amazonia (Sombroek 2001; Cheng et al. 2013) (Fig. 25.1b). The climatic pattern of western Amazonia is characterized by a low annual seasonality and high annual rainfall, which gradually changes towards Southern and eastern Amazonia, increasing in annual seasonality and decreasing in annual rainfall (Alvares et al. 2013) (Fig. 25.1b). It has been inferred that these two climatic regimes had distinct histories throughout the Quaternary, modifying the structure and affecting the stability of the forests in these regions (Cheng et al. 2013; Wang et al. 2017), and enabling periodical connections with other Neotropical rainforests (Cheng et al. 2013; Ledo and Colli 2017). Such dramatic cyclical changes in climatic patterns would have

Fig. 25.2 Generalized vertical profile and photographs of middle Tapajós River region, evidencing changes in landscape, terrain and vegetation structure. (**a**) Periodically flooded forest (*igapó*) from fluvial islands and banks of large rivers; (**b**) sand deposit (beach); (**c**) main course of the Tapajós River near the municipality of Itaituba, Pará state; (**d**) rocky outcrops of the crystalline basement; (**e**) non-flooded Dense Ombrophilous Forest; (**f**) alluvial Dense Ombrophilous Forest from banks of small streams evident in lighter green, varying in degrees of palm predominance; (**g**) non-flooded Open Ombrophilous Forest. Photographs by Dante Pavan (**a**, **d**, **f**), Elizângela Brito (**b**, **c**) and Leandro Moraes (**e**)

affected and altered the biota native of these climatic regions (Cheng et al. 2013; Wang et al. 2017).

Patterns of climatic and geological variation can also be observed in the current forest cover associated with the Tapajós River. On the left bank there is a predominance of Dense Ombrophylous Forest (Fig. 25.2e), an exuberant and diverse forest type, with complex stratification, massive presence of lianas and epiphytes, and high capacity of local climate regulation (Terezo et al. 1975; Coêlho et al. 1976; Loureiro et al. 1980), influenced by the more humid and less seasonal climate from the western Amazonian sedimentary basin. On the right bank of Tapajós River the Open Ombrophylous Forest dominates, a less-stratified forest type with lower canopy, greater presence of clearings, and more dense understory (Terezo et al. 1975; Coêlho et al. 1976; Loureiro et al. 1980), influenced by the drier and seasonal climate from eastern Amazonia (Brazilian Shield), and the contact between the Amazon-Cerrado biomes (Fig. 25.1d).

Based on these current patterns, it is evident that the Tapajós River basin, especially its upper and middle courses, is located in a climatic, geomorphological, and vegetational transition zone (Fig. 25.1). We propose that this ecotonal character be taken into account when investigating the spatial and temporal patterns of diversification of its associated biota (Moraes et al. 2016). Ecotonal regions are known to possess both unique and mixed ecological characteristics between adjacent environments (Smith et al. 1997; Werneck et al. 2012a). In Amazonia, forest ecotones present less obvious transitions (Cohn-Haft et al. 2007; Ortiz et al. 2018) when compared to closed-open environment contacts such as that between the Amazon and the Cerrado (Ackerly et al. 1989). Because the Tapajós River course crosses this ecotonal zone, acting as a physical barrier upon an environmental gradient (Moraes et al. 2016), its basin is a good model for studies of biotic responses to landscape evolution.

3 The Tapajós River Basin's Dynamic History

Unlike rivers running through sedimentary basins, where recent geomorphological dynamism is primarily driven by sedimentation patterns (Gascon et al. 2000), rivers running in the ancient terrain of the Brazilian Shield suffer mainly the effects of recent conformational changes associated with neotectonic events, generating course captures and recaptures (Rossetti and Valeriano 2007; Dagosta and de Pinna 2017). Geological maps show a high density of fractures and volcanic intrusions in the cratonic region of the Tapajós River basin, some with recent signs of reactivation (Santos et al. 1975). In addition to the geomorphological change that may have affected the conformation of the rivers inside the Tapajós River basin, recent climatic change likely modified the extent of available terrestrial ecosystems and the water level of the large rivers (Irion et al. 2006; Cheng et al. 2013; Wang et al. 2017), potentially promoting multiple periods of contact and allopatry between populations. Patterns of Amazonian fish diversity inside cratonic river basins also suggest a dynamism of river channels (Dagosta and de Pinna 2017), with the Tapajós River basin an example of a mixed assemblage where communities in the tributaries Jamanxim, Juruena and Teles Pires are more similar to other Brazilian Shield rivers, and those in the main course of the Tapajós River are more similar to rivers from the Guiana Shield and the sedimentary basin of Amazonas River (Dagosta and de Pinna 2017).

Brazilian Shield rivers have been associated with recent diversification events when compared to rivers that run in other geological formations, such as the Negro, Madeira and lower Amazonas (Ribas et al. 2012; Silva et al. 2019). This recent diversification was indirectly evident when all interfluves segregated by Xingu and Tocantins rivers were considered a single unit in former delimitations of Amazonian areas of endemism, mainly due to the fact that endemism was then mapped based on phenotypically-informed taxonomy only (Cracraft 1985). It was only after refined analyses of biotic diversity that the areas of endemism segregated by the Xingu and Tocantins rivers were recognized (da Silva et al. 2002). With the advent of molecular techniques, these three distinct areas of endemism delimited by Xingu and Tocantins rivers were also recovered, with most recent splits between taxa mostly associated with these rivers from the Brazilian Shield (Ribas et al. 2012; Silva et al. 2019).

These observations of shallow divergences across Brazilian Shield rivers appear counterintuitive, as those rivers run through older terrains—but this pattern seems recurrent as the volume of available data and analytical capacity increase (Silva et al. 2019). With longer time since their origins, rivers of the Brazilian Shield would have had an equally greater probability of experiencing course changes promoted by geomorphological and climatic dynamism over time (Irion et al. 2006; Wang et al. 2017), making them more permeable—or this region unsuitable for several Amazonian taxa until a recent period (Silva et al. 2019). These recent landscape disturbances or regional extinctions followed by reoccupation would have promoted equally recent splits between biological populations from Brazilian Shield (Silva et al. 2019).

Based on abiotic and biotic evidence for this region, one landscape evolution hypothesis for the Tapajós River basin includes the discharge of the ancient rivers (the proto-Teles Pires, proto-Juruena and proto-Jamanxim rivers) towards the lower Amazonian regions when the basin was craton-oriented (Hoorn et al. 2010a), or even more recently. This is consistent with the finding of Pleistocenic sediments from the Tapajós River basin near the region of Manaus (Caputo and Soares 2016; Fig. 25.3). During that craton-oriented Amazonian period, possible connections between sub-basins within the Brazilian Shield (caused by river captures due to neotectonic events) may have increased the similarity of the aquatic biota of these three proto-rivers (Dagosta and de Pinna 2017). Tectonic and geomorphological rearrangements derived from the positive inversion of the Cachimbo Graben between the proto-Teles Pires and the proto-Juruena rivers, in relation to the proto-Jamanxim River, resulted in a positive structural arch which divided the Amazonas and Solimões basins (Wanderley-Filho et al. 2010) (Fig. 25.3). This event, followed by the Andean uplift, the subsidence of the sub-Andean region, and the formation of the transcontinental Amazon River (Hoorn et al. 2010a; Caputo and Soares 2016), would have modified

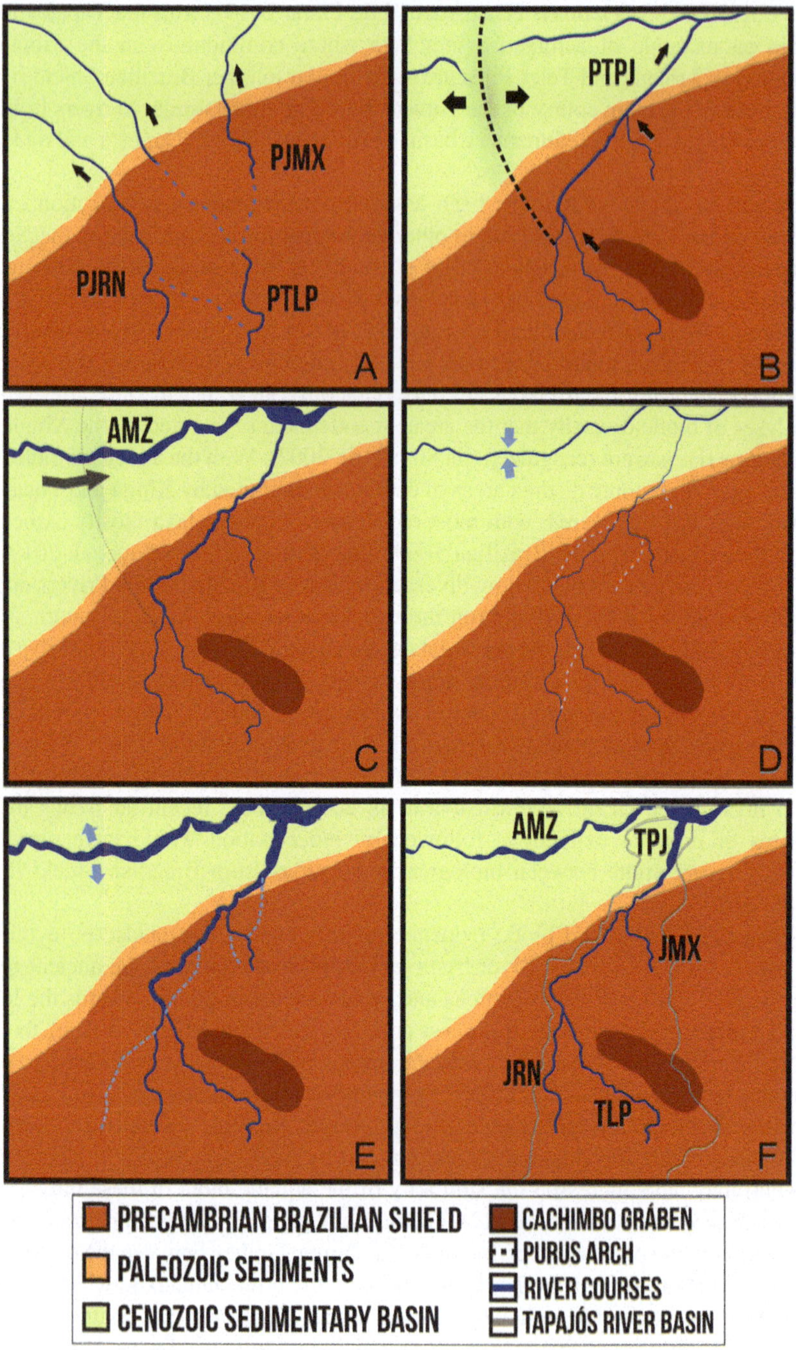

Fig. 25.3 Schematic hypothetic representation of the main spatial and temporal sequence of events hypothesized to have modified the Tapajós River basin landscape, consequently affecting its native biota. Accurate timing of such events remains unknown (see text for additional details and references). (**a**) Ancient discharge of Brazilian Shield rivers proto-Teles Pires (PTLP), proto-

the conformation and joined the proto-Teles Pires and proto-Juruena rivers. This connection of two large rivers would have discharged a large volume of water in a new channel that bypassed westward the Cachimbo Graben area, adjusted to a large fracture line (Tapajós line), and formed the current Tapajós River course (Fig. 25.3). This course also captured the proto-Jamanxim River and all smaller tributaries of its right bank that ran through the Brazilian Shield. Although this chronological sequence is plausible based on the evolution of the cratonic and sedimentary landscape of the Amazon basin (Hoorn et al. 2010a; Caputo and Soares 2016), the time periods in which these events would have occurred remain unknown (Fig. 25.3).

During the Pleistocene, the landscape in the basin would have been modified mainly by changes in the water level of the Tapajós River and its tributaries, due to glacial climatic variations, affecting river discharge and base level (Irion et al. 2006, 2010). The high sea level in interglacial periods may have led to an increase in the river base level within the Tapajós system, consequently increasing the barrier effect of the rivers to the surrounding biota and damming the lotic ria-lake system in its low course, as currently observed (Irion et al. 2006) (Fig. 25.3).

4 Amphibians and Reptiles of the Tapajós River Basin

The geographic distribution of the amphibians and reptiles of the Tapajós River basin appears to be strongly influenced by the marked environmental and historical heterogeneity described above. Although many generalist taxa are widely distributed across the basin (Duellman 1979), some are restricted to or are more abundant in distinct zones, causing striking shifts in beta diversity, especially between the

Fig. 25.3 (continued) Juruena (PJRN) and proto-Jamanxim (PJMX) in sedimentary basin, when the Amazon basin was craton-oriented, with the possible occurrence of riverine courses connections; (**b**) uplift of the Cachimbo Gráben resulting in geomorphological adjustments creating a central Amazonian watershed. These events may have connected the courses of proto-rivers Teles Pires and Juruena, discharging a large volume of water in the system, creating the proto-Tapajós River (PTPJ) embedded in geological faults, and connecting it to the proto-Jamanxim River. At this period, the discharge of Tapajós River basin was fully directed to the Amazonas sedimentary basin and eastern Amazonia; (**c**) The events arising from the Andean uplift and tectonic adjustments connected the western and eastern Amazonian sedimentary basins, generating the huge transcontinental Amazonas River. The Tapajós River basin becomes tributary to this system, (**d**, **e**) Intense climatic variations modified the sea level and favored periods of marine incursions and retractions, damming the ria-lake in the lower Tapajós River during higher sea-levels events. Such variation modified the amount of water discharging in the Tapajós River cratonic basin over time and its permeability to the native biota. The permeability is also affected by events of riverine course changes by neotectonic adjustments (dashed lines); (**f**) current conformation of the Tapajós River basin and its boundaries, with the main tributaries Teles Pires (TLP), Juruena (JRN), Jamanxim (JMX) and Tapajós (TPJ) rivers. For further details involving the current spatial variation of the landscape within this basin see Fig. 25.1

Brazilian Shield and the sedimentary basin. For example, the amphibian taxa *Boana punctata*, *B. raniceps*, *B. lanciformis*, *Lysapsus bolivianus*, *Scinax nebulosus*, *Rhinella* gr. *granulosa*, *Sphaenorhynchus* spp. *Scarthyla goinorum*, *Trachycephalus typhonius*, *Dendropsophus walfordi*, and the reptile taxa *Varzea bistriata*, *Anolis auratus*, *Gymnophthalmus* sp., *Crocodilurus amazonicus*, *Dracaena guianensis*, *Chironius carinatus* and *Crotalus durissus* are more abundant or only occur in the lower portion of the Tapajós River basin (sedimentary basin, Fig. 25.1a; Ribeiro-Júnior 2015a, b; Ribeiro Júnior and Amaral 2016, 2017; Frost 2019; Uetz et al. 2019), where the terrain is flatter and presents a well-developed alluvial plain, altering the flooded forest type and increasing its fertility, favoring the presence of aquatic macrophytes and the occurrence of water pools in the banks of large rivers (Sioli 1984; Piedade et al. 2010). Those taxa depend on these habitat characteristics throughout their life cycles. On the other hand, in the middle and upper portions of the basin, inserted in the Brazilian Shield (Fig. 25.1a), the more rugged terrain, with greater drainage capacity, lower development of alluvial plains, and flooded forests with lower fertility (Sioli 1984; Piedade et al. 2010), favors the occupation by other taxa, such as the amphibians *Adelphobates galactonotus*, *A. castaneoticus*, *Allobates crombiei*, *A. magnussoni*, *Ameerega munduruku*, *Proceratophrys concavitympanum*, and the reptiles *Gonatodes tapajonicus* and *Rondonops biscutatus* (Lima et al. 2014; Ribeiro-Júnior 2015a, b; Ribeiro Júnior and Amaral 2016, 2017; Frost 2019; Uetz et al. 2019). Other examples are detailed below, as the target region of this study is located on the northern edge of the Brazilian Shield terrain. The southern limits of the geographic distribution of many of these taxa typical from the Brazilian Shield are virtually unknown, being the environmental variation present in the Serra do Cachimbo and the transition with the Cerrado biome the most plausible determinants of their distribution boundaries (Fig. 25.1c). These environments at the upper portion of Tapajós River basin are then occupied by other unique taxa, typical from open habitats, such as the amphibians *Dendropsophus cachimbo*, *Adenomera martinezi*, *Scinax villasboasi*, and the reptiles *Hoplocercus spinosus*, *Tropidurus insulanus*, *Tupinambis matipu*, *Cercosaura eigenmanni*, *C. olivacea*, *Chironius flavolineatus* and *Lygophis meridionalis* (França et al. 2006; Ribeiro-Júnior 2015a, b; Ribeiro Júnior and Amaral 2016, 2017; Silva et al. 2018; Sturaro et al. 2018; Frost 2019; Uetz et al. 2019).

Focusing on the amphibian and reptile assemblages of the middle course of the Tapajós River, intensive standardized sampling in 11 localities from both banks of the river recorded a high diversity of species (193 taxa, Moraes et al. 2016) and challenged previously recognized biogeographic patterns (Jenkins et al. 2013; Moraes and Pavan 2017, 2018; Moraes et al. 2014, 2016, 2017a, 2019a, b). Since 2012, 10 amphibian and 3 reptile taxa recorded by Moraes et al. (2016) were formally described as new species (6% from total), 13 amphibian and 22 reptile taxa had their taxonomic status revised (18% from total), and about 30 amphibian and 9 reptile taxa still represent undescribed taxonomic entities (20% from total). Such expressive results highlight the importance and necessity of intensive faunal inventories, integrative data processing, and taxonomic refinement of databases to fill knowledge gaps in this Amazonian region and achieve greater accuracy in biogeographic analyses (Moraes et al. 2017b). High taxonomic diversity was also detected in this region for other vertebrate groups, such as mammals and birds (Oliveira et al. 2016; Maximiano et al. 2017).

Previous analyses of assemblage structure in the Tapajós river basin suggest that a dynamic equilibrium between regional historical events (i.e., riverine barriers) and local ecological conditions (i.e., flooding gradients) may have contributed (and still contribute) to the origin and maintenance of current species' geographic distribution patterns (Moraes et al. 2016). Specifically, they demonstrated changes in species composition along the flooding gradient, and differential effects of large rivers as barriers to dispersal in distinct zones of the flooding gradient (Moraes et al. 2016). Integration of biogeographic and ecological information shows that the permeability of a geographic barrier to Amazonian amphibians and reptiles unequally affects different taxonomic, functional and ecological groups, a result similarly found in other vertebrates (Gascon et al. 1998, 2000; Burney and Brumfield 2009; Fouquet et al. 2015; Naka and Brumfield 2018). The ecotonal zone crossed by the Tapajós River was also identified as a potential driver of geographic distribution patterns of amphibian and reptile assemblages on both banks of the middle Tapajós River (Moraes et al. 2016).

Building on the biogeographic patterns previously reported for this ecotonal landscape, and given the cratonic location of the basin (which is expected to result in a less dynamic geomorphological evolution), *we here implement a multi-taxa molecular study of amphibians and reptiles belonging to different functional groups, to (1) expand our knowledge about regional patterns of diversification and (2) to test the potential effect of the middle Tapajós River as a geographic barrier across space and time*. For that, we sequenced regions of the mitochondrial (mtDNA) genes *16S*, *ND4* and *COI* for 180 individuals representing taxa with distinct body sizes and ecological affinities, through methods described in the online Supplementary Material of this chapter. Comparative molecular data have become increasingly relevant to integrative studies of community assembly and evolutionary history by detecting shared spatial and temporal diversification patterns (Turchetto-Zolet et al. 2013; Baker et al. 2014), which are essential to identify the processes that generate and maintain current diversity patterns and landscape evolution (Bermingham and Moritz 1998). By investigating diversification of multiple co-distributed taxa, and evaluating spatial-temporal co-divergences, one can validate biogeographic models or hypotheses such as the relative effect of a river as a primary promoter or secondary contributor to diversification (Smith et al. 2014). To shed light on this issue, we compare the diversification patterns shown by 11 amphibian and five lizard taxa from the middle Tapajós River.

We focus on amphibians and reptiles because they are increasingly relevant model organisms in integrative studies of the Amazonian biota and landscape evolution (Marshall et al. 2018), given their sensitivity to climatic and environmental changes (Winter et al. 2016; Diele-Viegas et al. 2018) and relatively low individual mobility (Simões et al. 2014; Marshall et al. 2018). For the purposes of this study, we selected taxa present in high local abundances, and based on their distinct body sizes and ecological preferences, which may influence dispersal ability and hence genetic structure and diversification rates (Burney and Brumfield 2009; Cadena et al. 2011; Smith et al. 2014; Fouquet et al. 2015; Moraes et al. 2016; Harvey et al. 2017; Table 25.1, Figs. 25.4 and 25.5).

5 Amphibian and Reptile Diversification Across Space and Time

We used Bayesian phylogenetic gene trees to describe the spatial structure of infra-specific lineages in relation to the Tapajós River (see details in Supplementary Material in online version of this chapter). Haplotype sharing across river banks was observed in 31% of the analyzed taxa (5/16; the amphibians *B. tapajonica*, *L.* aff. *knudseni*, *O. taurinus* and *R.* gr. *margaritifera*, and the lizard *P. plica*), indicating the flow of individuals and genes between banks, or incomplete lineage sorting (Fig. 25.6). The remaining 69% taxa (11/16; the amphibians *A. andreae*, *A.* gr. *heyeri*, *A.* gr. *masniger*, *B. cinerascens*, *B.* gr. *albopunctata*, *P.* gr. *conspicillatus* and *R. magnussoni*, and the lizards *A. angulatus*, *A. reticulata*, *C.* gr. *ocellata* and *C. amazonicus*) have reciprocally monophyletic lineages on opposite banks of the Tapajós River (Figs. 25.6 and 25.7).

The lineages of *Adenomera andreae* present low values of genetic distance between opposite margins, suggesting relatively recent divergence (Fig. 25.7). In five amphibian taxa (*B. cinerascens*, *B.* gr. *albopunctata*, *A.* gr. *heyeri*, *A.* gr. *masniger*, *P.* gr. *conspicillatus*), the average genetic distance between banks varied from 4% to 11%, surpassing the threshold recognized for interspecific divergence (Fouquet et al. 2007a, b; Kok et al. 2018; Fig. 25.7). This suggests cryptic diversity, a common finding in Amazonian amphibians (Fouquet et al. 2007b; Vacher et al. 2017). Within those taxa, only five distinct lineages are formally described: *Boana leucocheila* and *Boana multifasciata* (from *B.* gr. *albopunctata*), *Allobates masniger* and *Allobates nunciatus* Moraes et al. 2019b (from *A.* gr. *masniger*), and *Pristimantis latro* (from *P. gr. conspicillatus*), while several other lineages remain unnamed, although they present clear phenotypic divergences in morphology and advertisement calls and are presently under description (authors, pers. obs.). Within the lizards, genetic distances between margins were also high; in three taxa (*A. reticulata*, *A. angulatus* and *C. amazonicus*) they surpass the proposed threshold of interspecific divergence based on the *ND4* gene (Kok et al. 2018; Fig. 25.7), ranging from 12% to 28% in average. This observation is consistent with the high levels of molecular differentiation associated with overall phenotypic conservatism reported for other lizard species in Amazonia (Geurgas and Rodrigues 2010; Nunes et al. 2012; Sturaro et al. 2018). Further analyses may revise the taxonomic status of these cryptic lizard lineages.

We did not detect high prevalence of riverine-related structure associated with any particular trait category regarding body size, forest type, or microhabitat use. Reciprocal monophyly across river banks was detected in taxa with small (e.g., *A.*

Table 25.1 Focal amphibian and lizard taxa from banks of Tapajós River, Pará State, Brazil, from which we investigated spatial and temporal diversification patterns with a molecular approach

Class	Order	Family	Taxa	Body size (mm)	Forest type	Microhabitat	N
Amphibia	Anura	Aromobatidae	*Allobates* gr. *masniger*	20	Non-flooded	Terrestrial	27
Amphibia	Anura	Craugastoridae	*Pristimantis* gr. *conspicillatus*	40	Non-flooded	Terrestrial	58
Amphibia	Anura	Leptodactylidae	*Adenomera andreae* (Müller, 1923)	20	Non-flooded	Terrestrial	22
Amphibia	Anura	Leptodactylidae	*Adenomera* gr. *heyeri*	25	Non-flooded	Terrestrial	11
Amphibia	Anura	Leptodactylidae	*Leptodacylus* aff. *knudseni*	140	Non-flooded	Terrestrial	8
Amphibia	Anura	Hylidae	*Boana cinerascens* (Spix, 1824)	35	Flooded	Arboreal	4
Amphibia	Anura	Hylidae	*Boana* gr. *albopunctata*	65	Flooded	Arboreal	7
Amphibia	Anura	Hylidae	*Osteocephalus taurinus* Steindachner, 1862	90	Generalist	Arboreal	13
Amphibia	Anura	Bufonidae	*Rhinella* gr. *margaritifera*	70	Non-flooded	Terrestrial	23
Amphibia	Anura	Bufonidae	*Rhinella magnussoni* Lima et al. 2007	45	Non-flooded	Terrestrial	16
Amphibia	Caudata	Plethodontidae	*Bolitoglossa tapajonica* Brcko et al. 2013	45	Non-flooded	Arboreal	5
Reptilia	Squamata	Gymnophthalmidae	*Arthrosaura reticulata* (O'Shaughnessy, 1881)	55	Non-flooded	Terrestrial	8
Reptilia	Squamata	Gymnophthalmidae	*Cercosaura* gr. *ocellata*	50	Non-flooded	Terrestrial	9
Reptilia	Squamata	Gymnophthalmidae	*Alopoglossus angulatus* (Linnaeus, 1758)	55	Non-flooded	Terrestrial	4
Reptilia	Squamata	Sphaerodactylidae	*Chatogekko amazonicus* (Andersson, 1918)	20	Non-flooded	Terrestrial	15
Reptilia	Squamata	Tropiduridae	*Plica plica* (Linnaeus, 1758)	145	Generalist	Arboreal	8

Body size (mean of snout-vent length based on field data), ecological preferences (typical forest types and microhabitats in which they are present), and number of specimens included in analyses (N), are detailed

Fig. 25.4 Focal amphibian taxa from the banks of the Tapajós River and studied here (**a**) *Allobates masniger* (Morales, 2002) (from *Allobates* gr. *masniger*); (**b**) *Pristimantis latro* (Oliveira et al. 2017) (from *Pristimantis* gr. *conspicillatus*); (**c**) *Adenomera andreae* (Müller 1923); (**d**) *Adenomera* gr. *heyeri*; (**e**) *Leptodactylus* aff. *knudseni*; (**f**) *Boana cinerascens* (Spix 1824); (**g**) *Boana*

gr. *masniger*, *C. amazonicus*) and medium (e.g., *B. cinerascens*) body-sizes, from flooded (e.g., *B. cinerascens*, *B.* gr. *albopunctata*) and non-flooded forests (e.g., *R. magnussoni*, *A.* gr. *heyeri*), and with terrestrial (e.g. *A. andreae*, *A. angulatus*) and arboreal habits (e.g., *B. cinerascens*, *B.* gr. *albopunctata*). The only taxa lacking examples of riverine-congruent divergences were those large-bodied (e.g. *O. taurinus*, *P. plica*), which may result from increased dispersal, and hence flow of individuals and genes, between banks. These results reinforce how idiosyncratic the diversification of distinct functional groups in association with Amazonian rivers can be (Fouquet et al. 2015; Moraes et al. 2016), especially in areas of highly dynamic landscape history. Still, additional sampling is needed to confirm if this observation is widespread in large-bodied organism.

All focal taxa with reciprocally monophyletic haplogroups on distinct river banks were submitted to temporal analyses, using published rates of substitution to time-calibrate the mitochondrial gene trees (see details in methods at Supplementary Material in online version of this chapter). We recovered a continuous range of riverine-associated divergence times, which we group in three main time periods for discussion purposes: (1) ancient divergences, dating from the middle Miocene (ca. 10 Ma; *C. amazonicus*); (2) divergences in the late Miocene-Pliocene (ca. 5–3.5 Ma; *A.* gr. *masniger*, *A.* gr. *heyeri*, *B.* gr. *albopunctata*, *P.* gr. *conspicillatus*, and *A. angulatus*), and (3) more recent divergences, dating to Pleistocene (ca. 2–1 Ma; *A. andreae*, *R. magnussoni* and *B. cinerascens*; Fig. 25.8).

6 Landscape Evolution and Diversification at the Tapajós River Basin

Our review described how the Tapajós River basin landscape would have varied through time, and our molecular results, allied to other published data, are consistent with a hypothesis that this dynamism shaped the multiple diversification histories of its native biota. This biological impact of landscape dynamism of the basin is also in agreement with its high number of endemic taxa, such as the amphibians *Allobates tapajos*, *A. magnussoni*, *B. tapajonica* and *R. magnussoni* (Lima et al. 2007, 2014, 2015; Brcko et al. 2013; Maia et al. 2017), several undescribed taxa (authors, pers. obs.), the lizard *G. tapajonicus*, the birds *L. vilasboasi* (Barrera-Guzmán et al. 2017; Dias et al. 2018) and *Thamnophilus nigrocinereus huberi* (Zimmer and Isler 2019), the monkey *Mico leucippe* (Pimenta and Silva Jr. 2005) and other mammals (Oliveira et al. 2016), besides several taxa of alluvial trees (Ferreira et al. 2013). In

Fig. 25.4 (continued) multifasciata (Günther 1859) (from gr. *albopunctata*); (**h**) *Osteocephalus taurinus* (Steindachner 1862). Photographs by José Cassimiro (**a**, **h**), Leandro Moraes (**b–f**) and Luis Storti (**g**)

Fig. 25.5 Focal amphibian and lizard taxa from the banks of Tapajós River, and studied here (**a**) *Rhinella* gr. *margaritifera*; (**b**) *Rhinella magnussoni* (Lima et al. 2007); (**c**) *Bolitoglossa tapajonica* (Brcko et al. 2013); (**d**) *Arthrosaura reticulata* (O'Shaughnessy 1881); (**e**) *Cercosaura* gr. *ocellata*;

addition, other bird, mammal, amphibian and reptile taxa have their current distributions delimited by the rivers within this basin (Cracraft 1985; Ayres and Clutton-Brock 1992; Ávila-Pires 1995; Silva et al. 2002; Moraes et al. 2016).

By analyzing the evolutionary histories of amphibians and lizards, we found a nearly continuous pattern of Plio-Pleistocene-centered divergence dates for taxa that currently occupy opposite margins of the Tapajós. These dates are older than most splits recorded for upland understory birds, which occurred mainly during the Pleistocene (Ribas et al. 2012; Silva et al. 2019). Amphibians and reptiles have lower dispersal abilities when compared to birds, present a strong association with landscape changes, and may retain older historical signals in their genetic profiles. The final uplift of the Brazilian Shield took place during the Late Miocene-Early Pliocene transition (7–5 Ma) (Del'Arco and Bezerra 1989), and this uplift is a known driver of diversification in amphibians and reptiles (Werneck et al. 2012b). We propose that the older splits inferred for some of the taxa analyzed here indicate changes in river courses that may have occurred by terrain adjustments during the uplift (10–3.5 Ma), segregating or gathering biological populations along the different areas of the flooding gradient created in their banks. Subsequent course rearrangements generated by neotectonic events, and changes in permeability due to climatic variations that altered water levels, could have maintained gene flow in some populations or extinguished some taxa due to stronger climatic instability in this region (Irion et al. 2006, 2010; Wang et al. 2017). Recolonizations from the climatically stable western Amazonia, and recent interruptions of gene flow associated with the higher water level of the system, may have driven the other, most recent splits detected in species of birds (Ribas et al. 2012; Silva et al. 2019), and may explain the more recent diversification events inferred here for some of the amphibians (2–1 Ma). Previous studies of vicariant diversification across the Tapajós River were similarly unable to detect synchronic responses to a single historical event—and found a signature of reticulated diversification in lineages isolated on opposite banks (Haffer 1997; Weir et al. 2015; Barrera-Guzmán et al. 2017; Dias et al. 2018; Pulido-Santacruz et al. 2018), with several periods of interruption and reconnection of gene flow (Naka and Brumfield 2018). As explained below, however, vicariant diversification is one of the several alternative hypotheses that may explain the genetic patterns we observed.

The fact that the Tapajós River runs through an ecotonal zone magnifies the landscape dynamism of the Tapajós River basin over time, certainly influencing the biotic diversification in this region (Moraes et al. 2016). Geologically and climatically dynamic zones are known to increase diversification through allopatric speciation and by presenting new opportunities for ecological adaptation (Hoorn et al. 2013; Tuomisto et al. 2016). Taxa with range edges in this ecotonal zone may be especially affected by the rise of a riverine geographical barrier (Moraes et al. 2016). However, recognizing diversification events determined by ecological barriers is a challenge, because such events generate evolutionary signatures with less evident

Fig. 25.5 (continued) (**f**) *Alopoglossus angulatus* (Linnaeus 1758); (**g**) *Chatogekko amazonicus* (Andersson 1918); (**h**) *Plica plica* (Linnaeus 1758). Photographs by Leandro Moraes

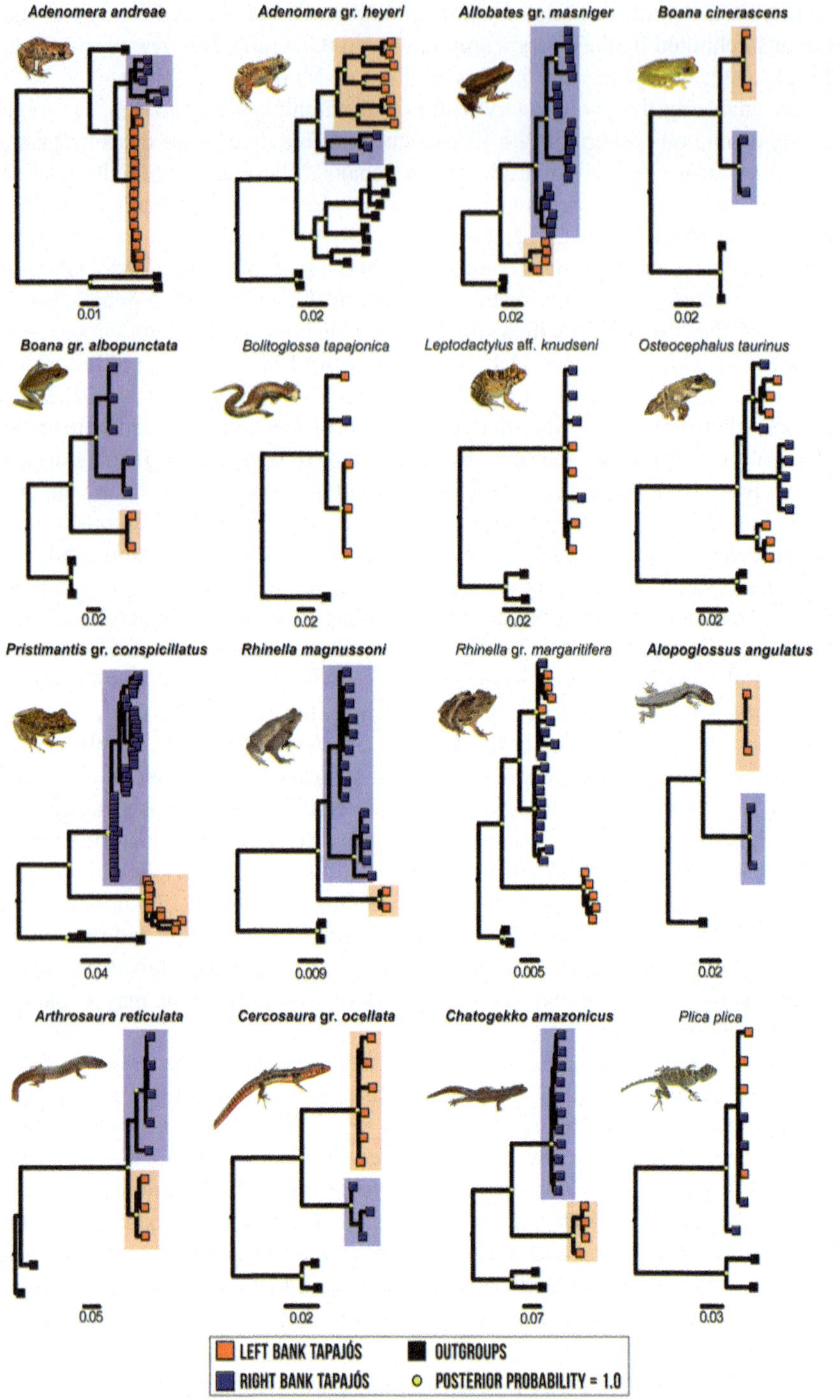

Fig. 25.6 Bayesian phylogenetic gene trees based on mtDNA variation in 16 focal taxa (*16S* + *COI* for amphibians and *16S* + *ND4* for lizards), indicating haplotype occurrence across distinct banks of

limits (Ortiz et al. 2018), compared to diversification determined by physical barriers. We propose that taxa with putative origin center centered in the western or the eastern Amazon, and which have their range edges in this ecotonal zone, may have been particularly impacted by ecological diversification combined to the emergence of the Tapajós River. Among the taxa analyzed here, *A.* gr. *masniger* is centered in the west (Tsuji-Nishikido et al. 2012), *B.* gr. *albopunctata* and *A.* gr. *heyeri*, are centered in the east (Fouquet et al. 2014; Pinheiro et al. 2018), and *C. amazonicus* has western and eastern divergent lineages (Geurgas and Rodrigues 2010) and may have been affected by the emergence of a geographical barrier in the ecotonal zone. Other examples may include taxa restricted to the middle Tapajós River, such as *L. vilasboasi* (Barrera-Guzmán et al. 2017), *G. tapajonicus* (Ribeiro-Júnior 2015b), and an undescribed taxa of *Cercosaura* Wagler, 1830 (Ribeiro-Júnior and Amaral 2017). Retractions of suitable habitats due to historical climatic variations, as pointed out in the evolutionary history of *L. vilasboasi*, may also have affected the diversification of these organisms in this ecotonal zone, by isolating its populations (Weir et al. 2015). With the expansion of suitable habitats, previously isolated populations would have been reconnected, especially in the basin headwaters, where hybridization zones are historically recognized (Haffer 1997; Weir et al. 2015).

Importantly, however, we cannot discard the occurrence of independent colonization events, or that the patterns we observe reflect secondary contact between formerly isolated lineages (Moritz et al. 2000; Naka and Brumfield 2018). Some taxa showing evidence of secondary contacts include non-sister lineages present in opposite banks of the Tapajós River, such as lizards in the genera *Alopoglossus* (Ribeiro-Júnior and Amaral 2017), *Cercosaura* (Sturaro et al. 2018) and *Chatogekko* (Geurgas and Rodrigues 2010). In fact, the lizard *Chatogekko amazonicus* showed the most ancient divergence across the Tapajós River; and previous studies already revealed a pattern of older split dates within this species and close relatives, compared to other lizards diversification (Geurgas et al. 2008; Gamble et al. 2011), as well as the occurrence of two very divergent lineages in western and eastern Amazonia, which may have the Tapajós River as their distributional boundary (Geurgas et al. 2008; Geurgas and Rodrigues 2010). Contacts between taxa typical of riparian forests, such as *B. cinerascens* and *B.* gr. *albopunctata*, may also be secondary, since the reduction of current gene flow is counterintuitive in these environments known for greater connectivity in relation to non-riparian forests (Moraes et al. 2016; Harvey et al. 2017). As abiotic information is sparse in this erosive region, and given the high dynamism to which the Tapajós river basin was subjected, a greater volume of biotic data and comparative approaches is needed to test for possible co-divergence events and further advance our understanding of the processes involved in the evolution of this landscape and its biota.

Fig. 25.6 (continued) the Tapajós River. Taxa in bold represent those with reciprocally monophyletic haplogroups on each river bank

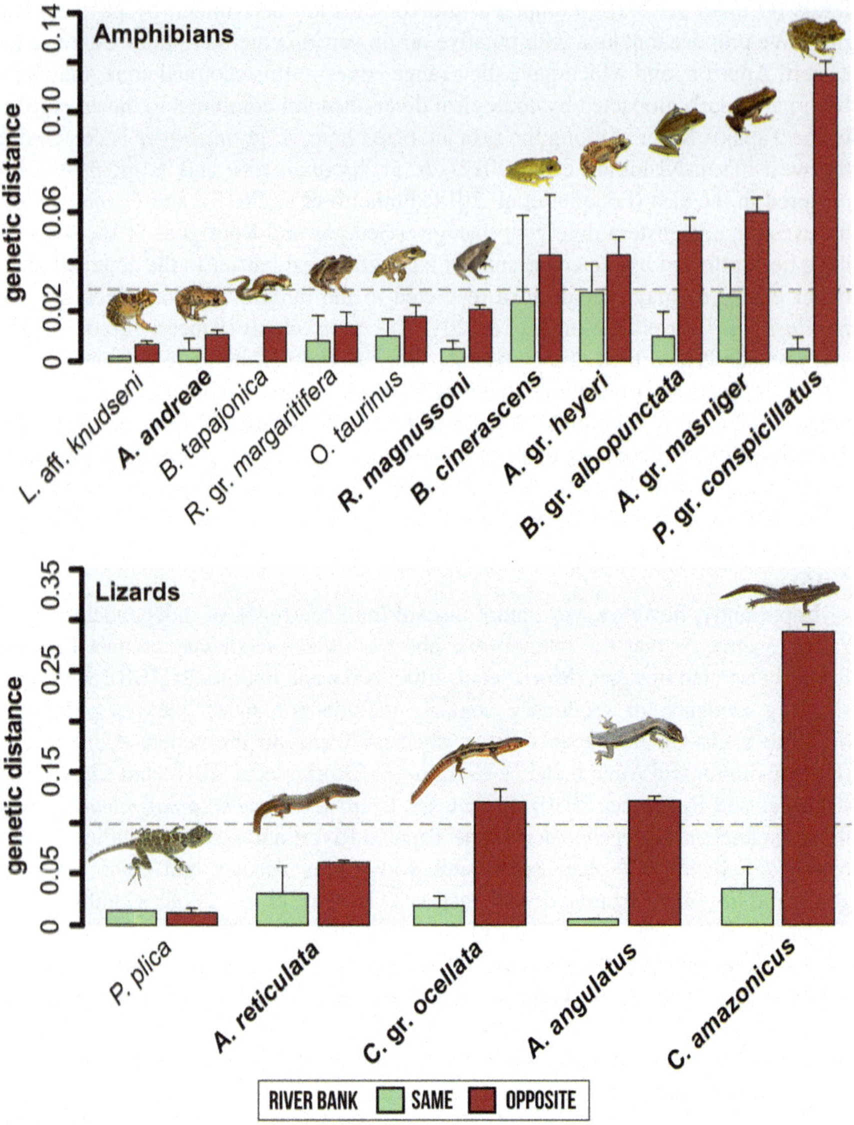

Fig. 25.7 Genetic uncorrected p-distances (mean ± standard deviation) between individuals of the 16 amphibian and lizard focal taxa from same and opposite Tapajós River bank (considering mtDNA gene *16S* for amphibians and *ND4* for lizards). Dashed lines represent known thresholds of interspecific distance for amphibian *16S* gene (Fouquet et al. 2007a, b) and lizard *ND4* gene (Kok et al. 2018). Taxa in bold represent those with reciprocally monophyletic haplogroups on each river bank

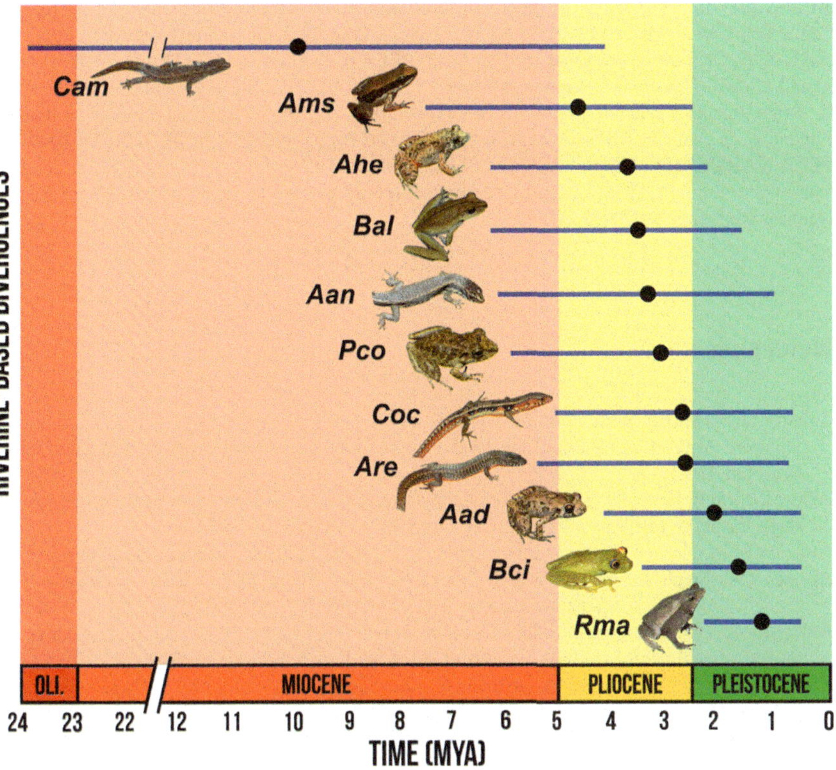

Fig. 25.8 Variation in temporal diversification of focal taxa with distinct lineages between opposite Tapajós River banks, showing divergence times of nodes (median ± 95% HPD) corresponding to current Tapajós River conformation, estimated with BEAST. Acronyms: (Oli.) Oligocene period; (Cam) *Chatogekko amazonicus*; (Ams) *Allobates* gr. *masniger*; (Ahe) *Adenomera* gr. *heyeri*; (Bal) *Boana* gr. *albopunctata*; (Aan) *Alopoglossus angulatus*; (Pco) *Pristimantis* gr. *conspicillatus*; (Coc) *Cercosaura* gr. *ocellata*; (Are) *Arthrosaura reticulata*; (Aad) *Adenomera andreae*; (Bci) *Boana cinerascens*; (Rma) *Rhinella magnussoni*

Acknowledgements Leandro J.C.L. Moraes is a member of Dra. Fernanda Werneck's lab at the Instituto Nacional de Pesquisas da Amazônia (INPA), and developed this work under a scholarship (131169/2013-0) and a research fellowship (313055/2015-7) from the Conselho Nacional de Desenvolvimento Científico e Tecnológico (CNPq) and a research fellowship (062.00962/2018) from the Fundação de Amparo à Pesquisa do Estado do Amazonas (FAPEAM). We thank Luis F. Storti, José Cassimiro, Jerriane Gomes, Mauro Hoffman, Tainá F.D. Rodrigues, José Mário B. Ghelere, Ana B. Barros, Elizângela S. Brito help in fieldwork sampling and to CNEC WorleyParsons Engenharia S.A. for fieldwork financial and logistical support. We are also grateful to José Cassimiro, Luis Storti and Elizângela S. Brito for contribution with landscape and specimen photographs; Ana Carolina Carnaval and an anonymous reviewer for the careful manuscript revision and editing; as well as to Erik Choueri, Mateus Ferreira, members of Thematic Laboratory of Molecular Biology from Instituto Nacional de Pesquisas da Amazônia (INPA) and Maria C. Barros' Molecular Biology lab from Universidade Estadual do Maranhão—campus Caxias (UEMA) for assistance with the molecular data survey and analyses. Fernanda P. Werneck thanks financial support from CNPq (projects 475559/2013-4 and 425571/2018-1; productivity fellowship

305535/2017-0), FAPEAM (projects 062.00665/2015, 062.01110/2017, and 062.00962/2018), Partnerships for Enhanced Engagement in Research from the U.S. National Academy of Sciences and U.S. Agency of International Development (PEER NAS/USAID AID-OAA-A-11-00012, cycle 3) and L'Oréal-Unesco For Women in Science Program (Brazil/ABC 2016 and IRT 2017). Camila C. Ribas thanks financial support from CNPq (308927/2016-8); National Science Foundation-Fundação de Amparo à Pesquisa do Estado de São Paulo (NSF-FAPESP) (2012/50260-6) and PEER NAS/USAID (AID-OAA-A-11-00012, cycle 5). The specimens were collected under collection permit number 066/2012 provided by Instituto Brasileiro do Meio Ambiente e dos Recursos Naturais Renováveis (IBAMA).

References

Ackerly DD, Thomas WW, Ferreira CAC, Pirani JR (1989) The forest-Cerrado transition zone in Southern Amazonia: results of the 1985 Projeto Flora Amazonica expedition to Mato Grosso. Brittonia 41:113–128

Albert JS, Val P, Hoorn C (2018) The changing course of the Amazon River in the neogene: center stage for neotropical diversification. Neotrop Ichthyol 16(3):e180033

Almeida FFM, Hasui I (1984) O Pré–Cambriano do Brasil. Edgard Blücher, São Paulo, p 378

Alvares CA, Stape JL, Sentelhas PC, Gonçalves JLM, Sparovek G (2013) Koppen's climate classification map for Brazil. Meteorol Z 22:711–728

Antonelli A, Nylander JAA, Persson C, Sanmartín I (2009) Tracing the impact of the Andean uplift on neotropical plant evolution. Proc Natl Acad Sci U S A 106(24):9749–9754

Antonelli A, Ariza M, Albert J, Andermann T, Azevedo J, Bacon C, Faurby S, Guedes T, Hoorn C, Lohmann L, Matos-Maraví P, Ritter C, Sanmartín I, Silvestro D, Tejedor M, ter Steege H, Tuomisto H, Werneck F, Zizka A, Edwards S (2018) Conceptual and empirical advances in neotropical biodiversity research. PeerJ 6:e5644

Araujo JFV, Montalvão RMG, Lima MIC, Fernandes ECA, Cunha FMB, Fernandes CAC, Basei MAS (1976) Geologia. In: RADAMBRASIL. Folha SA.21 Santarém, Projeto Radam Brasil, Brasília, pp 17–131

Arruda DM, Schaefer CEGR, Fonseca RS, Solar RRC, Fernandes-Filho EI (2018) Vegetation cover of Brazil in the last 21 ka: new insights into the Amazonian refugia and Pleistocenic arc hypotheses. Glob Ecol Biogeogr 27:47–56

Ávila-Pires TCS (1995) Lizards of Brazilian Amazonia (Reptilia: Squamata). Zoologische Verhandelingen 299:1–706

Ayres JM, Clutton-Brock TH (1992) River boundaries and species range size in Amazonian primates. Am Nat 14:531–537

Baker PA, Fritz SC, Dick CW, Eckert AJ, Horton BK, Manzoni S, Ribas CC, Garzione CN, Battisti DS (2014) The emerging field of geogenomics: constraining geological problems with genetic data. Earth Sci Rev 135:38–47

Barrera-Guzmán AO, Aleixo A, Shawkey MD, Weir JT (2017) Hybrid speciation leads to novel male secondary sexual ornamentation of an Amazonian bird. Proc Natl Acad Sci U S A 115(2): E218–E225

Batalha-Filho H, Fjeldsa J, Fabre PH, Miyaki CY (2012) Connections between the Atlantic and the Amazonian forest avifaunas represent distinct historical events. J Ornithol 154:41–50

Bermingham E, Moritz C (1998) Comparative phylogeography: concepts and applications. Mol Ecol 7:367–369

Brcko I, Hoogmoed MS, Neckel-Oliveira S (2013) Taxonomy and distribution of the salamander genus *Bolitoglossa* Duméril, Bibron & Duméril, 1854 (Amphibia, Caudata, Plethodontidae) in Brazilian Amazonia. Zootaxa 3686:401–431

Burney CW, Brumfield RT (2009) Ecology predicts levels of genetic differentiation in neotropical birds. Am Nat 174:358–368

Cadena CD, Gutiérrez-Pinto N, Dávila N, Chesser RT (2011) No population genetic structure in a widespread aquatic songbird from the neotropics. Mol Phylogenet Evol 58:540–545
Caputo MV, Soares EAA (2016) Eustatic and tectonic change effects in the reversion of the transcontinental Amazon River drainage system. Braz J Geol 46(2):301–328
Cheng H, Sinha A, Cruz FW, Wang X, Edwards RL, d'Horta FM, Ribas CC, Vuille M, Stott LD, Auler AS (2013) Climate change patterns in Amazonia and biodiversity. Nat Commun 4 (1):1411–1416
Clark SR (2018) Uncertainty in the breakup, spreading history, and velocity variations of Gondwana. Gondw Res 53:189–196
Coêlho FJF, Ferreira HC, Barros-Silva S, Ribeiro AG, Terezo EFM (1976) Vegetação. In: RADAMBRASIL. Folha SA.21 Santarém, Projeto Radam Brasil, Brasília, pp 309–414
Cohn-Haft M, Naka LN, Fernandaes AM (2007) Padrões de distribuição da avifauna da várzea dos rios Solimões e Amazonas. In: Albernaz AL (ed) Conservação da várzea: identificação e caracterização de regiões biogeográficas. Ibama/ProVárzea, pp 287–323
Cracraft J (1985) Historical biogeography and patterns of differentiation within the South American avifauna: areas of endemism. Ornithol Monogr 36:49–84
Dagosta FCP, de Pinna M (2017) Biogeography of Amazonian fishes: deconstructing river basins as biogeographic units. Neotrop Ichthyol 15:e170034
Del'Arco JO, Bezerra PEL (1989) Geologia. In: Fundação Instituto Brasileiro de Geografia e Estatística (ed) Geografia do Brasil. In: Região Centro-Oeste, vol 1. IBGE, Rio de Janeiro, pp 35–51
Dias C, Lima KA, Araripe J, Aleixo A, Vallinoto M, Sampaio I, Schneider A, Rêgo PS (2018) Mitochondrial introgression obscures phylogenetic relationships among manakins of the genus *Lepidothrix* (Aves: Pipridae). Mol Phylogenet Evol 126:314–320
Diele-Viegas L, Vitt LJ, Sinervo B, Colli GR, Werneck FP, Miles DB, Magnusson WE, Santos JC, Sette CM, Caetano GHO, Pontes E, Ávila-Pires TCS (2018) Thermal physiology of Amazonian lizards (Reptilia: squamata). PLoS One 13(3):e0192834
Duellman WE (1979) The South American herpetofauna: its origin, evolution, and dispersal. Museum of Natural History/University of Kansas, Kansas, p 504
Ferreira LV, Cunha DA, Chaves PP, Matos DCL, Parolin P (2013) Impacts of hydroelectric dams on alluvian riparian plant communities in eastern Brazilian Amazonian. An Acad Bras Cienc 85:1013–1023
Ferreira M, Aleixo A, Ribas CC, Santos MPD (2017) Biogeography of the neotropical genus *Malacoptila* (Aves: Bucconidae): the influence of the Andean orogeny, Amazonian drainage evolution and palaeoclimate. J Biogeogr 44:748–759
Fouquet A, Vences M, Salducci MD, Meyer A, Marty C, Blanc M, Gilles A (2007a) Revealing cryptic diversity using molecular phylogenetics and phylogeography in frogs of the *Scinax ruber* and *Rhinella margaritifera* species groups. Mol Phylogenet Evol 43:567–582
Fouquet A, Gilles A, Vences M, Marty C (2007b) Underestimation of species richness in neotropical frogs revealed by mtDNA analyses. PLoS One 2:e1109
Fouquet A, Cassini CS, Haddad CFB, Pech N, Rodrigues MT (2014) Species delimitation, patterns of diversification and historical biogeography of the Neotropical frog genus Adenomera (Anura, Leptodactylidae). J Biogeogr 41:855–870
Fouquet A, Courtois EA, Baudain D, Jucivaldo Dias Lima JD, Souza SM, Noonan BP, Rodrigues MT (2015) The trans–riverine genetic structure of 28 Amazonian frog species is dependent on life history. J Trop Ecol 31:361–373
França FGR, Mesquita DO, Colli G (2006) A checklist of snakes from Amazonian Savannas in Brazil, housed in the Coleção Herpetológica da Universidade de Brasília, with new distribution records. Occas Pap 17:1–13
Frost DR (2019) Amphibian species of the world: an online reference. Version 6.0. http://research.amnh.org/herpetology/amphibia/index.html
Gamble T, Daza JD, Colli GR, Vitt LJ, Bauer AM (2011) A new genus of miniaturized and pug-nosed gecko from South America (Sphaerodactylidae: Gekkota). Zool J Linnean Soc 163:1244–1266

Garzione CN, McQuarrie N, Perez ND, Ehlers TA, Beck SL, Kar N, Eichelberger N, Chapman AD, Ward KM, Ducea MN, Lease RO, Poulsen CJ, Wagner LS, Saylor JE, Zandt G, Horton BK (2017) Tectonic evolution of the Central Andean plateau and implications for the growth of plateaus. Annu Rev Earth Planet Sci 45:529–559

Gascon C, Lougheed SC, Bogart JP (1998) Patterns of genetic population differentiation in four species of Amazonian frogs: a test of the Riverine Barrier Hypothesis. Biotropica 30:104–119

Gascon C, Malcolm JR, Patton JL, Silva MNF, Bogart JP, Lougheed SC, Peres CA, Neckel S, Boag P (2000) Riverine barriers in the geographic distribution of Amazonian species. Proc Natl Acad Sci U S A 97:13672–13677

Geurgas S, Rodrigues MT (2010) The hidden diversity of *Coleodactylus amazonicus* (Sphaerodactylinae, Gekkota), revealed by molecular data. Mol Phylogenet Evol 54:583–593

Geurgas SR, Rodrigues MT, Moritz C (2008) The genus *Coleodactylus* (Sphaerodactylinae, Gekkota) revisited: a molecular phylogenetic perspective. Mol Phylogenet Evol 49:92–101

Godinho MBC, da Silva FR (2018) The influence of riverine barriers, climate, and topography on the biogeographic regionalization of Amazonian anurans. Sci Rep 8:3427

Günther ACLG (1859) Catalogue of the Batrachia Salientia in the collection of the British Museum. Taylor and Francis, London

Haffer J (1969) Speciation in Amazonian forest birds. Science 165:131–137

Haffer J (1997) Contact zones between birds of southern Amazonia. Ornithol Monogr 48:281–305

Harvey MG, Aleixo A, Ribas CC, Brumfield RT (2017) Habitat association predicts genetic diversity and population divergence in Amazonian birds. Am Nat 190:631–648

Hayakawa EH, Rossetti DF (2015) Late quaternary dynamics in the Madeira river basin, southern Amazonia (Brazil), as revealed by paleomorphological analysis. An Acad Bras Cienc 87:29–49

Hoorn C, Wesselingh FP, ter Steege H, Bermudez MA, Mora A, Sevink J, Sanmartín I, Sanchez-Meseguer A, Anderson CL, Figueiredo JP, Jaramillo C, Riff D, Negri FR, Hooghiemstra H, Lundberg J, Stadler T, Särkinen T, Antonelli A (2010a) Amazonia through time: Andean uplift, climate change, landscape evolution, and biodiversity. Science 330(6006):927–931

Hoorn C, Wesselingh FP, Hovikoski J, Guerrero J (2010b) The development of the Amazonian mega–wetland (Miocene; Brazil, Colombia, Peru, Bolivia). In: Hoorn C, Wesselingh FP (eds) Amazonia: landscape and species evolution. Wiley, Oxford, pp 123–142

Hoorn C, Mosbrugger V, Mulch A, Antonelli A (2013) Biodiversity from mountain building. Nat Geosci 6:154

Hoorn C, Bogota-A GR, Romero-Baez M, Lammertsma EI, Flantua SGA, Dantas EL, Dino R, do Carmo DA, Chemale F (2017) The Amazon at sea: onset and stages of the Amazon River from a marine record, with special reference to neogene plant turnover in the drainage basin. Glob Planet Chang 153:51–65

Horton BK, Parra M, Saylor J, Nie J, Mora A, Torres V, Stockli D, Strecker M (2010) Resolving uplift of northern Andes using detritical zircons age signatures. GSA Today 20(7):4–9

Irion G, Bush MB, de Mello JASN, Stüben D, Neumann T, Müller G, de Morais JO, Junk JW (2006) A multiproxy palaeoecological record of Holocene lake sediments from the Rio Tapajós, eastern Amazonia. Palaeogeogr Palaeoclimatol Palaeoecol 240:523–535

Irion G, de Mello JASN, Morais J, Piedade MTF, Junk WJ, Garming L (2010) Development of the Amazon valley during the Middle to Late Quaternary: sedimentological and climatological observations. In: Junk WJ, Piedade MTF, Wittmann F, Schöngart J, Parolin P (eds) Amazonian floodplain forests – ecophysiology, biodiversity and sustainable management. Springer, Berlin, pp 27–42

Jaramillo C, Romero IC, D'Apolito C, Bayona G, Duarte E, Louwye S, Escobar J, Luque J, Carrillo-Briceño JD, Zapata V, Mora JA, Schouten S, Zavada M, Harrington G, Ortiz J, Wesselingh FP (2017) Miocene flooding events of western Amazoni. Sci Adv 3(5):e1601693

Jenkins CN, Pimm SL, Joppa LN (2013) Global patterns of terrestrial vertebrate diversity and conservation. Proc Natl Acad Sci U S A 110:E2602–E2610

Junk WJ, Piedade MTF, Schöngart J, Cohn-Haft M, Adeney JM, Wittmann FA (2011) Classification of major naturally–occurring Amazonian lowland wetlands. Wetlands 31:623–640

Kok PJR, Bittenbinder MA, van den Berg JK, Marques Souza S, Nunes PMS, Laking AE, Teixeira M Jr, Fouquet A, Means DB, MacCulloch RD, Rodrigues MT (2018) Integrative taxonomy of the gymnophthalmid lizard *Neusticurus rudis* Boulenger, 1900 identifies a new species in the eastern Pantepui region, north-eastern South America. J Nat Hist 52:1029–1066

Latrubesse EM, Franzinelli E (2005) The Late quaternary evolution of the Negro River, Amazon, Brazil: implications for island and floodplain formation in large anabranching tropical systems. Geomorphology 70:372–397

Latrubesse EM, Kalicki T (2002) Late quaternary palaeohydrological changes in the upper Purus basin southwestern Amazônia, Brazil. Z Geomorphol 129:41–59

Latrubesse EM, Rancy A (1998) The late quaternary of the Upper Juruá River, Amazonia, Brazil. Q S Am Antarctic Peninsula 11:27–46

Latrubesse EM, Stevaux JC, Sinha R (2005) Tropical rivers. Geomorphology 70:187–206

Latrubesse EM, Silva SF, Cozzuol M, Absy ML (2007) Late Miocene continental sedimentation in the southwestern Amazonia and its regional significance: biotic and geological evidence. J S Am Earth Sci 23:61–80

Latrubesse EM, Cazzuol M, Silva-Caminha SAF, Rigsby CA, Absy ML, Jaramillo C (2010) The Late Miocene paleogeography of the Amazon Basin and the evolution of the Amazon River. Earth Sci Rev 99:99–124

Latrubesse EM, Arima EY, Dunne T, Park E, Baker VR, d'Horta FM, Wight C, Wittmann F, Zuanon J, Baker PA, Ribas CC, Norgaard RB, Filizola N, Ansar A, Flyvbjerg B, Stevaux JC (2017) Damming the rivers of the Amazon basin. Nature 546:363–369

Ledo RMD, Colli GR (2017) The historical connections between the Amazon and the Atlantic Forest revisited. J Biogeogr 44:2551–2563

Leite RN, Rogers DS (2013) Revisiting Amazonian phylogeography: insights into diversification hypotheses and novel perspectives. Org Divers Evol 13(4):639–664

Lima AP, Menin M, Araújo MC (2007) A new species of *Rhinella* (Anura: Bufonidae) from Brazilian Amazon. Zootaxa 1663:1–15

Lima AP, Simões PI, Kaefer IL (2014) A new species of *Allobates* (Anura: Aromobatidae) from the Tapajós River basin, Pará State, Brazil. Zootaxa 3889:355–387

Lima AP, Simões PI, Kaefer IL (2015) A new species of *Allobates* (Anura: Aromobatidae) from Parque Nacional da Amazônia, Pará State, Brazil. Zootaxa 3980:501–525

Loureiro RL, Dias AA, Magnago H (1980) Vegetação. In: RADAMBRASIL. Folha SC.21 Juruena, Projeto Radam Brasil, Brasília, pp 325–376

Maia GF, Lima AP, Kaefer IL (2017) Not just the river: genes, shapes, and sounds reveal population–structured diversification in the Amazonian frog *Allobates tapajos* (Dendrobatoidea). Biol J Linn Soc 121(1):95–108

Marshall JC, Bastiaans E, Caccone A, Camargo A, Morando M, Niemiller ML, Pabijan M, Russello MA, Sinervo B, Sites JW Jr, Vences M, Werneck FP, Valero KCW, Steinfartz S (2018) Mechanisms of speciation in reptiles and amphibians: a synopsis. PeerJ 6:e27279v1. https://doi.org/10.7287/peerj.preprints.27279v1

Maximiano MFA, d'Horta FM, Ribas CC (2017) Determinantes físicos e ecológicos que afetam as assembleias de aves no sudeste da Amazônia: o papel da história na distribuição das espécies. M.Sc. Dissertation, Instituto Nacional de Pesquisas da Amazônia, Manaus, AM

Mertes LAK, Dunne T, Martinelli LA (1996) Channel–floodplain geomorphology along the Solimoes-Amazon River, Brazil. Geol Soc Am Bull 108:1089–1107

Mora A, Baby P, Roddaz M, Parra M, Brusset S, Hermosa W, Espurt N (2010) Tectonic history of the Andes and sub-Andean zones: implication for the development of the Amazon drainage basin. In: Hoorn C, Wesselingh FP (eds) Amazonia, landscape and species evolution: a look into the past. Wiley, Chichester, pp 38–60

Moraes LJCL, Pavan D (2017) Another puzzle piece: new record of the Fringed Leaf Frog *Cruziohyla craspedopus* (Funkhouser 1957) (Amphibia: Anura: Phyllomedusidae) at eastern Amazonia. Check List 13(2):2095

Moraes LJCL, Pavan D (2018) On the occurrence of *Hemiphractus scutatus* (Spix, 1824) (Anura: Hemiphractidae) in eastern Amazonia. Amphibian Reptile Conserv 12:5–14

Moraes LJCL, Storti LF, Pavan D (2014) *Rhinella magnussoni*: Breeding site. Herpetol Rev 45:683–684

Moraes LJCL, Pavan D, Barros MC, Ribas CC (2016) The combined influence of riverine barriers and flooding gradients on biogeographical patterns for amphibians and squamates in south-eastern Amazonia. J Biogeogr 43(11):2113–2124

Moraes LJCL, Ribeiro-Júnior MA, Pavan D (2017a) *Tupinambis longilineus* Ávila-Pires, 1995 (Squamata, Teiidae): revised distribution of a rare Amazonian species. NW J Zool 13:364–368

Moraes LJCL, Almeida AP, Fraga R, Zamora RR, Pirani R, Silva AAA, Carvalho VT, Gordo M, Werneck FP (2017b) Integrative overview of the herpetofauna from Serra da Mocidade, a granitic mountain range in northern Brazil. Zookeys 715:103–159

Moraes LJCL, Werneck FP, Pavan D (2019a) New record and updated geographic distribution of *Anolis phyllorhinus* Myers & Carvalho, 1945. Herpetozoa 31(3/4):229–233

Moraes LJCL, Pavan D, Lima APL (2019b) A new nurse frog of *Allobates masniger-nidicola* complex (Anura, Aromobatidae) from the east bank of Tapajós River, eastern Amazonia. Zootaxa 4648:401–434

Morales VR (2002) Sistematica y biogeografía del grupo *trilineatus* (Amphibia, Anura, Dendrobatidae, *Colostethus*), con descripción de once nuevas especies. Publ Asoc Amigos Doñana 13:1–59

Moritz C, Patton JL, Schneider CJ, Smith TB (2000) Diversification of rainforest faunas: an integrated molecular approach. Annu Rev Ecol Evol Syst 31:533–563

Naka LN, Brumfield RT (2018) The dual role of Amazonian rivers in the generation and maintenance of avian diversity. Sci Adv 4(8):eaar8575

Nogueira ACR, Silveira R, Guimarães JTF (2013) Neogene-quaternary sedimentary and paleovegetation history of the eastern Solimões Basin, central Amazon region. J S Am Earth Sci 46:89–99

Nunes PMS, Fouquet A, Curcio FF, Kok PJR, Rodrigues MT (2012) Cryptic species in *Iphisa elegans* Gray, 1851 (Squamata: Gymnophthalmidae) revealed by hemipenial morphology and molecular data. Zool J Linnean Soc 166:361–376

Oliveira TG, Mazim FD, Vieira OQ, Barnett APA, Silva GN, Soares JBG, Santos VJP, Silva VF, Araújo PA, Tchaika L, Miranda CL (2016) Nonvolant mammal megadiversity and conservation issues in a threatened central Amazonian hotspot in Brazil. Trop Conserv Sci 9:1–16

Oliveira EA, Rodrigues LR, Kaefer IL, Pinto KC, Hernández-Ruz EJ (2017) A new species of *Pristimantis* from eastern Brazilian Amazonia (Anura, Craugastoridae). ZooKeys 687:101–129

Ortiz DA, Lima AP, Werneck FP (2018) Environmental transition zone and rivers shape intraspecific population structure and genetic diversity of an Amazonian rain forest tree frog. Evol Ecol 32(4):359–378

Piedade MTF, Junk W, D'Ângelo SA (2010) Aquatic herbaceous plants of the Amazon floodplains: state of the art and research needed. Acta Limnol Bras 22:165–178

Pimenta FE, Silva JS Jr (2005) An update on the distribution of primates of the Tapajós–Xingu interfluvium, Central Amazonia. Neotrop Primates 13(2):23–28

Pinheiro PDP, Cintra CED, Valdujo PH, Silva HLR, Martins IA, Silva NJ Jr, Garcia PCA (2018) A new species of the *Boana albopunctata* group (Anura: Hylidae) from the cerrado of Brazil. S Am J Herpetol 13:170–182

Pulido-Santacruz P, Aleixo A, Weir JT (2018) Morphologically cryptic Amazonian bird species pairs exhibit strong postzygotic reproductive isolation. Proc R Soc B Biol Sci 285:20172081

Ribas CC, Aleixo A, Nogueira ACR, Miyaki CY, Cracraft J (2012) A palaeobiogeographic model for biotic diversification within Amazonia over the past three million years. Proc R Soc B Biol Sci 279(1729):681–689

Ribas CC, Aleixo A, Gubili C, d'Horta FM, Brumfield RT, Cracraft J (2018) Biogeography and diversification of *Rhegmatorhina* (Aves: Thamnophilidae): implications for the evolution of Amazonian landscapes during the quaternary. J Biogeogr 45:917–928

Ribeiro-Júnior MA (2015a) Catalogue of distribution of lizards (Reptilia: Squamata) from the Brazilian Amazonia. I. Dactyloidae, Hoplocercidae, Iguanidae, Leiosauridae, Polychrotidae, Tropiduridae. Zootaxa 3983:1–110

Ribeiro-Júnior MA (2015b) Catalogue of distribution of lizards (Reptilia: Squamata) from the Brazilian Amazonia. II. Gekkonidae, Phyllodactylidae, Sphaerodactylidae. Zootaxa 3981 (1):1–55

Ribeiro-Júnior MA, Amaral S (2016) Catalogue of distribution of lizards (Reptilia: Squamata) from the Brazilian Amazonia. III. Anguidae, Mabuyidae, Teiidae. Zootaxa 5:401–430
Ribeiro-Júnior MA, Amaral S (2017) Catalogue of distribution of lizards (Reptilia: Squamata) from the Brazilian Amazonia. IV. Alopoglossidae, Gymnophthalmidae. Zootaxa 4269(2):151–196
Rossetti DF (2014) The role of tectonics in the late quaternary evolution of Brazil's Amazonian landscape. Earth Sci Rev 139:362–389
Rossetti DF, Valeriano MM (2007) Evolution of the lowest amazon basin modeled from the integration of geological and SRTM topographic data. Catena 70:253–265
Rossetti DF, Cohen MC, Bertani TC, Hayakawa EH, Paz JD, Castro DF, Friaes Y (2014a) Late quaternary fluvial terrace evolution in the main southern Amazonian tributary. Catena 116:19–37
Rossetti DF, Zani H, Cremon ÉH (2014b) Fossil megafans evidenced by remote sensing in the Amazonian wetlands. Z Geomorphol 58(2):145–161
Rossetti DF, Gribel R, Toledo PMD, Tatumi SH, Yee M, Tudela D, Munita CS, Coelho LS (2018) Unfolding long-term late Pleistocene-Holocene disturbances of forest communities in the southwestern Amazonian lowlands. Ecosphere 9:e02457
Santos DB, Fernandes PECA, Dreher AM, Cunha FMB, Basei MAS, Teixeira JBG (1975) Geologia. In: RADAMBRASIL. Folha SB.21 Tapajós, Projeto Radam Brasil, Brasília, pp 15–117
Shephard GE, Muller RD, Liu L, Gurnis M (2010) Miocene drainage reversal of the Amazon River driven by plate–mantle interaction. Nat Geosci 3:870–875
Silva JMC, Bates JM (2002) Biogeographic patterns and conservation in the South American Cerrado: a tropical savanna hotspot. Bioscience 52:225–233
Silva GH, Leal JWL, Montalvão RMG, Bezerra PEL, Pimenta ONS, Tassinari CCG, Fernandes CAC (1980) Geologia. In: RADAMBRASIL. Folha SC.21 Juruena, Projeto Radam Brasil, Brasília, pp 21–117
Silva JMC, Novaes FC, Oren DC (2002) Differentiation of *Xiphocolaptes* (Dendrocolaptidae) across the river Xingu, Brazilian Amazonia: recognition of a new phylogenetic species and biogeographic implications. Bull Br Ornithol Club 122:185–194
Silva MB, Ribeiro-Júnior MA, Ávila-Pires TC (2018) A new species of *Tupinambis* Daudin, 1802 (Squamata: Teiidae) from Central South America. J Herpetol 52(1):94–110
Silva SM, Peterson AT, Carneiro L, Burlamaqui TCT, Ribas CC, Sousa-Neves T, Miranda LS, Fernandes AM, d'Horta FN, Araújo-Silva LE, Batista R, Bandeira CHMM, Dantas SM, Ferreira M, Martins DM, Oliveira J, Rocha TC, Sardelli CH, Thom G, Rêgo PS, Santos MP, Sequeira F, Vallinoto M, Aleixo A (2019) A dynamic continental moisture gradient drove Amazonian bird diversification. Sci Adv 5(7):eaat5752
Simões PI, Stow A, Hödl W, Amézquita A, Farias IP, Lima AP (2014) The value of including intraspecific measures of biodiversity in environmental impact surveys is highlighted by the Amazonian brilliant–thighed frog (*Allobates femoralis*). Trop Conserv Sci 7:811–828
Sioli H (1968) Hydrochemistry and geology in the Brazilian Amazon Region. Amazoniana 3:267–277
Sioli H (1984) The Amazon: limnology and landscape ecology of a mighty tropical river and its basin. Junk, Dordretch, p 763
Smith TB, Wayne RK, Girman DJ, Bruford MW (1997) A role for ecotones in generating rainforest biodiversity. Science 276:1855–1857
Smith BT, McCormack JE, Cuervo AM, Hickerson MJ, Aleixo A, Cadena CD, Pérez-Emán J, Burney CW, Xie X, Harvey MG, Faircloth BC, Glenn TC, Derryberry EP, Prejean J, Fields S, Brumfield RT (2014) The drivers of tropical speciation. Nature 515:406–409
Soares EAA, Sonia HT, Riccomini C (2010) OSL age determinations of Pleistocene fluvial deposits in Central Amazonia. An Acad Bras Cienc 82:1–19
Soares EAA, D'Apolito C, Jaramillo C, Harrington G, Caputo MV, Barbosa RO, Santos EB, Dino R, Gonçalves AD (2017) Sedimentology and palynostratigraphy of a Pliocene–Pleistocene (Piacenzian to Gelasian) deposit in the lower Negro River: implications for the establishment of large rivers in Central Amazonia. J S Am Earth Sci 79:215–229

Sombroek W (2001) Spatial and temporal patterns of Amazon rainfall - consequences for the planning of agricultural occupation and the protection of primary forests. Ambio 30:388–396

Sturaro MJ, Rodrigues MT, Colli GR, Knowles LL, Ávila-Pires TCS (2018) Integrative taxonomy of the lizards *Cercosaura ocellata* species complex (Reptilia: Gymnophthalmidae). Zool Anz 275:37–65

Terezo EFM, Filho LG, Ferreira HC, Veloso HP, Fonseca WN, Silva SB (1975) Vegetação. In: RADAMBRASIL. Folha SB.21 Tapajós, Projeto Radam Brasil, Brasília, pp 287–385

Tsuji-Nishikido BM, Kaefer IL, Freitas FC, Menin M, Lima AP (2012) Significant but not diagnostic: differentiation through morphology and calls in the Amazonian frogs *Allobates nidicola* and *A. masniger*. Herpetol J 22:105–114

Tuomisto H, Ruokolainen K (1997) The role of ecological knowledge in explaining biogeography and biodiversity in Amazonia. Biodivers Conserv 6:347–357

Tuomisto H, Ruokolainen K, Kalliola K, Linna R, Danjoy W, Rodriguez Z (1995) Dissecting Amazonian biodiversity. Science 269:63–66

Tuomisto H, Moulatlet GM, Balslev H, Emilio T, Figueiredo FOG, Pedersen D, Ruokolainen K (2016) A compositional turnover zone of biogeographical magnitude within lowland Amazonia. J Biogeogr 43:2400–2411

Turchetto-Zolet AC, Pinheiro F, Salbueiro F, Palma-Silva C (2013) Phylogeographical patterns shed light on evolutionary process in South America. Mol Ecol 22:1193–1213

Uetz P, Freed P, Hošek J (2019) The reptile database. http://www.reptile-database.org

Vacher JP, Kok PJR, Rodrigues MT, Lima JD, Lorenzini A, Martinez Q, Fallet M, Courtois EA, Blanc M, Gaucher P, Dewynter M, Jairam R, Ouboter P, Thébaud C, Fouquet A (2017) Cryptic diversity in Amazonian frogs: integrative taxonomy of the genus *Anomaloglossus* (Amphibia: Anura: Aromobatidae) reveals a unique case of diversification within the Guiana Shield. Mol Phylogenet Evol 112:158–173

Valente CR, Latrubesse EM (2012) Fluvial archive of peculiar avulsive fluvial patterns in the largest quaternary intracratonic basin of tropical South America: the Bananal Basin, Central–Brazil. Palaeogeogr Palaeoclimatol Palaeoecol 356–357:62–74

Vanzolini PE, Williams EE (1970) South American anoles: the geographic differentiation and evolution of the *Anolis chrysolepis* species group (Sauria, Iguanidae). Arquivos de Zoologia 19:1–176

Wanderley-Filho JR, Eiras JF, Cunha PRC, van der Ven PH (2010) The Paleozoic Solimões and Amazonas basins and the Acre foreland basin of Brazil. In: Hoorn C, Wesselingh FP (eds) Amazonia, landscape and species evolution: a look into the past. Wiley, Chichester, pp 29–37

Wang X, Edwards RL, Auler AS, Cheng H, Kong X, Wang Y, Cruz FW, Dorale JA, Chiang HW (2017) Hydroclimate changes across the Amazon lowlands over the past 45,000 years. Nature 541(7636):204–207

Weir JT, Faccio MS, Pulido-Santacruz P, Barrera-Guzmán AO, Aleixo A (2015) Hybridization in headwater regions, and the role of rivers as drivers of speciation in Amazonian birds. Evolution 69:1823–1834

Werneck FP, Nogueira C, Colli GR, Sites JW, Costa GC (2012a) Climatic stability in the Brazilian Cerrado: implications for biogeographical connections of South American savannas, species richness and conservation in a biodiversity hotspot. J Biogeogr 39:1695–1706

Werneck FP, Gamble T, Colli GR, Rodrigues MT, Sites JW Jr (2012b) Deep diversification and long-term persistence in South American 'Dry Diagonal': integrating continent-wide phylogeography and distribution modeling of geckos. Evolution 66:3014–3034

Wesselingh FP, Salo JA (2006) Miocene perspective on the evolution of the Amazonian biota. Scr Geol 133:439–458

Winter M, Fiedler W, Hochachka WM, Koehncke A, Meiri S, de la Riva I (2016) Patterns and biases in climate change research on amphibians and reptiles: a systematic review. R Soc Open Sci 3:160158

Zimmer K, Isler ML (2019) Blackish–grey Antshrike (*Thamnophilus nigrocinereus*). In: del Hoyo J, Elliott A, Sargatal J, Christie DA, de Juana E (eds) Handbook of the birds of the world alive. Lynx Edicions, Barcelona. https://www.hbw.com/node/56672

Chapter 26
Diversification Processes in Lizards and Snakes from the Middle São Francisco River Dune Region, Brazil

Renato Sousa Recoder and Miguel Trefaut Rodrigues

Abstract In the middle course of the São Francisco River (SFR), inserted in the semiarid domain of the Caatinga, there is an extensive field of palaeodunes which represents an important testimony of former, drier climates in northeastern Brazil. The highly diverse sand-dwelling fauna of this dune region comprises several endemic reptiles with geographically structured distributions. Based on geomorphological, paleoclimatic, and distribution data of pairs of psammophilous reptiles, a paleolacustrine vicariant diversification scenario was proposed to explain the origin of this endemic fauna. In this model, after an endorheic phase during the last glacial maximum, when the SFR flowed into an interior lake, the SFR made its way out to the ocean through the dunes, separating ancestral populations of several groups of lizards and snakes on opposite margins. Phylogenetic data from endemic species are only partially concordant with this model, and highlight the contribution of both local and regional events as drivers of diversification. Estimates of divergence times suggest that speciation was not simultaneous, and that some events occurred prior to Pleistocene. Species that are widespread in the SFR dune region lack genetic structure between the two margins, contrasting with the pattern observed for endemic, habitat-specialist lineages. Both Pliocene and Pleistocene events appear to have promoted diversification in SFR dune endemic reptiles, and ecological interactions may have contributed to species turnover. The SFR may have acted both as a vicariant barrier and as a boundary delimiting secondary contact, reinforcing ongoing processes of speciation.

Keywords Caatinga · Endemism · Phylogeny · Squamata · Speciation

R. S. Recoder (✉) · M. T. Rodrigues
Instituto de Biociências, Universidade de São Paulo, São Paulo, Brazil
e-mail: mturodri@usp.br

V. Rull, A. C. Carnaval (eds.), *Neotropical Diversification: Patterns and Processes*,
Fascinating Life Sciences, https://doi.org/10.1007/978-3-030-31167-4_26

1 Introduction

In the middle course of the São Francisco River (SFR), a main stem river that crosses the semi-arid Caatinga of northeastern Brazil, lays the most extensive field of continental paleodunes of South America. The São Francisco sand dunes, currently inactive and stabilized by vegetation, stand out from the surrounding landscape and represent one of Brazil's most important testimonies of drier climates (Barreto et al. 2002). The distinctness of this dune field system readily catches the attention of scientists—both by the magnitude of the sand deposits (which in some sites may reach up to 100 m high), and by the permanent green appearance of its vegetation, which contrasts with the typical grayish look of the deciduous shrub-like caatingas in the dry season (Williams 1925; Ab'Sáber 2006).

This sandy ecosystem reveals a completely new and diverse fauna, sharply different from that of their surrounding Caatingas. In the last 30 years, several species and genera were discovered and described from the middle São Francisco dune region, including invertebrates (Xavier and Rocha 2001; Bravo et al. 2009), an echymid rodent (Rocha 1995), but, most remarkably, numerous species of Squamate reptiles—including new species of lizards, snakes, and amphisbaenians. Several of these species have striking morphological adaptations to live on sand, some of them previously unreported for South America (Rodrigues 1984, 1986, 1991a, b, c, d, 1992, 1993, 2002; Rodrigues et al. 1988; Vanzolini 1991a, b; Rodrigues and Juncá 2002). This reptile fauna reveals high levels of endemism and geographic structure, including several allopatric taxon pairs that provide interesting models to study diversification patterns (Rodrigues 1996, 2003).

Molecular data accumulated for lizards and snakes from the middle São Francisco dune region add a comparative framework to evaluate these patterns and speciation modes at a local scale, as well as to study the role of the São Francisco River as a historical barrier (Passoni et al. 2008; Siedchlag et al. 2010; Werneck et al. 2015). In this chapter, we (1) characterize and summarize current knowledge about the geological and paleoclimatic history of the São Francisco dune fields, (2) characterize its reptile fauna, with a focus on the **psammophilous**[1] species, highlighting a vicariant hypothesis for the origin of the endemic fauna, and (3) discuss the historical relationships of the endemic species, addressing the congruence between phylogenetic patterns and the geographic model of speciation proposed, using published molecular studies and novel data.

[1]Organisms that live in sandy environments.

2 Characterization of the Dune Fields

The São Francisco River dune region extends along ca. 7000 km^2 (roughly between 9°–11° S, 41°–44° W) in the middle course of the SFR, in the central-eastern portion of the state of Bahia, northeast Brazil (Fig. 26.1). It is inserted in the morphoclimatic domain of the Caatinga, in which a deciduous short forest and spiny shrub-like vegetation prevail (Ab'Sáber 1970). The climate is semi-arid, with low annual precipitation that reaches 700 mm, and irregular rains concentrated from April to October. A dry season, characterized by the complete lack of rain, extends from November to March. Temperatures are relatively constant throughout the year (24–28 °C), with absolute maxima near 40 °C and minima close to 10 °C (Nimer 1989).

The SFR dune system is composed of several dune fields and irregular sand banks forming patches of fine-sand habitats, isolated by typical Caatinga on rocky/clayish soils. The middle SFR dune region lays on the São Francisco depression, a large alluvial valley with extensive sedimentary deposition. To the west, it is bordered by the Serra do Estreito, which is a large and linear quartzitic outcrop that emerges from the alluvial plain on the left bank of the SFR to a height of 600 m above sea level.

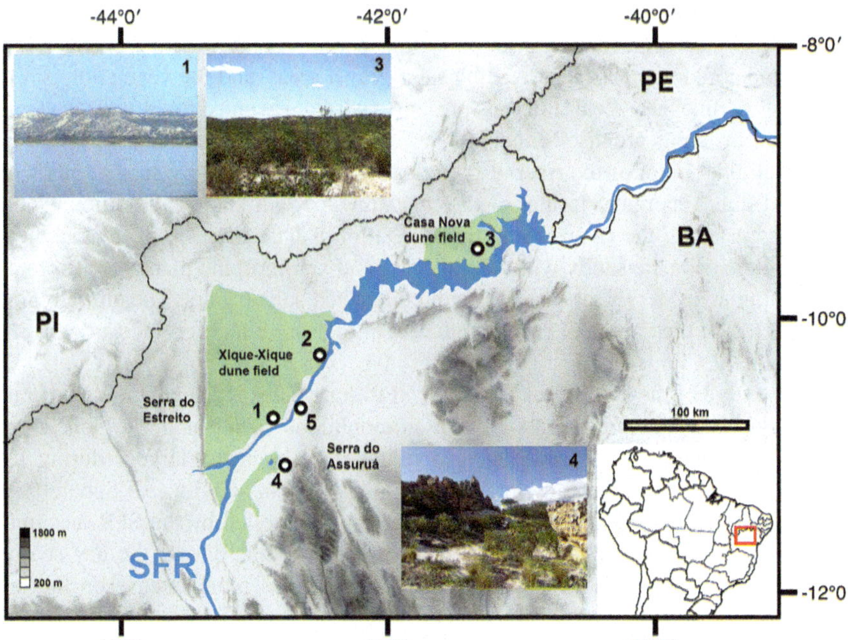

Fig. 26.1 Map of the São Francisco River dune region. Numbers refer to studied sites: 1. Ibiraba; 2. Queimadas; 3. Alagoado; 4. Santo Inácio; and 5. Vacaria. SFR—São Francisco River, BA—Bahia State, PI—Piauí State, PE—Pernambuco State. Map in detail delimits the location of the SFR dunes in South America

Situated between the Serra do Estreito and the river lies the Xique-Xique dune field, the largest portion of this dune system (Fig. 26.1). These white to yellowish eolian **parabolic dunes**[2] are the largest found in South America and are formed by fine eolic **dystrophic**[3] sands. Sand dunes are variable in elevation (20–100 m) and are held together by thickets of a thorny and xeromorphic vegetation. Dominant thickets are formed by associations of Cactaceae (*Opuntia* and *Cereus*) and Bromeliaceae (*Bromelia* sp.), with scattered shrubs of Euphorbiaceae, Mirtaceae, Bombacaceae, Celastraceae, Leguminosae, or Simaroubaceae. Some individuals of Olacaceae and Mirtaceae form isolated thickets. Large spaces of bare sand intercalate the clumps of vegetation, below which a leaf litter mat is also present. Only rarely do some of the shrubs reach 5 m. The only perennial river flowing across this area is the Icatu river, a small tributary of SFR with its headwaters in the Serra do Estreito (see Rodrigues 1996 and Rocha et al. 2004 for additional data on this area).

A smaller dune field occurs to the north, on the left bank: the Casa Nova dune field (Fig. 26.1). This dune field presents similar vegetation, soil and general physiognomy, but the relief is less accentuated with shorter and sometimes indistinct dunes. On the right bank, close to Santo Inácio village, a small indistinct dune represents a contact zone between the alluvial plain of the SFR and the terminal northwestern portion of the large Serra do Espinhaço range, known as Serra do Assuruá (Fig. 26.1). The Serra do Assuruá is a mountain range with typical "campos rupestres", an open vegetation characteristic of rocky substrates. Local habitats are a complex mosaic of several vegetation types: there are large outcrops of quartzitic rocks, and conglomerates separated by sand-gravel areas, and the vegetation is also clumped in thickets. Leaf litter is present: twisted savanna trees and scattered Velloziaceae plants are also found.

The alluvial plain of the river at Itaparica lake is dominated by sandy soils with scattered dunes identical to that of the opposite side, but shorter. Additional sand banks occur along the river, where the vegetation is characteristic of the caatingas although the soils are sandy with low, thorny, and xeromorphic plant species. In all intervening areas, a typical caatinga with compacted soils occurs. The entire region is subjected to the same general climatic regime, despite its high levels of habitat heterogeneity.

Knowledge of the geological history of SFR dune fields is relevant to understand the origins and diversification of the psammophilous fauna in the region. Dune formation in the middle São Francisco might have occurred largely during the Pleistocene, under a drier arid climate with intense eolian activity (Ab'Saber 1969; Tricart 1974). It is believed that sand deposits were supplied by the SFR, accumulated on its alluvial plain, and were transported by winds during the dry season (Williams 1925; Barreto 1996). Sand accumulation was more intense on the left

[2]Parabolic dunes are characterized by U-shaped mounds of sand produced by strong unidirectional winds.

[3]Poor in nutrients.

bank of the SFR, especially at the foot of the Serra do Estreito, in consequence of predominant E-SE to W-NW winds (Tricart 1974).

Although geomorphological studies indicate that sand dunes may have been present in the region since the Late Pliocene, the data that support the establishment of the current physiognomy of the dunes are limited to the Late Pleistocene. Furthermore, palynological evidence suggests the occurrence of abrupt climatic oscillations between climates drier and wetter than current in the middle SFR dune field region, with subsequent changes in vegetation cover in the Holocene. Together, these data highlight the complex history of this unique psammobiome.

Alternating sedimentary layers of clay and sand suggest a fluvio-lacustrian origins of deposits (Domingues 1948). According to Tricart (1974), the SFR had an **endorheic**[4] drainage during the Last Glacial Maximum (ca. 18,000 y bp), flowing into a continental paleolake. In this period, sand deposition and eolian transportation was intense in the region. After a wetter period with less wind activity, at the end of the last maximum glacial (12,000 y bp), the river made way towards the sea through the dune fields (Tricart 1974). Thus, the region experienced major geomorphological changes in recent geological time, triggered by paleoclimatic fluctuations. Accordingly, King (1956) pointed out the young geological age of the lower course of the SFR, admitting its middle drainage probably drained to the northwest (flowing to Tocantins or Parnaíba basins), being captured to current drainage only at Late Pleistocene.

The dunes are presently inactive, fixed in its current position by vegetation since at least 28,000 y bp, as inferred by thermoluminescence dating (Barreto 1996; Giannini et al. 2005). However, based on current sedimentology in the region and estimates of volume of eolian sand of the Xique-Xique dune field, Barreto (1996) concluded that at least 100,000 years would be necessary to accumulate all the sand present in the area. As presently admitted, the process of sand deposition and dune formation in the region is probably much older than Late Pleistocene, and can be tracked back to Pliocene, a period of intense denudation of sandstone highlands in Central/Eastern Brazil (King 1956). A long period of sand deposition is also suggested by the thickness of up to 150 m at some points in the dune field, relative to the bedrock (Barreto et al. 2002).

Currently, the predominant vegetation in the region is typical xerofilic caatinga (Rocha et al. 2004). Nevertheless, the intense paleoclimatic fluctuations that occurred in Northeastern Brazil during the Pliocene and Pleistocene are known to have produced significant changes in vegetation cover (Auler and Smart 2001; Oliveira et al. 2014). The best regional record comes from a palynological study that provides evidence for the presence of rainforest elements in the Icatu River valley during the Late Pleistocene/Holocene (Oliveira et al. 1999).

[4]Relative to closed drainages that lacks outflow to external water bodies such as rivers or the ocean.

3 Diversity and Distribution of the Squamate Fauna

Faunal studies of the middle São Francisco dune region have been concentrated in five sites in the state of Bahia (Fig. 26.1): Ibiraba and Queimadas (Xique-Xique dune field, left bank), Alagoado (Casa Nova dune field, left bank), Santo Inácio and Vacaria (right bank). These regions account for most of the environmental heterogeneity of the region. A total of 70 squamate reptile species were recorded in this SFR dune region, including 36 lizards, seven amphisbaenians, and 27 snakes (Appendix 1), representing roughly 35% of the species recorded in Caatinga (Rodrigues 2003; Guedes et al. 2014; Mesquita et al. 2017). Of these, 12 lizards, two amphisbaenians and six snakes are endemic to São Francisco dune region, which highlights its importance for the diversity of Caatinga reptiles (Rodrigues 2003; Guedes et al. 2014; Mesquita et al. 2017).

This squamate fauna of the SFR is very characteristic by presenting a high proportion of fossorial, nocturnal and psammophilous forms (Rodrigues 1996). Considering lizards and amphisbaenians, the typical caatinga forms are depauperate in the SFR dune region, being less abundant or rare relative to the endemic elements (Rodrigues 1996; Rocha and Rodrigues 2005). Furthermore, all endemic species of lizards are psammophilous and present some degree of morphological adaptation to live on sandy habitats; the endemic snake fauna also is entirely composed by fossorial species (Rodrigues 1996). This singular character of the SFR dune reptiles stands out when compared to other well-sampled regions of the Caatinga, especially regarding their lizards and amphisbaenians (Table 26.1).

Another remarkable feature of this reptile fauna is the geographically structured distribution of the species that inhabit the SFR dunes (Rodrigues 1996). At least three groups of psammophilous reptiles present pairs of taxa on opposite margins of the SFR, at relatively close localities (Rodrigues 1986, 1993, 1996): the tropidurid lizard genus *Eurolophosaurus,* the fossorial and limb-reduced lizards *Calyptommatus,* and the fossorial snake *Rodriguesophis* (Table 26.1). Some species pairs also occur on the left bank of São Francisco River, between Xique-Xique and Casa Nova dune fields. This is the case of the lizard genera *Ameivula* and *Procellosaurinus*. In general, there is high turnover in the composition of reptiles among dunes, which raises questions about whether the river may have acted as a historical barrier and structured those assemblages.

Considering the small dimensions of the middle SFR dune region, and its stranded condition in the caatingas, the mechanisms responsible for its high diversity and endemism are not immediately obvious. Similar patches of sandy habitats occur in other parts of the caatinga domain, although they usually contain caatinga species with no or only a few endemic species with psammophilic adaptations—in striking contrast with the situation observed in the SFR dune region (Table 26.1). Exceptions are Confusões/Capivara complex in the state of Piauí, The Raso da Catarina in the State of Bahia, and the Catimbau region in the state of Pernambuco, where a few endemic fossorial sand-dwelling reptiles are found (Vanzolini 1991b; Rodrigues et al. 2001; Rodrigues and dos Santos 2008; Table 26.1). The distinctness of the São

Table 26.1 Description of localities mentioned in the text, for which abundant data exist for lizards and amphisbaenians

	State	Relief	Vegetation	Soil	Area	Richness	Endemics (%)	Fossorial (%)	Refs.
Middle São Francisco dune region	Bahia	Lowland	Caatinga	Sand	7000 km^2	43	14 (33%)	17 (40%)	1, 2, 3
Serra das Confusões/Capivara complex	Piauí	Highland/ lowland	Caatinga	Sand/ clay	6000 km^2	31	6 (19%)	10 (32%)	4,5
Serra do Araripe bioregion	Ceará	Highland/ lowland	Caatinga/ forest	Sand/ clay	11,000 km^2	33	1 (3%)	10 (30%)	6
Raso da Catarina Ecological Station	Bahia	Lowland	Caatinga	Sand	1000 km^2	21	2 (10%)	5 (24%)	7
Serra Geral da Bahia, southern portion	Bahia/Minas Gerais	Highland	Savanna	Sand	15,000 km^2	28	5 (18%)	9 (32%)	8, 16
Central Chapada Diamantina range	Bahia	Highland	Caatinga/ forest	Sand/ rock	2000 km^2	31	7 (23%)	11 (35%)	9, 10
Catimbau National Park	Pernambuco	Highland/ lowland	Caatinga	Sand/ rock	620 km^2	25	2 (8%)	9 (36%)	11
Uruçuí-Una Ecological Station	Piauí	Highland/ lowland	Savanna	Sand/ rock	1350 km^2	23	1 (5%)	4 (17%)	12
Ibiapaba complex	Ceará	Highland	Forest	Clay/ rock	16,000 km^2	33	0	11 (33%)	13
Cabaceiras	Paraíba	Lowland	Caatinga	Clay/ rock	400 km^2	19	0	3 (15%)	1
Seridó Ecological Station	Rio Grande do Norte	Lowland	Caatinga	Clay/ rock	120 km^2	18	0	4 (22%)	15

(continued)

Table 26.1 (continued)

	State	Relief	Vegetation	Soil	Area	Richness	Endemics (%)	Fossorial (%)	Refs.
Exú	Pernambuco	Lowland	Caatinga	Clay	1400 km^2	14	0	4 (29%)	1, 14

Description includes dominant vegetation type, soil type, approximate area covered, total richness of lizards and amphisbaenians, and proportion of endemic and fossorial species. Data are based on references listed below the table. Approximate area was obtained from the literature or estimated based on the total size of municipalities included

1. Rodrigues (1996); 2. Rodrigues (2003); 3. Guedes et al. (2014); 4. Cavalcanti et al. (2014); 5. Dal Vechio et al. (2016); 6. Ribeiro et al. (2009); 7. Garda et al. (2013); 8. Recoder and Nogueira (2007); 9. Freitas et al. (2012); 10. Magalhães et al. (2015); 11. Pedrosa et al. (2014); 12. Dal Vechio et al. (2013); 13. Loebmann and Haddad (2010); 14. Vitt (1995); 15. Caldas et al. (2016); 16. Teixeira Jr. (2010)

Francisco dune fauna leads us to consider local scale processes for the diversification of sand-dwelling reptiles.

Although ecological causes can be invoked to explain the origin and coexistence of this highly diverse fauna – whose adaptations to psammophily and fossoriality are striking—the elevated number of endemic species and genera suggests we first consider historical reasons (Rodrigues 1996). Based on geomorphological and paleoclimatic data from the São Francisco sand dunes (Ab'Saber 1969; Tricart 1974), a model to explain the speciation of some of the lizards and snakes was proposed by Rodrigues (1986, 1996). This Paleolacustrine Vicariance Hypothesis proposed that the São Francisco river had an endorrheic pattern of drainage until the end of the last glacial period, flowing and carrying sands to nearby lacustrine or palustrine depressions. In a subsequent humid period, the river would have found its way out to the sea, isolating, on opposite margins, patches of sands of an otherwise continuous dune field. Psammophilous animals would have then evolved allopatrically to attain their present-day species status. The presence of pairs of taxa in the genera *Eurolophosaurus*, *Calyptommatus* and *Rodriguesophis*, occurring on opposite banks of the river, support this allopatric model of differentiation. Speciation would have occurred when the SFR attained full exorreism, dividing formerly continuous sands and synchronically isolating ancestral populations on opposite banks.

That scenario assumes that the period of speciation probably corresponded to the end of the Würm-Wisconsin glacial, roughly 12,000 y bp (Tricart 1974). Yet although the latter explains a final phase of speciation in the area, and a portion of the diversity, another sequence of events is necessary to account for the origin of adaptations to psammophily and fossoriality (Rodrigues 1996). Rodrigues (1993) hypothesized that the local fauna lived in typical caatinga depressions with inselbergs and isolated hills in a period previous to the origin of psammophilic adaptations. During a subsequent endorrheic phase, these habitats became submerged and surrounded by sediments accumulated from lacustrine deposition, isolating some elements of the fauna on disjunct hills above the flooded lowlands. Later, with the onset of a semiarid climate, the water level dropped—exposing sand deposits to sun and wind. These hills would have acted as vanishing faunal refugia in a sandy desert. Species with pre-adaptations for life in sand would have then expanded into the recent dune field whereas others became extinct. This last sequence of events corresponds, in essence, to the vanishing refuge ecogeographical model of speciation (Vanzolini and Williams 1981), which was recently revisited by Damasceno et al. (2014).

4 Phylogenetic Patterns

Molecular phylogenetic studies shed light on the origin and relationships of the endemic reptile fauna of the SFR dunes region. Assuming a hypothesis that sand-dwelling groups of squamate reptiles were already present in the region before the formation of the dune fields, preadaptations for a psammophilous and fossorial existence were likely present in those lineages (Rodrigues 1996). Molecular phylogenies are consistent with this hypothesis, showing that the exclusively psammophilous genera *Calyptommatus*, *Nothobachia*, *Procellosaurinus* and *Psilops* are part of a diverse gymnophthalmid radiation (Pellegrino et al. 2001) that includes other open-habitat genera such as *Gymnophthalmus*, *Micrablepharus* and *Vanzosaura*, which already show striking adaptations for a fossorial lifestyle such as the absence of eyelids, small body-size, and limb reduction (Rodrigues 1991c). The *Tropidurus torquatus* group, on the other hand, is a monophyletic group (Frost et al. 2001; Carvalho et al. 2016) in which several species are sandy-habitat specialists and show morphological adaptations associated with locomotion on sand (Rodrigues 1987; Kohlsdorf et al. 2004; Grizante et al. 2010). Relatives of the endemic species *T. psammonastes* and the sand-dwelling *T. cocorobensis* occur in distinct types of open habitats in Central and Northeastern Brazil (Carvalho et al. 2016). Whiptail lizards of the *Ameivula ocellifera* group and the genus *Glaucomastix* (Goicoechea et al. 2016) are present in the dry diagonal of open habitats of South America and coastal plains, and are also known to occur frequently or exclusively on sandy-soil in open habitats (Dias and Rocha 2007; Colli et al. 2009; Arias et al. 2011). Lastly, the Amphisbaenidae and Typhlopidae are known fossorial groups (Greene 1997; Pianka and Vitt 2003) and *Rodriguesophis* and *Apostolepis* are also two entirely fossorial genera of xenodontine snakes (Grazziotin et al. 2012). Therefore, the phylogenetic data lend support to the proposal that sand-dwelling forms, which account for the high species diversity of the SFD area, belong to groups of open-habitat species seemingly preadapted to a fossorial or psammophilous life.

The first population-level molecular study of psammophilous reptiles from the São Francisco dune region used segments of mitochondrial DNA to infer phylogenetic relationships and divergence times for lizards of the genus *Eurolophosaurus* (Passoni et al. 2008). Although molecular analyses supported the monophyly of the genus, the dune endemics *E. divaricatus* and *E. amathites* were not recovered as sister species. Instead, the later species was recovered as more closely related to saxicolous populations occurring in the Serra do Espinhaço mountain range. Such basal split between *E. divaricatus* (left bank) and the clade containing *E. amathites* (right bank) is congruent with the Paleolacustrine Vicariant Hypothesis. Nevertheless, estimates of divergence times based on a pre-defined sequence divergence rate of 2–2.8% per million years suggest that basal cladogenesis occurred between 5.4–9.5 My bp, well before the Pleistocene (Passoni et al. 2008).

Another molecular study focused on limb-reduced fossorial lizards of the genera *Calyptommatus* and *Nothobachia* (Siedchlag et al. 2010). Partial fragments of three mitochondrial genes were used to infer genetic structure and phylogenetic relationships among several populations. The *Calyptommatus* phylogeny recovered *C. sinebrachiatus* (Santo Inácio, right bank) as sister of a clade containing the remaining species. Within that clade, *C. confusionibus* (Confusões/Capivara complex- left bank) was recovered as sister to *C. leiolepis* (left bank) and *C. nicterus* (Vacaria, right bank). Approximate estimates of divergence times based on genetic distances of combined ntDNA markers, and considering an evolutionary rate of 2%

divergence per my, suggested that a basal split between *C. sinebrachiatus* and its congeners occurred ca. 4.4–6.5 my bp; *C. confusionibus* would have diverged from *C. leiolepis/nicterus* by 2.8–5.0 my bp, while *C. nicterus* would have separated from *C. leiolepis* 1.8–2.6 my bp (Siedchlag et al. 2010). This pattern inferred for *Calyptommatus* suggests the evolutionary history of *Calyptommatus* was characterized by alternate events of speciation involving sandy habitats on opposite margins of São Francisco River. Two events of vicariance are needed to admit that the SFR acted as a major isolation barrier for speciation of the endemic species of *Calyptommatus* in the dune region, in which one would have occurred during Miocene/Pliocene, and another in the Pleistocene. Although the Paleolacustrine Vicariance Hypothesis is partially congruent with this observed pattern of relationships, and could explain the most recent event of cladogenesis between *C. nicterus* and *C. leiolepis*, the role of sand islands within the SFR for differentiation, or even as transfer areas between opposite margins, cannot be dismissed.

For comparative purposes, we retrieved mtDNA and nuDNA sequences from Genbank for *Calyptommatus* and related Gymnophthalmini (Goicoechea et al. 2016), *Eurolophosaurus* and Tropidurinae (Passoni et al. 2008; Carvalho et al. 2016) and Pseudoboini (Grazziotin et al. 2012; Pyron et al. 2013), and added molecular data from the sand-dwelling snake genus *Rodriguesophis* (Rodrigues 1993) and *Eurolophosaurus* (Appendix 2). We applied a Bayesian approach to concurrently infer phylogenetic relationships and divergence times, using molecular clock rates previously estimated for Neotropical snakes and lizards (Daza et al. 2009; Olave et al. 2015). Protocols for obtaining sequences, and methods for Bayesian phylogenetic analysis are described in Appendix 2.

The phylogenetic tree inferred for *Calyptommatus* recovered the monophyly of the genus and presented a topology similar to previous molecular studies (Siedchlag et al. 2010; Goicoechea et al. 2016), with strong node support (Fig. 26.2a). Divergence time estimates agree with Siedchlag et al. (2010) in that a basal split in *Calyptommatus* may have occurred in the Miocene, and that only the split between *C. leiolepis* and *C. nicterus* is coincident with Pleistocene events (Fig. 26.2a, Table 26.2). Even assuming a scenario with faster clock rates (using the upper bound of rate variation), the basal split in *Calyptommatus* is inferred as older than the Pleistocene (Table 26.2).

The phylogeny inferred for *Eurolophosaurus* also agrees with the previously inferred topology, which was based on mtDNA (Passoni et al. 2008), with strong support for all nodes except the clade containing *E. amathites* and *E.* sp. (Fig. 26.2b). Estimates of divergence times recovered events older than previously estimated (Passoni et al. 2008), with most cladogenetic events occurring during the Miocene, and a split within the *E. divaricatus* group at the Pliocene (Fig. 26.2b, Table 26.2). Even assuming a fast-molecular clock, the events are estimated to have occurred before the Pleistocene (Table 26.2).

The phylogenetic tree inferred for Pseudoboini snakes supported the monophyly of *Rodriguesophis* but did not recover the two endemic species *R. chui* (right bank) and *R. scriptorcibatus* (left bank) as sister species. *Rodriguesophis scriporcibatus* was recovered as basal to a clade containing *R. iglesiasi* (left bank) and *R. chui*, with

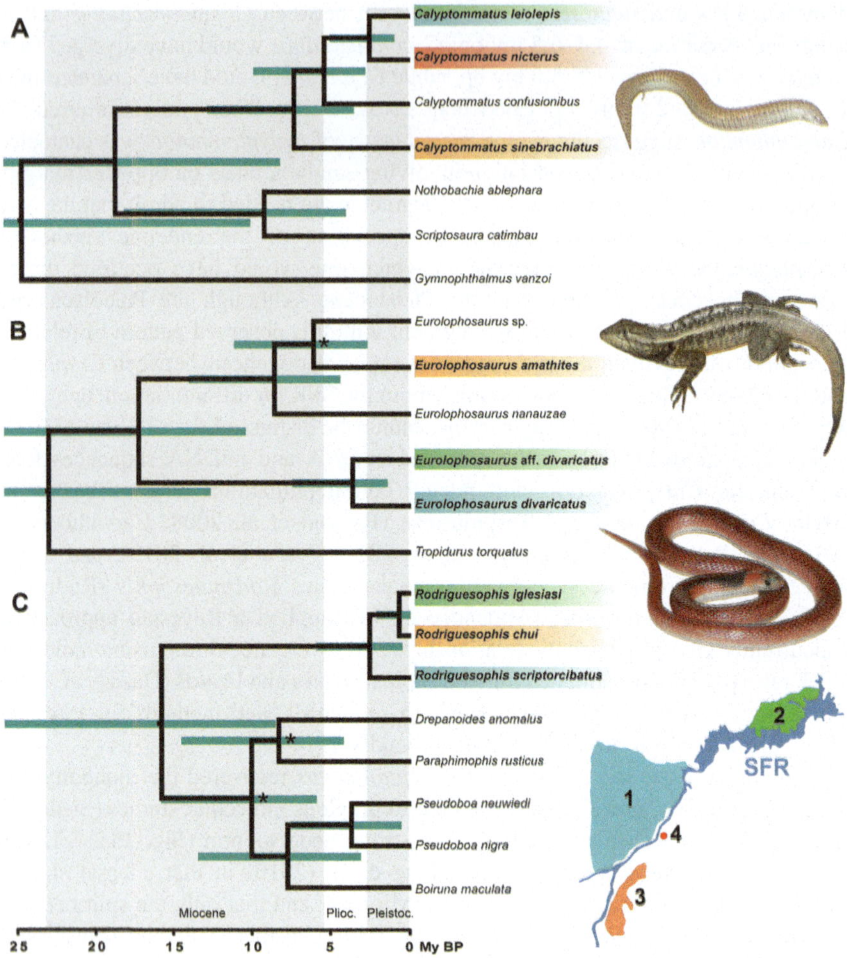

Fig. 26.2 Time-calibrated Bayesian phylogenies of three clades containing endemic reptiles from SFR dune region: (**a**) fossorial limb-reduced microteiid lizards, genera *Calyptommatus*, *Nothobachia* and *Scriptosaura* (Gymnophthalmini); (**b**) tropidurid lizards, genus *Eurolophosaurus*; (**c**) Pseudoboini snakes. Species occurring at SFD region are highlighted in bold, and color refer to localities on the map in detail: 1. Xique-Xique dunes (Ibiraba and Queimadas); 2. Casa Nova dunes (Alagoado); 3. Santo Inácio; and 4. Vacaria

strong support (Fig. 26.2c). Bayesian estimates of divergence times inferred that the basal diversification within *Rodriguesophis* may have occurred in the early Pleistocene, and that split between *R. iglesiasi* and *R. chui* may have occurred ca. 700,000 y bp, compatible with a Pleistocene scenario of cladogenesis (Fig. 26.2c, Table 26.2).

Despite the lack of more precise estimates of divergence times using fossil calibration, we observe that the divergence times within species pairs belonging to the three groups of endemic psammophilous reptiles from the SFR dunes vary

Table 26.2 Divergence times inferred for three clades containing endemic reptiles from the SFR dune region

	Divergence time estimates (Mya—95% HPD)	
Lineage pair	Mean rate	Upper rate
Calyptommatus		
C. sinebrachiatus × *Calyptommatus* spp.	7.3 (4.0–13.2)	5.8 (2.9–10.0)
C. confusionibus × *C. leiolepis*/*C. nicterus*	5.4 (2.7–9.7)	4.3 (1.9–7.4)
C. leiolepis × *C. nicterus*	2.5 (0.9–4.9)	2.0 (0.6–3.8)
Eurolophosaurus		
E. divaricatus group × *E. nanuzae* group	18.1 (10.4–28.7)	13.4 (7.1–20.6)
E. amathites × *Eurolophosaurus* sp.	6.5 (2.6–11.1)	4.9 (2.0–8.3)
E. divaricatus × *E.* aff. *divaricatus*	4.1 (1.4–7.3)	3.0 (1.5–3.4)
Rodriguesophis		
R. scriptorcibatus × *R. iglesiasi*/*R. chui*	2.6 (0.9–6.5)	–
R. iglesiasi × *R. chui*	0.7 (0.1–2.2)	–

Mean estimates of divergence times between clades occurring on distinct dunes are provided, with 95% high probability density given between brackets. For the lizards (*Calyptommatus* and *Eurolophosaurus*), estimates are based on the mean and upper bound of previously estimated molecular rates. For *Rodriguesophis* snakes, estimates are based only on mean rates (see Appendix 2 for details)

substantially, even when the data are analyzed under the same conditions. Thus, the comparative phylogenetic data do not seem to support a model where major speciation events occurred synchronically, although this hypothesis should be better tested against alternatives using a multi-locus model-based approach and additional lineages. Instead, the molecular estimates of divergence times suggest that the speciation of the local endemic fauna would have begun well before the fixation of the SFR dunes in the Late Pleistocene, with some events perhaps as early as the onset of the massive alluvial/eolian sand deposition that formed the dunes in the Late Pliocene and Early Pleistocene.

Taken together, the phylogenetic data from reptiles of the SFR suggest a complex pattern of diversification in which endemics can be either closely related to species that are geographically distant to them, such as the amphisbaenians (Teixeira Jr. et al. 2016) and tropidurids (Passoni et al. 2008; Carvalho et al. 2016), or closely related to endemic congeners, such as microteiids (Siedchlag et al. 2010) and the fossorial snakes *Rodriguesophis*. This indicates that the high diversity of psammophilous and fossorial reptiles in the SFR dunes might result from both the contribution of local events (speciation along the dune fields) and regional events (dispersal of preadapted lineages, following by speciation; Webb et al. 2002).

5 Geographic Structure within Species

The SFR has been considered a major historical barrier for gene flow at a regional scale, limiting migration and possibly playing a role in species diversification in small mammals, such as the mouse opossum *Gracilinanus agilis*, and punaré rats of the genus *Thrichomys* (Faria et al. 2013; Nascimento et al. 2013). The river may also have promoted deep genetic structuration among populations of the scansorial gecko *Phyllopezus pollicaris* and the saxicolous tropidurid lizard *Tropidurus semitaeniatus* (Werneck et al. 2012, 2015). Accordingly, the geographically structured composition of reptile fauna among dune fields suggests that the river effectively limits dispersal at a local scale (Rodrigues 1996). Many species of sand-dwelling endemic lizards and amphisbaenians that represent conspicuous components of local assemblages in the left bank dunes neither are present nor have close relatives on the opposite margin. This is the case of *Amphisbaena frontalis*, *Nothobachia ablephara*, *Procellosaurinus* spp., and *Tropidurus psammonastes* (Rodrigues 1996; Siedchlag et al. 2010), and the endemic psammophilous rodent *Proechimys yonenagae* (Rocha 1995). Accordingly, *Psilops paeminosus*, a widespread psammophilous lizard species that occurs in several sand banks along the middle and lower course of SFR, is always in the right margin (Delfim et al. 2006; Rodrigues et al. 2017). Given that the sandy habitats on opposite margins have been intensively and successively sampled, these absences seem to be factual. These patterns could be explained by extinctions and ecological factors, but more molecular data are necessary to investigate historical effects on species turnover.

One may further evaluate the contribution of SFR as a historical barrier for diversification of endemic reptiles by assessing genetic patterns in terrestrial reptiles that co-occur in the dune region. If the river represents a general historical barrier for gene flow in reptiles, then a significant signature may be found across local assemblages (Gascon et al. 2000; Naka and Brumfield 2018). Population-level genetic structure is hence expected to also be present in ecologically distinct species occurring in both margins of the river (Lougheed et al. 1999; Fouquet et al. 2015). We explored this hypothesis using fragments of mtDNA gene (*cyt b*) for two species of widespread Caatinga lizards occurring in sandy habitats at both margins: *Brasiliscincus heathi* and *Vanzosaura multiscutata*. Methods for molecular analysis are described in Appendix 2. Individuals of *B. heathi* presented genetic p-distances that varied between 2.3–10.7% (Fig. 26.3a). Genetic distances between individuals from dunes in opposite margins (Mocambo do Vento/Alagoado x Vacaria) were high (9.1–10.1%), but similar to the genetic distance between dune fields in the left bank (9.4%). Populations of *V. multiscutata* presented net between-group mean genetic distances ranging between 0.2–1.9% (Fig. 26.3b). Nevertheless, variation observed between populations on opposite margins was lower (0.2–0.9%). Analysis of molecular variance (AMOVA) do not provide support for a significant structure between populations of *V. multiscutata* from opposite margins of the SFR (Table 26.3).

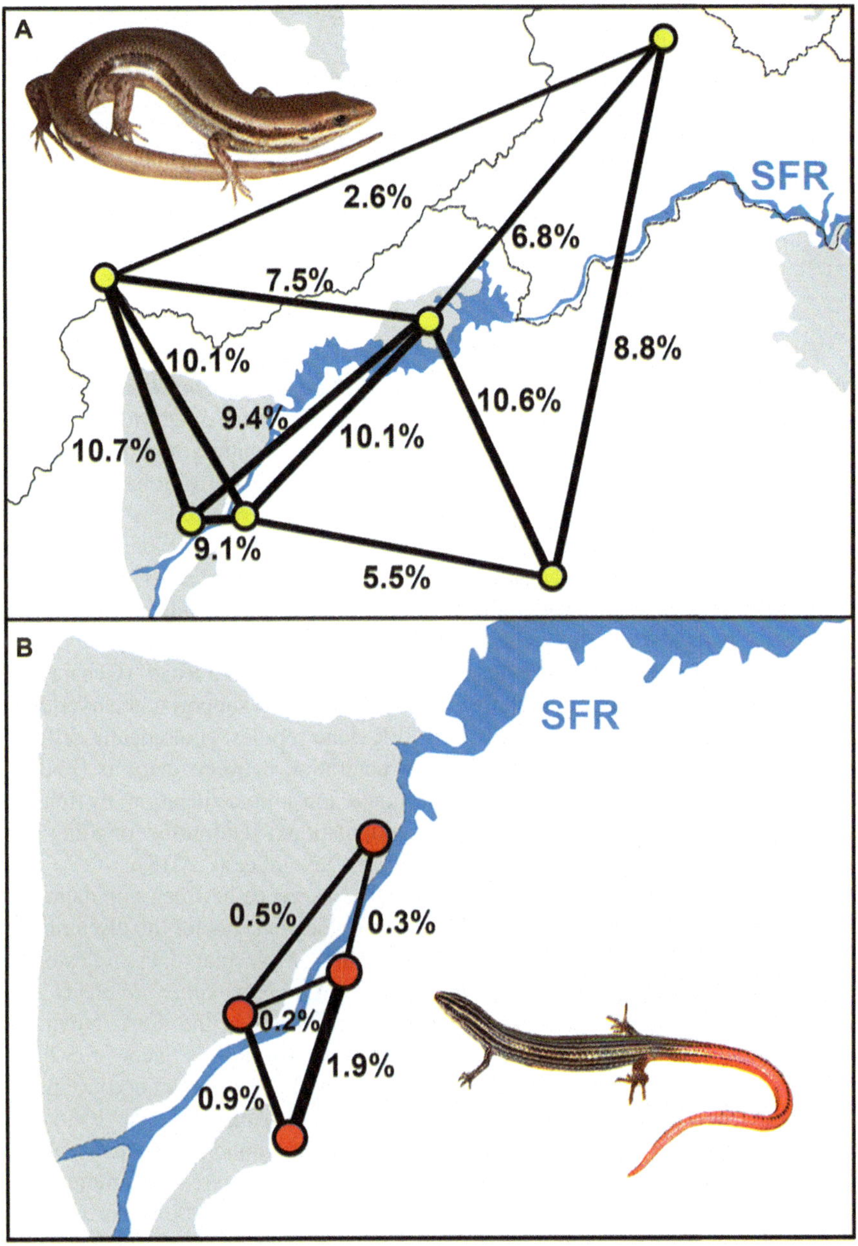

Fig. 26.3 Uncorrected pairwise genetic p-distances among samples of *Brasiliscincus heathi* (**a**) and net between-group mean distances among populations of *Vanzosaura multiscutata* (**b**) from both margins in the São Francisco River dune region

Table 26.3 AMOVA significance test for between-margin genetic structure in populations of *V. multiscutata* from the São Francisco dune region

Source of variation	d. f.	Sum of squares	Variance components	Percentage of variation	*P*-value
Among groups	1	1.025	0.0008	0.16	0.666
Among populations within groups	2	1.722	0.0839	16.75	0.010
Within populations	21	8.733	0.4159	83.09	0.000
Total	24	11.480	0.5005		

These patterns are consistent with the lack of genetic structure between river margins in the widespread Caatinga pitviper *Bothrops erythromelas*, in the SFR dune region (Machado et al. 2014). Taken together, the genetic patterns of differentiation observed for widespread, generalist species contrast sharply with the observed for endemic (Passoni et al. 2008; Siedchlag et al. 2010) or habitat specialists in the dune region (Werneck et al. 2015). This incongruence between endemic and widespread species is evident in *Vanzosaura multiscutata*, which despite its small-body size and close relationships with endemic psammophilous species, likely dispersed across the river through the alluvial plain and sand islands in a recent past, as attested by its low genetic variation in the region (Recoder et al. 2014; this study) and evidence of recent demographic expansion in Northeastern Brazil (Gehara et al. 2017). In other words, the comparative genetic data give no support to the hypothesis that the river is a vicariant barrier for *all* SFR dune reptiles: ecologically different species show distinct degrees of genetic structuration between margins (Fouquet et al. 2015). However, it may have promoted local diversification by limiting secondary contact between differentiating populations, reinforcing ongoing processes of speciation (Werneck et al. 2012, 2015; Oliveira et al. 2018).

Differences in species composition and high divergence between populations in dune fields on the left margin of the SFR also provide elements for the discussion of processes of diversification in the endemic fauna. Molecular data indicate that populations of *Eurolophosaurus divaricatus*, *Calyptommatus leiolepis* and *Nothobachia ablephara* present deep divergence between the Casa Nova and Xique-Xique dune fields, with genetic distances varying between 5.5–6.0%, 2.5–4.0% and 6.1–8.0% respectively, and supporting the presence of cryptic species (Passoni et al. 2008; Siedchlag et al. 2010). Currently isolated by ca. 100 km of caatinga vegetation on rocky and clayish soil, both sand dune fields were probably connected during the Late Pleistocene, when sand deposits were much more extensive than present and eolian activity was intense (Barreto et al. 2002). As discussed before, geomorphological data support major drainage changes in SFR during the Pleistocene, including a past northwest flow of its middle course to the Parnaíba basin (King 1956; Tricart 1974). Drainage changes left molecular evidences of diversification in both terrestrial and aquatic fauna (Nascimento et al. 2013; Costa et al. 2018) and may explain the presence of *Calyptommatus confusionibus* outside the SFR dune fields region. Alternating climatic fluctuations along the Pleistocene may have promoted phases of contact and isolation among sandy habitats and, consequently, dispersal and genetic differentiation in psammophilous species.

But why have those previously diverging populations failed to restore gene-flow, or coexisted locally, when in secondary contact? The current distribution of eyelidless lizards of the genus *Procellosaurinus* provides an interesting study case. The genus contains two species—*P. tetradactylus*, endemic to Alagoado, in the Casa Nova dune field, and *P. erythrocercus*, which occurs in the Xique-Xique dunes, the Confusões/Capivara complex, and several patches of sandy caatingas dispersed among these localities (Rodrigues 1991c). In the Casa Nova dune field, both species are tightly parapatric, with *P. erythrocercus* being rarer, but present at the edge of the range of *P. tetradactylus*, although no ecotonal change is evident along their contact zone. Similarly, at the Xique-Xique dune field, *Vanzosaura multiscutata* occurs only at the base of dunes, in the sandy alluvial plain, but never on dunes where *Procellosaurinus erythrocercus* occurs (Rodrigues 1996). The structured distribution of these sibling species readily suggests that competition, and probably density dependent advantage, would restrict coexistence of morphologically and ecologically similar species in the dunes (Bull 1991). Thus, even withstanding the above discussion on historical factors, some of the current patterns may be adequately explained by ecological interactions.

6 Conclusions

Molecular data gathered over the last years provided a historical framework and new interpretations for patterns of diversification of the psammophilous and fossorial reptile fauna of the SFR dune region. Data are available from only a few lineages yet, which limits the discussion. Nevertheless, comparative genetic information leads to some important conclusions: (1) vicariant events between lineages from opposite margins of the SFR were not simultaneous; (2) diversification of endemic reptiles occurred on a time-scale that extends back to the Late Miocene; (3) phylogenetic relationships among endemic species from distinct dune fields suggest a complex pattern of diversification with contribution of both local (speciation) and regional (dispersal) events; and (4) widespread species occurring in the SFR dune region lack genetic structure across margins, suggesting that the river likely limited gene-flow between populations of habitat specialist lineages, but not of generalist species.

More geologic, paleoclimatic and molecular studies are necessary for a clearer historical picture of the evolution of SFR dune fauna. From a biological perspective, we anticipate that phylogeographic analyses under denser genetic sampling will reveal a finer spatial scale of diversification patterns, leading to major advances in the field. Coalescent-based demographic inference can likewise help to test alternative scenarios for diversification at a local scale, as recently used for larger-scale analyses of Neotropical lizards and frogs (Carnaval et al. 2009; Thomé et al. 2016; Prates et al. 2016; Gehara et al. 2017).

Over the last decades, most discussions of the origins of Neotropical biodiversity focused on whether and how vegetation dynamics triggered by Quaternary climate changes, or landscape changes related to Tertiary tectonic events, led to diversification (Gentry 1982; Bush 1994; Haffer 1997). With the accumulation of molecular data, emergent patterns suggest that both processes are in charge of the large-scale patterns of diversification in multiple groups of tropical organisms (Moritz et al. 2000; Rull 2008; Turchetto-Zolet et al. 2013). Molecular data support this conclusion also in the endemic reptile fauna of the SFR dune region, which represents a remarkable empirical example of the synergy between geomorphological changes, paleoclimatic shifts, and ecological mechanisms for biotic diversification at a local scale.

Acknowledgements We thank Maria Lúcia Benozzati and Sabrina Baroni for generating sequences for this work and Ignacio de la Riva and Ana C. Carnaval for the critical reading and suggestions. Fundação de Amparo à Pesquisa do Estado de São Paulo (FAPESP) and Conselho Nacional de Desenvolvimento Científico e Tecnológico (CNPq) provided financial support.

Appendix 1: Updated list of Squamate Reptiles Occurring on the Middle São Francisco Dune Region

Species	Author	Habits	Endemic
SQUAMATA (LIZARDS $n = 36$)			
TROPIDURIDAE			
Eurolophosurus amathites	(Rodrigues, 1984)	Terrestrial[P]	X
Eurolophosaurus divaricatus	(Rodrigues, 1986)	Terrestrial[P]	X
Eurolophosaurus sp. n.		Terrestrial[P]	X
Tropidurus pinima	(Rodrigues, 1984)	Scansorial	
Tropidurus psammonastes	Rodrigues, 1988	Terrestrial[P]	X
Tropidurus hispidus	(Spix, 1825)	Scansorial	
Tropidurus cocorobensis	Rodrigues, 1987	Terrestrial[P]	
Tropidurus erythrocephalus	Rodrigues, 1987	Scansorial	
Tropidurus semitaeniatus	(Spix, 1825)	Scansorial	
HOPLOCERCIDAE			
Hoplocercus spinosus	Fitzinger, 1843	Terrestrial	
IGUANIDAE			
Iguana iguana	(Linnaeus, 1758)	Arboreal	
POLYCHROTIDAE			
Polychrus acutirostris	Spix, 1825	Arboreal	
GYMNOPHTHALMIDAE			
Acratosaura mentalis	(Amaral, 1933)	Terrestrial/Fossorial	
Calyptommatus leiolepis	Rodrigues, 1991a	Fossorial[P]	X

(continued)

Species	Author	Habits	Endemic
Calyptommatus nicterus	Rodrigues, 1991a	Fossorial[P]	X
Calyptommatus sinebrachiatus	Rodrigues, 1991a	Fossorial[P]	X
Nothobachia ablephara	Rodrigues, 1986	Fossorial[P]	X
Psilops paeminosus	(Rodrigues, 1991b)	Fossorial[P]	
Procellosaurinus erythrocercus	Rodrigues, 1991c	Fossorial[P]	X
Procellosaurinus tetradactylus	Rodrigues, 1991c	Fossorial[P]	
Micrablepharus maximiliani	(Reinhardt & Lütken, 1862)	Terrestrial/Fossorial	
Vanzosaura multiscutata	(Amaral, 1933)	Fossorial	
TEIIDAE			
Ameiva ameiva	(Linnaeus, 1758)	Terrestrial	
Ameiva sp. n.		Terrestrial[P]	X
Ameivula nigrigula	Arias, Carvalho, Rodrigues & Zaher, 2011	Terrestrial	
Ameivula sp. n. 1		Terrestrial[P]	X
Ameivula sp. n. 2		Terrestrial[P]	X
Glaucomastix cyanurus	Arias, Carvalho, Rodrigues & Zaher, 2011	Terrestrial	
Salvator merianae	(Duméril & Bibron, 1839)	Terrestrial	
SCINCIDAE			
Brasiliscincus heathi	(Schmidt & Inger, 1951)	Terrestrial	
PHYLLODACTYLIDAE			
Gymnodactylus geckoides	Spix, 1825	Terrestrial	
Phyllopezus pollicaris	(Spix, 1825)	Scansorial	
SPHAERODACTYLIDAE			
Coleodactylus meridionalis	(Boulenger, 1888)	Terrestrial	
GEKKONIDAE			
Hemidactylus brasilianus	(Amaral, 1935)	Scansorial	
Hemidactylus mabouia	(Moreau de Jonnès, 1818)	Scansorial	
Lygodactylus klugei	(Smith, Martin & Swain, 1977)	Arboreal	
SQUAMATA (AMPHISBAENIANS *n* = 7)			
AMPHISBAENIDAE			
Amphisbaena arda	Rodrigues, 2002	Fossorial[P]	
Amphisbaena frontalis	Vanzolini, 1991b	Fossorial[P]	
Amphisbena hastata	Vanzolini, 1991a	Fossorial[P]	X
Amphisbaena ignatiana	Vanzolini, 1991a	Fossorial[P]	X
Amphisbaena pretrei	Duméril & Bibron, 1839	Fossorial	
Amphisbaena vermicularis	Gans, 1963	Fossorial	
Leposternon polystegum	(Duméril, 1851)	Fossorial	

(continued)

Species	Author	Habits	Endemic
SQUAMATA (SNAKES *n* = 27)			
BOIDAE			
Boa constrictor	Linnaeus, 1758	Terrestrial/Arboreal	
Corallus hortulanus	(Linnaeus, 1758)	Arboreal	
COLUBRIDAE			
Apostolepis arenarius	Rodrigues, 1992	Fossorial^{p}	X
Apostolepis gaboi	Rodrigues, 1992	Fossorial^{p}	X
Apostolepis cf. *cearensis*		Fossorial^{p}	X
Boiruna sertaneja	Zaher, 1996	Terrestrial	
Drymoluber brazili	(Gomes, 1918)	Terrestrial	
Erythrolamprus poecilogyrus	(Wied-Neuwied, 1825)	Terrestrial	
Erythrolamprus viridis	(Günther, 1862)	Terrestrial	
Helicops leopardinus	(Schlegel, 1837)	Aquatic	
Lygophis dilepis	Cope, 1862	Terrestrial	
Mastigodryas bifossatus	(Raddi, 1820)	Terrestrial	
Oxybelis aeneus	(Wagler, 1824)	Arboreal	
Oxyrhopus trigeminus	Duméril, Bibron & Duméril, 1854	Terrestrial	
Philodryas nattereri	Steindachner, 1870	Terrestrial	
Rodriguesophis chui	(Rodrigues, 1993)	Fossorial^{p}	X
Rodriguesophis iglesiasi	(Gomes, 1915)	Fossorial^{p}	
Rodriguesophis scriptorcibatus	(Rodrigues, 1993)	Fossorial^{p}	X
Thamnodynastes sertanejo	Bailey, Thomas & da Silva, 2005	Terrestrial/arboreal	
Thamnodynastes phoenix	Franco, Trevine, Montingelli & Zaher, 2017	Terrestrial	
Xenodon merremii	(Wagler, 1824)	Terrestrial	
ELAPIDAE			
Micrurus ibiboboca	(Merrem, 1820)	Terrestrial/Fossorial	
VIPERIDAE			
Bothrops erythromelas	Amaral, 1923	Terrestrial	
Crotalus durissus	Linnaeus, 1758	Terrestrial	
LEPTOTYPHLOPIDAE			
Epictia borapeliotes	(Vanzolini, 1996)	Fossorial	
TYPHLOPIDAE			
Amerotyphlops yonenagae	(Rodrigues, 1991)	Fossorial^{p}	X
Amerotyphlops amoipira	(Rodrigues & Juncá, 2002)	Fossorial^{p}	

Endemic species are indicated by "X". Strictly psammophilous species are indicated by "^{p}" on "Habits" column

Appendix 2: Description of Methods Implemented for Molecular Analysis

Bayesian Phylogeny and Estimates of Divergence Times

We obtained partial sequences of the mitochondrial cytochrome b gene (*cyt b*, 281 bp) from one individual of *Rodriguesophis chui* (MTR 3535, Santo Inácio, BA) and one *R. scriptorcibatus* (MTR 3651, Mocambo do Vento, BA) following the protocol described in Siedchlag et al. (2010). Partial sequences of *cyt b* (398 bp) and sequences of the nuDNA exon prolactin receptor gene (*prlr*, 519 bp) were obtained from exemplars of *Eurolophosaurus* representing the five candidate species of Passoni et al. (2008): *E. amathites* (MTR 11348, Santo Inácio, BA), *E. nanuzae* (MTR 887662, Diamantina, MG), *E. divaricatus* (MTR 906421, Queimadas, BA), *E.* aff. *divaricatus* (MTR 11747, Alagoado, BA), and *E.* sp. (MTR 11073, Mucugê, BA). Primers and protocol for obtaining *prlr* sequences follows (Townsend et al. 2008); new sequences are deposited in Genbank (MK541043-MK541047; MK548582-MK548588)

We retrieved from Genbank sequences of mtDNA *cyt b* (692 bp) and nuDNA oocyte maturation factor gene (*C-mos*, 493 bp) for *Rodriguesophis iglesiasi* and outgroups of Pseudoboini (*Boiruna maculata*, *Drepanoides anomalus*, *Paraphimophis rusticus*, *Pseudoboa nigra*, *Pseudoboa neuwiedi*) (Grazziotin et al. 2012; Pyron et al. 2013); partial sequence of *cyt b* and *prlr* for one outgroup of *Eurolophosaurus* (*Tropidurus torquatus*; Carvalho et al. 2016); sequences of partial *cyt b* (712 bp) and nuDNA *C-mos* (390 bp) and dynein axonemal heavy chain 3 (*dnah3*, 691 bp) for representatives of Gymnophthalmini lizards (*Calyptommatus*, *Nothobachia*, *Scriptosaura*) and one outgroup (*Gymnophthalmus vanzoi*) (Goicoechea et al. 2016)

Sequences were aligned using MUSCLE (Edgar 2004). The best substitution model was selected under the Akaike's information criteria in jModelTest2 (Darriba et al. 2012): GTR+G for *cyt b* and HKY+I for *C-mos* and *dnah3* for the Gymnophthalmini alignment; GTR+I for *cyt b* and HKY for *prlr* for the *Eurolophosaurus* alignment; and HKY+I for *cyt b* and HKY for *C-mos* for the Pseudoboini alignment. Bayesian phylogenetic analysis and estimation of divergence times were performed using BEAST v.1.10 (Suchard et al. 2018). In all analyses, we implemented a Yule speciation prior, and implemented an uncorrelated relaxed-clock calibration using a lognormal distribution with previously estimated molecular clock rates for mtDNA and nuDNA genes (Daza et al. 2009; Olave et al. 2015). For Pseudoboini, we set the ucld.mean prior as the mean rate estimated for a Dipsadidae snake (Daza et al. 2009) for mtDNA (0.0134) and nuDNA (0.0014), based on *cyt b* and *C-mos*, respectively, with a SD of 0.1. For lizards, we used molecular clock rates estimated for Liolemidae species, based on several genes (Olave et al. 2015), implementing reported mean rates for *cyt b* (0.0194), *C-mos* (0.0009), *dnah3* (0.0008) and *prlr* (0.00133) for ucld.mean, with a SD of 0.1. We also simulated a scenario of faster molecular rates for lizards, using the upper bound of estimates for each marker: *cyt b* (0.0263), *C-mos* (0.0013), *dnah3* (0.0011) and *prlr* (0.00182). We simulated a faster clock rate considering the possibility that the small-bodied, highly abundant lizards from the Tropical SFR dunes, especially *Calyptommatus,* which presents early maturation and reproduces year-round (Ramiro et al. 2017), could have smaller generation times and higher metabolic rates than Liolemidae lizards, which in turn may influence molecular clock rates (Martin and Palumbi 1993). We implemented two independent runs, using four chains and 3×10^7 generations, and sampling each 3×10^3 generation. Stationarity, convergence, and high effective sample sizes (ESS > 200) were verified with Tracer 1.7 (Rambaut et al. 2018). Trees were summarized through maximum clade credibility (MCC) after removing the first 10% trees as burn-in with LogCombiner 1.10 (Suchard et al. 2018)

Genetic Distances

Uncorrected pairwise p-distances between samples was obtained with Arlequin 3.5 (Excoffier and Lischer 2010), based on the individual alignments of *Brasiliscincus heathi.* A final alignment (308 bp) was obtained with MUSCLE (Edgar 2004), and it consisted of six partial sequences of *cyt b* retrieved from GenBank (Whiting et al. 2006) and one novel sequence from Vacaria, Bahia (MTR 906801; Genbank MK548581), obtained as per Siedchlag et al. (2010). Net between-group mean distances were calculated for *Vanzosaura multiscutata* based on an alignment of partial *cyt b* sequences (735 bp) of 25 individuals collected in the São Francisco dune region (Recoder et al. 2014; Gehara et al. 2017). Individuals were assembled in four populations: Vacaria ($n = 10$); Ibiraba/Barra ($n = 6$); Queimadas ($n = 6$) and Santo Inácio ($n = 3$).

Statistical Test of Genetic Structure

The presence of significant genetic structure between populations of *Vanzosaura multiscutata* was tested with an analysis of molecular variance (AMOVA) implemented in Arlequin 3.5 (Excoffier and Lischer 2010). In it, we used opposite margins of the São Francisco River as an among-group factor, and the four populations occurring at the dune region (delimited as described above) as a within-group factor. The analysis was implemented with 100,000 replicates.

References

Ab'Sáber AN (1969) Participação das superfícies aplainadas nas paisagens do nordeste brasileiro. Geomorfologia 19:1–38

Ab'Sáber AN (1970) Províncias geológicas e domínios morfoclimáticos no Brasil. Geomorfologia 20:1–26

Ab'Sáber A (2006) O paleodeserto de Xique-Xique. Estudos Avançados 20:301–308

Arias F, Carvalho CM, Rodrigues MT, Zaher H (2011) Two new species of *Cnemidophorus* (Squamata: Teiidae) of the *C. ocellifer* group, from Bahia, Brazil. Zootaxa 3022:1–21

Auler AS, Smart PL (2001) Late quaternary paleoclimate in semiarid northeastern Brazil from U-series dating of travertine and water-table speleothems. Quat Res 55:159–167

Barreto AMF (1996) Interpretação paleoambiental do sistema de dunas fixadas do médio Rio São Francisco, Bahia. PhD thesis, Instituto de Geociências, Universidade de São Paulo, São Paulo

Barreto AMF, Suguio K, Oliveira PE, Tatumi SH (2002) Campo de Dunas Inativas do Médio Rio São Francisco, BA – Marcante registro de ambiente desértico do Quaternário brasileiro. In: Schobbenhaus C, Campos DA, Queiroz ET et al (eds) Sítios Geológicos e Paleontológicos do Brasil. Comissão Brasileira de Sítios Geológicos e Paleobiológicos (SIGEP), Brasília, pp 223–231

Bravo F, Pohl H, Silva-Neto A, Beutel RG (2009) Bahiaxenidae, a "living fossil" and a new family of Strepsiptera (Hexapoda) discovered in Brazil. Cladistics 25:614–623

Bull CM (1991) Ecology of parapatric distributions. Annu Rev Ecol Syst 22:19–36

Bush MB (1994) Amazonian speciation: a necessarily complex model. J Biogeogr 21:5–17

Caldas FLS, Costa TB, Laranjeiras DO et al (2016) Herpetofauna of protected areas in the Caatinga V: Seridó ecological station (Rio Grande do Norte, Brazil). Check List 12:1929

Carnaval AC, Hickerson MJ, Haddad CF et al (2009) Stability predicts genetic diversity in the Brazilian Atlantic forest hotspot. Science 323:785–789

Carvalho AL, Sena MA, Peloso PL et al (2016) A new *Tropidurus* (Tropiduridae) from the semiarid Brazilian Caatinga: evidence for conflicting signal between mitochondrial and nuclear loci affecting the phylogenetic reconstruction of south American collared lizards. Am Mus Novit 3852:1–68

Cavalcanti LBQ, Costa TB, Colli GR et al (2014) Herpetofauna of protected areas in the Caatinga II: Serra da Capivara National Park, Piauí, Brazil. Check List 10:18–27

Colli GR, Giugliano LG, Mesquita DO, França FG (2009) A new species of *Cnemidophorus* from the Jalapão region, in the central Brazilian Cerrado. Herpetologica 65:311–327

Costa WJ, Amorim PF, Mattos JLO (2018) Synchronic historical patterns of species diversification in seasonal aplocheiloid killifishes of the semi-arid Brazilian Caatinga. PLoS One 13:e0193021

Dal Vechio F, Recoder RS, Zaher H, Rodrigues MT (2013) The herpetofauna of the Estação Ecológica de Uruçuí-Una, state of Piauí, Brazil. Pap Avulsos Zool 53:225–243

Dal Vechio F, Teixeira M Jr, Recoder RS et al (2016) The herpetofauna of Parque Nacional da Serra das Confusões, state of Piauí, Brazil, with a regional species list from an ecotonal area of Cerrado and Caatinga. Biota Neotrop 16:e20150105

Damasceno R, Strangas ML, Carnaval AC et al (2014) Revisiting the vanishing refuge model of diversification. Front Genet 5:353

Darriba D, Taboada GL, Doallo R, Posada D (2012) jModelTest 2: more models, new heuristics and parallel computing. Nat Methods 9:772

Daza JM, Smith EN, Páez VP, Parkinson CL (2009) Complex evolution in the Neotropics: the origin and diversification of the widespread genus *Leptodeira* (Serpentes: Colubridae). Mol Phylogenet Evol 53:653–667

Delfim FR, Gonçalves EM, da Silva ST (2006) Squamata, Gymnophthalmidae, *Psilophthalmus paeminosus*: distribution extension, new state record. Check List 2:89–92

Dias EJDR, Rocha CFD (2007) Niche differences between two sympatric whiptail lizards (*Cnemidophorus abaetensis* and *C. ocellifer*, Teiidae) in the Restinga habitat of northeastern Brazil. Braz J Biol 67:41–46

Domingues AJP (1948) Contribuição à geologia do sudeste da Bahia. Rev Bras Geogr 10:255–289

Edgar RC (2004) MUSCLE: multiple sequence alignment with high accuracy and high throughput. Nucleic Acids Res 32:1792–1797

Excoffier L, Lischer HEL (2010) Arlequin suite ver 3.5: a new series of programs to perform population genetics analyses under Linux and windows. Mol Ecol Resour 10:564–567

Faria MB, Nascimento FF, de Oliveira JA, Bonvicino CR (2013) Biogeographic determinants of genetic diversification in the mouse opossum *Gracilinanus agilis* (Didelphimorphia: Didelphidae). J Hered 104:613–626

Fouquet A, Courtois EA, Baudain D et al (2015) The trans-riverine genetic structure of 28 Amazonian frog species is dependent on life history. J Trop Ecol 31:361–373

Freitas MA, Veríssimo D, Uhlig V (2012) Squamate reptiles of the Central Chapada Diamantina, with a focus on the municipality of Mucugê, state of Bahia, Brazil. Check List 8:16–22

Frost DR, Rodrigues MT, Grant T, Titus TA (2001) Phylogenetics of the lizard genus *Tropidurus* (Squamata: Tropiduridae: Tropidurinae): direct optimization, descriptive efficiency, and sensitivity analysis of congruence between molecular data and morphology. Mol Phylogenet Evol 21:352–371

Garda AA, Costa TB, Silva CRS et al (2013) Herpetofauna of protected areas in the Caatinga I: Raso da Catarina Ecological Station (Bahia, Brazil). Check List 9:405–414

Gascon C, Malcolm JR, Patton JL et al (2000) Riverine barriers and the geographic distribution of Amazonian species. Proc Natl Acad Sci 97:13672–13677

Gehara M, Garda AA, Werneck FP et al (2017) Estimating synchronous demographic changes across populations using hABC and its application for a herpetological community from northeastern Brazil. Mol Ecol 26:4756–4771

Gentry AH (1982) Neotropical floristic diversity: phytogeographical connections between central and South America, Pleistocene climatic fluctuations, or an accident of the Andean orogeny? Ann Mo Bot Gard 69:557–593

Giannini PCF, Assine ML, Barbosa LM et al (2005) Dunas e paleodunas eólicas costeiras e interiores. In: Souza CRG, Suguio K, Oliveira PE, Oliveira AMS (eds) Quaternário do Brasil. Holos Editora, Ribeirão Preto, pp 235–257

Goicoechea N, Frost DR, De la Riva I et al (2016) Molecular systematics of teioid lizards (Teioidea/Gymnophthalmoidea: Squamata) based on the analysis of 48 loci under tree-alignment and similarity-alignment. Cladistics 32:624–671

Grazziotin FG, Zaher H, Murphy RW et al (2012) Molecular phylogeny of the new world Dipsadidae (Serpentes: Colubroidea): a reappraisal. Cladistics 28:437–459

Greene HW (1997) Snakes: the evolution of mystery in nature. University of California Press, Berkeley, CA

Grizante MB, Navas CA, Garland T Jr, Kohlsdorf T (2010) Morphological evolution in Tropidurinae squamates: an integrated view along a continuum of ecological settings. J Evol Biol 23:98–111

Guedes TB, Nogueira C, Marques OA (2014) Diversity, natural history, and geographic distribution of snakes in the Caatinga, Northeastern Brazil. Zootaxa 3863:1–93

Haffer JR (1997) Alternative models of vertebrate speciation in Amazonia: an overview. Biodivers Conserv 6:451–476

King LG (1956) A Geomorfologia do Brasil Oriental. Rev Bras Geogr 18:147–265

Kohlsdorf T, James RS, Carvalho JE, Wilson RS, Dal Pai-Silva M, Navas CA (2004) Locomotor performance of closely related *Tropidurus* species: relationships with physiological parameters and ecological divergence. J Exp Biol 207:1183–1192

Loebmann D, Haddad CFB (2010) Amphibians and reptiles from a highly diverse area of the Caatinga domain: composition and conservation implications. Biota Neotrop 10:227–256

Lougheed SC, Gascon C, Jones DA et al (1999) Ridges and rivers: a test of competing hypotheses of Amazonian diversification using a dart-poison frog (*Epipedobates femoralis*). Proc R Soc Lond B Biol Sci 266:1829–1835

Machado T, Silva VX, Silva MJDJ (2014) Phylogenetic relationships within *Bothrops neuwiedi* group (Serpentes, Squamata): geographically highly-structured lineages, evidence of introgressive hybridization and Neogene/quaternary diversification. Mol Phylogenet Evol 71:1–14
Magalhães FM, Laranjeiras DO, Costa TB et al (2015) Herpetofauna of protected areas in the Caatinga IV: Chapada Diamantina National Park, Bahia, Brazil. Herpetol Notes 8:243–261
Martin AP, Palumbi SR (1993) Body size, metabolic rate, generation time, and the molecular clock. Proc Natl Acad Sci 90:4087–4091
Mesquita DO, Costa GC, Garda AA, Delfim FR (2017) Species composition, biogeography, and conservation of the Caatinga Lizards. In: da Silva JMC, Leal IR, Tabarelli M (eds) Caatinga: the largest tropical dry forest region in South America. Springer, Cham, pp 151–180
Moritz C, Patton JL, Schneider CJ, Smith TB (2000) Diversification of rainforest faunas: an integrated molecular approach. Annu Rev Ecol Syst 31:533–563
Nascimento FF, Lazar A, Menezes AN et al (2013) The role of historical barriers in the diversification processes in open vegetation formations during the Miocene/Pliocene using an ancient rodent lineage as a model. PLoS One 8:e61924
Naka LN, Brumfield RT (2018) The dual role of Amazonian rivers in the generation and maintenance of avian diversity. Sci Adv 4:eaar8575
Nimer E (1989) Climatologia do Brasil. IBGE, Rio de Janeiro
Olave M, Avila LJ, Sites JW Jr, Morando M (2015) Model-based approach to test hard polytomies in the *Eulaemus* clade of the most diverse south American lizard genus *Liolaemus* (Liolaemini, Squamata). Zool J Linnean Soc 174:169–184
Oliveira PE, Barreto AMF, Suguio K (1999) Late Pleistocene/Holocene climatic and Vegetational history of the Brazilian Caatinga: the fossil dunes of the middle São Francisco River. Paleogeogr Palaeoclimatol Palaeoecol 152:319–337
Oliveira PE, Pessenda LCR, Barreto AMF, Oliveira ES, Santos JC (2014) Paleoclimas da Caatinga brasileira durante o Quaternário tardio. In: de Souza Carvalho I, Garcia MJ, Lana CC, Stroschoen O (eds) Paleontologia: Cenários de Vida. Editora Interciência, Rio de Janeiro, pp 501–516
Oliveira EF, Martinez PA, São-Pedro VA et al (2018) Climatic suitability, isolation by distance and river resistance explain genetic variation in a Brazilian whiptail lizard. Heredity 120:251–265
Passoni JC, Benozzati ML, Rodrigues MT (2008) Phylogeny, species limits, and biogeography of the Brazilian lizards of the genus *Eurolophosaurus* (Squamata: Tropiduridae) as inferred from mitochondrial DNA sequences. Mol Phylogenet Evol 46:403–414
Pedrosa IMMC, Costa TB, Faria RG et al (2014) Herpetofauna of protected areas in the Caatinga III: the Catimbau National Park, Pernambuco, Brazil. Biota Neotrop 14:e20140046
Pellegrino KC, Rodrigues MT, Yonenaga-Yassuda Y, Sites JW Jr (2001) A molecular perspective on the evolution of microteiid lizards (Squamata, Gymnophthalmidae), and a new classification for the family. Biol J Linn Soc 74:315–338
Pianka ER, Vitt LJ (2003) Lizards: windows to the evolution of diversity. University of California Press, Berkeley, CA
Prates I, Xue AT, Brown JL et al (2016) Inferring responses to climate dynamics from historical demography in neotropical forest lizards. Proc Natl Acad Sci 113:7978–7985
Pyron RA, Burbrink FT, Wiens JJ (2013) A phylogeny and revised classification of Squamata, including 4161 species of lizards and snakes. BMC Evol Biol 13:93
Rambaut A, Drummond AJ, Xie D et al (2018) Posterior summarisation in Bayesian phylogenetics using tracer 1.7. Syst Biol 67:901–904
Ramiro CN, Teixeira M Jr, Rodrigues MT (2017) Reproductive biology of three sympatric species of gymnophthalmid lizards from the sand dunes of the middle São Francisco river, Bahia, Brazil. S Am J Herpetol 12:76–88
Recoder R, Nogueira C (2007) Composição e diversidade de répteis Squamata na região sul do Parque Nacional Grande Sertão Veredas, Brasil central. Biota Neotrop 7:267–278

Recoder RS, Werneck FP, Teixeira M Jr et al (2014) Geographic variation and systematic review of the lizard genus *Vanzosaura* (Squamata, Gymnophthalmidae), with the description of a new species. Zool J Linnean Soc 171:206–225

Ribeiro SC, Ferreira FS, Brito SV et al (2009) The squamata fauna of the Chapada do Araripe, northeastern Brazil. Cadernos de Cultura e Ciência 1:67–76

Rocha PLB (1995) *Proechimys yonenagae*, a new species of spiny rat (Rodentia: Echimyidae) from fossil sand dunes in the Brazilian Caatinga. Mammalia 59:537–550

Rocha PLB, Rodrigues MT (2005) Electivities and resource use by an assemblage of lizards endemic to the dunes of the São Francisco River, northeastern Brazil. Pap Avulsos Zool 45:261–284

Rocha PLB, Queiroz L, Pirani JR (2004) Plant species and habitat structure in a sand dune field in the Brazilian Caatinga: a homogeneous habitat harbouring an endemic biota. Rev Bras Bot 27:739–755

Rodrigues MT (1984) *Nothobachia ablephara*: novo gênero e espécie do nordeste do Brasil (Sauria: Teiidae). Pap Avulsos Zool 35:361–366

Rodrigues MT (1986) Um novo *Tropidurus* com crista dorsal do Brasil com comentários sôbre suas relações, distribuição e origem (Sauria; Iguanidae). Pap Avulsos Zool 36:171–179

Rodrigues MT (1987) Sistemática, ecologia e zoogeografia dos *Tropidurus* do grupo torquatus ao sul do Rio Amazonas (Sauria, Iguanidae). Arq Zool 31:105–230

Rodrigues MT (1991a) Herpetofauna das dunas interiores do rio São Francisco: Bahia: Brasil. I. Introdução à área e descrição de um novo gênero de microteiideos (*Calyptommatus*) com notas sobre sua ecologia, distribuição e especiação (Sauria, Teiidae). Papéis Avulsos de Zoologia, São Paulo 37:285–320

Rodrigues MT (1991b) Herpetofauna das dunas interiores do rio São Francisco: Bahia: Brasil. II. *Psilophthalmus*: um novo gênero de microteiideos sem pálpebra (Sauria, Teiidae). Papéis Avulsos de Zoologia, São Paulo 37:321–327

Rodrigues MT (1991c) Herpetofauna das dunas interiores do rio São Francisco: Bahia: Brasil. III. *Procellosaurinus*: um novo gênero de microteiideo sem pálpebra, com a redefinição do gênero *Gymnophthalmus* (Sauria, Teiidae). Papéis Avulsos de Zoologia, São Paulo 37:329–342

Rodrigues MT (1991d) Herpetofauna das dunas interiores do rio São Francisco: Bahia: Brasil. IV. Uma nova espécie de *Typhlops* (Ophidia, Typhlopidae). Papéis Avulsos de Zoologia, São Paulo 37:343–346

Rodrigues MT (1992) Herpetofauna das dunas interiores do rio São Francisco: Bahia: Brasil. V. Duas novas espécies de *Apostolepis* (Serpentes, Colubridae). Mem Inst Butantan 54:53–59

Rodrigues MT (1993) Herpetofauna of palaeoquaternary sand dunes of the middle São Francisco river: Bahia: Brazil. VI. Two new species of *Phimophis* (Serpentes: Colubridae) with notes on the origin of psammophilic adaptations. Papéis Avulsos de Zoologia, São Paulo 38:187–198

Rodrigues MT (1996) Lizards, snakes, and Amphisbaenians from the quaternary sand dunes of the middle Rio São Francisco, Bahia, Brazil. J Herpetol 30:513–523

Rodrigues MT (2002) Herpetofauna of the quaternary sand dunes of the middle Rio São Francisco, Bahia, Brazil: VIII. *Amphisbaena arda* sp. nov, a *fuliginosa*-like checkered patterned *Amphisbaena* (Squamata, Amphisbaenidae). Phyllomedusa 1:51–56

Rodrigues MT (2003) Herpetofauna da caatinga. In: Leal IR, Tabarelli M, Silva JMC (eds) Ecologia e Conservação da Caatinga. Universidade Federal de Pernambuco, Recife, pp 181–236

Rodrigues MT, Juncá FA (2002) Herpetofauna of the quaternary sand dunes of the middle Rio São Francisco: Bahia: Brazil. VII.: *Typhlops amoipira* sp. nov, a possible relative of *Typhlops yonenagae* (Serpentes, Typhlopidae). Papéis Avulsos de Zoologia, São Paulo 42:325–333

Rodrigues MT, dos Santos EM (2008) A new genus and species of eyelid-less and limb reduced gymnophthalmid lizard from northeastern Brazil (Squamata, Gymnophthalmidae). Zootaxa 1873:50–60

Rodrigues MT, Kasahara S, Yonenaga-Yassuda Y (1988) *Tropidurus psammonastes*: uma nova espécie do grupo torquatus com notas sôbre seu cariótipo e distribuição (Sauria, Iguanidae). Pap Avulsos Zool 36:307–313

Rodrigues MT, Zaher H, Curcio F (2001) A new species of lizard, genus *Calyptommatus*, from the caatingas of the state of Piauí, northeastern Brazil (Squamata, Gymnaphthalmidae). Pap Avulsos Zool 41:529–546

Rodrigues MT, Recoder R, Teixeira-Jr M et al (2017) A morphological and molecular study of *Psilops*, a replacement name for the Brazilian microteiid lizard genus *Psilophthalmus* Rodrigues 1991 (Squamata, Gymnophthalmidae), with the description of two new species. Zootaxa 4286:451–482

Rull V (2008) Speciation timing and neotropical biodiversity: the tertiary–quaternary debate in the light of molecular phylogenetic evidence. Mol Ecol 17:2722–2729

Siedchlag AC, Benozzati ML, Passoni JC, Rodrigues MT (2010) Genetic structure, phylogeny, and biogeography of Brazilian eyelid-less lizards of genera *Calyptommatus* and *Nothobachia* (Squamata, Gymnophthalmidae) as inferred from mitochondrial DNA sequences. Mol Phylogenet Evol 56:622–630

Suchard MA, Lemey P, Baele G et al (2018) Bayesian phylogenetic and phylodynamic data integration using BEAST 1.10 Virus Evolution 4:vey016

Teixeira Jr M (2010) Os lagartos do vale do Rio Peruaçu, MG, Brasil: Aspectos biogeográficos, história natural e implicações para a conservação. Dissertation, Universidade de São Paulo, São Paulo

Teixeira M Jr, Vechio FD, Rodrigues MT (2016) Diagnostic clarification, new morphological data and phylogenetic placement of *Amphisbaena arenaria* Vanzolini, 1991 (Amphisbaenia, Amphisbaenidae). Zootaxa 4205:293–296

Thomé MTC, Sequeira F, Brusquetti F et al (2016) Recurrent connections between Amazon and Atlantic forests shaped diversity in Caatinga four-eyed frogs. J Biogeogr 43:1045–1056

Townsend TM, Alegre RE, Kelley ST et al (2008) Rapid development of multiple nuclear loci for phylogenetic analysis using genomic resources: an example from squamate reptiles. Mol Phylogenet Evol 47:129–142

Tricart J (1974) Existence de Périodes Sèches au Quaternaire en Amazonie et dans les Regions Voisines. Revue de Geomorphologie Dynamique 23:145–158

Turchetto-Zolet AC, Pinheiro F, Salgueiro F, Palma-Silva C (2013) Phylogeographical patterns shed light on evolutionary process in South America. Mol Ecol 22:1193–1213

Vanzolini PE (1991a) Two new small species of *Amphisbaena* from fossil dune field of the middle Rio São Francisco, state of Bahia, Brasil (Reptilia, Amphisbaenia). Pap Avulsos Zool 37:259–276

Vanzolini PE (1991b) Two further new species of *Amphisbaena* from the semi-arid northeast of Brasil (Reptilia, Amphisbaenia). Pap Avulsos Zool 37:247–361

Vanzolini PE, Williams EE (1981) The vanishing refuge: a mechanism for ecogeographic speciation. Pap Avulsos Zool 34:251–255

Vitt LJ (1995) The ecology of tropical lizards in the Caatinga of Northeast Brazil. Occasional papers of the Oklahoma museum of. Nat Hist 1:1–29

Webb CO, Ackerly DD, McPeek MA, Donoghue MJ (2002) Phylogenies and community ecology. Annu Rev Ecol Syst 33:475–505

Werneck FP, Gamble T, Colli GR et al (2012) Deep diversification and long-term persistence in the south American 'dry diagonal': integrating continent-wide phylogeography and distribution modeling of geckos. Evolution 66:3014–3034

Werneck FP, Leite RN, Geurgas SR, Rodrigues MT (2015) Biogeographic history and cryptic diversity of saxicolous Tropiduridae lizards endemic to the semiarid Caatinga. BMC Evol Biol 15:94

Whiting AS, Sites JW Jr, Pellegrino KC, Rodrigues MT (2006) Comparing alignment methods for inferring the history of the new world lizard genus *Mabuya* (Squamata: Scincidae). Mol Phylogenet Evol 38:719–730

Williams HE (1925) Notas geológicas e econômicas sobre o vale do Rio São Francisco. Boletim do Serviço Geológico e Mineralógico 12:56

Xavier E, Rocha LS (2001) Autoecology and description of *Mummucia mauryi* (Solifugae, Mummuciidae), a new solifuge from Brazilian semi-arid Caatinga. J Arachnol 29:127–134

Chapter 27
Unexpected Beta-Diversity Radiations in Highland Clades of Andean Terraranae Frogs

Ignacio De la Riva

Abstract Several groups of direct-developing frogs of the Superfamily Brachycephaloidea (= Terraranae) are high Andean and occur in cloud forests, elfin forests, humid paramos, and grasslands, from Colombia to Bolivia.

They belong to different genera and experienced independent radiations, in some cases leading to stunning beta-diversity and high level of endemism.

Despite the remarkable accumulation of knowledge on the systematics and diversity of these frogs over the last decades, there has been no attempt to synthesize what is known about their morphology, life history, and evolutionary history. This is the goal of this contribution. Through this review, it becomes evident that similar life histories and ecological niches are associated with a particular, convergent morphology, here referred to as the "phrynopoid" ecomorph, which is also present in members of other groups that occupy those same habitats. It refers to small frogs with plump bodies, short legs, and simple digital tips, presumably adapted to a terrestrial life amidst the humid, mossed floor of paramos and upper cloud forests. Although it is possible to speculate on the mechanisms underlying the remarkable diversity of phrynopoid frogs, the geological and climatic processes that promoted the diversification and the present species diversity are still too poorly known to enable inferences about their evolutionary history. Much fieldwork and integrative studies are needed before we can attain a comprehensive knowledge of this important component of the Andean biota. Hopefully, this review provides a starting point.

Keywords Amphibia · Andes · Anura · Beta-diversity · Convergence · Evolutionary radiations · South America

I. De la Riva (✉)
Museo Nacional de Ciencias Naturales-CSIC, Madrid, Spain
e-mail: iriva@mncn.csic.es

V. Rull, A. C. Carnaval (eds.), *Neotropical Diversification: Patterns and Processes*, Fascinating Life Sciences, https://doi.org/10.1007/978-3-030-31167-4_27

1 Introduction

Almost half a century has elapsed since John Lynch published his extensive revision of the systematics and zoogeography of frogs of the family Leptodactylidae, a taxon at that time distributed from the Americas to South Africa and the Australo-Papuan region. He recognized seven subfamilies, 57 extant genera, and an inordinate number of species (Lynch 1971). In reality, for his doctoral dissertation, Lynch was initially seeking to study only one Neotropical genus, *Eleutherodactylus*, but the impossibility to put it in a proper evolutionary context led him to broaden the scope impressively. That, apparently, was not a deterrent to prevent him from tackling such monumental task.

Amphibian systematics has been in state of flux during the last decades, and the frogs that Lynch studied under the family Leptodactylidae are today split into 14 families, nearly 80 genera, and more than 1700 species (Frost 2019). For almost every group in the entire animal kingdom, species numbers are increasing incessantly, thanks to the effort of taxonomists. Yet, certainly, some of the frog groups studied by Lynch in the 1970s were already known to be highly diverse. However, even Lynch's predictions were quite conservative if we look at some figures today. He estimated that the genus *Eleutherodactylus*, for which ca. 300 species were named at that time, would probably contain 400. Today, the species of *Eleutherodactylus* sensu Lynch (1971) are distributed in at least 11 genera accounting for nearly 1000 species; 200 remain in *Eleutherodactylus*, while 548 belong to the genus *Pristimantis*, which is, furthermore, placed in a different family (Frost 2019). Thus, the split of *Eleutherodactylus*, a genus considered "large and unwieldy" (Lynch 1971: 148), has eventually led to even larger and, perhaps, more unwieldy groups.

Nevertheless, regarding Neotropical frogs, not everything needed splitting in the 1970s. Some lumping seemed to be necessary too. Stemming from previous work that examined all and every "leptodactylid" frog at hand, Lynch noticed that, in contrast to the expanded digital tips of *Eleutherodactylus* (supported by underlying T-shaped or knobbed terminal phalanges), a series of Andean species living in cloud forests and humid highland grasslands lacked expanded digital discs. From the synonymy of *Eusophus* (a genus from Chile and Argentina, now in the family Alsodidae), Lynch (1975) resurrected the genus *Phrynopus* (type species, *P. peruanus* Peters), placed the genera *Noblella* and *Niceforonia* in its synonymy, provided new combination names for eight species, and described five new ones. A brand-new taxonomic status and species composition was given for the genus *Phrynopus*, and the stage was set for further research. But its provisional character was obvious from the very start. In contrast to the hundreds of species of eleutherodactyline frogs that thrived along the piedmont and the tropical Andean slopes, only 14 species represented the genus *Phrynopus,* scattered here and there, along the immense span of the Andean cordillera from northern Colombia to western Bolivia. This ample distribution, and certain morphological oddities and inconsistences hidden under a general similar external morphology, did not pass unnoticed to

Lynch's acute taxonomic eye. For this reason, he recognized four species groups in the genus, and wrote:

> Some characteristics used to distinguish genera exhibit variation among the species of *Phrynopus*. Three species (*cophites*, *nanus* and *simonsii*) lack ears. Prevomerine teeth and odontophores are absent in the species of the *peruvianus* and *peruanus* groups as well as in *simmonsii*, but are uniformly present in species of the *flavomaculatus* group. One species (*cophites*) has a large frontoparietal fontanelle; one other species (*flavomaculatus*) has prominent cranial crests. *Phrynopus peruvianus* is unique in having a functionally firmisternal pectoral girdle.
>
> The distribution of *Phrynopus* is diperse (*sic*). A single case of sympatry is known (*cophites* and *peruvianus*). Only *flavomaculatus* is known to occur over any modest area; the other thirteen species occur at single localities or at a few clustered localities. A major distributional gap occurs from northern Perú to central Perú and many additional populations are expected in the 600 kilometer distributional hiatus.

These statements, read more than 40 years after they were written, sound really prophetic. Lynch was setting the basis for what might come afterwards; and it came. All the species mentioned in these two paragraphs are no longer in *Phrynopus*. The suspiciously ample distribution of the genus, in contrast to the restricted distribution of its species and the rare cases of sympatry, indicated that perhaps *Phrynopus* was not monophyletic and that many more species were likely involved. These facts also suggested that much more sampling and surveys were necessary to ascertain the nature—either real of artefactual—of the observed gaps. Sure enough, further analyses of these frogs' DNA, not accessible in Lynch's times, consistently tell us a history of convergence and different origins. Synchronously, the number of described species has augmented impressively (Fig. 27.1).

Unmasking convergence is one of the major challenges for systematists. By redefining *Phrynopus*, what Lynch (1975) actually did was to cluster under this name 14 representatives of a single ecomorph (sensu Williams 1972). Although in

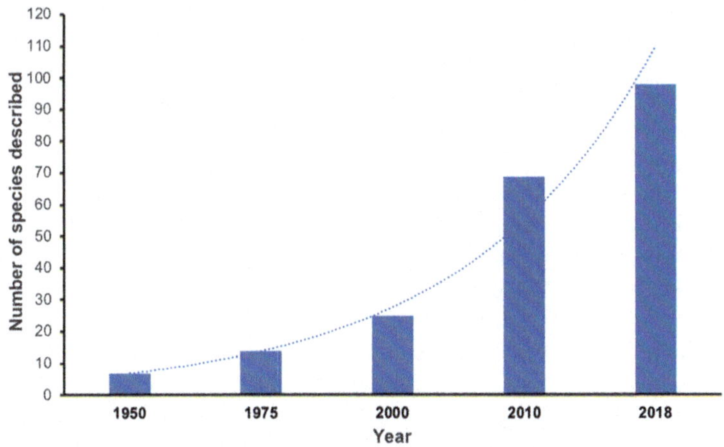

Fig. 27.1 Bar graph and trendline showing the increase in the cumulative number of described species of high Andean phrynopoid frogs for selected years from 1950 to 2018

some cases they showed obvious osteological differences, externally they looked more or less the same: small frogs with plump bodies, short legs, and simple digital tips, presumably adapted to a terrestrial life amidst the humid, mossed floor of paramos and upper cloud forests. As the years passed and the evolutionary relationships and actual species diversity of members of this ecomorph were progressively unraveled, researchers identified and described some outliers—species departing from the general "phrynopoid" morphology. They also described some intruders—species from other groups where most lineages are not high Andean, but which have adapted to the highlands and adopted such morphology.

This chapter is an essay that provides an integrative approach to the phrynopoid phenotype, including a brief historical review about the taxonomy of the groups that share it, pertinent information on the diversity of morphological and life history traits in phrynopoid frogs, and available data on the evolutionary and biogeographical processes that led to them.

2 Historical Account and the State of the Art

The genus *Phrynopus*, as recognized by Lynch (1975) had its northern distribution limit in the Colombian department of Norte de Santander, at a latitude of ca. 7°15′N (near the Venezuelan border), and its southern limit in western central Bolivia, at 16°18′S, encompassing an approximate distance of 3400 km. Despite the taxonomic turmoil of the last decades, little has changed in this sense: the current distribution of the taxa recognized today does not surpass the northern limit as formerly identified, and only in the south it has been extended by a little more than one degree of latitude and 380 km to the southeast. However, where there used to be a few scattered dots along four Andean countries, there is now a plethora of points; in most cases, each dot represents a single species, for many of the species are known from the type locality only (Fig. 27.2).

During 33 years, next to nothing changed substantially in the taxonomy of *Phrynopus*, except for the intermittent addition of more and more new species (Fig. 27.1), and the almost anecdotic departure of a couple of species that, due to their expanded digit pads, were exiled to the genus *Eleutherodactylus* by Lehr (2005). But as soon as DNA was interrogated about internal *Phrynopus'* relationships, things started to change dramatically. In 2005, the first molecular analysis of the genus was published, based on 12S and 16S mitochondrial rDNA sequences of 15 out of 34 species then recognized. In that work, some taxonomic foundations of *Phrynopus* were shaken and strongly questioned for the first time (Lehr et al. 2005). As a consequence, *P. simmonsii* was transferred to *Eleutherodactylus*, *P. brunneus* exited the genus pending its possible inclusion in *Eleutherodactylus* as well, and the bizarre *P. carpish* was transferred to a new genus in another, distantly related family (see below). Furthermore, Lehr et al. (2005) commented on the unclear relationships between *Phyllonastes* (another genus of small direct-developing frogs, nowadays known as *Noblella*, with representatives in the lowlands and in the Andes) and

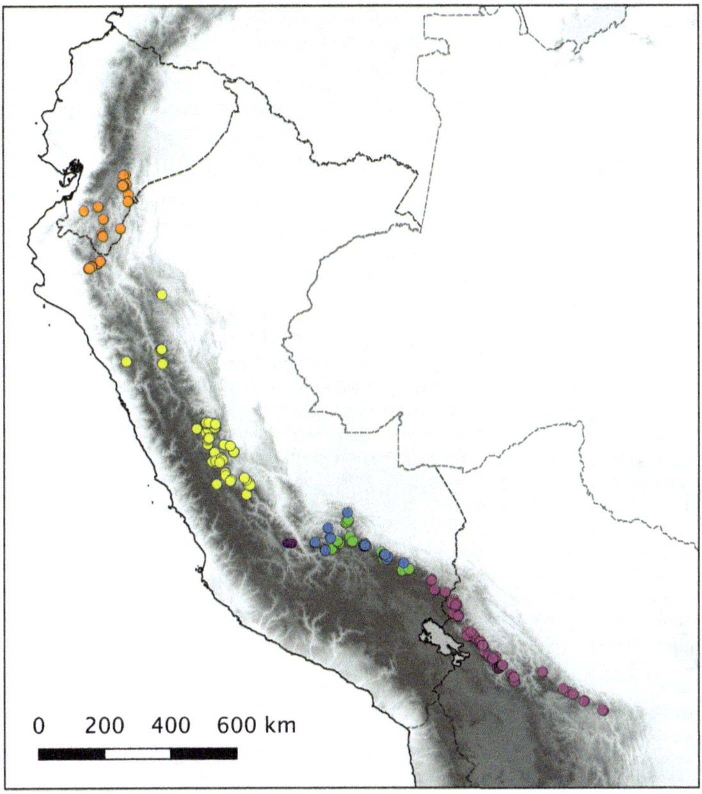

Fig. 27.2 Map of Ecuador, Peru and Bolivia showing the distribution of currently known species (some undescribed) of five genera of phrynopoid frogs: *Lynchius* (orange), *Phrynopus* (yellow), the "intruder" (see text) *Oreobates ayacucho* (purple), *Bryophryne* (green), *Psychrophrynella* (blue), and *Microkayla* (pink). A map by Lynch (1975: 28) of the same region contained only 12 dots

Phrynopus, and they found that central Peruvian species of *Phrynopus* and Bolivian ones were in well-supported and reciprocally monophyletic groups. Later, it was suggested that *Phrynopus* species north of the Huancabamba depression should be placed in other genera (Lehr 2006), to ensure that the taxonomy reflected the biogeographic pattern. A comprehensive revision of the genus in Bolivia was then published (De la Riva 2007), including the description of 12 new species, and raising the total number for that country to 17. Based on the diversity pattern found—a new micro-endemic species appeared in almost each suitable Andean valley surveyed—De la Riva (2007) suggested, for the first time, that the genus *Phrynopus* was likely composed of a vast number of species still unknown, well beyond any figure imagined before. The author advocated that it may represent a poorly studied case of an extraordinary vertebrate evolutionary radiation, unprecedented for the high Andes.

By 2008, the genus contained already 53 species, mostly thanks to the studies of Edgar Lehr and collaborators in Peru and Ignacio De la Riva and collaborators in Bolivia. And then, that year, everything changed abruptly when Hedges et al. (2008a) published a seminal paper that totally altered the taxonomy of American direct-developing frogs, which they grouped under the unranked taxon Terrarana [later emended as Terraranae by Duellman et al. (2016)]. This massive group, comprising 882 described species at that time, had been previously unified under the family Brachycephalidae by Frost et al. (2006), based on a limited taxon sampling which included only two representatives of *Phrynopus*. Hedges et al. (2008a) restricted Brachycephalidae to only two genera from southeastern Brazil, recognized the family Eleutherodactylidae, with 199 species (mostly in Middle America and the Caribbean region), and erected the new families Craugastoridae, with 113 species (mostly in Middle America, except for two species in Southeastern Brazil) and the large Strabomantidae, with 493 species in ten genera (mostly in the northern Andean region and the Amazon Basin). Although they recognized a poor representation of *Phrynopus* diversity in their taxon sampling of Strabomantidae, 23 samples of former *Phrynopus* species were enough for them to demonstrate the non-monophyletic nature of the genus. Consequently, they erected four new genera belonging to two new subfamilies, Holoadeninae and Strabomantinae. Holoadeninae contained the new genera *Bryophryne* (southern Peru) and *Psychrophrynella* (southern Peru to central Bolivia), while Strabomantinae contained *Isodactylus* (central Colombia to central Peru), and *Lynchius* (southern Ecuador and northern Peru), as well as the resurrected Colombian genus *Niceforonia*, and *Phrynopus*, which was redefined and restricted to central Peru; *Isodactylus* was later emended as *Hypodactylus* by Hedges et al. (2008b). Thus, the convergence of characters in *Phrynopus* was finally unraveled and the old taxonomy was dismantled by splitting it into six different genera and two subfamilies.

Few truly important novelties have been proposed since 2008, except for those regarding the unstable inferences on the internal phylogenetic relationships of certain groups, including changes at family and subfamily levels (Hedges et al. 2008a; Pyron and Wiens 2011; Padial et al. 2014; Motta et al. 2016; De la Riva et al. 2017a; Heinicke et al. 2018; Venegas et al. 2018). It is out of the scope of this chapter to make a comprehensive, historical taxonomic revision of these groups, describing or discussing all these changes. However, there have been two recent important steps for a better knowledge of the evolutionary history of the Andean phrynopoid ecomorph: (1) the erection of the new genus *Microkayla* to accommodate all the species representing the spectacular radiation of these frogs in Bolivia and southeastern Peru (De la Riva et al. 2017a); *Microkayla* (25 species) is the sister group of *Psychrophrynella* (four); and (2), the inclusion of *Niceforonia* in a molecular analysis for the first time, which rendered *Hypodactylus* a junior synonym (Acosta-Galvis et al. 2018).

The continuous shuffling of internal topologies for certain Terraranae groups indicates that still some relationships remain unclear, but the progress achieved since Lynch's proposal in 1975 is impressive, and the genera described are clearly supported. We have now an acceptable understanding of the evolutionary complexity of these groups, and the uncertainties and problems to be solved are well identified. While all these taxonomic rearrangements take place, the pace of description of new species remains incessant. For example, 11 years after the erection of *Bryophryne* by Hedges et al. (2008a) to accommodate the single species *B. cophites*, the number of species in this genus has risen to 14, the same figure as Lynch's original *Phrynopus*. The total number of species in the six genera containing phrynopoid frogs is now 99 species (Frost 2019)... and counting. All researchers studying these anurans have specimens of several new species accumulated in their cabinets, awaiting to be formally described.

In the present chapter I deal mostly with currently recognized genera of high Andean direct-developing frogs present solely in Peru and Bolivia, because they are the only ones that comply with three facts: (1) almost all their members match the phrynopoid ecomorph; (2) their species have small distributions (found in a single locality or a few clustered localities); and (3) they experienced a (previously unexpected) spectacular species radiation. These three features combined exclude members of the genera *Niceforonia* and *Lynchius*. *Niceforonia* (15 species) occurs in Colombia, Ecuador and Peru between 1200 and 3850 m a.s.l. (the taxonomically unclear species *N. columbiana*, might occur between 1000 and 1300, but the holotype and only known specimen is lost); only those forms inhabiting the highlands exhibit phrynopoid morphology (e.g., *N. brunnea*, *N. lucida*, *N. nana* and *N. peraccai*). *Lynchius* (seven species) occurs in Ecuador and Peru between 1300 and 3300 m (Motta et al. 2016), and the species have in general phrynopoid morphology except the lowland form *L. simmonsi* [which is so divergent that it was formerly considered an *Oreobates* (Padial et al. 2012)] and the bizarre *L. megacephalus* (Sánchez-Nivicela et al. 2019); some species have moderately broad distributions (e.g., *L. flavomaculatus*). Most comments and analyses herein, if not stated otherwise, will be referred to the genera *Bryophryne*, *Microkayla*, *Phrynopus,* and *Psychrophrynella*.

3 Morphological Diversity and Ecological Correlates

What defines the phrynopoid ecomorph are three main external characteristics, namely: (1) stocky body, from moderately to markedly robust; (2) legs short in relation to body; and (3) tips of digits rounded, not expanded in discs or only slightly knobbed, lacking circumferential grooves. The conditions of robust body and short limbs clearly denote a terrestrial life with a gait that commonly does not include saltatorial behavior, except in a few reported cases; also, short laps have been observed in certain species when animals are disturbed (De la Riva et al. 2017a). The proportion between limbs and body, as measured by the ratio tibia + foot length/snout-vent length (SVL) for the 25 described species of *Microkayla* only exceed slightly 80% in three species, being more often below 75%, and in several cases even below 65%. For comparison purposes within Holoadeninae, this ratio in, for example, regular *Oreobates* species, is always >100% (De la Riva, unpubl.); it reaches

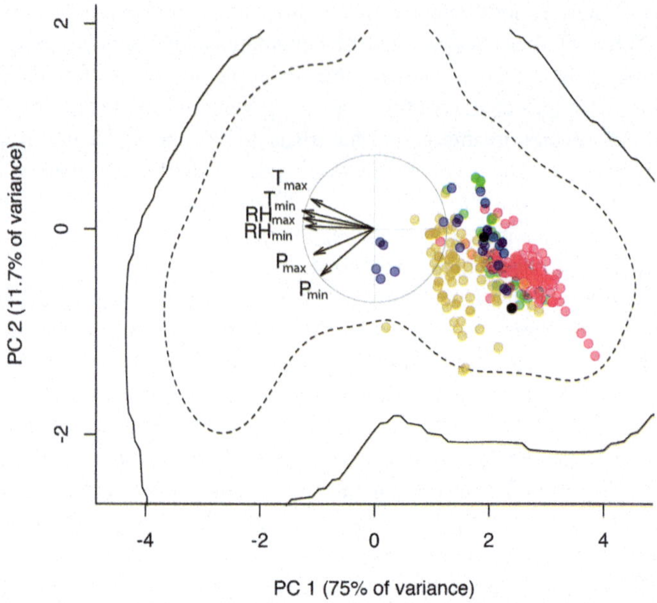

Fig. 27.3 Overlap of climatic conditions shared by species of *Phrynopus* (yellow points), *Bryophryne* (green), *Psychrophrynella* (blue), *Microkayla* (pink) and *Oreobates ayacucho* (black). The solid and dashed contour lines represent respectively, 100 % and 50 % of the available climatic space comprised between the parallels 4°S and 22°S and the meridians 82°W and 60°W (calculated from 1000 background points and presence records). This space is defined from a Principal Component Analysis resulting from six climatic variables whose loadings are represented in the inset arrow graph, where the radius of the circle equals a unit in each of the axes. Climate data were extracted from WorldClim version 2 (Fick and Hijmans 2017; Tmax: Maximum temperature of the warmest month; Tmin: minimum temperature of the coldest month; Pmax: precipitation of the wettest month; Pmin: precipitation of the driest month; RHmax: minimum relative humidity of the wettest month; RHmin: minimum relative humidity of the driest month). Relative humidity values were calculated from the monthly partial water vapor pressure layers and monthly maximum temperature by means of the function wvapsat() in the package insol (Corripio 2014)

almost 150% in great jumpers like some forms of African *Ptychadena* frogs [data from Bwong et al. (2009)].

The three characteristics mentioned above are considered adaptations to a life in the floor in humid highlands covered my mosses, ferns, and herbaceous vegetation, either in open formations like grasslands—wet puna from northern Peru to central Bolivia, and paramos and subparamos from Ecuador to Venezuela—or in elfin forests and cloud forests. In fact, many species are found in both habitats, although usually one of them predominates. However, for these small, terrestrial, nocturnal secretive frogs with low vagility, the concept of habitat (e.g., open vs. closed) has to be interpreted at a very small, microhabitat scale. Overall, members of the four genera herein considered live under approximately the same climatic conditions (Fig. 27.3).

Within this general morphological scheme, there is considerable variation regarding size, shape, skin texture, presence of middle ear and of some secondary sexual characters, and, of course, color pattern. Phrynopoid frogs are small, and females are larger than males; their SVL ranges from 13.4 mm in the smallest species, *Phrynopus auriculatus* (Duellman and Hedges 2008) to 56.4 mm in the largest, *P. kauneorum* (Lehr et al. 2002a). It has been suggested that Bergmann's rule applies to amphibians, although evidence is somehow contradictory (Ashton 2002; Adams and Church 2007). Von May et al. (2018) found that *Phrynopus* species living at higher elevations are larger than those from lower altitudes, although this rule does not necessarily apply to every Andean group of frogs, and maybe not even to other phrynopoid taxa. Perhaps not coincidentally, *Psychrophrynella usurpator* reaches 30.5 mm SVL and an elevation of 3539 m a.s.l., and is the largest of the four species of the genus described hitherto, only second to *P. chirihampatu* (27.5 mm SVL and 3180 m); the two other species, *P. bagrecito* and *P. glauca* are tiny (19.0 and 19.8 mm SVL, respectively) and inhabit lower elevations (1830–2740 m and 2225 m, respectively) (Catenazzi and Ttito 2016, 2018). But *Microkayla* frogs do not seem to follow this pattern. Von May et al. (2018) also found a relationship between size and habitat, the larger species occurring in humid grasslands and paramos, and the smaller in forests. It could be argued that the ultimate driver promoting larger size is still obscured by the correlation existing between altitude and vegetation type.

Andean phrynopoid frogs vary between two morphological extremes. On the one hand, there are large species having moderately longer and acute snouts, with a more or less marked canthus rostralis, and a robust but moderately elongated body and not extremely short limbs or digits. On the other hand, there are tiny species with rounded head, globular body, and very short limbs and digits (tibia length + foot length < 70% SVL). These two extremes are illustrated by, for example, *Microkayla wettsteini* and *M. kempffi*, respectively (Fig. 27.4); many intermediate combinations of size and shape are found. There is inconclusive evidence regarding the relation of these forms with phylogenetic relatedness, and most probably one and the other have evolved independently several times. For example, all species of *Microkayla* east of the Cordillera Quimsacruz in Bolivia (*M. adenopleura*, *M. iatamasi*, *M. harveyi*, *M. kempffi* and two undescribed species) which form a clade (De la Riva et al. 2017a) belong to the second type, but the same morphology is scattered in other areas of the Bolivian Cordillera Oriental, and there are cases in which a species with a given morphology can be embedded in a clade in which all members have the other morphology. However, any conclusion is uncertain until more complete geographical and phylogenetic sampling are done. Von May et al. (2018) found that species of *Phrynopus* from the high humid grasslands are less variable in morphology than those that inhabit cloud forests or both kinds of habitats, and that the combination of shorter limbs and short, rounded head, is also more common in species living at higher elevations.

A character subjected to variation is the presence of an external ear. *Microkayla* species lack a tympanic membrane and the tympanic annulus is, with variable degrees of distinctiveness, visible under the skin, either totally [condition 2 of

Fig. 27.4 Examples of convergence and variation in morphology and color pattern in phrynopoid Andean frogs. (**a**, **b**) Two different color morphs of an undescribed species of *Microkayla* from Utururo, Department of Cochabamba, Bolivia; (**c**, **d**) extreme interspecific differences in morphology as shown by a large, moderately long-legged species, *Microkayla wettsteini* from Aceramarca river valley, La Paz (**c**), and a tiny, short-legged species, *M. kempffi* from La Siberia, Cochabamba (**d**); (**e**) calling male of *Microkayla saltator*, a divergent species with a huge vocal sac, saltatorial behavior, and arboreal habits; (**f**) an "intruder" phrynopoid species, *Pristimantis ginesi*, from Laguna de Mucubají (Sierra Nevada, Mérida, Venezuela; photo by César Barrio); (**g**) *Oreobates ayacucho* from between Punki and Anco, 3850 m (Ayachucho, Peru; photo by José M. Padial); note the short legs and stout complexion when compared to congeneric *O. sanderi* (H) from Bilunto Mountains, 1800 m (La Paz, Bolivia)

Lynch and Duellman (1997) and Duellman and Lehr (2009)] or only its lower, ventral or anteroventral part (condition 3). There can be some intraspecific variation, with some individuals clearly showing the annulus while in others it is not visible at all. Thus, the actual specific condition is not easy to ascertain unless a good sample of individuals is examined. Members of *Psychrophrynella* mostly show condition 3. In *Phrynopus*, most species are earless (condition 4: tympanum and tympanic annulus absent) and, consequently, these species do not emit advertisement calls, but there are four species with a well-marked tympanum and tympanic annulus (Condition 1): the pair of sister species *P. auriculatus* and *P. mariellaleo*, and *P. peruanus* and its sister undescribed species (which together form the most primitive clade in the genus) (Venegas et al. 2018; Von May et al. 2018). *Bryophryne* species seem to have a range of conditions from 2 to 4, with at least some species emitting advertisement calls [e.g., *B. flammiventris*, *B. gymnotis*; Lehr and Catenazzi (2010)].

Morphology apart, phrynopoid species present intraspecific variation in color pattern, often broader than interspecific variation, as is the case in many groups of anurans (De la Riva 2007). Thus, members of species not closely related and living in different geographic areas may be externally undistinguishable (sometimes making the writing of comparative diagnoses a true nightmare), while individuals of the same species found under the same stone may appear to belong to species completely different (Fig. 27.4) [see also, for example, the ample variation in *Microkayla illimani* shown by Willaert et al. (2016)]. Likewise, it is common some ontogenetic change in color pattern, juveniles often being more patterned that adults.

As mentioned above, the general phrynopoid morphology represents an example of convergent evolution towards an ecomorph oriented to fill an ecological niche available in the Andean humid highlands. This trend is obvious in the genera split from the former genus *Phrynopus* sensu Lynch (1975), but the selective forces at play also acted on particular members of other anuran groups primarily adapted to other kinds of habitats. Among these "intruders", it is streaking the case of a member of the genus *Oreobates*, *O. ayacucho* (Fig. 27.4), that only after molecular analyses was discovered to belong to *Oreobates* instead of *Phrynopus* (Padial et al. 2012). Members of *Oreobates* are terrestrial inhabitants of cloud forests, lowland rainforests, and even dry forests, extending to central Brazil, being *O. ayacucho* the only representative found in high Andean open grasslands, well above the tree line, at 3850 m. Standard *Oreobates* frogs are more or less slender jumpers and they have long limbs and digits with expanded tips, and particularly long feet (Fig. 27.4), which is not the case of *P. ayacucho*. It is highly interesting the fact that there is a gap of ca. 300 km between the distribution of the southernmost species of *Phrynopus (P. chaparroi)* and the northernmost species of *Bryophryne* (*B. flammiventris*), and this gap is perhaps not coincidentally filled by *O. ayacucho* (Fig. 27.2), which is found ca. 60 km (airline) to the west of *B. flammiventris* (Lehr and Catenazzi 2010). Whether this gap is an artefact due to low sampling or a real void in phrynopoid diversity is a matter of further research.

To the north of the Huancabamba depression in northern Peru, certain representatives of the genera *Lynchius* and *Niceforonia* have phrynopoid morphology and they fill the corresponding highland niches, but their species diversity is low compared to that of the southern phrynopoid genera. This paucity of species is partially compensated by species of *Pristimantis*, in some cases belonging to different clades of this primarily lowland and montane genus (see González-Durán et al. 2017). Highland phrynopoid *Pristimantis* are found in Peru (e.g., members of the likely non-monophyletic *P. orestes* group: *P. atrabracus, P. chimu, P. cordovae, P. melanogaster, P. pataikos, P. pinguis, P. seorsus, P. simonsii,* etc.), Ecuador (e.g., *P. myersi, P. ocreatus* [*P. myersi* group], and *P. orestes, P. simonbolivari* [*P. orestes* group]), Colombia (*P. affinis, P. elegans, P. nervicus, P. nicefori*), and even Venezuela (*P. ginesi* [Fig. 27.4], *P. lancinii, P. paramerus*). No phrynopoid *Pristimantis* have been reported for Bolivia.

The most bizarre case of convergence to the phrynopoid ecomorph cheating herpetologists is that of the species initially described as *Phrynopus carpish* by Lehr et al. (2002b). The new species was vaguely similar to members of the genus *Telmatobius*, had long limbs (for phrynopoid standards), and the only female collected had small ovaric eggs, suggesting non-direct development. Not surprisingly, when the species was included in the first molecular phylogeny of *Phrynopus* (Lehr et al. 2005), it was nested within the outgroup, somewhere between some leptodactylids and one bufonid! Eventually, further osteological studies led to the conclusion that it was a member of the family Microhylidae, and a new genus, *Melanophryne*, was erected to accommodate this species and another newly described (Lehr and Trueb 2007). Ulterior genetic studies would tentatively place *Melanophryne* in the synonymy of *Ctenophryne* (de Sá et al. 2012). The history of *P. carpish* as an unnoticed phrynopoid intruder was short but picturesque.

On the opposite side of the intruders, are the outliers: true members of genuine phrynopoid genera that depart from the phrynopoid morphology. The northernmost species of *Phrynopus, P. mariellaleo,* occurs in cloud forests at relatively low elevation (2575 m), 170 km north of the nearest species (*P. capitalis, P. dumicola,* and *P. personatus*), indicating a gap where new species of *Phrynopus* would be expected. It has a rather *Niceforonia*-like morphology, with large size (up to 39.7 mm SVL), complete dorsolateral folds, and long legs (tibia length + foot length > 100% SVL) (Venegas et al. 2018). Interestingly, the phylogenetically closest species, *P. auriculatus*, occurs 540 km to the south, and both species share the plesiomorphic condition of having a distinct tympanum. Another peculiar species is *Microkayla saltator* (Fig. 27.4), which occurs at relatively low elevation (2500 m) in semi-humid forests of the Bolivian Cordillera de Apolobamba (De la Riva 2007). It is a small species (maximum SVL = 19.0 mm) but possesses relatively long legs (tibia length + foot length 70–80% SVL), a huge vocal sac, a complex, divergent advertisement call, and saltatorial behavior, being partially arboreal, like *Phrynopus horstpauli* (Lehr et al. 2000).

4 Life History Variation

The unranked American group Terraranae (Hedges et al. 2008a), or superfamily Brachycephaloidea (Padial et al. 2014), comprises 1153 species of frogs described hitherto (Frost 2019). The main unifying character of the whole group is that they have direct development, following reproductive modes 17, 20 and 28 of Duellman and Trueb (1986) [or their equivalents 23, 27 and 38 of Haddad and Prado (2005)]. These frogs lay large terrestrial eggs, few in number (when compared with those in other, more generalist reproductive modes), deposited on the floor, underground nests, bromeliads, caves, etc., with embryos experiencing intra-egg direct development, with no tadpole stage. Different degrees or purported parental care have been reported, most commonly male or female attendance of eggs. For some species the reproduction has been well studied, but entire groups are poorly known. For example, despite the huge number of species of *Pristimantis*, clutches have been observed only for a handful of species (Duellman and Lehr 2009). In Andean phrynopoid frogs, eggs are usually laid in burrows under rocks or mosses, where parental care takes place. Clutch size in observed nests varies between 8 eggs in *Microkayla wettsteini* (Ergueta 1993) and *M. illampu* (De la Riva 2007) to 41 in *M. teqta*; such large clutches might correspond to more than one female (De la Riva and Burrowes 2014). Males beside clutches of eggs have been reported in *Microkayla illimani* (Willaert et al. 2016) and *M. teqta* (De la Riva and Burrowes 2014), and purported females in the cases of *M. wettsteini* (Ergueta 1993) and *Bryophryne cophites* (Catenazzi 2006). Next to nothing is known regarding time of development, but it takes at least more than 28 days in *M. illimani* (Willaert et al. 2016). The only specific study of life history of a phrynopoid species, *Microkayla wettsteini*, described its reproduction and diet—small arthropods and their larvae—(Ergueta 1993).

Sexual dimorphism is not marked in high Andean phrynopoids. Regarding coloration, sexual dimorphism has been reported only for *Phrynopus peruanus* and *P. pesantesi* (see Duellman and Lehr 2009). Females are usually slightly larger than males, which, in those species with external ear, usually possess vocal slits and a subgular vocal sac, commonly discrete in size, but exceptionally large and expanded in *Microkayla saltator* (Fig. 27.4); the gular region sometimes has a different color, e.g., dark in males of *M. huayna* and *M. wettsteini* (De la Riva et al. 2017b), or orange in females of *Bryophryne hanssaueri* (Lehr and Catenazzi 2009). Nuptial pads are present in the thumbs of males of certain species [e.g., *Bryophryne cophites, Phrynopus horstpauli, P. montium* (Lehr 2006; Duellman and Lehr 2009)].

Amplexus in phrynopoid frogs have been rarely observed. While most Terraranae have axillary amplexus, it is inguinal in at least *Phrynopus horstspauli* (Lehr et al. 2000), and axillar in an undescribed species of *Psychrophrynella* (L. Mamani, pers. comm.). To the best of my knowledge, amplexus has never been observed in any species of *Bryophryne* or *Microkayla*.

Phrynopoid frogs that emit calls usually do it mostly at dusk and night, when they become active, but, depending on local climatic conditions and, likely, species-specific behavior, calls can be heard at any time of day. Advertisement calls have been thoroughly described for several phrynopoid species. *Phrynopus* species with external ear emit calls (Von May et al. 2018) but these have not been quantitatively described. In *Bryophryne*, vocalizations have been described as a short whistle in *B. bustamantei* (Chaparro et al. 2007), and were well-studied in *B. gymnotis*, whose call consists of a single tonal note with two weakly modulated parts, repeated at irregular intervals (Lehr and Catenazzi 2009). The call of *B. mancoinca* consists of two or three tonal notes (Mamani et al. 2017). All *Psychrophrynella* species have a call consisting in a trill of multiple, short notes (Catenazzi and Ttito 2016, 2018). In *Microkayla*, in which presumably all the species vocalize, the standard call seems to be a simple, short whistle-like tonal note, emitted at irregular intervals by day or night. This is the case for *M. ankohuma, M. boettgeri, M. chacaltaya, M. chapi, M. iatamasi,* and *M. kempffi*. In other species, vocalizations are arranged in call groups, like the many short, repeated calls of *M. illimani, M. saltator* and *M. wettsteini* (De la Riva 2007; Willaert et al. 2016), or the few calls of *P. huayna* (De la Riva et al. 2017b). The pulsed call of *P. teqta* is unique in the genus (De la Riva and Burrowes 2014). *Microkayla huayna, M. teqta* and *M. wettsteini* are closely related, and *M. saltator* is the sister species of *M. illampu* (De la Riva et al. 2017a), a species that, like *M. katantika*, emits short series of three tonal, simple calls (De la Riva 2007). Due to the many species remaining to be discovered and described, the knowledge on internal phylogenetic relationships in *Microkayla* is still too incomplete as to study intrageneric evolutionary trends in traits of the advertisement calls. Finally, the call of *Oreobates ayacucho* consists of groups of 3–5 pulsed notes, a spectral structure rather different from that found in *Microkayla* species (Padial et al. 2012).

5 Genetics, Biogeographical Patterns, and Species Diversity: Why So Many Species?

Phylogenetic studies of phrynopoid frogs mentioned above have provided data on genetic distances (based on the 16S mitochondrial gene) for the four Peruvian and Bolivian genera considered in this review. Because of the small distributions of these frogs and the fact that conspecific samples are often collected at the same locality or very close, intraspecific distances found are usually 0 or very low, but reach 1% between sympatric samples of *M. boettgeri* (De la Riva et al. 2017a, b) and 2.3 between localities (~20 km airline) of *Phrynopus inti* (Lehr et al. 2017). Interspecific genetic distances in these genera are considerably high. In *Bryophryne*, they range from 5.4 (*B. bakersfield-B. bustamantei*) to 10.9 (*B. phuyuhampatu-B. tocra*) (Catenazzi and Ttito 2018), and in *Psychrophrynella* they span from 7.9 (*P. chirihampatu-P. usurpator*) to 12.7 (*P. glauca-P. chirihampatu*) (Catenazzi

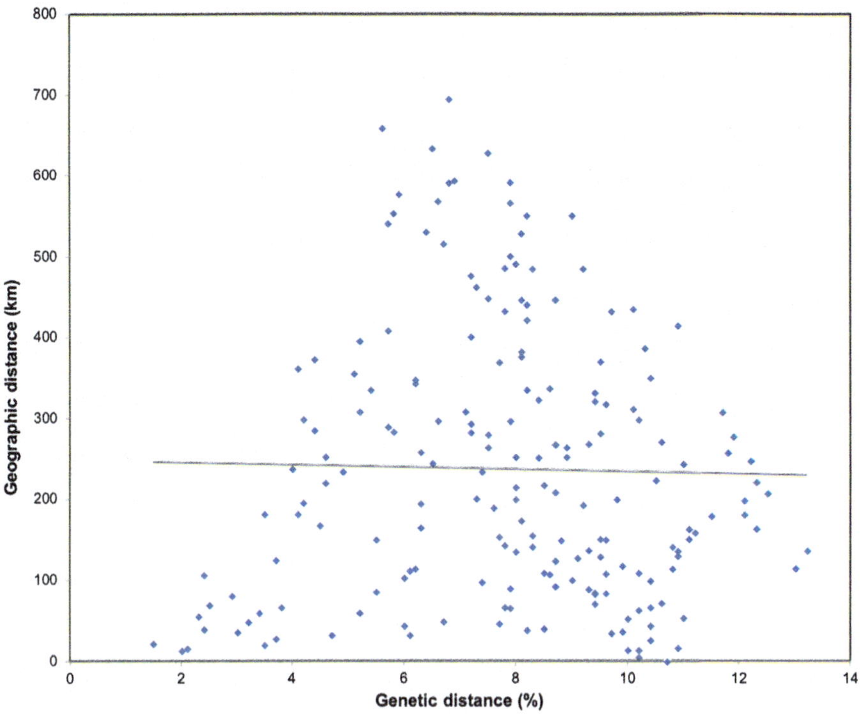

Fig. 27.5 Mantel test of correlation between genetic and geographic distances of 20 species of *Microkayla* based on pairwise comparison of homologous sequences of 532 bp of the 16S mitochondrial gene (De la Riva et al. 2017a). Geographic distances between the type localities of the species were obtained from Google Earth following the contour of the Andes when the shortest distance crossed lowlands unsuitable for these frogs. The clear absence of correlation (Rxy = –0.022; p = 0.410) is corroborated by the fact that sister species recovered by phylogenetic analyses are not necessarily those from adjacent or close valleys

and Ttito 2018). In *Phrynopus*, the lowest interspecific distance found is 1.5 (between an undescribed species and *P. inti*), and the highest is 14.7 (*P. bufoides-P. heimorum*) (Lehr et al. 2017). Finally, in *Microkayla*, genetic distances show a broad range, being the lowest 1.1 between two undescribed forms from Central Bolivia and the highest 13.2 (*M. huayna-M. saltator*). Although species with low genetic divergences are often found in close geographic proximity, the results of a Mantel test for 20 species of *Microkayla* were not significant, suggesting absence of a linear relationship (Fig. 27.5). For example, the geographically most distant species, *M. boettgeri* and *M. kempffi* (695 km) (Fig. 27.6), show an intermediate genetic distance (6.8%), while the highest genetic distance (13.2%), between *M. huayna* and *M. saltator*, corresponds to an intermediate geographic distance (136 km) (De la Riva et al. 2017a). *Microkayla guillei* and *M. saltator* occupy basically the same point in the map (Fig. 27.6), albeit separated by ca. 1000 m in elevation, and their genetic distance is 10.7. Also, Lehr et al. (2005) observed that in

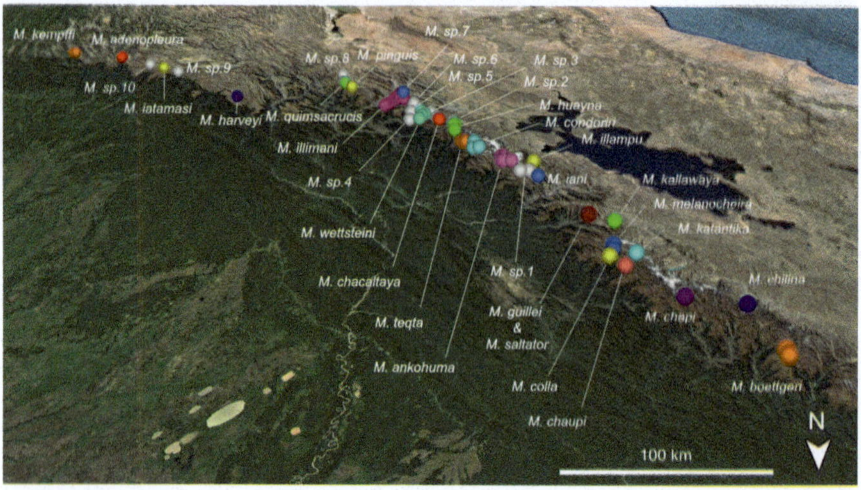

Fig. 27.6 Google Earth's aerial map of the Cordillera Oriental of southern Peru and Central Bolivia, showing the distribution of all known (ten undescribed) species of *Microkayla*. From left to right: the Beni savannas, the Amazonian rainforest piedmont, the Andes, the Altiplano with Lake Titicaca, and the Pacific Ocean, can be appreciated. The Peruvian-Bolivian border crosses between *M. chapi* and *M. chaupi*. The broad valley of Sorata (between *M. guillei*-*M. saltator* and *M. iani*), separates the Cordillera de Apolobamba from the Cordillera Real, and the valley of La Paz (between *M. illimani* and *M. pinguis*), separates the Cordillera Real from Cordillera de Quimsacruz. Few surveys have been done between Cordillera Quimsacruz and Cordillera del Tunari, where *M. harveyi* occurs

the three cases of sympatry in *Phrynopus*, the sympatric species were not sister species. All these results indicate old divergence times for the different subclades, with widespread distributions of common ancestors pre-dating the main orogenic and geologic phenomena that shaped the Andean landscape as seen today. A good example is the phylogenetic position of *M. illimani* from the slopes of mount Illimani, which marks the easternmost limit of the Cordillera Real (Fig. 27.6); this species is consistently recovered as the sister species of a clade formed by *M. quimsacrucis* and an undescribed species, both occurring in the Cordillera Quimsacruz. Mount Illimani and the Cordillera Quimsacruz are separated by the broad, deep and arid valley of La Paz, today representing an unsurmountable barrier for any craugastorid frog. A similar example is that of *M. illampu* and *M. saltator*, sister species found in two different mountain ranges, Cordillera Real and Cordillera de Apolobamba, separated by the large valley of Sorata (De la Riva et al. 2017a) (Fig. 27.6).

Species of phrynopoid frogs have, in general, remarkably small distributions, sometimes apparently restricted to a single valley within a more or less narrow altitudinal belt, resulting in a high geographic turnover in species composition (De la Riva 2007). There are few cases in which continuity of appropriate habitat and lack of barriers allow for broader distributions, comprising contiguous valleys, and spanning over moderately long distances, even a few dozens of kilometers (e.g,

Phrynopus inti, *P. tribulosus* and, perhaps, *Bryophryne cophites*). But, why are there so many species and such high beta-diversity of high Andean phrynopoid frogs? There have been some attempts to explain the incontestable evolutionary success of Terraranae frogs. Hedges et al. (2008a) suggested direct development as a key factor because it implies less barriers for dispersal than reproduction tied to aquatic environments, but this does not explain the high beta-diversity observed in some groups. Rapid chromosomal changes in Terraranae and the unusual number of different karyotypes present in this group might have contributed to intense speciation processes [see Schmid et al. (2010)]. Also, events of depredation on eggs of direct developing anurans will often affect entire clutches, and this could accelerate speciation via fixation of particular alleles (Dubois 2005). However, all these factors would affect all Terraranae more or less equally, and they do not explain the particular question posed above. For Andean phrynopoid frogs, the speciation model proposed by Lynch and Duellman (1997) offers a better explanation for the occurrence of so many species. In this model, dynamic processes involving Andean orogeny, erosion, and climatic changes during the Cenozoic would provide multiple opportunities for allopatric speciation. González-Voyer et al. (2011) support this model, but suggested that divergent selection must be also involved to explain the different degrees of diversification in clades of similar age occupying the same areas.

Because a given phrynopoid species may occur either in humid grassland, in cloud forest, or in both, and altitudinal segregation is generally observed when more than one species occurs in the same valley, ecological speciation may have played a role in the diversification of these frogs during climatic fluctuations. Niche conservatism would have kept species tied to suitable habitats during shifts of vegetation belts, favoring allopatric or peripatric speciation (Wiens 2004). Also, the so-called "vanishing refuge", which acts as a mechanism of ecogeographic speciation (Vanzolini and Williams 1981; Damasceno et al. 2014), might have played a role; nonetheless, a moderately widespread distribution for the adapted species would perhaps be rare in these frogs. The extremely complex topography of the Andes, and the many different outcomes of climatic fluctuations (depending on topography, orientation, degree and type of change, duration, etc.), have led to an intricate history of ecosystem shuffling, affecting organisms in different ways that are not easy to interpret. Von May et al. (2018) found evidence of some cases in which a *Phrynopus* species is a grassland inhabitant and the sister species occurs in cloud forests, but this pattern is not necessarily common. Lynch and Suárez-Mayorga (2002) suggested that speciation processes in some groups of anurans in the high paramos of Colombia have taken place within the same vegetation formation, not having their ancestors in other formations or thermal belts. In summary, ecological radiations to occupy the wet highland ecological niche by Andean phrynopoid frogs, was followed by intense allopatric speciation driven by geology—Andean uplift and erosion—and altitudinal changes in vegetation caused by Pleistocene climatic oscillations, leading to the currently observed beta-diversity (De la Riva et al. 2017a); this has been the general process for many other components of the Andean biota [see Hazzi et al. (2018) and references therein].

Hedges et al. (2008a) suggested that northern clades of phrynopoid frogs—*Phrynopus*, together with *Niceforonia* (as *Hypodactylus*), *Lynchius*, and *Oreobates*—evolved in the northern Andes during in the Pliocene, 5.3–1.8 Ma. However, *Oreobates* is widespread in the lowlands, reaching even central Brazil, and high Andean *Oreobates* (*O. ayacucho* and its sister species *O. pereger*) evolved from forest Andean ancestors (Padial et al. 2012). Southern clades—*Bryophryne* and *Psychrophrynella*, now including *Microkayla*—would have arisen from a southern Holoadeninae stock around 23 million years ago (Hedges et al. 2008a). This old origin complicates making correlations between cladogenetic events and the corresponding geologic and climatic conditions that led to the current diversity and biogeographic patterns. Despite the Andean orogeny having proceeded in the central Andes from south-to-north (Gregory-Wodzicki 2000), older clades of *Microkayla* are not necessarily from the southern part of the genus' range, and it seems that the most recent clades in the genus comprise the southernmost species, from central Bolivia (De la Riva et al. 2017a) (Fig. 27.6). This general pattern is also found in Andean lizards of the genus *Proctoporus* (Goicoechea et al. 2012), which are often found in micro-sympatry with species of *Bryophryne* and *Microkayla*.

In summary, the diversity of phrynopoid frogs in the high Andes of Peru (south of Huancabamba depression) and Bolivia is made up by four major radiations (De la Riva et al. 2017a) which, overall, have a span of ca. 2000 km (following the contour of the Andes) from the northernmost species, *Phrynopus mariellaleo*, to the southernmost, *Microkayla kempffi* (Fig. 27.2). The span of *Phrynopus* distribution comprises 740 km, with two large internal gaps that most likely represent sampling artefacts. *Phrynopus thompsoni* is the only species from the Cordillera Occidental. There is a still poorly surveyed area of ca. 280 km between the distributions of *Phrynopus* and *Bryophryne* (which is partially filled by the intruder *Oreobates ayacucho*). The deep Apurimac river valley has been proposed as a barrier separating both genera (Lehr and Catenazzi 2008), as for other organisms (Hazzi et al. 2018); thus, reported *Phrynopus* species from the isolated Cordillera de Vilcabamba (Icochea et al. 2001; Rodríguez 2001) are, most likely, part of *Bryophryne* or *Psychrophrynella*, or both. The distributions of these two genera span over ca. 315 km and they overlap broadly in the department of Cusco, with species of *Psychrophrynella* generally occupying lower elevations. A few *Bryophryne* species enter Puno department, where the ca. 75 km gap between this genus and *Microkayla* will likely be filled by species of one or the other genus when extensive surveys are done. As already mentioned above, the span of *Microkayla* distribution is 695 km from northwest to southeast (Fig. 27.6). Altitudinal distributions are generally lower in *Psychrophrynella* (2225 m in *P. glauca*; Catenazzi and Ttito 2018), with only *P. usurpator* found above 3200 m. Reported altitudinal ranges in the other three genera are broad, from 2466 m in *Microkayla colla* (De la Riva and Aparicio 2016) to 4490 m in *Phrynopus chaparroi* (Mamani and Malqui 2014), albeit an undescribed species of Bolivian *Microkayla* from the Illimani slopes has the absolute record, reaching almost 4700 m (De la Riva and Burrowes, unpubl.).

6 Future Research and Conservation

We have still to understand why northern Andean phrynopoid genera of Holoadeninae in Ecuador and Colombia—e.g., *Lynchius* and *Niceforonia*—have not experienced explosive radiations like those of their southern counterparts (instead, those corresponding niches have been partially filled by members of the genus *Pristimantis*). One reason may be that the high elevations and multiple glacial valleys of the Cordillera Oriental of Peru and Bolivia create a dramatically complex geographical pattern made up of suitable habitats and barriers, favoring vicariant processes. This type of orography is not as apparent in the northern Andes, which in general are lower in elevation, and thus may facilitate continuity and connections between paramo and subparamo habitats in certain areas. The fact that the phrynopoid ecomorph is absent from many suitable habitats of the northern Andes is most likely a consequence of historical, ecological, and biogeographic factors. The likelihood of a given niche being occupied by many similar species over extensive areas as a result of multiple events of convergent evolution, is probably much lower than that of achieving the same result by repeated vicariant events affecting an adapted ancestor. This hypothesis can be tested when more comprehensive information is attained on species numbers, distributions, and phylogenetic relationships of all phrynopoid frogs, including highland *Pristimantis*, from Venezuela to Bolivia.

Taking into consideration the impressive progress made in the knowledge of phrynopoid diversity during the current century, and the still many suitable and not surveyed Andean areas, it is pertinent to wonder how many species in the four genera herein considered remain to be discovered and described. Ongoing research based on niche models and empirical data indicates that projections on plausible species numbers will surprise everybody (De la Riva et al., unpubl.).

However, we might be running a race against time. Amphibians as a group are subject to an extensive number of different threats. Virtually, almost all major causes of biodiversity loss affect strongly amphibians, but there is one totally specific and dramatic: chytridiomycosis, caused by pathogenic strains of the chytrid fungus *Batrachochytrium dendrobatidis* (*Bd*). Several species of phrynopoid frogs have been found to be infected by *Bd*, although no sick frogs or die-offs have been reported (De la Riva and Burrowes 2011, 2014; Catenazzi et al. 2011, 2014). Climatic warming can pose a threat to amphibians as well. There are no studies on home range in phrynopoid frogs but, presumably, individuals are mostly linked to their retreat sites under mosses, stones and forest litter, from which males do call, moving only very short distances. Such low vagility limits their dispersal capacity, perhaps trapping the species if rapid climatic shifts take place. However, studies on thermal conditions of amphibians at microhabitat scales have shown these microhabitats to be buffered in relation to general climatic conditions in the surrounding environment (e.g., Storlie et al. 2014). Thermal ecology studies on Andean anurans indicate that they are active well below their preferred temperatures and that climate warming would not pose a threat per se, but it might have a detrimental effect if it implies less humidity, which is the important limiting factor for direct-developing frogs (Catenazzi et al. 2014; Von May et al. 2017; Burrowes et al. 2020). Despite nowadays *Bd* and climate warming do not seem to represent impending, serious threats for phrynopoid frogs, it must be considered that synergetic interactions between both stressors may exacerbate their detrimental effect on amphibians (Longo and Burrowes 2010).

Habitat loss, mostly due to agriculture, is a real or potential threat for many species of high Andean phrynopoid frogs, but, overall, for the moment, most species seem to be safe in their restricted areas of distribution. The fact that certain protected areas (e.g., Yanachaga Chemillén National Park in Perú, or Apolobamba Integrated Management Natural Area in Bolivia) hold several endemic species, must be emphasized and seriously considered by authorities in charge of those areas (De la Riva and Aparicio 2016). More importantly, it should be studied how much of this astonishing beta-diversity falls outside any protected area and goes mostly unnoticed. More field surveys are needed to fill the many existing distribution gaps, as well as appropriate integrative taxonomic investigations, in order to obtain more accurate knowledge of the actual diversity, evolutionary history, and conservation status of these interesting creatures.

In any case, when looking backwards, it becomes obvious that the history of research on phrynopoid frogs, since the descriptions of the first species but, especially, since Lynch's (1975) synthesis, is an excellent example of how field work, exploration, study, development and application of new techniques, integration of different sources of evidence, and, ultimately, human curiosity, can translate into impressive progress in the knowledge of otherwise hidden natural treasures and histories that must be preserved.

Acknowledgements I am grateful to Octavio Jiménez-Robles (MNCN) for his invaluable help making of modifying some of the figures; to Violeta López Márquez and Annie Machordom (MNCN) for their assistance with the Mantel test; to Patricia A. Burrowes (University of Puerto Rico) for her valuable comments and corrections, help in many ways, and patience while I was finishing the first draft of this chapter; to Santiago Castroviejo-Fisher for his critical reading of the manuscript and the highly pertinent advice for improvement; to César Barrio for the picture of *Pristimantis ginesi* and José M. Padial for that of *Oreobates ayacucho*; and, last but not least, to James Aparicio (Colección Boliviana de Fauna, La Paz) and the Museo Nacional de Historia Natural (La Paz, Bolivia) for the continuous help through many years of fieldwork in Bolivia. This work has been funded by projects CGL2005-03156/BOS, CGL2008-04164/BOS, CGL2011-30393, and CGL2014-56160-P of the Spanish Government (PI, I. De la Riva). I dedicate this paper to my post-doctoral advisor (in the 1990s), friend, and colleague William E. Duellman (The University of Kansas) as an homage for his outstanding and impressive contribution to the knowledge of Neotropical amphibians in general and Andean Terraranan frogs in particular (Bill, I'm sorry; despite all your efforts, a huge bunch of species still remain undescribed!).

References

Acosta-Galvis AR, Streicher JW, Manuelli L, Cuddy T, De Sá RO (2018) Molecular insights into the phylogenetic placement of the poorly known genus *Niceforonia* Goin & Cochran, 1963 (Anura: Brachycepahloidea). Zootaxa 4514(4):487–500

Adams DC, Church JO (2007) Amphibians do not follow Bergmann's rule. Evolution 62 (2):413–420

Ashton KG (2002) Do amphibians follow Bergmann's rule? Can J Zool 80:708–716

Burrowes PA, Navas CA, Jiménez-Robles O, Delgado P, De la Riva I (2020) Climatic heterogeneity in the Bolivian Andes: are frogs trapped? S Am J Herpetol (in press)

Bwong BA, Chira R, Schick S, Veith M, Lötters S (2009) Diversity of Ridged Frogs (Ptychadenidae: *Ptychadena*) in the easternmost remnant of the Guineo-Congolian rain forest: an analysis using morphology, bioacoustics and molecular genetics. Salamandra 45(3):129–146

Catenazzi A (2006) *Phrynopus cophites* (Cuzco Andes Frog). Reproduction. Herp Rev 37(2):206

Catenazzi A, Ttito A (2016) A new species of *Psychrophrynella* (Amphibia, Anura, Craugastoridae) from the humid montane forests of Cusco, eastern slopes of the Peruvian Andes. PeerJ 4:e1807

Catenazzi A, Ttito A (2018) *Psychrophrynella glauca* sp. n., a new species of terrestrial-breeding frogs (Amphibia, Anura, Strabomantidae) from the montane forests of the Amazonian Andes of Puno, Peru. PeerJ 6:e4444

Catenazzi A, Lehr E, Rodríguez LO, Vredenburg VT (2011) *Batrachochytrium dendrobatidis* and the collapse of anuran species richness and abundance in the upper Manu National Park, southeastern Peru. Conserv Biol 25(2):382–391

Catenazzi A, Lehr E, Vredenburg VT (2014) Thermal physiology, disease, and amphibian declines on the eastern slopes of the Andes. Conserv Biol 28:509–517

Chaparro JC, De la Riva I, Padial JM, Ochoa JA, Lehr E (2007) A new species of *Phrynopus* from Departamento Cusco, southern Peru (Anura: Brachycephalidae). Zootaxa 1618:61–68

Corripio JG (2014) insol: Solar Radiation. R package version 1.1.1. https://CRAN.R-project.org/package=insol

Damasceno R, Strangas ML, Carnaval AC, Rodrigues MT, Moritz C (2014) Revisiting the vanishing refuge model of diversification. Front Genet 5:353

De la Riva I (2007) Bolivian frogs of the genus *Phrynopus* with the description of twelve new species (Anura: Brachycephalidae). Herpetol Monogr 21:242–278

De la Riva I, Burrowes PA (2011) Rapid assessment of the presence of *Batrachochytrium dendrobatidis* in Bolivian Andean frogs. Herpetol Rev 42(3):372–375

De la Riva I, Burrowes PA (2014) A new species of *Psychrophrynella* (Anura: Craugastoridae) from the Cordillera Real, Department La Paz, Bolivia. Zootaxa 3887(3):459–470

De la Riva I, Aparicio J (2016) Three new Bolivian species of *Psychrophrynella* (Anura: Craugastoridae), and comments on the amphibian fauna of the Cordillera de Apolobamba. Salamandra 52:283–292

De la Riva I, Chaparro JC, Castroviejo-Fisher S, Padial JM (2017a) Underestimated anuran radiations in the high Andes: five new species, a new genus, and their phylogenetic relationships (Anura: Craugastoridae). Zool J Linn Soc 182:129–172

De la Riva I, Cortez C, Burrowes PA (2017b) A new species of *Microkayla* (Anura: Craugastoridae: Holoadeninae) from Department La Paz, Bolivia. Zootaxa 4363(3):350–360

de Sá RO, Streicher JW, Sekonyela R, Forlani MC, Loader SP, Greenbaum E, Richards SJ, Haddad CFB (2012) Molecular phylogeny of microhylid frogs (Anura: Microhylidae) with emphasis on relationships among New World genera. BMC Evol Biol 12(241):1–21

Dubois A (2005) Development pathway, speciation and supraspecific taxonomy in amphibians. 1. Why are there so many frog species in Sri Lanka? Alytes 22:19–37

Duellman WE, Trueb L (1986) Biology of amphibians. McGraw Hill, New York

Duellman WE, Hedges SB (2008) Two new minute species of *Phrynopus* (Lissamphibia, Anura) from the Cordillera oriental in Peru. Zootaxa 1675:59–65

Duellman WE, Lehr E (2009) Terrestrial-breeding frogs (Strabomantidae) in Peru. Nature und Tier Verlag, Münster

Duellman WE, Marion AB, Hedges SB (2016) Phylogenetics, classification, and biogeography of treefrogs (Amphibia: Anura: Arboranae). Zootaxa 4104:1–109

Ergueta P (1993) Aspectos de la biología y ecología de *Phrynopus lapalacai* (Anura: Leptodactylidae) en un bosque nublado de altura de yungas (La Paz, Bolivia). Ecol Bol 21:19–29

Fick SE, Hijmans RJ (2017) WorldClim 2: new 1-km spatial resolution climate surfaces for global land areas. Int J Climatol 37:4302–4315

Frost DR (2019) Amphibian species of the world: an online reference. Version 6.0 (accessed 9 December 2019). Electronic Database accessible at http://research.amnh.org/herpetology/amphibia/index.html. American Museum of Natural History, New York, USA

Frost DR, Grant T, Faivovich J, Bain RH, Haas A, Haddad CFB, de Sá RO, Channing A, Wilkinson M, Donnellan SC, Raxworthy CJ, Campbell JA, Blotto BL, Moler P, Drewes RC, Nussbaum RA, Lynch JD, Green DM, Wheeler WC (2006) The amphibian tree of life. Bull Am Mus Nat Hist 297:1–370

Goicoechea N, Padial JM, Chaparro JC, Castroviejo-Fisher S, De la Riva I (2012) Molecular phylogenetics, species diversity, and biogeography of the Andean lizards of the genus *Proctoporus* (Squamata: Gymnophthalmidae). Mol Phyl Evol 65:953–964

González-Voyer A, Padial JM, Castroviejo-Fisher S, De la Riva I, Vilà C (2011) Correlates of species richness in the largest Neotropical amphibian radiation. J Evol Biol 24:931–942

González-Durán GA, Targino M, Rada M, Grant T (2017) Phylogenetic relationships and morphology of the *Pristimantis leptolophus* species group (Amphibia: Anura: Brachycephaloidea), with the recognition of a new species group in *Pristimantis* Jiménez de la Espada, 1870. Zootaxa 4243(1):42–74

Gregory-Wodzicki KM (2000) Uplift history of the central and northern Andes: a review. GSA Bull 112:1091–1105

Haddad CFB, Prado CPA (2005) Reproductive modes in frogs and their unexpected diversity in the Atlantic forest of Brazil. BioScience 55(3):207–217

Hazzi NA, Moreno JS, Ortiz-Movliav C, Palacio RD (2018) Biogeographic regions and events of isolation and diversification of the endemic biota of the tropical Andes. PNAS. https://doi.org/10.1073/pnas.1803908115

Hedges BS, Duellman WE, Heinicke MP (2008a) New World direct-developing frogs (Anura: Terrarana): molecular phylogeny, classification, biogeography, and conservation. Zootaxa 1737 (1):182

Hedges BS, Duellman WE, Heinicke MP (2008b) A replacement name for *Isodactylus* Hedges, Duellman, and Heinicke, 2008. Zootaxa 1795:67–68

Heinicke MP, Lemmon AR, Lemmon EM, McGrath K, Hedges SB (2018) Phylogenomic support for evolutionary relationships of New World direct-developing frogs (Anura: Terraranae). Mol Phyl Evol 118:145–155

Icochea J, Quispitupac E, Portilla A, Ponce E (2001) Amphibians and reptiles of the southern Vilcabamba region, Peru. In: Alonso LE, Alonso A, Schulenberg TS, Dallmeier F (eds) Biological and social assessments of the Cordillera de Vilcabamba, Peru. RAP working papers 12 and SI/MAB series 6, Conservation International, pp 131–137

Lehr E (2005) A new species of the *Eleutherodactylus nigrovittatus* group (Anura: Leptodactylidae) from Andean Peru. Herpetologica 61:199–208

Lehr E (2006) Taxonomic status of some species of Peruvian *Phrynopus* (Anura: Leptodactylidae), with the description of a new species from the Andes of southern Peru. Herpetologica 62 (3):331–347

Lehr E, Catenazzi A (2008) A new species of *Bryophryne* (Anura: Strabomantidae) from southern Peru. Zootaxa 1784(1):10

Lehr E, Catenazzi A (2009) Three new species of *Bryophryne* (Anura: Strabomantidae) from the Region of Cusco, Peru. S Am J Herpetol 4(2):125–138

Lehr E, Catenazzi A (2010) Two new species of *Bryophryne* (Anura: Strabomantidae) from high elevations in southern Peru (Region of Cusco). Herpetologica 66(3):308–319

Lehr E, Trueb L (2007) Diversity among New World microhylid frogs (Anura: Microhylidae): morphological and osteological comparisons between *Nelsonophryne* (Günther 1901) and a new genus from Peru. Zool J Linn Soc 149:583–609

Lehr E, Köhler G, Ponce E (2000) A new species of *Phrynopus* from Peru (Amphibia, Anura, Leptodactylidae). Senck Biol 80(1/2):205–212

Lehr E, Aguilar C, Córdova JH (2002a) Morphological and ecological remarks on *Phrynopus kauneorum* (Amphibia, Anura, Leptodactylidae). Zool Abh 52:71–75

Lehr E, Rodríguez D, Córdova JH (2002b) A new species of *Phrynopus* (Amphibia, Anura, Leptodactylidae) from the Cordillera de Carpish (Departamento Huánuco, Peru). Zool Abh 52:65–70

Lehr E, Fritzsch G, Müller A (2005) Analysis of Andes frogs (*Phrynopus,* Leptodactylidae, Anura) phylogeny based on 12S and 16S mitochondrial rDNA sequences. Zool Scr 34(6):593–603

Lehr E, Von May R, Moravec J, Cusi JC (2017) A new species of *Phrynopus* (Amphibia, Anura, Craugastoridae) from upper montane forests and high Andean grasslands of the Pui Pui Protected Forest in central Peru. ZooKeys 713:131–157

Longo AV, Burrowes PA (2010) Persistence with chytridiomycosis does not assure survival of direct-developing frogs. EcoHealth 7:185–195

Lynch JD (1971) Evolutionary relationships, osteology, and zoogeography of leptodactyloid frogs. Univ Kans Mus Nat Hist Misc Publ 53:1–238

Lynch JD (1975) A review of the Andean leptodactylid frog genus *Phrynopus*. Occ Pap Mus Nat Hist Univ Kans 35:1–51

Lynch JD, Duellman WE (1997) Frogs of the genus *Eleutherodactylus* in Western Ecuador. Systematics, Ecology, and Biogeography. Univ Kans Nat Hist Mus Spec Publ 23:1–236

Lynch JD, Suárez-Mayorga AM (2002) Análisis biogeográfico de los anfibios paramunos. Caldasia 24(2):471–480

Mamani L, Malqui S (2014) A new species of *Phrynopus* (Anura: Craugastoridae) from the central Peruvian Andes. Zootaxa 3838:207–214

Mamani L, Catenazzi A, Ttito A, Mallqui S, Chaparro JC (2017) A new species of *Bryophryne* (Anura: Strabomantidae) from the Cordillera de Vilcabamba, southeastern Peruvian Andes. Phyllomedusa 16(2):129–141

Motta AP, Chaparro JC, Pombal JP et al (2016) Molecular phylogenetics and taxonomy of the Andean genus *Lynchius* Hedges, Heinicke and Duellman, 2008 (Anura: Craugastoridae). Herpetol Monogr 30:119–142

Padial JM, Chaparro JC, Castroviejo-Fisher S, Guayasamin JM, Lehr E, Delgado AJ, Vaira M, Teixeira M Jr, Aguayo R, De la Riva I (2012) A revision of species diversity in the Neotropical genus *Oreobates* (Anura: Strabomantidae), with the description of three new species from the Amazonian slopes of the Andes, and the proposal of candidate species. Am Mus Nov 3752:1–55

Padial JM, Grant T, Frost DR (2014) Molecular systematics of terraranas (Anura: Brachycephaloidea) with an assessment on the effects of alignment and optimality criteria. Zootaxa 3825:1–132

Pyron RA, Wiens JJ (2011) A large-scale phylogeny of Amphibia including over 2,800 species, and a revised classification of extant frogs, salamanders, and cecilians. Mol Phyl Evol 61:543–583

Rodríguez LO (2001) The herpetofauna of the northern Cordillera de Vilcabamba, Peru. In: Alonso LE, Alonso A, Schulenberg TS, Dallmeier F (eds) Biological and social assessments of the Cordillera de Vilcabamba, Peru. RAP working papers 12 and SI/MAB series 6, Conservation International, pp 127–130

Sánchez-Nivicela JC, Urgiles VL, Navarrete MJ, Yánez-Muñoz MH, Ron S (2019) A bizarre new species of *Lynchius* (Amphibia; Anura, Strabomantidae) from the Andes of Ecuador and first report of *Lynchius parkeri* in Ecuador. Zootaxa 4567(1):001–024

Schmid M, Bogart JP, Hedges SB (eds) (2010) The chromosomes of terraranan frogs. Cytogenet Genome Res 130–131(1–8):1–568

Storlie C, Merino-Viteri A, Phillips B, VanDerWal J, Welbergen J, Williams S (2014) Stepping inside the niche: microclimate data are critical for accurate assessment of species' vulnerability to climate change. Biol Lett 10:20140576

Vanzolini PE, Williams EE (1981) The vanishing refuge: a mechanism for ecogeographic speciation. Pap Avuls Zool 34(23):251–255

Venegas P, Barboza AC, De la Riva I, Padial JM (2018) A new species of *Phrynopus* from the northeastern Andes of Peru, its phylogenetic position, and notes on the relationships of Holoadeninae (Anura: Craugastoridae). Zootaxa 4446(4):501–524

von May R, Catenazzi A, Corl A, Santa-Cruz R, Carnaval AC, Moritz C (2017) Divergence of thermal physiological traits in terrestrial breeding frogs along a tropical elevational gradient. Ecol Evol 7:3257–3267

von May R, Lehr E, Rabosky DL (2018) Evolutionary radiation of earless frogs in the Andes: molecular phylogenetics and habitat shifts in high-elevation terrestrial breeding frogs. PeerJ 6: e4313

Wiens JJ (2004) Speciation and ecology revisited: phylogenetic niche conservatism and the origin of species. Evolution 58(1):193–197

Willaert B, Reichle S, Stegen G, Martel A, Barrón S, Sánchez de Lozada N, Greenhawk NA, Agostini G, Muñoz A (2016) Distribution, ecology, and conservation of the critically endangered frog *Psychrophrynella illimani* (Anura: Craugastoridae) with description of its call. Salamandra 52(4):317–327

Williams EE (1972) The origin of faunas. Evolution of lizard congeners in a complex island fauna: a trial analysis. In: Dobzhansky T, Hetch MK, Steere WC (eds) Evolutionary biology, vol 6. Meredith, New York, pp 47–89

Chapter 28
Disparate Patterns of Diversification Within Liolaemini Lizards

Melisa Olave, Andrea González Marín, Luciano J. Avila, Jack W. Sites Jr., and Mariana Morando

Abstract Lizards are a major component of temperate-to-tropical terrestrial vertebrate biotas, and have played a central role as model systems for evolutionary and ecological research. The most diverse lizard group of the southern half of South America is the clade Liolaemini (=family Liolaemidae), which includes three genera characterized by large differences in species richness, as well as many other aspects of their biology. At one extreme is the monotypic genus *Ctenoblepharys*, restricted to sandy beaches and dunes in the coastal desert of Peru, oviparous and insectivorous. At the other extreme, *Liolaemus* is the world's richest temperate zone amniote genus of the temperate zone, with 262 described species. *Liolaemus* is widely distributed across southern South America, from north-central Peru to Tierra del Fuego, inhabiting climatic regimes extending from sea level to 5176 m in Bolivia, and exhibiting great diversity in biological features such as body size, color pattern, diet, reproductive mode (viviparous, oviparous, and one parthenogenetic species)

Electronic supplementary material The online version of this chapter (https://doi.org/10.1007/978-3-030-31167-4_28) contains supplementary material, which is available to authorized users.

M. Olave (✉)
Department of Biology, University of Konstanz, Konstanz, Germany

Instituto Argentino de Investigaciones de Zonas Arídas, Consejo Nacional de Investigaciones Cientificas y Técnicas (IADIZA-CONICET), Mendoza, Argentina

A. G. Marín · L. J. Avila
Instituto Patagónico para el Estudio de los Ecosistemas Continentales, Consejo Nacional de Investigaciones Científicas y Técnicas (IPEEC-CONICET), Puerto Madryn, Chubut, Argentina

J. W. Sites Jr
Department of Biology and M.L. Bean Life Science Museum, Brigham Young University (BYU), Provo, UT, USA

Department of Biology, Austin Peay State University, Clarksville, TN, USA

M. Morando
Instituto Patagónico para el Estudio de los Ecosistemas Continentales, Consejo Nacional de Investigaciones Científicas y Técnicas (IPEEC-CONICET), Puerto Madryn, Chubut, Argentina

Universidad Nacional de la Patagonia San Juan Bosco, Sede Puerto Madryn, Puerto Madryn, Chubut, Argentina

V. Rull, A. C. Carnaval (eds.), *Neotropical Diversification: Patterns and Processes*, Fascinating Life Sciences, https://doi.org/10.1007/978-3-030-31167-4_28

and karyotype. The third clade is the genus *Phymaturus*, which includes 44 described species distributed along the eastern and western Andean slopes in Argentina and Chile and through Patagonia. All *Phymaturus* species are viviparous, primarily herbivorous, strictly saxicolous and restricted to volcanic plateaus and peaks. Here, we contrast diversification patterns between the more specialized *Phymaturus* and *Ctenoblepharys,* with the more generalist *Liolaemus*. We found disparate patterns of diversification among the three genera, with *Liolaemus* showing the highest net diversification rate and, surprisingly, *Phymaturus* showing the highest speciation rates. The lower species diversity in *Phymaturus*, however, appears to be due to a high extinction rate, while the extraordinary species richness in *Liolaemus* is likely due to a lower extinction rate. The monotypic *Ctenoblepharys* is characterized by a negative net diversification rate, highlighting its vulnerability. We also found evidence of selection acting on the body sizes of Liolaemini species, in the form of a positive correlation between body size evolution and net diversification, speciation and extinction rates in *Phymaturus*, and a clear slowdown of morphological evolution in the *Phymaturus patagonicus* clade. We discuss the advantages and disadvantages of generalist vs. specialist life histories in Liolaemini, and provide recommendations for their conservation based on our findings.

Keywords Liolaemini · Lizards · South America · Diversification · Speciation · Extinction · BAMM · Macroevolution · Biogeography

1 Introduction

Since Darwin's revolutionary book (1859), biologists have been fascinated by highly diversified lineages (Simpson 1953; Losos and Ricklefs 2009). Rapid evolutionary radiations are thought to be responsible for a great deal of the extant and extinct organismal diversity on Earth, and they have served as model systems to explore mechanisms of clade diversification (Simpson 1953; Schluter 2000; Mayr 2001; Coyne and Orr 2004; Berner and Salzburger 2015). In the last two decades, however, a wealth of new molecular data, improved phylogenetic algorithms, and new methods for quantifying and interpreting species diversification patterns and phenotypic change (e.g. Harmon et al. 2003; Rabosky 2014), coupled with the incorporation of ecological data (Schluter 2000), have permitted evolutionary biologists to address long-standing macroevolutionary questions (Stroud and Losos 2016). These approaches are shedding insights into many conceptual aspects of what promotes clade diversification in space and time (e.g. Rabosky and Lovette 2008; Morlon et al. 2010; Rabosky and Glor 2010; Li and Wiens 2019). Specifically, rates of speciation and extinction vary through time and among clades (Etienne and Haegeman 2012; Jetz et al. 2012; Moen and Morlon 2014; Alfaro et al. 2018), contributing to dramatic heterogeneity in species richness across the Tree of Life (Alfaro et al. 2009; Jetz et al. 2012; Barker et al. 2013). By characterizing variation in rates of speciation and extinction, we can better understand the dynamics of biological diversity through time, across geographic and environmental gradients (Mittelbach et al. 2007; Silvestro et al. 2011; Rabosky et al. 2015), and in relation to traits and key innovations (Near et al. 2012; Rabosky et al. 2014; Beaulieu and O'Meara 2016).

The Neotropical biogeographic region holds one of the most biodiverse biota of the world and reflects the main global latitudinal biodiversity gradient, with higher species richness in the tropics decreasing towards temperate areas (Lomolino et al. 2010). Although greater species richness is well explained by an increase in speciation rates in some taxonomic groups (e.g. Pyron 2014), a recent example focused on marine fishes have shown an opposite pattern: an inverse latitudinal gradient of speciation rates (Rabosky et al. 2018). It is still highly debated how environmental drivers and evolutionary and ecological process are involved in diversification rates, mainly across this latitudinal gradient (e.g. Mittelbach et al. 2007), and a general integrated theory is still lacking (Rull 2018). To contribute to a unified general theory of Neotropical diversification, it is essential to study different empirical examples, and counter examples, in order to identify causal factors that may be related to the origin and maintenance of diversity under different evolutionary and ecological conditions. One counter example is the clade Liolaemini (=Liolaemidae), which includes the world's richest genus of extant amniotes in the temperate zone.

Lizards are a major component of temperate-to-tropical terrestrial vertebrate biotas, and have played a central role as model systems for evolutionary and ecological research (Camargo et al. 2010; Vitt and Pianka 2014; Losos and Ricklefs 2009). The clade Liolaemini (=Liolaemidae family; Fig. 28.1) is the most diverse lizard group of the southern half of South America, comprising three genera with large differences in species richness, as well as many other aspects of their biology. At one extreme, *Ctenoblepharys* is monotypic and restricted to the coastal desert of Peru, diurnal, oviparous and feed mostly on arthropods (Fig. 28.1). On the other hand, *Liolaemus* is the world's richest genus of extant amniotes in the temperate zone (Olave et al. 2018). It includes 262 described species (Reptile Database 2 February 2019) and, for the last two decades, an average of five species has been described each year (Lobo et al. 2010). *Liolaemus* has an extensive distribution in southern South America, ranging from north-central Peru to Tierra del Fuego (14.5°–52.5°S; Fig. 28.1), from sea level to 5176 m (in Bolivia; Aparicio and Ocampo 2010), and inhabits many climatic regimes (Etheridge and Espinoza 2000). This clade is characterized by a great diversity in biological features such as body size (Pincheira-Donoso et al. 2015), color pattern (Abdala and Quinteros 2014), diet (Espinoza et al. 2004; Pincheira-Donoso 2012), karyotype (Morando 2004; Aiassa et al. 2005) and parity mode. *Liolaemus* lizards reach sexual maturity in only 1 or 2 years, and have independently transitioned between reproductive modes (oviparity-viviparity) in several lineages (Schulte et al. 2000; Pincheira-Donoso et al. 2013; Esquerré et al. 2019). In addition, it also includes the only parthenogenetic species within higher-level clade Pleurodonta (Abdala et al. 2016). There are two major clades (subgenera) within *Liolaemus*: *Liolaemus* sensu stricto

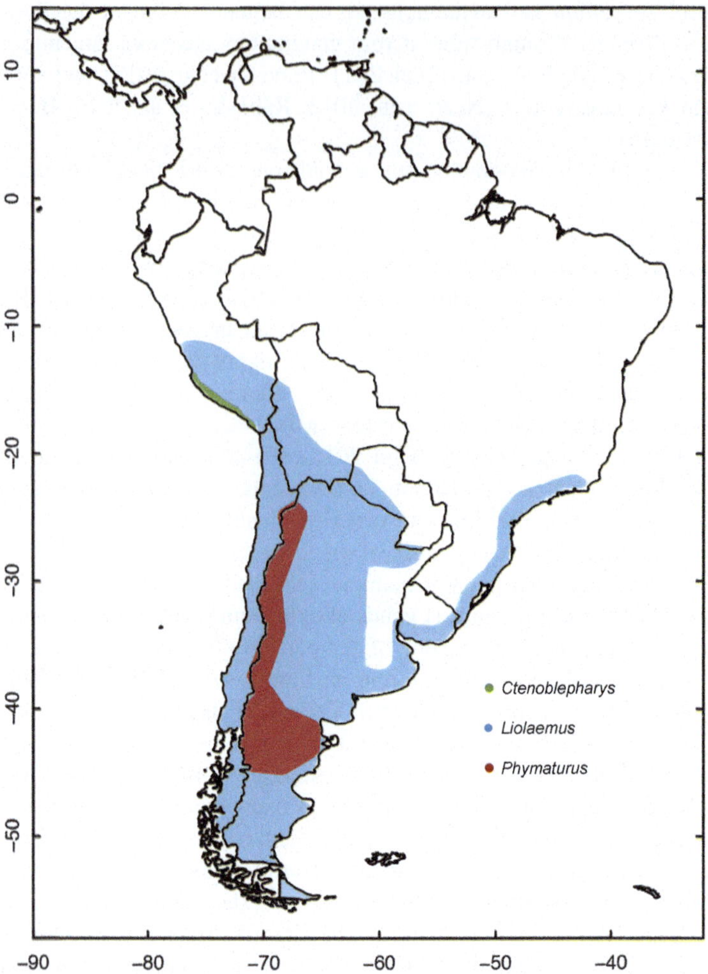

Fig. 28.1 Map showing the distribution range of the three genera, *Liolaemus* in blue, *Phymaturus* in red and the monotypic *Ctenoblepharys* in green

(primarily distributed in Chile; also called the "chiliensis" clade), and *Eulaemus* (mainly in Argentina; also called the "argentinus" clade). During the last decade, *Liolaemus* has emerged as a model system for phylogenetic, phylogeographic, and ecological studies (e.g. Espinoza et al. 2004; Camargo et al. 2010; Breitman et al. 2013). Further, some authors have considered it an example of a continental-scale adaptive radiation (Pincheira-Donoso et al. 2015), although not conforming to the strict definition of this term (Soulebeau et al. 2015; Simões et al. 2016). The third genus, *Phymaturus*, includes 44 described species (Reptile Database; 2 February 2019), distributed along both the eastern and western Andean slopes in Argentina and Chile (*P. palluma* clade, northern distribution) and through Patagonia (*P. patagonicus* clade, southern distribution). *Phymaturus* lizards reach sexual maturity at ~7–8 years (Piantoni et al. 2006), all species are viviparous (Schulte et al. 2000; Cei et al. 2003), primarily herbivorous (Espinoza et al. 2004) and strictly saxicolous, highly specialized to rock crevices on rocky formations of volcanic plateaus and mountain peaks (Cei 1986). These lizards use rock crevices to hide from predators (Pietrek et al. 2009) and for thermoregulation (Ibargüengoytía et al. 2008; Vidal et al. 2010).

The clade Liolaemini is distributed over a very large heterogeneous landscape characterized by a complex geo-climatic history that includes extensive tectonic uplift, volcanism, and many cycles of glacial advance and retreat (Ponce et al. 2011; Rabassa 2008; Rabassa et al. 2005). The uplift of the Andes has long been proposed as a major influence in shaping and promoting biodiversity (e.g. Hazzi et al. 2018). Hellmich (1951) hypothesized that climate change driven by the Andean orogeny is causally linked to the diversification of *Liolaemus*. More recently, Esquerré et al. (2019) proposed that the Andean Cordillera promoted diversification in Liolaemini by acting as a species pump into surrounding biomes, and increased speciation via allopatric fragmentation (but see Olave et al. 2019).

Here, we investigate contrasting diversification patterns within the Liolaemini by comparing the more specialized, ecologically constrained *Phymaturus* (Marín et al. 2018) and *Ctenoblepharys* (Etheridge 1995) genera against the ecologically diverse and species-rich *Liolaemus* (Abdala and Quinteros 2014). At a macroscale, *Liolaemus* has been considered a "generalist" genus (Olave et al. 2018), since the majority of these species have functionally intermediate morphology (Arnold 1998) that Tulli et al. (2012) described as "jack of all trades and master of none"; this generalized morphology could represent a morphological optimum (Tulli et al. 2016) that allows them to disperse and survive on a variety of habitats. Thus, *Liolaemus* clearly contrasts with the highly specialized species of *Phymaturus* and *Ctenoblepharys*. This could be an advantage for the genus because it may be more tolerant to environmental changes and higher dispersal capacity (Jansson and Dynesius 2002; Kassen 2002; Lanchier and Neuhauser 2006). An intuitive hypothesis is that *Phymaturus* and *Ctenoblepharys*, the two ecologically constrained genera with a smaller number of species, will have lower diversification rates than the ecologically broader and highly diverse genus *Liolaemus*.

Here, we examine the ability of each genus to produce new species and persist through time, by comparing speciation and extinction rates based on sequences from one mitochondrial and 12 nuclear genes and the only known *Liolaemus* fossil. In *Liolaemus*, phenotypic disparity, as measured by body size, has been shown to be high and overlapped among its various clades (Harmon et al. 2003; Pincheira-Donoso et al. 2015) and under stabilizing selection (Tulli et al. 2012; Olave et al. 2017). Body size has been related to thermoregulation in *Liolaemus* (e.g. Labra et al. 2009; Moreno Azócar et al. 2016), a central trait in ectotherms, particularly for those inhabiting extreme harsh environments like the Andes or the Patagonian Steppe. In *Phymaturus*, however, lower body size disparity, with two adaptive peaks, have been detected along with evidence for stabilizing selection as the possible mechanism underlying niche conservatism (Marín et al. 2018; Reaney et al. 2018). Thus, we have here a unique model system with the Liolaemidae family, as its two main genera have had the same time to diversify, yet they have very distinct morphologies and ecological characteristics. On the one hand, we expect to find higher rates of evolution in morphological evolution in *Liolaemus* than in *Phymaturus*. Additionally, given the importance of body size in the evolution of these lizards, we also expect to detect a link between diversification rates and body size evolution rate in each genus. To address the later, we evaluate the possible influence of body size evolution upon diversification patterns in Liolaemini.

2 Materials and Methods

2.1 DNA Sequences

We obtained DNA sequences from 215 species of Liolaemini (~75% coverage), ~65% coverage for *Liolaemus* (171 species), ~98% for *Phymaturus* (43 species), and 100% for *Ctenoblepharys* (1 species; used to root the tree). The majority of DNA sequences were obtained from GenBank and we included five new cyt-b sequences (accession numbers presented in Table S1). We constructed a DNA matrix of 13 loci (1 mitochondrial gene and 12 nuclear genes; Table 28.1; alignment available on File S1). Our selection was made on the basis of gene nature (mitochondrial and nuclear), and sought to minimize the amount of missing data. We selected both presumably unlinked nuclear protein-coding loci, with lower mutation rates, and anonymous

Table 28.1 Details of loci used for species tree estimation, substitution model selected and priors used for divergence time estimation

Gene	Model	Lognormal hyperprior mean	Lognormal hyperprior s.d.
cyt-b	GTR+I+G	0.01935500	1.2
CMOS	HKY+G	0.000879215	1.2
A12D	HKY+G	0.00076162	1.2
A1D	HKY+G	0.001765	1.2
A4B	HKY+G	0.0035965	1.2
A9C	HKY+G	0.0017753	1.2
PRLR	HKY	0.00132228	1.2
ND2	GTR+I+G	0.01	1.2
Phy38	HKY	0.001	1.2
Phy41	HKY	0.001	1.2
Phy60	HKY+G	0.001	1.2
Phy64	HKY	0.001	1.2
Phy84	HKY	0.001	1.2

nuclear loci, with higher expected mutation rates, with the aim of reconstructing relationships across all levels of lineage divergence.

2.2 Time-Calibrated Phylogenetic Tree

We used BEAST 1.8.3 (Suchard et al. 2018) to estimate a time-calibrated phylogenetic tree. Following Fontanella et al. (2012) and Olave et al. (2015), *Eulaemus* origin was calibrated using the fossil reported by Albino (2008), using a lognormal distribution with mean in real space equal to 20 mya and standard deviation 0.1 as a hyperprior (Table 28.1).

Using an uncorrelated lognormal clock model, we used the Olave et al. (2015) mutation rate estimations to inform the mutation hyperpriors for the cyt-b, CMOS, A12D, A1D, A4B, A9C and PRLR loci (see Table 28.1 for details). We set a wide hyperprior distribution for those genes with previous unpublished mutation rates, following previous estimations for mitochondrial and nuclear genes in Liolaemini. Specifically, we set the ND2 hyperprior to 0.01 (s.d. 1.2) and the nuclear genes Phy38, Phy41, Phy60, Phy64 and Phy84 hyperpriors to 0.001 (s.d. 1.2). Note that BEAST estimates mutation rates from the data (see Table S2 for all gene mutation rate estimates by BEAST). Two independent analyses were run for 1×10^9 MCMC generations, sampling every 1×10^5 steps and with 25% burn-in.

2.3 Net Diversification, Speciation and Extinction Rates

We estimated the net diversification, speciation and extinction rates based on the estimated phylogenetic tree (described above). Time-calibrated phylogenetic trees can be used to estimate these rates given the observed branching pattern per time interval (Nee 2001; Ricklefs 2007). One of the simplest and most widely applied models is the speciation–extinction process (birth-death), this model assigns both rates instantaneous probabilities (λ and μ, respectively; with units 1/time). The intervals between the formation of a new lineage by speciation and subsequent lineage splitting or termination are exponentially distributed, with average times $1/\lambda$ and $1/\mu$, respectively (Ricklefs 2007). We used BAMM 2.5 (Rabosky et al. 2014) to estimate speciation and extinction rates. BAMM is a Bayesian approach that uses a rjMCMC algorithm to estimate lineage-specific speciation and extinction rates, and rates of phenotypic change. Because the method estimates rates per linage, it allows us to compare changes of these rates among clades and nested clades (species groups) of interest (i.e. tips). As in most other similar models, BAMM assumes the given topology is the true phylogenetic tree. However, to account for topological uncertainty, we ran the analysis based on each of the 500 trees sampled from the posterior during the BEAST MCMC search. We incorporated the proportion of missing taxa using *globalSamplingFraction* = 0.75, which accounts for the missing tips (i.e. 75% coverage). Priors were generated using *setBAMMpriors* in BAMMtools (Rabosky et al. 2014), and we used all 500 obtained means for target groups (genus, subgenus, clades) to construct final distributions for all downstream comparisons. All BAMM analyses were run for 5×10^6 generations sampling every 1000 generations and with a 25% burn-in.

2.4 Body Size Evolution

Phenotypic data consisted of snout-vent length (SVL) for 195 species, collected from the literature (Table S3). We removed tips (terminal branches) from the phylogenetic tree with missing data for these analyses using the drop.tip() function of the ape package in R (Paradis and Schliep 2018). We separated sexes for these analyses, and repeated all analyses with sexes combined, taking the average between them or using general values from the literature. Because all results were very similar, we only present here the general results with sexes combined. Rates of body size evolution were estimated using BAMM 2.5, with the similar approach described above, by considering each of the last 500 trees inferred during the MCMC of BEAST. All BAMM analyses were run for 5×10^6 generations, sampling every 1000 generations and with a 25% burn-in.

To obtain insight about the process of character evolution and the type of selection impacting body size, we used the GEIGER package (Harmon et al. 2007) to test three different models of morphological evolution: Brownian motion (BM), Ornstein-Uhlenbeck (OU) and Early-burst (EB). The BM model represents random walks through time (i.e. neutral evolution), while OU and EB include natural selection. Specifically, the OU model represents a tendency of the trait (in this case, body size, represented by SVL) to move towards a long-term mean (e.g. stabilizing selection), which may capture the importance of constraints on evolution. In contrast, the EB model allows rates of evolution to slow through time, with highest rate at the root of the tree, corresponding to one classical idea of adaptive radiation. We used the Akaike Information Criterion corrected for small samples (AICc) to select the best-fit model.

2.5 Statistical Comparisons

Focusing on rate estimations obtained from the last 500 phylogenetic trees sampled from the posterior of BEAST (as described above), we constructed distributions that capture uncertainty in the reconstruction of phylogenetic relationships. We calculated summary statistics using R (mean, standard deviation and quartiles), and evaluated differences among specific target clades with ANOVA tests using the R function aov().

To test the possible link of body size evolution and diversification, we extracted species-specific rates from the tree tips and constructed regressions between net diversification, speciation and extinction rates against the body size evolution rates. We used simple linear models using the R function lm(), as well as phylogenetically controlled for historical relationships, given that rates are not independent of the phylogeny. For this last case, a phylogenetically independent contrast model was applied by obtaining a contrast matrix with the pic() function of the ape R package (Paradis and Schliep 2018) and its implementation into the gls() function of the nlme package (Pinheiro et al. 2012) as a correlation matrix. For *Phymaturus* lizards, we also fitted a non-linear model using the nls() function. We tried different formulas and selected the one that maximized the R^2: $y \sim a * \beta/(b + \beta)$, where β is the rate of body size evolution, a is the origin and b is the decay rate of the curve. We selected starting values $a = 1$ and $b = 1$, and compared the linear and non-linear models using an ANOVA test in *R*.

3 Results

3.1 Liolaemini Phylogenetic Relationships

The BEAST phylogenetic tree resolved the monophyly of both multi-species Liolaemini genera with strong support (PP = 1; Figs. S1 and S2; phylogenetic trees on Files S2 and S3). The origin of the clade Liolaemini family is dated at 62.33 mya (95% HPD = 43.74–83.76 mya), with the first split separating the monotypic *Ctenoblepharys*, and inferring the origins of *Liolaemus* and *Phymaturus* at 30.3 mya (95% HPD = 23.9–38.61 mya) and 22.48 mya (95% HPD = 15.2–31.36 mya), respectively. The sub-genera *Liolaemus* sensu stricto [s.s.] and *Eulaemus* are strongly inferred as monophyletic as well (PP = 1), with crown nodes origins estimated at 25.16 mya (95% HPD = 19.3–31.99 mya) and 21.21 mya (95% HPD = 17.44–25.42 mya), respectively. The *P. palluma* clade is inferred with a younger coalescence at 4.12 mya (95% HPD = 2.71–6.11 mya); a slightly older date at 5.06 mya (95% HPD = 3.49–6.98 mya) was inferred for the *P. patagonicus* clade.

3.2 Net Diversification, Speciation and Extinction Rates

Overall Liolaemini net diversification rate increases through time from the origin of the clade (60 mya) to the present (Fig. 28.2c), with an inflection point at 30 mya. Speciation and extinction rates have experienced a maximum rate about 20 mya, followed by slowdown around 10 mya, and accelerating again in the past 5 mya (Fig. 28.2d–e). All estimated rates show a disparate pattern among the three genera of Liolaemini (summary statistics in Table S4). At one extreme, the phylogenetic tree reflecting the lineage-specific net diversification rates (Fig. 28.2a) displays a

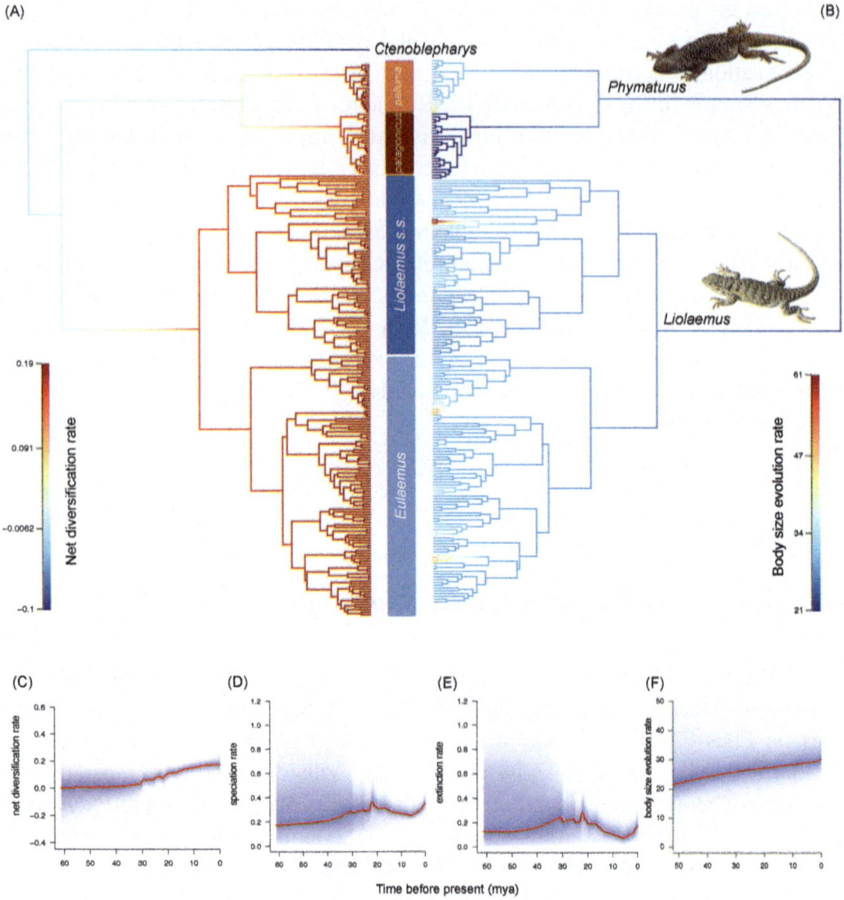

Fig. 28.2 Liolaemini evolution. (**a**) Diversification rates and (**b**) morphological evolution rate projected in color gradient on the phylogenetic tree. Note that some tips were removed for the estimation of phenotypic rates, due to missing data. (**c–f**) Rates through time (in millions of years), with the median (in red) and the 95% credibility interval (blue shadows)

clear diversification slowdown in *Ctenoblepharys*, whereas there is a significant increase in speciation rates in *Liolaemus* and *Phymaturus* (Fig. 28.3a; $P < 0.001$; $F = 7759.6$; $DF = 2$; see Table S7 for ANOVA test details). Not surprisingly, *Ctenoblepharys* shows a negative diversification rate and the lowest of the three genera (Fig. 28.3a), *Phymaturus* has an intermediate net diversification rate, while *Liolaemus* has the highest. However, the greatest net diversification rate of *Liolaemus* is not explained by a high speciation rate but, instead by the lowest extinction rate (Fig. 28.3b, c and S3). Another interesting result is the distinctively high speciation (Figs. 28.3b and S3) and extinction rates (Figs. 28.3c and S4) of *Phymaturus*. Results are consistent when considering the different main clades

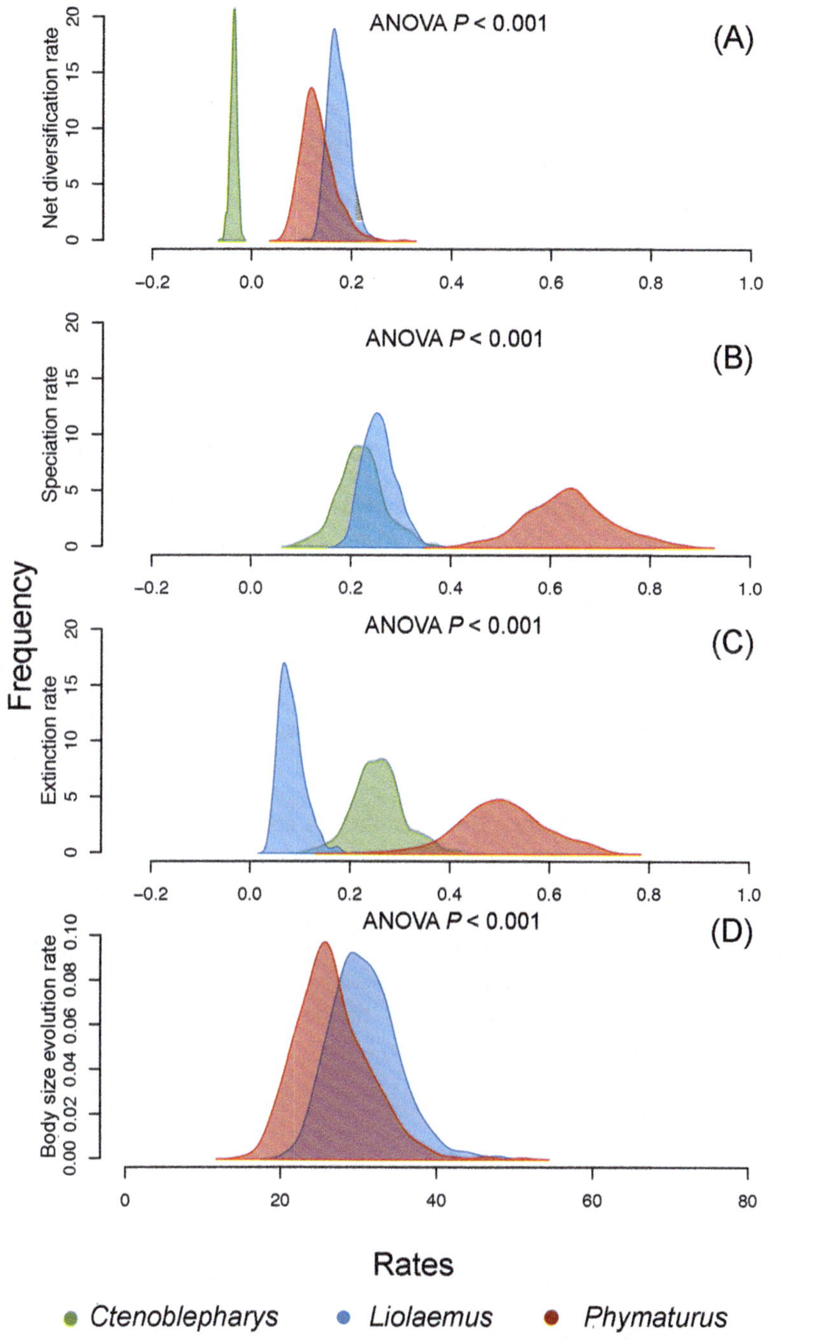

Fig. 28.3 Net diversification, speciation, extinction and morphological evolution rates obtained for *Ctenoblepharys* (green), *Liolaemus* (blue) and *Phymaturus* (red). Note that *Ctenoblepharys* density plot for the net diversification rate was re-scaled (y-axis divided by 3) in order to get a better observation of all three genera distributions. The density plots are constructed considering the mean

within each genus (i.e. *Eulaemus* and *Liolaemus* sensu stricto [s.s.], plus *P. palluma* and *P. patagonicus* clades; Fig. 28.4a–c; Table S7).

Because *Liolaemus* comprises a very large number of species with different morphologies and ecological requirements, we also explored rates among clades within this genus (Figs. S5 and S6). The net diversification rate differed significantly among *Liolaemus* clades ($P < 0.001$; $F = 11.773$; $DF = 13$; Table S7), with the *lemniscatus* group being the only clade different from the others (Figs. S5A and S6A). This is explained by a small decrease in speciation rate (Figs. S5B and S6B), while extinction rates did not differ among *Liolaemus* clades (Figs. S5C and S6C; $P = 0.9403$; $F = 0.4734$; $DF = 13$; Table S7).

3.3 *Body Size Evolution*

Liolaemini lizards in general display a monotonic increment in the rate of body size evolution through time (Fig. 28.2f). *Phymaturus* has a slower rate of body size evolution than *Liolaemus* (Fig. 28.3d; $P < 0.001$; $F = 194.18$; $DF = 1$; Table S7) and its two main clades differ from each other: the southernmost *P. patagonicus* clade has a lower mean rate of body size change, while the northern *P. palluma* clade has a higher mean than rates for the *Liolaemus* subgenera (Fig. 28.4d; Tables S4 and S7). The slowdown in the *P. patagonicus* clade rate is also very prominent in the color-coded phylogenetic tree (Fig. 28.2b). We found significant differences in rates of body size evolution between *Liolaemus* clades (Figs. S5D and S6D; $P < 0.001$; $F = 250.15$; $DF = 13$; Table S7), with the *elongatus*, *lemniscatus*, *montanus* and *rothi* clades having high rates. However, there is not a clear latitudinal pattern as in *Phymaturus* (i.e. southern clades having a lower rate). Instead, the "high rate" *Liolaemus* clades are from different regions: the *montanus* group has a northern distribution (central-northern Argentina to Peru), the *elongatus* and *lemniscatus* groups are from central Argentina-Chile, and *rothi* lizards are from central Patagonia. The model-selection analysis of the evolution of SVL indicated that the Ornstein-Uhlenbeck (OU) model better fit the data than the Brownian Motion (BM) or Early Burst (EB) models (Table 28.2).

Fig. 28.3 (continued) obtained from each of the last 500 trees of BEAST analysis. The *p*-value corresponds to an ANOVA test comparing distributions (see Table S7 for ANOVA details)

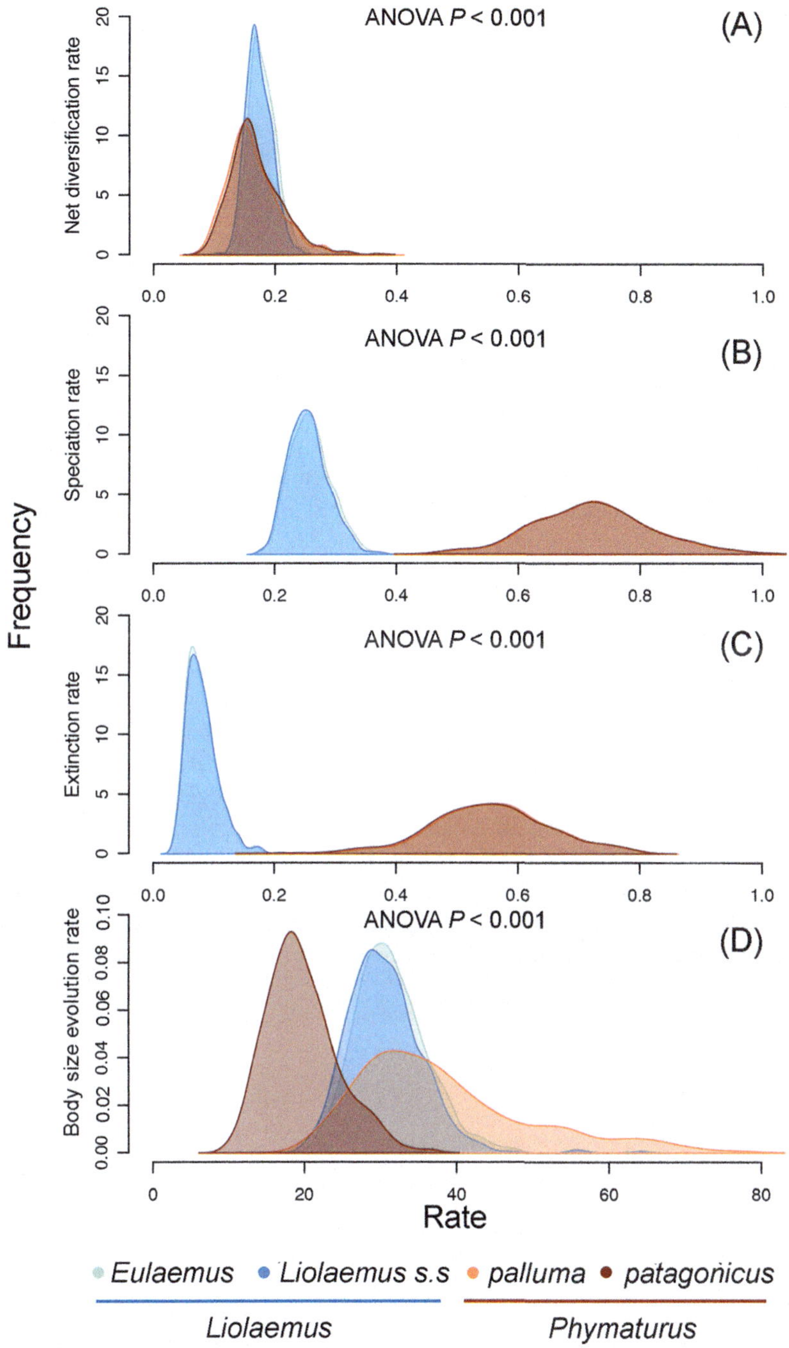

Fig. 28.4 Net diversification, speciation, extinction and morphological evolution rates obtained for *Liolaemus* subgenera (blue scale): *Eulaemus* and *Liolaemus sensu stricto* (s.s.); and clades within *Phymaturus* (red scale): *palluma* and *patagonicus*. The density plots are constructed considering the

Table 28.2 Akaike Criterion (corrected) calculation for the fit of the three models of morphological evolution OU, EB and BM. Selected model for all cases corresponds to the OU with the lowest AICc value

	AICc		
Model	Liolaemini	*Liolaemus*	*Phymaturus*
Ornstein-Uhlenbeck (OU)	**1472.02**	**1173.41**	**281.78**
Brownian Motion (BM)	1503.92	1211.47	294.54
Early Burst (EB)	1506.01	1213.58	296.97

3.4 Link Between Diversification Rates and Body Size Evolution Rates

Linear regressions between diversification rates and body size evolution were not significant for any of the *Liolaemus* clades with phylogenetically controlled rates (Fig. 28.5; Table S6). Although we found a significant correlation between rates using the phylogenetically independent contrast approach, the fit of the model remains poor ($R^2 < 0.1$ for all cases; Fig. 28.5; Table S6). In contrast, for *Phymaturus* while we found a significant link between diversification rates and body size evolution when considering phylogenetically uncontrolled rates, the signal is lost when rates are phylogenetically controlled (Fig. 28.6; Table S6). The best-fit model selected for uncontrolled rates is a non-linear model ($P < 0.001$ in all cases; Table S6) with an increment of the R^2 (Table S6). In this case, an incremental increase in the rate of body size evolution is linked to incremental increases in net diversification, speciation and extinction rates, in a log-like relationship. When considering the *P. palluma* clade alone and using the phylogenetically uncontrolled rates, we found a similar log-like trend for the net diversification and speciation rates as function of the rate of body size evolution, but a negative linear regression for the extinction rate (Fig. S7; Table S6). We found a weaker correlation for the *P. patagonicus* clade, with a non-significant correlation for the extinction rate in both, phylogenetically controlled and uncontrolled rates (Fig. S8; Table S6).

Fig. 28.4 (continued) mean obtained from each of the last 500 trees of BEAST analysis. The *p*-value corresponds to an ANOVA test comparing distributions (see Table S7 for ANOVA details)

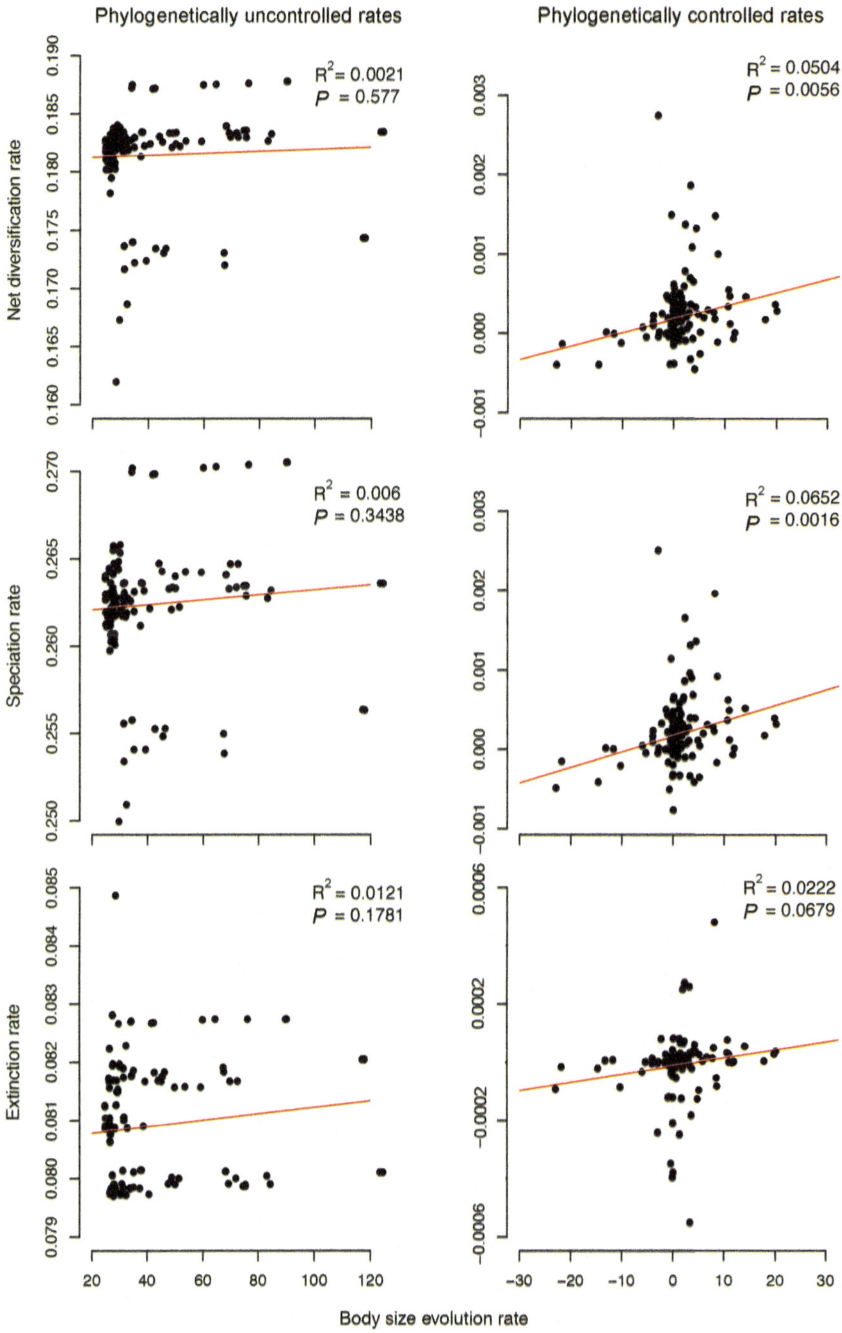

Fig. 28.5 Phylogenetically uncorrected (left) and corrected (right) linear regressions between net diversification, speciation and extinction rates as function of the morphological evolution rate in *Liolaemus*. Each point is the rate obtained per species (see Table S6 for details of regressions)

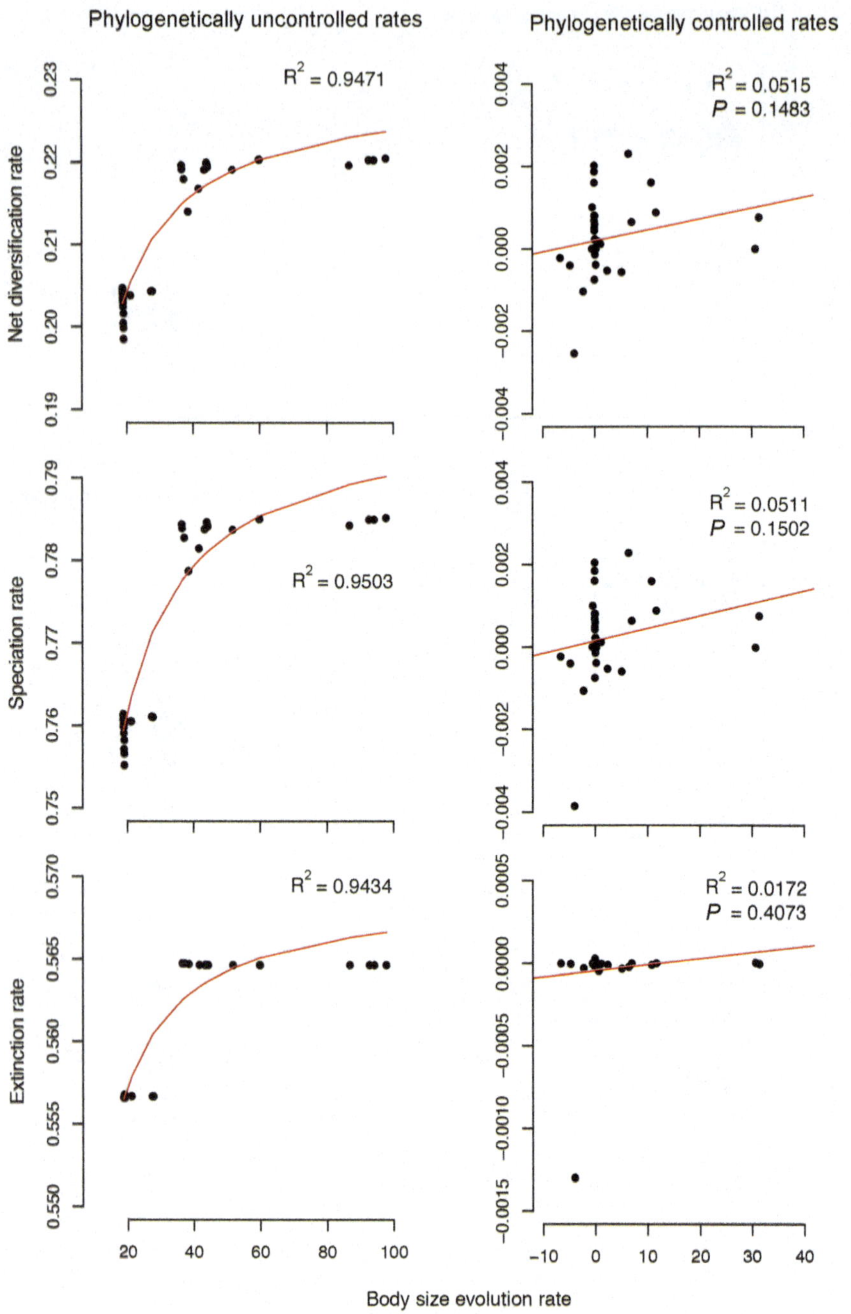

Fig. 28.6 Phylogenetically uncorrected non-linear regressions (left) and corrected linear regressions (right) between net diversification, speciation and extinction rates as function of the morphological evolution rate in *Phymaturus*. Each point corresponds to the rate obtained per species (see Table S6 for details of regressions)

4 Discussion

4.1 *Disparate Patterns of Diversification in Liolaemini Lizards*

The influence of specialization on macroevolutionary potential has been debated for over a century (Day et al. 2016), with a focus on two alternative interpretations. One argument suggests that specialization is a potential evolutionary "dead end", because it constrains capacity for future evolutionary change or diversification, due to narrowing resource use which may limit adaptation to environment changes, ultimately leading to higher extinction rates (e.g. Cope 1896; Haldane 1951; Kelley and Farrell 1998; Nosil and Mooers 2005; Vamosi et al. 2014; Day et al. 2016). The alternative view holds that specialization has seems to have increased diversification rates in a range of taxa, by allowing more species to be "packed" in an area (e.g. Price et al. 2012; Salisbury et al. 2012; Ebel et al. 2015; Brodersen et al. 2018; reviewed by Day et al. 2016). Empirical support for both hypotheses suggests that there is not a general rule, and that each study system has its own unique evolutionary history, and that specific environments must be considered for each case (Vamosi et al. 2014; Day et al. 2016). Specialists are often hypothesized to do a better job in an "homogenous" niche, and "generalism" may then be an ephemeral state (Kassen 2002; Janz and Nylin 2008). However, morphological generalists are considered to have advantages over specialists in fluctuating and heterogeneous environments, because the former are more tolerant to climatic/ecological changes and more likely to find suitable habitats (Jansson and Dynesius 2002; Kassen 2002; Lanchier and Neuhauser 2006).

Here, we found very interesting disparate patterns of diversification among the three genera within Liolaemini (Fig. 28.3). The split between *Ctenoblepharys* and (*Liolaemus* + *Phymaturus*) has been inferred to be ~62 (44–84) Mya, placing it in the Paleocene -Eocene (Cenozoic), and the split between *Liolaemus* and *Phymaturus* at ~55 (72–40) Mya (Fig. S1). The ancestral area of *Liolaemus* + *Phymaturus* has been inferred in the Cordillera Andina and Valle Central (Díaz Gómez 2011), which correspond to 32–36 S, where the initial uplift of the Andes started at early Cenozoic, congruent with the inferred dates. The sequence uplift of the Andes at this region that started in the Miocene, at 20 Mya with the Cordillera Principal (the political division between Argentina and Chile), lasted until 8 Mya (Giambiagi et al. 2016). During this period intense tectonic as well as volcanism and climatic changes continuously shaped the environment (Gregory-Wodzicki 2000; Garzione et al. 2008); and more recently glaciation cycles also contributed to the formation of heterogeneous landscapes (Ponce et al. 2011). Even though *Liolaemus* has an extraordinary species richness and the highest net diversification rate relative to the other Liolaemini genera (Fig. 28.3), we inferred higher speciation rates in its sister group, the ecologically highly specialized *Phymaturus* lizards (Figs. 28.3 and 28.4). *Phymaturus* species only occur in isolated patches of rock outcrops; therefore, migration between populations is likely severely limited relative to *Liolaemus*

(Vicenzi et al. 2017). The strong fidelity of *Phymaturus* species to specific microhabitats, "islands" of big boulders with deep crevices in volcanic cliffs, peaks and plateaus (Cei 1986), might have promoted fragmentation and speciation over short periods of time, as is the case in other lizard species, either in wildlife as well as in urban populations (e.g. Templeton et al. 2001; Thompson et al. 2018). *Phymaturus* specialization may be advantageous if it results in efficient selection for adaptation to a stable and narrow niche, reducing the cost of trade-offs by abandoning traits needed to utilize a wider range of resources (Futuyma and Moreno 1988). However, these shorter term microevolutionary benefits may come at the cost of longer term macroevolutionary success (e.g. Agnarsson et al. 2006; Anacker et al. 2011; Forister et al. 2012; Armbruster 2014). Thus, the same factors promoting speciation in *Phymaturus* might explain its high extinction rate as a trade-off (Fig. 28.3c).

In contrast, *Liolaemus* species occur in a wide diversity of environments, even extreme ones, which track the complex geoclimatic history of southern South America. Here, we have shown that the extraordinary species richness of *Liolaemus* seems to be a product of a lower extinction rate rather than a high speciation rate (Fig. 28.3a–c). Thus, the versatility of the genus to colonize new environments and to persist through environmental changes seems to be key for explaining its higher net diversification rate (Jansson and Dynesius 2002; Kassen 2002; Lanchier and Neuhauser 2006).

The disparate diversity and biology of the three genera within Liolaemini makes this group an exciting system to approach questions in macroevolution. A recent study focused on general patterns of Liolaemini diversification has proposed an important role for the Andean Cordillera promoting lineage diversification and acting as a species pump into the surrounding biomes (Esquerré et al. 2019). Here, we revisited the diversification of Liolaemini lizards, but we focused on a more inclusive level by treating each genus as a different group. Given our novel findings, it is clear that an overly general estimation of rates will likely confound results or hide interesting patterns. For example, while there is an increment through time in net diversification, speciation and extinction rates for the Liolaemini clade as a whole (Fig. 28.2c–e), the monotypic *Ctenoblepharys* has experienced a decrease in both speciation and extinction rates (Figs. 28.2a, S3 and S4). Further, the overall mean speciation and extinction rates based on the single (*Phymaturus* + *Liolaemus*) clade is a meaningless dispersion statistic, due to the clearly distinct rate distributions of each genus (Fig. 28.3). Here, we do not find trends for increased diversification rates in *Liolaemus* groups distributed in or near the Andes area (Fig. S6), and we argue that the higher Andean speciation rates reported by Esquerré et al. (2019) are simply detecting a dominant "*Phymaturus* signal". Our findings highlight the importance of considering clade-specific biology in macroevolutionary analyses (see also Olave et al. 2019).

4.2 The Link Between Body Size Evolution and Diversification

Body size is a fundamental character in animal biology that may covary with ecological, physiological or other life history traits, such as fecundity, age at maturity, metabolic efficiency and thermoregulation (Sookias et al. 2012). Thus, body size has complex interactions with fitness (LaBarbera 1989). For example, larger body size could increase success in obtaining food, defending a home range, and reproductive success (Hone and Benton 2005), against trade-offs with cuckoldry or predation. Here, we found stabilizing selection acting on the evolution of body size (SVL) in each genus (Table 28.2). Specifically, an OU model was selected, indicating a tendency to a central value, consistent with previous results (Olave et al. 2017; Reaney et al. 2018; Marín et al. 2018). This result was expected, given the observed increment through time shown in Fig. 28.2f, with a more recent accumulation of change.

We found a positive log-like correlation between the phylogenetically uncontrolled rate of SVL evolution and net diversification, speciation and extinction rates in *Phymaturus* lizards, although this signal is lost when rates are phylogenetically controlled (Fig. 28.6; Table S6). A plausible interpretation for the uncontrolled rates results is that an increase of the rate of body size evolution promotes diversification, but this is also not a straightforward conclusion from this analysis. Instead, we found a clear slow-down in the rate of body size evolution in the *Phymaturus patagonicus* clade (Figs. 28.2b and 28.4d), which could reflect a response to the more homogeneous habitats used by these species over the Patagonian steppe, compared to a more heterogeneous and wider latitudinal range for the *P. palluma* clade. *Phymaturus palluma* species seem to be more flexible in the rock-type association, relative to a sympatric species of the *P. patagonicus* clade that are only observed on two types of rocks (Corbalán et al. 2013). In addition, some *P. palluma* lizards tolerate wider temperature ranges (Corbalán and Debandi 2013) and higher altitudinal ranges (1200–4500 m) than species of the *P. patagonicus* clade (2700 m maximum; Debandi et al. 2012). Further, in one of the few areas were species of these two main clades of *Phymaturus* are in sympatry, it was reported that their diets and foraging schedules are different (Corbalán and Debandi 2014).

The greater specialization of the *Phymaturus patagonicus* clade may have constrained phenotypic evolution to an optimum and promoted their apparent phenotypic stasis (see also Marín et al. 2018). A more heterogeneous environment in the central-north of Argentina might explain the higher rate of body size evolution associated with an acceleration in net diversification rate in the *Phymaturus palluma* clade (Fig. 28.2a). Specifically, we found a linear reduction in the extinction rate associated with an increase in the rate of body size evolution, in addition to a positive log-like correlation with the speciation rate (Fig. S7). Considering that there is little support for a universal and consistent mechanism of climate-driven size evolution within Squamates (Slavenko et al. 2019), further studies are needed to investigate the role of temperature and thermoregulation on the evolution of body size and diversification of *Phymaturus*.

The genus *Liolaemus* extends to the southernmost distribution of any reptile (Breitman et al. 2011), but we did not find the same pattern of reduced body size evolution in southern *Liolaemus* clades as demonstrated in *Phymaturus* (Fig. 28.2a). *Liolaemus* seems to be independent of the rate of body size evolution, given that we found no significant regressions under the phylogenetically uncontrolled rates and the phylogenetically controlled model with significant result had a very poor fit ($R^2 < 0.1$; Fig. 28.5; Table S6). For example, the *L. lineomaculatus* clade includes the southernmost distributed species of any *Liolaemus* species and shows no differences in rates of body size evolution with respect to any of the other clades (Figs. S5 and S6). We acknowledge that factors that contribute to the diversification rate of *Liolaemus* likely correspond to several axes of morphological evolution, given the biological variability within the *Liolaemus* clades. Further, hybridization seems to be a common phenomenon in the genus, and it has been hypothesized as contributing to the evolution of a successful generalist morphology by preventing specialization (Olave et al. 2018). How *Liolaemus* has maintained its low extinction rate is a very interesting and complex question that deserves further study.

4.3 Implications for Conservation

Traditional conservation efforts have been focused on few of the multiple facets of biodiversity, particularly focusing on species richness, detection of hotspots (Myers et al. 2000), and threat status as defined by the World Conservation Union Red List criteria (https://www.iucnredlist.org). Several metrics and approaches have been developed to take into account various other aspects related to conservation, with some incorporated into the IUCN criteria; these include genetic variability (reviewed in Diniz 2004), connectivity (e.g. Lowe and Allendorf 2010), geospacial data (Bachman et al. 2011), and phylogenetic history (Faith 1992, 2018). But phylogenetic diversity itself has three distinct facets: lineage diversification, character divergence, and survival time (Kling et al. 2018). Our example allows us to elaborate some on general points that contribute to this discussion while taking into account comparative diversification results.

For the monotypic genus *Ctenoblepharys* we inferred a negative diversification rate (Figs. 28.2a and 28.3a), i.e. its extinction rate is higher than the speciation rate (Fig. 28.2b–c); thus the expectation for the long-term persistence of the genus through time is doubtful. *Ctenoblepharys adspersa* is considered a vulnerable species by the IUCN Red List (www.iucnredlist.org), and is of conservation interest in Peru (Peru's National Red List; http://www.serfor.gob.pe). Its distribution coincides with several protected areas, including the Reserva Nacional de Paracas and Lomas de Lachay. However, little is known about its life history, except that it runs swiftly over sand and that its color pattern is cryptic on this substrate (Etheridge

1995). We encourage further ecological studies and active conservation actions to preserve this unique lineage.

The *Phymaturus* clade is characterized by an alarmingly high extinction rate (Fig. 28.3c), and at least some species are likely highly vulnerable to climate change (Sinervo et al. 2010; Vicenzi et al. 2017). We recommend immediate conservation of all volcanic habitats, as these are essential for persistence of all *Phymaturus* species and can be implemented with little economic cost.

Finally, the *Liolaemus* clade seems to have the highest probability of persisting via adaptation to future environmental changes. However, at the species level, most *Liolaemus* are reported to be currently unprotected by legislation (Corbalán et al. 2011), and that anthropogenic actions are negatively affecting the abundance and distributions of some species (e.g. Vega et al. 2000; Pelegrin and Bucher 2012; Stellatelli et al. 2015; Abdala et al. 2017). There is also a "taxonomic impediment" to implementing long-term conservation actions for this clade; we are still in an active phase of species discovery and large remote parts of the clade's distribution have yet to be extensively explored. For most of their evolutionary history, these lizards have been highly successful in surviving changing and extreme natural environments, but current climate change and, in some places, urbanization, are likely to present major challenges to the conservation of some species. We strongly encourage both intensive field work in poorly-known regions, especially large sections of the Andes Cordillera, and implementation of management plans on existing natural areas (Corbalán et al. 2011).

Acknowledgments We thank Guarino R. Colli for useful comments and suggestions made on the first version of this chapter. We thank all members of the Grupo de Herpetología Patagónica (IPEEC-CONICET) for continuing support. Financial support was provided by ANPCYT-FONCYT 1252/2015 (MM), and a postdoctoral fellowship (AGM) from CONICET, Argentina and a postdoctoral fellowship (MO) from the Alexander von Humboldt Foundation at Meyer Lab, Konstanz, Germany.

References

Abdala CS, Quinteros AS (2014) Los últimos 30 años de estudios de la familia de lagartijas más diversa de Argentina: Actualización taxonómica y sistemática de Liolaemidae. Cuad Herpetol 28(2):55–82

Abdala CS, Baldo D, Juárez RA et al (2016) The first parthenogenetic Pleurodont Iguanian: a new all-female *Liolaemus* (Squamata: Liolaemidae) from western Argentina. Copeia 104 (2):487–497

Abdala CS, Semhan RV, Laspiur A et al (2017) Rediscovery of *Liolaemus rabinoi* (Iguania: Liolaemidae) after 35 years: redescription, biological and phylogenetic information, and conservation challenges. Salamandra 53(1):14–125

Agnarsson I, Avilés L, Coddington JA et al (2006) Sociality in theridiid spiders: repeated origins of an evolutionary dead end. Evolution 60(11):2342–2351

Aiassa D, Gorla N, Martori R (2005) Citogenética de los lagartos del género *Liolaemus* (Iguania: Liolaemidae) de América del Sur. Cuad Herpetol 18:23–36

Albino AM (2008) Iguanian lizards from the Colhuehuapian (early Miocene) of Gaiman, (Chubut province, Argentina). Ameghiniana 45(4):775–782

Alfaro ME, Santini F, Brock C et al (2009) Nine exceptional radiations plus high turnover explain species diversity in jawed vertebrates. Proc Natl Acad Sci U S A 106:13410–13414

Alfaro ME, Faircloth BC, Harrington RC et al (2018) Explosive diversification of marine fishes at the cretaceous–Palaeogene boundary. Nat Ecol Evol 2:688–696

Anacker BL, Whittall JB, Goldberg EE et al (2011) Origins and consequences of serpentine endemism in the California flora. Evolution 65(2):365–376

Aparicio J, Ocampo M (2010) *Liolaemus* grupo *montanus* Etheridge 1995 (Iguania: Liolaemidae). Cuad Herpetol 24:133–135

Armbruster WS (2014) Floral specialization and angiosperm diversity: phenotypic divergence, fitness trade-offs and realized pollination accuracy. AoB Plants 6. https://doi.org/10.1093/aobpla/plu003

Arnold EN (1998) Structural niche, limb morphology and locomotion in lacertid lizards (Squamata, Lacertidae); a preliminary survey. Bull Br Mus Nat Hist (Zool) 64:63–89

Bachman S, Moat J, Hill AW et al (2011) Supporting red list threat assessments with GeoCAT: geospatial conservation assessment tool. ZooKeys 150:117–126

Barker FK, Burns KJ, Klicka J et al (2013) Going to extremes: contrasting rates of diversification in a recent radiation of new world passerine birds. Syst Biol 62:298–320

Beaulieu JM, O'meara BC (2016) Detecting hidden diversification shifts in models of trait-dependent speciation and extinction. Syst Biol 65(4):583–601

Berner D, Salzburger W (2015) The genomics of organismal diversification illuminated by adaptive radiations. Trends Genet 31(9):491–499

Breitman MF, Avila LJ, Sites JW Jr et al (2011) Lizards from the end of the world: phylogenetic relationships of the *Liolaemus lineomaculatus* section (Squamata: Iguania: Liolaemini). Mol Phylogenet Evol 59(2):364–376

Breitman MF, Morando M, Avila LJ (2013) Past and present taxonomy of the *Liolaemus lineomaculatus* section (Liolaemidae): is the morphological arrangement hypothesis valid? Zool J Linnean Soc 168(3):612–668

Brodersen J, Post DM, Seehausen O (2018) Upward adaptive radiation cascades: predator diversification induced by prey diversification. Trends Ecol Evol 33(1):59–70

Camargo A, Sinervo B, Sites JW Jr (2010) Lizards as model organisms for linking phylogeographic and speciation studies. Mol Ecol 19(16):3250–3270

Cei JM (1986) Reptiles del centro, centro-oeste y sur de la Argentina: Herpetofauna de las zonas áridas y semiáridas. Museo Regionale di Scienze Naturali, Torino, p 527

Cei JM, Videla F, Vicente L (2003) From oviparity to viviparity: a preliminary note on the morphometric differentiation between oviparous and viviparous species assigned to the genus *Liolaemus* (Reptilia, Squamata, Liolaemidae). J Zool Syst Evol Res 41(3):152–156

Cope ED (1896) The primary factors of organic evolution. The Open Court Publishing Company, Chicago, IL

Corbalán V, Debandi G (2013) Basking behaviour in two sympatric herbivorous lizards (Liolaemidae: *Phymaturus*) from the Payunia volcanic region of Argentina. J Nat Hist 47 (19–20):1365–1378

Corbalán V, Debandi G (2014) Resource segregation in two herbivorous species of mountain lizards from Argentina. Herpetol J 24(4):201–208

Corbalán V, Tognelli MF, Scolaro JA et al (2011) Lizards as conservation targets in Argentinean Patagonia. J Nat Conserv 19(1):60–67

Corbalán V, Debandi G, Kubisch E (2013) Thermal ecology of two sympatric saxicolous lizards of the genus *Phymaturus* from the Payunia region (Argentina). J Therm Biol 38(7):384–389

Coyne JA, Orr HA (2004) Speciation. Sinauer Associates, Sunderland, MA

Day EH, Hua X, Bromham L (2016) Is specialization an evolutionary dead end? Testing for differences in speciation, extinction and trait transition rates across diverse phylogenies of specialists and generalists. J Evol Biol 29(6):1257–1267

Debandi G, Corbalan V, Scolaro JA et al (2012) Predicting the environmental niche of the genus *Phymaturus*: are *palluma* and *patagonicus* groups ecologically differentiated? Austral Ecol 37 (3):392–400
Díaz Gómez JM (2011) Estimating ancestral ranges: testing methods with a clade of Neotropical lizards (Iguania: Liolaemidae). PLoS One 6(10):e26412
Diniz JAF (2004) Phylogenetic diversity and conservation priorities under distinct models of phenotypic evolution. Conserv Biol 18:698–704
Ebel ER, DaCosta JM, Sorenson MD et al (2015) Rapid diversification associated with ecological specialization in Neotropical *Adelpha butterflies*. Mol Ecol 24(10):2392–2405
Espinoza RE, Wiens JJ, Tracy CR (2004) Recurrent evolution of herbivory in small, cold-climate lizards: breaking the ecophysiological rules of reptilian herbivory. Proc Natl Acad Sci U S A 101(48):16819–16824
Esquerré D, Brennan IG, Catullo RA et al (2019) How mountains shape biodiversity: the role of the Andes in biogeography, diversification, and reproductive biology in South America's most species-rich lizard radiation (Squamata: Liolaemidae). Evolution 73(2):214–230
Etheridge RE (1995) Redescription of *Ctenoblepharys adspersa* Tschudi, 1845, and the taxonomy of Liolaeminae (Reptilia: Squamata: Tropiduridae). Am Mus Novit 3142:1–34
Etheridge R, Espinoza RE (2000) Taxonomy of the Liolaeminae (Squamata: Iguania: Tropiduridae) and a semi-annotated bibliography. Smithson Herpetol Inf Serv 126(1):64
Etienne RS, Haegeman B (2012) A conceptual and statistical framework for adaptive radiations with a key role for diversity dependence. Am Nat 180:E75–E89
Faith DP (1992) Conservation evaluation and phylogenetic diversity. Biol Conserv 61:1–10
Faith DP (2018) Phylogenetic diversity and conservation evaluation: perspectives on multiple values, indices, and scales of application. In: Scherson R, Faith DP (eds) Phylogenetic diversity: applications and challenges in biodiversity science. Springer, Cham, pp 1–26
Fontanella FM, Olave M, Avila LJ et al (2012) Molecular dating and diversification of the South American lizard genus *Liolaemus* (subgenus Eulaemus) based on nuclear and mitochondrial DNA sequences. Zool J Linnean Soc 164(4):825–835
Forister ML, Dyer LA, Singer MS et al (2012) Revisiting the evolution of ecological specialization, with emphasis on insect–plant interactions. Ecology 93(5):981–991
Futuyma DJ, Moreno G (1988) The evolution of ecological specialization. Annu Rev Ecol Syst 19 (1):207–233
Garzione CN, Hoke GD, Libarkin JC et al (2008) Rise of the Andes. Science 320(5881):1304–1307
Giambiagi L, Mescua J, Bechis F et al (2016) Cenozoic Orogenic evolution of the southern Central Andes (32–36°S). In: Folguera A (ed) Growth of the southern Andes. Springer, Cham, pp 63–98
Gregory-Wodzicki KM (2000) Uplift history of the central and northern Andes: a review. Geol Soc Am Bull 112(7):1091–1105
Haldane JBS (1951) Everything has a history. Allen & Unwin, London
Harmon LJ, Schulte JA, Larson A et al (2003) Tempo and mode of evolutionary radiation in Iguanian lizards. Science 301(5635):961–964
Harmon LJ, Weir JT, Brock CD et al (2007) GEIGER: investigating evolutionary radiations. Bioinformatics 24(1):129–131
Hazzi NA, Moreno JS, Ortiz-Movliav C et al (2018) Biogeographic regions and events of isolation and diversification of the endemic biota of the tropical Andes. Proc Natl Acad Sci U S A 115 (31):7985–7990
Hellmich WC (1951) On ecotypic and autotypic characters, a contribution to the knowledge of the evolution of the genus *Liolaemus* (Iguanidae). Evolution 5(4):359–369
Hone DW, Benton MJ (2005) The evolution of large size: how does Cope's rule work? Trends Ecol Evol 20(1):4–6
Ibargüengoytía NR, Acosta JC, Boretto JM, Villavicencio HJ, Marinero JA, Krenz JD (2008) Field thermal biology in Phymaturus lizards: comparisons from the Andes to the Patagonian steppe in Argentina. J Arid Environ 72(9):1620–1630

Jansson R, Dynesius M (2002) The fate of clades in a world of recurrent climatic change: Milankovitch oscillations and evolution. Annu Rev Ecol Syst 33(1):741–777

Janz N, Nylin S (2008) The oscillation hypothesis of host-plant range and speciation. In: Tilmon KJ (ed) Specialization, speciation, and radiation: the evolutionary biology of herbivorous insects. University of California Press, Los Angeles, CA, pp 203–215

Jetz W, Thomas GH, Joy JB et al (2012) The global diversity of birds in space and time. Nature 491:444–448

Kassen R (2002) The experimental evolution of specialists, generalists, and the maintenance of diversity. J Evol Biol 15(2):173–190

Kelley ST, Farrell BD (1998) Is specialization a dead end? The phylogeny of host use in *Dendroctonus* bark beetles (Scolytidae). Evolution 52(6):1731–1743

Kling MM, Mishler BD, Thornhill AH et al (2018) Facets of phylodiversity: evolutionary diversification, divergence and survival as conservation targets. Philos Trans R Soc B 374 (1763):20170397

LaBarbera M (1989) Analyzing body size as a factor in ecology and evolution. Annu Rev Ecol Syst 20(1):97–117

Labra A, Pienaar J, Hansen TF (2009) Evolution of thermal physiology in *Liolaemus* lizards: adaptation, phylogenetic inertia, and niche tracking. Am Nat 174(2):204–220

Lanchier N, Neuhauser C (2006) A spatially explicit model for competition among specialists and generalists in a heterogeneous environment. Ann Appl Probab 16(3):1385–1410

Li H, Wiens JJ (2019) Time explains regional richness patterns within clades more often than diversification rates or area. Am Nat 193(4):514–529

Lobo F, Slodki D, Valdecantos S (2010) Two new species of lizards of the *Liolaemus montanus* group (Iguania: Liolaemidae) from the northwestern uplands of Argentina. J Herpetol 44:279–293

Lomolino MV, Riddle BR, Whittaker RJ et al (2010) Biogeography. Sinauer, Sunderland, MA

Losos JB, Ricklefs RE (2009) Adaptation and diversification on islands. Nature 457 (7231):830–836

Lowe WH, Allendorf FW (2010) What can genetics tell us about population connectivity? Mol Ecol 19:3038–3051

Marín AG, Olave M, Avila LJ et al (2018) Evidence of body size and shape stasis driven by selection in Patagonian lizards of the *Phymaturus patagonicus* clade (Squamata: Liolaemini). Mol Phylogenet Evol 129:226–241

Mayr E (2001) What evolution is. Science Masters Series. Basic Books, London

Mittelbach GG, Schemske DW, Cornell HV et al (2007) Evolution and the latitudinal diversity gradient: speciation, extinction and biogeography. Ecol Lett 10:315–331

Moen D, Morlon H (2014) Why does diversification slow down? Trends Ecol Evol 29:190–197

Morando M (2004) Sistemática y filogenia de grupos de especies de los géneros *Phymaturus* y *Liolaemus* (Squamata: Tropiduridae: Liolaeminae) del oeste y sur de Argentina. Unpublished D. Phil. Thesis, Universidad Nacional de Tucumán, Argentina

Moreno Azócar DL, Bonino MF, Perotti MG et al (2016) Effect of body mass and melanism on heat balance in *Liolaemus* lizards of the *goetschi* clade. J Exp Biol 219:1162–1171

Morlon H, Potts MD, Plotkin JB (2010) Inferring the dynamics of diversification: a coalescent approach. PLoS Biol 8(9):e1000493

Myers N, Mittermeier RA, Mittermeier CG et al (2000) Biodiversity hotspots for conservation priorities. Nature 403:853–858

Near TJ, Dornburg A, Kuhn KL et al (2012) Ancient climate change, antifreeze, and the evolutionary diversification of Antarctic fishes. Proc Natl Acad Sci U S A 109:3434–3439

Nee S (2001) Inferring speciation rates from phylogenies. Evolution 55(4):661–668

Nosil P, Mooers AØ (2005) Testing hypotheses about ecological specialization using phylogenetic trees. Evolution 59(10):2256–2263

Olave M, Avila LJ, Sites JW Jr et al (2015) Model-based approach to test hard polytomies in the *Eulaemus* clade of the most diverse south American lizard genus *Liolaemus* (Liolaemini, Squamata). Zool J Linnean Soc 174(1):169–184

Olave M, Avila LJ, Sites JW Jr et al (2017) Hidden diversity within the lizard genus *Liolaemus*: genetic vs morphological divergence in the *L. rothi* complex (Squamata: Liolaeminae). Mol Phylogenet Evol 107:56–63
Olave M, Avila LJ, Sites JW Jr et al (2018) Hybridization could be a common phenomenon within the highly diverse lizard genus *Liolaemus*. J Evol Biol 31(6):893–903
Olave M, Avila LJ, Sites JW Jr et al (2019) How important is it to consider lineage diversification heterogeneity in macroevolutionary studies: lessons from the lizard family Liolaemidae. bioRxiv. https://doi.org/10.1101/563635
Paradis E, Schliep K (2018) Ape 5.0: an environment for modern phylogenetics and evolutionary analyses in R. Bioinformatics 35(3):526–528
Pelegrin N, Bucher EH (2012) Effects of habitat degradation on the lizard assemblage in the arid Chaco, Central Argentina. J Arid Environ 79:13–19
Piantoni C, Ibargüengoytía NR, Cussac VE (2006) Age and growth of the Patagonian lizard *Phymaturus patagonicus*. Amphibia-Reptilia 27(3):385–392
Pietrek AG, Walker RS, Novaro AJ (2009) Susceptibility of lizards to predation under two levels of vegetative cover. J Arid Environ 73(4–5):574–577
Pincheira-Donoso D (2012) Intraspecific predation in the *Liolaemus* lizard radiation: a primer. Anim Biol 62:277–287
Pincheira-Donoso D, Tregenza T, Witt MJ et al (2013) The evolution of viviparity opens opportunities for lizard radiation but drives it into a climatic cul-de-sac. Glob Ecol Biogeogr 22 (7):857–867
Pincheira-Donoso D, Harvey LP, Ruta M (2015) What defines an adaptive radiation? Macroevolutionary diversification dynamics of an exceptionally species-rich continental lizard radiation. BMC Evol Biol 15(1):153
Pinheiro J, Bates D, DebRoy S et al (2012) Nlme: linear and nonlinear mixed effects models. R Package Version 3:1–117
Ponce JF, Rabassa J, Coronato A et al (2011) Palaeogeographical evolution of the Atlantic coast of Pampa and Patagonia from the last glacial maximum to the middle Holocene. Biol J Linn Soc 103(2):363–379
Price SA, Hopkins SS, Smith KK et al (2012) Tempo of trophic evolution and its impact on mammalian diversification. Proc Natl Acad Sci U S A 109(18):7008–7012
Pyron RA (2014) Temperate extinction in squamate reptiles and the roots of latitudinal diversity gradients. Glob Ecol Biogeogr 23(10):1126–1134
Rabassa J (2008) The late Cenozoic of Patagonia and Tierra del Fuego. In: van der Meer JJM (ed) Developments in quaternary sciences, vol 11. Elsevier, Amsterdam
Rabassa J, Coronato AM, Salemme M (2005) Chronology of the late Cenozoic Patagonian glaciations and their correlation with biostratigraphic units of the Pampean region (Argentina). J S Am Earth Sci 20(1–2):81–103
Rabosky DL (2014) Automatic detection of key innovations, rate shifts, and diversity-dependence on phylogenetic trees. PLoS One 9(2):e89543
Rabosky DL, Glor RE (2010) Equilibrium speciation dynamics in a model adaptive radiation of island lizards. Proc Natl Acad Sci U S A 107:22178–22183
Rabosky DL, Lovette IJ (2008) Explosive evolutionary radiations: decreasing speciation or increasing extinction through time? Evolution 62(8):1866–1875
Rabosky DL, Grundler M, Anderson C et al (2014) BAMM tools: an R package for the analysis of evolutionary dynamics on phylogenetic trees. Methods Ecol Evol 5(7):701–707
Rabosky DL, Title PO, Huang H (2015) Minimal effects of latitude on present-day speciation rates in New World birds. Proc R Soc Lond Ser B Biol Sci 282:20142889
Rabosky DL, Chang J, Title PO et al (2018) An inverse latitudinal gradient in speciation rate for marine fishes. Nature 559(7714):392–395
Reaney AM, Saldarriaga-Córdoba M, Pincheira-Donoso D (2018) Macroevolutionary diversification with limited niche disparity in a species-rich lineage of cold-climate lizards. BMC Evol Biol 18(1):16

Ricklefs RE (2007) Estimating diversification rates from phylogenetic information. Trends Ecol Evol 22(11):601–610

Rull V (2018) Neotrophical diversification: historical overview and conceptual insights. PeerJ Preprints e27294v1

Salisbury CL, Seddon N, Cooney CR et al (2012) The latitudinal gradient in dispersal constraints: ecological specialisation drives diversification in tropical birds. Ecol Lett 15(8):847–855

Schluter D (2000) The ecology of adaptive radiation. OUP, Oxford, p 296

Schulte JA, Macey JR, Espinoza RE et al (2000) Phylogenetic relationships in the iguanid lizard genus *Liolaemus*: multiple origins of viviparous reproduction and evidence for recurring Andean vicariance and dispersal. Biol J Linn Soc 69(1):75–102

Silvestro D, Schnitzler J, Zizka G (2011) A Bayesian framework to estimate diversification rates and their variation through time and space. BMC Evol Biol 11:311

Simões M, Breitkreuz L, Alvarado M et al (2016) The evolving theory of evolutionary radiations. Trends Ecol Evol 31(1):27–34

Simpson GG (1953) The Baldwin effect. Evolution 7(2):110–117

Sinervo B, Mendez de la Cruz F, Miles DB et al (2010) Erosion of lizard diversity by climate change and altered thermal niches. Science 328(5980):894–899

Slavenko A, Feldman A, Allison A et al (2019) Global patterns of body size evolution in Squamate reptiles are not driven by climate. Glob Ecol Biogeogr 28(4):471–483

Sookias RB, Butler RJ, Benson RB (2012) Rise of dinosaurs reveals major body-size transitions are driven by passive processes of trait evolution. Proc R Soc Lond B Biol Sci 279 (1736):2180–2187

Soulebeau A, Aubriot X, Gaudeul M et al (2015) The hypothesis of adaptive radiation in evolutionary biology: hard facts about a hazy concept. Org Divers Evol 15(4):747–761

Stellatelli OA, Block C, Vega LE et al (2015) Responses of two sympatric sand lizards to exotic forestations in the coastal dunes of Argentina: some implications for conservation. Wildl Res 41 (6):480–489

Stroud JT, Losos JB (2016) Ecological opportunity and adaptive radiation. Annu Rev Ecol Evol Syst 47:507–532

Suchard MA, Lemey P, Baele G, et al (2018) Bayesian phylogenetic and phylodynamic data integration using BEAST 1.10 Virus Evolution 4, vey016

Templeton AR, Robertson RJ, Brisson J et al (2001) Disrupting evolutionary processes: the effect of habitat fragmentation on collared lizards in the Missouri Ozarks. Proc Natl Acad Sci U S A 98 (10):5426–5432

Thompson KA, Rieseberg LH, Schulter D (2018) Speciation and the city. Trends Ecol Evol 33:815–826

Tulli MJ, Abdala V, Cruz FB (2012) Effects of different substrates on the sprint performance of lizards. J Exp Biol 215(5):774–784

Tulli MJ, Cruz FB, Kohlsdorf T et al (2016) When a general morphology allows many habitat uses. Integr Zool 11(6):483–499

Vamosi JC, Armbruster WS, Renner SS (2014) Evolutionary ecology of specialization: insights from phylogenetic analysis. Proc R Soc B 281:20142004

Vega LE, Bellagamba PJ, Fitzgerald LA (2000) Long-term effects of anthropogenic habitat disturbance on a lizard assemblage inhabiting coastal dunes in Argentina. Can J Zool 78 (9):1653–1660

Vicenzi N, Corbalán V, Miles D et al (2017) Range increment or range detriment? Predicting potential changes in distribution caused by climate change for the endemic high-Andean lizard *Phymaturus palluma*. Biol Conserv 206:151–160

Vidal M, Habit E, González-Gajardo A et al (2010) Thermoregulation and activity pattern of the high-mountain lizard *Phymaturus palluma* (Tropiduridae) in Chile. Fortschr Zool 27(1). https://doi.org/10.1590/S1984-46702010000100003

Vitt LJ, Pianka ER (2014) Lizard ecology: historical and experimental perspectives. Princeton University Press, Princeton, NJ, 403pp

Chapter 29
Diversification History of Neotropical Lecythidaceae, an Ecologically Dominant Tree Family of Amazon Rain Forest

Oscar M. Vargas and Christopher W. Dick

Abstract The Neotropical subfamily of Lecythidaceae (Lecythidoideae) is a clade of 10 genera with an estimated number of 232 species. Lecythidaceae is the third most abundant family of trees in Amazon forests, and its most diverse genus, *Eschweilera* (ca. 100 species) is the most abundant genus of Amazon trees. In this chapter we explore the diversification history of the Lecythidoideae through space and time in the Neotropics. We inferred a time-calibrated phylogeny of 118 species, which we used to reconstruct the biogeographic origins of Lecythidoideae and its main clades. To test for significant changes of speciation rates in the subfamily, we performed a diversification analysis. Our analysis dated the crown clade of Lecythidoideae at 46 Ma (95% CI = 36.5–55.9 Ma) and the stem age at 62.7 Ma (95% CI = 56.7–68.9 Ma), suggesting dispersal from the paleotropics long after the Gondwana breakup. Most major crown clades in the Lecythidoideae (*Grias*, *Gustavia*, *Eschweilera*, *Couroupita*, *Couratari*, and all *Lecythis* and *Eschweilera* subclades) differentiated during the Miocene (ca. 5.3–23 Ma). The Guayana floristic region (Guiana Shield + north-central Amazon) is the inferred ancestral range for 8 out of the 18 Lecythidoideae clades (129 species, ~55%), highlighting the region's evolutionary importance, especially for the species-rich Bertholletia clade, which includes the genera *Eschweilera*, *Lecythis*, *Corythophora* and *Bertholletia*. Our results indicate that the Bertholletia clade colonized the Trans-Andean region at least three times in the last 10 Ma. We found no significant changes in the rate of diversification inside Lecythidoideae over the Cenozoic, and found no evidence of increased speciation during the Pleistocene. Lecythidoideae has diversified not in

O. M. Vargas (✉)
Department of Ecology and Evolutionary Biology and Herbarium, University of Michigan, Ann Arbor, MI, USA

Department of Ecology and Evolutionary Biology, University of California, Santa Cruz, CA, USA

C. W. Dick (✉)
Department of Ecology and Evolutionary Biology and Herbarium, University of Michigan, Ann Arbor, MI, USA

Smithsonian Tropical Research Institute, Panama City, Republic of Panama
e-mail: cwdick@umich.edu

V. Rull, A. C. Carnaval (eds.), *Neotropical Diversification: Patterns and Processes*, Fascinating Life Sciences, https://doi.org/10.1007/978-3-030-31167-4_29

pulses, but in a pattern of steady accumulation, akin to a museum model of diversification.

Keywords Ericales · Brazil nut · Speciation · Boreotropics · Long-distance dispersal · Phylogeny

1 Introduction

Lecythidaceae is an important family of woody plants in the Amazon forests, both in terms of species richness and abundance. As such, its diversification history can greatly illuminate our understanding of the assembly of biological communities in the Neotropics. Lecythidaceae is a pantropical family of trees in the order Ericales often referred to include either five (APG IV 2016) or three subfamilies (Huang et al. 2015). In the latter classification, Napoleonaeoideae and Scytopetaloideae are excluded from Lecythidaceae, leaving three core Lecythidaceae subfamilies Foetidioideae (in Madagascar), Planchonioideae (in Asia and Africa), and the Lecythidoideae, which is restricted to the Neotropics (Mori et al. 2017). The Lecythidoideae contains ca. 232 (Mori 2017) of the ca. 278 known species in the family (Mori et al. 2007, 2017; Huang et al. 2015; Mori 2017). An enigmatic species, *Asteranthos brasiliensis* Desf., is the single Neotropical representative of the West African family Scytopetalaceae, which is sometimes included in the Lecythidaceae (subfamily Scytopetaloideae; Mori et al. 2017).

Neotropical Lecythidaceae (excluding *A. brasiliensis*, and henceforth called "Lecythidoideae") are understory, canopy, or emergent trees with distinctive floral morphology, fibrous bark, and woody fruit capsules. Although the subfamily is distributed from Mexico to Southern Brazil, its epicenter of species diversity is Amazonia. Lecythidaceae is the third most abundant family of trees in Amazon forests, ranking only behind Fabaceae and Sapotaceae (ter Steege et al. 2013). The most species-rich genus, *Eschweilera* with ca. 100 species (Mori 2017), is also the most abundant Amazonian tree genus (ter Steege et al. 2013). The canopy species *Eschweilera coriacea* is the most common tree in much of Amazonia, contributing substantially to the biomass and carbon sequestration of Amazon forests (ter Steege et al. 2013). Other notable Lecythidaceae include the Brazil nut tree, *Bertholletia excelsa*, which has been a keystone food source for Amazon peoples for millennia (Shepherd and Ramirez 2011); *Cariniana micrantha*, the monkey nut tree (tauarí), which is the oldest documented Amazon tree (carbon dated at >1400 years; Chambers et al. 1998); and *Couroupita guianensis* (cannonball tree), a cauliflorous tree with cannonball-sized fruits that is grown in tropical botanical gardens around the world. Lecythidaceae includes important timber species, such as *Cariniana legalis*, known from Brazil's Atlantic Forest. Most of the species are found in lowland rain forest, although some species are adapted to tropical montane forest, flooded forest (igapó and várzea), and savanna. Bees are the main pollinators for Lecythidoideae, with some taxa showing specialization toward carpenter and euglossine bees (Huang

2010) and even bats (Prance and Mori 1979). Fruits in Lecythidoideae are woody and can be dehiscent or indehiscent. Although there is much to be learned about seed dispersal in the family, preliminary studies and field observations have suggested that water (e.g. some species of *Allantoma*), wind (*Cariniana* and *Couratari*), parrots and macaws (some *Eschweilera* spp.), fish (*Gustavia* spp.), rodents (including agoutis, which gnaw through the Brazil nut fruit wall), monkeys (some *Lecythis* spp.), and bats (*Lecythis* spp.) act as dispersal agents (Prance and Mori 1979; Mori and Prance 1990).

Phylogenetic analyses of Lecythidaceae, using chloroplast markers and the internal transcribed spacer (ITS), identified the Neotropical subfamily Lecythidoideae as monophyletic and sister to the Paleotropical clade Foetidioideae + Planchonioideae (Mori et al. 2007). These analyses also identified the ecologically important and diverse Bertholletia clade, comprised of four genera with zygomorphic flowers (*Eschweilera*, *Lecythis*, *Corythophora*, and *Bertholletia*) (Mori et al. 2007; Huang et al. 2015). The Bertholletia clade is nested within a paraphyletic grade of actinomorphic-flowered (*Grias*, *Gustavia*, and *Allantoma*) and zygomorphic-flowered genera (*Cariniana*, *Couroupita*, and *Couratari*; Huang et al. 2015; Mori et al. 2017). The two most species-rich genera in the Bertholletia clade, *Lecythis* and *Eschweilera*, are not monophyletic in these phylogenetic reconstructions. These groups may be paraphyletic. Alternatively, the lack of monophyly may result from poor resolution of deeper nodes using ITS and a handful of chloroplast markers.

Given that Lecythidoideae is nested in a grade comprising predominantly Paleotropical clades with actinomorphic flowers (Foetidioideae, Planchonioideae, Napoleonaceae and Scytopetalaceae), the leading hypotheses to explain the dispersal of Lecythidoideae from the Paleotropics are (1) a vicariance event after the breakup of Gondwana, (2) migration using a high latitude land bridge(s) (e.g. North Atlantic land bridge or Beringia, Graham 2018), and (3) a long-distance dispersal event across the Atlantic (Mori et al. 2017). In a phylogenetic analysis of Ericales, the crown and stem ages of Lecythidoideae have been estimated to be 43 Ma and 53 Ma, respectively, with Lecythidoideae sister to Southeast Asian Planchonioideae (Rose et al. 2018). This time frame post-dates the breakup of Gondwana (90–100 Ma) by tens of millions of years, supporting a land-bridge migration or a long-distance dispersal.

After its establishment in the Neotropics, Lecythidoideae evolved into a speciose clade with ca. 232 species. This species richness has been partially attributed to Pleistocene refugia (Haffer 1969) in *Cariniana*, *Couratari*, and *Eschweilera* (Prance 1973, 1978), but a holistic investigation about Lecythidoideae diversification is missing. Considering the Eocene arrival of the family to the Neotropics (Rose et al. 2018), it is also possible that Lecythidoideae have accumulated species steadily akin to a museum model (Stebbins 1974).

In this chapter, we present the first chronogram and ancestral range reconstructions for the Lecythidoideae. Our phylogenetic inference combines published data (Mori et al. 2007; Huang et al. 2015) and DNA sequences of 13 highly informative regions extracted from 26 plastomes (Thomson et al. 2018; Yu et al. 2017). Our main goals were to (1) identify the geographic origins and divergence times of the primary clades of Lecythidoideae, (2) address the hypotheses (land-bridge migration *vs.* oceanic dispersal) for the origin of Lecythidaceae in the Neotropics, and (3) test for heterogeneity in diversification rates across the tree to address the Pleistocene refuge *vs.* the museum hypotheses.

2 Methods

2.1 Tree Building

Lecythidaceae genetic data were obtained from GenBank using PYPHLAWD (Smith and Brown 2018). We focused on the Lecythidoideae and the outgroup genus *Barringtonia*, which was identified as a close relative (Mori et al. 2007). We retained DNA markers represented in >50 species (i.e. ITS, *ndhF*, *psbA-trnH*, and *trnL-F*) and species with least two DNA regions (we noted inconsistencies in the placement of species represented by only one marker). We combined the PYPHLAWD-generated matrix with a matrix containing 13 plastome regions identified by Thomson et al. (2018) as phylogenetically informative for the Lecythidoideae. The 13-region matrix was extracted from an alignment of 26 plastomes that contained all Lecythidoideae genera and three *Barringtonia* species (Thomson et al. 2018; Yu et al. 2017). Our strategy of combining a species-rich matrix (PYPHLAWD-generated) with a sequence-rich one (13-region) aimed to create a comprehensive Lecythidoideae tree with a strong backbone. The combined matrix contained a total of 118 species (including 8 outgroups) and 16 markers (PYPHLAWD-mined: ITS, *ndhF*, and *trnL-F*; 13 highly-informative plastome regions: *petN-trnD*, *psaJ-rps18*, *psbM-trnD*, *psbZ-trnfM*, *rps15-ycf1*, *trnE-trnT*, *trnfM-psbA*, *trnK-rps16*, *trnT-psbD*, *trnV-atpE*, *ycf1*(1), *ycf1*(2); *psbA-trnH* was shared by both marker-sets). We used MAFFT v7.310 (Katoh and Standley 2013) to align individual regions before concatenation and searched for the best maximum likelihood (ML) tree in RAxML v. 8.2.11 (Stamatakis 2014) using the option "-f a", which performs rapid bootstraps (200) and then searches for the best-scoring ML topology. Every region was treated by RAxML as an independent partition using a GTRGAMMA model of molecular evolution (Stamatakis 2015). The best-scoring ML tree was subsequently time-calibrated using BEAST v.2.5 (Bouckaert et al. 2014). We employed two fossils to calibrate our phylogeny: (1) a seed fossil, *Lecythidospermum* (Pons and Rica 1983), assigned to the crown node containing *Allantoma*, *Cariniana*, *Couratari*, and the Bertholletia clade, with a lognormal prior distribution with a median of 24.3 Ma and a 95% confidence interval (CI) of 20.0–29.2 Ma; (2) a wood fossil, *Barringtonioxylon deccanense* (Shallom 1960; Srivastava et al. 2009), to the stem node of *Barringtonia*, with a lognormal prior distribution with a median of 66.0 Ma and a 95% CI of 60.3–72.0 Ma. We ran three independent BEAST analyses of ten million generation sampling every 4000; these results were combined to calculate a chronogram using a 0.25 burnin fraction in LogCombiner v.2.5 and TreeAnnotator v.2.5 (Bouckaert et al. 2014). We checked for a minimum effective sample size >200 for the estimated parameters with Tracer v.1.6 (Rambaut et al. 2018).

2.2 Ancestral Range Reconstruction

To infer the geographic history of Lecythidoideae in the Neotropics and identify areas of importance for its diversification, we performed an ancestral range reconstruction on the chronogram, using a dispersal-extinction-cladogenesis model (DEC, Ree and Smith 2008) in BioGeoBEARS (Matzke 2013). Given a phylogeny and the presence or absence of extant taxa in a set of geographical regions, DEC infers the geographical ranges of internal nodes by modeling dispersal, extinction, and cladogenesis (speciation). We did not include a founder speciation event parameter "J" (Matzke 2014), given the caveats noted by Ree and Sanmartín (2018). Although the models DIVA (Ronquist 1997) and BayArea (Landis et al. 2013) are also implemented in BioGeoBEARS and typically run in conjunction with DEC, we opted not employ them in our analysis because DIVA does not model some cladogenetic processes (i.g. parapatric speciation; Kodandaramaiah 2010)—a process expected to have occurred in the Lecythidoideae given the numerous species found in Amazon forests—and BayArea is optimized for taxa distributed on numerous geographic regions like islands systems (Landis et al. 2013)—Lecythidoideae are found in contiguous low and middle elevations forests throughout tropical Central and South America.

We divided the Neotropical region in eight areas, based on a previous bioregionalization (Morrone 2014; Löwenberg-Neto 2014) modified in QGIS v2.18 (QGIS Development Team 2005). Presence of species in our regions was codified from a database of specimens that included data from GBIF (https://www.gbif.org) and the New York Botanical Garden. Duplicates and unlikely data points (e.g. maritime areas) were filtered out.

2.3 Diversification Analyses

We performed an analysis of diversification on our chronogram using BAMM v2.5 (Rabosky 2014). BAMM uses reversible-jump Markov chain Monte Carlo to infer diversification processes using phylogenies. We ran BAMM with four chains for ten million generations, sampling every 1000, with a sampling correction that considers missing taxa in the phylogeny, and based on the number of species sampled in each one of the main Lecythidoideae clades described in Mori et al. (2017) (Table 29.1). We enforced an effective sample size >200 and used BAMMtools (Rabosky et al. 2014) to overlay speciation rates onto our phylogeny and to graph average speciation rates over time.

Table 29.1 Crown clade ages, with 95% confidence interval (CI), most probable origin with its probability, speciation rates (Lambda), and percentage sampled for main Lecythidoideae clades

Clade/*genus*	Age	95% CI	Most probable origin	Prob.	Species	Lambda	Sampled (%)
Lecythidoideae	46.1	36.5–55.9	W Amazonia	0.036	232	0.159	47
Grias	7.0	2.4–12.8	Transandean, Andean, N&W Amazonia	0.239	11	0.156	27
Gustavia	20.4	11.2–30.9	Transandean	0.109	45	0.163	20
Couroupita	8.8	3.9–14.2	Transandean, N&W Amazonia, Guayana	0.106	3	0.149	100
Allantoma	4.8	2.1–8.0	Amazonia	0.302	8	0.151	38
Cariniana	23.2	15.3–30.2	W Amazonia	0.710	9	0.151	33
Echinata (*Couratari*)	17.2	9.8–25.7	Transandean, W Amazonia	0.187	6	0.152	50
Guianensis (*Couratari*)	13.3	6.8–20.5	Guayana	0.319	13	0.151	31
Bertholletia clade	28.6	22.4–34.7	Guayana	0.822	136	0.163	62
Chartacea (*Lecythis*)	16.3	11.8–21.5	Guayana	0.956	14	0.154	71
Integrifolia (*Eschweilera*)	12.2	7.8–16.5	Guayana	0.165	22	0.156	77
Tetrapetala (*Eschweilera*)	11.4	6.3–17.2	Cerrado & Caatinga	0.337	8	0.152	38
Ollaria (*Lecythis*)	13.8	8.5–19.3	Transandean, N Amazonia	0.464	3	0.153	100
Poiteaui (*Lecythis*)	17.0	11.2–22.8	Guayana	0.799	6	0.153	83
Pisonis (*Lecythis*)	9.0	5.2–13.2	Amazonia, Cerrado & Caatinga, Mata Atlantica	0.106	5	0.154	80
Corythophora	12.5	7.8–17.5	Guayana	0.990	4	0.154	100
Corrugata (*Lecythis*)	8.2	4.6–12.2	Guayana	0.856	5	0.164	100
Parvifolia (*Eschweilera*)	13.6	10.2–17.1	Guayana	0.600	65	0.186	46

All GenBank accessions, DNA alignments, locality data, bioregionalization map, control files, scripts, and additional results and figures used for or generated by this study can be found at https://bitbucket.org/oscarvargash/lecythidaceae_diversification

3 Results

Our matrix comprised 118 species and 16 DNA regions, containing 12,726 aligned nucleotides. Overall matrix gene-occupancy was 39%, with ITS = 84%, *ndhF* = 97%, *psbA-trnH* = 83%, *trnL-F* = 95%, and the remaining 12 highly-informative regions = 22%. Nearly half (48%) of the nodes in our ML phylogeny had high ($\geq$80%) bootstrap support (BS), while 14% had moderate BS (60–79) and 38% had low BS (<60). Our tree had more supported nodes than the most recently published comprehensive molecular tree for the Lecythidoideae (Huang et al. 2015) which contained 24% nodes with high BS. Our topology largely agrees with that of Huang et al. (2015) and recovers all of the clades detected by their study: Bertholletia clade (*Eschweilera* + *Lecythis* + *Corythophora* + *Bertholletia*); the *Lecythis* clades Ollaria, Pisonis, Corrugata, Poiteaui, and Chartacea; the *Eschweilera* clades Integrifolia, Tetrapetala, and Parvifolia; *Corythophora*, and the species *Bertholletia excelsa* (Fig. 29.1). There are three main differences in our tree when compared against previous phylogenies (Mori et al. 2007; Huang et al. 2015). First, our phylogeny recovers *Couratari* as polyphyletic (with moderate BS), while previous studies recovered it as monophyletic. Second, our phylogeny suggests that *Bertholletia excelsa* is sister to the Chartacea clade and the Integrifolia clade, while the phylogeny of Huang et al. (2015) placed it as sister of the Integrifolia clade. Finally, in our tree, *Eschweilera amazoniciformis* is sister to a clade comprising the Pisonis clade, *Corythophora*, the Corrugata clade, and the Parvifolia clade, while in previous phylogenies this species was inferred to be nested within the Integrifolia clade (Huang et al. 2015). The positions of *Bertholletia excelsa* and *Eschweilera amazoniciformis*, however, have low support.

When comparing prior vs. posterior distributions for the nodes calibrated with our Bayesian analysis, we found no difference for the node calibrated with *Barringtonioxylum*, while the node dated with *Lecythidospermum* showed a posterior distribution slightly shifted towards older ages. The latter suggests that the sequence data used in this study provide information for the estimation of a posterior estimate of *Lecythidospermum*. The chronogram dates the crown clade for the Lecythidoideae at 46 Ma (95% CI = 36.5–56 Ma) (Fig. 29.1, Table 29.1) and its stem age at 62.71 Ma (95% CI = 56.7–68.92 Ma). Bertholletia clade's stem and crown ages date back to the Oligocene. In the Bertholletia clade, stem and crown ages of main clades (e.g. Parvifolia, Chartaceae, Huang et al. 2015) fall within the Miocene, with only the stem age of the Tetrapetala clade falling within the Oligocene.

Our biogeographic analysis shows a preliminary historical range reconstruction for Lecythidoideae (Fig. 29.2, Table 29.1). While many of the range reconstructions for early ancestors in the subfamily are ambiguous (e.g. the ancestor for the Lecythidoideae), the reconstruction for many ancestors in the Bertholletia clade is dominated by the Guayana area. Furthermore, Guayana is reconstructed as the most probable ancestral region for eight diverse clades (comprising ~129 species), suggesting it as the most important geographic region for neotropical diversification in the subfamily. Despite ambiguous reconstructions for ancestors of *Grias* and

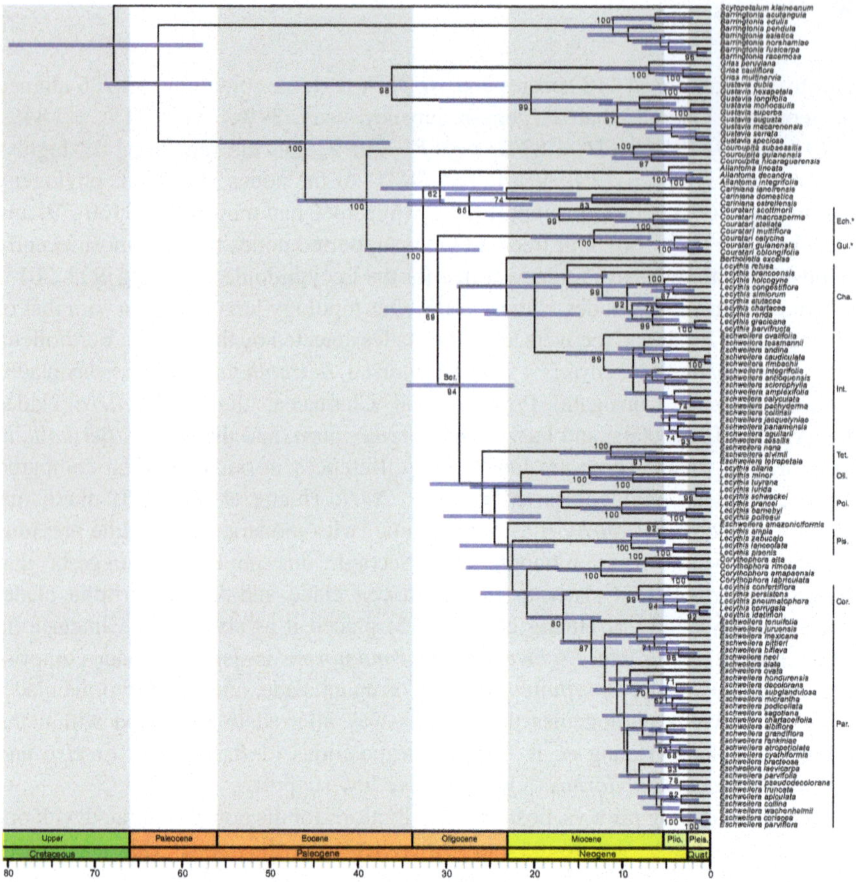

Fig. 29.1 Lecythidoideae chronogram. Blue bars at nodes indicate the 95% confidence intervals for the age of a given node. Numbers at nodes indicate bootstrap support, not shown for values <60. Main clades are indicated as: Ber. = Bertholletia, Ech. = Echinata, Gui. = Guianensis, Cha. = Chartacea, Int. = Integrifolia, Tet. = Tetrapetala, Oll. = Ollaria, Poi. = Poiteaui, Pis. = Pisonis, Cor. = Corrugata, Par. = Parvifolia. Stars indicate newly defined clades in this study

Gustavia, the Trans-Andean region seems central to their diversification (Fig. 29.2, Table 29.1). Similarly, the Western Amazon area seems to have played an important role as the setting for diversification of the clade comprised by *Allantoma* + Echinata (*Couratari*) clade + *Cariniana*. Finally, our reconstruction also suggests at least three instances of dispersal into the Trans-Andean region within the Bertholletia clade, whose center of diversification appears to be the Guayana region, in the last 10 Ma.

We found no evidence for shifts in the diversification dynamics of the Lecythidoideae through the Cenozoic: a scenario of no significant changes in diversification rates was found to have the highest posterior probability (0.62). The

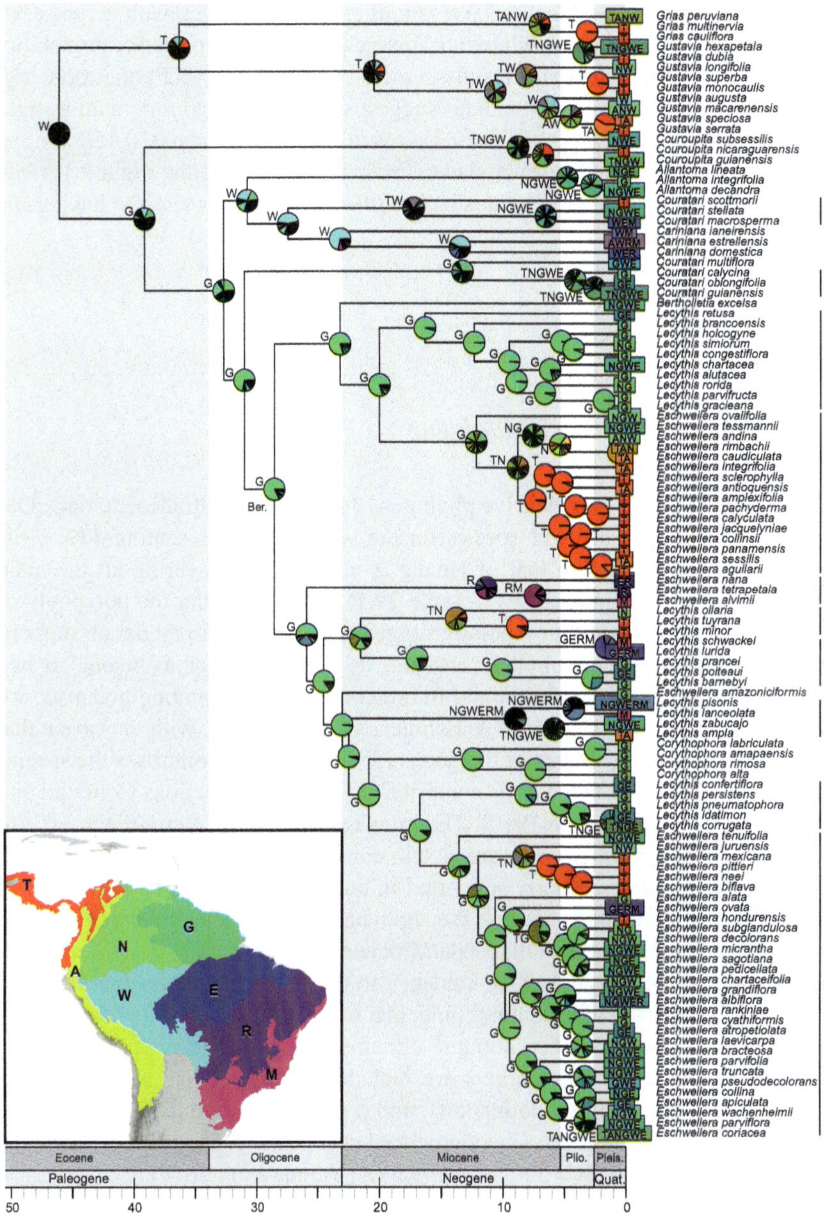

Fig. 29.2 Biogeographic ancestral reconstruction for Lecythidoideae. Percent probabilities are shown as a pie charts, the most probable range is indicated on top of the pie chart. Notice that ancestral ranges comprising multiple areas are represented by a combination of colors/letters of the individual areas. Main clades are indicated as in Fig. 29.1

next probable scenario is that of one significant change, yet with a posterior probability of 0.27; scenarios with two or more changes have a posterior probability of <0.09 (see additional BAMM results in the on-line repository). Furthermore, our estimations of speciation rates (lambda) suggest stability through time with a slight increase towards the present (Fig. 29.3); overall mean lambda is 0.156 for the Lecythidoideae with the Parvifolia clade (*Eschweilera*) having the highest lambda of 0.186 (this higher rate is not statistically different), and *Couroupita* having the lowest lambda of 0.149 (Table 29.1).

4 Discussion

4.1 Systematics of Lecythidoideae

We present the most comprehensive phylogeny for the Lecythidoideae to date. Our phylogeny, which contains 111 species for the subfamily (representing 47% of its species), largely agrees with that of Huang et al. (2015), recovering all the main clades described by them (Fig. 29.1, Table 29.1), and supporting the polyphyly of *Eschweilera* and *Lecythis*. Our topology suggests, however, with moderate support, that *Couratari* is not monophyletic; instead, its species appear to belong to two clades. To facilitate the discussion and future communication among botanists, we named the two clades of *Couratari* as Echinata and Guianensis, with the caveat that we only sampled seven species of the genus (37%). Echinata comprises the species of section Echinata, and Guianensis comprises species of the sections Couratari and Microcarpa (Mori and Prance 1990). The inferred positions of *Bertholletia excelsa* and *Eschweilera amazoniciformis* are also in conflict with the phylogeny of Huang et al. (2015), but they are poorly supported in both phylogenies. The placement of these two taxa is problematic because they both bear long branches and seem to be of old age (both dating from the Oligocene/Miocene boundary).

While our phylogeny adds more evidence to the polyphyly of *Eschweilera* and *Lecythis* and suggests for the first time the non-monophyly of *Couratari*, we maintain a conservative position about re-circumscribing these genera until a larger number of unlinked nuclear markers are included in the analysis. Our matrix is primarily composed of plastome markers, and preliminary phylogenomic evidence for Lecythidoideae suggest deep incongruence between the plastome phylogeny and a species tree inferred with >300 nuclear genes (Vargas et al. 2019).

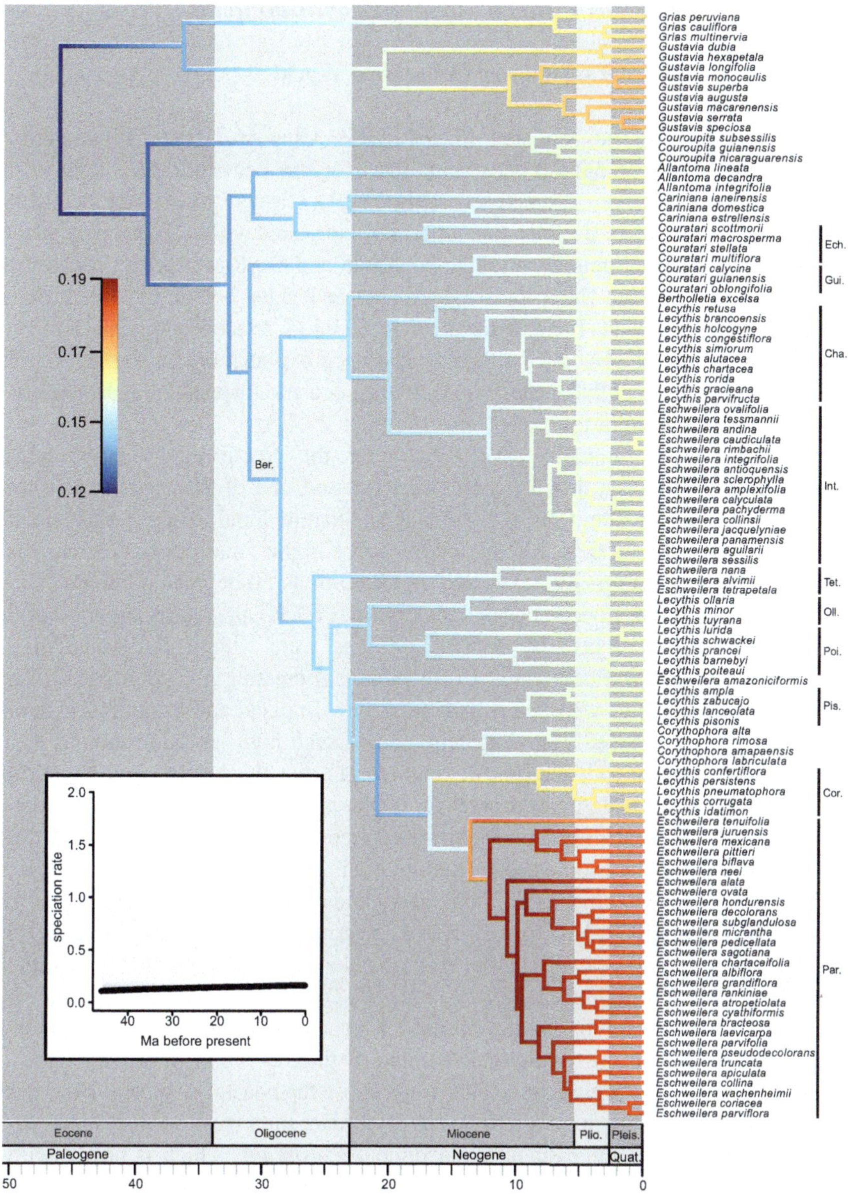

Fig. 29.3 Model-averaged speciation rates plotted on the Lecythidoideae phylogeny and speciation rates through time (inset). Main clades are indicated as in Fig. 29.1

4.2 *Spatiotemporal Patterns in the Lecythidoideae*

4.2.1 Dispersal of Lecythidaceae to the New World

Our chronogram suggests that stem (62.7 Ma 95% CI = 56.7–68.9 Ma) and crown (46 Ma 95% CI = 36.5–55.9 Ma) clade ages for the Lecythidoideae date to the Paleocene and Eocene, respectively (Fig. 29.1), firmly within the Cenozoic (i.e. within past 66 Ma). Because Lecythidoideae is nested within Neotropical clades (Foetidioideae, Planchonioideae, Napoleonaceae and Scytopetalaceae; Mori et al. 2017), we can infer that Neotropical Lecythidaceae has its roots in the Paleotropics. Given that the Gondwana breakup occurred in the mid-Cretaceous (e.g. 80–100 Ma), our age estimates rule out a Gondwana vicariance explanation for the pantropical disjunction of the Lecythidaceae, supporting an oceanic dispersal (Fig. 29.4a) or a land-bridge migration in the early Cenozoic (Fig. 29.4b).

The land-bridge scenario (Fig. 29.4b) for the Neotropical distribution of Lecythidaceae proposes that early members crossed one of the high latitude land bridges (e.g. Bering Land Bridge, or North Atlantic Land Bridge) early in the Cenozoic, when Earth surface temperatures were higher, and cold-intolerant thermophilic ("tropical") lineages extended as far north as 50 degrees (Graham 2011). During the Paleocene-Eocene thermal maximum (PETM), the warmest period of the Cenozoic, some plant taxa of tropical affinities were widely distributed in the Northern hemisphere (Eurasia and North America) creating a boreotropical distribution—palms and figs were left as fossils in New England and London. Lavin and Luckow (1993) proposed a phylogenetic test of such high latitude (boreotropical) dispersal for Neotropical lineages. It requires (1) that the Neotropical lineage be nested within a tropical Asian or African clade, and (2) existence of high latitude fossil records. Several important Neotropical tree clades are shown to meet these criteria, including Melastomataceae (Renner et al. 2001), Burseraceae (Weeks et al. 2005), Meliaceae (Muellner et al. 2006), and some clades of Lauraceae (Chanderbali and van der Werff 2001; Huang et al. 2016) and Annonaceae (Couvreur et al. 2011). Land-bridge migration can also occur from the Neotropics to the Paleotropics: there is evidence that Malpighiaceae (Davis et al. 2002) originated in the Neotropics but obtained its pantropical distribution through boreotropical or "Laurasian" dispersal.

Because there are no fossils of either the distinctive wood or woody fruits of Lecythidaceae in any high latitude site, we propose the oceanic dispersal hypothesis as the most plausible to explain the Lecythidaceae arrival to the Neotropics (Fig. 29.4a). As with the disjunct Amazonian *Asteranthos*, which is nested within west African Scytopetalaceae, the ancestor of many Amazon tree lineages likely arrived by oceanic dispersal (Pennington and Dick 2004). This conclusion stands in contrast to that of Rose et al. (2018), who estimated similar crown/stem ages for Lecythidoideae but accepted a Gondwana vicariance explanation. We believe that the disagreements between our study and that of Rose et al. (2018) are based on the evidence and analyses taken into consideration to draw interpretations. While we take fossil evidence (absence of Lecythidaceae in North America) and

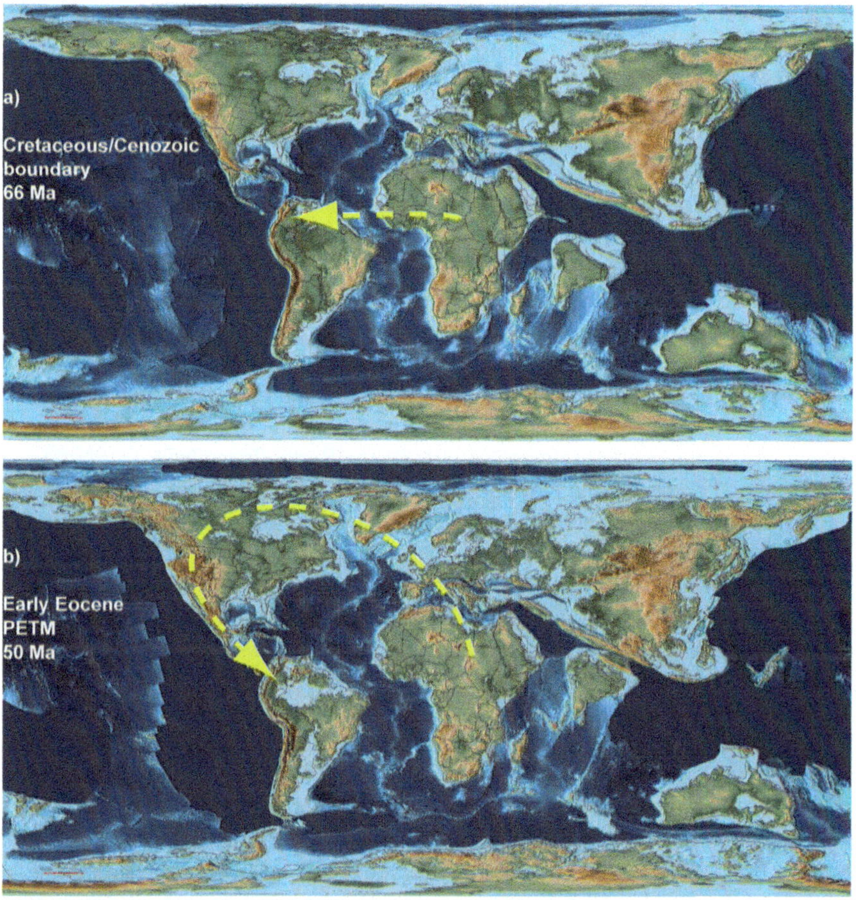

Fig. 29.4 Competing hypothesis about the origin of the Lecythidoideae. Arrows represent colonization routes. (**a**) Long dispersal hypothesis (supported by the present study) represented on the Cretaceous-Tertiary boundary earth tectonic reconstruction 66 Ma, which is close to the Lecythidoideae stem age 62.7 Ma (95% confidence interval = 56.7–68.9 Ma). (**b**) Boreotropical colonization hypothesis represented on the Earth tectonic reconstruction during the Paleocene-Eocene thermal maximum 50 Ma. Figures modified from Scotese (2001)

paleogeography into consideration (stem and crown Lecythidoideae ages postdate the break up Gondwana), the conclusions of Rose et al. appear to be based solely on an ancestral biogeographic reconstruction that suggests a vicariance event, in which the parental node of the Neotropical Lecythidoideae, Paleotropical Foetidioideae, and Paleotropical Planchonioideae is reconstructed as being distributed in both the Neotropics and the Afrotropics with a moderate probability of 0.61 (Rose et al. 2018, p. 67).

4.2.2 Diversification Within the Neotropics

Mori et al. (2017) suggested that the first Lecythidaceae arrived in Central America and first diverged into the *Gustavia* + *Grias* clade. While our phylogeny supports the *Gustavia* + *Grias* clade as sister to the rest of Lecythidoideae, we cannot pinpoint the arrival location to the Neotropics from the Paleotropics due to the ambiguity of our reconstruction for the deepest nodes in the subfamily (Fig. 29.2).

Our analyses suggest that the Bertholletia clade diversified in the transition of the Oligocene to the Miocene at 28.6 Ma (95%CI = 22.4–34.7), with a high probability of having originated in the Guayana region (0.82) (Fig. 29.1, Table 29.1). All the clades that make up the Bertholletia clade *sensu* Huang et al. (2015) (e.g. Chartacea, Parvifolia) appear to have diverged during the Miocene, with six clades having Guayana as the most probable center of origin (Chartacea, Integrifolia, Poiteaui, Corythophora, Guianensis, and Parvifolia). These results suggest that the Guayana region has been a center of diversification for the Bertholletia clade.

We hypothesize that the elevated terrain of the Guayana shield, which predates the origin of the Lecythidaceae, has provided constant non-flooded *terra firme* for the family to diversify over time. In contrast to Guayana, Northern Amazonia and Western Amazonia (following our bioregionalization, Fig. 29.2) experienced marine incursions and wetland systems during the Miocene (Hoorn et al. 2010), making these areas improbable for Lecythidaceae speciation. It has been documented that Lecythidoideae is more abundant in *terra-firme*, with near 70% of the Neotropical species inhabiting non-flooded forest (Mori et al. 2017). The preference for *terra-firme* predicts that other areas, i.e. the proto-Andean Cordillera, which could have sustained *terra-firme* forest in the Eocene, could also be a center of diversification for the family. Estimates indicate that the proto-Andean cordilleras began their uplift during the Cretaceous (Gregory-Wodzicki 2000), likely providing low elevation mountains (<1000 m) but with enough elevation for a *terra-firme* forest during the Eocene—when the Lecythidoideae started to diversify. In fact, current distributions of *Grias*, *Gustavia*, and the Integrifolia clade show higher concentration of species in the hills or in the surroundings of the Andes Cordillera (Mori et al. 2017). There is little support for the proto-Andean hypothesis in our biogeographic analysis (Fig. 29.2, Table 29.1), but the absence of the Andean region in our ancestral reconstruction could be an artifact of the (perhaps too stringent) threshold used in our analysis to score Andean species (>1000 m), and our biogeographic modeling failing to include Andean paleoelevation estimates that could allow for an appropriate inference of the distribution of ancestors in the early Andean Cordillera.

4.3 Diversification Rates

Our estimations of diversification rates in the Lecythidoideae shows that the subfamily has diversified relatively constantly over time, with a slight tendency to increase speciation rates towards the present (Fig. 29.3). Our analysis does not show significant increases of speciation rates for the Pleistocene (inset Fig. 29.3), as in a refuge hypothesis (Haffer 1969), which was previously a favored explanation for the sympatric occurrence of closely related *Eschweilera* species in central Amazon forests (Prance 1978). Instead, our results favor a museum hypothesis (Stebbins 1974) with most nodes (stem and crown) leading to the main Lecythidoideae crown clades being older than 10 Ma. Nevertheless, there are species that diverged recently during the Pleistocene (e.g. some species of *Lecythis* and *Eschweilera*) and our phylogeny is missing 53% of Lecythidoideae species. Beyond these classical hypotheses to explain the Amazon biodiversity, our results show that Lecythidaceae arrived to the Amazon forests after its formation (Graham 2011) and that Lecythidoideae speciated at a steady rate until the present. While our results support the museum hypothesis, we believe there is much to learn from Amazon speciation beyond binary categorizations (i.e. cradle vs. museum). In particular, we believe that causes for speciation are pivotal for understanding biological diversification in the Amazon region. Lineage splitting processes remain a mystery and should be studied to, for example, explain the presence of numerous closely-related Lecythidaceae species in the Guayana floristic region.

4.4 Caveats of Our Study

Despite the fact that we presented the most comprehensive phylogeny of Lecythidoideae to date, we sampled only 47% of its current diversity, with the lowest sampled genera being *Grias* and *Gustavia* with 27% and 20%, respectively. The gaps in our sampling, specifically in *Grias* and *Gustavia*, affected our biogeographic reconstruction (inconclusive for some of the ancestors of these two sister taxa) and diversification analysis; a denser taxon sampling may increase diversification rates in main clades especially towards the Pleistocene. Sampling efforts are unequal across the Amazon basin as major Lecythidaceae collectors, Mori and Prance, focused primarily in the central Amazon around Manaus and the Guayana region. It is possible that these collections biased our sampling and our estimation of distributional ranges for our biogeographic analysis.

Our dating analysis is based solely on two fossils, with only one of them, *Lecythidospermum*, located in the ingroup. Our diversification analysis is based solely on tree shape and no fossils were used to estimate diversification rates. While additional New World fossils have been reported (Mori et al. 2017), these need to be revised in terms of age or taxonomic identity for accurate utilization in the Lecythidoideae phylogeny (C. Martinez, pers. comm.).

Finally, our model for reconstructing ancestral ranges does not consider historical geology and climate, meaning that our reconstruction assumes no change in landscape and forest coverage through time. Northwestern South America was considerably different before the middle Miocene because major uplift of the Central and Northern Andes occurred in the last 10 Ma (Gregory-Wodzicki 2000), meaning that

Andean paleoelevation during the Eocene was much lower. Additionally, the isthmus of Panama remained opened until at least 15 Ma (Montes et al. 2015) and it is known that wetlands existed during the Miocene in western Amazonia. These historical landscape changes, not accounted in this study, most certainly affected the coverage of Amazon forests and hence the past distribution of the Lecythidoideae.

4.5 Conclusions and Future Directions

Based on the most comprehensive Lecythidoideae phylogeny to date, we inferred spatiotemporal diversification patterns of this subfamily during the Cenozoic. Our results support an oceanic dispersal event from Africa as early as the late Cretaceous and early divergence of the Lecythidoideae during the Eocene. We found that its most important center of diversification is the Guayana region, which was likely the origin of the Bertholletia clade that includes the most speciose genera in the family, *Eschweilera* and *Lecythis*. Since its arrival to the Neotropics, Lecythidoideae has speciated at a steady rate, with a slight but non-significant increase in speciation rates towards the present. To corroborate some of new hypotheses stated in this study it will be necessary to increase the number of taxa included in the phylogenetic reconstruction of the subfamily, especially for the non-Bertholletia genera *Allantoma*, *Couratari*, *Grias*, and *Gustavia*. The inclusion of nuclear data will be pivotal in obtaining a robust species tree in the future, and a comprehensive study and reassessment of the fossil record should provide further insight into the history of the family. Comparative phylogenetic approaches, like the reconstruction of historical niches, have the promise of complementing the preliminary spatiotemporal inferences presented here.

Acknowledgements The National Science Foundation (DEB 1240869 and FESD Type I 1338694 to CWD and Stephen Smith) and the University of Michigan (Associate Professor Award to CWD) provided financial support for this work. We thank the members of the Stephen Smith lab, Scott Mori, Ana Carnaval, Lucia Lohmann, João Capurucho, and Giorgia Auteri for useful comments and support. Camila Martinez kindly provided expert advice about fossil utilization in the dating analysis. We are grateful to Scott Mori for sharing many insights into the family, and for providing access to botanical collections and DNA samples through the New York Botanical Garden.

References

APG IV (2016) An update of the Angiosperm Phylogeny Group classification for the orders and families of flowering plants: APG IV. Bot J Linn Soc 181:1–20. https://doi.org/10.1111/boj.12385

Bouckaert R, Heled J, Kühnert D et al (2014) BEAST 2: a software platform for Bayesian evolutionary analysis. PLoS Comput Biol 10:1–6. https://doi.org/10.1371/journal.pcbi.1003537

Chambers JQ, Higuchi N, Schimel JP (1998) Ancient trees in Amazonia. Nature 391:135–136. https://doi.org/10.1038/34325

Chanderbali PAS, van der Werff H (2001) Phylogeny and historical biogeography of Lauraceae: evidence from the chloroplast and nuclear genomes. Ann Missouri Bot Gard 88:104–134

Couvreur TLP, Pirie MD, Chatrou LW et al (2011) Early evolutionary history of the flowering plant family Annonaceae: steady diversification and boreotropical geodispersal. J Biogeogr 38:664–680. https://doi.org/10.1111/j.1365-2699.2010.02434.x

Davis CC, Bell CD, Mathews S, Donoghue MJ (2002) Laurasia migration explains Gondwanan disjuctions: evidence from Malpighiaceae. Proc Natl Acad Sci U S A 99:6833–6837. https://doi.org/10.1073/pnas.102175899

Graham A (2011) A natural history of the New World, the ecology and evolution of plants in the Americas. University of Chicago Press, Chicago

Graham A (2018) Land bridges: ancient environments, plant migrations, and new world connections. University of Chicago Press, Chicago

Gregory-Wodzicki KM (2000) Uplift history of the Central and Northern Andes: a review. Geol Soc Am Bull 112:1091–1105. https://doi.org/10.1130/0016-7606(2000)112<1091

Haffer J (1969) Speciation in Amazonian forest birds. Science 165:131–137

Hoorn C, Wesselingh FP, ter Steege H et al (2010) Amazonia through time: Andean uplift, climate change, landscape evolution, and biodiversity. Science 330:927–931. https://doi.org/10.1126/science.1194585

Huang YY (2010) Systematics of Lecythidoideae (Lecythidaceae) with emphasis on *Bertholletia, Corythophora, Eschweilera*, and *Lecythis*. A dissertation submitted to the Graduate Faculty in Biology in partial fulfillment of the requirements for the degree of Doctor of Philosophy, The City University of New York

Huang YY, Mori SA, Kelly LM (2015) Toward a phylogenetic-based generic classification of neotropical Lecythidaceae–I. Status of *Bertholletia*, *Corythophora*, *Eschweilera* and *Lecythis*. Phytotaxa 203:85–121. https://doi.org/10.11646/phytotaxa.203.2.1

Huang JF, Li L, van der Werff H et al (2016) Origins and evolution of cinnamon and camphor: a phylogenetic and historical biogeographical analysis of the Cinnamomum group (Lauraceae). Mol Phylogenet Evol 96:33–44. https://doi.org/10.1016/j.ympev.2015.12.007

Katoh K, Standley DM (2013) MAFFT multiple sequence alignment software version 7: improvements in performance and usability. Mol Biol Evol 30:772–780. https://doi.org/10.1093/molbev/mst010

Kodandaramaiah U (2010) Use of dispersal – vicariance analysis in biogeography – a critique. J Biogeogr 37:3–11. https://doi.org/10.1111/j.1365-2699.2009.02221.x

Landis MJ, Matzke NJ, Moore BR, Huelsenbeck JP (2013) Bayesian analysis of biogeography when the number of areas is large. Syst Biol 62:789–804. https://doi.org/10.1093/sysbio/syt040

Lavin M, Luckow M (1993) Origins and relationships of tropical North America in the context of the Boreotropics hypothesis. Am J Bot 80:1–14

Löwenberg-Neto P (2014) Neotropical region: a shapefile of Morrone's (2014) biogeographical regionalisation. Zootaxa 3802:300. https://doi.org/10.11646/zootaxa.3802.2.12

Matzke NJ (2013) BioGeoBEARS: Biogeography with Bayesian (and likelihood) evolutionary analysis in R scripts, CRAN: The Comprehensive R Archive Network, Vienna, Austria. http://cran.r-project.org/package=BioGeoBEARS

Matzke NJ (2014) Model selection in historical biogeography reveals that founder-event speciation is a crucial process in island clades. Syst Biol 63:951–970. https://doi.org/10.1093/sysbio/syu056

Montes C, Cardona A, Jaramillo C et al (2015) Middle Miocene closure of the Central American Seaway. Science 348:226–229. https://doi.org/10.1126/science.aaa2815

Mori SA (2017) The Lecythidaceae pages. http://sweetgum.nybg.org/science/projects/lp/

Mori SA, Prance GT (1990) Lecythidaceae. Part II. The zygomorphic-flowered New World genera (*Couroupita, Corythophora, Bertholletia, Couratari, Eschweilera, & Lecythis*). Flora Neotrop Monogr 21:1–376

Mori SA, Tsou CH, Wu CC et al (2007) Evolution of Lecythidaceae with an emphasis on the circumscription of neotropical genera: information from combined ndhF and trnL-F sequence data. Am J Bot 94:289–301. https://doi.org/10.3732/ajb.94.3.289

Mori SA, Kiernan EA, Smith NP et al (2017) Observations on the phytogeography of the Lecythidaceae clade (Brazil nut family). Phytoneuron 30:1–85

Morrone JJ (2014) Biogeographical regionalisation of the neotropical region. Zootaxa 3782:1–110. https://doi.org/10.11646/zootaxa.3782.1.1

Muellner AN, Savolainen V, Samuel R, Chase MW (2006) The mahogany family "out-of-Africa": divergence time estimation, global biogeographic patterns inferred from plastid rbcL DNA sequences, extant, and fossil distribution of diversity. Mol Phylogenet Evol 40:236–250. https://doi.org/10.1016/j.ympev.2006.03.001

Pennington RT, Dick CW (2004) The role of immigrants in the assembly of the South American rainforest tree flora. Philos Trans R Soc Lond B Biol Sci 359:1611–1622. https://doi.org/10.1098/rstb.2004.1532

Pons D, Rica P (1983) *Lecythidospermum bolivarensis* (Berry) nov. comb., graine fossile de Lecythidaceae (Angiospermae) du Tertiaire de la Colombie. Ann Paléontologie 69:1–12

Prance GT (1973) Phytogeographic support the theory of Pleistocene forest refuges in the Amazon Basin, based on evidence from distribution patterns in Caryocaraceae, Chrysobalanaceae, Dichapetalaceae and Lecythidaceae. Acta Amaz 3:5–28

Prance GT (1978) The origin and evolution of the Amazon flora. Interciencia 3:207–222

Prance GT, Mori SA (1979) Lecythidaceae. Part I. The actinomorphic-flowered New World Lecythidaceae (*Asteranthos*, *Gustavia*, *Grias*, *Allantoma*, & *Cariniana*). Flora Neotrop Monogr 21:1–270

QGIS Development Team (2005) QGIS geographic information system. Open source geospatial foundation project. http://qgis.osgeo.org

Rabosky DL (2014) Automatic detection of key innovations, rate shifts, and diversity-dependence on phylogenetic trees. PLoS One 9:e89543. https://doi.org/10.1371/journal.pone.0089543

Rabosky DL, Grundler M, Anderson C et al (2014) BAMMtools: an R package for the analysis of evolutionary dynamics on phylogenetic trees. Methods Ecol Evol 5:701–707. https://doi.org/10.1111/2041-210X.12199

Rambaut A, Drummond AJ, Xie D et al (2018) Posterior summarization in Bayesian phylogenetics using Tracer 1.7. Syst Biol 67:901–904. https://doi.org/10.1093/sysbio/syy032

Ree RH, Sanmartín I (2018) Conceptual and statistical problems with the DEC+J model of founder-event speciation and its comparison with DEC via model selection. J Biogeogr 45:741–749. https://doi.org/10.1111/jbi.13173

Ree RH, Smith SA (2008) Maximum likelihood inference of geographic range evolution by dispersal, local extinction, and cladogenesis. Syst Biol 57:4–14. https://doi.org/10.1080/10635150701883881

Renner SS, Clausing G, Meyer K (2001) Historical biogeography of Melastomataceae: the roles of tertiary migration and long-distance dispersal. Am J Bot 88:1290–1300

Ronquist F (1997) Dispersal-vicariance analysis: a new approach to the quantification of historical biogeography. Syst Biol 46:195–203. https://doi.org/10.1093/sysbio/46.1.195

Rose JP, Kleist TJ, Löfstrand SD et al (2018) Phylogeny, historical biogeography, and diversification of angiosperm order Ericales suggest ancient Neotropical and East Asian connections. Mol Phylogenet Evol 122:59–79. https://doi.org/10.1016/j.ympev.2018.01.014

Scotese CR (2001) Atlas of earth history, vol 1, Paleogeography, PALEOMAP Project. Arlington, TX

Shallom LJ (1960) Fossil dicotyledonous wood of Lecythidaceae from the Deccan Intertrappean beds of Mahurzari. J Ind Bot Soc 39:198–203

Shepard GH, Ramirez H (2011) "Made in Brazil": human dispersal of the Brazil nut (*Bertholletia excelsa*, Lecythidaceae) in ancient Amazonia. Econ Bot 65:44–65. https://doi.org/10.1007/s12231-011-9151-6

Smith SA, Brown JW (2018) Constructing a broadly inclusive seed plant phylogeny. Am J Bot 105:1–13. https://doi.org/10.1002/ajb2.1019

Srivastava R, Kapgate DK, Chatterjee S (2009) Permineralized fungal remains in the fossil wood of *Barringtonia* from the Deccan Intertrappean sediments of Yavatmal District, Maharashtra, India. The Palaeobotanist 58:11–19

Stamatakis A (2014) RAxML version 8: a tool for phylogenetic analysis and post-analysis of large phylogenies. Bioinformatics 30:1312–1313. https://doi.org/10.1093/bioinformatics/btu033

Stamatakis A (2015) The RAxML v8.1.X manual. http://sco.h-its.org/exelixis/web/software/raxml/index.html

Stebbins GL (1974) Flowering plants: evolution above the species level. Belknap Press, Cambridge

ter Steege H, Pitman NCA, Sabatier D et al (2013) Hyperdominance in the Amazonian tree flora. Science 342:325–342. https://doi.org/10.1126/science.1243092

Thomson AM, Vargas OM, Dick CW (2018) Complete plastome sequences from *Bertholletia excelsa* and 23 related species yield informative markers for Lecythidaceae. Appl Plant Sci 6: e1151. https://doi.org/10.1002/aps3.1151

Vargas OM, Heuertz M, Smith SA, Dick CW (2019) Target sequence capture in the Brazil nut family (Lecythidaceae): marker selection and in silico capture from genome skimming data. Mol Phylogenet Evol 135:98–104. https://doi.org/10.1016/j.ympev.2019.02.020

Weeks A, Daly DC, Simpson BB (2005) The phylogenetic history and biogeography of the frankincense and myrrh family (Burseraceae) based on nuclear and chloroplast sequence data. Mol Phylogenet Evol 35:85–101. https://doi.org/10.1016/j.ympev.2004.12.021

Yu XQ, Gao LM, Soltis DE et al (2017) Insights into the historical assembly of East Asian subtropical evergreen broadleaved forests revealed by the temporal history of the tea family. New Phytol 215:1235–1248. https://doi.org/10.1111/nph.14683

Part IV
Epilogue

Chapter 30
Conservation in the Neotropics: A Final Reflection

Ana Carolina Carnaval

When Valentí Rull first contacted me with a proposal to co-edit this book, it was clear that we needed to seek for, and welcome, different kinds of contributions. Reviews of the literature would have a place here—but also new data, novel methods, study cases, as well as established, controversial, and challenging perspectives. We wanted to provide a forum to contrast and integrate information streams from diverse, yet complementary, disciplines: paleoclimatology, geology, molecular phylogenetics, biogeography, historical demography, Earth and atmospheric sciences, paleontology, community and macroecology, geography... Because convergence across scientific fields is needed to shed full light on the topic of Neotropical Diversification, we felt that merging distinct streams of information was particularly relevant (Rull 2020). Of me, it was asked to assist with the editorial process—but also to contribute with a final chapter about the conservation implications of what we have learned.

This joint effort, driven by the work of the multiple contributors to the book, resulted in a diverse collection of ecosystem- and organismal-focused essays that speak to (or about) conservation in multiple, different ways. Here we find reviews of spatial distribution patterns of biodiversity in target Neotropical regions; see for instance Señaris and Rojas-Runjaic for the Guayana (2020), and Reginato and Michelangeli (2020) and Peres et al. (2020) for the Atlantic Forest, which are discussed under the light of biogeography, and current and past environments. Comprehensive and integrative discussions of the processes underscoring diversification are provided for megadiverse systems such as the mountains east of the Andes (Guedes et al. 2020) and the Cerrado savannahs (Azevedo et al. 2020), based on the compilation of data from many of the disciplines cited above. Other chapters present novel data and analyses to understand diversification and biodiversity patterns in

A. C. Carnaval (✉)
Department of Biology, City College of New York, New York, NY, USA

Biology Program, Graduate Center of CUNY, New York, NY, USA

V. Rull, A. C. Carnaval (eds.), *Neotropical Diversification: Patterns and Processes*, Fascinating Life Sciences, https://doi.org/10.1007/978-3-030-31167-4_30

regions as varied and understudied as the Amazonian White Sands Ecosystem (Capurucho et al. 2020) and the dunes system in the São Francisco region (Recoder and Rodrigues 2020), integrating information from molecular biology, inventory efforts, and environmental description. Some present new information on particularly diverse groups of Neotropical organisms, to address hypotheses about the processes underscoring diversification at local and regional scales. For instance, Olave et al. (2020) study the influence of ecological and morphological traits on the diversification of lizards in southern South America, whereas Cadena and Céspedes (2020) propose a mechanism to explain elevational replacements in Andean birds. Focusing on broader spatial scales, both Luebert et al. (2020) and Maestri and Duarte (2020) employ novel methods that combine geographical, niche and molecular analyses to investigate former connectivity between Neotropical forests. The examples go on.

How does one, then, leverage this large, heterogeneous collective knowledge of the processes behind Neotropical diversification (or, more broadly, the data generated by studies of Neotropical biodiversity patterns and their underscoring mechanisms) to guide conservation in the tropical Americas?

First and foremost, all field work needed for the compilation of occurrence, voucher, and molecular data discussed here inform conservation by reporting back on the *status* of biological diversity today. The consistent message across many of the chapters of this book is worrisome. It suggests widespread negative impacts of human-mediated landscape change and anthropogenic climate change on the abundance and diversity of the fauna and flora of Central and South America [e.g. Nores (2020), de La Riva (2020), Señaris and Rojas-Runjaic (2020)].

Studies of Neotropical diversification also guide conservation by identifying geographical *patterns* of biodiversity distribution that help to prioritize areas or taxa for management. Maps of diversity accumulation, endemism, and turnover have traditionally been used to this end (Myers et al. 2000; Ceballos and Ehrlich 2006). More recently, they have incorporated phylogenetic information to identify areas that contain high or unique evolutionary history for conservation planning (e.g. Thomassen et al. 2011). Several chapters of this book summarize our understanding of the spatial structuring of biological diversity in the Neotropics by identifying centers of endemism and areas of turnover of species, lineages, or ecological functions (traits). By doing so, they provide the very foundation for strategic conservation plans. As an example, both Reginato and Michelangeli (2020), and Peres et al. (2020) locate and discuss unique areas of species composition in the Atlantic Forest. Given the demonstrated importance of biogeographical connections as drivers of present-day diversity patterns (e.g. Maestri and Duarte 2020), it may be appropriate to develop conservation strategies tailored to each of these unique regions—particularly if they contain pools of largely co-existing, or co-evolved, taxa. While several conservation priorization schemes for Neotropical regions treat each domain as a single entity (e.g. MMA—Ministério do Meio Ambiente 2002), an understanding of the distinct biogeographical regions within each one of them, as well as of the environmental and historical constraints at play, will result in more effective strategies for conservation. In the Atlantic Forest case, for instance, a conservation plan of montane assemblages (e.g. the Serra do Mar region identified in Reginato and Michelangeli 2020 and Peres et al. 2020) should address threats and risks very distinct from those relevant to the conservation of northern, lowland assemblages (e.g. Bahia coastal communities). Because the historical and contemporary drivers of biodiversity patterns differ among these regions, so should their conservation strategies.

Some of the chapters of this book inform conservation by pinpointing ecological and evolutionary *correlates* of diversity (e.g. Villalobos et al. 2020). Given the high rates of discovery of unique evolutionary lineages in the Neotropics (e.g. Peloso et al. 2011), these studies are particularly interesting by identifying biodiverse yet undersampled regions for inventory effort—something particularly relevant in the face of the rampant habitat changes faced by South and Centro-America (Ribeiro et al. 2011). This exercise has been done at broad spatial scales (Albuquerque and Beier 2015; Holt et al. 2017; Antonelli et al. 2018), and combined with projections of climate change to predict biological shifts under future conditions (Brown et al. 2016). However, it can still be improved by linking in situ biodiversity data obtained through inventory work (as opposed to correlative models) and descriptors of climatic conditions experienced locally, and measured through sensors or remote sensing. By combining field-based data on community composition and environmental conditions (Scholes et al. 2008) with frequent monitoring of land-use shifts (Shimabukuro et al. 2012, www.globalforestwatch.org), conservation biologists may soon be able to report and model biodiversity loss at infra-annual scales (Turner et al. 2015), and keep a closer eye on biodiversity erosion. Further integration of these data with knowledge of phylogenetic history and ecological traits will allow us to closely monitor and report the change of taxonomic, evolutionary, and functional diversity in these megadiverse countries (Brum et al. 2017), in near-real time. Imagine the potential implications of this exercise—particularly for public outreach, management, and policy change.

By forcing us to look into the past, studies of Neotropical diversification also shed light on the historical (geological, climatic, evolutionary) and the ecological *processes* that result in present-day species ranges. As such, they can be used to identify at-risk taxa and ecosystems (Akçakaya et al. 2006) and directly feed into conservation practices. Señaris and Rojas-Runjaic (2020), for instance, call attention to the close relationship between threat category and narrow endemism in amphibians and reptiles of the Venezuelan Guayanas. By doing so, they identify something common to all naturally fragmented systems in the Neotropics: the combined role of environmental history, lineage history, and topography in restricting the range of many local species, which increases vulnerability today and under future global change. Incorporating these and other elements in simulations of species ranges is possible now, and evolutionary data can be added to enable more mechanistic predictions of biodiversity accumulation based on biological processes that are modeled as a function of environmental conditions available in space and over time (Rangel et al. 2018). While this exercise has been used to test our ability to reconstruct contemporary patterns of diversity in geographic space, it may likewise be utilized to flag lineages or regions under potential threat.

Another way by which Neotropical diversification studies contribute to conservation is by illuminating our understanding of the biological *mechanisms* underscoring biodiversity generation. Both Olave et al. (2020) and Vargas and Dick (2020) estimate and compare diversification rates across Neotropical organisms. On the other hand, Recoder and Rodrigues (2020) and Cadena and Céspedes (2020) combine datasets to identify the processes likely driving diversification in their study taxa. It has been argued that both taxon-specific and process-related data like those listed above may be useful in long-term planning and conservation, particularly if one aims to ensure that local populations maintain evolutionary potential over time (Moritz and Agudo 2013).

However, any backward- or forward-looking projections will need to acknowledge a human footprint that did not exist in deep history; this footprint almost certainly altered the distribution of biodiversity in the recent past, and will keep on doing so in the future. Montoya et al. (2020) review a particularly relevant and timely topic: the influence of human societies, and the role of fire, on the diversity and distribution of the Amazonia biota. As we finalize the edition of this book in 2019, citizens and governments worldwide turn their attention to Amazonia and the large extensions of forests burning in Brazil, Bolivia, Paraguay, and Peru. Knowledge about the Amazonian environment and its biota—now and in the past—is relevant to place the potential biological impacts of these fires in a broader perspective.

For instance, the paleorecords reviewed and discussed by Montoya and collaborators (2020) indicate the occasional presence of charcoal deposits at deeper time scales, demonstrating that the Amazon is not a complete stranger to fire. Specifically, existing data link the occurrence of historical fires with the presence of Pre-Columbian societies and human-driven burning (see references in Montoya et al. 2020)—in fact, the authors argue that indigenous peoples may have been altering Amazonian plant diversity for more than 13,000 years, including man-made fires. Understanding this impact is relevant to explain current patterns of plant diversity in the system, and may inform models predicting future trajectories of the local fauna and flora. That being said, the authors remind us that not all fire records in the history of this ecosystem should be attributed to human influence. The southwestern portion of Amazonia, for instance, experienced a significantly direr climate in the mid-Holocene; it has been argued that increased fire frequency in that region may have been caused by more flammable dry forests and savannahs (Mayle and Power 2008).

Yet what Montoya and collaborators emphasize (2020) is that fire is not a natural element in today's Amazonia—instead, it is associated with human-induced burning following deforestation. Importantly, the scale of habitat alternation now afforded by technologies unavailable to pre-Columbian societies is changing the structure, the composition, and the functioning of the Amazonian forest at a dramatically fast pace (Carmenta et al. 2018; Longo et al. 2018). As the scientific community seeks to better understand and document the ecological and evolutionary impacts of burning in the Amazonian system (e.g. Barlow and Peres 2008), this book argues that the paleoecological reconstructions and future predictions of Amazonian fires, as well as of their biological impacts, must incorporate both environmental and anthropogenic influences. Transferring knowledge about the processes of biodiversity build-up and erosion across temporal scales into conservation is challenging. In the case of the Neotropics, this has to be thought of, and implemented, in a way to include the human footprint that exists, and will keep existing, in the region (Peres et al. 2010; Lewis et al. 2015).

Lastly, integrative studies of the assembly of biotas also inform conservation by providing insight about their *behavior* in response to environmental shifts over time (Devictor and Robert 2009). By merging geological, geographical, paleoclimatological and biological data from White Sands Ecosystems in Amazonia, for instance, Capurucho et al. (2020) demonstrate this habitat's low resilience to disturbance. Knowledge of the historical ability of Neotropical domains to bounce back is central to the identification of ecological tipping points (Hughes et al. 2013; Lenton and Williams 2013), and one more example of how evolutionary, ecological, and Earth observation data ought to be combined to guide management and policy in the face of global change.

This compilation of examples is by no way comprehensive; the ways by which integrative studies of Neotropical diversification inform conservation go well beyond those addressed in this book. Studies of population dynamics in biodiversity refugia or microrefugia (Rull 2010), for instance, can guide planning for the preservation of evolutionary potential over time. Insights into environmental correlates or processes leading to adaptive genetic change (Hoffmann and Sgro 2011; Razgour et al. 2019), which were also not mentioned here, can inform assisted migration and priorization of areas for conservation (MacDonald et al. 2017). Technological advances and genome-level sampling of wild populations already allow for the estimation of ecological parameters from molecular data and their inclusion in predictions of shifts in the distribution of genetic diversity under future climates (Prates et al. 2016). In addition, studies of co-evolutionary dynamics, and how species interactions shift in space and time, are particularly relevant to conservation and the maintenance of ecological and evolutionary processes (Koskella et al. 2012).

So where to, from here? Valentí and I hope that this book will demonstrate the power of combining and integrating data across multiple scientific fields to understand the processes and outcomes of diversification in the Neotropics. It also documents our collective concern about how to preserve these very same biodiversity patterns and processes. More than an interdisciplinary approach within STEM fields, we need a trans-disciplinary effort to integrate these data, and the insights our analyses provide, with the social and economical structure of the biodiverse areas we study and love (Griggs et al. 2013). Any attempt to preserve the ecological and evolutionary legacy discussed at length within this book will need to go beyond pure scientific silos and reach out to the humanities (Aronson 2011) to translate the relevance of our object of study into societal and policy change. While we are now able to generate models of environmental and biodiversity shifts under future conditions for several Neotropical areas, the conservation solutions to be implemented on the ground ought to be region- and people-specific (Kasecker et al. 2018). By summarizing the many unique biogeographical regions, patterns of biodiversity, and processes of diversity generation, maintenance, and erosion in the Neotropics, this book gives one important step in this direction. Ahead of us, though, lies the challenging task of building a path to sustainability that takes into account the biological diversity, the history—both deep and recent—, and the human component of the Neotropics.

References

Akçakaya H, Butchart SHM, Mace GM, Stuart SN, Hilton-Taylor C (2006) Use and misuse of the IUCN Red List Criteria in projecting climate change impacts on biodiversity. Glob Chang Biol 12:2037–2043

Albuquerque F, Beier P (2015) Global patterns and environmental correlates of high-priority conservation areas for vertebrates. J Biogeogr 42:1397–1405

Antonelli A et al (2018) Geological and climatic influences on mountain biodiversity. Nat Geosci 11(10):718–725

Aronson J (2011) Sustainability science demands that we define our terms across diverse disciplines. Landsc Ecol 26:457–460

Azevedo JAR, Collevatti RG, Jaramillo CA, Strömberg CAE, Guedes TB, Matos-Maraví P, Bacon CD, Carrillo JD, Faurby S, Antonelli A (2020) On the young savannas in the land of ancient forests. In: Rull V, Carnaval AC (eds) Neotropical diversification. Springer, Berlin

Barlow J, Peres CA (2008) Fire-mediated dieback and compositional cascade in an Amazonian forest. Philos Trans R Soc B Biol Sci 363:1787–1794

Brown JL, Weber JJ, Alvarado-Serrano DF, Hickerson MJ, Franks SJ, Carnaval AC (2016) Predicting the genetic consequences of future climate change: the power of coupling spatial demography, the coalescent, and historical landscape changes. Am J Bot 103:153–163

Brum FT, Graham CH, Costa GC, Hedges SB, Penone C, Radeloff VC, Rondinini C, Loyola R, Davidson AD (2017) Global priorities for conservation across multiple dimensions of mammalian diversity. Proc Natl Acad Sci U S A 114(29):7641–7646

Cadena CD, Céspedes LN (2020) Origin of elevational replacements in a clade of nearly flightless birds – most diversity in tropical mountains accumulates via secondary contact following allopatric speciation. In: Rull V, Carnaval AC (eds) Neotropical diversification. Springer, Berlin

Capurucho JMG, Borges SH, Cornelius C, Vicentini A, Prata EMB, Costa FM, Campos P, Sawakuchi AO, Rodrigues F, Zular A, Aleixo A, Bates JM, Ribas CC (2020) Patterns and processes of diversification in Amazonian white sand ecosystems: insights from birds and plants. In: Rull V, Carnaval AC (eds) Neotropical diversification. Springer, Berlin

Carmenta R, Coudel E, Stewards AM (2018) Forbidden fire: does criminalising fire hinder conservation efforts in swidden landscapes of the Brazilian Amazon? Geogr J 184:1–15

Ceballos G, Ehrlich P (2006) Global mammal distributions, biodiversity hotspots, and conservation. Proc Natl Acad Sci U S A 103(51):19374–19379

De La Riva I (2020) Unexpected beta-diversity radiations in highland clades of Andean Terraranae frogs. In: Rull V, Carnaval AC (eds) Neotropical diversification. Springer, Berlin

Devictor V, Robert A (2009) Measuring community responses to large-scale disturbance in conservation biogeography. Divers Distrib 15(1):122–130

Griggs D et al (2013) Sustainable development goals for people and planet. Nature 495:305–307

Guedes TB, Azevedo JAR, Bacon CD, Provete DB, Antonelli A (2020) Diversity, endemism, and evolutionary history of mountain biotas outside the Andean region. In: Rull V, Carnaval AC (eds) Neotropical diversification. Springer, Berlin

Hoffmann AA, Sgrò CM (2011) Climate change and evolutionary adaptation. Nature 470:479–485

Holt BG, Costa GC, Penone C, Lessard J-P, Brooks TM, Davidson AD, Hedges SB, Radeloff VC, Rahbek C, Rondinini C, Graham CH (2017) Environmental variation is a major predictor of global trait turnover in mammals. J Biogeogr 2017:1–13
Hughes TP, Carpenter S, Rockstrom J, Scheffer M, Walker B (2013) Multiscale regime shifts and planetary boundaries Trends Ecol. Evolution 28:389–395
Kasecker TP, Ramos-Neto MB, Silva JMC, Scarano FR (2018) Ecosystem-based adaptation to climate change: defining hotspot municipalities for policy design and implementation in Brazil. Mitig Adapt Strateg Glob Chang 23:981–993
Koskella B, Lin DM, Buckling A, Thompson JN (2012) The costs of evolving resistance in heterogeneous parasite environments. Proc R Soc B Biol Sci 279:1896–1903
Lenton TM, Williams HTP (2013) On the origin of planetary-scale tipping points. Trends Ecol Evol 28:380–382
Lewis SL, Edwards DP, Galbraith D (2015) Increasing human dominance of tropical forests. Science 349:827–832
Longo M, Knox RG, Levine NM et al (2018) Ecosystem heterogeneity and diversity mitigate Amazon forest resilience to frequent extreme droughts. New Phytol 219:914–931
Luebert F, Lörch M, Acuña R, Mello-Silva R, Weigend M, Mutke J (2020) Clade-specific biogeographic history and climatic niche shifts of the southern Andean-southern Brazilian disjunction in plants. In: Rull V, Carnaval AC (eds) Neotropical diversification. Springer, Berlin
MacDonald SL, Llewelyn J, Moritz C, Phillips BL (2017) Peripheral isolates as sources of adaptive diversity under climate change. Front Ecol Evol 5:88
Maestri R, Duarte L (2020) Evolutionary imprints on species distribution patterns across the Neotropics. In: Rull V, Carnaval AC (eds) Neotropical diversification. Springer, Berlin
Mayle FE, Power MJ (2008) Impact of a drier Early-Mid-Holocene climate upon Amazonian forests. Philos Trans R Soc B 363:1829–1838
MMA – Ministério do Meio Ambiente (2002) Biodiversidade Brasileira: Avaliação e Identificação das áreas e ações prioritárias para a Conservação, Utilização Sustentável e Repartição dos Benefícios da Biodiversidade nos Biomas Brasileiros. Ministério do Meio Ambiente, Brasília
Montoya E, Lombardo U, Levis C, Aymard GA, Mayle FE (2020) Human contribution to Amazonian plant diversity: legacy of pre-Columbian land use in modern plant communities. In: Rull V, Carnaval AC (eds) Neotropical diversification. Springer, Berlin
Moritz C, Agudo R (2013) The future of species under climate change: resilience or decline? Science 341:504–508
Myers N, Mittermeier RA, Mittermeier CG, da Fonseca GAB, Kent J (2000) Biodiversity hotspots for conservation priorities. Nature 403:853–859
Nores M (2020) Avian diversity in humid tropical and subtropical South American forests, with a discussion about their related climatic and geological underpinnings. In: Rull V, Carnaval AC (eds) Neotropical diversification. Springer, Berlin
Olave M, González MA, Avila L, Sites J Jr, Morando M (2020) Disparate patterns of diversification within Liolaemini lizards. In: Rull V, Carnaval AC (eds) Neotropical diversification. Springer, Berlin
Peloso PLV, Pellegrino KCM, Rodrigues MT, Avila-Pires TCS (2011) Description and phylogenetic relationships of a new genus and species of lizard (Squamata, Gymnophthalmidae) from the Amazonian rainforest of northern Brazil. Am Mus Novit 3713:1–24
Peres CA, Gardner TA, Barlow J, Zuanon J, Michalski F, Lees AC, ICG V, Moreira FM, Sand Feeley KJ (2010) Biodiversity conservation in human-modified Amazonian forest landscapes. Biol Conserv 143:2314–2327
Peres EA, Pinto-da-Rocha R, Lohmann LG, Michelangeli FC, Miyaki CY, Carnaval AC (2020) Patterns of species and lineage diversity in the Atlantic rainforest of Brazil. In: Rull V, Carnaval AC (eds) Neotropical diversification. Springer, Berlin
Prates I, Xue A, Brown JL, Alvarado-Serrano DF, Rodrigues MT, Hickerson MJ, Carnaval AC (2016) Inferring responses to climate dynamics from historical demography in Neotropical forest lizards. Proc Natl Acad Sci U S A 113(29):7978–7985

Rangel TF, Edwards NR, Holden PB, Diniz-Filho JAF, Gosling WD, Coelho MTP, Cassemiro FAS, Rahbek C, Colwell RK (2018) Modeling the ecology and evolution of biodiversity: biogeographical cradles, museums, and graves. Science 361:eaar5452

Razgour O, Forester B, Taggart JB, Bekaert M, Juste J, Ibáñez C, Puechmaille SJ, Novella-Fernandez R, Alberdi A, Manel IS (2019) Considering adaptive genetic variation in climate change vulnerability assessment reduces species range loss projections. Proc Natl Acad Sci U S A 116(21):10418–10423

Recoder RS, Rodrigues MT (2020) Diversification processes in lizards from the middle São Francisco River dune region. In: Rull V, Carnaval AC (eds) Neotropical diversification. Springer, Berlin

Reginato M, Michelangeli FC (2020) Bioregions of eastern Brazil based on vascular plant occurrence data. In: Rull V, Carnaval AC (eds) Neotropical diversification. Springer, Berlin

Ribeiro MC, Martensen AC, Metzger JP, Tabarelli M, Scarano FR, Fortin M-J (2011) The Brazilian Atlantic forest: a shrinking biodiversity hotspot. In: Zachos FE, Habel JC (eds) Biodiversity hotspots: distribution and protection of conservation priority areas. Springer, Berlin, pp 405–434

Rull V (2010) On microrefugia and cryptic refugia. J Biogeogr 37:1623–1625

Rull V (2020) Neotropical diversification: historical overview and conceptual insights. In: Rull V, Carnaval AC (eds) Neotropical diversification. Springer, Berlin

Scholes RJ et al (2008) Ecology: toward a global biodiversity observing system. Science 321 (5892):1044–1045

Señaris JC, Rojas-Runjaic FJM (2020) Amphibians and reptiles of Venezuelan Guayana: diversity, biogeography and conservation. In: Rull V, Carnaval AC (eds) Neotropical diversification. Springer, Berlin

Shimabukuro YE, Santos JR, Formaggio AR, Duarte V, Rudorff BFT (2012) The Brazilian Amazon monitoring program: PRODES and DETER projects. In: Achard F, Hansen MC (eds) Global forest monitoring from earth observation. CRC Press, New York, pp 153–169

Thomassen HA et al (2011) Mapping evolutionary process: a multi-taxa approach to conservation prioritization. Evol Appl 4:397–413

Turner W et al (2015) Free and open-access satellite data are key to biodiversity conservation. Biol Conserv 182:173–176

Vargas OM, Dick CW (2020) Diversification history of Neotropical Lecythidaceae, an ecologically dominant tree family of Amazon rain forest. In: Rull V, Carnaval AC (eds) Neotropical diversification. Springer, Berlin

Villalobos F, Pinto-Ledezma J, Diniz-Filho JAF (2020) Evolutionary macroecology and the geographical patterns of Neotropical diversification. In: Rull V, Carnaval AC (eds) Neotropical diversification. Springer, Berlin

Correction to: Neotropical Diversification: Patterns and Processes

Valentí Rull and Ana Carolina Carnaval

Correction to:
V. Rull, A. C. Carnaval (eds.), *Neotropical Diversification: Patterns and Processes*, Fascinating Life Sciences https://doi.org/10.1007/978-3-030-31167-4

The original version of the book was inadvertently published with an incorrect fore name of the author in chapters 10 and 11 as C. Camila Ribas. The name has been now corrected as Camila C. Ribas.

The updated online versions of the chapters can be found at
https://doi.org/10.1007/978-3-030-31167-4_10
https://doi.org/10.1007/978-3-030-31167-4_11

V. Rull, A. C. Carnaval (eds.), *Neotropical Diversification: Patterns and Processes*,
Fascinating Life Sciences, https://doi.org/10.1007/978-3-030-31167-4_31

Correction to: Neotropical Diversification: Patterns and Processes

The manufacturer's authorised representative in the EU is Springer Nature Customer Service Centre GmbH, Europaplatz 3, 69115 Heidelberg, Germany. If you have any concerns regarding our products, please contact ProductSafety@springernature.com

Printed and bound by CPI Group (UK) Ltd, Croydon, CR0 4YY
30/09/2026
02225454-0017